Mathematics in Action

Algebraic, Graphical, and Trigonometric Problem Solving

Sixth Edition

The Consortium for Foundation Mathematics

Ralph Bertelle — *Columbia-Greene Community College*
Judith Bloch — *University of Rochester*
Roy Cameron — *SUNY Cobleskill*
Carolyn Curley — *Erie Community College—South Campus*
Ernie Danforth — *Corning Community College*
Brian Gray — *Howard Community College*
Arlene Kleinstein — *SUNY Farmingdale*
Kathleen Milligan — *Monroe Community College*
Patricia Pacitti — *SUNY Oswego*
Renan Sezer — *Ankara University, Turkey*
Patricia Shuart — *Polk State College—Winter Haven, Florida*
Sylvia Svitak — *Queensborough Community College*

VP, Portfolio Management: Chris Hoag
Director, Portfolio Management: Michael Hirsch
Courseware Portfolio Manager: Matt Summers and Karen Montgomery
Courseware Portfolio Management Assistant: Kayla Shearns
Content Producer: Kathleen A. Manley
Managing Producer: Karen Wernholm
Producer: Marielle Guiney
TestGen Content Manager: Rajinder Singh
Manager, Content Development: Eric Gregg
Senior Product Marketing Manager: Alicia Frankel
Marketing Assistant: Brooke Imbornone
Senior Author Support/Technology Specialist: Joe Vetere
Composition: Pearson CSC
Cover and Interior Design: Pearson CSC
Cover Image: lzf/Shutterstock

Cataloging-in-Publication Data on file with the Library of Congress

ISBN 13: 978-0-13-498930-3
ISBN 10: 0-13-498930-9
ISBN 13: 978-0-13-511561-9
ISBN 10: 0-13-511561-2

In Memoriam

In the spring of 2018, two of our fellow authors, Brian Gray and Carolyn Curley lost long battles with cancer. They have been an integral part of the team from the beginning of this project, which began as a National Science Foundation project in the 1990's.

Brian was a steadying influence on the team. His enthusiasm never got too high and never too low. Brian was always able to focus on what needed to be done and accomplish it with quality in a timely fashion. Brian was knowledgeable in many fields. This gave him a broad perspective on life and the lives of our students as well.

Carolyn was our "Joy Girl"! Carolyn loved life, and she loved to have fun. Her motto seemed to be, "If it is not fun, why are we doing it." She always encouraged and inspired the team and was incredibly insightful and creative. She was totally dedicated to the book and to our team. Even during her final months, she was constantly asking, "what can I do?" Carolyn had a great feel for the needs of the students as well as the instructor. She was instrumental in creating our Instructor Resource Manuals.

To paraphrase from a Broadway musical, Brian and Carolyn, because we knew you we have been changed. We have been changed for good! May you both rest in peace.

CONTENTS

APPENDICES

PREFACE

Our Vision

Mathematics in Action: Algebraic, Graphical, and Trigonometric Problem Solving, Sixth Edition, is intended to help college mathematics students gain mathematical literacy in the real world and simultaneously help them build a solid foundation for future study in mathematics and other disciplines.

Our team of twelve faculty, primarily from the State University of New York and the City University of New York systems, used the AMATYC *Crossroads* Standards to develop this *Mathematics in Action* series to serve a very large population of college students who, for whatever reason, have not yet succeeded in learning mathematics. It became apparent to us that teaching the same content in the same way to students who have not previously comprehended it is not effective, and this realization motivated us to develop a new approach.

Mathematics in Action is based on the principle that students learn mathematics best by doing mathematics within a meaningful context. In keeping with this premise, students solve problems in a series of realistic situations from which the crucial need for mathematics arises. *Mathematics in Action* guides students toward developing a sense of independence and taking responsibility for their own learning. Students are encouraged to construct, reflect on, apply, and describe their own mathematical models, which they use to solve meaningful problems. We see this as the key to bridging the gap between abstraction and application and as the basis for transfer learning. Appropriate technology is integrated throughout the books, allowing students to interpret real-life data verbally, numerically, symbolically, and graphically.

We expect that by using the *Mathematics in Action* series, all students will be able to achieve the following goals:

- Develop mathematical intuition and a relevant base of mathematical knowledge.
- Gain experiences that connect classroom learning with real-world applications.
- Prepare effectively for further college work in mathematics and related disciplines.
- Learn to work in groups as well as independently.
- Increase knowledge of mathematics through explorations with appropriate technology.
- Develop a positive attitude about learning and using mathematics.
- Build techniques of reasoning for effective problem solving.
- Learn to apply and display knowledge through alternative means of assessment, such as mathematical portfolios and journal writing.

Our vision for you is to join the growing number of students using our approaches who discover that mathematics is an essential and learnable survival skill for the 21st century.

Pedagogical Features

The pedagogical core of *Mathematics in Action* is a series of guided-discovery activities in which students work in groups to discover mathematical principles embedded in realistic situations. The key principles of each activity are highlighted and summarized at the activity's conclusion. Each activity is followed by exercises that reinforce the concepts and skills revealed in the activity.

The activities are clustered within each chapter. Each cluster contains regular activities along with project and lab activities that relate to particular topics. The lab activities require more than just paper, pencil, and calculator; they also require measurements and data collection and are ideal for in-class group work. The project activities are designed to allow students to explore specific topics in greater depth, either individually or in groups. These activities are usually self-contained and have no accompanying exercises. For specific suggestions on how to use the three types of activities, we strongly encourage instructors to refer to the *Instructor's Resource Manual with Tests* that accompanies this text. New PowerPoints have been created to support instructors looking to implement this contextual approach to algebra.

Each cluster concludes with two sections: "What Have I Learned?" and "How Can I Practice?" The "What Have I Learned?" exercises are designed to help students pull together the key concepts of the cluster. The "How Can I Practice?" exercises are designed primarily to provide additional work with the numeric, algebraic, and graphing skills of the cluster. Taken as a whole, these exercises give students the tools they need to bridge the gaps between abstraction, skills, and application.

Each chapter ends with a Summary containing a brief description of the concepts and skills discussed in the chapter, plus examples illustrating these concepts and skills. The concepts and skills are also referenced to the activity in which they appear, making the format easier to follow for those students who are unfamiliar with our approach. Each chapter also ends with a Gateway Review, providing students with an opportunity to check their understanding of the chapter's concepts and skills.

What's New in the Sixth Edition

The Sixth Edition retains all the features of the previous edition, with the following content changes:

- All the data-based activities and exercises have been updated to reflect the most recent information and/or replaced with more relevant topics.
- The introductory scenarios in several activities have been replaced with more robust, up-to-date situations.
- Several new real-world exercises have been added throughout.
- The exposition and treatment of topics has been carefully reviewed and revised/rewritten where necessary to provide students with a more clear and easy to understand presentation.
- New PowerPoint presentations have been developed to support instructors looking to implement the contextual approach to algebra.
- Learning Catalytics questions have been developed for nearly every activity, providing an opportunity for instructors to quickly assess the progress on a given concept and give students an opportunity to use technology as an interactive learning tool.
- For ease in course planning, the mathematical concept explored in each activity is now included in the activity head.

MyLab Math Changes in the Sixth Edition

- Exercise coverage has been enhanced to ensure better conceptual flow, encourage conceptual thinking about math topics, and balance out the coverage of skills related questions.
- A new video program built around the Consortium approach will provide additional multimedia support.
- New to this edition, Integrated Review in MyLab Math provides additional assignments and study aids for select prerequisite topics for students who will benefit from remediation. Integrated Review materials can be used to help underprepared students get up to speed, or for a corequisite course model.
- Learning Catalytics questions are now premade to complement teaching with Mathematics in Action. Learning Catalytics allows students to use their own mobile devices in the classroom for real-time engagement.
- Skill Builder assignments provide just-in-time adaptive practice at the exercise level, delivering questions personalized to each student with the goal of enabling them to better complete their homework assignment.

Acknowledgments

The Consortium would like to acknowledge and thank the following people for their invaluable assistance in reviewing and testing material for this text in the past and current editions:

Mark Alexander, *Kapi'olani Community College*

Kathleen Bavelas, *Manchester Community College*

Shirley J. Beil, *Normandale Community College*

Carol Bellisio, *Monmouth University*

Ann Boehmer, *East Central College*

Barbara Burke, *Hawai'i Pacific University*

San Dong Chung, *Kapi'olani Community College*

Marjorie Deutsch, *Queensboro Community College*

Jennifer Dollar, *Grand Rapids Community College*

Irene Duranczyk, *University of Minnesota*

Kristy Eisenhart, *Western Michigan University*

Mary Esteban, *Kapiolani Community College*

Brian J. Garant, *Morton College*

Thomas Grogan, *Cincinnati State Technical and Community College*

Maryann Justinger, *Erie Community College—South Campus*

Brian Karasek, *South Mountain Community College*

Jim Larson, *Lake Michigan College*

Miriam Long, *Madonna University*

Ellen Musen, *Brookdale Community College*

Roberta Pardo, *Chandler-Gilbert Community College*

Kathy Potter, *St. Ambrose University*

Cindy Pulley, *Heartland Community College*

Robbie Ray, *Sul Ross State University*

Janice Roy, *Montcalm Community College*

Andrew S. Russell, *Queensborough Community Collge*

Amy C. Salvati, *Adirondack Community College*

Philomena Sawyer, *Manchester Community College*

Brenda Shepard, *Lake Michigan College*

Karen Smith, *US Blue Ash*

Kurt Verderber, *SUNY Cobleskill*

Our sincere thanks goes to these dedicated individuals at Pearson who worked to make this revision a success: Michael Hirsch, Matt Summers, Karen Montgomery, Alicia Frankel, Kathy Manley, Marielle Guiney, Eric Gregg, Kayla Shearns and the entire salesforce.

We would also like to thank our accuracy checkers, Rick Ponticelli and Lauri Semarne.

Finally, a special thank you to our families for their unwavering support and sacrifice, which enabled us to make this text a reality.

The Consortium for Foundation Mathematics

Resources for Success

Pearson MyLab

Get the Most Out of MyLab Math for *Mathematics in Action: Algebraic, Graphical, and Trigonometric Problem Solving*, Sixth Edition by the Consortium for Foundation Mathematics

The active learning at the heart of the Mathematics in Action series is complemented by a full suite of resources in the MyLab course created specifically for the unique Consortium approach. Brand new and expanded MyLab resources ensure more than ever that students have a consistent experience from the text to MyLab. The integration of Consortium's activity-based learning with the #1 choice in digital learning for developmental math ensures that students build a solid conceptual understanding of topics.

A new video program for support where and when students need it

A completely new video series offers students support in and out of the classroom. Videos are available as short, objective-level videos or longer section lecture videos. This program is created specifically for the Consortium series and its unique approach—so students can be assured that terminology and problem solving techniques in the videos are completely consistent with what they experience in the classroom.

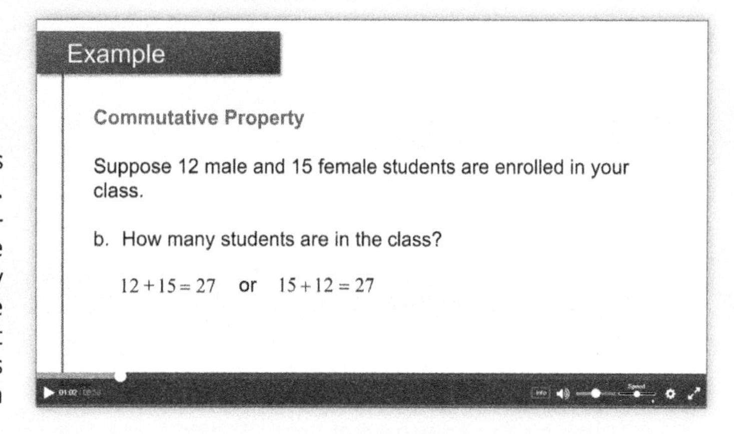

A new PowerPoint program

PowerPoints are available for the first time with this title! Presenting an overview of key concepts from each activity, these can be used by students as study or review aids, and can also be used by instructors to help structure class-time. Accessible, screen-reader friendly versions of the PowerPoints are available in the MyLab course.

Get Students Prepared

New! Integrated Review content ensures students are caught up on prior skills, or can be used for a corequisite course.

Integrated Review provides embedded and personalized review of prerequisite topics within relevant chapters. Students can check their prerequisite skills, and receive personalized practice on the topics they need to focus on, with study aids like worksheets and videos also available to help.

Integrated Review assignments are premade and available to assign in the Assignment Manager.

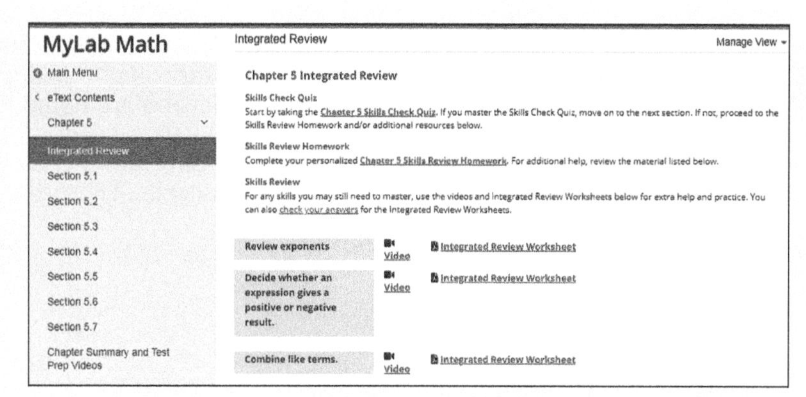

pearson.com/mylab/math

Support College Success

A new **Mindset module** is available in the course, with mindset-focused videos and exercises that encourage students to maintain a positive attitude about learning, value their own ability to grow, and view mistakes as a learning opportunity.

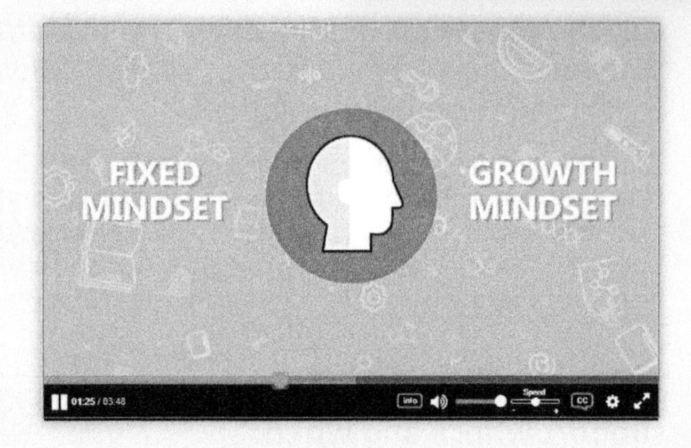

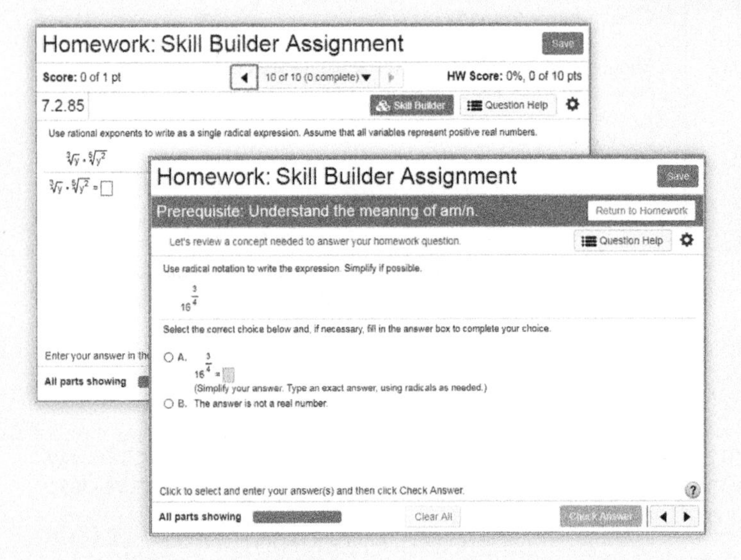

Personalize Learning

New! Skill Builder exercises offer just-in-time additional adaptive practice. The adaptive engine tracks student performance and delivers questions to each individual that adapt to his or her level of understanding. This new feature allows instructors to assign fewer questions for homework, allowing students to complete as many or as few questions as they need.

Get Students Engaged

New! Learning Catalytics questions specific to Consortium's content are pre-built and available through MyLab Math. Learning Catalytics is an interactive student response tool that uses students' smartphones, tablets, or laptops to engage them in more sophisticated tasks and thinking. Consortium-specific questions have been pre-made, and are noted in the Annotated Instructor's Edition at point-of-use when relevant for a particular section's objective. Search for MathInActionGREEN#, where # is the chapter number, in Learning Catalytics to begin using the Consortium questions with your students!

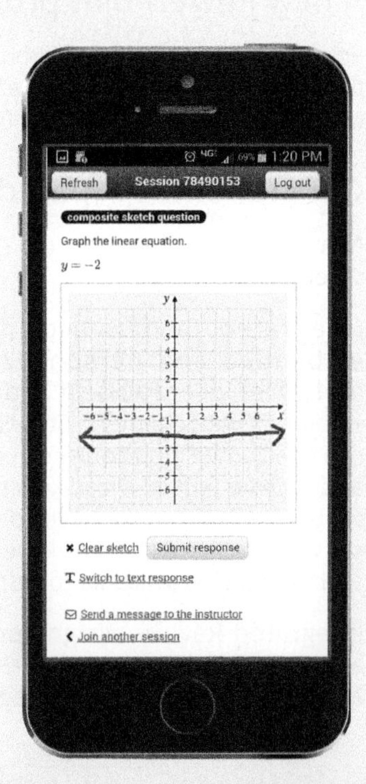

pearson.com/mylab/math

Resources for Success

Whether you are using Mathematics in Action for the first time or the tenth time, we know that having a full suite of resources to support teaching and learning is essential to implementing this unique approach. All resources are built specifically for each Consortium title, giving students and instructors resources that match and complement the main text and MyLab.

Instructor Resources

Annotated Instructor's Edition

Contains all the content found in the student edition, plus answers to all exercises directly beneath each problem and Learning Catalytics instructor annotations.

The following instructor resources are available to download through Pearson's Instructor Resource Center, or from MyLab Math.

Instructor's Resource Manual with Tests

This resource includes:

- Sample syllabi suggesting ways to structure the course around core and supplemental activities.
- Sample course outlines with timelines for covering topics.
- Teaching notes for each chapter—ideal for using the text for the first time!
- Extra skills practice worksheets for difficult topics.
- Sample chapter tests and final exams.
- Information about incorporating technology in the classroom, such as graphing calculators.

TestGen®

Enables instructors to build, edit, print, and administer tests using a computerized bank of questions developed to cover all the Objectives of the text. TestGen is algorithmically based, allowing instructors to create multiple but equivalent versions of the same question or test with the click of a button. Instructors can also modify test bank questions or add new questions.

Instructor Training Videos

From author Ernie Danforth, the videos provide instructors with advice ranging from the Consortium teaching philosophy to tips for implementing group-work.

New! PowerPoint Lecture Slides

These slides present key concepts and definitions from the text. These have been created to support instructors looking to implement this contextual approach in the classroom, and can also be used as a student study aid.

Student Resources

Worksheets for Classroom or Lab Practice with Integrated Review

Provide extra practice to ensure that students have many opportunities to work problems related to the concepts learned in every activity. Concept Connections, a feature unique to these worksheets, offer students an opportunity to show in words that they understand the mathematical concepts they have just practiced.

TO THE STUDENT

The book in your hands is most likely very different from any mathematics textbook you have seen before. In this book, you will take an active role in developing the important ideas of arithmetic and beginning algebra. You will be expected to add your own words to the text. This will be part of your daily work, both in and out of class. It is the belief of the authors that students learn mathematics best when they are actively involved in solving problems that are meaningful to them.

The text is primarily a collection of situations drawn from real life. Each situation leads to one or more problems. By answering a series of questions and solving each part of the problem, you will be led to use one or more ideas of introductory college mathematics. Sometimes, these will be basic skills that build on your knowledge of arithmetic. Other times, they will be new concepts that are more general and far reaching. The important point is that you won't be asked to master a skill until you see a real need for that skill as part of solving a realistic application.

Another important aspect of this text and the course you are taking is the benefit gained by collaborating with your classmates. Much of your work in class will result from being a member of a team. Working in small groups, you will help each other work through a problem situation. While you may feel uncomfortable working this way at first, there are several reasons we believe it is appropriate in this course. First, it is part of the learning-by-doing philosophy. You will be talking about mathematics, needing to express your thoughts in words. This is a key to learning. Secondly, you will be developing skills that will be very valuable when you leave the classroom. Currently, many jobs and careers require the ability to collaborate within a team environment. Your instructor will provide you with more specific information about this collaboration.

One more fundamental part of this course is that you will have access to appropriate technology at all times. You will have access to calculators and some form of graphics tool—either a calculator or computer. Technology is a part of our modern world, and learning to use technology goes hand in hand with learning mathematics. Your work in this course will help prepare you for whatever you pursue in your working life.

This course will help you develop both the mathematical and general skills necessary in today's workplace, such as organization, problem solving, communication, and collaborative skills. By keeping up with your work and following the suggested organization of the text, you will gain a valuable resource that will serve you well in the future. With hard work and dedication you will be ready for the next step.

The Consortium for Foundation Mathematics

CHAPTER 1

Function Sense

CLUSTER 1

Modeling with Functions

ACTIVITY 1.1

Parking Problems

Functions; Function Notation and Terminology

OBJECTIVES

1. Identify the input and output in situations involving two variable quantities.
2. Identify a functional relationship between two variables.
3. Identify the independent and dependent variables.
4. Use a table to numerically represent a functional relationship between two variables.
5. Write a function using function notation.
6. Identify the input and output values in an ordered pair.

Introduction

A key step in the problem-solving process is to look for relationships and connections between the variable quantities in a given situation. Problems encountered in the world around us, including the environment, medicine, economics, and the Internet, are often very complicated and contain several variables. In this text, you will deal primarily with situations that contain two variables. In many of these situations, the variables will have a special relationship called a **function**.

Function

Did you have trouble finding a parking space this morning? Was the time that you arrived on campus a factor? As part of a reconstruction project at a small community college, the number of cars in the parking lot was counted each hour from 7 A.M. to 10 P.M. on a particular day. The results are shown in the following table.

TIME OF DAY	NUMBER OF CARS
7 A.M.	24
8 A.M.	212
9 A.M.	384
10 A.M.	426
11 A.M.	538
12 P.M.	497
1 P.M.	384
2 P.M.	337
3 P.M.	285
4 P.M.	278
5 P.M.	302
6 P.M.	427
7 P.M.	384
8 P.M.	315
9 P.M.	187
10 P.M.	56

This situation involves two variables, the time of day and the number of cars in the parking lot. A **variable**, usually represented by a letter, is a quantity that may change in value from one instance to another. Typically, one variable is designated as the **input** and the other is called the **output**. An output value corresponds to, or is determined by, a given input value.

For the data in the preceding table, let the time of day be the input variable and the number of cars be the output variable.

1. a. For an input of 10 A.M., how many cars are in the parking lot (output)?

426

b. For an input of 5 P.M., how many cars are in the parking lot (output)?

302

c. For each value of input (time of day), how many different outputs (number of cars) are there?

1

The set of data in the table is an example of a mathematical function.

DEFINITION

A **function** is a correspondence between an input variable and an output variable that assigns a single output value to each input value. Therefore, for a function, any given input value has exactly one corresponding output value. If x represents the input variable and y represents the output variable, then the function assigns a single, unique y-value to each x-value.

2. Use the results from Problem 1 to explain how the data in the table fits the description of a function.

At each time of day (input), there is only one number of cars (output).

A functional relationship is stated as follows: "The output variable is a function of the input variable." Using x for the input variable and y for the output variable, the functional relationship is stated "y is a function of x." Because the input for the parking lot function is the time of day and the output is the number of cars in the lot at that time, you write that the number of cars in the parking lot is a function of the time of day. Note that the input follows the word "of."

EXAMPLE 1

Consider the following table listing the official high temperature (in degrees Fahrenheit) in the village of Lake Placid, New York, during the first week of January. Note that the date has been designated the input and the high temperature on that date the output. Is the high temperature a function of the date?

Date (Input)	1	2	3	4	5	6	7
Temperature (Output)	25	30	32	24	23	27	30

SOLUTION

From this table, you observe that the high temperature is a function of the date. For each date, there is exactly one high temperature. The relationships in this example can be visualized as follows:

DATE (INPUT)	TEMPERATURE (OUTPUT)
1 →	25
2 →	30
3 →	32
4 →	24
5 →	23
6 →	27
7 →	30

If d represents the input (date) and T represents the output (temperature), then T is a function of d.

If the input and output in Example 1 are switched (see table below), the daily high temperature becomes the input and the date becomes the output. The date is not a function of the high temperature. The input value 30 has two output values, 2 and 7.

TEMPERATURE (INPUT)	DATE (OUTPUT)
25 →	1
30 →	2, 7
32 →	3
24 →	4
23 →	5
27 →	6

3. Interchange the input and the output in the parking lot situation. Let the number of cars in the lot be the input and the time of day be the output. Is the time of day a function of the number of cars in the lot? Write a sentence explaining why this switch does or does not fit the description of a function.

No, it is not a function because the input 384 has three corresponding output values: 7 P.M., 1 P.M., and 9 A.M.

EXAMPLE 2 ***Determine whether the following situation describes a function. Give a reason for your answer.***

The amount of postage for a letter is a function of the weight of the letter.

SOLUTION

Yes, this statement does describe a function. The weight of the letter is the input, and the amount of postage is the output. Each letter has one weight. This weight determines the postage necessary for the letter. There is only one amount of postage for each letter. Therefore, for each value of input (weight of the letter), there is one output (postage). Note that if w represents the input (weight of the letter) and p represents the output (postage), then p is a function of w.

4. Determine whether each situation describes a function. Give a reason for your answer.

a. For a part-time student, the amount of tuition you have to pay is a function of the number of credit hours taken.

This is a function. For a given number of credit hours (input), there is exactly one corresponding amount of tuition (output).

b. The amount of calories you burn while spinning at your local gym is a function of the time you spin.

This is not a function. For any given time you spin (input), different amounts of calories can be burned (output) depending, for example, on how fast you spin or how humid it is.

c. The amount of current interest owed on a college loan is a function of the amount borrowed.

This is a function. For any given amount borrowed (input), there is exactly one corresponding amount of interest (output).

d. The value of a used pickup truck is a function of its age.

This is not a function because the value of a 5-year-old (input) pickup truck does not have a unique value (output).

DEFINITION

If the relationship between two variables is a function, the input variable is called the **independent** variable and the output variable is called the **dependent** variable. If x represents the input variable and y represents the output variable, then x is the independent variable and y is the dependent variable.

5. The independent variable in Example 1 is the date. The dependent variable is the temperature. Identify the independent and dependent variables in Example 2.

The weight of the letter is the independent variable. The amount of postage is the dependent variable.

Defining Functions Numerically

The input–output pairing in the parking lot function on page 1 is presented as a **table of matched pairs**. In such a situation, the function is defined **numerically**. Another way to define a function numerically is as a set of ordered pairs.

DEFINITION

An **ordered pair** of numbers consists of two numbers written in the form

(input value, output value).

The order in which they are listed is significant.

EXAMPLE 3

The ordered pair (3, 4) is distinct from the ordered pair (4, 3). In the ordered pair (3, 4), 3 is the input and 4 is the output. In the ordered pair (4, 3), 4 is the input and 3 is the output.

DEFINITION

A function may be defined **numerically** as a set of ordered pairs in which the first number of each pair represents the input value and the second number represents the corresponding output value. To be a function, no two ordered pairs have the same input with a different output value.

EXAMPLE 4 ***(9 A.M., 384)** or **(0900, 384)** (using a 24-hour clock) is an ordered pair that is part of the parking lot function.*

6. Using a 24-hour clock, write three other ordered pairs for the parking lot function.

   (1000, 426), (1300, 384), (1800, 427) Other pairs are acceptable.

7. In Example 1, the high temperature in Lake Placid is a function of the date.

   Convert to ordered pairs all the values in the table on page 2.

   (1, 25), (2, 30), (3, 32), (4, 24), (5, 23), (6, 27), (7, 30)

Function Notation

There is a special notation for functions in which the function itself is represented by a name or letter. For example, the function that relates the time of day to the number of cars can be represented by the letter f. Let t represent time, the input variable, and let c represent the number of cars, the output variable. The following simplification (really an abbreviation) is now possible.

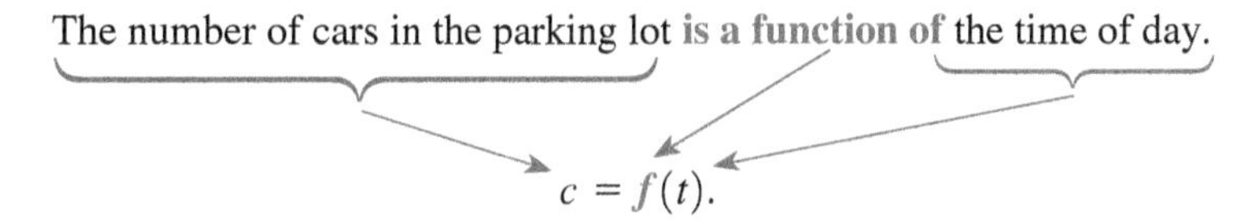

The final function notation is read "c equals f of t."

Note: $f(t)$ does not mean "f times t."

Notice that the output c (the number of cars) is equal to $f(t)$. So $f(t)$ is the output of f when the input is t. For example, $f(1400)$ represents the number of cars (the output of f) when the input is 1400 (at 2 P.M.).

In general, function notation is written as follows:

$$\textit{output variable} = \text{name of function } (\textit{input variable})$$

The input variable or input value is also called the **argument** of the function.

EXAMPLE 5 ***Values from the table or ordered pairs for the parking lot function can be written as follows using function notation.***

$$212 = f(800) \qquad 302 = f(1700) \qquad f(2100) = 187$$

8. a. Rewrite the three examples in Example 5 as three ordered pairs. Pay attention to which is the input value and which is the output value.

   (800, 212) (1700, 302) (2100, 187)

b. Write a sentence explaining the meaning of $f(1600) = 278$ in the parking lot situation.

There are 278 cars in the lot at 1600 hours (4 P.M.).

LC LEARNING CATALYTICS

Suppose the function, f, is defined by the following table;

x	1	2	3	4	5	5
$f(x)$	7	6	5	3	0	−4

determine $f(3)$.

9. a. Referring to the table in Example 1, determine $g(3)$, where g is the name of the temperature function.

$g(3) = 32$

b. Write a sentence explaining the meaning of $g(5) = 23$.

The high temperature in Lake Placid on January 5 was 23 degrees.

10. If you work for an hourly wage, your gross pay is a function of the number of hours you work.

a. Identify the input and output.

The input is the number of hours you work, and the output is your gross pay.

b. If you earn $10.50 per hour, complete the following table.

Number of Hours	0	3	5	7	10	12
Gross Pay	0	31.50	52.50	73.50	105	126

c. Let n represent the number of hours worked and $f(n)$ represent the gross pay. Use the table to determine $f(5)$.

$f(5) = 52.50$

d. Write a sentence explaining the meaning of $f(10) = 105$.

The gross pay for 10 hours worked is $105.

11. a. The value of your car, in dollars, is a function, f, of the number of miles driven. Explain the meaning of $f(52{,}000)$.

$f(52{,}000)$ is the value of the car after you've driven it 52,000 miles.

b. At a particular coffeehouse, the cost, in dollars, of a cup of coffee is a function, g, of the size of the cup, in ounces. Explain the meaning of $g(12)$.

$g(12)$ is the cost of a 12-ounce cup of coffee.

SUMMARY Activity 1.1

1. A **variable**, usually represented by a letter, is a quantity that may change in value from one instance to another.
2. In a situation involving two variables, one variable is called the **input** and the other the **output.** The output is the value that corresponds to or is determined by the given input value.
3. A **function** is a rule relating an input variable (sometimes called the argument) and an output variable so that a single output value is assigned to each input value. In such a case, you state that the output variable is a function of the input variable.
4. **Independent variable** is another name for the input variable of a function.
5. **Dependent variable** is another name for the output variable of a function.

6. An **ordered pair** of numbers consists of two numbers written in the form

$$(input\ value,\ output\ value).$$

The order in which they are listed is significant.

7. Functions may be defined **numerically** using ordered pairs of numbers. These can be displayed as a table of values or points on a graph. For each input value, there is one and only one corresponding output value.

8. The function relationship is often defined using function notation:

$$\text{output variable} = \text{name of function(input variable)}$$

If y represents the output variable, f is the name of the function, and x represents the input variable, then

$y = f(x)$ is read "y equals f of x."

EXERCISES Activity 1.1

1. The weights and heights of six mathematics students are given in the following table:

WEIGHT (lb.)	HEIGHT (in.)
165	67
123	61
212	71
175	69
165	64
147	65

a. In the statement "Height is a function of weight," which variable is the input and which is the output?

Weight is the input, and height is the output.

b. Is height a function of weight for the six students? Explain using the definition of function.

No, height is not a function of weight because two students weigh 165 pounds (input) but have different heights (output). So there is not a single value of output for the input of 165.

c. In the statement "Weight is a function of height," which variable is the input and which is the output?

Height is the input, and weight is the output.

d. Is weight a function of height for the six students? Explain using the definition of function.

Yes. For each value of input (height), there is one output (weight).

e. For all students, is weight a function of height? Explain.

No, many students with the same height (input) have different weights (output).

2. a. At a particular gas station, the cost of a gallon of gas, in dollars, is a function of its octane. Explain the meaning of $f(89)$.

$f(89)$ is the cost, in dollars, of one gallon of 89-octane gasoline.

b. When you are driving to school, the distance traveled, in miles, is a function of time, in minutes. Explain the meaning of $s(28)$.

$s(28)$ is the number of miles traveled in 28 minutes.

For Exercises 3–7, determine whether each of the situations describes a function. Give a reason for your answer.

3. a. The letter grade in this course is a function of your numerical grade.

Yes. For each numerical grade, there will correspond only one letter grade.

b. The numerical grade in this course is a function of the letter grade.

No. For each letter grade, there may correspond several different numerical grades. For example, an A may be based on 92% or 94%.

c. The score on the next math exam is a function of the number of hours studied for the exam.

This is not a function. For each number of hours studied for the exam, there may correspond several different scores. For example, a study time of 3 hours could result in a score of 71, 82, or 94.

d. The number of tweets the President posts every day on Twitter is a function of the date.

Yes, this is a function. For each date, there is just one corresponding number of tweets.

e. The sales tax on a purchased item is a function of the final selling price.

This situation represents a function. For any given final selling price, there is just one corresponding amount of tax.

4. a. The input is any number, and the output is the square of the number.

Yes, each input has one output.

b. The square of a number is the input, and the output is the number.

No, 4 can have 2 and -2 as outputs. Other answers are possible.

5. a. In the following table, elevation is the input and amount of snowfall is the output.

ELEVATION (ft.)	SNOWFALL (in.)
2000	4
3000	6
4000	9
5000	12

Yes. In this table, each elevation is paired with only one amount of snowfall.

b. In the preceding table, snowfall is the input and elevation is the output.

Yes. In this table, each quantity of snow is paired with one elevation.

6. Number of hours using the Internet is the input, and the monthly cost for the Internet service is \$39.95.

NUMBER OF HOURS	MONTHLY COST
10	\$39.95
50	\$39.95
75	\$39.95
100	\$39.95

Yes. For each input, there is one output.

7. a. $(2, 5), (-3, 5), (10, 5), (\pi, 5)$

Yes, each input value has only one output value.

b. $(5, 2), (5, -3), (5, 10), (5, \pi)$

No, the one input 5 is paired with four different outputs.

8. In Exercise 5, the amount of snowfall is a function of the elevation.

a. Let x represent the elevation in feet and $a(x)$ represent the amount of snowfall in inches. Determine $a(4000)$.

$a(4000) = 9$

b. Write $a(5000) = 12$ as an ordered pair.

(5000, 12)

c. Write a sentence explaining the meaning of $a(5000) = 12$.

There were 12 inches of snowfall at an elevation of 5000 feet.

9. Identify the input, the output, and the name of the function. For each of the functions, write in words the equation as you would say it.

a. $y = g(x)$

The input is x.

The output is $g(x)$ or y.

The function name is g.

y equals g of x.

b. $h(a) = b$

The input is a.

The output is b or $h(a)$.

The name is h.

h of a equals b.

c. $f(6) = 3.527$

The input is 6.

The output is 3.527.

The name is f.

f of 6 equals 3.527.

d. $520 = g(t)$

The input is t.

The output is 520.

The name is g.

520 equals g of t.

e. sales tax $= T(\text{price})$

The input is price.

The output is sales tax.

The name is T.

Sales tax is a function of price, or sales tax equals T of price.

10. Your college community service organization has volunteered to help with Spring Cleanup Day at a youth summer camp. You have been assigned the job of supplying paint for the exterior of the bunkhouses. You discover that 1 gallon of paint will cover 400 square feet of flat surface.

 a. If n represents the number of gallons of paint you supply and s represents the number of square feet you can cover with the paint, complete the following table.

n, Number of Gallons of Paint	1	2	4	6
s, Square Feet Covered by the Paint	400	800	1600	2400

 b. Let s be represented by $f(n)$, where f is the name of the function. Determine $f(6)$.

 $f(6) = 2400$

 c. Write a sentence explaining the meaning of $f(4) = 1600$.

 Four gallons of paint will cover 1600 square feet.

11. a. Give an example of a function that you may encounter in your daily life or that describes something about the world around you.

 i. Identify the input and the output variables.

 ii. Write the function in the form "output is a function of the input."

 iii. Explain how the example fits the definition of a function.

 (Answers will vary.) The letter grade on a true/false quiz is a function of the number of questions answered correctly.

 i. The input is the number of questions answered correctly. The output is the grade received.

 ii. The letter grade is a function of the number of questions answered correctly.

 iii. There will be one grade for each number of questions answered correctly.

 b. Switch the input and the output of the function you determined in part a.

 i. Identify the input and the output.

 ii. Explain how the example does or does not fit the definition of a function.

 (Answers will vary.)

 i. The number of questions answered correctly is the output. The grade received is the input.

 ii. This is probably not a function because the same letter grade could have resulted from different numbers of questions answered correctly.

 c. Write the function you listed in part a in function notation. Represent the input variable, the output variable, and the function itself by letters.

 Let n represent the number of questions answered correctly and g represent the grade. Then $g = f(n)$, where f is the name of the function.

ACTIVITY
1.2

Fill'er Up
Defining Functions by a Symbolic Rule (Equation)

OBJECTIVES

1. Determine the equation (symbolic representation) that defines a function.
2. Determine the domain and range of a function.

You probably need to fill your car with gas more often than you would like. You commute to college each day and to a part-time job each weekend. Your car gets good gas mileage, but the recent dramatic fluctuation in gas prices has wreaked havoc on your budget.

1. Two input variables determine the cost (output) of a fill-up. What are they? Be specific.

The number of gallons pumped and the price per gallon of the gas are the two variables.

2. Assume that you need 12.6 gallons to fill up your car. Now one of the input variables in Problem 1 will become a constant. The value of a constant will not vary throughout the problem. The cost of a fill-up is now dependent on only one variable, the price per gallon.

a. Complete the following table.

Price per Gallon	\$2.00	\$2.50	\$3.00	\$3.50	\$4.00
Cost of Fill-Up	\$25.20	\$31.50	\$37.80	\$44.10	\$50.40

b. Is the cost of a fill-up a function of the price per gallon? Explain.

Yes, for each value of input (price per gallon), there is one value of output (cost).

3. a. Write a verbal statement that describes how the cost of a fill-up is determined.

The cost is the price per gallon multiplied by 12.6.

b. Let p represent the price of a gallon of gasoline pumped (input) and c represent the cost of the fill-up (output). Translate the verbal statement in part a into a symbolic statement (an equation) that expresses c in terms of p.

$c = 12.6p$

Defining Functions by a Symbolic Rule (Equation)

The symbolic rule (equation) $c = 12.6p$ is an example of a second method of defining a function. Recall that the first method is numerical (tables and ordered pairs).

4. a. Use the given equation to determine the cost of a fill-up at a price of \$2.60 per gallon.

$c = 12.6(\$2.60) = \$32.76$

b. Explain the steps that you used to determine the cost in part a.

I multiplied 12.6 by \$2.60 to determine the total cost of \$32.76.

Recall from Activity 1.1 that function notation is an efficient and convenient way of representing the output variable. The equation $c = 12.6p$ may be written using the function notation by replacing c with $f(p)$ as follows:

$$f(p) = 12.6p$$

Now, if the price per gallon is \$2.60, then the cost of a fill-up can be represented by $f(2.60)$. To evaluate $f(2.60)$, substitute 2.60 for p in $f(p) = 12p$ as follows:

$$f(2.60) = 12.6(2.60) = 32.76$$

The results can be written as $f(2.60) = 32.76$ or as the ordered pair (2.60, 32.76). Therefore, at a price of \$2.60 per gallon, the cost of filling your car with 12.6 gallons of gas is \$32.76.

5. a. Using function notation, write the cost if the price is \$3.85 per gallon and evaluate. Write the result as an ordered pair.

$f(3.85) = 12.6(3.85) = 48.51$; (3.85, 48.51)

LC LEARNING CATALYTICS

Suppose the function, g, is defined by $g(x) = 3x - 4$. Determine $g(2)$.

b. Use the equation for the cost-of-fill-up function to evaluate $f(2.95)$, and write a sentence describing its meaning. Write the result as an ordered pair.

$f(2.95) = 12.6(2.95) = \$37.17$

If the price of a gallon of gas is \$2.95, then the cost of a fill-up is \$37.17.

Real Numbers

The numbers that you will be using as input and output values in this text will be **real numbers**. A real number is any rational or irrational number.

A **rational number** is any number that can be expressed as the quotient of two integers (negative and positive counting numbers as well as zero) such that the division is not by zero.

EXAMPLE 1 ***Rational numbers include the following:***

$$\frac{3}{4}; -\frac{7}{8}; 2\frac{1}{3} = \frac{7}{3}; 5 = \frac{5}{1}; 0 = \frac{0}{1}; -3\frac{1}{4} = -3.25; \frac{2}{3} = 0.666\ldots = 0.\overline{6}$$

An **irrational number** is defined as one that cannot be written exactly as a quotient of two integers. The decimal representation of an irrational number never ends and never repeats. We use symbols to represent the exact values of irrational numbers.

EXAMPLE 2 ***Irrational numbers include $\sqrt{2}, -\sqrt{7}, \sqrt[3]{5}, \pi$.***

All of the numbers in Examples 1 and 2 are real numbers. A real number can be represented as a point on the number line.

$-3\frac{1}{4}$ $-\sqrt{7}$ $-\frac{7}{8}$ 0 $\frac{2}{3}$ $\sqrt{2}$ π

-4 -3 -2 -1 0 1 2 3 4

Domain and Range

You are training to run your first marathon. You have been training hard and believe you can run the entire 26 miles at an average speed of 6.5 mi./hr. Once the race starts, your distance from the finish line d can be approximated by $d = 26 - 6.5t$, where t is the time measured in hours.

6. Can any number be substituted for the input variable t in the marathon function? Describe values of t that make sense and explain why they do.

The only values of t that make sense are values between 0 and 4. Negative times don't make sense because they are before the race starts, and values greater than 4 do not make sense because you finish the marathon in 4 hours.

DEFINITION

Given a function, the collection of all possible values of the input or independent variable is called the **domain** of the function. The **practical domain** is the collection of replacement values of the input variable that makes practical sense in the context of the situation.

7. a. Determine the practical domain of the cost-of-fill-up function. Refer to Problem 2.

(Answers will vary.) The practical domain is the collection of rational numbers rounded to two decimal places from $2 to $4.

b. Determine the domain for the general function defined by $c = f(p) = 12.6p$, with no connection to the context of the situation.

The domain is the set of real numbers because the expression $12.6p$ is defined for any real number replacement for the variable p.

DEFINITION

Given a function, the collection of all possible values of the output or dependent variable is the **range** of the function. The **practical range** corresponds to the practical domain.

8. a. What is the practical range for the cost function defined by $f(p) = 12.6p$ if the practical domain is 2 to 4?

The practical range of this function is all rational numbers $f(p)$ from $25.20 to $50.40.

$25.20 is the cost at $2 per gallon; $50.40 is the fill-up cost at $4 per gallon.

b. What is the range of this function if it has no connection to the context of the situation?

The range of the general function is all real numbers.

EXAMPLE 3 *Consider the following table that gives the percentage of mothers in the workforce from 2011 to 2016 with children under the age of 6:*

YEAR	PERCENTAGE
2011	64.2
2012	64.7
2013	64.7
2014	64.3
2015	62.6
2016	63.3

Source: U.S. Bureau of Labor Statistics.

The six pairs of values given in the table represent a function. The input or independent variable is the year, and the output or dependent variable is the percentage. The domain of the function is {2011, 2012, 2013, 2014, 2015, 2016} because this is the set of all the input values. The range of the function is {62.6, 64.2, 63.3, 64.3, 64.7} because this is the set of all the output values. Note that although 64.7 occurs twice in the table as an output value, it is listed only once in the range.

Constructing Tables of Input–Output Values

9. Use the symbolic form of the gas cost-of-fill-up function, $f(p) = 12.6p$, to evaluate $f(2)$, $f(2.50)$, $f(3)$, $f(3.50)$, and $f(4)$, and complete the following table. Note that the input variable p increases by 0.50 unit. In such a case, you say that the input increases by an **increment** of 0.50 unit.

Price per Gallon, p	$2.00	$2.50	$3.00	$3.50	$4.00
Cost of Fill-Up, $f(p)$	$25.20	$31.50	$37.80	$44.10	$50.40

A numerical form of the cost-of-fill-up function is a table or a collection of ordered pairs. When a function is defined in symbolic form, you can use technology to generate the table. The TI-84 Plus C calculator is a function grapher. The y variables Y_1, Y_2, etc., represent function output (dependent) variables. The input, or independent variable, is x. The steps to build tables with the TI-84 Plus C can be found in Appendix C.

10. Use your graphing calculator to generate a table of values for the function represented by $f(p) = 12.6p$ to check your values in the table in Problem 9. The screens on your graphing calculator should appear as follows.

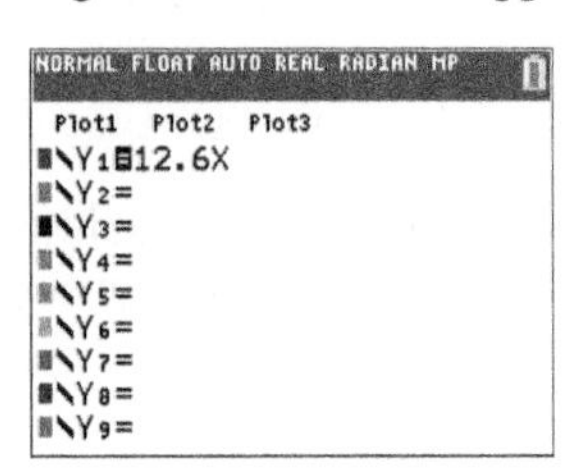

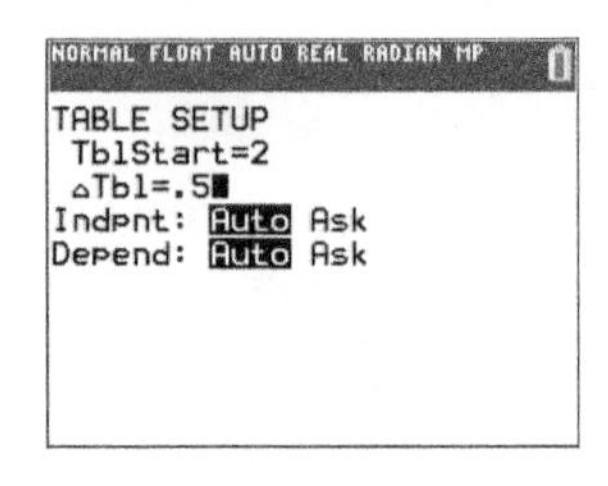

NORMAL FLOAT AUTO REAL RADIAN MP
PRESS + FOR ΔTbl

X	Y_1
2	25.2
2.5	31.5
3	37.8
3.5	44.1
4	50.4
4.5	56.7
5	63
5.5	69.3
6	75.6
6.5	81.9
7	88.2

X=2

Recall from Activity 1.1 that if you work for an hourly wage, your gross pay is a function of the number of hours you work.

11. a. You are working a part-time job. You work between 0 and 25 hours per week. If you earn $10.50 per hour, write an equation to determine the gross pay, g, for working h hours.

$g = 10.50h$

b. What is the independent variable? What is the dependent variable?

The number of hours worked is the independent variable. The gross pay is the dependent variable.

c. Complete the following table using your graphing calculator:

Number of Hours, h	0	5	10	15	20	25
Gross Pay, g ($)	0	52.50	105	157.50	210	262.50

d. Verify your results in part c by using the table feature of your graphing calculator.

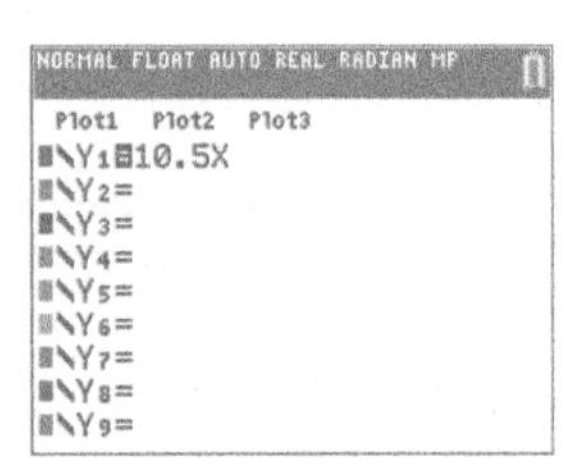

NORMAL FLOAT AUTO REAL RADIAN MP
PRESS + FOR ΔTbl

X	Y_1
0	0
5	52.5
10	105
15	157.5
20	210
25	262.5
30	315
35	367.5
40	420
45	472.5
50	525

X=0

e. Using f for the name of the function, the output variable g can be written as $g = f(h)$.

Rewrite the equation in part a using the function notation $f(h)$ for gross pay.

$f(h) = 10.50h$

f. What are the practical domain and the practical range of the function? Explain.

The practical domain is 0 to 25. The practical range is 0 to 262.5.

g. Evaluate $f(14)$ and write a sentence describing its meaning.

$f(14) = \$147$ Your gross pay is $147 for working 14 hours.

SUMMARY Activity 1.2

1. Given a function, the collection of all possible replacement values for the independent or input variable is called the **domain of the function**. The **practical domain** is the collection of replacement values of the input variable that makes practical sense in the context of the situation.

2. Given a function, the collection of all possible replacement values for the dependent or output variable is called the **range of the function**. When a function describes a real situation or phenomenon, its range is often called the **practical range** of the function.

3. When a function is represented by an equation, the function may also be written in function notation. For example, given $y = 2x + 3$, you can replace y with $f(x)$ and rewrite the equation as $f(x) = 2x + 3$.

EXERCISES Activity 1.2

In Exercises 1 and 2,

a. *identify the independent and dependent variables.*

b. *let x represent the input variable. Use function notation to write the function in symbolic form.*

1. Sales tax is a function of the price of an item. The amount of sales tax is 0.08 times the price of the item. Use h to represent the function.

   a. independent (input) price of an item
   dependent (output) sales tax

   b. $h(x) = 0.08x$

2. The Fahrenheit measure of temperature is a function of the Celsius measure. The Fahrenheit measure is 32 more than 9/5 times the Celsius measure. Use g to represent the function.

   a. independent (input) Celsius temperature
   dependent (output) Fahrenheit temperature

   b. $g(x) = \frac{9}{5}x + 32$

For each function in Exercises 3–5, evaluate $f(2), f(-3.2)$, *and* $f(a)$.

3. $f(d) = 2d - 5$
   $f(2) = -1$
   $f(-3.2) = -11.4$
   $f(a) = 2a - 5$

4. $f(t) = -16t^2 + 7.8t + 12$
   $f(2) = -36.4$
   $f(-3.2) = -176.8$
   $f(a) = -16a^2 + 7.8a + 12$

5. $f(x) = 4$
   $f(2) = 4$
   $f(-3.2) = 4$
   $f(a) = 4$

In Exercises 6–8, construct a table of values of four ordered pairs for the given function. Check your results using the table feature of your graphing calculator.

6. $g(x) = x^2$. Start the inputs at 3, and use an increment of 2.

x	3	5	7	9
g(x)	9	25	49	81

Exercise numbers appearing in color are answered in the Selected Answers appendix.

7. $h(x) = \frac{1}{x}$. Start the inputs at 10, and use an increment of 10.

x	10	20	30	40
$h(x)$	0.1	0.05	0.033	0.025

8. $f(x) = 3.5x + 6$. Start the inputs at 0, and use an increment of 5.

x	0	5	10	15
$f(x)$	6	23.5	41	58.5

9. a. The distance you travel while hiking is a function of how fast you hike and how long you hike at this rate. You usually maintain a speed of 2 miles per hour while hiking. Write a verbal statement that describes how the distance you travel is determined.

   The distance traveled is 2 times the number of hours I hiked.

   b. Identify the input and output variables of this function.

   The input is hours. The output is distance.

   c. Write the verbal statement in part a using function notation. Let t represent the input variable, h represent the function, and $h(t)$ represent the output variable.

   $h(t) = 2t$

   d. Which variable is the dependent variable? Explain.

   $h(t)$ is the dependent variable because distance is the output.

   e. Use the equation from part c to determine the distance traveled in 4 hours.

   $h(4) = 2(4) = 8$ miles

   f. Evaluate $h(7)$ and write a sentence describing its meaning. Write the results as an ordered pair.

   $h(7) = 2(7) = 14$. If I hike for 7 hours, I expect to hike 14 miles. (7, 14)

   g. Determine the domain and range of the general function.

   The domain is all real numbers. The range is all real numbers.

   h. Determine the practical domain and the practical range for the function.

   The practical domain depends on the individual and in this situation is probably real numbers from 0 to about 8. The practical range would be real numbers from 0 to about 16.

   i. Use your calculator to generate a table of values beginning at zero with an increment of 0.5.

t	0	0.5	1	1.5	2	2.5	3
$h(t)$	0	1	2	3	4	5	6

10. Determine the domain and range of each function.

   a. $\{(-2, 4), (0, 3), (5, 8), (8, 11)\}$

   domain $\{-2, 0, 5, 8\}$; range $\{3, 4, 8, 11\}$

   b. $\{(-6, 5), (-2, 5), (0, 5), (3, 5)\}$

   domain $\{-6, -2, 0, 3\}$; range $\{5\}$

Note that the graph of the parking lot function consists of 16 distinct points that are not connected. The input variable (time of day) is defined only for the car counts in the parking lot for each hour from 7 A.M. to 10 P.M. The parking lot function is said to be **discrete** because it is defined only at isolated, distinct input values (practical domain). The function is not defined for input values between these particular values.

Caution: To use the graph for a relationship such as the parking lot situation to make predictions or to recognize patterns, it is convenient to connect the points with line segments. This creates a type of continuous graph. This changes the domain shown in the graph from "some values" to "all values." Therefore, you need to be cautious. Connecting data points may cause confusion when working with real-world situations.

3. a. In Example 1 on page 2 of Activity 1.1, the high temperature in Lake Placid is a function of the date. Plot each ordered pair as a point on an appropriately scaled and labeled set of coordinate axes.

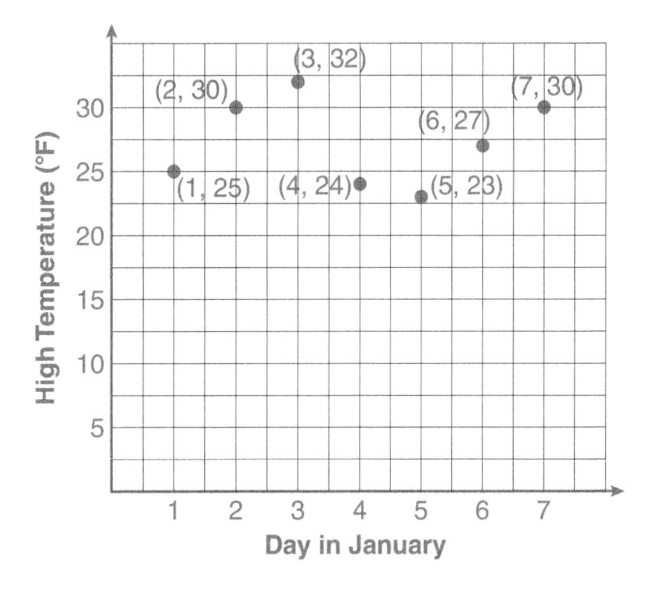

b. Determine the practical domain of the temperature function.

The practical domain is $\{1, 2, 3, 4, 5, 6, 7\}$.

c. Determine the practical range of the function.

$\{23, 24, 25, 27, 30, 32\}$.

d. Is this function discrete?

Yes. Because a particular day of the week is the input, the only input values that are defined are integers from 1 through 7. Although you can have a fraction of a day, this function is not defined at such values.

4. The marathon function in Activity 1.2 was defined by the equation $d = 26 - 6.5t$, where d represents the distance from the finish line and t represents the number of hours you have run.

a. What is the practical domain of this function? Refer to Problem 6 in Activity 1.2.

The practical domain is all numbers from 0 to 4.

b. List five ordered pairs of the marathon function in the following table.

Number of Hours Run, t	0	1	2	3	4
Distance from the Finish Line, d	26	19.5	13	6.5	0

(Answers will vary.)

c. Sketch a graph of the marathon function by first plotting the five points from part b on properly scaled and labeled coordinate axes.

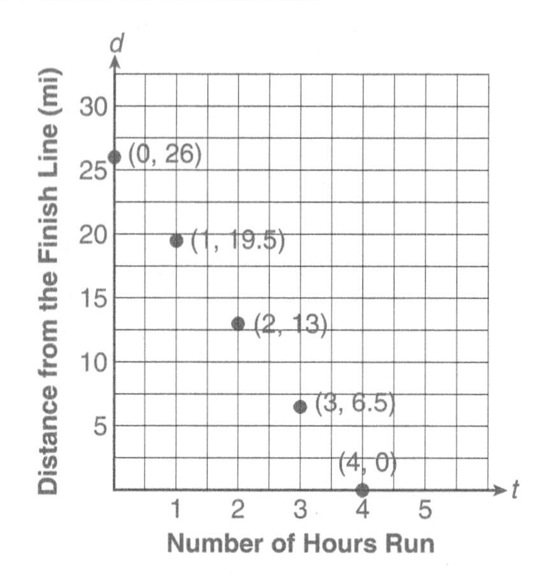

d. Does the graph of the function consist of just the five points from part b? Explain.

No, the hours run can take on any value from 0 to 4.

e. Describe any patterns or trends in the graph.

The graph falls to the right at a steady rate and appears to be a straight line.

The marathon function is defined for all input values in the practical domain in Problem 4a. The five points determined in part b can be connected to form a **continuous** graph. Such a graph is said to be **continuous** over its practical domain.

5. a. Consider the gross pay function defined by $g = 10.50h$ in Problem 11 of Activity 1.2. Plot the ordered pairs determined in Problem 11c, page 14.

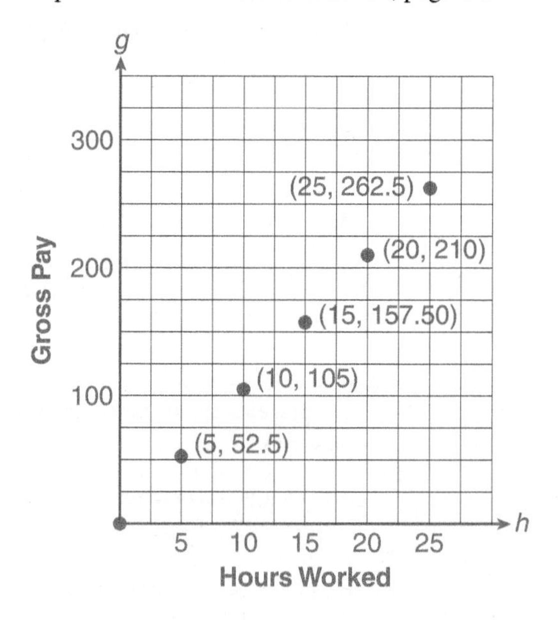

b. Is the gross pay function discrete or continuous? Explain.

Continuous. It is appropriate to connect the data points in the graph in part a because the domain is all real numbers from 0 to 25.

Graphing Functions Using Technology

Following is the graph of the cost-of-fill-up function defined by $c = 12.6p$ over its practical domain.

Graph 1

LC LEARNING CATALYTICS

You plot a graph from data collected at isolated distinct input values, will the graph be discrete or continuous?

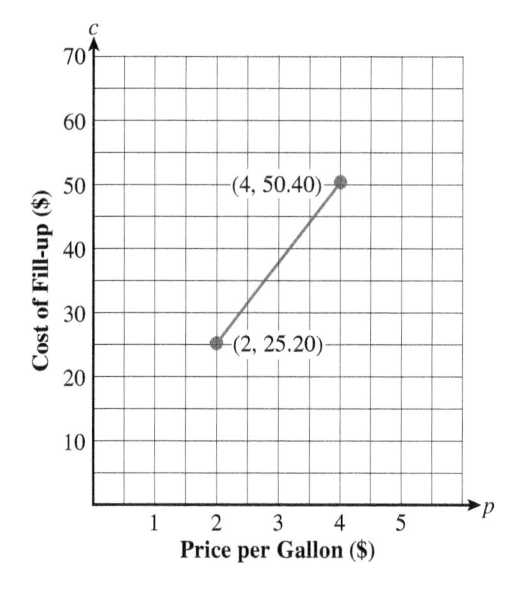

The domain for the general function defined by $c = 12.6p$, with no connection to the context of the situation, is the set of all real numbers because any real number can be substituted for p in $12.6p$. Following is a graph of $c = 12.6p$ for any real number p.

Graph 2

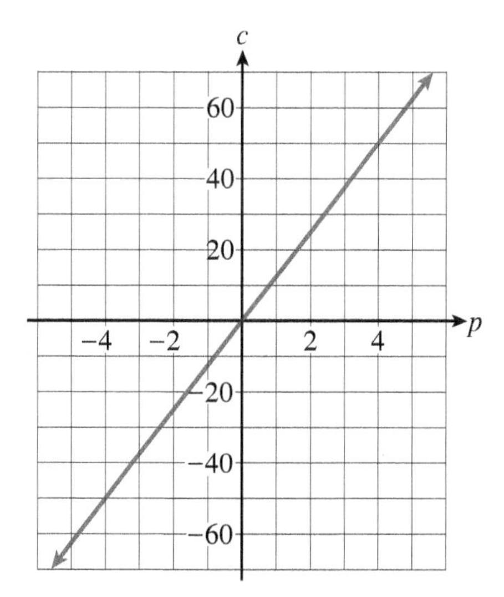

Each of these graphs can be obtained using your graphing calculator as demonstrated in Problems 6 and 7.

Recall that the independent variable in your graphing calculator is represented by x and the dependent variable is represented by y. Therefore, the cost-of-fill-up equation $c = 12.6p$ needs to be keyed in the "Y=" menu as $y = 12.6x$. The procedure for graphing a function using the TI-84 Plus C calculator can be found in Appendix C.

6. a. The viewing window is the portion of the rectangular coordinate system that is displayed when you graph a function. Use the practical domain and practical range of the cost-of-fill-up function to determine X_{min}, X_{max}, Y_{min}, and Y_{max} in the window screen. Key these values into your calculator. Your screens should appear as follows:

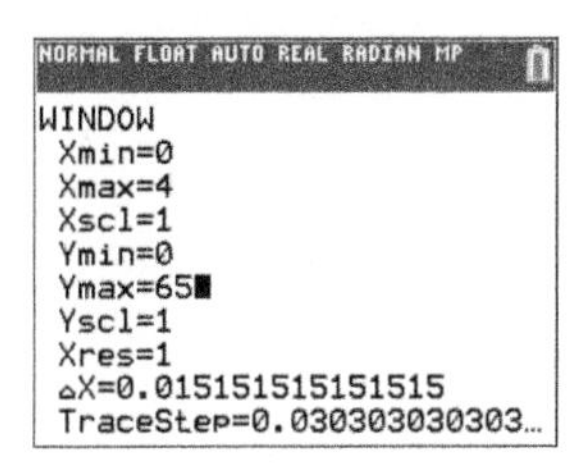

b. Graph the function. Your screen should appear as follows:

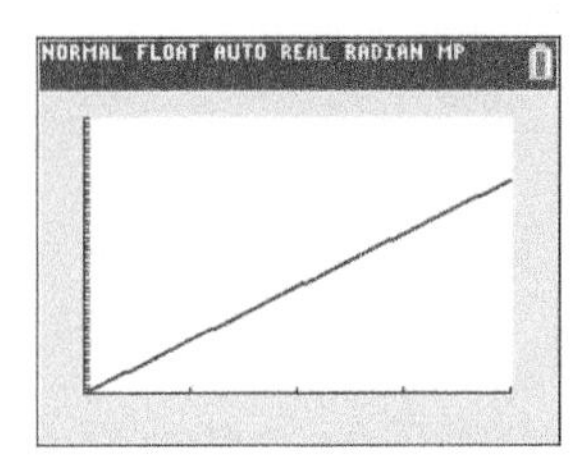

7. a. Determine the window settings, X_{min}, X_{max}, Y_{min}, and Y_{max}, to obtain the graph of the general function $c = 12.6p$ on the interval $-10 \le p \le 10$.

$X_{min} = -10$, $X_{max} = 10$, $Y_{min} = -25$, and $Y_{max} = 25$ (Answers will vary.)

b. Type in the values determined in part a, and graph the function. The screens should appear as follows:

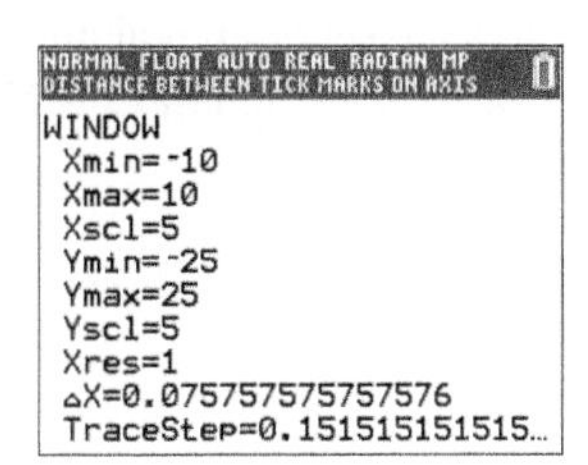

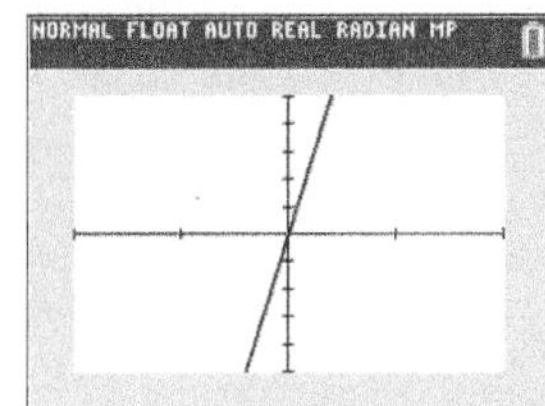

c. How does the graph in part b compare with graph 2 on the previous page?

The graphs are very similar. (Answers will vary.)

Defining Functions: A Summary

A function can be defined by a written statement (verbally), symbolically, numerically, and graphically. The following example illustrates the different ways the gross pay function can be defined.

EXAMPLE 1 ***If you work for an hourly wage, then your gross pay is a function of the number of hours worked. If you earn \$10.50 per hour, define the gross pay function verbally, symbolically, numerically, and graphically.***

SOLUTION

Verbal Definition: A Statement of the Definition of the Function:

The gross pay will be the number of hours worked multiplied by \$10.50.

Symbolic Definition: If g represents the gross pay and h represents the number of hours worked, then

$$g = 10.50h.$$

Numerical Definition: The gross pay is represented by the following table:

Number of Hours Worked, h	0	10	15	20	30
Gross Pay, g ($)	0.00	105.00	157.50	210.00	315.00

Graphical Definition: The gross pay is represented by the following graph:

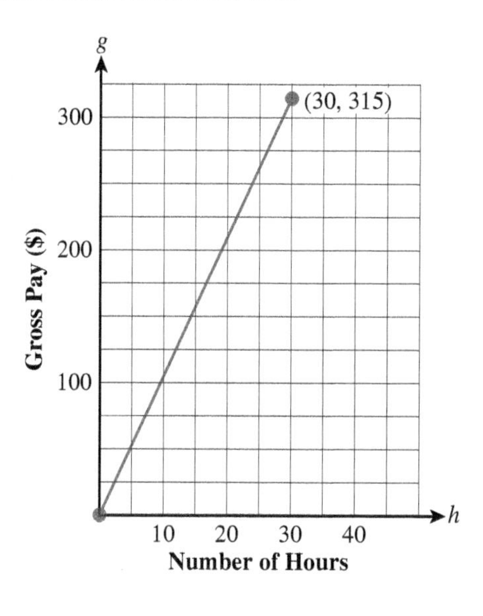

8. Sales tax is a function of the cost of any item. If the sales tax in your area is 6%, define the cost function in four different ways: verbally, symbolically, numerically, and graphically.

A Statement Definition of the Function: Sales tax is a function of the cost of an item. The sales tax rate is 6%. The sales tax on an item is calculated by multiplying the cost of the item by 0.06.

Symbolic Definition: If t represents the sales tax on an item that costs C dollars, then

$$t = 0.06C.$$

Numerical Definition:

Cost of Item ($)	10	20	30	40	50	100	150	200	300	500	1000
Sales Tax ($)	0.60	1.20	1.80	2.40	3.00	6.00	9.00	12.00	18.00	30.00	60.00

Graphical Definition:

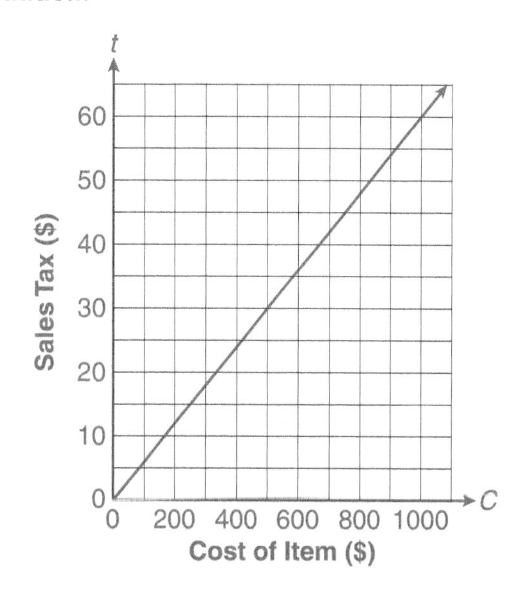

SUMMARY Activity 1.3

1. When a function is defined **graphically**, the input variable will be represented on the horizontal axis and the output on the vertical axis.
2. Functions are **discrete** if they are defined only at isolated input values and do not make sense or are not defined for input values between those values. Discrete functions can be counted.
3. Functions are **continuous** if they are defined for entire intervals of input values between any consecutive input values. The outputs of continuous functions include entire intervals.

EXERCISES Activity 1.3

1. In the following table, the amount of snowfall is a function of the elevation:

ELEVATION (ft.)	SNOWFALL (in.)
2000	4
3000	6
4000	9
5000	12

a. Plot the ordered pairs on an appropriately scaled and labeled set of coordinate axes.

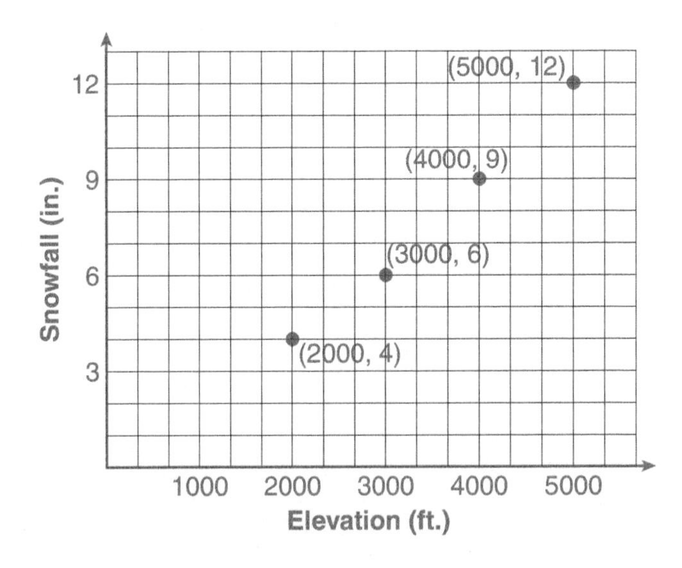

b. Would you consider this a discrete situation (consisting of separate, isolated points) or a continuous situation? Explain.

Continuous because the amount of snowfall is defined at any elevation.

2. Plot the ordered pairs of each of the following functions on an appropriately scaled and labeled set of axes. Then determine whether a continuous or discrete graph is more appropriate for the situation.

a. In a science experiment, the amount of water displaced in a graduated cylinder is a function of the number of marbles placed in the cylinder. The results of an experiment are recorded in the following table:

Number of Marbles	1	2	3	4	5
Volume of Water Displaced (mL)	20	40	60	80	100

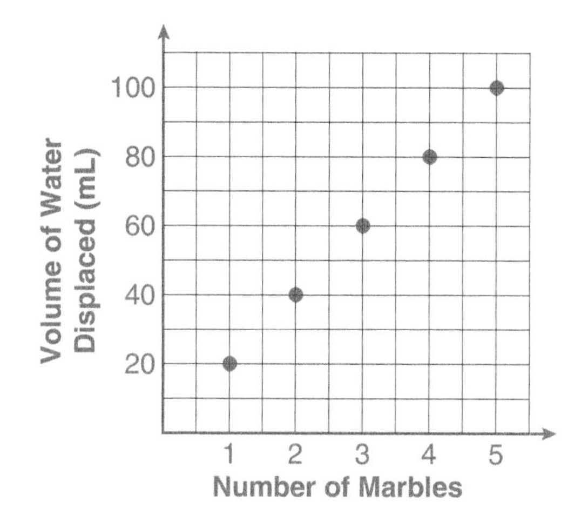

Because a fraction of a marble does not make sense in this situation, this is a discrete function.

b. Forensic anthropologists estimate the height of a male based on the length of his femur. The following table demonstrates this functional relationship:

Length of Femur (in.)	10	12	14	16	18
Height (in.)	51	55	59	62	66

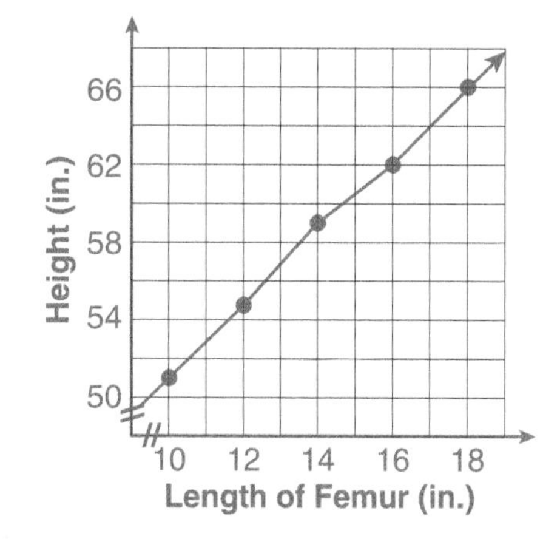

Because the function is defined for all values between the input values in the table, it is appropriate to connect the data points with a smooth, continuous curve. The graph appears to be a line. (The // marks on the axes indicate that some numbers between 0 and the number on the first tick mark have been skipped.)

A primary objective of this textbook is to help you develop a familiarity with graphs, equations, and properties of a variety of functions. In Chapter 4, you will study a type of function called a **quadratic function**.

In Exercises 3–6, sketch a graph of each quadratic function in the standard window of a graphing calculator. Then match each graph with the corresponding graphs in parts a–d.

3. $y = x^2$ 4. $y = -1x^2$ 5. $y = x^2 + 2$ 6. $y = x^2 - 2$

a.

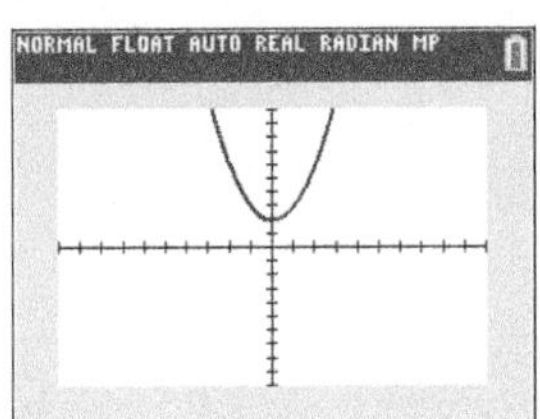

b.

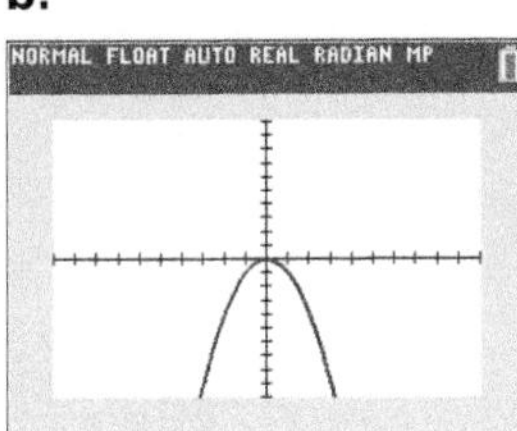

c.

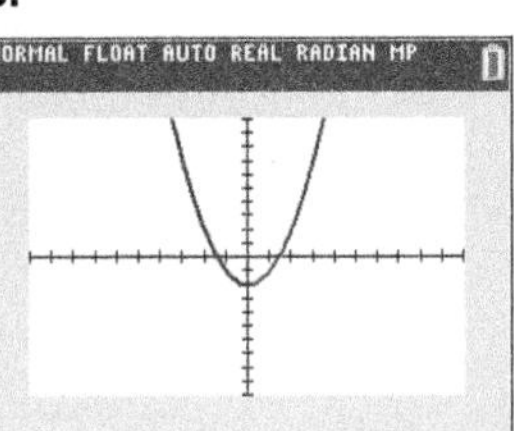

d.

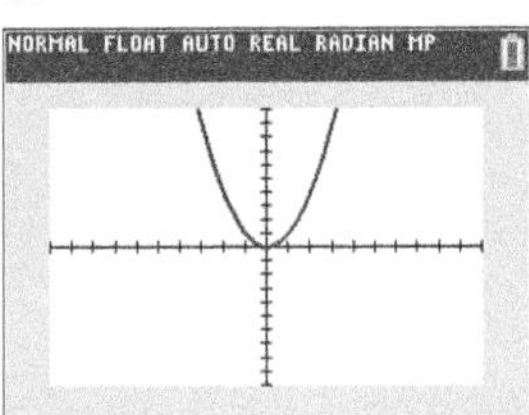

Equation 3 matches graph d. Equation 4 matches graph b. Equation 5 matches graph a. Equation 6 matches graph c.

7. a. Sketch a graph of the quadratic function defined by $y = x^2 + 15$ in the standard window of your graphing calculator. What do you observe?

   No graph appears in the screen.

   b. Use the table feature of your graphing calculator to complete the following table:

x	−3	−2	−1	0	1	2	3
y	24	19	16	15	16	19	24

   c. Describe how you would use the results in part b to help select an appropriate viewing window.

   The y-values are increasing from a minimum of 15. Therefore, you need to have Y_{max} of at least 30.

   d. Sketch a graph of the function. Your screen should appear as follows:

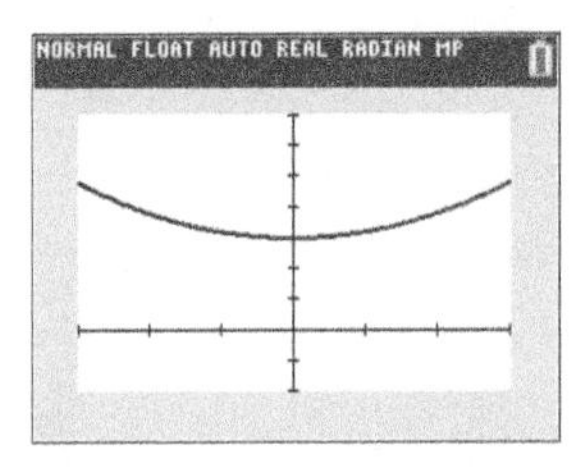

8. In Exercise 10 on page 10 of Activity 1.1, you were assigned the job of supplying paint for the exterior of the bunk houses at a youth summer camp. You found that 1 gallon of paint will cover 400 square feet of flat surface. The number of square feet you can cover with the paint is a function of the number of gallons of paint used. Define the paint coverage function in four different ways: verbally, symbolically, numerically, and graphically.

   Verbal Definition: The number of square feet of coverage is 400 times the number of gallons of paint used.

   Symbolic Definition: If s represents the number of square feet I can cover with the paint and n represents the number of gallons of paint used, then $s = 400n$.

Numerical Definition:

n, Number of Gallons of Paint	1	2	4	6
s, Square Feet Covered by the Paint	400	800	1600	2400

Graphical Definition:

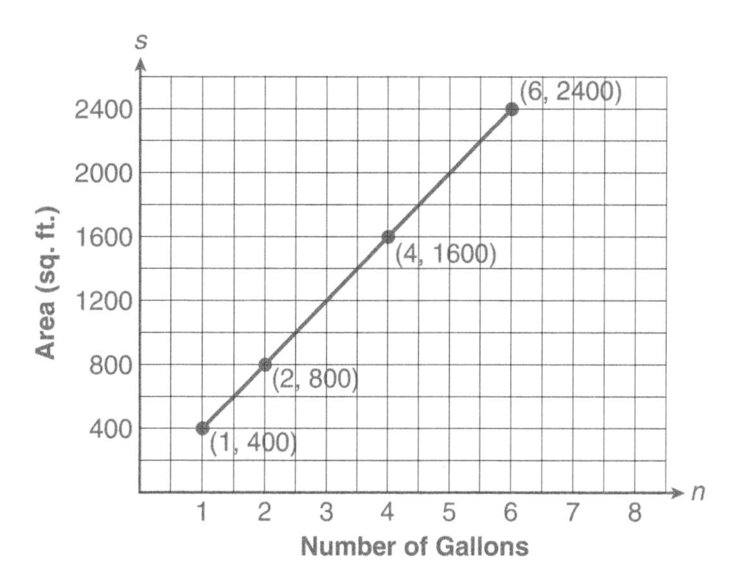

ACTIVITY 1.4

Stopping Short

Functions as Mathematical Models

OBJECTIVES

1. Use a function as a mathematical model.
2. Determine when a function is increasing, decreasing, or constant.
3. Use the vertical line test to determine whether a graph represents a function.

When you translate a situation or problem into mathematical language, you develop a mathematical model of the situation. Information about a situation containing variable quantities is often given by a collection of data. Very often, a table of data in combination with a graph of the data reveals a pattern that can be modeled or represented by an equation. Such mathematical models can be used to answer questions involving the situation, make predictions, and/or gain a better understanding of the situation.

Braking Distance

After an automobile accident, the investigating police officers often estimate the speed of the vehicle by measuring the length of the skid distance. The following table gives the average skid distances for an automobile with good tires on dry pavement:

SKID DISTANCE (ft.)	SPEED (mph)
28	25
54	35
89	45
132	55
184	65
244	75
313	85
390	95

1. Does the table data define speed as a function of skid distance? Explain using the definition of *function*.

   Yes. For every skid distance, there is one and only one speed. Thus, it is a function.

2. Identify the independent variable and the dependent variable.

   The length of the skid distance is the independent variable. Speed is the dependent variable.

3. Plot the input–output ordered pairs on an appropriately scaled and labeled set of coordinate axes. Remember, the input values appear along the horizontal axis, and the output values appear along the vertical axis.

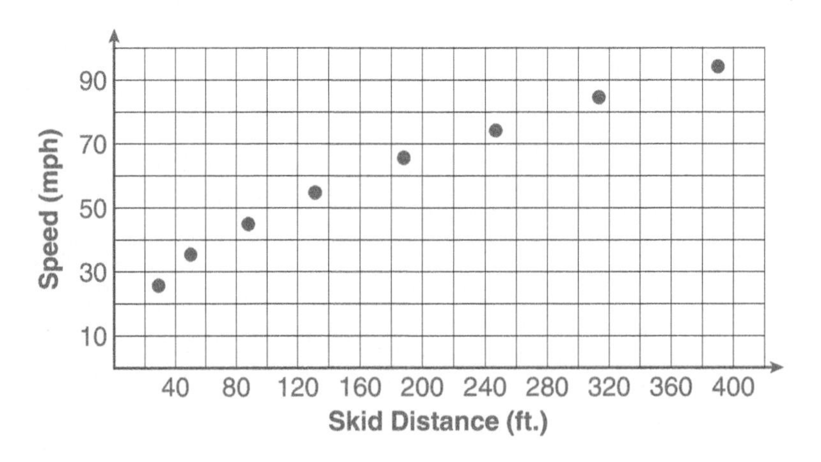

After a particular accident, a skid distance was measured to be 200 feet. From the table, the investigating officer knows that the speed of the vehicle was between 65 and 75 miles per hour. She would, however, like to be more precise in reporting the speed. One way of getting values that are not listed in the table is to use a graph or an equation of a function that best fits the actual data.

DEFINITION

A **mathematical model** is an equation or a graph that fits or approximates the actual data. The model can be used to predict output values for input values not in the table.

Note that the points on the graph in Problem 3 do not lie exactly on a specific curve. However, calculators can produce an equation that best models actual data. From the data in the table on page 28, the TI-84 Plus C can be used to generate the model

$$f(x) = -0.00029x^2 + 0.31x + 18.6 \text{ (coefficients are rounded)},$$

where x represents the length of skid distances in feet and $f(x)$ represents the speed in miles per hour. The process for generating these equations is covered in later activities.

4. **a.** Enter the function equation above into your calculator. For help with the TI-84 Plus C, see Appendix C. The screen should appear as follows:

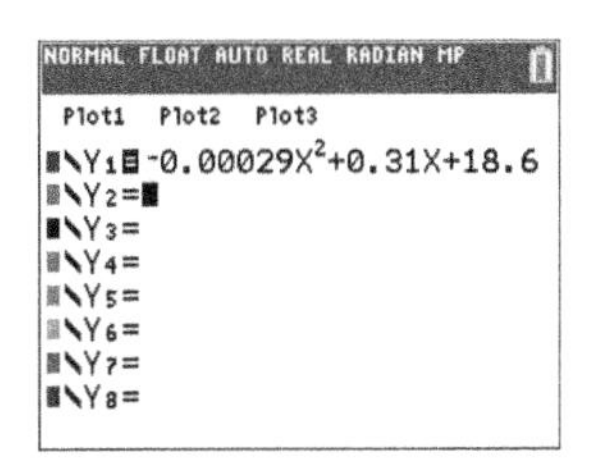

b. The values in the table on page 28 can help you set appropriate window values in your calculator to view the graph. Starting at a minimum skid distance of 0, a reasonable maximum value of the skid distance in this situation would be 450 feet. Beginning at a minimum speed of 0, a reasonable maximum speed value would be 100 mph. Using these settings, the window screen should appear as follows:

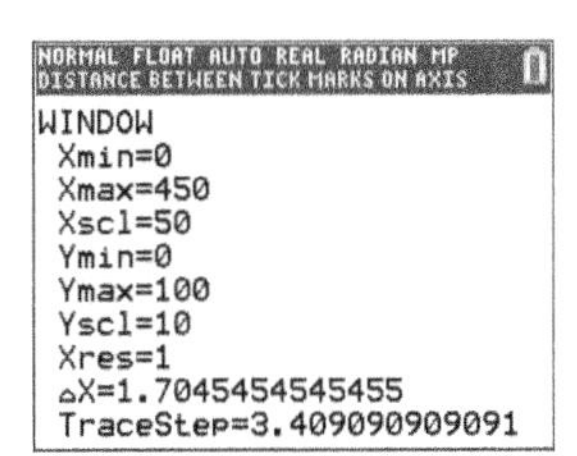

c. Display your graph using the window settings from part b. Your graph should resemble the following:

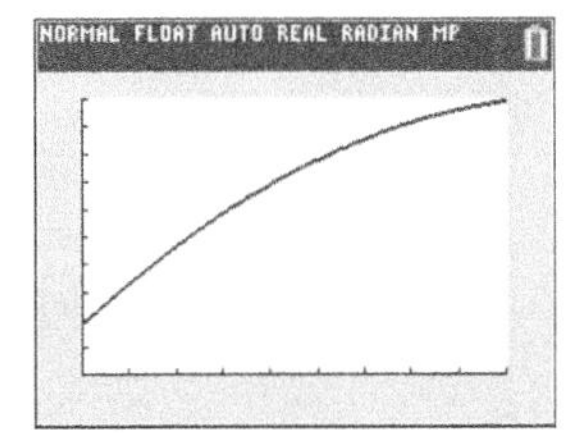

5. a. By pressing the trace button and the left and right arrow keys, you can display the x-y values of points on the graph on the bottom of the display. Use the trace feature of your calculator to approximate the speed of the car when the length of the skid distance is 200 feet. You can obtain the exact value for any x-value between X_{min} and X_{max} by entering the x-value while in trace mode and pressing ENTER. Your screen should appear as follows:

69 mph

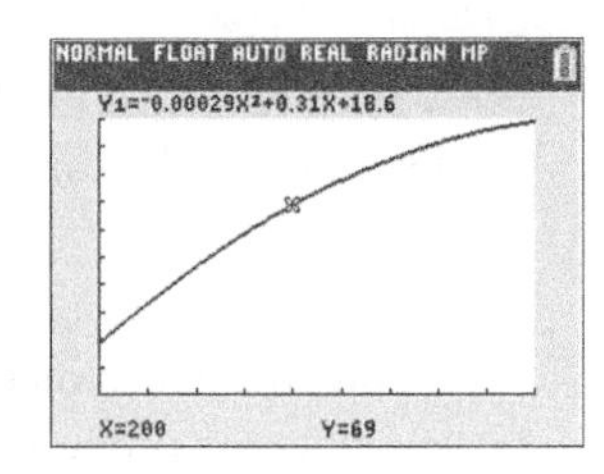

b. Use your calculator to verify the result in part a by evaluating $-0.00029(200)^2 + 0.31(200) + 18.6$.

There is another way to evaluate $f(200)$ when the function is entered in the calculator as a Y-variable ($Y1$, $Y2$, $Y3$, . . .). Because the skid distance function is in $Y1$, you can evaluate $f(200)$ by entering $Y1(200)$ in the home screen. Enter $Y1$ in the home screen by pressing ALPHA – TRACE to access the f4 menu. Then press 1 or ENTER to select $Y1$ and place it at the cursor. Enter (200) ENTER to display the result.

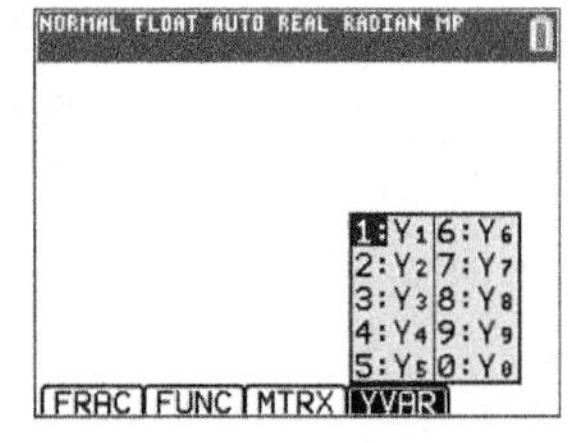

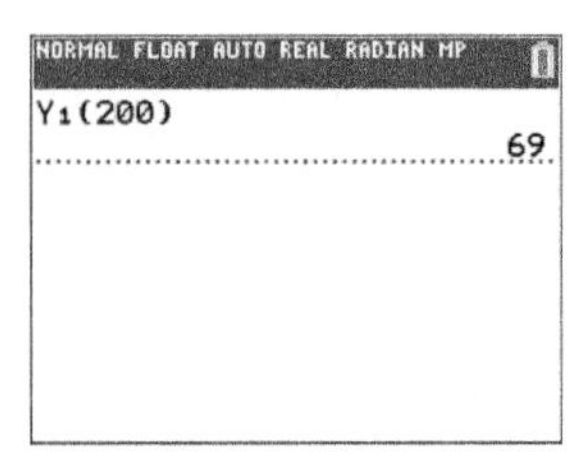

Increasing and Decreasing Functions

There are many advantages to having function models represented in graphical form. For example, you are often interested in determining how the output values change as the input values increase.

DEFINITION

A function is **increasing** if its graph goes up to the right, **decreasing** if its graph goes down to the right, and **constant** if its graph is horizontal. In each case, you are viewing the graph as a point moves along the curve from left to right, that is, as the input values increase.

EXAMPLE 1

a. ***The graph of the function $y = 3x + 1$ is always increasing. The graph of this function goes up from left to right.***

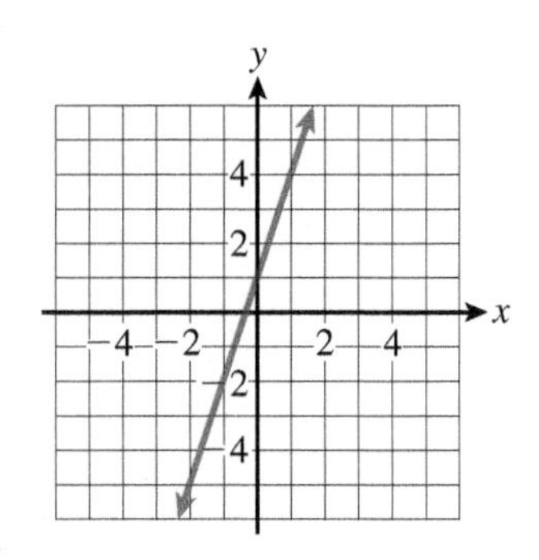

b. ***The graph of the function $f(x) = -x + 2$ is decreasing because its graph goes down from left to right.***

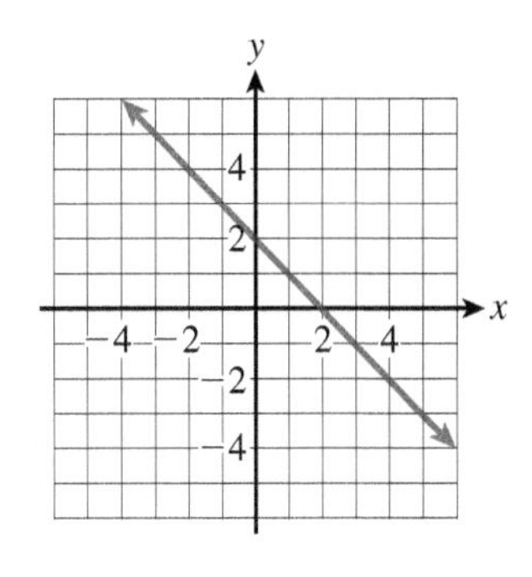

c. ***The graph of $y = 3$ is constant because the graph goes neither up nor down. The output value is always 3 no matter what the input value is.***

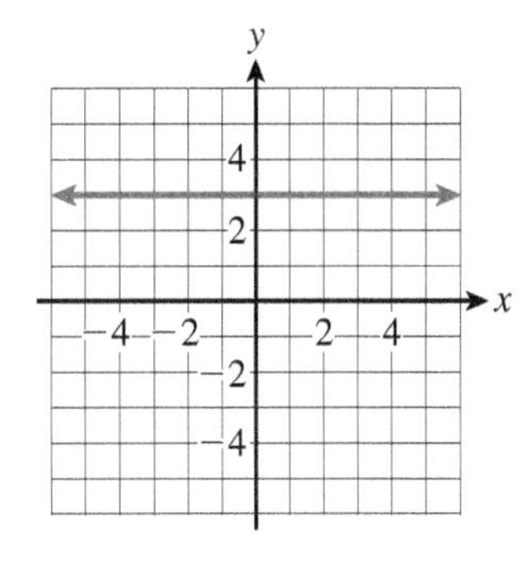

6. Is the function $f(x) = -0.00029x^2 + 0.31x + 18.6$ in Problem 4 increasing, decreasing, or constant over the domain displayed in the window?

The function is increasing.

Graphs of Functions: A Second Look

7. To cover your weekly expenses while going to college, you earn $100 each week as a part-time aide in your college's health center. The following table represents a typical month:

x, Weekly Salary ($)	100	100	100	100
y, Weekly Expenses ($)	50	70	90	60

a. Plot the data points on the following grid:

b. Are weekly expenses a function of the weekly salary? Explain using the definition of function.

No. For a weekly salary of $100 (input), there is more than one corresponding weekly expense (output).

c. Do all four points lie on the same straight line? Yes.

d. Is the line horizontal, vertical, or slanted?

The line is vertical. All points have the same input value.

e. How can you determine from the graph that it does not represent a function?

More than one point on the graph has an input value of 100. A vertical line drawn through one of the points on the graph will intersect other points on the graph.

8. Notice that all the input values of points on the vertical line in Problem 7 are the same (100 in this case). A vertical line can be thought of as the graphical representation of an input value.

a. Draw a line whose points all have an input value of 2.

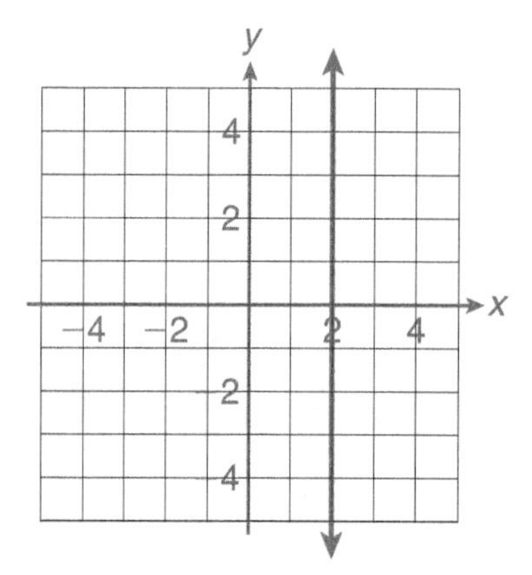

b. Draw a line whose points all have an input value of −3.

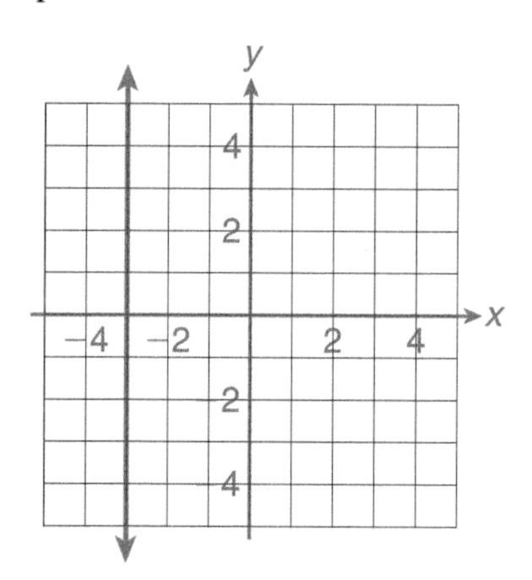

9. A unit circle is a circle having a center at the origin and a radius of 1. Such circles are used in the study of trigonometry. Following is the graph of a unit circle:

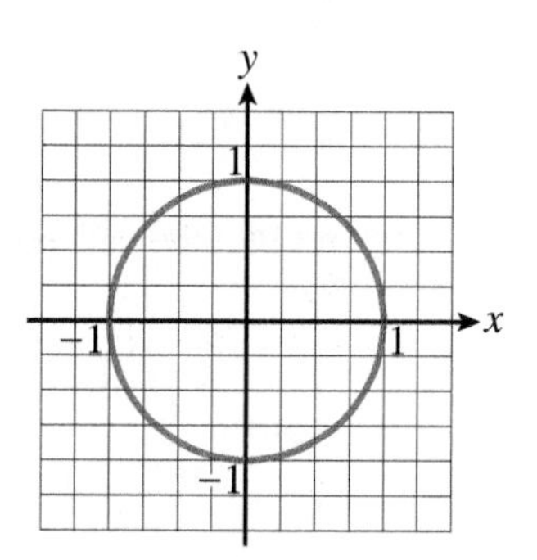

Does the graph of a unit circle represent a function? Explain how you can determine from the graph whether it represents a function.

No, a vertical line drawn through any one of the points on the circle [except $(\pm 1, 0)$] will intersect the graph in two places.

Problem 9 illustrates what is known as the vertical line test. You know that a function can have only one output value for each input value. Because any two points with the same first coordinate lie on a vertical line, a graph defines a function if any vertical line drawn through the graph intersects the graph no more than once. This is called the **vertical line test**.

10. Use the vertical line test to verify whether the graph displayed in Problem 4 defines a function. The graph in Problem 4 represents a function.

LC LEARNING CATALYTICS

The graph of f is crossed by a vertical line in more than one place at x_0. Does this graph represent a function?

11. Use the vertical line test to determine which of the following graphs represent functions:

a. yes

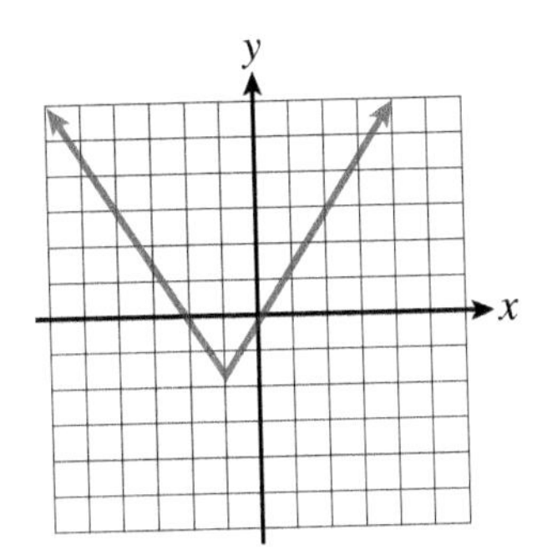

b. no

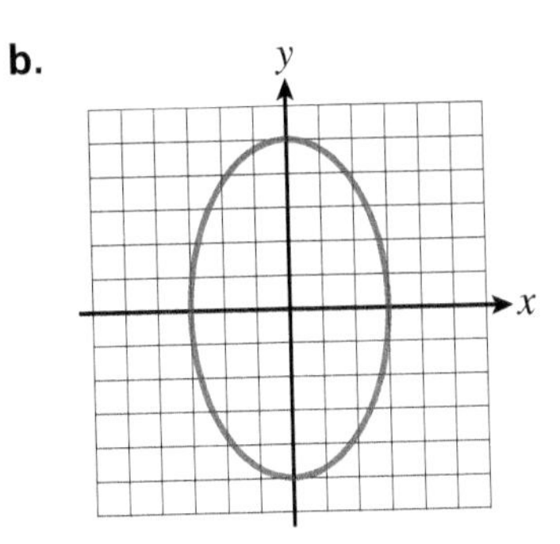

12. If a graph is increasing everywhere, does the graph represent a function? Explain.

Yes. For every value of input, there will be only one value of output.

SUMMARY Activity 1.4

1. A **mathematical model** is an equation or a graph that fits or approximates the actual data. The model can be used to predict output values for input values not in the data set.

2. A function is **increasing** if its graph goes up to the right, **decreasing** if its graph goes down to the right, and **constant** if its graph is horizontal.

3. In the **vertical line test**, a graph defines a function if any vertical line drawn through the graph intersects the graph no more than once.

EXERCISES Activity 1.4

1. The following table defines snowfall as a function of elevation for a recent snowstorm in upstate New York:

ELEVATION (ft.), x	SNOWFALL (in.), $f(x)$
1000	4
2000	6
3000	9
4000	12

A function that closely models the data in the table is

$$f(x) = 0.0027x + 1.$$

Enter this function into a Y variable on your calculator.

a. Complete the following table using the function f. Verify your answers using the table feature of your graphing calculator.

ELEVATION	1000	2000	3000	4000
$f(x)$	3.7	6.4	9.1	11.8

Exercise numbers appearing in color are answered in the Selected Answers appendix.

As you read the graph in Example 1 from left to right, it shows how your distance from your car changes as time passes. One possible scenario this graph describes is as follows:

You leave your car and walk at a steady pace toward the library. You meet some friends and stop to chat for a while. You realize that you forgot something and quickly return to your car. After rummaging around for a while, you hurry off to the library.

How did anyone come up with this from the graph? Look at the graph in sections.

a. The first increasing line segment indicates that you are moving away from your car because the time and the distance are increasing.

b. The first horizontal section indicates that your distance from the car is constant, so you are standing still.

c. The decreasing line segment indicates that your distance from the car is decreasing. When it reaches the horizontal axis, it tells you that you are back at your car.

d. The second horizontal segment indicates that you stay at your car for a time.

e. The final increasing segment is steeper and longer than the first, so you are moving away from the car faster and farther than in the first segment.

Graphs to Stories

The graphs in Problems 1–4 present visual images of several situations. Each graph shows how the y-values change in relation to the x-values. In each situation, identify the independent variable and the dependent variable. Then interpret the situation; that is, describe in words what the graph is telling you about the situation. Indicate whether the graph rises, falls, or is constant and whether the graph reaches either a minimum (smallest) or maximum (largest) y-value.

1. A person's core body temperature (in degrees Fahrenheit) in relation to time of day

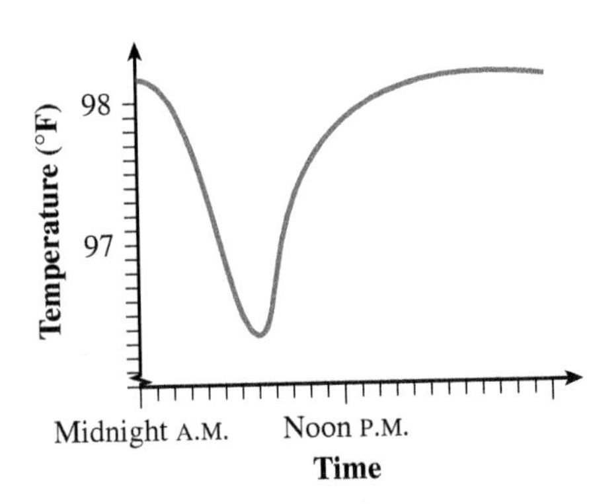

a. Independent: time of day   Dependent: temperature

b. Interpretation:

This person's body temperature was normal at midnight and then dropped below normal as he or she slept, reaching a minimum of approximately 96.3°F at 7 A.M. During the day, the person's temperature returned to normal and remained at that level.

LC LEARNING CATALYTICS

Alpha Company's stock declines a great deal during the first month of the quarter. It then steadies during the second month before gradually regaining about half of its early losses during the third month of the quarter. Which month could be represented graphically by a horizontal line?

a. the first month
b. the second month
c. the third month
d. none of these

2. Performance of a simple task in relation to anxiety level

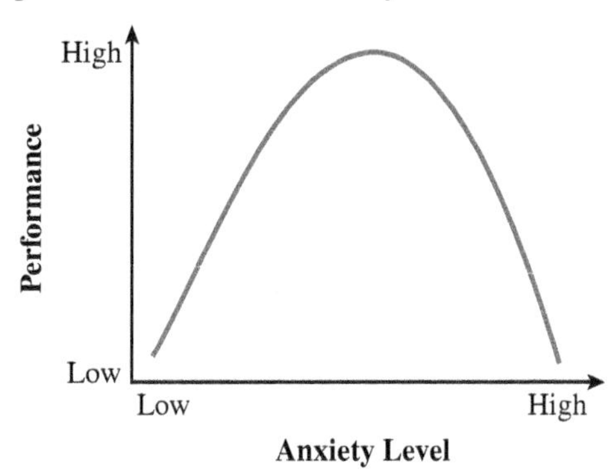

a. Independent: anxiety level Dependent: performance

b. Interpretation:

At first, performance increases as anxiety level increases. As anxiety level reaches a certain point, the performance reaches a maximum. At that point, the performance level drops because anxiety becomes an obsession. Too much anxiety is a detriment to high performance.

3. Net profit of a particular business in relation to time given quarterly

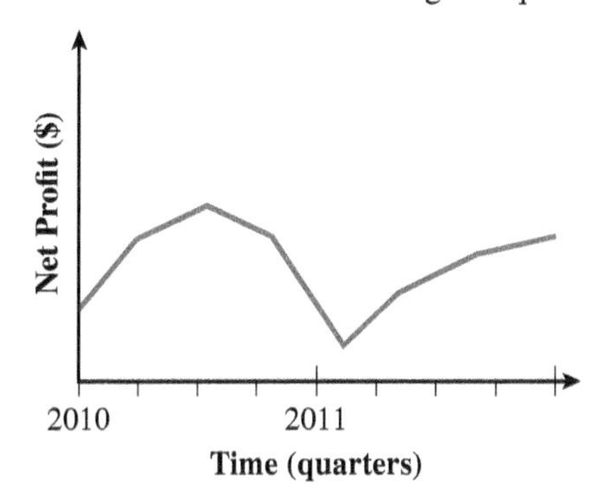

a. Independent: time Dependent: net profit

b. Interpretation:

Profit increases during the first quarter of 2010 and then increases more slowly to mid-year, when the profit reaches a maximum. During the third quarter, profit decreases at a constant rate and then drops sharply. Profit begins to increase during 2011 (but at a slower rate each quarter) until the profit reaches the same level as in the third quarter of 2010.

Possibly, this business sells or services summer merchandise.

4. Annual gross income in relation to number of years

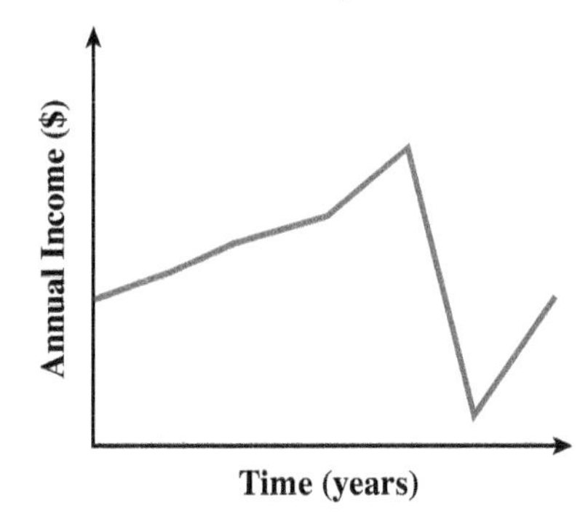

a. Independent: time in years Dependent: annual income

b. Interpretation:

For a few years, income rises at about the same rate. Then it rises more sharply, perhaps due to a raise, and reaches a maximum income. Income drops rather quickly, perhaps due to layoff, and reaches a minimum. When work begins again, income begins to rise.

Stories to Graphs

In Problems 5–9, sketch a graph that best represents the given situation. Many times you won't know the actual values, so estimate what seems reasonable to you. Your graphs of these problems will be more qualitative than quantitative. Be sure to label your axes, with the input variable always on the horizontal axis. Provide numerical scales when appropriate.

5. You leave home on Friday afternoon for your weekend getaway. Heavy traffic slows you down for the first half of your trip, but you make good time by the end. Express your distance from home as a function of time.

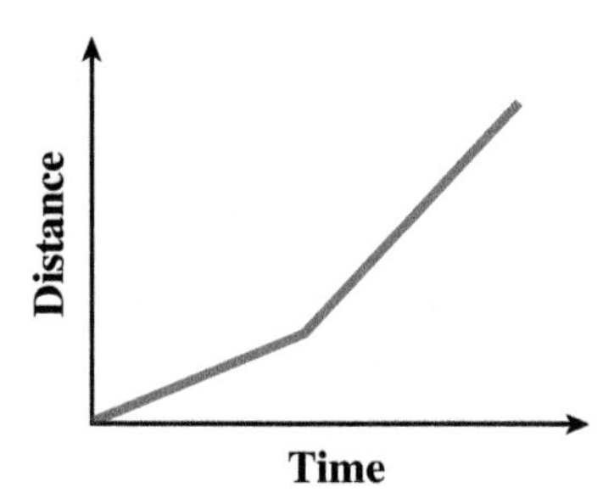

6. You just started a new job that pays \$16 per hour, with a raise of \$2 per hour every 6 months. After one and a half years, you receive a promotion that gives you a wage increase of \$5 per hour, but your next raise won't come for another year. Sketch a graph of your wage over your first *two and a half* years.

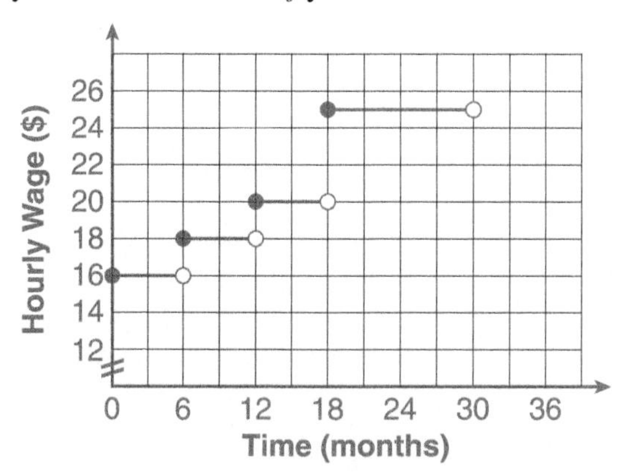

7. Your small business started slowly, losing money in its first 2 years, then breaking even in year 3. By the fourth year, you made as much as you lost in the first year and then doubled your profits each of the next 2 years. Graph your profit as the output and time as the input.

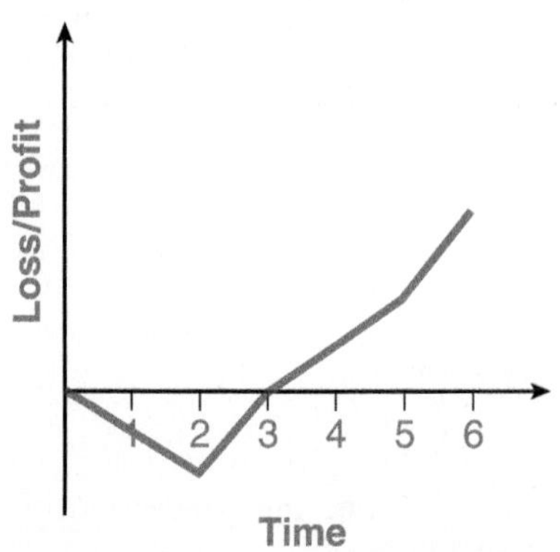

8. Hair grows at a steady rate. Suppose you get your hair cut every month. Measuring the longest hair on your head, graph your hair length over the course of 6 months.

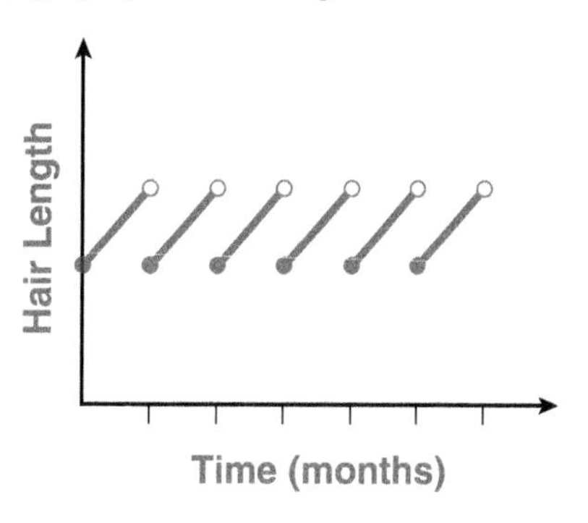

9. The distance traveled is a function of speed in a fixed time interval.

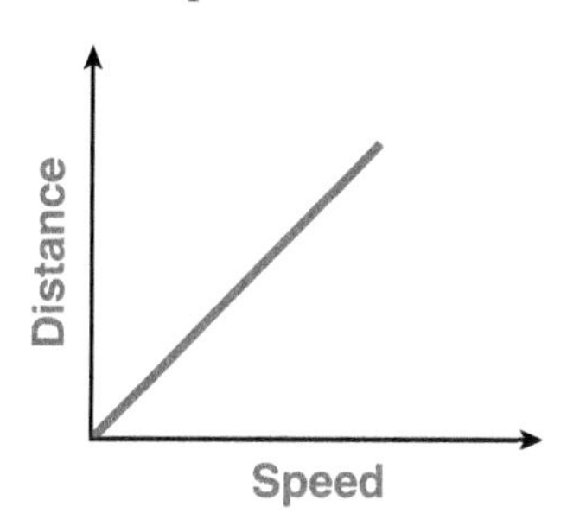

SUMMARY Activity 1.5

If a function increases and then decreases, the point where the graph changes from rising to falling is called a **maximum point**. The y-value of this point is called a **local maximum value**. If a function decreases and then increases, the point where the graph changes from falling to rising is called a **minimum point**. The y-value of this point is called a **local minimum value**.

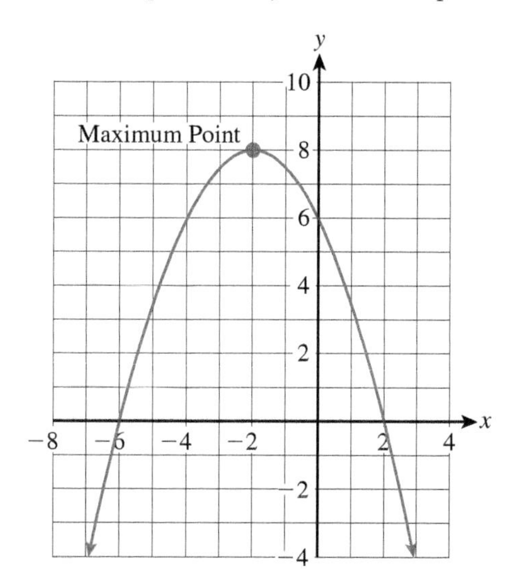

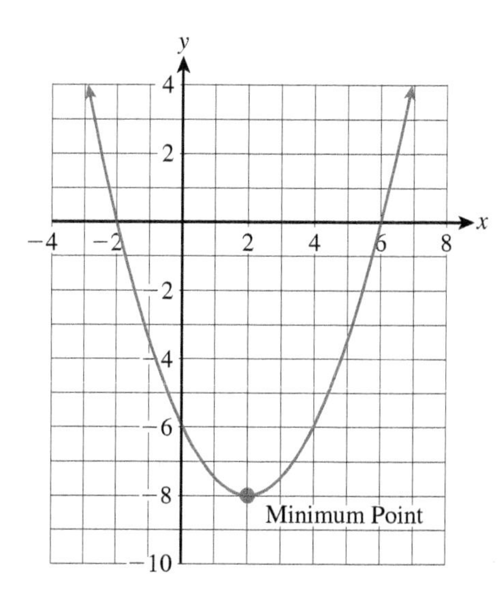

EXERCISES Activity 1.5

1. You are a technician at the local power plant, and you have been asked to prepare a report that compares the output and efficiency of the six generators in your sector. Each generator has a graph that shows output of the generator as a function of time over the previous week, Monday through Sunday. You take all the paperwork home for the night (your supervisor wants this report on his desk at 7 A.M.), and to your dismay, your cat scatters your pile of papers out of the neat order in which you left them. Unfortunately, the graphs for generators A through F were not labeled (you will know better next time!). You recall some information and find evidence elsewhere for the following facts:

- Generators A and D were the only ones that maintained a fairly steady output.
- Generator B was shut down for a little more than two days during midweek.
- Generator C experienceed a slow decrease in output during the entire week.
- On Tuesday morning, there was a problem with generator E that was corrected in a few hours.
- Generator D was the most productive one over the entire week.

Match each graph with its corresponding generator. Explain in complete sentences how you arrive at your answers.

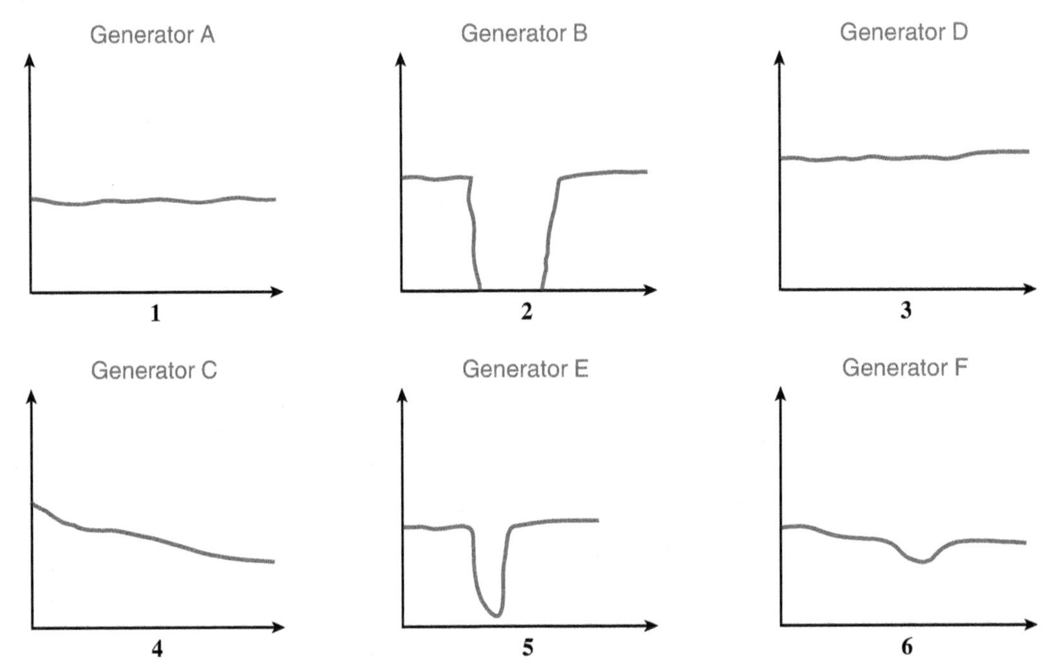

Generators A and D were fairly steady, so graphs 1 and 3 would be appropriate for them. Because D was most productive, it follows that graph 3 represents the output of generator D and that graph 1 represents the output of generator A.

Graph 2 is the only one that shows a shutdown, so it represents the output of generator B.

The sharp drop shown in graph 5 represents the problem with generator E on Tuesday.

Generator C experienced a slow decrease in output. Graph 4 represents C as it steadily decreases.

The output of generator F is represented by graph 6 by elimination.

Exercise numbers appearing in color are answered in the Selected Answers appendix.

In Exercises 2 and 3, identify the independent variable and the dependent variable. Then interpret the situation being represented. Indicate whether the graph rises, falls, or is constant and whether the graph reaches either a minimum (smallest) or maximum (largest) output value.

2. Time required to complete a task in relation to number of times the task is attempted

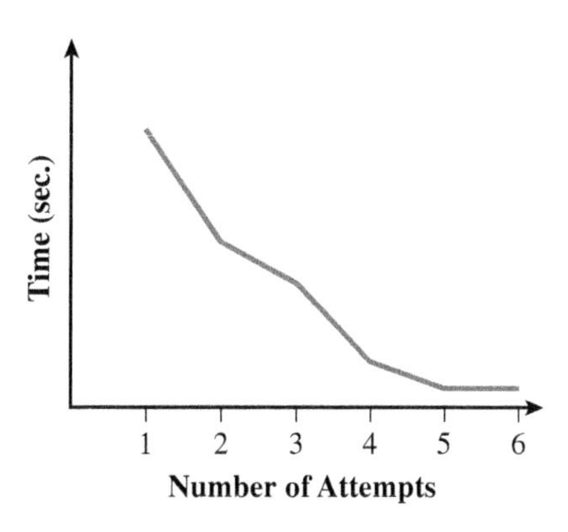

a. Independent: number of attempts Dependent: time in seconds

b. Interpretation:

The time required to complete a task decreases as the number of attempts increases. As a person attempts a task more times, the task takes less time to complete. After five attempts, no further improvement is made.

3. Number of units sold in relation to selling price

a. Independent: selling price Dependent: number of units sold

b. Interpretation:

As the selling price increases, the number of units sold increases slightly at first, reaches a maximum, and then declines until none are sold.

In Exercises 4 and 5, sketch a graph that represents the situation. Be sure to label your axes.

4. The sale price of a computer is a function of the percent of discount.

5. The area of a square is a function of the length of one side of the square.

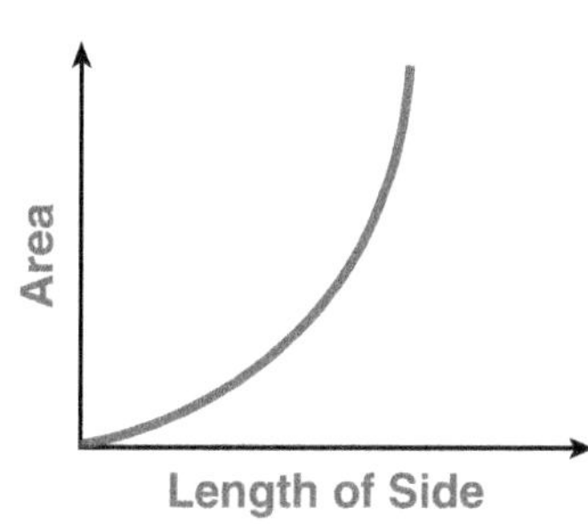

b. Use $f(h)$ to represent the cost, and rewrite the equation in part a using function notation.

$f(h) = 245h$

c. Complete the following table:

h	2	4	7	8	11
$f(h)$	490	980	1715	1960	2695

d. Evaluate $f(3)$ and write a sentence describing its meaning. Write the results as an ordered pair.

$f(3) = 245(3) = 735$ The tuition cost for 3 hours of coursework is \$735; $(3, 735)$.

e. Given $f(h) = \$1225$, determine the value of h.

$1225 = 245h$, $h = 5$ hr.

f. Which variable is the output? Explain.

$f(h)$ or c is the output variable. This is the variable that depends on the number of credits taken.

g. Which is the independent variable? Explain.

h is the independent variable. It is the input variable.

h. Explain (using the table in part c) how you know that the data represent a function.

For each value of input, there is one value of output.

i. What is the practical domain for this function?

Assuming that there are no half-credit courses, the practical domain is all whole numbers from 0 to 11.

j. Plot the ordered pairs on an appropriately scaled and labeled coordinate system. Which axis represents the input values?

The horizontal axis represents the input.

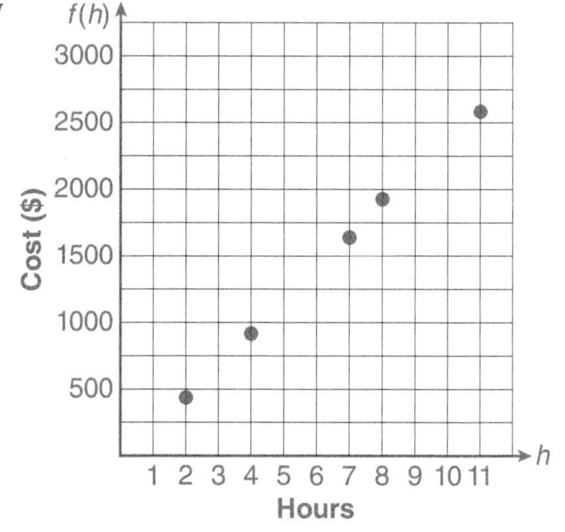

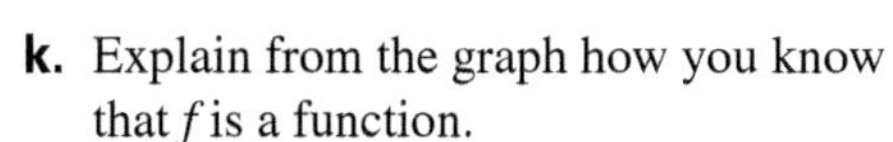

k. Explain from the graph how you know that f is a function.

f is a function because the graph passes the vertical line test.

l. Use your graphing calculator to verify your answers to parts c and j.

m. Use the trace and table features to determine the cost of 9 credit hours.

The cost of 9 credit hours is \$2205.

12. Given $p(x) = 2x + 7$, determine each of the following:

a. $p(3)$

$p(3) = 13$

b. $p(-4)$

$p(-4) = -1$

c. $p\left(\frac{1}{2}\right)$

$p\left(\frac{1}{2}\right) = 8$

d. $p(0)$

$p(0) = 7$

13. Given $t(z) = 2z^2 - 3z - 5$, determine each of the following:

a. $t(2)$

$t(2) = -3$

b. $t(-3)$

$t(-3) = 18 + 9 - 5 = 22$

14. According to the U.S. Centers for Disease Control and Prevention, the average life expectancy from birth for males in the United States may be modeled by the function $f(x) = 0.19x + 65.0$, where x is the number of years since 1950.

a. Use your calculator to complete the following table. Round your results to the nearest tenth.

	YEAR						
	1950	1960	1970	1985	2000	2010	2020
x, Years since 1950	0	10	20	35	50	60	70
$f(x)$, Life Expectancy	65.0	66.9	68.8	71.7	74.5	76.4	78.3

b. Evaluate $f(30)$ and explain its practical meaning in this situation.

$f(30) = 0.19(30) + 65.0 = 70.7$

The life expectancy for a male born in 1980 is 70.7 years.

c. Use the table values to set appropriate window values to view the graph of f. Graph the function on your calculator. Identify the window used.

$X_{min} = 0$ $X_{max} = 75$

$Y_{min} = 60$ $Y_{max} = 80$

(Answers may vary.)

d. Is the graph increasing, decreasing, or constant? Explain.

The graph is increasing. It is rising to the right.

e. Use the trace feature of your calculator to determine $f(30)$. Compare your answer with your result in part b.

They are the same.

15. Determine the domain and the range of each of the following functions:

a. $\{(3, 5), (4, 5), (5, 8), (6, 10)\}$

domain $\{3, 4, 5, 6\}$; range $\{5, 8, 10\}$

b.

GLOBAL REVENUE FROM DIGITAL MUSIC (billions of dollars)						
2008	2009	2010	2011	2012	2013	2014
4.0	4.4	4.7	5.3	6.0	6.4	6.9

Source: International Federation of the Phonographic Industry (IFPI).

domain {2008, 2009, 2010, 2011, 2012, 2013, 2014}; range {4.0, 4.4, 4.7, 5.3, 6.0, 6.4, 6.9}

c.

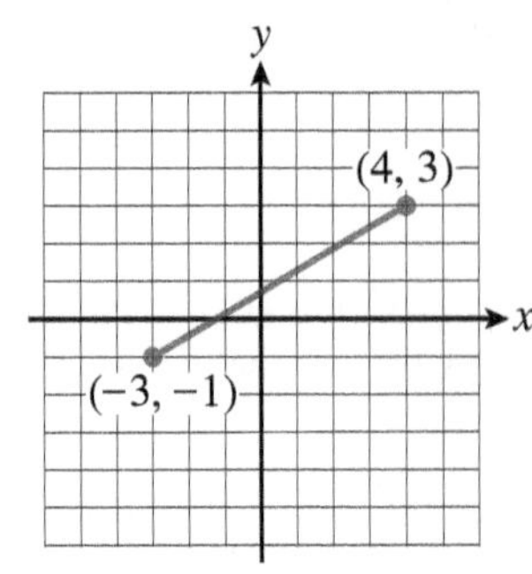

domain $-3 \le x \le 4$

range $-1 \le y \le 3$

d.

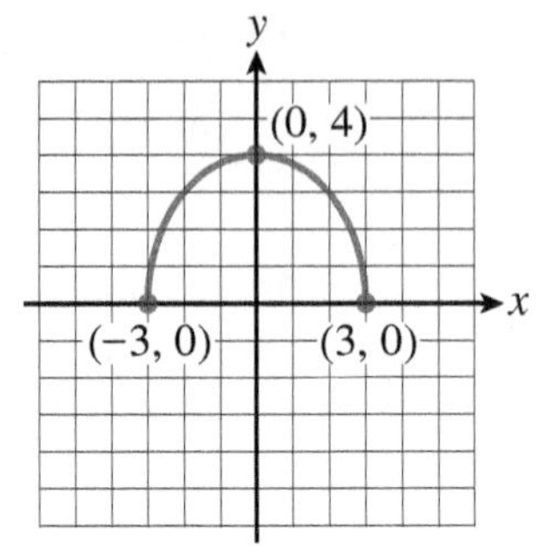

domain $-3 \le x \le 3$

range $0 \le y \le 4$

e. $y = 3x$

Domain is all real numbers; range is all real numbers.

16. Each of the following graphs shows how the inputs and outputs change in relation to each other. Describe in words what the graph is telling you about the situation. Provide a reasonable explanation for the behavior you describe.

a.

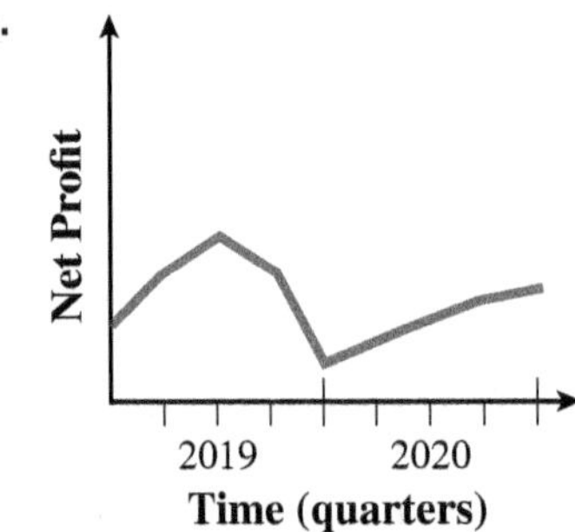

b.

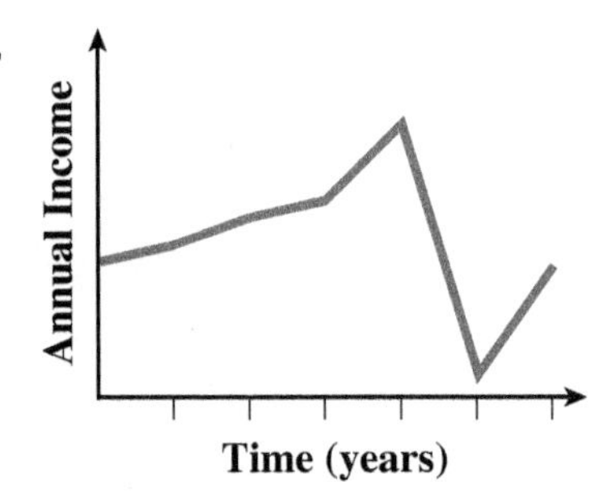

a. The net profit increases during the first two quarters of 2019. The net profit then decreases for about two quarters, and then it increases through 2020.

b. The annual income rises rather steadily for 3 years; in the fourth year, it rises sharply. Then it suffers a sharp decline during the next year. During the last year, the income recovers to about the point it was originally.

17. Sketch a graph representing the fact that the hours of daylight depend on the day of the year in the Northern Hemisphere.

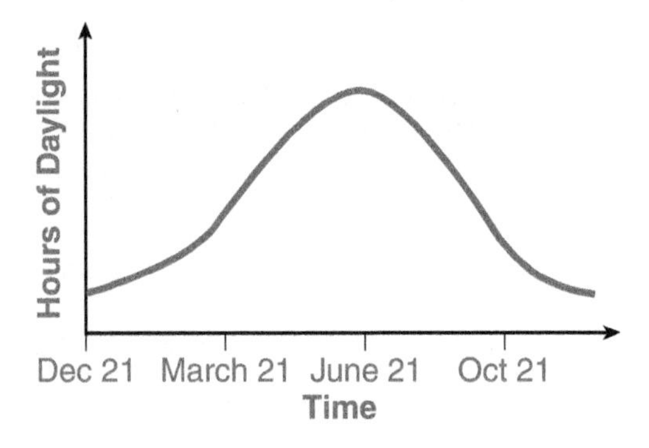

CLUSTER 2 Linear Functions

ACTIVITY 1.6

Walking for Fitness

Average Rate of Change

OBJECTIVE

Determine the average rate of change.

Suppose you are a member of a health and fitness club. Your personal trainer has developed a special diet and exercise program for you. At the beginning of the program and once a week thereafter, you are tested on the treadmill. The test consists of how many minutes it takes you to walk, jog, or run 3 miles on the treadmill. The following data gives your time, t, over an 8-week period:

End of Week, w	0	1	2	3	4	5	6	7	8
Time, t (Min.)	45	42	40	39	38	38	37	39	36

Note that $w = 0$ corresponds to the first time on the treadmill, $w = 1$ is the end of the first week, $w = 2$ is the end of the second week, etc.

1. **a.** Is time, t, a function of weeks, w? If so, what are the input and output variables?

   Yes, time is a function of weeks. w is the input variable. t is the output variable.

   b. Plot the data points using ordered pairs of the form (w, t).

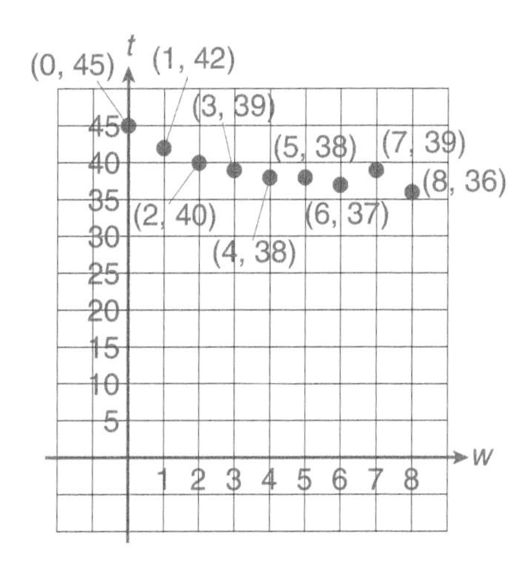

DEFINITION

A set of points in the plane whose coordinate pairs represent input–output pairs of a data set is called a **scatterplot**.

EXAMPLE 1

The points plotted in Problem 1b are a scatterplot of the treadmill data. Your graphing calculator can generate a scatterplot of data points. See Appendix C for instructions. The final screen should appear as follows:

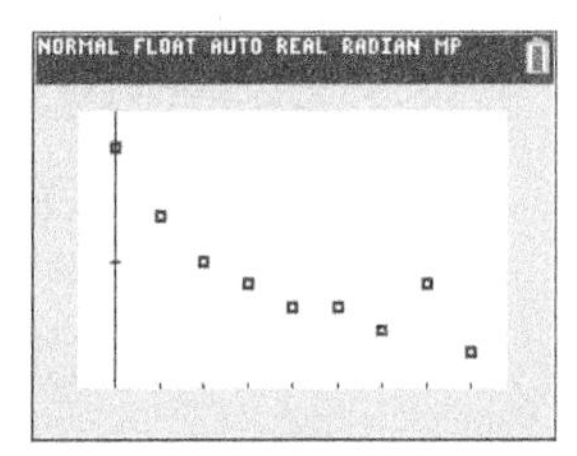

2. a. What was your treadmill time at the beginning of the program?

At the beginning of the program, my treadmill time was 45 minutes.

b. What was your treadmill time at the end of the first week?

My time at the end of the first week was 42 minutes.

An important question in this situation is how your time changed from one week to the next.

3. a. During which week(s) did your time increase?

My time increased only in week 7.

b. During which week(s) did your time decrease?

My time decreased in weeks 1, 2, 3, 4, 6, and 8.

c. During which week(s) did your time remain unchanged?

My time remained constant during week 5.

PROCEDURE

Determining Total Change

The change in time, t, is represented by the symbol Δt. The symbol Δ (delta) is used to represent "change in." You generally calculate the change in time, t, from a first (initial) value to a second (final) value of t. The first time is represented by t_1 (read "t sub 1"), and the second time is represented by t_2 (read "t sub 2"). The change in t is then calculated by subtracting the first (initial) value from the second (final) value. This is symbolically represented by

$$\Delta t = t_2 - t_1 \quad \text{or} \quad \Delta t = \text{final time} - \text{initial time}.$$

Because t is the output variable, Δt is the change in output.

Similarly, Δw represents the change in weeks, w; w_1 represents the first (initial) value of w; and w_2 represents the second (final) value of w. Symbolically,

$$\Delta w = w_2 - w_1 \quad \text{or} \quad \Delta w = \text{final week} - \text{initial week}.$$

Because w is the input variable, Δw is the change in input.

4. Your time decreased during each week of the first 4 weeks of the program.

a. Determine the total change in time, t, during the first 4 weeks of the program (i.e., from $t = 45$ to $t = 38$). Why should your answer contain a negative sign? Explain.

$38 - 45 = -7$.

The negative makes sense because my time decreased.

b. Determine the change in weeks, w, during this period (i.e., from $w = 0$ to $w = 4$).

$4 - 0 = 4$

5. Use the Δ notation to express your results in Problems 4a and 4b.

$\Delta t = -7$, $\Delta w = 4$

Average Rate of Change

Neither the change in treadmill time nor the change in the number of weeks completely describes your progress during the first 4 weeks. The ratio of the change in t, Δt, to the change in w, Δw, written $\frac{\Delta t}{\Delta w}$, provides more relevant information about the effect of the exercise program over time. This ratio, $\frac{\Delta t}{\Delta w}$, shows how the time changed on average over the 4-week period.

6. Use your results from Problem 5 to determine the ratio $\frac{\Delta t}{\Delta w}$ during the first 4-week period. Interpret your answer.

$\frac{\Delta t}{\Delta w} = -\frac{7}{4} = -1.75$. Your time decreased by an average of 1.75 min./week.

7. a. What are the units of measurement of the ratio determined in Problem 6?

minutes per week

b. On your graph from Problem 1, connect the points (0, 45) and (4, 38) with a line segment. Does the output increase, decrease, or remain unchanged over the interval?

The output decreases over the interval.

The ratio $\frac{\Delta t}{\Delta w}$ is called the **average rate of change** of time, t, with respect to weeks, w.

DEFINITION

If x represents the input and y represents the output, then the quotient

$$\frac{\text{change in output}}{\text{change in input}} = \frac{\Delta y}{\Delta x} = \frac{y_2 - y_1}{x_2 - x_1}$$

is called the average rate of change of y with respect to x over the x-interval from x_1 to x_2.

The units of measurement of the quantity $\frac{\Delta y}{\Delta x}$ are y-units per x-unit.

8. a. Determine the average rate of change of t with respect to w during the sixth and seventh weeks (from the point where $w = 5$ to the point where $w = 7$).

$\frac{\Delta t}{\Delta w} = \frac{39 - 38}{7 - 5} = \frac{1}{2}$ min./week

b. What is the significance of the positive sign of the average rate of change in this situation?

The output, time, increased over this time period.

c. Connect the data points (5, 38) and (7, 39) on your graph from Problem 1 using a line segment. Is the output increasing, decreasing, or constant on the interval?

The output is increasing.

LC LEARNING CATALYTICS

Consider the points (6, 15) and (10, 10). What is the average rate of change from 6 to 10? What is the significance of the sign of your result?

a. the slope is $-5/4$. the negative sign indicates a decrease in the output values as the input increases.
b. the slope is $-5/4$. The negative sign indicates a decrease in the input values as the output increases.
c. the slope is $-4/5$. The negative sign indicates that the output values decreased as the input values increased.
d. The slope is $-4/5$. The negative sign indicates that both the input and output values are decreasing.

9. a. At what average rate did your time change during the fifth week (from $w = 4$ to $w = 5$)?

$\frac{\Delta t}{\Delta w} = \frac{38 - 38}{5 - 4} = \frac{0}{1} = 0$ min./week

b. Interpret your answer in this situation.

My time did not change this week.

c. Connect the data points (4, 38) and (5, 38) on the graph from Problem 1 using a line segment. Is the output increasing, decreasing, or constant over this interval?

The output is constant.

10. a. At what average rate is your time changing as w increases from $w = 3$ to $w = 7$?

$$\frac{\Delta t}{\Delta w} = \frac{39 - 39}{7 - 3} = \frac{0}{4} = 0 \text{ min./week}$$

b. Does your answer mean that your time did not change in this 4-week period? Interpret your answer in this situation.

This means that the final output was the same as the initial output. It says nothing about what happened in the meantime.

11. As part of your special diet and exercise program, you record your weight at the beginning of the program and each week thereafter. The following data give your weight, w, over a 5-week period:

Weeks, *w*	0	1	2	3	4	5
Weight, *y* (lb.)	196	183	180	177	174	171

a. Determine the average rate of change of your weight during the first 3 weeks.

$$\frac{177 - 196}{3 - 0} = -\frac{19}{3} \approx -6.3 \text{ lb./week}$$

b. Determine the average rate of change during the 5-week period.

$$\frac{171 - 196}{5 - 0} = -\frac{25}{5} = -5 \text{ lb./week}$$

c. Determine the change in weight during each week of your exercise program.

first week change = −13 pounds, second week change = −3 pounds, third week change = −3 pounds, fourth week change = −3 pounds, and fifth week change = −3 pounds

d. What are the units of measure of the average rate of change?

lb./week

e. What is the practical meaning of the average rate of change in this situation?

The average rate of change indicates the rate at which I am losing weight every week.

f. What can you say about the average rate of change of weight during any time interval in this situation?

Weight is always decreasing because the average rate is always negative.

SUMMARY Activity 1.6

1. Let y_1 represent the corresponding output value for the x_1, and let y_2 represent the corresponding output value for the x_2. As the variable x changes in value from x_1 to x_2,

a. the change in x is represented by $\Delta x = x_2 - x_1$.

b. the change in y is represented by $\Delta y = y_2 - y_1$.

2. The quotient $\frac{\Delta y}{\Delta x} = \frac{y_2 - y_1}{x_2 - x_1}$ is called the **average rate of change** of y with respect to x over the x-interval from x_1 to x_2. The units of measurement of the quantity $\frac{\Delta y}{\Delta x}$ are y-units per x-unit.

3. The rate of change measures the change in output for a 1-unit increase in the input.

4. The line segment connecting the points (x_1, y_1) and (x_2, y_2)

   a. increases from left to right if $\frac{\Delta y}{\Delta x} > 0$.

   b. decreases from left to right if $\frac{\Delta y}{\Delta x} < 0$.

   c. remains level if $\frac{\Delta y}{\Delta x} = 0$.

5. The average rate of change indicates how much and in which direction the output changes when the input increases by a single unit. It measures how the output changes on average.

EXERCISES Activity 1.6

The following table of data from the U.S. Bureau of the Census gives the median age of an American man at the time of his first marriage:

Year	1920	1930	1940	1950	1960	1970	1980	1990	2000	2010	2015
Median Age	24.6	24.3	24.3	22.8	22.8	23.2	24.7	26.1	26.8	28.2	29.2

Use this data to answer Exercises 1–6.

1. a. Determine the average rate of change in median age per year from 1950 to 2010.

   $$\frac{28.2 - 22.8}{2010 - 1950} = \frac{5.4}{60} = 0.09 \text{ years of age/yr.}$$

   b. Describe what the average rate of change in part a represents in this situation.

   The median age of a man at the time of his first marriage is increasing at an average rate of 0.09 years of age/yr.

2. Determine the average rate of change in median age per year from 1930 to 1960.

   $$\frac{22.8 - 24.3}{1960 - 1930} = \frac{-1.5}{30} = -0.05 \text{ years of age/yr.}$$

3. What is the average rate of change over the 95-year period described in the table?

   $$\frac{29.2 - 24.6}{95} \approx 0.048 \text{ years of age/yr.}$$

4. During what 10-year period did the average age increase the most?

   From 1970 to 1980, the median age of a man at the time of his first marriage increased at an average rate of 0.15 years of age/yr.

5. a. What does it mean in this situation if the average rate of change is negative?

   It means that the median age of a man at the time of his first marriage is decreasing.

b. Determine at least one 10-year period when the average rate of change is negative.

1920–1930, or 1940–1950

c. What trend would you observe in the graph of median age if the average rate of change were negative? That is, would the graph rise, fall, or remain horizontal?

The graph would fall to the right.

6. a. Is the average rate of change zero over any 10-year period? If so, when?

yes, 1930–1940 and 1950–1960

b. What does a rate of change of zero mean in this situation?

The median age of a man at the time of his first marriage is on average constant.

c. What trend would you observe in the graph during this period? That is, would the graph rise, fall, or remain horizontal?

The graph would be horizontal.

7. The total amount of rainfall in a given community can vary widely from year to year. The following table gives information on the total rainfall over a recent 7-year period for Corning, New York:

	ANNUAL RAINFALL IN CORNING, NEW YORK						
Year, *t*	1	2	3	4	5	6	7
Rainfall, *r* (in.)	45.49	41.88	39.63	32.91	37.47	50.08	37.54

a. Plot the data points using ordered pairs of the form (t, r)

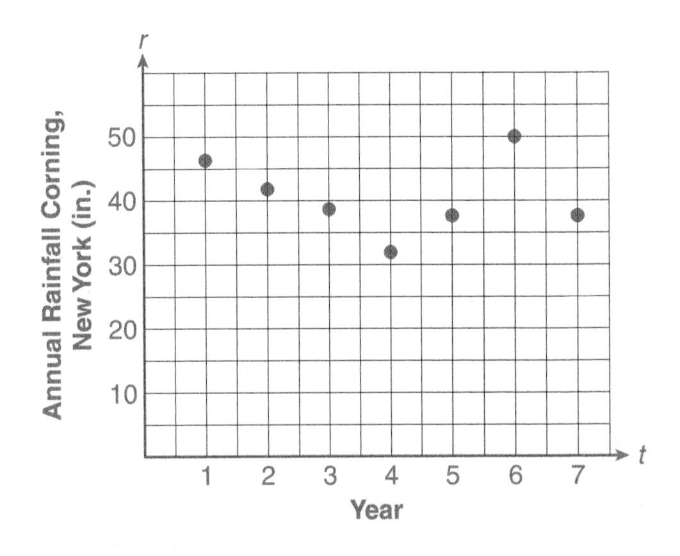

b. Determine the average rate of change of rainfall in Corning, New York, from year 4 to year 7.

$$\frac{37.54 - 32.91}{7 - 4} = \frac{4.63}{3} \approx 1.54 \text{ in./yr.}$$

c. Determine the average rate of change of rainfall in Corning, New York, from year 1 to year 4.

$$\frac{32.91 - 45.49}{4 - 1} = \frac{-12.58}{3} \approx -4.19 \text{ in./yr.}$$

d. Compare the average rate of change from year 1 to year 4 with the average rate of change from year 4 to year 7.

The annual rainfall in Corning, New York, decreased from year 1 to year 4. On average, the annual rainfall in Corning, New York, increased from year 4 to year 7.

e. When the average rate of change is negative, what trend do you observe in the graph? What does that mean in this situation?

The graph is going down to the right. The annual rainfall in Corning, New York, is decreasing.

8. Between 1960 and 2010, the size and shape of automobiles in the United States have changed almost annually. The fuel consumed by these vehicles has also changed. The following table describes the average fuel consumed per year per passenger car in gallons of gasoline:

Year, t	1960	1970	1980	1990	1995	2000	2005	2010	2015
Gallons Consumed per Passenger Car (average), g	668	760	576	520	530	547	567	453	480

Source: U.S. Department of Transportation.

a. Determine the average rate of change, in gallons of fuel used per passenger car, from 1960 to 1970.

$$\frac{\Delta g}{\Delta t} = \frac{760 - 668}{1970 - 1960} = \frac{92}{10} = 9.2 \text{ gal./yr.}$$

b. Determine the average rate of change, in gallons of gas per year, from 1960 to 1990.

$$\frac{520 - 668}{1990 - 1960} = \frac{-148}{30} \approx -4.93 \text{ gal./yr.}$$

c. Determine the average rate of change, in gallons of gas per year, from 1995 to 2005.

$$\frac{567 - 530}{2005 - 1995} = \frac{37}{10} = 3.7 \text{ gal./yr.}$$

d. Determine the average rate of change, in gallons of gas per year, between 1960 and 2015.

$$\frac{480 - 668}{2015 - 1960} = \frac{-188}{55} \approx -3.4 \text{ gal./yr.}$$

e. What does the result in part d mean in this situation?

It means that from 1960 to 2015, the average fuel consumption per year of a passenger car in the United States decreased by about 3.4 gal./yr.

ACTIVITY 1.7

Depreciation

Slope-Intercept Form of an Equation of a Line

OBJECTIVES

1. Interpret slope as an average rate of change.
2. Use the formula to determine slope.
3. Discover the practical meaning of vertical and horizontal intercepts.
4. Develop the slope-intercept form of an equation of a line.
5. Use the slope-intercept formula to determine vertical and horizontal intercepts.
6. Determine the zeros of a function.

You have decided to buy a new Honda Civic, but you are concerned about the value of the car depreciating over time. You search the Internet and obtain the following information:

- Suggested retail price: $22,905
- Depreciation per year: $1750 (assume constant)

1. a. Complete the following table in which V represents the value of the car after n years of ownership:

n, YEARS	V, VALUE IN DOLLARS
0	22,905
1	21,155
2	19,405
3	17,655
5	14,155
8	8,905

b. Is the value of the car a function of the number of years of ownership? Explain.

Yes. For each year of ownership, n, there is only one value, V.

c. What is the input? What is the output?

The input is the number of years of ownership, n. The output is the value, V.

2. a. Select two ordered pairs of the form (n, V) from the table in Problem 1, and determine the average rate of change.

(Chosen values may vary.) $\frac{21{,}155 - 22{,}905}{1 - 0} = -1750$

b. What are the units of measure of the average rate of change?

dollars/year

c. What is the practical meaning of the sign of the average rate of change?

The value of the car is decreasing.

d. Select two different ordered pairs, and compute the average rate of change.

(Chosen values may vary.) $\frac{17{,}655 - 19{,}405}{3 - 2} = -1750$

e. Select two ordered pairs not used in parts a or d, and compute the average rate of change.

(Chosen values may vary.) $\frac{8905 - 14{,}155}{8 - 5} = -1750$

f. Using the results in parts a, d, and e, what can you infer about the average rate of change over any interval of time?

The average rate of change is constant at $-\$1750/\text{yr}$.

If the computation of the average rate of change using any two ordered pairs yields the same result, the average rate of change is said to be constant.

DEFINITION

Any function in which the average rate of change, $\frac{\text{change in output}}{\text{change in input}}$, between any pair of points is constant is called a **linear function**.

3. **a.** Is the value, V, of the car a linear function of the number of years, n, of ownership? Explain using the definition of linear function.

   Yes, because the average rate of change is a constant −$1750/yr.

   b. Is this function increasing, decreasing, or constant?

   The function is decreasing.

Graph of a Linear Function

4. Consider the ordered pairs of the form (n, V), and plot each ordered pair in Problem 1 on an appropriately scaled and labeled set of axes. Connect the points to see if there is a pattern.

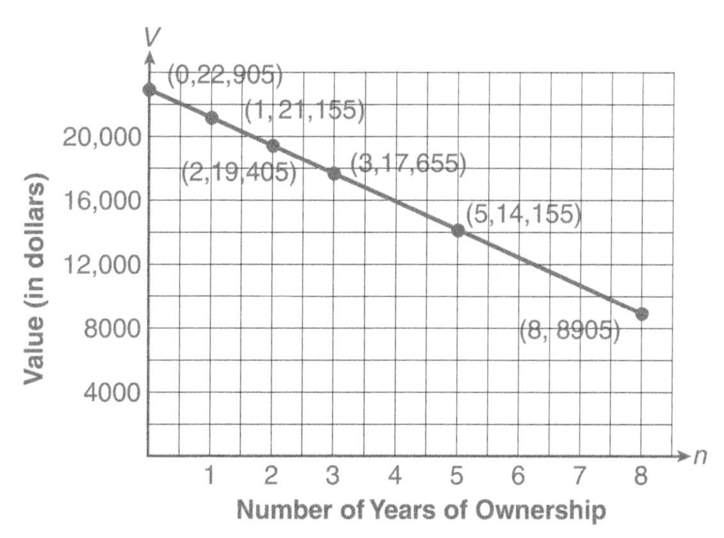

The graph of a linear function is a nonvertical line. The constant average rate of change is called the **slope** of the line and is denoted by the letter m.

DEFINITION

If x represents the input variable and y represents the output variable of a linear function, then the **slope m** is given by

$$m = \frac{\Delta y}{\Delta x} = \frac{y_2 - y_1}{x_2 - x_1}, \text{ where } x_1 \neq x_2.$$

5. a. What is the slope of the line graphed in Problem 4?

The slope is −1750.

b. What is the relationship between the slope of the line and the average rate of change?

They are the same.

c. What is the practical meaning of slope in this situation?

The value of the car decreases (depreciates) at a constant rate of \$1750/yr.

Vertical Intercept

DEFINITION

The **vertical intercept** is the point where the graph of a function crosses, or intercepts, the vertical axis. The input value of a vertical intercept is always zero. If the output variable is represented by y, the vertical intercept is referred to as the ***y*-intercept**. The vertical intercept is written in the form of $(0, b)$ or $(0, y)$.

6. a. Using the table of data in Problem 1 or the graph in Problem 4, determine the vertical intercept (V-intercept).

(0, 22,905)

b. What is the practical meaning of the vertical intercept in this situation? Include units.

The retail (initial) price was \$22,905.

Slope-Intercept Form of a Linear Equation

7. a. Review how you determined the value, V, of the car in Problem 1 for a given number of years, n, of ownership. Write an equation for V in terms of n.

$V = 22{,}905 - 1750n$

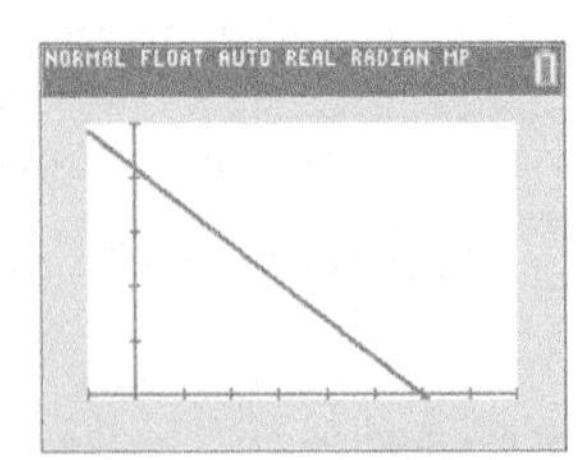

b. Use your graphing calculator to sketch a graph of this equation. Use the window $X_{min} = -2$, $X_{max} = 16$, $Y_{min} = -500$, and $Y_{max} = 25{,}000$.

c. How does this graph compare with your graph in Problem 4?

It is the same.

8. Recall that the slope of your line is $m = -1750$ and that the vertical intercept is (0, 22,905). How is this information contained in the equation of the line you determined in Problem 7a?

The slope is the coefficient of n. The constant term is the output value of the vertical intercept.

DEFINITION

The coordinates of all points (x, y) on the line with slope m and vertical intercept $(0, b)$ satisfy the equation

$$y = mx + b \quad \text{or} \quad y = b + mx.$$

This is called the **slope-intercept form** of the equation of a line.

Note that the coefficient of x, which is m, is the *slope* of the line. The constant term, b, is the *y-coordinate of the vertical intercept*. If $f(x)$ replaces y, the equation $y = mx + b$ can be written as

$$f(x) = mx + b.$$

EXAMPLE 1 ***The slope-intercept form of the equation of the line with slope 3 and vertical intercept $(0, -6)$ is $y = 3x - 6$. Using function notation and replacing y with $f(x)$, the equation becomes $f(x) = 3x - 6$.***

LC LEARNING CATALYTICS

Identify the slope of the line defined by $y = -3x + 4$.

9. Identify the slope and vertical intercept of the line whose equation is given. Write the vertical intercept as an ordered pair.

a. $y = -2x + 5$

slope $= -2$

vertical intercept $(0, 5)$

b. $s = \frac{3}{4}t + 2$

slope $= \frac{3}{4}$

vertical intercept $(0, 2)$

c. $q = 2 - r$

slope $= -1$

vertical intercept $(0, 2)$

d. $y = \frac{5}{6} + \frac{x}{3}$

slope $= \frac{1}{3}$

vertical intercept $\left(0, \frac{5}{6}\right)$

EXAMPLE 2 ***In the equation $f(x) = 3x - 6$ in Example 1, the slope is 3 and the vertical intercept is $(0, -6)$. You can now sketch the graph of this function by plotting one point and using the slope to obtain additional points on the line.***

Plot the vertical intercept $(0, -6)$. To use the slope $(m = 3)$ to obtain additional points, write the slope as a fraction, $= 3 = \frac{3}{1} = \frac{\Delta y}{\Delta x}$. Starting with the vertical intercept, $(0, -6)$, the slope tells you to go up 3 units $(\Delta y = 3)$ and to the right 1 unit $(\Delta x = 1)$ to obtain another point $(1, -3)$. Now, use $(1, -3)$ and the slope to obtain another point in the same way, $(2, 0)$. Connect the points to graph the line. Another option is to write the slope as $= 3 = \frac{-3}{-1} = \frac{\Delta y}{\Delta x}$. Now starting at the y-intercept, $(0, -6)$, go down 3 $(\Delta y = -3)$ and left 1 $(\Delta x = -1)$ to obtain another point $(-1, -9)$.

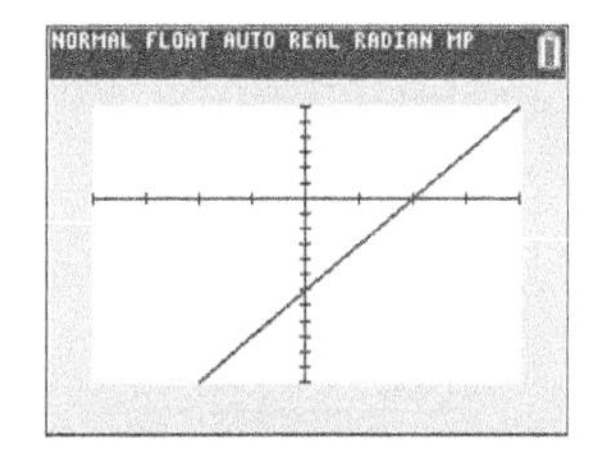

10. Graph each of the following from Problem 9 using the vertical intercept and the slope of the line:

a. $y = -2x + 5$

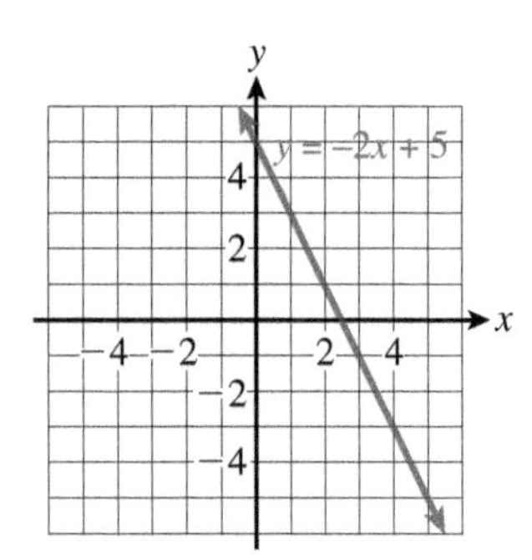

b. $s = \frac{3}{4}t + 2$

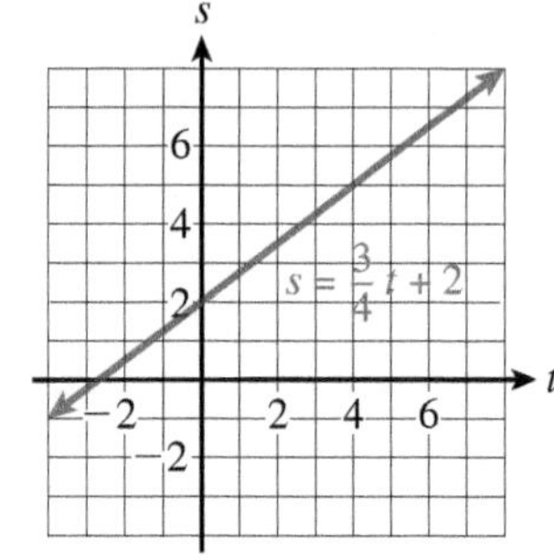

c. $q = 2 - r$

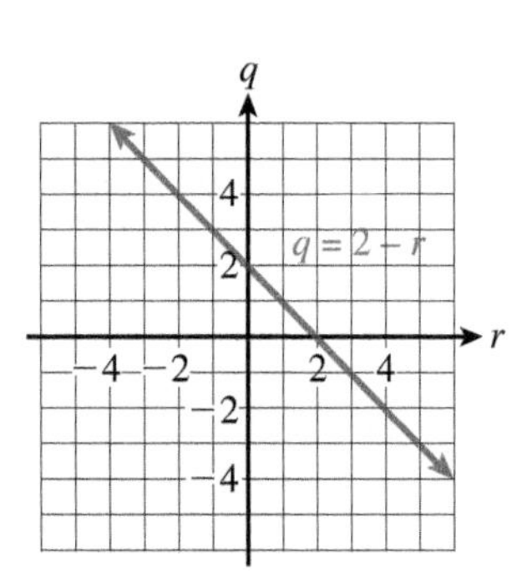

d. $y = \frac{5}{6} + \frac{x}{3}$

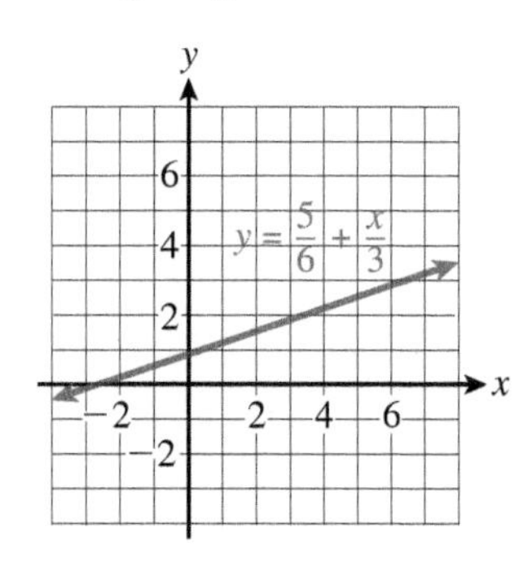

Horizontal Intercepts

DEFINITION

A **horizontal intercept** of a graph is a point where the graph meets or crosses the horizontal axis. The output value of a horizontal intercept is always zero. If the input variable is represented by x, the horizontal intercept is referred to as the ***x*-intercept**. The horizontal intercept is written in the form of $(a, 0)$ or $(x, 0)$.

EXAMPLE 3 ***Consider the equation $y = 2x - 10$.***

a. The vertical intercept (y-intercept) occurs where the line crosses the vertical axis, that is, where $x = 0$. Letting $x = 0$, $y = 2(0) - 10$, or $y = -10$. The vertical intercept is $(0, -10)$.

b. The horizontal intercept (x-intercept) occurs where the line crosses the horizontal axis, that is, where $y = 0$.

Letting $y = 0$,

$$0 = 2x - 10$$

$$\underline{+10 \qquad + 10} \qquad \text{Add 10 to each side.}$$

$$10 = 2x$$

$$\frac{10}{2} = \frac{2x}{2} \qquad \text{Divide each side by 2.}$$

$$5 = x$$

Therefore, the horizontal intercept is $(5, 0)$.

c. You can now sketch a graph of $y = 2x - 10$ by plotting the horizontal and vertical intercepts and connecting the points.

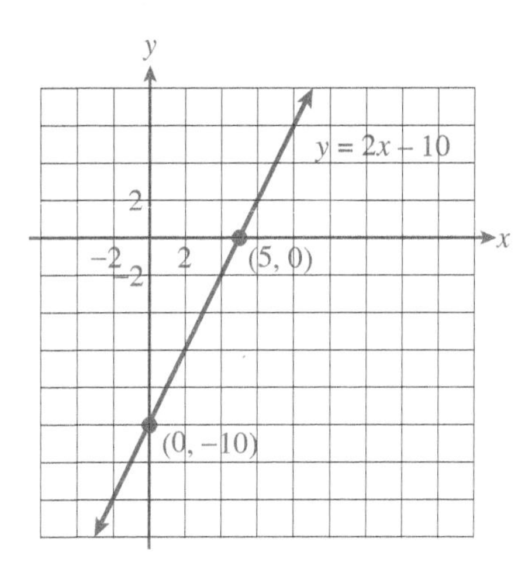

11. Determine the vertical and horizontal intercepts of the graph of the given linear equation. Write the intercepts as ordered pairs and use them to graph the line.

a. $y = 2x - 6$

(0, –6) (3, 0)

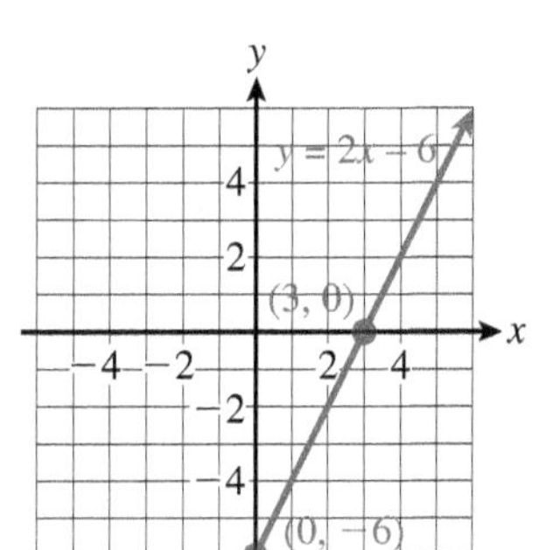

b. $y = 4 - x$

(0, 4) (4, 0)

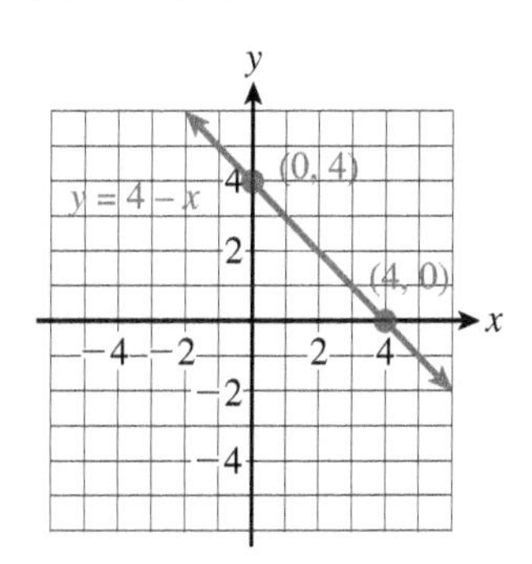

Converting Celsius to Fahrenheit

12. The temperature on a warm summer day may be 27°. Your refrigerator is set at 5°. You set your oven to 180° to cook a turkey. If these temperatures seem strange, it is because they are measured on the Celsius scale. If the TV weather forecaster says that it is going to be 65°F tomorrow, you know what that temperature will feel like and you can dress accordingly. How should you dress if the reporter says that it will be 10°C? To answer such a question, you convert Celsius to Fahrenheit using the equation

$$F = 1.8C + 32,$$

where C is the Celsius measure and F is the Fahrenheit measure.

a. Use the equation to complete the following table:

C	−10	0	5	10	27	100
F	14	32	41	50	80.6	212

b. Sketch a graph of the equation $F = 1.8C + 32$ by plotting the ordered pairs in part a.

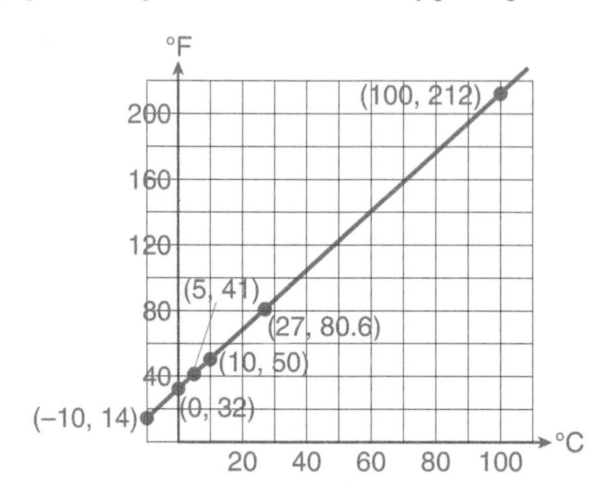

c. Determine the average rate of change of F as C increases from 0 to 100.

1.8°F/°C

d. What is the slope of the line? How does the slope compare to the average rate of change in part c?

$m = 1.8$°F/°C. They are the same.

e. Is the function increasing, decreasing, or constant?

increasing

f. Determine the vertical and horizontal intercepts of the graph using the equation of the line. Verify your results using the graph of the line.

vertical intercept (0, 32); horizontal intercept (−17.8, 0)

g. What is the practical significance of each intercept in this situation?

When $C = 0$, $F = 32$, and when $C = -17.8$, $F = 0$.

h. When does $F = C$?

when $F = C = -40°$

Zeros of a Function

A value, a, is a zero of a function f if $f(a) = 0$. In other words, a zero of a function is an input value that makes the output zero.

EXAMPLE 4 *If $f(x) = 2x - 6$, then $f(3) = 2(3) - 6 = 0$. So 3 is a zero of f.*

13. If the value, 3, is a zero of a function f, then $f(3) = 0$.

a. Write the ordered pair corresponding to $f(3) = 0$.

(3, 0)

b. Describe the location of the point whose ordered pair is given in part a.

(3, 0) is on the x-axis.

Zeros and x-intercepts

If a value, a, is a zero of a function f, then $f(a) = 0$ and $(a, 0)$ is an x-intercept of the graph of $y = f(x)$.

14. a. Determine the zero of $f(x) = 3x + 12$ by setting $f(x) = 0$ and solving for x.

$3x + 12 = 0, 3x = -12, x = -4.$

b. What is the x-intercept of the graph of $f(x) = 3x + 12$?

The x-intercept is $(-4, 0)$.

SUMMARY Activity 1.7

1. A function for which the average rate of change between any pair of points remains constant is called a **linear function**.

2. The graph of a linear function is a nonvertical line. The constant average rate of change is called the **slope** and is denoted by the letter m.

3. The **slope** of a line segment joining two points (x_1, y_1) and (x_2, y_2) is denoted by m and is given by $m = \frac{y_2 - y_1}{x_2 - x_1}$, where $x_1 \neq x_2$. The slope gives the change in output for a 1-unit increase in the input.

4. The **vertical intercept** $(0, b)$ of a graph is the point where the graph crosses the vertical axis. The **horizontal intercept** $(a, 0)$ of a graph is the point where the graph crosses the horizontal axis.

5. The **slope-intercept form** of the equation of a line is $\mathbf{y = mx + b}$.

6. To determine the b-value of the vertical intercept $(0, b)$ from $y = mx + b$, set $x = 0$ and solve for y.

7. Determine the zero of $f(x) = mx + b$ by setting $f(x) = 0$. If the value, a, is a zero of the function, then $f(a) = 0$ and $(a, 0)$ is an x-intercept of the graph of $y = f(x)$.

8. To determine the a-value of the horizontal intercept $(a, 0)$ from $y = mx + b$, set $y = 0$ and solve for x.

9. If the **slope** of a linear function is **positive**, the graph of the function rises to the right.

10. If the **slope** of a linear function is **negative**, the graph of the function falls to the right.

11. There are three methods to graph a linear equation. You may choose to plot points, plot one point and use the slope to plot additional points, or plot the intercepts.

EXERCISES Activity 1.7

1. Determine whether the following functions are linear. That is, determine whether the average rate of change is constant.

   a.

x	y
0	−1
1	9
2	19

   Yes, linear; the constant rate of change is 10.

   b.

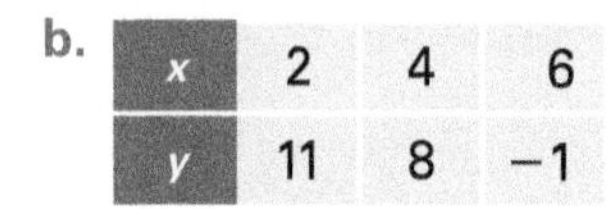

x	2	4	6
y	11	8	−1

   No, not linear.

   c. $\{(-2, 18), (2, 9), (6, 0)\}$

   Yes, linear; the constant rate of change is $\frac{-9}{4}$.

   d.

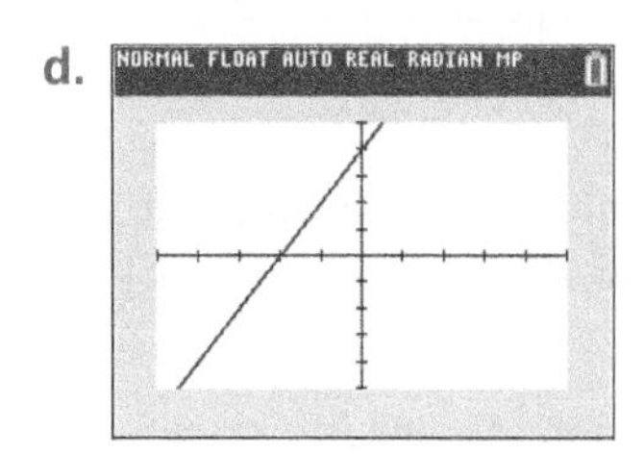

   Yes, linear; the constant rate of change is 2, assuming that each tic mark represents 1 unit.

   e.

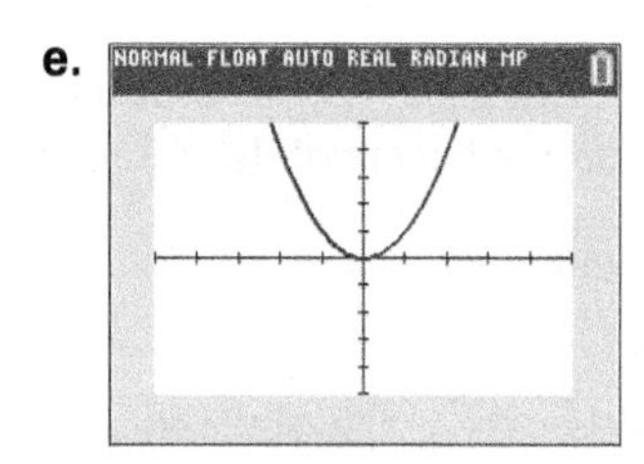

   No, not linear; the rate of change is not constant.

2. Determine whether the following tables contain data that represent a linear function. Assume that the first row of the table is the input and the second row is the output. Explain your reasoning.

   a. Your uncle helped you financially your first semester at the community college. As such, you owe him \$1000. The conditions of the loan are that you must pay him back the whole amount in one payment using a simple interest rate of 2% per year. He doesn't care in which year you

Exercise numbers appearing in color are answered in the Selected Answers appendix.

pay him. The table contains input and output values to represent how much money you will owe your uncle one, two, three, or four years later.

Year	1	2	3	4
Amount Owed ($)	1020	1040	1060	1080

Yes, the average rate of change is a constant $20/yr.

b. You are driving along the highway, and just before you reach the crest of a hill, you notice a sign that indicates the elevation at the crest is 2250 feet. As you proceed down the hill, your elevation is as given by the following table:

Distance Traveled from the Crest of the Hill (ft.)	0	1000	3000	6000	10,000
Elevation (ft.)	2250	2180	2040	1830	1550

Yes, this data is linear. The constant average rate of change is −0.07.

c. You decide to invest $1000 of your 401k funds into an account that pays 3.5% interest compounded continuously. The table contains input and output values that represent the amount an initial investment of $1000 is worth at the end of each year.

Year	1	2	3	4	5
Total Investment ($)	1036	1073	1111	1150	1191

No, the rate of change is $37/yr. during the first year and $41/yr. during the fourth year.

d. For a fee of $40 per month, you may have breakfast (all you can eat) at the college snack bar each day. The table contains input and output values that represent the total number of breakfasts consumed each month and the amount you pay each month.

Number of Breakfasts	10	22	16	13
Cost ($)	40	40	40	40

Yes. The output is constant, so the slope is a constant 0.

3. You belong to a health and fitness center. You and your friends are enrolled in the center's weight-loss program. The charts contain input and output values (assume that week is input and weight is output) that represent the weight over a 4-week period for you and your two friends. Determine which charts contain data that is linear and explain why.

a.

Week	1	2	3	4
Weight (lb.)	150	147	144	141

Yes, the average rate of change is a constant −3.

b.

Week	1	2	3	4
Weight (lb.)	183	178	174	171

No. Between weeks 1 and 2, the slope is −5. Between weeks 2 and 3, the slope is −4.

c.

Week	1	2	3	4
Weight (lb.)	160	160	160	160

Yes, the slope is 0 for all pairs of points.

3. a. What is the slope, m, of the payment function?

$m = 40$

b. What is the initial value, b, of the payment function?

$b = 240$

c. Write the linear equation that gives the amount paid, A, as a function of the number of payments made, n. Note that the equation will have the general form $A = mn + b$.

$A = 40n + 240$

d. Graph the equation from part c for values of n between 0 and 24.

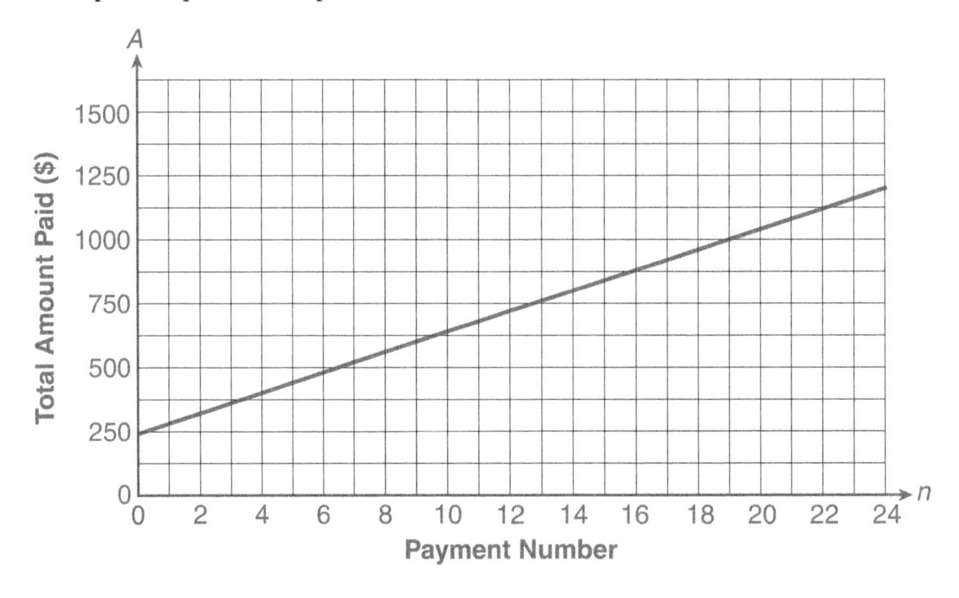

e. What is the slope of the line graphed in part d? What is the practical meaning of the slope in this situation?

slope, $m = 40$; each payment is $40.

f. What is the vertical intercept of the line graphed in part d (write the answer as an ordered pair)? What is the practical meaning of the vertical intercept in this situation?

(0, 240) The down payment is $240; that is, $240 is the amount paid with $n = 0$.

4. The number of students at a community college was 1100 in 1994 and 1940 in 2018. The number of students increased at a constant rate over this 24-year period. Let t represent the number of years after 1994 and N represent the number of students at the community college in a given year.

a. Use the given enrollment figures at the community college to write two ordered pairs of the form (t, N). Note that 1994 is $t = 0$.

(0, 1100) *and* (24, 1940)

b. Use the result from part a to complete the following table:

YEARS AFTER 1994, t	NUMBER OF STUDENTS, N
0	1100
24	1940

c. What is the initial value of this function? (Remember, the initial value is the output when the input is zero.)

The initial value is 1100 students.

d. What is the average rate of change of this function?

The average rate of change is 35 students per year.

e. Is the average rate of change constant? Explain.

Yes, the number of students increases at a constant average rate of 35 students per year.

f. Write a linear equation that gives the number N of community college students as a function in terms of t, the number of years after 1994 $(N = mt + b)$.

$N = 35t + 1100$

Point-Slope Form of an Equation of a Line

In Problem 4 you were given two points, one of which had an input value of zero. You determined the slope and produced a linear equation in slope-intercept form. In the next situation, you are given two points, neither of which has an input value of zero, and are asked to determine a linear equation.

5. The cost for a part-time student is determined by a fixed activity fee plus a fixed tuition amount per credit. The cost for a student taking 6 credits is $1520. The cost for a student taking 9 credits is $2255. The total cost, C, is a function of the number of credits taken, n.

a. Write two ordered pairs of the form (n, C) for the college cost function.

(6, 1520) and (9, 2255)

b. Use the result from a part a to complete the following table:

NUMBER OF CREDITS, n	TOTAL COST, C ($)
6	1520
9	2255

c. Determine the average rate of change for the college cost function.

$$\frac{\Delta C}{\Delta n} = \frac{2255 - 1520}{9 - 6} = \frac{735}{3} = 245$$

d. Interpret the average rate of change in the context of the college costs.

Each credit costs $245.

e. Is the average rate of change constant? Explain.

Yes, the cost increases by $245 for each additional credit.

The cost, C, is a linear function of the number of credits, n. Therefore, the relationship between C and n can be modeled by an equation written in the slope-intercept form, $C = b + mn$.

6. a. What is the value of m?

The value of m is 245, the average rate of change.

b. Write the equation $C = b + mn$, replacing m with its value.

$C = b + 245n$

c. Use the ordered pair (6, 1520) from Problem 5a, and rewrite the equation in part b by replacing C and n with the values in the ordered pair.

Using (6, 1520), $1520 = b + 245(6)$.

d. Solve the equation in part c for b.

$1520 = b + 245(6)$

$1520 = b + 1470$

$50 = b$

e. Interpret the value of b in the college cost function.

The $50 represents some fee, like an activity fee or a parking fee. b is also the vertical intercept of the graph of the college cost function.

f. Finally, rewrite the equation $C = b + mn$, replacing b and m with their respective values.

$C = 245n + 50$

In Problems 5 and 6, you determined a linear equation in slope-intercept form, using two ordered pairs that satisfy the equation. You determined the slope in Problem 5c and the vertical intercept in Problem 6d. There is a general method for writing a linear equation from two ordered pairs, based on the method of Problems 5 and 6.

Let (x_0, y_0) be a known point on a line, and let (x, y) be any other point on the line. The slope of the line is determined by the two ordered pairs, (x_0, y_0) and (x, y).

$$m = \frac{\Delta y}{\Delta x} = \frac{y - y_0}{x - x_0}, \text{ so}$$

$$y - y_0 = m(x - x_0) \quad \text{or} \quad y = y_0 + m(x - x_0)$$

The last equation, $y = y_0 + m(x - x_0)$, gives a linear equation in **point-slope form.**

EXAMPLE 1 ***Write an equation of the linear function whose graph is a line containing the points $(-1, 3)$ and $(2, 9)$.***

SOLUTION

Step 1. Determine the slope of the line.

$$m = \frac{9 - 3}{2 - (-1)} = \frac{6}{3} = 2$$

Step 2. Choose any point on the line, and substitute for x_0 and y_0. Using the point-slope form $y = y_0 + m(x - x_0)$ with $m = 2$ and $(x_0, y_0) = (-1, 3)$, you have

$$y = 3 + 2(x - (-1))$$
$$y = 3 + 2(x + 1)$$
$$y = 3 + 2x + 2, \text{ or } y = 2x + 5.$$

Note that in Problem 5, you determined two ordered pairs, $(6, 1520)$ and $(9, 2255)$, for the college cost function and then used the two ordered pairs to determine the slope, $m = 245$. Because the input and output variables are n and C, respectively, the point-slope form is $C = C_0 + m(n - n_0)$.

7. a. Use the ordered pair $(6, 1520)$ for (n_0, C_0) and $m = 245$ to write the point-slope form of the college cost function.

$C = 1520 + 245(n - 6)$

LC LEARNING CATALYTICS

Write the equation of the line that passes through $(-1, 4)$ and $(3, -4)$ in slope-intercept form.

b. Simplify the right side of the point-slope form of the equation, solve for C, and compare the result to the slope-intercept form in Problem 6f.

$C = 1520 + 245n - 1470$, or $C = 245n + 50$

The result is the same as the result of Problem 6f.

c. Repeat Problems 7a and 7b using the ordered pair $(9, 2255)$ for (n_0, C_0).

$2255 + 245(n - 9)$

$2255 + 245n - 2205$ or $C = 245n + 50$

The result is the same as the result of Problem 6f.

Problem 7 demonstrates that you obtain the same equation of the line regardless of which point you use for (n_0, C_0).

8. Follow the steps below to write the point-slope form of the equation of the line containing the points $(2, 4)$ and $(5, -2)$.

a. Use the given points to determine the slope, m.

$m = (-2 - 4) \div (5 - 2) = -6 \div 3 = -2$

b. Use the point $(2, 4)$ for (x_0, y_0) to write the point-slope form of the equation.

$y = 4 - 2(x - 2)$

Parallel Lines

Recall in Problems 1 through 3 that you developed the equation $A = 40n + 240$, where A represented the amount paid toward the camera and lens after n monthly payments. Suppose you are considering ordering a tripod. The cost for the tripod is \$160. You do not want to increase your monthly payments, so you want to pay for the tripod by adding its cost to the down payment.

9. a. What will be the amount of the down payment if you add the cost of the tripod?

$240 + 160 = \$400$

b. Write the linear equation that gives the amount paid, A, as a function of the number of payments made, n, with the new down payment amount. Remember, the payment amount is unchanged at \$40 per month.

$A = 400 + 40n$

c. Graph the equation from part b on the grid from Problem 3b copied below.

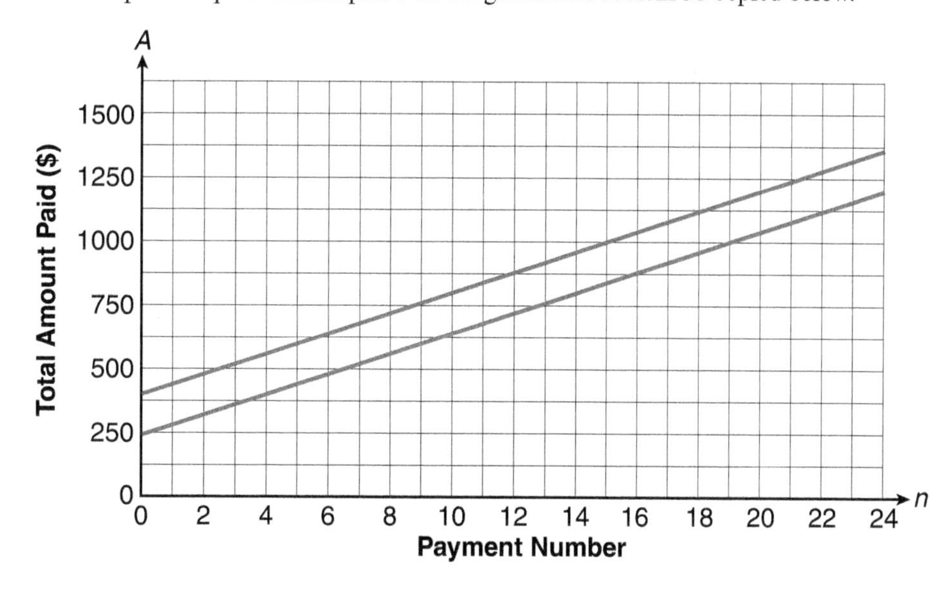

d. What can you say about the graphs of the two lines?

The lines are parallel.

e. Compare the equations of the two lines. What is same, and what is different?

The slopes are the same and the vertical intercepts are different.

DEFINITION

Two lines are **parallel** if the lines have equal slopes but different vertical intercepts.

10. Write an equation of a line parallel to the line whose equation is $y = 0.5x + 4$.

$y = 0.5x + b$, where b is any number except 4

SUMMARY Activity 1.8

1. To determine the equation of a line when two points on the line are known:

Step 1. Determine the **slope** of the line: $m = \dfrac{\Delta \text{ output}}{\Delta \text{ input}}$.

Step 2. If the input value of one of the points is zero, then b is the output value and you can write the equation in **slope-intercept form,** $y = mx + b$.

Step 3. If neither point has an input value of zero, then choose one of the points as (x_0, y_0) and write the equation in **point-slope form,** $y = y_0 + m(x - x_0)$.

2. Parallel lines have the same slope with different y-intercepts.

EXERCISES Activity 1.8

1. Write an equation of the line to satisfy the given conditions. The final equation should be solved for the ouput y.

a. The slope is $\frac{1}{2}$; the vertical intercept is -1.

$y = \frac{1}{2}x - 1$

b. The line contains the points $(-3, 5)$ and $(0, 1)$.

$y = -\frac{4}{3}x + 1$

c. The slope is -3, and the line contains the point $(-2, 1)$.

$y = 1 + (-3)(x + 2)$, or $y = 1 - 3x - 6$, or $y = -3x - 5$

Exercise numbers appearing in color are answered in the Selected Answers appendix.

d. The line contains the points $(-4, -3)$ and $(2, 6)$.

$m = \frac{6-(-3)}{2-(-4)} = \frac{6+3}{2+4} = \frac{9}{6} = \frac{3}{2}$, $y = 6 + \frac{3}{2}(x-2)$ or $y = \frac{3}{2}x + 3$

e. The line is parallel to the line $y = 3x - 7$ and passes through the point $(2, -5)$. Recall that parallel lines have the same slope.

$y = -5 + 3(x-2)$, $y = -5 + 3x - 6$ or $y = 3x - 11$

2. The following table gives the cost, c, of a car rental as a function of miles driven, x:

X (mi.)	C ($)
0	35
100	40

Assume that the function is linear and that you want to determine the equation of the line from the table.

a. What is the vertical intercept? How do you know this?

(0, 35) The vertical intercept occurs where the input $x = 0$.

b. Calculate the slope $m = \frac{\Delta c}{\Delta x}$ from the data in the table. What is the practical meaning of this slope?

$m = \frac{40-35}{100-0} = \frac{5}{100} = 0.05$

The mileage charges are $0.05 per mile.

c. Use your results from parts a and b to write the equation of the line in slope-intercept form.

$c = 0.05x + 35$

3. The following table gives the distance, d, in miles of a boat from a marina as a linear function of time, t, in hours:

t (hr.)	d (mi.)
2	75
4	145

a. Determine the slope of the line. What is the practical meaning of slope in this situation?

$m = \frac{145-75}{4-2} = \frac{70}{2} = 35$ mph. This represents the average rate of change or the average speed of the boat from $t = 2$ to $t = 4$.

b. Write the equation of the line in slope-intercept form.

$d = 75 + 35(t-2)$

$d = 35t + 5$

4. For each graph below,

i. determine the slope of the graph.

ii. determine the y-intercept of each graph.

iii. write the equation of the line whose graph is given.

Each tick mark denotes 1 unit.

a.

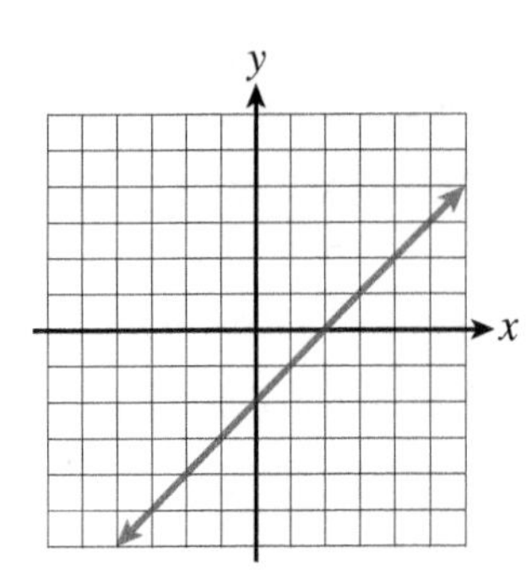

i. $m = 1$

ii. $(0, -2)$

iii. $y = x - 2$

b.

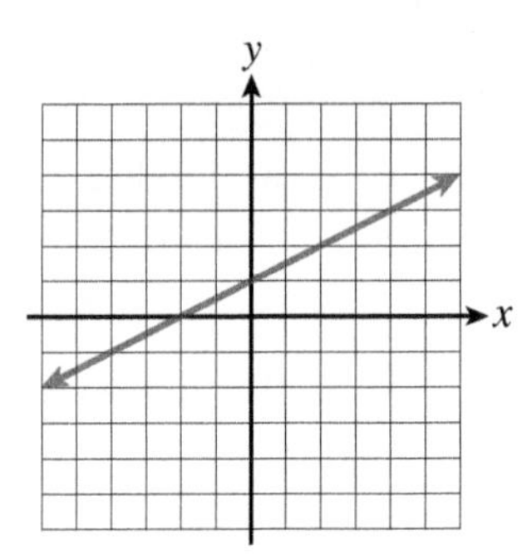

i. $m = \frac{1}{2}$

ii. $(0, 1)$

iii. $y = \frac{1}{2}x + 1$

c.

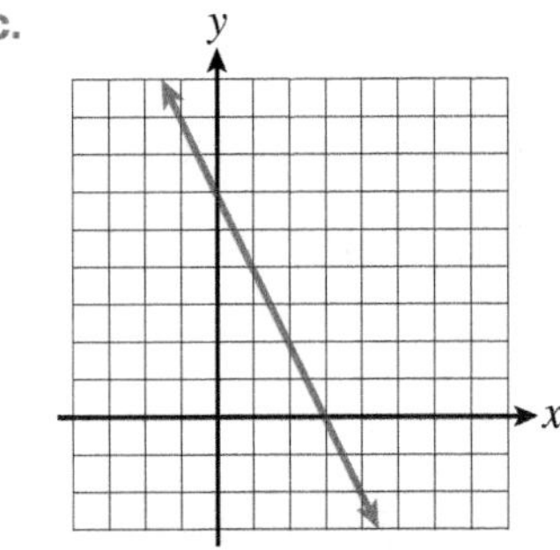

i. $m = -\frac{6}{3} = -2$

ii. $(0, 6)$

iii. $y = -2x + 6$

d.

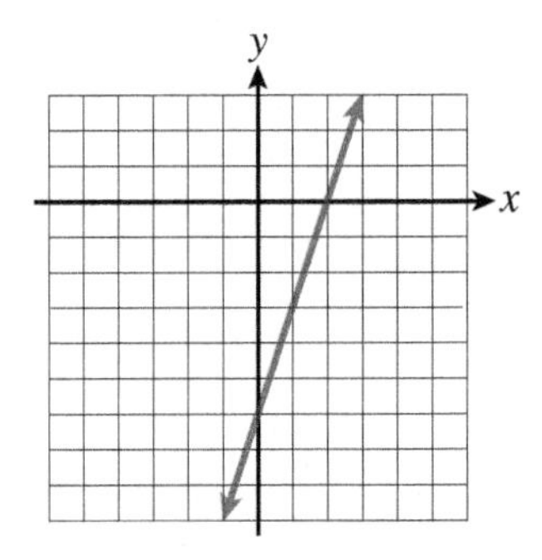

i. $m = \frac{6}{2} = 3$

ii. $(0, -6)$

iii. $y = 3x - 6$

5. The following graph represents the distance from home of a car as a function of time (in hours):

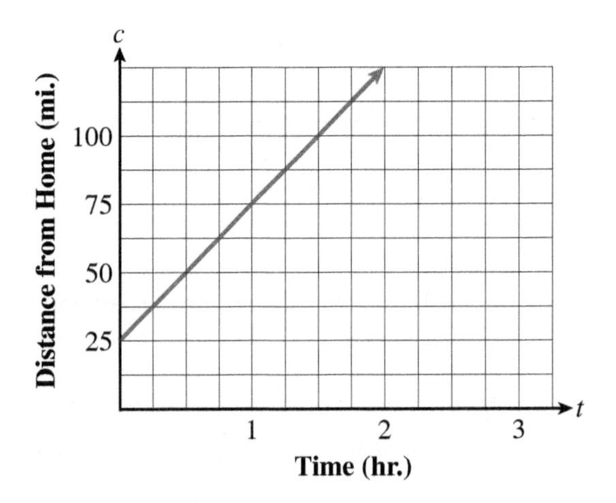

a. How fast is the car traveling? 50 mph

b. Determine the vertical intercept. $(0, 25)$

c. Write the equation of the line in slope-intercept form. $y = 50t + 25$

6. a. Graph a line with a slope of 4 that goes through the point (1, 5). Write the equation of the line.

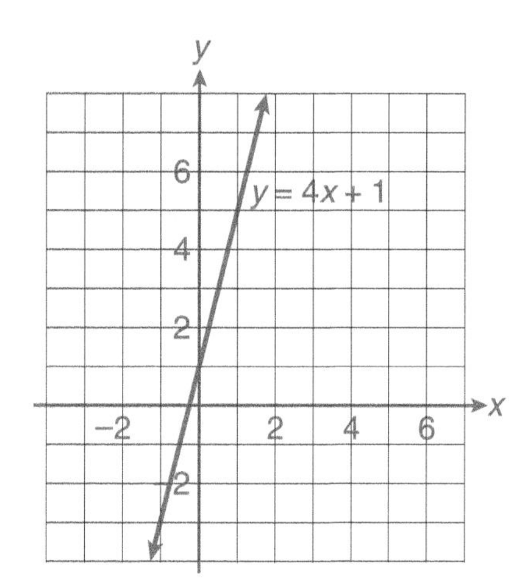

$y = 5 + 4(x - 1)$

$y = 4x + 1$

b. Graph a line with a slope of $-\frac{1}{2}$ that goes through the point $(-2, 3)$. Write the equation of the line.

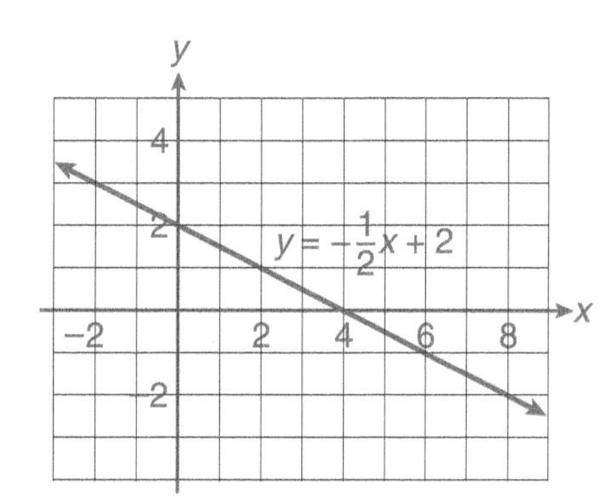

$y = 3 - \frac{1}{2}(x - (-2))$

$y = 3 - \frac{1}{2}(x + 2)$

$y = 3 - \frac{1}{2}x - 1$

$y = -\frac{1}{2}x + 2$

7. The data in the following table show the circumference of a circle as a function of its radius:

r (radius in ft.)	0	5	10	15
C (circumference in ft.)	0	31.42	62.83	94.25

a. Assuming that this function is linear, determine the slope to the nearest hundredth place.

$$m = \frac{31.42 - 0}{5 - 0} = 6.28$$

b. From the table, determine the vertical intercept.

(0, 0)

c. Write the equation of the linear function.

$C = 6.28r$

d. What is the formula of the circumference of a circle?

$C = 2\pi r$

e. Does your equation approximate this formula? Explain.

Yes, π is approximately 3.14, so 2π is approximately 6.28.

8. You own a kayak company and it is open only during the summer months. You discover that if you sell a certain type of kayak for $400, your sales per day average $5200. If you raise the price of the kayak to $450, the sales fall to approximately $3600 per day.

a. Assume that the sales per day is a function of the price of the kayak. Write two ordered pairs that describe this situation.

(400, 5200)

(450, 3600)

b. Assume that the sales per day is a linear function of the price of the kayak. Write an equation describing this relationship.

$$m = \frac{3600 - 5200}{450 - 400} = \frac{-1600}{50} = -32$$

$y = 5200 - 32(x - 400)$

$y = 5200 - 32x + 12{,}800$

$y = -32x + 18{,}000$

c. You cannot make enough profit if you sell the kayak for less than $375. What will be the average sales per day if you change the price to $375?

$y = -32(375) + 18{,}000 = 6000$

9. Your architect will charge you a flat fee of $5000 for the plans for your home. The cost of your home is estimated by the square footage. The following table gives the total estimated cost of your home, *c*, including the architect's fees, as a function of the square footage, *h*. Assume that the total cost is a linear function of square footage.

TOTAL SQUARE FEET, *h*	TOTAL COST, *c* ($)
0	5000
3000	380,000

a. What is the vertical intercept of the line containing these points? Explain how you determined this intercept.

(0, 5000) The vertical intercept occurs where the input value is 0.

b. Using the data in the table and the formula $m = \frac{\Delta c}{\Delta h}$, calculate the slope.

What is the practical meaning of the slope in this situation?

$$\frac{\Delta c}{\Delta h} = \frac{380{,}000 - 5000}{3000} = \frac{375{,}000}{3000} = 125$$

The building cost is $125 per square foot.

c. Use the results from parts a and b to write the equation of the line in slope-intercept form that can be used to determine the cost for any given square footage.

$c = 125h + 5000$

d. You decide that you cannot afford a house with 3000 square feet. Using the equation from part c, determine the cost of your home if you decrease its size to 2500 square feet.

$c = 125(2500) + 5000 = 317{,}500$

10. Straight-line depreciation helps spread the cost of new equipment over a number of years. The value of your company's copy machine after 1 year will be \$14,700 and after 4 years will be \$4800.

a. Write a linear function that will determine the value of the copy machine for any specified year.

$$m = \frac{4800 - 14{,}700}{4 - 1} = \frac{-9900}{3} = -3300$$

$y = 14{,}700 - 3300(x - 1)$

$y = 14{,}700 - 3300x + 3300$

$y = -3300x + 18{,}000$

b. The salvage value is the value of the equipment when it gets replaced. What will be the salvage value of the copier if you plan to replace it after 5 years?

$y = -3300(5) + 18{,}000 = 1500$

11. The basal energy requirement is the daily number of calories that a person needs to maintain basic life processes. For a 20-year-old male who weighs 75 kilograms and is 190.5 centimeters tall, the basal energy requirement is 1952 calories. If his weight increases to 95 kilograms, he will require 2226 calories.

The given information is summarized in the following table:

20-YEAR-OLD MALE, 190.5 CENTIMETERS TALL		
w, Weight (kg)	75	95
B, Basal Energy Requirement (cal.)	1952	2226

a. Assume that the basal energy requirement, *B*, is a linear function of weight, *w* for a 20-year-old male who is 190.5 centimeters tall. Determine the slope of the line containing the two points indicated in the above table.

$$\text{slope} = \frac{2226 - 1952}{95 - 75} = \frac{274}{20} = 13.7$$

b. What is the practical meaning of the slope in the context of this situation?

For each additional 1-kilogram increase in weight, a 20-year-old, 190.5-centimeter-tall male experiences an increase of 13.7 additional calories in his basal energy requirement.

c. Using the point-slope form, determine an equation that expresses *B* in terms of *w* for a 20-year-old, 190.5-centimeter-tall male.

$B = k + m(w - h)$

$B = 1952 + 13.7(w - 75)$

$B = 13.7w - 1027.5 + 1952$

$B = 13.7w + 924.5$

The equation $B = 13.7w + 924.5$ expresses the basal energy rate B for a 20-year-old, 190.5-centimeter-tall male in terms of his weight w.

d. Does the *B*-intercept have any practial meaning in this situation? Determine the practical domain of the basal energy function.

The *B*-intercept has no practical meaning in this situation because it would indicate a weight of 0 kilograms. A possible practical domain is a set of weights from 55 to 182 kilograms.

ACTIVITY 1.9

Skateboard Heaven

General Form of an Equation of a Line

OBJECTIVES

1. Write an equation of a line in general form $Ax + By = C$.
2. Write the slope-intercept form of a linear equation given the general form.
3. Determine the equation of a horizontal line.
4. Determine the equation of a vertical line.
5. Graph equations using intercepts.

Your town has just authorized funding to build a new ramp and pathways for skateboarding. For security, the ramp and pathways must have a rectangular fence surrounding them. The money allocated in the budget for fencing will be enough to purchase 350 feet of fence. The only stipulation is that the width must be between 35 and 60 feet to properly enclose the new ramp. The length will depend on the width you choose. Your task is to determine the length and width of the rectangular region so that you use all of the fencing.

1. a. What does the value of 350 represent with regard to the rectangular region?

   350 represents the perimeter of the rectangle.

   b. Using x to represent the width and y to represent the length, write an equation for the perimeter of this rectangular region.

   $2x + 2y = 350$

   c. The linear equation in part b should be in the form $Ax + By = C$. Identify the values of the constants A, B, and C in the equation.

   $A = 2, B = 2, C = 350$

DEFINITION

When a linear equation is written in the form $Ax + By = C$, it is said to be in **general form**, where A, B, and C represent constants and A and B cannot both be zero.

EXAMPLE 1

Sometimes it is advantageous to rewrite a linear equation given in general form as its equivalent slope-intercept form. For example, consider the equation 3x + 7y = 5. To write this equation in slope-intercept form, you need to solve for y as follows:

$$3x + 7y = 5$$

$$-3x \qquad -3x$$ Add $-3x$ to each side of the equation.

$$7y = -3x + 5$$

$$\frac{7y}{7} = \frac{-3x}{7} + \frac{5}{7}$$ Divide each side of the equation by 7, the coefficient of y.

$$y = -\frac{3}{7}x + \frac{5}{7}$$

2. a. Rewrite the equation from Problem 1b in slope-intercept form by solving the equation for y in terms of x.

$y = 175 - x$

b. What is the slope of the line represented by the equation you wrote in part a? What is the practical meaning of the slope in this situation?

The slope is -1. This means that as the width increases by 1 foot, the length decreases by 1 foot.

c. What is the vertical intercept of the line represented by the equation you determined in part a? What is the practical meaning of the vertical intercept in this situation?

The vertical intercept is (0, 175). This means that if the length is 175, the width is 0, and there is no rectangle.

d. What is the practical domain and range of this function?

$35 \leq x \leq 60$; $115 \leq y \leq 140$

3. a. Determine the corresponding y-value for $x = 35$ using $2x + 2y = 350$ (general form of the equation of a line).

$$2(35) + 2y = 350$$
$$2y = 280$$
$$y = 140$$

b. Determine the corresponding y-value for $x = 35$ using $y = -x + 175$ (slope-intercept form of the equation of a line).

$y = -35 + 175 = 140$

c. Was it more convenient to use the standard form or the slope-intercept form to determine a y-value given $x = 35$? Explain.

Slope-intercept form is easier because it involves just substituting and simplifying; no solving for variables is required.

d. Complete the table to determine some possible lengths and widths for the rectangular region.

x	35	40	50	60
y	140	135	125	115

4. Use function notation to write the length as a function of the width, letting f represent the function.

$f(x) = 175 - x$

5. You are working in the purchasing department for an electronics retailer. Your job this month is to stock up on virtual reality (VR) headsets and tablets. Your supervisor informs you that you have a budget of $12,000 this month. You know that the average wholesale cost of a VR headset system is $125 and the average wholesale cost of a tablet is $400.

a. If g represents the number of VR headsets you can purchase, write an expression that represents the amount you can spend on VR headsets.

$125g$

b. If c represents the number of tablets you can purchase, write an expression that represents the amount you can spend on tablets.

$400c$

c. Write a linear equation in general form that relates the number of VR headsets and tablets you can expect to purchase with your budget.

$125g + 400c = 12{,}000$

d. Solve your equation in part c for c. In other words, express the number of tablets you can purchase as a function of the number of VR headsets you can purchase.

$$c = \frac{12{,}000 - 125g}{400} = 30 - 0.3125g$$

e. What is the horizontal intercept for this function? What is its meaning in this situation?

(96, 0) The maximum number of VR headsets I can purchase is 96.

f. What is the vertical intercept for this function? What is its meaning in this situation?

(0, 30) The maximum number of tablets I can purchase is 30.

g. What is the slope of this function? What is the significance of the slope in this situation?

The slope is -0.3125, or $-\frac{5}{16}$. This means for every 16 VR headsets I purchase, I can purchase 5 fewer tablets.

h. What are the practical domain and the practical range in this situation?

The practical domain is whole numbers from 0 to 96. The practical range is whole numbers from 0 to 30.

Equations of a Horizontal Line

If either A, the coefficient of x, or B, the coefficient of y, in the general form $Ax + By = C$ equals zero, a special situation arises. You will explore this in the following problems.

6. a. When the ramp for skateboarding is complete, you and your friends will be able to pay a monthly fee of \$12.50 to use the ramp for as many hours as you wish. Complete the table of values below, where x is the number of hours per month that each person who pays the fee uses the ramp and y is the total cost per month for each person.

x, Time (hr.)	0	5	10	15	20
y, Cost (\$)	12.50	12.50	12.50	12.50	12.50

b. Use the data points from the table to sketch the graph.

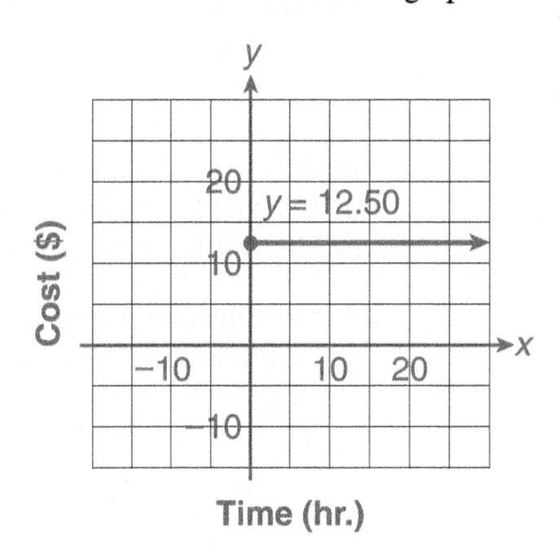

c. Describe the graph in words.

The graph is a horizontal line through (0, 12.50).

d. Choose two ordered pairs from part a, and determine the slope, m, of the line. What is the practical meaning of the slope in this situation?

$m = 0$ The cost does not change.

e. What are the intercepts (vertical and horizontal) of the line (if they exist)?

vertical intercept: (0, 12.50); horizontal intercept: none

f. Write the equation of this line in slope-intercept form, $y = mx + b$.

$y = 0x + 12.50$

DEFINITION

A graph in which the output y is a constant, or equivalently $f(x)$ is a constant, is a **horizontal line**. The equation of a horizontal line is $y = c$ or $f(x) = c$, where c is some fixed real number. The slope of any horizontal line is zero.

7. Determine three ordered pairs that satisfy each of the following equations, and then sketch each graph on the same coordinate axes.

a. $y = -2$

$(0, -2), (1, -2), (3, -2)$

b. $f(x) = 1$

$(0, 1), (1, 1), (2, 1)$

c. $g(x) = \frac{5}{2}$

$\left(0, \frac{5}{2}\right), \left(-1, \frac{5}{2}\right), \left(2, \frac{5}{2}\right)$

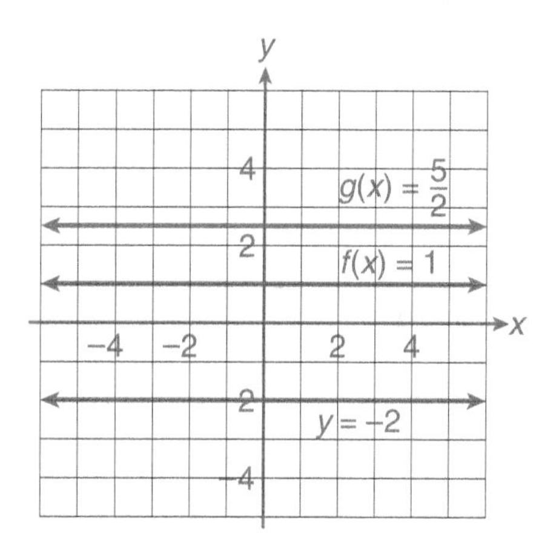

Equation of a Vertical Line

8. You apply for a part-time job at the skateboarding rink to help cover your weekly expenses while going to college. The following table gives your weekly salary, x, and corresponding weekly expenses, y, for a typical month:

x, Weekly Salary ($)	70	70	70	70
y, Weekly Expenses ($)	45	35	50	60

LC LEARNING CATALYTICS

Determine the equation of the vertical line through the point $(-2, 15)$.

a. Sketch a graph of the given data points.

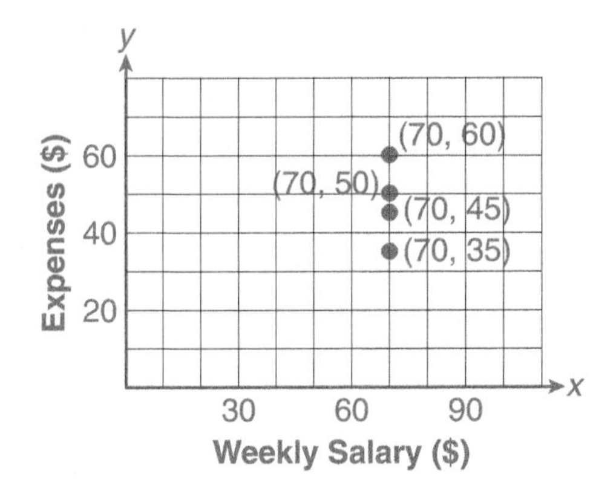

b. Describe the graph in words.

The graph is a vertical line.

c. Choose two ordered pairs from the table, and determine the slope, m.

$$m = \frac{45 - 35}{70 - 70} = \frac{10}{0}, \text{ which is undefined}$$

d. What are the intercepts (vertical and horizontal) of the line (if they exist)?

no vertical; horizontal (70, 0)

e. Is y a function of x? Explain using the definition of a function.

No. For the x-value 70, there are multiple outputs.

f. Do the ordered pairs in the table satisfy the equation $x = 70$? Explain.

Yes, $70 = 70$ is always true.

g. Can you graph the equation $x = 70$ using your graphing calculator? Explain.

No, because I can't solve for y.

DEFINITION

A graph in which x is a constant is a **vertical line**. The equation of a vertical line is $x = a$, where a is some fixed real number. The slope of a vertical line is undefined. The graph does not represent a function.

9. Determine three ordered pairs that satisfy each of the following equations, and then sketch each graph. (Answers will vary.)

a. $x = -2$

$(-2, 1), (-2, 0), (-2, 3)$

b. $x = 4$

$(4, 0), (4, -2), (4, 3)$

c. $x = \frac{5}{2}$

$\left(\frac{5}{2}, 0\right), \left(\frac{5}{2}, -5\right), \left(\frac{5}{2}, 10\right)$

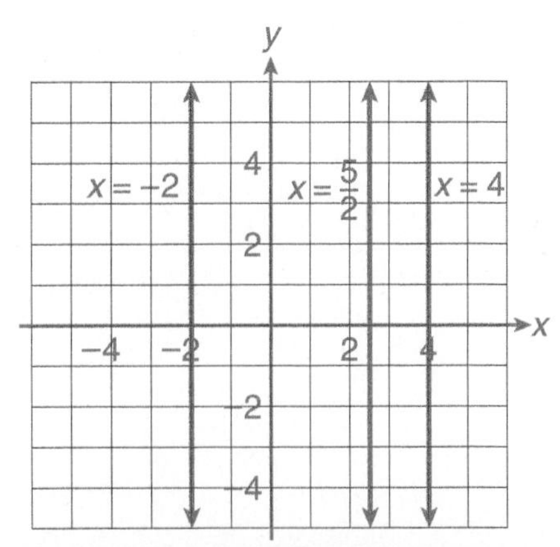

SUMMARY Activity 1.9

1. The **general form** of a linear equation is $Ax + By = C$, where A, B, and C are constants and A and B are not both zero.

2. The graph of $y = c$ or $f(x) = c$ is a **horizontal line**. In this case, f is called a constant function. Every point on this line has a y-coordinate equal to c. A horizontal line has slope of zero.

3. The graph of $x = a$ is a **vertical line**. Every point on this line has an x-coordinate equal to a. The slope of a vertical line is undefined. The graph does not represent a function.

EXERCISES Activity 1.9

1. Write the following linear equations in slope-intercept form. Determine the slope and vertical intercept of each line.

   a. $2x - y = 3$

   $y = 2x - 3, m = 2, (0, -3)$

   b. $x + y = -2$

   $y = -x - 2, m = -1, (0, -2)$

   c. $2x - 3y = 7$

   $y = \frac{2}{3}x - \frac{7}{3}, m = \frac{2}{3}, \left(0, -\frac{7}{3}\right)$

   d. $-x + 2y = 4$

   $y = \frac{1}{2}x + 2, m = \frac{1}{2}, (0, 2)$

   e. $0x + 3y = 12$

   $y = 0x + 4, m = 0, (0, 4)$

2. **a.** Sketch the graph of the horizontal line through the point $(-2, 3)$.

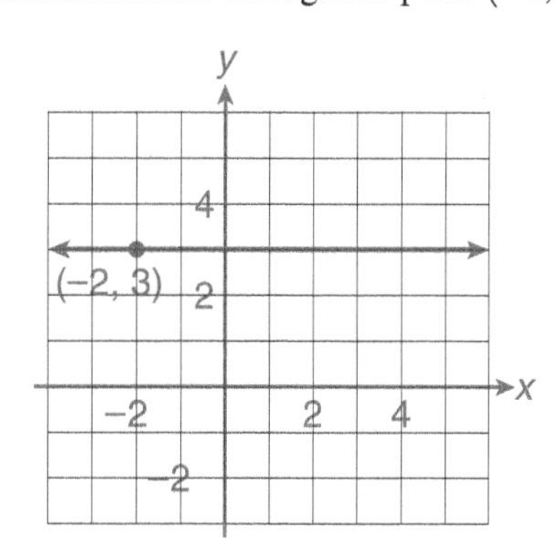

   b. Write the equation of a horizontal line through the point $(-2, 3)$.

   $y = 3$

   c. What is the slope of the line?

   $m = 0$

d. What are the vertical and horizontal intercepts of the line?

vertical: (0, 3); horizontal: none

e. Does the graph represent a function? Explain.

Yes. For every x-value of input, there is exactly one y-value of output.

3. a. Sketch the graph of the vertical line through $(-2, 3)$.

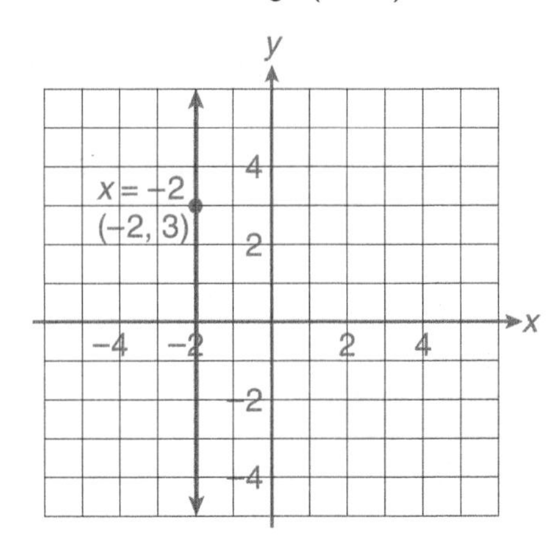

b. Does the graph represent a function? Explain.

This is not a function. It does not pass the vertical line test.

c. Write the equation of a vertical line through the point $(-2, 3)$.

$x = -2$

d. What is the slope of the line?

The slope is undefined.

e. What are the vertical and horizontal intercepts of the line?

vertical: none; horizontal: (−2, 0)

4. Explain the difference between a line with a zero slope and a line with an undefined slope.

A line with a zero slope is a horizontal line. It does represent a function. A line with an undefined slope is a vertical line. It does not represent a function.

5. You are retained as a consultant for a major computer company. You receive $2000 per month as a fee no matter how many hours you work.

a. Using x to represent the number of hours you work each month, write a function, f, in symbolic form to represent the total amount received from the company each month.

$f(x) = 2000$

b. Complete the following table of values.

Hours Worked per Month	15	25	35
Fee per Month ($)	2000	2000	2000

c. Use your graphing calculator to sketch the graph of this function.

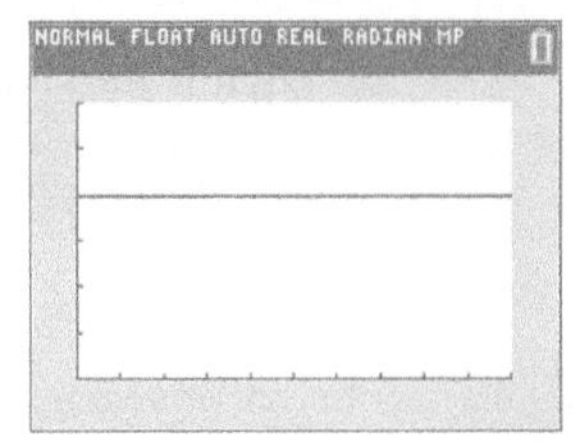

d. What is the slope of the line? What is the practical meaning of the slope in this situation?

The slope is zero. This means that the fee does not change.

e. Describe the graph of the function.

The graph is a horizontal line through (0, 2000).

6. You work in the purchasing department of an appliance retailer. This month you are stocking up on washers and dryers. Your supervisor informs you that your budget this month is $10,000. You know that the average wholesale cost of a washer over the past year has been $250, whereas the average wholesale cost of a dryer has been $200.

a. If w represents the number of washers you can purchase, write an expression that represents the amount you can spend on washers.

$250w$

b. If d represents the number of dryers you can purchase, write an expression that represents the amount you can spend on dryers.

$200d$

c. Write a linear equation in general form that relates the number of washers and dryers you can expect to purchase with your budget.

$250w + 200d = 10{,}000$

d. Solve your equation in part c for d. In other words, express the number of dryers you can expect to purchase as a function of the number of washers you can expect to purchase.

$$d = \frac{10{,}000 - 250w}{200} = 50 - \frac{5}{4}w$$

e. What is the horizontal intercept for this function? What is its practical meaning in this situation?

(40, 0) The maximum number of washers I can purchase is 40.

f. What is the vertical intercept for this function? What is its practical meaning in this situation?

(0, 50) The maximum number of dryers I can purchase is 50.

g. What is the slope of this function? What is its significance in this situation?

The slope is $-\frac{5}{4}$. This means that for every 4 washers I purchase, I can purchase 5 fewer dryers.

h. Use your graphing calculator to graph the function in part d. What part of this graph is relevant to this situation?

The only relevant part is where both w and d are positive (quadrant I).

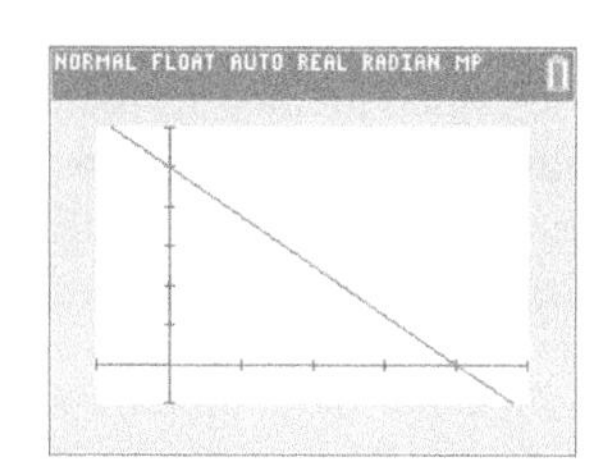

i. What are the practical domain and range of this situation?

domain: $0 \leq w \leq 40$; range: $0 \leq d \leq 50$

ACTIVITY 1.10

College Tuition

Modeling Data with Linear Regression Equations

OBJECTIVES

1. Recognize when patterns of points in a scatterplot have a linear form.
2. Recognize when the pattern in the scatterplot shows that the two variables are positively related or negatively related.
3. Estimate and draw a line of best fit through a set of points in a scatterplot.
4. Use a graphing calculator to determine the equation of the line of best fit.
5. Measure the strength of the correlation (association) by a correlation coefficient.
6. Recognize that a strong correlation does not necessarily imply a linear or a cause-and-effect relationship.

The following table contains the average tuition and room and board for full-time matriculated students at private nonprofit 4-year colleges as published by the College Board:

COLLEGE COSTS					
Year Ending	1981	1991	2001	2011	2017
Years since 1980	1	11	21	31	37
Cost in 2016 dollars	$10,525	$17,237	$22,382	$29,545	$33,479

Ending year means the year in which the school year ended. That is, the 2000–2001 school year is represented by 2001.

1. Let t, the number of years since 1980, represent your input variable and c, the average cost, your output variable. Determine an appropriate scale, and create a scatterplot from the table above.

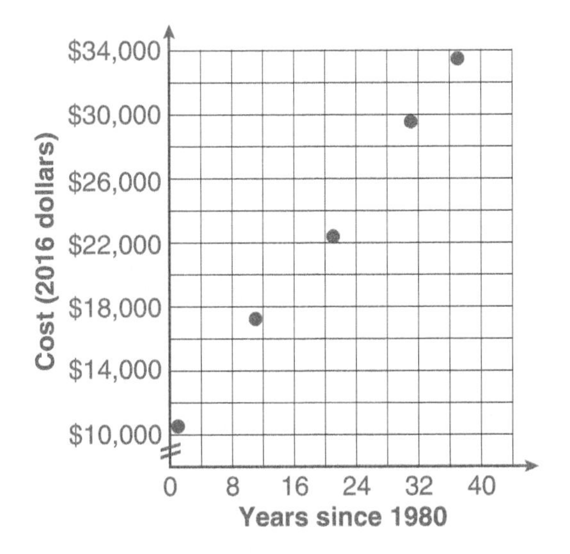

2. **a.** Does there appear to be a linear relationship between the years since 1981 and the cost of tuition?

   The graph appears to be somewhat linear.

   b. As the years since 1981 increase, what is the general trend in tuition costs?

   In general, the values of tuition cost increase as time increases. The two variables are said to be positively related.

3. Using a straightedge (a taut string or dry strand of spaghetti is helpful in positioning the line), draw a line that will have as many data points as close to the line as possible. The line you are drawing is called the **line of best fit**.

The accepted method for determining the line of best fit to a set of data is called the **method of least squares**. This method is used to calculate a slope and y-intercept. The line with this slope and y-intercept is called the **regression line** for your data. Your graphing calculator is programmed with an algorithm that determines the equation of the regression line.

4. **a.** Enter the tuition data into your calculator by pressing STAT and choosing EDIT.

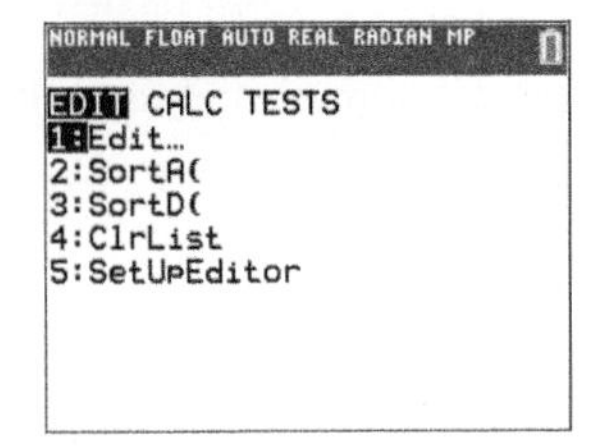

NORMAL FLOAT AUTO REAL RADIAN MP
L1 L2 L3 L4 L5 1
L1(1)=

b. Enter the year data in L1 and the tuition data in L2.

c. Determine the regression line by pressing STAT and choosing CALC and option 4:LinReg $(ax + b)$. Your screens should appear as follows:

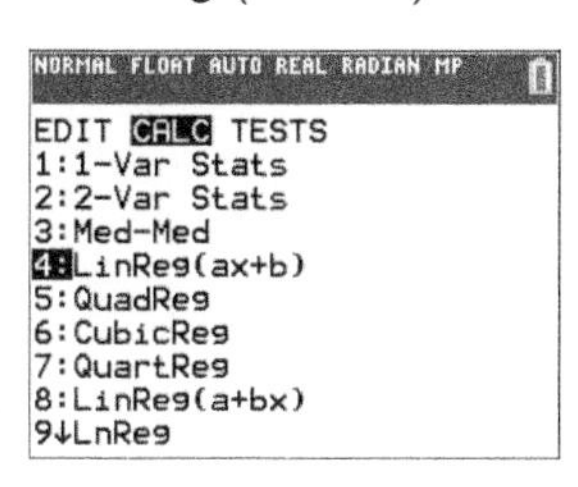

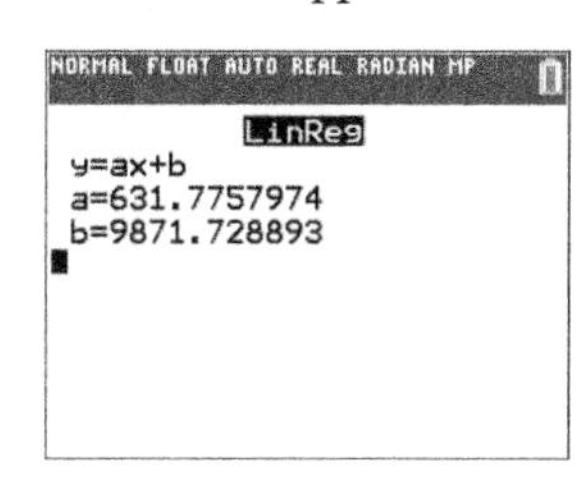

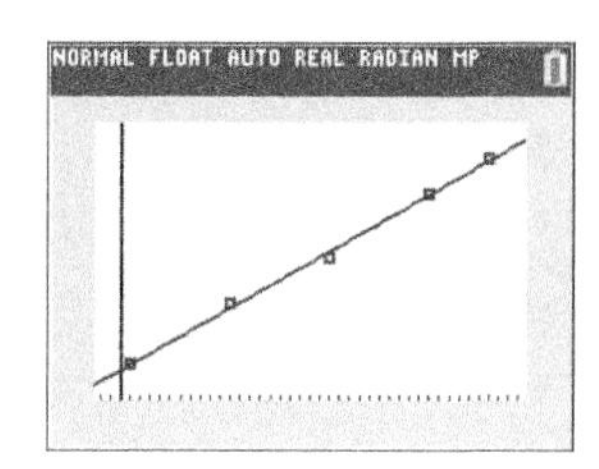

d. Write the result, rounding the coefficients to two decimal places if necessary. See Appendix C for help operating the TI-84 Plus C graphing calculator.

$c = 631.78t + 9871.73$

LC LEARNING CATALYTICS

For the following data

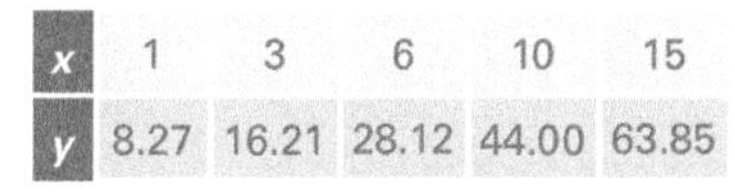

x	1	3	6	10	15
y	8.27	16.21	28.12	44.00	63.85

determine the linear regression equation.

5. Now use the regression equation you determined in Problem 4 to model the tuition data. Use the Table feature on your graphing calculator to determine the model's predicted output in the following table:

INPUT, t	ACTUAL OUTPUT, c	MODEL'S PREDICTED OUTPUT
1	$10,525	$10,504
11	$17,237	$16,821
21	$22,382	$23,139
31	$29,545	$29,457
37	$33,479	$33,248

6. a. Use the regression equation to predict the average tuition and fees at four-year colleges ending in 2004 $(t = 24)$.

$c = 631.78(24) + 9871.73 = \$25{,}034$

DEFINITION

Using a regression model to predict an output within the boundaries of the input values of the given data is called **interpolation**. Using a regression model to predict an output outside the boundaries of the input values of the given data is called **extrapolation**. In general, interpolation is more reliable than extrapolation.

b. Use the regression equation to predict the average tuition and room and board at four-year colleges in the years ending in 2020 and 2025.

$c = 631.78(40) + 9871.73 = \$35{,}143$

$c = 631.78(45) + 9871.73 = \$38{,}302$

c. Which prediction do you believe would be more accurate? Explain.

The prediction for the year ending 2020 is more believable because the input is closer to the given data.

d. Use the regression equation to estimate the year in which average tuition and room and board will be at least $45,000.

The graph of the regression line indicates that tuition and room and board will reach $45,000 55.6 years after 1980. I will not use partial years, so the regression line indicates that tuition and room and board will be at least $45,000 in the year 2036.

7. The National Center for Education Statistics projects that the number of bachelor's degrees earned by women in the United States will grow steadily. The following table contains the projections for selected years after 2019. The number of degrees is measured in thousands. The input, t, represents the number of years since 2019.

BACHELOR'S DEGREES PROJECTED TO BE EARNED BY WOMEN					
	2019	2021	2023	2025	2027
t, number of years since 2019	0	2	4	6	8
$f(t)$, number of bachelor's degrees (thousands)	1077	1118	1152	1181	1207

a. Sketch a scatterplot of the given data.

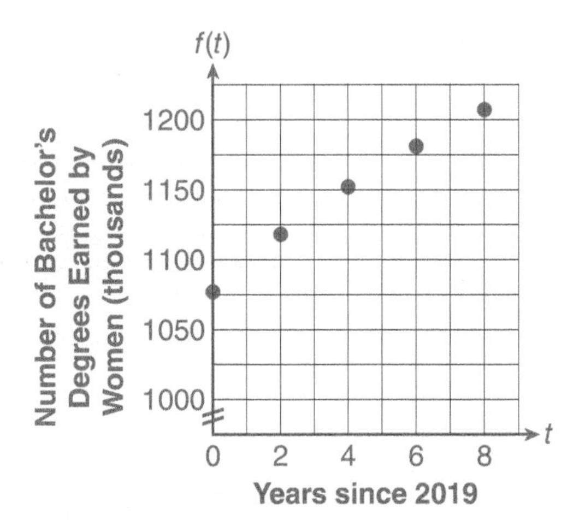

b. Enter the data into your graphing calculator, and determine a linear regression model to represent the data. Write the result here.

$y = 16.15t + 1082.4$

c. What is the slope of the line? What is the practical meaning of the slope in this situation?

$m = 16.15$

The projected number of bachelor's degrees earned by women after 2019 will increase by an average of 16.15 thousand per year.

d. Use the regression model to predict the number of bachelor's degrees that will be granted in the year 2030.

$y = 16.15(11) + 1082.4 = 1260.05$ thousand or 1,260,050 bachelor's degrees

Linear Correlation

When a linear pattern is evident in a scatterplot, there is said to be a linear correlation between the two variables. (When the word *correlation* appears by itself, it usually means **linear** correlation.) Your calculator window in Problem 4c should have included a value for r, called the **correlation coefficient**. (If it didn't, see Appendix C to enter DiagnosticOn.)

C Appendix

$$r = 0.9957570126 \quad \text{or} \quad r = 1.00$$

Note that the correlation coefficient is generally rounded to the nearest hundredth.

The value of the correlation coefficient indicates how strong a linear relationship exists between the two variables under consideration. The value of r ranges between -1 and 1. When the value of r is closer to zero, you would conclude that there is little or no linear correlation. The closer r is to either 1 or -1, the stronger the linear correlation between the two variables.

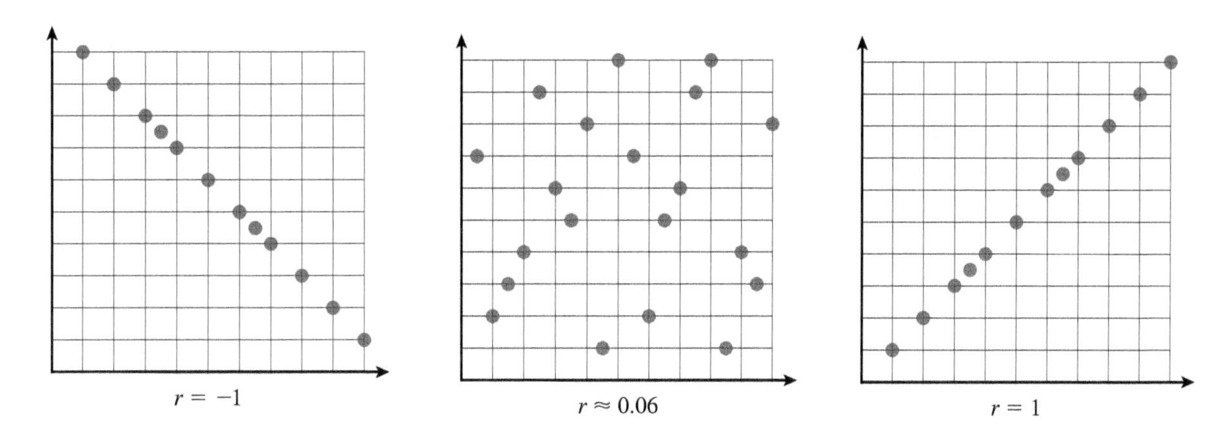

If $r = 1$, then all the data points lie exactly on a straight line with positive slope. In this case, there is a perfect positive correlation. If $r = -1$, then all the decreasing data points lie exactly on a straight line with negative slope. In this case, there is a perfect negative correlation. The strength of the correlation is generally best described by both the correlation coefficient and a visual interpretation of the scatterplot. The following scatterplots with their respective correlation coefficients should help make this clear:

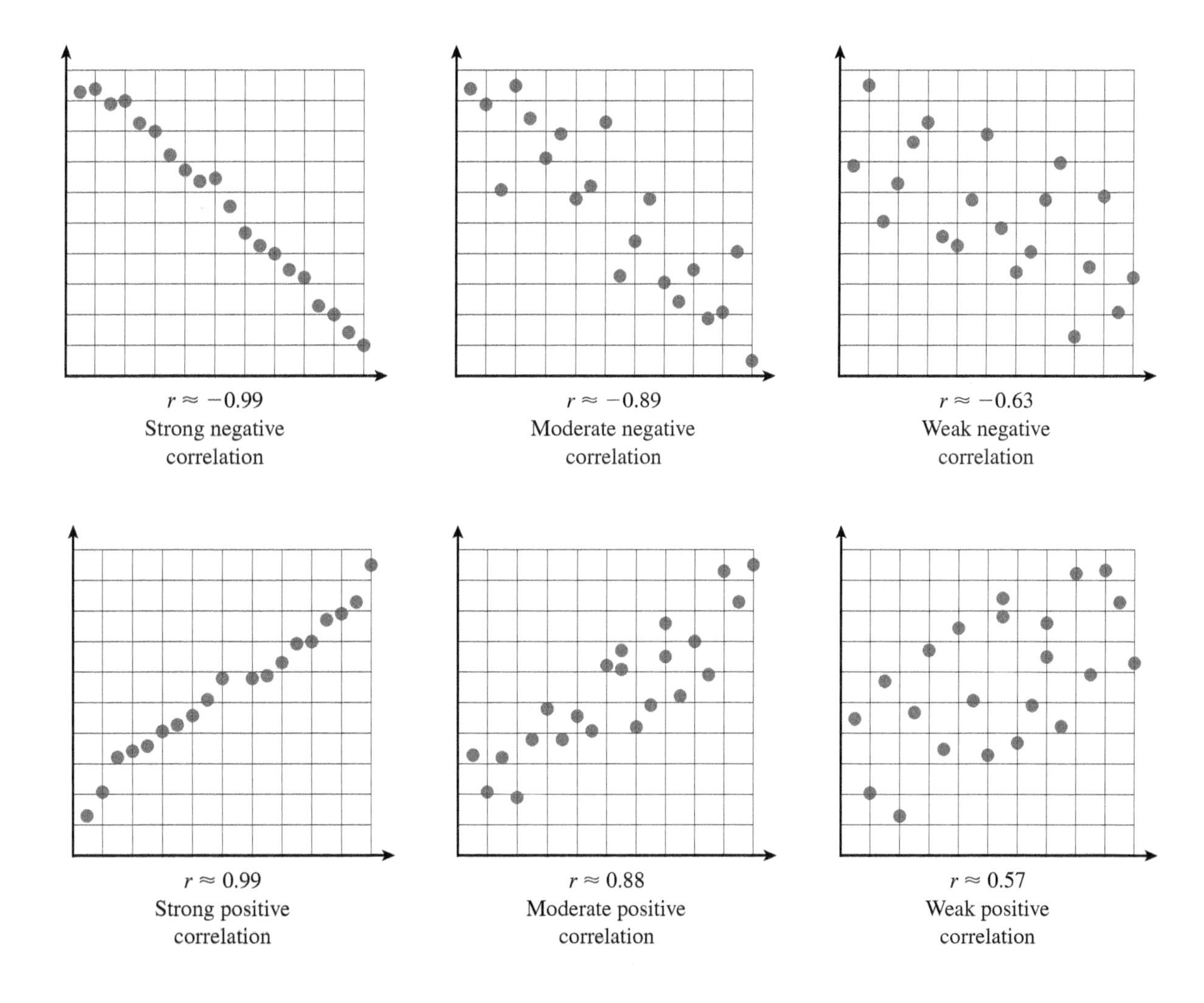

8. For each of the following, match the correlation coefficient r with the corresponding scatterplot of data points:

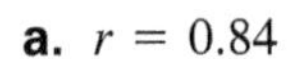

a. $r = 0.84$ **b.** $r = 0.53$ **c.** $r = -0.45$

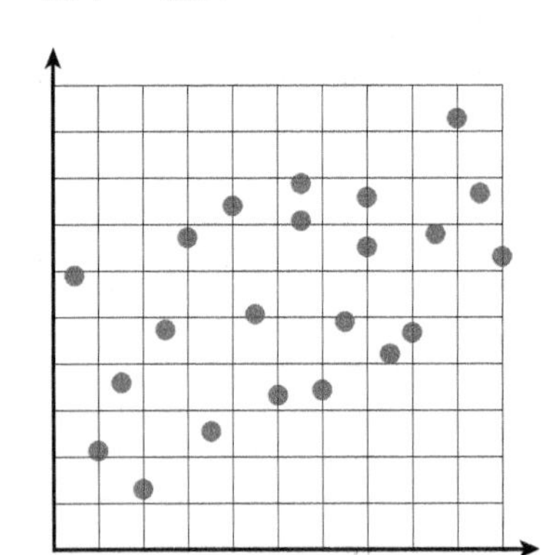

b.

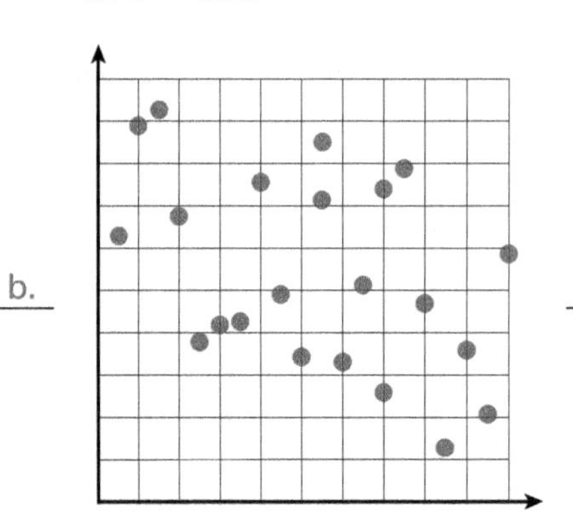

c.

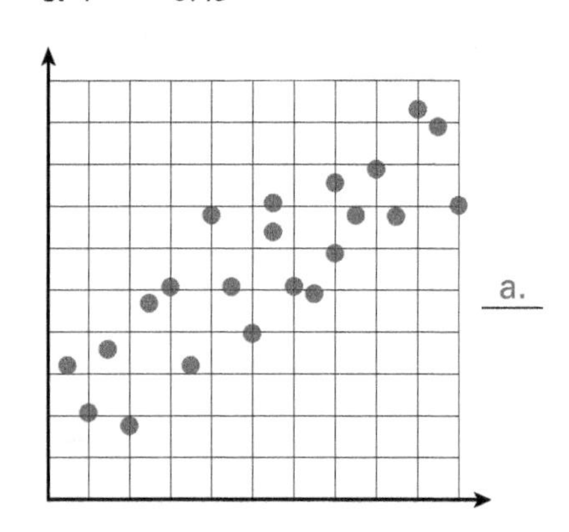

a.

A correlation coefficient value of $r = 0.85$ suggests a moderately strong linear association between two variables. However, as demonstrated in Problem 9, a line may not be the best model or representation of the data.

9. a. Produce a scatterplot for the following table of values. How would you describe the association between x and y?

x	0.5	0.5	1	1	1.5	2	2.5	3	3.5	4	4.5	5	5	5.5	5.5	6	6.5	7	7	7.5	8	8
y	3.2	2.9	2.5	2.1	2.6	1.8	2.6	2.2	3.2	2.9	3.1	3.6	4	4.1	3.1	3.9	4.9	4.6	5.3	6.3	7.4	6.2

There appears to be a moderately strong positive correlation.

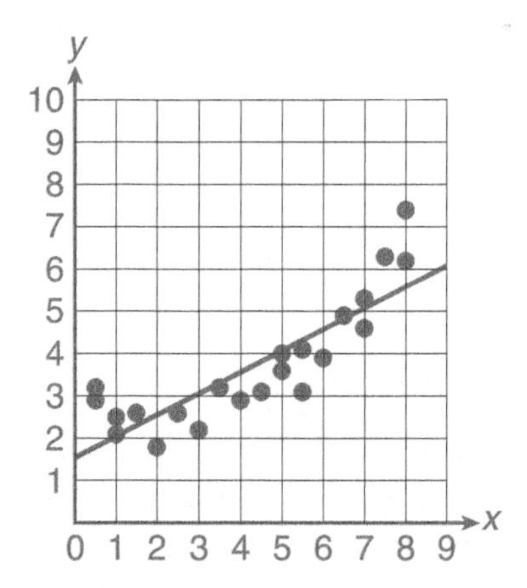

b. Determine the regression line equation and correlation coefficient. Does the coefficient confirm what you said in part a?

$y = 0.502x + 1.58$ and $r = 0.85$

This value for r confirms that there is a moderately strong linear correlation.

c. Graph the regression line with the scatterplot. Do you think a different graph might better describe the pattern shown in the scatterplot?

It looks as though a curved graph would better fit the data. Even though approximately half the data points are above and half are below the regression line, those points above the line are to the extreme left and right, whereas those points below the line are in the middle.

Cause-and-Effect Relationships

A strong correlation between two variables does not necessarily mean that a cause-and-effect relationship exists between the two variables. For example, there is a strong positive correlation between height and weight. However, an increase in height does not necessarily cause an increase in weight. Many other variables could cause the increase in weight, such as gender, age, or body type. Such variables are often called **lurking variables**, hidden in the background and not explicitly measured.

10. a. Suppose your home is heated with natural gas. There is a strong positive correlation between the amount of natural gas usage and the outside temperature. Is there a cause-and-effect relationship between these two variables? Explain.

Yes, cold weather causes an increase in the use of natural gas to keep a house warm. But be careful: The two variables play different roles. You would not say that natural gas usage causes the cold weather.

b. In a 2002 publication *The Natural History of the Rich*, the author claimed a positive correlation between people with two cars and a longer life span. Is there a cause-and-effect relationship between the number of cars you own and length of life? Explain.

No, owning several cars is a sign of affluence. Well-to-do people often own more cars but tend to live longer because they are better educated, take better care of themselves, and have better medical care.

SUMMARY Activity 1.10

1. The **linear regression equation** is the linear equation that "best fits" a set of data.
2. The **regression line** is a mathematical model for the data.
3. **Interpolation** is the process of using a regression equation to predict a value of output for an input value that lies within the boundaries of the given data.
4. **Extrapolation** is the process of using a regression equation to predict a value of output for an input value that lies outside the boundaries of the given data.
5. A **correlation coefficient, r,** measures how strongly two related variables follow a linear pattern. The value of r ranges between -1 and 1. When the value of r is closer to zero, you would conclude that there is little or no linear correlation. The closer r is to either 1 or -1, the stronger the linear correlation between the two variables. If $r < 0$, a negative correlation, then the linear pattern follows a negative slope. If $r > 0$, a positive correlation, then the linear pattern follows a positive slope.

EXERCISES Activity 1.10

1. a. Plot the following data:

x	0	3	6	9	12	15	18
$f(x)$	−0.8	6.3	13.1	19.6	27.0	33.5	40.8

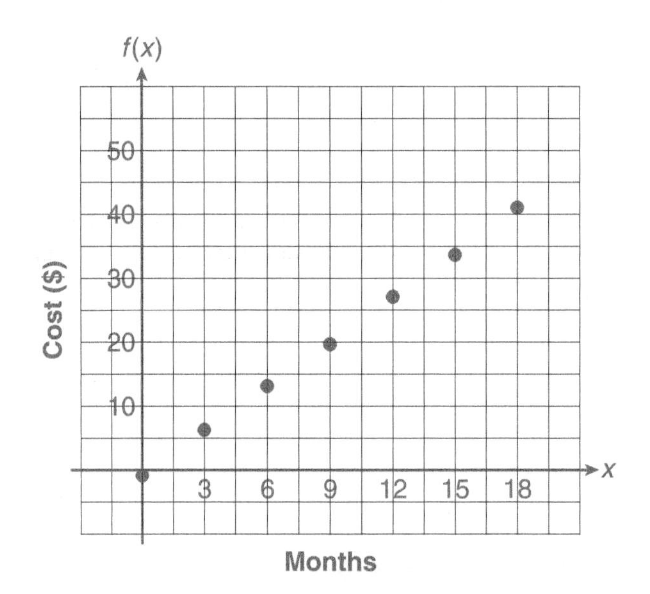

b. With a straightedge, draw a line that you think looks like the line of best fit. Does this data appear to be linear?

(Answers will vary.) Yes, the points are very close to a line.

c. Use your graphing calculator to determine the equation of the regression line. Write the result below.

$f(x) = 2.299x - 0.761$

d. Use your equation from part c to predict the value of y when $x = 10$.

The regression equation yields 22.229.

e. Use your equation from part c to predict the value of y when $x = 25$.

The regression equation yields 56.714.

f. Which prediction, $f(10)$ or $f(25)$, would be more accurate? Explain.

$f(10)$ is probably more accurate. 10 is within the given data; 25 is not.

$f(10)$ uses interpolation; $f(25)$ uses extrapolation.

2. Worldwide PC shipments have declined sinces 2011 as mobile technology has become more and more accessible. The following table gives worldwide PC shipments (in millions of units) for various years since 2011. $t = 0$ corresponds to the year 2011.

t (years since 2011)	0	1	2	3	4	5
c (millions of units)	365.3	351.0	316.3	302.6	266.9	250.1

a. Plot the data.

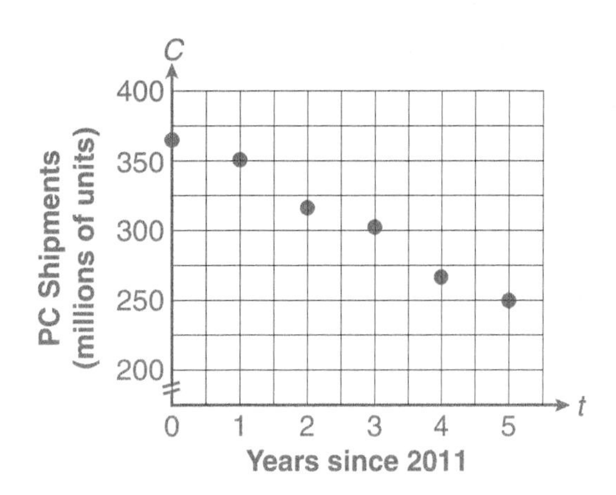

b. Use your graphing calculator to determine the equation of the regression line. Write the result here. (Round your results to two decimal places, if necessary.)

$c = f(t) = -24.06t + 368.84$

c. What is the slope of the line? What is the practical meaning of slope in this situation?

The slope of the line is –24.06. This means that the number of PC shipped worldwide since 2011 is declining at an average rate of 24,060,000 units per year.

d. Use your regression line to determine the number of PCs shipped in 2013 ($t = 2$). Compare your result with the actual value of 316.3 in 2013.

The regression line predicts the number of PC shipments to be 320.72. This is 4.42 million units above the actual, but the error is only about 1.4%.

e. Use the regression equation to predict the number of PC shipments in 2020.

The regression equation predicts PC shipments in 2020 to be 152.3 million units.

f. What process is used to make the prediction in part e?

extrapolation

g. What is the correlation coefficient for these data? Discuss the meaning of this coefficient.

$r = -0.9931127822$. There is a strong negative correlation between these two variables.

3. In 1966, the U.S. Surgeon General's health warnings began appearing on cigarette packages. The following data seem to demonstrate that public awareness of the health hazards of smoking has had some effect on consumption of cigarettes.

	YEAR							
	2001	2003	2005	2007	2009	2011	2014	2015
% of Total Population 18 and Older Who Smoke, P	22.8	21.6	20.9	19.8	20.6	19.0	16.8	15.0

Source: U.S. National Center for Health Statistics.

a. Plot the given data as ordered pairs of the form (t, P), where t is the number of years since 2000 and P is the percent of the total population (18 and older) who smoke. Appropriately scale and label the coordinate axes.

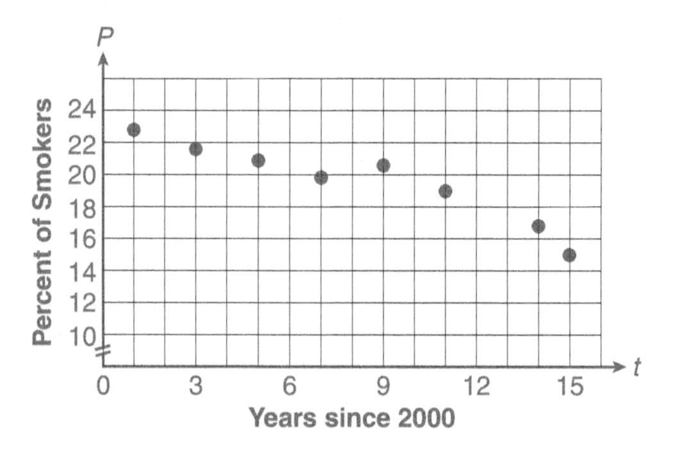

b. Determine the equation of the regression line that best represents the data. (Round your results to two decimal places, if necessary.)

The regression equation is $P = -0.48t + 23.48$.

c. Use the equation to predict the percent of the total population 18 and older who will smoke in 2025.

$2025 - 2000 = 25$

$P = -0.48(25) + 23.48 = 11.48$

According to the model, approximately 11.5% of the total population 18 years of age and older will smoke in the year 2025.

4. Per capita personal income is calculated by taking the total income of a population and dividing it by the total number of people in that population. The following table shows the per capita income of the United States as reported by the Department of Commerce for 2001–2015:

YEAR	PER CAPITA INCOME
2001	$30,574
2003	$31,484
2005	$34,757
2007	$38,611
2009	$38,846
2011	$41,633
2013	$41,706
2015	$43,925

a. Plot the data points on appropriately scaled and labeled axes. Let x represent the number of years since 2001.

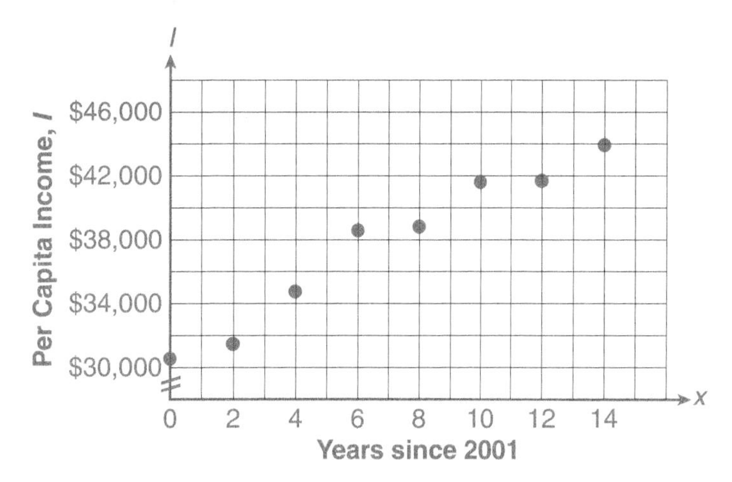

b. Use your graphing calculator's statistics menu (STAT) to determine the equation of the regression line. Round your results to two decimal places.

$I = 984.70x + 30{,}799.08$

c. What is the slope of the line in part b? What is the practical meaning of the slope?

The slope is 984.70. This means that on average per capita income rose by $984.70 every year from 2001 through 2015.

d. Use the linear model from part b to determine when the per capita income will reach $50,000. How confident are you in the prediction?

According to the model, the per capita income will reach $50,000 for the first time 20 years after 2001 or 2021. I am not very confident because this prediction is extrapolating into the future 6 years past my last data point.

Collecting and Analyzing Data

5. Several articles were written in the early 2010s that indicated marriage rates in the United States were at historic lows. Your mathematics professor has asked the class to investigate trends in marriage rates in the United States from 1980 to the present day.

Each member of the group should submit a report of the group's findings. The report should contain the following information:

a. the source(s) of the data

b. a description of the dependent and independent variables

c. a table and scatterplot of the data

d. identification of any patterns in the scatterplot

e. a linear regression equation and correlation coefficient for the data

f. a prediction of the marriage rate per thousand population in the United States in 2020 along with a description of the group's level of confidence in its prediction

g. a conclusion based on the data analysis process

e.

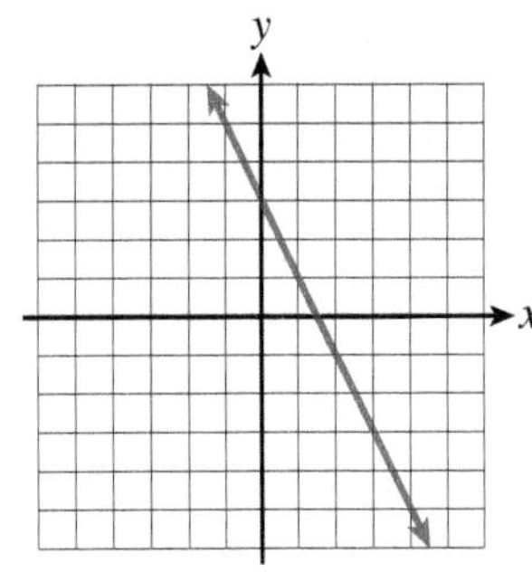

$f(x) = -2x + 3$

f. 

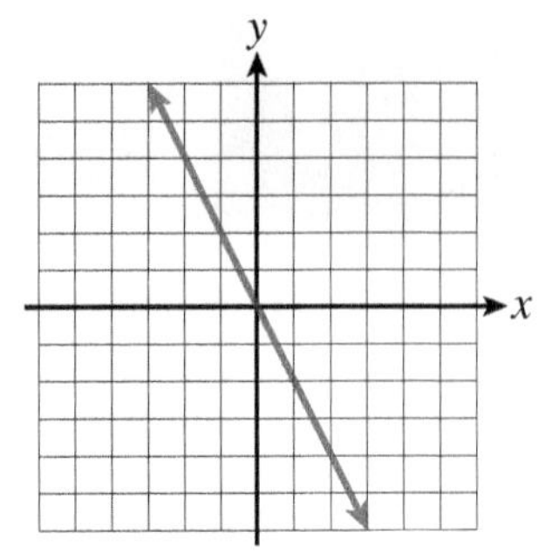

$y = -2x$

g.

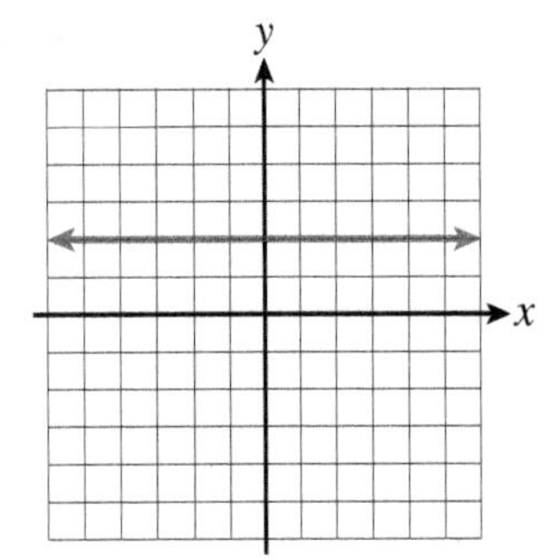

$y = 2$

h.

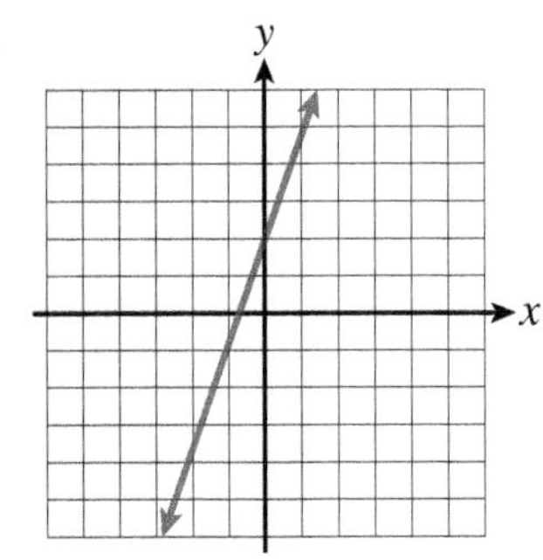

none

i.

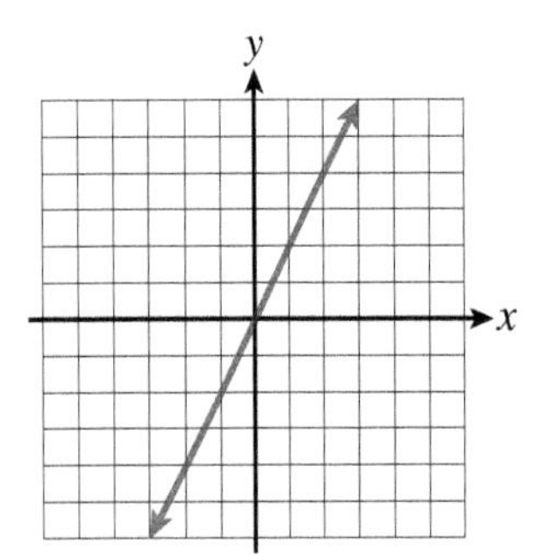

none

3. The cost of renting a graphing calculator from the bookstore is \$20, plus \$4 per month for as long as you want to rent.

a. Complete the table.

Months, m	2	5	8	10	12
Cost In \$, c	28	40	52	60	68

b. Is the cost, c, a linear function of months, m?

Yes, it is linear.

c. Determine the slope.

The slope is 4.

d. Write the equation for the function.

$c = f(m) = 4m + 20$

e. Graph this function, and compare the results from your graphing calculator.

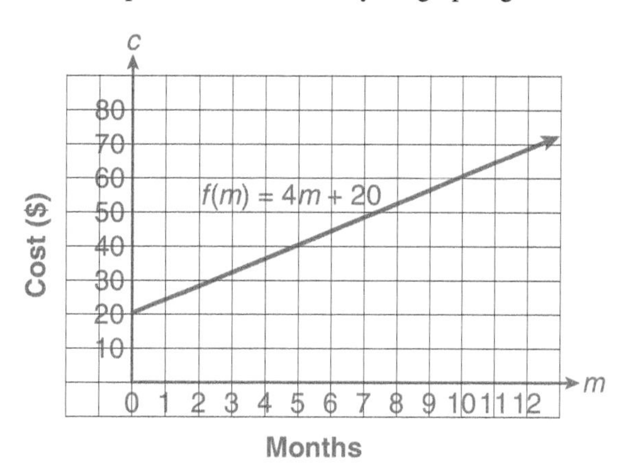

f. What is the practical meaning of the slope in this situation?

This represents the monthly charge.

g. Determine the vertical intercept of the graph of the function. What is the practical meaning of the vertical intercept in this situation?

The vertical intercept is (0, 20). It indicates that the initial rental cost is $20.

h. Determine the horizontal intercept of the graph of the function. What practical meaning does this have in this situation?

The horizontal intercept is (−5, 0). It has no practical meaning in this situation.

i. Approximately how many months will you be able to keep your graphing calculator if you have $65 budgeted for this expense?

$65 = 4m + 20$, or $45 = 4m$, or $m = 11.25$. I can keep the graphing calculator for 11 months.

4. Refer to the following table of values:

t	10	20	40
$s(t)$	0	15	45

a. Determine the average rate of change from $t = 10$ to $t = 20$. 1.5

b. Determine the average rate of change from $t = 20$ to $t = 40$. 1.5

c. Is s a linear function? Explain.

It is a linear function because the rate of change is constant.

5. Determine the slope of the line through the points (3, 8) and (−5, 12).

$$m = \frac{12 - 8}{-5 - 3} = -\frac{1}{2}$$

6. Determine the slope of the line represented by the equation $y = -4x + 2$.

$m = -4$

7. Determine the slope of the line having equation $2x - 5y = 9$.

$$m = \frac{2}{5}$$

8. Write an equation of a line with a slope of −7 and a vertical intercept of (0, 4).

$y = -7x + 4$

9. Write an equation of a line that has a slope of 2 and goes through the point (0, 10).

$y = 2x + 10$

10. Write an equation of a line that has a slope of 0 and goes through the point $(-4, 5)$.

$y = 5$

11. Write an equation of a vertical line that goes through the point $(-3, -5)$.

$x = -3$

12. Write an equation of a line that goes through the points $(2, -3)$ and $(-4, 0)$.

$y = -\frac{1}{2}x - 2$

13. Write an equation of a line that is parallel to $y = \frac{1}{3}x - 7$ and goes through the point $(6, -1)$.

$y = \frac{1}{3}x - 3$

14. Sketch a graph of the line through $(2, -3)$ with a slope of $\frac{1}{2}$.

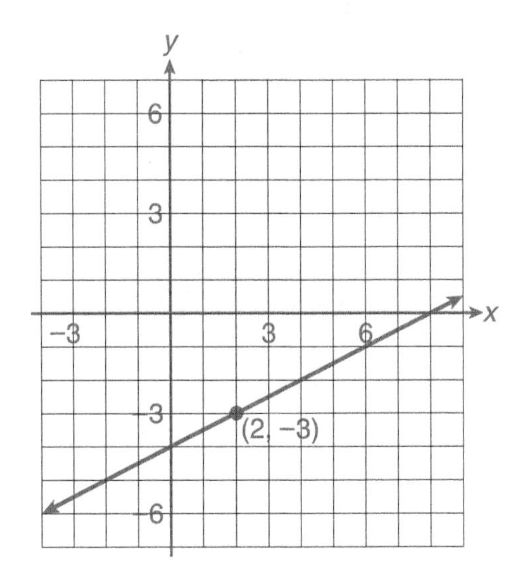

In Exercises 15–17, graph each of the functions by hand. Then compare your results using your graphing calculator.

15. $y = \frac{4}{5}x - 3$

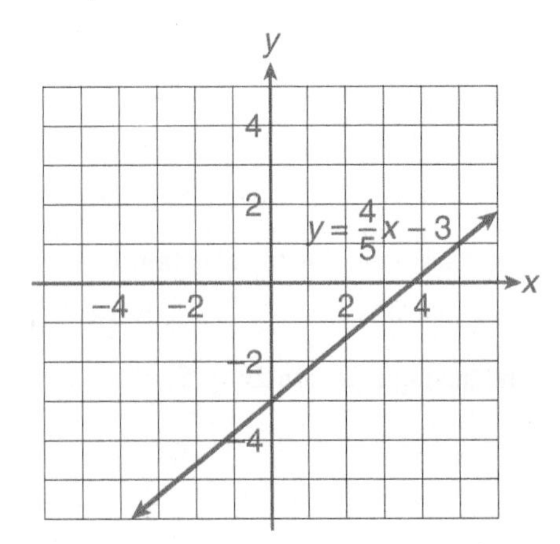

16. $4x - 5y = 20$

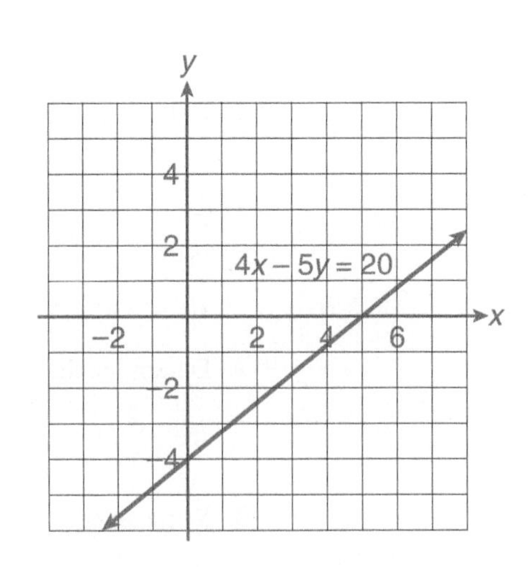

17. $y = -4$

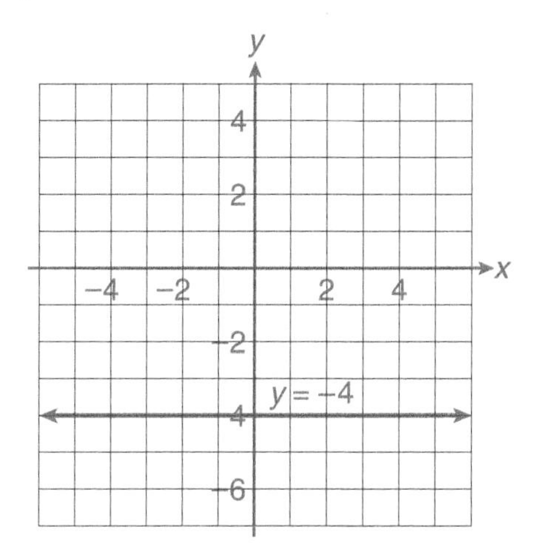

18. a. Plot the data on the following grid using an appropriate scale:

x	0	2	4	6	8	10	12
y	23.76	24.78	25.93	26.24	26.93	27.04	27.93

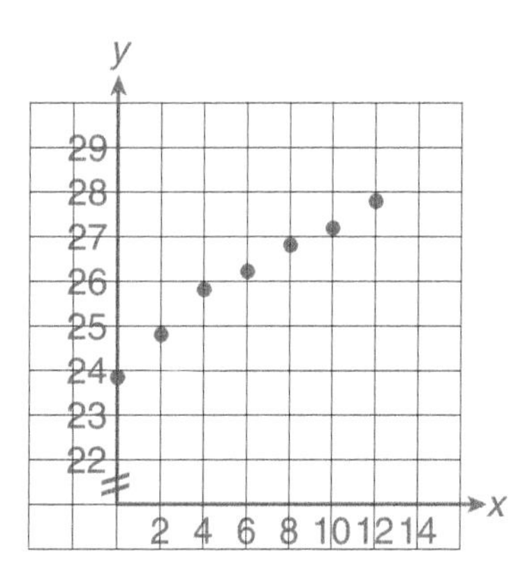

b. Use your graphing calculator to determine the equation of the regression line and the correlation coefficient. Write the result.

$y = 0.322x + 24.155, \quad r = 0.976$

c. Use the result from part b to predict the value of y when $x = 9$.

$y = 27.053$

d. Use the result from part b to predict the value of y when $x = 20$.

$y = 30.595$

19. The following table shows the percentage of seats in the U.S. Congress held by women in selected years from 1991 to 2017, as tabulated by the Center for American Women and Politics:

Year	1991	1995	2001	2006	2011	2013	2017
Percentage of Congressional Seats Held by Women	6.0	10.3	13.6	15.0	16.8	18.3	19.3

Let x represent the number of years from 1991 and y represent the percentage of women in the U.S. Congress.

a. Plot the data from the table on the following grid:

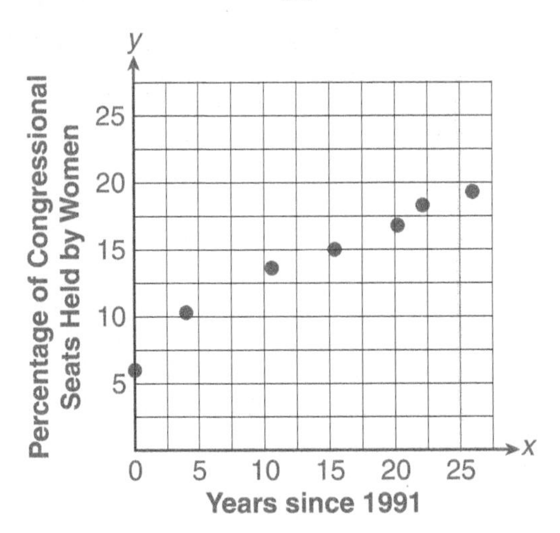

b. Use your graphing calculator to produce the linear regression equation for the data in the graph, and record it here. Round the results to three decimal digits.

regression equation $y = 0.479x + 7.552$

c. Use the equation from part b to estimate the percentage of women in the U.S. Congress in 2008 and 2025.

Year	Percent
2008 ($x = 17$)	15.69
2025 ($x = 34$)	23.83

d. In which of the values found in part c do you have more confidence? Explain.

The year 2008 is within the given data (interpolation); therefore, I have confidence in the result 15.69%. 2025 is well outside of the given data (extrapolation); therefore, I have little or no confidence in the result of 23.83%.

CLUSTER 3 Systems of Linear Equations, Inequalities, and Absolute Value Functions

ACTIVITY 1.11

Moving Out
Systems of Linear Equations in Two Variables

OBJECTIVES

1. Solve a system of 2×2 linear equations numerically and graphically.
2. Solve a system of 2×2 linear equations using the substitution method.
3. Solve an equation of the form $ax + b = cx + d$ for x.

Finals are over, and you are moving back home for the summer. You need to rent a truck to move your possessions from the college residence hall back to your home. You contact two local rental companies and obtain the following information for the 1-day cost of renting a truck:

Company 1: $60.00 per day plus $0.75 per mile

Company 2: $30.00 per day plus $1.00 per mile

The total cost of renting a truck for 1 day is a function of the number of miles driven.

1. Identify the input and output in this situation.

The input variable is the number of miles driven. The output variable is the total cost of renting a truck.

2. Let n represent the total number of miles driven in 1 day.

a. Write an equation to determine the total cost C of renting a truck for 1 day from company 1.

$C = 60 + 0.75n$

b. Write an equation to determine the total cost C of renting a truck for 1 day from company 2.

$C = 30 + 1.00n$

3. a. Complete the following table to compare the total cost of renting the vehicle for the day. Verify your results using the Table feature of your graphing calculator.

n, NUMBER OF MILES DRIVEN	TOTAL COST, C, COMPANY 1 ($)	TOTAL COST, C, COMPANY 2 ($)
0	60	30
20	75	50
40	90	70
60	105	90
80	120	110
100	135	130
120	150	150
140	165	170
160	180	190

b. For what mileage is the 1-day rental cost the same?

$n = 120$ miles

c. Which company should you choose if you intend to drive less than 120 miles?

I should choose company 2 if my miles will be less than 120.

d. Which company should you choose if you intend to drive more than 120 miles?

I should choose company 1 if my miles will be greater than 120.

4. a. Graph the two cost functions, for n between 0 and 160 miles, on the same coordinate axes.

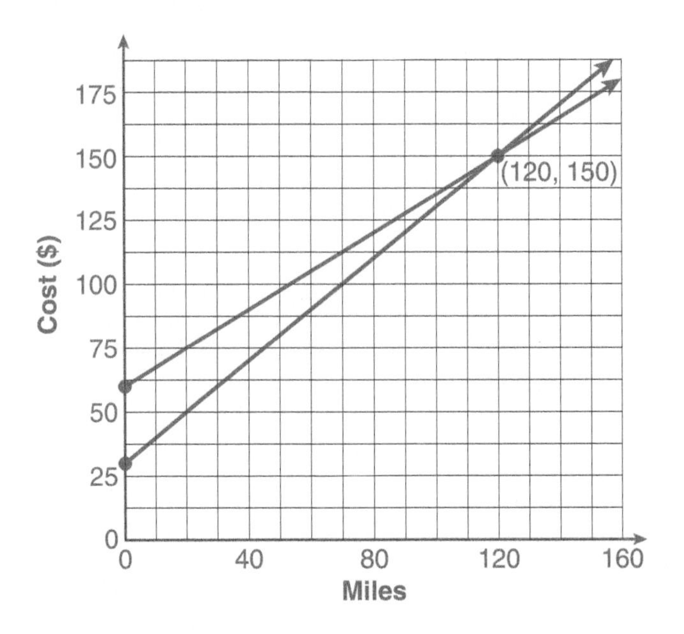

b. Use the table from Problem 3a to determine the point where the lines in part a intersect. What is the significance of the point in this situation?

The point is (120, 150); at 120 miles the rental cost of both companies is \$150.

c. Verify your results from part b using your graphing calculator. Use the INTERSECT feature of your graphing calculator. See Appendix C for the procedure for the TI-84 Plus C. Your final screens should appear as follows:

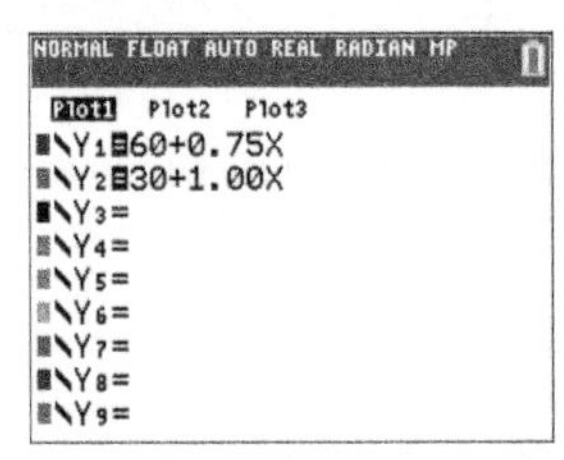

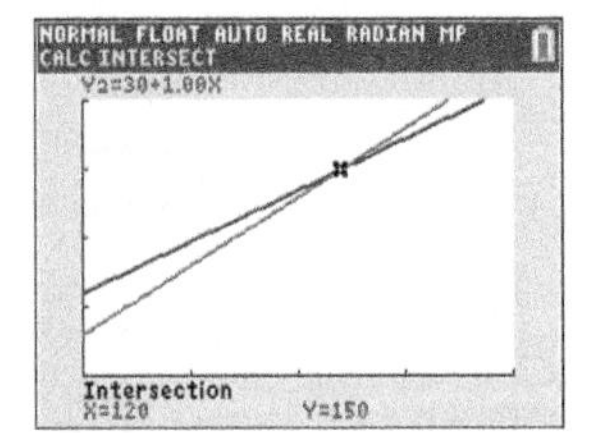

DEFINITION

The two equations in Problem 2 are an example of a 2×2 **system of linear equations.** Such a linear system of two equations in two variables can be written

$$y = ax + b$$

$$y = cx + d.$$

A **solution of the system**, if it exists, is the ordered pair of numbers (x, y) that makes both equations true. If the system has exactly one solution, the system is called **consistent**.

The system in the truck rental situation can be written

$$C = 60 + 0.75n$$
$$C = 30 + 1.00n$$

You first solved the system **numerically** in Problem 3 by completing the table and noting the value of the input that produced the same output. You then solved this system **graphically** in Problem 4 by locating the point of intersection of the two lines.

Substitution Method

You can also use algebraic methods to solve systems of equations. One algebraic method is the **substitution** method. To use the substitution method, you solve one of the equations for one of the variables and then substitute for that variable in the other equation. The following example illustrates this method.

EXAMPLE 1 ***Solve the following system algebraically using the substitution method:***

$$y = 3x - 10 \qquad (1)$$
$$y = 5x + 14 \qquad (2)$$

SOLUTION

Because equation (1) is solved for y, substitute $3x - 10$ for y into equation (2) to get

$$3x - 10 = 5x + 14.$$

To solve the equation for x, you need to isolate the variable. Rewrite the equation so that all terms involving x are on one side and all other terms are on the other side. This is generally accomplished by adding and/or subtracting the appropriate terms to/from each side of the equation.

$$\begin{array}{rcl} 3x - 10 &=& 5x + 14 \\ -5x \quad\;\; && -5x \\ \hline -2x - 10 &=& 14 \end{array}$$ Subtract $5x$ from both sides, and collect like terms.

$$\begin{array}{rcl} -2x - 10 &=& 14 \\ +10 && +10 \\ \hline \end{array}$$ Add 10 to each side, and collect like terms.

$$\frac{-2x}{-2} = \frac{24}{-2}$$ Divide each side by -2, the coefficient of the variable x.

$$x = -12$$

To determine the corresponding y-value, substitute -12 for x into either of the original equations and solve for y. Substituting into equation (1),

$$y = 3(-12) - 10 = -36 - 10 = -46.$$

Remember that the ordered pair $(-12, -46)$ represents the point of intersection of the two lines as well as the solution to the system. This is verified by the following screen:

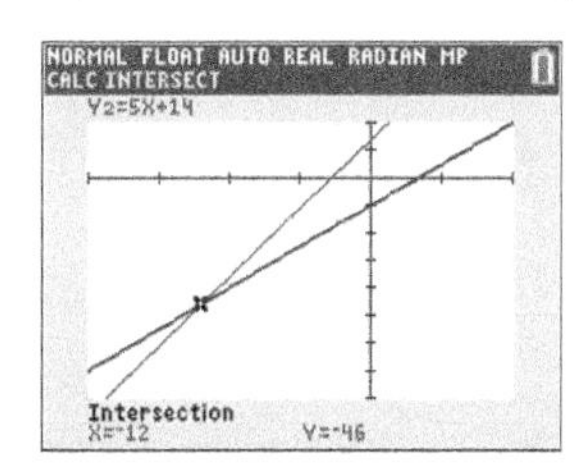

For the truck rental problem, you want C (total cost) to be the same for both functions. Because each expression is solved for C, set the expressions equal to each other. You are substituting an expression for C from one equation into the other equation.

5. a. Solve the introductory problem involving the 1-day cost of renting a truck by solving the following system for n using the substitution method:

$$C = 60.00 + 0.75n$$
$$C = 30.00 + 1.00n$$
$$60.00 + 0.75n = 30.00 + 1.00n$$
$$30.00 = 0.25n$$
$$n = 120$$

b. What does the value of n represent in part a?

n represents the number of miles for which the rental cost for the two companies is the same.

c. Determine the value of C by substituting the value of n from part a into one of the original equations and solving for C.

$C = 60.00 + 0.75(120) = 150$

d. Check to see if your solution satisfies both equations.

$C = 30.00 + 1.00(120) = 150$

This is the same result as that from equation (1).

6. You are going to graduate and are interested in purchasing a new car. You have narrowed the choice to a Honda Accord LX and a Passat. You are concerned about the value of the car depreciating over time. You search the Internet and obtain the following information. The depreciation per year is the amount by which the value of the car will decrease each year.

MODEL	SUGGESTED RETAIL PRICE ($)	DEPRECIATION PER YEAR (ASSUME CONSTANT) ($)
Accord LX	$23,130	$1385
Passat	$26,370	$1790

a. Complete the following table:

YEARS THE CAR IS OWNED	VALUE OF ACCORD LX ($)	VALUE OF PASSAT ($)
0	23,130	26,370
1	21,745	24,580
2	20,360	22,790
3	18,975	21,000
4	17,590	19,210

b. Will the value of the Passat ever be lower than the value of the Accord LX? Explain.

Yes, the Passat decreases in value faster than the Accord.

c. Let V represent the value of the car after x years of ownership. Write an equation to determine V in terms of x for the Accord LX. Write another equation to determine V in terms of x for the Passat.

$V = 23{,}130 - 1385x$

$V = 26{,}370 - 1790x$

d. The two equations in part c form a 2×2 system of linear equations. Solve this system using the substitution method.

$V = 23{,}130 - 1385x$

$V = 26{,}370 - 1790x$

$$\begin{array}{rcl} 23{,}130 - 1385x & = & 26{,}370 - 1790x \\ + 1790x & & + 1790x \\ \hline 23{,}130 + 405x & = & 26{,}370 \\ -23{,}130 & & -23{,}130 \\ \hline 405x & = & 3240 \\ x & = & 8 \end{array}$$

$V = 23{,}130 - 1385(8)$

$V = 12{,}050$ dollars

e. Use your graphing calculator to solve this system graphically. How does your solution compare with your solution in part d?

They are the same.

f. The solution to this system is an ordered pair of the form (x, V). What do the values x and V represent in this solution?

x represents the number of years until the values of the cars are equal.

V represents the value when they are equal.

g. Which car has the better resale value? Explain.

It depends on how long you want to keep the car. The Passat has greater resale dollar value up to 8 years. After 8 years, the Accord has the greater resale dollar value.

Types of Linear Systems

7. You and your friend are traveling in the same direction but in different cars on the New York State Thruway.

a. Let d represent the distance you are from the common starting point. Write an equation for d in terms of time, t, if you are traveling 65 miles per hour.

$d = 65t$

b. Write another equation to determine the distance, d, in terms of time, t, that your friend is from the starting point. She is traveling 60 miles per hour and has a 30-mile head start.

$d = 60t + 30$

8. The linear equations in Problem 7 form a 2×2 system of linear equations.

$$d = 65t$$
$$d = 30 + 60t$$

a. Solve the given system using each of the following methods:

Numerically

t (IN HOURS)	YOUR DISTANCE (MILES)	YOUR FRIEND'S DISTANCE (MILES)
0	0	30
1	65	90
2	130	150
3	195	210
4	260	270
5	325	330
6	390	390

Graphically

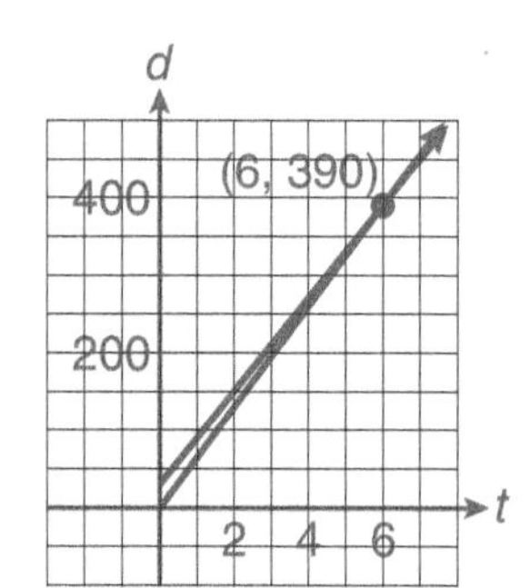

Algebraically

$$d = 65t$$
$$d = 60t + 30$$
$$65t = 60t + 30$$
$$-60t \quad -60t$$
$$\frac{5t}{5} = \frac{30}{5}$$
$$t = 6$$
$$d = 65(6) = 390 \text{ miles}$$

b. What is the solution of this system? What is the meaning of this solution in the context of this problem?

The solution is (6, 390). It means that after 6 hours, you catch up to your friend 390 miles from your starting point.

The linear system in Problem 8 has exactly one solution. Such a system is said to be **consistent**.

Do all systems have exactly one solution? Consider the following.

9. a. You and your friend are traveling again. She still has a 30-mile head start, but this time both of you are traveling 60 miles per hour. When will you catch up with your friend? Explain.

Never, our speeds are the same. I will always be 30 miles behind.

b. The situation just described can be represented by the system

$$d = 60t + 30$$
$$d = 60t.$$

Try to solve this system graphically. Do the lines intersect? What is the solution to the system?

No, there is none. The lines are parallel.

c. Try to solve the system algebraically using the substitution method. What type of equation do you obtain?

$$60t + 30 = 60t$$
$$30 = 0$$

This is a false equation, not true for any value of t.

The linear system in Problem 9 is said to be **inconsistent**. There is no solution because the lines never intersect. Graphically, the slopes of the lines are equal, but the vertical intercepts are different. Therefore, the graphs are parallel lines. Solving such an equation algebraically results in a false equation such as $30 = 0$.

LC LEARNING CATALYTICS

Consider the three systems

a. $y = 3x + 1$
$y = -3x - 1$

b. $y = 3x + 1$
$y = 3x - 3$

c. $y = 3x + 1$
$2y = 6x + 2$.

Which of the three systems is inconsistent?

10. a. You and your friend are taking one more trip. This time she does not have a head start. You both leave from your house, both travel in the same direction, and both travel 60 miles per hour. When will you both be at the same point?

We will always be together.

b. The system for this situation is

$$d = 60t$$
$$d = 60t.$$

Try to solve this system graphically. Do the lines intersect? What is the solution to the system?

The lines are the same line. All points on the line represent solutions.

c. Try to solve this system algebraically. What type of equation do you obtain?

$60t = 60t$

$0 = 0$

The result will be a true equation for any value of t.

The system in Problem 10 is an example of a **dependent** system. Graphically, in such a system, both equations represent the same line. The system has an infinite number of solutions. Solving a dependent system algebraically results in an equation that is true for all t, such as $0 = 0$.

SUMMARY Activity 1.11

1. A 2×2 system of linear equations consists of two equations with two variables. The graph of each equation is a line.
2. A **solution** to a 2×2 system of equations is an ordered pair that satisfies both equations of the system.
3. Solutions can be found in three different ways:
   - **Numerically**, by examining tables of values for both functions.
   - **Graphically**, by graphing each equation and finding the point of intersection.
   - **Algebraically**, by combining the two equations to form a single equation in one variable, which can then be solved. This is called the substitution method.
4. A linear system is **consistent** if there is at least one solution, the points of intersection of the graphs.
5. A linear system is **inconsistent** if there is no solution; the lines are parallel.
6. A linear system is **dependent** if there are infinitely many solutions. The equations represent the same line.

EXERCISES Activity 1.11

Solve the following systems of linear equations numerically, graphically, and algebraically (substitution method).

1. **a.** $y = 2x + 3$
$y = -x + 6$

Numerically

x	y_1	y_2
−2	−1	8
−1	1	7
0	3	6
1	5	5
2	7	4
3	9	3

Graphically

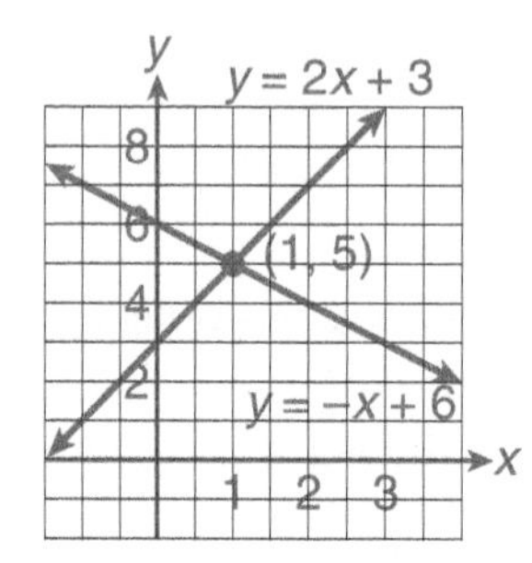

Algebraically (substitution method)

$y = 2x + 3 \quad y = -x + 6$

$$2x + 3 = -x + 6$$
$$+x \qquad +x$$
$$3x + 3 = 6$$
$$-3 \quad -3$$
$$\frac{3x}{3} = \frac{3}{3}$$
$$x = 1$$
$$y = -1 + 6$$
$$y = 5$$

The answer is (1, 5).

b. $y = 0.5x + 9$
$y = 4.5x + 17$

Numerically

x	y_1	y_2
−2	8	8
−1	8.5	12.5
0	9	17
1	9.5	21.5
2	10	26
3	10.5	30.5

Graphically

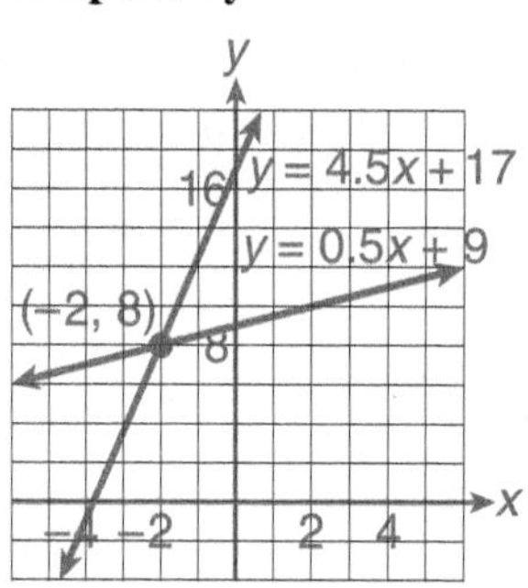
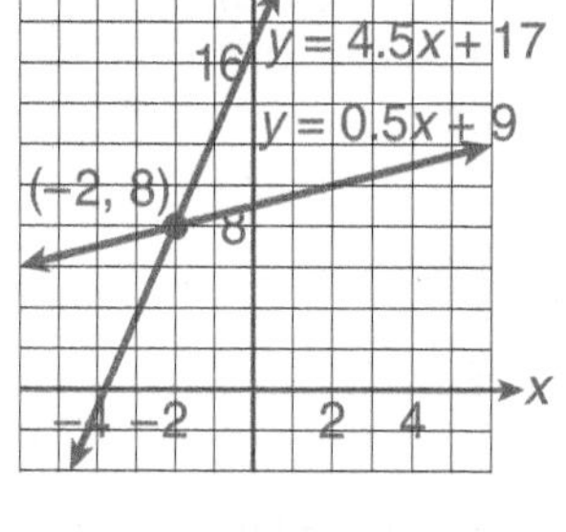

Algebraically (substitution method)

$y = 0.5x + 9 \quad y = 4.5x + 17$

$$0.5x + 9 = 4.5x + 17$$
$$-0.5x \qquad -.5x$$
$$9 = 4x + 17$$
$$-17 = \quad -17$$
$$\frac{-8}{4} = \frac{4x}{4}$$
$$x = -2$$
$$y = 0.5(-2) + 9$$
$$y = -1 + 9 = 8$$

The answer is (−2, 8).

c. $y = 5x - 3$
$y = 5x + 7$

Numerically

x	y_1	y_2
0	−3	7
1	2	12
2	7	17
3	12	22
4	17	27

Graphically

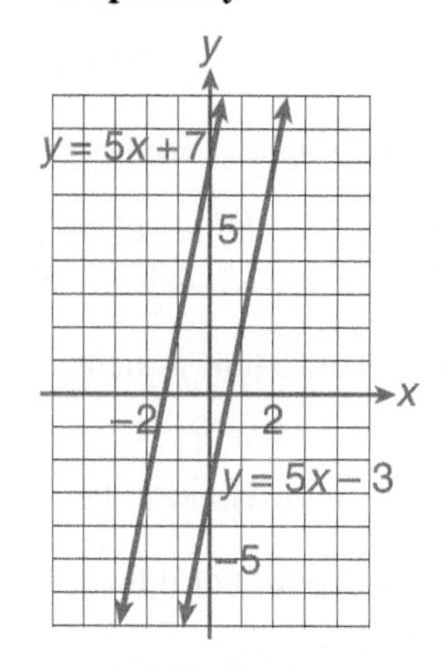

Algebraically (substitution method)

$y = 5x - 3 \quad y = 5x + 7$

$$5x - 3 = 5x + 7$$
$$-5x \qquad -5x$$
$$-3 = 7 \text{ false}$$

There is no solution.

Exercise numbers appearing in color are answered in the Selected Answers appendix.

2. Two companies sell software products. In 2019, company A had total sales of $17.2 million. Its marketing department projects that sales will increase by $1.5 million per year for the next several years. Company B had total sales of $9.6 million of software products in 2019 and projects that its sales will increase by $2.3 million each year.

   Let n represent the number of years since 2019.

   a. Write an equation that represents the total sales, s, of company A since 2019.

      $s = 17.2 + 1.5n$

   b. Write an equation that represents the total sales, s, of company B since 2019.

      $s = 9.6 + 2.3n$

   c. The two equations in parts a and b form a system. Solve this system to determine the year in which the total sales of both companies will be the same.

      $$17.2 + 1.5n = 9.6 + 2.3n$$
      $$17.2 = 9.6 + 0.8n$$
      $$7.6 = 0.8n$$
      $$9.5 = n; \quad \text{in the year 2028}$$

3. You are considering installing a security system in your new house. You gather the following information from two local home security dealers for similar security systems:

   Dealer 1: $3560 to install and $15 per month monitoring fee

   Dealer 2: $2850 to install and $28 per month monitoring fee

   Although the initial fee of dealer 1 is much higher than that of dealer 2, dealer 1's monitoring fee is lower.

   Let n represent the number of months you have the security system.

   a. Write an equation that represents the total cost, c, of the system with dealer 1.

      $c = 3560 + 15n$

   b. Write an equation that represents the total cost, c, of the system with dealer 2.

      $c = 2850 + 28n$

   c. Solve the system of equations that results from parts a and b to determine the number of months for which the total cost of the systems will be equal.

      $$3560 + 15n = 2850 + 28n$$
      $$710 = 13n$$
      $$n = 54.6, \text{ or 55 months}$$

   d. If you plan to live in the house and use the system for 10 years, which system would be less expensive?

      dealer 1's system

4. You can run a 400-meter race at an average rate of 6 meters per second. Your friend can run the race at a rate of 5 meters per second. You give your friend a 40-meter head start. She then runs 360 meters.

   a. Write an equation for your distance, in meters, from the starting point as a function of time, in seconds.

      $d = 6t$

   b. Write an equation for your friend's distance, in meters, from the starting point as a function of time, in seconds.

      $d = 5t + 40$

c. How long does it take you to catch up with your friend?

$6t = 5t + 40$

$t = 40$ sec.

d. How far from the finish line do you meet?

$d = 6(40) = 240$ m

You meet 160 meters from the finish line.

5. For many years, the life expectancy for women has been longer than the life expectancy for men. In the past few years, the life expectancy for men has been increasing at a faster rate than that for women. Using data from the Centers for Disease Control and Prevention and the National Center for Health Statistics, the life expectancy, E, for men and women in the United States can be modeled as follows:

Women: $E = 0.109t + 77.5$

Men: $E = 0.192t + 70.0,$

where t represents the number of years since 1980

a. Solve the system numerically by completing the following table. Approximate the value of t to the nearest year. Use your calculator.

t, NUMBER OF YEARS SINCE 1980	LIFE EXPECTANCY FOR WOMEN	LIFE EXPECTANCY FOR MEN
0	77.5	70
25	80.2	74.8
50	83	79.6
100	88.4	89.2
85	86.8	86.3
88	87.1	86.9
90	87.3	87.3

b. What does the solution to this system represent in the context of this problem?

Ninety years after 1980, in the year 2070, the life expectancy for both men and women will be 87.3 years.

c. Solve this problem graphically using your calculator. Use the window $X_{min} = 0, X_{max} = 125, Y_{min} = 60$, and $Y_{max} = 100$.

(90.4, 87.3)

d. Solve the system algebraically using the substitution method.

$0.192t + 70 = 0.109t + 77.5$

$0.083t = 7.5$

$t = 90.36$

$E = 0.192(90.36) + 70$

$E = 87.34$

6. You are the manager of a small company producing interlocking paving pieces, called pavers, for driveways. You sell the pavers in bundles that cost \$200; each bundle contains 144 pavers. The total cost in dollars, $C(x)$, of producing x bundles of pavers is modeled by $C(x) = 160x + 1000$.

a. The revenue is the amount of money collected from the sale of the product. Write an equation for the revenue function in dollars, R, from the sale of x bundles of pavers.

$R = 200x$

b. Determine the slope and the C-intercept of the given cost function. Explain the practical meaning of each in this situation.

The slope of the cost function is 160. This represents a cost of \$160 to produce each bundle of pavers. The C-intercept is (0, 1000). Fixed costs of \$1000 are incurred even when no pavers are produced.

c. Determine the slope and the R-intercept of the revenue function. Explain the practical meaning of each in this situation.

The slope of the revenue function is \$200, the income for each bundle sold. The R-intercept is (0, 0). If no pavers are sold, there is no income.

d. Graph the cost and revenue functions on the same set of axes where the number of bundles varies from 0 to 50. A company will break even when its revenue, R, exactly equals its cost, C. Estimate the break-even point from the graph. Express your answer as an ordered pair, giving units. Check your estimate of the break-even point by graphing the two functions on your graphing calculator.

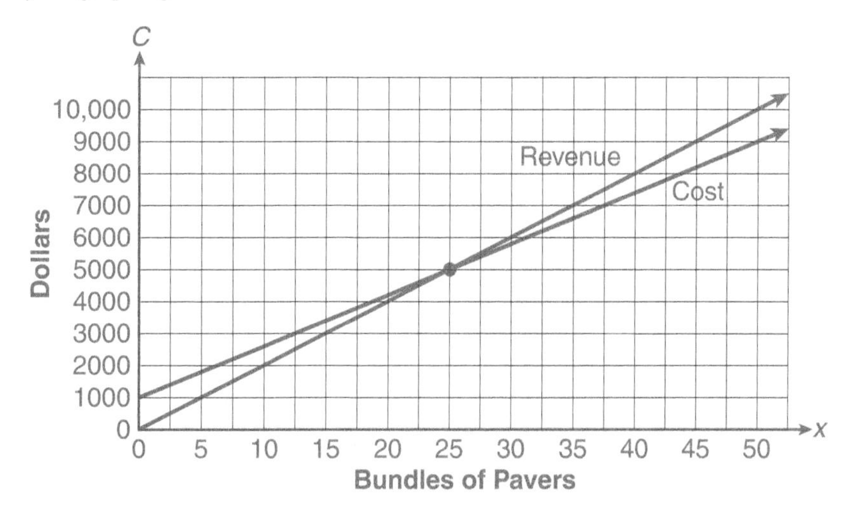

From the graph, the break-even point is (25 bundles, \$5000).

e. Determine the exact break-even point algebraically. If your algebraic solution does not approximate your answer from the graph in part d, explain why.

$R = C$ at the break-even point

$$200x = 160x + 1000$$
$$40x = 1000$$
$$x = 25$$
$$R(25) = 200(25)$$
$$= \$5000$$

The break-even point is (25, 5000), which agrees with the answer obtained from the graph.

f. How many bundles of pavers must the company sell to break even?

The company must sell 25 bundles of pavers to break even.

g. What is the total cost to the company when you break even? Verify that the cost and revenue values are equal at the break-even point.

$C(25) = 160(25) + 1000 = \$5000$

Total cost is \$5000 at the break-even point.

h. For what values of x will your revenue exceed your cost?

Revenues will exceed cost when more than 25 bundles are sold.

i. If you knew you could sell only 30 bundles of pavers, would you make them? Consider how much it would cost you and how much you would make. What if you could sell only 20?

The profit is only \$200 on 30 bundles of pavers, so I might or might not make them.

$C(30) = 160(30) + 1000 = \$5800$

$R(30) = 200(30) = \$6000$

There is a loss of \$200 if only 20 bundles are sold. I would not make them.

$C(20) = 160(20) + 1000 = \$4200$

$R(20) = 200(20) = \$4000$

Collecting and Analyzing Data

7. The women's world record time in the 400-meter run is decreasing at a faster rate than the men's record time. This is true of world record times in many events, including men's and women's Olympic 500-meter speed skating and the 1500-meter run.

Choose one of these competitions, and use linear functions to model the data for the women's event and the men's event. Use the models to estimate when the women's record time will equal the men's record time.

Work in a group, and be prepared to give a presentation to the class. The presentation should include visual displays showing the tables, scatterplots, and equations used in the problem-solving process. Be sure to identify the source of your data.

ACTIVITY

1.12

Fireworks

Solving Systems of Linear Equations using Algebraic Methods

OBJECTIVES

1. Solve a 2×2 linear system algebraically using the substitution method and the addition method.
2. Solve equations containing parentheses.

You have volunteered to serve on the committee formed by local service organizations that will be responsible for the Fourth of July fireworks in your town. The first thing you learn about fireworks is some terminology. A cake is an item that has a single fuse used to light several tubes in sequence. Cakes can have a variety of intricate aerial effects, including spinners, fish, flower bouquets, comets, and other elements. Cakes are the most popular consumer fireworks item outside of sparklers and firecrackers. A peony is an aerial effect that looks like a spherical ball of colored lights in the sky.

You are put in charge of purchasing cakes and peonies for the first 10 minutes of the show. You determine that you need a total of 30 cakes and peonies to cover your time frame. You discover that you can purchase peonies for $44 each and a group of cakes for $47 each. You have a budget of $1350. You need to determine how many cakes and how many peonies you can purchase and still fill up your required time for the show.

Let c represent the number of cakes you will purchase. Let p represent the number of peonies you will purchase.

1. Write an equation that relates c and p to the number of fireworks you need to cover your 10 minutes of the show.

   $c + p = 30$

2. Write an equation that relates c and p to your budget of $1350.

   $47c + 44p = 1350$

   The situation just described can be represented by the following system:

$$\begin{aligned} c + p &= 30 \\ 47c + 44p &= 1350 \end{aligned}$$

   Note that each equation is in general form. One approach to solving this system is to solve each equation for one variable in terms of the other and then use the substitution method.

3. **a.** Solve each of the equations above for p.

   $p = 30 - c \qquad 44p = 1350 - 47c$

   $p = \dfrac{1350 - 47c}{44}$

   b. Solve the system using the substitution method.

   $30 - c = \dfrac{1350 - 47c}{44}$

   $1320 - 44c = 1350 - 47c$

   $3c = 30$, or $c = 10$; $p = 30 - 10 = 20$ You purchase 10 cakes and 20 peonies.

Addition Method

Sometimes it is more convenient to leave each equation in the linear system in general form $(Ax + By = C)$ rather than solving for one variable in terms of the other. Look at the same system:

$$\begin{aligned} c + p &= 30 \qquad (1) \\ 47c + 44p &= 1350 \qquad (2) \end{aligned}$$

If you apply the addition principle of algebra by adding the two equations (left side to left side and right side to right side), you may be able to obtain a single equation containing only one variable.

$$\begin{aligned} c + p &= 30 \quad (1) \\ \underline{47c + 44p} &\underline{= 1350} \quad (2) \\ 48c + 45p &= 1380 \end{aligned}$$

In this case, adding the equations does not eliminate a variable. But if you multiply both sides of equation (1) by -47, the coefficients of c will be opposites, and the variable c can be eliminated.

$$\begin{aligned} -47(c + p) &= -47(30) \rightarrow -47c - 47p = -1410 \\ 47c + 44p &= 1350 \qquad\quad \underline{47c + 44p = \ \ 1350} \\ & \qquad\qquad\qquad\quad 0c - \ \ 3p = -60 \end{aligned}$$

Add the corresponding sides of the two equations.

Solving the resulting equation, $-3p = -60$ for p,

$$\begin{aligned} -3p &= -60 \\ p &= 20 \end{aligned}$$

Substituting for p in $c + p = 30$, you have

$$\begin{aligned} c + 20 &= 30 \\ c &= 10 \end{aligned}$$

This method of solving systems algebraically is called the **addition method**.

To solve a 2×2 linear system by the addition method,

Step 1. Line up the like terms in each equation vertically.

Step 2. If necessary, multiply one or both equations by constants so that the coefficients of one of the variables are opposites.

Step 3. Add the corresponding sides of the equations and solve the resulting equation for the remaining variable.

Step 4. Substitute the value of the variable from step 3 into one of the original equations and solve for the second variable.

Step 5. Check the solution in the original equation not used in step 4.

4. Solve the same system again using the addition method. Multiply the appropriate equation by the appropriate factor to eliminate p, and solve for c first.

$$\begin{aligned} c + p &= 30 \\ 47c + 44p &= 1350 \end{aligned}$$

$$-44(c + p) = -44(30) \qquad -44c - 44p = -1320 \qquad 10 + p = 30$$

$$47c + 44p = 1350 \qquad \underline{47c + 44p = 1350} \qquad p = 20$$

$$3c = 30 \qquad (10, 20)$$

$$c = 10$$

Not all systems have convenient coefficients, and you may need to multiply one or both equations by a factor that will produce coefficients of the same variable that are additive inverses, or opposites.

5. Consider solving the following system using the addition method:

$$-2x + 5y = -16$$
$$3x + 2y = 5$$

a. Identify which variable you want to eliminate. Multiply the appropriate equations by the appropriate factors so that the coefficients of your chosen variable are opposites. Show the two equations after you multiply by the factor(s). (Remember to multiply both sides of the equation by the factor.)

You decide to eliminate the x-variable. Use the factors 3 and 2 respectively.

$$3(-2x + 5y) = 3(-16) \qquad -6x + 15y = -48$$
$$2(3x + 2y) = 2(5) \qquad 6x + 4y = 10$$

b. Add the two equations to eliminate the chosen variable.

$$19y = -38$$

c. Solve the resulting linear equation.

$$\frac{19y}{19} = \frac{-38y}{19}$$
$$y = -2$$

d. Determine the complete solution. Remember to check by substituting into both of the original equations.

$$-2x + 5(-2) = -16$$
$$-2x - 10 = -16$$
$$-2x = -6$$
$$x = 3 \quad (3, -2)$$

$$-2(3) + 5(-2) = -16 \qquad 3(3) + 2(-2) = 5$$
$$-6 - 10 = -16 \qquad 9 - 4 = 5$$
$$-16 = -16 \qquad 5 = 5$$

Substitution Method Revisited

The substitution method of solving a 2×2 system of linear equations is generally used when at least one of the equations in the system is solved for one variable in terms of the other. For example, solve the following system:

$$y = -x - 1 \qquad (1)$$
$$4x - 3y = 24 \qquad (2)$$

Using the substitution method, substitute the expression $-x - 1$ from equation (1) for y in equation (2) as follows:

$$y = -x - 1 \qquad 4x - 3y = 24$$

The resulting equation, $4x - 3(-x - 1) = 24$, contains just one variable, namely, x. The variable y has been eliminated. Now solving for x, you have

$$4x - 3(-x - 1) = 24 \qquad \text{Apply the distributive property.}$$
$$4x + 3x + 3 = 24 \qquad \text{Combine like terms on the left-hand side.}$$
$$7x + 3 = 24 \qquad \text{Add the opposite of 3 to each side to isolate the variable term.}$$
$$7x = 21 \qquad \text{Divide each side by 7.}$$
$$x = 3, \ y = -(3) - 1 = -4$$

6. a. To solve the following system using the substitution method, you need to solve either of the equations for one variable in terms of the other. Solve equation (1) for x.

$$x - 2y = 5 \quad (1)$$
$$5x - y = -2 \quad (2)$$

$x - 2y = 5;\ x = 2y + 5$

b. Substitute the expression obtained for x in part a into equation (2).

$5(2y + 5) - y = -2$

c. Solve the equation in part b for y.

$$5(2y + 5) - y = -2$$
$$10y + 25 - y = -2$$
$$9y + 25 = -2$$
$$9y = -27$$
$$y = -3$$

d. Determine the value of x that satisfies the system.

$x = 2(-3) + 5 = -6 + 5 = -1$

EXAMPLE 1 ***In the fireworks system***

$$c + p = 30 \quad (1)$$
$$47c + 44p = 1350 \quad (2)$$

You could solve equation (1) for c and substitute for c in equation (2) as follows.

Step 1. Solve $c + p = 30$ for c.

$$c = 30 - p$$

Step 2. Substitute $30 - p$ for c in equation (2)

$$47(30 - p) + 44p = 1350$$

Step 3. Solve the resulting equation for c.

$47(30 - p) + 44p = 1350$	Remove the parentheses by applying the distributive property.
$1410 - 47p + 44p = 1350$	Collect like terms on the same side.
$1410 - 3p = 1350$	Add the opposite of 1410 to both sides.
$-3p = -60$	Divide each side by -3
$p = 20$	

Step 4. From equation (1), $c = 10$ as before.

LC LEARNING CATALYTICS

Consider the two following systems

a. $y = 1 - 2x$
$3x + 2y = 4$

b. $-2x + y = -9$
$2x + 3y = 5$

Which system would most likely be easier to solve using substitution?

7. a. Solve the following linear system using the substitution method in which you solve only one of the equations for a variable:

$$x - y = 5$$
$$4x + 5y = -7$$

$$x = y + 5$$
$$4(y + 5) + 5y = -7$$
$$4y + 20 + 5y = -7$$
$$9y + 20 = -7$$
$$9y = -27$$
$$y = -3$$
$$x = -3 + 5 = 2 \quad (2, -3)$$

b. Check your answer in part a by solving the system using the addition method.

$$5(x - y) = 5(5) \qquad 5x - 5y = 25 \qquad 2 - y = 5$$
$$4x + 5y = -7 \qquad \underline{4x + 5y = -7} \qquad -y = 3$$
$$\frac{9x}{9} = \frac{18}{9} \qquad y = -3$$
$$x = 2 \quad (2, -3)$$

SUMMARY Activity 1.12

1. There are two methods for solving a 2 × 2 system of linear equations algebraically:

a. the substitution method

b. the addition method.

2. To solve linear systems by **substitution**,

Step 1. Solve one equation for a variable.

Step 2. Substitute the expression from step 1 for that variable in the other equation.

Step 3. Solve the resulting equation for the remaining variable.

Step 4. Substitute the value from step 3 into one of the original equations, and solve for the other variable.

3. To solve linear systems by **addition** with equations written in the form $Ax + By = C$,

Step 1. Multiply one equation or both equations by the number(s) that will make the coefficients of one of the variables opposites.

Step 2. Add the two equations to eliminate one variable and solve the resulting equation for the remaining variable.

Step 3. Substitute the value of the variable from step 2 into one of the original equations and solve for the second variable.

Step 4. Check the solution in the original equation not used in step 3.

EXERCISES Activity 1.12

Solve by substitution.

1. $y = x - 2$

$2x - y = 6$

$$2x - (x - 2) = 6$$
$$2x - x + 2 = 6$$
$$x = 4$$
$$y = (4) - 2 = 2$$
$$(x, y) = (4, 2)$$

2. $y = 1 - 2x$

$3x + 2y = 4$

$$3x + 2(1 - 2x) = 4$$
$$3x + 2 - 4x = 4$$
$$-x = 2$$
$$x = -2$$
$$y = 1 - 2(-2) = 5$$
$$(x, y) = (-2, 5)$$

3. $x = 2 - y$

$2x - 4y = 13$

$$2(2 - y) - 4y = 13$$
$$4 - 2y - 4y = 13$$
$$-6y = 9$$
$$y = -\frac{9}{6} = -\frac{3}{2}$$
$$x = 2 - \left(-\frac{3}{2}\right) = \frac{7}{2}$$
$$(x, y) = \left(\frac{7}{2}, -\frac{3}{2}\right) = (3.5, -1.5)$$

4. $x = 2y + 1$

$4x - 3y = 2$

$$4(2y + 1) - 3y = 2$$
$$8y + 4 - 3y = 2$$
$$5y = -2$$
$$y = -\frac{2}{5}$$
$$x = 2\left(-\frac{2}{5}\right) + 1 = -\frac{4}{5} + 1 = \frac{1}{5}$$
$$(x, y) = \left(\frac{1}{5}, -\frac{2}{5}\right) = (0.2, -0.4)$$

Solve for y in the first equation, then substitute for y in the second equation.

5. $x - y = 2$

$3x + 2y = 1$

$$-y = 2 - x$$
$$y = -2 + x = x - 2$$
$$3x + 2(x - 2) = 1$$
$$3x + 2x - 4 = 1$$
$$5x = 5$$
$$x = 1$$
$$y = (1) - 2 = -1$$
$$(x, y) = (1, -1)$$

6. $2x + y = 4$

$4x - 3y = 5$

$$y = 4 - 2x$$
$$4x - 3(4 - 2x) = 5$$
$$4x - 12 + 6x = 5$$
$$10x = 17$$
$$x = 1.7$$
$$y = 4 - 2(1.7) = 0.6$$
$$(x, y) = (1.7, 0.6)$$

Solve for x in the first equation and then substitute for x in the second equation.

7. $x + y = 3$

$2x + 3y = 1$

$$x = 3 - y$$
$$2(3 - y) + 3y = 1$$
$$6 - 2y + 3y = 1$$
$$y = -5$$
$$x = 3 - (-5) = 8$$
$$(x, y) = (8, -5)$$

8. $-x + 2y = 1$

$-3x + 4y = 2$

$$-x = 1 - 2y$$
$$x = -1 + 2y = 2y - 1$$
$$-3(2y - 1) + 4y = 2$$
$$-6y + 3 + 4y = 2$$
$$-2y = -1$$
$$y = \frac{1}{2}$$
$$x = 2\left(\frac{1}{2}\right) - 1 = 0$$
$$(x, y) = \left(0, \frac{1}{2}\right) = (0, 0.5)$$

Solve one equation for x or y; then solve the system using substitution.

9. $x + y = 2$
$2x + 3y = 4$

$$x = 2 - y$$
$$2(2 - y) + 3y = 4$$
$$4 - 2y + 3y = 4$$
$$y = 0$$
$$x = 2 - (0) = 2$$
$$(x, y) = (2, 0)$$

10. $-2x + y = -9$
$2x + 3y = 5$

$$y = 2x - 9$$
$$2x + 3(2x - 9) = 5$$
$$2x + 6x - 27 = 5$$
$$8x = 32$$
$$x = 4$$
$$y = 2(4) - 9 = -1$$
$$(x, y) = (4, -1)$$

11. $x + 4y = 2$
$-2x + 3y = 7$

$$x = 2 - 4y$$
$$-2(2 - 4y) + 3y = 7$$
$$-4 + 8y + 3y = 7$$
$$11y = 11$$
$$y = 1$$
$$x = 2 - 4(1) = -2$$
$$(x, y) = (-2, 1)$$

12. $x - 2y = 4$
$3x - 4y = -1$

$$x = 2y + 4$$
$$3(2y + 4) - 4y = -1$$
$$6y + 12 - 4y = -1$$
$$2y = -13$$
$$y = -\frac{13}{2}$$
$$x = 2\left(-\frac{13}{2}\right) + 4 = -13 + 4 = -9$$
$$(x, y) = \left(-9, -\frac{13}{2}\right) = (-9, -6.5)$$

Solve using the addition method.

13. $2x - y = 7$
$3x + 2y = 7$

$$4x - 2y = 14$$
$$3x + 2y = 7$$
$$7x = 21$$
$$x = 3$$
$$2(3) - y = 7$$
$$-y = 1$$
$$y = -1$$
$$(x, y) = (3, -1)$$

14. $x + 2y = 12$
$4x + 3y = 23$

$$-4x + -8y = -48$$
$$4x + 3y = 23$$
$$-5y = -25$$
$$y = 5$$
$$x + 2(5) = 12$$
$$x = 2$$
$$(x, y) = (2, 5)$$

15. $2x + 5y = 1$
$4x + 3y = -5$

$$-4x - 10y = -2$$
$$4x + 3y = -5$$
$$-7y = -7$$
$$y = 1$$
$$2x + 5(1) = 1$$
$$2x = -4$$
$$x = -2$$
$$(x, y) = (-2, 1)$$

16. $3x + 4y = 2$
$2x + 5y = 6$

$$-6x - 8y = -4$$
$$6x + 15y = 18$$
$$7y = 14$$
$$y = 2$$
$$3x + 4(2) = 2$$
$$3x = -6$$
$$x = -2$$
$$(x, y) = (-2, 2)$$

17. A catering service placed an order for eight centerpieces and five glasses, and the bill was $106. For the wedding reception, they were short one centerpiece and six glasses and had to reorder. This order came to $24. Let x represent the cost of one centerpiece, and let y represent the cost of one glass.

a. Write a system of equations that represents both orders.

$$8x + 5y = 106$$
$$x + 6y = 24$$

b. Solve the system using the substitution method. Interpret your solution.

$$x = -6y + 24$$
$$8(-6y + 24) + 5y = 106$$
$$-48y + 192 + 5y = 106$$
$$-43y = -86$$
$$y = 2$$
$$x = -6(2) + 24 = 12$$

(12, 2)

Each centerpiece costs $12, and each glass costs $2.

c. Check your result in part b using the addition method.

$$8x + 5y = 106 \qquad 8x + 5y = 106$$
$$-8(x + 6y) = -8(24) \qquad \underline{-8x - 48y = -192}$$
$$-43y = -86$$
$$y = 2$$
$$x + 6(2) = 24$$
$$x + 12 = 24$$
$$x = 12$$

d. Use your graphing calculator to solve the system.

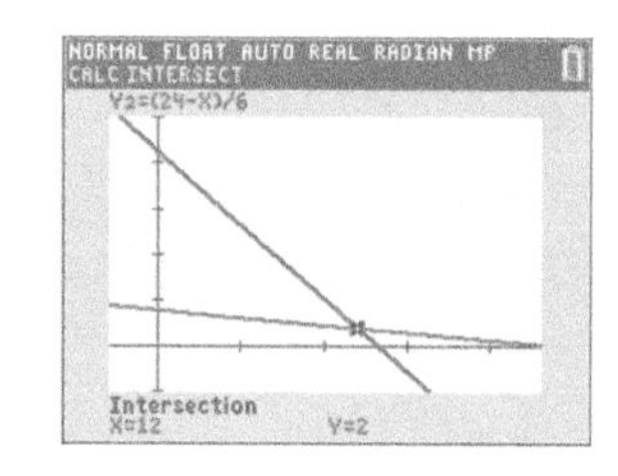

ACTIVITY 1.13

Manufacturing Pewter Oil Lamps

Solving Systems of Linear Equations in Three Variables

OBJECTIVE

Solve a 3×3 linear system of equations.

You recently accepted a job as the production manager for a small company that manufactures handmade pewter products. During the next week, you have to schedule the manufacturing of three oil lamps: a small basic version, a larger basic version, and a deluxe limited edition version.

Your entire staff is trained in the steps of the process: spinning the base, polishing the base, and assembling the lamp. You have determined the following for the week to come: a total of 260 hours for spinning, 170 hours for polishing, and 120 hours for the assembly of the lamps.

The time allotted for each task on each lamp is summarized in the following table:

	SMALL BASIC	LARGE BASIC	LIMITED EDITION
Spinning	1 hr.	3 hr.	4 hr.
Polishing	1 hr.	2 hr.	2 hr.
Assembly	1 hr.	1 hr.	2 hr.

Use x, y, and z to represent the number of each type of lamp you are to build this week, with x representing the number of small basic lamps, y the number of large basic lamps, and z the number of limited edition lamps.

1. Write an equation for the total hours spent on spinning the lamp bases each week.

   $x + 3y + 4z = 260$

2. Write an equation for the total hours spent on polishing lamp bases each week.

   $x + 2y + 2z = 170$

3. Write an equation for the total hours spent on assembling lamps each week.

   $x + y + 2z = 120$

Taken together, the equations in Problems 1–3 form a **3×3 system of linear equations.**

The solution to this system is the ordered triple of numbers (x, y, z) that satisfies all three equations. The strategy for solving such a system is typically to reduce the system to a 2×2 linear system and then proceed to solve this smaller system.

4. Select two equations from Problems 1–3, and use substitution or addition to eliminate one of the variables.

   Using the equations in Problems 1 and 2 to eliminate the variable x, you have

$$\begin{array}{r} x + 3y + 4z = 260 \\ \underline{-x - 2y - 2z = -170} \\ y + 2z = 90 \end{array}$$

5. Select a different pair of equations from Problems 1–3, and eliminate the same variable.

$$\begin{array}{r} x + 3y + 4z = 260 \\ \underline{-x - y - 2z = -120} \\ 2y + 2z = 140 \\ y + z = 70 \end{array}$$

6. The equations from Problems 4 and 5 form a 2×2 system. Solve this new 2×2 system.

$$\begin{array}{r} y + 2z = 90 \\ \underline{-y - z = -70} \\ z = 20 \end{array}$$

   $y + 20 = 70$ or $y = 50$

7. Substitute the solutions from Problem 6 into one of the original three equations. Now solve for the third variable.

Substituting 50 for y and 20 for z into the equation for Problem 3,

$$x + 50 + 2(20) = 120$$
$$x = 30$$

8. How many of each type of oil lamp should you manufacture this week to make optimal use of your available time? Make sure your solution agrees with each of the three original assumptions.

30 small basic lamps, 50 large basic lamps, and 20 limited edition lamps

9. Explain why it is not possible to solve this 3×3 system by graphing on your calculator.

The graphing calculator will only graph functions of two variables, not three.

All the equations in the 3×3 systems in this activity are called *linear equations* even though they cannot all be graphed as single lines. In this case, linearity refers to each variable being linear, that is, raised to the first power.

10. Solve the following 3×3 linear system:

$$x - 2y + z = -5 \qquad (1)$$
$$2x + y - z = 6 \qquad (2)$$
$$3x + 3y - z = 11 \qquad (3)$$

If you are not sure where to start, follow these steps.

Step 1. Is it possible to add two of the equations (right side to right side and left side to left side) so that one of the variables is eliminated? [Add equation (1) to equation (2).] Yes.

$$\begin{array}{r} x - 2y + z = -5 \\ \underline{2x + y - z = 6} \\ 3x - y = 1 \end{array}$$

Step 2. Is it possible to add a different pair of equations to eliminate the same variable? [Add equation (1) to equation (3).] Yes.

$$\begin{array}{r} x - 2y + z = -5 \\ \underline{3x + 3y - z = 11} \\ 4x + y = 6 \end{array}$$

Step 3. Notice that your equations from parts a and b form a 2×2 linear system. Solve this 2×2 system.

Solve $4x + y = 6$ for $y: y = 6 - 4x$.

Substitute $6 - 4x$ for y into $3x - y = 1$.

$$3x - (6 - 4x) = 1$$
$$3x - 6 + 4x = 1$$
$$7x = 7$$
$$x = 1;\ y = 2$$

Step 4. Substitute your solution from step 3 into any one of the three original equations, and solve the resulting equation for the remaining variable.

$$1 - 4 + z = -5$$
$$z = -2$$

so $(x, y, z) = (1, 2, -2)$

Step 5. The final step is to substitute your potential solution into each of the three original equations. This is the only way you can be confident that your solution is correct.

Check: Equation (1): $1 - 4 - 2 = -5$
Equation (2): $2 + 2 + 2 = 6$
Equation (3): $3 + 6 + 2 = 11$

Most 3×3 systems will not have coefficients that are as convenient as the ones you just encountered. The following application provides a case in point.

11. In your job as a buyer for The Better Cup, a nationwide coffee bar, you need to buy three grades of coffee that sell at different price points. Currently, the three grades sell for $2.00, $2.80, and $3.50 per pound. You need to determine the number of pounds of each grade of coffee to buy.

LC LEARNING CATALYTICS

Solve the following linear system.
$x + y + z = 3$
$2x - y + 2z = 3$
$3x + 2y - z = 0$

Let x = the number of pounds of $2.00 per pound coffee.
Let y = the number of pounds of $2.80 per pound coffee.
Let z = the number of pounds of $3.50 per pound coffee.

Write the equation that corresponds to each of the following assumptions:

a. You need to buy a total of 12,000 pounds of coffee.

$x + y + z = 12{,}000$

b. You need to spend a total of $30,020.

$2.00x + 2.80y + 3.50z = 30{,}020$

c. You need to buy 800 more pounds of the $2.00 per pound coffee than the other two grades combined.

$x = y + z + 800$ or $x - y - z = 800$

12. Solve the 3×3 system of equations determined in Problem 11. Verify that all three assumptions are satisfied and state your solution in terms of the context of this problem.

Equation (1): $x + y + z = 12{,}000$
Equation (2): $2x + 2.8y + 3.5z = 30{,}020$
Equation (3): $x = y + z + 800$

Substitute $y + z + 800$ for x in equations (1) and (2).

Equation (1): $(y + z + 800) + y + z = 12{,}000$
$2y + 2z = 11{,}200$
$y + z = 5600$
Equation (2): $2(y + z + 800) + 2.8y + 3.5z = 30{,}020$
$4.8y + 5.5z = 28{,}420$

Solve the 2×2 system of equations for y and z.

Equation (4): $y + z = 5600$
Equation (5): $4.8y + 5.5z = 28{,}420$
$-4.8y - 4.8z = -26{,}880$
$4.8y + 5.5z = 28{,}420$
$0.7z = 1540$
$z = 2200$
$y = 5600 - z = 5600 - 2200 = 3400$
$x = y + z + 800 = 3400 + 2200 + 800 = 6400$

You need to buy 6400 pounds of $2.00 per pound coffee, 3400 pounds of $2.80 per pound coffee, and 2200 pounds of $3.50 per pound coffee.

Further examples and practice in solving 3×3 linear systems of equations can be found in Appendix A.

SUMMARY Activity 1.13

- A **3×3 system of linear equations** consists of three linear equations involving three variables.
- A **linear equation in the three variables** x, y, and z is of the form $Ax + By + Cz = D$, where A, B, C, and D are any constants. The equation is linear because the variables are all raised to the first power.
- To solve a 3×3 system algebraically,

  1. reduce the system in size to a 2×2 system.

  2. solve the 2×2 system.

  3. substitute the 2×2 solution into any original equation to solve for the third unknown.

  4. check the solution by substituting into all three of the original equations.
- Note that 3×3 systems may also be **inconsistent** (have no solution) or **dependent** (have infinitely many solutions).

EXERCISES Activity 1.13

In Exercises 1–4, solve the 3×3 linear systems. Be sure to check your solution in all three of the original equations.

1. $x + y - z = -8$

$-x + y + z = 2$

$2x - y + z = 8$

$2y = -6, \quad y = -3$

$3x = 0, \quad x = 0$

$(0, -3, 5)$

2. $2x - 3y + z = 7 \quad (1)$

$x + 2y - 2z = -5 \quad (2)$

$-2x + y + z = -1 \quad (3)$

$(1) + (3) \quad -2y + 2z = 6 \quad (4)$

$2(2) + (3) \quad 5y - 3z = -11 \quad (5)$

$3(4) \quad -6y + 6z = 18$

$2(5) \quad 10y - 6z = -22$

$4y = -4$

$y = -1; z = 2; x = 1 \ (1, -1, 2)$

3. $x + 2y - z = 0$ (1)

$3x + 2y + z = -8$ (2)

$2x + 3y + z = 0$ (3)

$$\begin{array}{lrl}
(1)+(2) & 4x + 4y = -8 & (4) \\
(1)+(3) & 3x + 5y = 0 & (5) \\
3(4) & 12x + 12y = -24 & \\
-4(5) & -12x - 20y = 0 & \\
 & -8y = -24 & \\
 & y = 3 & \\
 & 4x + 12 = -8 & \\
 & 4x = -20 & \\
 & x = -5 & \\
 & 2(-5) + 3(3) + z = 0 & \\
 & -10 + 9 + z = 0 & \\
 & z = 1 &
\end{array}$$

$(-5, 3, 1)$

4. $x + 2y - 3z = 5$ (1)

$-x + y + 2z = 0$ (2)

$2x - y + z = -1$ (3)

$$\begin{array}{lrl}
(1) & x + 2y - 3z = 5 & \\
(2) & -x + y + 2z = 0 & \\
 & 3y - z = 5 & (4) \\
2(2) & -2x + 2y + 4z = 0 & \\
(3) & 2x - y + z = -1 & \\
 & y + 5z = -1 & (5) \\
 & 3y - z = 5 & \\
-3(5) & -3y - 15z = 3 & \\
 & -16z = 8 & \\
 & \overline{-16} \quad \overline{-16} &
\end{array}$$

$z = \frac{-1}{2}, y = \frac{3}{2}, x = \frac{1}{2}$ $\left(\frac{1}{2}, \frac{3}{2}, -\frac{1}{2}\right)$

5. Recall that some 2×2 linear systems do not have unique solutions. Solve these 3×3 linear systems. Identify each system as either dependent or inconsistent.

a. $x + 2y + z = 4$ (1)

$2x - y + 3z = 2$ (2)

$3x + y + 4z = 6$ (3)

$$\begin{array}{rl}
x + 2y + z = 4 & (1) \\
4x - 2y + 6z = 4 & 2(2) \\
5x + 7z = 8 & (4) = (1) + 2(2) \\
5x + 7z = 8 & (5) = (2) + (3) \\
0 = 0 & (4) - (5)
\end{array}$$

dependent

b. $2x - y + 3z = 3$ (1)

$-x + 2y - z = 1$ (2)

$x + y + 2z = 2$ (3)

$$\begin{array}{rl}
4x - 2y + 6z = 6 & (4) = 2(1) \\
-x + 2y - z = 1 & (2) \\
3x + 5z = 7 & (5) = (4) + (2) \\
3x + 5z = 5 & (6) = (1) + (3) \\
0 = 2 &
\end{array}$$

inconsistent

6. You are responsible for buying parts from a wholesale distributor. Three types of comparable switches are needed. The cost per switch is \$1.20, \$1.90, and \$2.30. You need all three types and will place your order as dictated by the following facts:

i. You need a total of 12,000 switches in this order.

ii. Your budget will allow an expenditure of \$23,400.

iii. You need 3 times as many of the most expensive switches as the least expensive switches.

Let x, y, and z represent the number of the first type, second type, and third type of switch, respectively.

a. Write a 3×3 linear system of equations to model this problem.

$$\begin{array}{rl}
x + y + z = 12{,}000 & (1) \\
1.2x + 1.9y + 2.3z = 23{,}400 & (2) \\
3x - z = 0 & (3)
\end{array}$$

b. Solve the system. Be sure to check your solution.

$$
\begin{array}{rl}
-19x - 19y - 19z = -228{,}000 & (4) = -19(1) \\
\underline{12x + 19y + 23z = 234{,}000} & (5) = 10(2) \\
-7x + 4z = 6000 & (6) = (4) + (5) \\
\underline{12x - 4z = 0} & (7) = 4(3) \\
5x = 6000 & (8) = (6) + (7)
\end{array}
$$

$x = 1200$; $y = 7200$; $z = 3600$

ACTIVITY
1.14

Earth Week
Using Matrices to Solve Systems of Linear Equations

OBJECTIVE

Solve a linear system of equations using matrices.

Earth Day in your community has been expanded to Earth Week. As part of the festivities, groups of local middle school, high school, and community college students will participate in three projects: cleaning up roadsides, planting trees, and helping to collect electronic equipment for recycling.

For the week, a total of 2150 pounds of trash was collected along roadsides, 525 new trees were planted in the community, and 9750 pounds of electronic equipment was recycled. The average production of the three groups is summarized in the following table:

	MIDDLE SCHOOL	HIGH SCHOOL	COMMUNITY COLLEGE
Highway Cleanup	10 lb./student	20 lb./student	30 lb./student
Trees Planted	3 trees/student	4 trees/student	7 trees/student
Electronics Recycled	50 lb./student	75 lb./student	150 lb./student

1. Let x represent the number of middle school student participants, y represent the number of high school student participants, and z represent the number of community college student participants. Write a 3×3 linear system that describes this situation.

$$\begin{aligned} 10x + 20y + 30z &= 2150 \\ 3x + 4y + 7z &= 525 \\ 50x + 75y + 150z &= 9750 \end{aligned}$$

Suppose you solve this system algebraically. Do the variables actually enter into the calculations? Not really. The variables are important in setting up (placement or alignment of) the system. But once the system is set up, the variables represent a position within the system. It is the coefficients and the constants that are important.

Using just the coefficients and constants, any system of linear equations can be represented by a matrix.

DEFINITION

Any rectangular array of numbers or symbols is called a **matrix**. That is, a matrix is a systematic arrangement of numbers or symbols in rows and columns.

Matrices are useful for displaying information. The following example demonstrates how a system of linear equations can be represented by a matrix.

EXAMPLE 1 ***Consider the following* 3×3 *linear system:***

$$\begin{aligned} 3x + 2y - 3z &= -2 \\ 2x - 5y + 2z &= -2 \\ 4x - 3y + 4z &= 10 \end{aligned}$$

Note that the variables x, y, and z hold a particular position in the system of equations. The system can be viewed as an array of numbers.

The matrix

$$\begin{bmatrix} 3 & 2 & -3 \\ 2 & -5 & 2 \\ 4 & -3 & 4 \end{bmatrix}$$

derived from the linear system is called the **matrix of coefficients**. Note that the first column contains the coefficients of the variable x in the system. The second and third columns are the coefficients of the y and z variables, respectively.

The matrix

$$\begin{bmatrix} 3 & 2 & -3 & -2 \\ 2 & -5 & 2 & -2 \\ 4 & -3 & 4 & 10 \end{bmatrix},$$

also derived from the same linear system, is called the **augmented matrix**. The fourth column contains the constants on the right side of the system.

To solve a system using matrices, you use a combination of operations on the rows of the augmented matrix to obtain new matrices that give the same solution as the original system.

Elementary Row Operations

a. Interchange two rows.

b. Multiply one row by a nonzero constant.

c. Add a multiple of one row to another row, and replace the second row.

Note that these row operations are the same operations you use on a system of linear equations to eliminate variables and reduce the system to a simpler system. For a 3×3 linear system, your goal is to apply a sequence of row operations to obtain an augmented matrix of the form

$$\begin{bmatrix} 1 & 0 & 0 & c_1 \\ 0 & 1 & 0 & c_2 \\ 0 & 0 & 1 & c_3 \end{bmatrix}.$$

Such a matrix is said to be in reduced row echelon form.

DEFINITION

A matrix is said to be in **reduced row echelon form** if the matrix has 1s down the main diagonal and 0s above and below each 1.

The following example demonstrates how to solve a 2×2 linear system using matrices. In this situation, your goal is to obtain a matrix in the reduced row echelon form:

$$\begin{bmatrix} 1 & 0 & c_1 \\ 0 & 1 & c_2 \end{bmatrix}.$$

EXAMPLE 2 ***Solve the following system using an augmented matrix and row operations:***

$$8x + 3y = 23$$
$$3x - 9y = 12$$

SOLUTION

Step 1. Write the system as an augmented matrix.

The augmented matrix for this system is

$$\begin{bmatrix} 8 & 3 & 23 \\ 3 & -9 & 12 \end{bmatrix}.$$

Step 2. Use row operations to reduce the augmented matrix to the reduced row echelon form

$$\begin{bmatrix} 1 & 0 & c_1 \\ 0 & 1 & c_2 \end{bmatrix}.$$

In general, your strategy is first to change the number in the first column, first row, to a 1.

1. Interchange the rows to obtain

$$\begin{bmatrix} 3 & -9 & 12 \\ 8 & 3 & 23 \end{bmatrix}.$$

2. Divide the first row by 3.

$$\begin{bmatrix} 1 & -3 & 4 \\ 8 & 3 & 23 \end{bmatrix}$$

Next, you want to change the number in the first column, second row, to a 0.

3. Multiply the first row by -8, and add the results to the second row to obtain

$$\begin{bmatrix} 1 & -3 & 4 \\ 0 & 27 & -9 \end{bmatrix}.$$

Now change the number in the second column, second row, to a 1.

4. Divide the second row by 27.

$$\begin{bmatrix} 1 & -3 & 4 \\ 0 & 1 & -\frac{1}{3} \end{bmatrix}$$

Finally, change the number in the second column, first row, to a 0.

5. Multiply the second row by 3 and add to the first.

$$\begin{bmatrix} 1 & 0 & 3 \\ 0 & 1 & -\frac{1}{3} \end{bmatrix}$$

This augmented matrix is in reduced row echelon form and is equivalent to the following system:

$$1x + 0y = 3 \quad \text{or} \quad x = 3$$

$$0x + 1y = -\frac{1}{3} \qquad y = -\frac{1}{3}$$

Therefore, the solution to the system is $x = 3$ and $y = -\frac{1}{3}$ or $\left(3, -\frac{1}{3}\right)$.

2. Why would you want to rewrite an augmented matrix in reduced row echelon form?

When an augmented matrix is rewritten in reduced row echelon form, the solution can be read from the final column.

You can use your calculator to determine reduced row echelon form. Problem 3 demonstrates the steps in solving the system in Example 2 using the TI-84 Plus C.

3. a. Choose the MATRIX option, and choose the EDIT menu.

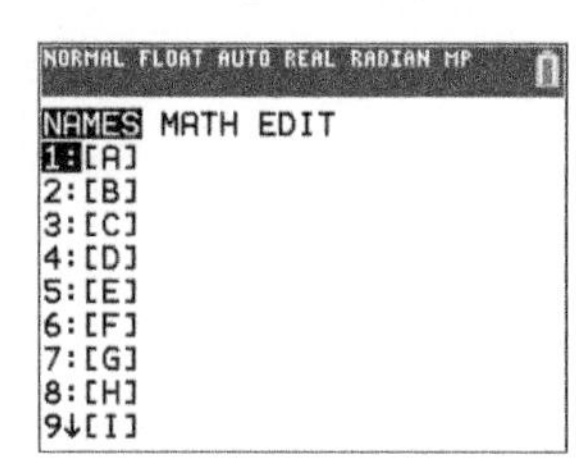

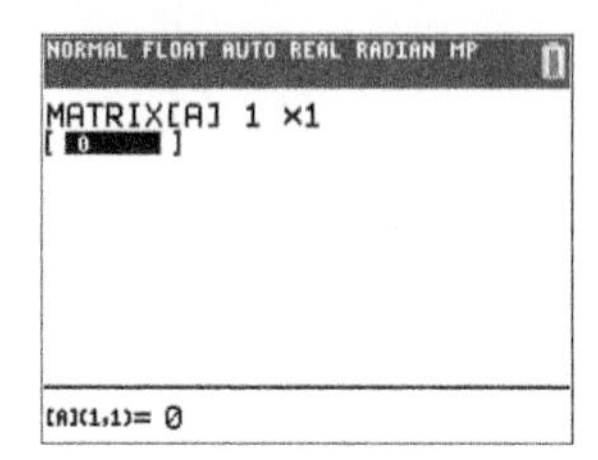

b. For the system

$$8x + 3y = 23$$

$$3x - 9y = 12,$$

enter 2×3 for the dimension and enter the elements of the matrix into MATRIX [A].

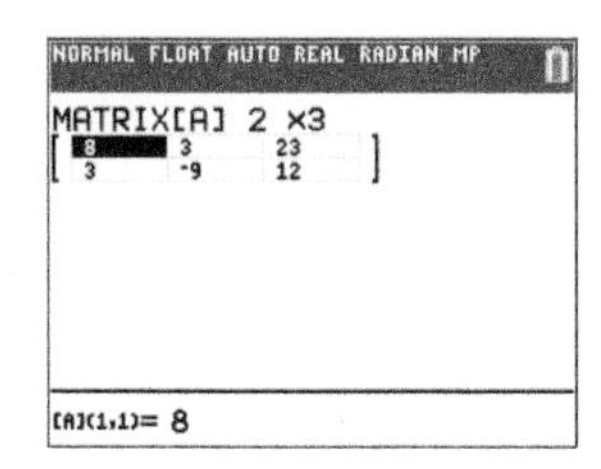

c. Return to the home screen; then choose the MATRIX option again. This time select the MATH menu.

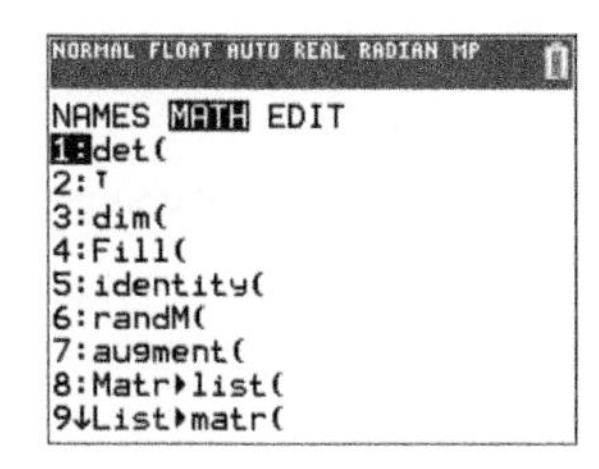

d. Choose option B, the reduced row echelon form command rref. Once it has been selected, you are back at the home screen, as in screen capture 2.

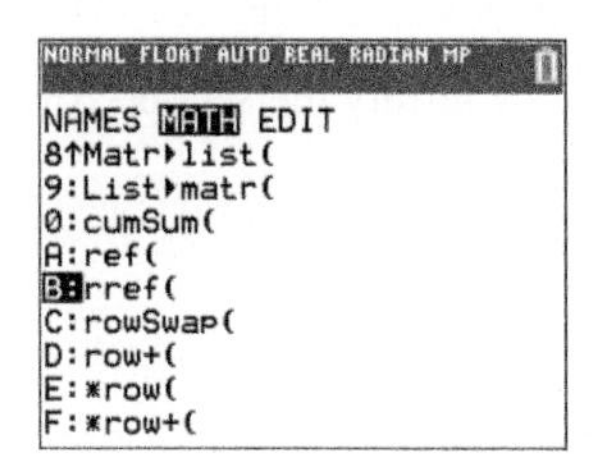

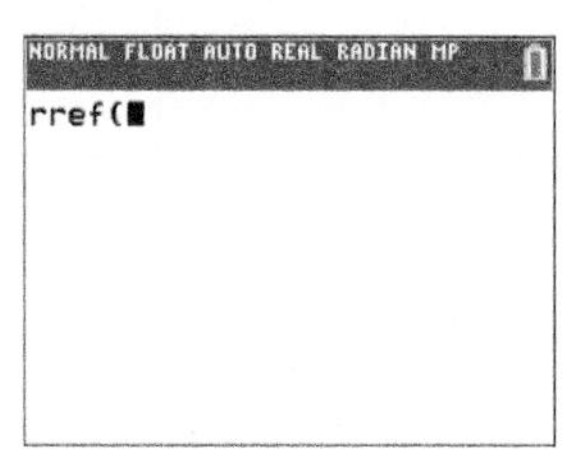

e. Choose the MATRIX option again. The name of your matrix is [A] or number 1. Select 1: [A] and enter. Your home screen will now look like this.

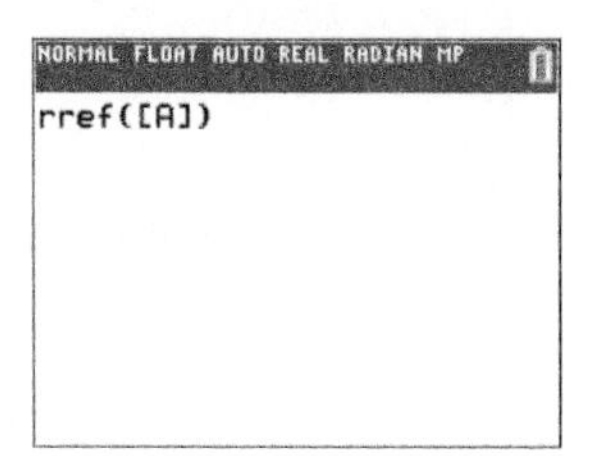

LC LEARNING CATALYTICS

For the linear system

$$2x + 4y - z = -2$$
$$x - 2y + z = -5$$
$$-2x + y + 2z = 7$$

determine the reduced row echelon form of the augmented matrix for this system.

f. Press ENTER.

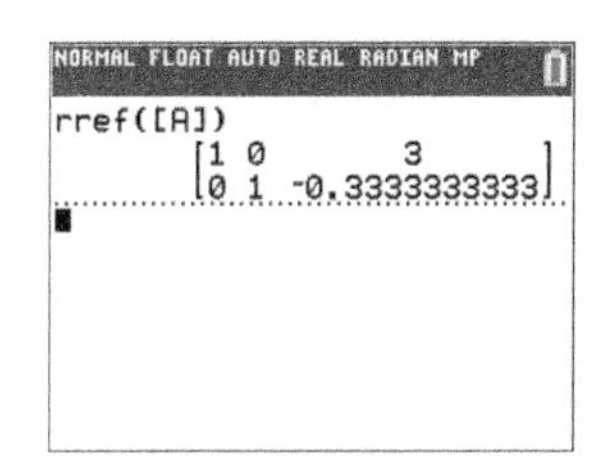

g. Read the solution to the system from the matrix.

$$x = 3 \qquad y = -\frac{1}{3}$$

4. Solve the 3×3 linear system from Example 1.

$$3x + 2y - 3z = -2$$
$$2x - 5y + 2z = -2$$
$$4x - 3y + 4z = 10$$

using your calculator to determine the reduced row echelon form. Your screens should appear as follows:

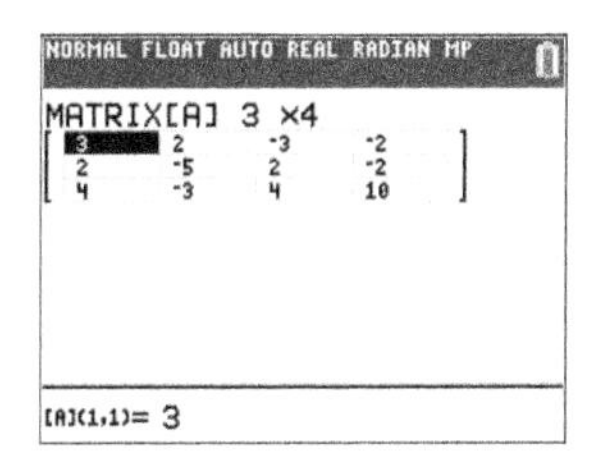

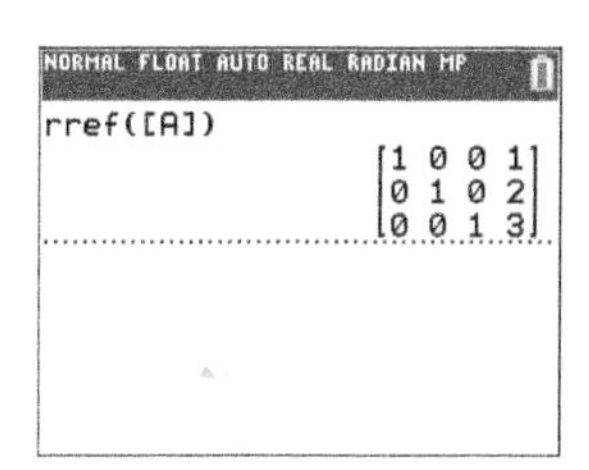

The screens indicate a solution of $x = 1$, $y = 2$, and $z = 3$.

5. a. Use matrices to solve the system from Problem 1.

The augmented matrix and the reduced row echelon form are

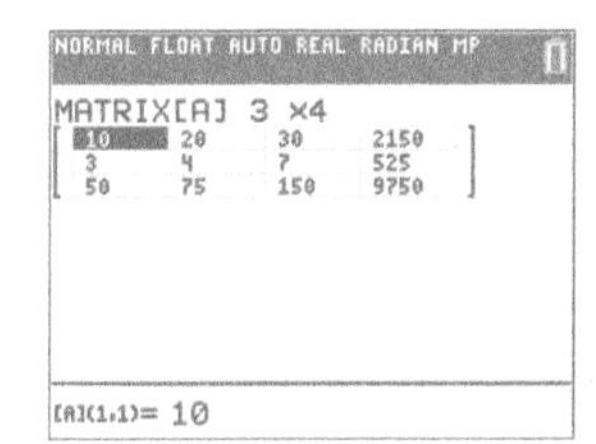

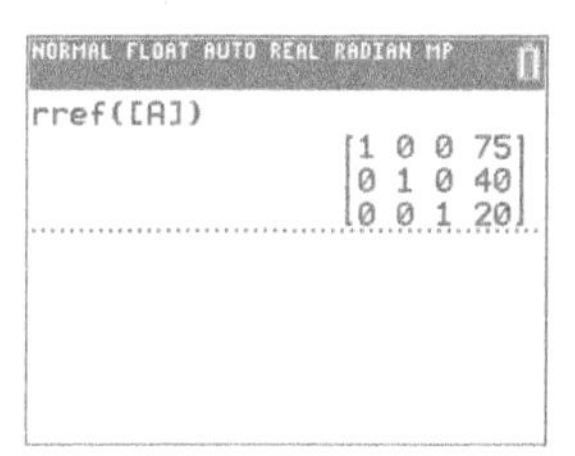

The solution is $x = 75$, $y = 40$, and $z = 20$.

b. Interpret your solution in part a in the context of the original Earth Day situation.

This means that 75 middle school students, 40 high school students, and 20 community college students participated in the Earth Week activities.

SUMMARY Activity 1.14

1. Any rectangular array of numbers or symbols is called a **matrix**.
2. The **matrix of coefficients of a linear system** is the matrix that contains only the coefficients of the variables of the system. The variables must be in the same order in each equation. For any missing term, a zero is entered for the coefficient.
3. The **augmented matrix** is the matrix of coefficients with one column added to the right-hand end of the matrix. This column contains the constant terms from each equation in the system.
4. Matrix **elementary row operations** are as follows:
   a. Interchange two rows.
   b. Multiply one row by a nonzero constant.
   c. Add a multiple of one row to another row and replace the second row.
5. To solve a system, use elementary row operations to rewrite the augmented matrix in **reduced row echelon form.**
6. You can use your calculator to determine reduced row echelon form.

   Step 1. Choose the MATRIX option, and choose the EDIT menu.

   Step 2. Enter the dimension, and enter the elements of the matrix into MATRIX [A].

   Step 3. Return to the home screen; then choose the MATRIX option again. This time select the MATH menu.

   Step 4. Choose option B, the reduced row echelon form command rref. Once it has been selected, you will be back at your home screen.

   Step 5. Choose the MATRIX option again. The name of your matrix is [A] or number 1. Select 1:[A] and enter. Again, you will be back at your home screen.

   Step 6. Read the solution to the system from the matrix.

EXERCISES Activity 1.14

Write the augmented matrix for each system of linear equations.

1. $4x + 3y = -1$
   $2x - y = -13$

$$\begin{bmatrix} 4 & 3 & -1 \\ 2 & -1 & -13 \end{bmatrix}$$

2. $3x - 2y + 5z = 31$
   $x + 3y - 3z = -12$
   $-2x - 5y + 3z = 11$

$$\begin{bmatrix} 3 & -2 & 5 & 31 \\ 1 & 3 & -3 & -12 \\ -2 & -5 & 3 & 11 \end{bmatrix}$$

3. $4x - 2y + z = 15$
   $3x + 2y - 2z = -4$
   $x + z = 5$

$$\begin{bmatrix} 4 & -2 & 1 & 15 \\ 3 & 2 & -2 & -4 \\ 1 & 0 & 1 & 5 \end{bmatrix}$$

4. $x - 2y + 3z = 9$
   $y + 3z = 5$
   $z = 2$

$$\begin{bmatrix} 1 & -2 & 3 & 9 \\ 0 & 1 & 3 & 5 \\ 0 & 0 & 1 & 2 \end{bmatrix}$$

Exercise numbers appearing in color are answered in the Selected Answers appendix.

Write the system of l inear equations represented by the augmented matrix. Use variables of x, y, and z when appropriate.

5. $\begin{bmatrix} 4 & 3 & 15 \\ 2 & -5 & 1 \end{bmatrix}$

$4x + 3y = 15$

$2x - 5y = 1$

6. $\begin{bmatrix} 2 & -6 & 4 & 10 \\ 1 & 5 & -5 & 0 \\ 3 & 0 & 4 & 7 \end{bmatrix}$

$2x - 6y + 4z = 10$

$x + 5y - 5z = 0$

$3x + 4z = 7$

7. The given system of equations

$$\begin{aligned} 2x - y + 3z &= 9 \\ x + 2y - z &= -3 \\ 3x + y + z &= 4 \end{aligned}$$

has the following row-reduced augmented matrix (use your calculator to check).

$$\begin{bmatrix} 1 & 0 & 0 & 1 \\ 0 & 1 & 0 & -1 \\ 0 & 0 & 1 & 2 \end{bmatrix}$$

What is the solution to the system?

$(x, y, z) = (1, -1, 2)$

8. The senior class will sell pizzas as their fund-raiser this year. The medium cheese and pepperoni pizza will sell for \$10.95; the large, for \$14.95. On Valentine's Day, a total of 49 pizzas were sold for a total of \$620.55. The class president asked the class treasurer how many medium and how many large pizzas were sold. Help the treasurer figure this out using matrices.

Twenty-eight medium pizzas and 21 large pizzas were sold.

9. In a recent game at your community college, the men's basketball team made 45 baskets from the field; some were 2-pointers and some were 3-pointers. In total, 101 points were made from the field. The coach wanted to know how many were 2-pointers and how many were 3-pointers. Use matrices to determine the answer, and give the information to your coach.

Thirty-four 2-pointers and eleven 3-pointers were made for a total of 101 points.

10. Your college offers a degree in cosmetology. You are interested in the number of men (over the age of 25), young adults (25 and under), and women (over the age of 25) who take advantage of this program. For 1 week, you keep track of the number of hours spent working in each area, as shown in the table below. Students spend a maximum of 31 hours giving haircuts, 19 hours giving permanents, and 25 hours applying hair color. If all of the available time must be used, how many young adults, men, and women will be serviced at your school during that week?

	MEN	YOUNG ADULTS	WOMEN
Haircuts	3 hours	1 hour	2 hours
Permanents	1 hour	1 hour	2 hours
Hair Color	1 hour	3 hours	2 hours

Six men, 3 young adults, and 5 women will take advantage of the services during that week.

11. A car dealer needs to buy luxury sedans, small hatchbacks, and hybrid models. The dealer needs a total of 28 cars and has a budget of $692,000. Luxury sedans cost $36,000 each, small hatchbacks cost $18,000 each, and hybrids cost $28,000 each. The dealer needs as many small hatchbacks as the number of luxury sedans and hybrids combined. How many of each model should the dealer buy?

6 luxury sedans

14 small hatchbacks

8 hybrids

12. Your doctor tells you that your diet should consist of 2100 calories per day. Your doctor also recommends that you consume a total of 437.5 grams of protein, carbohydrates, and fat. Protein and carbohydrates have 4 calories per gram, and fat has 9 calories per gram. To maintain the correct proportions, the number of grams of carbohydrates is 140 grams more than the sum of the grams of protein and fat. How many grams of protein, carbohydrates, and fat should you consume daily?

78.75 grams of protein (315 calories), 288.75 grams of carbohydrates (1155 calories), and 70 grams of fat (630 calories)

ACTIVITY 1.15

How Long Can You Live?

Linear Inequalities; Compound Inequalities

OBJECTIVES

1. Solve linear inequalities in one variable numerically and graphically.
2. Use properties of inequalities to solve linear inequalities in one variable algebraically.
3. Solve compound inequalities algebraically.
4. Use interval notation to represent a set of real numbers described by an inequality.

Life expectancy in the United States is steadily increasing, and the number of Americans aged 100 and older will exceed 850,000 by the middle of this century. Medical advancements have been a primary reason for Americans living longer. Another factor has been the increased awareness of maintaining a healthy lifestyle.

The life expectancies at birth for men and women born after 1980 in the United States can be modeled by the following functions:

$$W(x) = 0.109x + 77.5$$

$$M(x) = 0.192x + 70.0,$$

where $W(x)$ represents the life expectancy for women, $M(x)$ represents the life expectancy for men, and x represents the number of years since 1980 that the person was born. That is, $x = 0$ corresponds to the year 1980, $x = 5$ corresponds to 1985, etc.

1. a. Complete the following table:

	YEAR								
	1980	1985	1990	1995	2000	2005	2010	2015	2020
x, years since 1980	0	5	10	15	20	25	30	35	40
$W(x)$	77.5	78.0	78.6	79.1	79.7	80.2	80.8	81.3	81.9
$M(x)$	70.0	71.0	71.9	72.9	73.8	74.8	75.8	76.7	77.7

b. For people born between 1980 and 2020, do men or women have the greater life expectancy?

women

c. Is the life expectancy of men or women increasing more rapidly? Explain using slope.

Men. The slope for men is 0.192; the slope for women is 0.109.

You would like to determine in what birth years the life expectancy of men is greater than that of women. The phrase "greater than" indicates a mathematical relationship called an **inequality**. Symbolically, the relationship can be represented by

$$\underbrace{M(x)}_{\text{Life expectancy for men}} \quad \underbrace{>}_{\text{is greater than}} \quad \underbrace{W(x).}_{\text{life expectancy for women.}}$$

Other commonly used phrases that indicate inequalities are given in the following example.

EXAMPLE 1

STATEMENT, WHERE x REPRESENTS A REAL NUMBER	TRANSLATION TO AN INEQUALITY
x is greater than 10.	$x > 10$ or $10 < x$
x is less than 10.	$x < 10$ or $10 > x$
x is at least 10.	$x \geq 10$ (also read "x is greater than or equal to 10")
x is at most 10.	$x \leq 10$ (also read "x is less than or equal to 10")

2. Substitute the appropriate expressions for $M(x)$ and $W(x)$ to obtain an inequality involving x that can be used to determine the birth years for which the life expectancy of men is greater than that of women.

$0.192x + 70.0 > 0.109x + 77.5$

Solving Inequalities in One Variable Numerically and Graphically

DEFINITION

Solving an inequality in one variable is the process of determining the values of the variable that make the inequality a true statement. These values are called the **solutions** of the inequality.

3. Solve the inequality in Problem 2 numerically. That is, continue to construct a table of values (see Problem 1) until you determine the values of the years x (inputs) for which $0.192x + 70.0 > 0.109x + 77.5$. Use the Table feature of your graphing calculator.

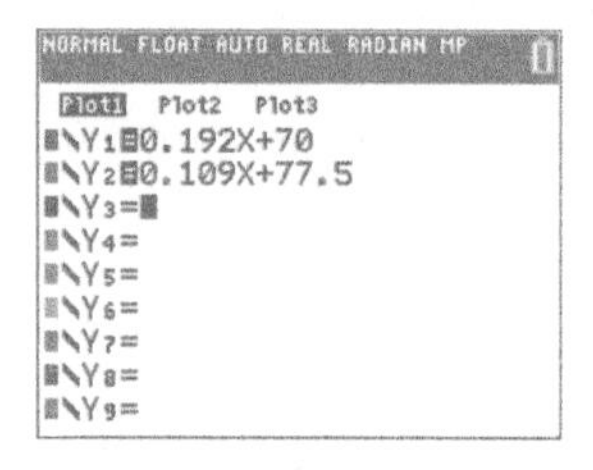

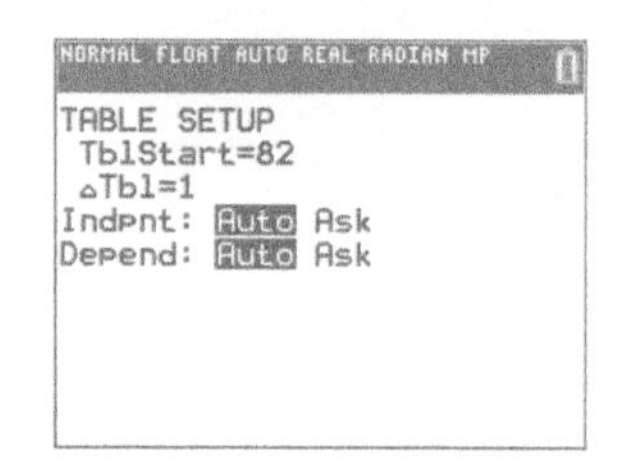

NORMAL FLOAT AUTO REAL RADIAN MP
PRESS + FOR ΔTbl

X	Y1	Y2
82	85.744	86.438
83	85.936	86.547
84	86.128	86.656
85	86.32	86.765
86	86.512	86.874
87	86.704	86.983
88	86.896	87.092
89	87.088	87.201
90	87.28	87.31
91	87.472	87.419
92	87.664	87.528

X=82

Therefore, if the trends given by the equations for $M(x)$ and $W(x)$ continue, the approximate solution to the inequality $M(x) > W(x)$ is $x > 90$. That is, according to the models, after the year 2070, men will live longer than women.

4. Now solve the inequality $0.192x + 70.0 > 0.109x + 77.5$ graphically.

a. Use your graphing calculator to sketch a graph of $M(x) = 0.192x + 70.0$ and $W(x) = 0.109x + 77.5$ on the same coordinate axis.

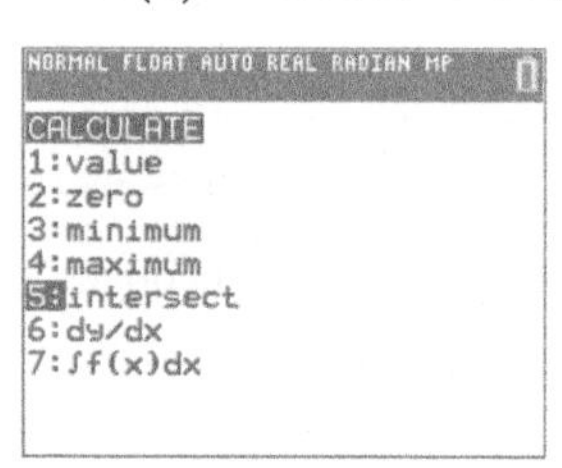

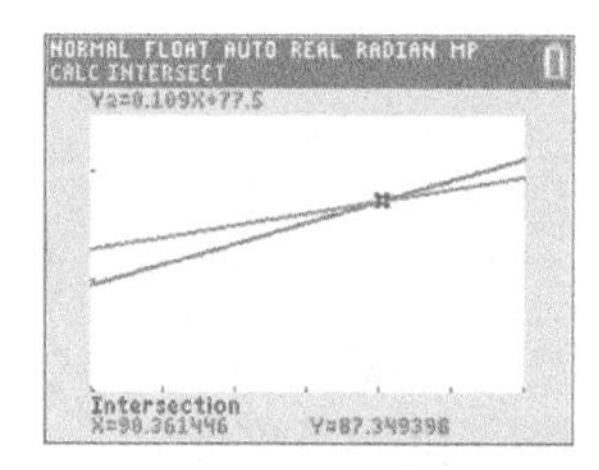

b. Determine the point of intersection of the two graphs using the intersect feature of your graphing calculator. What does the point represent in this situation?

(90.36, 87.35). The point of intersection says that during the ninetieth year after 1980, the year 2070, the life expectancy for men will be the same as the life expectancy for women, 87.35 years.

To solve the inequality $M(x) > W(x)$ graphically, you need to determine the values of x for which the graph of $M(x) = 0.192x + 70.0$ is above the graph of $W(x) = 0.109x + 77.5$.

c. Use the graph to solve $M(x) > W(x)$. How does your solution compare to the solution in Problem 3?

The solution is approximately the same, $x > 90.36$. Men born after mid 2070 will have a longer life expectancy than women.

5. a. Write an inequality to determine the birth years of women whose life expectancy is at least 85.

$0.109x + 77.5 \geq 85$

b. Solve the inequality numerically, using the Table feature of your graphing calculator.

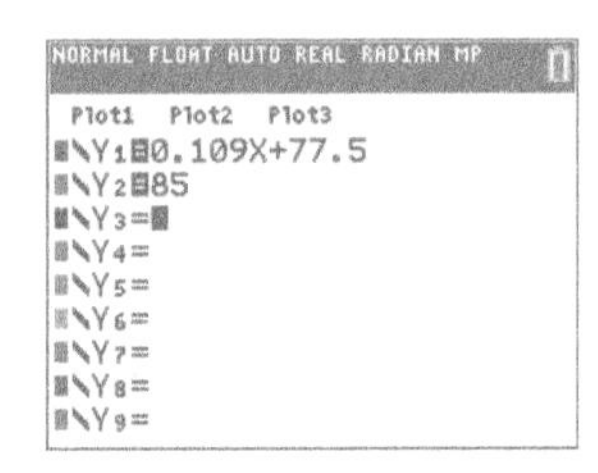

NORMAL FLOAT AUTO REAL RADIAN MP
PRESS + FOR ΔTbl

X	Y1	Y2
65	84.585	85
66	84.694	85
67	84.803	85
68	84.912	85
69	85.021	85
70	85.13	85
71	85.239	85
72	85.348	85
73	85.457	85
74	85.566	85
75	85.675	85

X=65

The birth years in which women's life expectancy is at least 85 start 69 years after 1980, or all years after 2049.

c. Use your graphing calculator to solve this inequality graphically.

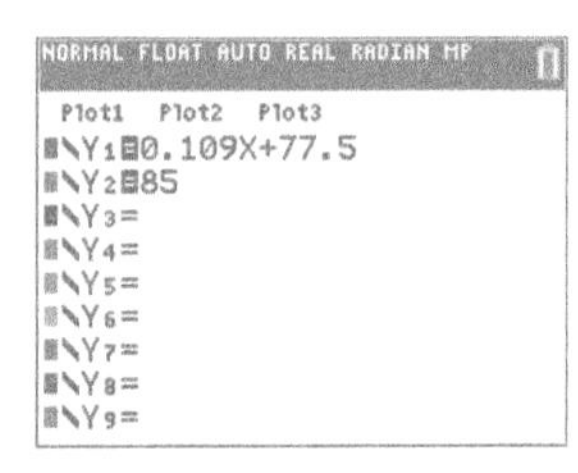

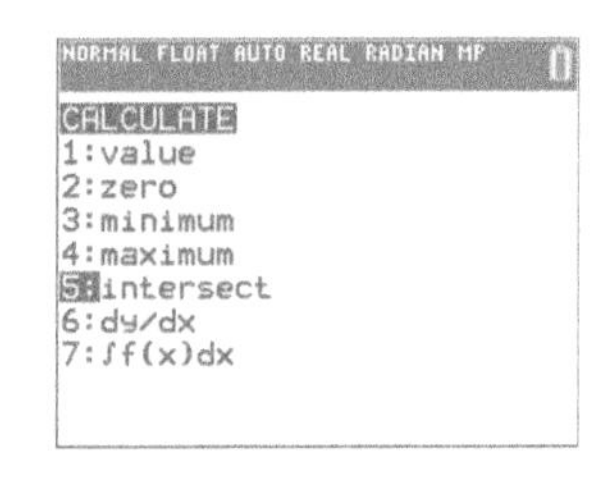

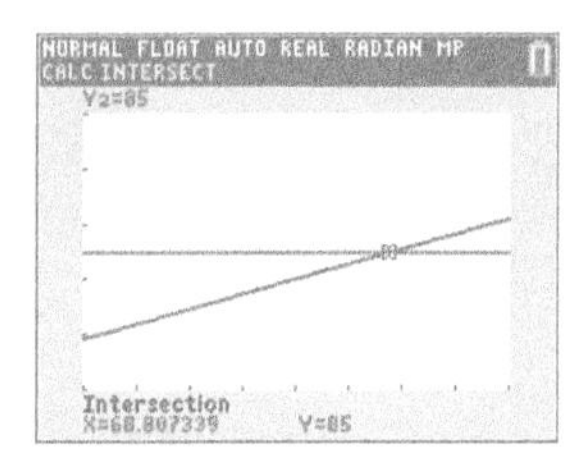

Solving Inequalities in One Variable Algebraically

The process of solving an inequality in one variable algebraically is very similar to solving an equation in one variable algebraically. Your goal is to isolate the variable on one side of the inequality symbol. You isolate the variable in an **equation** by performing the same operations to both sides of the equation so as not to upset the **balance**. You isolate the variable in an **inequality** by performing the same operations to both sides so as not to upset the **imbalance**.

6. a. Write the statement "15 is greater than 6" as an inequality.

$15 > 6$

b. Add 5 to each side of $15 > 6$. Is the resulting inequality a true statement? (That is, is the left side still greater than the right side?)

$20 > 11$; yes

c. Subtract 10 from each side of $15 > 6$. Is the resulting inequality a true statement?

$5 > -4$; yes

d. Multiply each side of $15 > 6$ by 4. Is the resulting inequality true?

$60 > 24$; yes

e. Multiply each side of $15 > 6$ by -2. Is the left side still greater than the right side?

$-30 > -12$; no

f. Reverse the direction of the inequality symbol in part e. Is the new inequality a true statement?

$-30 < -12$; yes

Problem 6 demonstrates two very important properties of inequalities.

Property 1 If $a < b$ represents a true inequality, then if

i. the same quantity is added to or subtracted from both sides or

ii. both sides are multiplied or divided by the same *positive number,* the resulting inequality remains a true statement and the direction of the inequality symbol remains the same

For example, because $-4 < 10$, then

i. $-4 + 5 < 10 + 5$, or $1 < 15$, is true;
$-4 - 3 < 10 - 3$, or $-7 < 7$, is true.

ii. $-4(6) < 10(6)$, or $-24 < 60$, is true;
$\frac{-4}{2} < \frac{10}{2}$, or $-2 < 5$, is true.

Property 2

If $a < b$ represents a true inequality, then if both sides are multiplied or divided by the same *negative number*, the inequality symbol in the resulting inequality statement must be reversed ($<$ to $>$ or $>$ to $<$) for the resulting statement to be true.

For example, because $-4 < 10$, then $-4(-5) > 10(-5)$, or $20 > -50$. Because $-4 < 10$, then $\frac{-4}{-2} > \frac{10}{-2}$, or $2 > -5$.

Properties 1 and 2 will be true if $a < b$ is replaced with $a \le b$, $a > b$, or $a \ge b$.

The following example demonstrates how properties of inequalities can be used to solve an inequality algebraically.

EXAMPLE 2 ***Solve* $3(x - 4) > 5(x - 2) - 8$.**

SOLUTION

$3(x-4) > 5(x-2) - 8$	Apply the distributive property.
$3x - 12 > 5x - 10 - 8$	Combine like terms on the right side.
$3x - 12 > 5x - 18$	
$-5x \quad\quad - 5x$	Subtract $5x$ from both sides; the direction of the inequality symbol remains the same.
$-2x - 12 > -18$	
$+12 \quad\quad +12$	Add 12 to both sides; the direction of the inequality does not change.
$\frac{-2x}{-2} < \frac{-6}{-2}$	Divide both sides by -2; the direction is reversed!
$x < 3$	

Therefore, from Example 2, any number less than 3 is a solution to the inequality $3(x - 4) > 5(x - 2) - 8$. The solution set can be represented on a number line by shading all points to the left of 3.

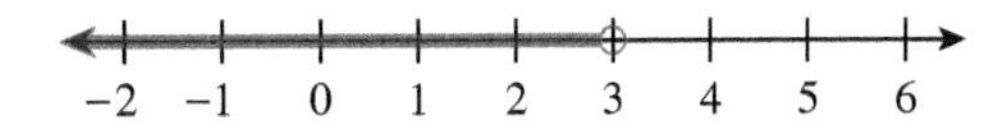

The open circle at 3 indicates that 3 is *not* a solution. A closed circle indicates that the number beneath the closed circle *is* a solution. The arrow shows that the solutions extend indefinitely to the left.

7. Solve the inequality $0.192x + 70 > 0.109x + 77.5$ algebraically to determine the birth years in which men will be expected to live longer than women. How does your solution compare with the solutions determined numerically and graphically in Problems 3 and 4c?

$$0.192x + 70 > 0.109x + 77.5$$
$$0.192x - 0.109x > 77.5 - 70$$
$$0.083x > 7.5$$
$$x > 90.36$$

The solution is the same.

Compound Inequality

You have joined a health and fitness club. Your aerobics instructor recommends that to achieve the most cardiovascular benefit from your workout, you should maintain your pulse rate between a lower and upper range of values. These values depend on your age.

8. If the variable a represents your age, then the lower and upper values for your pulse rate are determined by the following:

lower value: $0.72(220 - a)$

upper value: $0.87(220 - a)$

a. Determine your lower value.

(Answers vary.) For a 20-year-old, the lower value is $0.72(200) = 144$.

b. Determine your upper value.

(Answers vary.) For a 20-year-old, the upper value is $0.87(200) = 174$.

For the most cardiovascular benefit, a 20-year-old's pulse rate should be between 144 and 174. The phrase "between 144 and 174" means that the pulse rate should be greater than 144 *and* less than 174. Symbolically, this combination or **compound inequality** is written as

$144 <$ pulse rate *and* pulse rate < 174.

This statement is written more compactly as $144 <$ pulse rate < 174.

The numbers that satisfy this compound inequality can be represented on a number line as follows:

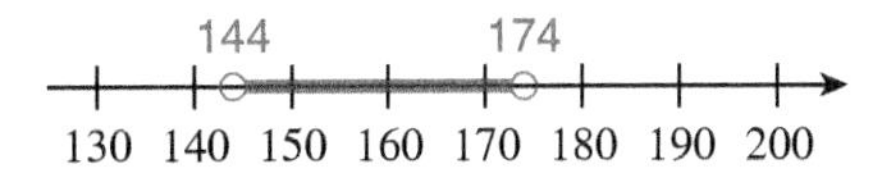

Other commonly used phrases that indicate compound inequalities involving the word *and* are given in the following example.

EXAMPLE 3

STATEMENT, WHERE x REPRESENTS A REAL NUMBER	TRANSLATION TO A COMPOUND INEQUALITY
x is greater than or equal to 10 and less than 20.	$10 \le x < 20$
x is greater than 10 and less than or equal to 20.	$10 < x \le 20$
x is from 10 to 20 inclusive.	$10 \le x \le 20$

9. Recall that the life expectancy for men is given by the expression $0.192x + 70$, where x represents the number of years since 1980. Use this expression to write a compound inequality that can be used to determine in what birth years men will be expected to live into their eighties.

$80 \le 0.192x + 70 < 90$

The following example demonstrates how to solve a compound linear inequality algebraically and graphically.

EXAMPLE 4 ***Solve $-4 < 3x + 5 \le 11$ using an algebraic approach.***

SOLUTION

Note that the compound inequality has three parts: left: -4, middle: $3x + 5$, and right: 11. To solve this inequality, isolate the variable in the middle part.

$$\begin{array}{rcl} -4 < 3x + 5 \le 11 & & \\ \underline{-5 \qquad -5 \quad -5} & & \text{Subtract 5 from each part.} \\ -9 < 3x \le 6 & & \\ -\frac{9}{3} < \frac{3x}{3} \le \frac{6}{3} & & \text{Divide each part by 3.} \\ -3 < x \le 2 & & \end{array}$$

The solution can be represented on a number line as follows:

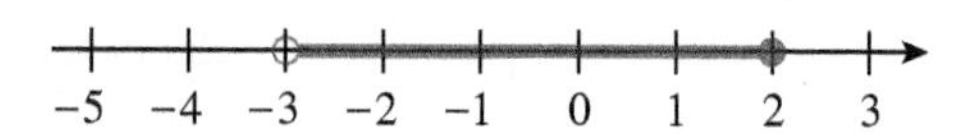

LEARNING CATALYTICS

Solve $-1 < 2x + 5 < 7$ algebraically or graphically. Which of the following is the solution?

a. $-6 < x < 2$
b. $-3 < x < 2$
c. $-3 < x < 1$
d. None of these

10. Solve the compound inequality $80 \le 0.192x + 70 < 90$ from Problem 9 to determine in what birth years men will be expected to live into their eighties.

$$\begin{array}{c} 80 \le 0.192x + 70 < 90 \\ \underline{-70 \qquad\qquad -70 \;-70} \\ \frac{10}{0.192} \le \frac{0.192x}{0.192} < \frac{20}{0.192} \\ 52.1 \le x < 104.2 \end{array}$$

Men born between 2032 and 2084 should live into their eighties.

Interval Notation

Interval notation is an alternative method to represent a set of real numbers described by an inequality. The **closed interval** $[-3, 4]$ represents all real numbers x for which $-3 \le x \le 4$.

The square brackets [ ] indicate that the endpoints of the interval are included. The **open interval** $(-3, 4)$ represents all real numbers x for which $-3 < x < 4$. Note that the parentheses ( ) indicate that the endpoints are not included. The interval $(-3, 4]$ is said to be **half-open** or **half-closed**. The interval is open at -3 (endpoint not included) and closed at 4 (endpoint included).

Suppose you want to represent the set of real numbers x for which x is greater than 3. The symbol $+\infty$ (positive infinity) is used to indicate **unboundedness** in the positive direction. Therefore, the interval $(3, +\infty)$ represents all real numbers x for which $x > 3$. Note that $+\infty$ is always open.

The symbol $-\infty$ (negative infinity) is used to represent unboundedness in the negative direction. Therefore, the interval $(-\infty, 5]$ represents all real numbers x for which $x \leq 5$.

11. In parts a–d, express each inequality in interval notation.

a. $-5 \leq x \leq 10$ $[-5, 10]$

b. $4 \leq x < 8.5$ $[4, 8.5)$

c. $x > -2$ $(-2, +\infty)$

d. $x \leq 3.75$ $(-\infty, 3.75]$

In parts e–h, express each of the following using inequalities:

e. $(-6, 4]$ $-6 < x \leq 4$

f. $(-\infty, 1.5]$ $x \leq 1.5$

g. $(-2, 2)$ $-2 < x < 2$

h. $(-3, +\infty)$ $x > -3$

SUMMARY Activity 1.15

1. The **solution set** of an inequality is the set of all values of the variable that satisfy the inequality.

2. The direction of an inequality is not changed when

i. the same quantity is added to or subtracted from both sides of the inequality. Stated symbolically, if $a < b$, then $a + c < b + c$ and $a - c < b - c$.

ii. the same positive quantity is multiplied or divided on both sides of the inequality. Stated symbolically, if $a < b$, then $ac < bc$, and $\frac{a}{c} < \frac{b}{c}$, whenever $c > 0$.

3. The direction of an inequality is reversed if both sides of an inequality are multiplied by or divided by the same negative number. These properties can be written symbolically as

i. if $a < b$, then $ac > bc$, where $c < 0$.

ii. if $a < b$, then $\frac{a}{c} > \frac{b}{c}$, where $c < 0$.

The two properties of inequalities above (items 2 and 3) will still be true if $a < b$ is replaced by $a \leq b$, $a > b$, or $a \geq b$.

4. Inequalities such as $f(x) < g(x)$ can be solved using three different methods.

i. a **numerical approach**, in which a table of input–output pairs is used to determine values of x for which $f(x) < g(x)$

ii. a **graphical approach**, in which values of x are located so that the graph of f is below the graph of g

iii. an **algebraic approach**, in which the properties of inequalities are used to isolate the variable

Similar statements can be made for solving inequalities of the form $f(x) \leq g(x)$, $f(x) > g(x)$, and $f(x) \geq g(x)$.

EXERCISES Activity 1.15

In Exercises 1–6, translate the given statement into an algebraic inequality or a compound inequality.

1. To avoid an additional charge, the sum of the length, l, width, w, and depth, d, of a piece of luggage to be checked on a commercial airline can be at most 61 inches.

   $l + w + d \leq 61$

2. A PG-13 movie rating means that your age, a, must be at least 13 years for you to view the movie.

   $a \geq 13$

3. The cost, $C(A)$, of renting a car from company A is less expensive than the cost, $C(B)$, of renting from company B.

   $C(A) < C(B)$

4. The label on a bottle of film developer states that the temperature, t, of the contents must be kept between 68°F and 77°F.

   $68°F < t < 77°F$

5. You are in a certain tax bracket if your taxable income, i, is over \$24,650 but not over \$59,750.

   $\$24{,}650 < i \leq \$59{,}750$

6. The temperature, t, on Mars as measured by the rovers *Spirit* and *Opportunity* has ranged from 30°C down to –110°C. See https://mars.nasa.gov/mer/spotlight/20070612.html.

   $-110°C \leq t \leq 30°C$

Solve Exercises 7–14 graphically and algebraically.

7. $3x > -6$

   $x > -2$

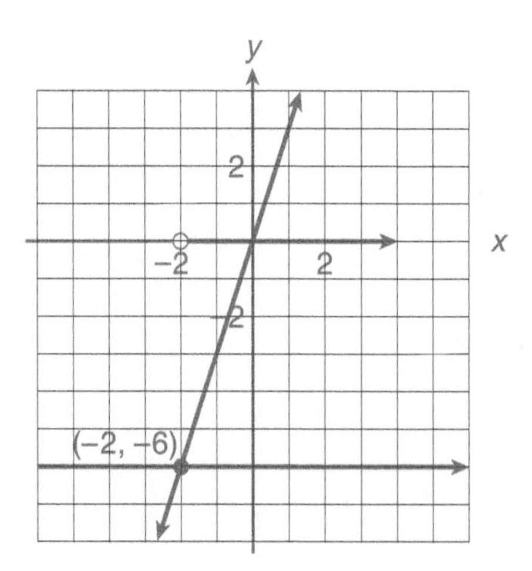

8. $3 - 2x \leq 5$

   $-2x \leq 2$

   $x \geq -1$

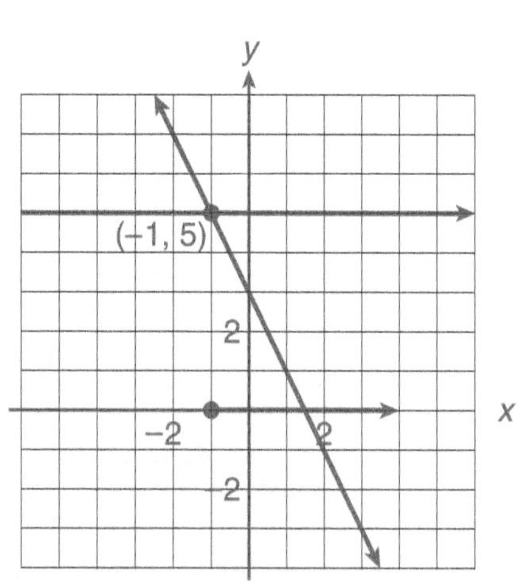

9. $x + 2 > 3x - 8$

   $-2x > -10$

   $x < 5$

10. $5x - 1 < 2x + 11$

    $3x < 12$

    $x < 4$

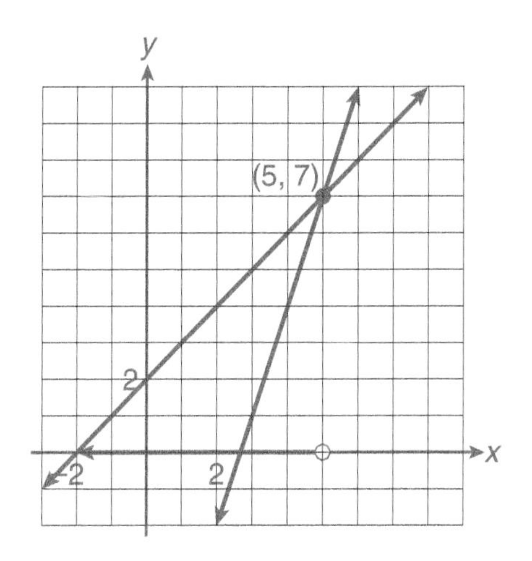

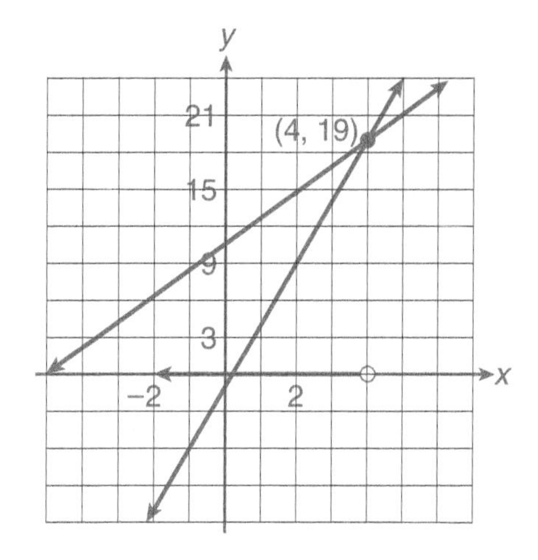

11. $8 - x \geq 5(8 - x)$

$8 - x \geq 40 - 5x$

$4x \geq 32$

$x \geq 8$

12. $5 - x < 2(x - 3) + 5$

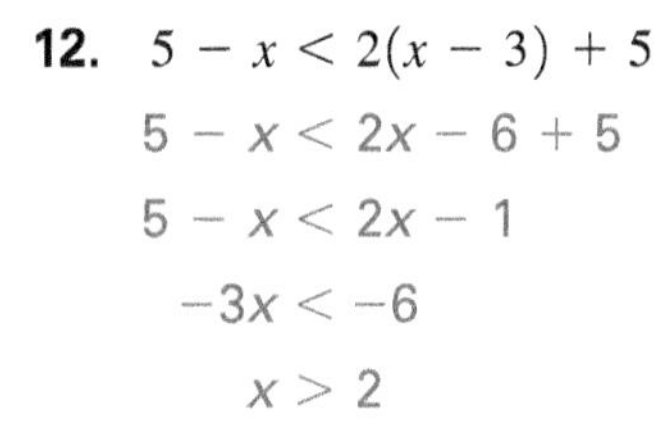

$5 - x < 2x - 6 + 5$

$5 - x < 2x - 1$

$-3x < -6$

$x > 2$

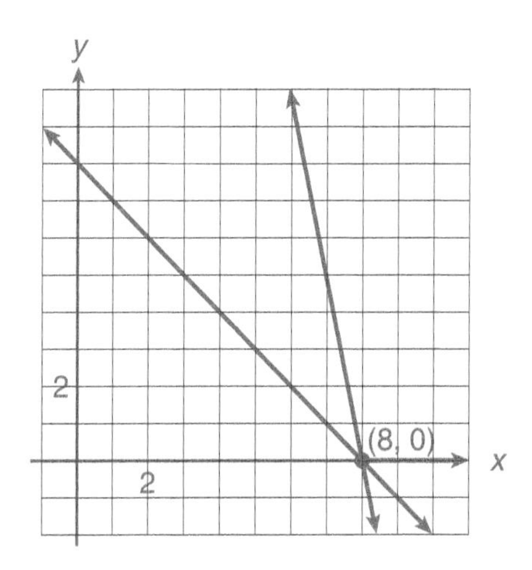

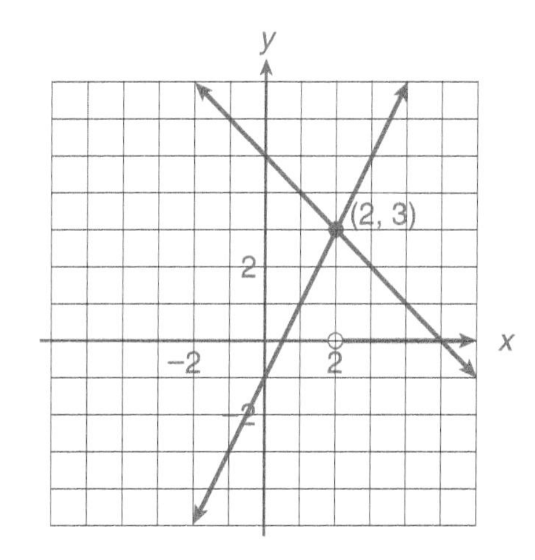

13. $\frac{x}{2} + 1 \leq 3x + 2$

$-2.5x \leq 1$

$x \geq -0.4$

14. $0.5x + 3 \geq 2x - 1.5$

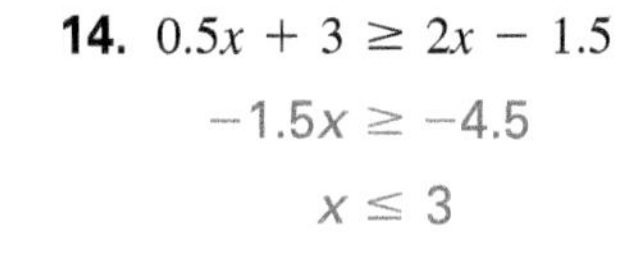

$-1.5x \geq -4.5$

$x \leq 3$

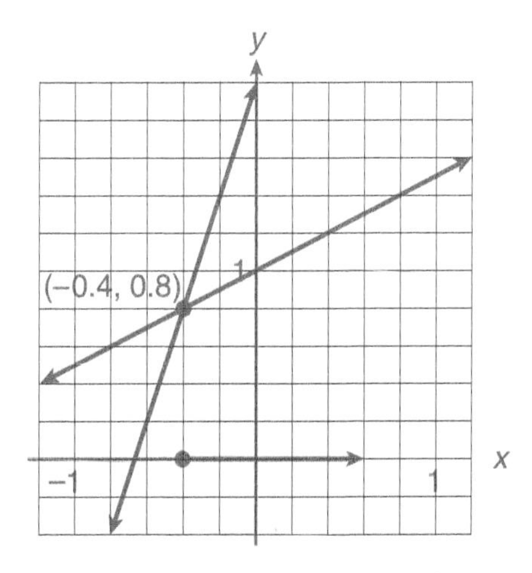

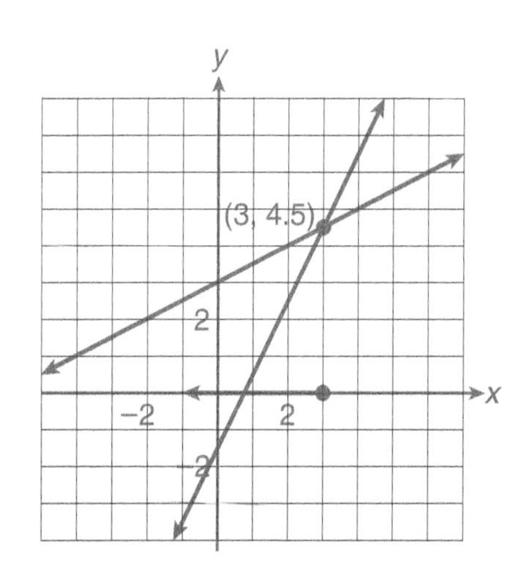

Solve Exercises 15 and 16 algebraically.

15. $1 < 3x - 2 < 4$

$3 < 3x < 6$

$1 < x < 2$

16. $-2 < \frac{x}{3} + 1 < 5$

$-6 < x + 3 < 15$

$-9 < x < 12$

17. The consumption of cigarettes in the United States is declining. If t represents the number of years since 2000, then the consumption, C, is modeled by

$$C = -13.4t + 445.2$$

where C represents the number of billons of cigarettes smoked per year.

a. Write an inequality that can be used to determine the first year in which cigarette consumption is less than 200 billion cigarettes per year.

$-13.4t + 445.2 < 200$

b. Solve the inequality in part a using both algebraic and graphical approaches.

$-13.4t + 445.2 < 200$

$-13.4t < -245.2$

$t > 18.3$ years from 2000 or 2018.

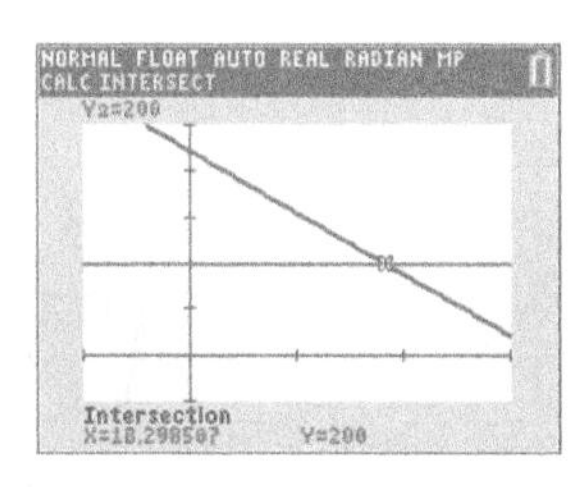

18. In Activity 1.11, you contacted two local rental companies and obtained the following information for the 1-day cost of renting a truck:

Company 1: \$60.00 per day plus \$0.75 per mile

Company 2: \$30.00 per day plus \$1.00 per mile

Let n represent the total number of miles driven in 1 day.

a. Write an expression to determine the total cost, C, of renting a truck for 1 day from company 1.

$C = 60 + 0.75n$

b. Write an expression to determine the total cost, C, of renting a truck for 1 day from company 2.

$C = 30 + 1.00n$

c. Use the expressions in parts a and b to write an inequality that can be used to determine for what number of miles it is less expensive to rent the truck from company 2.

$30 + 1.00n < 60 + 0.75n$

d. Solve the inequality.

$0.25n < 30$

$n < 120$

19. The sign on the elevator in a seven-story building on campus states that the maximum weight it can carry is 1200 pounds. As part of your work-study program, you need to move a large shipment of books to the sixth floor. Each box weighs 60 pounds.

a. Let n represent the number of boxes placed in the elevator. Assuming that you weigh 150 pounds, write an expression that represents the total weight in the elevator. Assume that only you and the boxes are in the elevator.

$150 + 60n$

b. Using the expression in part a, write an inequality that can be used to determine the maximum number of boxes you can place in the elevator at one time.

$150 + 60n \leq 1200$

c. Solve the inequality.

$60n \leq 1050$

$n \leq 17.5$

The maximum number of boxes that can be placed in the elevator is 17.

20. The following equation is used in meteorology to determine the temperature humidity index, T:

$$T = \frac{2}{5}(w + 80) + 15,$$

where w represents the wet-bulb thermometer reading. For what values of w would T range from 70 to 75?

$$70 \leq \frac{2}{5}(w + 80) + 15 \leq 75$$

$$55 \leq \frac{2}{5}(w + 80) \leq 60$$

$137.5 \leq w + 80 \leq 150$

$57.5 \leq w \leq 70$

21. The temperature readings in the United States have ranged from a record low of −79.8°F (Alaska, January 23, 1971) to a record high of 134°F (California, July 10, 1913).

a. If F represents the Fahrenheit temperature, write a compound inequality that represents the interval of temperatures (in degrees Farenheit) in the United States.

$-79.8 \leq F \leq 134$

b. Recall that Fahrenheit and Celsius temperatures are related by the formula

$$F = 1.8C + 32.$$

Rewrite the compound inequality in part a to determine the temperature range in degrees Celsius.

$-79.8 \leq 1.8C + 32 \leq 134$

c. Solve the compound inequality.

$-111.8 \leq 1.8C \leq 102$

$-62.1 \leq C \leq 56.7$

22. You are enrolled in a wellness course at your college. You achieved grades of 70, 86, 81, and 83 on the first four exams. The final exam counts the same as an exam given during the semester.

a. If x represents the grade on the final exam, write an expression that represents your course average (arithmetic mean).

$$\frac{70 + 86 + 81 + 83 + x}{5}$$

b. If your average is greater than or equal to 80 and less than 90, you will earn a B in the course. Using the expression from part a for your course average, write a compound inequality that must be satisfied to earn a B.

$$80 \le \frac{70 + 86 + 81 + 83 + x}{5} < 90$$

c. Solve the inequality.

$$400 \le 320 + x < 450$$

$$80 \le x < 130$$

ACTIVITY
1.16

Working Overtime
Piecewise Linear Functions

OBJECTIVES

1. Graph a piecewise linear function.
2. Write a piecewise linear function to represent a given situation.
3. Graph a function defined by $y = |x - c|$.

According to federal overtime pay law, overtime pay is extra cash compensation for the number of hours nonexempt (eligible) employees work in excess of 40 in one workweek. The current overtime pay rate for eligible employees is one and one-half (1.5) times their regular rates of pay (also known as time and a half).

1. a. If your regular hourly rate of pay is \$9.00 per hour, then what is your overtime rate?

Your overtime rate is $1.5 \cdot 9.00 = \$13.50$ per hour.

b. If your regular hourly rate of pay is \$14.00 per hour, then what is your overtime rate?

Your overtime rate is $1.5 \cdot 14.00 = \$21.00$ per hour.

2. a. Suppose your regular hourly rate of pay is \$12.00 per hour. Complete the following table:

Number of Hours Worked	25	35	45
Gross Pay	\$300	\$420	\$570

b. Is gross pay a function of the number of hours worked? Explain.

Yes, each volume of hours worked (input) has exactly one corresponding value of gross pay (output).

The method used to determine your gross pay for 45 hours is different from the method used for 25 or 35 hours.

When the output value of a function is calculated differently depending on the input value, the function is called a **piecewise function.**

3. Let h represent the number of hours worked in 1 week, and let P represent the gross pay. Assume your regular hourly rate of pay is \$12.00 per hour.

a. Write an equation for P if h is less than or equal to 40 hours.

If h is less than or equal to 40 hours, $P = 12.00h$.

b. Write an equation for P if h is *greater than* 40 hours.

If h is *greater than or equal* to 40 hours, $P = 480 + 18.00(h - 40)$.

The equations from Problems 3a and 3b are two pieces of one function, the overtime function. This piecewise function is written in the following form:

$$P = f(h) = \begin{cases} 12h, & 0 \le h \le 40 \\ 480 + 18(h - 40), & h > 40 \end{cases}$$

4. a. Use the piecewise function to verify the results from Problem 2a.

b. Graph the gross pay function for $0 \le h \le 60$.

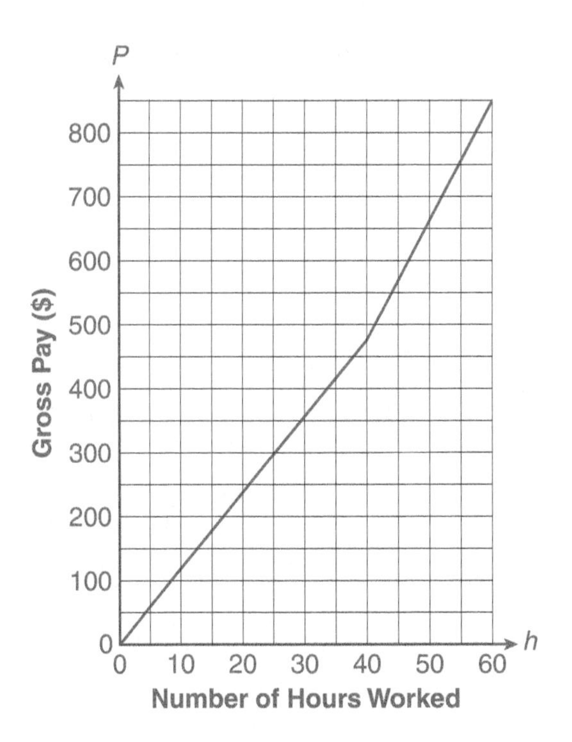

Solid Waste

5. The county solid waste facility charges a tipping fee for waste disposal. There is a minimum charge of \$14.70 for up to 280 pounds of waste. The charge for waste over 280 pounds is \$50 per ton. Note: The \$50 rate only applies to the amount of waste above 280 pounds.

a. What is the charge per pound of waste at the rate of \$50 per ton? **Note:** 1 ton = 2000 pounds.

$\frac{\$50}{2000} = \$0.025$ per pound

b. Complete the following table:

Pounds of Waste	100	200	400	600
Cost	\$14.70	\$14.70	\$17.70	\$22.70

\$14.70; \$14.70; $14.70 + 0.025(400 - 280) = \$17.70$;
$14.70 + 0.025(600 - 280) = \$22.70$

6. Let w represent the pounds of waste and C represent the cost of disposing of w pounds of waste. The cost, C, is a function of the pounds of waste, w, $C = f(w)$.

a. Write the equation for $f(w)$ if w is less than or equal to 280 pounds.

$f(w) = 14.70$

b. Write the equation for $f(w)$ if w is greater than 280 pounds. Check your equation by substituting 400 and 600 for w, and compare the result with your answer to Problem 5b.

$f(w) = 14.70 + 0.025(w - 280) = 0.025w + 7.70$

c. Combine your answers to parts a and b to write C as a piecewise function of w.

$$C = f(w) = \begin{cases} 14.70 & 0 < w \leq 280 \\ 0.025w + 7.70 & w > 280 \end{cases}$$

d. Sketch a graph of the solid waste cost function for $0 < w \leq 600$ pounds.

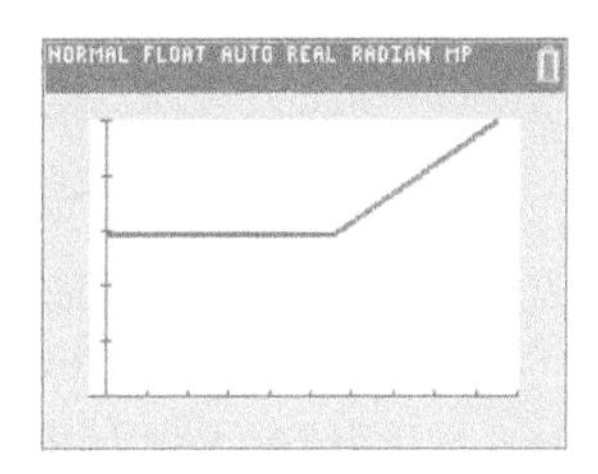

Commission

As an incentive to obtain more sales, a salesperson is often paid on a commission basis. A common commission scheme is the **accumulative plateau method**. Using this method, a salesperson is paid commission at a greater rate for a greater level of sales.

A particular company pays commission as follows:

5% on sales up to $10,000

7.5% on the next $10,000 in sales

10% on all sales over $20,000

For example,

i. The commission on $8000 in sales is 5% of $8000.

$$0.05 \cdot 8000 = \$400$$

ii. The commission on $12,000 in sales is 5% of the first $10,000 plus 7.5% of the sales over $10,000.

$$0.05 \cdot 10{,}000 + 0.075 \cdot 2000 = 500 + 150 = \$650$$

iii. The commission on $27,000 in sales is 5% of the first $10,000 plus 7.5% of the second $10,000 plus 10% of the sales over $20,000.

$$500 + 750 + 0.10(7000) = 1250 + 700 = \$1950$$

7. a. How much will a salesperson earn in commission on $6000 in sales?

The commission on $6000 in sales will be $0.05 \cdot 6000 = \$300$.

b. How much will a salesperson earn in commission on $15,000 in sales?
Note: The 7.5% rate only applies to sales above $10,000.

The commission on $15,000 in sales will be $500 + 0.075 \cdot 5000 = \$875$.

c. How much will a salesperson earn in commission on $24,000 in sales?
Note: The 10% rate only applies to sales above $20,000.

The commission on $24,000 in sales will be $1250 + 0.10 \cdot 4000 = \$1650$.

8. Is the commission earned a function of the amount of the sales? Explain.

Yes, the input is the sales amount, and the output is the commission.

Note that when using the accumulation plateau method, the commission earned is calculated differently depending on the level (plateau) of sales.

9. a. Refer to Problem 7a, and write an equation that gives the commission, C, as a function of sales, s, if the sales amount is less than or equal to $10,000.

If the sales are less than or equal to $10,000, the commission is $C = 0.05s$.

b. Refer to Problem 7b and write an equation that gives the commission, C, as a function of sales, s, if the sales amount is greater than \$10,000 but less than or equal to \$20,000.

If the sales are greater than \$10,000 but less than or equal to \$20,000, the commission is $C = 500 + 0.075(s - 10{,}000)$.

c. Refer to Problem 7c and write an equation that gives the commission, C, as a function of sales, s, if the sales amount is greater than \$20,000.

If the sales are greater than \$20,000, the commission is $C = 1250 + 0.10(s - 20{,}000)$.

10. Combine the three equations from Problem 9 into one piecewise function with three pieces.

$$C = f(s) = \begin{cases} 0.05s & 0 \leq s \leq 10{,}000 \\ 500 + 0.075(s - 10{,}000) & 10{,}000 < s \leq 20{,}000 \\ 1250 + 0.10(s - 20{,}000) & s > 20{,}000 \end{cases}$$

LC LEARNING CATALYTICS

Suppose f is defined by

$$f(x) = \begin{cases} 2x - 4, & x < -1 \\ -2x + 5, & x > 1 \end{cases}.$$

What is the value of $f(-1)$?

a. -6

b. 1

c. $f(-1)$ is undefined

d. none of these

11. Use the piecewise function from Problem 10 to evaluate the following. Interpret the results.

a. $f(6000)$

$f(6000) = 0.05 \cdot 6000 = 300$. The commission on \$6000 in sales is \$300.

b. $f(16{,}000)$

$f(16{,}000) = 500 + 0.075(16{,}000 - 10{,}000) = 950$. The commission on \$16,000 in sales is \$950.

c. $f(24{,}000)$

$f(24{,}000) = 1250 + 0.10(24{,}000 - 20{,}000) = 1650$. The commission on \$24,000 in sales is \$1650.

12. Sketch a graph of the commission function over the domain 0 to 30,000. Be sure to use an appropriate scale for the horizontal and vertical axes.

Absolute Value Function

13. Consider the piecewise function defined by

$$f(x) = \begin{cases} -x & \text{if } x < 0 \\ x & \text{if } x \geq 0 \end{cases}.$$

a. What is the domain of the function f?

all real numbers

b. Complete the following table:

x	−4	−3	−2	−1	0	1	2	3	4
$f(x)$	4	3	2	1	0	1	2	3	4

c. Sketch a graph of the function f. Use your graphing calculator to verify the graph.

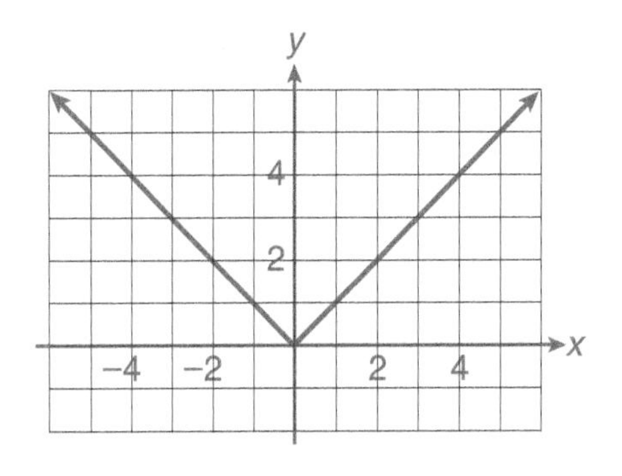

d. Describe the shape of the graph of function f.

The graph is V-shaped with the vertex at the origin.

e. What is the range of the function?

$y \geq 0$

The function f defined in Problem 13 may look familiar. This function is the piecewise definition of the absolute value function defined by $f(x) = |x|$.

14. a. Use your graphing calculator to obtain the graph of $y = |x|$. The absolute value function is located in the Math menu under the Num submenu.

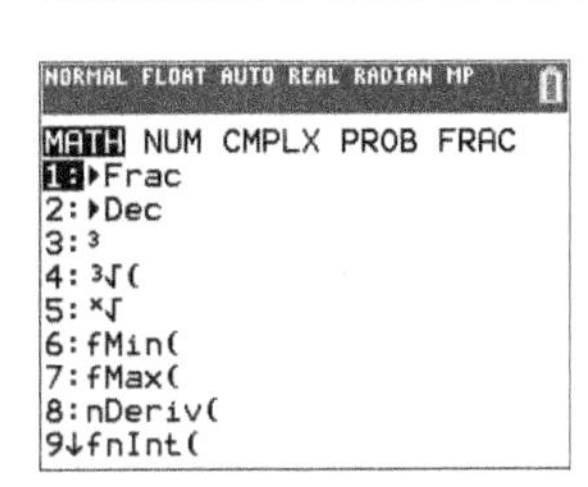

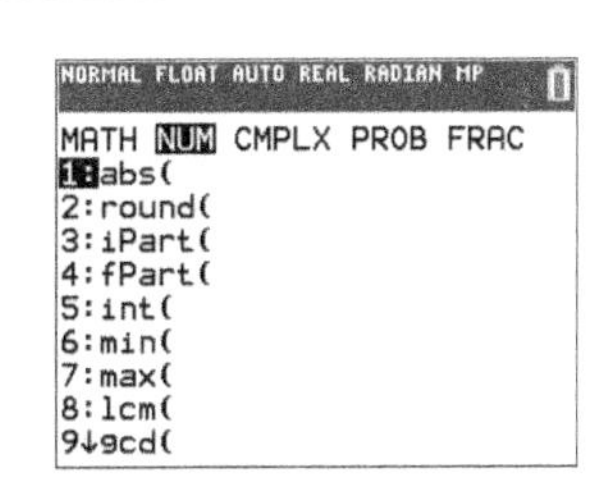

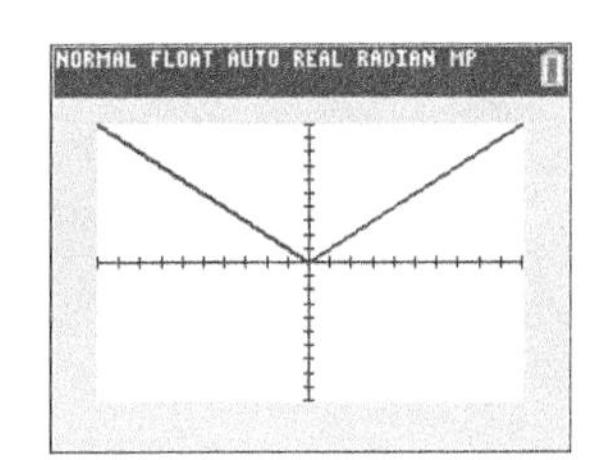

b. How does the graph compare with the graph in Problem 13?

They are exactly the same.

If $|x| = f(x) = \begin{cases} -x, & x < 0 \\ x, & x \geq 0 \end{cases}$, then

$$g(x) = |x - 2| = f(x - 2) = \begin{cases} -(x - 2), & x - 2 < 0 \\ x - 2, & x - 2 \geq 0 \end{cases} = \begin{cases} -x + 2, & x < 2 \\ x - 2, & x \geq 2 \end{cases}.$$

15. Use the piecewise function for $g(x) = |x - 2|$ to evaluate the following:

a. $g(0)$

$g(0) = -(0) + 2 = 2$

b. $g(5)$

$g(5) = (5) - 2 = 3$

c. $g(2)$

$g(2) = (2) - 2 = 0$

16. a. Use your graphing calculator to sketch a graph of $y = |x - 2|$.

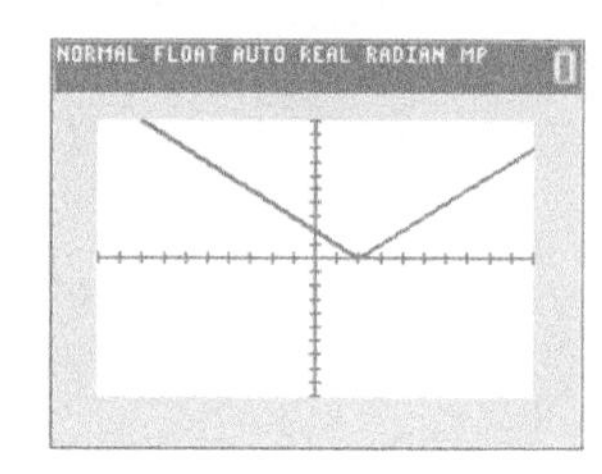

b. Describe how the graph of $y = |x - 2|$ can be obtained from the graph of $y = |x|$.

The graph of $y = |x - 2|$ is the same as the graph of $y = |x|$ shifted 2 units to the right.

SUMMARY Activity 1.16

1. A **piecewise** function is a function that is defined differently for certain "pieces" of its domain.
2. The **absolute value** function is a special piecewise function defined by

$$|x| = \begin{cases} x & \text{if } x \geq 0 \\ -x & \text{if } x < 0 \end{cases}.$$

3. The absolute value of the linear function $g(x) = |x - c|$ always has a V-shaped graph with a vertex at $(c, 0)$.

EXERCISES Activity 1.16

1. A certain airport taxicab company charges \$5.00 for the first mile or less of travel. After the first mile, the charge is an additional \$0.90 per mile for all travel outside the city limits.

a. Complete the following table:

x, Number of Miles Outside City Limits	0	0.5	1	2	5	10
C, Total Cost (\$)	5.00	5.00	5.00	5.90	8.60	13.10

b. Write an equation for the total cost, C, of the first mile or less of travel.

$C = 5.00$

c. Write an equation that gives the total cost, C, if you travel more than 1 mile, all outside city limits.

$C = 5.00 + 0.90(x - 1)$

d. Write a piecewise function for the total cost, C.

$$C = \begin{cases} 5.00 & 0 \leq x \leq 1 \\ 5.00 + 0.90(x - 1) & x > 1 \end{cases}$$

e. Sketch a graph of the cost function over the domain 0 to 10 miles.

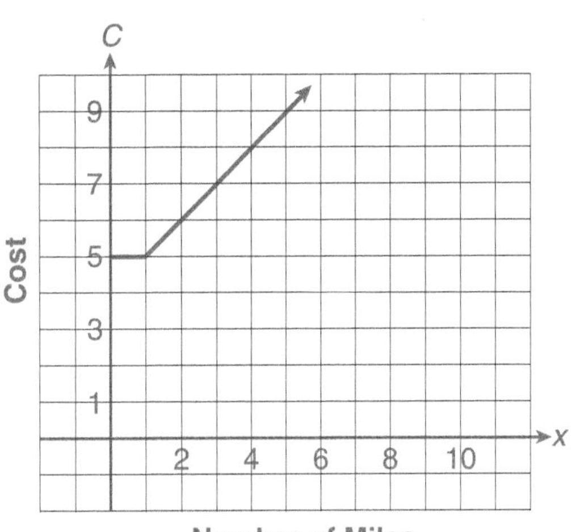

f. Determine the cost of a 12-mile taxi ride, all outside city limits.

$5.00 + .90(11) = \$14.90$

2. Sketch the graph of the piecewise function. You are given the following piecewise function:

$$H(x) = \begin{cases} -2x + 3 & x < -2 \\ 4 & -2 \le x < 1 \\ x - 1 & x \ge 1 \end{cases}$$

a. Evaluate $H(-4)$.

$H(-4) = -2(-4) + 3 = 11$

b. Evaluate $H(0)$.

$H(0) = 4$

c. Evaluate $H(3)$.

$H(3) = (3) - 1 = 2$

d. Sketch the graph of the piecewise function, $H(x)$.

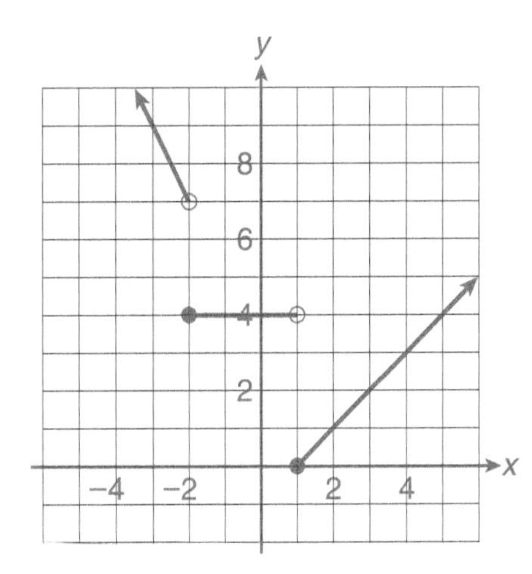

3. You are an author about to publish your first novel, *Hockey Today*. The book will sell for \$25. You will be paid royalties of 10% on the first 15,000 copies sold, 12% on the next 6000 copies, and 16% on any additional copies.

a. Write a piecewise function, *f*, that specifies the total royalties if *x* copies are sold.

$$f(x) = \begin{cases} 2.5x & x \le 15{,}000 \\ 37{,}500 + 3(x - 15{,}000) & 15{,}000 < x \le 21{,}000 \\ 55{,}500 + 4(x - 21{,}000) & x > 21{,}000 \end{cases}$$

b. Use your function from part a to determine whether the royalty is $46,500 when 18,000 books are sold. Is the royalty $71,500 when 25,000 copies are sold? If you do not obtain these answers when substituting into the function, then discover where you went wrong.

c. With the royalties from your book, you would like to pay for your advanced degree in journalism. It will cost $65,000. How many books must sell to cover the cost of the degree?

$65{,}000 = 55{,}500 + 4x - 84{,}000$

$65{,}000 = 4x - 28{,}500$

$93{,}500 = 4x$

$23{,}375 = x$

23,375 books

4. You receive a bill each month for your credit-card use. The bill indicates the minimum amount that is due by a certain date. The minimum amount due depends on your unpaid balance. One credit-card company uses the following criteria to determine your bill:

- The entire amount is due if the balance is less than $25.
- A minimum of $25 is due if the balance is $25 or more but less than $1000.
- A minimum of $30 is due if the balance is $1000 or more but less than $1250.
- A minimum of $50 is due if the balance is $1250 or more but less than $1500.
- A minimum of 5% of the balance is due if the balance is $1500 or more.

a. Let x represent the dollar amount of the unpaid balance. Complete the following table:

Unpaid Balance ($)	0	15	100	500	1000	1300	1500	2000
Minimum Amount Due, M ($)	0	15	25	25	30	50	75	100

b. Write a piecewise function for the minimum payment due.

$$M(x) = \begin{cases} x & x < 25 \\ 25 & 25 \le x < 1000 \\ 30 & 1000 \le x < 1250 \\ 50 & 1250 \le x < 1500 \\ 0.05x & x \ge 1500 \end{cases}$$

c. Sketch a graph of the minimum payment function

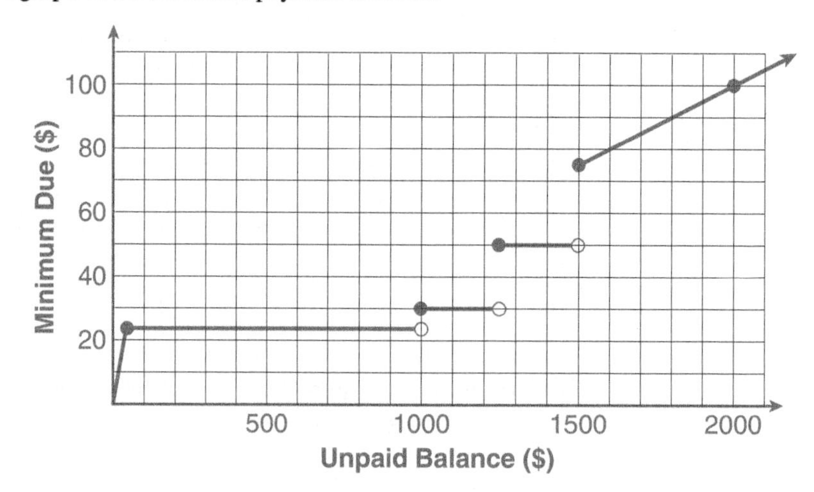

5. a. Complete the following table, where $f(x) = |x - 3|$ and $g(x) = |x + 3|$:

x	−5	−4	−3	−2	−1	0	1	2	3	4	5
$f(x)$	8	7	6	5	4	3	2	1	0	1	2
$g(x)$	2	1	0	1	2	3	4	5	6	7	8

b. Sketch a graph of each of the functions f and g. Verify using your graphing calculator.

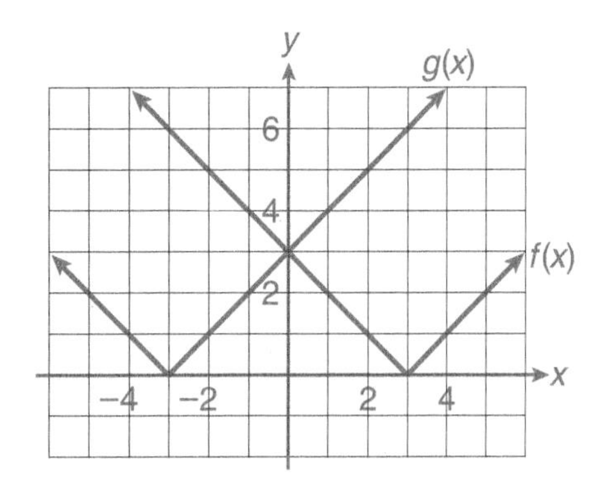

c. Describe how the graphs of f and g can be obtained from the graph of $y = |x|$.

f is the same as $y = |x|$, only it is shifted 3 units to the right. g is the same as $y = |x|$, only it is shifted 3 units to the left.

d. Write $f(x) = |x - 3|$ as a piecewise function.

$$f(x) = \begin{cases} -x + 3 & \text{if } x \leq 3 \\ x - 3 & \text{if } x > 3 \end{cases}$$

e. What is the domain of $f(x)$? What is the domain of $g(x)$?

The domains of f and g are all real numbers.

f. What is the range of $f(x)$? What is the range of $g(x)$?

The ranges of f and g are $y \geq 0$.

6. a. Without using a table of values or a graphing calculator, sketch the graph of $f(x) = |x - 5|$.

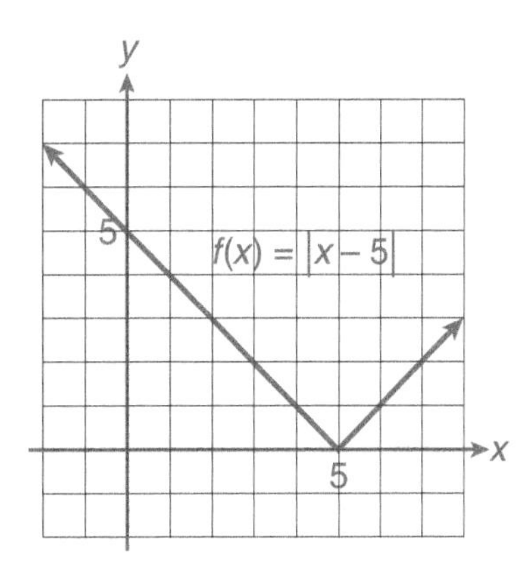

b. Describe the location and shape of the graph of $g(x) = |x + 5|$.

The graph of g is the same as the graph of $y = |x|$ shifted 5 units to the left.

c. Sketch the graph of $h(x) = |x| + 5$ without a graphing calculator or a table of values.

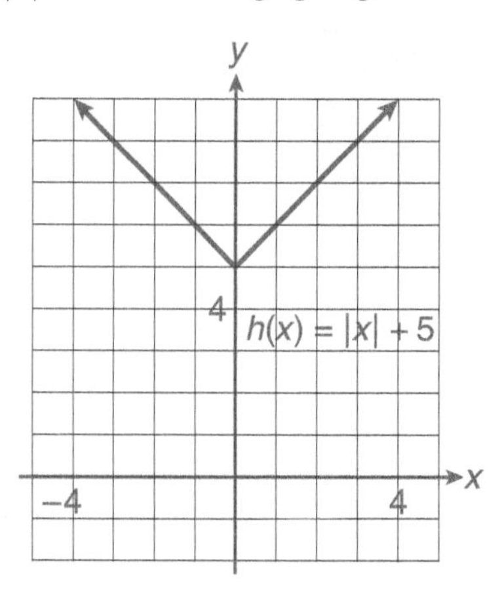

d. Describe the similarities and differences in the graphs in parts b and c. Include shape, location, and intercepts.

The graphs have the same shape of $y = |x|$. Where g is shifted 5 units left, h is shifted 5 units up. The x-intercept of g is $(-5, 0)$, and the y-intercept is $(0, 5)$. The y-intercept of h is $(0, 5)$, and there is no x-intercept.

7. Tax brackets for federal income tax are described by a piecewise function that gives the tax (output) for a given taxable income (input). Go to the IRS website and access the current tax rate schedules. Write a piecewise function for the federal income tax for single people. The function has multiple pieces, one piece for each tax bracket.

CLUSTER 3 What Have I Learned?

1. You are given two linear equations in slope-intercept form. How can you tell by inspection if the system is consistent, inconsistent, or dependent? Give examples.

   If the slopes are different, such as $y = 2x + 1$ and $y = 3x + 2$, then I can say immediately that the system is consistent and that it has one solution. If the system is inconsistent, the slopes will be equal but the lines will possess different y-intercepts (e.g., $y = 2x + 1$ and $y = 2x + 3$.) If the system is dependent, then the equations will be exactly the same.

2. In this cluster, you solved 2×2 linear systems four ways. List them. Give an advantage of each approach.

   The four ways are graphically; numerically; and algebraically by two methods, substitution and addition. The advantage to the graphical approach is that it is visual. The disadvantage is that if the solutions are not integers, they may be hard to read accurately. Numeric solutions can be relatively easy to find using the table feature of the calculator if the solution is integral. But if it is not, the numeric approach can be cumbersome and time-consuming. The advantage of an algebraic approach is accuracy. The disadvantage is that it can be cumbersome.

3. Describe a procedure that will combine the following two linear equations in three variables into a single linear equation in two variables:

   $$2x + 3y - 5z = 10$$
   $$3x - 2y + 2z = 4$$

   Multiply the first equation by 2 and the second by 3. The result will be two equations in which the y-coefficients will be opposites. Adding the two equations will eliminate the y-variable from the system. (Other answers are possible.)

4. What number is its own opposite?

   Zero is its own opposite.

5. The graphs of the absolute value functions in this cluster look like a V. What are the coordinates of the point of the V of the graph of $f(x) = |x - 10|$?

   The coordinates of the vertex of the V in the graph of $f(x) = |x - 10|$ are (10, 0).

6. Explain when the addition method would be more efficient to use than the substitution method as you solve a system of linear equations algebraically.

   If the equations are written in general form and the coefficients of a variable are opposites or are easily made opposites by application of the multiplication principle, then the addition method would be more efficient than the substitution method.

7. In solving an inequality, explain when you would change the direction of the inequality symbol.

   When multiplying or dividing each side of an inequality by a negative number, I would reverse direction of an inequality symbol.

CLUSTER 3 How Can I Practice?

1. Solve the following systems both graphically and algebraically.

a. $x + y = -3$

$y = x - 5$

$x + y = -3$ $\quad x + x - 5 = -3$

$y = x - 5$ $\quad 2x = 2$, or $x = 1$

$y = 1 - 5$, or $y = -4$

$(1, -4)$

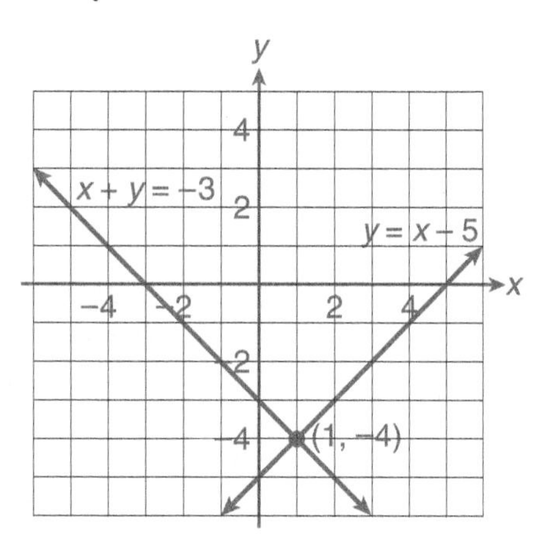

b. $x - 2y = -1$

$4x - 3y = 6$

$-4x + 8y = 4$ $\quad 5y = 10$, or $y = 2$

$4x - 3y = 6$ $\quad x - 4 = -1$, or $x = 3$

$(3, 2)$

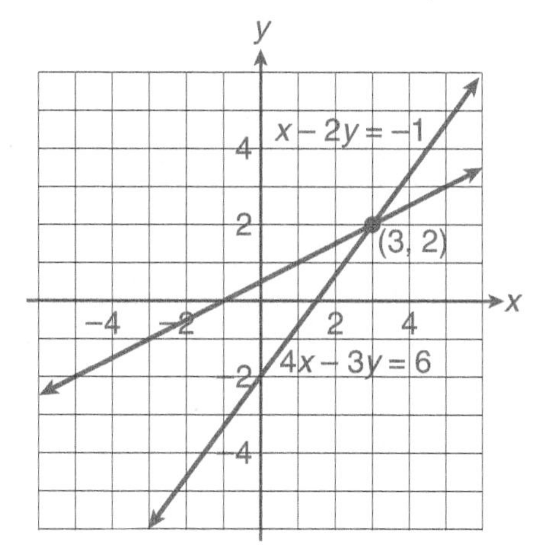

c. $2x - 3y = 7$

$5x - 4y = 0$

$8x - 12y = 28$

$-15x + 12y = 0$

$-7x = 28$, or $x = -4$

$-8 - 3y = 7$, or $y = -5$

$(-4, -5)$

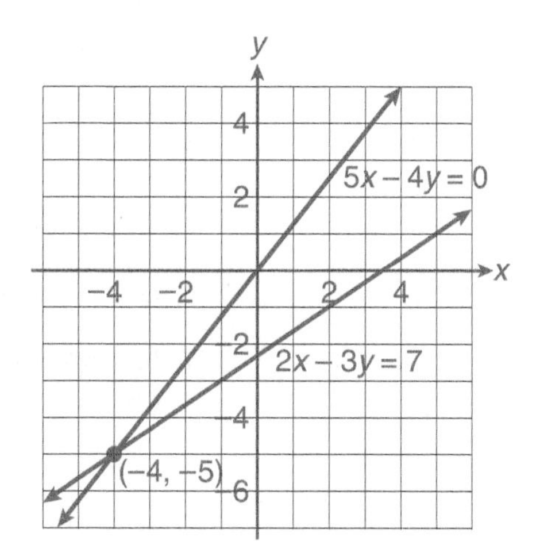

Answers to all How Can I Practice exercises are included in the Selected Answers appendix.

d. $x - y = 6$

$y = x + 2$

$x - (x + 2) = 6$

$-2 = 6$

inconsistent; no solution

2. Rewrite the systems in Exercise 1 in the form

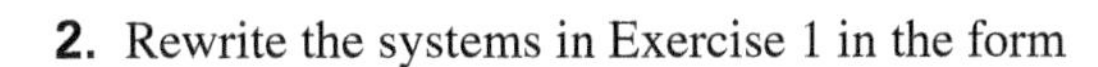

$$y = ax + b$$
$$y = cx + d,$$

and check your solutions numerically using the table feature of your graphing calculator.

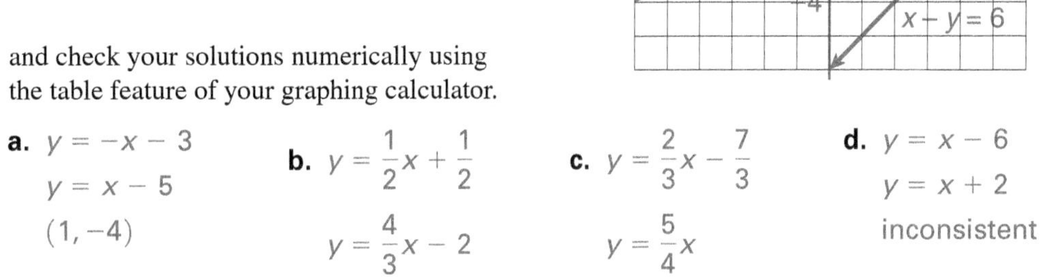

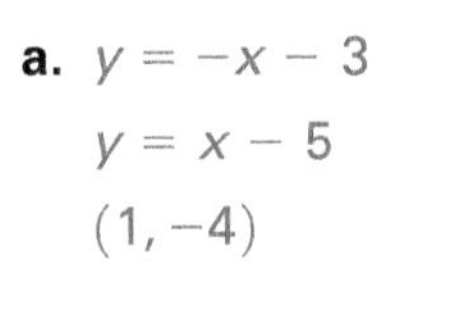

a. $y = -x - 3$

$y = x - 5$

$(1, -4)$

b. $y = \frac{1}{2}x + \frac{1}{2}$

$y = \frac{4}{3}x - 2$

$(3, 2)$

c. $y = \frac{2}{3}x - \frac{7}{3}$

$y = \frac{5}{4}x$

$(-4, -5)$

d. $y = x - 6$

$y = x + 2$

inconsistent

3. Solve the 3×3 system algebraically, or using matrix techniques.

$2x - y + z = -5$

$x - 2y + 2z = -13$

$3x + y - 2z = 12$

$$\begin{array}{l} 2x - y + z = -5 \\ 3x + y - 2z = 12 \\ \hline 5x \quad - z = 7 \end{array} \qquad \begin{array}{l} x - 2y + 2z = -13 \\ 6x + 2y - 4z = 24 \\ \hline 7x \quad - 2z = 11 \end{array} \qquad \begin{array}{l} 7x - 2z = 11 \\ -10x + 2z = -14 \\ \hline -3x = -3;\ x = 1 \end{array}$$

$5(1) - z = 7;\ z = -2$

$3(1) + y - 2(-2) = 12;\ y = 5$

$(1, 5, -2)$

4. Solve the following inequalities algebraically. Check your solutions graphically in parts a and b.

a. $2.5x + 9.8 \geq 14.3$

$2.5x \geq 4.5$

$x \geq 1.8$

b. $-3x + 14 < 32$

$-3x < 18$

$x > -6$

c. $-5 \leq 3x - 8 < 7$

$3 \leq 3x < 15$

$1 \leq x < 5$

5. You are going to help your grandmother plant a garden of tulips and daffodils. She has space for approximately 80 bulbs. The florist tells you that tulips cost \$0.50 per bulb and daffodils cost \$0.75 per bulb. How many of each can you purchase if your grandmother's budget is \$52?

a. Write the system of equations.

$t + d = 80$

$0.50t + 0.75d = 52$

b. Solve the system algebraically.

$0.50(80 - d) + 0.75d = 52$

$40 - 0.50d + 0.75d = 52$

$40 + 0.25d = 52$

$0.25d = 12$

$d = 48,\ t = 32$

c. Check your solution graphically using your graphing calculator.

6. You need some repair work done on your truck. Towne Truck charges $80 just to examine the truck and $70 per hour for labor costs. World Transport Co. charges $50 for the initial exam and $80 per hour for the labor.

a. Write a cost equation for each company. Use y to represent the total cost of doing the work and x to represent the number of hours of labor.

$y = 80 + 70x$

$y = 50 + 80x$

b. Complete the table of values for the cost function

x (NUMBER OF HOURS)	y, TOWNE TRUCK COST ($)	y, WORLD TRANSPORT COST ($)
2	220	210
4	360	370
6	500	530
8	640	690

c. Graph the functions.

d. From the graph, determine after how many hours the costs will be equal. What will be the cost?

after 3 hours; the cost will be $290.

e. Check your solution in part d by solving the system algebraically.

$80 + 70x = 50 + 80x$

$30 = 10x$

$x = 3; y = 80 + 70(3) = 290$

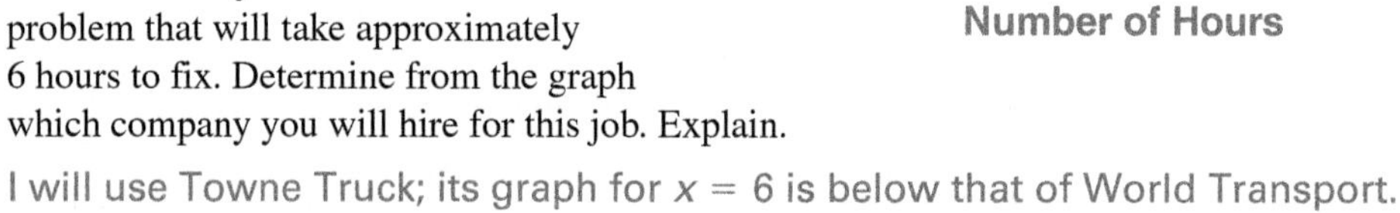

f. You think that you have a transmission problem that will take approximately 6 hours to fix. Determine from the graph which company you will hire for this job. Explain.

I will use Towne Truck; its graph for $x = 6$ is below that of World Transport.

7. Your friend tells you that he has 27 coins. Some coins are nickels, some are dimes, and the rest are quarters. The total value is $3.25. When your friend gives you the last clue by saying that he has twice as many dimes as nickels, you can easily solve the system to tell him how many of each coin he has.

a. Write the system of equations.

$x + y + z = 27$

$0.05x + 0.10y + 0.25z = 3.25$

$y = 2x$

b. Solve the system algebraically.

$x + 2x + z = 27$

$0.05x + 0.20x + 0.25z = 3.25$

$z = 27 - 3x$

$0.25x + 0.25(27 - 3x) = 3.25$

$0.25x + 6.75 - 0.75x = 3.25$

$3.5 = 0.5x$

$x = 7, y = 14, z = 6$

c. Check your solution.

It checks.

8. Translate each of the following into an inequality statement:

a. x is greater than -5 and at most 6.

$-5 < x \leq 6$

b. x is greater than or equal to -3 and less than 4.

$-3 \leq x < 4$

9. You own a hot dog cart in New York City. Your monthly profit is determined by the expression $1.50x - 50$, where x represents the number of hot dogs sold each month. The number 1.50 in the expression is the profit for each hot dog. The cost of leasing the hot dog stand is the number 50 in the expression $1.50x - 50$.

a. To ensure a profit of at least \$4000 per month, how many hot dogs must you sell? Write the inequality and solve.

$1.50x - 50 \geq 4000$

$1.50x \geq 4050$

$x \geq 2700$

b. Your profit has been fluctuating between \$3500 and \$5000 per month. Determine approximately between what two values your hot dog sales must be to realize this range of profit. Write the inequality and solve.

$3500 \leq 1.50x - 50 \leq 5000$

$3550 \leq 1.50x \leq 5050$

$2367 \leq x \leq 3367$

10. The first three brackets for a single individual for the 2018 New York State income tax are as follows:

Taxable Income	Tax
up to \$8,500	4% of the income
\$8,500 to \$11,700	\$340 plus 4.5% of the income over \$8,500.
\$11,700 to \$13,900	\$484 plus 5.25% of the income over \$11,700

Write a piecewise function for the income tax for a single individual on taxable income, x, up to \$13,900.

$$Tax = f(x) = \begin{cases} 0.04x, & 0 \leq x \leq 8500 \\ 340 + 0.045(x - 8500), & 8500 < x \leq 11{,}700 \\ 484 + 0.0525(x - 11{,}700), & 11{,}700 < x \leq 13{,}900 \end{cases}$$

11. Write $f(x) = |x + 2|$ as a piecewise function.

$$f(x) = \begin{cases} -(x+2) & x + 2 \leq 0 \\ x + 2 & x + 2 > 0 \end{cases} = \begin{cases} -x - 2, & x \leq -2 \\ x + 2, & x > -2 \end{cases}$$

CHAPTER 1 Summary

The bracketed numbers following each concept indicate the activity in which the concept is discussed.

CONCEPT/SKILL	DESCRIPTION	EXAMPLE
Variable [1.1]	A variable, usually represented by a letter, is a quantity or quality that may change in value from one instance to another.	In a survey of your class, an individual's height, weight, and gender are all variables.
Input variable [1.1]	The input variable is the value given first in a relationship.	In the relationship for computing the perimeter given the side of a square, $P = 4s$, s is the input variable.
Output variable [1.1]	The output is the value that corresponds to or is determined by the given input value.	In the relationship for computing the perimeter given the side of a square, $P = 4s$, P is the output variable.
Function [1.1]	A function is a correspondence between an input variable and an output variable that assigns a single output value to each input value.	See Example 1 in Activity 1.1 (pages 2 and 3).
Ordered pair [1.1]	An ordered pair of numbers consists of two numbers written in the form (input value, output value). The order in which they are listed is-significant.	(2, 3) is an ordered pair. In this pair, 2 is the input and 3 is the output.
Verbally defined function [1.3]	A function is defined verbally when it is defined using words.	The high temperature in Albany, New York, is a function of the day of the year because for each day, there is one high temperature.
Numerically defined function [1.1]	A function is defined numerically using ordered pairs.	The table on page 1 in Activity 1.1 is a numerically defined function because for each hour, there is only one value for the number of cars in the parking lot.
Function notation [1.1]	output variable = name of function (input variable). $y = f(t)$ is read "y equals f of t."	See Example 4 of Activity 1.1 (page 5).
Independent variable [1.1]	*Independent variable* is another name for the input variable of a function.	See Problem 5 of Activity 1.1 (page 4).
Dependent variable [1.1]	*Dependent variable* is another name for the output variable of a function.	See Problem 5 of Activity 1.1 (page 4).

CONCEPT/SKILL	DESCRIPTION	EXAMPLE
Domain of a function [1.2]	The domain of the function is the collection of all replacement values for the independent or input variable.	See Example 3 in Activity 1.2 (page 13).
Practical domain of a function [1.2]	The practical domain is the collection of replacement values of the input variable that make practical sense in the context of the situation.	See Example 3 in Activity 1.2 (page 13).
Range of a function [1.2]	The range of a function is the collection of all output values of a function.	See Problem 8 in Activity 1.2 (page 13).
Practical range of a function [1.2]	The practical range is the collection of all output values that make practical sense in the context of the situation.	See Problem 8 in Activity 1.2 (page 13).
Graphically defined function [1.3]	A function is defined graphically when the input variable is represented on the horizontal axis and the output variable on the vertical axis.	The graph drawn on page 18 in Problem 1 of Activity 1.3 defines a function. See Vertical Line Text
Mathematical model [1.4]	A function can be used as a mathematical model that best fits the actual data and can be used to predict output values for input values not in a table.	See Problem 4 in Activity 1.4 (page 29).
Increasing function [1.4]	A function is increasing if its graph goes up to the right.	See Example 1 in Activity 1.4 (page 30).
Decreasing function [1.4]	A function is decreasing if its graph goes down to the right.	See Example 1 in Activity 1.4 (page 30).
Constant function [1.4]	A function is constant if its graph is horizontal.	See Example 1 in Activity 1.4 (page 31).
Vertical line test [1.4]	A graph defines a function only if any vertical line intersects the graph no more than once. This is called the vertical line test.	No circle can represent a function because a vertical line through the center will pass through the circle twice, indicating that there is at least one input value paired with two different output values.
Average rate of change [1.6]	The average rate of change of a function over a specified input interval is the ratio $\frac{\text{change in output}}{\text{change in input}}$.	See Problem 8 in Activity 1.6 (page 53).
Linear function [1.7]	A function for which the rate of change between any pair of points remains constant is called a linear function.	$f(x) = 2x + 1$ defines a linear function. Its rate of change stays equal to 2.

CONCEPT/SKILL	DESCRIPTION	EXAMPLE
Slope [1.7]	The slope of a line segment joining two points (x_1, y_1) and (x_2, y_2) is denoted by m and defined by $m = \frac{y_2 - y_1}{x_2 - x_1}$.	The slope of the line segment joining $(2, -1)$ and $(5, 2)$ is given by $m = \frac{2 - (-1)}{5 - 2} = \frac{3}{3} = 1$.
Vertical intercept [1.7]	The vertical intercept $(0, b)$ of a graph is the point where the graph crosses the vertical axis.	The vertical intercept of $f(x) = 2x + 1$ is $(0, 1)$.
Horizontal intercept [1.7]	The horizontal intercept $(a, 0)$ of a graph is the point where the graph crosses the horizontal axis.	The horizontal intercept of $f(x) = 2x + 1$ is $\left(-\frac{1}{2}, 0\right)$.
Slope-intercept form [1.7]	The slope-intercept form of the equation of a line is $f(x) = mx + b$.	$f(x) = 2x + 1$ is a linear function in slope-intercept form.
Parallel lines [1.8]	The graphs of linear functions with the same slope but different y-intercepts are parallel lines.	The graphs of $f(x) = 2x + 1$ and $g(x) = 2x - 3$ are parallel lines.
Point-slope form [1.8]	The point-slope form of the equation of a line is $y = y_0 + m(x - x_0)$	The point-slope form of the equation of the line through $(2,-1)$ with slope 4 is $y = -1 + 4(x - 2)$
General form of a linear equation [1.9]	A linear function whose equation is in the form $Ax + By = C$, where A, B, and C are constants, is said to be written in general form.	$2x + 3y = 6$ is an equation of a linear function written in general form.
Horizontal line [1.9]	The graph of $y = c$ or $f(x) = c$ is a horizontal line.	The graph of $y = 3$ is a horizontal line 3 units above the x-axis.
Vertical line [1.9]	The graph of $x = a$ is a vertical line.	The graph of $x = 2$ is a vertical line 2 units to the right of the y-axis.
Linear regression equation [1.10]	The linear regression equation is the linear equation that best fits a set of data.	See Problem 4 in Activity 1.10 (pages 89–90).
Interpolation [1.10]	Interpolation is the process of using a regression equation to predict a value of output for an input value that lies within the range of the original input data.	See Problem 6 in Activity 1.10 (pages 90–91).
Extrapolation [1.10]	Extrapolation is the process of using a regression equation to predict a value of output for an input value that lies outside the range of the original input data.	See Problem 6 in Activity 1.10 (pages 90–91).

CONCEPT/SKILL	DESCRIPTION	EXAMPLE
2×2 system of linear equations [1.11]	A 2×2 system of linear equations consists of two linear equations with two variables.	$y = 3x - 10$ $y = 5x + 14$
Solution to a 2×2 linear system [1.11]	A solution to a 2×2 linear system is an ordered pair that solves both equations of the system.	$(-12, -46)$ is a solution to $y = 3x - 10$ $y = 5x + 14$.
Consistent system [1.11]	A linear system is consistent if there is at least one solution.	$y = 3x - 10$ $y = 5x + 14$ is a consistent system.
Inconsistent system [1.11]	A linear system is inconsistent if there is no solution. The lines are parallel.	$y = 2x + 1$ $y = 2x - 3$
Dependent system [1.11]	A linear system is dependent if there are infinitely many solutions. The equations represent the same line.	$2x - 3y = 6$ $4x - 6y = 12$
3×3 system of linear equations [1.13]	A 3×3 system of linear equations consists of three equations with a total of three variables.	$x + y - z = 8$ $-x + y + z = 2$ $2x - y + z = 8$ is a system of three linear equations.
Linear equation in three variables [1.13]	A linear equation in three variables x, y, and z is of the form $Ax + By + Cz = D$, where A, B, C, and D are any constants.	$2x + 3y - 7z = 23$ is a linear equation in three variables.
Matrix [1.14]	Any rectangular array of numbers or symbols is called a matrix. That is, a matrix is a systematic arrangement of numbers or symbols in rows and columns.	$\begin{bmatrix} 1 & 2 & 3 \\ -1 & 5 & 0 \end{bmatrix}$ is an example of a 2×3 matrix—two rows and three columns.
Matrix of coefficients [1.14]	A matrix of coefficients of a linear system is a matrix containing the coefficients of the variables of the system.	For the system $2x + 3y = 1$ $-x - 2y = 4$, $\begin{bmatrix} 2 & 3 \\ -1 & -2 \end{bmatrix}$ is the matrix of coefficients.
Augmented matrix [1.14]	An augmented matrix of a linear system is the matrix of coefficients with the constant terms included.	For the system $2x + 3y = 1$ $-x - 2y = 4$, $\begin{bmatrix} 2 & 3 & 1 \\ -1 & -2 & 4 \end{bmatrix}$ is the augmented matrix.

CONCEPT/SKILL	DESCRIPTION	EXAMPLE
Reduced row echelon form [1.14]	A matrix is said to be in reduced row echelon form if the matrix has 1s down the main diagonal and 0s above and below each 1.	The matrix $\begin{bmatrix} 1 & 0 & 3 \\ 0 & 1 & -2 \end{bmatrix}$ is in reduced row echelon form.
Interval notation [1.15]	Interval notation is an alternative method used to represent sets of real numbers described by inequalities.	See Problem 11 of Activity 1.15 (page 147).
A compound inequality [1.15]	A compound inequality is a statement that involves more than one inequality symbol: $<, >, \leq, \geq$, or $\neq$.	$-3 < x + 7 \leq 10$
Piecewise function [1.16]	A piecewise function is a function that is defined differently for certain "pieces" of its domain.	$f(x) = \begin{cases} x & \text{if} \quad x \leq 2 \\ -x + 1 & \text{if} \quad x > 2 \end{cases}$
Absolute value function [1.16]	The absolute value function is the function defined by $\lvert x \rvert = \begin{cases} x & \text{if} \quad x \geq 0 \\ -x & \text{if} \quad x < 0 \end{cases}$.	$\lvert x \rvert = f(x) = \begin{cases} x & \text{if} \quad x \geq 0 \\ -x & \text{if} \quad x < 0 \end{cases}$
The graph of the absolute value of a linear function defined by $g(x) = \lvert x - c \rvert$ [1.16]	The absolute value of a linear function $g(x) = \lvert x - c \rvert$ has a V-shaped graph with a vertex at $(c, 0)$.	$g(x) = \lvert x + 2 \rvert$ has a vertex at $(-2, 0)$.

CHAPTER 1 Gateway Review

1. Determine whether each of the following is a function:

a. The distance traveled by a walker in 1 hour is a function of her average speed.

Yes, it is a function. For each value of input (average speed), there is one output (distance traveled).

b. 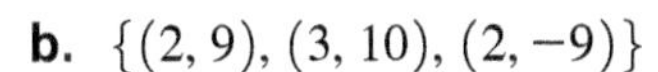
$\{(2, 9), (3, 10), (2, -9)\}$

No, it is not a function. Two different outputs are paired with 2.

c.

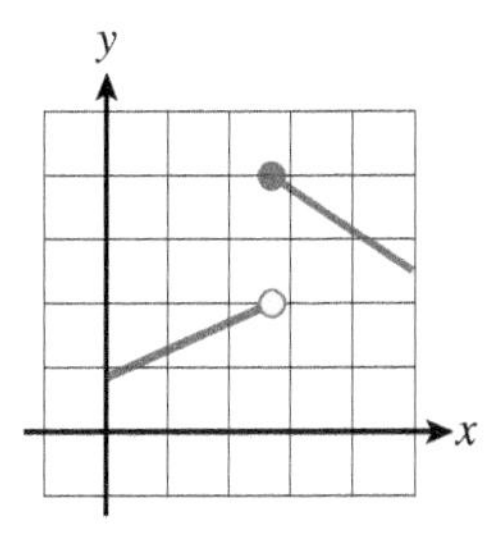

Yes, it is a function.

2. For a certain yard, the fertilizer costs \$20. You charge \$8 per hour to do yard work. If x represents the number of hours worked on the yard and $f(x)$ represents the total cost, including fertilizer, complete the following table:

x	0	2	3	5	7
$f(x)$	20	36	44	60	76

a. Is the total cost a function of the hours worked? Explain.

Yes. For each input, there is one output.

b. Which variable represents the input?

The input is x, the number of hours worked.

c. Which is the dependent variable?

The dependent variable is $f(x)$, the total cost.

d. Which value(s) of the domain would not be realistic for this situation? Explain.

Negative values would not be realistic domain values. A negative number of hours worked does not make sense.

e. What is the average rate of change from 0 to 3?

The average rate of change is \$8 per hour.

f. What is the average rate of change from 5 to 7?

The average rate of change is \$8 per hour.

g. What can you say about the rate of change between any two of the points?

The rate of change between any two points is \$8 per hour.

h. What kind of relationship exists between the two variables?

The relationship is linear.

i. Write this relationship in the form $f(x) = mx + b$.

$f(x) = 8x + 20$

Answers to all Gateway exercises are included in the Selected Answers appendix.

j. What is the practical meaning of the slope in this situation?

The slope is the hourly rate I charge, $8 per hour.

k. What is the vertical intercept? What is the practical meaning of this point?

(0, 20) is the vertical intercept. The 20 represents the fertilizer cost.

l. Determine $f(4)$.

$f(4) = 8(4) + 20 = 52$

m. For what value(s) of x does $f(x) = 92$? Interpret your answer in the context of the situation.

$8x + 20 = 92$, or $8x = 72$, or $x = 9$

I need to work 9 hours for the cost to equal exactly $92.

3. Let $f(x) = x^2 - 5x$, and let $g(x) = -3x + 4$. Evaluate each of the following:

a. $f(-2)$ and $g(-2)$

$f(-2) = 14$, $g(-2) = 10$

b. $f(3) + g(3)$

$-6 + (-5) = -11$

c. $f(-3) - g(-4)$

$24 - 16 = 8$

d. $f(-4) \cdot g(2)$

$36(-2) = -72$

4. Which of the following sets of data represent a linear function?

a.

x	0	2	4	6	8
$f(x)$	14	22	30	38	46

This represents a linear function. The slope is 4.

b.

x	5	10	15	20	25
y	4	2	0	−2	−4

This represents a linear function; $m = \frac{-2}{5}$.

c.

x	1	3	4	6	7
$g(x)$	10	20	30	40	50

This does not represent a linear function.

d.

t	0	10	20	30	40
d	143	250	357	464	571

This represents a linear function; $m = 10.7$.

5. a. Determine the slope of the line through the points $(5, -3)$ and $(-4, 9)$.

$$m = \frac{9 + 3}{-4 - 5} = \frac{12}{-9} = \frac{-4}{3}$$

b. From the equation $3x - 7y = 21$, determine the slope.

$$7y = 3x - 21$$

$$y = \frac{3}{7}x - 3$$

$$m = \frac{3}{7}$$

c. Determine the slope of the line from its graph. $m = \frac{1}{2}$

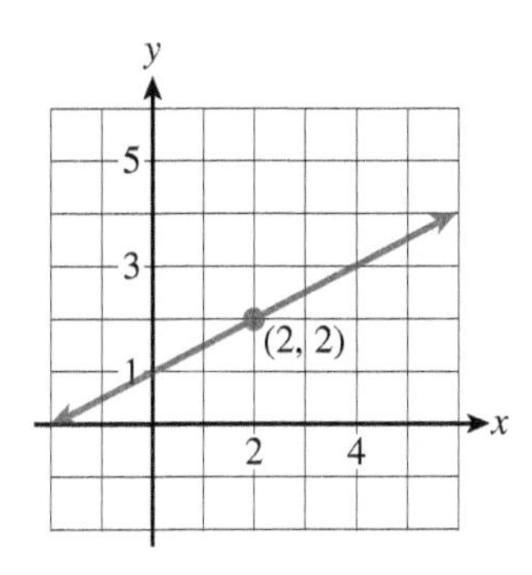

6. Write the equation of the line described in each of the following:

a. a slope of 0 and passing through the point (2, 4)

$y = 4$

b. a slope of 2 and a vertical intercept of (0, 5)

$y = 2x + 5$

c. a slope of -3 and passing through the point $(6, -14)$

$-14 = -3(6) + b, b = 4, y = -3x + 4$

d. a slope of 2 and passing through the point $(7, -2)$

$-2 = 2(7) + b, b = -16, y = 2x - 16$

e. a line with no slope passing through the point $(2, -3)$

$x = 2$

f. a slope of -5 and a horizontal intercept of $(4, 0)$

$0 = -5(4) + b, b = 20, y = -5x + 20$

g. a line passing through the points $(-3, -4)$ and $(2, 16)$

$$m = \frac{16 + 4}{2 + 3} = \frac{20}{5} = 4;$$

$16 = 4(2) + b, b = 8, y = 4x + 8$

h. a line parallel to $y = \frac{-1}{2}x$ and passing through the point (0, 5)

$$y = \frac{-1}{2}x + 5$$

7. Given the following graph of the linear function, determine the equation of the line:

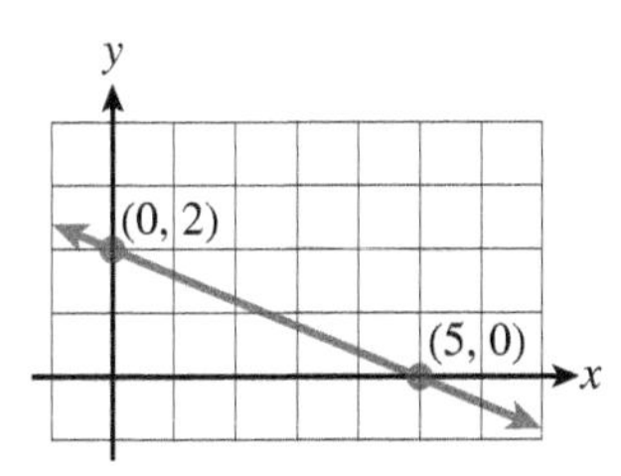

$$y = \frac{-2}{5}x + 2$$

8. a. The building where your computer graphics store is located is 10 years old and has a value of \$200,000. When the building was 1 year old, its value was \$290,000. Assuming that the building's depreciation is linear, express the value of the building as a function, f, of its age, x, in years.

$f(x) = 300{,}000 - 10{,}000x$

b. What is the slope of the line? What is the practical meaning of the slope in this situation?

$m = -10{,}000$. The building depreciates \$10,000 per year.

c. What is the vertical intercept? What is the practical meaning of the vertical intercept in this situation?

(0, 300,000). The original value is \$300,000.

d. What is the horizontal intercept? What is the practical meaning of the horizontal intercept in this situation?

(30, 0). It takes 30 years for the building to fully depreciate.

9. Determine the vertical intercept of the following functions. Solve for y if necessary.

a. $y = 2x - 3$

$(0, -3)$

b. $y = -3$

$(0, -3)$

c. $x - y = 3$

$(0, -3)$

d. What relationship do the graphs of these functions have to one another?

The graphs all intersect at the point $(0, -3)$.

e. Use your graphing calculator to graph the functions in parts a–c on the same coordinate axes. Compare your results with those of part d.

The results are the same.

10. Determine the slopes and y-intercepts of each of the following functions. Solve for y if necessary.

a. $y = -2x + 1$

$m = -2$; $(0, 1)$

b. $2x + y = -1$

$m = -2$; $(0, -1)$

c. $-4x - 2y = 6$

$m = -2$; $(0, -3)$

d. What relationship do the graphs of these functions have to one another?

The graphs are parallel lines.

e. Use your graphing calculator to graph the functions in parts a–c on the same coordinate axes. Compare your results with those of part d.

The results are the same.

11. Determine the slopes and y-intercepts of each of the following functions. Solve for y if necessary.

a. $y = -3x + 2$

$m = -3$; $(0, 2)$

b. $3x + y = 2$

$m = -3$; $(0, 2)$

c. $6x + 2y = 4$

$m = -3$; $(0, 2)$

d. What relationship do the graphs of these functions have to one another?

The graphs are all the same.

e. For two lines to be parallel to each other, what must be the same?

the slopes

f. For two lines to lie on top of each other (coincide), what must be the same?

the slopes and the y-intercepts

g. Use your graphing calculator to graph the functions in parts a–c on the same coordinate axes. Compare your results with those of part d.

The results are the same.

12. a. Graph the function defined by $y = -2x + 150$. Indicate the vertical and horizontal intercepts. Make sure you include some negative values of x.

(0, 150), (75, 0)

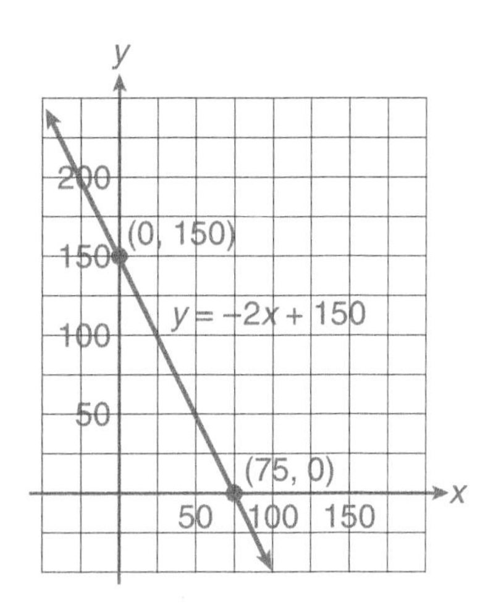

b. Using your graphing calculator, verify the graph you drew in part a.

c. Using the graph, determine the domain and range of the function.

The domain and range are all real numbers.

d. Assume that a 150-pound person starts a diet and loses 2 pounds per week for 15 weeks. Write the equation modeling this situation.

$w(t) = -2t + 150$

e. Compare the equation you found in part d with the one given in part a.

They are essentially the same.

f. What is the practical meaning of the vertical and horizontal intercepts you found in part a?

The vertical intercept is (0, 150). It indicates the person's initial weight of 150 pounds.

The horizontal intercept (75, 0) indicates that after 75 weeks of weight loss, the person weighs nothing.

g. What is the practical domain and range of this function for the situation given in part d?

The practical domain is $0 \leq t \leq 15$. The practical range is $120 \leq w(t) \leq 150$.

13. a. You pay a flat fee of $25 per month for your trash to be picked up, and it doesn't matter how many bags of trash you have. Use x to represent the number of bags of trash, and write a function, f, in symbolic form to represent the total cost of your trash for the month.

$f(x) = 25$

b. Sketch the graph of this function.

horizontal line starting at (0, 25)

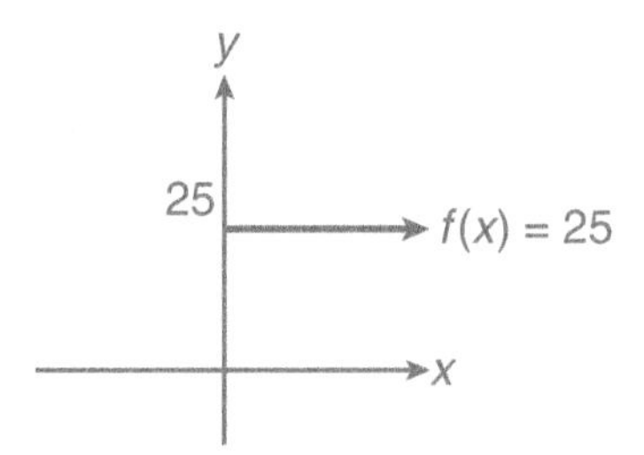

c. What is the slope of the line?

The slope is 0.

14. You work as a special events salesperson for your city-owned golf course. Your salary is based on the following. You receive a flat salary of \$1500 per month for sales of \$10,000 or less; for the next \$30,000 of sales, you receive your salary plus 2% of the sales over \$10,000 and up to \$40,000; and for any sales exceeding \$40,000, you receive your salary and commission of 4% of sales over \$40,000.

a. Write a piecewise function, *f*, that specifies the total monthly salary when *x* represents the amount of sales for the month.

$$f(x) = \begin{cases} 1500 & \text{if } x \leq 10{,}000 \\ 1500 + 0.02(x - 10{,}000) & \text{if } 10{,}000 < x \leq 40{,}000 \\ 2100 + 0.04(x - 40{,}000) & \text{if } x > 40{,}000 \end{cases}$$

b. Graph the function.

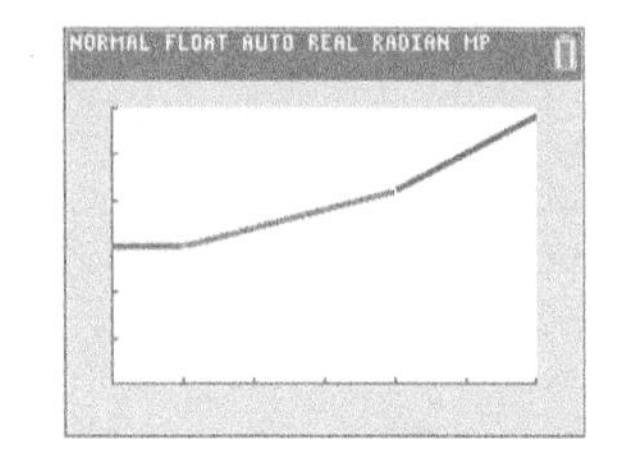

c. What is your salary if your sales are \$25,000?

$f(25{,}000) = 1500 + 0.02(15{,}000) = 1800$

d. You need to make \$3150 to cover your expenses this month. What must your sales be for your salary to be that amount?

$2100 + 0.04(x - 40{,}000) = 3150$

$x = 66{,}250$

15. During the years 2014–2018, the number of finishers in a large marathon increased. The following table gives the total number of finishers (to the nearest hundred) each year, where *t* represents the number of years after 2014.

Years after 2014, *t*	0	1	2	3	4
Number of Finishers, *n*	7800	9100	10,000	10,900	12,100

a. Enter the data from your table into your calculator. Determine the linear regression equation model, and write the result.

$y = 1040x + 7900$ or $t(n) = 1040n + 7900$

b. What is the slope of the regression line? What is the practical meaning of the slope in this situation?

1040. The number of finishers increased at a rate of 1040 per year.

c. What is the vertical intercept? What is the practical meaning of the vertical intercept in this situation?

(0, 7900). The model indicates that there were 7900 finishers in 2014.

d. Use your graphing calculator to graph the regression line in the same screen as the scatterplot. How well do you think the line fits the data?

fairly well

e. Use your regression model to determine the number of finishers in 2020.

14,140

f. Did you use interpolation or extrapolation to determine your result in part e? Explain.

I used extrapolation because I am predicting outside the original data.

g. Do you think that the prediction for the year 2020 will be as accurate as that in 2014? Explain.

No, 2020 is farther from the data than 2014. The farther removed we are from the data, the more likely our prediction is incorrect.

16. Solve the systems of equations. Solve at least one algebraically and at least one graphically.

a. $3x - y = 10$

$5x + 2y = 13$

$3x - 10 = y$

$5x + 2(3x - 10) = 13$

$5x + 6x - 20 = 13$

$11x = 33, x = 3$

$y = 9 - 10 = -1$

$(3, -1)$

b. $4x + 2y = 8$

$x - 3y = -19$

$x = -19 + 3y$

$4(-19 + 3y) + 2y = 8$

$-76 + 12y + 2y = 8$

$14y = 84, y = 6$

$x = -19 + 18 = -1$

$(-1, 6)$

c. $2x + y = 10$

$y = -2x + 13$

d. $2x + 6y = 4$

$x + 3y = 2$

$x = -3y + 2$

$2(-3y + 2) + 6y = 4$

$-6y + 4 + 6y = 4$

$4 = 4$

This is a dependent system. Any pair of numbers that satisfies one equation satisfies both equations.

c.

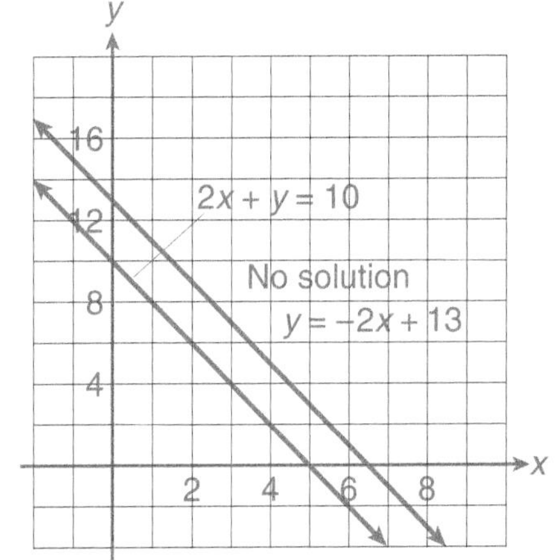

17. You need to replace the heating system in your house. A conventional heating system will cost \$5000 with a yearly fuel cost of \$5400. A modern heating system will cost \$8000 with a yearly fuel cost of \$4500.

a. Write an equation for the total cost, C, of a conventional system for t years. The total cost is the cost of the system plus the fuel cost.

$C = 5000 + 5400t$

b. Write an equation for the total cost, C, of a modern system for t years.

$C = 8000 + 4500t$

c. Solve the system of equations from parts a and b to find the number of years it takes for the total cost of the conventional system to equal the total cost of the modern system.

$C = 5000 + 5400t$

$C = 8000 + 4500t$

$5000 + 5400t = 8000 + 4500t$

$900t = 3000$

$t = 3.33$ yr. $C = \$22,982$

CHAPTER 2

The Algebra of Functions

CLUSTER 1

Addition, Subtraction, and Multiplication of Polynomial Functions

ACTIVITY 2.1

Spending and Earning Money

Polynomial Functions

OBJECTIVES

1. Identify a polynomial expression.
2. Identify a polynomial function.
3. Add and subtract polynomial expressions.
4. Add and subtract polynomial functions.

The Kennel

You are the owner of a small pet kennel. You are looking to expand your facilities since business is picking up. You locate a 2-acre piece of property that will give you enough room to build your kennel and the exercise area. The property includes a small building that you cannot use, but its cost is only $15,000, so you jump on the opportunity.

1. You find out you need a permit to tear down the old structure as well as a building permit for the new kennel. The demolition permit is $200, and the building permit is $0.80 per square foot for your new building.

   a. Complete the following table. $P(x)$ represent the cost of the permits in terms of the total square footage of the kennel, x.

Total Square Footage of the Kennel, x	1200	1400	1600	1800
Cost of the Permits, $P(x)$ ($)	$1160	$1320	$1480	$1640

   b. Write an equation for the cost, $P(x)$, of the permits for the kennel project in terms of the square footage of the kennel, x.

   $P(x) = 200 + 0.80x$

2. You also learn that the construction costs for your kennel will average $35.00 per square foot. Your in-laws are donating the fence and posts for the exercise area, so your costs will be the cost of the land, the cost of the permit, and the cost of the construction.

   a. Complete the following table for the construction costs of the kennel, where $C(x)$ represents the cost of construction and x represents the square footage of the kennel:

Total Square Footage of the Kennel, x	1200	1400	1600	1800
Construction Cost of the Kennel, $C(x)$ ($)	$42,000	$49,000	$56,000	$63,000

   b. Write an equation for the construction costs of the new kennel, $C(x)$, in terms of the square footage of the kennel, x.

   $C(x) = 35x$

3. The total cost of the kennel project is a function of the square footage of the new kennel, x.

Complete the following table, where $T(x)$ is the total cost of the kennel project, $P(x)$ is the cost of the permits, and $C(x)$ is the construction cost:

x, SQUARE FOOTAGE OF THE KENNEL	PRICE OF THE LAND	$P(x)$, COST OF THE PERMITS	$C(x)$, CONSTRUCTION COSTS	$T(x)$, TOTAL COST OF THE KENNEL PROJECT
1200	15,000	1160	42,000	$58,160
1400	15,000	1320	49,000	$65,320
1600	15,000	1480	56,000	$72,480
1800	15,000	1640	63,000	$79,640

b. What is the relationship among the price of the land, the cost of the permits, the construction costs, and the total cost of the kennel project?

The total cost is the sum of the other three.

c. Add 15,000, the expression for $P(x)$, from Problem 1 and $C(x)$ from Problem 2 to define the total cost of the project, $T(x)$, as a function of the square footage of the new kennel, x.

$T(x) = 15{,}200 + 35.80x$

DEFINITION

The kennel situation involves the addition of functions. The total cost function determined in Problem 3 is called a **sum function**. The notation is

$$T(x) = 15{,}000 + P(x) + C(x).$$

Note: when you add functions, you add the output values. The input value is unchanged.

EXAMPLE 1 ***Functions f and g are defined by the following tables:***

x	−2	0	2	4	6	8
$f(x)$	0	2	6	20	42	72

x	−2	0	2	4	6	8
$g(x)$	7	3	7	19	39	67

Complete the table below for the sum of f and g.

SOLUTION

For any given x, when functions are added, the outputs are added.

x	−2	0	2	4	6	8
$f(x) + g(x)$	$0 + 7 = 7$	$2 + 3 = 5$	$6 + 7 = 13$	$20 + 19 = 39$	$42 + 39 = 81$	$72 + 67 = 139$

4. Enter the constant function, $F(x) = 15{,}000$; the permit cost function, P; the construction cost function, C; and the total cost function, T, into your graphing calculator. Use the table feature in ASK mode to complete the table for the four input values given.

x	$F(x)$	$P(x)$	$C(x)$	$T(x)$
1250	15,000	1200	43,750	59,950
1500	15,000	1400	52,500	68,900
1725	15,000	1580	60,375	76,955
2050	15,000	1840	71,750	88,590

Note: The outputs of the sum function, T, are the sums of the output of the functions, F, P, and C.

Subtraction of Functions

The kennel has been built and you are back in business. Your new kennel can accommodate at most 25 dogs. Your current charge for boarding a dog is \$35 per day. The local utility bills you approximately \$15 per day. The cost of feeding each dog, cleaning its pen, and exercising the dog is approximately \$15.15 per day.

5. **a.** Suppose you let the input variable d represent the number of dogs boarding on a given day. Determine an equation that expresses the total revenue, $R(d)$, as a function of the number of dogs, d, boarding on a given day. $R(d) = 35d$

b. Complete the following input-output table for the revenue function, R:

d	0	5	10	15	20	25
$R(d)$ (\$)	0	175	350	525	700	875

6. **a.** The total daily cost of operating the kennel is a function of the number of dogs boarding on a given day. If $C(d)$ represents the cost, write an equation for $C(d)$ in terms of d.
$C(d) = 15 + 15.15d$

b. Complete the following table for the daily cost function, C.

d	0	5	10	15	20	25
$C(d)$ (\$)	15	90.75	166.50	242.25	318	393.75

7. The results from Problems 5 and 6 can be used to determine the profit, $P(d)$, in terms of the number of boarded dogs, d.

a. Use the output values from Problems 5b and 6b to complete the following table for the profit. Recall that profit = revenue − cost.

d	0	5	10	15	20	25
$P(d)$ (\$)	−15	84.25	183.50	282.75	382	481.25

b. Using the equations for revenue, $R(d)$, and cost, $C(d)$, in Problems 5a and 6a, determine an equation for the profit, $P(d)$, as a function of d. Use the new equation to verify some of the entries in the table in part a.

$P(d) = 35d - (15 + 15.15d) = 19.85d - 15$

DEFINITION

The kennel situation involves the subtraction of functions. The profit function determined in Problem 7b is called a **difference function**. The notation is

$$P(d) = R(d) - C(d).$$

Note: When you subtract functions, you subtract the output values. The input value is unchanged.

EXAMPLE 2 ***Functions f and g are defined by the following tables:***

x	−2	0	2	4	6	8
$f(x)$	2	0	6	20	42	72

x	−2	0	2	4	6	8
$g(x)$	7	3	7	19	39	67

Complete the table below for the difference of f and g.

SOLUTION

For any given x, when functions are subtracted, the outputs are subtracted.

x	−2	0	2	4	6	8
$f(x) - g(x)$	$2 - 7 = -5$	$0 - 3 = -3$	$6 - 7 = -1$	$20 - 19 = 1$	$42 - 39 = 3$	$72 - 67 = 5$

8. a. Sketch the graphs of the revenue function, R; the cost function, C; and the profit function, P, on the following grid. Label the axes with the appropriate scales.

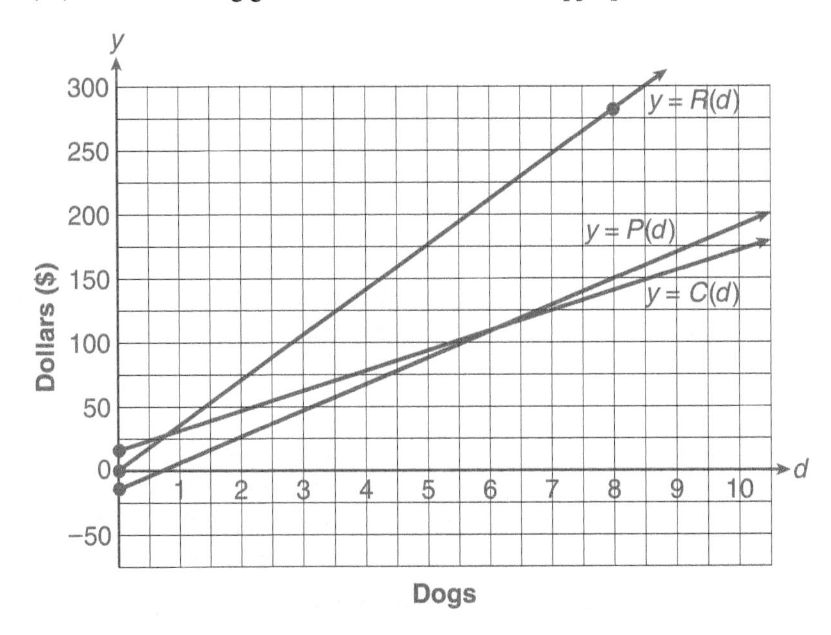

Note: The practical domain of these functions consists of whole numbers. While it is common practice to draw the graphs with continuous lines, the functions are *discrete*.

b. Use the graphs from part a to determine the break-even point for your pet kennel business. That is, determine the number of dogs necessary for the profit to equal $0. Explain how you determined your answer.

The graph of the profit function crosses the x-axis at a point just less than $x = 1$. So, If you board 1 dog you will break even.

c. Write and solve an equation to determine the break-even point.

$19.85d - 15 = 0$

$19.85d = 15$

$d = 15/19.85 \approx 0.76$

You break even if you board 1 dog.

d. For you to obtain a profit of at least $150, how many dogs must you board?

$19.85d - 15 \geq 150$

$19.85\,d \geq 165$

$d \geq 165/19.85 \text{ approximately} = 8.3$

You will need to board at least 9 dogs to obtain a profit of at least $150.

Polynomial Expressions, Functions, and Terminology

The functions encountered in this activity, such as the total cost of running a kennel function defined by $C(d) = 15 + 15.15d$ are examples of a special category of functions called **polynomial functions**. Such functions are defined by equations of the form $y = p(x)$ where x is the input variable and $p(x)$ is a polynomial expression involving the input variable. Therefore, to identify a polynomial function, you must be able to identify a polynomial expression.

DEFINITION

Any expression that is formed by adding one or more terms of the form ax^n, where a is a real number and n is a nonnegative integer, is called a **polynomial expression** in x.

EXAMPLE 3

POLYNOMIAL EXPRESSIONS	EXPRESSIONS THAT ARE NOT POLYNOMIALS
$10, 5x, -3x^2 + 2,$ $4x^3 + 7x^2 - 3$, and $\frac{5}{3}x^4$	$\frac{3}{x}$, $\sqrt{x}$, and $2x^3 + \frac{1}{x^3}$

The polynomial expression $4x^3 + 7x^2 - 3$ is said to be written in **descending order** because the term having the largest exponent is written first, the term having the next largest exponent is written second, etc. The same polynomial written in **ascending order** is

$$-3 + 7x^2 + 4x^3.$$

The expressions $\frac{3}{x}$ and $\sqrt{x}$ are not polynomial expressions in x because the variable, x, appears in a denominator and under a radical, respectively. Later in this chapter, you will learn that $\frac{3}{x} = 3x^{-1}$ and $\sqrt{x} = x^{1/2}$. Variables in polynomial expressions can only have nonnegative integer exponents.

Polynomial expressions can be classified by the number of terms that are contained in the expression.

Terminology

- A **monomial** is a single term that consists of a constant or a constant times a variable or variables raised to nonnegative integer powers, such as -3, $2x^4$, and $\frac{1}{2}s^2$.
- A **binomial** is a polynomial that has two nonzero terms, such as $4x^3 + 2x$ and $3t - 4$.
- A **trinomial** is a polynomial that has three terms, such as $3x^4 - 5x^2 + 10$ and $5x^2 + 3x - 4$.

9. a. Write yes if the expression is a polynomial in x. Write no if the expression is not a polynomial.

$3x - 5$	yes; binomial	$5x^3 - 2x + 7$	yes; trinomial
$\frac{5x}{3}$	yes; monomial	$\sqrt{x + 10}$	no
$\frac{2}{x^2} - 8$	no		

b. In part a, classify any polynomial expressions as a monomial, binomial, or trinomial.

EXAMPLE 4 ***Examples of polynomial and nonpolynomial functions are given in the following table:***

POLYNOMIAL FUNCTIONS	NONPOLYNOMIAL FUNCTIONS
$y = 10, y = 5x$	$f(x) = \frac{3}{x}$
$f(x) = -3x^2 + 2$	$g(x) = \sqrt{x}$
$y = 4x^3 + 7x - 3$	$h(x) = 2x^3 + \frac{1}{x^3}$
$g(x) = \frac{5}{3}x^4$	$y = 3x^{-2}$

10. Write yes if the equation defines a polynomial function. Write no if it does not.

a. $y = 5x^2 + 2x - 1$

yes

b. $f(x) = 3x + \frac{1}{x}$

no

c. $g(m) = 1.75m - 7$

yes

d. $R(t) = \sqrt{t} + 7$

no

Addition and Subtraction of Polynomial Expressions

Operations with polynomial functions (such as addition and subtraction) involve operations with polynomial expressions. Example 5 demonstrates how to perform these operations.

EXAMPLE 5 ***Consider the polynomial functions f and g defined by***

$$f(x) = 2x^2 + 3x - 5 \quad \text{and} \quad g(x) = -x^2 + 5x + 1.$$

Determine each of the following:

a. $f(x) + g(x) = (2x^2 + 3x - 5) + (-x^2 + 5x + 1)$ Remove parentheses.

$= 2x^2 + 3x - 5 - x^2 + 5x + 1$ Combine like terms.

$= x^2 + 8x - 4$

b. $f(x) - g(x) = (2x^2 + 3x - 5) - (-x^2 + 5x + 1)$ Change the sign of each term of the polynomial being subtracted.

$= 2x^2 + 3x - 5 + x^2 - 5x - 1$

$= 3x^2 - 2x - 6$

c. $-5f(x) = -5(2x^2 + 3x - 5)$ Apply the distributive property.

$= -10x^2 - 15x + 25$

LC LEARNING CATALYTICS

Given $f(x) = 3x^2 - 5x + 7$ and $g(x) = -2x^2 + 6x - 9$, determine $f(x) + g(x)$.

11. Given the polynomial functions g and h, defined by

$$g(x) = 4x^2 - 3x + 10, \qquad h(x) = -3x^2 + 5x - 2,$$

determine each of the following:

a. $g(x) + h(x)$ $g(x) + h(x) = x^2 + 2x + 8$

b. $g(x) - h(x)$ $g(x) - h(x) = 7x^2 - 8x + 12$

c. $-2g(x)$ $-2g(x) = -8x^2 + 6x - 20$

SUMMARY Activity 2.1

1. Given two functions, f and g, the **sum function**, s, is defined by

$$s(x) = f(x) + g(x)$$

and the **difference function**, d, is defined by

$$d(x) = f(x) - g(x).$$

2. Any expression that is formed by adding terms of the form ax^n, where a is a real number and n is a nonnegative integer, is called a **polynomial expression** in x.

3. A **monomial** is a polynomial with one term. A **binomial** is a polynomial with two terms. A **trinomial** is a polynomial with three terms. Polynomials having more than three terms are not given special names.

4. A **polynomial function** is any function defined by an equation of the form $y = f(x)$, where $f(x)$ is a polynomial expression in x. For example,

$$y = \underbrace{2x^3 + 5x^2 - x + 1}_{f(x),\text{ a polynomial}}.$$

EXERCISES Activity 2.1

1. You are a financial planner. In an effort to attract new customers, you sponsor a dinner at a local restaurant. The restaurant will charge \$100.00 for the banquet room, plus \$12.50 per person for each meal. You will pay these expenses yourself. From past experience, you can expect to make sales to approximately 15% of the people attending. You also know that the average sales you can expect from each new client is \$750, for which you receive a 13% commission. It is clear that your personal financial success from the event depends on how many people you can attract to this dinner.

   a. Complete the following table:

x, NUMBER OF PEOPLE ATTENDING	COST OF THE BANQUET HALL ($)	TOTAL MEAL COST ($)	TOTAL COST ($)
20	100	250	350
40	100	500	600
60	100	750	850
80	100	1000	1100
100	100	1250	1350

Exercise numbers appearing in color are answered in the Selected Answers appendix.

b. Determine a formula for the total cost of restaurant expenses as a function of x, the number of attendees. Represent the total cost by $C(x)$.

$C(x) = 12.50x + 100$

c. If 20 people attend, determine the number of new customers.

$0.15(20) = 3$

d. Determine the total sales of the new customers in part c.

$3(750) = \$2250$

e. What is your commission on the sales in part d?

$0.13(2250) = \$292.50$

f. Complete the following table. The results from parts c, d, and e are recorded.

x, NUMBER OF PEOPLE ATTENDING	NUMBER OF NEW CUSTOMERS	TOTAL SALES ($)	YOUR COMMISSION ($)
20	3	2250	292.50
40	6	4500	585.00
60	9	6750	877.50
80	12	9000	1170.00
100	15	11,250	1462.50

g. Determine a formula for the total commission that you can expect to generate from this dinner as a function of x, the number of attendees. Represent the total commission by $T(x)$.

$$T(x) = \underbrace{0.13(750)}_{\text{commission}}\underbrace{(0.15x)}_{\text{new customers}} = 14.625x$$

h. Combine the formulas in parts b and g to define a new function for the profit, P, that you can expect from your dinner. A basic business equation is

$$\text{profit} = \text{revenue} - \text{cost}.$$

In this situation, the revenue is the total commission.

$$P(x) = T(x) - C(x) = 14.625x - (12.50x + 100)$$
$$= 2.125x - 100$$

i. What is the practical domain of this new function?

$0 \le x \le$ the number of people the banquet room will accommodate.

j. Use your new function to determine how many people must attend for you to break even. Explain how you arrive at your decision.

Set $P(x) = 0$ and solve for x.

$$0 = 2.125x - 100$$
$$100 = 2.125x$$
$$x \approx 47.06$$

48 people must attend.

k. You hope to make a profit of $500 on the dinner. How many people must attend for you to meet this goal? Explain. Write an equation that can be solved to answer this question. Then show how to solve the equation.

Set the profit equal to 500 and solve for x.

$500 = 2.125x - 100$

$600 = 2.125x$

$x \approx 282.35$

283 people must attend.

2. Write yes if the expression is a polynomial in x. Write no if the expression is not a polynomial. If the expression is a polynomial, classify it as a monomial, binomial, or trinomial.

a. $5x^{-3} + 4$

no

b. $-3x^{10} - 2x^2 - 1$

yes, trinomial

c. $x^{1/2}$

no

d. x

yes, monomial

e. $\frac{5}{4x} - 8$

no

3. Suppose that f and g are defined by the following tables:

x	0	2	4	6	8	10
$f(x)$	3	−5	0	7	−1	4

x	0	2	4	6	8	10
$g(x)$	1	−1	1	−1	3	4

Complete the following table:

x	0	2	4	6	8	10
$f(x) + g(x)$	4	−6	1	6	2	8
$f(x) - g(x)$	2	−4	−1	8	−4	0

4. a. Suppose that $f(x) = 4x + 1$ and $g(x) = -2x + 4$. Determine an algebraic expression for $f(x) - g(x)$ by subtracting $g(x)$ from $f(x)$ and combining like terms.

$f(x) - g(x) = (4x + 1) - (-2x + 4) = 4x + 1 + 2x - 4 = 6x - 3$

b. Complete the following table, using f and g from part a:

x	$f(x)$	$g(x)$	$f(x) - g(x)$
0	1	4	−3
2	9	0	9
4	17	−4	21
6	25	−8	33

c. Use the table feature of your graphing calculator to complete the following table for four input values not used in part b.

(Answers will vary depending on the choices of x.)

x	$f(x)$	$g(x)$	$f(x) - g(x)$
1	5	2	3
−1	−3	6	−9
2.5	11	−1	12
3.2	13.8	−2.4	16.2

d. Do the results in part c agree with your understanding of the difference of two functions? Explain.

Yes. When two functions are subtracted, the outputs are subtracted.

5. The algebraic skills necessary for determining the algebraic form of the new functions are those of simplifying expressions and combining like terms. Simplify the following:

a. $(2x + 3) + (3x - 5)$

$5x - 2$

b. $(2x^2 - 3x + 1) - (x^2 - 6x + 9)$

$x^2 + 3x - 8$

c. $2(x + 9) - 3(x - 4)$

$-x + 30$

d. $14x - 9 - 3(x^2 + 2x - 2)$

$-3x^2 + 8x - 3$

e. $4(3x - 2) - (7 - 3x)$

$15x - 15$

f. $6x + 5 + 3(2 - 2x)$

11

g. $2x^2 + 5x - 3(3 - x^2)$

$5x^2 + 5x - 9$

h. $(5x - 2) - 2(3x^2 - 5x + 1)$

$-6x^2 + 15x - 4$

i. $2x + 5 - [3x - 4(5 - x)]$

$-5x + 25$

j. $7x + 2[3x - 2(4 - 5x)] + 6$

$33x - 10$

6. Given $f(x) = 3x - 5$ and $g(x) = -x^2 + 2x - 3$, determine a formula, in simplest form, for each of the following:

a. $f(x) + g(x)$

$-x^2 + 5x - 8$

b. $f(x) - g(x)$

$x^2 + x - 2$

c. $2f(x) + 3g(x)$

$-3x^2 + 12x - 19$

d. $f(x) - 2g(x)$

$2x^2 - x + 1$

7. Given $h(x) = 6$, $p(x) = 3 - 4x$, and $r(x) = 4x^2 - x - 6$, determine an expression, in simplest form, for each of the following:

a. $r(x) + h(x)$

$4x^2 - x$

b. $p(x) + r(x) - h(x)$

$4x^2 - 5x - 9$

c. $h(x) - p(x)$

$3 + 4x$

d. $r(x) + p(x) + h(x)$

$4x^2 - 5x + 3$

8. Given $f(x) = x^2 - 2$ and $g(x) = x + 4$, determine a value for each of the following:

a. $f(2) + g(2)$

$2 + 6 = 8$

b. $g(3) - f(3)$

$7 - 7 = 0$

c. $f(-5) + g(-5)$

$23 + (-1) = 22$

d. $f(-2) - g(-2)$

$2 - 2 = 0$

9. Suppose f and g are defined by the following tables:

x	-6	-4	-2	0	2	4
$f(x)$	50	16	-2	-4	10	40

x	-6	-4	-2	0	2	4
$g(x)$	49	25	9	1	1	9

a. Complete the following table:

x	-6	-4	-2	0	2	4
$f(x) - g(x)$	1	-9	-11	-5	9	31

b. If $f(x) = 2x^2 + 3x - 4$ and $g(x) = x^2 - 2x + 1$, determine an algebraic expression for $f(x) - g(x)$.

$f(x) - g(x) = x^2 + 5x - 5$

c. Check your answers in the table in part a by using the function you determined in part b.

The answers check.

ACTIVITY 2.2

The Dormitory Parking Lot

Multiplication of Polynomials; Multiplicative Properties of Exponents

OBJECTIVES

1. Multiply two binomials using the FOIL Method.
2. Multiply two polynomial functions.
3. Apply the property of exponents to multiply powers having the same base.

Your dormitory was recently renovated to bring it up to twenty-first-century standards, but the parking lot is still inadequate. The current lot has two rows with six parking spaces in each row.

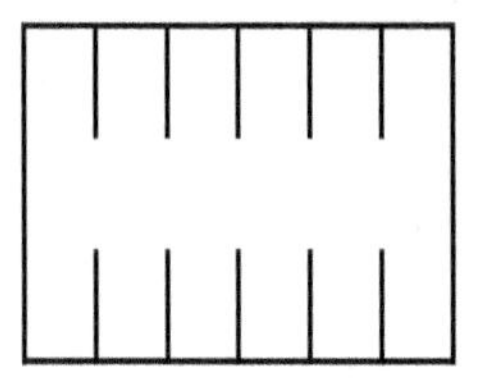

1. What is the current number of parking spaces? 12

Fortunately, the college has decided to expand this lot. It has also been determined that to maintain the relative shape of the lot, for every four cars added to a row, the college will add another row to the lot.

Starting with a geometric representation of the original lot, extend the length by four cars and the width by one row to obtain a geometric model of the new parking lot.

Number of Parking Spaces per Row

Number of Rows	6	+ 4
2	12	8
+ 1	6	4

2. Determine the number of parking spaces in each of the four sections of this diagram of the parking lot; then record the total in the appropriate place in the diagram.

3. Determine the total number of parking spaces in two different ways.

   a. Sum the parking spaces of the sections.

   $12 + 8 + 6 + 4 = 30$ parking spaces

   b. Multiply the number of cars in each row by the total number of rows.

   $10(3) = 30$ parking spaces

4. Let x represent the number of new rows added to the original parking lot.

   a. Determine an expression for the total number of rows when x new rows are added.

   $2 + x$

   b. Determine an expression for the number of cars in each row when x new rows are added. Remember, each new row requires an additional four cars to the length of each row.

   $6 + 4x$

5. Starting with a geometric representation of the current parking lot, extend the parking lot by x rows to obtain a geometric model of the new lot.

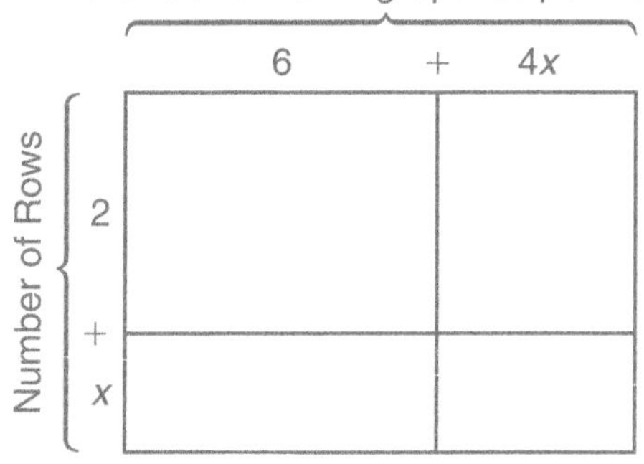

6. a. The number of parking spaces in each row in Problem 5 is a function of x. Write an equation for the number of parking spaces per row, $s(x)$, as a function of x.

$s(x) = 6 + 4x$

b. The number of rows in the parking lot is also a function of x. Write an equation for the number of rows, $n(x)$, in terms of x.

$n(x) = 2 + x$

c. The total number of parking spaces is a function of x. If $T(x)$ represents the total number of parking spaces in the new lot, use the results in parts a and b to write an equation that defines $T(x)$ as a function of x. Do not simplify.

$T(x) = (6 + 4x)(2 + x)$

DEFINITION

The total parking spaces function T in Problem 6c is called a **product function** because $T(x)$ is determined by the multiplication or product of two functions. The notation for a product function is

$$T(x) = s(x) \cdot n(x).$$

Note: when you determine the product of two functions, you multiply the outputs. The input is unchanged.

d. Complete the following table using the results from parts a–c:

x	$s(x)$	$n(x)$	$T(x)$
2	14	4	56
5	26	7	182
8	38	10	380

Multiplication of Binomials

The total parking space function is defined by $T(x) = s(x) \cdot n(x) = (6 + 4x)(2 + x)$. The product of the binomials $6 + 4x$ and $2 + x$ can be determined using a geometric model.

7. a. Return to the geometric model of the parking lot from Problem 5. Determine the number of parking spaces in each section, and fill in that number in each section of the geometric model.

Number of Parking Spaces per Row: $6 + 4x$; Number of Rows: $2 + x$

	6	4x
2	12	8x
x	6x	$4x^2$

b. Sum the parking spaces in all of the sections in the geometric model, and simplify the expression. What does this algebraic expression represent?

$12 + 8x + 6x + 4x^2 = 12 + 14x + 4x^2$

The expression represents the total number of parking spaces in the new parking lot when x rows are added to the lot.

As a result of Problem 7, you now know that

$$T(x) = (6 + 4x)(2 + x) = 12 + 14x + 4x^2.$$

The geometric model can be used to develop an algorithm (a process or procedure) for determining the product of two binomials such as $6 + 4x$ and $2 + x$. The sum of the parking spaces in Problem 7 is

$$2 \cdot 6 + 2 \cdot 4x + x \cdot 6 + x \cdot 4x.$$

This sum can be obtained from the terms of the binomial factors $6 + 4x$ and $2 + x$ as follows:

$$(6 + 4x) \cdot (2 + x) = \overbrace{6 \cdot 2}^{F} + \overbrace{6 \cdot x}^{O} + \overbrace{4x \cdot 2}^{I} + \overbrace{4x \cdot x}^{L}$$
$$= 12 + 6x + 8x + 4x^2$$

Combining like terms, you have $12 + 14x + 4x^2$. This procedure is called the FOIL Method. FOIL is an acronym for the sum of the products of the first, outer, inner, and last terms of the binomials. Essentially, you multiply each term of the first binomial by each term of the second binomial.

EXAMPLE 1 ***If $f(x) = 3x + 2$ and $g(x) = 2x - 5$, determine $f(x) \cdot g(x)$.***

SOLUTION

$$f(x) \cdot g(x) = (3x + 2)(2x - 5)$$

(First, Outer, Inner, Last)

$= 3x \cdot 2x + 3x(-5) + 2 \cdot 2x + 2(-5)$

$= 6x^2 - 15x + 4x - 10$

$= 6x^2 - 11x - 10$

Note that the product $(3x + 2)(2x - 5)$ can be represented by the following diagram:

	$2x$	-5
$3x$ +	$6x^2$	$-15x$
2	$4x$	-10

Therefore, $(3x + 2)(2x - 5) = 6x^2 - 15x + 4x - 10 = 6x^2 - 11x - 10$.

8. a. Given $f(x) = x + 7$ and $g(x) = x + 5$, determine a single polynomial expression for $f(x) \cdot g(x)$ by multiplying $(x + 7)(x + 5)$. Write your answer as a sum of terms.

$f(x) \cdot g(x) = (x + 7)(x + 5) = x^2 + 12x + 35$

b. Using the functions defined in part a, complete the following table:

x	$f(x)$	$g(x)$	$f(x) \cdot g(x)$
0	7	5	35
1	8	6	48
2	9	7	63
3	10	8	80
4	11	9	99

c. Use the table feature of your graphing calculator to complete the following table for four input values not used in part b.

(Answers will vary depending on choices of x.)

x	$f(x)$	$g(x)$	$f(x) \cdot g(x)$
5	12	10	120
6	13	11	143
7	14	12	168
8	15	13	195

d. Are the results found in part c consistent with your algebraic solutions in part b? Explain.

Yes. When functions are multiplied, the outputs are multiplied.

e. When you determine the product function, what do you multiply: domain values, range values, or both? Explain.

I multiply the range values for any given domain value, x.

Multiply Powers Having the Same Base

The volume, v (in cubic feet), of a partially cylindrical storage tank of liquid fertilizer is represented by the formula

$$v(r) = r^2(2.1r + 37.7),$$

where r is the radius (in feet) of the cylindrical part of the tank.

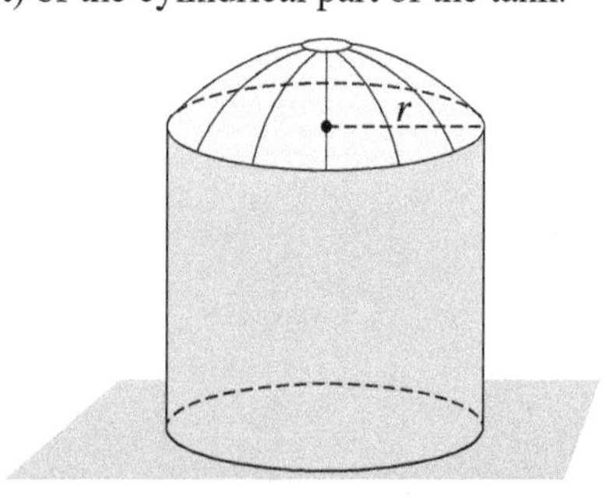

9. Determine the volume of the tank if its radius is 3 feet.

$v = 3^2(2.1(3) + 37.7) = 9(6.3 + 37.7) = 9(44) = 396$ cubic feet

Suppose you were asked to write the expression $r^2(2.1r + 37.7)$ as an equivalent expression without parentheses. Using the distributive property, you would multiply each term within the parentheses by r^2. The first product is $r^2(2.1r)$. What is r^2 times r?

Recall that in the expression r^2, the exponent 2 tells you that the base r is used as a factor two times. In the expression r, the exponent, 1, tells you that the base r is used as a factor once.

$$r^2 \cdot r = \underbrace{(r \cdot r) \cdot r}_{} = r^3$$

base r is used as a factor three times.

10. a. Complete the following table:

INPUT r	OUTPUT FOR $r^2 \cdot r$	OUTPUT FOR r^3
2	$4(2) = 8$	8
4	$16(4) = 64$	64
5	$25(5) = 125$	125

b. How does the table demonstrate that $r^2 \cdot r$ is equivalent to r^3?

The output for a specific input value is the same.

c. Consider the following products:

i. $x \cdot x^4 = x^5$ **ii.** $w^2 \cdot w^5 = w^7$ **iii.** $a^2 \cdot a^3 \cdot a^4 = a^9$

What pattern do you observe?

You add the exponents when you are multiplying, as long as the bases are the same.

Multiplication Property of Exponents

Let m and n be rational numbers.

To multiply powers of the same base, keep the base and add the exponents.

$$a^m \cdot a^n = a^{m+n}$$

11. Expressions for $p(x)$ and $q(x)$ are given in the following table. Fill in the last column of the table using a single power of x.

$p(x)$	$q(x)$	$p(x) \cdot q(x)$
x^2	x^4	x^6
$2x^3$	x	$2x^4$
$-3x^5$	$4x^2$	$-12x^7$
$5x^4$	$3x^4$	$15x^8$

12. Multiply $(-2a^5)(8b^3)(3a^2b)$. Explain the steps you used to determine this product.

$-48a^7b^4$. Multiply the coefficients -2, 8, and 3, and add the exponents on the same bases.

13. If $f(x) = 2x^2 + 3$ and $g(x) = 5x^3 - 2$, determine $f(x) \cdot g(x)$.

$f(x) \cdot g(x) = 10x^5 + 15x^3 - 4x^2 - 6$

When multiplying polynomials with more than two terms, FOIL cannot be applied. However, the geometric principles behind the FOIL Method can still be applied, or you can simply multiply each term of the first polynomial by each term of the second and collect like terms.

14. a. Multiply $(4x + 2)(x^2 - 4x + 3)$. Determine the appropriate products to complete the table below. Combine like terms, and write the final answer for the product in descending order of the exponents.

	x^2	$-4x$	$+3$
$4x$	$4x^3$	$-16x^2$	$12x$
2	$2x^2$	$-8x$	6

$4x^3 - 14x^2 + 4x + 6$

b. Multiply $(4x + 2)(x^2 - 4x + 3)$ by multiplying each term of the first polynomial by each term of the second. Combine like terms, and write the final answer for the product in descending order of the exponents.

$4x^3 - 16x^2 + 12x + 2x^2 - 8x + 6$

$4x^3 - 14x^2 + 4x + 6$

c. How do the final answers in parts a and b compare?

They are the same.

Special Products

15. Use FOIL or the rectangle method to determine the product $(x + 4)(x - 4)$.

$(x + 4)(x - 4) = x^2 - 4x + 4x - 16 = x^2 - 16$

Notice that the outer and inner products subtract out. The general form for this product is the following identity:

$$(a + b)(a - b) = a^2 - b^2$$

This expression, $a^2 - b^2$, is the **difference of the squares** of the binomial terms. For Problem 15, $a = x$ and $b = 4$, so $a^2 = x^2$ and $b^2 = 4^2 = 16$. Then $a^2 - b^2 = x^2 - 16$.

16. To use the identity $(a + b)(a - b) = a^2 - b^2$ to determine the product $(2x + 3)(2x - 3)$, complete the following steps:

a. Identify a and b.

$a = 2x$, $b = 3$

b. Identify a^2 and b^2.

$a^2 = (2x)^2 = 4x^2$,

$b^2 = 3^2 = 9$

c. Write the product as $a^2 - b^2$.

$(2x+3)(2x-3) = 4x^2 - 9$

17. Use FOIL or the rectangle method to determine the product $(x - 4)^2 = (x - 4)(x - 4)$.

$(x - 4)^2 = (x - 4)(x - 4) = x^2 - 4x - 4x + 16 = x^2 - 8x + 16$

Notice that the outer and inner products are equal. The general form for this product is the following identity:

$$(a + b)^2 = a^2 + 2ab + b^2$$

The expression $(a + b)^2$ is the **square of a binomial**. For Problem 17, $a = x$ and $b = -4$, so $a^2 = x^2$ and $b^2 = (-4)^2 = 16$. Then $(a + b)^2 = a^2 + 2ab + b^2 = x^2 + 2 \cdot x \cdot (-4) + 4^2 = x^2 - 8x + 16$.

18. To use the identity $(a + b)^2 = a^2 + 2ab + b^2$ to determine the square of the binomial $(2x + 3)^2$, complete the following steps:

a. Identify a and b.

$a = 2x$, $b = 3$

b. Identify a^2 and b^2.

$a^2 = (2x)^2 = 4x^2$, $b^2 = 3^2 = 9$

c. Write the product as $a^2 + 2ab + b^2$.

$(2x + 3)^2 = 4x^2 + 12x + 9$

19. Use the identity $(a + b)(a - b) = a^2 - b^2$ or $(a + b)^2 = a^2 + 2ab + b^2$ to determine the following products:

a. $(x + 6)(x - 6)$

$x^2 - 36$

b. $(x - 6)^2$

$x^2 - 12x + 36$

c. $(4x + 1)(4x - 1)$

$16x^2 - 1$

d. $(3x + 5)^2$

$9x^2 + 30x + 25$

LC LEARNING CATALYTICS

Given $f(x) = -2x + 7$ and $g(x) = 3x - 1$, determine $f(x) \cdot g(x)$.

SUMMARY Activity 2.2

1. To multiply any two polynomials, multiply each term of the first by each term of the second.

2. A common method used to multiply two binomials is the **FOIL Method**.

Step 1. Multiply the FIRST terms in each binomial.

Step 2. Multiply the OUTER terms.

Step 3. Multiply the INNER terms.

Step 4. Multiply the LAST terms.

Step 5. Sum the products in steps 1–4.

3. Given two functions, f and g, the **product function** is defined by $y = f(x) \cdot g(x)$.

4. To multiply powers of the same base, keep the base and add the exponents. Symbolically, this property of exponents is written as $a^m \cdot a^n = a^{m+n}$, where m and n are rational numbers.

5. Special products:

a. $(a + b)(a - b) = a^2 - b^2$

b. $(a + b)^2 = a^2 + 2ab + b^2$

EXERCISES Activity 2.2

1. a. You are drawing up plans to enlarge your square patio. You want to triple the length of one side and double the length of the other side. If x represents a side of your square patio, write an expression for the new area in terms of x.

   $(3x)(2x) = 6x^2$

   b. You discover from the plan that after doubling one side of the patio, you must cut off 3 feet from that side to clear a shrub. Write an expression in terms of x to represent the length of this side.

   $2x - 3$

   c. Use the result from part b to write an expression without parentheses to represent the new area of the patio. Remember that the length of the other side of the original square patio was tripled.

   $(2x - 3)(3x) = 6x^2 - 9x$

2. A rectangular bin has the following dimensions:

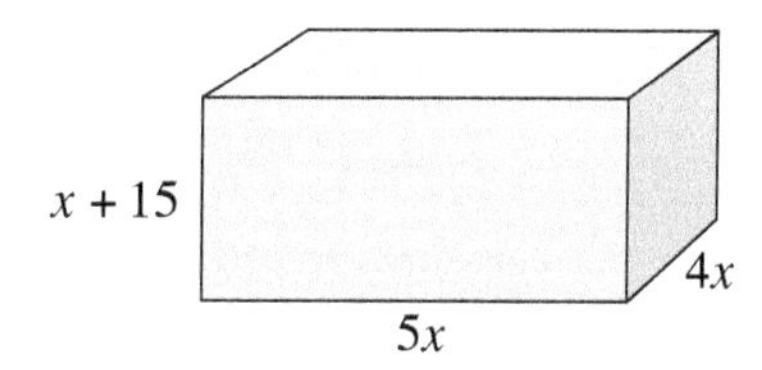

   a. Write an expression that represents the area of the base of the bin.

   $5x(4x) = 20x^2$

   b. Using the result from part a, write an expression that represents the volume of the bin. (Note that the volume is computed by multiplying the area of the base by the height.)

   $20x^2(x + 15) = 20x^3 + 300x^2$

3. You are working for a concert promoter, and she has assigned you the task of setting the ticket prices. She knows from experience that you will sell 3000 tickets if you price them at $40 each. She also knows that you will sell 100 more tickets for every dollar that you reduce the ticket price. Your job is to determine the ticket price that will maximize the revenue for the concert.

   a. Let x represent the number of $1 reductions in the price of the tickets. Write an equation for the price of each ticket, $P(x)$.

   $P(x) = 40 - x$

   b. Write an equation for the number of tickets sold, $N(x)$.

   $N(x) = 3000 + 100x$

   c. The revenue is the total amount of money collected. In this case, the revenue will be determined by multiplying the number of tickets sold by the cost of each ticket. Determine an equation for the total revenue, $R(x)$, as a function of the number of $1 reductions in price, x.

   $R(x) = (40 - x)(3000 + 100x)$

   d. What is the domain of the revenue function if the promoter has informed you that she will not sell the tickets for less than $30 each?

   The domain is $0 \le x \le 10$.

Exercise numbers appearing in color are answered in the Selected Answers appendix.

e. Complete the following table:

Number of \$1 Reductions, x	0	2	4	6	8	10
Price per Ticket, $P(x)$ (\$)	40	38	36	34	32	30
Number of Tickets Sold, $N(x)$	3000	3200	3400	3600	3800	4000
Total Revenue, $R(x)$ (\$)	120,000	121,600	122,400	122,400	121,600	120,000

f. How do the values in the fourth row of the table in part e relate to the values in the second and third rows?

The values in the fourth row are the product of the values in the second and third rows.

g. Using the table in part e, determine an appropriate window and graph the total revenue function on your calculator.

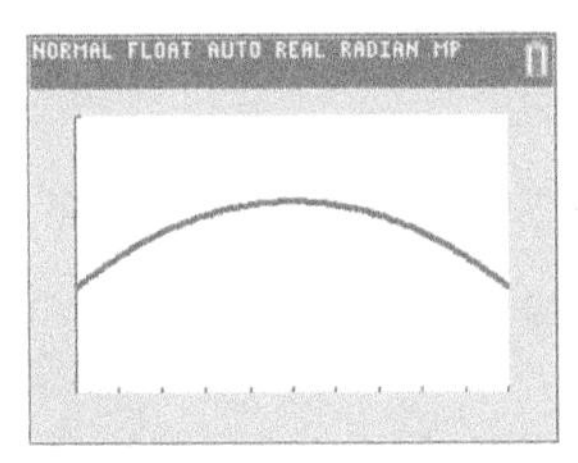

$X_{min} = 0$ $X_{max} = 10$ $Y_{min} = 117{,}000$ $Y_{max} = 125{,}000$

h. Use your calculator to estimate the price that will produce the maximum revenue as well as the maximum revenue.

The maximum point of the graph is (5, 122,500). This means the price that produces the maximum revenue is 40 – 5 or \$35. The maximum revenue is \$122,500.

i. Rewrite the revenue function by multiplying the factors and then combining like terms.

$R(x) = (40 - x)(3000 + 100x) = 120{,}000 + 4000x - 3000x - 100x^2$
$= -100x^2 + 1000x + 120{,}000$

j. Graph the new function from part i in the same window as part g. What do you see? Compare this graph with the graph in part g.

There is only one graph. The graphs are the same.

4. Use the property of exponents $a^m \cdot a^n = a^{m+n}$ to determine the following products:

a. $3^5 \cdot 3^7$

3^{12}

b. $t^4 \cdot t$

t^5

c. x^2y^5

x^2y^5

d. $(2z^4)(3z^8)$

$6z^{12}$

e. $(-2x)(3x^2)(-5x^3)$

$30x^6$

f. $(a^2b^2)(a^3b^4)$

a^5b^6

g. $x^{2n} \cdot x^n$

x^{3n}

5. Multiply $(x + 3)(x^2 + 3x - 5)$. Determine the appropriate products to complete the chart. Combine like terms, and write the final answer for this multiplication in descending order of the exponents.

	x^2	$3x$	-5
x	x^3	$3x^2$	$-5x$
3	$3x^2$	$9x$	-15

$(x + 3)(x^2 + 3x - 5) = x^3 + 3x^2 - 5x + 3x^2 + 9x - 15 = x^3 + 6x^2 + 4x - 15$

6. Multiply $(x^2 + 2x - 3)(2x^2 + 3x - 4)$. Determine the appropriate products to complete the chart. Combine like terms, and write the final answer for this multiplication in descending order of the exponents.

	$2x^2$	$3x$	-4
x^2	$2x^4$	$3x^3$	$-4x^2$
$2x$	$4x^3$	$6x^2$	$-8x$
-3	$-6x^2$	$-9x$	12

$2x^4 + 7x^3 - 4x^2 - 17x + 12$

7. Determine each product, and simplify the result.

a. $(3x + 2)(2x + 5)$

$6x^2 + 19x + 10$

b. $(3x - 2)(2x - 5)$

$6x^2 - 19x + 10$

c. $(x + 2)(4x - 3)$

$4x^2 + 5x - 6$

d. $(x - 2)(4x + 3)$

$4x^2 - 5x - 6$

8. Determine the following products, and simplify the results:

a. $(2x + 5)(x - 3)$

$2x^2 - x - 15$

b. $(4x + 3)(3x - 2)$

$12x^2 + x - 6$

c. $(x + 2)(x^2 + 4x - 3)$

$x^3 + 6x^2 + 5x - 6$

d. $(x^2 - 3x + 4)(2x^2 + x)$

$2x^4 - 5x^3 + 5x^2 + 4x$

e. $(x - 3)(2x^2 - 5x + 1)$

$2x^3 - 11x^2 + 16x - 3$

f. $(2x^2 + 5x)(2x - 6)$

$4x^3 - 2x^2 - 30x$

g. $(x^2 - 3x + 1)(3x^2 - 5x + 2)$

$3x^4 - 14x^3 + 20x^2 - 11x + 2$

h. $(2x^2 + 5x)(6 - 2x)$

$-4x^3 + 2x^2 + 30x$

i. $(x - 2)(x^2 + 2x + 4)$

$x^3 - 8$

9. a. Expand $(2x + 3)^2$.

$4x^2 + 12x + 9$

b. Multiply $(3x - 2)^2$.

$9x^2 - 12x + 4$

c. Multiply $(5x + 2)(5x - 2)$.

$25x^2 - 4$

d. Multiply $(x^2 + 5)(x^2 - 5)$.

$x^4 - 25$

e. Expand $(x + 4)^3$. Note that $(x + 4)^3 = (x + 4)(x + 4)^2$.

$x^3 + 12x^2 + 48x + 64$

f. After simplifying in parts c and d, the product contains only two terms. Explain why. (*Hint:* Compare the first terms with each other and the second terms with each other.)

The outer product and inner product are opposites. Their sum is 0.

10. a. Given $f(x) = x + 1$ and $g(x) = 2x - 3$, determine $f(x) \cdot g(x)$ by multiplying and combining like terms.

$f(x) \cdot g(x) = 2x^2 - x - 3$

b. Use f and g as defined in part a to complete the following table:

x	$f(x)$	$g(x)$	$f(x) \cdot g(x)$
0	1	−3	−3
1	2	−1	−2
2	3	1	3
3	4	3	12
4	5	5	25

c. Use your graphing calculator to plot all three functions, f, g, and the product of f and g. Use the trace or table feature to complete the following table for four input values not used in part b:

x	$f(x)$	$g(x)$	$f(x) \cdot g(x)$
6	7	9	63
7	8	11	88
8	9	13	117
9	10	15	150

(Answers will vary depending on the choices of x.)

ACTIVITY 2.3

Stargazing

Scientific Notation; Additional Properties and Definitions of Exponents

OBJECTIVES

1. Convert scientific notation to decimal notation.
2. Convert decimal notation to scientific notation.
3. Apply the property of exponents to divide powers having the same base.
4. Apply the definition of exponents $a^0 = 1$, where $a \neq 0$.
5. Apply the definition of exponents $a^{-n} = \frac{1}{a^n}$, where $a \neq 0$ and n is any positive integer.

On any clear evening, the sky is filled with millions of stars. Some of these are closer to Earth than others. Some are large. Some are small. All of them send light to us. The speed of light through the universe is constant. Light travels at a speed of approximately 300,000 kilometers per second.

1. How many kilometers does light travel in 1 minute?

   300,000(60) = 18,000,000 km

2. How many kilometers does light travel in 1 hour?

   18,000,000(60) = 1,080,000,000 km

3. How many kilometers does light travel in 1 day?

   1,080,000,000(24) = 25,920,000,000 km

The result of Problem 3 displayed on the TI-84 Plus C is 2.592E10. This is the way your calculator displays a very large number in **scientific notation**.

DEFINITION

Scientific notation is a convenient way to write a very large (or small) number. A positive number is written in scientific notation as a number (the base) between 1 and 10 times a power of 10. A negative number is written as a number (the base) between -10 and -1 times a power of 10.

EXAMPLE 1 ***Convert 3.2×10^3 and -9.8×10^7 from scientific notation to decimal notation.***

SOLUTION

$3.2 \times 10^3 = 3.2 \times 1000 = 3200$

$-9.8 \times 10^7 = -9.8 \times 10{,}000{,}000 = -98{,}000{,}000$

Notice that to convert a number whose absolute value is greater than 1 from scientific notation to decimal notation, you move the decimal point of the base to the right the number of decimal places being indicated by the exponent of the power of 10.

4. Convert the following numbers from scientific notation to decimal notation:

   a. 2.23×10^4 = 22,300

   b. -4.78×10^6 = −4,780,000

   c. 8.37×10^{12} = 8,370,000,000,000

You can use the TI-84 Plus C to check your answers. To input 2.23×10^4, type in the base, 2.23, press [2nd][,] to access the EE command, and then press [4][ENTER]. Your display should look like this:

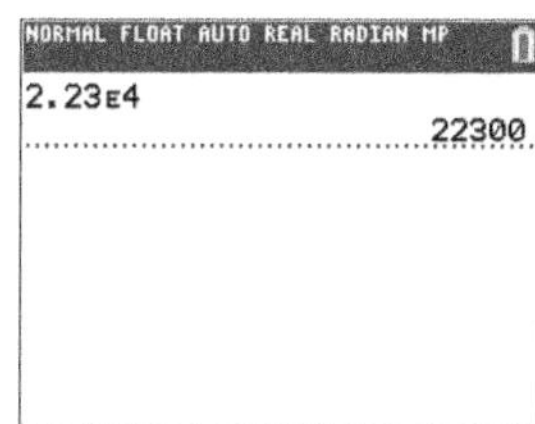

The EE button, which displays as E on the screen, is followed by the exponent of the power of 10.

The calculator display for converting -4.78×10^6 is

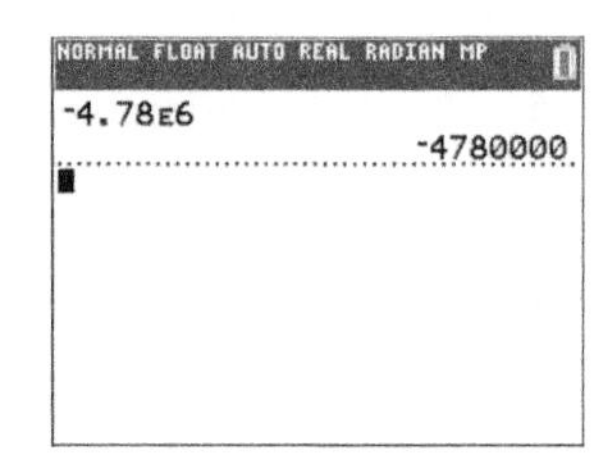

Because the calculator displays 10 digits at most, it will not convert 8.37×10^{12} to a decimal number. Try it.

Decimal Notation to Scientific Notation

In English units, light travels approximately 5,880,000,000,000 miles in 1 year. This distance is called the light-year. If you enter 5,880,000,000,000 into the calculator, it converts the number to scientific notation and returns 5.88E12.

5. a. Write the number 5,880,000,000,000 in scientific notation.

5.88×10^{12}

b. Describe the process the calculator used to convert 5,880,000,000,000 into 5.88E12.

The decimal point was moved 12 places to the left, then the calculator attached E12.

In general, to convert a number whose absolute value is greater than 1 from decimal notation to scientific notation, move the decimal point to the immediate right of the first nonzero digit. Count the decimal places moved. This number is the exponent of the power of 10.

EXAMPLE 2 ***Convert* 345,000,000 *into scientific notation.***

SOLUTION

Because the first nonzero digit is 3, the decimal part of the number is 3.45. Because the decimal point needs to be moved eight places to produce 3.45, the exponent on the 10 is 8. Therefore, $345{,}000{,}000 = 3.45 \times 10^8$ in scientific notation.

6. Convert the following decimal numbers into scientific notation:

a. 7,605,000,000,000

7.605×10^{12}

b. $-98{,}300{,}000$

-9.83×10^7

Division Property of Exponents

Light travels at a speed of approximately 300,000 kilometers per second or approximately 25,920,000,000 kilometers per day. The nearest star to Earth other than the Sun is Proxima Centauri, which is approximately 39,740,000,000,000 kilometers from Earth. How long does it take light from Proxima Centauri to reach Earth?

The answer to the question is determined by dividing 39,740,000,000,000 kilometers by 25,920,000,000 kilometers per day. If you do this with your calculator, it will produce the following screen.

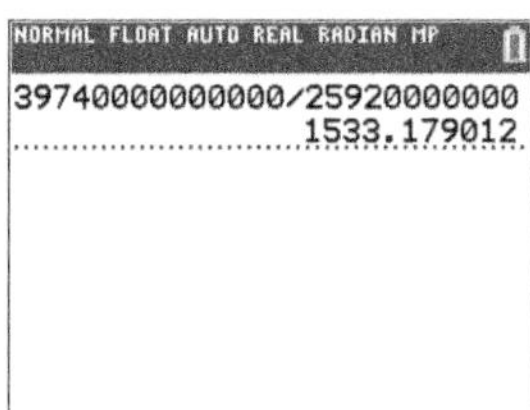

7. Change the mode of the calculator to Scientific Notation (Sci). Now divide 39,740,000,000,000 kilometers per day by 25,920,000,000 kilometers by entering both numbers into your calculator in scientific notation. How does your result compare with the result performed in Normal mode?

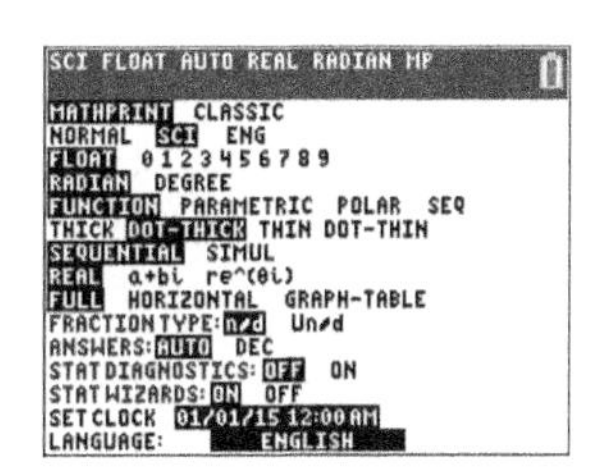

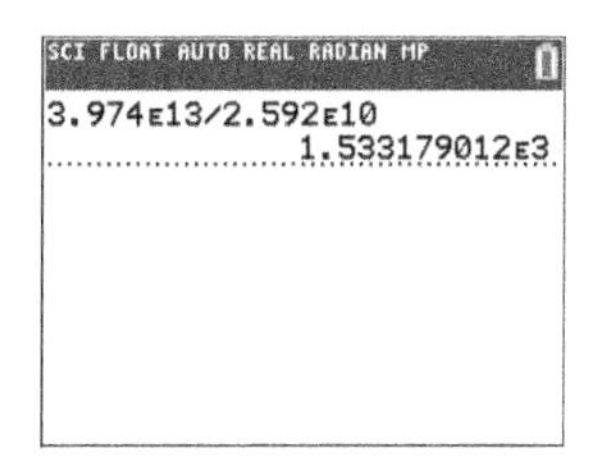

The result is equivalent, but it is in scientific notation.

8. Look at this calculation in scientific notation more closely.

$$\frac{3.974 \times 10^{13}}{2.592 \times 10^{10}} = 1.533179012 \times 10^3$$

a. The 1.533179012 is a result of dividing 3.974 by 2.592. Check this result with your calculator.

It checks.

b. The 10^3 is a result of dividing 10^{13} by 10^{10}. This means that $\frac{10^{13}}{10^{10}} = 10^3$. Note that

$$\frac{10^{13}}{10^{10}} = \frac{10 \times 10 \times 10 \times 10 \times 10 \times 10 \times 10 \times 10 \times 10 \times 10 \times 10 \times 10 \times 10}{10 \times 10 \times 10 \times 10 \times 10 \times 10 \times 10 \times 10 \times 10 \times 10} = 10^3.$$

Rather than expanding the powers and dividing out the common factors, you can obtain the exponent 3 from the exponents of the powers you are dividing. You simply subtract the exponent in the denominator, 10, from the exponent in the numerator, 13.

Problem 8b demonstrates another important property of exponents.

Division Property of Exponents

Let m and n be rational numbers. To divide powers of the same base, keep the base and subtract the exponents

$$\frac{a^m}{a^n} = a^{m-n}, \text{ where } a \neq 0.$$

EXAMPLE 3

a. $\frac{3^5}{3^2} = 3^{5-2} = 3^3$

b. $\frac{x^{15}}{x^9} = x^{15-9} = x^6$

c. $\frac{15t^6}{3t^2} = \frac{15}{3} \cdot \frac{t^6}{t^2} = 5 \cdot t^{6-2} = 5t^4$

d. $\frac{-4a^7}{2a^5} = -2a^2$

e. $\frac{x^9}{y^5}$ ***cannot be simplified because the bases x and y are different.***

LC LEARNING CATALYTICS

Use the division property to simplify $\frac{14t^{13}}{2t^7}$

9. Use the division property to simplify the following expressions:

a. $\frac{8^9}{8^4} = 8^5$ **b.** $\frac{x^6}{x} = x^5$ **c.** $\frac{10w^8}{4w^5} = \frac{5w^3}{2}$

d. $\frac{6t^{13}}{2t^7} = 3t^6$ **e.** $\frac{5^2}{5^2} = 1$

The result of Problem 9e using the division property is $\frac{5^2}{5^2} = 5^{2-2} = 5^0$. If you did the same problem by first writing 5^2 as 25, the result would be $\frac{5^2}{5^2} = \frac{25}{25} = 1$. Therefore, it must be true that $5^0 = 1$. In the same way, it can be shown that $2^0 = 1$, $10^0 = 1$, etc. This leads to the following definition.

DEFINITION

Zero Exponents

$a^0 = 1$ *if* $a \neq 0$.

EXAMPLE 4

a. $16^0 = 1$ **b.** $\left(\frac{3}{x}\right)^0 = 1, x \neq 0$

c. $(3x)^0 = 1$ provided that $x \neq 0$. Note that $3x^0 = 3 \cdot 1 = 3$.

d. $5(x + 3)^0 = 5, \ x \neq -3$

10. Simplify the following expressions. Assume that $x \neq 0$.

a. 7^0

1

b. $2x^0$

2

c. $(5x)^0$

1

d. $\left(\frac{4}{x}\right)^0$

1

e. $-3(x^2 + 4)^0$

-3

Negative Integer Exponents

Light travels at a rate of 25,920,000,000 kilometers per day. The Sun is 149,600,000 kilometers from Earth. You have determined that it takes 1533 days for light to travel from the second-nearest star, Proxima Centauri, to Earth. How many days does it take for light from the nearest star, the Sun, to travel to Earth?

To answer that question, divide 149,600,000 kilometers by 25,920,000,000 kilometers per day. If you convert both numbers to scientific notation and work in Sci mode on your calculator, your results should resemble the following:

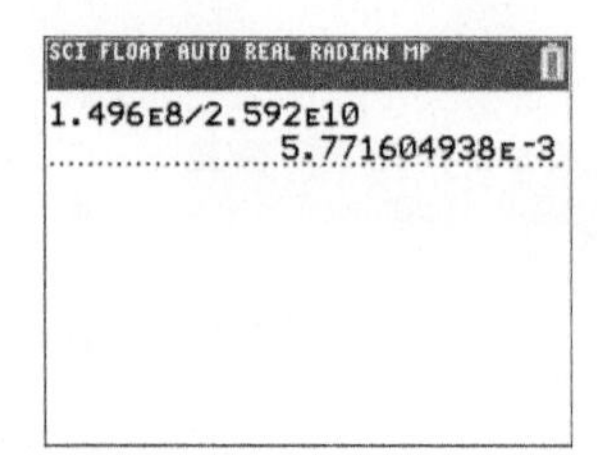

This says that light travels from the Sun to Earth in 5.772×10^{-3} days. If you perform the same calculation in Normal mode, your results should resemble the following:

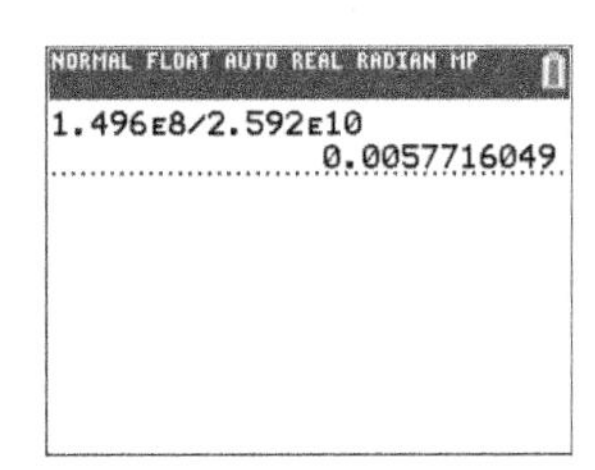

Therefore, you know that 5.772×10^{-3} days $\approx$ 0.005772 days.

11. Describe in your own words how to convert a number such as 5.772×10^{-3} (written in scientific notation) to its equivalent representation 0.005772 (written in decimal notation).

(Answers will vary.) The digits of the number will not change. The decimal point will be moved the same number of places indicated by the exponent on 10. If the exponent is positive, the decimal point is moved to the right. If the exponent is negative, the decimal point is moved to the left.

Therefore, to convert a number in scientific notation with a negative exponent, $-n$, to decimal notation, move the decimal point n places to the left.

EXAMPLE 5 ***Convert 6.3×10^{-3} and -17.7×10^{-4} to decimal notation.***

SOLUTION

a. $6.3 \times 10^{-3} = 0.0063$

b. $-17.7 \times 10^{-4} = -0.00177$

You can check these results by entering the numbers into your calculator in scientific notation as long as your calculator is in Normal mode.

12. Convert the following to decimal notation:

a. $5.61 \times 10^{-5} = 0.0000561$

b. $9.071 \times 10^{-7} = 0.0000009071$

Look again at 5.772×10^{-3} days $=$ 0.005772 days. You know that

$$\begin{aligned} 0.005772 &= 5.772 \times 0.001 \\ &= 5.772 \times \frac{1}{1000} \\ &= 5.772 \times \frac{1}{10^3}. \end{aligned}$$

Therefore, $5.772 \times 10^{-3} = 5.772 \times \frac{1}{10^3}$. It follows that $10^{-3} = \frac{1}{10^3}$. This leads to the following definition of negative exponents.

DEFINITION

Negative Exponents

If $a \neq 0$ and n is a rational number, then $a^{-n} = \left(\frac{1}{a}\right)^n = \frac{1}{a^n}$.

EXAMPLE 6 *Rewrite the following expressions using only positive exponents:*

a. $3^{-4} = \frac{1}{3^4} = \frac{1}{81}$

b. $(2x)^{-3} = \frac{1}{(2x)^3} = \frac{1}{8x^3}$

c. $x^{-1} = \frac{1}{x}$

d. $\frac{1}{x^{-4}} = x^4$

e. $3y^{-2} = \frac{3}{y^2}$

f. $\frac{-2a^{-3}}{b^{-2}} = \frac{-2b^2}{a^3}$

g. $x^{-3} \cdot x^{-5}$

Method 1. Apply the multiplication property of exponents first.

$x^{-3} \cdot x^{-5} = x^{-8} = \frac{1}{x^8}$

Method 2. Apply the definition of negative exponents first.

$x^{-3} \cdot x^{-5} = \frac{1}{x^3} \cdot \frac{1}{x^5} = \frac{1}{x^8}$

h. $\frac{x^{-3}}{x^4}$

Method 1. Apply the division property of exponents first.

$\frac{x^{-3}}{x^4} = x^{-3-4} = x^{-7} = \frac{1}{x^7}$

Method 2. Apply the definition of negative exponents first.

$\frac{x^{-3}}{x^4} = \frac{1}{x^3} \cdot \frac{1}{x^4} = \frac{1}{x^7}$

13. Rewrite the following expressions using positive exponents only:

a. 5^{-3} $\frac{1}{5^3}$

b. $(2z)^{-4}$ $\frac{1}{16z^4}$

c. $6y^{-5}$ $\frac{6}{y^5}$

d. $\frac{4}{x^{-1}}$ $4x$

e. $\left(\frac{x}{y}\right)^{-3}$ $\frac{y^3}{x^3}$

f. $\frac{x^3}{y^{-4}}$ x^3y^4

g. $x^{-4} \cdot x^{-2}$ $\frac{1}{x^6}$

h. $\frac{a^{-2}}{a^{-5}}$ a^3

Additional examples and exercises involving properties of exponents in Activity 2.3 are given in Appendix A.

SUMMARY Activity 2.3

1. In **scientific notation**, a positive number is written as a number (the base) between 1 and 10 times a power of 10. A negative number is written as a number (the base) between −10 and −1 times a power of 10.

2. To convert a number whose absolute value is greater than 1 from scientific notation to decimal notation, move the decimal point of the base to the right the number of decimal places given by the exponent on 10.

3. To convert a number whose absolute value is greater than 1 from decimal notation to scientific notation, move the decimal point to the immediate right of the first nonzero digit. Count the decimal places moved. This number is the exponent on 10.

4. To divide powers of the same base, keep the base and subtract the exponents: $\frac{a^m}{a^n} = a^{m-n}$, where $a \neq 0$ and m and n are rational numbers.

5. $a^0 = 1$, where $a \neq 0$.

6. To convert a number in scientific notation with a negative exponent, $-n$, to decimal notation, move the decimal point n places to the left.

7. If $a \neq 0$ and n is a real number, then $a^{-n} = \left(\frac{1}{a}\right)^n = \frac{1}{a^n}$.

EXERCISES Activity 2.3

1. According to the U.S. Department of the Treasury, in September 2017, the total currency estimated in circulation in the United States was $1,580,000,000,000. Write this number in scientific notation. Check your result by entering the number in your calculator.

   1.58×10^{12} 1.58E12

2. According to HIS Markit, an information services company based in London, in December 2016, there were an estimated 1.76×10^{10} active devices.

a. Write this number in standard notation.

17,600,000,000 active devices

b. According to DOMO, a computer software company based in Utah, 22,000,000,000 text messages are sent every day.

2.2×10^{10} text messages every day

c. Use scientific notation on your calculator to determine the average number of daily text messages per active device.

$\frac{2.2 \times 10^{10}}{1.76 \times 10^{10}} = 1.25$ text messages per device

3. a. A sextillion has 21 zeros. Write 3 sextillion in scientific notation.

3×10^{21}

b. The number 45,000,000,000,000,000 is read 45 quadrillion. Write this number in scientific notation.

4.5×10^{16}

c. Write the number 9×10^{27} in standard notation. The number will be read 9 octillion.

9,000,000,000,000,000,000,000,000,000

d. Use scientific notation to divide 9 octillion by 45 quadrillion. Use the properties of exponents.

$\frac{9 \times 10^{27}}{4.5 \times 10^{16}} = 2 \times 10^{11}$

Exercise numbers appearing in color are answered in the Selected Answers appendix.

4. a. One square inch is equivalent to approximately 0.000000159423 acre. Write this number in scientific notation.

1.59423×10^{-7}

b. 5.78704×10^{-4} cubic feet is equivalent to 1 cubic inch. Write this number in standard notation.

0.000578704 cubic feet

5. The amount of federal acreage in the United States is approximately 6.35×10^{8}. The total acreage in the United States is approximately 2.27×10^{9}.

a. Use scientific notation to write the ratio of the federal acreage to the total acreage.

$$\frac{6.35 \times 10^{8}}{2.27 \times 10^{9}}$$

b. Use the properties of exponents to simplify your answer.

2.797×10^{-1}

In Exercises 6–20, use the properties of exponents to simplify the following, where $x \neq 0$. Write your results with positive exponents only.

6. $\frac{3^5}{3^2}$

$3^3 = 27$

7. $\left(\frac{6}{x}\right)^0$

1

8. 2^{-5}

$\frac{1}{32}$

9. $10x^0$

10

10. $\left(\frac{1}{x}\right)^{-2}$

x^2

11. $4x^{-4}$

$\frac{4}{x^4}$

12. $(2x)^{-3}$

$\frac{1}{8x^3}$

13. $\frac{6x^8}{3x}$

$2x^7$

14. $\frac{9x^8}{3x^{12}}$

$\frac{3}{x^4}$

15. $\frac{6x^3y^5z^2}{10x^7yz^2}$

$\frac{3y^4}{5x^4}$

16. $x^{-3} \cdot x$

$\frac{1}{x^2}$

17. $(3x^{-2})(x^{-3})x^2$

$3x^{-3} = \frac{3}{x^3}$

18. $\frac{10x^4}{5x^{-3}}$

$2x^7$

19. $\frac{4a^0b^{-4}}{-8a^2b^{-1}}$

$\frac{-1}{2a^2b^3}$

20. $a^{-3}(4a^{-1})(-5a^7)$

$-20a^3$

ACTIVITY
2.4

The Cube of a Square
Rational Exponents and *n*th Roots

OBJECTIVES

1. Apply the property of exponents to simplify an expression involving a power to a power.
2. Apply the property of exponents to expand the power of a product.
3. Determine the *n*th root of a real number.
4. Write a radical as a base to a rational exponent, and write a base to a rational exponent as a radical.

Suppose that you are in charge of decorations for a Monte Carlo Night benefit for the math club. Some of the decorations will be in the form of dice (cubes) that you will make out of foam material. You decide that if the length of each edge of the small die is x units, you want the edges of each large die to be x^2 units.

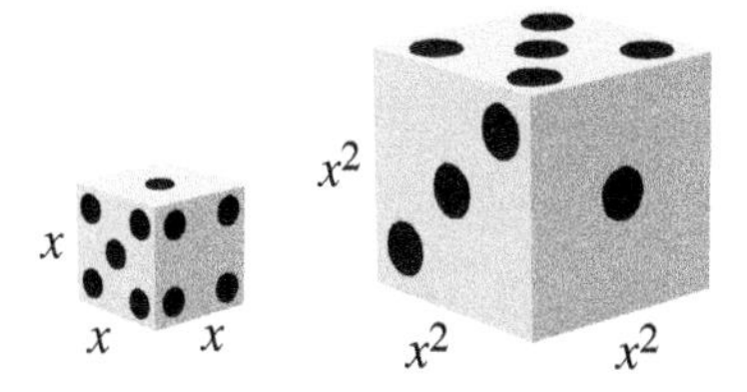

1. **a.** Determine the volume V of a die for the edge lengths given in the following table. Also evaluate the expression in the fourth column for the given lengths. The first row is done for you.

LENGTH OF EDGE, x (in.)	SMALL DIE, $V = x^3$ (Cu. in.)	LARGE DIE, $V = (x^2)^3$ (Cu. in.)	x^6
2	8	$(2^2)^3 = 4^3 = 64$	$2^6 = 64$
3	27	$(3^2)^3 = 9^3 = 729$	$3^6 = 729$
4	64	$(4^2)^3 = 16^3 = 4096$	$4^6 = 4096$

b. The volume of the large die is the cube of a square function defined by $f(x) = (x^2)^3$.

Use your graphing calculator to sketch a graph of $f(x) = (x^2)^3$ and $g(x) = x^6$.

c. How do these graphs compare?

The graphs are identical.

d. What do the results of parts a and c demonstrate about the relationship between $(x^2)^3$ and x^6?

The expressions $(x^2)^3$ and x^6 are equivalent.

The results of Problem 1 suggest that the expression $(x^2)^3$ can be written as x^6. Note also that $(x^2)^3$ indicates the base x^2 is used as a factor three times. Therefore,

$$(x^2)^3 = \underbrace{x^2 \cdot x^2 \cdot x^2}_{\text{The base } x^2 \text{ is used as a factor 3 times.}} = \underbrace{x^{2+2+2}}_{\text{Property 1 of exponents}} = (x^{2 \cdot 3}) = x^6$$

The simplified expression x^6 can be obtained from $(x^2)^3$ by multiplying the exponents 2 and 3. This leads to the following property of exponents.

Property of Exponents (Power to a Power)

If a is a nonzero real number and m and n are rational numbers, then

$$(a^n)^m = (a^m)^n = a^{mn}.$$

EXAMPLE 1

a. $(3^2)^3 = 3^{2 \cdot 3} = 3^6 = 729$; $(3^3)^2 = 3^{3 \cdot 2} = 3^6 = 729$

b. $(x^3)^5 = x^{3 \cdot 5} = x^{15}$; $(x^5)^3 = x^{5 \cdot 3} = x^{15}$

c. $(a^{-2})^3 = a^{-6} = \dfrac{1}{a^6}$

2. Use the properties of exponents to simplify each of the following:

a. $(t^3)^5$
t^{15}

b. $(y^2)^4$
y^8

c. $(3^2)^4$
$3^8 = 6561$

d. $2(a^5)^3$
$2a^{15}$

e. $x(x^2)^3$
$x(x^6) = x^7$

f. $-3(t^2)^4$
$-3t^8$

g. $(5xy^2)(3x^4y^5)$
$15x^5y^7$

EXAMPLE 2 ***Multiply the factors $3x^4(x^2)^3 \cdot 2x^3$.***

SOLUTION

$3x^4(x^2)^3 \cdot 2x^3$

$= 3x^4x^6 \cdot 2x^3$ Remove parentheses by applying a property of exponents (power to power).

$= 6x^4x^6x^3$ Multiply the numerical coefficients.

$= 6x^{13}$ Apply the multiplicative property of exponents to variable factors that have the same base.

3. Multiply the series of factors $2a^3(a^2)^4 7a^5$.

$2a^3 \cdot a^8 \cdot 7a^5$

$= 14a^{16}$

The Power of a Product

Consider the square of the expression $4x^3$, written as $(4x^3)^2$. Because the base for the squaring is $4x^3$, you can write $(4x^3)^2$ as

$$(4x^3)^2 = 4x^3 \cdot 4x^3 = 16x^6. \qquad (1)$$

↑
base

Note that $4x^3 \cdot 4x^3$ can be written equivalently as

$$4x^3 \cdot 4x^3 = 4 \cdot 4 \cdot x^3 \cdot x^3 = 4^2(x^3)^2. \qquad (2)$$

Comparing the results on lines 1 and 2, the expression $(4x^3)^2$ can be written as

$$(4x^3)^2 = 4^2 \cdot (x^3)^2.$$

Note that each factor in the base $4x^3$ (namely, 4 and x^3) can be raised to the second power. This illustrates another important property of exponents.

Property of Exponents (Power of a Product)

If a and b are real numbers and n is a rational number, then

$$(ab)^n = a^n b^n.$$

EXAMPLE 3

a. $(x^2y^3)^4 = (x^2)^4(y^3)^4 = x^8y^{12}$

b. $(-2a^5)^3 = (-2)^3(a^5)^3 = -8a^{15}$

c. $(3x^{-3})^4 = 3^4(x^{-3})^4 = 81x^{-12} = \dfrac{81}{x^{12}}$

4. Use the properties of exponents to simplify the following:

a. $(2x^4)^3$
$2^3(x^4)^3 = 8x^{12}$

b. $(-3x^3)^2$
$(-3)^2(x^3)^2 = 9x^6$

c. $(a^3b^{-3})^2$
$(a^3)^2(b^{-3})^2 = a^6b^{-6}$
$= \dfrac{a^6}{b^6}$

d. $x^2(x^4x^5)^2$
$x^2(x^9)^2 = x^2 \cdot x^{18} = x^{20}$

e. $3x(-2x^3)^4$
$3x(-2)^4(x^3)^4 = 3x(16)x^{12}$
$= 48x^{13}$

LC LEARNING CATALYTICS
Use the properties of exponents to simplify $4x(-2x^2)^3$.

5. Apply the properties of exponents to simplify the following:

i. $y = (5x^3)^2$
$y = 25x^6$

ii. $y = (-x^4)^3$
$y = -x^{12}$

Fractional Exponents

What does $a^{1/2}$ represent? The properties of exponents allow you to adopt a reasonable definition for rational (fractional) exponents such as $\frac{1}{2}$. Let us begin by reviewing the definition of square root.

DEFINITION

Let a represent a nonnegative real number, symbolically written as $a \geq 0$. The principal **square root** of a, denoted by $\sqrt{a}$, is defined as the nonnegative number that, when squared, produces a.

EXAMPLE 4

a. $\sqrt{9} = 3$ *because* $3^2 = 9$

b. $\sqrt{100} = 10$ *because* $10^2 = 100$

Note that because $\sqrt{9} = 3$ and $3^2 = 9$, it follows that $(\sqrt{9})^2 = 9$.

In general, $(\sqrt{a})^2 = a$ if $a \geq 0$.

Is there a relationship between $\sqrt{a}$ and $a^{1/2}$? To answer this question, you need to assume that there is an exponent, m, such that

$$a^m = \sqrt{a}, \qquad a \geq 0.$$

Squaring both sides of this equation, you have

$$(a^m)^2 = (\sqrt{a})^2$$
$$a^{2m} = a^1.$$

Because the bases are equal, the exponents must be equal. Therefore,

$$2m = 1$$
$$m = \frac{1}{2}.$$

From this result, the following definition is obtained.

DEFINITION

$a^{1/2} = \sqrt{a}$, where a is a real number and $a \geq 0$.

6. Evaluate each of the following if possible, and check the answer using your graphing calculator:

a. $36^{1/2}$

6

b. $-9^{1/2}$

-3

c. $(-9)^{1/2}$

is not a real number

d. $0^{1/2}$

0

Cube Roots

7. The volume of the following cube is 64 cubic inches:

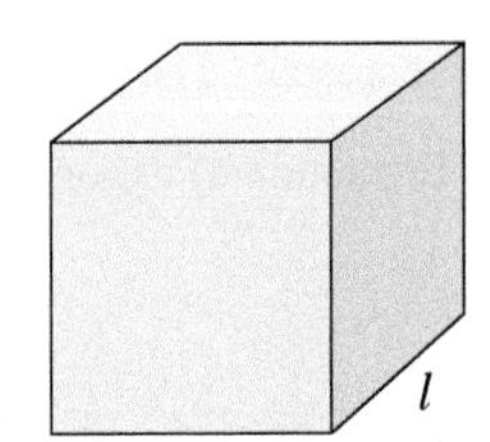

a. Determine the length, l, of one side (edge) of this cube.

4 in.

b. Explain how you obtained your answer.

The volume is the length of the edge cubed. The number that is cubed to produce 64 is 4.

The answer in Problem 7a is called the **cube root** of 64.

DEFINITION

The **cube root** of any real number a, denoted by $\sqrt[3]{a}$, is defined as the number that, when cubed, gives a.

EXAMPLE 5

a. $\sqrt[3]{8} = 2$ *because* $2^3 = 8$

b. $\sqrt[3]{125} = 5$ *because* $5^3 = 125$

c. $\sqrt[3]{-1000} = -10$ *because* $(-10)^3 = -1000$

Numbers such as 8, 125, and −1000 that have exact cube roots are called perfect cubes.

8. Evaluate each of the following, and check the answer using your graphing calculator. See Appendix C for help in determining cube roots on the TI-84 Plus C.

a. $\sqrt[3]{1000}$

10

b. $\sqrt[3]{0}$

0

c. $\sqrt[3]{-8}$

−2

d. $\sqrt[3]{100}$ (nearest tenth)

4.6

Just as $\sqrt{a}$, where $a \geq 0$, can be written equivalently as $a^{1/2}$, the cube root of a real number a can be written as $a^{1/3}$.

Consider the following argument. If $\sqrt[3]{a}$ can be written in exponential form,

$a^m = \sqrt[3]{a}$ — Raise both sides to the third power.

$(a^m)^3 = (\sqrt[3]{a})^3$ — Simplify both sides.

$a^{3m} = a^1$ — Because the bases are equal, the exponents must be equal as well.

$3m = 1$

$m = \frac{1}{3}$

Therefore, $\sqrt[3]{a} = a^{1/3}$, the cube root of a real number a.

Similarly, $\sqrt[4]{a} = a^{1/4}$, the fourth root of a real number, a, such that $a \geq 0$, *and*

$\sqrt[5]{a} = a^{1/5}$, the fifth root of a real number a.

DEFINITION

In general, $\sqrt[n]{a} = a^{1/n}$, the nth root of a. The number a, called the **radicand**, must be nonnegative if n, called the **index**, is even.

9. Calculate each of the following, and then verify your answer using your graphing calculator. See Appendix C for determining nth roots on the TI-84 Plus C.

a. $\sqrt[4]{81}$

3

b. $32^{1/5}$

2

c. $\sqrt[5]{-32}$

−2

d. $-225^{1/4}$

−3.9

10. a. Try to compute $\sqrt[4]{-81}$ using your graphing calculator. Explain what happens.

It tells me to quit. There is no real number n such that $n^4 = -81$.

b. Explain why the value of a in $\sqrt[n]{a}$, where n is even, cannot be negative.

There is no real number that can be raised to an even power and be negative.

11. Yachts that compete in the America's Cup must satisfy the International America's Cup Class rule that requires

$$L + 1.25\sqrt{S} - 9.8\sqrt[3]{D} \leq 16.296 \text{ meters,}$$

where L represents the yacht's length in meters,
S represents the rated sail area, in square meters, and
D represents the water displacement, in cubic meters.

a. Is a yacht with length 21.85 meters, sail area 305.5 square meters, and displacement 21.85 cubic meters eligible to compete? Explain.

$21.85 + 1.25\sqrt{305.5} - 9.8\sqrt[3]{21.85} = 16.301 > 16.296$

The yacht is not eligible to compete.

b. Explain why the units of your numerical answer in part a are meters.

L is in meters, the square root of square meters is meters, and the cube root of cubic meters is meters. Meters + meters + meters = meters.

Rational Exponents

The properties of exponents can be expanded to include rational exponents where the numerator is different from 1. For example,

$$\begin{aligned} 8^{2/3} &= 8^{2\cdot(1/3)} && \text{Apply the property of exponents.} \\ &= (8^2)^{1/3} && \text{Apply the definition of the 1/3 exponent.} \\ &= \sqrt[3]{8^2} \\ &= \sqrt[3]{64} \\ &= 4 \end{aligned}$$

In a similar fashion,

$$8^{2/3} = 8^{(1/3)\cdot 2} = (8^{1/3})^2 = (\sqrt[3]{8})^2 = 2^2 = 4.$$

Therefore, $8^{2/3}$ can be written equivalently as $\sqrt[3]{8^2}$ or $(\sqrt[3]{8})^2$. Note that 3, the denominator of the rational exponent 2/3, is the index. The numerator 2 indicates the power.

DEFINITION

$a^{p/q} = \sqrt[q]{a^p}$ or $a^{p/q} = (\sqrt[q]{a})^p$, where $a \geq 0$ if q is even and p and q are integers.

EXAMPLE 6

a. $(-27)^{2/3} = (\sqrt[3]{-27})^2 = (-3)^2 = 9$

b. $16^{3/4} = (\sqrt[4]{16})^3 = 2^3 = 8$

c. $8^{-2/3} = \dfrac{1}{(8^{\frac{1}{3}})^2} = \dfrac{1}{(\sqrt[3]{8})^2} = \dfrac{1}{2^2} = \dfrac{1}{4}$

12. Compute each of the following, and then verify the answer using your graphing calculator:

a. $25^{3/2}$
125

b. $(-8)^{2/3}$
4

c. $32^{4/5}$
16

d. $-16^{3/4}$
-8

e. $243^{2/5}$
9

f. $(-16)^{3/4}$
not a real number

13. Compute $7^{2/3}$ on your calculator, and explain why your answer is reasonable.

3.659 It is reasonable because if I raise it to the 3/2 power, the result is 7.

14. Write each of the following using fractional exponents:

a. $\sqrt[3]{x}$

$x^{1/3}$

b. $\sqrt[5]{x^3}$

$x^{3/5}$

c. $\sqrt[4]{x+1}$

$(x+1)^{1/4}$

d. $\sqrt{xy}$

$(xy)^{1/2}$

15. Perform the indicated operations by applying the appropriate property of exponents.

a. $x^{1/2}x^{2/3}$

$x^{1/2+2/3} = x^{7/6} = \sqrt[6]{x^7}$

b. $(x^3)^{3/4}$

$x^{3\cdot(3/4)} = x^{9/4} = \sqrt[4]{x^9}$

c. $\dfrac{x^3}{x^{1/3}}$

$x^{3-1/3} = x^{8/3} = \sqrt[3]{x^8}$

16. Determine the domain of each of the following:

a. $f(x) = \sqrt{x}$

$x \geq 0$

b. $g(x) = \sqrt[3]{x}$

all real numbers

17. If $f(x) = \sqrt{x+2}$, then determine each of the following:

a. $f(-2)$

0

b. $f(0)$

$\sqrt{2} \approx 1.414$

c. $f(7)$

3

d. $f(-6)$

not a real number

18. If $g(x) = \sqrt[3]{x-5}$, then determine each of the following:

a. $g(5)$

0

b. $g(13)$

2

c. $g(-3)$

-2

19. The area of the base of a cube is related to the volume of the cube by the formula

$$A = V^{2/3}.$$

Determine the base area of a cube having volume 216 cubic inches.

$A = (216)^{2/3} = 36$ square inches

Additional examples and exercises involving properties of exponents are given in Appendix A.

SUMMARY Activity 2.4

1. If a is a real number and m and n are rational numbers, then $(a^m)^n = a^{mn}$.
2. If a and b are real numbers and n is a rational number, then $(ab)^n = a^n b^n$.
3. Let a represent a nonnegative number, symbolically written as $a \geq 0$. The **principal square root** of a, denoted by $\sqrt{a}$, is defined as the nonnegative number that, when squared, produces a.

4. $(\sqrt{a})^2 = a$, if $a \geq 0$.

5. $a^{1/2} = \sqrt{a}$, where $a \geq 0$.

6. $\sqrt[n]{a} = a^{1/n}$, the nth root of a. The number a, called the **radicand**, must be nonnegative if n, called the **index**, is even.

7. $a^{p/q} = \sqrt[q]{a^p}$ or $a^{p/q} = (\sqrt[q]{a})^p$, where $a \geq 0$ if q is even and p is a positive integer.

EXERCISES Activity 2.4

1. Simplify each expression by applying the properties of exponents. Write your results with positive exponents only.

a. $(x^3)^6$
x^{18}

b. $(2x^5)^2$
$4x^{10}$

c. $(-3x^2)^3$
$-27x^6$

d. $(x^4)^{-3}$
$x^{-12} = \dfrac{1}{x^{12}}$

e. $(-4a^{-5})^2$
$(-4)^2(a^{-5})^2 = \dfrac{16}{a^{10}}$

f. $(a^2b^{-2}c^3)^{-4}$
$a^{-8}b^8c^{-12} = \dfrac{b^8}{a^8c^{12}}$

g. $-(x^6)^3$
$-x^{18}$

h. $(-x^3)^6$
x^{18}

2. Compute each of the following quantities:

a. $100^{1/2}$
10

b. $144^{1/2}$
12

c. $64^{1/3}$
4

d. $64^{4/3}$
256

e. $5^{2/5}$
1.903654

f. $(-8)^{2/3}$
4

g. $25^{-1/2}$
$\dfrac{1}{\sqrt{25}} = \dfrac{1}{5}$

h. $27^{-2/3}$
$\dfrac{1}{(\sqrt[3]{27})^2} = \dfrac{1}{3^2} = \dfrac{1}{9}$

3. Simplify the following:

a. $x^{1/3} \cdot x^{3/4}$
$x^{1/3 + 3/4} = x^{13/12} = \sqrt[12]{x^{13}}$

b. $(x^{-1/3})^{-1/2}$
$x^{1/6} = \sqrt[6]{x}$

c. $\dfrac{x^{4/5}}{x^{1/3}}$
$x^{4/5 - 1/3} = x^{12/15 - 5/15} = x^{7/15} = \sqrt[15]{x^7}$

4. Write each of the following using fractional exponents:

a. $\sqrt{x}$

$x^{1/2}$

b. $\sqrt[4]{x^3}$

$x^{3/4}$

c. $\sqrt[3]{x + y}$

$(x + y)^{1/3}$

d. $\sqrt[5]{a^2b^3}$

$a^{2/5}\,b^{3/5}$

5. If $f(x) = \sqrt{x - 3}$, determine each of the following:

a. $f(28)$

5

b. $f(3)$

0

c. $f(-1)$

not a real number

6. If $g(x) = \sqrt[3]{x + 10}$, determine each of the following:

a. $g(-74)$

-4

b. $g(-10)$

0

c. $g(17)$

3

7. The length of time, t (in seconds), it takes the pendulum of a clock to swing through one complete cycle is a function of the length of the pendulum in feet. The relationship is defined by

$$t = f(L) = 2\pi\sqrt{\frac{L}{32}},$$

where L is the length of the pendulum.

a. Rewrite the formula using fractional exponents.

$$t = f(L) = 2\pi\left(\frac{L}{32}\right)^{1/2}$$

b. Determine the length of the cycle in time if the pendulum is 4 feet long.

$$t = 6.28\sqrt{\frac{4}{32}} \approx 2.22 \text{ sec.}$$

CLUSTER 1 What Have I Learned?

1. For any two functions f and g,

i. determine which of the following equations are true and explain why.

ii. determine which of the following equations are not always true and give an example to show why not.

a. $f(x) + g(x) = g(x) + f(x)$

(i) Equation a is true because addition of real numbers is commutative.

b. $f(x) - g(x) = g(x) - f(x)$

(ii) Let $f(x) = 2x^2 + 1$ and $g(x) = x - 1$. I will use these functions to show that the statement in part b is false.

$f(x) - g(x) = 2x^2 + 1 - (x - 1) = 2x^2 - x + 2$

$g(x) - f(x) = x - 1 - (2x^2 + 1) = -2x^2 + x - 2$

Clearly, these are not equal.

(Answers will vary depending on choice of f and g.)

c. $f(x) \cdot g(x) = g(x) \cdot f(x)$

(i) Equation c is true because multiplication of real numbers is commutative.

2. Given $f(x) = 2x - 3$ and $g(x) = 4$, for what values of x is $f(x) + g(x) = f(x) \cdot g(x)$?

$f(x) + g(x) = 2x + 1$ and $f(x) \cdot g(x) = 8x - 12$

$2x + 1 = 8x - 12$

$13 = 6x$

$x = \frac{13}{6}$

3. Given the defining equations for two functions, describe how to determine the output of the product function for a particular input value.

For any given x-value, multiply the values of the outputs of $f(x)$ and $g(x)$ to determine the product $f(x) \cdot g(x)$.

4. Explain the difference between 3^4 and 3^{-4}.

$3^4 = 81$ and $3^{-4} = 1/81$. The answers are reciprocals.

5. What will be the sign of the answer if you raise a positive base to a negative exponent?

The sign of the answer will be positive.

CLUSTER 2 How Can I Practice?

1. **a.** The College International Club is planning a holiday dinner. A banquet room has been reserved, and catering arrangements have been made for a total of $600, a fixed fee independent of the number of people who attend. The planning committee has decided that a price of $20 per couple is the most it will charge for tickets. What is the *least* number of tickets the committee needs to sell to break even?

   $20x - 600 = 0$

   $20x = 600$

   $x = 30$

   b. The committee also decides that once it has met its expenses, it will reduce the ticket charge by $0.50 per couple for each additional ticket (couple) above the break-even point you determined in part a. Let t represent the *additional* tickets sold. Write an equation to show the total number, N, of couples attending the banquet (i.e., the number of tickets sold) as a function of t. Call this function f. $N = f(t) = 30 + t$

   c. The charge per ticket can be represented by the function $C = g(t)$. Write an equation for the charge per ticket, C. $C = g(t) = 20 - 0.5t$

   d. Complete the following table for each of the functions $N = f(t)$ and $C = g(t)$:

t	0	2	4	6	8	10
$N = f(t)$	30	32	34	36	38	40
$C = g(t)$	20	19	18	17	16	15

   e. The total revenue obtained from the ticket sales is the total number of tickets sold multiplied by the charge per ticket. Use the output values from part d to complete the following table for the total revenue function, $R(t) = N \cdot C = f(t) \cdot g(t)$:

t	0	2	4	6	8	10
$R(t)$	600	608	612	612	608	600

   f. Use the results from parts b and c to determine a symbolic rule for $R(t)$.

   $R(t) = f(t) \cdot g(t) = (30 + t)(20 - 0.5t) = -0.5t^2 + 5t + 600$

   g. Use your graphing calculator to graph the total revenue function, $R(t)$, on the accompanying grid. What window values—Xmin, Xmax, Ymin, Ymax—do you use?

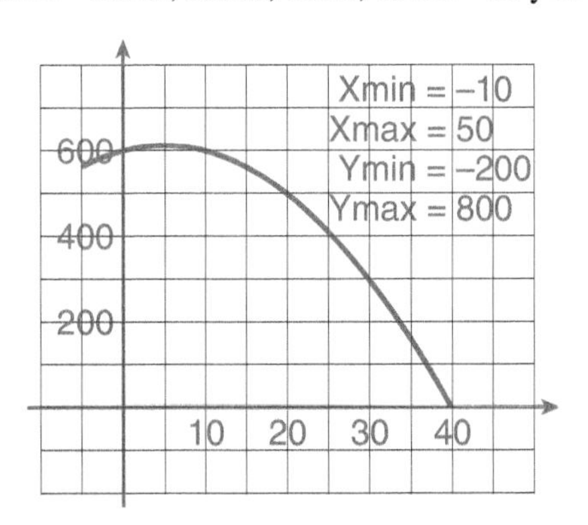

Answers to all How Can I Practice exercises are included in the Selected Answers appendix.

h. Determine the maximum revenue that can be obtained.

The maximum revenue is \$612.50 if 35 couples attend.

i. What is the total number of tickets that must be sold to obtain this maximum revenue? (Be careful.)

35 tickets must be sold to obtain the maximum revenue.

2. If $f(x) = x + 2$ and $g(x) = 2x - 3$, determine the following:

a. $f(x) + g(x)$

$3x - 1$

b. $f(x) - g(x)$

$-x + 5$

c. $f(x) \cdot g(x)$

$2x^2 + x - 6$

d. $f(3) - g(3)$

2

e. $f(-2) \cdot g(-2)$

0

f. $3 \cdot f(x)$

$3x + 6$

3. If $f(x) = x^2 - 2x + 1$ and $g(x) = x^2 + x - 4$, determine each of the following:

a. $f(x) - g(x)$

$-3x + 5$

b. $f(x) \cdot g(x)$

$x^4 - x^3 - 5x^2 + 9x - 4$

c. $f(-1) + g(-1)$

0

d. $2f(x) - 3g(x)$

$-x^2 - 7x + 14$

4. Perform the indicated operations and simplify.

a. $(4x + 5) + (x - 7)$

$5x - 2$

b. $(x^2 - 3x + 1) + (x^2 + x - 9)$

$2x^2 - 2x - 8$

c. $(x + 4) - (3x - 8)$

$-2x + 12$

d. $(3x^2 - 4x - 5) - (x^2 + 9x + 3)$

$2x^2 - 13x - 8$

e. $(5x^2 + 6x - 1) - (7x - 3)$

$5x^2 - x + 2$

5. Perform the indicated operations and simplify.

a. $x \cdot x^3$

x^4

b. $x^4 \cdot x^5$

x^9

c. $(2x^6)(3x^2)$

$6x^8$

d. $(xy^3)(x^4y^3z)$

x^5y^6z

e. $(2x^4y^5z^7)(5x^2z)$

$10x^6y^5z^8$

f. $(-3a^2b)(-2a)(-5a^2b^2)$

$-30a^5b^3$

6. Perform the indicated operations and simplify.

a. $(x - 2)(x - 5)$

$x^2 - 7x + 10$

b. $(4x - 3)(x + 7)$

$4x^2 + 25x - 21$

c. $(2x - 3)(2x + 3)$

$4x^2 - 9$

d. $(x - 2)(x^2 + 3x - 5)$

$x^3 + x^2 - 11x + 10$

e. $(2x + 1)(x^2 - x + 2)$

$2x^3 - x^2 + 3x + 2$

f. $3(x - 7) - 2(x^2 + 4x)$

$-2x^2 - 5x - 21$

g. $2x(x + 5) - 3x(4 - 3x)$

$11x^2 - 2x$

h. $3x^2 - (x^3 + 1) - x(x^4 - 2)$

$-x^5 - x^3 + 3x^2 + 2x - 1$

i. $(3x + 5)^2$

$9x^2 + 30x + 25$

j. $(2x - 7)^2$

$4x^2 - 28x + 49$

k. $(x + 4)^3$

$x^3 + 12x^2 + 48x + 64$

l. $(5x - 7)(5x + 7)$

$25x^2 - 49$

7. The tuition at a local state college was $5636 for the academic year ending in 2016. Since then tuition has increased approximately $160 per year. Let t represent the years since 2016 and $f(t)$ represent the tuition for any given year. At the same college, the cost for room and board was $12,660 for the year ending in 2016. And it has increased at a rate of $430 per year since. Let $g(t)$ represent the cost of room and board for any given year since 2016. The college fees have not changed over the years. Let $h(t)$ represent the college fees. Assume that the college fees remain at $650 per year.

a. Complete the following input–output table for the tuition function, $f(t)$:

t, Years since 2016	0	5	10	15	20
f(*t*), Cost of Tuition ($)	$5636	$6436	$7236	$8036	$8836

b. Write an equation that will give the cost of tuition for any year beginning in 2016.

$f(t) = 5636 + 160t$

c. Complete the following input–output table for the room and board, $g(t)$:

t, Years since 2016	0	5	10	15	20
g(*t*), Cost of Room and Board ($)	$12,660	$14,810	$16,960	$19,110	$21,260

d. Write an equation that will give the cost of room and board for any year beginning in 2016.

$g(t) = 12{,}660 + 430t$

e. Complete the following input–output table for the fees, $h(t)$:

t, Years since 2016	0	5	10	15	20
h(*t*), Cost of Fees ($)	$650	$650	$650	$650	$650

f. Write an equation that will give the cost of fees for any year beginning in 2016.

$h(t) = 650$

g. Use the equations in parts b, d, and f to determine a function, k, that will give the total cost of attending this college for any year after 2016. Write the function in simplest form.

$k(t) = 5636 + 160t + 12{,}660 + 430t + 650$ or $k(t) = 18{,}946 + 590t$

h. Use your graphing calculator to graph functions *f*, *g*, *h*, and *k*. Use the trace or table feature to complete the following table for four input values:

t, YEARS SINCE 2016	$f(t)$	$g(t)$	$h(t)$	$k(t)$
3	6116	13,950	650	20,716
12	7556	17,820	650	26,026
18	8516	20,400	650	29,566
25	9636	23,410	650	33,696

i. If the increase continues at the same rate, in what year will the total cost first equal or exceed $25,000?

$$18{,}946 + 590t > 25{,}000$$
$$590t > 6054$$
$$t > \frac{6054}{590} \approx 10.26$$

The cost will exceed $25,000 approximately eleven years after 2016 or 2027.

8. Functions *f* and *g* are defined by the following tables:

x	−4	−2	0	2	4	6
$f(x)$	49	9	−7	1	33	89

x	−4	−2	0	2	4	6
$g(x)$	32	12	0	−4	0	12

a. Complete the following table:

x	−4	−2	0	2	4	6
$f(x) - g(x)$	17	−3	−7	5	33	77

b. If $f(x) = 3x^2 - 2x - 7$ and $g(x) = x^2 - 4x$, determine an algebraic expression for $f(x) - g(x)$.

$f(x) - g(x) = 2x^2 + 2x - 7$

c. Check your answers in the table in part a by using the function you found in part b.

The answers check.

9. Use the properties of exponents to simplify the following, where $x \neq 0$. Write your results with positive exponents only.

a. $2x^{-3}$

$\frac{2}{x^3}$

b. $(-3x)^2$

$9x^2$

c. 3^{-4}

$\frac{1}{3^4}$

d. $\left(\frac{x}{5}\right)^0$

1

e. $\frac{4x^9}{2x^2}$

$2x^7$

f. $(-2x^3y^6)(3xy^2)$

$-6x^4y^8$

g. $4x^0$

4

h. $\frac{8x}{12x^5}$

$\frac{2}{3x^4}$

i. $\frac{10xy^3z^4}{2x^5yz^4}$

$\frac{5y^2}{x^4}$

j. $(3x^{-2})(-5x^{-4})(x)$

$\frac{-15}{x^5}$

k. $\frac{-4x^3}{2x^{-3}}$

$-2x^6$

l. $\frac{a^4b^{-2}c^{-5}}{a^0b^2}$

$\frac{a^4}{b^4c^5}$

10. Simplify the following by applying the appropriate properties of exponents:

a. $(x^2)^4$

x^8

b. $(xy)^3$

x^3y^3

c. $(2x^4y)^5$

$32x^{20}y^5$

d. $(3x^2)^3(2xy)^4$

$432x^{10}y^4$

e. $(-5)^2$

25

f. -5^2

-25

g. $(-2x)^2$

$4x^2$

h. $(-2x)^3$

$-8x^3$

i. $(x^2)^5(-2x^4)^3$

$-8x^{22}$

j. $25^{3/2}$

125

k. $64^{2/3}$

16

l. $(-27)^{1/3}$

-3

m. $\left(\frac{9}{16}\right)^{3/2}$

$\frac{27}{64}$

n. $4^{1/2}$

2

o. $(x^3y^6)^{2/3}$

x^2y^4

p. $x^{2/3}x^{1/2}$

$x^{7/6} = \sqrt[6]{x^7}$

q. $(8xy)^{1/3}$

$2x^{1/3}y^{1/3}$

11. a. The *Exxon Valdez* spilled oil in Prince William Sound, Alaska, when it ran aground in March 1989. It is estimated that 1.008×10^7 gallons of oil spilled. The British Petroleum–leased undersea well in the Gulf of Mexico began leaking oil on April 20, 2010. Approximately 18 times more crude oil spilled into the Gulf of Mexico than spilled into Prince William Sound. How much crude oil is estimated to have spilled into the Gulf of Mexico? Write your answer in standard form.

181,440,000 gal. $1.008 \times 18 \times 10^7 = 18.144 \times 10^7 = 1.8144 \times 10^8$

b. For insurance purposes, you need to determine how many square miles of hunting land you own. One square foot is equivalent to 0.00000003587006 square mile. If your hunting land measures approximately 6,000,000 square feet, how many square miles is this? Use scientific notation and the rules of exponents to determine how many square miles of land you own.

I own approximately 0.2 of a square mile.

c. The mean radius of the largest planet, Jupiter, is 43,441 miles. The formula for determining the volume of a sphere, V, is

$$V = \frac{4}{3}\pi r^3.$$

Determine the approximate volume, in cubic miles, of Jupiter. Express your answer in scientific notation rounded to four decimal places.

3.4339×10^{14} cubic miles

12. According to the U.S. population clock at the Census Bureau, the population estimate for December 30, 2017; is 326,963,000. The world population according to the population clock for that same day is 7,591,584,000. Write both of these numbers in scientific notation.

3.26963×10^8 7.591584×10^9

13. According to their respective websites, in September 2016, Princeton University's endowment was approximately $22,710,000,000 and the University of Chicago's endowment was approximately $7,550,000,000. Show that Princeton's endowment is three times the endowment of the University of Chicago using scientific notation. Which property of exponents did you use in your explanation?

$$\frac{2.27 \times 10^{10}}{7.55 \times 10^{9}} \approx 0.3 \times 10^{1} = 3$$

To divide powers with the same base, I kept the base and subtracted the exponents.

14. You plan to put in a rectangular patio that measures 5 feet by 7 feet. However, you neglected to include enough seating room around your patio table. Let x be the number of additional feet you will extend your plan in each direction.

a. Determine a formula for the area of the extended patio.

$A = 35 + 12x + x^2$

b. If x is 4 feet, by how much have you increased the area of the patio from that of the original plan?

$(7 + 4)(5 + 4) = 99$ or $35 + 12(4) + 4^2 = 99$; $99 - 35 = 64$ sq. ft.

c. Your patio table is round with a radius of r feet. You need to purchase an umbrella for the table with an overhang of 2 feet all around. Write an expression for the area the umbrella will cover.

$A = \pi r^2$ For the umbrella, $A = \pi(r + 2)^2$ or $A = \pi r^2 + 4\pi r + 4\pi$.

CLUSTER 2 Composition and Inverses of Functions

ACTIVITY 2.5

Inflated Balloons

Composite Functions

OBJECTIVES

1. Determine the composition of two functions.
2. Explore the relationship between $f(g(x))$ and $g(f(x))$.

The volume of an inflated balloon increases as the air temperature rises. The following table shows the data from experimental measurements for a particular balloon:

Temperature (°F)	32	39	42	45	50	58	63	68
Volume (cu. in.)	35.1	36.5	37.1	37.7	38.7	40.3	41.3	42.3

1. Treating volume as a function of temperature, is this relationship a linear function? Describe how you determined your answer.

   $\frac{37.1 - 35.1}{42 - 32} = \frac{2}{10} = 0.2, \frac{38.7 - 37.7}{50 - 45} = \frac{1}{5} = 0.2, \frac{42.3 - 40.3}{68 - 58} = \frac{2}{10} = 0.2$

   The relationship appears to be linear because the average rate of change appears to be a constant. The slope of the line is 0.2.

2. Use your graphing calculator to sketch a scatterplot of these data points. Refer to Appendix C for procedures to plot a set of data on the TI-84 Plus C. Your screen should resemble the following:

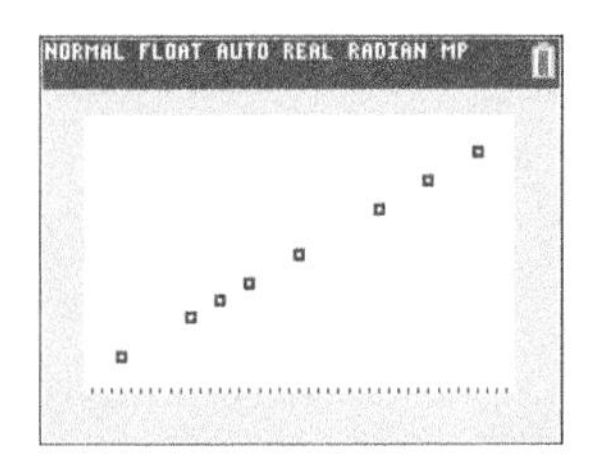

   Does your plot verify your answer to Problem 1? Yes

3. **a.** Use the table of values and/or your graphing calculator to determine the equation for this function. Use V for the output variable and F for the input variable. Call the function g so that $V = g(F)$.

   $V = g(F) = 0.2F + 28.7$

   b. Use your graphing calculator to sketch the graph of the volume function, g. Explain how you can be reasonably sure that your function is correct.

   I can check that the data points are close to the graph.

   c. Use the volume function, g, to determine the volume of the balloon when the temperature is 55°F.

   $V = 0.2(55) + 28.7 = 39.7$ cubic inches

4. Suppose you only have a Celsius thermometer and you want to know the volume of the balloon when the temperature is 10°C. To use the volume function from Problem 3, you must first convert degrees Celsius to degrees Fahrenheit.

   a. The formula $F = 32 + 1.8C$ is used to convert from degrees Celsius, C, to degrees Fahrenheit, F. Note that the formula defines F as a function of C. Therefore, F can be written as $h(C)$, where h is the name of the function. Use the given formula to determine the Fahrenheit temperature equivalent to 10°C. That is, determine $h(10)$.

   $h(10) = 32 + 1.8(10) = 32 + 18 = 50°F$

   b. Now use the result from part a to determine the volume for 10°C.

   $V = 0.2(50) + 28.7 = 10 + 28.7 = 38.7$ cubic inches

5. You have two functions. For the first function defined by $V = g(F) = 0.2F + 28.7$, F is the input and $V = g(F)$ is the output.
In the second function defined by $F = h(C) = 32 + 1.8C$, C is the input and $F = h(C)$ is the output.

Complete the following table using a combination of these functions. (Note that the temperature is given in Celsius units.)

Temperature (°C)	0	10	20	30	40
Volume (cu. in.)	35.1	38.7	42.3	45.9	49.5

In calculating the volumes in Problem 5, you followed a two-step calculation.

Step 1. You used the degrees Celsius, C, as the input to $F = g(C) = 32 + 1.8C$ to convert the temperature to degrees Fahrenheit.

Step 2. You used this output, degrees Fahrenheit, as the input to the **function** $V = f(F) = 0.2F + 28.7$ to obtain the volume, V.

To shorten this calculation, you can combine these two functions in a special way as described in Problem 6.

6. a. Because $F = 32 + 1.8C$, substitute the expression $32 + 1.8C$ for F in $V = 0.2F + 28.7$ and simplify. You have just determined an equation for V as a function of C.

$$V = 0.2F + 28.7$$
$$= 0.2(32 + 1.8C) + 28.7$$
$$= 6.4 + 0.36C + 28.7$$
$$= 0.36C + 35.1$$

b. Use the equation in part a and your graphing calculator to verify the table of values in Problem 5.

The results are the same.

Using function notation to describe the procedures in Problem 6a, you started with $F = h(C)$. Then substituting $h(C)$ for F in the second function, $V = g(F)$, you have

$$V = g(F) = g(h(C)).$$

This function of a function is called the **composition of g and h**.

EXAMPLE 1 ***Given $f(x) = 2x + 3$ and $g(x) = 4x - 1$, determine $g(f(x))$.***

SOLUTION

Substitute $2x + 3$ for $f(x)$ in $g(f(x))$.

$g(f(x)) = g(2x + 3)$ Replace x in the function rule for $g(x)$ with the expression $2x + 3$.

$= 4(2x + 3) - 1$
$= 8x + 12 - 1$ Simplify.
$= 8x + 11$

Therefore, $g(f(x)) = 8x + 11$.

7. If $f(x) = 2x + 5$, determine each of the following:

a. $f(3) = 2(3) + 5$
$= 6 + 5 = 11$

b. $f(a) = 2(a) + 5 = 2a + 5$

c. $f(\text{KATE}) = 2(\text{KATE}) + 5$

d. $f(\square) = 2(\square) + 5 = 2\square + 5$

e. $f(g(x)) = 2(g(x)) + 5$

8. If $g(x) = 3x^2 + 2x - 4$, determine each of the following:

a. $g(2) = 3(2)^2 + 2(2) - 4$
$= 3 \cdot 4 + 4 - 4$
$= 12$

b. $g(\text{KATE}) = 3(\text{KATE})^2 + 2(\text{KATE}) - 4$

c. $g(h(x)) = 3(h(x))^2 + 2(h(x)) - 4$

9. a. If $f(x) = 2x + 1$ and $g(x) = 4x - 3$, determine an equation for $f(g(x))$.

$f(g(x)) = 2(4x - 3) + 1$
$= 8x - 6 + 1$
$= 8x - 5$

b. Use the result from part a to determine $f(g(3))$.

$f(g(3)) = 8(3) - 5 = 24 - 5 = 19$

c. Determine an equation for $g(f(x))$.

$g(f(x)) = g(2x + 1) = 4(2x + 1) - 3$
$= 8x + 4 - 3$
$= 8x + 1$

d. Does $f(g(x)) = g(f(x))$?

no

LC LEARNING CATALYTICS

Given $f(3) = 2, f(-1) = 3, g(2) = 7$, and $g(5) = -1$, determine $f(g(5))$.

SUMMARY Activity 2.5

1. If x is the input of a function g, the output is $g(x)$. If $g(x)$ is then used as the input of a function f, the output is $f(g(x))$. The result is a function h, defined by $h(x) = f(g(x))$. The function h is the **composition** of the functions f and g.

2. In general, $f(g(x)) \neq g(f(x))$.

EXERCISES Activity 2.5

1. Oil is leaking from a tanker and is spreading outward in the shape of a circle. The area, A, of the oil slick is a function of radius, r (in feet), and is given by $A = f(r) = \pi r^2$. The input, r, for the area function is itself a function of time, t (in hours), since the oil began leaking.

   a. If $r = g(t) = 100t$, determine $g(2)$ and interpret its meaning in this situation.

   $g(2) = 200$. The radius of the slick is 200 feet 2 hours after the spill.

   b. Determine $f(g(2))$, and interpret its meaning in this situation.

   $f(g(2)) = f(200) = \pi(200)^2 = 40{,}000\pi \approx 125{,}664$. The area of the oil slick is 125,664 square feet 2 hours after the spill.

   c. Determine $f(g(10))$, and interpret its meaning in this situation.

   $f(g(10)) = f(1000) = \pi(1000)^2 = 1{,}000{,}000\pi$ square feet. This is the area of the slick after 10 hours.

   d. Determine a general expression for $f(g(t))$.

   $f(g(t)) = f(100t) = \pi(100t)^2 = 10{,}000\pi t^2$

   e. Determine the area of the circular oil spill after 10 hours using the new composite area function found in part d, and compare the result with your answer in part c.

   $f(g(10)) = 10{,}000\pi(10)^2 = 1{,}000{,}000\pi$. The results are the same.

2. If $A = g(r) = 4\pi r^2$ and $r = f(t) = t + 1$, determine $g(f(t))$.

   $g(f(t)) = g(t + 1) = 4\pi(t + 1)^2 = 4\pi(t^2 + 2t + 1)$

3. If $s = u(t) = -2t^2 + 2t + 1$ and $t = v(x) = 3x - 1$, determine each of the following:

   a. $u(v(x))$

   $$\begin{aligned} u(v(x)) &= u(3x - 1) = -2(3x - 1)^2 + 2(3x - 1) + 1 \\ &= -2(9x^2 - 6x + 1) + 6x - 2 + 1 = -18x^2 + 12x - 2 + 6x - 1 \\ &= -18x^2 + 18x - 3 \end{aligned}$$

   b. $v(u(t))$

   $$\begin{aligned} v(u(t)) &= v(-2t^2 + 2t + 1) = 3(-2t^2 + 2t + 1) - 1 \\ &= -6t^2 + 6t + 3 - 1 \\ &= -6t^2 + 6t + 2 \end{aligned}$$

Exercise numbers appearing in color are answered in the Selected Answers appendix.

4. In parts a and b, use the first two tables to complete the third.

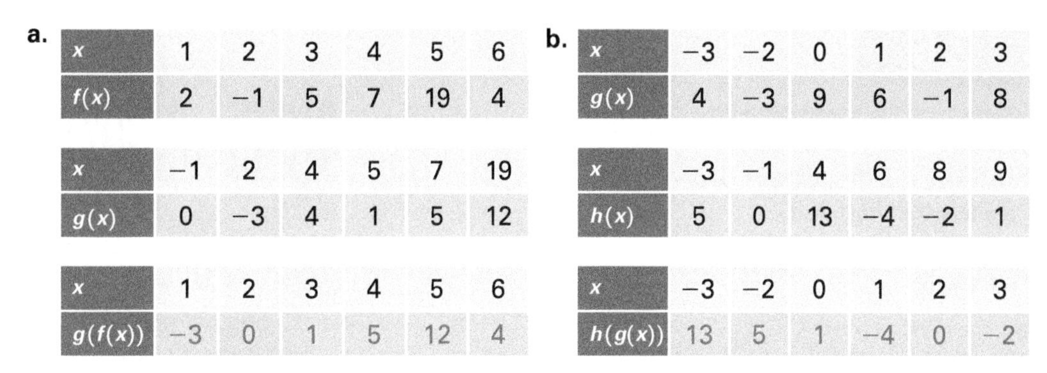

a.

x	1	2	3	4	5	6
$f(x)$	2	−1	5	7	19	4

x	−1	2	4	5	7	19
$g(x)$	0	−3	4	1	5	12

x	1	2	3	4	5	6
$g(f(x))$	−3	0	1	5	12	4

b.

x	−3	−2	0	1	2	3
$g(x)$	4	−3	9	6	−1	8

x	−3	−1	4	6	8	9
$h(x)$	5	0	13	−4	−2	1

x	−3	−2	0	1	2	3
$h(g(x))$	13	5	1	−4	0	−2

5. You read about the safety features on your brand-new car. According to *Consumer Reports*, in a 30-mile-per-hour collision, the seat belt locks properly about 99% of the time. In approximately 90% of these collisions, the air bag will successfully deploy. Let x represent the number of 30-mile-per-hour collisions in cars of the same year and model that you purchased.

a. Write an equation for $L(x)$ that represents the number of collisions in which the seat belt locks.

$L(x) = 0.99x$

b. Write an equation for $D(x)$ that represents the number of collisions in which the air bag deploys.

$D(x) = 0.90x$

c. If you suppose that every occupant is uninjured only when everything works as it should in the 30-mile-per-hour collision—that is, the seat belts lock and air bags deploy—write an equation for $S(x)$ that represents the number of times no one is injured out of x collisions. (*Hint:* Use the results of parts a and b to write a composite function.)

$S(x) = L(D(x)) = L(0.90x) = 0.99(0.90x) = 0.891x$

d. Evaluate $S(500)$.

$S(500) = 0.891(500) \approx 446$

e. Examine $L(x)$ and $D(x)$, and decide which of the safety features should be improved immediately to increase the number of survivors, $S(x)$.

$D(x)$ needs to be improved because the air bags fail 10% of the time, whereas the seat belts fail only 1% of the time.

6. A quality control inspector at a bottler of carbon-filtered drinking water notes that the first six bottles processed each day are not acceptable because they are not properly labeled. After those first six bottles, the labeler is warmed up and works just fine for the rest of the shifts. After labeling, the bottles are filled and caps are put in place. Caps are properly applied approximately 99% of the time. After capping, the bottles are inspected. Let x represent the number of bottles processed in a day.

a. Write an equation for $f(x)$ that represents the number of bottles that are properly labeled.

$f(x) = x - 6$

b. Write an equation for $g(x)$ that represents the number of bottles that are satisfactorily capped.

$g(x) = 0.99x$

c. Now determine $g(f(10{,}000))$, and interpret its meaning in this situation.

$g(f(10{,}000)) = g(9994) = 0.99(9994) = 9894.06$ or 9894

If 10,000 bottles are processed, 9894 of them should be properly labeled and capped.

ACTIVITY 2.6

Finding a Bargain

Problem Solving and Using Composite Functions

OBJECTIVE

Solve problems using the composition of functions.

You have been waiting for the best price for a winter coat. You see the following advertisement:

This is it! The time is right! Last week, the coat was on sale for 25% off the original price, and now you can get the coat for 40% less than last week's sale price.

1. a. Complete the following table:

ORIGINAL COST ($) OF THE COAT, x	LAST WEEK'S PRICE ($)	TODAY'S PRICE ($)
80	60	36
100	75	45
120	90	54
140	105	63

b. What percent of the original price would you have paid for the coat during last week's sale?

75%

c. Write an equation that gives last week's sale price, y, as a function of the original price, x. Therefore, y represents the sale price after the first reduction.

$y = f(x) = 0.75x$

2. The ad indicates that you can now save an additional 40%.

a. What percent of last week's sale price would you now have to pay?

60%

b. Write an equation that gives this week's sale price, z, as a function of last week's sale price, y. Therefore, z represents the sale price after the second reduction.

$z = g(y) = 0.6y$

3. Determining the final sale price of the coat requires a sequence of two calculations:

First: $y = f(x) = 0.75x$ (With 25% off, you pay 75% of the cost.)

Second: $z = g(y) = 0.60y$ (With 40% off, you now pay 60% of the reduced price.)

a. Substitute the expression $0.75x$ for y into the second equation to determine the single-step equation for this composition function.

$z = 0.60(0.75x) = 0.45x$

b. Your answer to part a should imply a savings of 55% from the original price. Does it? Explain.

Yes. $0.45x$ means that I pay 45% of the original price, a savings of 55%.

c. Using function notation, you can determine the equation of the composition function as follows:

$$z = g(y) = g(f(x))$$

Determine $g(f(x))$. How does this compare with the final equation in part a?

$z = g(0.75x) = 0.60(0.75x) = 0.45x$

They are the same.

LC LEARNING CATALYTICS

A store has a 30% off sale. You have a 20% off coupon. How much will you pay for an item regularly priced at $60?

d. Use the single-step equation to determine the final price of a coat having the original price

i. $80 **ii.** $100 **iii.** $120 **iv.** $140

i. $36 ii. $45 iii. $54 iv. $63

SUMMARY Activity 2.6

Taking a discount of a discount is an example of **composition of functions** denoted by $g(f(x))$, where

1. f is the function that calculates the first reduction.
2. x is the original price and $f(x)$ is the price after the first reduction.
3. g is the function that calculates the second reduction.
4. $g(f(x))$ is the price after the second reduction has been applied to the reduced price, $f(x)$.

EXERCISES Activity 2.6

1. In this activity, the coat was on sale for 25% off the original price, followed by an additional 40% off the sale price. Suppose the salesclerk entered the discount as a 65% reduction.

   a. Why do you think the clerk took a 65% reduction?

   The clerk is thinking 25% + 40% = 65%.

   b. Is this okay with you? Explain.

   Yes, because I am getting a bigger discount than I should.

   c. Would this be okay with the manager of the department? Explain.

   No, because I should be getting a 55% discount.

2. Function f gives the approximate percent increase in harmful ultraviolet rays for an x percent decrease in the thickness of the ozone layer.

x (%)	0	1	2	3	4	5	6
$f(x)$ (%)	0	1.5	3.0	4.5	6.0	7.5	9.0

Exercise numbers appearing in color are answered in the Selected Answers appendix.

Function g gives the expected percent increase in cases of skin cancer for a p percent increase in ultraviolet radiation.

p (%)	0	1.5	3.0	4.5	6.0	7.5	9.0
$g(p)$ (%)	0	5.25	10.5	15.75	21.0	26.25	31.5

In Exercises a–d, determine the output value and interpret your answer.

a. $f(3)$

4.5

A 3% decrease in the ozone layer produces a 4.5% increase in ultraviolet rays.

b. $g(4.5)$

15.75

A 4.5% increase in ultraviolet radiation produces a 15.75% increase in skin cancer.

c. $g(f(3))$

$g(4.5) = 15.75$

A 3% decrease in the ozone layer produces a 15.75% increase in skin cancer.

d. $g(f(6))$

$g(9.0) = 31.5$

A 6% decrease in the ozone layer produces a 31.5% increase in skin cancer.

e. What does $g(f(x))$ determine? That is, what does the output for $g(f(x))$ represent in this situation?

$g(f(x))$ determines the expected percent increase of skin cancer for an x percent decrease in the thickness of the ozone layer.

f. Complete the following table:

x (%)	0	1	2	3	4	5	6
$g(f(x))$ (%)	0	5.25	10.5	15.75	21.0	26.25	31.5

3. A car dealership advertises a factory rebate of \$1500, followed by a 10% discount.

a. Let x represent the price of the car. Let $f(x)$ represent the price of the car after the rebate. Determine a rule for $f(x)$.

$f(x) = x - 1500$

b. If $g(x)$ represents the price of the car after the 10% discount, determine a rule for $g(x)$.

$g(x) = 0.9x$

c. Determine $g(f(20{,}000))$, and interpret your answer.

$g(f(20{,}000)) = g(20{,}000 - 1500) = g(18{,}500) = 0.9(18{,}500) = 16{,}650$

The price of a \$20,000 car with a \$1500 rebate taken first and then followed by a 10% discount is \$16,650.

d. Suppose the price of a car is \$20,000. Determine $f(g(20{,}000))$, and interpret your answer.

$f(g(20{,}000)) = f(0.9(20{,}000)) = f(18{,}000) = 18{,}000 - 1500 = 16{,}500$

The price of a \$20,000 car with a 10% discount taken first and then followed by a \$1500 rebate is \$16,500.

e. Compare the sale price obtained by subtracting the rebate first and then taking the discount with the sale price obtained by taking the discount first and then subtracting the rebate.

Taking the discount first is better for the consumer. Taking the rebate first is better for the dealership.

4. You drop a pebble off a bridge. Ripples move out from the point of impact as concentric circles. The radius (in feet) of the outer ripple is given by

$$R = f(t) = 0.5t,$$

where t is the number of seconds after the pebble hits the water. The area, A, of a circle is a function of its radius and is given by

$$A = g(r) = \pi r^2.$$

a. Determine a formula for $g(f(t))$.

$g(f(t)) = g(0.5t) = \pi(0.5t)^2 = 0.25\pi t^2$

b. What are the input and output for the function defined in part a?

The input is the number of seconds after the pebble hits the water. The output is the area enclosed by the outer ripple in square feet.

5. The following table gives some conversions from U.S. dollars to Canadian dollars in October 2017:

U.S. Dollars	50	100	250	500	1000
Canadian Dollars	63.13	126.27	315.66	631.33	1262.65

The next table gives some conversions from Canadian dollars to euros on the same day.

Canadian Dollars	63.13	126.27	315.66	631.33	1262.65
Euros	42.42	84.85	212.12	424.26	848.50

Use the two previous tables to complete the following table for converting U.S. dollars to euros on the same day in October 2017:

U.S. Dollars	50	100	250	500	1000
Euros	42.42	84.85	212.12	424.26	848.50

ACTIVITY

2.7

Study Time

Inverse Functions

OBJECTIVES

1. Determine the inverse of a function represented by a table of values.
2. Use the notation f^{-1} to represent an inverse function.
3. Use the property $f(f^{-1}(x)) = f^{-1}(f(x)) = x$ to recognize inverse functions.
4. Determine the domain and range of a function and its inverse.

You are interested in taking 16 credits this semester to complete your program, but you are concerned about the amount of time you will need for studying. The Academic Advising Center provides you with the information listed in the following table:

CREDITS TAKEN (INPUT)	HOURS OF STUDY PER WEEK (OUTPUT)
12	22
13	24
14	26
15	28
16	30
17	32
18	34

Notice that the number of hours of study time is a function of the number of credits taken. Call this function h.

1. Determine $h(14)$ and explain its meaning in practical terms.

$h(14) = 26$

The Advising Center recommends 26 hours of study time per week if your credit hour load is 14 hours.

2. Use the information from the preceding table to construct another table in which the number of hours of study time is the input and the number of credits taken is the output.

HOURS OF STUDY PER WEEK (INPUT)	CREDITS TAKEN (OUTPUT)
22	12
24	13
26	14
28	15
30	16
32	17
34	18

3. a. Using the input–output values from the second table, determine whether the number of credits taken is a function of the number of hours of study time. Explain.

Yes. For each value of input, there is only one value of output.

b. Call this new function c, and determine $c(30)$. Explain its meaning in practical terms.

$c(30) = 16$

If you have 30 hours per week to study, you can handle 16 credit hours.

4. a. What are the domain and the range of function h, where the number of credits is the input and the number of hours of study per week is the output?

$D = \{12, 13, 14, 15, 16, 17, 18\}$

$R = \{22, 24, 26, 28, 30, 32, 34\}$

b. What are the domain and the range of function c, where the number of hours of study time is the input and the number of credits taken is the output?

$D = \{22, 24, 26, 28, 30, 32, 34\}$

$R = \{12, 13, 14, 15, 16, 17, 18\}$

c. How are the domain and range of function h related to the domain and range of function c?

The domains and ranges are interchanged.

5. Determine each of the following. Refer to the appropriate table.

a. $h(15)$

28

b. $c(28)$

15

Notice that in Problem 5, the output of h—namely, 28—in part a was used as the input of function c in part b. Recall that this is the composition of h and c and can be written as $c(h(15))$.

6. Determine each of the following. Use the tables given at the beginning of this activity.

a. $c(h(18))$

$c(h(18)) = c(34) = 18$

b. $h(c(34))$

34

c. $c(h(13))$

13

d. $h(c(32))$

32

7. a. Let x represent the input, number of credits taken, for the function h. What is the result of the composition $c(h(x))$?

$c(h(x)) = x$

b. Let x represent the input, number of hours of study, for the function c. What is the result of the composition $h(c(x))$?

$h(c(x)) = x$

Inverse Functions

DEFINITION

When the output of the composition of two functions is always the same as the input of the composition, the two functions are **inverses** of each other. Each function "undoes" the other. Symbolically, if functions f and g are inverses, then $f(g(x)) = x$ and $g(f(x)) = x$.

EXAMPLE 1 ***Show that f and g, defined by $f(x) = 3x - 1$ and $g(x) = \frac{x+1}{3}$, are inverses.***

SOLUTION

First, check the composition $f(g(x))$.

$$f(g(x)) = f\left(\frac{x+1}{3}\right) = 3\left(\frac{x+1}{3}\right) - 1 = (x+1) - 1 = x$$

This verifies the first equation. Now check $g(f(x))$.

$$g(f(x)) = g(3x-1) = \frac{(3x-1)+1}{3} = \frac{3x}{3} = x$$

Because $f(g(x)) = g(f(x)) = x$, f and g are inverses.

8. Are the functions h and c in Problem 7 inverses? Explain.

Yes, because $c(h(x)) = h(c(x)) = x$.

In general, the inverse of a function f is written f^{-1}. Using this notation, $h = c^{-1}$ and $c = h^{-1}$.

Inverse Functions

- The two functions f and g are inverses if $f(g(x)) = x$ and $g(f(x)) = x$.
- The notation for the inverse of f is f^{-1}.
- The domain of f is the range of f^{-1}, and the range of f is the domain of f^{-1}.

Important Note about Notation

The notation for the inverse function of f is potentially confusing. The -1 in $f^{-1}(x)$ is not an exponent! The notation is derived from the fact that the inverse function undoes the arithmetic operations of the original function. You may have seen -1 used as an exponent to denote the reciprocal of a number: $2^{-1} = \frac{1}{2}$; however, $f^{-1}(x) \neq \frac{1}{f(x)}$.

An important property of inverse functions is that the domain and range values are interchanged. For example, from the first table in this section,

$$h = \{(12, 22), (13, 24), (14, 26), \text{etc.}\}.$$

If the input and output values of each ordered pair are interchanged, you have

$$\{(22, 12), (24, 13), (26, 14), \text{etc.}\}.$$

These ordered pairs match the (input, output) pairs of the function c, the inverse of h. See the second table in this section.

9. Given the function $p = \{(2, 4), (-5, 6), (0, 1), (7, 8)\}$, determine the following.

a. The inverse function, p^{-1}

$p^{-1} = \{(4, 2), (6, -5), (1, 0), (8, 7)\}$

b. $p(2)$

4

c. $p^{-1}(4)$

2

d. $p(p^{-1}(4))$

4

e. $p^{-1}(p(2))$

2

f. $p(p^{-1}(x))$

x

g. $p^{-1}(p(x))$

x

10. The function q is defined by the following table, where $s = q(t)$:

t (INPUT)	s (OUTPUT)
1	2
2	3
3	5
4	3

a. Interchange the input and output values, and record the results in the following table:

s (INPUT)	t (OUTPUT)
2	1
3	2
5	3
3	4

b. Does the table in part a represent a function? Explain.

No, because the input value 3 is paired with two different output values, 2 and 4.

LC LEARNING CATALYTICS

If f and g are inverse functions defined for all real numbers, what is the value of $f(g(2))$? A. 22 B. 1/2 C. 2 D. none of these

c. As a result of part b, the function q does not have an inverse. Could you have predicted that the function q does not have an inverse from the original table? Explain.

Yes, because an output value 3 was repeated with different input values.

SUMMARY Activity 2.7

1. The two functions f and g are **inverses** if $f(g(x)) = x$ and $g(f(x)) = x$.
2. The notation for the inverse of f is f^{-1}.
3. The domain of f is the range of f^{-1}, and the range of f is the domain of f^{-1}.
4. $f^{-1}(x) \neq \dfrac{1}{f(x)}$.

EXERCISES Activity 2.7

1. The functions f and g are defined by the following tables:

x	y = f(x)
2	3
4	5
6	7
8	9

x	y = g(x)
3	2
5	4
7	6
9	8

Determine each of the following:

a. $f(g(7))$

7

b. $g(f(4))$:

4

c. $f(g(x))$

x

d. $g(f(x))$

x

Exercise numbers appearing in color are answered in the Selected Answers appendix.

2. The function h is defined by the following set of ordered pairs:

$$\{(2, 3), (3, 4), (4, 5), (5, 6)\}$$

a. Write h^{-1} as a set of ordered pairs.

$h^{-1} = \{(3, 2), (4, 3), (5, 4), (6, 5)\}$

b. Determine $h(3)$ and $h^{-1}(h(3))$.

$h(3) = 4$; $h^{-1}(h(3)) = h^{-1}(4) = 3$

c. Determine $h^{-1}(5)$ and $h(h^{-1}(5))$.

$h^{-1}(5) = 4$; $h(h^{-1}(5)) = h(4) = 5$

3. The function r is defined by the table to the right.

x	$r(x)$
0	2
1	3
2	4
3	2

Does the function r have an inverse that is also a function? Explain.

No, the interchange of the input and output values does not result in a function.

4. You are planning a trip to Canada and want to exchange some U.S. currency for Canadian money. The following table will help you with this conversion:

Function, *f*

Amount in U.S. Dollars (Input)	50	100	250	500	1000
Amount in Canadian Money (Output)	63.13	126.27	315.66	631.33	1262.65

Source: The Money Converter, October 2017.

a. Determine $f(100)$ and determine its meaning in practical terms.

$f(100) = \$126.27$. \$100 in U.S. currency is equivalent to \$126.27 in Canadian money.

b. Use the information from the preceding table to construct another table in which the amount of Canadian money is the input and the U.S. dollar amount is the output.

Amount in Canadian Money (Input)	63.13	126.27	315.66	631.33	1262.65
Amount in U.S. Dollars (Output)	50	100	250	500	1000

c. Use the information from the table in part b to determine whether the amount in U.S. dollars is a function of the amount in Canadian money. Explain.

The amount in U.S. dollars is a function of Canadian dollars because there is exactly one output for each input.

d. Let g represent the function in the table in part b. Determine $g(315.66)$. Explain its meaning in this practical situation.

$g(315.66) = 250$. \$315.66 in Canadian currency is equivalent to \$250.00 in U.S. dollars.

e. Determine the domain of f and the range of f.

Domain of f: $\{50, 100, 250, 500, 1000\}$

Range of f: $\{63.13, 126.27, 315.66, 631.33, 1262.65\}$

f. Determine the domain of g and the range of g.

Domain of g: $\{63.13, 126.27, 315.66, 631.33, 1262.65\}$

Range of g: $\{50, 100, 250, 500, 1000\}$

g. How are the domain and range of f related to the domain and range of g?

The domain of f is the range of g, and the range of f is the domain of g.

h. Determine $g(f(1000))$ using the two preceding tables. Explain your answer.

$g(f(1000)) = 1000$. If I exchange 1000 U.S. dollars for Canadian currency, I will have $1262.65 in Canadian money, and if I exchange $1262.65 in Canadian money, I will have $1000 U.S. currency.

i. Determine $f(g(63.13))$ using the two preceding tables.

$f(g(63.13)) = 50.00$. If I exchange $63.13 in Canadian currency, I will have 50 U.S. dollars, and if I exchange $50 U.S. currency, I will again have $63.13 in Canadian currency.

5. You are now familiar with the conversion tables in Exercise 4, but you would like to be able to take any amount of money and convert between U.S. and Canadian currencies.

First, verify that the functions f and g in Exercise 4 are linear. Then refer to the data in the tables to answer the following.

a. Determine the average rate of change (slope) of function f. Round to four decimal places.

$$m = \frac{126.27 - 63.13}{100 - 50} = 1.2628$$

b. What is the practical meaning of the slope in this situation?

You have 1.2628 Canadian dollars for every U.S. dollar.

c. Use x to represent the input (amount in U.S. dollars) to write the linear equation of the function f.

$f(x) = 1.2628x$

d. Use the equation to determine $f(3000)$. Explain the practical meaning in this situation.

$f(3000) = 1.2628(3000) = 3788.40$. If I have 3000 U.S. dollars, then I can exchange them for $3788.40 in Canadian currency.

e. Determine the average rate of change (slope) for function g. Round to four decimal places.

$$m = \frac{100 - 50}{126.27 - 53.13} \approx 0.7919$$

f. What is the practical meaning of the slope in function g?

You have 0.7919 U.S. dollar for 1 Canadian dollar.

g. Use x to represent the input to write the linear equation of the function g.

$g(x) = 0.7919x$

h. Use the equation to determine $g(6000)$. Explain the meaning in this situation.

$g(6000) = 0.7919(6000) = 4751.40$. If I have 6000 Canadian dollars, then I can exchange them for 4751.40 U.S. dollars.

i. Prove that the two functions are inverse functions by showing that $f(g(x)) = x$ and $g(f(x)) = x$. Round your calculations to the nearest whole number.

$f(g(x)) = f(0.7919x) = 1.2628(0.7919x) = x$

$g(f(x)) = g(1.2628x) = 0.7919(1.2628x) = x$

ACTIVITY

2.8

Temperature Conversions

Equations and Graphs of Inverse Functions

OBJECTIVES

1. Determine the equation of the inverse of a function represented by an equation.
2. Describe the relationship between the graphs of inverse functions.
3. Determine the graph of the inverse of a function represented by a graph.
4. Use the graphing calculator to produce graphs of inverse functions.

In Activity 2.5, you used the function defined by $F = 32 + 1.8C$ to convert from a temperature measured in degrees Celsius to a temperature measured in degrees Fahrenheit. Call this function T.

1. a. Identify the input and output variables for the conversion function T.

The input is C. The output is F or $T(C)$.

b. Determine $T(-5)$ and explain its meaning in this situation.

$T(-5) = 32 - 9 = 23$

A Celsius temperature of $-5°$ is equivalent to a Fahrenheit temperature of 23°.

2. When the temperature is 70°F, what is the temperature in degrees Celsius?

$70 = 32 + 1.8C$

$38 = 1.8C$

$C \approx 21.1$

70°F is equivalent to approximately 21.1°C.

3. If you need to determine the temperature in Celsius for several Fahrenheit temperatures, it is easier to have a single formula (in fact, a new function) in which Celsius is the output and Fahrenheit is the input. Solve $F = 32 + 1.8C$ for C to determine this new function. Call this function H.

$$H(F) = C = \frac{F - 32}{1.8}$$

4. a. Identify the input and output variables for the function H.

The input is F. The output is C.

b. Determine $H(62)$ and explain its meaning in this situation.

$$H(62) = \frac{62 - 32}{1.8} = \frac{30}{1.8} \approx 16.7$$

A Fahrenheit temperature of 62° is equivalent to a Celsius temperature of approximately 16.7°.

5. Determine each of the following:

a. $T(10)$

$T(10) = 32 + 1.8(10) = 50$

b. $H(T(10))$

10

6. Determine each of the following:

a. $T(H(95))$

95

b. $H(T(0))$

0

c. $T(H(212))$

212

7. a. Write a general expression for $H(T(C))$, and simplify the result.

$$H(T(C)) = H(32 + 1.8C) = \frac{(32 + 1.8C) - 32}{1.8} = \frac{1.8C}{1.8} = C$$

b. Write a general expression for $T(H(F))$, and simplify the result.

$$T(H(F)) = 32 + 1.8\left(\frac{F - 32}{1.8}\right) = 32 + (F - 32) = F$$

c. Based on the results of parts a and b, what can you conclude about the functions H and T?

H and T are inverses. $T(C) = H^{-1}(C)$, and $H(F) = T^{-1}(F)$.

Inverse Function Algorithm

Given y as a function of x, follow the steps below to determine the inverse function.

Step 1. Write the equation for y in terms of x.

Step 2. Solve for x to get $x = f^{-1}(y)$.

Step 3. If f is a noncontextual (abstract) function, you can interchange x and y to get $y = f^{-1}(x)$.

EXAMPLE 1 *Given $f(x) = 7 + 5x$, determine the equation for the inverse function f^{-1}.*

SOLUTION

Step 1. Write the equation for y in terms of x: $y = 7 + 5x$.

Step 2. Solve for x to get $x = f^{-1}(y)$: $5x = y - 7$, $x = \frac{y-7}{5}$, $f^{-1}(y) = \frac{y-7}{5}$.

Step 3. Because f is a noncontextual function, we can interchange x and y to get $y = f^{-1}(x) = \frac{x-7}{5}$.

8. Given $y = f(x) = 5 - 3x$, determine an equation of the inverse function f^{-1}.

$3x = 5 - y$

$x = \frac{5-y}{3}$

$y = f^{-1}(x) = \frac{5-x}{3}$

Graphs of Inverse Functions

9. a. Consider the graphs of the functions $f(x) = 7 + 5x$ and $f^{-1}(x) = \frac{x-7}{5}$ from Example 1. Graph the two functions on the same coordinate system.

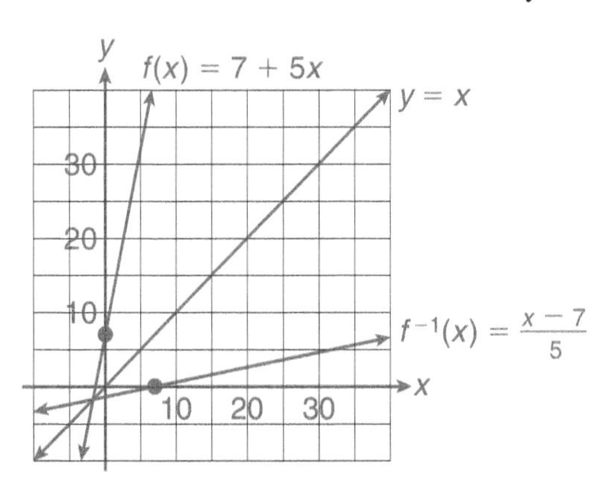

b. Draw the graph of $y = x$ on the grid in part a. Describe the relationship among the graphs of $f(x) = 7 + 5x, f^{-1}(x) = \frac{x - 7}{5}$, and $y = x$.

The graphs of $f(x) = 7 + 5x$ and $f^{-1}(x) = \frac{x - 7}{5}$ are symmetric about the line $y = x$.

Graphs of Inverse Functions

The graphs of inverse functions are reflections about the line $y = x$.

You can draw the inverse of a function with your TI-84 Plus C without determining the equation. Turn off the graph of f^{-1} from Problem 9, and keep the graphs of f and $y = x$. Use the window Xmin $= -9$; Xmax $= 9$, Ymin $= -6$; and Ymax $= 6$. Your screen should resemble the following:

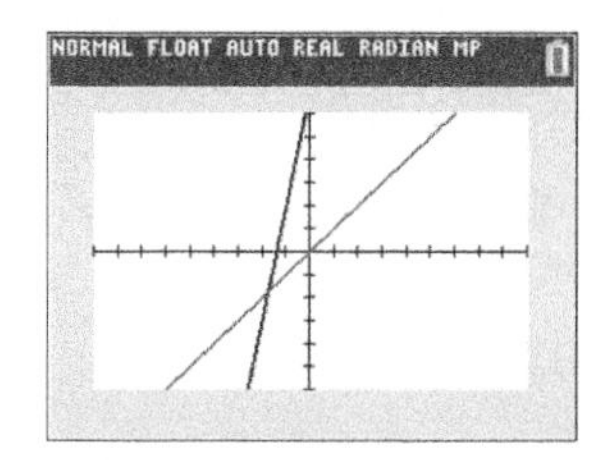

Access the Draw menu by pressing (2nd)(PRGM). Toggle down to option 8, DrawInv.

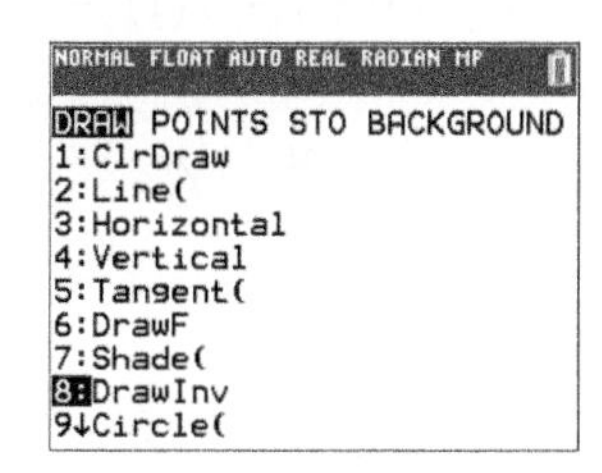

Pressing (ENTER) will place the DrawInv command in the home screen. Now press (VARS) and then (Y-VARS) and (1: Functions) to access the following screen:

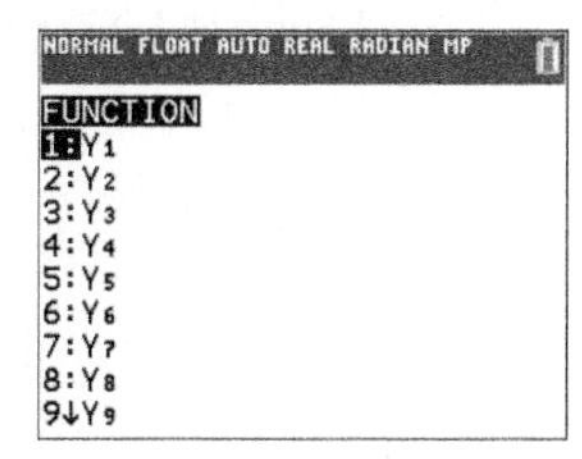

Choose the variable that corresponds to your function T (most likely Y1), and press (ENTER) again. Your screen should resemble the following:

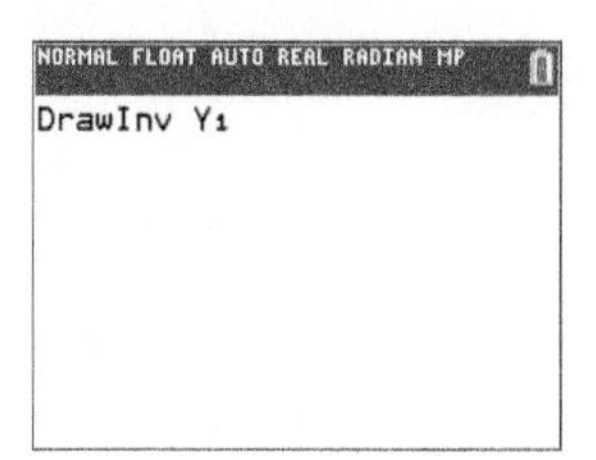

Press (ENTER). This adds the graph of the inverse to the graph screen.

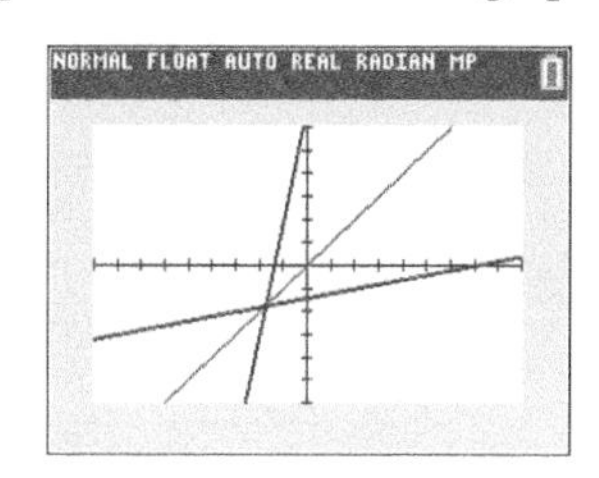

How does this compare with the graph of f^{-1} in Problem 9a? Note that the graph of f^{-1} in the graph screen of the TI-84 Plus C is not active. It cannot be traced, nor can values be viewed in the table.

10. Determine the point of intersection of the graphs of f and f^{-1}. What is the significance of this point in this situation?

The point of intersection is $(-1.75, -1.75)$. $f(-1.75) = f^{-1}(-1.75) = -1.75$

11. a. Complete the following table for the graphs of f and f^{-1} given in Problem 9:

	VERTICAL INTERCEPT	HORIZONTAL INTERCEPT
$f(x) = 7 + 5x$	$(0, 7)$	$\left(-\frac{7}{5}, 0\right)$
$f^{-1}(x) = \frac{x - 7}{5}$	$\left(0, -\frac{7}{5}\right)$	$(7, 0)$

b. Write a sentence that describes the relationship of the intercepts of the graphs of f and f^{-1}.

The coordinates of the vertical intercept of f are the coordinates of the horizontal intercept of f^{-1} in reverse order. Also, the coordinates of the vertical intercept of f^{-1} are the coordinates of the horizontal intercept of f in reverse order.

LC LEARNING CATALYTICS

If $f(x) = 2x + 6$, then $f^{-1}(x) =$ ____.

12. a. Determine the slopes of the graphs of f and f^{-1} given in Problem 9.

The slope of the graph of $f(x) = 7 + 5x$ is $m = 5$. The slope of the graph of $f^{-1}(x) = \frac{x - 7}{5}$ is $m = \frac{1}{5}$.

b. Write a sentence that describes the relationship of the slope of f and the slope of f^{-1}.

The slopes of f and f^{-1} are reciprocals.

13. Sketch a graph of f^{-1} on the same coordinate system as the graph of f for the functions that follow.

a.

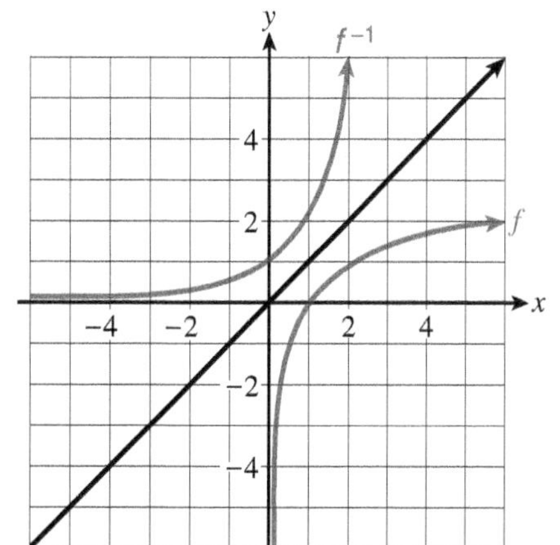

b.

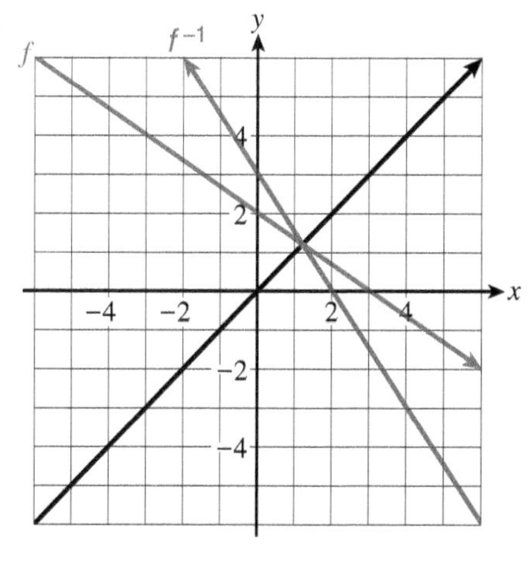

c.

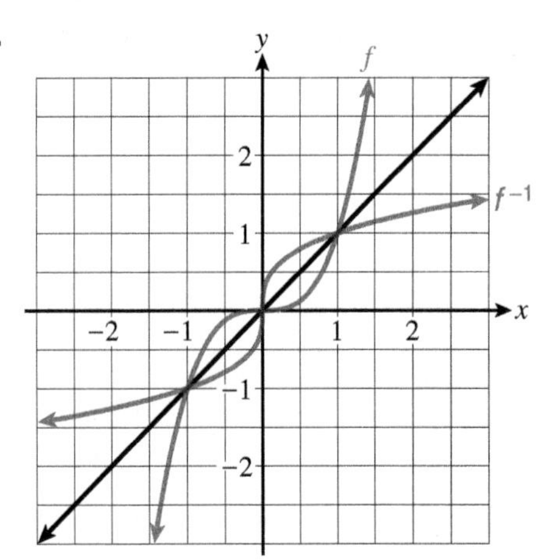

SUMMARY Activity 2.8

1. To determine the equation of the inverse of a function defined by $y = f(x)$,

   Step 1. Write the equation for y in terms of x.

   Step 2. Solve for x to get $x = f^{-1}(y)$.

   Step 3. If f is noncontextual, interchange x and y to get $y = f^{-1}(x)$.

2. The graphs of inverse functions are reflections about the line $y = x$.

3. If two linear functions are inverses, the slopes of the graphs of the lines are reciprocals.

EXERCISES Activity 2.8

1. As a sales representative, you are paid a base salary of $500 a week, plus a 5% commission.

   a. Determine a function, f, for your weekly gross pay, P (before taxes), as a function of S, your weekly sales in dollars, $P = f(S)$.

   $P = f(S) = 0.05S + 500$

   b. Determine $f(6000)$ and interpret its meaning in this situation.

   $f(6000) = 0.05(6000) + 500 = 800$. In a week when you sell $6000 worth of merchandise, your gross salary is $800.

   c. Solve the equation $P = 0.05S + 500$ for S to determine the equation for a new function, $g(P)$, whose input is P and whose output is S.

   $$S = g(P) = \frac{P - 500}{0.05}$$

Exercise numbers appearing in color are answered in the Selected Answers appendix.

d. Determine $g(600)$ and interpret its meaning in this situation.

$g(600) = \frac{600 - 500}{0.05} = 2000$. A weekly salary of \$600 means that I sold \$2000 worth of merchandise.

e. Determine $g(f(8000))$.

$g(f(8000)) = 8000$

2. a. Complete the following table for the function $y = f(x) = 3x^2 - 2$:

x	−2	−1	0	1	2
y	10	1	−2	1	10

b. Describe how you know that f is a function by examining the table in part a.

f is a function because for each input value, there is only one output value.

c. Use the table of values from part a to determine f^{-1} if possible. If it is not possible, explain why.

f^{-1} is not a function because two different x-values are paired with the y-value 10.

3. Determine the equation of the inverse of the given function.

a. $y = f(x) = 3x - 4$

$y + 4 = 3x$ or $x = f^{-1}(y) = \frac{y + 4}{3}$ or $y = f^{-1}(x) = \frac{x + 4}{3}$

b. $w = g(z) = \frac{z - 4}{2}$

$2w = z - 4$ or $z = f^{-1}(w) = 2w + 4$ or $w = g^{-1}(z) = 2z + 4$

c. $s = \frac{5}{t}$

$t = \frac{5}{s}$ or $s = \frac{5}{t}$

4. a. Given $y = f(x) = 2x + 6$, determine f^{-1}. Write f^{-1} as a function of x.

$f^{-1}(x) = \frac{x - 6}{2}$

b. Sketch the graphs f and f^{-1} on the same coordinate system.

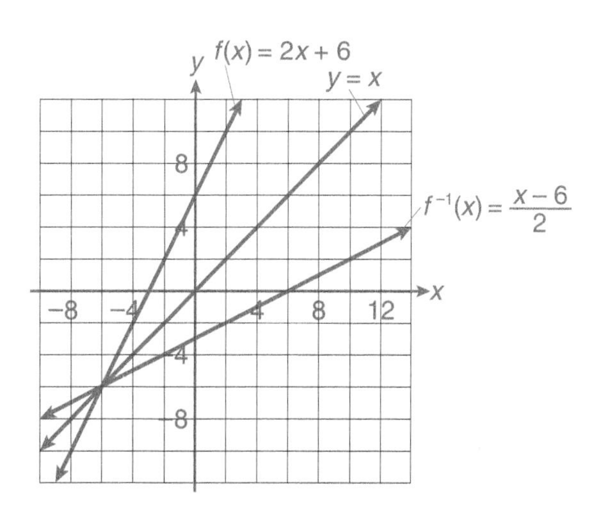

c. Add the sketch of the graph of $y = x$ to the sketches of f and f^{-1} in part b. Describe any symmetry in the graphs of f and f^{-1}.

The graphs of f and f^{-1} are symmetric with respect to the line $y = x$.

d. Determine the horizontal and vertical intercepts of the graphs of f and of f^{-1}. Explain the relationship between the intercepts of f and the intercepts of f^{-1}.

The intercepts of f are $(0, 6)$ and $(-3, 0)$. The intercepts of f^{-1} are $(6, 0)$ and $(0, -3)$. The intercepts of the inverse are obtained by interchanging the coordinates of the intercepts of f.

e. Determine the slope of the graph of f and the slope of the graph of f^{-1}. What is the relationship between the slope of f and the slope of f^{-1}?

The slope of the graph of f is 2; the slope of the inverse is 1/2. The slopes are reciprocals.

5. Consider the graphs of functions g and h. Are the functions inverses of each other? Explain using a symmetry argument.

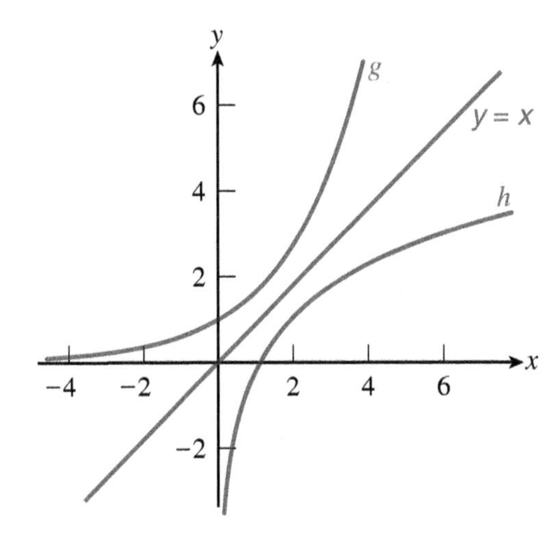

g and h are inverses. The graphs of g and h are symmetric with respect to the line $y = x$.

6. Consider the functions $f(x) = 3x + 6$ and $g(x) = \frac{1}{3}x - 2$.

a. Determine $f(g(x))$.

$$f\left(\frac{1}{3}x - 2\right) = 3\left(\frac{1}{3}x - 2\right) + 6 = x$$

b. Determine $g(f(x))$.

$$g(3x + 6) = \frac{1}{3}(3x + 6) - 2 = x$$

c. Are f and g inverse functions? Explain.

Yes, because $g(f(x)) = f(g(x)) = x$.

d. Complete the following tables for the given functions:

x	$f(x)$
−2	0
0	6
2	12
4	18

x	$g(x)$
0	−2
6	0
12	2
18	4

e. What do you notice about the ordered pairs of *f* and *g*?

The inputs and outputs have been interchanged.

f. Graph *f* and *g* on the following grid:

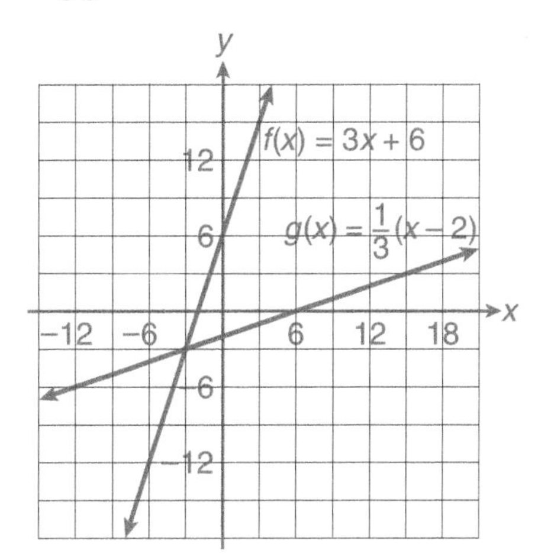

g. What can you say about the graphs of *f* and *g* with respect to the graph of $y = x$?

The graphs are reflections of each other in the line $y = x$.

7. **a.** Given $g(x) = \dfrac{6 + 4x}{3}$, determine an equation for $g^{-1}(x)$.

$$g^{-1}(x) = \frac{3x - 6}{4}$$

b. Sketch the graphs of g and g^{-1} on the following grid:

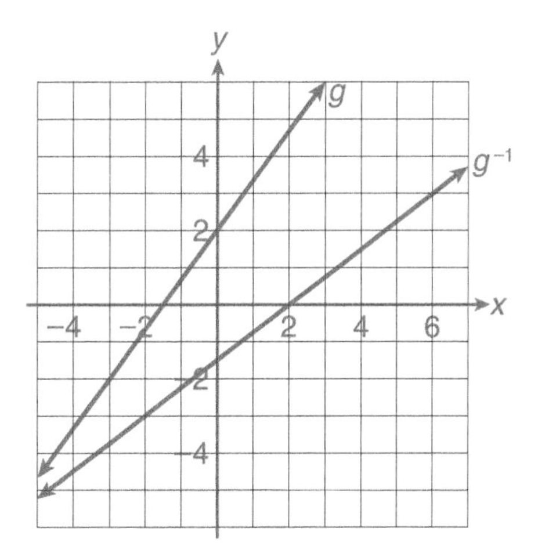

c. Do you believe that your equation for g^{-1} in part a is correct? Explain.

Yes, because the graphs are reflections in the line $y = x$.

d. Determine $g^{-1}(g(x))$. Does your result support your answer in part c? Explain.

$$g^{-1}\left(\frac{6 + 4x}{3}\right) = \frac{3\left(\frac{6 + 4x}{3}\right) - 6}{4} = \frac{4x}{4} = x.$$ Yes, because $g^{-1}(g(x)) = x$.

The functions in Problems 8 and 9 model contextual situations. The input and output represent real-world quantities. In general, you should not graph the function and its inverse on the same set of axes because the input and output quantities are different. However, we do here to emphasize the geometric relationship between the graph of a function and its inverse.

8. One inch measures 2.54 centimeters. To convert inches to centimeters, use the equation $C(x) = 2.54x$, where x is the number of inches.

a. Using the graph of C, draw the graph of the inverse function C^{-1}. Ignore the context of the situation. Graph the inverse of the general function. (*Hint:* First draw the graph of $y = x$ on the same axes.)

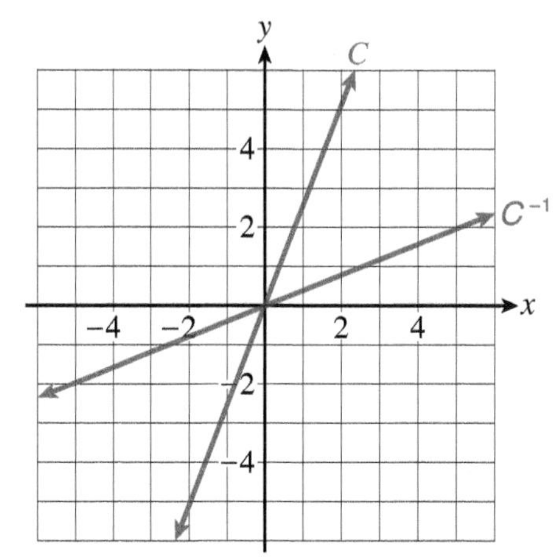

b. Locate $C^{-1}(5)$ on the graph, and interpret its meaning in this situation.

$C^{-1}(5) = 1.97$. 5 centimeters is approximately 1.97 inches. (Answers may vary.)

9. Your new kitchen has a square island that needs a granite top. The side of the largest square top you can purchase measures $S = f(a) = \sqrt{a}$, where S is the length of the square in feet and a is the area of the square.

a. Using the graph of S, sketch S^{-1}. Do this for the general function, ignoring the context of the situation.

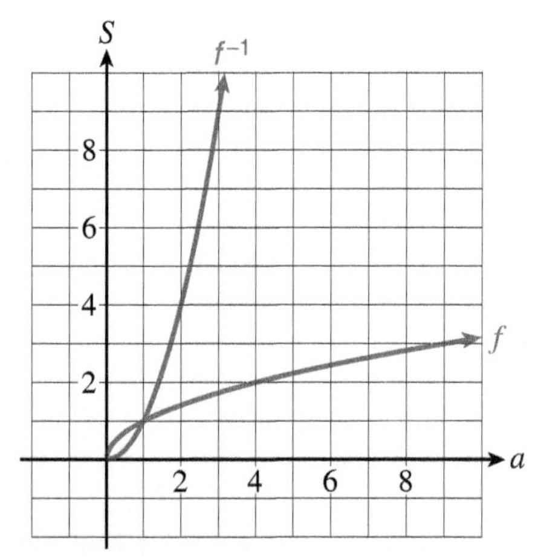

b. Check your graph of f^{-1} using the DrawInv command.

9.b.

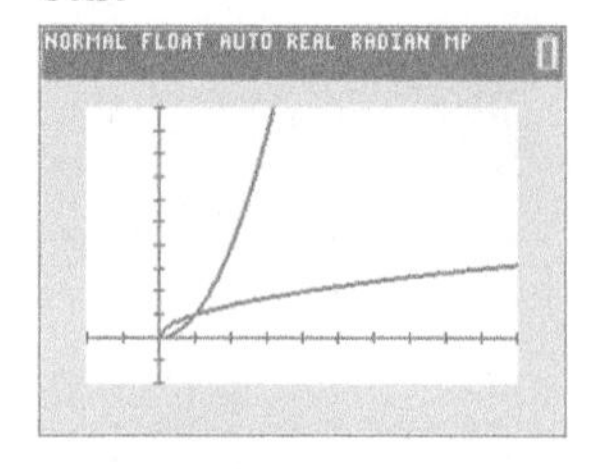

c. What are input and output variables for f?

The area of the square is the input. The length of the side of the square is the output.

d. What are the input and output variables for f^{-1}?

The length of the side of the square is the input. The area of the square is the output.

e. Would f^{-1} ever be useful in this situation? Explain.

Given the length of the side, we can determine the area of the granite top.

CLUSTER 2 What Have I Learned?

1. For any two functions f and g,

i. determine which of the following equations are true and explain why.

ii. determine which of the following equations are not always true and give an example to show why not.

a. $f(g(x)) = g(f(x))$

(ii) Let $f(x) = 2x^2 + 1$ and $g(x) = x - 1$. I will use these functions to show that the statement in part a is false.

$$f(g(x)) = f(x - 1) = 2(x-1)^2 + 1 = 2(x^2 - 2x + 1) + 1 = 2x^2 - 4x + 3$$

and

$$g(f(x)) = g(2x^2 + 1) = 2x^2 + 1 - 1 = 2x^2$$

Again, these outputs are clearly not equal. (Answers will vary depending on the choice of f and g.)

b. $f(f^{-1}(x)) = f^{-1}(f(x))$

Equation b is true because it is one of the properties of inverse functions.

$f(f^{-1}(x)) = f^{-1}(f(x)) = x$

2. Describe how you would determine whether two functions are inverses of each other.

Given f and g, I would evaluate $f(g(x))$ and $g(f(x))$. If both expressions simplify to x, I would conclude that they are inverses. If they do not, I would conclude that they are not inverses. (Geometric or numeric arguments could also be valid here.)

3. Suppose f is a nonconstant linear function.

a. Will f always have an inverse function?

Yes, it will.

b. What will always be true about the slopes of the lines representing the graphs of f and f^{-1}?

f and f^{-1} will have reciprocal slopes.

4. If $f(g(x)) = x^6$, determine at least three different ways to define f and g.

$f(x) = x$, $g(x) = x^6$ or $f(x) = x^2$, $g(x) = x^3$ or $f(x) = x^3$, $g(x) = x^2$

Other answers are possible.

5. It is possible to compose more than two functions. If $f(x) = 2x - 1$, $g(x) = 5 - 3x$, and $h(x) = 2x^2$, determine each of the following:

a. $f(g(h(1)))$

$f(g(h(1))) = f(g(2)) = f(-1) = -3$

b. $g(f(h(1)))$

$g(f(h(1)) = g(f(2)) = g(3) = -4$

c. $h(g(f(1)))$

$h(g(f(1))) = h(g(1)) = h(2) = 8$

d. $h(f(g(1)))$

$h(f(g(1))) = h(f(2)) = h(3) = 18$

CLUSTER 2 How Can I Practice?

1. Given $f(x) = x^2 - 4$ and $g(x) = x + 2$, determine each of the following:

   a. $f(g(-3))$
   $f(-1) = -3$

   b. $g(f(-3))$
   $g(5) = 7$

   c. $f(g(x))$
   $f(x + 2) = (x + 2)^2 - 4 = x^2 + 4x$

   d. $g(f(x))$
   $g(x^2 - 4) = x^2 - 4 + 2 = x^2 - 2$

   e. $f(f(x))$
   $f(x^2 - 4) = (x^2 - 4)^2 - 4$
   $= x^4 - 8x^2 + 12$

   f. $g^{-1}(x)$
   $y = x + 2$ or $x = y - 2$ or
   $y = g^{-1}(x) = x - 2$

2. Given $f(x) = x - 4$ and $g(x) = 4 + x - x^2$, determine each of the following.

   a. $f(g(3))$
   $f(-2) = -6$

   b. $g(f(3))$
   $g(-1) = 2$

   c. $g(g(3))$
   $g(-2) = -2$

   d. $f(g(x))$
   $f(4 + x - x^2) = (4 + x - x^2) - 4$
   $= x - x^2$

   e. $g(f(x))$
   $g(x - 4) = 4 + (x - 4) - (x - 4)^2$
   $= x - (x^2 - 8x + 16)$
   $= -x^2 + 9x - 16$

   f. $f^{-1}(x)$
   $y = x - 4$ or $x = y + 4$ or
   $y = f^{-1}(x) = x + 4$

3. Given the following two tables, complete the third table:

x	0	1	2	3	4
$f(x)$	6	−1	−3	0	2

x	−3	−1	0	2	6
$g(x)$	2	3	1	0	−1

x	0	1	2	3	4
$g(f(x))$	−1	3	2	1	0

4. Given $f(x) = 3x^2$ and $g(x) = -2x^3$, determine each of the following:

   a. $g(f(x))$
   $g(3x^2) = -2(3x^2)^3 = -54x^6$

   b. $f(g(x))$
   $f(-2x^3) = 3(-2x^3)^2 = 12x^6$

   c. $g(f(-4))$
   $g(48) = -2(48)^3 = -221{,}184$

Answers to all How Can I Practice exercises are included in the Selected Answers appendix.

5. Given $s(x) = x^2 + 4x - 1$ and $t(x) = 4x - 1$, determine each of the following:

a. $s(t(x))$

$s(4x - 1) = (4x - 1)^2 + 4(4x - 1) - 1 =$

$16x^2 - 8x + 1 + 16x - 4 - 1 = 16x^2 + 8x - 4$

b. $t(s(x))$

$t(x^2 + 4x - 1) = 4(x^2 + 4x - 1) - 1 = 4x^2 + 16x - 5$

c. $t^{-1}(x)$

$y = 4x - 1$ or $4x = y + 1$ or $x = \frac{y+1}{4}$ or $y = t^{-1}(x) = \frac{x+1}{4}$

6. Given $p(x) = \frac{1}{x}$ and $c(x) = \sqrt{x + 2}$, determine each of the following:

a. $p(c(x))$

$p(\sqrt{x+2}) = \frac{1}{\sqrt{x+2}}$

b. $c(p(x))$

$c\left(\frac{1}{x}\right) = \sqrt{\frac{1}{x} + 2}$

c. $p^{-1}(x)$

$x = \frac{1}{y}$ or

$y = p^{-1}(x) = \frac{1}{x}$

7. The function q is defined by the following set of ordered pairs:

$$\{(4, 6), (7, -9), (-2, 1), (0, 0)\}$$

Determine q^{-1} as a set of ordered pairs.

$\{(6, 4), (-9, 7), (1, -2), (0, 0)\}$

8. Show that $f(x) = 2x - 3$ and $g(x) = \frac{x + 3}{2}$ are inverses of each other.

$f(g(x)) = f\left(\frac{x+3}{2}\right) = 2\left(\frac{x+3}{2}\right) - 3 = x$

$g(f(x)) = g(2x - 3) = \frac{(2x - 3) + 3}{2} = x$

Because $f(g(x)) = g(f(x)) = x$, f and g are inverse functions.

9. a. Determine the equation of the inverse function of $f(x) = 4x + 3$.

$y = 4x + 3$; $x = f^{1}(y) = \frac{y-3}{4}$; $y = f^{-1}(x) = \frac{x-3}{4}$

b. Sketch the graph of the function f and its inverse, f^{-1}, on the same coordinate system.

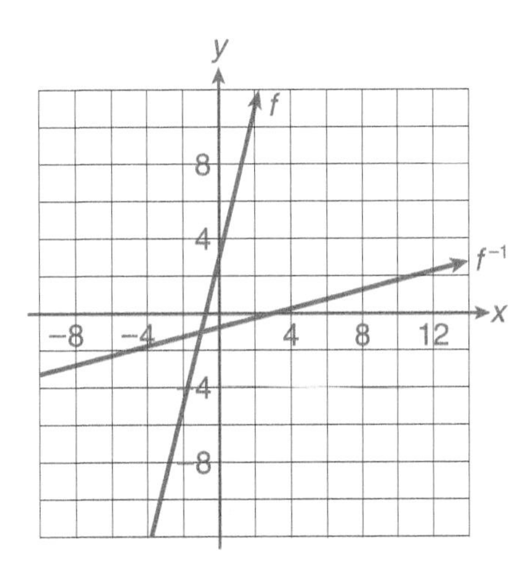

c. Determine the horizontal and vertical intercepts of the graph of f and of f^{-1}.

The intercepts of f are $(0, 3)$ and $\left(-\frac{3}{4}, 0\right)$. The intercepts of f^{-1} are $\left(0, -\frac{3}{4}\right)$ and $(3, 0)$.

d. Determine the slopes of the graphs of f and of f^{-1}.

The slope of the graph of f is 4. The slope of the graph of the inverse is $\frac{1}{4}$.

e. Sketch a graph of the line $y = x$ on the same coordinate system.

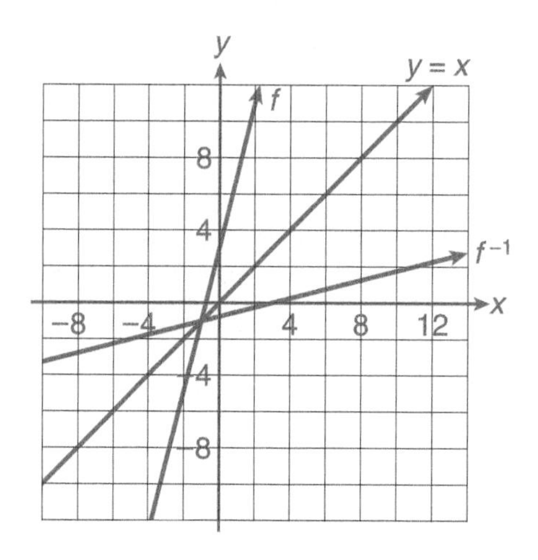

10. Cigarette smoking among adults in the United States has been decreasing steadily since 1990. The percent of the U.S. adult population who smokes is given by $P = f(x) = -0.39x + 26.4$, where x represents the number of years after 1990.

Source: Centers for Disease Control and Prevention.

a. Complete the following table:

x, Years since 1990	0	5	10	15	20	25	30	35
p, Percent of the Adult Population That Smokes	26.4	24.5	22.5	20.6	18.6	16.7	14.7	10.8

b. Interchange the input and output data.

p, Percent of the Adult Population That Smokes	26.4	24.5	22.5	20.6	18.6	16.7	14.7	10.8
x, Years since 1990	0	5	10	15	20	25	30	35

c. Determine the equation of the inverse function, and use this equation to check the table values in part b.

$$x = f^{-1}(p) = \frac{26.4 - p}{0.39}$$

d. Ignoring the context of the situation, graph the function and its inverse using your graphing calculator. Use the window Xmin $= -10$, Xmax $= 50$, Ymin $= -10$, Ymax $= 50$. Use Zoom, ZSquare to view the graphs in a square window. Use the appropriate graph to determine when the percent of the adult U.S. population who smokes drops to 10%.

According to the models,10% of the population will smoke 42 years after 1990, or in the year 2032.

e. Add the graph of $y = x$ to the graphs of the function and its inverse. Describe the symmetry in the graphs.

The graphs are symmetric in the line $y = x$.

f. Determine the horizontal and vertical intercepts of the function and its inverse. Explain the relationship between the intercepts of the two functions.

	HORIZONTAL INTERCEPT	VERTICAL INTERCEPT
$f(x) = -0.39x + 26.4$	(67.7, 0)	(0, 26.4)
$f^{-1}(x) = \frac{26.4 - x}{0.39}$	(26.4, 0)	(0, 67.7)

The horizontal intercept of the function interchanged is the vertical intercept of the inverse. The vertical intercept of the function interchanged is the horizontal intercept of its inverse.

g. Determine the slopes of the graphs of the function and its inverse. What is the relationship between the slope of the function and the slope of the inverse?

Function: $m = -0.39$

Inverse function: $m = \frac{-1}{0.39} \approx -2.56$

The slopes are reciprocals.

h. According to the function, when will there be no smoking adults in the United States?

68 years after 1990, or 2058.

11. A Japanese student is coming to the United States as an exchange student. She has saved 60,000 yen for the trip. She realizes that because she is going to stop briefly in Europe, she can convert her money to euros. She uses the Internet on January 2, 2018, to find that the conversion factor to convert yen to euros is 0.00740. Let x represent the amount in Japanese currency and $f(x)$ the amount in euros.

a. Write a conversion function to convert her Japanese yen to euros.

$f(x) = 0.00740x$

b. Use the function to determine how many euros she will receive for her 60,000 yen.

444 euros

c. She then comes to the United States with her euros. To convert her money to U.S. dollars, she finds another conversion factor on the Internet. The function is $g(E) = 1.20400E$, with E representing the amount in euros and $g(E)$ representing the U.S. dollar amount. How many dollars will she have if she has 350 euros?

$421.40

d. If the Japanese student does not go to Europe, but flies directly to the United States, write a new function that would tell her how much her 60,000 yen would yield in U.S. dollars. (*Hint*: Determine $g(f(x))$.)

$g(f(x)) = g(0.00740x) = 1.20400(0.00740x) = 0.00891x$

e. Use the function you found in part d to determine what her 60,000 yen would be worth in U.S. dollars.

$g(f(60{,}000)) = 0.00891(60{,}000) = \$534.60$

12. Given $f(x) = 4x - 9$ and $g(x) = 10 - 3x$, what input to the composition function $f(g(x))$ will result in an output of 15?

$f(g(x)) = 15$ or $4(10 - 3x) - 9 = 15$

$$40 - 12x - 9 = 15$$

$$31 - 12x = 15$$

$$-12x = -16$$

$$x = \frac{4}{3}$$

13. The volume of a rectangular box is equal to the area of the base times the height. Suppose a particular box has a height of 10 inches and a square base.

a. Using V for the volume of the box, b for the area of the base, and x for the length of one side of the square base, express b as a function of x and V as a function of b.

$b = f(x) = x^2$; $V = g(b) = 10b$

b. Show how the volume is the composition of these two functions.

$V = g(f(x)) = g(x^2) = 10x^2$

CHAPTER 2 Summary

The bracketed numbers following each concept indicate the activity in which the concept is discussed.

CONCEPT/SKILL	DESCRIPTION	EXAMPLE
The sum function $f + g$ [2.1]	Given two functions, f and g, the sum function is defined by $y = f(x) + g(x)$.	See Example 1, Activity 2.1.
The difference function, $f - g$ [2.1]	Given two functions, f and g, the difference function is defined by $y = f(x) - g(x)$.	See Example 1, Activity 2.1.
Polynomial expression [2.1]	Any expression that is formed by adding or subtracting terms of the form ax^n, where a is a real number and n is a non-negative integer, is called a polynomial expression in x.	$3x^3 - 2x^2 + 6x - 7$
Monomial [2.1]	A monomial is a polynomial with one term.	$13x^5$
Binomial [2.1]	A binomial is a polynomial with two terms.	$14x^4 - 3x$
Trinomial [2.1]	A trinomial is a polynomial with three terms.	$5x^4 - 7x + 13$
Polynomial function [2.1]	A polynomial function is any function defined by an equation of the form $y = f(x)$, where $f(x)$ is a polynomial expression.	$f(x) = 3x^3 - 2x - 7$
The product of two polynomials [2.2]	To multiply any two polynomials, multiply each term of the first by each term of the second.	$(x^2 + 1)(3x^2 + 6x - 2)$ $= 3x^4 + 6x^3 - 2x^2 + 3x^2 + 6x - 2$ $= 3x^4 + 6x^3 + x^2 + 6x - 2$
FOIL Method [2.2]	FOIL is a common method used to multiply two binomials.	See Example 1, Activity 2.2.
The product of two functions [2.2]	Given two functions, f and g, the product function is defined by $y = f(x) \cdot g(x)$.	See Problem 8, Activity 2.2.
$a^m \cdot a^n$ [2.2]	To multiply powers of the same base, keep the base and add the exponents. Symbolically, this property of exponents is written as $a^m \cdot a^n = a^{m+n}$.	$x^4 \cdot x^3 = x^{4+3} = x^7$
Zero exponents [2.3]	$a^0 = 1$, where $a \neq 0$	$\left(\frac{2x}{4y}\right)^0 = 1 \quad x \text{ and } y \neq 0$

CONCEPT/SKILL	DESCRIPTION	EXAMPLE
Negative exponents [2.3]	$a^{-n} = \frac{1}{a^n}$ if $a \neq 0$ and n is a rational number	$\left(\frac{2}{3}\right)^{-3} = \left(\frac{3}{2}\right)^3 = \frac{27}{8}$
$(a^m)^n$ [2.4]	If a is a real number and m and n are rational numbers, then $(a^m)^n = a^{mn}$.	$(x^2)^5 = x^{2 \cdot 5} = x^{10}$
$(ab)^n$ [2.4]	If a and b are real numbers and n is a rational number, then $(ab)^n = a^n b^n$.	$(2x)^4 = 2^4 x^4 = 16x^4$
The principal square root of a, $\sqrt{a}$ [2.4]	Let a represent a nonnegative real number, symbolically written as $a \geq 0$. The principal square root of a, denoted by $\sqrt{a}$, is defined as the nonnegative real number that, when squared, produces a.	$\sqrt{16} = 4$ because $4^2 = 16$.
Fractional exponents [2.4]	$a^{1/2} = \sqrt{a}$, where $a \geq 0$	$13^{1/2} = \sqrt{13}$
$\sqrt[n]{a}$ [2.4]	$\sqrt[n]{a} = a^{1/n}$, the nth root of a. The real number a, called the radicand, must be nonnegative if n, called the index, is even.	$\sqrt[3]{36} = 36^{1/3}$
$a^{p/q}$ [2.4]	For positive integers p and q, $a^{p/q} = \sqrt[q]{a^p}$ or $a^{p/q} = (\sqrt[q]{a})^p$, where $a \geq 0$ if q is even.	$16^{3/4} = (\sqrt[4]{16})^3 = 2^3 = 8$
Composition of functions [2.5, 2.6]	The composition of the functions f and g is a function, h, defined by $h(x) = f(g(x))$.	If $f(x) = 2x + 1$ and $g(x) = 3x - 2$, $f(g(x)) = f(3x - 2)$ $= 2(3x - 2) + 1 = 6x - 3$.
Noncommutativity of composition [2.5]	In general, $f(g(x)) \neq g(f(x))$.	If $f(x) = 2x + 1$ and $g(x) = 3x - 2$, $f(g(x)) = 6x - 3$, $g(f(x)) = g(2x + 1)$ $= 3(2x + 1) - 2 = 6x + 1$.
Inverse functions [2.7]	The two functions f and g are inverses if $f(g(x)) = x$ and $g(f(x)) = x$.	See Example 1, Activity 2.7.
Domain and range of inverse functions [2.7]	The domain of f is the range of f^{-1}, and the range of f is the domain of f^{-1}.	See the tables at Problems 1 and 2, Activity 2.7.
Graphs of inverse functions [2.8]	The graphs of inverse functions are reflections about the line $y = x$.	See the graph before Problem 9, Activity 2.8.
Slopes of inverse linear functions [2.8]	If two linear functions are inverses, the slopes of the graphs of the lines are reciprocals.	The slope of $f(x) = 3x + 1$ is 3. The inverse, defined by $f^{-1}(x) = \frac{1}{3}x - \frac{1}{3}$, has a slope of $\frac{1}{3}$.

CHAPTER 2 Gateway Review

1. Simplify the following:

a. $(x + 6) + (2x^2 - 3x - 7)$

$2x^2 - 2x - 1$

b. $(x^2 + 4x - 3) - (2x^2 - x + 1)$

$-x^2 + 5x - 4$

c. $(x - 3)(4x - 1)$

$4x^2 - 13x + 3$

d. $(x - 5)(x^2 - 2x + 3)$

$x^3 - 7x^2 + 13x - 15$

e. $4(x + 2) - 3(5x - 1)$

$-11x + 11$

f. $(2x^2 + x - 1)(x^2 - 3x + 4)$

$2x^4 - 5x^3 + 4x^2 + 7x - 4$

2. Simplify the following. Write all of your results with positive exponents only. Assume that all variables in denominators do not equal 0.

a. $(3x^3)(2x^5)$

$6x^8$

b. $(4x^3y)^2$

$16x^6y^2$

c. $(xy)^2(-2x^3y)$

$-2x^5y^3$

d. $(5x^3y^4z)(-2x^2yz^3)$

$-10x^5y^5z^4$

e. $(3x^2y)^0(3x^3y)^2$

$9x^6y^2$

f. $(-5xy)^3$

$-125x^3y^3$

g. $\dfrac{6x^4}{3x}$

$2x^3$

h. $2x^0$

2

i. $\dfrac{3^3}{3^3}$

1

j. $\dfrac{6xy^4z^2}{4xyz^5}$

$\dfrac{3y^3}{2z^3}$

k. $(-5x^{-3})(x^{-5})$

$-5x^{-8} = \dfrac{-5}{x^8}$

l. $\dfrac{8x^{-4}}{-2x^{-6}}$

$-4x^2$

m. $(-5x^{-3})^3$

$(-5)^3(x^{-3})^3 = -125x^{-9} = \dfrac{-125}{x^9}$

n. $x^{4/5} \cdot x^{1/2}$

$x^{4/5+1/2} = x^{8/10+5/10} = x^{13/10}$

o. $(x^{2/3})^3$

$x^{(2/3)\cdot 3} = x^2$

3. Given $f(x) = 6x - 2$ and $g(x) = -2x + 3$, determine each of the following:

a. $f(-3)$

-20

b. $f(x) + g(x)$

$4x + 1$

c. $f(3) - g(3)$

$f(3) - g(3) = 16 - (-3) = 19$

d. $f(x) \cdot g(x)$

$(6x - 2)(-2x + 3)$

$= -12x^2 + 22x - 6$

e. $f(g(x))$

$f(-2x + 3) = 6(-2x + 3) - 2$

$= -12x + 16$

f. $g(f(2))$

$g(10) = -2(10) + 3 = -17$

g. f^{-1} (Determine the inverse of f.)

$y = 6x - 2$ or $\dfrac{y + 2}{6} = x$ or $y = f^{-1}(x) = \dfrac{x + 2}{6}$

Answers to all Gateway exercises are included in the Selected Answers appendix.

4. Given $f(x) = x^2 - x + 3$ and $g(x) = 3x - 2$, determine each of the following:

a. $f(x) - g(x)$

$x^2 - 4x + 5$

b. $f(x) \cdot g(x)$

$3x^3 - 3x^2 + 9x - 2x^2 + 2x - 6$

$= 3x^3 - 5x^2 + 11x - 6$

c. $f(g(x))$

$f(3x - 2) = (3x - 2)^2 - (3x - 2) + 3$

$= 9x^2 - 12x + 4 - 3x + 2 + 3$

$= 9x^2 - 15x + 9$

d. $g(f(2))$

$g(5) = 3(5) - 2 = 13$

5. Determine the value of each of the following:

a. $49^{1/2}$

7

b. $32^{2/5}$

4

c. $(-27)^{4/3}$

81

d. $7^{3/5}$

3.214

e. $\sqrt[3]{27^2}$

9

f. $\sqrt[4]{16^5}$

32

g. 4^{-2}

$\frac{1}{16}$

6. To ship the mail-order ceramic figures that you produce, you need to make square-bottomed boxes. For the size of the box to be proportional to the figurines, the height of the box must be three times longer than the width. The cost of the material to make the top and bottom of the box molded to fit the figurine is \$0.01 per square inch, and the cost of the material for the sides of the box sells for \$0.004 per square inch.

a. Write a function, f, to represent the cost of producing the top and the bottom of a box. Use x to represent the width of the bottom of the box in inches.

$f(x) = 2(0.01)x^2 = 0.02x^2$

b. Write a function, g, to represent the cost of producing the sides of the box.

$g(x) = 4(0.004)(x)(3x) = 0.048x^2$

c. Combine the functions in parts a and b to write one function, T, that represents the total cost of making the box. Write the equation in simplest form.

$T(x) = f(x) + g(x) = 0.02x^2 + 0.048x^2 = 0.068x^2$

d. Using $f(x)$, $g(x)$, and $T(x) = f(x) + g(x)$ as defined in parts a, b, and c, complete the following table:

x (in.)	f(x)(\$)	g(x)(\$)	T(x) = f(x) + g(x)(\$)
2	0.08	0.19	0.27
4	0.32	0.77	1.09
6	0.72	1.73	2.45
8	1.28	3.07	4.35
10	2.00	4.80	6.80

7. You have a knitting machine in your home, and your business is making ski hats. The fixed cost to run your knitting company is \$300 per month, and the cost to produce each hat averages approximately \$12. The hats will sell for \$25.95.

a. Write a function C to represent the cost of making the hats. Use x to represent the number of hats made per month.

$C(x) = 12x + 300$

b. Write a function R to represent the revenue from the sale of the hats.

$R(x) = 25.95x$

c. Write a function p to represent the profit for the month. Express this function in simplest form.

$p(x) = 25.95x - (12x + 300) = 13.95x - 300$

d. Graph the three functions on your graphing calculator. How many hats must be sold in 1 month to break even?

22 hats must be sold because 21 hats is not quite enough. The solution was obtained graphically.

e. Determine the value of $C(50)$, $R(50)$, and $p(50)$. Explain the practical meaning of the values that you find.

$C(50) = 900$; the cost of producing 50 hats is \$900. $R(50) = 1297.50$; the revenue from 50 hats is \$1297.50. $p(50) = 397.50$; the profit from selling 50 hats is \$397.50.

f. Explain how the difference function $y = f(x) - g(x)$ pertains to this situation.

The profit function is the difference between the revenue and cost functions.

8. The manufacturer of a certain brand of computer printer sells her printers at a wholesale price of \$110 per printer, based on selling 60 printers. Because the warehouse is overstocked and new high-speed printers are arriving, the manufacturer decides to make the following one-time-only offer. The price of each printer ordered in addition to the basic 60 will be reduced by \$2 per extra printer ordered. You may not, however, purchase more than 90 printers.

a. Let x represent the number of printers in excess of 60 that you will purchase. Write a function for the total cost, C, of the printers as a function of the number of printers, x. Call this function f.

$f(x) = 60(110) + x(110 - 2x)$

b. Rewrite the cost function by multiplying the factors and combining like terms.

$f(x) = 6600 + 110x - 2x^2$

c. What is the domain of the cost function?

integers $0 \leq x \leq 30$

d. If you decide that it is to your advantage to purchase 75 printers rather than 60 printers, explain the cost savings to you and your business.

$f(15) = 7800$. At the regular price, the cost is \$8250, so the savings are \$450.

9. Functions f and g are defined by the following tables:

x	−1	0	1	2	4	5	8
$f(x)$	8	5	2	−1	−7	−10	−19

x	−1	0	1	2	4	5	8
$g(x)$	5	1	−1	−1	5	11	41

Determine the values for each of the following:

a. $f(g(4))$

−10

b. $g(f(-1))$

41

c. $f(g(0))$

2

d. $g(f(2))$

5

10. You manufacture snowboards. You cannot produce more than 30 boards per day. The cost, C, of producing x boards is represented by the function

$$C = f(x) = 150x - 0.9x^2.$$

a. What is the practical domain of the function?

integers $0 \leq x \leq 30$

b. What is the cost if 22 boards are produced?

$f(22) = \$2864.40$

If 22 snowboards are produced, the cost is \$2864.40.

c. The number of snowboards that can be produced in t hours is represented by the function $g(t) = 3.75t$. Determine $f(g(t))$.

$f(3.75t) = 150(3.75t) - 0.9(3.75t)^2 = 562.50t - 12.65625t^2$

d. What is the input variable in the composition of the functions in part c?

The input variable is t.

e. Because of a blizzard, your employees work only 4 hours on a certain day. Determine the production cost for that day.

$f(g(4)) = \$2047.50$

f. The company prefers to keep production costs at approximately \$3500. How many hours each day must the company operate to maintain this production cost? Your employees do not work more than 8 hours per day.

$3500 = 562.50t - 12.65625t^2$

t is about 7.5 hours (determined graphically).

11. a. Determine the equation of the inverse of the function $f(x) = \dfrac{2x - 3}{5}$.

$x = \dfrac{2y - 3}{5}$, $y = \dfrac{5x + 3}{2}$, or $f^{-1}(x) = \dfrac{5x + 3}{2}$

b. What is the slope of the line for each function? What is the relationship between the slopes of the two functions?

The slope of f is $\frac{2}{5}$. The slope of f^{-1} is $\frac{5}{2}$. The slopes are reciprocals.

12. a. Show that $f(x) = -2x + 1$ and $g(x) = \dfrac{1 - x}{2}$ are inverse functions of each other.

$f(g(x)) = f\left(\dfrac{1 - x}{2}\right) = -2\left(\dfrac{1 - x}{2}\right) + 1 = x$

$g(f(x)) = g(-2x + 1) = \dfrac{1 - (-2x + 1)}{2} = \dfrac{2x}{2} = x$

Because $f(g(x)) = g(f(x)) = x$, f and g are inverses.

b. Sketch the graphs of the functions in part a on the same axis, and check the result with your graphing calculator.

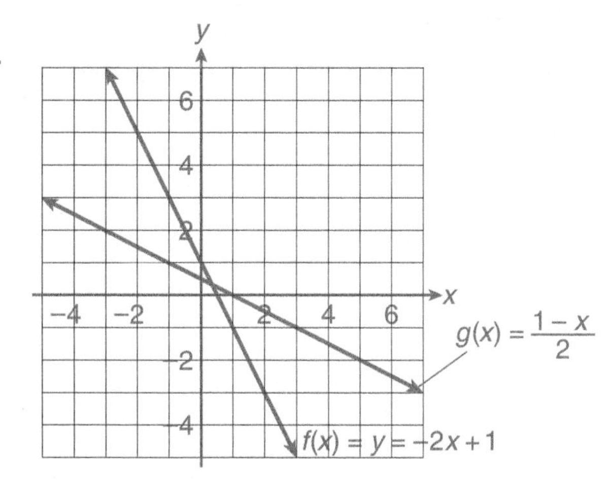

c. What can you say about the graphs of f and g with respect to the graph of $y = x$?

f and g are symmetric with respect to the line $y = x$.

13. You work in the box office of a movie theater. The system for purchasing tickets is automated. You enter the number of tickets you need for each category (adult, child, senior), and the output is the total cost. The following table is a sample from a computer screen that shows the cost for adult tickets only. The number of tickets represents the input, and the total cost represents the output. Let c represent the total cost of the tickets purchased, and let n represent the number of tickets purchased.

NUMBER OF ADULT TICKETS (n)	TOTAL COST (c)
2	\$17.70
5	\$44.25
7	\$61.95
12	\$106.20

a. Does this table represent a linear function? Explain.

Yes, the ratio (change in cost)/(change in number of tickets) is constant.

b. Write a function, f, to represent the total cost, c, as a function of the number of tickets, n, that are purchased.

$f(n) = 8.85n$

c. Determine the cost of one ticket. What does this value represent in your function?

The cost of one ticket is \$8.85. This represents the slope.

Often the customer approaches the window with the exact amount of cash for the tickets. In that case, you enter the total amount into the computer and press adult ticket, and out comes the number of tickets. In this case, the total cost is the input and the number of tickets is the output.

d. Fill in the table showing this situation. (*Hint:* Consider the table above.)

TOTAL COST (c)	NUMBER OF ADULT TICKETS (n)
\$17.70	2
\$44.25	5
\$61.95	7
\$106.20	12

e. Write a function g that represents the total number of tickets, n, purchased as a function of the total cost, c.

$n = g(c) = \frac{1}{8.85}c = \frac{20}{177}c$

f. What is the slope of this line? What is the relationship between the slopes of the two functions f and g?

The slope is $\frac{20}{177}$. The slopes are reciprocals.

g. Determine $f(g(c))$ and $g(f(n))$. Are the functions inverses of each other? Explain.

$f(g(c)) = f\left(\frac{20}{177}c\right) = 8.85\left(\frac{20}{177}c\right) = c$; $g(f(n)) = g(8.85n) = \frac{20}{177}(8.85n) = n$

The functions are inverses because they undo each other.

CHAPTER 3

Exponential and Logarithmic Functions

CLUSTER 1 Exponential Functions

ACTIVITY 3.1

Princess Charlotte and Dracula

Increasing Exponential Functions

OBJECTIVES

1. Determine the growth factor of an exponential function.
2. Identify the properties of the graph of an exponential function defined by $y = b^x$, where $b > 1$.
3. Graph an increasing exponential function.

The birth of Princess Charlotte in May 2015 sparked interest in investigating her family tree. It has been reported that through Queen Mary, the wife of King George V, Princess Charlotte's line goes back to the princes of Romania. Among those ancestors is a man known as Vlad the Impaler, who became Bram Stoker's inspiration for *Dracula*. The line stretches back about 20 generations, which means that Princess Charlotte may have more than 1 million grandparents.

Although you are still somewhat skeptical about the new heir being a descendant of Dracula, the 1 million possible grandparents fascinates you.

1. You start by creating and completing the following table:

GENERATION (input)	NUMBER OF PARENTS/GRANDPARENTS (output)
1	2
2	4
3	$8 = 2^3$
4	$16 = 2^4$
5	$32 = 2^5$
6	$64 = 2^6$
7	$128 = 2^7$

2. Do you notice a pattern in the output values? Describe how you determine the number of parents/grandparents in any given generation, knowing the number in the previous generation.

I multiply the number in the previous generation by 2.

3. Use the pattern you discovered to determine the number of grandparents in the eighth generation.

$2 \cdot 128 = 256$

4. a. The number of possible parents/grandparents in any given generation can be written as a power of 2. Write each parent/grandparent entry in the output column of Problem 1 as a power of 2 (e.g., $2 = 2^1$ and $4 = 2^2$).

b. Let n represent the number of generations. Write an equation for the number of parents/grandparents, P, as a function of n, the number of the generation.

$P = 2^n$

c. Use the equation from part b to determine the number of possible different grandparents in the 20th generation back.

$2^{20} = 1{,}048{,}576$

d. Was the claim of 1 million parents or grandparents exaggerated? Explain.

There are two acceptable answers here. One is no, based on the powers of 2. The other is yes, if students discuss the likelihood of intermarriage that could reduce the number of ancestors.

5. a. What is the practical domain (inputs that make sense for the situation) of the function defined by $P(n) = 2^n$?

The input, n, is the number of generations, so the practical domain is whole numbers from $n = 1$ to 20.

b. Sketch a scatterplot of ordered pairs of the form $(n, P(n))$ for the first 10 generations on an appropriately scaled and labeled axis.

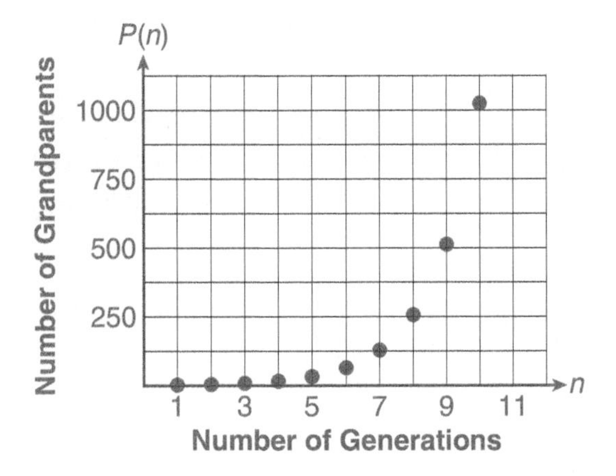

c. Is the function discrete or continuous?

The function is discrete because the function is not defined for real numbers between 1 and 2, 2 and 3, etc.

The function defined by $P(n) = 2^n$ gives the relationship between the number of parents/grandparents $P(n)$, and the given generation, n. This function belongs to a family of functions called **exponential functions**.

Some **exponential functions** can be defined by equations of the form $y = b^x$, where the base b is a constant such that $b > 1$. Such functions are called **exponential** functions because the independent variable (input) x is the exponent.

EXAMPLE 1 ***Examples of exponential functions are***

i. $g(x) = 10^x$, where $b = 10$

ii. $h(x) = 1.08^x$, where $b = 1.08$

Graphs of Increasing Exponential Functions

Because n in $P(n)$ (the Princess Charlotte and Dracula situation) represents the generation, the practical domain (whole numbers from 1 to 20) limits the investigation of the exponential function.

6. a.

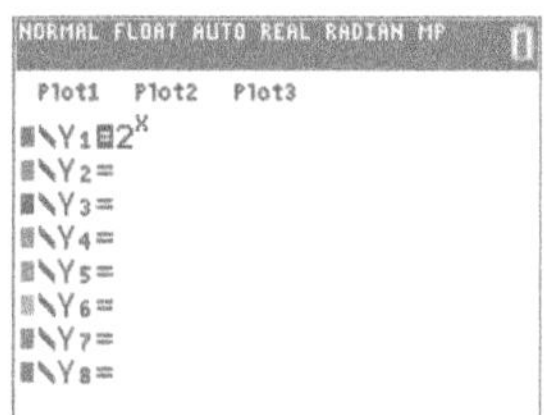

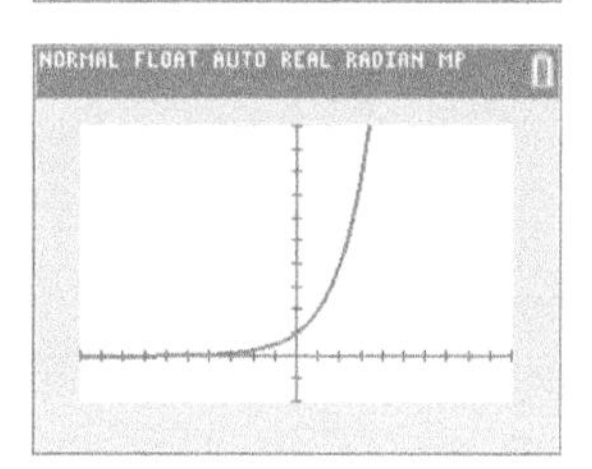

6. a. Consider the general function defined by $f(x) = 2^x$. Use your graphing calculator to sketch a graph of this function. Use the window Xmin $= -10$, Xmax $= 10$, Ymin $= -2$, and Ymax $= 10$.

b. Because the graph of the general function $f(x) = 2^x$ is continuous (it has no holes or breaks), what appears to be the domain of the function f? What is the range of the function f?

The domain of f is the set of all real numbers. The range of f is the set of all positive real numbers.

c. Determine the y-intercept of the graph of f by substituting 0 for x in the equation $y = 2^x$ and solving for y.

$y = 2^0 = 1$ Therefore, the y-intercept is (0, 1).

d. Is the function f increasing or decreasing?

increasing

DEFINITION

If the base b of an exponential function defined by $y = b^x$ is greater than 1, then b is the **growth factor.** The graph of $y = b^x$ is increasing if $b > 1$. For each increase of 1 of the value of the input, the output increases by a factor of b.

EXAMPLE 2 ***The base 2 of $f(x) = 2^x$ is the growth factor because each time the input, x, is increased by 1, the output is multiplied by 2.***

LC LEARNING CATALYTICS

Identify the growth factor for the function defined by $g(x) = 3.7^x$.

7. Identify the growth factor, if any, for the function defined by the given equation.

a. $g(x) = 10^x$

The growth factor is 10.

b. $y = 1.08^x$

The growth factor is 1.08.

c. $y = \left(\frac{4}{3}\right)^x$

The growth factor is $\frac{4}{3}$.

d. $h(x) = 0.8^x$

The base 0.8 is less than 1; therefore, it is not a growth factor.

e. $y = 8x$

This is not an exponential function; therefore, there is no growth factor.

There is a special relationship between the graph of $f(x) = 2^x$ and the x-axis when the input x becomes more negative. Problem 8 investigates this relationship.

8. Return to the graph of $f(x) = 2^x$.

a. Does the graph of $f(x) = 2^x$ appear to have an x-intercept?

No, the graph never touches the x-axis.

b. Use your calculator to complete the following table:

x	−1	−2	−4	−6	−8	−10
$f(x) = 2^x$	0.5	0.25	0.0625	0.0156	0.00391	0.000977

Note: $2^{-10} = \frac{1}{2^{10}} \approx 0.000977$.

c. As the values of the input variable, x, decrease (become more negative), what happens to the output values?

The output values get closer to zero.

d. Use the trace feature of your graphing calculator to trace the graph of $f(x) = 2^x$ for $x < 0$. What appears to be the relationship between the graph of $y = 2^x$ and the x-axis when x decreases (becomes more negative)?

The graph of $f(x) = 2^x$ gets closer to the x-axis as x becomes more negative.

DEFINITION

A horizontal axis having equation $y = 0$ is called a **horizontal asymptote** of the graph of a function defined by $y = b^x$, where $b > 1$. The graph of the function gets closer to the x-axis $(y = 0)$ as the input gets farther from the origin, in the negative direction.

EXAMPLE 3 ***The x-axis is the horizontal asymptote of $y = 7^x$ because as x gets more negative, the graph gets closer and closer to the x-axis. See the following graph:***

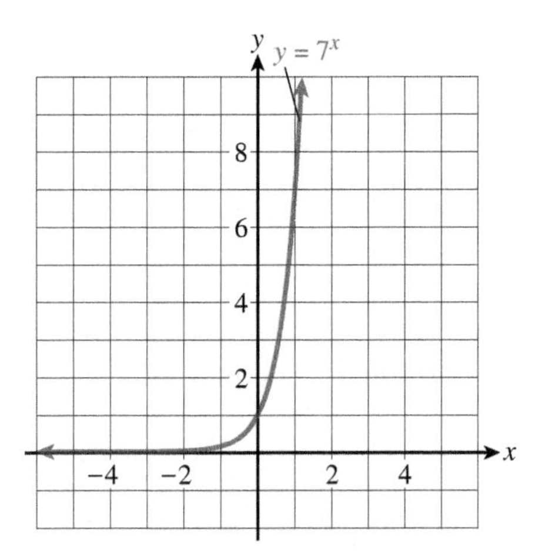

9. a. Complete the following table:

x	−3	−2	−1	0	1	2	3	4	5
$f(x) = 2^x$	0.125	0.25	0.5	1	2	4	8	16	32
$g(x) = 10^x$	0.001	0.01	0.1	1	10	100	1000	10,000	100,000

9. b.

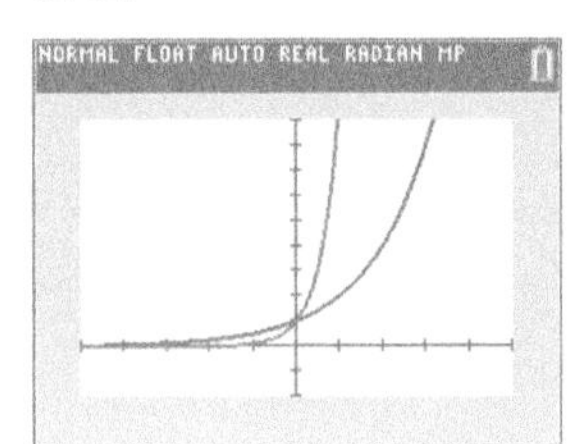

b. Sketch the graph of the functions f and g on your graphing calculator. Use the window Xmin = −5, Xmax = 5, Ymin = −2, and Ymax = 9.

c. Use the results from parts a and b to describe how the graphs of $f(x) = 2^x$ and $g(x) = 10^x$ are similar and how they are different. Be sure to include domain, range, growth factor, x- and y-intercepts, and horizontal asymptotes. Also discuss whether the graph of g increases faster or slower than the graph of f.

The graphs of f and g are both increasing. The domain of each function is the set of all real numbers, and the range is the set of all positive real numbers. The y-intercept is (0, 1) for each graph, and there is no x-intercept. The x-axis is a horizontal asymptote. The graph of g increases at a faster rate than the graph of f because its growth factor, 10, is larger than 2.

10. Examine the output pattern to determine which of the following data sets is linear and which is exponential. For the linear set, determine the slope. For the exponential set, determine the growth factor.

a.

x	−2	−1	0	1	2	3	4
y	−8	−4	0	4	8	12	16

This data is linear; the slope is 4.

b.

x	−2	−1	0	1	2	3	4
y	$\frac{1}{16}$	$\frac{1}{4}$	1	4	16	64	256

This data exponential; the growth factor is 4.

SUMMARY Activity 3.1

Functions defined by equations of the form $y = b^x$, where $b > 1$, are called **exponential functions** and have the following properties:

1. The domain is all real numbers.
2. The range is $y > 0$.
3. If $b > 1$, the function is increasing and has the following general shape:

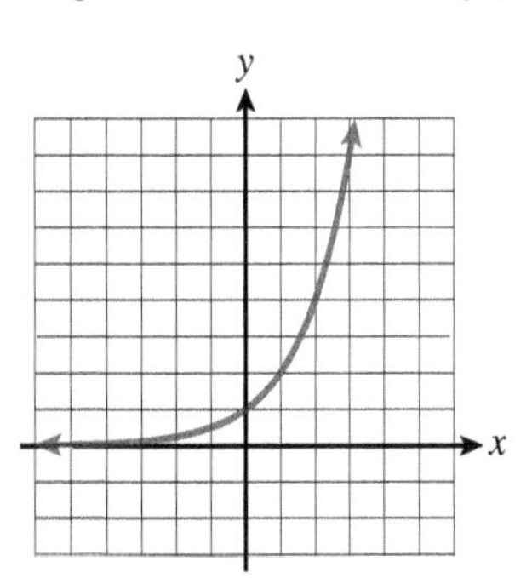

In this case, b is called the **growth factor**.

4. The vertical intercept (y-intercept) is (0, 1).

5. The graph does not intersect the horizontal axis. There is no x-intercept.

6. The line $y = 0$ (the x-axis) is a **horizontal asymptote** to the graph.

7. The function is continuous.

EXERCISES Activity 3.1

1. Your neighbor's son will be attending college in the fall, majoring in mathematics. On July 1, he comes to your house looking for summer work to help pay for his college expenses. You are interested in hiring him to do some odd jobs, but you don't have much extra money to pay him. He can start right away and will work all day, July 1, for 2 cents. This gets your attention, but you wonder whether there is a catch. He says that he will work on July 2 for 4 cents, July 3 for 8 cents, July 4 for 16 cents, etc., for every day in July. Do you think you should hire him?

   a. Complete the following table:

DAY IN JULY (input)	PAY IN CENTS (output)
1	2
2	4
3	8
4	16
5	32
6	64
7	128
8	256

   b. The pay on any given day can be written as a power of 2. Write each pay entry in the output column of the table in part a as a power of 2.

   $2 = 2^1, 4 = 2^2, 8 = 2^3, 16 = 2^4, 32 = 2^5, 64 = 2^6, 128 = 2^7, 256 = 2^8$

   c. Let n represent the number of days worked. Write an equation for the daily pay, $P(n)$ (in cents), as a function of n, the number of days worked. Note that the number of days worked is the same as the July date.

   $P(n) = 2^n$

   d. How much will he earn on July 31? Be sure to indicate the units of your answer. Was it a good idea to hire him? Explain.

   $P(31)$ = 2,147,483,648 cents = $21,474,836.48. No, I could not afford him.

   e. Is the function linear? Explain.

   The function is not linear. The average rate of change is not constant.

f. What is the practical domain of the function defined by $P(n) = 2^n$?

The input, n, is the July date, so the practical domain is whole numbers $n = 1$ to 31.

g. Sketch a scatterplot of ordered pairs of the form $(n, P(n))$ from July 1 to July 8 on an appropriately scaled and labeled axis.

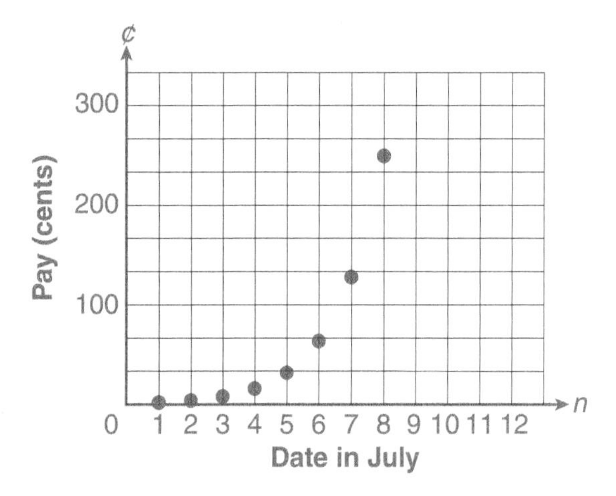

h. Is the function discrete or continuous? Explain.

The function is discrete because the function is not defined for real numbers between 1 and 2, 2 and 3, etc.

i. There is an interesting story about the inventor of chess and grains of wheat. Use the Internet to look up the wheat and chessboard problem and answer the following question:

> If a chessboard were to have wheat placed upon each square such that one grain were placed on the first square, two on the second, four on the third, etc. (doubling the number of grains on each subsequent square), how many grains of wheat would be on the last square of the chessboard at the finish?

$2^{63} = 9.223372036854775808 \times 10^{18} = 9{,}223{,}372{,}036{,}854{,}775{,}808$

9 quintillion 223 quadrillion 372 trillion 36 billion 854 million 775 thousand 808

2. a. Use your graphing calculator to graph the general exponential function defined by $f(x) = 2^x$. Use the window Xmin $= -10$, Xmax $= 10$, Ymin $= -2$, and Ymax $= 10$.

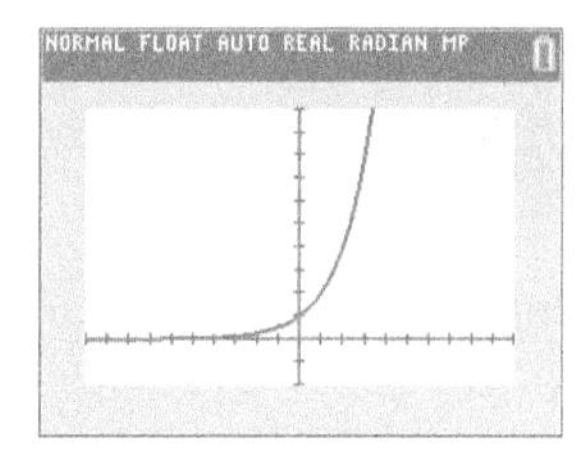

b. Is the general function $f(x) = 2^x$ discrete or continuous? Explain.

The general function is continuous. It has no holes or breaks.

c. What is the domain of the general function f? What is the range of the general function f?

The domain of f is the set of all real numbers. The range of f is the set of all positive real numbers.

d. Determine the y-intercept of the graph of f by substituting zero for x in the equation $y = 2^x$ and solving for y.

$y = 2^0 = 1$. The y-intercept is $(0, 1)$.

e. Is the function increasing or decreasing?

The function is increasing.

3. a. Complete the following tables:

x	−3	−2	−1	0	1	2	3
$h(x) = 5^x$	0.008	0.04	0.2	1	5	25	125

x	−3	−2	−1	0	1	2	3
$g(x) = 2.65^x$	0.0537	0.1424	0.3774	1	2.65	7.0225	18.61

b. Sketch graphs of h and g on the following grid:

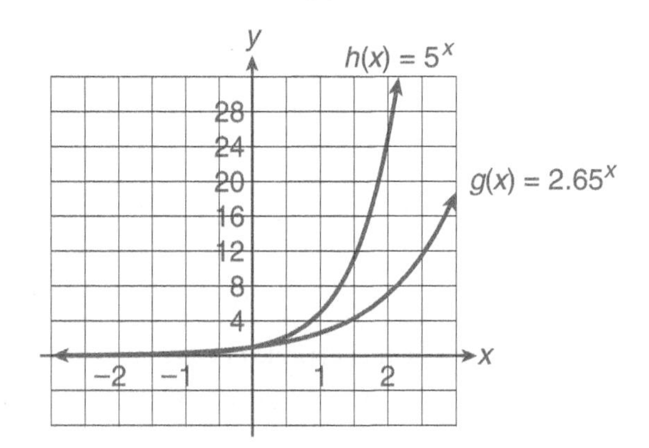

c. Use the tables and graphs in parts a and b to complete the following table:

FUNCTION	BASE, b	GROWTH FACTOR	x-INTERCEPT	y-INTERCEPT	HORIZONTAL ASYMPTOTE	INCREASING OR DECREASING
$h(x) = 5^x$	5	5	none	(0, 1)	$y = 0$	increasing
$g(x) = 2.65^x$	2.65	2.65	none	(0, 1)	$y = 0$	increasing

4. a. Complete the following table:

x	−3	−2	−1	0	1	2	3	4
$f(x) = 3^x$	0.037	0.111	0.333	1	3	9	27	81
$g(x) = x^3$	−27	−8	−1	0	1	8	27	64
$h(x) = 3x$	−9	−6	−3	0	3	6	9	12

b. Sketch a graph of each of the given functions f, g, and h.

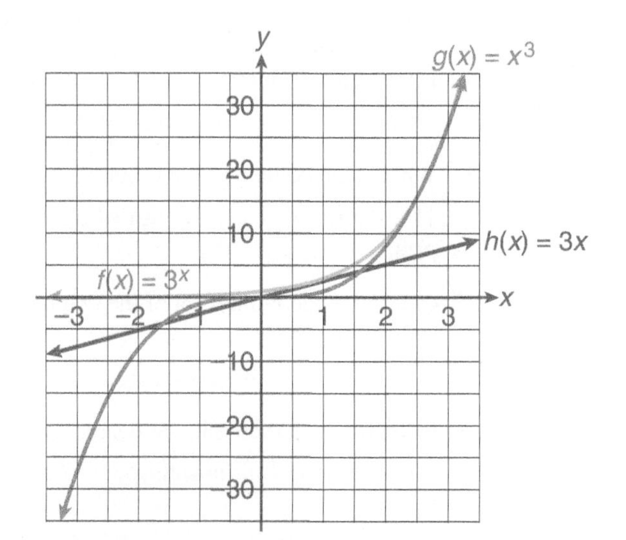

c. Describe any similarities or differences that you observe in the functions. Use the table and the graphs.

All the graphs are increasing. All have the domain of all real numbers. $f(x) = 3^x$ increases fastest for $x > 3$. $f(x) = 3^x$ has a horizontal asymptote ($y = 0$).

5. Determine which of the following data sets are linear and which are exponential. For the linear sets, determine the slope. For the exponential sets, determine the growth factor.

a.

x	−2	−1	0	1	2	3	4
y	$\frac{1}{9}$	$\frac{1}{3}$	1	3	9	27	81

This data is exponential with a growth factor of 3.

b.

x	−2	−1	0	1	2	3	4
y	2	2.5	3	3.5	4	4.5	5

This data is linear with a slope of 0.5.

c.

x	−2	−1	0	1	2	3	4
y	0.75	1.5	3	6	12	24	48

This data is exponential with a growth factor of 2.

6. Would you expect $f(x) = 3^x$ to increase more quickly or more slowly than $g(x) = 2.5^x$ for $x > 0$? Explain. (*Hint:* You may want to use your graphing calculator for help.)

The function f will increase more quickly than g because its growth factor is larger.

7. Take a sheet of paper from your notebook. Let x represent the number of times you fold the paper in half and $f(x)$ represent the number of sections the paper is divided into after the folding.

a. Complete the table of values.

x	0	1	2	3	4	5
$f(x)$	1	2	4	8	16	32

b. If you could fold the paper eight times, how many individual sections will there be on the paper?

256

c. Does this data represent an exponential function? Explain.

The data is exponential with a growth factor of 2.

d. What is the practical domain and range in this situation?

(Answers will vary.) The practical domain is the set of nonnegative integers from 0 to 10. The practical range is the set of whole-number powers of 2 up to 2^{10}.

ACTIVITY

3.2

Half-Life of Drugs or Medications

Decreasing Exponential Functions

OBJECTIVES

1. Determine the decay factor of an exponential function.
2. Graph a decreasing exponential function.
3. Identify the properties of an exponential functions defined by $y = b^x$, where $b > 0$ and $b \neq 1$.

The term half-life refers to the amount of time it takes a person's body to break down and eliminate half of the initial dose of any drug or medication. This is important information to know because some drugs stay in the system longer than others. This may present a risk because although people may not still feel the effects of a certain medication, they may take something that interacts with it. For this reason, doctors and pharmacies are vigilant about obtaining and recording all the drugs an individual is taking and noting the timing of when the ingestion of any drug or medication should take place. The half-life of most cholesterol medications, like Lipitor or Atorvastatin, is approximately 24 hours, so the timing of the ingestion is not an issue if it is consistent from one day to the next.

1. a. What fraction of a dose of Lipitor is left in your body after 1 day?

$\frac{1}{2}$ of the dose remains after 1 day.

b. What fraction of the dose of Lipitor is left in your body after 2 days? (This is one-half the result from part a.)

$\frac{1}{4}$ of the dose remains after 2 days.

c. Complete the following table. Let t represent the number of days since a dose of Lipitor is taken, and let Q represent the fraction of the dose of Lipitor remaining in your body.

t, DAYS	Q
0	1
1	$\frac{1}{2}$
2	$\frac{1}{4}$
3	$\frac{1}{8}$
4	$\frac{1}{16}$

2. a. The values of the fraction of the dosage, Q, can be written as powers of $\frac{1}{2}$. For example, $1 = \left(\frac{1}{2}\right)^0, \frac{1}{2} = \left(\frac{1}{2}\right)^1$, etc. Complete the following table by writing each value of Q in the above table as a power of $\frac{1}{2}$. The values for 1 and $\frac{1}{2}$ have already been entered.

t, DAYS	Q
0	$\left(\frac{1}{2}\right)^0$
1	$\left(\frac{1}{2}\right)^1$
2	$\left(\frac{1}{2}\right)^2$
3	$\left(\frac{1}{2}\right)^3$
4	$\left(\frac{1}{2}\right)^4$

b. Use the result of part a to write an equation for Q in terms of t.

$Q = \left(\frac{1}{2}\right)^t$, or $Q = 0.5^t$

c. What is the practical domain of the half-life function?

(Answers will vary.) The practical domain is $0 \le t \le 10$.

d. Sketch a scatterplot of the data in part a on an appropriately scaled and labeled coordinate axis.

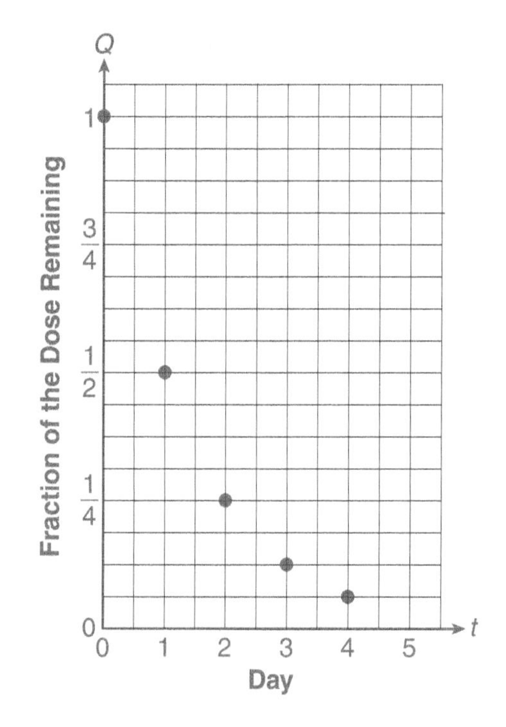

e. Is this function discrete or continuous?

The function is continuous because the function is defined for all values in the practical domain and the graph will not have any holes or breaks.

Notice that the equation $Q = \left(\frac{1}{2}\right)^t$ fits the equation form of an exponential function, $y = b^x$, given in the previous activity. However, the value of b for this exponential function is $\frac{1}{2}$, which is not greater than 1. Therefore, the base $\frac{1}{2}$ is not a growth factor.

When the base, b, of an exponential function is between 0 and 1, the base is called a **decay factor**. The result is a **decreasing exponential function.**

Graphs of Decreasing Exponential Functions

Because t in the equation $Q = \left(\frac{1}{2}\right)^t$ represents time after the drug has been taken, the function is not defined for negative values of t and therefore limits the investigation of the function. Therefore, consider the general function defined by $g(x) = \left(\frac{1}{2}\right)^x$.

3. a. Complete the following table:

x	−3	−2	−1	0	1	2	3	4	5
$g(x) = \left(\frac{1}{2}\right)^x$	8	4	2	1	0.5	0.25	0.125	0.0625	0.03125

b. Describe how you can obtain the output value for $x = 6$, using the output value for $x = 5$.

Multiply $g(5)$ by $\frac{1}{2}$ to get $g(6)$.

c. Sketch the graph of $g(x) = \left(\frac{1}{2}\right)^x$ for $-6 \le x \le 6$ and $-4 \le y \le 10$.

Verify your sketch using your graphing calculator.

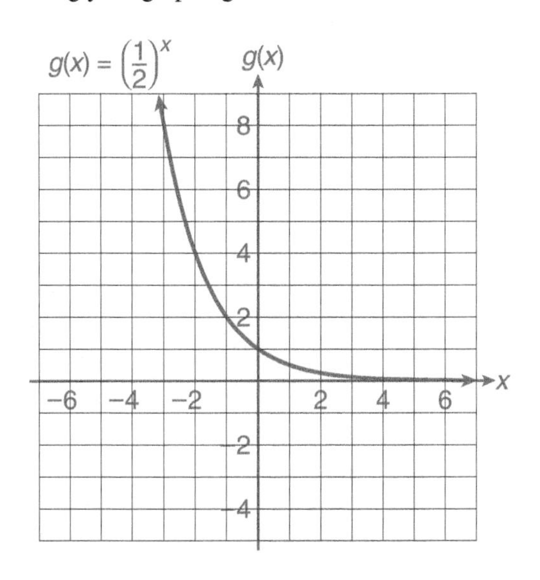

d. What are the domain and range of the function g?

The domain is the set of all real numbers, and the range is the set of all positive real numbers.

e. Determine the vertical intercept of the graph of g.

(0, 1)

f. Is the function g increasing or decreasing?

decreasing

DEFINITION

If the base b of an exponential function $y = b^x$ is between 0 and 1, then b is the **decay factor**. The graph of $y = b^x$ is decreasing if $0 < b < 1$. For each increase of 1 in the value of the input, the output decreases by a factor of b.

EXAMPLE 1 ***The base $\frac{1}{2}$ in the function $g(x) = \left(\frac{1}{2}\right)^x$ is the decay factor because each time x is increased by 1, the output value is multiplied by $\frac{1}{2}$.***

4. Identify the decay factor, if any, for the given function.

a. $g(x) = \left(\frac{2}{7}\right)^x$

The decay factor is $\frac{2}{7}$.

b. $y = 0.98^x$

The decay factor is 0.98.

c. $h(x) = 1.8^x$

The base 1.8 is greater than 1, so it is a growth factor.

d. $y = 0.8x$

$y = 0.8x$ defines a linear function, so there is no decay factor.

5. Return to the graph of $g(x) = \left(\frac{1}{2}\right)^x$.

a. Does the graph of $g(x) = \left(\frac{1}{2}\right)^x$ have an x-intercept?

no

b. Complete the following tables:

x	1	3	5	7	10
$g(x) = \left(\frac{1}{2}\right)^x$	0.5	0.125	0.03125	0.00781	0.000977

c. As the values of the input variable x get larger, what happens to the output values?

The output values get closer to zero.

d. Does the graph of g have a horizontal asymptote? Explain.

Yes, the horizontal asymptote is the x-axis ($y = 0$).

6. a. For each of the following exponential functions, identify the base, b, and determine whether the base is a growth or decay factor. Graph each function on your graphing calculator, and complete the table below.

FUNCTION	BASE, b	GROWTH OR DECAY FACTOR	x-INTERCEPT	y-INTERCEPT	HORIZONTAL ASYMPTOTE	INCREASING OR DECREASING
$h(x) = 1.08^x$	1.08	growth	none	(0, 1)	$y = 0$	increasing
$T(x) = 0.75^x$	0.75	decay	none	(0, 1)	$y = 0$	decreasing
$f(x) = 3.2^x$	3.2	growth	none	(0, 1)	$y = 0$	increasing
$r(x) = \left(\frac{1}{4}\right)^x$	$\frac{1}{4}$	decay	none	(0, 1)	$y = 0$	decreasing

b. Without graphing, how might you determine which of the functions in part a increase and which decrease? Explain.

If the base is greater than 1, the function is increasing. If the base is between 0 and 1, the function is decreasing.

7. Examine the output pattern to determine which of the following data sets is linear and which is exponential. For the linear set, determine the slope. For the exponential set, determine the growth or decay factor.

a.

x	−2	−1	0	1	2	3	4
y	−6	−3	0	3	6	9	12

This data is linear; the slope is 3.

b.

x	−2	−1	0	1	2	3	4
y	9	3	1	$\frac{1}{3}$	$\frac{1}{9}$	$\frac{1}{27}$	$\frac{1}{81}$

This data is exponential; the decay factor is $\frac{1}{3}$.

LC LEARNING CATALYTICS

Determine the decay factor of the function represented by the following table:

x	-1	0	1	2	3
y	27	9	3	1	$\frac{1}{3}$

8. Determine the decay factor of the function represented by the data, and complete the table.

x	-2	-1	0	1	2
$f(x)$	16	4	1	0.25	0.0625

The decay factor is 0.25.

SUMMARY Activity 3.2

Functions defined by equations of the form $y = b^x$, where $b > 0$ and $b \neq 1$, are called **exponential functions** and have the following properties:

1. The domain is all real numbers.

2. The range is $y > 0$.

3. If $0 < b < 1$, the function is decreasing and has the general shape below.

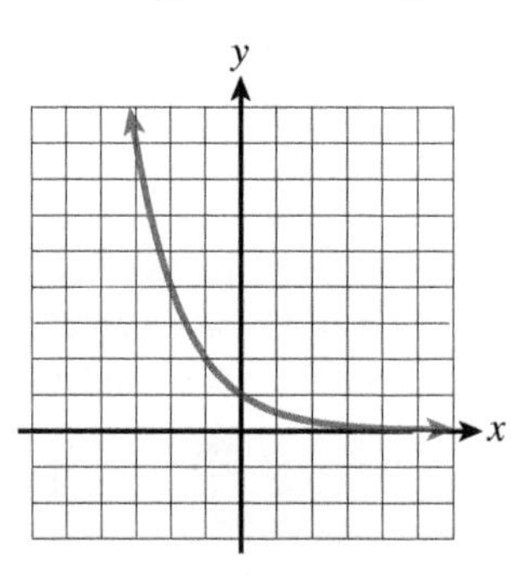

In this case, b is called the **decay factor**.

4. If $b > 1$, the function is increasing and has the general shape below.

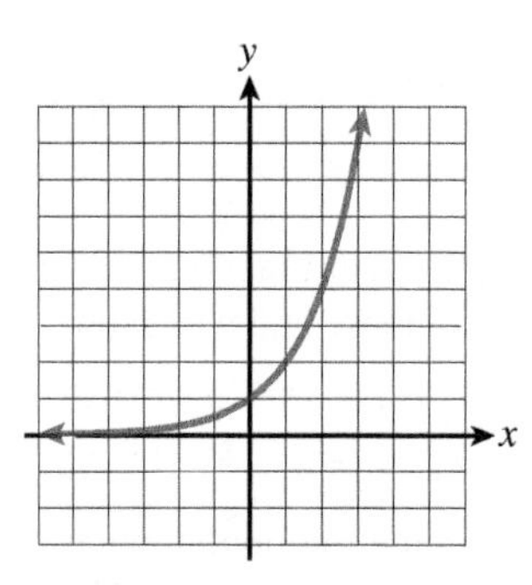

In this case, b is called the **growth factor**.

5. The vertical intercept (y-intercept) is $(0, 1)$.

6. The graph does not intersect the horizontal axis. There is no x-intercept.

7. The line $y = 0$ (the x-axis) is a **horizontal asymptote to the graph.**

8. The function is continuous.

EXERCISES Activity 3.2

1. a. Complete the following tables:

x	−3	−2	−1	0	1	2	3
$h(x) = 0.35^x$	23.32	8.16	2.86	1	0.35	0.1225	0.043

x	−3	−2	−1	0	1	2	3
$g(x) = \left(\frac{1}{5}\right)^x$	125	25	5	1	0.2	0.04	0.008

b. Sketch graphs of h and g on the following grid:

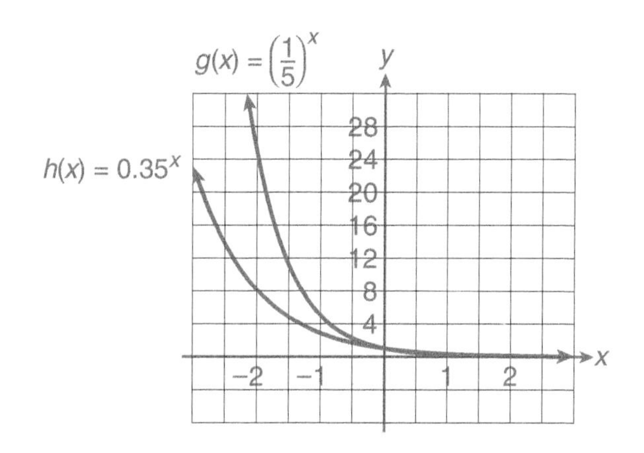

c. Use the tables and graphs in parts a and b to complete the following table:

FUNCTION	BASE, b	GROWTH OR DECAY FACTOR	x-INTERCEPT	y-INTERCEPT	HORIZONTAL ASYMPTOTE	INCREASING OR DECREASING
$h(x) = 0.35^x$	0.35	decay	none	(0, 1)	$y = 0$	decreasing
$g(x) = \left(\frac{1}{5}\right)^x$	$\frac{1}{5}$	decay	none	(0, 1)	$y = 0$	decreasing

2. Using your graphing calculator, investigate the graphs of the following pairs of functions. Describe any relationships within each pair, including domain and range, growth and decay factors, vertical and horizontal intercepts, and asymptotes. Identify the functions as increasing or decreasing.

a. $f(x) = \left(\frac{3}{4}\right)^x, g(x) = \left(\frac{4}{3}\right)^x$

The graph of f is a decreasing exponential function with a decay factor of $\frac{3}{4}$.
The graph of g is an increasing exponential function with a growth factor of $\frac{4}{3}$.
The graphs of f and g are reflections in the y-axis. Both functions have a vertical intercept of (0, 1). Both functions have a domain of all real numbers and a range of all positive numbers.

b. $f(x) = 10^x, g(x) = -10^x$

The graph of *f* is an increasing exponential function with a growth factor of 10. The graph of *f* has a vertical intercept of (0, 1). The graph of *g* has a vertical intercept of (0, −1). Both have a domain of all real numbers. The range of *f* is all positive real numbers. The range of *g* is all negative real numbers. Both functions have a horizontal asymptote of *y* = 0.

c. $f(x) = 3^x, g(x) = \left(\frac{1}{3}\right)^x$

The graph of *f* is an increasing exponential function with a growth factor of 3. The graph of *g* is a decreasing exponential function with a decay factor of $\frac{1}{3}$. The graphs of *f* and *g* are reflections in the *y*-axis. Both functions have a vertical intercept of (0, 1). Both functions have a horizontal asymptote of *y* = 0. Both functions have a domain of all real numbers and a range of all positive real numbers.

3. Determine which of the following data sets are linear and which are exponential. For the linear sets, determine the slope. For the exponential sets, determine the growth factor or the decay factor.

a.

x	−2	−1	0	1	2	3	4
y	0.5	1	1.5	2	2.5	3	3.5

This data is linear with a slope of 0.5.

b.

x	−2	−1	0	1	2	3	4
y	0.50	2	8	32	128	512	2048

This data is exponential with a growth factor of 4.

c.

x	−2	−1	0	1	2	3	4
y	6.25	2.5	1	0.4	0.16	0.064	0.0256

This data is exponential with a decay factor of 0.4.

4. Assume that *y* is an exponential function of *x*.

a. If the growth factor is 1.25, then complete the following table:

x	0	1	2	3
y	10.5	13.125	16.406	20.508

b. If the decay factor is 0.75, then complete the following table:

x	0	1	2	3
y	10	7.5	5.625	4.21875

5. a. Would you expect $f(x) = 5^x$ to increase more quickly or more slowly than $g(x) = 7^x$ for $x > 0$? Explain. (*Hint:* You may want to use your graphing calculator for help.)

The function *f* will increase more slowly than *g* because its growth factor is smaller.

b. Would you expect $f(x) = \left(\frac{1}{2}\right)^x$ to decrease more quickly or more slowly than $g(x) = (0.70)^x$ for $x > 0$? Explain.

The function *f* will decrease more quickly than *g* because its decay factor is smaller, cutting the outputs by 50% while *g* cuts the outputs by 30%.

6. Determine the domain and range of each of the following exponential functions:

a.

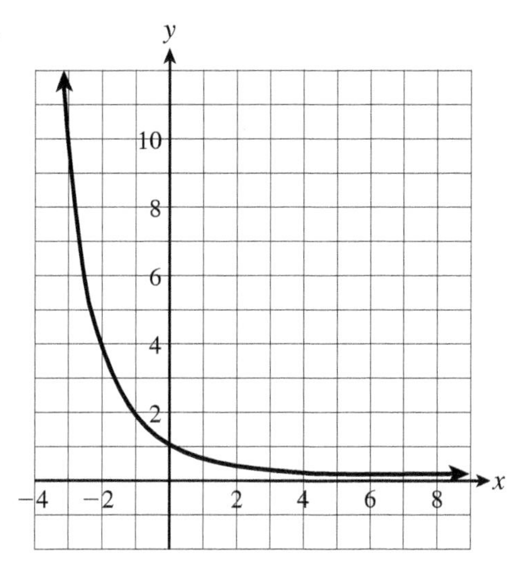

The domain is all real numbers; the range is $y > 0$.

b.

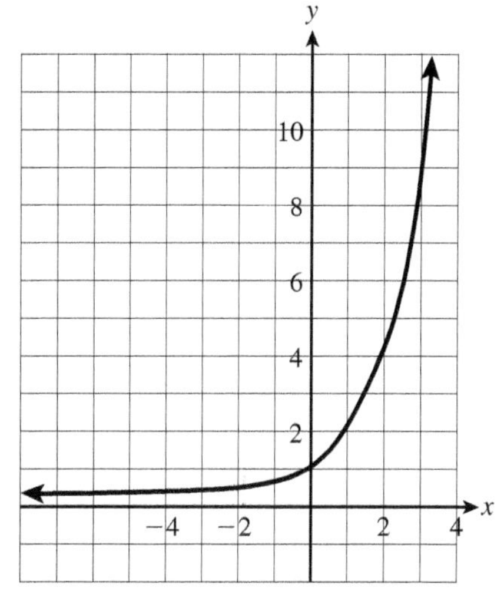

The domain is all real numbers; the range is $y > 0$.

ACTIVITY 3.3

Spotify
More Growth and Decay Factors

OBJECTIVES

1. Determine the growth and decay factor for an exponential function represented by a table of values or an equation.
2. Graph exponential functions defined by $y = ab^x$, where $b > 0$ and $b \neq 1$, $a \neq 0$.
3. Determine the doubling or halving time of an exponential function.

Spotify is a music, podcast, and video streaming service that originated and is still based in Stockholm, Sweden. Since 2012, the number of active users worldwide has increased dramatically.

The following table summarizes the number of Spotify users from June 2012 to June 2017:

YEARS SINCE JUNE 2012, t	NUMBER OF ACTIVE SPOTIFY USERS, n (in millions)
0	15
1	24
2	40
3	75
4	100
5	140

1. Does the table represent a linear function? How do you know?

   The table does not represent a linear function. The rate of change is not constant.

2. a. Use the data in the table above to evaluate the following ratios to complete the table:

USERS IN 2013 / USERS IN 2012	USERS IN 2014 / USERS IN 2013	USERS IN 2015 / USERS IN 2014	USERS IN 2016 / USERS IN 2015	USERS IN 2017 / USERS IN 2016
1.6	1.67	1.88	1.33	1.4

   b. What do you notice about all of the values in the table?

   The values are actually rather close together. They vary from 1.33 to 1.88.

   c. Compute the average of the table values in the second row of part a.

   1.58

In an exponential function with base b, equally spaced input values yield output values whose successive ratios are constant. If the input values increase by increments of 1, the common ratio is the base b. If $b > 1$, b is the growth factor: if $0 < b < 1$, b is the decay factor.

The ratios computed in Problem 2a are not exactly the same, but as noted in 2b, they are relatively close in value. The value computed in 2c may be a good approximation of a growth factor for the Spotify users function.

Once you know the growth factor, b, you can create an equation that will model the Spotify users function. The output will be the number of Spotify users worldwide measured in millions, and the input will be the number of years, t, since 2012. Note that $t = 0$ corresponds to June 2012, $t = 1$ corresponds to June 2013, etc.

3. a. Using 1.58 as the value for b, complete the following table:

t	CALCULATION OF THE NUMBER OF SPOTIFY USERS (millions)	EXPONENTIAL FORM	NUMBER OF SPOTIFY USERS (millions)
0	15	$15(1.58)^0$	15
1	$15 \cdot 1.58$	$15 \cdot 1.58^1$	23.7
2	$15 \cdot 1.58 \cdot 1.58$	$15 \cdot 1.58^2$	37.4
3	$15 \cdot 1.58 \cdot 1.58 \cdot 1.58$	$15 \cdot 1.58^3$	59.2
4	$15 \cdot 1.58 \cdot 1.58 \cdot 1.58 \cdot 1.58$	$15 \cdot 1.58^4$	93.5
5	$15 \cdot 1.58 \cdot 1.58 \cdot 1.58 \cdot 1.58 \cdot 1.58$	$15 \cdot 1.58^5$	147.7

b. Use the pattern of the preceding table to help you write an equation of the form $n(t) = a \cdot b^t$ to model the Spotify users function represented in the table on page 284.

$n(t) = 15 \cdot 1.58^t$

c. What is the practical domain of this function?

The practical domain is the whole numbers from 0 to 5

4. a. Use your graphing calculator to create a scatterplot of the original data given for the function.

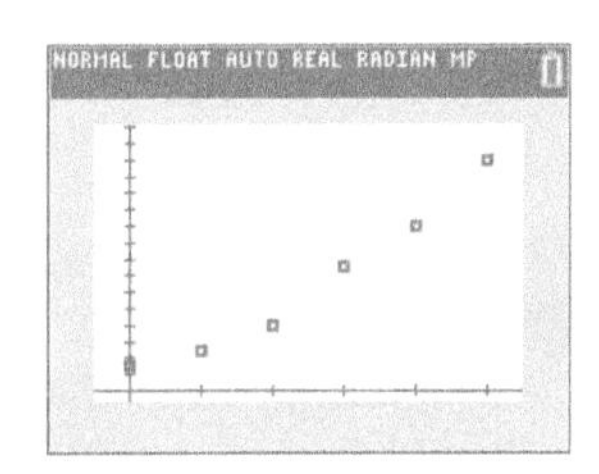

b. Add the model from Problem 3b to the scatterplot on your calculator.

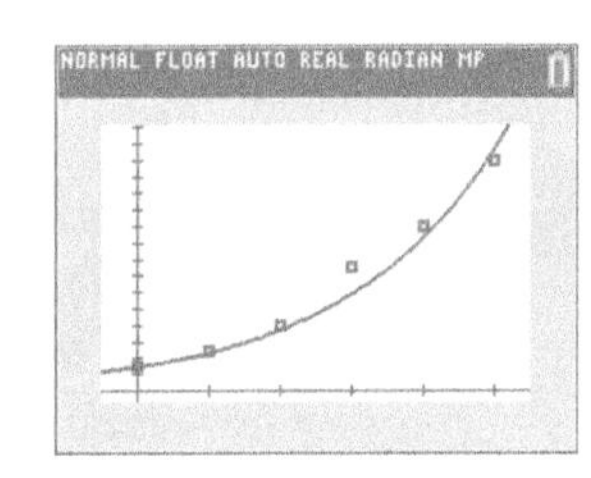

c. Do you believe the function is a reasonable model for the Spotify users function? Explain.

The model passes quite closely to the points of the scatterplot, so yes, the function is a reasonable model on the practical domain.

d. Determine the vertical intercept of the model by substituting 0 for the value of t. What is the practical meaning of the vertical intercept in this situation?

(0, 15). There were 15 million Spotify users worldwide in 2012.

DEFINITION

Many exponential functions can be represented symbolically by $f(t) = a \cdot b^t$, where a is the value of f when $t = 0$ and b is the growth or decay factor. If the input, t, of $y = a \cdot b^t$ represents time, then the coefficient a is called the initial value.

EXAMPLE 1 ***The exponential function defined by $f(x) = 5 \cdot 2^x$ has y-intercept $(0, 5)$ and growth factor $b = 2$. The exponential function defined by $h(x) = \frac{1}{2}(0.75)^x$ has y-intercept $\left(0, \frac{1}{2}\right)$ and decay factor $b = 0.75$.***

5. Use the Spotify function defined by $n(t) = 15(1.58)^t$ to estimate the number of Spotify users in 2020 ($t = 8$). Do you think this is a good estimate? Explain.

$n(8) = 15(1.58)^8$ approximately $= 582.57$. The function predicts the number of Spotify users in 2020 to be 582,570,000. This is not reliable because $t = 8$ is well outside of the original data.

6. a. Use the graph of the Spotify function $n(t) = 15(1.58)^t$ and the graph or table feature of your graphing calculator to estimate the number of years it takes for the number of users to double from 15 million to 30 million.

The number of Spotify user will double in about 1.5 years.

b. Estimate the number of years necessary for the number of Spotify users to double from 30 million to 60 million. Verify your estimate using your calculator.

The number of Spotify users will double in 1.5 years.

c. How long will it take for the number of Spotify users to double, in general?

1.5 years

DEFINITION

The **doubling time** of an exponential function is the time it takes an output to double. The doubling time is determined by the growth factor and remains the same for all output values.

EXAMPLE 2 ***The balance $B(t)$, in dollars, of an investment account is defined by $B(t) = 5500(1.12)^t$, where t is the number of years. The initial value for this function is \$5500. Determine the value of t when the balance is doubled or equal to \$11,000.***

LC LEARNING CATALYTICS

The following table represents an exponential function;

x	-1	0	1	2	3
y	$\frac{4}{3}$	2	3	4.5	6.75

Determine the growth factor.

SOLUTION

Solve $5500(1.12)^t = 11000$ for t. $t \approx 6.1$. It takes about 6.1 years for the balance to double, as indicated by the following table and graph:

NORMAL FLOAT AUTO REAL DEGREE MP
PRESS + FOR △Tbl

X	Y1
5.8	10613
5.9	10734
6	10856
6.1	10980
6.2	11105
6.3	11231
6.4	11359
6.5	11489
6.6	11620
6.7	11752
6.8	11886

X=6.1

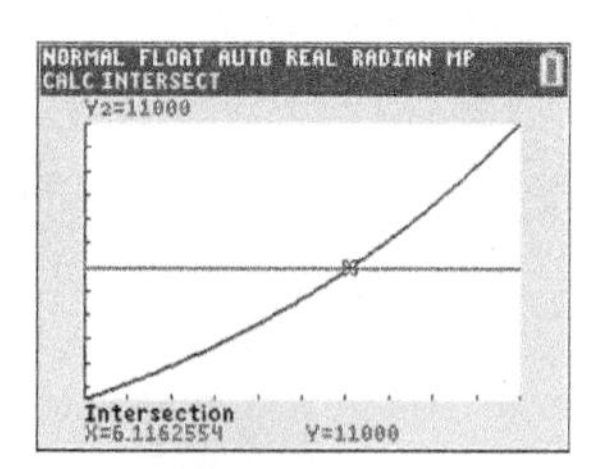

Decreasing Exponential Functions, Decay Factor, and Halving Time

You just purchased a new automobile for \$22,000. Much to your dismay, you learned that you should expect the value of your car to depreciate by 30% per year! The following table shows the book value of the car for the next several years, where V is the value in thousands of dollars.

DEPRECIATION: TAKING ITS TOLL

t (year)	0	1	2	3	4
V(*t*) (thousands of dollars)	22	15.4	10.8	7.5	5.3

When a quantity is increased or decreased by a constant percent rate, it can be modeled by an exponential function. In this situation, the car value is decreased by 30% per year. A decreasing exponential function has a decay factor, b, with $0 < b < 1$. For consecutive values of the input, an output value is determined by multiplying the previous output value by b.

7. As the input, t, increases from 0 to 1, the output, V, decreases from 22 to 15.4 (in \$1000).

a. Determine the value that 22 is multiplied by to get 15.4.

$15.4 \div 22 = 0.7$

b. Use the result from part a to complete the following table:

t	CALCULATION OF THE VALUE OF THE CAR	EXPONENTIAL FORM	VALUE (in \$1000)
0	22	$22(0.7)^0$	22
1	22(0.7)	$22(0.7)^1$	15.4
2	22(0.7)(0.7)	$22(0.7)^2$	10.8
3	22(0.7)(0.7)(0.7)	$22(0.7)^3$	7.5
4	22(0.7)(0.7)(0.7)(0.7)	$22(0.7)^4$	5.3

c. Use the pattern in the preceding table to write an equation in the form $V = a \cdot b^t$ that gives the car value, V, as a function years, t.

$V = 22(0.7)^t$

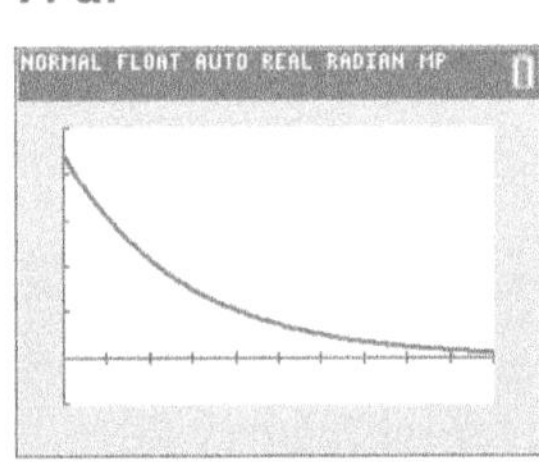

d. Input the function into Y_1 on your calculator, and graph it in an appropriate window.

e. Determine the value of the vertical intercept of the graph. Write the result as an ordered pair.

(0, 22)

f. What is the practical meaning of the vertical intercept in this situation?

The car cost \$22,000 when it was new ($t = 0$).

8. **a.** Estimate the number of years, t, it takes the value of the car to be \$11,000, half the original value. (*Hint:* Put 11,000 in Y_2, and find the point of intersection.)

   $t \approx 1.94$ years

   b. How many years will it take the car value to be halved again, that is, from \$11,000 to \$5500?

   another 1.94 years

DEFINITION

The **half-life** of an exponential function is the time it takes an output to decay by one-half. The half-life is determined by the decay factor and remains the same for all output values.

EXAMPLE 3

The population of Detroit, Michigan, can be modeled by the function $D(t) = 1022.8(0.983)^t$ where t represents the number of years from 1990 and $D(t)$ is the population of Detroit in thousands. If the population of Detroit continues to decline at the same rate, determine the value of t so that the population of Detroit will be one-half the 1990 population.

SOLUTION

The function indicates that the 1990 Detroit population was 1022.8 thousand or 1,022,800 people ($D(0) = 1022.8(0.983)^0 = 1022.8$). We need to solve the equation $1022.8(0.983)^t = \frac{1}{2}(1022.8) = 511.4$ for t. Using the table feature of the graphing calculator, we see the value of t is approximately 40.4 years from 1990 (2030). The intersection of the graphs confirms this result.

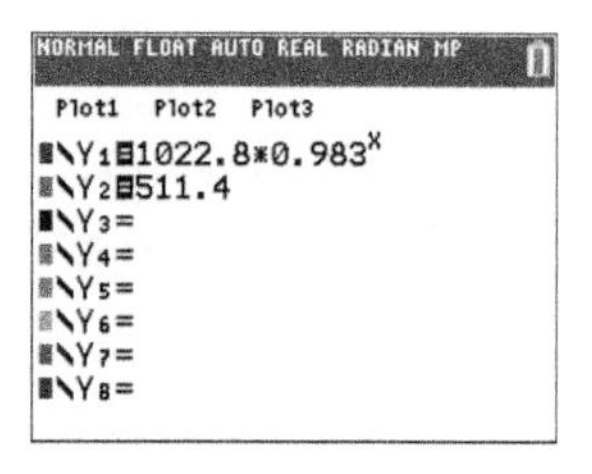

NORMAL FLOAT AUTO REAL RADIAN MP
PRESS + FOR ΔTbl

X	Y1	Y2
39.8	516.92	511.4
39.9	516.03	511.4
40	515.15	511.4
40.1	514.26	511.4
40.2	513.38	511.4
40.3	512.5	511.4
40.4	511.63	511.4
40.5	510.75	511.4
40.6	509.87	511.4
40.7	509	511.4
40.8	508.13	511.4

X=40.4

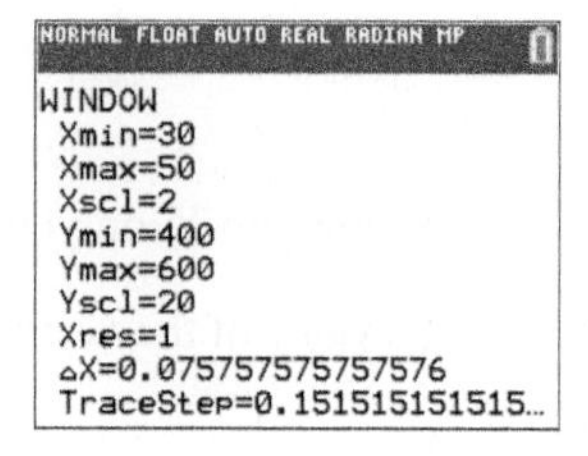

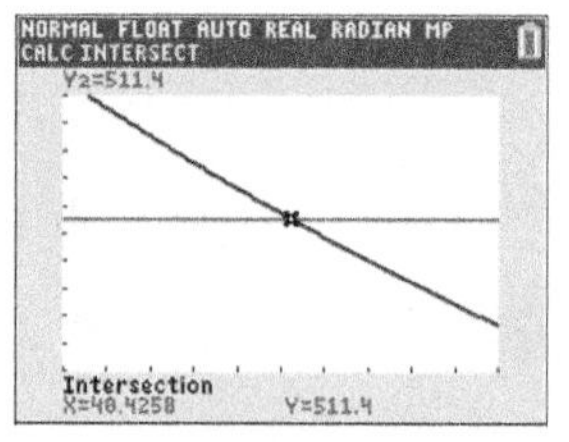

9. Inflation means that a current dollar will buy less in the future. According to the U.S. Consumer Price Index, the inflation rate for the 12 months from January 2016 to January 2017 was 2.5%. This means that a 1-pound loaf of white bread that cost a dollar in January 2016 cost \$1.03 (rounded) in January 2017. The change in price is usually expressed as an annual percentage rate, known as the inflation rate.

a. If you assume that the inflation rate remains at 2.5% per year for the next decade, you can calculate the cost of a pizza that currently costs $12 for each year over the next decade. Complete the following table where t represents the number of years from January 2020 and C is the price of a pizza, in dollars. Round your results to the nearest cent.

t	0	1	2	3	4	5	6	7	8	9	10
C	12.00	12.30	12.61	12.92	13.25	13.58	13.92	14.26	14.62	14.99	15.36

b. What is the initial value?

The point (0, 12) indicates an initial value of $12.

c. Determine the growth factor by dividing one output value by the previous output value.

The growth factor is 1.025

d. Determine the exponential function in the form $C(t) = a \cdot b^t$ that represents the cost of a $12 pizza t years from January 2020.

$C(t) = 12(1.025)^t$

e. Explain the meaning of the values a and b in this situation.

$a = 12$, the cost of a pizza in January 2020 ($t = 0$) was $12.00;
$b = 1.025$, the growth factor corresponding to an annual growth rate of 2.5%.

f. Determine $C(5)$ and interpret the result.

$C(5) = 12(1.025)^5 = 13.58$ (rounded). The cost of a pizza 5 years from January 2020 (2025) is $13.58.

g. Graph the function $C(t) = 12(1.025)^t$ on your graphing calculator with $-20 \le t \le 20$ and $0 \le C(t) \le 20$.

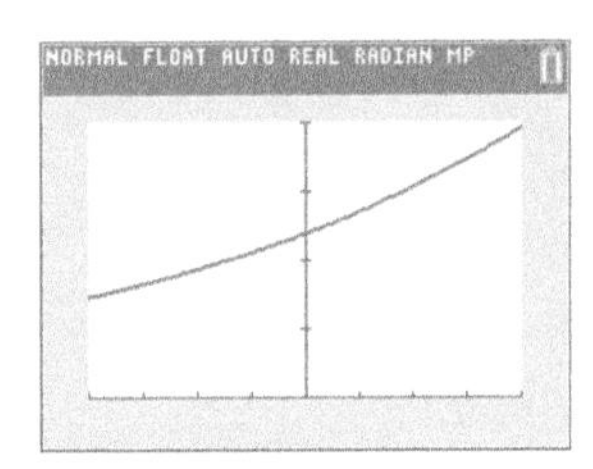

h. Is the entire graph in part g relevant to the original problem? Explain.

No. We assumed the 2.5% inflation rate was from January 2020 onward; therefore, the practical domain is only nonnegative values.

i. Resize the window to include only the first quadrant with $0 \le t \le 20$ and $0 \le C(t) \le 20$. Graph the function with the same window.

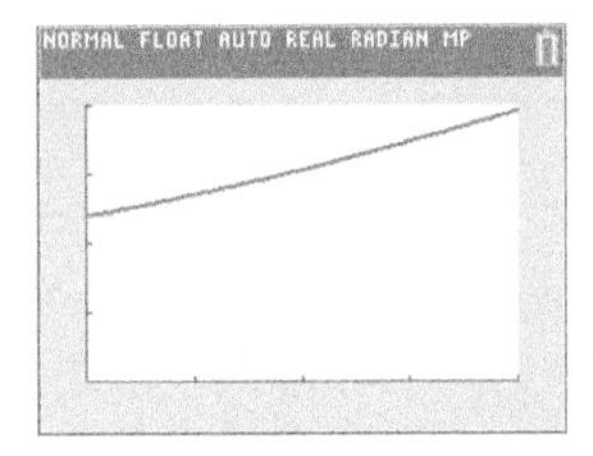

j. How many years will it take the cost of the pizza to double?

Solve $12(1.025)^t = 24$ for t as follows:

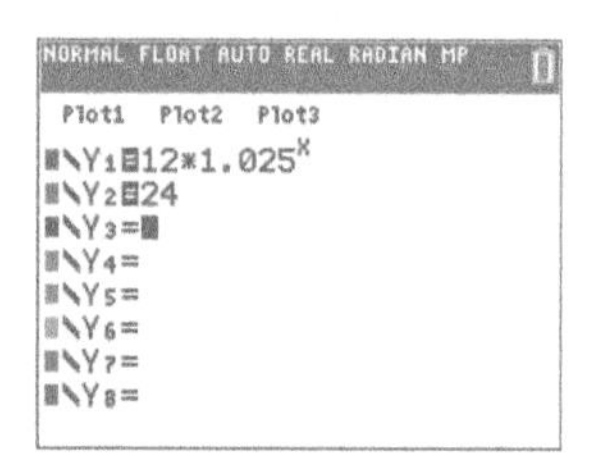

X	Y1	Y2
22	20.659	24
23	21.175	24
24	21.705	24
25	22.247	24
26	22.804	24
27	23.374	24
28	23.958	24
29	24.557	24
30	25.171	24
31	25.8	24
32	26.445	24

X=22

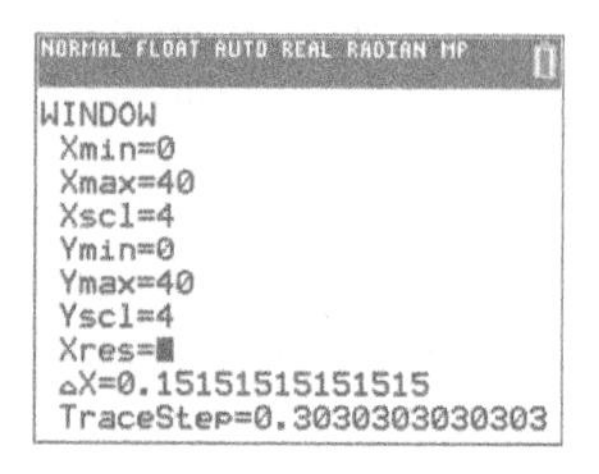

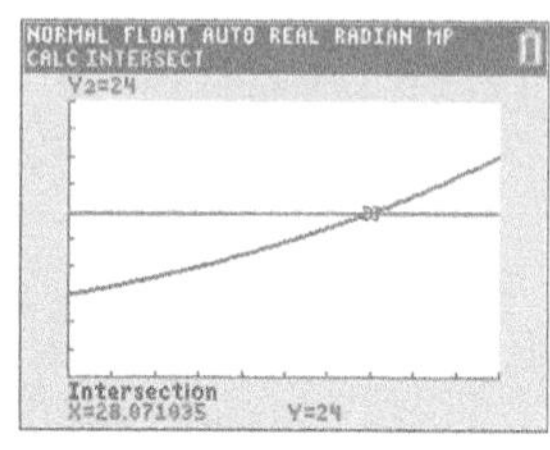

It will take 28 years for the cost of a pizza to double with an inflation rate of 2.5%.

SUMMARY Activity 3.3

1. For **exponential functions** defined by $f(x) = ab^x$, a is the value of f when $x = 0$ (sometimes called the initial value) and b is the growth or decay factor.
2. The vertical intercept of these functions is $(0, a)$.
3. In an exponential function, equally spaced input values yield output values whose successive ratios are constant. If the input values increase by 1 unit, then
   - **a.** the constant ratio is the **growth factor** if the output values are increasing.
   - **b.** the constant ratio is the **decay factor** if the output values are decreasing.
4. The **doubling time** of an increasing exponential function is the time it takes an output to double. The doubling time is set by the growth factor and remains the same for all output values.
5. The **half-life** of a decreasing exponential function is the time it takes an output to decay by one-half. The half-life is determined by the decay factor and remains the same for all output values.

EXERCISES Activity 3.3

1. The population of Buffalo, New York, in selected years can be approximated by the following table:

Year	1990	1994	2002	2003	2008	2012	2016
Population (thousands)	328.1	317.4	287.1	284.6	271.2	260.1	256.9

a. Let 1990 correspond to $t = 0$. Let b be the ratio between the population of Buffalo in 2003 and 2002. Determine an exponential function of the form $y = a \cdot b^t$ to represent the population of Buffalo symbolically. Round to three decimal places.

$P = 328.1(0.991)^t$

b. Does the function in part a give an accurate value of the population of Buffalo in 2016? Explain.

Close; substituting 26 for t yields $P = 328.1(0.991)^{26} = 259.4$ thousand.

c. Use your model in part a to predict the population of Buffalo in 2025.

$P = 328.1(0.991)^{35} = 239.1$ thousand

d. How much confidence do you have in the projection in part c? Explain.

Not too much confidence in the projection. In 10 years, a lot of factors can enter into the increase or decrease of a population.

2. Without using your graphing calculator, match each graph with its equation. Then check your answer using your graphing calculator.

a. $f(x) = 0.5(0.73)^x$

b. $g(x) = 3(1.73)^x$

i.

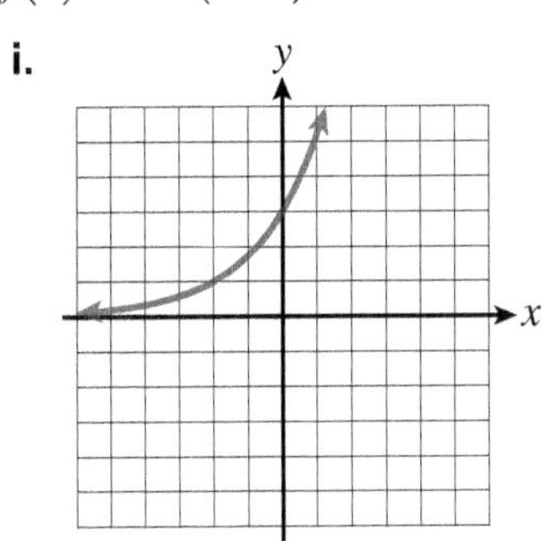

ii.

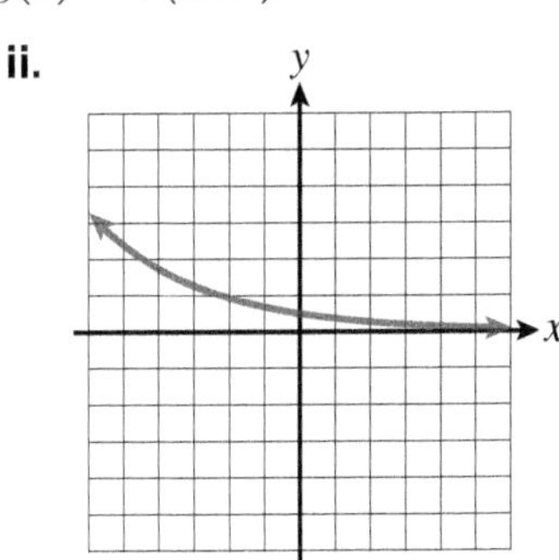

a. ii

b. i

3. Which of the following tables represent exponential functions? Indicate the growth or decay factor for the data that exponential.

a.

x	0	1	2	3	4
y	0	2	16	54	128

This data is not exponential. The successive ratios are not constant.

b.

x	0	1	2	3	4
y	1	4	16	64	256

This data is exponential. The growth factor is $b = 4$.

c.

x	1	2	3	4	5
y	1750	858	420	206	101

This data is exponential. The decay factor is $b \approx 0.49$.

4. An exponential function may be increasing or decreasing. Determine which is the case for each of the following functions. Explain how you determined each answer.

a. $y = 5^x$

b. $y = \left(\frac{1}{2}\right)^x$

a. The function is increasing. The base 5 is greater than 1 and is a growth factor.

b. The function is decreasing. The base $\frac{1}{2}$ is less than 1 and is a decay factor.

c. $y = 1.5^t$

d. $y = 0.2^p$

c. The function is increasing. The base 1.5 is a growth factor (greater than 1).

d. The function is decreasing. The base 0.2 is a decay factor (less than 1).

5. a. Evaluate the functions in the following table for the input values x:

input x	0	1	2	3	4	5
$g(x) = 3x$	0	3	6	9	12	15
$f(x) = 3^x$	1	3	9	27	81	243

b. Compare the functions $f(x) = 3^x$ and $g(x) = 3x$ from $x = 0$ to $x = 5$ by looking at the output values in the table.

$f(x) = 3^x$ has larger output values.

c. Compare the graphs of the functions f and g that are shown in the following window. Approximate the interval in which the exponential function f grows more slowly than the linear function g and the interval in which it grows more quickly.

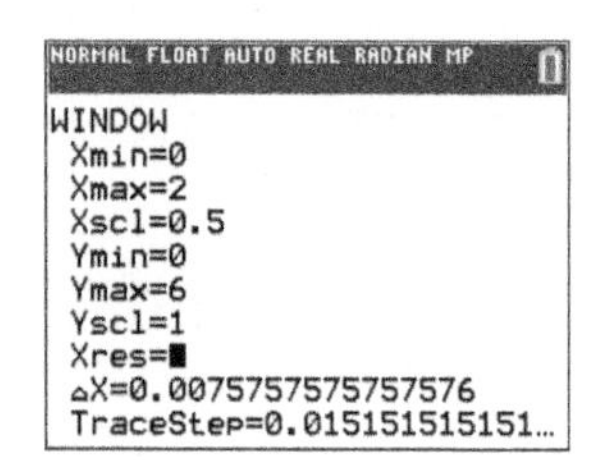

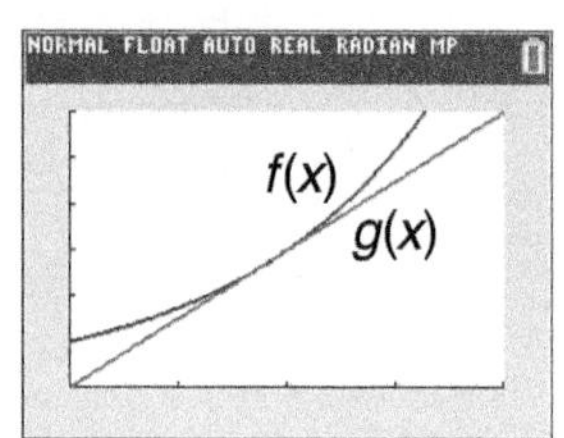

From $x = 0$ to $x = 1$, g is increasing more quickly than f. From $x = 1$ onward, f increases more quickly.

d. Compare the rate of increase of the function $f(x) = 3^x$ and $g(x) = 3x$ from $x = 0$ to $x = 5$ by calculating the average rate of change for each function from $x = 0$ to $x = 5$. Determine which function grows more quickly, on average, in the given interval.

$$\frac{\Delta f}{\Delta x} = \frac{3^5 - 3^0}{5 - 0} = \frac{243 - 1}{5} = \frac{242}{5} = 48.4$$

$$\frac{\Delta g}{\Delta x} = \frac{3(5) - 3(0)}{5 - 0} = \frac{15 - 0}{5} = \frac{15}{5} = 3$$

$f(x) = 3^x$ increases at a faster rate, on average, than does $g(x) = 3x$.

In this interval, $f(x)$ is increasing about 16 times more quickly.

6. If $f(x) = 3 \cdot 4^x$, determine the exact value of each of the following when possible. Otherwise, use your calculator to approximate the value to the nearest hundredth.

a. $f(-2)$

$f(-2) = 3 \cdot 4^{-2} = 3 \cdot \frac{1}{4^2} = \frac{3}{16}$

b. $f\left(\frac{1}{2}\right)$

$f\left(\frac{1}{2}\right) = 3 \cdot 4^{1/2} = 3 \cdot 2 = 6$

c. $f(2)$

$f(2) = 3 \cdot 4^2 = 3 \cdot 16 = 48$

d. $f(1.3)$

$f(1.3) = 3 \cdot 4^{1.3} \approx 18.19$

7. According to industry reports, the global sales of smartphones can be modeled by the equation $P(t) = 1428.6 \cdot 1.034^t$, where t represents the number of years since 2014 and $P(t)$ represents global smartphone sales measured in millions of units.

a. Complete the following table:

t, Number of Years since 2014	0	1	2	3	4	5
$P(t)$, Global Smartphone Sales (millions)	1428.6	1477.2	1527.4	1579.3	1633.0	1688.5

b. Determine the growth factor of smartphone sales.

The growth factor is 1.034.

c. Sketch a graph of this exponential function. Use $0 \le t \le 5$ and $1400 \le P(t) \le 1900$

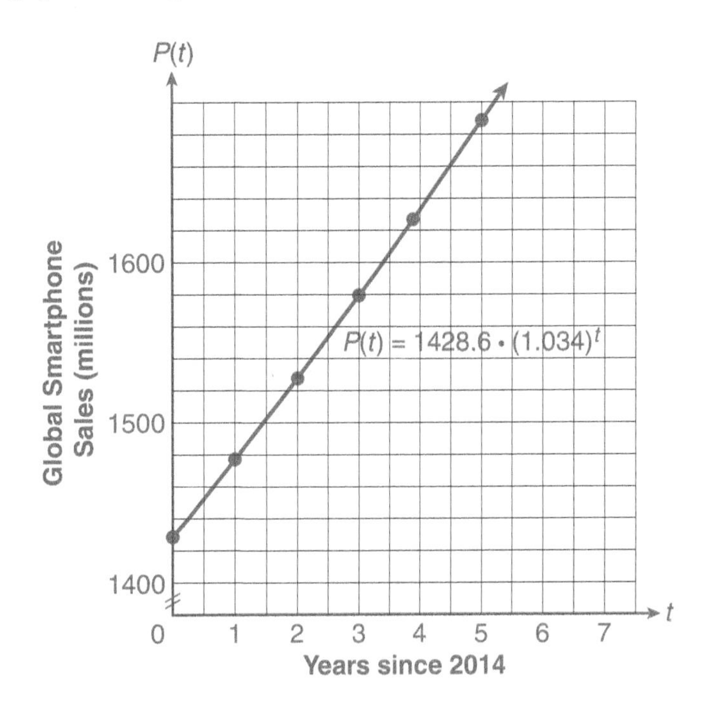

d. Use the equation to determine the projected smartphone sales in 2025. Do you believe your projection is reasonable?

In 2025, $t = 11$, so $P(11) = 1428.6 \cdot 1.034^{11} = 2063.7$ million smartphones. This represents relatively slow growth; thus, it is reasonable.

e. Based on the data plotted in part c, do you believe a model other than an exponential one might be reasonable in this case? Explain.

The growth is so slow that a linear model might work just as well, but without specific data this is difficult to determine.

8. Chlorine is used to disinfect swimming pools. The chlorine concentration should be between 1.5 and 2.5 parts per million (ppm). On sunny, hot days, 30% of the chlorine dissipates into the air or combines with other chemicals. Therefore, chlorine concentration, $A(x)$ (in parts per million), in a pool after x sunny days can be modeled by

$$A(x) = 2.5(0.7)^x.$$

a. What is the initial concentration of chlorine in the pool?

2.5 ppm

b. Complete the following table:

x	0	1	2	3	4	5
$A(x)$	2.5	1.75	1.225	0.8575	0.6003	0.4202

c. Sketch the graph of the chlorine function.

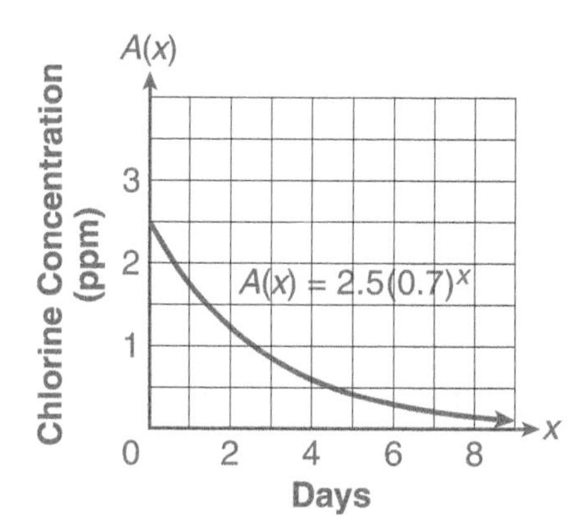

d. What is the chlorine concentration in the pool after 3 days?

$A(3) = 2.5(0.70)^3 = 0.8575$ ppm

e. Approximate graphically and numerically the number of days before chlorine should be added.

Chlorine should be added in 1.4 days.

9. The population of Ft. Myers, Florida, from 2010 to 2016 is approximated in the following table:

Year	2010	2011	2012	2013	2014	2015	2016
Population (thousands)	62.4	63.7	65.6	67.9	70.6	73.8	77.1

a. Does the relationship in the table seem to represent an exponential function? Explain.

Possibly, the consecutive ratios are approximately constant.

b. Approximate the growth factor for the entire period by using the growth factor from 2012 to 2013. Record that factor below. Use three decimal places.

The growth factor is $b = 1.035$.

c. Determine the exponential equation that gives the population of Ft. Myers, N, in thousands as a function of t, the number of years since 2010. Note that $t = 0$ corresponds to 2010, etc.

$N = 62.4 \cdot 1.035^t$

d. Graph the function.

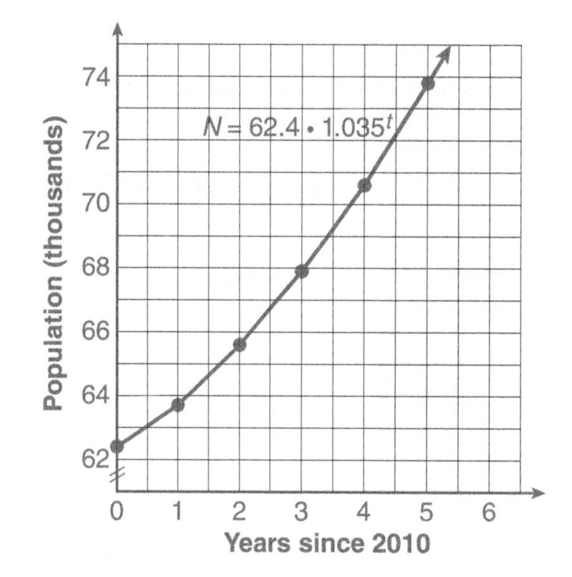

e. What is the vertical intercept? What is the practical meaning of the intercept in this situation?

The vertical intercept is (0, 62.4). The population of Ft. Myers, Florida, in 2010 was 62.4 thousand people.

f. Use the equation to estimate the population of Ft. Myers, Florida, in 2028. Do you believe this is a good estimate? Explain.

2028 is 18 years after 2010, so $t = 18$. $N = 62.4 \cdot 1.035^{18} \approx 115.9$ thousand people. This probably is not a good estimate because $t = 18$ is extrapolating too far into the future.

g. Use the exponential function of the population and the graph or table feature of your graphing calculator to estimate the number of years it will take for the population of Ft. Myers to double from 62.4 thousand to 124.8 thousand.

According to the model, the population of Ft. Myers will double in 20.1 years.

10. As a radiology specialist, you use the radioactive substance iodine-131 to diagnose conditions of the thyroid gland. The amount of radioactive substance decays with time. Your hospital currently has a 20-gram supply of iodine-131. The following table gives the number of grams remaining after a specified number of days:

t, Number of Days Starting from a 20-Gram Supply of Iodine-131	0	1	2	3	4	5	6
N, Number of Grams of Iodine-131 Remaining from a 20-Gram Supply	20.00	18.34	16.82	15.42	14.14	12.97	11.89

a. Does the relationship represent an exponential function? Explain.

Yes, the constant ratio is approximately 0.917.

b. What is the decay factor?

The decay factor is approximately 0.917.

c. Write an exponential decay formula for *N*, the number of grams of iodine-131 remaining, in terms of *t*, the number of days from the current supply of 20 grams.

The formula is $N = 20(0.917)^t$.

d. Determine the number of grams of iodine-131 remaining from a 20-gram supply after 2 months (60 days).

$N = 20(0.917)^{60} \approx 0.11$. After 2 months, approximately a 10th of a gram of iodine-131 remains from a 20-gram supply.

e. Graph the decay formula for iodine-131, $N = 20(0.917)^t$, as a function of the time *t* (days). Use appropriate scales and labels on the following grid or an appropriate window on a graphing calculator.

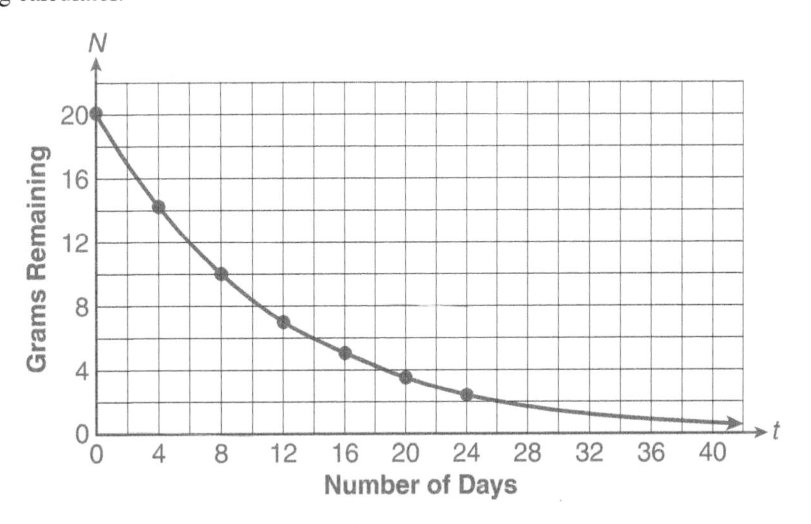

f. How long will it take iodine-131 to decrease to half its original value? Explain how you determined your answer.

It will take about 8 days for iodine-131 to decay from 20 grams to 10 grams.

(Explanations will vary.)

I determined the result by graphing $y = 10$, which is half the amount, on the same axes and found the intersection.

ACTIVITY

3.4

Population Growth

Growth and Decay Rates; More Graphing

OBJECTIVES

1. Determine the annual percentage growth or decay rate of an exponential function represented by a table of values or an equation.
2. Graph an exponential function having equation $y = a(1 + r)^x$.

1. According to the U.S. Census Bureau, in 2016 the city of Bellevue, Washington, was one of the fastest-growing cities in the entire nation. The population of Bellevue in 2016 was 141,400.

a. Assuming that the population increases at a constant percent rate of 1.3 %, determine the population of Bellevue (in thousands) in 2017.

In 2017, the population will be $141{,}400 + 0.013(141{,}400) = 141{,}400 + 1838 = 143{,}238$ or 143.2 thousand people.

b. Determine the population of Bellevue (in thousands) in 2018.

In 2018, the population will be $143.2 + 0.013(143.2) = 143.2 + 1.9 = 145.1$ thousand.

c. Divide the population in 2017 by the population in 2016 and record this ratio.

$$\frac{143.2}{141.4} = 1.013$$

d. Divide the population in 2018 by the population in 2017 and record this ratio.

$$\frac{145.1}{143.2} = 1.013$$

e. What do you notice about the ratios in parts c and d? What do these ratios represent?

The ratios are the same; they represent the annual growth factor, 1.013.

Linear functions represent quantities that change at a constant average rate (slope). Exponential functions represent quantities that change at a constant percent rate.

EXAMPLE 1

Population growth, sales and advertising trends, compound interest, spread of disease, and concentration of a drug in the blood are examples of quantities that increase or decrease at a constant percent rate.

2. a. Let t represent the number of years since 2016 ($t = 0$ corresponds to 2016). Use the results from Problem 1 to complete the following table:

t, Years (since 2016)	0	1	2	3	4	5
P, Population (thousands)	141.4	143.2	145.1	147.0	148.9	150.8

The table above will differ depending if you use $b = 1.01$ or $b = 1.013$. I used 1.013. Once you know the growth factor, b, and the initial value, a, you can write the exponential equation. In this situation, the initial value is the population in thousands in 2016 ($t = 0$), and the growth factor is $b = 1.013$.

b. Write the exponential equation, $P = a \cdot b^t$, for the population of Bellevue, Washington.

$P = 141.4(1.013)^t$

The growth factor is $b = 1.013$, and the growth rate is $r = 1.3\%$.

3. a. Write the growth rate, $r = 1.3\%$, as a decimal.

$r = 0.013$

b. Add 1 to the decimal form of the growth rate, r.

$1 + r = 1 + .013 = 1.013$

The growth factor, b, is determined from the growth rate, r, by writing r in decimal form and adding 1:

$$b = 1 + r.$$

EXAMPLE 2 ***Determine the growth factor, b, for a growth rate of r = 8%.***

SOLUTION

$$r = 8\% = 0.08, b = 1 + r = 1 + 0.08 = 1.08$$

c. Solve the equation for the growth factor, $b = 1 + r$, for r.

$r = b - 1$

The growth rate, r, is determined from the growth factor, b, by subtracting I from b and writing the result in percent form.

EXAMPLE 3 ***Determine the growth rate, r, for a growth factor of b = 1.054.***

SOLUTION

$$r = b - 1 = 1.054 - 1 = 0.054 = 5.4\%$$

The growth rate is 5.4%.

4. a. Complete the following table for the population of Bellevue, Washington:

t	CALCULATION FOR POPULATION (THOUSANDS)	EXPONENTIAL FORM	$P(t)$, POPULATION IN THOUSANDS
0	141.4	$141.4 \cdot (1.013)^0$	141.4
1	(141.4)1.013	$141.4 \cdot (1.013)^1$	143.2
2	(141.4)(1.013)(1.013)	$141.4 \cdot (1.013)^2$	145.1
3	(1414)(1.013)(1.013)(1.013)	$141.4 \cdot (1.013)^3$	147.0

b. Use the pattern in the table in part a to help you write the equation for $P(t)$, the population of Bellevue, Washington (in thousands), using t, the number of years since 2016, as the input value. How does your result compare with the equation obtained in Problem 2?

$P(t) = 141.4 \cdot (1.013)^t$; the equations are the same.

EXAMPLE 4 **a.** ***Determine the growth factor and the growth rate of the function defined by $f(x) = 250(1.7)^x$.***

SOLUTION

The growth factor $1 + r$ is the base 1.7. To determine the growth rate, solve the equation $1 + r = 1.7$ for r.

$$r = 0.7, \text{ or } 70\%$$

b. ***If the growth rate of a function is 5%, determine the growth factor.***

SOLUTION

If $r = 5\%$, or 0.05, the growth factor is $1 + r = 1 + 0.05 = 1.05$.

5. a. In the Bellevue, Washington, population function $P(t) = 141.4(1.013)^t$, determine the growth factor.

The growth factor is 1.013.

b. Determine the growth rate. Express your answer as a percent.

The growth rate is 0.013 = 1.3%.

6. a. Using the function defined by $P(t) = 141.4(1.013)^t$, determine the population of Bellevue in 2025. That is, determine $P(t)$ when $t = 9$.

In 2025, $P(9) = 141.4(1.013)^9 \approx 158.8$ thousand people.

b. Graph the function with your graphing calculator. Use the window Xmin = 0, Xmax = 100, Ymin = 0, Ymax = 600. Your graph should resemble the following:

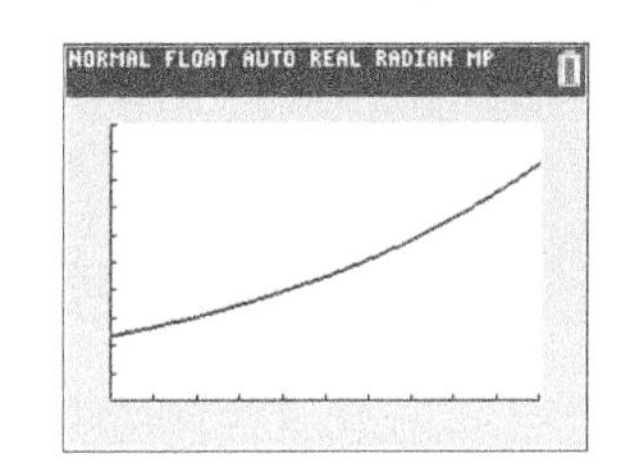

c. Determine $P(0)$. What is the graphical and practical meaning of $P(0)$?

$P(0) = 141.4$ (0, 141.4) is the vertical intercept of the graph. 141.4 thousand is the population of Bellevue, Washington, in 2016 (when $t = 0$).

7. a. Use your prediction to predict the population of Bellevue, Washington, in 2031.

$2031 - 2016 = 15; P(15) = 141.4(1.013)^{15} \approx 171.6$ thousand

b. Verify your prediction graphically.

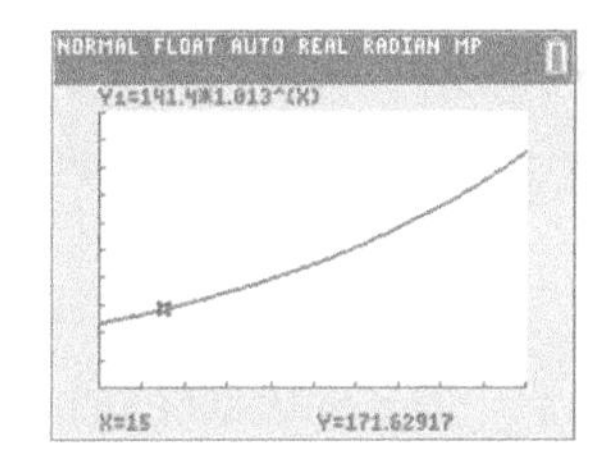

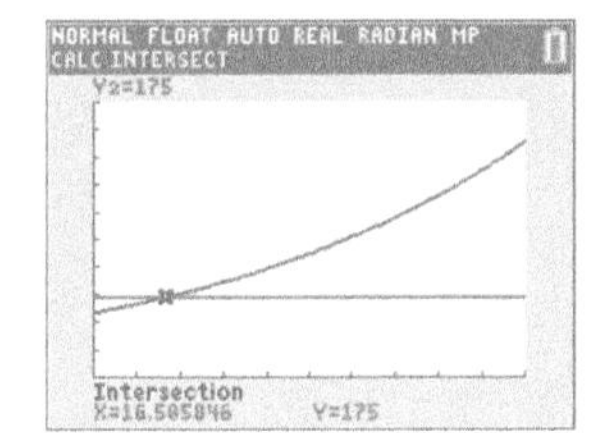

8. a. Use the graph to estimate when Bellevue's population will reach 175,000. Assume that the population continues to grow at the same rate. Remember that $P(t)$ is measured in thousands.

$P = 175$ when $t \approx 16.5$, or midway through the year 2032.

b. Evaluate $P(32)$ and describe what it means.

$P(32) \approx 213.8$ If Bellevue, Washington, continues to grow at the same rate, the population will be 213,800 in the year 2048.

9. Use the model to predict the population of Bellevue in 2019 and 2039. In which prediction would you be more confident? Why?

2019: $P(3) = 141.4(1.013)^3 \approx 147.0$ thousand

2039: $P(23) = 141.4(1.013)^{23} \approx 190.3$ thousand

I am more confident in the estimate for 2019. The growth rate will probably not remain constant for 23 years.

NORMAL FLOAT AUTO REAL RADIAN MP
CALC INTERSECT
$Y_2=282.8$
Intersection
X=53.664041 Y=282.8

10. a. Assuming the growth rate remains constant, how long will it take the population of Bellevue, Washington, to double its 2016 population?

53.7 yr.

b. Explain how you reached your conclusion in part a.

(Answers may vary.) I added $Y_2 = 282.8$ to the equation editor and found the intersection point.

Wastewater Treatment Facility

You are working at a wastewater treatment facility. You are presently treating water contaminated with 18 micrograms (μg) of pollutant per liter. Your process is designed to remove 20% of the pollutant during each treatment. Your goal is to reduce the pollutant to less than 3 micrograms per liter.

11. a. What percent of pollutant present at the start of a treatment remains at the end of the treatment?

Twenty percent of the pollutant is removed, so 80% remains.

b. The concentration of pollutants is 18 micrograms per liter at the start of the first treatment. Use the result of part a to determine the concentration of pollutant at the end of the first treatment.

The concentration at the end of the first treatment is

$18 \cdot 0.80 = 14.4$ micrograms per liter.

c. Complete the following table. Round the results to the nearest 10th.

n, Number of Treatments	0	1	2	3	4	5
C(*n*), Concentration of Pollutant, in μg/L, at the End of the *n*th Treatment	18	14.4	11.5	9.2	7.4	5.9

d. Write an equation for the concentration, $C(n)$, of the pollutant as a function of the number of treatments, n.

$C(n) = 18(0.80)^n$

The equation $C(n) = 18(0.80)^n$ has the general form $C = C_0(1 - r)^n$, where r is the **decay rate**, $(1 - r)$ is the **decay factor** or the base of the exponential function, n is the numer of treatments, and C_0 is the initial value, the concentration when $n = 0$.

EXAMPLE 5

a. ***Determine the decay factor and the decay rate of the function defined by $h(x) = 123(0.43)^x$.***

SOLUTION

The decay factor $1 - r$ is the base, 0.43. To determine the decay rate, solve the equation $1 - r = 0.43$ for r.

$$r = 0.57, \quad \text{or} \quad 57\%$$

b. ***If the decay rate of a function is 5%, determine the decay factor.***

SOLUTION

If $r = 5\%$, or 0.05, the decay factor is $1 - r = 1 - 0.05 = 0.95$.

LC LEARNING CATALYTICS

Given the function defined by $g(x) = 67.5(0.977)^x$ is this a growth or decay function?

12. a. If the decay rate is 2.5%, what is the decay factor?

The decay factor is $b = 1 - 0.025 = 0.975$.

b. If the decay factor is 0.76, what is the decay rate?

The decay rate is $1 - 0.76 = 0.24$, or 24%.

13. a. Use the function defined by $C(n) = 18(0.8)^n$ to predict the concentration of contaminants at the wastewater treatment facility after seven treatments.

After seven treatments, the concentration of the contaminants will be $C(7) = 18 \cdot 0.80^7 \approx 3.77$ micrograms per liter.

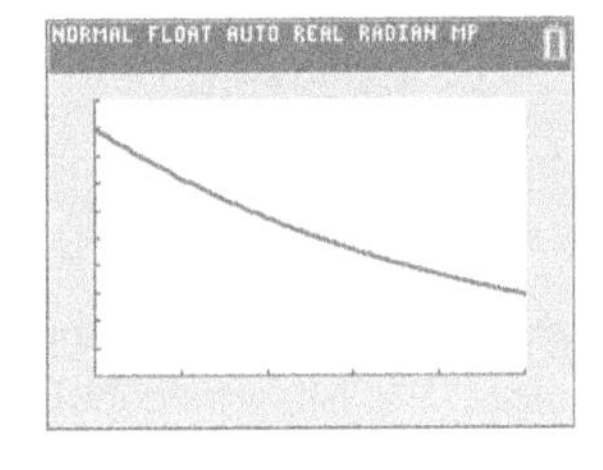

b. Sketch a graph of the concentration function on your graphing calculator. Use the table in Problem 11c to set a window. Does the graph look like you expected it would? Explain.

The graph, to the left, looks like a decreasing exponential function.

c. What is the vertical intercept? What is the practical meaning of the intercept in this situation?

The vertical intercept is $C(0) = 18(0.80)^0 = 18$. The initial concentration is 18 micrograms per liter.

d. Reset the window of your graphing calculator to Xmin = −5, Xmax = 15, Ymin = −10, and Ymax = 50. Does the graph have a horizontal asymptote? Explain what this means in this situation.

The horizontal asymptote is $C = 0$. The concentration of contaminants approaches 0 as the number of treatments increases.

14. Use the table or trace feature of your graphing calculator to estimate the number of treatments necessary to bring the concentration of pollutant below 3 micrograms per liter.

The concentration will be below 3 micrograms per liter after the ninth treatment.

SUMMARY Activity 3.4

1. **Exponential functions** are used to describe phenomena that grow or decay by a constant percent rate per unit time.

2. If r represents the **growth rate**, the exponential function that models the quantity, P, can be written as

$$P(t) = P_0(1 + r)^t,$$

where P_0 is the initial amount, t represents the amount of elapsed time, and $1 + r$ is the growth factor.

3. If r represents the **decay rate**, the exponential function that models the amount remaining can be written as

$$P(t) = P_0(1 - r)^t,$$

where $1 - r$ is the decay factor.

EXERCISES Activity 3.4

1. Determine the growth and decay factors and growth and decay rates in the following tables:

GROWTH FACTOR	GROWTH RATE
1.02	2%
1.029	2.9%
2.23	123%
1.34	34%
1.0002	0.02%

DECAY FACTOR	DECAY RATE
0.77	23%
0.32	68%
0.953	4.7%
0.803	19.7%
0.9948	0.52%

2. In 2010, the U.S. Census Bureau estimated the population of Miami, Florida, as 400.9 thousand people and the population of Baltimore, Maryland, as 642.0 thousand people. Since 2010, Miami's population has been increasing at approximately 1.36% per year. Baltimore's population has been decreasing at approximately 0.52% per year. Assume that the growth and decay rates remain constant.

   a. Let $M(t)$ represent the population of Miami, and $B(t)$ represent the population of Baltimore, t years after 2010. Determine the exponential functions that model the population (in thousands) of both cities.

   Miami: $M(t) = 400.9(1.0136)^t$ Baltimore: $B(t) = 642.0(0.9948)^t$

   b. Use the models from part a to predict the population of both cities in 2020.

   2020 is 10 years from 2010, so $t = 10$. Miami: $M(10) = 400.9(1.0136)^{10} = 458.9$ thousand

   Baltimore: $B(10) = 642.0(0.9948)^{10} = 609.4$ thousand

   c. Estimate the number of years it will take for the population of Baltimore to be cut in half.

   Solve $642.0(0.9948)^t = \frac{1}{2}642.0 = 321.0$ for t.

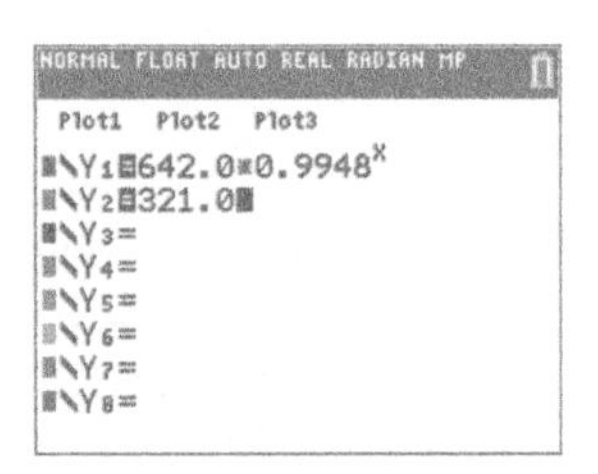

NORMAL FLOAT AUTO REAL RADIAN MP
PRESS + FOR ΔTbl

X	Y1	Y2
128	329.39	321
129	327.68	321
130	325.98	321
131	324.28	321
132	322.59	321
133	320.92	321
134	319.25	321
135	317.59	321
136	315.94	321
137	314.29	321
138	312.66	321

X=133

NORMAL FLOAT AUTO REAL RADIAN MP
DISTANCE BETWEEN TICK MARKS ON AXIS
WINDOW
Xmin=100
Xmax=150
Xscl=10
Ymin=300
Ymax=400
Yscl=10
Xres=1
ΔX=0.18939393939394
TraceStep=0.378787878787...

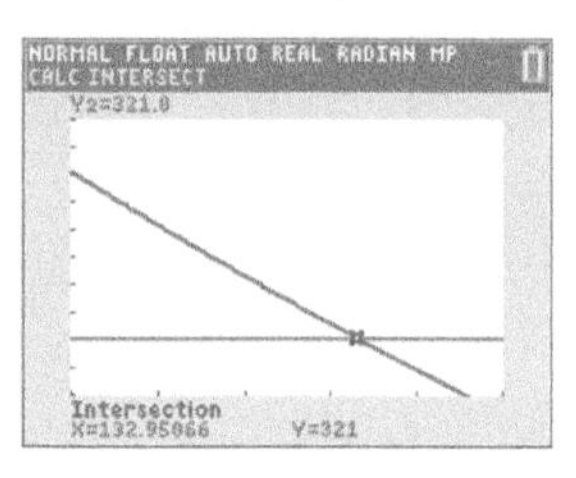

   The population of Baltimore will halve 133 years from 2010.

   d. Use the table and/or graphs of these functions to predict when the populations of Miami and Baltimore will be equal.

   Solve $400.9(1.0136)^t = 642.0(0.9948)^t$ for t.

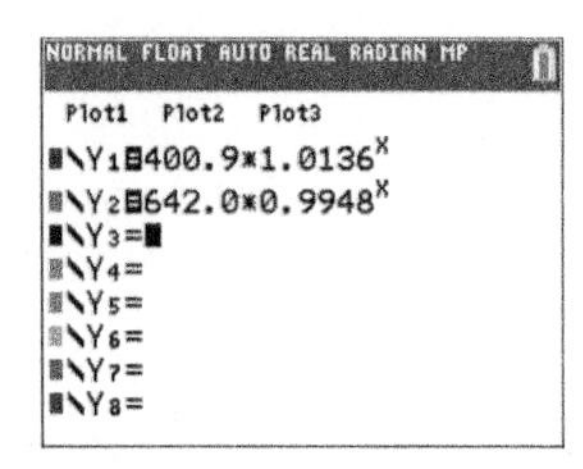

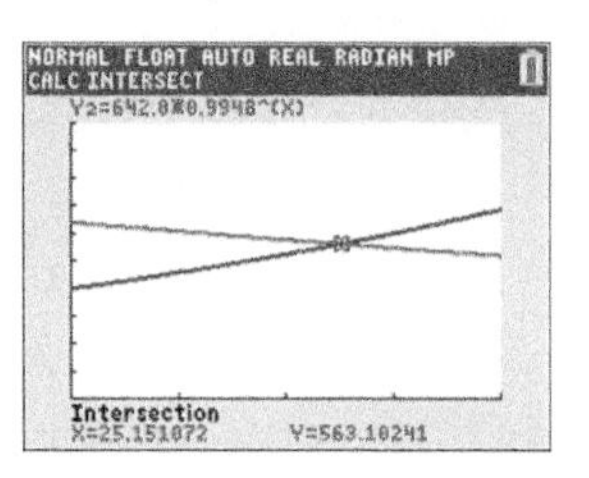

   The populations will be equal approximately 25 years from 2010 or 2035.

3. You have just taken over as the manager of a small city. The personnel expenses were \$8,500,000 in 2018. Over the previous 5 years, the personnel expenses increased at a rate of 3.2% annually.

a. Assuming that this rate continues, write an equation describing personnel costs, $C(t)$, in millions of dollars, where $t = 0$ corresponds to 2018.

personnel costs $C(t) = 8.5(1.032)^t$ million

b. Sketch a graph of this function up to the year 2033 ($t = 15$).

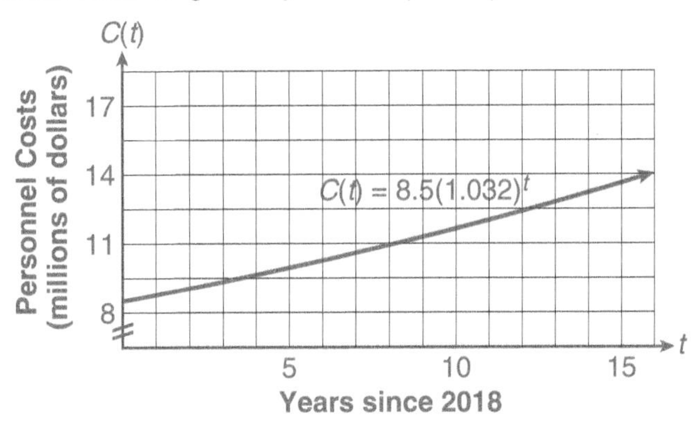

c. What is the vertical intercept of the graph? What is the practical meaning of the intercept in this situation?

The vertical intercept is (0.8.5). The 2018 ($t = 0$) personnel costs were 8.5 million dollars.

d. What are the projected personal costs in the year 2024?

2024 is 6 years from 2018, so $t = 6$. $C(6) = 8.5(1.032)^6 = 10.268$ million dollars.

e. In what year will the personnel expenses be double the 2018 personnel expenses?

Solve the equation $8.5(1.032)^t = 2 \cdot 8.5 = 17$ for t.

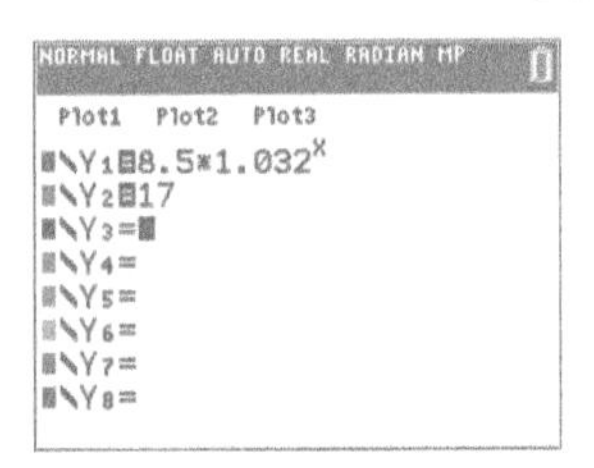

NORMAL FLOAT AUTO REAL RADIAN MP
PRESS + FOR ΔTbl

X	Y1	Y2
15	13.634	17
16	14.07	17
17	14.52	17
18	14.985	17
19	15.464	17
20	15.959	17
21	16.47	17
22	16.997	17
23	17.541	17
24	18.102	17
25	18.681	17

X=22

NORMAL FLOAT AUTO REAL RADIAN MP
WINDOW
Xmin=0
Xmax=30
Xscl=5
Ymin=8
Ymax=20
Yscl=2
Xres=1
ΔX=0.11363636363636
TraceStep=0.227272727272...

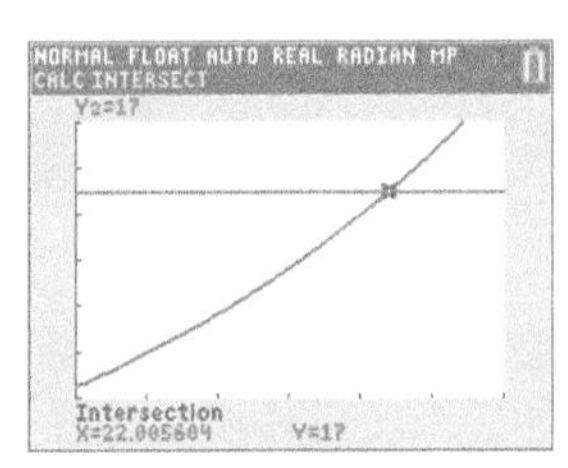

The personnel expenses will double 22 years from 2018, in 2040.

4. According to the U.S. Census Bureau, the population of the United States (in millions) can be modeled by $P(t) = 123.1 \cdot 1.0116^t$, where t represents the number of years since 1930.

a. Sketch a graph of the U.S. population model from 1930 to 2010.

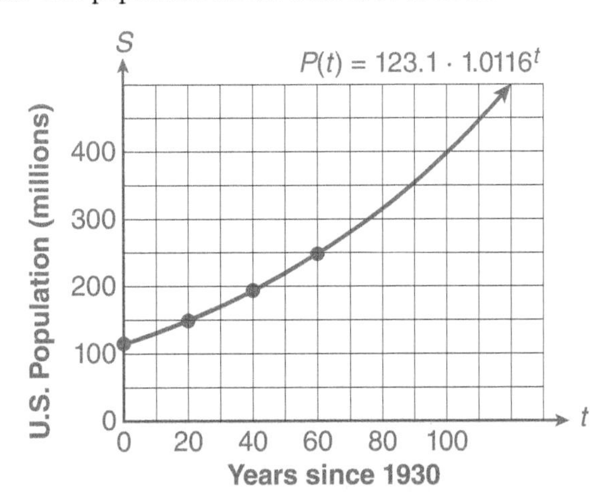

b. Determine the annual growth rate and the growth factor from the equation.

The growth factor is $b = 1.0116$; the growth rate is 0.0116, or 1.16%.

c. Use the model to determine the population (in millions) of the United States in 2015. How does your answer compare with the actual population of 320.9 million?

$P(85) = 123.1 \cdot 1.0116^{85} \approx 328.1$ million. The model is a little higher than the actual population.

5. You recently purchased a new car for $25,000 by arranging financing for the next 5 years. You are curious to know what your new car will be worth when the loan is paid off.

a. Assuming that the value depreciates at a constant rate of 15%, write an equation that represents the value, $V(t)$, of the car t years from now.

$V(t) = 25{,}000(0.85)^t$

b. What is the decay factor in this situation?

The decay factor is $b = 0.85$.

c. What is the decay rate in this situation?

The decay rate is $r = 1 - b = 1 - 0.85 = 0.15 = 15\%$.

d. Use the equation from part a to estimate the value of your car 5 years from now.

$V(5) = 25000(0.85)^5 = \$11{,}092.63$

e. Use the trace and table features of your graphing calculator to check your results in part d.

```
NORMAL FLOAT AUTO REAL RADIAN MP
 Plot1  Plot2  Plot3
■\Y1■25000*0.85^X
■\Y2=■
■\Y3=
■\Y4=
■\Y5=
■\Y6=
■\Y7=
■\Y8=
```

NORMAL FLOAT AUTO REAL RADIAN MP
PRESS + FOR ΔTbl

X	Y1
0	25000
1	21250
2	18063
3	15353
4	13050
5	11093
6	9428.7
7	8014.4
8	6812.3
9	5790.4
10	4921.9

X=5

```
NORMAL FLOAT AUTO REAL RADIAN MP
FREE TRACE VALUES
WINDOW
 Xmin=0
 Xmax=10
 Xscl=1
 Ymin=0
 Ymax=25000
 Yscl=2500
 Xres=1
 ΔX=0.037878787878788
 TraceStep=0.075757575757...
```

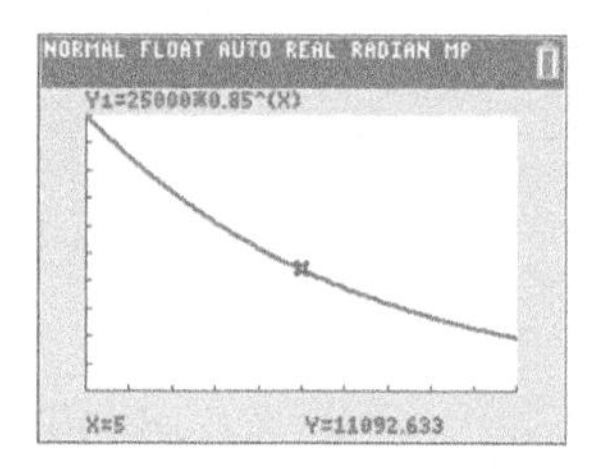

f. Use the trace or table features of your graphing calculator to determine when your car will be worth $15,000.

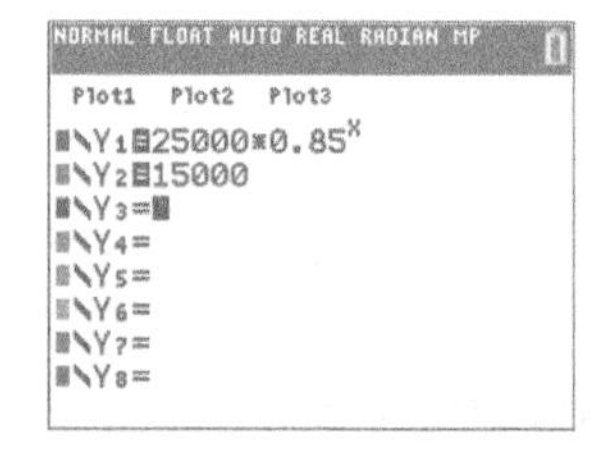

NORMAL FLOAT AUTO REAL RADIAN MP
PRESS + FOR ΔTbl

X	Y1	Y2
3	15353	15000
3.1	15106	15000
3.2	14862	15000
3.3	14623	15000
3.4	14387	15000
3.5	14155	15000
3.6	13927	15000
3.7	13702	15000
3.8	13481	15000
3.9	13264	15000
4	13050	15000

X=3.1

```
NORMAL FLOAT AUTO REAL RADIAN MP
FREE TRACE VALUES
WINDOW
 Xmin=0
 Xmax=10
 Xscl=1
 Ymin=1000
 Ymax=25000
 Yscl=2500
 Xres=1
 ΔX=■.037878787878788
 TraceStep=0.075757575757...
```

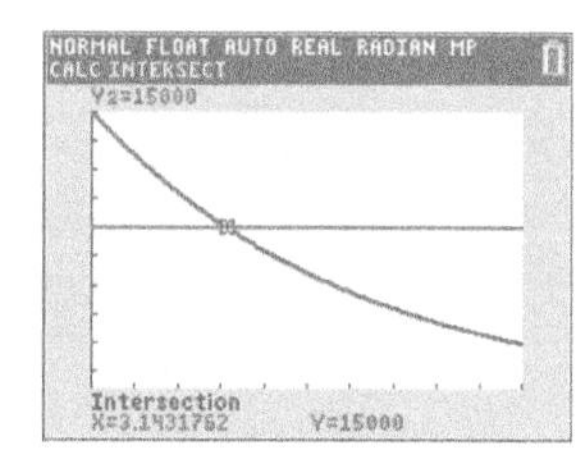

The value of the car will be $15,000 in about 3.1 years.

6. Suppose the inflation rate is 2% per year and remains the same for the next 7 years.

a. Determine the annual growth factor for a 2% inflation rate.

The growth factor is $1 + 0.02 = 1.02$.

b. If the yearly inflation rate remains at 2%, what exponential function would you use to determine the cost of $65 cross-training shoes after t years.

$c(t) = 65(1.02)^t$

c. Use the function in part b to determine the cost of the cross-training shoes in 10 years. What assumption are you making regarding the inflation rate?

$79.23, assuming that the inflation rate remains constant for 10 years

d. Complete the following table for a pair of cross-training shoes that cost $65 now. Round to the nearest cent.

t, Years from Now	0	1	2	3	4	5	6	7
$c(t)$, Cost of Sneakers ($)	65	66.30	67.63	68.98	70.36	71.77	73.20	74.66

ACTIVITY 3.5

Time Is Money
Compound Interest and Continuous Compounding

OBJECTIVE

Apply the compound interest and continuous compounding formulas to a given situation.

Congratulations, you have inherited \$20,000! Your grandparents suggest that you use half of the inheritance to start a retirement fund. Your grandfather claims that an investment of \$10,000 could grow to over half a million dollars by the time you retire. You are intrigued by this statement and decide to investigate whether this can happen.

1. **a.** Suppose the \$10,000 is deposited in an investment account at a 3.5% annual interest. What is the interest earned after 1 year?

   $10{,}000 \cdot 0.035 = \$350$

   b. Suppose you left the money in the account for 10 years. Can the total amount of interest on your investment be calculated by multiplying your answer in part a by 10? What assumption are you making if you said yes?

   It would be simple annual interest.

Recall that for **simple interest** on an investment, the interest earned during the first period does not earn interest for the rest of the life of the investment. Therefore, at 3.5% simple annual interest, the \$10,000 would earn a total interest of $10{,}000(0.035)(10) = \$3500$.

The interest paid on savings accounts in most banks is **compound interest**. The interest earned for each period is added to the previous principal before the next interest calculation is made. Simply stated, interest earns interest.

For example, if you deposit \$10,000 in the bank at 3.5%, the balance after 1 year is

$$10{,}000 + 0.035(10{,}000) = 10{,}000 + 350 = \$10{,}350.$$

The interest, \$350, earned during the year becomes part of the new balance. At the end of the second year, your balance is

$$10{,}350 + 0.035(10{,}350) = 10{,}350 + 362.25 = \$10{,}712.25.$$

Note that you made interest on the original deposit, plus interest on the first year's interest. In this situation, we say that interest is compounded. Usually, the compounding occurs at fixed intervals (typically at the end of every year, quarter, month, or day). In this example, interest is compounded annually.

If interest is compounded, then the current balance is given by the formula

$$A = P\left(1 + \frac{r}{n}\right)^{nt},$$

where A is the current balance, or compound amount in the account.

P is the principal (the original amount deposited),
r is the annual interest rate (in decimal form),
n is the number of times per year that interest is compounded, and
t is the time in years the money has been invested.

The given formula for the compound amount A is called the **compound interest formula**.

EXAMPLE 1 ***You invest \$100 at 4% compounded quarterly. How much money do you have after 5 years?***

SOLUTION

The principal is \$100, so $P = 100$. The annual interest rate is 4%, so $r = 0.04$. Interest is compounded quarterly (i.e., four times per year), so $n = 4$. The money is invested for 5 years, so $t = 5$. Substituting the values for the P, r, n, and t in the compound interest formula, you have

$$A = 100\left(1 + \frac{0.04}{4}\right)^{4 \cdot 5} \approx \$122.02.$$

2. a. Suppose you deposit \$10,000 in an account that has a 3.5% annual interest rate (usually referred to as APR, for annual percentage rate) and whose interest is compounded annually ($n = 1$). Substitute the appropriate values for P, n, and r into the compound interest formula to get the balance, A, as a function of time, t.

$$A = 10{,}000\left(1 + \frac{0.035}{1}\right)^{1 \cdot t}$$

b. Use the compound interest formula from part a to determine your balance, A, at the end of the first year ($t = 1$).

$$A = 10{,}000\left(1 + \frac{0.035}{1}\right)^{1 \cdot 1} = \$10{,}350$$

c. What will be the amount of interest earned in the first year?

\$350

d. Use the compound interest formula developed in part a to complete the following table:

t, Year	0	1	2	3	4
A, Balance	10,000.00	10,350.00	10,712.25	11,087.18	11,475.23

e. The compound interest formula in part a defines A as an exponential function of t. Identify the base.

1.035

f. Is the base a growth or decay factor? Explain.

Because the base is greater than 1, it is a growth factor.

3. a. Suppose you deposit the \$10,000 in an account that has the same interest rate (APR) of 3.5%, with compounding quarterly ($n = 4$) rather than annually ($n = 1$). Write a new formula for your balance, A, as a function of time.

$$A = 10{,}000\left(1 + \frac{0.035}{4}\right)^{4t} = 10{,}000(1.00875)^{4t}$$

b. What would your balance be after the first year?

$A = 10{,}000(1.00875)^{4 \cdot 1} \approx \$10{,}354.62$

c. Enter the function from Problem 3a into your calculator. Use the table feature to determine the balance every 5 years for 40 years. Record the values in the table in Problem 4 under $n = 4$ (compounded quarterly).

d. Write the function $A = 10{,}000 \cdot (1.00875)^{4t}$ in the form $A = ab^t$:

$A = 10{,}000(1.03546)^t$

e. What is the base of this exponential function from Problem 3d?

$(1.00875)^4 \approx 1.03546$

4. Now deposit your \$10,000 in a 3.5% APR account with *monthly* compounding ($n = 12$) and then an account with *daily* compounding ($n = 365$). Use your graphing calculator and the appropriate formula to complete the table on the next page:

COMPARISON OF $10,000 PRINCIPAL IN 3.5% APR ACCOUNTS WITH VARYING COMPOUNDING PERIODS			
t	$n = 4$	$n = 12$	$n = 365$
0	10,000	10,000	10,000
5	11,903	11,909	11,912
10	14,169	14,183	14,190
15	16,866	16,892	16,904
20	20,076	20,117	20,137
25	23,898	23,958	23,988
30	28,446	28,533	28,575
35	33,861	33,981	34,040
40	40,306	40,469	40,549

5. In Problem 4, you calculated the balance on a deposit of $10,000 at an annual interest rate of 3.5% that was compounded at different intervals. After 40 years, which account has the highest balance? Does this seem reasonable? Explain.

Compounding daily results in the largest balance after 40 years. The balance is greater where the number of compounding periods per year is larger because you compute interest on interest more frequently.

Continuous Compounding

You could extend this problem so that interest is compounded every hour or every minute or even every second. However, compounding more frequently than every hour does not increase the balance very much.

To discover why this happens, take a closer look at the exponential functions from Problems 1–3.

$$n = 1 \qquad A = 10{,}000\left(1 + \frac{0.035}{1}\right)^{1\cdot t} = \underline{10{,}000\left[(1 + 0.035)^{1}\right]^{t}}$$

$$n = 4 \qquad A = 10{,}000\left(1 + \frac{0.035}{4}\right)^{4\cdot t} = \underline{10{,}000\left[\left(1 + \frac{0.035}{4}\right)^{4}\right]^{t}}$$

$$n = 12 \qquad A = 10{,}000\left(1 + \frac{0.035}{12}\right)^{12\cdot t} = \underline{10{,}000\left[\left(1 + \frac{0.035}{12}\right)^{12}\right]^{t}}$$

$$n = 365 \qquad A = 10{,}000\left(1 + \frac{0.035}{365}\right)^{365\cdot t} = \underline{10{,}000\left[\left(1 + \frac{0.035}{365}\right)^{365}\right]^{t}}$$

Can you discover a pattern in the form of the underlined expressions?

Each formula can be expressed as $A = 10{,}000b^t$, where $b = \left(1 + \frac{0.035}{n}\right)^n$ for $n = 1, 4, 12$, and 365. The number b is called the **growth** factor, and n is the number of compounding periods per year.

EXAMPLE 2

If $n = 4$ *in the formula* $b = \left(1 + \frac{0.035}{n}\right)^n$, *then* $b = \left(1 + \frac{0.035}{4}\right)^4 \approx 1.03546$. *The number* **1.03546** *is the growth factor.*

6. Determine the value of b in the following table, where $b = \left(1 + \frac{0.035}{n}\right)^n$. Round to five decimal places.

n, Number of Compounding Periods	1	4	12	365
b, Growth Factor	1.035	1.03546	1.03557	1.03562

The growth rate is the percentage by which the balance grows in 1 year. It is called the **effective yield, r_e**. Notice that as the number of compounding periods increases, the effective yield increases. This means that with the same annual interest rate (APR), your investment will earn more with more compounding periods.

PROCEDURE

To calculate the effective yield

1. Determine the growth factor $b = \left(1 + \frac{r}{n}\right)^n$.

2. Subtract 1 from b and write the result as a decimal.

$$r_e = b - 1 = \left(1 + \frac{r}{n}\right)^n - 1$$

EXAMPLE 3 ***Determine the effective yield for an APR of 4.5% compounded monthly.***

SOLUTION

$r = 4.5\% = 0.045,\ n = 12$

$$r_e = b - 1 = \left(1 + \frac{r}{n}\right)^n - 1 = \left(1 + \frac{0.045}{12}\right)^{12} - 1 \approx 1.04594 - 1 = 0.04594$$

$r_e = 4.594\%$

7. a. If interest is compounded hourly, then $n = 365 \cdot 24 = 8760$. Compute the growth factor, b, for compounding hourly, using an APR of 3.5%.

$$b = \left(1 + \frac{0.035}{8760}\right)^{8760} \approx 1.03562$$

b. Determine the effective yield associated with each of the growth factors in the table from Problem 6.

n	1	4	12	365
Growth Factor, *b*	1.035	1.03546	1.03557	1.03562
Effective Yield, r_e	3.5%	3.546%	3.557%	3.562%

c. Write a sentence comparing the growth factor b for compounding hourly, $n = 8760$, with that for compounding daily, $n = 365$.

The growth factors are the same up to five decimal places.

If the compounding periods become shorter and shorter (compounding every hour, every minute, every second), n gets larger and larger. If you consider the period to be so short that it's essentially an instant in time, you have what is called **continuous compounding**. Some banks use this method for compounding interest.

The compound interest formula $A = P\left(1 + \frac{r}{n}\right)^{nt}$ is no longer used when interest is compounded continuously. The following steps develop a formula for continuous compounding.

Step 1. Rewrite the given formula as indicated using properties of exponents.

$$A = P\left(1 + \frac{r}{n}\right)^{nt} = P\left[\left(1 + \frac{r}{n}\right)^{n/r}\right]^{rt} \text{ because } \frac{n}{r} \cdot rt = nt$$

Step 2. Let $\frac{n}{r} = x$. It follows that $\frac{r}{n} = \frac{1}{x}$. Note that as n gets very large, the value of x also gets very large.

Step 3. Substituting x for $\frac{n}{r}$ and $\frac{1}{x}$ for $\frac{r}{n}$, in the rewritten formula in step 1, you have

$$A = P\left[\left(1 + \frac{r}{n}\right)^{n/r}\right]^{rt} = P\left[\left(1 + \frac{1}{x}\right)^{x}\right]^{rt}.$$

8. a. Now take a closer look at the expression $\left(1 + \frac{1}{x}\right)^x$. Enter $\left(1 + \frac{1}{x}\right)^x$ into your calculator as a function of x. Display a table that starts at 0 and is incremented by 100. The results are displayed below.

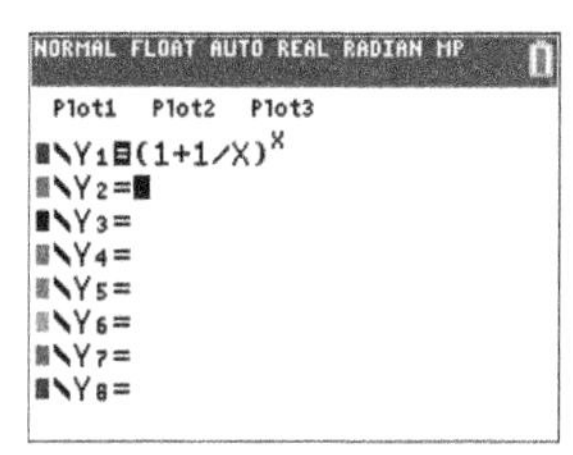

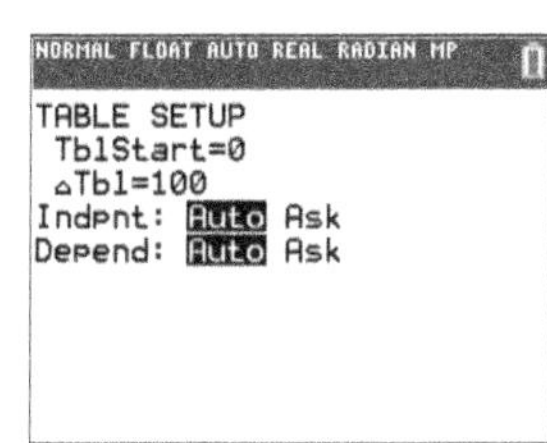

NORMAL FLOAT AUTO REAL RADIAN MP
PRESS + FOR ΔTbl

X	Y1
0	ERROR
100	2.7048
200	2.7115
300	2.7138
400	2.7149
500	2.7156
600	2.716
700	2.7163
800	2.7166
900	2.7168
1000	2.7169

X=0

b. In the table of values, why is there an error at $x = 0$?

There is an error at $x = 0$ because $\frac{1}{0}$ is undefined.

c. Determine the output for $x = 4000$, 5000, and 10,500. What happens to the output, $\left(1 + \frac{1}{x}\right)^x$, as the input, x, gets very large?

As x gets very large, $\left(1 + \frac{1}{x}\right)^x$ does not change very much and approaches 2.7182.

The letter e is used to represent the number that $\left(1 + \frac{1}{x}\right)^x$ approaches as x gets very large. This notation was devised by mathematician Leonhard Euler (pronounced oiler) (1707–1783). Euler used the letter e to denote this number. The number is irrational, so its decimal representation never ends and never repeats.

9. The number e is a very important number in mathematics. Find it on your calculator, and write its decimal approximation below. How does this approximation compare with the result in Problem 8a?

$e \approx 2.718281828$; very close

You are now ready to complete the compound interest formula for continuous compounding. Substituting e for $\left(1 + \frac{1}{x}\right)^x$ in $A = P\left[\left(1 + \frac{1}{x}\right)^x\right]^{rt}$, you obtain the following continuous compounding formula:

$$A = Pe^{rt},$$

where A is the current amount, or balance, in the account;

P is the principal;
r is the annual interest rate (annual percentage rate in decimal form);
t is the time in years that your money has been invested; and
e is the base of the continuously compounded exponential function.

EXAMPLE 4 ***You invest \$100 at a rate of 4% compounded continuously. How much money will you have after 5 years?***

SOLUTION

The principal is \$100, so $P = 100$. The annual interest rate is 4%, so $r = 0.04$. The money is invested for 5 years, so $t = 5$. Because interest is compounded continuously, you use the formula for continuous compounding as follows:

$$A = 100e^{0.04 \cdot 5} \approx \$122.14$$

10. a. Calculate the balance of your \$10,000 investment in 10 years with an annual interest rate of 3.5% compounded continuously.

$A = 10{,}000e^{0.035 \cdot 10} \approx \$14{,}190.68$

The formula used for the preceding result was $A = 10{,}000e^{0.035t}$. Comparing $A = 10{,}000e^{0.035t}$ with $A = 10{,}000b^t$ shows that the growth factor is $b = e^{0.035}$.

b. Determine the growth factor in this situation.

$b \approx 1.03562$

c. What is the effective yield of an annual interest rate of 3.5% compounded continuously?

$r_e = b - 1 = 1.03562 - 1 = 0.03562 = 3.562\%$

11. Beginning in 2008, the Federal Reserve, the nation's central bank, has kept its lending rate to member banks very low. This has resulted in low interest rates on both deposits and loans. In 2013, the interest rate paid on a "premium" savings account was 0.85%.

a. Determine the interest created by a \$10,000 deposit after 1 year if the money was deposited at 0.85% compounded quarterly.

$A = 10{,}000 \cdot \left(1 + \dfrac{0.0085}{4}\right)^{4 \cdot 1} \approx 10{,}085.27.$ The interest created was \$85.27.

b. Determine the value of a \$10,000 deposit after 5 years if the money was deposited at 0.85% compounded monthly.

$A = 10{,}000 \cdot \left(1 + \dfrac{0.0085}{12}\right)^{12 \cdot 5} \approx \$10{,}434.00$

c. Determine the value of a \$10,000 deposit after 10 years if the money was deposited at 0.85% compounded continuously.

$A = 10{,}000 \cdot e^{0.0085 \cdot 10} \approx \$10{,}887.17$

d. Do you believe that keeping interest rates low encourages savings? Explain.

No. The money is growing slowly, so why bother?

LC LEARNING CATALYTICS

You have \$30,000 that is in an account that earns 4% annually compounded monthly. What will the value of the account be after 5 years?

12. a. Historically, investments in the stock market have yielded an average rate of 11.7% per year. Suppose you invest \$10,000 in an account at an 11% annual interest rate that compounds continuously. Use the formula $A = Pe^{rt}$ to determine the balance after 35 years.

$A = 10{,}000e^{0.11 \cdot 35} \approx \$469{,}930.63$

b. What is the balance after 40 years?

$A = 10{,}000e^{0.11 \cdot 40} \approx \$814{,}508.69$

c. Your grandfather claimed that \$10,000 could grow to more than half a million dollars by the time you retire (40 years). Is your grandfather correct in his claim?

yes

SUMMARY Activity 3.5

1. The formula for **compounding interest** is $A = P\left(1 + \frac{r}{n}\right)^{nt}$.

2. The formula for **continuous compounding** is $A = Pe^{rt}$.

3. If the number of **compounding periods** is large, $A = P\left(1 + \frac{r}{n}\right)^{nt}$ is approximated by $A = Pe^{rt}$.

EXERCISES Activity 3.5

1. You inherit \$25,000 and deposit it in an investment account that earns 4.5% annual interest compounded quarterly.

   a. Write an equation that gives the amount of money in the account after t years.

   $A = 25{,}000\left(1 + \frac{0.045}{4}\right)^{4t}$

   b. How much money will be in the account after 10 years?

   $A = 25{,}000\left(1 + \frac{0.045}{4}\right)^{4 \cdot 10} \approx \$39{,}109.42$

   c. You want to have approximately \$65,000 in an investment account when your first child begins college. Use your graphing calculator to determine in how many years you will reach this goal.

   approximately 21.4 years

   d. If the interest were to be compounded continuously at 4.5%, how much money would be in the account after 10 years?

   $A = 25{,}000e^{0.045 \cdot 10} \approx \$39{,}207.80$

   e. Use your graphing calculator to determine in how many years you would reach your goal of \$65,000 if the interest is compounded continuously.

   approximately 21.2 years

   f. Should you look for an investment account that will be compounded continuously?

   Yes, I should.

Exercise numbers appearing in color are answered in the Selected Answers appendix.

2. You deposit \$2000 in an account that earns 5% annual interest compounded monthly.

a. What will your balance be after 2 years?

$$A = 2000\left(1 + \frac{0.05}{12}\right)^{12\cdot 2} \approx \$2209.88$$

b. Estimate how long it would take your investment to double.

approximately 13.9 years

c. Identify the annual growth rate and the growth factor.

The growth factor is $b = \left(1 + \frac{0.05}{12}\right)^{12} \approx 1.0512$. The growth rate is 0.0512, or 5.12%.

3. Your friend deposits \$1900 in an account that earns 3% compounded continuously.

a. What will her balance be after 2 years?

$A = 1900e^{0.03\cdot 2} \approx \$2017.49$

b. Estimate how long it will take your friend's investment to double.

approximately 23.1 years

4. You are 25 years old and begin to work for a large company that offers you two supplemental retirement options. Assume that you work for the company at least 10 years.

Option 1. A lump sum of \$2000 for each year you work for the company will be paid when you retire.

Option 2. The company will deposit \$10,000 in an account that pays 5% annual interest compounded monthly, guaranteed. When you retire, the money in the account will be given to you.

Let A represent the amount of money you will be given at retirement t years from now.

a. Write an equation that represents option 1.

$A = 2000t$

b. Write an equation that represents option 2.

$$A = 10{,}000 \cdot \left(1 + \frac{0.05}{12}\right)^{12t}$$

c. Use your graphing calculator to sketch a graph of the two functions. Use the window Xmin = 0, Xmax = 50, Ymin = 0, Ymax = 120,000.

d. If you plan to retire at age 65, which is the better plan?

After 40 years, option 1 will yield \$80,000. After 40 years, option 2 will yield \$73,584.17. So option 1 is the better plan.

e. If you plan to retire at age 70, which is the better plan?

After 45 years, option 1 will yield \$90,000. After 45 years, option 2 will yield \$94,434.89. So option 2 is the better plan.

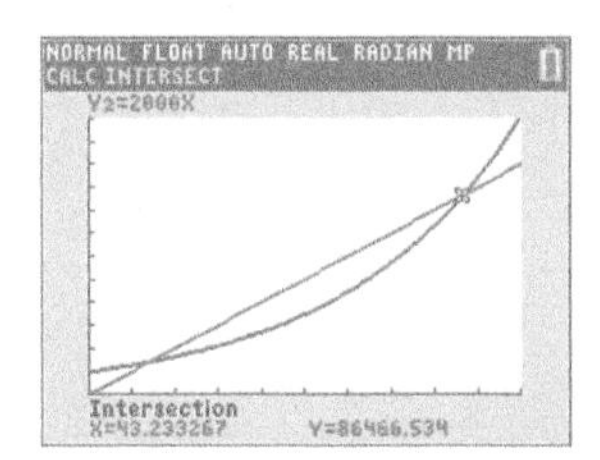

f. Use your graphing calculator to determine at what age it will not make a difference which plan you choose.

If I work 43.2 years or if you are 25 + 43.2 = 68.2 years old, both plans will yield \$86,466.53.

5. The compound interest formula that gives the balance in an account with a principal of $1500 that earns interest at the rate of 4.8% compounded monthly is $A = 1500\left(1 + \frac{0.048}{12}\right)^{12t}$.

Compare this formula with the exponential equation $A = 1500 \cdot b^t$.

a. What takes the place of b in the compound interest formula?

$b = \left(1 + \frac{0.048}{12}\right)^{12}$

b. What is the value of the growth factor b?

$b \approx 1.04907$

c. What is the effective yield?

$r_e = 4.907\%$

6. The compound interest formula that gives the balance in an account with a principal of $1500 that earns interest at the rate of 4.8% compounded continuously is $A = 1500 \cdot e^{0.048t}$. Compare this formula with the exponential equation $A = 1500 \cdot b^t$.

a. What takes the place of b in the compound interest formula?

$b = e^{0.048}$

b. What is the value of the growth factor b?

$b \approx 1.04917$

c. What is the effective yield?

$r_e = 4.917\%$

ACTIVITY 3.6

Continuous Growth and Decay

Problem Solving with Continuous Growth and Decay Models

OBJECTIVES

1. Discover the relationship between the equations of exponential functions defined by $y = ab^t$ and the equations of continuous growth and decay exponential functions defined by $y = ae^{kt}$.
2. Solve problems involving continuous growth and decay models.
3. Graph base *e* exponential functions.
4. Determine the domain of an exponential function.
5. Determine the range of an exponential function.

Instructor's note: This is a more complicated activity. It can be skipped with no loss of continuity.

The U.S. Census Bureau estimated that the U.S. population was 323,756,681 in September 28, 2016. The population grew to 325,973,284 in September 28, 2017.

1. Assuming exponential growth, the U.S. population, *P*, can be modeled by the equation $P = a \cdot b^t$, where *t* is the number of years since September 28, 2016. Therefore, $t = 0$ corresponds to September 28, 2016.

   a. What is the initial value, *a*?

   $a = 323{,}756{,}681$

   b. Determine the annual growth factor, *b*, for the U.S. population.

   $b = 325{,}973{,}284 \div 323{,}756{,}681 \quad b = 1.007$

   c. What is the annual growth rate? Write the result as a percent.

   $r = b - 1 \quad r = 1.007 - 1 \quad r = 0.007 \quad r = 0.7\%$

   d. Write the equation for the U.S. population, *P*, as a function of *t*.

   $P(t) = 323{,}756{,}681\,(1.007)^t$

The U.S. population did not remain constant at 323,756,681 from September 28, 2016 to September 27, 2017 and suddenly increase to 325,973,284 on September 28, 2016. The population grew continuously throughout the year. The exponential function used to model continuous growth is the same type of function used to model continuous compounding for an investment.

Recall from Activity 3.5 that the formula for continuous compounding is $A = Pe^{rt}$, where *A* (output) is the amount of the investment, *P* is the initial principal, *r* is the compounding rate, *t* (input) is time, and *e* is the constant irrational number. When this function is used more generally, it is written as

$$y = ae^{kt},$$

where *A* has been replaced with *y*, the output;
P has been replaced with *a*, the initial value; and
r has been replaced with *k*, the continuous growth rate.

Now the exponential growth model for the U.S. population determined in Problem 1d, written in the form $y = ab^t$, can be rewritten equivalently in the continuous growth form, $y = ae^{kt}$. Because *y* represents the same output value in each case, $ab^t = ae^{kt}$. Because *a* represents the same initial value in each model, it follows that $b^t = e^{kt}$.

2. **a.** Notice that e^{kt} can be written as $(e^k)^t$. How are *b* and e^k related?

   $b = e^k$

   b. Set the value of *b* determined in Problem 1 ($b = 1.007$) equal to e^k, and solve for *k*, the continuous growth rate. Solve the equation $e^k = 1.007$ graphically by entering e^x into Y_1 and 1.007 into Y_2. Use the window Xmin = 0, Xmax = 0.02, Ymin = 1, and Ymax = 1.02. Round the results to five decimal digits.

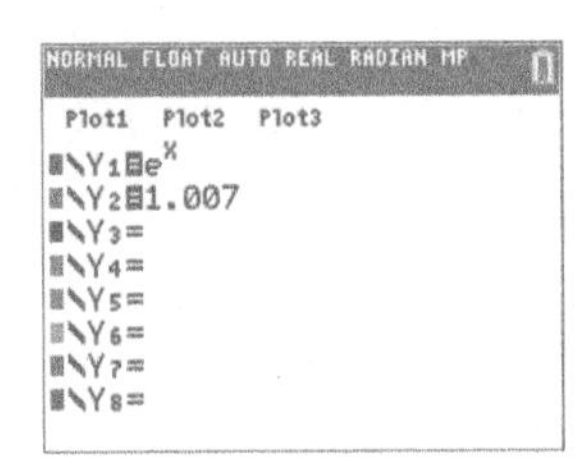

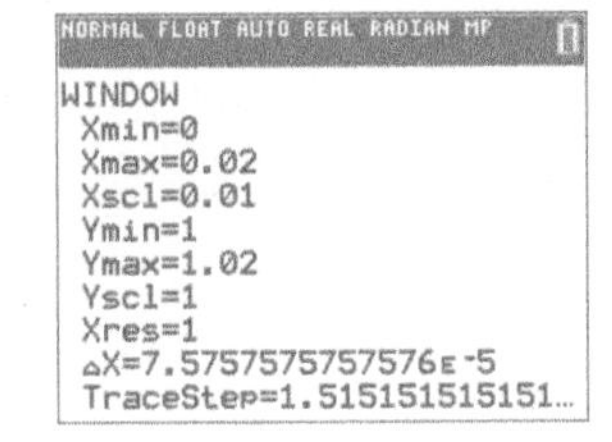

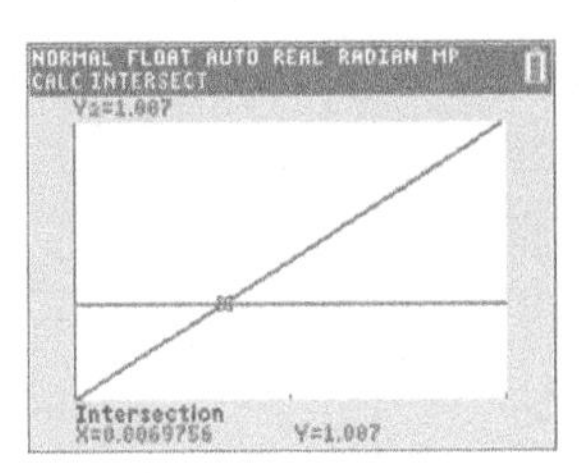

   $k = 0.00698$

c. Rewrite the U.S. population function $P = 323{,}756{,}681(1.007)^t$ in the continuous growth form.

$P = 323{,}756{,}681e^{0.00698t}$

Notice that an annual growth rate of $r = 0.7\%$ is equivalent to a continuous growth rate of $k = 0.00698 = 0.698\%$.

The exponential model used to describe continuous growth at a constant percent rate is $y = ae^{kt}$, where k is the constant continuous growth rate, a is the amount present initially (when $t = 0$), and e is the constant irrational number approximately equal to 2.718.

EXAMPLE 1

LC LEARNING CATALYTICS

Rewrite the equation $y = 16(2.31)^t$ as a continuous growth equation of the form $y = ae^{kt}$.

a. Rewrite the equation $y = 42(1.23)^t$ as a continuous growth equation of the form $y = ae^{kt}$.

SOLUTION

Set $1.23 = e^k$. Then $k \approx 0.207$, and $y = 42e^{0.207t}$

b. What is the continuous growth rate?

SOLUTION

$$0.207 = 20.7\%$$

c. What is the initial amount present (when $t = 0$)?

SOLUTION

$$42(1.23)^0 = 42$$

Now consider a situation that involves continuous decay at a constant percentage rate. Tylenol (acetaminophen) is metabolized in your body and eliminated at the rate of 24% per hour. You take two Tylenol tablets (1000 milligrams) at noon.

3. Assume that the amount of Tylenol in your body can be modeled by an exponential function $Q = ab^t$, where t is the number of hours from noon.

a. What is the initial value, a, in this situation?

$a = 1000$

b. Determine the decay factor, b, for the amount of Tylenol in your body.

$b = 1 - r = 1 - 0.24 = 0.76$

c. Write the amount of Tylenol in your body as an exponential function of t.

$Q(t) = 1000(0.76)^t$

Of course, the amount of Tylenol in your body does not decrease suddenly by 24% at the end of each hour; it is metabolized and eliminated continuously. The equation $y = ae^{kt}$ can also be used to model a quantity that decreases at a continuous rate.

Recall that in Problem 2a, you compared $y = ab^t$ with $y = ae^{kt}$ and established that $b = e^k$.

4. **a.** The value of b for the Tylenol equation is 0.76. Set $b = 0.76$ equal to e^k, and solve for k graphically as in Problem 2b. Use the window Xmin $= -1$, Xmax $= 0$, Ymin $= 0$, and Ymax $= 1$.

   $k \approx -0.274$

   b. Write the equation for the amount of Tylenol in your body, $Q = 1000(0.76)^t$, in the form $Q = ae^{kt}$.

   $Q = 1000e^{-0.274t}$

Notice that the value of k in Problem 4 is negative. Whenever $0 < b < 1$, then b is a decay factor and the value of k will be negative. A decreasing exponential function written in the form $y = ae^{kt}$ will have $k < 0$, and $|k|$ is the continuous rate of decrease.

For exponential decrease (decay) at a continuous constant percent rate, the model $y = ae^{kt}$ is used, where $k < 0$, $|k|$ is the constant continuous decrease (decay) rate, a is the amount present initially (when $t = 0$), and e is the constant irrational number approximately equal to 2.718.

EXAMPLE 2

a. Rewrite the decay equation $y = 12.5(0.83)^t$ in the form $y = ae^{kt}$.

SOLUTION

$$0.83 = e^k. \quad \text{Then } k = -0.186 \text{ and } y = 12.5e^{-0.186t}.$$

b. What is the continuous percentage rate of decay?

SOLUTION

$$0.186 = 18.6\%$$

c. What is the initial amount present when $t = 0$?

SOLUTION

12.5

Graphs of Exponential Functions Having Base *e*

5. Consider the exponential function defined by $y = e^x$.

   a. What is the domain of the function?

   all real numbers

   b. Is this function increasing or decreasing? Explain.

   The function is increasing because the base $b = e$ and $e > 1$.

   c. Complete the following table. If necessary, round the y-values to the nearest two decimal places.

x	−3	−2	−1	0	1	2	3	4	5
$y = e^x$	0.05	0.14	0.37	1	2.72	7.39	20.09	54.60	148.41

d. Sketch a graph of $y = e^x$. Verify using a graphing calculator.

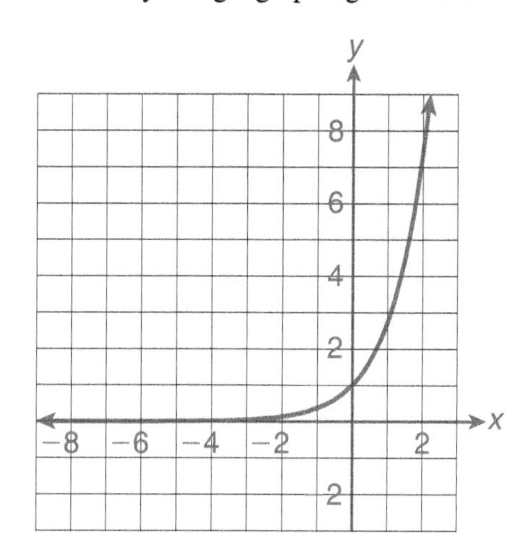

e. What is the horizontal asymptote?

$y = 0$, the x-axis

f. What is the range of the function?

The range is the set of all positive real numbers $y > 0$.

g. What are the intercepts of the graph?

y-intercept: (0, 1); x-intercept: none

h. Is the function continuous over its domain?

Yes, the graph is totally connected for all real values of x.

6. A vertical shift of the graph of $y = e^x$ changes the horizontal asymptote.

a. What is the equation of the horizontal asymptote of the graph of $y = e^x + 2$?

$y = 2$

b. What is the equation of the horizontal asymptote of the graph of $y = e^x - 2$?

$y = -2$

7. Identify the given exponential function as increasing or decreasing. In each case, give the initial value and rate of increase or decrease.

a. $P = 2500e^{0.04t}$

The initial value is 2500, and P is increasing at the continuous rate of 4%.

b. $Q = 400(0.86)^t$

The initial value is 400, and Q is decreasing at the rate of 14%.

c. $A = 75(1.032)^t$

The initial value is 75, and A is increasing at the rate of 3.2%.

d. $R = 12e^{-0.12t}$

The initial value is 12, and R is decreasing at the continuous rate of 12%.

SUMMARY Activity 3.6

1. When a quantity increases or decreases continuously at a constant percent rate, the amount present at time t can be modeled by $y = ae^{kt}$, where a is the initial quantity at $t = 0$. If the quantity is increasing, then $k > 0$ and k is the continuous rate of increase. If the quantity is decreasing, then $k < 0$ and $|k|$ is the continuous rate of decrease.

2. The graph of an increasing exponential function of the form $y = ab^x$ or $y = ae^{kt}$, where $a > 0$ and $b > 1$ or $k > 0$, will be shaped as follows:

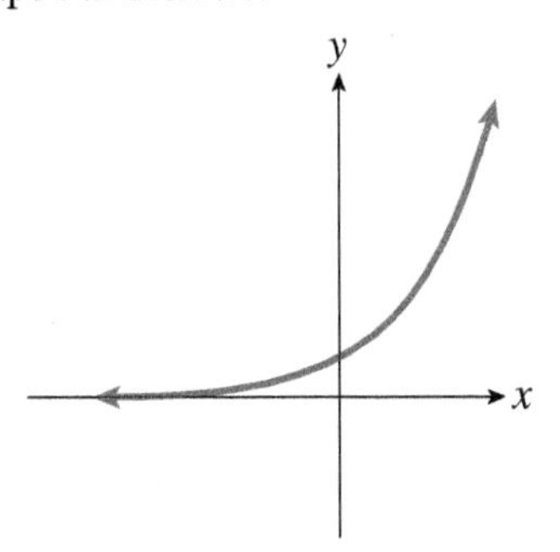

3. The graph of a decreasing exponential function of the form $y = ab^x$ or $y = ae^{kt}$, where $a > 0$ and $0 < b < 1$, or $k < 0$, will be shaped as follows:

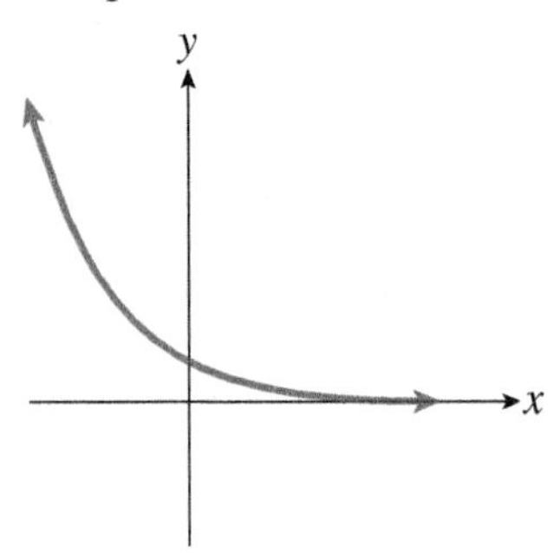

4. The domain of an exponential function of the form $y = ab^x$ or $y = ae^{kx}$, where $a > 0$ is all real numbers.

5. The range of an exponential function of the form $y = ab^x$ or $y = ae^{kx}$, where $a > 0$, is $y > 0$.

Notation

The function $y = ae^{kt}$ may be written in other forms, such as $y = y_0e^{kt}$, where the initial value is y_0 (y sub zero) or $Q = Q_0e^{kt}$, where the initial value is Q_0 (Q sub zero).

EXERCISES Activity 3.6

1. In 2016, Orem, Utah, was one of the fastest-growing cities in the United States. The population of Orem had been growing steadily, and the exponential function $f(t) = 88.5(1.015)^t$ models the population from 2010 to 2016, where t represents the number of years from 2010 and $f(t)$ represents the population of Orem measured in thousands of people.

   a. Is this an increasing or decreasing exponential function? Explain.

   Increasing. $b = 1.015$ is greater than 1, so it is a growth factor.

Exercise numbers appearing in color are answered in the Selected Answers appendix.

b. Determine the annual growth or decay rate from the model.

The annual growth rate is $r = b - 1 = 1.015 - 1 = 0.015 = 1.5\%$.

c. According to the model, what is the initial value? What does the value mean in this situation?

The initial value is $f(0) = 88.5(1.015)^0 = 88.5$. The 2010 ($t = 0$) population of Orem, Utah, was 88.5 thousand (88,500) people.

d. The equation for continuous growth is $y = ae^{kt}$. Set the value of b in your model equal to e^k (e^x in the calculator) and use your graphing calculator to determine the value for k graphically. This will be the continuous growth or decay model.

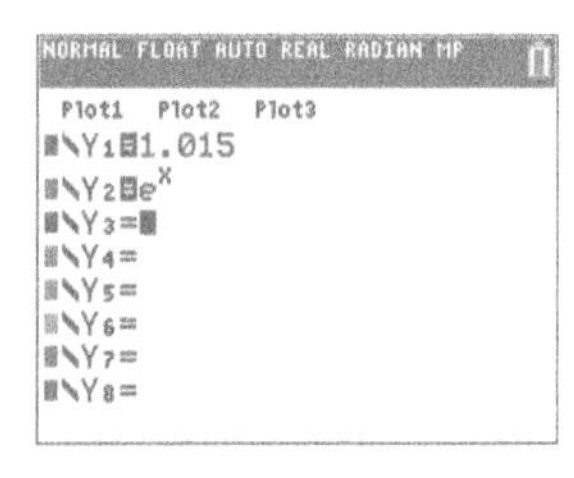

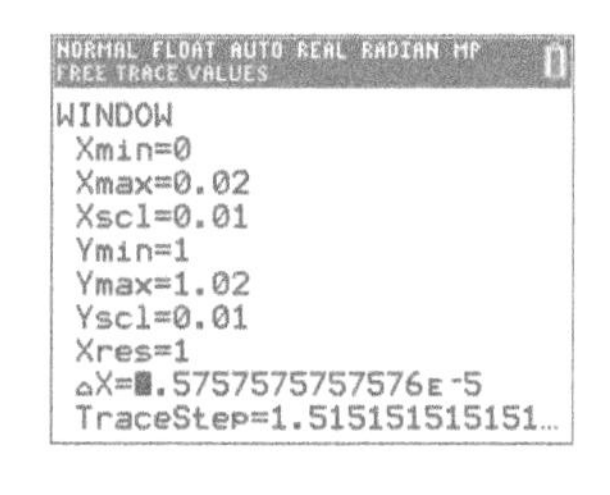

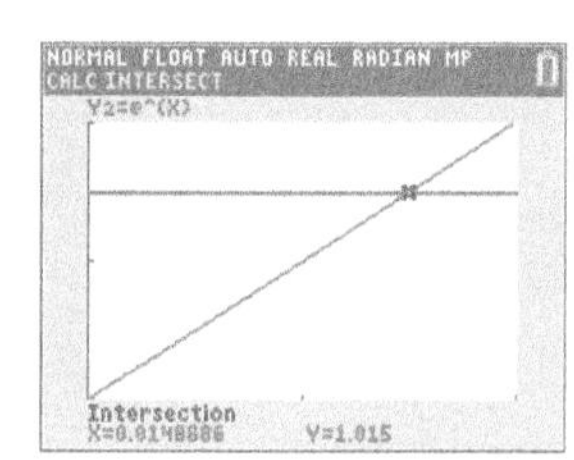

$k \approx 0.0149$

e. Rewrite the population function in the form $f(t) = ae^{kt}$.

$f(t) = 88.5e^{0.0149}$

f. Use the function in part e to predict the population of Orem in 2025.

2025 corresponds to $t = 15$. $f(15) = 88.5e^{0.0149 \cdot 15} = 110.664$. If the population continues to grow at the same rate, the 2025 population will be 110,664 people.

2. The table below shows the smoking prevalence among U.S. adults (18 years and older) as a percent of the population.

SMOKING PREVALENCE AMONG U.S. ADULTS							
Year	1970	1980	1990	2000	2005	2010	2015
Years since 1970	0	10	20	30	35	40	45
Percent of the Population	37.4	33.2	25.5	23.3	20.9	19.3	15.1

Source: The U.S. Centers for Disease Control and Prevention.

This data can be modeled by an exponential function $p(t) = 38.7(0.981^t)$, where t represents the number of years since 1970 and p is the smoking prevalence among adults as a percent of the U.S. population.

a. Is this an increasing or decreasing exponential function? Explain.

The function is decreasing because the base of the model is 0.981, which is greater than 0, but less than 1.

b. Use your graphing calculator to create a scatterplot of the data and then add the graph of the function to the graph. Does the graph reinforce your answer in part a?

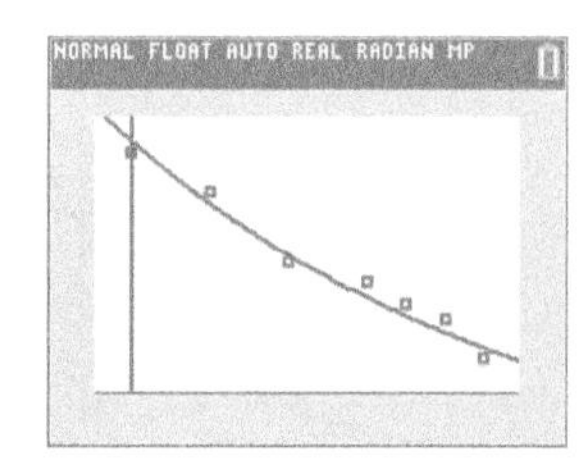

Yes, the percent of the population decreases as the years since 1970 increase.

c. Determine the growth or decreasing (decay) rate from the model.

It is a decreasing (decay) rate $1 - 0.981 = 0.019$ or 1.9%.

d. According to the model, what is the initial value?

Based on the model, the initial value was 38.7%. This value differs from the value in the table because the model only approximates the given data.

e. The equation for the continuous growth or decay is $y = ae^{kt}$. Set the value of b in your model equal to e^k, and use your graphing calculator to determine the value for k graphically. This will be the continuous growth or decay rate.

$k \approx -0.0192$; the continuous decay rate is 0.0192 or 1.92%.

f. Rewrite the smoking function in the form $y = ae^{kt}$.

$y = 38.7 \cdot e^{-0.0192t}$

g. Use the function from part f to determine the percent of adults in the U.S. population who will be smoking in 2025.

Approximately 13.5% of adults in the U.S. population will be smokers in 2025 if the decline continues at the same rate.

3. Identify the given exponential function as increasing or decreasing. In each case, give the initial value and rate of increase or decrease.

a. $R(t) = 33(1.097)^t$

The initial value is 33, and R is increasing at the rate of 9.7%.

b. $f(x) = 97.8e^{-0.23x}$

The initial value is 97.8, and $f(x)$ is decreasing at the continuous rate of 23%.

c. $S = 3250(0.73)^t$

The initial value is 3250, and S is decreasing at the rate of 27%.

d. $B = 0.987e^{0.076t}$

The initial value is 0.987, and B is increasing at the continuous rate of 7.6%.

4. Sketch a graph of each of the following, and verify using your graphing calculator.

a. $y = 20e^{0.08x}$

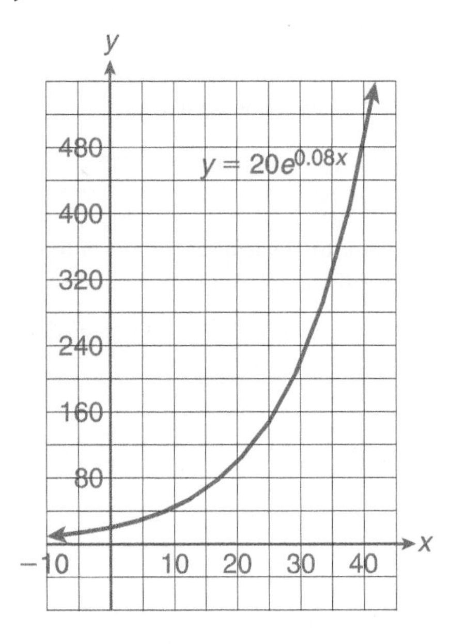

b. $f(x) = 10e^{-0.3x}$

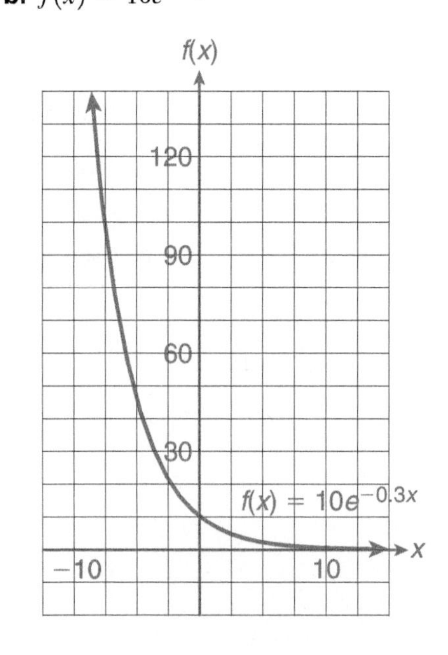

c. Compare the two graphs. Include shape, direction, and initial values.

Both graphs are exponential curves. The first is increasing. The second is decreasing. The initial value for the first is 20. The initial value for the second is 10.

5. Strontium 90 is a radioactive material that decays according to the function defined by $y = y_0e^{-0.0244t}$, where y_0 is the amount present initially and t is time in years.

a. If 20 grams of strontium 90 are present today, how much will be present in 20 years?

$y = 20e^{-0.0244 \cdot 20} \approx 12.277$ g

b. Use the graph of the function to approximate how long it will take 20 grams to decay to 10 grams, 10 grams to decay to 5 grams, and 5 grams to decay to 2.5 grams. The length of time is called the *half-life*. In general, a half-life is the time required for half of a radioactive substance to decay.

The half-life is approximately 28.4 years.

c. Identify the annual decay rate and the decay factor.

The decay factor is $b = e^{-0.0244} \approx 0.9759$. The decay rate is 0.0241, or 2.41%.

6. When drugs are administered into the bloodstream, the amount present decreases continuously at a constant rate. The amount of a certain drug in the bloodstream is modeled by the function $y = y_0e^{-0.35t}$, where y_0 is the amount of the drug injected (in milligrams) and t is time (in hours).

a. Suppose that 10 milligrams are injected at 10:30 A.M. How much of the drug is still in the bloodstream at 2 P.M.?

At 2 P.M., $t = 3.5$, so $y = 10e^{-0.35 \cdot 3.5} \approx 2.94$ milligrams.

b. If another dose needs to be administered when 1 milligram of the drug is present in the bloodstream, approximately when should the next dose be given (to the nearest quarter hour)?

In about 6.6 hours, at about 5 P.M., the next dose should be given.

7. As everyone knows, Amazon has become a dominant force in the retail world. Amazon's annual net revenue since 2007 can be modeled by $R = f(t) = 15.75e^{0.247t}$, where R is measured in billions of U.S. dollars and t is the number of years since 2007.

a. According to the model, what was Amazon's net revenue in 2012?

2012 is 5 years from 2007, so $t = 5$. $f(5) = 15.75e^{0.247 \cdot 5} = 54.154$ billion. Amazon's 2012 revenue was $54,154,000,000.

b. Sketch a graph of Amazon's net revenue function.

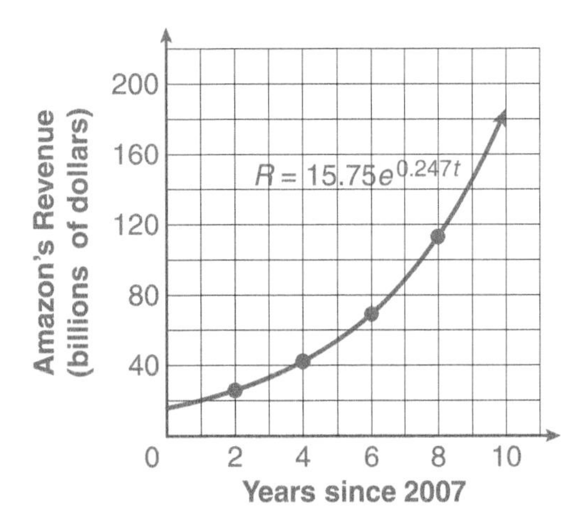

c. What is the vertical intercept of the graph? What is the practical meaning of the intercept in this situation?

The vertical intercept of the graph is (0, 15.75). Amazon's 2007 revenue was $15,750,000,000.

d. Determine graphically and/or numerically the year when Amazon's net revenue first reached $100 billion.

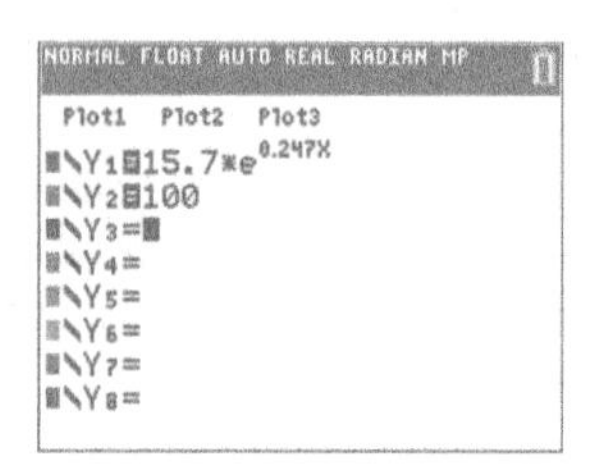

NORMAL FLOAT AUTO REAL RADIAN MP
PRESS + FOR △Tbl

X	Y1	Y2
0	15.7	100
1	20.099	100
2	25.73	100
3	32.939	100
4	42.168	100
5	53.983	100
6	69.107	100
7	88.47	100
8	113.26	100
9	144.99	100
10	185.61	100

X=7

NORMAL FLOAT AUTO REAL RADIAN MP
FREE TRACE VALUES
WINDOW
Xmin=0
Xmax=10
Xscl=1
Ymin=0
Ymax=200
Yscl=20
Xres=1
△X=0.037878787878788
TraceStep=0.075757575757...

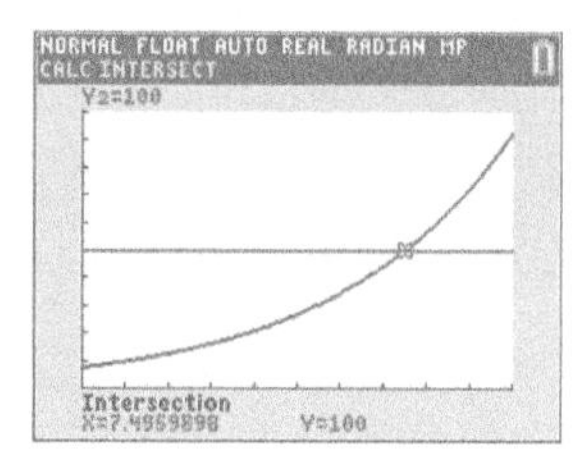

Amazon's revenue reached $100 billion about 7.5 years from 2007 or mid-2014.

e. What is the doubling time?

Solve the equation $15.75e^{0.247t} = 2 \cdot 15.75 = 31.50$.

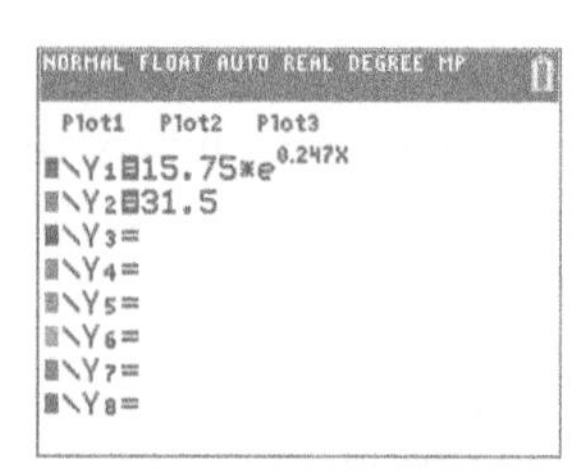

NORMAL FLOAT AUTO REAL DEGREE MP
PRESS + FOR △Tbl

X	Y1	Y2
2	25.812	31.5
2.1	26.458	31.5
2.2	27.119	31.5
2.3	27.797	31.5
2.4	28.492	31.5
2.5	29.205	31.5
2.6	29.935	31.5
2.7	30.684	31.5
2.8	31.451	31.5
2.9	32.238	31.5
3	33.044	31.5

X=2.8

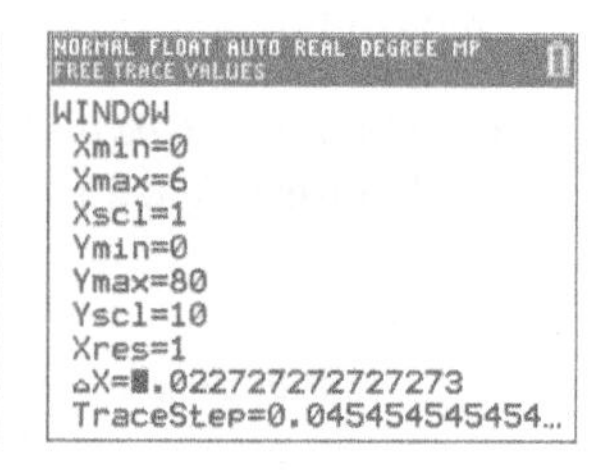

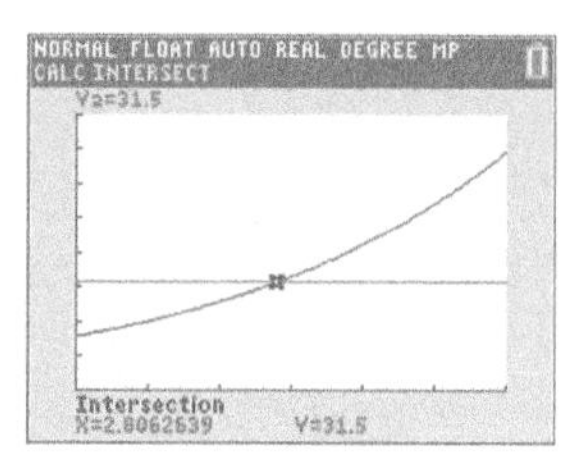

The doubling time is approximately 2.8 years.

8. *E. coli* bacteria are capable of very rapid growth, doubling in number approximately every 49.5 minutes. The number, N, of *E. coli* bacteria per milliliter after x minutes can be modeled by the function

$$N = 500{,}000e^{0.014x}.$$

a. What is the initial number of bacteria per milliliter?

500,000

b. How many *E. coli* bacteria would you expect after 99 minutes? (*Hint:* There will be two doublings.) Verify your estimate using the equation.

500,000 doubled twice is 2,000,000 per milliliter.

c. Use a graphing or numerical approach to determine the elapsed time when there will be 20,000,000 *E. coli* bacteria per milliliter.

There will be 20,000,000 bacteria when $x \approx 264$ minutes, or 4 hours 24 minutes.

ACTIVITY 3.7

Ebola
Modeling Data with Exponential Regression Equations

OBJECTIVES

1. Determine the regression equation of an exponential function that best fits the given data.
2. Make predictions using an exponential regression equation.
3. Determine whether a linear or exponential model best fits the data.

In May 2017, health officials in the Democratic Republic of Congo reported three cases of Ebola. Because of an earlier outbreak of Ebola in West Africa, health officials took this news very seriously. Left unchecked, the Ebola virus is capable of sparking a worldwide pandemic. The infection rate (the number of people that any single infected person will infect) and the incubation period (the time between exposure and the development of symptoms) of this virus are known to vary.

A very conservative infection rate (without treatment) would be 2, and the incubation period is between 2 and 21 days. We will assume an incubation period of 15 days or ½ of a month. This means that the first person could be expected to infect 2 people in about 15 days. After 15 days, the person cannot infect anyone else. This assumes that the spread of the virus is not checked by inoculation or vaccination.

Thus, the total number of infected people 0.5 month after the first person was infected would be 3, the sum of the original infected person and the 2 newly infected people. During the second half of the month, the 2 newly infected people would infect $2 \times 2 = 4$ new people. This means you have 4 people added to the 3 people previously infected, or approximately 7 people infected with Ebola at the end of the first month.

1. The following table represents the total number of people who could be infected with the Ebola virus over a period of 5 months. Complete the table.

Months since the First Person Was Infected	0	0.5	1	1.5	2	2.5	3	3.5	4	4.5	5
Number of Newly Infected		2	4	8	16	32	64	128	256	512	1024
Total Number of People Infected	1	3	7	15	31	63	127	255	511	1023	2047

2. Let t represent the number of months since the first person was infected, and T represent the total number of people infected by the Ebola virus. Use your graphing calculator to create a scatterplot of the given data.

3. Does the scatterplot indicate a linear relationship between t and T? Explain.

No. The relationship does not appear to be linear because the data points do not lie on a straight line.

4. a.

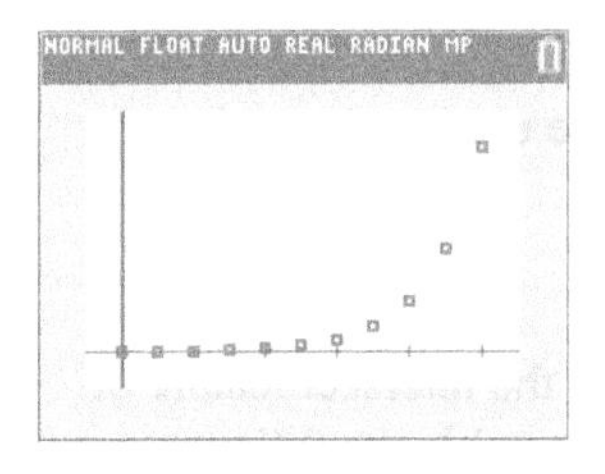

4. a. Use your calculator to model the data with an exponential function. Use option 0:ExpReg in the STAT CALC menu to determine an exponential function that best fits the given data. Record the regression equation of the exponential model below. Round a and b to the nearest 0.001. Your screen should appears as follows:

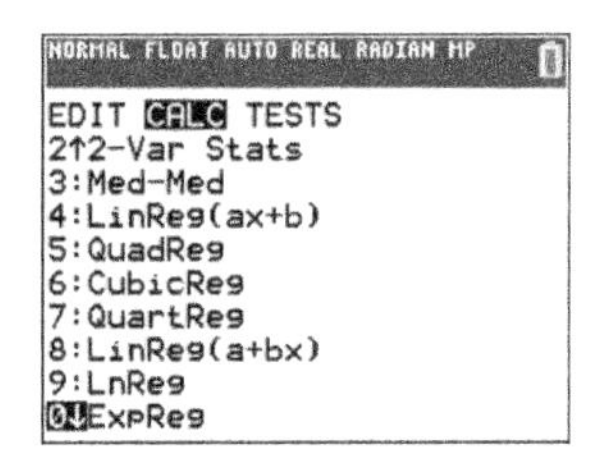

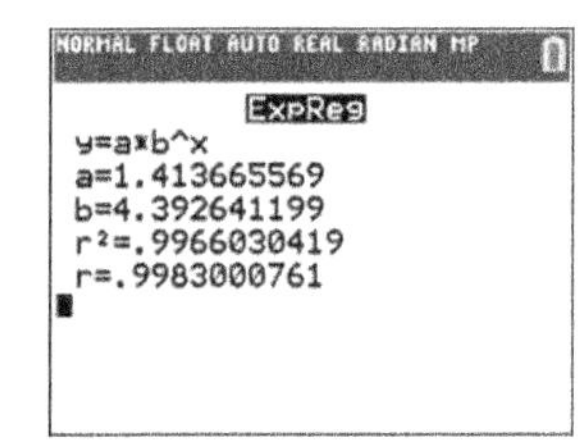

$Y = T = 1.414(4.393)^t$

c. Solve the equation in part b using a graphing approach. Use the intersect feature of your graphing calculator; the screen containing the solution should appear as follows. How confident are you in your prediction?

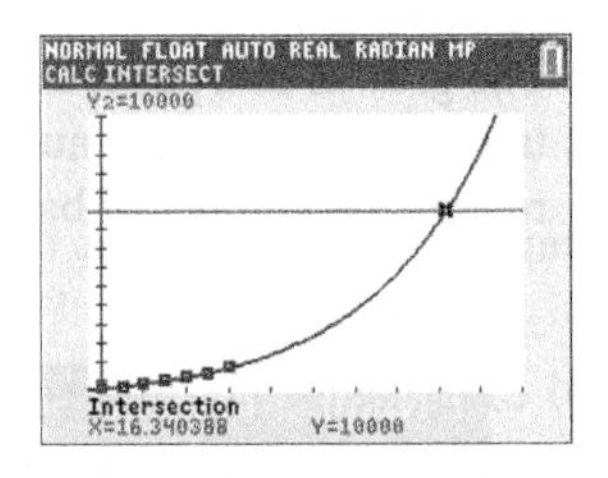

$x \approx 16.3$. I am not very confident. The results are far from the original data.

11. What is the doubling time for the exponential model? That is, approximately how long does it take the number of bacteria to double?

It takes 4 days. This answer can be read from the original table.

Decreasing Exponential Model

The price of 32" flat screen television sets has been decreasing since they were first becoming popular in 2005. Data for various years from 2005 to 2018 is found in the following table:

Year	2005	2008	2012	2018
Average Selling Price of a 32" Television Set	\$911	\$649	\$435	\$204

12. a. Use your graphing calculator to determine the regression equation of an exponential function that models the given data. Let your input, t, represent the number of years since 2005. Round your values to three decimal places.

$P = 924.01 \cdot 0.892^t$

b. Sketch a graph of the exponential model using your graphing calculator.

12. b.

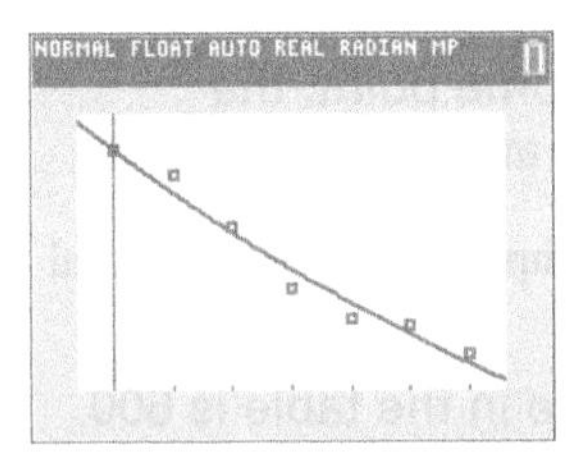

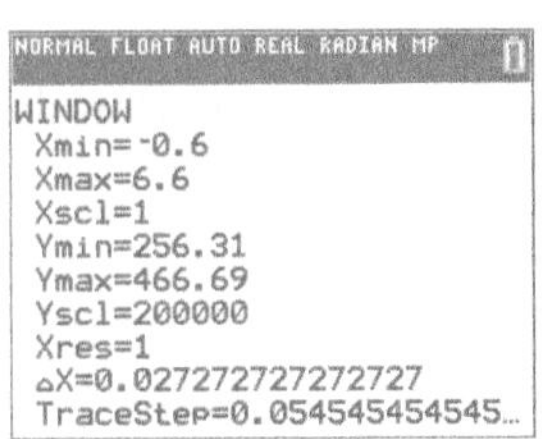

c. What is the base of the exponential model? Is the base a growth or decay factor? How do you know?

The base of the model is 0.892. The base is less than 1, so it is a decay function.

d. What is the annual decay rate?

$r = 1 - 0.892 = 0.108 = 10.8\%$

e. Does the graph have a horizontal asymptote? What is the practical meaning of this asymptote in the context of the situation?

The horizontal asymptote is $P = 0$. There is no practical meaning for $P = 0$, the cost of the tv will not be \$0.

EXERCISES Activity 3.7

1. The total amount of money spent on healthcare in the United States is increasing at an alarming rate. The following table gives the total national healthcare expenditures, in billions of dollars, for selected years from 1980 through 2016.

Year	1980	1990	2000	2010	2012	2014	2016
Total Spent, C (billions of dollars)	253	714	1354	2595	2800	3093	3458

a. Sketch the scatterplot of this data.

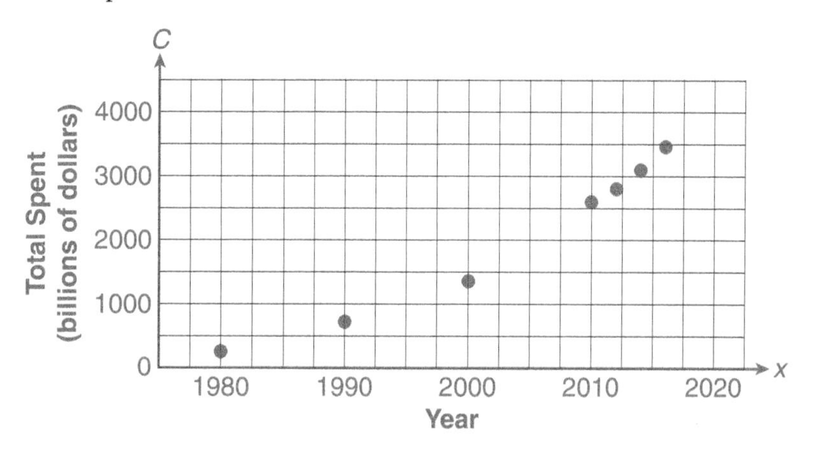

b. Use your graphing calculator to determine the exponential regression equation that best fits the healthcare data in the table. Let your input, t, represent the number of years since 1980.

$C(t) = 299.73(1.0728)^t$

c. Using the regression equation from part b, determine the predicted total healthcare expenditures for 2005.

2005 is 25 years form 1980, so $t = 25$. $C(25) = 299.73(1.0728)^{25} = 1736.598$. The 2005 estimated expenditure is $1,736,598,000,000.

d. According to the exponential model, what is the growth factor for total healthcare costs per year?

The growth factor is $b = 1.0728$.

e. What is the growth rate?

The growth rate is $r = 1.0728 - 1 = 0.0728 = 7.28\%$.

f. According to the exponential model, in what year did total healthcare costs first exceed $1.5 trillion?

1.5 trillion = 1500 billion. The table indicates that healthcare costs will exceed $1.5 trillion 23 years from 1980 or in the year 2003.

NORMAL FLOAT AUTO REAL RADIAN MP
PRESS + FOR △Tbl

X	Y1
20	1221.3
21	1310.2
22	1405.5
23	1507.8
24	1617.5
25	1735.2
26	1861.5
27	1997
28	2142.3
29	2298.1
30	2465.4

X=23

Exercise numbers appearing in color are answered in the Selected Answers appendix.

g. What is the doubling time for your exponential model?

Solve the equation $299.73(1.0728)^t = 2 \cdot 299.73 = 599.46$ for t.

NORMAL FLOAT AUTO REAL RADIAN MP
Plot1 Plot2 Plot3
\Y1=299.73017265678*1.072
\Y2=599.46
\Y3=
\Y4=
\Y5=
\Y6=
\Y7=
\Y8=

NORMAL FLOAT AUTO REAL RADIAN MP
PRESS + FOR ΔTbl

X	Y1	Y2
9.3	576.01	599.46
9.4	580.07	599.46
9.5	584.16	599.46
9.6	588.27	599.46
9.7	592.42	599.46
9.8	596.6	599.46
9.9	600.8	599.46
10	605.04	599.46
10.1	609.3	599.46
10.2	613.6	599.46
10.3	617.92	599.46

X=9.9

NORMAL FLOAT AUTO REAL RADIAN MP
WINDOW
Xmin=0
Xmax=20
Xscl=2
Ymin=300
Ymax=1200
Yscl=100
Xres=1
ΔX=0.075757575757576
TraceStep=0.151515151515...

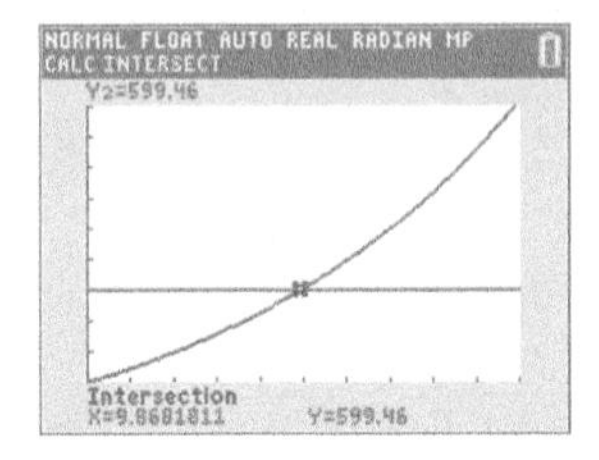

The healthcare costs will double in about 9.9 years.

2. a. Consider the following data set for the variables x and y:

x	5	8	11	15	20
y	70.2	50.7	35.1	22.6	9.5

Plot these points on the following grid:

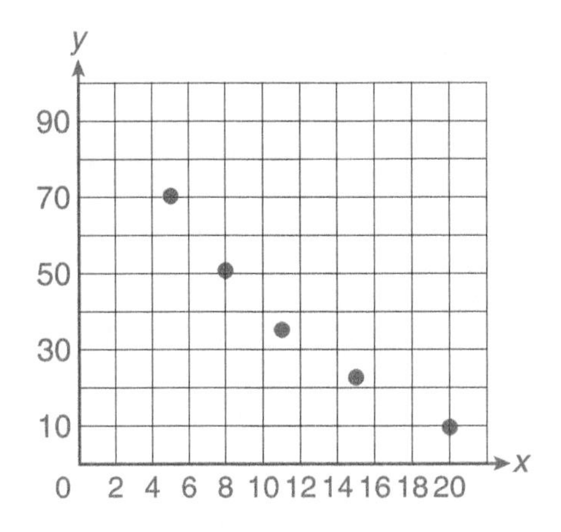

b. Use your graphing calculator to determine both a linear regression and an exponential regression model of the data. Record the equations for these models here.

linear: $y = -3.95x + 84.20$

exponential: $y = 144.21(0.877)^x$

c. Use the graph of each equation in part b to determine which model appears to fit the data better? Explain.

The graph of the exponential model is closer to the points on the scatterplot.

d. Use the better model to determine y when $x = 13$ and y when $x = 25$.

$y = 144.21(0.877)^{13} \approx 26.18$

$y = 144.21(0.877)^{25} \approx 5.42$

e. For the exponential model, what is the decay factor?

0.877

f. What does it mean that the decay factor is between 0 and 1?

As the input increases, the output decreases.

g. What is the half-life for the exponential model?

approximately 5.28

3. Use the graph of $y = 5(2)^x$ as a model, and summarize the properties of the exponential function defined by $y = ab^x$, where $a > 0$.

a. What is the domain?

the set of all real numbers

b. What is the range?

the set of all positive real numbers

c. For what value of x is $y = ab^x$ positive?

$y = a \cdot b^x$ is positive for all values of x.

d. For what value of x is $y = ab^x$ negative?

$y = a \cdot b^x$ is never negative.

e. What is the vertical intercept of the graph of $y = ab^x$?

$(0, a)$

4. According to GfK, Germany's largest market research institute, the total sales value of smartphones in China from 2013 through 2017 is as given in the following table. The sales are measured in billions of dollars.

Year	2013	2014	2015	2016	2017
Smartphone Sales in China (billions of dollars)	90.1	99.0	116.2	133.5	148.3

a. Use your graphing calculator to determine an exponential regression model for the data in the table. Let the independent variable, x, represent the number of years since 2013.

$y = 89.105(1.138)^x$

b. Assuming the rate of growth continues, approximate the value of smartphone sales in China in 2025.

For 2025, $x = 12$. So $y = 89.105(1.138)^{12} \approx 420.3$ billion dollars in smartphone sales.

Collecting and Analyzing Data

5. You need to find some real-world data that appears to be increasing exponentially. Newspapers, magazines, scientific journals, almanacs, and the Internet are good resources. Once you obtain the data, model the data using an exponential function defined by $y = ab^x$. Explain why an exponential function would best represent the data. Be sure to describe the meaning of a and b in the exponential model in terms of the situation. Make a prediction about the dependent variable, y, for a specific value of the independent variable, x. Describe the reliability of this prediction.

CLUSTER 1 What Have I Learned?

1. Consider a linear function defined by $g(x) = mx + b, m \neq 0$, and an exponential function defined by $f(x) = ab^x$. Explain how you can determine from the equation whether the function is increasing or decreasing. Assume $a > 0$.

   The function g is increasing if $m > 0$. f is increasing if $b > 1$. g is decreasing if $m < 0$. f is decreasing if $0 < b < 1$.

2. Suppose you have an exponential function of the form $f(x) = ab^x$, where $a > 0$ and $b > 0$ and $b \neq 1$. By inspecting the graph of f, can you determine if $b > 1$ or if $0 < b < 1$? Explain.

   Yes. If the function is increasing, $b > 1$. If the function is decreasing, $0 < b < 1$.

3. You are given a function defined by a table, and the input values are in increments of 1. By looking at the table, can you determine whether the function can be approximated by an exponential model? Explain.

   Yes. If the ratio between consecutive output values is roughly constant, the function can be approximated by an exponential function.

4. Explain the difference between growth rate and growth factor.

   If the growth rate is r, the growth factor is $1 + r$. The growth rate is the percentage increase in output per unit increase in input. The growth factor is the common factor multiplier that produces the next output (per unit increase of input).

5. An exponential function $y = ab^x$ passes through the point $(0, 2.6)$. What can you conclude about the values of a and b?

   $a = 2.6$ Nothing can be concluded about the value of b without further information.

6. You just received a substantial tax refund of P dollars. You decide to invest the money in an investment account for 2 years and narrow your choices to two companies. Company A will give you 6.75% interest compounded quarterly. Company B offers you 6.50% compounded continuously. Where do you deposit your money? Explain.

   The growth factor for company A is $\left(1 + \frac{0.0675}{4}\right)^4 = 1.06923$. The growth factor for company B is $e^{0.065} = 1.06716$. Therefore, I would deposit the money in company A because the growth factor is larger.

7. Explain why the base in an exponential function cannot equal 1.

   If $b = 1$, the function will be a constant function because $1^n = 1$ for all values of n.

CLUSTER 1 How Can I Practice?

1. You are planning to purchase a new car and have your eye on a specific model. You know that new car prices are projected to increase at a rate of 4% per year for the next few years.

   a. Write an equation that represents the projected cost, C, of your dream car t years in the future, given that it costs $23,000 today.

   $C = 23{,}000(1.04)^t$

   b. Identify the growth rate and the growth factor.

   The growth rate is 0.04; the growth factor is 1.04.

   c. Use your equation in part a to project the cost of your car 3 years from now.

   $C = 23{,}000(1.04)^3 \approx \$25{,}871.87$

   d. Use your graphing calculator to approximate how long it will take your dream car to cost $35,000 if the price continues to increase at 4% per year.

   10.7 yr.

2. Without using your graphing calculator, match the graph with its equation.

   a. $g(x) = 2.5(0.47)^x$

   b. $h(x) = 1.5(1.47)^x$

i.

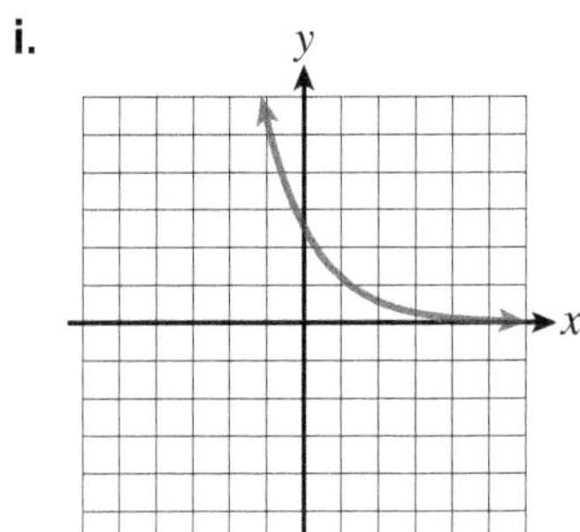

Graph i is function g.

ii.

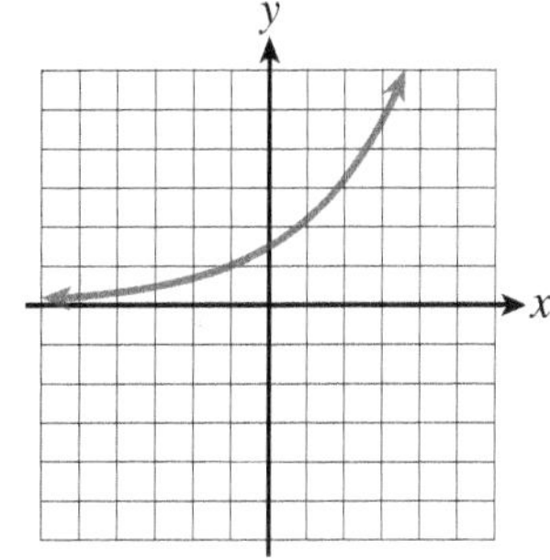

Graph ii is function h.

3. Explain the reasons for your choices in Problem 2.

   Graph i is function g because it is decreasing with a decay factor of 0.47, which is between 0 and 1.

   Graph ii is function h because it is increasing with the growth factor of 1.47.

4. Complete the following tables representing exponential functions. Round calculations to two decimal places whenever necessary.

   a.

x	0	1	2	3	4
y	2.00	5.10	13.01	33.18	84.61

   b.

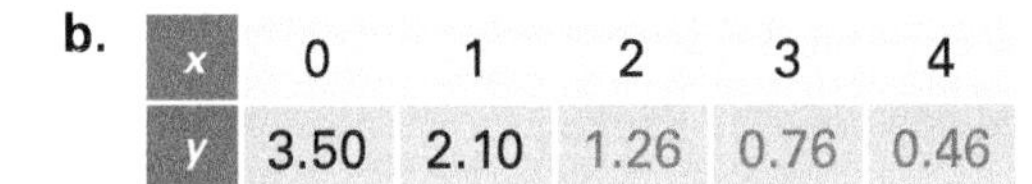

x	0	1	2	3	4
y	3.50	2.10	1.26	0.76	0.46

Answers to all How Can I Practice exercises are included in the Selected Answers appendix.

11. You are a college freshman and have a credit card. You immediately purchase a refurbished tablet for \$415. Your credit limit is \$500. Assume that you make no payments and purchase nothing more and that there are no other fees. The monthly interest rate is 1.18%.

a. What is your initial credit card balance?

\$415

b. What is the growth rate of your credit card balance?

0.0118, or 1.18% per month

c. What is the monthly growth factor of your credit card balance?

1.0118

d. Write an exponential function to determine how much you will owe [represented by $f(x)$] after x months with no more purchases or payments.

$f(x) = 415(1.0118)^x$

e. Use your graphing calculator to graph this function. What is the vertical intercept?

(0, 415)

f. What is the practical meaning of this intercept in this situation?

This represents the initial balance on the card.

g. How much will you owe after 10 months? Use the table feature on your graphing calculator to determine the solution.

\$466.65

h. When you reach your credit limit of \$500, the bank will expect a payment. How long do you have before you must start paying back the money? Use the trace feature on your grapher to approximate the solution.

With no payments, I exceed my credit limit during the sixteenth month.

12. You are working part-time while going to college. The following table shows the hourly wage, $w(t)$, in dollars, that you earn as a function of time, t. Time is measured in years since the beginning of 2015 when you started working.

Time, t, Years, since 2015	0	1	2	3	4	5
Hourly Wage, $w(t)$ (\$)	12.50	12.75	13.01	13.27	13.53	13.81

a. Calculate the ratios of the outputs to determine whether the data in the table is exponential. Round each ratio to the nearest hundredth.

The ratios are all approximately 1.02, so the data is exponential.

b. What is the growth factor?

1.02

c. Write an exponential model for the data in the table.

$w(t) = 12.50(1.02)^t$

d. What percent raise did you receive each year?

2%

e. If you continue to work for this company, what can you expect your hourly wage to be in 2025?

\$15.24

f. For approximately how many years must you work for the company for your hourly wage to double? (Assume that you will receive the same percentage increase each year.)

about 35 years

13. You deposited \$10,000 in an account that pays 3% annual interest compounded monthly.

a. Write an equation to determine the amount, A, you will have in t years.

$$A = 10{,}000\left(1 + \frac{0.03}{12}\right)^{12t}$$

$$A = 10{,}000(1.0025)^{12t}$$

b. How much will you have in 5 years?

approximately \$11,616.17

c. Use your graphing calculator to determine in how many years your investment will double.

approximately 23 years

d. Write an equation to determine the amount, A, you will have in t years if the interest is compounded continuously.

$$A = 10{,}000e^{0.03t}$$

e. Use the equation in part d to determine how much you will have in 5 years. Compare your answer with your answer in part b.

\$11,618.34; \$2.17 more than in part b

14. The number of farms in the United States has declined from 1950 to 2016, as the data in the following table shows. The data is estimated from the National Agricultural Statistics Service, U.S. Department of Agriculture.

Year	1950	1960	1970	1980	1990	2000	2010	2016
Number of Farms, in Millions, N	5.8	4	3	2.5	2.2	2.2	2.17	2.06

a. Make a scatterplot of this data.

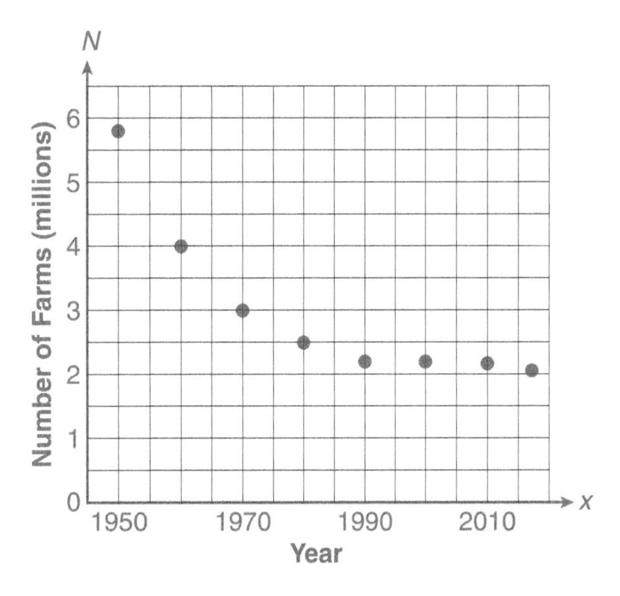

b. Does the scatterplot show that the data would be better modeled by a linear model or by an exponential model? Explain.

An exponential decay function would better model this data. The rate of change does not appear to be constant.

c. Use your graphing calculator to determine the exponential regression equation that best fits the U.S. farm data. Let x represent the number of years since 1950.

$N(x) = 4.5569(0.98597)^x$

d. Use the regression equation to predict the total number of farms in the United States in 2020.

2020 is 70 years from 1950, so $x = 70$. $N(70) = 4.5569(0.98597)^{70} \approx 1.695$.

The number of farms in 2020 in predicted to be 1,695,000.

e. According to your exponential model, what is the decay factor for the total number of farms in the United States?

The decay factor is $b = 0.98597$.

f. What is the decay rate?

The decay rate is $r = 1 - 0.98597 = 0.01403 = 1.403\%$.

g. Explain the meaning of the decay rate determined in part f of this situation.

Over the years from 1950 to 2016, the number of farms in the United States has been decreasing at a rate of 1.403% per year.

h. Use your graphing calculator to determine the half-life of the exponential model in this situation.

Solve the equation $N(x) = 4.5569(0.98597)^x = \frac{1}{2} \cdot 4.5569 = 2.27845$.

NORMAL FLOAT AUTO REAL RADIAN MP
Plot1 Plot2 Plot3
Y1=4.5569*0.98597^X
Y2=2.2785
Y3=
Y4=
Y5=
Y6=
Y7=
Y8=

NORMAL FLOAT AUTO REAL RADIAN MP
PRESS + FOR ΔTbl

X	Y1	Y2
45	2.4129	2.2785
46	2.379	2.2785
47	2.3457	2.2785
48	2.3127	2.2785
49	2.2803	2.2785
50	2.2483	2.2785
51	2.2168	2.2785
52	2.1857	2.2785
53	2.155	2.2785
54	2.1248	2.2785
55	2.0949	2.2785

X=49

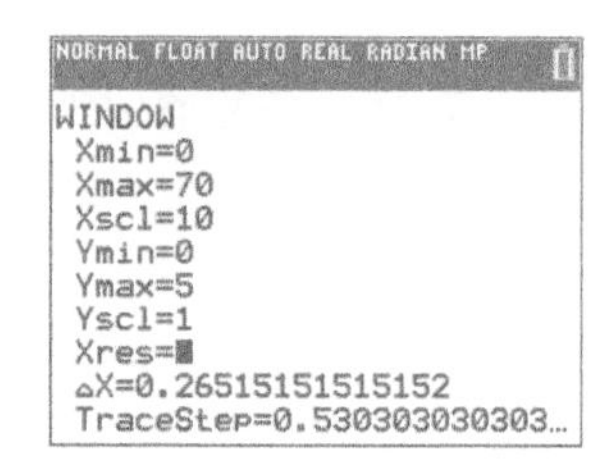

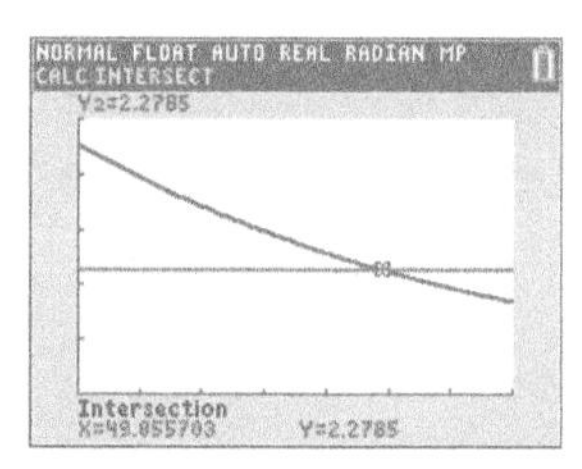

The half-life is approximately 49 years.

CLUSTER 2 Logarithmic Functions

ACTIVITY 3.8

The Diameter of Spheres

Logarithmic and Exponential Forms

OBJECTIVES

1. Define *logarithm*.
2. Write an exponential statement in logarithmic form.
3. Write a logarithmic statement in exponential form.
4. Determine log and ln values using a calculator.

Spheres are all around you (pardon the pun). You play sports with spheres such as baseballs, basketballs, and golf balls. You live on a sphere. Earth is a big ball in space, as are the other planets, the Sun, and the Moon. All spheres have properties in common. For example, the formula for the volume, V, of any sphere is $V = \frac{4}{3}\pi r^3$, and the formula for the surface area, S, of any sphere is $S = 4\pi r^2$, where r represents the radius of the sphere.

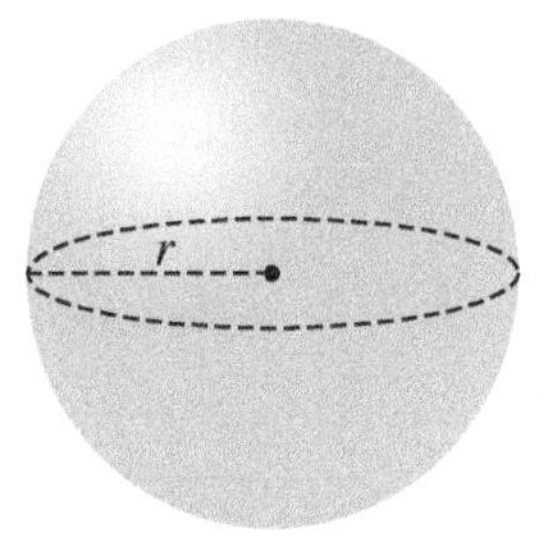

However, not all spheres are the same size. The following table gives the diameter, d, of some spheres you know. Recall that the diameter, d, of a sphere is twice the radius, r.

SPHERE	DIAMETER, *d*(m)
Golf ball	0.043
Baseball	0.075
Basketball	0.239
Moon	3,476,000
Earth	12,756,000
Jupiter	142,984,000

If you want to determine either the volume or surface area of any of the spheres in the preceding table, the diameter of the given sphere would be the input value and would be referenced on the horizontal axis. But how would you scale this axis?

1. a. Plot the values in the first three rows of the table. Scale the axis starting at 0 and incrementing by 0.02 meter.

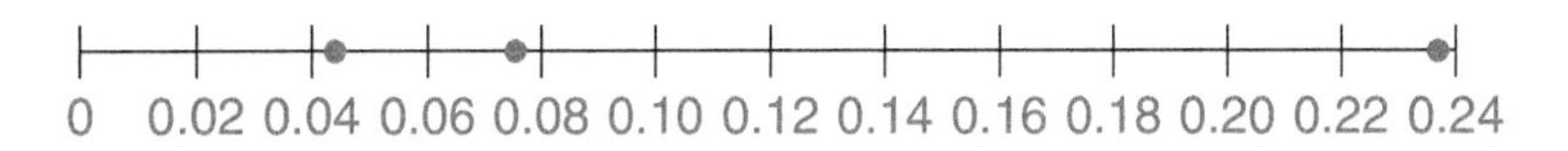

b. Can you plot the values in the last three rows of the table on the same axis? Explain.

No, the scale is so small I run out of paper.

2. a. Plot the values in the last three rows of the table on a different axis. Scale the axis starting at 0 and incrementing by 10,000,000 meters.

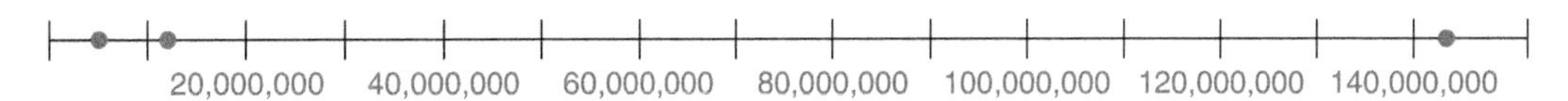

b. Can you plot the values in the first three rows of the table on the axis in part a? Explain.

Not very easily. The scale is too big.

Logarithmic Scale

3. There is a way to scale the axis so that you can plot all the values in the table on the same axis.

a. Starting with the leftmost tick mark, give the first tick mark a value of 0.01 meter. Write 0.01 as a power of 10 as follows: $0.01 = \frac{1}{100} = \frac{1}{10^2} = 10^{-2}$ meters. Give the next tick mark a value of 0.1, written as 10^{-1} meters. Continue in this way by giving each consecutive tick mark a value that is one power of 10 greater than the preceding tick mark.

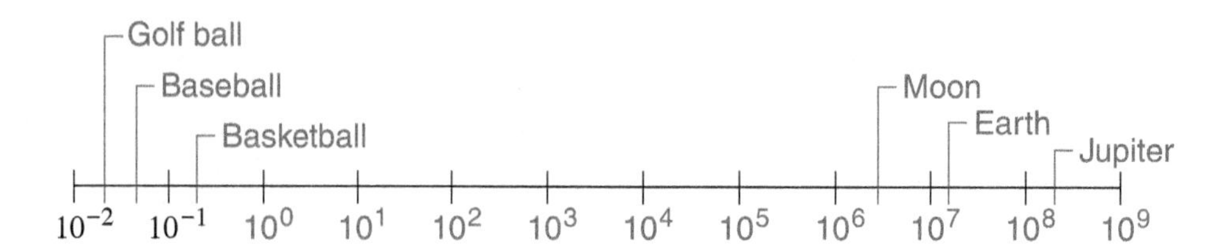

b. Complete the following table by writing all of the diameters from the preceding table in scientific notation:

SPHERE	DIAMETER, d(m)	d, IN SCIENTIFIC NOTATION
Golf ball	0.043	4.3×10^{-2}
Baseball	0.075	7.5×10^{-2}
Basketball	0.239	2.39×10^{-1}
Moon	3,476,000	3.476×10^{6}
Earth	12,756,000	1.2756×10^{7}
Jupiter	142,984,000	1.4298×10^{8}

c. To plot the diameter of a golf ball, notice that 0.043 meter is between $10^{-2} = 0.01$ meter and $10^{-1} = 0.1$ meter. Now using the axis in part a, plot 0.043 meter between the tick mark labeled 10^{-2} and 10^{-1} meters, closer to the tick mark labeled 10^{-2} meters.

d. To plot the diameter of Earth, notice that 12,756,000 meters is between $10^7 = 10{,}000{,}000$ meters and $10^8 = 100{,}000{,}000$ meters. Now plot 12,756,000 meters between the tick marks labeled 10^7 and 10^8 meters, closer to the tick mark labeled 10^7.

e. Plot the remaining data in the same way by first determining between which two powers of 10 the number lies.

The scale you used to plot the diameter values is a *logarithmic*, or *log scale*. The tick marks on a logarithmic scale are usually labeled with just the exponent of the powers of 10.

4. a. Rewrite the axis from Problem 3a by labeling the tick marks with just the exponents of the powers of 10.

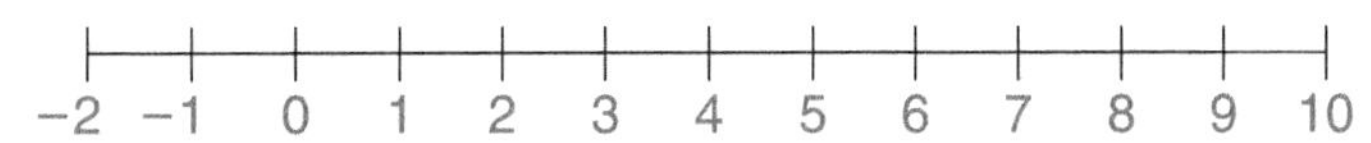

b. The axis looks like a standard axis with tick marks labeled $-1, -2, 0, 1$, etc. However, it is quite different. Describe the difference between this log scale and a standard axis labeled in the same way. Focus on the values between consecutive tick marks.

Each unit increases the exponent on base 10 by 1 rather than an addend of 1.

DEFINITION

The exponents used to label the tick marks of the preceding axis are **logarithms** or simply **logs**. Since these are exponents of powers of 10, the exponents are logs **base** 10, known as **common logarithms** or common logs.

EXAMPLE 1

a. ***The common logarithm of*** 10^3 ***is the exponent to which 10 must be raised to obtain a result of*** 10^3***. Therefore, the common log of*** 10^3 ***is 3.***

b. ***The common log of*** 10^{-2} ***is*** -2.

c. ***The common log of*** $100 = 10^2$ ***is 2.***

5. Determine the common log of each of the following:

a. 10^{-1}
-1

b. 10^4
4

c. 1000
3

d. 100,000
5

e. 0.0001
-4

Logarithmic Notation

Remember that **a logarithm is an exponent**. The common log of x is an exponent, y, to which the base, 10, must be raised to get result x. That is, in the equation $10^y = x$, y is the logarithm. Using log notation, $\log_{10} x = y$. Therefore, $\log_{10} 10{,}000 = \log_{10} 10^4 = 4$.

EXAMPLE 2

x, THE NUMBER	y, THE EXPONENT (LOGARITHM) TO WHICH THE BASE, 10, MUST BE RAISED TO GET x	LOG NOTATION $\log_{10} x = y$
10^3	3	$\log_{10} 10^3 = 3$
10^{-2}	-2	$\log_{10} 10^{-2} = -2$
100	2	$\log_{10} 100 = 2$

When using logs base 10, the notation $\log_{10}$ is shortened by dropping the 10. Therefore,

$$\log_{10} 10^3 = \log 10^3 = 3; \log_{10} 100 = \log 100 = 2.$$

6. Determine each of the following. Compare your result with those from Problem 5.

a. $\log 10^{-1}$

−1

b. $\log 10^4$

4

c. $\log 1000$

3

d. $\log 100{,}000$

5

e. $\log 0.0001$

−4

The results are the same as those in Problem 5.

Bases for Logarithms

The logarithmic scale for the diameter of spheres situation was labeled with the exponents of powers of 10. Using 10 as the base for logarithms is common since the number 10 is the base of our number system. However, other numbers could be used as the base for logs. For example, you could use exponents of powers of 5 or exponents of powers of 2.

EXAMPLE 3 ***Base-5 logarithms: The log base 5 of a number, x, is the exponent to which the base, 5, must be raised to obtain x. For example,***

a. $\log_5 5^4 = 4$, or in words, log base 5 of 5 to the fourth power equals 4.

b. $\log_5 125 = \log_5 5^3 = 3.$

c. $\log_5 \frac{1}{25} = \log_5 5^{-2} = -2.$

Base-2 logarithms: The log base 2 of a number, x, is the exponent to which 2 must be raised to obtain x. For example,

a. $\log_2 2^5 = 5$, or log base 2 of 2 to the fifth power equals 5.

b. $\log_2 16 = \log_2 2^4 = 4.$

c. $\log_2 \frac{1}{8} = \log_2 2^{-3} = -3.$

In general, a statement in logarithmic form is $\log_b x = y$, where b is the base of the logarithm, x is a power of b, and y is the exponent. The base b for a logarithm can be any positive number except 1.

7. Determine each of the following:

a. $\log_4 64$

3

b. $\log_2 \frac{1}{16}$

−4

c. $\log_3 9$

2

d. $\log_3 \frac{1}{27}$

−3

The examples and problems so far in this activity demonstrate the following property of logarithms.

Property of Logarithms

In general, $\log_b b^n = n$, where $b > 0$ and $b \neq 1$.

8. Determine each of the following:

a. $\log 1$

0

b. $\log_5 1$

0

c. $\log_{\frac{1}{2}} 1$

0

d. $\log 10$

1

e. $\log_5 5$

1

f. $\log_{1/2}\left(\frac{1}{2}\right)$

1

9. a. Referring to Problems 8a–c, write a general rule for $\log_b 1$.

$\log_b 1 = 0$

b. Referring to Problems 8d–f, write a general rule for $\log_b b$.

$\log_b b = 1$

Property of Logarithms

In general, $\log_b 1 = 0$ and $\log_b b = 1$, where $b > 0, b \neq 1$.

Natural Logarithms

Because the base of a log can be any positive number except 1, the base can be the number e. Many applications involve the use of log base e. Log base e is called the **natural** log and has the following special notation.

$\log_e x$ is written as $\ln x$, read simply as el-n-x.

EXAMPLE 4

a. $\ln e^2 = \log_e e^2 = 2$

b. $\ln \frac{1}{e^4} = \ln e^{-4} = -4$

10. Evaluate the following:

a. $\ln e^7$

7

b. $\ln\left(\frac{1}{e^3}\right)$

-3

c. $\ln 1$

0

d. $\ln e$

1

e. $\ln\sqrt{e}$

$\frac{1}{2}$

Logarithmic and Exponential Forms

Because logarithms are exponents, logarithmic statements can be written as exponential statements and exponential statements can be written as logarithmic statements.

For example, in the statement $3 = \log_5 125$, the base is 5, the exponent (logarithm) is 3, and the result is 125. This relationship can also be written as the equation $5^3 = 125$.

In general, the logarithmic equation $y = \log_b x$ is equivalent to the exponential equation $b^y = x$.

EXAMPLE 5 ***Rewrite the exponential equation $e^{0.5} = x$ as an equivalent logarithmic equation.***

SOLUTION

In the equation $e^{0.5} = x$, the base is e, the result is x, and the exponent (logarithm) is 0.5. Therefore, the equivalent logarithmic equation is $0.5 = \log_e x$, or $0.5 = \ln x$.

LC LEARNING CATALYTICS

Rewrite $7^4 = 2401$ in logarithmic form.

11. Rewrite each exponential equation as a logarithmic equation and each log equation as an exponential equation.

a. $3 = \log_2 8$

$2^3 = 8$

b. $\ln e^3 = 3$

$e^3 = e^3$

c. $\log_2 \frac{1}{16} = -4$

$2^{-4} = \frac{1}{16}$

d. $6^3 = 216$

$\log_6 216 = 3$

e. $e^1 = e$

$\ln e = 1$

f. $3^{-2} = \frac{1}{9}$

$\log_3 \frac{1}{9} = -2$

Logarithms and the Calculator

The numbers whose logarithms you have been working with have been exact powers of the base. However, in many situations, you must evaluate a logarithm where the number is not an exact power of the base. For example, what is log 20 or ln 15? Fortunately, the common log (base 10) and the natural log (base e) are functions on your calculator.

12. Use your calculator to evaluate the following:

a. log 20

1.3010

b. ln 15

2.7081

c. $\ln \frac{1}{2}$

−0.6931

d. log 0.02

−1.6990

e. Use your calculator to check your answers to Problems 6 and 10.

They check.

13. a. Use your calculator to complete the following table:

SPHERE	DIAMETER, d(m)	d, IN SCIENTIFIC NOTATION	log(d)
Golf ball	0.043	4.3×10^{-2}	−1.3665
Baseball	0.075	7.5×10^{-2}	−1.1249
Basketball	0.239	2.39×10^{-1}	−0.6216
Moon	3,476,000	3.476×10^{6}	6.5411
Earth	12,756,000	1.2756×10^{7}	7.1057
Jupiter	142,984,000	1.4298×10^{8}	8.1553

b. Plot the values from the log column in the preceding table on the following axis:

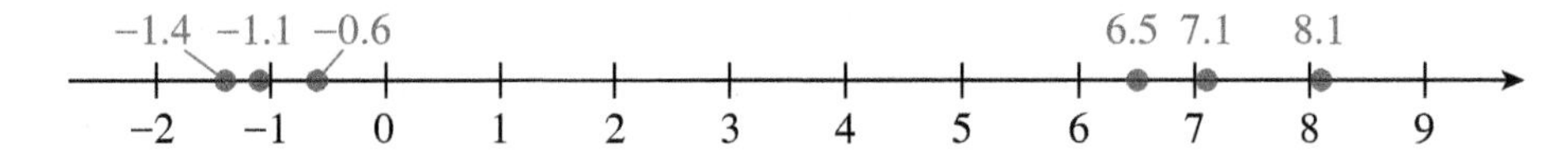

c. Compare the preceding plot with the plot on the log-scaled axis in Problem 3a and comment.

Answers will vary. The locations of the points are the same. The scale is units of 1 rather than powers of 10. However, each successive tick mark in the graph above represents the corresponding power of 10.

SUMMARY Activity 3.8

1. The notation for logarithms is $\log_b x = y$, where b is the base of the log, x is the resulting power of b, and y is the exponent. The base, b, can be any positive number except 1; x can be any positive number. The range of y values includes all real numbers.
2. The notation for the **common logarithm**, or base-10 logarithm, is $\log_{10} x = \log x$.
3. The notation for the **natural logarithm**, or base e logarithm, is $\log_e x = \ln x$.
4. The **logarithmic equation** $y = \log_b x$ is equivalent to the **exponential equation** $b^y = x$.
5. If $b > 0$ and $b \neq 1$,

   a. $\log_b 1 = 0$.

   b. $\log_b b = 1$.

   c. $\log_b b^n = n$.

EXERCISES Activity 3.8

1. Use the definition of logarithm to determine the exact value of each of the following:

   a. $\log_2 32$
   5

   b. $\log_3 27$
   3

   c. $\log 0.1$
   -1

   d. $\log_2 \left(\frac{1}{64}\right)$
   -6

   e. $\log_5 1$
   0

   f. $\log_{1/2} \left(\frac{1}{4}\right)$
   2

   g. $\log_7 \sqrt{7}$
   (*Hint:* $\sqrt{7} = 7^{1/2}$.)
   $\frac{1}{2}$

   h. $\log_{100} 10$
   $\frac{1}{2}$

   i. $\log 1$
   0

   j. $\log_2 1$
   0

   k. $\ln e^5$
   5

   l. $\ln \left(\frac{1}{e^2}\right)$
   -2

   m. $\ln 1$
   0

2. Evaluate each common logarithm without the use of a calculator.

   a. $\log \left(\frac{1}{1000}\right) =$ -3

   b. $\log \left(\frac{1}{100}\right) =$ -2

   c. $\log \left(\frac{1}{10}\right) =$ -1

   d. $\log 1 =$ 0

   e. $\log 10 =$ 1

   f. $\log 100 =$ 2

   g. $\log 1000 =$ 3

   h. $\log \sqrt{10} =$ $\frac{1}{2}$

3. Rewrite the following equations in logarithmic form:

a. $3^2 = 9$
$\log_3 9 = 2$

b. $\sqrt{121} = 11$ (*Hint:* First rewrite $\sqrt{121}$ in exponential form.)
$\log_{121} 11 = \frac{1}{2}$

c. $4^t = 27$
$\log_4 27 = t$

d. $b^3 = 19$
$\log_b 19 = 3$

4. Rewrite the following equations in exponential form:

a. $\log_3 81 = 4$
$3^4 = 81$

b. $\frac{1}{2} = \log_{100} 10$
$100^{1/2} = 10$

c. $\log_9 N = 12$
$9^{12} = N$

d. $y = \log_7 x$
$7^y = x$

e. $\ln \sqrt{e} = \frac{1}{2}$
$e^{1/2} = \sqrt{e}$

f. $\ln \left(\frac{1}{e^2}\right) = -2$
$e^{-2} = \frac{1}{e^2}$

5. Estimate between what two integers the solutions for the following equations fall. Then solve each equation exactly by changing it to log form. Use your calculator to approximate your answer to three decimal places.

a. $10^x = 3.25$
$0 < x < 1$
$x = \log 3.25$
$x = 0.512$

b. $10^x = 590$
$2 < x < 3$
$x = \log 590$
$x = 2.771$

c. $10^x = 0.0000045$
$-6 < x < -5$
$x = \log 0.0000045$
$x = -5.347$

ACTIVITY
3.9

Walking Speed of Pedestrians
Logarithmic Functions

OBJECTIVES

1. Determine the inverse of the exponential function.
2. Identify the properties of the graph of a logarithmic function.
3. Graph the natural logarithmic function.

On a recent visit to Boston, you notice that people seem rushed as they move about the city. Upon returning to college, you mention this observation to your psychology instructor. The instructor refers you to a psychology study that investigates the relationship between the average walking speed of pedestrians and the population of the city. The study cites statistics presented graphically as follows:

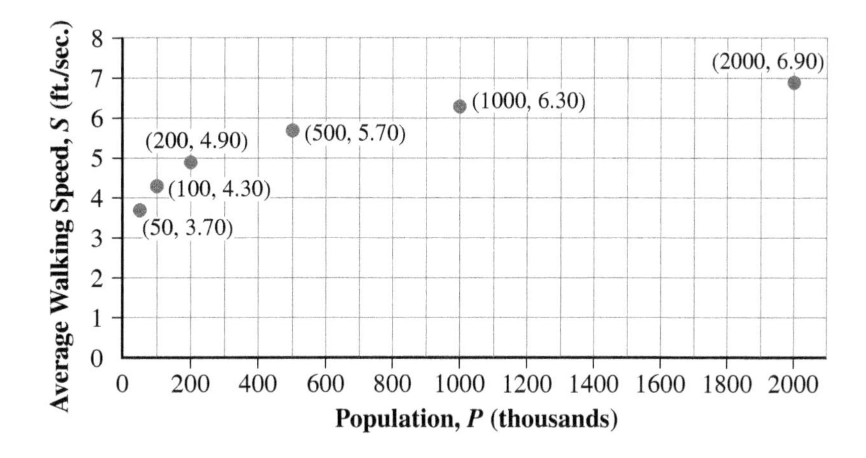

1. a. Does the data appear to be linear? Explain.

No, the rate of change is not constant.

b. Do the data appear to be exponential? Explain.

No, it is growing slower rather than faster as x increases.

This data is actually logarithmic. Situations that can be modeled by logarithmic functions will be the focus of this and the following activity.

Introduction to the Logarithmic Function

The logarithmic function base b is symbolized by $y = \log_b x$, where

i. b represents the base of the logarithmic function $(b > 0, b \neq 1)$.

ii. x is the input and represents a power of the base b (x is also called the argument) and y is the output and is the exponent needed on the base b to obtain x.

LC LEARNING CATALYTICS

What is the domain of the function $y = \log x$?
a. All real numbers.
b. $x > 0$.
c. All real numbers except 0.
d. None of these.

2. a. Evaluate $\log_{10}(-100)$ using your calculator. What do you observe? Does it seem reasonable? Explain.

I get an error that is not a real answer. Yes, it seems reasonable because 10 raised to any power will never be negative.

b. Is it possible to determine $\log(0)$? Explain.

No, because 10 raised to any power will never be zero.

c. What is the domain for the function defined by $y = \log x$?

$x > 0$

d. What is the range? Remember, the output y is an exponent.

The range is all real numbers because 10^y is defined for any y.

3. The exponential function defined by $f(x) = 10^x$ has a special relationship with the corresponding logarithmic function defined by $g(x) = \log_{10} x = \log x$.

a. Complete the following tables for $f(x) = 10^x$ and $g(x) = \log x$:

x	$f(x) = 10^x$
−2	0.01
−1	0.1
0	1
1	10
2	100

x	$g(x) = \log x$
0.01	−2
0.1	−1
1	0
10	1
100	2

b. Compare the input and output values for functions f and g.

The inputs and outputs are interchanged.

c. Sketch the graphs of $Y_1 = 10^x$ and $Y_2 = \log_{10} x$ using your graphing calculator. For the window, use the window Xmin = −4, Xmax = 4, Ymin = −3, and Ymax = 3. Your screen should appear as follows:

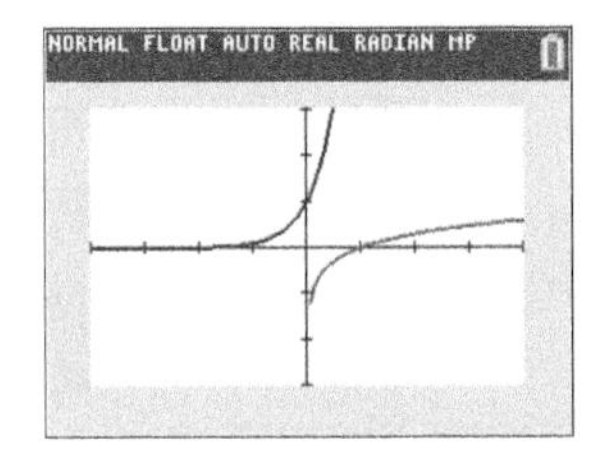

d. Graph $y = x$ on the same coordinate axes as functions f and g. Describe in a sentence or two the symmetry you observe in the graphs of f and g.

The graphs are reflected over the line $y = x$.

Recall the concept of an inverse function from Chapter 2. The inverse function interchanges the domain and range of the original function. Also, the graph of an inverse function is the reflection of the original function about the line $y = x$. Therefore, the results in Problem 3 demonstrate that $f(x) = 10^x$ and $g(x) = \log x$ are inverse functions.

You can determine the equation of the inverse function by solving the defining equation for the input (x-value) and then interchanging the input (x-value) and the output (y-value).

EXAMPLE 1 ***Determine the equation of the inverse of the function defined by $y = 5^x$.***

SOLUTION

Step 1: Solve the equation for x by writing the statement in logarithmic notation. $x = \log_5 y$

Step 2: Interchange the x and y variables. $y = \log_5 x$

4. Use the algebraic approach demonstrated in Example 1 to verify that $y = \log x$ is the inverse of $y = 10^x$.

Solve for x by writing $y = \log x$ in exponential form: $10^y = x$. Then interchange x and y to obtain $y = 10^x$.

Problems 2, 3, and 4 illustrate the following properties of the common logarithmic function:

Properties of the Common Logarithmic Function Defined by $f(x) = \log x$

1. The domain of f is the set of all positive real numbers ($x > 0$).
2. The range of f is all real numbers.
3. f is the inverse of the function defined by $g(x) = 10^x$.

The Graph of the Natural Logarithmic Function

5. a. Using your calculator, complete the following table. Round your answers to three decimal places.

x	0.1	0.5	1	5	10	20	50
$y = \ln x$	−2.303	−0.693	0	1.609	2.303	2.996	3.912

b. Sketch a graph of $y = \ln x$.

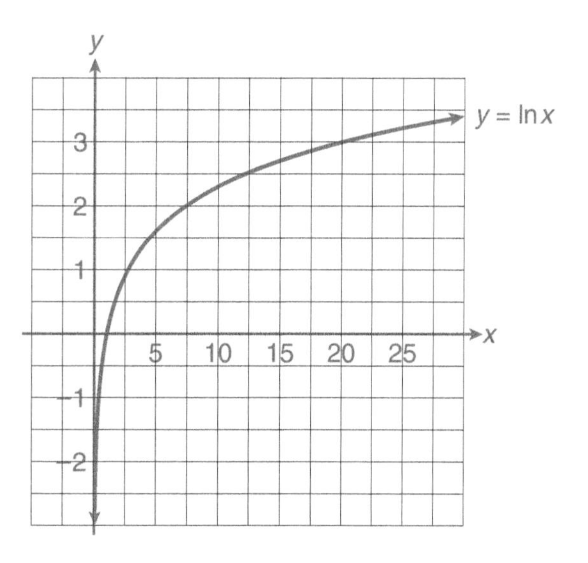

c. Verify your graph in part b using your graphing calculator. Using the window Xmin $= -1$, Xmax $= 4$, Ymin $= -2.5$, and Ymax $= 2.5$, your screen should appear as follows:

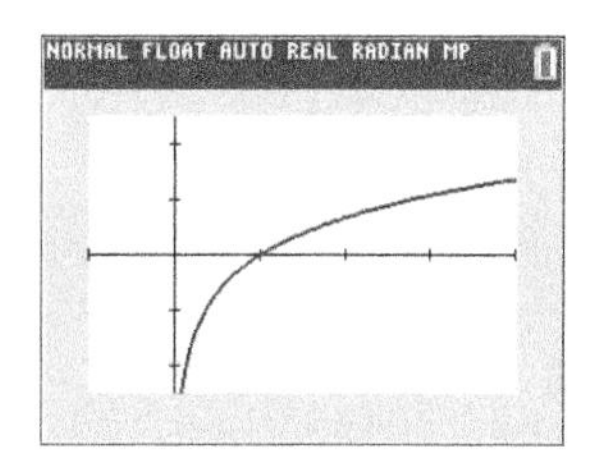

d. What are the domain and range of the function defined by $y = \ln x$?

The domain is $x > 0$. The range is all real numbers.

e. Determine the intercepts of the graph.

(1, 0), no y-intercept

f. Does the graph of $y = \ln x$ have a horizontal asymptote? Explain.

No. As x increases in value, the output continues to increase slowly. The output values eventually get infinitely large.

g. Complete the following table using your calculator. Round your answers to the nearest tenth.

x	1	0.5	0.25	0.1	0.01	0.001
$y = \ln x$	0	−0.7	−1.4	−2.3	−4.6	−6.9

h. As the input values take on values closer and closer to zero, what happens to the corresponding output values?

The output values become very large negative numbers.

The y-axis (the line $x = 0$) is a vertical asymptote of the graph of $y = \ln x$.

DEFINITION

A **vertical asymptote** is a vertical line, $x = a$, that the graph of a function becomes very close to but never touches. As the input values get closer to $x = a$, the output values get larger in magnitude. That is, the output values become very large positive or very large negative values.

EXAMPLE 2

The vertical asymptote of the graphs of $y = \log x$ and $y = \ln x$ is the vertical line $x = 0$ (the y-axis).

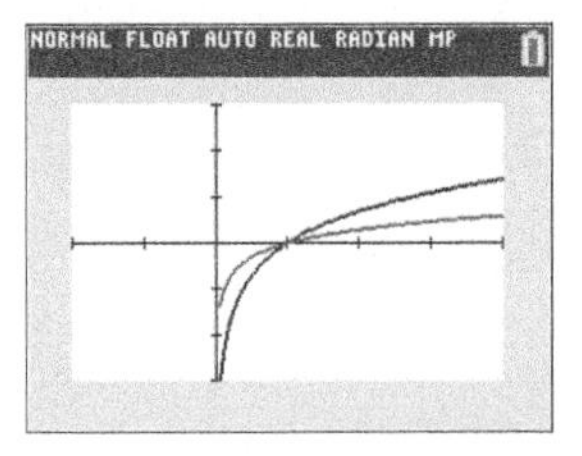

6. a. Graph $y = e^x$, $y = \ln x$, and $y = x$ on the same set of coordinate axes using the window Xmin = −7.5, Xmax = 7.5, Ymin = −5, and Ymax = 5. Describe the symmetry that you observe.

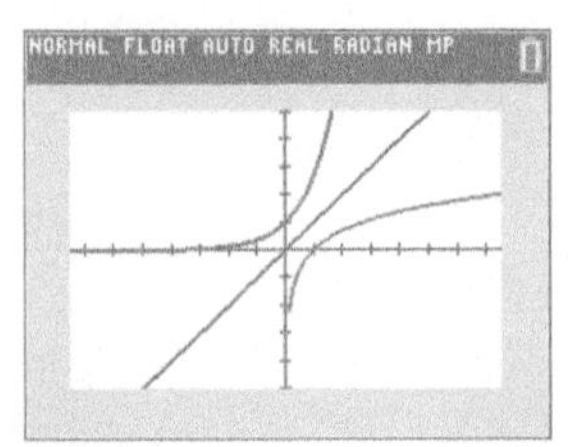

The graphs are symmetric in the line $y = x$.

b. Use an algebraic approach to determine the inverse of the exponential function defined by $y = e^x$.

Write in logarithmic form: $x = \ln y$

Interchange x and y: $y = \ln x$

7. A horizontal shift of the graph of $y = \ln x$ changes the vertical asymptote.

a. What is the equation of the vertical asymptote of the graph of $y = \ln (x + 3)$?

$x = -3$

b. What is the equation of the vertical asymptote of the graph of $y = \ln (x - 2)$?

$x = 2$

SUMMARY Activity 3.9

1. Properties of the logarithmic function defined by $y = \log_b x$, where $b > 1$.

a. The domain of f is $x > 0$.

b. The range of f is all real numbers.

c. f is the inverse of the function defined by $g(x) = b^x$.

2. The graph of a logarithmic function defined by $y = \log_b x$, where $b > 1$,

a. is increasing for all $x > 0$.

b. has an x-intercept of $(1,0)$.

c. has no y-intercept.

d. has a vertical asymptote of $x = 0$, the y-axis.

e. resembles the following graph:

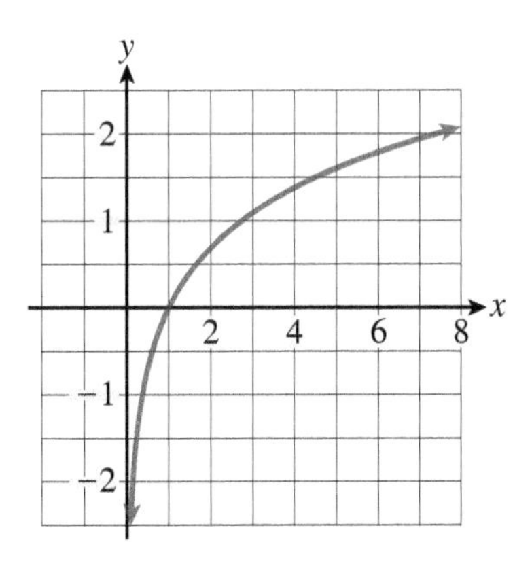

3. The **common logarithmic function** is defined by

$$y = \log x = \log_{10} x.$$

4. The **natural logarithmic function** is defined by

$$y = \ln x = \log_e x.$$

5. Because the base of the natural log function is $e > 1$, the graph of the natural logarithmic function $y = \ln x$

a. is increasing for all $x > 0$.

b. has an x-intercept of $(1, 0)$.

c. has a vertical asymptote of $x = 0$, the y-axis.

6. The logarithmic function defined by $y = \log_b x$, where $b > 1$, is continuous over its domain, $x > 0$.

7. You can determine the equation of the inverse of the function by solving the equation of the function for x and then interchanging the input (x-values) and the output (y-values) in the new equation.

EXERCISES Activity 3.9

1. Using the graph of $y = \log x$ as a check, summarize the following properties of the common logarithmic function:

   a. What is the domain?

   $x > 0$

   b. What is the range?

   the set of all real numbers

   c. For what values of x is $\log x$ positive?

   $x > 1$

   d. For what values of x is $\log x$ negative?

   $0 < x < 1$

   e. For what values of x does $\log x = 0$?

   $x = 1$

   f. For what values of x does $\log x = 1$?

   $x = 10$

2. a. Complete the following table using your calculator. Round your answers to the nearest 10th.

x	0.001	0.01	0.1	0.25	0.5	1
$y = \log x$	−3	−2	−1	−0.6	−0.3	0

   b. As the positive input values take on values closer to 0, what happens to the corresponding output values?

   The output values become very large negative numbers (large in absolute value).

   c. Determine the vertical asymptote of the graph of $y = \log x$.

   $x = 0$, the y-axis

3. The exponential function defined by $y = 2^x$ has an inverse. Determine the equation of the inverse function. Write your answer in logarithmic form.

   Start with $y = 2^x$. Rewrite in log form.

   $x = \log_2 y$. Interchange x and y.

   $y = \log_2 x$

4. Using the graph of $y = \ln x$ as a check, summarize the following properties of the natural logarithmic function:

   a. What is the domain?

   $x > 0$

   b. What is the range?

   the set of all real numbers

c. For what values of x is $\ln x$ positive?

$x > 1$

d. For what values of x is $\ln x$ negative?

$0 < x < 1$

e. For what values of x does $\ln x = 0$?

$x = 1$

f. For what values of x does $\ln x = 1$?

$x = e$

5. The life expectancy of a piece of equipment is the number, n, of years for the equipment to depreciate to a known salvage value, V. The life expectancy, n, is given by the formula

$$n = \frac{\log V - \log C}{\log (1 - r)},$$

where C is the initial cost of the piece of equipment and r is the annual rate of depreciation expressed as a decimal. If a backhoe costs \$45,000 and has a salvage value of \$2500, what is the life expectancy if the annual rate of depreciation is 40%?

$$n = \frac{\log (2500) - \log (45{,}000)}{\log (1 - 0.40)} \approx 5.7 \text{ years}$$

6. The Pew Research Center's Internet & American Life Project Surveys gathered information that included the percent of Americans, 18 years old and older, who use the Internet. Some of that information is summarized in the following table:

Years since 1994, t	1	3	5	7	9	11	13	15	18	21
Percent of U.S. Adults Using the Internet, P	14	30	40	61	63	70	72	79	85	85

a. Use your graphing calculator to produce a scatterplot of the data. Do you believe that the data can be best modeled by a linear, exponential, or logarithmic model?

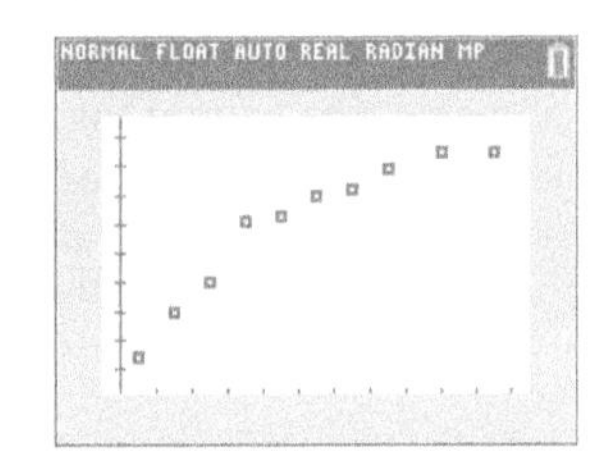

The data can be best modeled by a logarithmic model because in general, the output grows over time but the rate of growth slows over time.

The data can be modeled by $P = 7.77 + 25.48 \ln(t)$

b. Use the model to predict the percent of Americans 18 years and older who will be using the Internet in 2025.

$t = 2025 - 1994 = 31.$ $P(31) = 7.77 + 25.48 \ln (31) = 95.3.$

The model predicts that 95.3% of Americans 18 years and older will be using the Internet in 2025.

c. Use your model from part b to predict the percent of Americans 18 years old and older who will be using the Internet in 2040. Do you have confidence in this result? Explain.

$t = 2040 - 1994 = 46.$ $P(46) = 7.77 + 25.48 \ln(46)$; $P(46) = 105.3\%$. The model predicts that 105.3% of Americans 18 years and older will be using the Internet in 2040. This is not possible. I have no confidence in the result because I am extrapolating too far from the data.

ACTIVITY

3.10

Walking Speed of Pedestrians, continued

Modeling Data with Logarithmic Regression Equations

OBJECTIVES

1. Compare the average rate of change of increasing logarithmic, linear, and exponential functions.
2. Determine the regression equation of a natural logarithmic function having equation $y = a + b \ln x$ that best fits a set of data.

In Activity 3.9, you looked at a psychology study that investigated the relationship between the average walking speed of pedestrians and the population of the city. Graphically, the data were presented as follows:

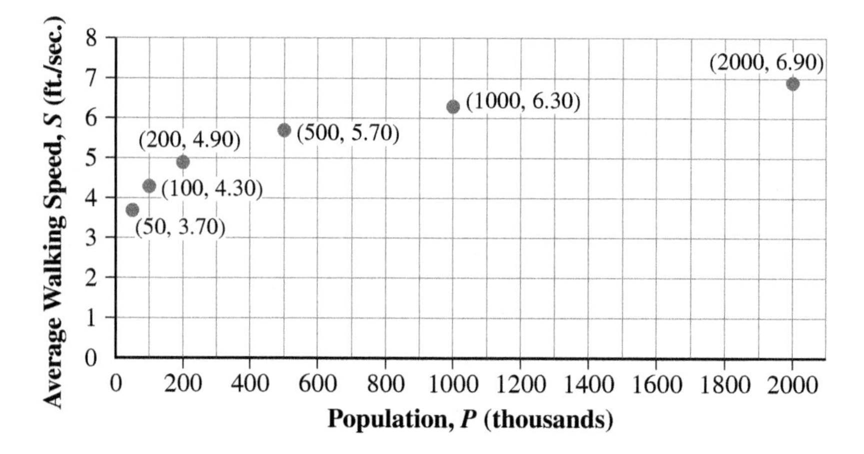

1. a. Do the data appear to be logarithmic? Explain.

Yes. It grows rather quickly at first, but for larger values of input, the growth slows a great deal.

b. Use the data in the graph to complete the following table:

Population, *P* (thousands)	50	100	200	500	1000	2000
Average Walking Speed, *S*	3.70	4.30	4.90	5.70	6.30	6.90

The natural logarithmic function can be used to model a variety of scientific and natural phenomena. The natural logarithmic function is so prevalent that on most graphing calculators, it has its own built-in regression finder.

2. a. Use your graphing calculator and the table in Problem 1b to produce a scatterplot of the average walking speed data.

2. a.

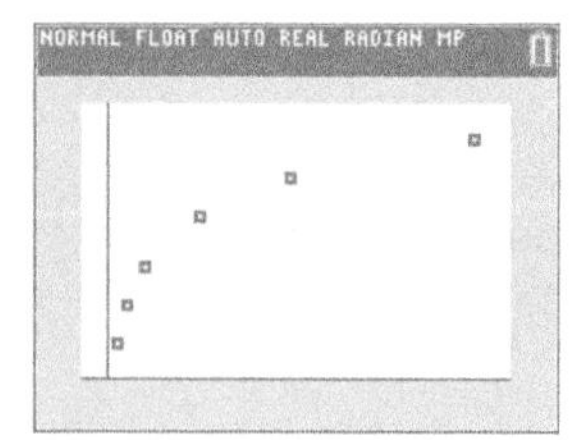

b. Use the regression feature of your calculator to produce a natural logarithmic curve that approximates the data in the table. Use option 9 from the STAT CALC menu.

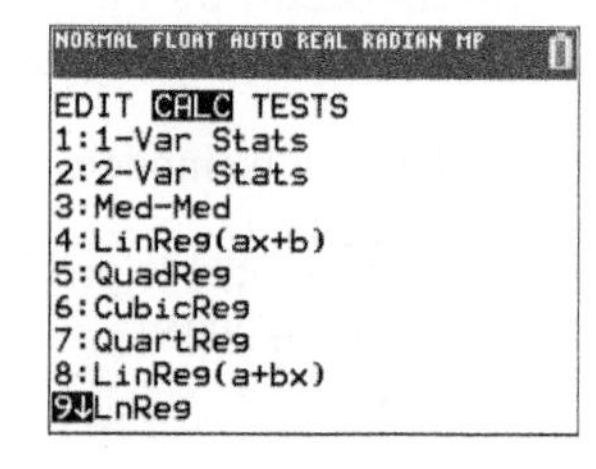

The LnReg option will generate a regression equation of the form $y = a + b \ln x$. Round a and b to the nearest thousandth, and record the function below.

$S = 0.303 + 0.868 \ln P$

2. c.

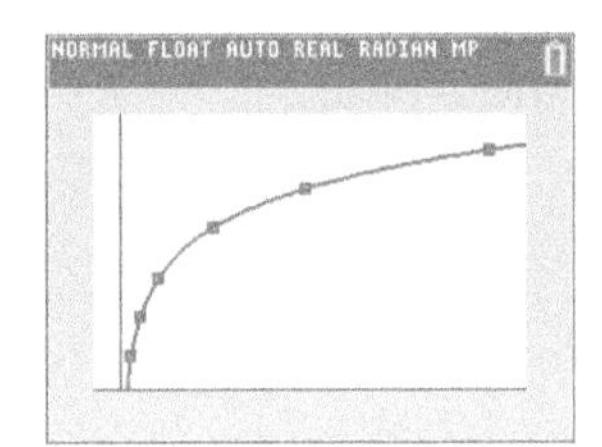

c. Enter the function from part b into your graphing calculator. Verify visually that this function is a good model for your data.

d. What is the practical domain of this function?

The domain is all real numbers greater than 0 and less than 9000. (Answers will vary.)

e. Use the function from part b to predict the average walking speed in Boston, population 589,121. [**Note:** P is in thousands (589.121 thousands).]

$S = 0.303 + 0.868 \ln(589.121) \approx 5.84$ feet per second

f. Use the model to predict the average walking speed in New York City, population 8,008,278.

$S = 0.303 + 0.868 \ln(8008.278) \approx 8.10$ feet per second

3. a. If the average walking speed in a certain city is 5.2 feet per second, write an equation that can be used to estimate the population P of the city.

$5.2 = 0.303 + 0.868 \ln P$

b. Solve the equation using a graphical approach.

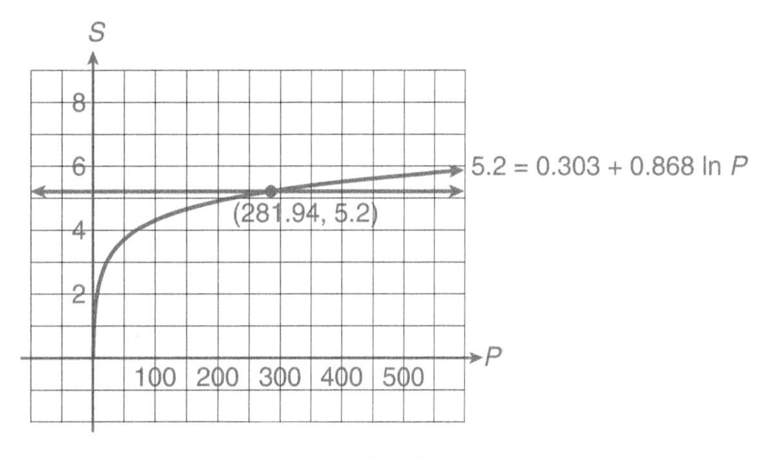

The population is approximately 282,000.

Comparing the Average Rate of Change of Logarithmic, Linear, and Exponential Functions

4. a. Complete the following table using the function defined by $S = 0.303 + 0.868 \ln P$:

P, Population (thousands)	10	20	150	250
S, Average Walking Speed (ft./sec.)	2.30	2.90	4.65	5.10

b. Determine the average rate of change of S as the population increases from

i. 10 to 20 thousand.

$$\frac{2.90 - 2.30}{10} = 0.06 \text{ feet per second per thousand}$$

ii. 20 to 150 thousand.

$$\frac{4.65 - 2.90}{150 - 20} = \frac{1.75}{130} = 0.013 \text{ feet per second per thousand}$$

iii. 150 to 250 thousand.

$$\frac{5.10 - 4.65}{250 - 150} = \frac{0.45}{100} = 0.0045 \text{ feet per second per thousand}$$

c. What can you say, in general, about the average rate of change in the walking speed as the population increases?

The average rate of change in the walking speed decreases as the population increases.

You should have discovered that the average rate of change in this situation is always positive. This means that the walking speed increases as the population increases. Nevertheless, in general, the increase gets smaller as the population increases. This is characteristic of logarithmic functions.

As the input of a logarithmic function with $b > 1$ increases, the output increases at a slower rate (the graph becomes less steep).

LC LEARNING CATALYTICS

Which of the following statements is true?

a. An increasing logarithmic function increases at a constant rate.
b. An increasing logarithmic function increases at an increasing rate.
c. An increasing logarithmic function increases at a decreasing rate.
d. None of these.

5. Complete the following statements by describing the rate at which the output values change:

a. For an increasing linear function, as the input variable increases, the output <u>increases at a constant rate.</u>.

b. For an increasing exponential function, as the input increases, the output <u>increases at an increasing rate.</u>.

c. For an increasing logarithmic function, as the input increases, the output <u>increases at a decreasing rate.</u>.

6. Consider the graphs of

i. $f(x) = e^x$ **ii.** $h(x) = x$ **iii.** $g(x) = \ln x$

using the window Xmin $= -7.5$, Xmax $= 7.5$, Ymin $= -5$, and Ymax $= 5$.

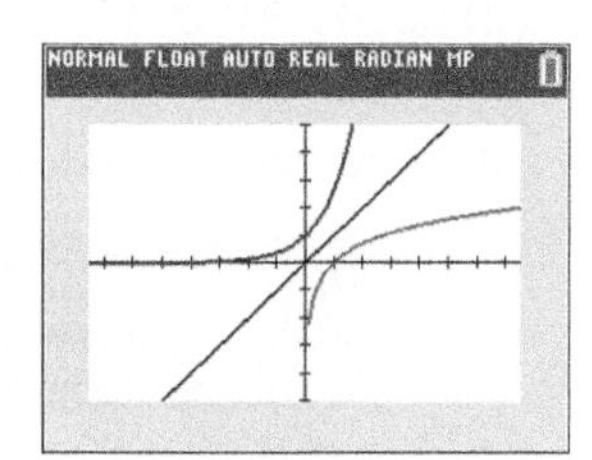

a. Which of the functions are increasing?

All three are increasing.

b. Which of the functions are decreasing?

None are decreasing.

c. As the input values get larger, which of the functions grows fastest?

$f(x) = e^x$

d. As the input values get larger, which of the functions grows most slowly?

$g(x) = \ln x$

e. Do any of these functions have a horizontal asymptote?

$f(x) = e^x$ has the horizontal asymptote $y = 0$.

f. Do any of these functions have a vertical asymptote?

$g(x) = \ln x$ has the vertical asymptote $x = 0$.

g. Compare the domains of these functions.

The domain of $f(x) = e^x$ and $h(x) = x$ is all real numbers. The domain of $g(x) = \ln x$ is $x > 0$.

h. Compare the ranges of these functions.

The range of $g(x) = \ln x$ and $h(x) = x$ is all real numbers. The range of $f(x) = e^x$ is $y > 0$.

Problem 6 illustrates some of the relationships between $f(x) = b^x$, where $b > 1$; $g(x) = \log_b x$, where $b > 1$; and $y = mx + b$, where $m > 0$.

Application

7. You are working on the development of an "elastic" ball for the IBF Toy Company. The question you are investigating is, If the ball is launched straight up, how far has it traveled vertically when it hits the ground for the 10th time?

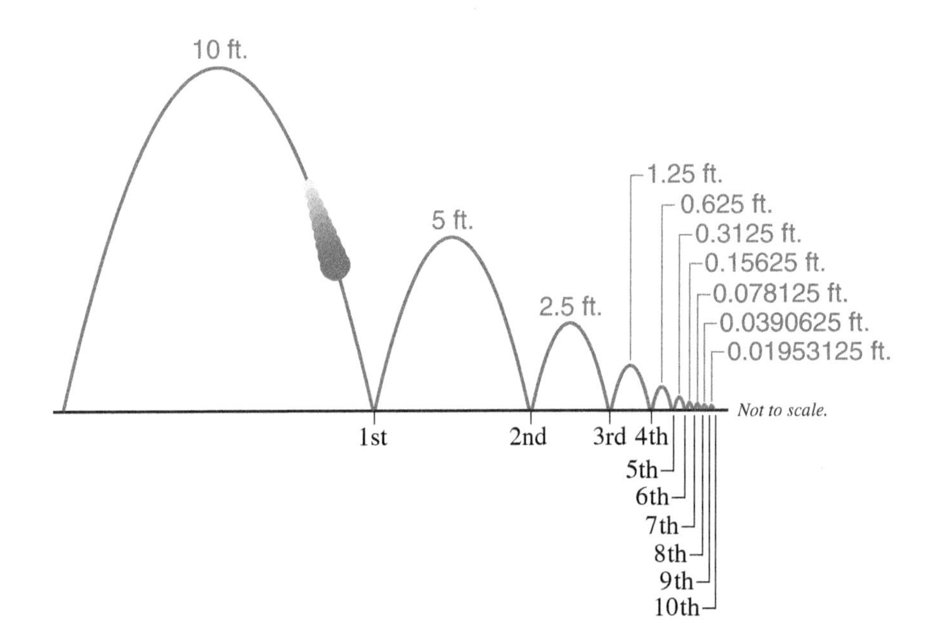

Your launcher will project the ball 10 feet into the air. This means that it will travel 20 feet (10 feet up and 10 feet down) before it hits the ground the first time. Assuming that the ball returns to 50% of its previous height, it will rebound 5 feet and travel 10 feet before it hits the ground again. The following table summarizes this situation:

N, Times the Ball Hits the Ground	1	2	3	4	5	6
Distance Traveled since Last Time (ft.)	20	10	5	2.5	1.25	0.625
T, Total Distance Traveled (ft.)	20	30	35	37.5	38.75	39.375

a. Using the window Xmin = 0, Xmax = 7, Ymin = 0, and Ymax = 45, a plot of N versus T should resemble the following:

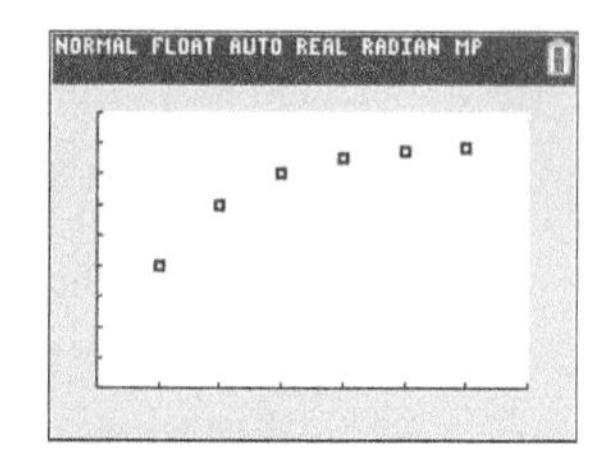

b. Do the table and scatterplot indicate that the data is linear, exponential, or logarithmic?

The rate of change appears to be slowing, so the data should be approximately logarithmic.

c. Use your graphing calculator to produce linear, exponential, and natural log regression equations for the given data.

linear: $T \approx 3.589N + 20.875$

exponential: $T \approx 21.372(1.128)^N$

natural log: $T \approx 21.373 + 11.002 \ln N$

d. Graph each equation, and visually determine which of the regression models best fits the data.

The natural log fits best.

e. Use the equation of best fit to predict the total distance the ball traveled when it hits the ground for the 10th time.

$T(10) = 21.373 + 11.002 \ln(10) \approx 46.706$ ft.

SUMMARY Activity 3.10

1. As the input of a logarithmic function increases, the output increases at a slower rate (the graph becomes less steep).

2. The relationships among the graphs of $f(x) = b^x$, where $b > 1$, $g(x) = \log_b x$, where $b > 1$, and $y = mx + b$, where $m > 0$, are identified in the following table:

FUNCTION	INCREASING OR DECREASING	GROWTH RATE	HORIZONTAL OR VERTICAL ASYMPTOTE	DOMAIN	RANGE
$f(x) = b^x$, $b > 1$	increasing	fastest	horizontal asymptote	all real numbers	$y > 0$
$g(x) = \log_b x$, $b > 1$	increasing	slowest	vertical asymptote	$x > 0$	all real numbers
$y = mx + b$, $m > 0$	increasing	constant	none	all real numbers	all real numbers

EXERCISES Activity 3.10

1. According to IDC, International Data Corporation, the global shipment of smartphones has risen and will continue to rise at least through 2021. The following data shows global smartphone shipments and their forecast from 2012 to 2021:

Year	2012	2013	2014	2015	2016	2017	2018	2019	2020	2021
Global Smartphone Shipments (millions of units)	725.3	1018.7	1301.7	1437.2	1470.6	1517.0	1585.5	1642.3	1697.0	1744.6

a. Use your graphing calculator to create a scatterplot of the data using t, the number of years since 2011, as the input variable and $g(t)$, the global smartphone shipments measured in millions of units, as the output variable.

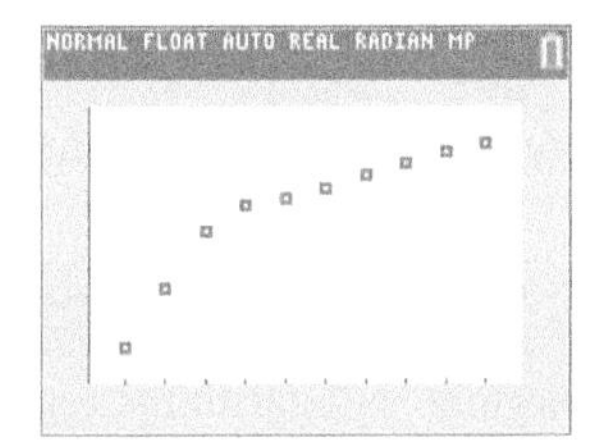

b. Does the scatterplot indicate that the data may be logarithmic? Explain.

The data could very well be logarithmic. The data tends to increase more slowly as the input increases.

c. Determine the natural log regression equation. Round your values to two decimal places. Record the regression equation below, and add the curve to your scatterplot in part a.

$g(t) = 757.43 + 434.68 \ln(t)$

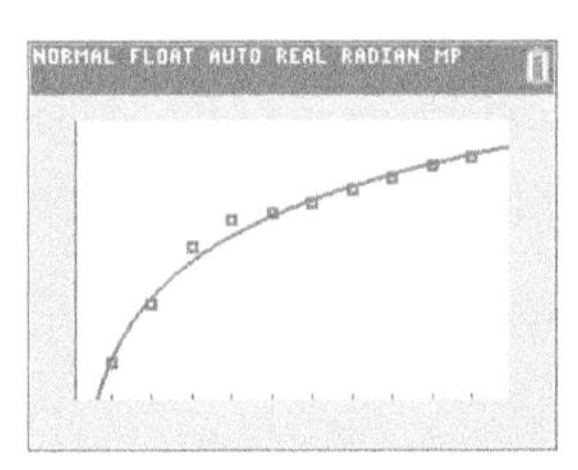

d. Does this appear to be a good fit? Explain.

Yes, it is a good fit.

e. Use your model to predict the number of smartphones that will ship globally in the year 2025.

$T = 14$; $g(14) = 757.43 + 434.68 \ln(14) \approx 1904.6$. The model predicts that in 2025, there will be 1904.6 million smartphones shipped globally.

2. a. Consider a data set for the variables x and y.

x	1	4	7	10	13
$f(x)$	3.0	4.5	5.0	5.2	5.8

Plot these points on the following grid:

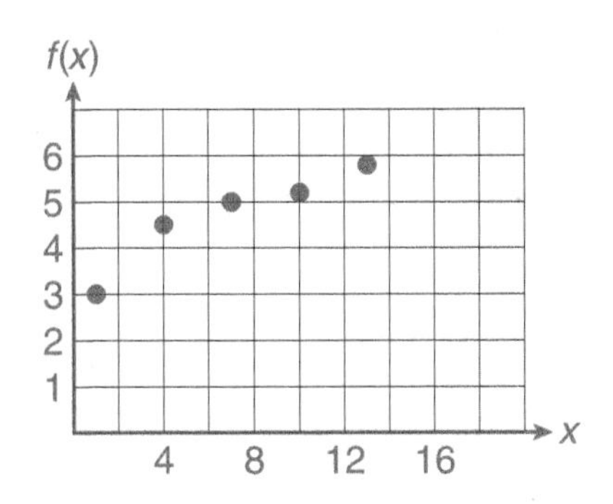

b. Does the scatterplot indicate that the data is more likely linear, exponential, or logarithmic? Explain.

logarithmic because it is increasing, but at continually lesser rates

c. Use your graphing calculator to determine a logarithmic regression model that represents this data.

$y = 3.006 + 1.033 \ln x$

d. Use your model to determine $f(11)$ and $f(20)$.

$f(11) \approx 5.483$; $f(20) \approx 6.101$

3. The barometric pressure, P, in inches of mercury at a distance x miles from the eye of a moderate hurricane can be modeled by

$$P = f(x) = 0.48 \ln(x + 1) + 27.$$

a. Determine $f(0)$. What is the practical meaning of the value in this situation?

$f(0) = 27$ in. This is the pressure in the eye of the storm.

b. Sketch a graph of this function.

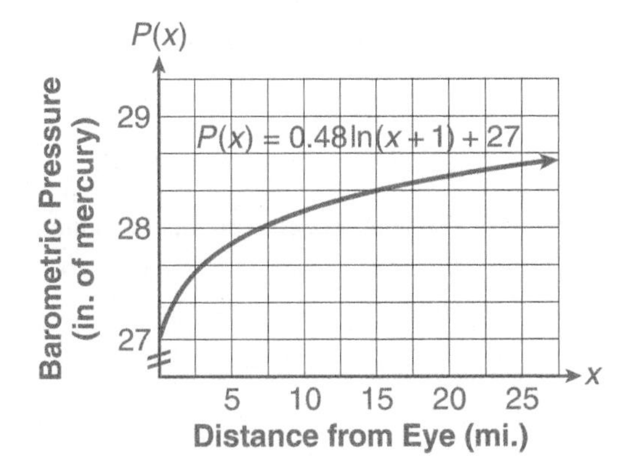

c. Describe how air pressure changes as one moves away from the eye of the hurricane.

As one moves away from the hurricane's eye, the pressure increases quickly at first and then more slowly.

4. The following data was collected during an experiment in science class. The table shows the yield y of a substance (in milligrams) after x minutes of a chemical reaction.

Minutes, x	1	2	3	4	5	6	7	8
Yield, y (mg)	1.5	7.4	10.2	13.4	15.8	16.3	18.2	18.3

a. Produce a scatterplot of the chemical reaction data.

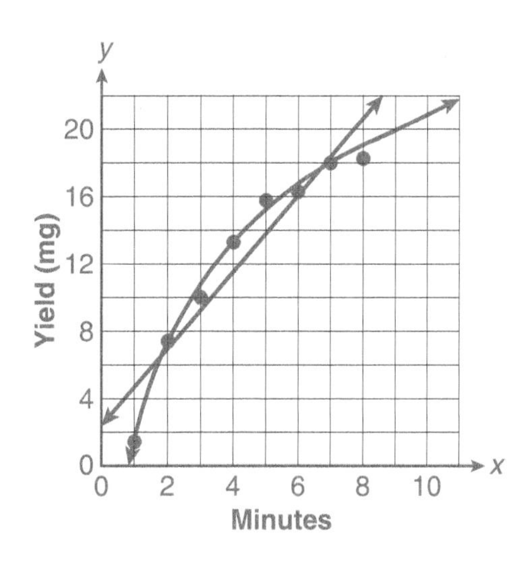

b. Use the regression feature of a calculator to obtain a natural logarithmic model and a linear model that approximates the data in the table. Graph each equation on the scatterplot in part a.

logarithmic model: $y = 1.538 + 8.373 \ln x$

linear model: $y = 2.29x + 2.34$

c. Determine which model best fits the data.

The best-fitting model is the logarithmic model. Its y-values are closest to the actual yield values in the table.

5. The formula $R = 80.4 - 11 \ln x$ is used to approximate the minimum required ventilation rate, R, as a function of the air space per child in a public school classroom. The rate R is measured in cubic feet per minute, and x is measured in cubic feet.

a. Sketch a graph of the rate function for $100 \leq x \leq 1500$.

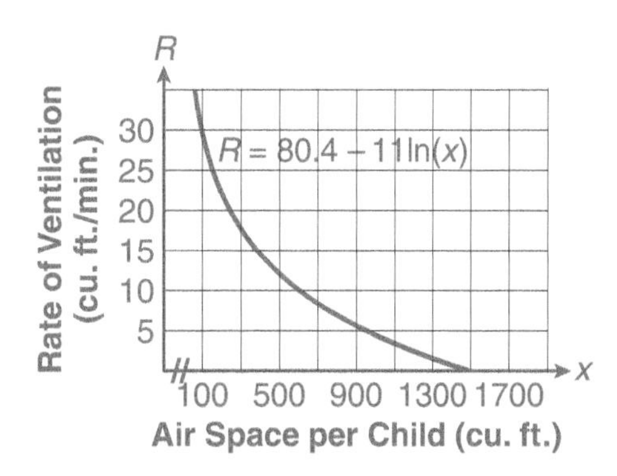

b. Determine the required ventilation rate if the air space per child is 300 cubic feet.

$R = 80.4 - 11 \ln(300) \approx 17.66$ cubic feet per minute

6. You have recently accepted a job working in the coroner's office of a large city. Because of the large numbers of homicides, it has been difficult for the coroners to complete all of their work. In part, your job is to assist them with the paperwork. On one particular day, you work on a case in which you are attempting to establish the time of death.

The coroner tells you that to establish the time of death, he uses the formula

$$t = 4 \ln\left(\frac{98.6 - T_s}{T_b - T_s}\right),$$

where t is the number of hours the victim has been dead,

T_b represents the temperature of the body when discovered, and

T_s represents the temperature of his surroundings.

The coroner also tells you that the thermostat was set at 68°F in the apartment in which the body was found and that the victim's body temperature was 78°F.

a. Using the preceding formula, determine the number of hours the victim has been deceased. Use your calculator to approximate your answer to one decimal place.

4.5 hr.

b. If the body was discovered at 10:07 P.M., what do you estimate for the time of death?

5:37 P.M.

7. The following formula can be used to determine the time, t, it takes for an investment to double or triple, and so forth in value:

$$t = \frac{\ln m}{n \ln\left(1 + \frac{r}{n}\right)},$$

where m represents the number of times the investment is to grow in value ($m = 2$ is double, $m = 3$ is triple, etc.),

r is the annual interest rate expressed as a decimal,

n is the number of corresponding compounding periods per year.

a. How many years will it take an investment to double if you are receiving an annual rate of 5.5% compounded quarterly ($n = 4$)?

$$t = \frac{\ln 2}{4 \ln\left(1 + \frac{0.055}{4}\right)} \approx 12.689 \text{ yr.}$$

b. How many years will it take the investment in part a to triple in value?

$$t = \frac{\ln 3}{4 \ln\left(1 + \frac{0.055}{4}\right)} \approx 20.11 \text{ yr.}$$

c. Suppose the interest on the investment in part a was compounded monthly ($n = 12$). How long will it take the value to double?

$$t = \frac{\ln 2}{12 \ln\left(1 + \frac{0.055}{12}\right)} \approx 12.632 \text{ yr.}$$

ACTIVITY 3.11

The Elastic Ball

Properties of Logarithms

OBJECTIVES

1. Apply the log of a product property.
2. Apply the log of a quotient property.
3. Apply the log of a power property.
4. Discover the change-of-base formula.

You are continuing your work on the development of the elastic ball. You are still investigating the following question: If the ball is launched straight up, how far has it traveled vertically when it hits the ground for the 10th time? However, your supervisor tells you that you cannot count the initial launch distance. You must calculate only the rebound distance.

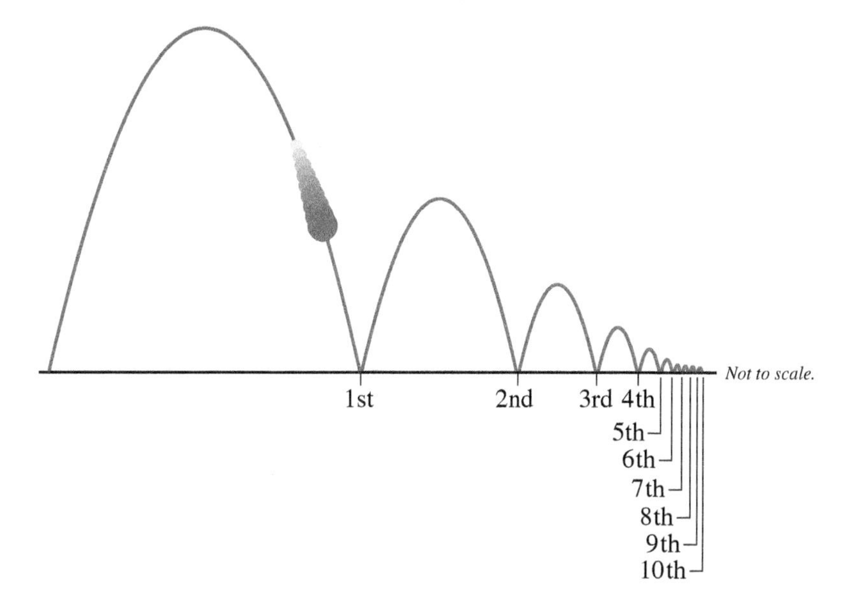

Using some physical properties, timers, and your calculator, you collect the following data:

N, Number of Times the Ball Hits the Ground	1	2	3	4	5	6
T, Total Rebound Distance (ft.)	0	9.0	13.5	16.3	18.7	21.0

1. Does the data seem reasonable? Explain.

   Yes. The total rebound distance increases, but more slowly as N increases.

2. Use your graphing calculator to construct a scatterplot of the data with N as the input and T as the output. Using a window of Xmin = 0, Xmax = 7, Ymin = 0, and Ymax = 25, your graph should resemble the following:

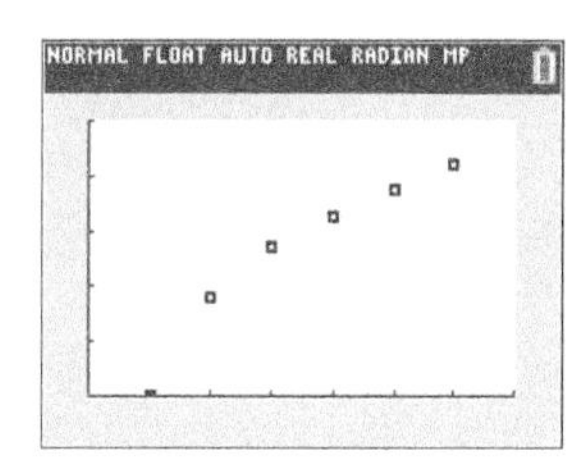

3. Do you believe that the data can be modeled by a logarithmic function? Explain.

   Yes, because it is increasing, but more slowly as N increases.

4. This data can be modeled by $T = 26.75 \log N$. Use your graphing calculator to verify visually that this is a reasonable model for the given data.

4.

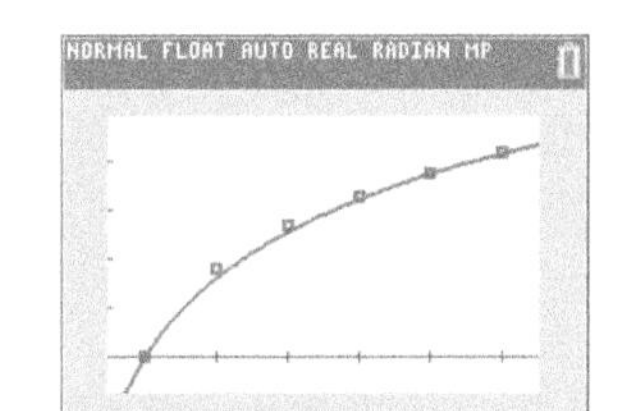

5. **a.** Using the log model, complete the following table. Round values to the nearest hundredth.

N	2	5	10
$T = 26.75 \log N$	8.05	18.70	26.75

b. How are the T-values for $N = 2$ and $N = 5$ related to the T-value for $N = 10$?

T evaluated at 2 plus T evaluated at 5 equals T evaluated at 10.

c. Using the results from part b, how could you determine the total rebound distance after 10 bounces?

Add the result after 2 bounces and the result after 5 bounces to give the result after 10 bounces.

The results from Problem 5 can be written as follows:

$$\underbrace{26.75 \log 10}_{26.75} = \underbrace{26.75 \log 2}_{8.05} + \underbrace{26.75 \log 5}_{18.70}$$

$$26.75 \log (2 \cdot 5) = 26.75 \log 2 + 26.75 \log 5$$

Dividing both sides by 26.75, you have

$$\log (2 \cdot 5) = \log 2 + \log 5.$$

This result illustrates an important property of logarithms.

Property of the Logarithm of a Product
If $A > 0$, $B > 0$, then $\log_b (A \cdot B) = \log_b A + \log_b B$, where $b > 0$, $b \neq 1$. Expressed verbally, this property states that the logarithm of a product is the sum of the individual logarithms.

EXAMPLE 1

a. $\log_2 32 = \log_2 (4 \cdot 8) = \log_2 4 + \log_2 8 = 2 + 3 = 5$

b. $\log (5st) = \log 5 + \log s + \log t$

c. $\ln (xy) = \ln x + \ln y$

6. Use the property of the logarithm of a product to write the following as the sum of two or more logarithms:

a. $\log_b (7 \cdot 13)$

$\log_b 7 + \log_b 13$

b. $\log_3 (xyz)$

$\log_3 x + \log_3 y + \log_3 z$

c. $\log 15$

$\log 3 + \log 5$

d. $\ln (3xy)$

$\ln 3 + \ln x + \ln y$

7. Write the following as the logarithm of a single expression:

a. $\ln a + \ln b + \ln c$

$\ln (abc)$

b. $\log_4 3 + \log_4 9$

$\log_4 27$

Logarithm of a Quotient

Consider the following table from Problem 5:

N	2	5	10
$T = 26.75 \log N$	8.05	18.70	26.75

This table also indicates that the rebound distance after the ball has hit the floor twice (8.05 feet) is the total rebound distance when the ball has hit the ground 10 times (26.75 feet) minus the total rebound distance when the ball has hit the ground 5 times (18.70 feet).

This can be written as

$$\underbrace{8.05}_{26.75 \log 2} = \underbrace{26.75}_{26.75 \log 10} - \underbrace{18.70}_{26.75 \log 5}$$

$$\log 2 = \log 10 - \log 5.$$

Substituting $\log\left(\frac{10}{5}\right)$ for log 2, you have

$$\log\left(\frac{10}{5}\right) = \log 10 - \log 5.$$

This suggests another important property of logarithms. The property is demonstrated further in Problem 8.

8. a. Complete the following table. Round your answers to the nearest thousandth.

x	$Y1 = \log\left(\frac{x}{4}\right)$	$Y2 = \log x - \log 4$
1	−0.602	−0.602
5	0.097	0.097
10	0.398	0.398
23	0.760	0.760

b. Is the expression $\log\left(\frac{x}{4}\right)$ equivalent to $\log x - \log 4$? Explain.

It would appear that the expressions are equivalent. In part a, for each x, the two expressions produce equal outputs.

c. Sketch the graph of $y = \log\left(\frac{x}{4}\right)$ and $y = \log x - \log 4$ using your graphing calculator. What do the graphs suggest about the relationship between $\log\left(\frac{x}{4}\right)$ and $\log x - \log 4$?

The graphs are the same, indicating that the expressions are equivalent.

Property of the Logarithm of a Quotient

If $A > 0$, $B > 0$, then $\log_b\left(\frac{A}{B}\right) = \log_b A - \log_b B$, where $b > 0$, $b \neq 1$. Expressed verbally, this property states that the logarithm of a quotient is the difference of the logarithm of the numerator and the logarithm of the denominator.

EXAMPLE 2

a. $\log_3\left(\frac{81}{27}\right) = \log_3 81 - \log_3 27 = 4 - 3 = 1$

Note that $\log_3\left(\frac{81}{27}\right) = \log_3 3 = 1.$

b. $\log\left(\frac{2x}{y}\right) = \log 2x - \log y = \log 2 + \log x - \log y$

c. $\ln\left(\frac{x^2}{5}\right) = \ln x^2 - \ln 5$

9. Use the properties of logarithms to write the following as the sum or difference of logarithms:

a. $\log_6 \frac{17}{3}$

$\log_6 17 - \log_6 3$

b. $\ln \frac{x}{23}$

$\ln x - \ln 23$

c. $\log_3 \frac{2x}{y}$

$\log_3 2 + \log_3 x - \log_3 y$

d. $\log \frac{3}{2z}$

$\log 3 - \log 2 - \log z$

10. Write the following expressions as the logarithm of a single expression:

a. $\log x - \log 4 + \log z$

$\log \frac{xz}{4}$

b. $\log x - (\log 4 + \log z)$

$\log \frac{x}{4z}$

11.

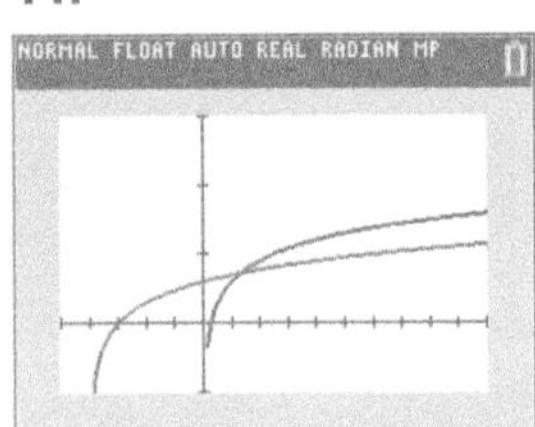

11. a. Use your graphing calculator to sketch the graphs of $y = \log x + \log 4$ and $y = \log(x + 4)$.

b. How do these graphs compare?

The graphs are not the same. The first is a vertical shift of $y = \log x$. The second is a horizontal shift of $y = \log x$.

c. What do the graphs suggest about the relationship between $\log(A + B)$ and $\log A + \log B$?

$\log(A + B) \neq \log A + \log B$

Logarithm of a Power

Before calculators, logarithms were used to help in computing products and quotients of numbers. More important, logarithms were used to compute powers such as 734.21^3 and $\sqrt{0.0761} = (0.0761)^{1/2}$. In such a case, the first step was to take the logarithm of the power and rewrite the resulting expression. To determine how to rewrite $\log 734.21^3$, you can investigate the expression $\log x^3$.

12. b.

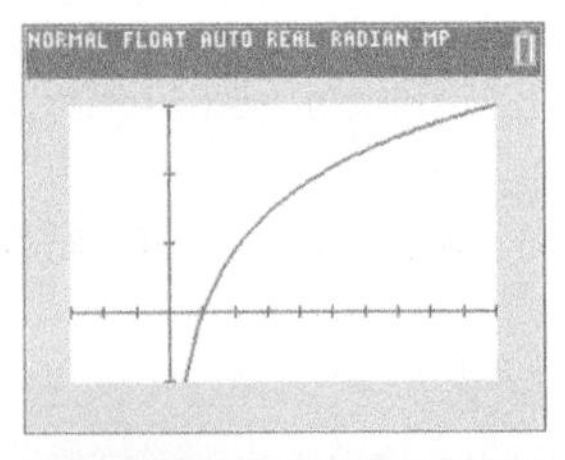

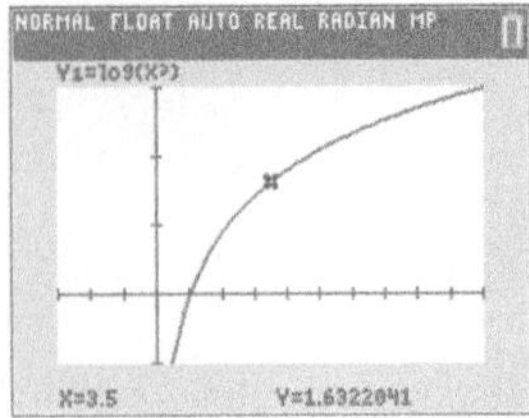

12. a. Complete the following table. Round to the nearest thousandth.

x	Y1 = $\log x^3$	Y2 = $3 \log x$
2	0.903	0.903
7	2.535	2.535
15	3.528	3.528

b. Sketch the graphs of $y = \log x^3$ and $y = 3 \log x$ using your graphing calculator.

c. What do the results of part a and part b demonstrate about the relationship between $\log x^3$ and $3 \log x$?

$\log x^3 = 3 \log x$

The results in Problem 12 illustrate another property of logarithms.

Property of the Logarithm of a Power

If $A > 0$ and p is any real number, then $\log_b A^p = p \cdot \log_b A$, where $b > 0, b \neq 1$. In words, the property states that the logarithm of a power is equivalent to the exponent times the logarithm of the base.

EXAMPLE 3

a. $\log_3 9^2 = 2\log_3 9 = 2 \cdot 2 = 4$

b. $\log_5 x^4 = 4\log_5 x$

c. $\ln (xy)^7 = 7\ln (xy)$

d. $\ln x^{1/4} = \frac{1}{4}\ln x$

e. $\log \sqrt{63} = \log 63^{1/2} = \frac{1}{2}\log 63$

LC LEARNING CATALYTICS

Which of the following represents the expansion of the expression $\log_4(\frac{x^6}{y^4})$ as much as possible?

a. $\log_4(x^6) + \log_4(y^4)$.
b. $6\log_4(x) + 4\log_4(y)$.
c. $24\log_4 x - 24\log_4 y$.
d. None of these.

13. Use the properties of logarithms to write the given logarithms as the sum or difference of two or more logarithms or as the product of a real number and a logarithm. All variables represent positive numbers.

a. $\log_3 x^{1/2}$

$\frac{1}{2}\log_3 x$

b. $\log_5 x^3$

$3\log_5 x$

c. $\ln t^2$

$2\ln t$

d. $\log \sqrt[3]{50}$ (*Hint:* $\sqrt[3]{50} = 50^{1/3}$.)

$\frac{1}{3}\log 50$

e. $\log_5 \frac{x^2y^3}{z}$

$2\log_5 x + 3\log_5 y - \log_5 z$

f. $\log_3 \frac{3x^2}{y^3}$

$\log_3 3 + 2\log_3 x - 3\log_3 y$ or $1 + 2\log_3 x - 3\log_3 y$

14. Write each of the following as the logarithm of a single expression with coefficient 1:

a. $2\log_3 5 + 3\log_3 2$

$\log_3(25 \cdot 8) = \log_3 200$

b. $\frac{1}{2}\log x^4 - \frac{1}{2}\log y^5$

$\log\sqrt{\frac{x^4}{y^5}}$

c. $3\log_b 10 - 4\log_b 5 + 2\log_b 3$

$\log_b \frac{5^3 \cdot 2^3 \cdot 9}{5^4} = \log_b \frac{72}{5}$

d. $3\ln 4 - (4\ln 5 + 2\ln 3)$

$\ln(4^3) - \ln(5^4 \cdot 3^2) = \ln \frac{4^3}{5^4 \cdot 3^2}$

Using the properties of logarithms to solve exponential equations algebraically will be investigated in the next activity.

Change-of-Base Formula

Because the T1-84 Plus C has only the log base 10 (log) and the log base e (ln) keys, you cannot graph a logarithmic function such as $y = \log_2 x$ directly. Consider the following argument to rewrite the expression $\log_2 x$ as an equivalent expression using log base 10.

By definition of logs, $y = \log_2 x$ is the same as $x = 2^y$. Taking the log base 10 of both sides of the second equation, $x = 2^y$, you have

$$\log x = \log 2^y.$$

Using the property of the log of a power, $\log x = y\log 2$. Solving for y, you have

$$y = \frac{\log x}{\log 2}.$$

Therefore, the equation $y = \log_2 x$ is equivalent to $y = \frac{\log x}{\log 2}$.

15. To graph $y = \log_2 x$, enter log(X)/log(2) for Y_1 in your TI-84 Plus C calculator. Your graph should resemble the following:

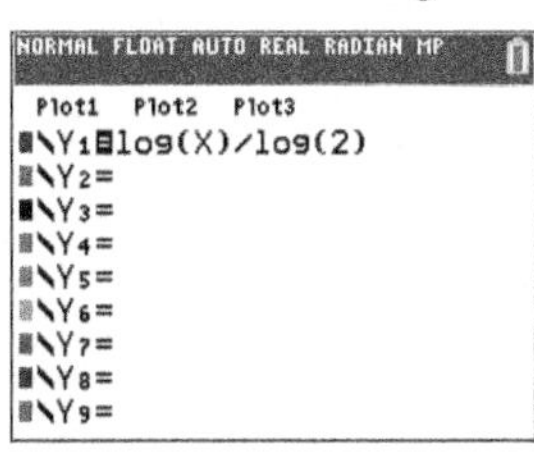

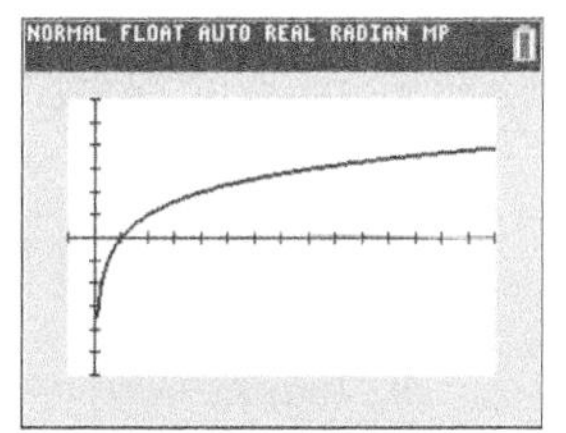

16. a. Write $y = \log_6 x$ as an equivalent equation using base 10.

$$y = \frac{\log x}{\log 6}$$

b. Use the result from part a to graph $y = \log_6 x$.

16. b.

NORMAL FLOAT AUTO REAL RADIAN MP
Y1=log(X)/log(6)
X=7 Y=1.0860331

c. What is the domain of the function?

$x > 0$

d. What is the x-intercept of the graph?

(1, 0)

The formula you used in Problems 15 and 16 for graphing log functions of different bases is a special case of the formula

$$\log_b x = \frac{\log_a x}{\log_a b}, \text{ where } a > 0, a \neq 1, b > 0, \text{ and } b \neq 1.$$

This is often called the **change-of-base formula.**

In the change-of-base formula, $\log_b x$ has base b and argument x. To change the base, divide the log of the argument by the log of the base.

Because most calculators have log base 10 (log) and log base e (ln) keys, you usually convert to one of those bases. For those bases,

$$\log_b x = \frac{\log x}{\log b} \quad \text{or} \quad \log_b x = \frac{\ln x}{\ln b}.$$

EXAMPLE 4 ***Change the equation $y = \log_5 x$ to an equivalent equation in base 10 and/or base e.***

$$y = \log_5 x = \frac{\log x}{\log 5} \quad \text{or} \quad y = \log_5 x = \frac{\ln x}{\ln 5}$$

17. Use each of the change-of-base formulas to determine $\log_4 1024$.

a. Using base 10:

$$\frac{\log 1024}{\log 4} = 5$$

b. Using base e:

$$\frac{\ln 1024}{\ln 4} = 5$$

c. How do the results in parts a and b compare?

The results are the same.

SUMMARY Activity 3.11

Properties of the Logarithmic Function

If $A > 0, B > 0, b > 0$, and $b \neq 1$, then

1. $\log_b (A \cdot B) = \log_b A + \log_b B.$

2. $\log_b \left(\frac{A}{B}\right) = \log_b A - \log_b B.$

3. $\log_b (x + y) \neq \log_b x + \log_b y.$

4. $\log_b A^p = p \log_b A$

5. You can use the calculator to change logarithms in base b to common or natural logarithms by

$$\log_b x = \frac{\log x}{\log b} \quad \text{or} \quad \log_b x = \frac{\ln x}{\ln b}.$$

EXERCISES Activity 3.11

1. Use the properties of logarithms to write the following as a sum or difference of two or more logarithms:

a. $\log_b (3 \cdot 7)$

$\log_b 3 + \log_b 7$

b. $\log_3 (3 \cdot 13)$

$\log_3 3 + \log_3 13 = 1 + \log_3 13$

c. $\log_7 \frac{13}{17}$

$\log_7 13 - \log_7 17$

d. $\log_3 \frac{xy}{3}$

$\log_3 x + \log_3 y - \log_3 3$
$= \log_3 x + \log_3 y - 1$

2. Write the following expressions as the logarithm of a single number:

a. $\log_3 5 + \log_3 3$

$\log_3 15$

b. $\log 25 - \log 17$

$\log \frac{25}{17}$

c. $\log_5 x - \log_5 5 + \log_5 7$

$\log_5 \frac{7x}{5}$

d. $\ln (x + 7) - \ln x$

$\ln \frac{x + 7}{x}$

3.

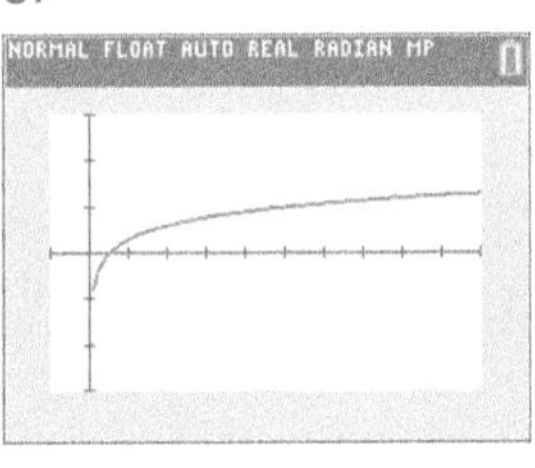

3. a. Sketch the graphs of $y = \log (2x)$ and $y = \log x + \log 2$ on your graphing calculator.

b. Are you surprised by the results? Explain.

The graphs are the same. This is not surprising because the log of a product is the sum of the logs.

4. a. Sketch the graphs of $y = \log \left(\frac{3}{x}\right)$ and $y = \log x - \log 3$ on your graphing calculator.

4.

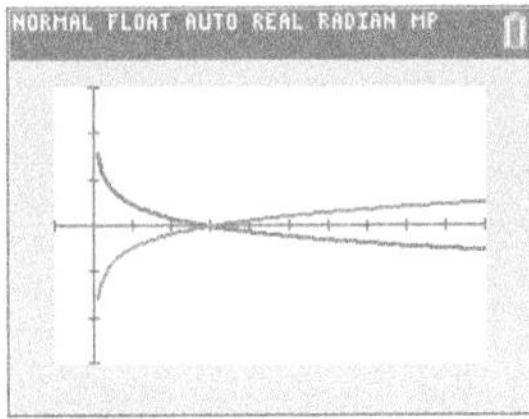

b. Are you surprised by your results? Explain.

Not really. The graphs are reflections in the x-axis.

c. If your graphs in part a are not identical, can you modify the second function to make the graphs identical? Explain.

To make them identical, I must change all the signs: $y = \log 3 - \log x$.

5. You have been hired to handle the local newspaper advertising for a large used car dealership in your community. The owner tells you that your predecessor in this position used the formula

$$N(A) = 7.4 \log A$$

to decide how much to spend on newspaper advertising over a 2-week period. The owner admits that he doesn't know much about the formula except that $N(A)$ represents the number of cars that the owner can expect to sell and A is the amount of money spent on local newspaper advertising. He also indicates that the formula seems to work well. You can purchase small ads in the local paper for \$15 per day, larger ads for \$50 per day, and giant ads for \$750 per day.

a. How many cars do you expect to sell if you purchase one small ad?

$7.4 \log 15 \approx 8.7$, or 9 cars

b. To understand the relationship between the amount spent on advertising and the number of cars sold, you set up a table. Complete the following table:

AD COST, A	EXPECTED CAR SALES, $N(A)$
15	9
50	13
750	21

c. How do the expected car sales from one small ad and one larger ad compare with the expected car sales from just one giant ad?

The sum of the sales from the smaller ads exceeds the sales from the larger ad by 1.

d. Are the results in the table in part b consistent with what you know about the properties of logarithms? Explain.

Fairly close. 15 times 50 equals 750, so I would have expected the sum of the sales from the smaller ads to equal the sales from the largest. The error is due to rounding.

e. What are you going to advise the owner regarding the purchase of a giant ad?

Forget about the giant ad. It is a waste of money.

6. Use the properties of logarithms to write the given logarithms as the sum or difference of two or more logarithms or as the product of a real number and a logarithm. Simplify if possible. All variables represent positive numbers.

a. $\log_3 3^5$

$5 \log_3 3 = 5$

b. $\log_2 2^x$

$x \log_2 2 = x$

c. $\log_b \dfrac{x^3}{y^4}$

$3 \log_b x - 4 \log_b y$

d. $\ln \dfrac{\sqrt[3]{x}\sqrt[4]{y}}{z^2}$

$\dfrac{1}{3} \ln x + \dfrac{1}{4} \ln y - 2 \ln z$

e. $\log_3 (2x + y)$

$\log_3 (2x + y)$, it does not simplify

7. Write each of the following as the logarithm of a single expression with coefficient 1:

a. $2 \log_2 7 + \log_2 5$

$\log_2 245$

b. $\dfrac{1}{4} \log x^3 - \dfrac{1}{4} \log z^5$

$\log \sqrt[4]{\dfrac{x^3}{z^5}}$

c. $2 \ln 10 - 3 \ln 5 + 4 \ln z$

$\ln \dfrac{2^2 5^2 z^4}{5^3} = \ln \dfrac{4z^4}{5}$

d. $\log_5 (x + 2) + \log_5 (x + 1) - 2 \log_5 (x + 3)$

$\log_5 \dfrac{x^2 + 3x + 2}{x^2 + 6x + 9}$

8. Given that $\log_a x = 6$ and that $\log_a y = 25$, determine the numeric value of each of the following:

a. $\log_a \sqrt{y}$

12.5

b. $\log_a x^3$

18

c. $3 + \log_a x^2$

15

d. $\log_a \frac{x^2 y}{a}$

$2 \log_a x + \log_a y - \log_a a$

$= 12 + 25 - 1 = 36$

9. Use the change-of-base formula and your calculator to determine a decimal approximation of each of the following to the nearest ten thousandth:

a. $\log_7 5$

0.8271

b. $\log_6 \sqrt{15}$

0.7557

c. $\log_{13} 47$

1.5011

d. $\log_5 \sqrt[3]{31}$

0.7112

10. The formula

$$P = 95 - 30 \log_2 t$$

gives the percentage, P, of students who could recall the important content of a classroom presentation as a function of time, t, where t is the number of days that have passed since the presentation was given.

a. Sketch a graph of the function.

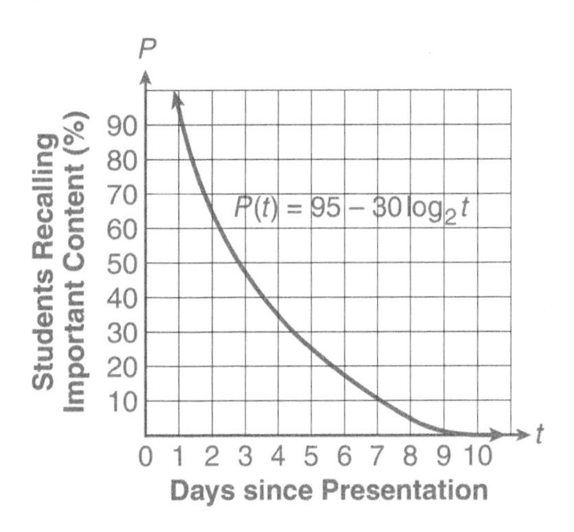

b. After 3 days, what percentage of the students will remember the important content of the presentation?

47.5%

c. According to the model, after how many days do only half $(P = 50)$ of the students remember the important features of the presentation? Use a graphing approach.

2.83 days

ACTIVITY
3.12

Changing Demographics
Solving Exponential Equations

OBJECTIVE

Solve exponential equations both graphically and algebraically.

The number of people in the United States who claim Hispanic origins has been increasing rapidly since 1970. The following table gives U.S. Census Bureau data on the Hispanic population:

Year	1970	1980	1990	2000	2010	2014
Hispanic Population (millions of people)	9.1	14.6	22.4	35.3	50.5	54.3

1. Do you believe that the data is better modeled by a linear or exponential function?

The data is not linear because the rate of change is not constant. But all the ratios of the consecutive output are between 1.4 and 1.6, so it could be modeled fairly well by an exponential function.

2. Let t represent the number of years since 1970. Use your graphing calculator to produce a scatterplot of the U.S. Hispanic population. Your screen should appear as follows:

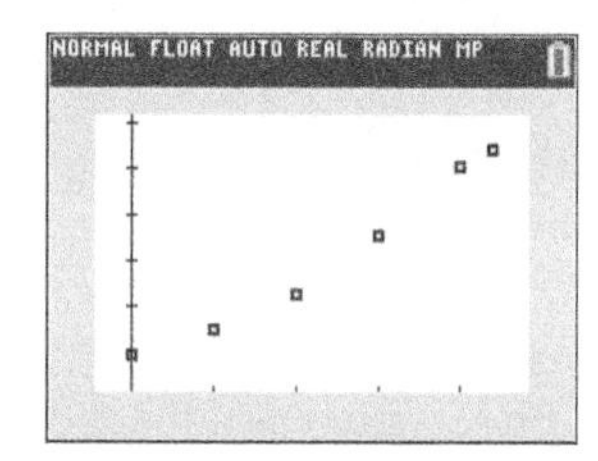

3. Use your graphing calculator to determine the regression equation of the exponential model that best represents the Hispanic population data. Remember that the input variable, t, is the number of years since 1970. In your regression equation, $P = a \cdot b^t$, round the value for a to two decimal places and the value of b to three decimal places. Record your model below.

$P = 9.55 \cdot 1.042^t$

4. Use your graphing calculator to visually check how well the equation in Problem 3 fits the data. Your graph should resemble the following:

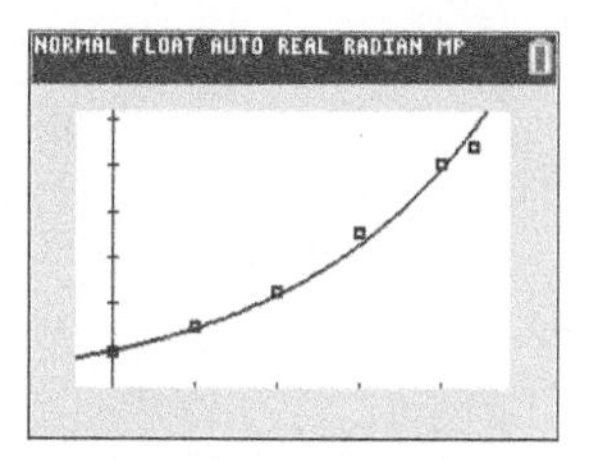

5. Use the exponential model from Problem 3 to determine what the projected Hispanic population will be in the United States in 2025.

$2025 - 1970 = 55$

$9.55(1.042)^{55} = 91.8$. Therefore, the model predicts that the U.S. Hispanic population will be 91.8 million in 2025.

6. a. Using your model from Problem 3, write an equation that can be used to determine the year in which the Hispanic population in the United States will first reach 75 million.

$9.55 \cdot 1.042^t = 75$

b. Solve this equation using a graphing approach. Your screen should resemble the following. What is the equation of the horizontal line in the graph?

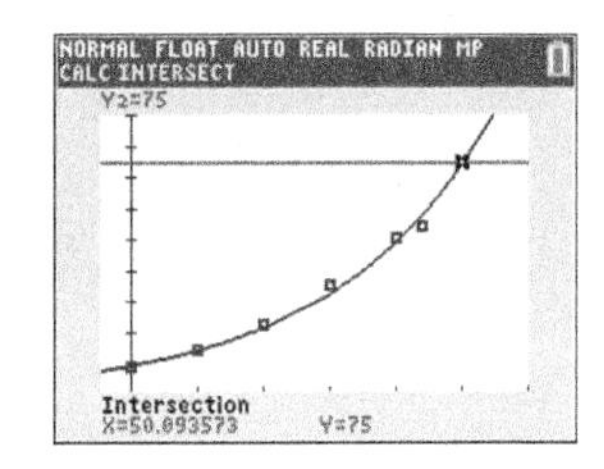

The equation of the horizontal line is $y = 75$. In the year $1970 + 50 = 2020$, the U.S. Hispanic population will first reach 75 million people.

To solve the equation $9.55 \cdot (1.042)^t = 75$ for t using an algebraic approach, you need to remove t as an exponent. The following problem guides you through this process. As you will discover, logarithms are essential in this algebraic approach.

7. Solve $9.55(1.042)^t = 75$ for t using an algebraic approach.

a. Isolate the exponential factor 1.042^t on one side of the equation.

$9.55(1.042)^t = 75$

$1.042^t = \dfrac{75}{9.55}$

b. Take the log (or ln) of each side of the equation in part a.

$\ln(1.042^t) = \ln\left(\dfrac{75}{9.55}\right)$

c. Apply the appropriate property of logarithms on the left side of the equation to "remove" t as an exponent.

$t \cdot \ln(1.042) = \ln\left(\dfrac{75}{9.55}\right)$

d. Solve the resulting equation in part c for t.

$t = \dfrac{\ln\left(\frac{75}{9.55}\right)}{\ln(1.042)} = 50.09$. The number of Hispanics in the United States will reach 75,000,000 in 2020 (50 years from 1970).

e. How does your solution in part d compare with the estimate you obtained graphically in Problem 6b?

The solutions are identical.

8. As the Hispanic population of the country has grown, so has the number of Hispanic voters in presidential elections. The following table shows the number of Hispanic voters in presidential elections from 1992 to 2016:

Year	1992	1996	2000	2004	2008	2012	2016
Number of Hispanic Voters (millions)	4.1	5.0	5.0	7.6	9.7	11.2	13.1

a. Determine an exponential model for the number of Hispanic voters, V, in presidential elections. Let the variable, t, represent the number of years since 1992. In the regression equation $V = a \cdot b^t$, round the value of a to one decimal place and the value of b to three decimal places.

$V(t) = 3.9(1.053)^t$

b. Use the exponential model to predict the number of Hispanic voters in the 2020 presidential election.

$P(28) = 3.9(1.053)^{28} = 16.6$. The model predicts that 16.6 million Hispanics will vote in the 2020 presidential election.

c. Write an equation that can be used to predict the year in which the number of Hispanic voters in a presidential election will first exceed 20 million.

$3.9 \cdot 1.053^t = 20$

d. Solve the equation in part c using an algebraic approach. Keep in mind that presidential elections take place every 4 years.

$$1.053^t = \frac{20}{3.9}$$

$$\ln(1.053^t) = \ln\frac{20}{3.9}$$

$$t\ln(1.053) = \ln\frac{20}{3.9}$$

$$t = \frac{\ln\left(\frac{20}{3.9}\right)}{\ln(1.053)} \approx 31.7$$

The election 32 years after 1992, or 2024, is predicted to be the first time there will be at least 20 million Hispanic voters.

Radioactive Decay

Radioactive substances such as uranium-235, strontium-90, iodine-131, and carbon-14 decay continuously with time. If P_0 represents the original amount of a radioactive substance, then the amount P present after a time t (usually measured in years) is modeled by

$$P = P_0 e^{-kt},$$

where k represents the rate of continuous decay.

EXAMPLE 1 ***One type of uranium decays at a rate of 0.35% per day. If 40 pounds of this uranium are available today, how much will be available after 90 days?***

SOLUTION

The uranium decays at a constant rate of $0.35\% = 0.0035$ per day. The initial amount, the amount available on the first day, is 40 pounds, so the equation for the amount available after t days is

$$P = 40e^{-0.0035t}.$$

To determine the amount available after 90 days, let $t = 90$. The amount available 90 days from now is

$$P = 40e^{-0.0035(90)} \approx 29.2 \text{ lb.}$$

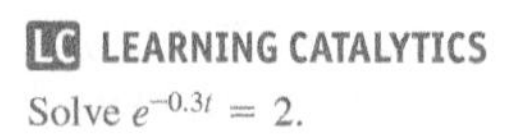

Solve $e^{-0.3t} = 2$.

9. Strontium-90 decays continuously at a constant rate of 2.4% per year. Therefore, the equation for the amount P of strontium-90 after t years is

$$P = P_0 e^{-0.024t}.$$

a. If 10 grams of strontium-90 are present initially, determine the number of grams present after 20 years.

$P = 10e^{-0.024(20)} = 6.2$ g

b. How long will it take the given quantity to decay to 2 grams?

$$\frac{2}{10} = \frac{10e^{-0.024t}}{10}$$
$$0.2 = e^{-0.024t}$$
$$\ln(0.2) = -0.024t$$
$$t = \frac{\ln(0.2)}{-0.024}$$
$$t \approx 67 \text{ yr.}$$

c. How long will it take the given amount of strontium-90 to decay to one-half of its original size (called its half-life)? Round to the nearest whole number.

$$\frac{5}{10} = \frac{10}{10}e^{-0.024t}$$
$$0.5 = e^{-0.024t}$$
$$\ln(0.5) = -0.024t$$
$$\frac{\ln(0.5)}{-0.024} = t$$
$$t \approx 29 \text{ yr.}$$

d. Do you think that the half-life of strontium-90 is 29 years regardless of the initial amount? Answer part c using P_0 as the initial amount. (*Hint:* Find t when $P = \frac{1}{2}P_0$.)

Yes, $\frac{1}{2}P_0 = P_0e^{-0.024t}$ or $\ln(0.5) = -0.024t$ or $t \approx 28.9$ years.

SUMMARY Activity 3.12

To solve exponential equations of the form $ab^x = c$, where $a > 0$, $b > 0$, $b \neq 1$, and $c > 0$,

1. isolate the exponential factor on one side of the equation.
2. take the log (or ln) of each side of the equation.
3. apply the property $\log b^x = x \log b$ to remove the variable x as an exponent.
4. solve the resulting equation for the variable.

EXERCISES Activity 3.12

1. The amount of monetary damage caused by reported cyber-crime to the Internet Crime Complaint Center, IC3, for various years from 2001 to 2016 (in millions of U.S. dollars) is given in the following table:

Year	2001	2004	2007	2009	2012	2014	2016
Cyber-Crime Damage, D (millions of dollars)	17.8	68.1	239.1	559.7	581.44	800.49	1330.0

   a. Use your graphing calculator to determine the exponential regression model $D = a \cdot b^t$, where t represents the number of years since 2000. Round the value of a to 1 decimal place and b to 3 decimal places.

   $D = 23.3 \cdot 1.312^t$

Exercise numbers appearing in color are answered in the Selected Answers appendix.

b. According to the model, what was the monetary damage caused by cyber-crime in 2013?

$D = 23.3 \cdot 1.312^{13} \approx 795.2$ million dollars

c. According to the model, when will the cyber-crime damage reach 2 billion dollars for the first time?

$$23.3 \cdot 1.312^t = 2000$$

$$t\ln(1.312) = \ln\left(\frac{2000}{23.3}\right)$$

$$t = \frac{\ln\left(\frac{2000}{23.3}\right)}{\ln(1.312)} \approx 16.4$$

According to the model, the amount of cyber-crime will first reach 2 billion dollars 16.4 years after 2000 or rounding up to the next year, 2017.

2. The U.S. Department of Transportation recommended that states adopt a 0.08% blood-alcohol concentration as the legal measure of drunk driving. Medical research has shown that as the concentration of alcohol in the blood increases, the risk of having a car accident increases exponentially. The risk, R, expressed as a percentage, is modeled by

$$R(x) = 6e^{12.77x},$$

where x is the blood-alcohol concentration, expressed as a percent.

a. What is a driver's risk of having a car accident if his or her blood-alcohol concentration is 0.08% $(x = 0.08)$?

$R(0.08) = 6e^{12.77(0.08)} \approx 17$

There is a 17% risk of a car accident for a driver whose blood-alcohol concentration is 0.08.

b. What blood-alcohol concentration has a corresponding 25% risk of a car accident?

$$6e^{12.77x} = 25$$

$$e^{12.77x} = \frac{25}{6}$$

$$12.77x = \ln\left(\frac{25}{6}\right)$$

$$x = \frac{\ln\left(\frac{25}{6}\right)}{12.77} \approx 0.11\%$$

3. In 1990, the International Panel on Climate Change projected the following future amounts of carbon dioxide (in parts per million or ppm) in the atmosphere:

Year	1990	2000	2075	2175	2275
Amount of Carbon Dioxide (ppm)	353	375	590	1090	2000

3. a.

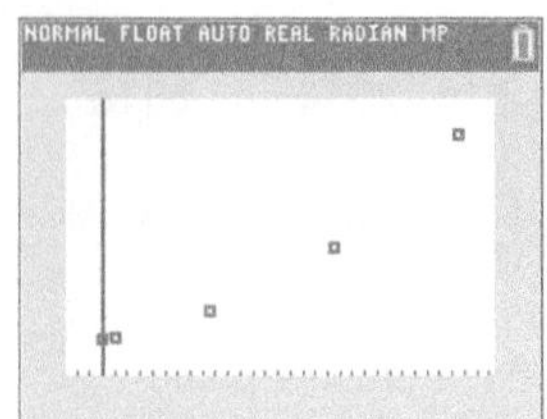

a. Use your graphing calculator to create a scatterplot of the data. Let t represent the number of years since 1990 and $A(t)$ represent the amount of carbon dioxide (in ppm) in the atmosphere. Do the carbon dioxide levels appear to be growing exponentially?

Yes

b. Use your graphing calculator to determine the regression equation of an exponential model that best fits the data.

$A(t) = 352.65(1.006)^t$

c. Use the model in part b to determine in what year the 1990 carbon dioxide level is expected to double.

$$705.3 = 352.65(1.006)^t$$

$$1.006^t = \frac{705.3}{352.65}$$

$$t\ln(1.006) = \ln\left(\frac{705.3}{352.65}\right)$$

$$t = \frac{\ln\left(\frac{705.3}{352.65}\right)}{\ln(1.006)} \approx 116$$

116 years after 1990 would be the year 2106. The carbon dioxide level is expected to double about 2106.

d. Verify your result in part c graphically.

3. d.

NORMAL FLOAT AUTO REAL RADIAN MP
CALC INTERSECT
Y2=705.3
Intersection
X=115.87876 Y=705.3

In Exercises 4–9, solve each equation using an algebraic approach. Verify your answers graphically.

4. $2^x = 14$

$$x\ln 2 = \ln 14$$

$$x = \frac{\ln 14}{\ln 2} \approx 3.81$$

5. $3^{2x} = 8$

$$2x\ln 3 = \ln 8$$

$$x = \frac{\ln 8}{2\ln 3} \approx 0.95$$

6. $1000 = 500(1.04)^t$

$$2 = 1.04^t$$

$$t = \frac{\ln 2}{\ln(1.04)} \approx 17.7$$

7. $e^{0.05t} = 2$ (*Hint:* Take the natural log of both sides.)

$$0.05t = \ln 2$$

$$t \approx 13.86$$

8. $2^{3x+1} = 100$

$$3x + 1 = \frac{\ln(100)}{\ln(2)}$$

$$3x = \frac{\ln(100)}{\ln(2)} - 1$$

$$3x \approx 5.6439;$$

$$x \approx 1.881$$

9. $e^{-0.3t} = 2$

$$-0.3t = \ln 2$$

$$t \approx -2.31$$

10. a. Iodine-131 disintegrates at a continuous constant rate of 8.6% per day. Determine its half-life. Use the model

$$P = P_0e^{-0.086t},$$

where t is measured in days. Round your answer to the nearest whole number.

$$\ln(0.5) = -0.086t$$

$$t \approx 8 \text{ days}$$

b. If dairy cows eat hay containing too much iodine-131, their milk will be unsafe to drink. Suppose that hay contains five times the safe level of iodine-131. How many days should the hay be stored before it can be fed to dairy cows?

(*Hint:* Find t when $P = \frac{1}{5}P_0$.)

$$\frac{1}{5}P_0 = P_0e^{-0.086t} \text{ or } t = \frac{\ln(0.2)}{-0.086} \approx 19 \text{ days}$$

11. a. In 1969, a report written by the National Academy of Sciences (U.S.) estimated that Earth could reasonably support a maximum world population of 10 billion. The world's population was approximately 3.6 billion and growing continuously at 2% per year. If this growth rate remained constant, in what year would the world population reach 10 billion, referred to as Earth's carrying capacity? Use the model

$$P = P_0e^{kt},$$

where P is the population (in billions), $P_0 = 3.6$, $k = 0.02$, and t is the number of years since 1969.

$10 = 3.6e^{0.02t}$

$$t = \frac{\ln\left(\frac{10}{3.6}\right)}{0.02} \approx 51 \text{ yr.}$$

$1969 + 51 = 2020$

b. According to your growth model, when would this 1969 population double?

$e^{0.02t} = 2$ or $t = \frac{\ln 2}{0.02} \approx 34.7$

sometime in the year 2003

c. The world population in 1995 was approximately 5.7 billion. How does this compare with the population predicted by your growth model in part a?

$1995 - 1969 = 26$ yr.

$P(26) = 3.6e^{0.02(26)} \approx 6.06$ billion

The actual population was approximately 360 million below the prediction.

d. The growth rate in 1995 was 1.5%. Assuming this growth rate remains constant, determine when Earth's carrying capacity will be reached. Use the model $P = P_0e^{kt}$.

$10 = 5.7e^{0.015t}$

$$t = \frac{\ln\left(\frac{10}{5.7}\right)}{0.015} \approx 37.5.$$ The year would be 2032.

Collecting and Analyzing Data

12. The combined populations of China and India currently represent over 38% of the world's population. Go to the Internet to obtain the population of each country for every 5 years from 1960 to 2010.

a. Beginning with 1960, make a scatterplot of the data from each country's population. Let the independent variable represent the number of years since 1960 and the dependent variable represent the population of the country in billions. Describe any patterns you observe.

b. Determine whether a linear, quadratic, or exponential function best fits each set of data. Explain.

c. Determine a regression equation for each set of population data. Which country has the largest rate of increase in population?

d. Sketch the graph of each regression equation on the appropriate scatterplot in part a.

e. Predict the population of each country in 2015.

f. Determine the year in which the population of each country should reach 1.5 billion.

g. Use the intersect feature of your graphing calculator to estimate the year when the populations of China and India will be equal.

CLUSTER 2 What Have I Learned?

1. A logarithm is an exponent. Explain how this fact relates to the following properties of logarithms:

a. $\log_b (x \cdot y) = \log_b x + \log_b y$

Multiplication of powers with the same base is accomplished by adding the exponents.

b. $\log_b \frac{x}{y} = \log_b x - \log_b y$

Division of powers of the same base is accomplished by subtracting the exponents.

c. $\log_b x^n = n \cdot \log_b x$

Exponentiation of a power is accomplished by multiplying the exponents.

2. You have $20,000 to invest. Your broker tells you that the value of shares of mutual fund A has been growing exponentially for the past 2 years and that shares of mutual fund B have been growing logarithmically over the same period. If you make your decision based solely on the past performances of the funds, in which fund would you choose to invest? Explain.

I will choose fund A because exponential growth results in more rapid growth over time. Logarithmic growth results in slower growth over time.

3. Study the following graphs, which show various types of functions you have encountered in this course:

a.

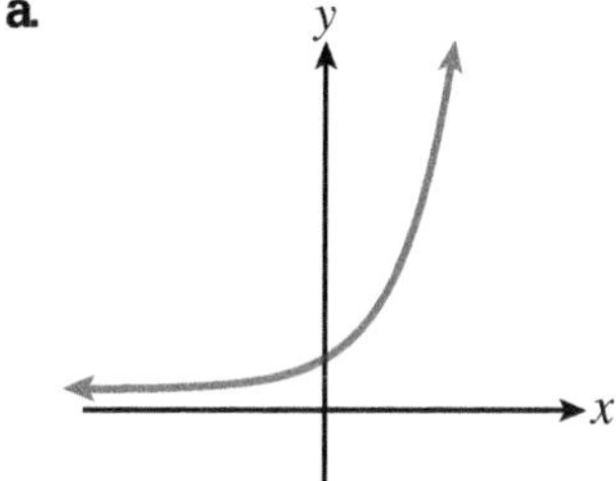

b.

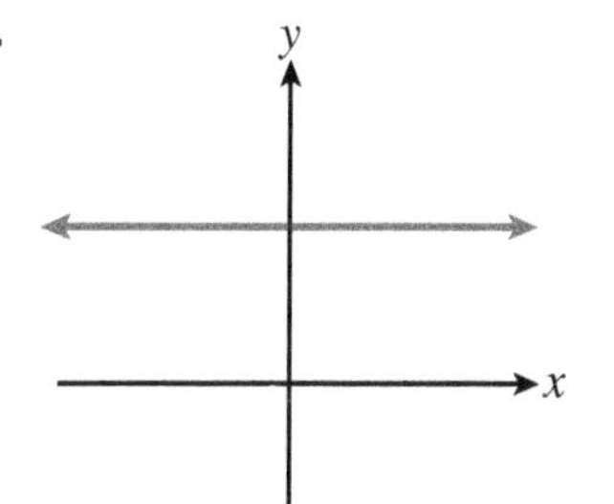

c.

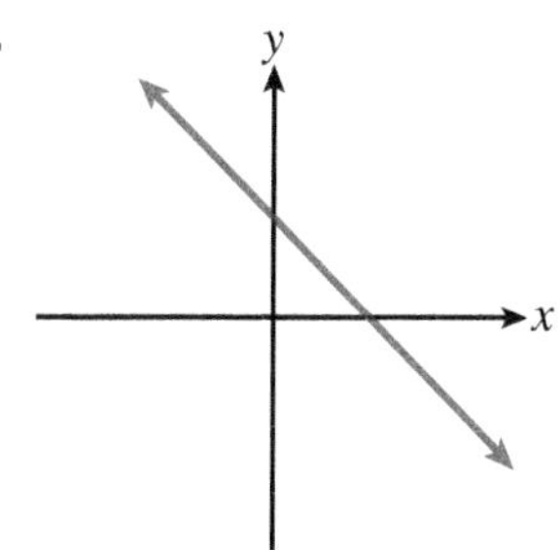

d.

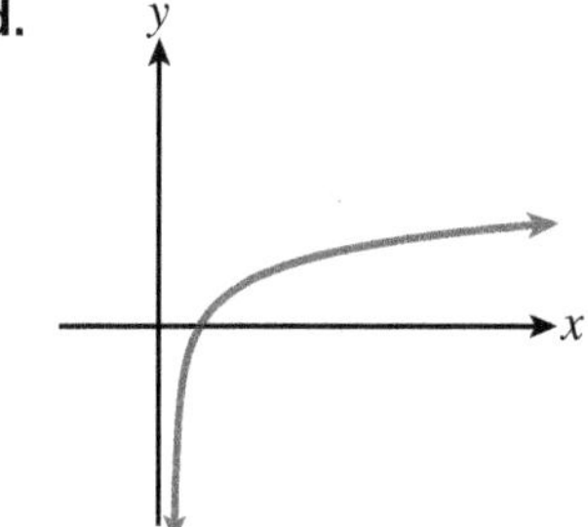

e.

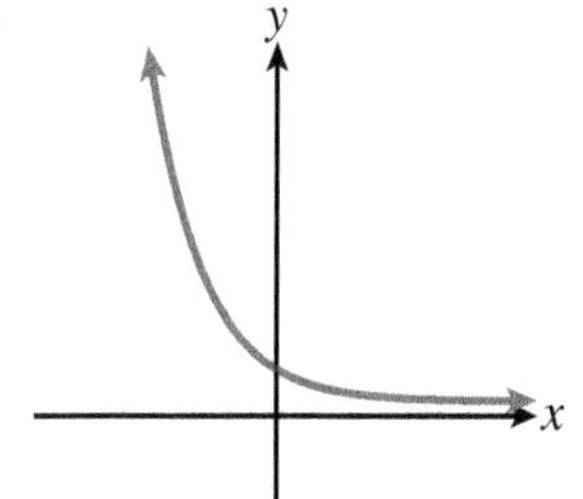

f.

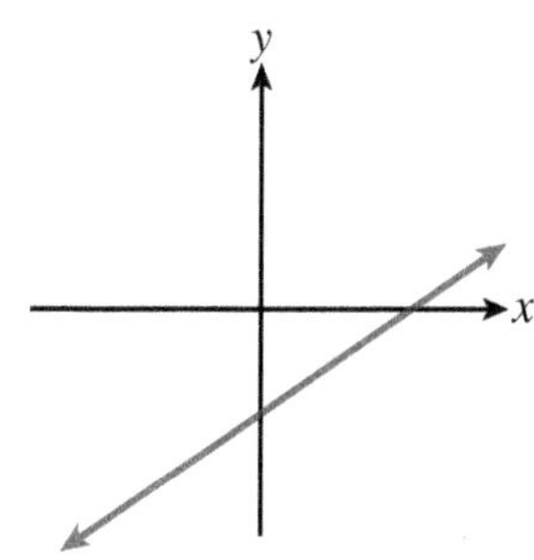

Complete the following table with respect to the preceding graphs:

DESCRIPTION	GRAPH LETTER	GENERAL EQUATION
Constant function	b	$y = a$
Linearly decreasing function	c	$y = mx + b; m < 0$
Logarithmically increasing function	d	$y = \log_b x; b > 1$
Exponentially decreasing function	e	$y = b^x; 0 < b < 1$
Exponentially increasing function	a	$y = b^x; b > 1$
Linearly increasing function	f	$y = mx + b; m > 0$

4. The graph of $y = \log_b x$ will never be located in the second or third quadrants. Explain.

The domain of $y = \log_b x$ is $x > 0$ because b^y cannot be negative.

5. What function would you enter into Y_1 on your graphing calculator to graph the function $y = \log_4 x$?

$y = \dfrac{\log x}{\log 4}$ or $\dfrac{\ln x}{\ln (4)}$

6. What values of x cannot be inputs in the function $y = \log_b (3x - 2)$?

$3x - 2 \le 0$ or $x \le \dfrac{2}{3}$

7. What is the relationship between the functions $y = \log x$ and $y = 10^x$? How are the graphs related?

The functions are inverses. The graphs are symmetric about the line $y = x$.

CLUSTER 2 How Can I Practice?

1. Write each equation in logarithmic form.

a. $4^2 = 16$

$\log_4 16 = 2$

b. $0.0001 = 10^{-4}$

$\log_{10}(0.0001) = -4$

c. $3^{-4} = \frac{1}{81}$

$\log_3\left(\frac{1}{81}\right) = -4$

2. Write each equation in exponential form.

a. $\log_2 32 = 5$

$2^5 = 32$

b. $\log_5 1 = 0$

$5^0 = 1$

c. $\log_{10} 0.001 = -3$

$10^{-3} = 0.001$

d. $\ln e = 1$

$e^1 = e$

3. Solve each equation for the unknown variable.

a. $\log_4 x = -3$

$x = 4^{-3} = \frac{1}{64}$

b. $\log_b 32 = 5$

$b^5 = 32 = 2^5$

$b = 2$

c. $\log_5 125 = y$

$5^y = 125 = 5^3$

$y = 3$

4. a. Complete the table of values for the function $f(x) = \log_4 x$.

x	0.25	0.5	1	4	16	64
$f(x)$	−1	−0.5	0	1	2	3

b. Sketch a graph of the function f.

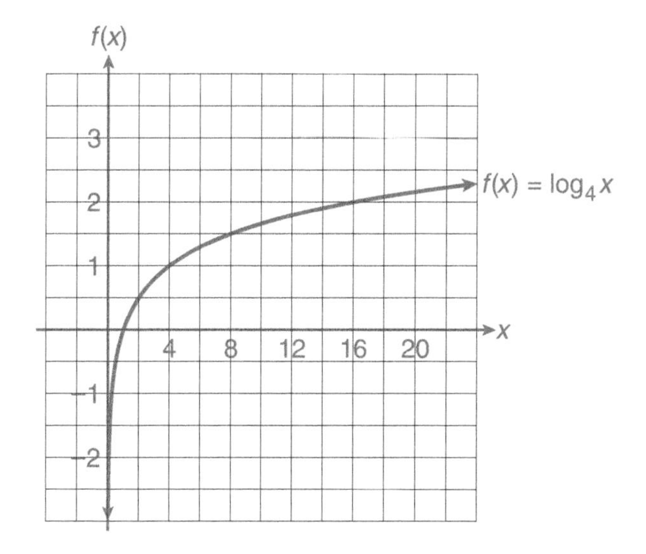

c. Use your graphing calculator to check your result in parts a and b.

d. Determine the x-intercept.

(1, 0)

e. What is the domain of the function?

$x > 0$

f. What is the range?

all real numbers

g. Is the function increasing or decreasing?

increasing

h. Does the graph have a vertical or horizontal asymptote?

The y-axis ($x = 0$) is a vertical asymptote.

i. Use your graphing calculator to determine $f(32)$.

$f(32) = 2.5$

j. Use your graphing calculator to determine x when $f(x) = 3.25$.

$x \approx 90.5$

5. Write each of the following as a sum, difference, or multiple of logarithms. Assume that x, y, and z represent positive numbers.

a. $\log_b \frac{xy^2}{z}$

$\log_b x + 2\log_b y - \log_b z$

b. $\log_3 \frac{\sqrt{x^3 y}}{z}$

$\frac{3}{2}\log_3 x + \frac{1}{2}\log_3 y - \log_3 z$

c. $\log_5 (x\sqrt{x^2 + 4})$

$\log_5 x + \frac{1}{2}\log_5(x^2 + 4)$

d. $\log_4 \sqrt[3]{\frac{xy^2}{z^2}}$

$\frac{1}{3}\log_4 x + \frac{2}{3}\log_4 y - \frac{2}{3}\log_4 z$

6. Rewrite the following as the logarithm of a single quantity:

a. $\log x + \frac{1}{3}\log y - \frac{1}{2}\log z$

$\log \frac{x\sqrt[3]{y}}{\sqrt{z}}$

b. $3\log_3 (x + 3) + 2\log_3 z$

$\log_3 (x + 3)^3 z^2$

c. $\frac{1}{3}\log_3 x - \frac{2}{3}\log_3 y - \frac{4}{3}\log_3 z$

$\log_3 \sqrt[3]{\frac{x}{y^2 z^4}}$

7. Use the change-of-base formula and your calculator to approximate the following:

a. $\log_5 17$

$\frac{\log 17}{\log 5} = 1.76$

b. $\log_{13} \sqrt[3]{41}$

$\frac{1}{3} \cdot \frac{\log 41}{\log 13} = 0.4826$

8. Solve each of the following using an algebraic approach:

a. $25 + 3\ln x = 10$

$3\ln x = -15$

$\ln x = -5$

$x = e^{-5}$

$x \approx 0.0067$

b. $1.5\log_4 (x - 1) = 7$

$x - 1 = 4^{7/1.5}$

$x = 4^{7/1.5} + 1$

$x = 646.08$

9. Solve the following algebraically. Check your solutions using graphs or tables.

a. $3^x = 17$

$$x = \frac{\log 17}{\log 3} \approx 2.5789$$

b. $42 = 3e^{1.7x}$

$$14 = e^{1.7x}$$

$$1.7x = \ln 14$$

$$x = \frac{\ln 14}{1.7} \approx 1.55$$

10. You have invested $10,000 in a money market account that will pay you 4% interest compounded continuously.

a. Write an equation that relates the current value of the account, V, to the number of years you have held the account, t.

$V = 10{,}000 \cdot e^{0.04t}$

10. c.

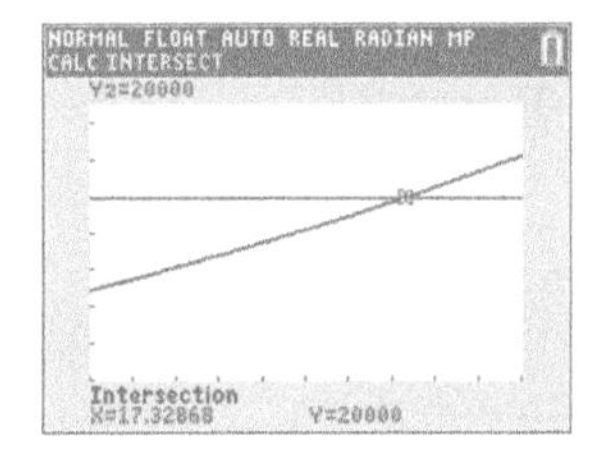

b. Use the equation in part a to determine algebraically the number of years it will take the value of the account to double.

$$10{,}000 \cdot e^{0.04t} = 20{,}000$$

$$0.04t = \ln(2)$$

$t \approx 17.3$. It will take 17.3 years for the account to double.

c. Verify your result in part b graphically.

11. Data collected from over 100 countries showed that the relationship between per capita health-care expenditures, H, in dollars and average life expectancy, E, could be modeled by the formula $E = 0.035 + 9.669 \ln(H)$, where $0 < H \le 4500$.

a. Sketch a graph of the healthcare/life expectancy model using the domain $0 < H \le 4500$.

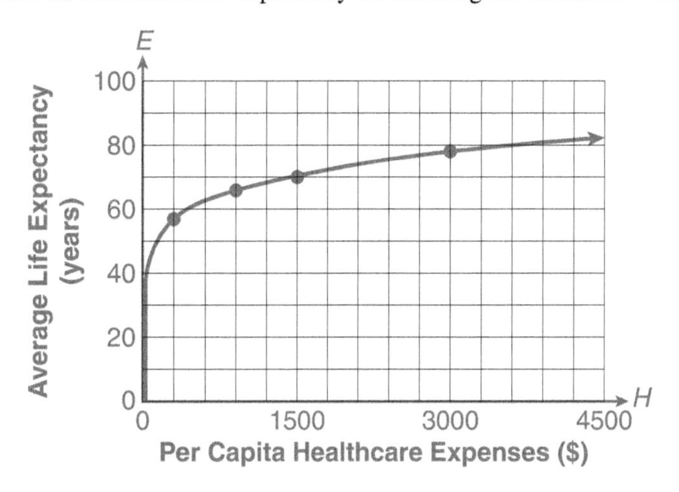

b. Use this model to predict the average life expectancy in a country whose per capita healthcare expenditure is $1500 per year.

$E = 0.035 + 9.669 \ln(1500) \approx 70.7$ yr.

c. Use the model and your graphing calculator to predict per capita healthcare expenditures in a country whose average life expectancy is 77 years.

$H \approx \$2864$

CHAPTER 3 Summary

The bracketed numbers following each concept indicate the activity in which the concept is discussed.

CONCEPT/SKILL	DESCRIPTION	EXAMPLE
Exponential functions [3.1]	The exponential functions can be defined by $y = b^x, b > 0, b \neq 1$.	$y = 3^x$
Growth factor of an exponential function [3.1]	If $b > 1$, the function $y = b^x$ is increasing and b is called the growth factor.	The exponential function $y = 3^x$ has a growth factor of 3.
Vertical intercept of an exponential function [3.1]	The vertical intercept (y-intercept) of an exponential function $y = b^x$ is $(0, 1)$.	The graph of $y = 2^x$ passes through the point (0, 1).
Horizontal asymptote of an exponential function [3.1]	The line $y = 0$ is a horizontal asymptote of an exponential function $y = b^x$.	As x gets more negative, the output values of $y = 3^x$ approach 0.
Decay factor of an exponential function [3.2]	If $0 < b < 1$, the function $y = b^x$ is decreasing and b is called the decay factor.	The exponential function $y = \left(\frac{1}{2}\right)^x$ has a decay factor of $\frac{1}{2}$.
Doubling time [3.3]	The doubling time of an increasing exponential function is the time it takes an output to double. The doubling time is set by the growth factor and remains the same for all output values.	Example 2, Activity 3.3; see page 286.
Half-life [3.3]	The half-life of a decreasing exponential function is the time it takes an output to be cut in half. The half life is determined by the decay factor and remains the same for all output values.	Example 3, Activity 3.3; see page 288.
Growth model [3.4]	If r represents the annual percentage growth rate, the exponential function that models the quantity P can be written as $P(t) = P_0(1 + r)^t$, where P_0 is the initial amount, t represents the number of elapsed years, and $1 + r$ is the growth factor.	Example 4, Activity 3.4; see page 298.
Decay model [3.4]	If r represents the annual percentage decay rate, the exponential function that models the amount remaining can be written as $P(t) = P_0(1 - r)^t$, where $1 - r$ is the decay factor.	Example 5, Activity 3.4; see page 300.
Compound interest [3.5]	The formula for compound interest is $A = P(1 + \frac{r}{n})^{nt}$.	Example 1, Activity 3.5; see page 305.

CONCEPT/SKILL	DESCRIPTION	EXAMPLE
Continuous compounding [3.5]	The formula for continuous compounding is $A = Pe^{rt}$.	Example 4, Activity 3.5; see page 310.
Continuous growth at a constant percentage rate [3.5], [3.6]	Whenever growth is continuous at a constant percentage rate, the exponential model used is $y = y_0e^{rt}$.	Problem 10, Activity 3.5; see page 310.
Continuous decay at a constant percentage rate [3.6]	Whenever decay is continuous at a constant percentage rate, the model used is $y = y_0e^{-rt}$.	Example 2, Activity 3.6; see page 316.
Logarithm [3.8]	In the equation $y = b^x$, where $b > 0$ and $b \neq 1$, x is called a logarithm, or log, base b.	For the equation $3^4 = 81$, 4 is the logarithm of 81, base 3.
Notation for logarithms [3.8]	The notation for logarithms is $\log_b x = y$, where b is the base of the log, x (a positive number) is the power of b, and y is the exponent.	In the equation $\log_2 16 = 4$, 2 is the base, 4 is the log or exponent, and 16 is the power of 2.
Common logarithm [3.8]	A common logarithm is a base-10 logarithm. The notation is $\log_{10} x = \log x$.	$1000 = 10^3$. The common logarithm of 10^3 is 3 (i.e., $\log 1000 = 3$).
Natural logarithm [3.8]	A natural logarithm is a base-e logarithm. The notation is $\log_e x = \ln x$.	$\log_e e^3 = \ln e^3 = 3$
Logarithmic equation [3.8]	The logarithmic equation $y = \log_b x$ is equivalent to the exponential equation $b^y = x$.	The equations $6 = \log_4 x$ and $x = 4^6$ are equivalent.
Basic properties of logarithms [3.8]	If $b > 0$ and $b \neq 1$, $\log_b 1 = 0$, $\log_b b = 1$, and $\log_b b^n = n$.	$\log_4 1 = 0$, $\log_7 7 = 1$, $\log_6 6^4 = 4$
Logarithmic function [3.9]	If $b > 0$ and $b \neq 1$, the logarithmic function is defined by $y = \log_b x$.	$y = \log_4 x$
Graphs of the logarithmic functions [3.9]	The graph of $y = \log_b x$, where $b > 1$, is increasing for all $x > 0$, has an x-intercept of $(1, 0)$ and has a vertical asymptote of $x = 0$, the y-axis.	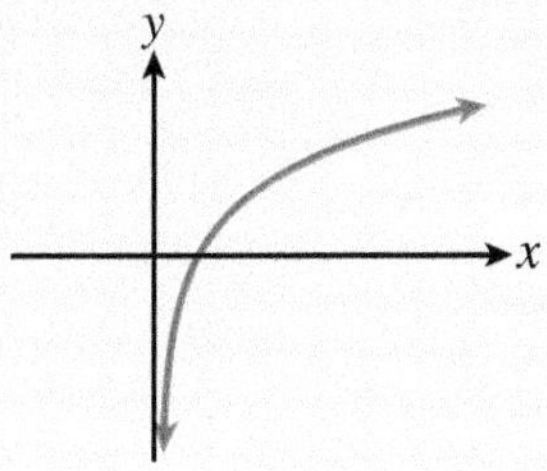
Comparison of the graphs of $f(x) = b^x$, where $b > 1$, and $g(x) = \log_b x$, where $b > 1$ [3.10]	Both graphs increase. The exponential function increases more quickly as x increases; the log function increases more slowly as x increases. The domain of the exponential function is the range of the log, which is all real numbers; the range of the exponential function is the domain of the log, which is the interval $(x > 0)$ $f(x)$ and $g(x)$ are inverse functions.	Problem 6, Activity 3.10; see pages 354–355.

CONCEPT/SKILL	DESCRIPTION	EXAMPLE
Log of a Product: If $A > 0, B > 0, b > 0$, and $b \neq 1$, then $\log_b (A \cdot B) = \log_b A + \log_b B$. [3.11]	The logarithm of a product is the sum of the logarithms.	$\log_2 4 \cdot 8 = \log_2 4 + \log_2 8$ $= 2 + 3 = 5$
Log of a Quotient: If $A > 0, B > 0, b > 0$, and $b \neq 1$, then $\log_b\left(\frac{A}{B}\right) = \log_b A - \log_b B$. [3.11]	The logarithm of a quotient is the difference of the logarithms.	$\log_3\left(\frac{81}{27}\right) = \log_3 81 - \log_3 27$ $= 4 - 3 = 1$
Log of a Sum: If $A > 0, B > 0, b > 0$, and $b \neq 1$, then $\log_b (A + B) \neq \log_b (A) + \log_b (B)$. [3.11]	The logarithm of a sum is not the sum of the logarithms.	$\log 2 + \log 3 \approx$ $0.3010 + 0.4771 = 0.7781$ $\log(2 + 3) = \log 5 \approx 0.6990$
Log of a Power: If $A > 0$, p is a real number, $b > 0$, and $b \neq 1$, then $\log_b A^p = p \log_b A$. [3.11]	The logarithm of a power of A is the exponent times the logarithm of A.	$\log_5 x^4 = 4 \log_5 x$ $\log_3 \sqrt{x} = \frac{1}{2} \log_3 x$
Change-of-base formula [3.11]	The logarithm of any positive number x to any base can be found using the formula $\log_b x = \frac{\log x}{\log b}$ or $\log_b x = \frac{\ln x}{\ln b}$.	$\log_2 2.5 = \frac{\log 2.5}{\log 2} \approx 1.3219$

CHAPTER 3 Gateway Review

1. Tuition will increase 3% each year.

a. If tuition is \$300 per credit now, how much will it be in 5 years? In 10 years?

The growth factor is $100\% + 3\% = 103\% = 1.03$. Let T represent the tuition.

$T = 300(1.03)^t$

$T = 300(1.03)^5 \approx 347.78$

In 5 years, the tuition will be \$347.78 per credit.

$T = 300(1.03)^{10} \approx 403.17$

In 10 years, the tuition will be \$403.17 per credit.

b. Calculate the average rate of change in tuition over the next 5 years.

$$\frac{347.78 - 300}{5} = \frac{47.78}{5} \approx 9.56$$

Tuition increases \$9.56 per credit per year.

c. Calculate the average rate of change in tuition over the next 10 years.

$$\frac{403.17 - 300}{10} = \frac{103.17}{10} \approx 10.32$$

Tuition increases \$10.32 per credit per year.

d. If the rate of increase of tuition stays at 3%, approximately how long will it take tuition to double?

Tuition will double in approximately 23.4 years if the rate of increase stays at 3%.

2. a. Determine some of the output values for the function $f(x) = 8^x$ by completing the following table:

x	-1	$-\frac{1}{3}$	0	1	$\frac{4}{3}$	2	3
$f(x) = 8^x$	$\frac{1}{8}$	$\frac{1}{2}$	1	8	16	64	512

b. Sketch the graph of the function f.

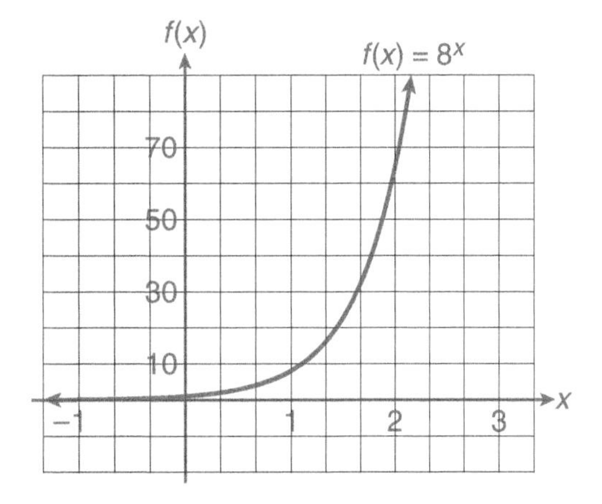

Answers to all Gateway exercises are included in the Selected Answers appendix.

c. Is this function increasing or decreasing? Explain how you know this by looking at the equation of the function.

The function is increasing because $b = 8 > 1$.

d. What is the domain?

all real numbers

e. What is the range?

$y > 0$

f. What are the x- and y-intercepts?

There is no x-intercept. The y-intercept is (0, 1).

g. Are there any asymptotes? If so, write the equations of the asymptotes.

There is one horizontal asymptote, the x-axis, $y = 0$.

h. Compare the graph of f with the graph of $g(x) = \left(\frac{1}{8}\right)^x$. What are the similarities and the differences?

The domain and range are the same. The graphs are reflections in the y-axis.

f is increasing; g is decreasing.

i. In what way does the graph of $h(x) = 8^x + 5$ differ from that of $f(x) = 8^x$?

f is moved vertically upward 5 units to obtain h.

j. Write the equation of the function that is the inverse of the function $f(x)$.

Solve $y = 8^x$ for x: $x = \log_8 y$. Interchange x and y: $y = \log_8 x$.

3. Complete the table for each exponential function. Use your graphing calculator to check your work.

FUNCTION	BASE, b	GROWTH OR DECAY FACTOR	x-INTERCEPT	y-INTERCEPT	HORIZONTAL ASYMPTOTE	INCREASING OR DECREASING
$h(x) = 6^x$	6	growth	none	(0, 1)	$y = 0$	increasing
$g(x) = \left(\frac{1}{3}\right)^x$	$\frac{1}{3}$	decay	none	(0, 1)	$y = 0$	decreasing
$p(x) = 5(2.34)^x$	2.34	growth	none	(0, 5)	$y = 0$	increasing
$q(x) = 3(0.78)^x$	0.78	decay	none	(0, 3)	$y = 0$	decreasing
$r(x) = 2^x - 4$	2	growth	(2, 0)	(0, −3)	$y = -4$	increasing

4. Use your graphing calculator to help you determine the domain and range for each function.

Function	$f(x) = 0.8^x$	$h(x) = 6^x + 2$	$t(x) = 3^x - 5$	$q(x) = \log_4 x$	$r(x) = \ln(x - 3)$
Domain	all reals	all reals	all reals	$x > 0$	$x > 3$
Range	$y > 0$	$y > 2$	$y > -5$	all reals	all reals

5. a. Given the following table, determine whether the given data can be approximately modeled by an exponential function. If it can, what is the growth or decay factor?

x	0	1	2	3	4
y	10	15.5	24	36	55.5

The table is approximately exponential. The growth factor is about 1.55.

b. Determine an exponential equation that models this data.

$y = 10 \cdot 1.55^x$

6. Complete the following tables, which represent exponential functions. Round calculations to two decimal places whenever necessary after each calculation.

a.

x	0	1	2	3	4
y	3.00	6.12	12.48	25.46	51.94

b.

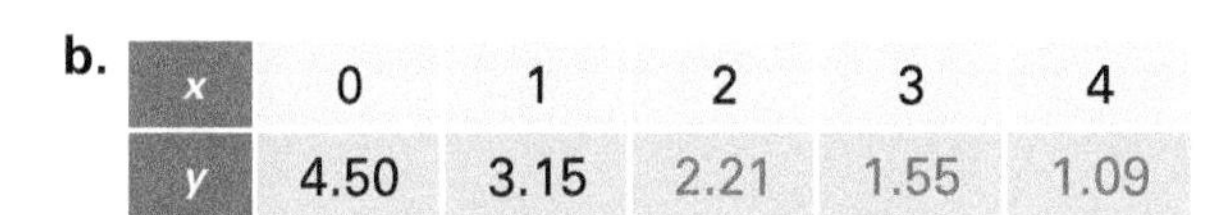

x	0	1	2	3	4
y	4.50	3.15	2.21	1.55	1.09

c.

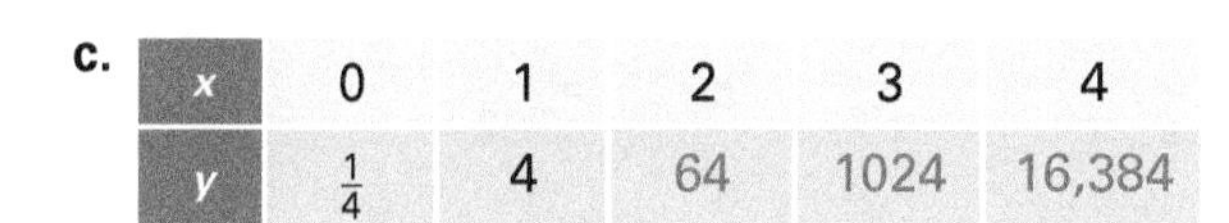

x	0	1	2	3	4
y	$\frac{1}{4}$	4	64	1024	16,384

d. Write the equation of the exponential function that represents the data in each table in parts a, b, and c.

i. $y = 3.00(2.04)^x$ **ii.** $y = 4.50(0.7)^x$ **iii.** $y = 0.25(16)^x$

7. a. Your salary has increased at the rate of 1.5% annually for the past 5 years, and your boss projects this rate will remain unchanged for the next 5 years. You were making $35,000 annually in 2015. Complete the following table:

2015	2016	2017	2018	2019	2020
$35,000	$35,525	$36,058	$36,599	$37,148	$37,705

b. Write the exponential growth function that models your annual salary during this period of time. Let x represent the number of years since 2015.

$y = 35{,}000\,(1.015)^x$

c. If your increase in salary continues at this rate, how much will you make in 2023? Is this realistic?

$x = 8$; $y = \$39{,}427$. This is realistic if you assume that $35,000 is a reasonable starting salary and that the 1.5% salary increase per year remains constant.

d. You would like to double your salary. How many years must you work before your salary will be twice the salary you made in 2015?

$70{,}000 = 35{,}000\,(1.015)^x;\quad 2 = 1.015^x;\quad x = \dfrac{\ln 2}{\ln 1.015} \approx 46.6 \text{ yr.}$

8. a. You just inherited $10,000. You can invest the money at a rate of 4% compounded continuously. In 8 years, your oldest child will be going to college. How much will be in the bank for her education? Use the equation $A = A_0e^{rt}$.

$A = 10{,}000 \cdot e^{0.04(8)}\quad A = \$13{,}771.28$

b. You actually will need $16,000 for your child's first year of college. For how many years must you leave the money in the bank to accumulate the $16,000?

$16{,}000 = 10{,}000 \cdot e^{(0.04 \cdot t)}$

$0.04t = \ln(8/5)$

$t = 11.75 \text{ yr.}$

9. Determine the value of each of the following without using your calculator:

a. $25^{3/2}$

125

b. $81^{3/4}$

27

c. $64^{-5/6}$

$\frac{1}{32}$

d. $\sqrt[3]{125^2}$

25

e. $\log_3 \frac{1}{9}$

-2

f. $\log_5 625$

4

g. $\log 0.001$

-3

h. $\ln e^2$

2

10. Write each equation in logarithmic form.

a. $6^2 = 36$

$\log_6 36 = 2$

b. $0.000001 = 10^{-6}$

$\log_{10} 0.000001 = -6$
or $\log 0.000001 = -6$

c. $2^{-5} = \frac{1}{32}$

$\log_2 \frac{1}{32} = -5$

11. Write each equation in exponential form.

a. $\log_3 81 = 4$

$3^4 = 81$

b. $\log_7 1 = 0$

$7^0 = 1$

c. $\log_{10} 0.0001 = -4$

$10^{-4} = 0.0001$

d. $\ln e = 1$

$e^1 = e$

e. $\log_q y = b$

$q^b = y$

12. Solve each equation for the unknown variable.

a. $\log_5 x = -3$

$5^{-3} = x$

$x = \frac{1}{125}$

b. $\log_b 256 = 4$

$b^4 = 256$

$b = 4$

c. $\log_2 64 = y$

$2^y = 64$

$y = 6$

d. $\log_4 x = \frac{3}{2}$

$x = 4^{3/2}$

$x = 2^3 = 8$

13. a. Complete the table of values for the function $f(x) = \log_5 x$.

x	0.008	0.04	0.2	1	5	25
$f(x)$	-3	-2	-1	0	1	2

b. Sketch a graph of the function.

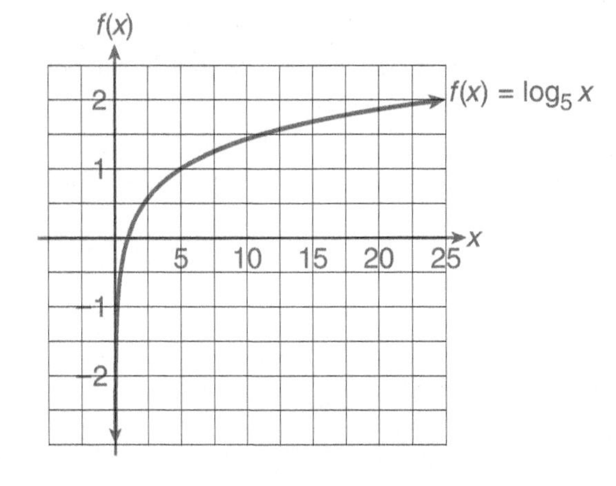

13. c

c. Use your graphing calculator to check your result in parts a and b.

d. Determine the x-intercept.

(1, 0)

e. What is the domain of the function?

$x > 0$

f. What is the range?

all real numbers

g. Does the graph have a vertical or horizontal asymptote?

It has a vertical asymptote at $x = 0$. The function gets closer and closer to the y-axis but does not cross it.

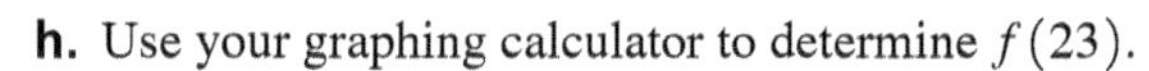

h. Use your graphing calculator to determine $f(23)$.

$f(23) \approx 1.948$

i. Use your graphing calculator to determine x when $f(x) = 2.46$.

$x \approx 52.416$

14. Use the change-of-base formula and your calculator to approximate the following:

a. $\log_7 21$

$\frac{\log 21}{\log 7} \approx 1.56$

b. $\log_{15} \frac{8}{9}$

$\frac{\log\left(\frac{8}{9}\right)}{\log 15} \approx -0.0435$

15. Write each of the following as a sum, difference, or multiple of logarithms. Assume that x, y, and z are all greater than 0.

a. $\log_2 \frac{x^3 y}{z^{1/2}}$

$3 \log_2 x + \log_2 y - \frac{1}{2} \log_2 z$

b. $\log \sqrt[3]{\frac{x^4 y^3}{z}}$

$\frac{1}{3}(4 \log x + 3 \log y - \log z)$

16. Rewrite the following as the logarithm of a single quantity:

a. $\log x + \frac{1}{4} \log y - 3 \log z$

$\log \frac{x\sqrt[4]{y}}{z^3}$

b. $\frac{1}{3}(\log x - 2 \log y - \log z)$

$\log \sqrt[3]{\frac{x}{y^2 z}}$

17. Solve the following algebraically:

a. $3^{3+x} = 7$

$3 + x = \frac{\log 7}{\log 3}; x \approx -1.23$

b. $\log_2 (4x + 9) = 4$

$4x + 9 = 2^4; x = 1.75$

c. $50 + 6 \ln x = 85$

$6 \ln x = 35$

$\ln x = \frac{35}{6}$

$x = e^{35/6}$

$x \approx 341.5$

18. a. Sketch the graph of the function using the data from the given table.

x	0.1	0.5	1	2	4	16
$f(x)$	−1.66	−0.5	0	.5	1	2

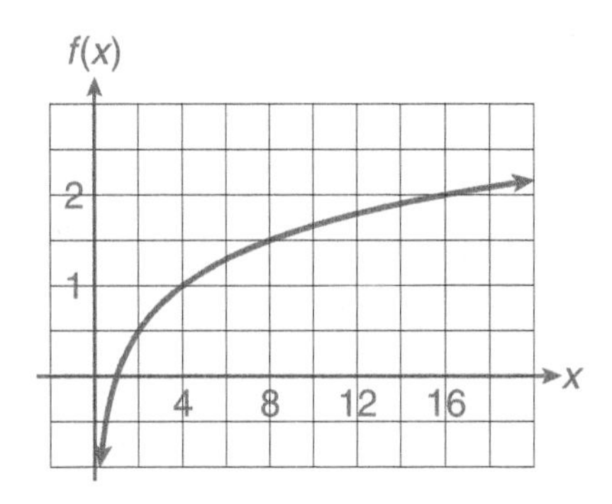

b. Use the table and the graphing feature of your calculator to verify that the equation that defines function f is $f(x) = 0.5 \log_2 x$.

c. Use the function to determine the value of $f(54)$.

2.87744

d. If $f(x) = 2.319$, determine the value of x.

$2.319 = \frac{\log x}{2 \log 2}; x \approx 24.9$

e. Use your graphing calculator to verify that the function $g(x) = 4^x$ is the inverse of f.

19. The populations of Ohio and Georgia (in millions) can be modeled by the following:

Ohio	$P_O = 11.53 \cdot e^{0.0013t}$
Georgia	$P_G = 9.71 \cdot e^{0.008t}$

where t represents years since 2010.

a. Determine the populations of Ohio and Georgia in 2010.

Ohio: 11.53 million; Georgia: 9.71 million

b. Sketch a graph of each function on the same set of coordinate axes. Let t vary from 0 to 40.

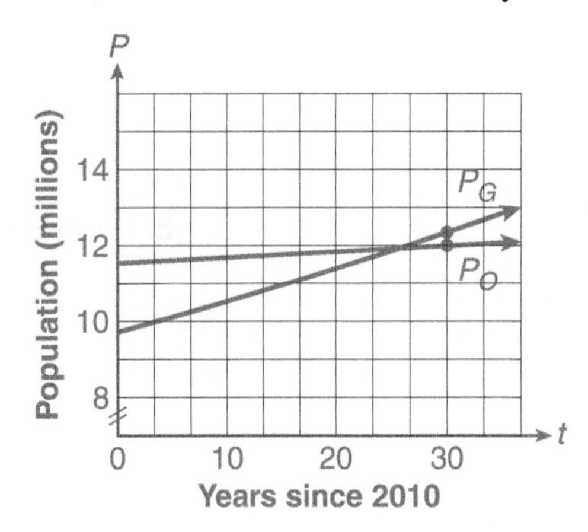

c. Determine graphically the year when the population of Georgia will be equal to the population of Ohio.

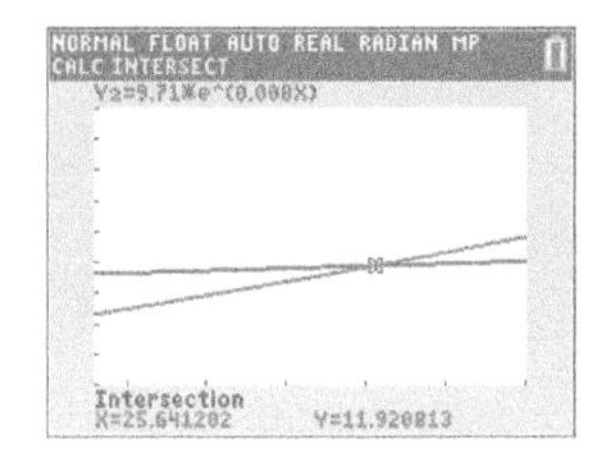

The models indicate that the populations of Ohio and Georgia will be equal 25.6 years after 2010 or sometime in the year 2035.

d. Determine algebraically the year that the population of Georgia first exceeds 13 million people.

$$9.71 \cdot e^{0.008x} > 13$$

$$e^{0.008x} > \frac{13}{9.71}$$

$$0.008x > \ln\left(\frac{13}{9.71}\right)$$

$$x > \frac{\ln\left(\frac{13}{9.71}\right)}{0.008}$$

$x > 36.5$. According to the model, sometime during the year 2046, the population of Georgia will first exceed 13 million people.

20. Atmospheric pressure decreases with increasing altitude. The following data was collected during a science experiment to investigate the relationship between the height of an object above ground level and the pressure exerted on the object.

Atmospheric Pressure, x (mm of mercury)	760	740	725	700	650	630	600	580	550
Height, y (km)	0	0.184	0.328	0.565	1.079	1.291	1.634	1.862	2.235

a. Use a graphing calculator to determine a logarithmic regression model that best fits the data.

$y = 45.786 - 6.903 \ln x$

b. Use the log equation in part a to predict the height of the object if the atmospheric pressure is 500 millimeters of mercury.

$y = 45.786 - 6.903 \ln(500) \approx 2.887$ km

CHAPTER 4

Quadratic and Higher-Order Polynomial Functions

CLUSTER 1 Introduction to Quadratic Functions

ACTIVITY 4.1

Baseball and the Willis Tower

Equations Defining Quadratic Functions

OBJECTIVES

1. Identify functions of the form $f(x) = ax^2 + bx + c$ as quadratic functions.
2. Explore the role of c as it relates to the graph of $f(x) = ax^2 + bx + c$.
3. Explore the role of a as it relates to the graph of $f(x) = ax^2 + bx + c$.
4. Explore the role of b as it relates to the graph of $f(x) = ax^2 + bx + c$.

Note: $a \neq 0$ in Objectives 1–4.

Imagine yourself standing on the roof of the 1450-foot-high Willis Tower (formerly called the Sears Tower) in Chicago. When you release and drop a baseball from the roof of the tower, *the ball's height above the ground*, H (in feet), can be described as a function of the time, t (in seconds), since it was dropped. This height function is defined by

$$H(t) = -16t^2 + 1450.$$

1. Sketch a diagram illustrating the Willis Tower and the path of the baseball as it falls to the ground.

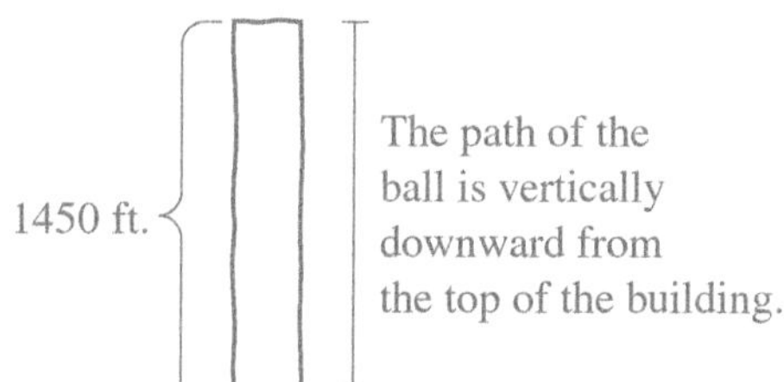

2. a. Complete the following table:

TIME, t (sec.)	$H(t) = -16t^2 + 1450$
0	1450
1	1434
2	1386
3	1306
4	1194
5	1050
6	874
7	666
8	426
9	154
10	−150

b. How far does the baseball fall during the first second?

1450 − 1434 = 16 ft.

c. How far does it fall during the interval from 1 to 3 seconds?

1434 − 1306 = 128 ft.

3. Use the height function defined by, $H(t) = -16t^2 + 1450$, to determine the average rate of change of H with respect to t over the given interval. Remember:

$$\text{average rate of change} = \frac{\text{change in output}}{\text{change in input}}.$$

a. $0 \le t \le 1$

$-\frac{16}{1} = -16$ feet per second

b. $1 \le t \le 3$

$-\frac{128}{2} = -64$ feet per second

c. Based on the results of parts a and b, do you believe that $H(t) = -16t^2 + 1450$ defines linear function? Explain.

No, because the rate of change is not constant.

4. a. What is the value of H when the baseball strikes the ground? Use the table in Problem 2a to estimate the time when the ball is at ground level.

The value of H is 0 when the ball strikes the ground. It appears from the table that the object hits the ground in approximately 9.5 seconds.

b. What is the practical domain of the height function?

$0 \le t \le 9.5$

c. Determine the practical range of the height function.

$0 \le H \le 1450$

d. On the following grid, plot the points in Problem 2a that satisfy part b (practical domain) and sketch a curve representing the height function:

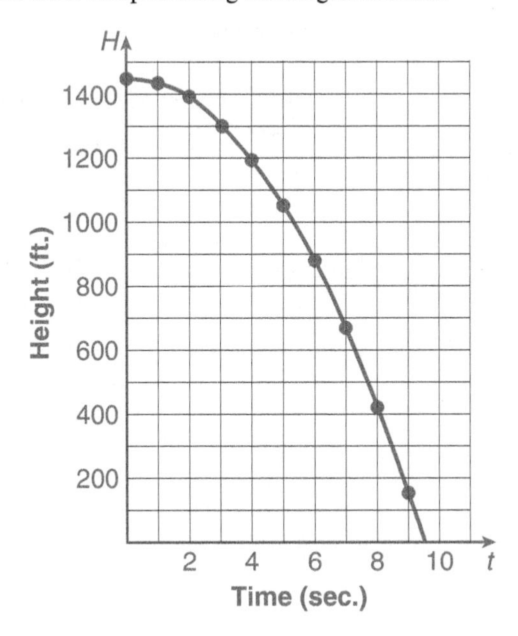

e. Is the graph of the height function in part d the actual path of the object (see Problem 1)? Explain.

No. The actual path of the ball is a vertical path downward, as drawn in Problem 1. The graph shows how the height of the baseball changes with respect to time.

Some interesting properties of the function defined by $H(t) = -16t^2 + 1450$ arise when you ignore the falling object context. Replace $H(t)$ with y and t with x, and consider the general function defined by $y = -16x^2 + 1450$.

5. a. Graph the function defined by $y = -16x^2 + 1450$, setting the window parameters at Xmin $= -10$ and Xmax $= 10$ for the input and Ymin $= -50$ and Ymax $= 1500$ for the output. Your graph should appear as follows:

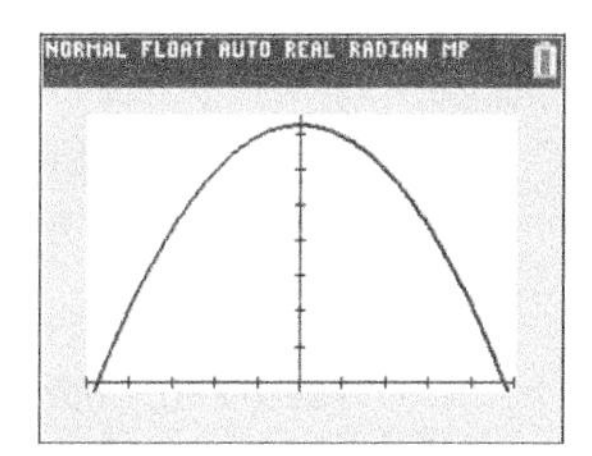

b. Describe the important features of the graph of $y = -16x^2 + 1450$. Discuss the shape, symmetry, and intercepts.

The graph is an upside-down U-shape that is symmetrical with respect to the y-axis. The graph opens downward, resulting in a maximum point. The vertical intercept is (0, 1450). The horizontal intercepts are approximately $(\pm 9.52, 0)$.

Quadratic Functions

The graph of the function defined by $y = -16x^2 + 1450$ is a parabola. The graph of a **parabola** is a U-shaped figure that opens upward, ⋃, or downward, ⋂. Parabolas are graphs of a special category of functions called quadratic functions.

DEFINITION

Any function defined by an equation of the form $y = ax^2 + bx + c$ or $f(x) = ax^2 + bx + c$, where a, b, and c represent real numbers and $a \neq 0$, is called a **quadratic function**. The output variable y is defined by an expression having three terms: the **quadratic term**, ax^2; the **linear term**, bx; and the **constant term**, c. The numerical factors of the quadratic and linear terms, a and b, are called the **coefficients** of the terms.

EXAMPLE 1

$H(t) = -16t^2 + 1450$ defines a quadratic function. The quadratic term is $-16t^2$. The linear term is $0t$, although it is not written as part of the expression defining $H(t)$. The constant term is 1450. The numbers -16 and 0 are the coefficients of the quadratic and linear terms, respectively. Therefore, $a = -16$, $b = 0$, and $c = 1450$.

6. For each of the following quadratic functions, identify the value of a, b, and c.

QUADRATIC FUNCTION	a	b	c
$y = 3x^2$	3	0	0
$y = -2x^2 + 3$	−2	0	3
$y = x^2 + 2x - 1$	1	2	−1
$y = -x^2 + 4x$	−1	4	0

The Constant Term *c*: A Closer Look

Consider once again the height function defined by $H(t) = -16t^2 + 1450$ from the beginning of the activity.

7. a. What is the vertical intercept of the graph? Explain how you obtained the results.

(0, 1450); substituting 0 for *t*, $H(0) = -16(0)^2 + 1450 = 1450$.

b. What is the practical meaning of the vertical intercept in this situation?

The vertical intercept represents the height of the release point at the top of the tower.

c. Predict what the graph of $H(t) = -16t^2 + 1450$ would look like if the constant term 1450 were changed to 800. That is, the baseball is dropped from a height of 800 feet rather than 1450 feet. Verify your prediction by graphing $H(t) = -16t^2 + 800$. What does the constant term tell you about the graph of the parabola?

The graph would be the same shape, but the vertical intercept would be (0, 800). The constant term represents the *H*-value of the vertical intercept. The graph of $H(t) = -16t^2 + 800$ is a shift downward of the graph of $H(t) = -16t^2 + 1450$ by 650 units.

The constant term *c* of a quadratic function $f(x) = ax^2 + bx + c$ *always* indicates the vertical intercept of the parabola. The vertical intercept of any quadratic function is (0, *c*) because $f(0) = a(0)^2 + b(0) + c = c$.

8.

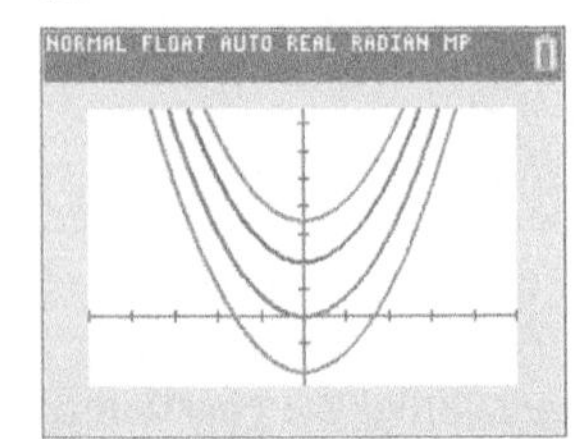

8. Graph the parabolas defined by the following quadratic equations. Note the similarities and differences among the graphs, especially the vertical intercepts. Be careful in your choice of a window.

a. $f(x) = 1.5x^2$

vertical intercept: (0, 0)

b. $g(x) = 1.5x^2 + 7$

vertical intercept: (0, 7); the graph of *g* is the graph of *f* shifted up 7 units.

c. $q(x) = 1.5x^2 + 4$

vertical intercept: (0, 4); the graph of *q* is the graph of *f* shifted up 4 units.

d. $s(x) = 1.5x^2 - 4$

vertical intercept: (0, −4); the graph of *s* is the graph of *f* shifted down 4 units.

All four graphs open upward and are symmetric with respect to the *y*-axis.

The Effects of the Coefficient *a* on the Graph of $y = ax^2 + bx + c$

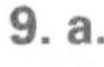

9. a.

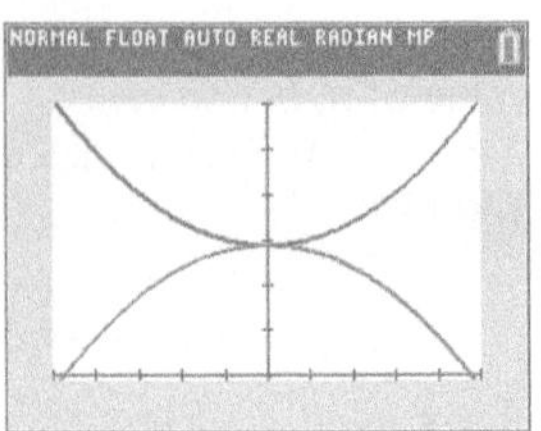

9. a. Graph the quadratic function defined by $y_1 = 16x^2 + 1450$ on the same screen as $y_2 = -16x^2 + 1450$. Use the window settings Xmin $= -10$, Xmax $= 10$, Ymin $= -50$, and Ymax $= 3000$.

b. What effect does the sign of the coefficient of x^2 appear to have on the graph of the parabola?

The sign of the coefficient of x^2 determines whether the parabola opens upward or downward. If the sign is positive, the parabola opens upward. If the sign is negative, the parabola opens downward.

10.

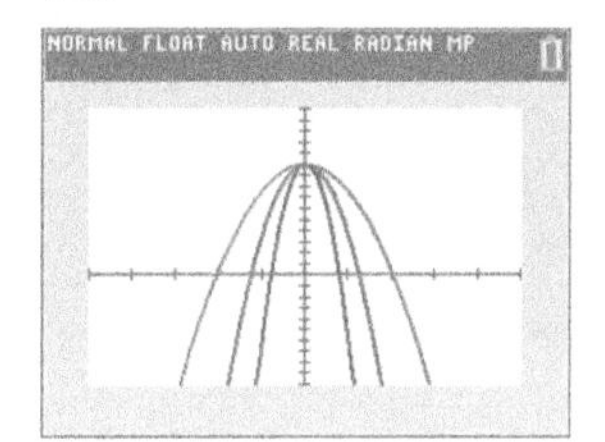

10. Graph the functions $y_3 = -16t^2 + 100, y_4 = -6t^2 + 100, y_5 = -40t^2 + 100$ in the same window. What effect does the magnitude of the coefficients of x^2 (namely, $|-16| = 16, |-6| = 6$, and $|-40| = 40$) appear to have on the graph of that particular parabola? Use window settings Xmin $= -15$, Xmax $= 15$, Ymin $= -200$, and Ymax $= 200$.

As the absolute value of *a* increases, the parabola gets narrower.

The results from Problems 9 and 10 regarding the effects of the coefficient a can be summarized as follows:

The graph of a quadratic function defined by $f(x) = ax^2 + bx + c$ is called a parabola.

- If $a > 0$, the parabola opens upward.
- If $a < 0$, the parabola opens downward.
- The magnitude of a affects the width of the parabola. The larger the absolute value of a, the narrower the parabola.

11. a. Is the graph of $h(x) = 0.3x^2$ wider or narrower than the graph of $f(x) = x^2$?

The graph of *h* is wider than the graph of *f*.

b. How do the output values of h and the output values of f compare for the same input value?

The output values of *h* are 0.3 times the output values of *f* for the same input values.

c. Is the graph of $g(x) = 3x^2$ wider or narrower than the graph of $f(x) = x^2$?

The graph of *g* is narrower than the graph of *f*.

d. How do the output values of g and f compare for the same input value?

The output values of *g* are 3 times the output values of *f* for the same input values.

e. Describe the effect of the magnitude of the coefficient a on the width of the graph of the parabola.

The larger the magnitude of *a*, the narrower the parabola.

f. Describe the effect of the magnitude of the coefficient of a on the output value for a given value of input.

For $a > 0$, the larger the value of *a*, the larger the output value. For $a < 0$, the larger the magnitude of *a*, the more negative the output value.

The Effects of the Coefficient *b* on the Turning Point

Assume for the time being that you are back on the roof of the 1450-foot Willis Tower. Instead of merely releasing the ball, suppose you *throw it down* with an initial velocity of 40 feet per second. Then the function describing its height above ground as a function of time is modeled by

$$H_{\text{down}}(t) = -16t^2 - 40t + 1450.$$

If you tossed the ball straight up with an initial velocity of 40 feet per second, then the function describing its height above ground as a function of time would be modeled by

$$H_{\text{up}}(t) = -16t^2 + 40t + 1450.$$

12. Predict what features the graphs of H_{down} and H_{up} have in common with

$$H(t) = -16t^2 + 1450.$$

The graphs have the same vertical intercept. All of the graphs open downward.

13. a.

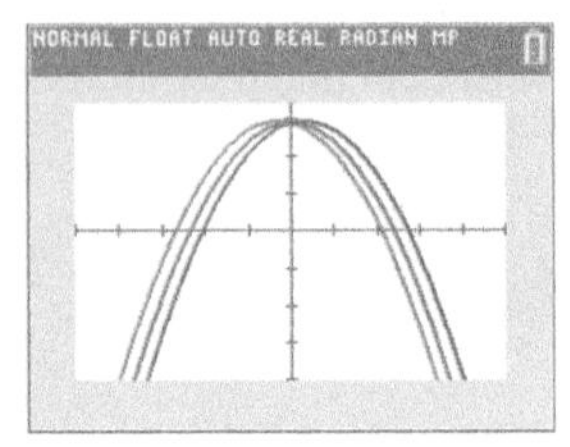

13. a. Graph the three functions $H(t)$, $H_{down}(t)$, and $H_{up}(t)$ using the same window settings given in Problem 5a.

b. What effect do the $-40t$ and $40t$ terms seem to have upon the turning point of the graphs?

Answers will vary. The $-40t$ term moves the turning point of the graph of H to the left of the y-axis. The $40t$ term moves the turning point of the graph of H to the right of the y-axis.

If $b = 0$, the turning point of the parabola is located on the vertical axis. If $b \neq 0$, the turning point will not be on the vertical axis.

LC LEARNING CATALYTICS

If $a > 0$, the graph of the function defined by $f(x) = ax^2 + bx + c$

a. opens upward
b. opens downward
c. opens upward and the y-intercept is on the positive y-axis
d. There is not enough information to determine any of these.

14. For each of the following quadratic functions, identify the value of b and then, without graphing, determine whether the turning point is on the y-axis. Verify your conclusion by graphing the given function. Set the window of your calculator to Xmin $= -8$, Xmax $= 8$, Ymin $= -20$, and Ymax $= 20$.

a. $y = x^2$

$b = 0$; the turning point is on the y-axis.

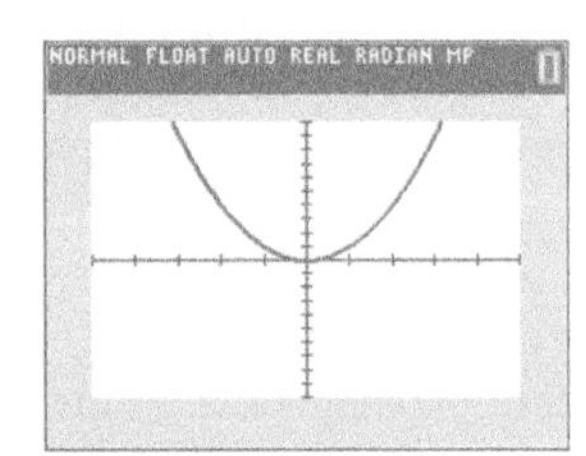

b. $y = x^2 - 4x$

$b = -4$; the turning point is not on the y-axis.

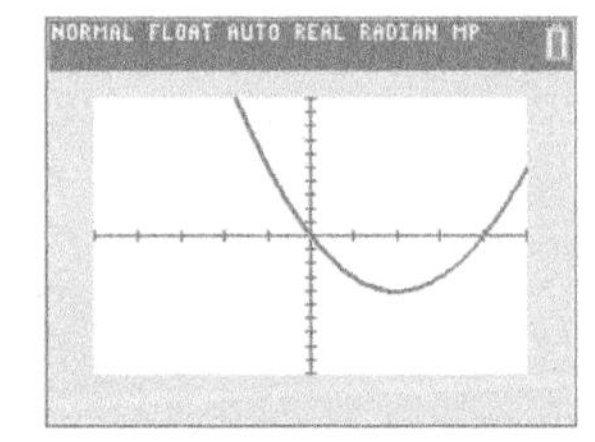

c. $y = x^2 + 4$

$b = 0$; the turning point is on the y-axis.

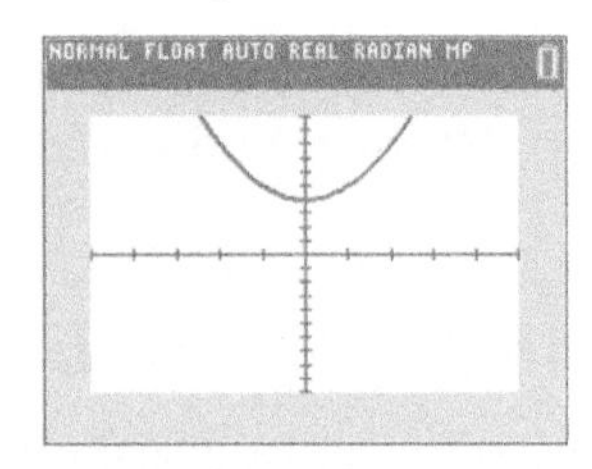

d. $y = x^2 + x$

$b = 1$; the turning point is not on the y-axis.

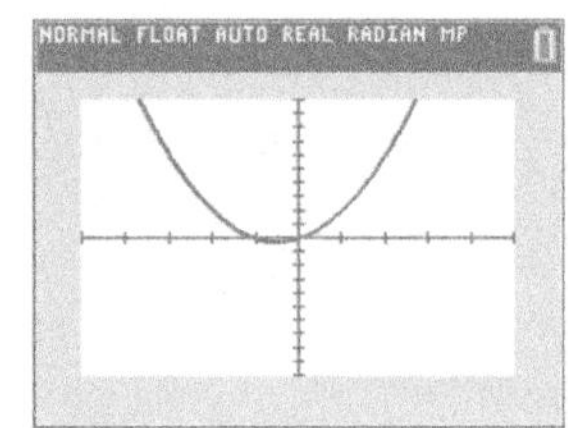

e. $y = x^2 - 3$

$b = 0$; the turning point is on the y-axis.

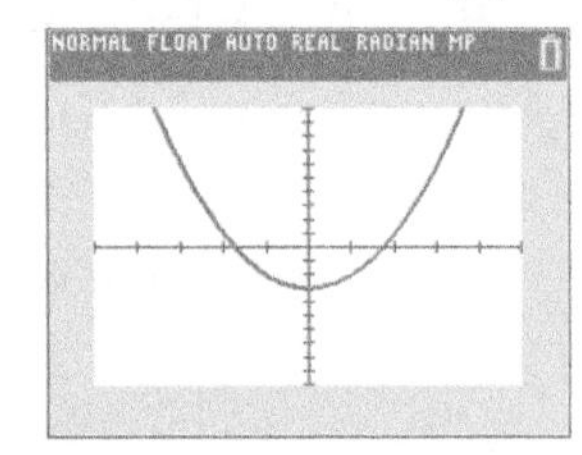

15. Match each function with its corresponding graph below, and then verify using your graphing calculator.

a. $f(x) = x^2 + 4x + 4$ **b.** $g(x) = 0.2x^2 + 4$ **c.** $h(x) = -x^2 + 3x$

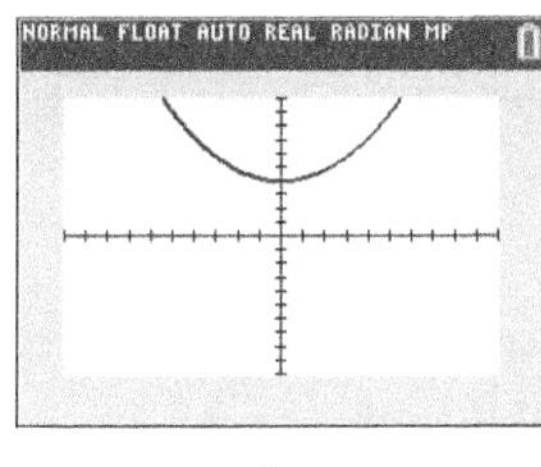

i.

The graph of **b** is **i**.

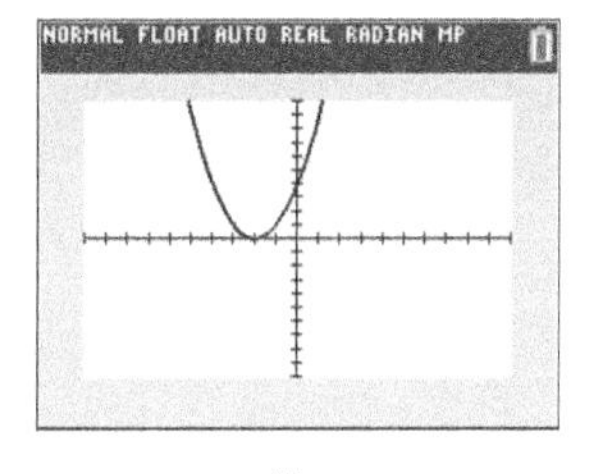

ii.

The graph of **a** is **ii**.

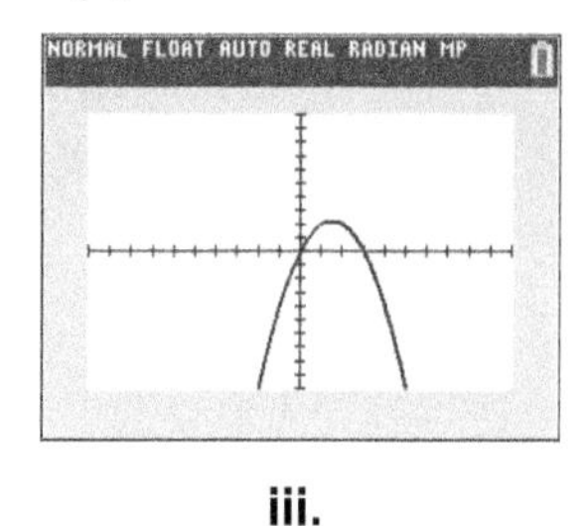

iii.

The graph of **c** is **iii**.

SUMMARY Activity 4.1

1. The equation of a **quadratic function** with x as the input variable and y as the output variable has the standard form

$$y = ax^2 + bx + c,$$

where a, b, and c are real numbers and $a \neq 0$.

2. The graph of a quadratic function is called a **parabola**.

3. For the quadratic function defined by $f(x) = ax^2 + bx + c$

- if $a > 0$, the parabola opens upward.
- if $a < 0$, the parabola opens downward.

The magnitude of a affects the width of the parabola. The larger the absolute value of a, the narrower the parabola.

4. If $b = 0$, the turning point of the parabola is located on the vertical axis. If $b \neq 0$, the turning point will not be on the vertical axis.

5. The constant term, c, of a quadratic function $f(x) = ax^2 + bx + c$ always indicates the vertical intercept of the parabola. The vertical intercept of the graph of any quadratic function is $(0, c)$.

EXERCISES Activity 4.1

1. a. Complete the following table for $f(x) = x^2$:

x	-3	-2	-1	0	1	2	3
$f(x) = x^2$	9	4	1	0	1	4	9

Exercise numbers appearing in color are answered in the Selected Answers appendix.

b. Use the results of part a to sketch a graph $y = x^2$. Verify using a graphing calculator.

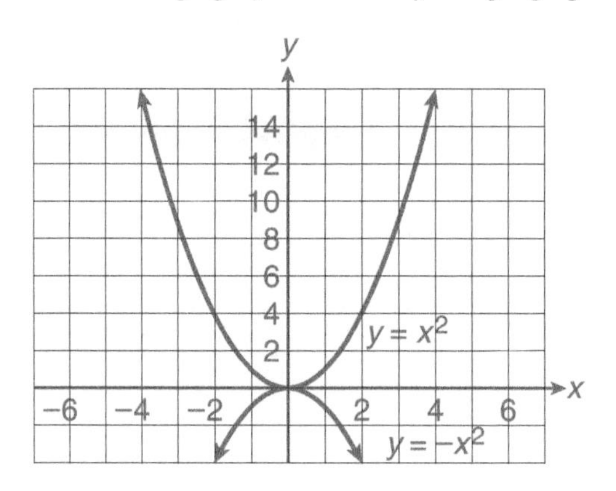

c. What is the coefficient of the term x^2?

1

d. From the graph, determine the domain and range of the function.

The domain is all real numbers. The range is all real numbers greater than or equal to zero.

e. Create a table similar to the one in Exercise 1a to show the output for $g(x) = -x^2$.

x	−3	−2	−1	0	1	2	3
$g(x) = -x^2$	−9	−4	−1	0	−1	−4	−9

f. Sketch the graph of $g(x) = -x^2$ on the same coordinate axes in part a. Verify using a graphing calculator.

g. What is the coefficient of the term $-x^2$?

−1

h. How can the graph of $y = -x^2$ be obtained from the graph of $y = x^2$?

The graph of $y = -x^2$ is a mirror image of $y = x^2$ over the x-axis.

2. In each of the following functions defined by an equation of the form $y = ax^2 + bx + c$, identify the value of a, b, and c:

a. $y = -2x^2$

$a = -2, b = 0, c = 0$

b. $y = \frac{2}{5}x^2 + 3$

$a = \frac{2}{5}, b = 0, c = 3$

c. $y = -x^2 + 5x$

$a = -1, b = 5, c = 0$

d. $y = 5x^2 + 2x - 1$

$a = 5, b = 2, c = -1$

3. Predict how the graph of each of the following quadratic functions will look. Use your graphing calculator to verify your prediction.

a. $f(x) = 3x^2 + 5$

Because $a = 3 > 0$, the graph is U-shaped; vertical intercept: (0, 5); symmetric with respect to the y-axis

b. $g(x) = -2x^2 + 1$

Because $a = -2 < 0$, the graph is ∩-shaped; vertical intercept: (0, 1); symmetric with respect to the y-axis

c. $h(x) = 0.5x^2 - 3$

Because $a = 0.5 > 0$, the graph is U-shaped; vertical intercept: $(0, -3)$; symmetric with respect to the y-axis

4. Graph the following pairs of functions, and describe any similarities as well as differences that you observe in the graphs:

a. $f(x) = 3x^2, g(x) = -3x^2$

f opens upward; *g* opens downward; both pass through (0, 0).

b. $h(x) = \frac{1}{2}x^2, f(x) = 2x^2$

Both *f* and *h* open upward; *h* is wider than *f*.

c. $g(x) = 5x^2, h(x) = 5x^2 + 2$

h is *g* shifted up 2 units; both open upward.

d. $f(x) = 4x^2 - 3, g(x) = 4x^2 + 3$

Both *f* and *g* open upward.

The low point of *f* is 3 units below the x-axis; the low point of *g* is 3 units above the x-axis.

e. $f(x) = 6x^2 + 1, h(x) = -6x^2 - 1$

f opens upward with a vertical intercept at (0, 1);

h opens downward with a vertical intercept at $(0, -1)$;

both are symmetric with respect to the y-axis.

5. Use your graphing calculator to graph the two functions $y_1 = 3x^2$ and $y_2 = 3x^2 + 2x - 2$.

a. What is the vertical intercept of the graph of each function?

y_1: (0, 0)

y_2: $(0, -2)$

b. Compare the two graphs to determine the effect of the linear term $2x$ and the constant term -2 on the graph of $y_1 = 3x^2$.

The $2x$ and the -2 make the graph shift to the left and down.

For Exercises 6–10, determine

a. *whether the parabola opens upward or downward.*

b. *the vertical intercept.*

6. $f(x) = -5x^2 + 2x - 4$

a. downward

b. $(0, -4)$

7. $g(t) = \frac{1}{2}t^2 + t$

a. upward

b. (0, 0)

8. $h(v) = 2v^2 + v + 3$

a. upward

b. (0, 3)

9. $r(t) = 3t^2 + 10$

a. upward

b. (0, 10)

10. $f(x) = -x^2 + 6x - 7$

a. downward

b. $(0, -7)$

11. Does the graph of $y = -2x^2 + 3x - 4$ have any horizontal intercepts? Explain.

The graph does not cross the x-axis; therefore, there are no horizontal intercepts.

12. **a.** Is the graph of $y = \frac{3}{5}x^2$ wider or narrower than the graph of $y = x^2$?

The graph of $y = \frac{3}{5}x^2$ is wider than the graph of $y = x^2$.

b. For the same input value, which graph would have a larger output value?

The graph of $y = x^2$ would have a greater output value for any input x except $x = 0$.

13. Put the following in order from narrowest to widest:

a. $y = 0.5x^2$ **b.** $y = 8x^2$ **c.** $y = -2.3x^2$

b, *c*, *a* because $|0.5| < |-2.3| < |8|$

ACTIVITY
4.2

The Shot Put
Properties of Graphs of Quadratic Functions

OBJECTIVES

1. Determine the vertex, or turning point, of a parabola.
2. Identify the vertex as a maximum or minimum.
3. Determine the axis of symmetry of a parabola.
4. Identify the domain and range.
5. Determine the *y*-intercept of a parabola.
6. Determine the *x*-intercept(s) of a parabola using technology.
7. Interpret the practical meaning of the vertex and intercepts in a given problem.

Parabolas are good models for a variety of situations that you encounter in everyday life. Examples include the path of a golf ball after it is struck, the arch (cable system) of a bridge, the path of a baseball thrown from the outfield to home plate, the stream of water from a drinking fountain, and the path of a cliff diver.

At the Rio Olympics in 2016, USA's Ryan Crouser won the men's shot put event for the first time. His winning throw traveled 73.88 ft., an Olympic record. The path of the throw can be modeled by the quadratic function defined by

$$y = H(x) = -0.014630x^2 + x + 6$$

where x is the horizontal distance from the point of release, in feet, and y is the vertical height of the shot above the ground, in feet.

1. a. After inspecting the equation for the path of the winning throw, which way do you expect the parabola to open? Explain.

The parabola will open downward. The coefficient $a = -0.014630$ is negative.

b. What is the vertical intercept of the graph of the parabola? What practical meaning does this intercept have in this situation?

The vertical intercept is (0, 6). Just before the shot was released, the shot was 6 feet above the ground.

2. Use your graphing calculator to produce a plot of the path of the winning throw. Be sure to adjust your window settings so that all of the important features of the parabola (including x-intercepts) appear on the screen. Your graph should resemble the following:

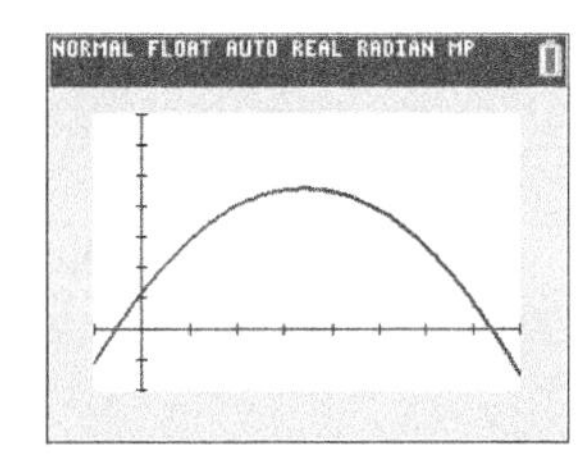

3. a. Use the graph to estimate the practical domain of the function.

$0 \leq x \leq 73.9$

b. What does the practical domain mean in the shot put situation?

It gives the horizontal position of the shot put in feet from the initial release to the landing of the shot put.

c. Use the graph to estimate the practical range of the function.

$0 \leq H(x) \leq 23.1$

d. What does the practical range mean in the shot put situation?

It gives the vertical height in feet of the shot put from the initial release to the landing. The highest point was about 23.1 feet above the ground.

4. a. Use the table feature of your graphing calculator to complete the following table:

x	10	20	30	40	50
$H(x)$	14.5	20.1	22.8	22.6	19.4

b. Use the table to estimate the horizontal distance from the point of release when the shot put reaches its maximum height above the ground.

approximately 34.2 feet

Vertex of a Parabola

An important feature of the *graph* of any quadratic function defined by $f(x) = ax^2 + bx + c$ is its **turning point**, also called the **vertex**. The turning point of a parabola that opens downward or upward is the point at which the parabola changes direction from increasing to decreasing or decreasing to increasing, respectively.

5. Use the results from Problems 3c and 4b to approximate the coordinates of the vertex of the shot put function H.

(Answers will vary.) One possibility is (34.2, 23.1).

6. The vertex is often very important in a situation. What is the significance of the coordinates of the turning point in this problem?

When the shot was displaced approximately 34.2 feet horizontally from the thrower, it was at its maximum height, approximately 23.1 feet.

The coordinates of the vertex of a parabola having equation $y = ax^2 + bx + c$ can be determined from the values of a and b in the equation.

DEFINITION

The **vertex** is the turning point of a parabola. The vertex of a parabola with the equation $y = f(x) = ax^2 + bx + c$ has coordinates

$$\left(-\frac{b}{2a}, f\left(-\frac{b}{2a}\right)\right),$$

where a is the coefficient of the x^2 term and b is the coefficient of the x term.

Note that the y-coordinate (output) of the vertex is determined by substituting the x-coordinate of the vertex into the equation of the parabola and evaluating the resulting expression.

EXAMPLE 1 ***Determine the vertex of the parabola defined by the equation*** $\boldsymbol{y = -3x^2 + 12x + 5}$.

SOLUTION

Step 1. Determine the x-coordinate of the vertex by substituting the values of a and b into the formula $x = \frac{-b}{2a}$.

Because $a = -3$ and $b = 12$, you have

$$x = \frac{-(12)}{2(-3)} = \frac{-12}{-6} = 2.$$

Step 2. The y-value of the vertex is the corresponding output value for $x = 2$. Substituting 2 for x in the equation, you have

$$y = -3(2)^2 + 12(2) + 5 = 17.$$

Therefore, the vertex is (2, 17).

Because the parabola in Example 1 opens downward ($a = -3 < 0$), the vertex is the high point (maximum) of the parabola, as demonstrated by the following graph of the parabola defined by $y = -3x^2 + 12x + 5$:

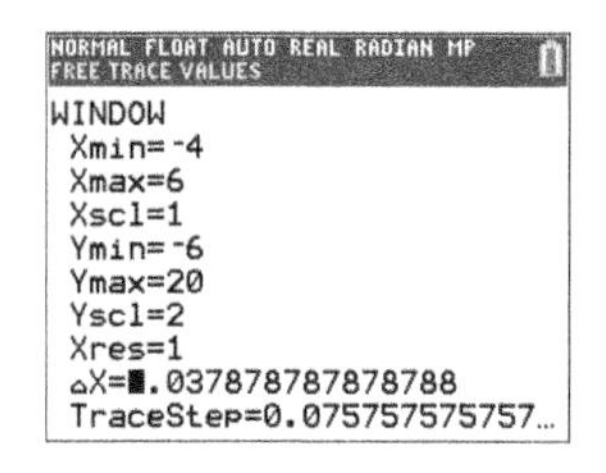

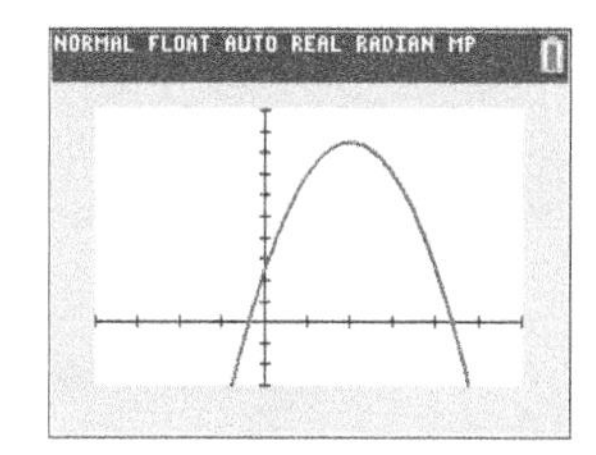

7. Use the method demonstrated in Example 1 to determine the vertex of the parabola defined by $y = -0.014630x^2 + x + 6$ (the shot put function).

Because $a = -0.014630$ and $b = 1$, $x = -\dfrac{b}{2a} = -\dfrac{1}{2(-0.014630)} \approx 34.18$.

Then $y = -0.014630(34.18)^2 + 34.18 + 6 \approx 23.09$.

The vertex is (34.18, 23.09).

You can also determine the vertex of a parabola by selecting the maximum (or minimum) option in the CALC menu of your graphing calculator. Follow the prompts to obtain the coordinates of the maximum point (vertex) of $y = -3x^2 + 12x + 5$.

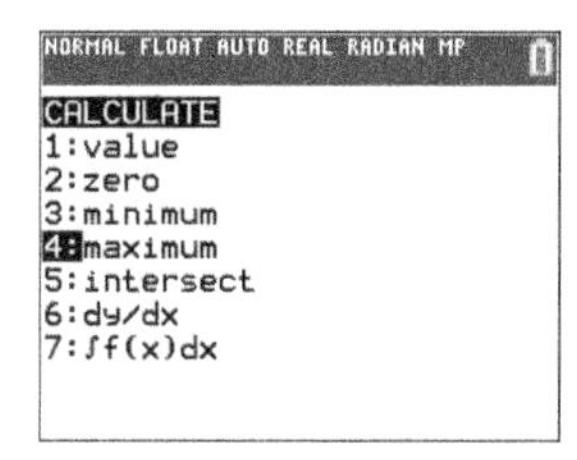

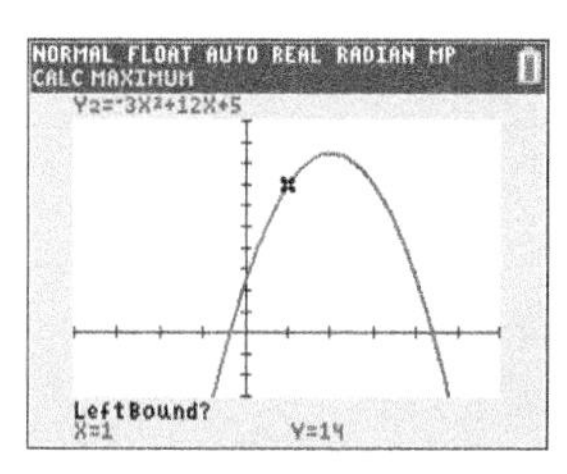

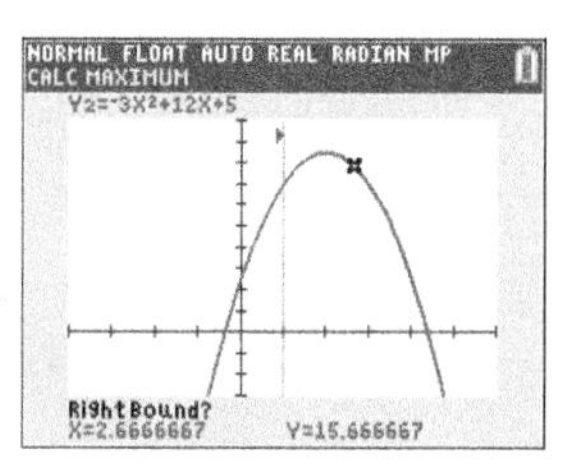

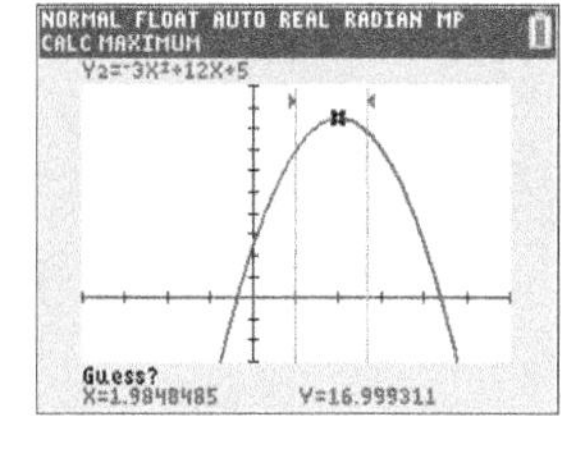

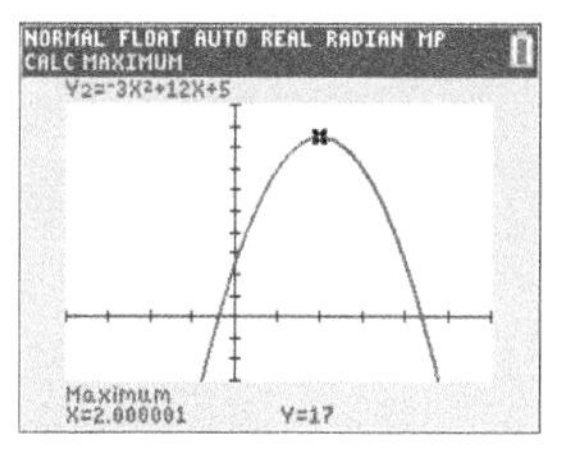

For further help with the TI-84 Plus C calculator, see Appendix C.

8. a. Use your graphing calculator to determine the vertex of the parabola having equation $y = -0.014630x^2 + x + 6$ (the shot put function).

(34.18, 23.09)

b. How do the coordinates you determined using your graphing calculator compare with your results in Problem 7?

They are the same.

c. What is the practical meaning of the coordinates of the vertex in this situation?

The coordinates represent the highest point of the shot during its flight.

Axis of Symmetry of a Parabola

DEFINITION

The **axis of symmetry** of a parabola is a vertical line that divides the parabola into two symmetric parts that are mirror images about the line.

EXAMPLE 2 ***Consider the parabola having equation $y = -3x^2 + 12x + 5$ from Example 1. The axis of symmetry of the parabola is $x = 2$. Note that the line of symmetry passes through the vertex of the parabola.***

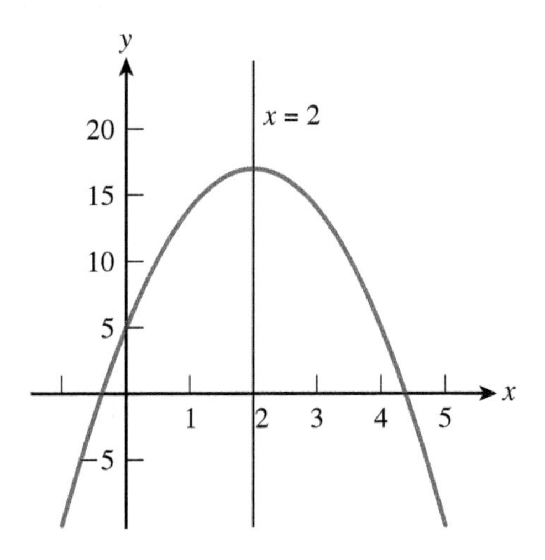

Because the vertex (turning point) of a parabola lies on the axis of symmetry, the equation of the axis of symmetry is

$$x = \frac{-b}{2a}.$$

9. What is the axis of symmetry of the shot put function, H?

$x = \dfrac{-b}{2a}$; $x = \dfrac{-1}{2(-0.014630)} \approx 34.17635$

Intercepts of the Graph of a Parabola

The y-intercept (vertical intercept) of the graph of the parabola defined by $y = -3x^2 + 12x + 5$ (see Example 1) can be determined directly from the equation. If $x = 0$, then

$$y = -3(0)^2 + 12(0) + 5 = 5,$$

and the y-intercept is $(0, 5)$.

In general, the y-intercept of the parabola defined by $y = ax^2 + bx + c$ is $(0, c)$.

In Example 1, it was determined that (2,17) is the vertex of the parabola having the equation $y = -3x^2 + 12x + 5$. The point (2, 17) is above the x-axis (the y-coordinate is positive) and the parabola opens downward. The parabola must intersect the horizontal axis in two places. This is verified by the following graph:

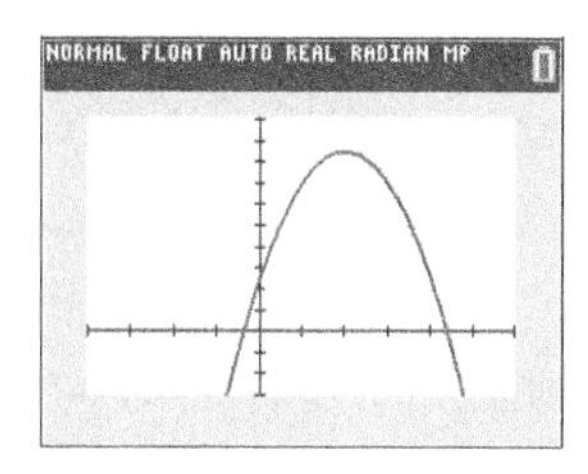

The x-intercepts of $y = -3x^2 + 12x + 5$ can be determined using the zero option in the CALC menu of your graphing calculator. Follow the prompts to obtain one x-intercept at a time. The screens should appear as follows:

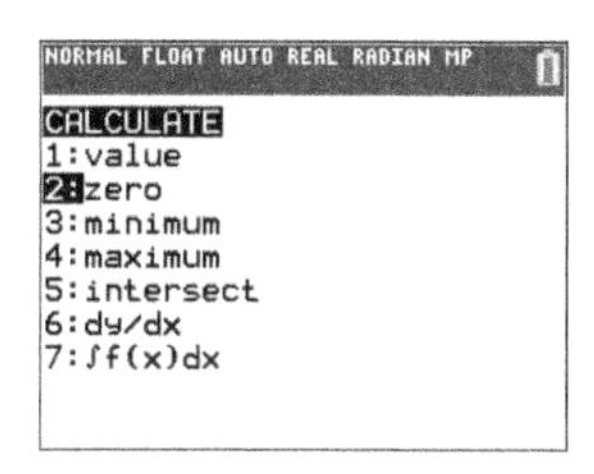

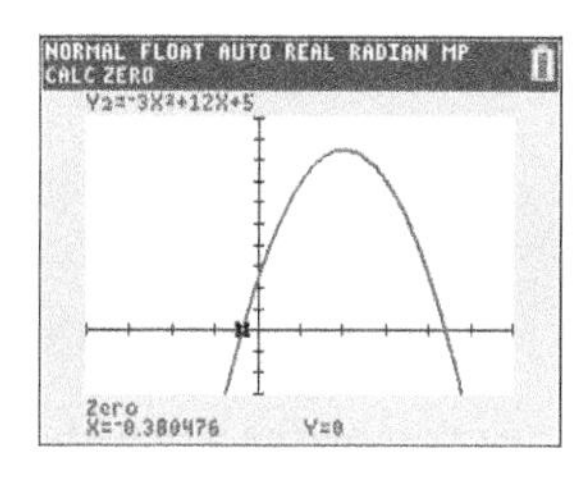

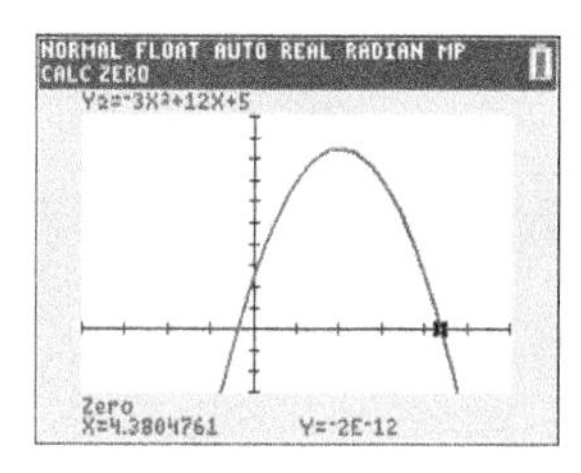

For further help with the TI-84 Plus C, see Appendix C.

10. a. Use the zero option of your graphing calculator to approximate the x-intercept(s) for the shot put function having equation $y = -0.014630x^2 + x + 6$. The rightmost intercept appears in the following screen:

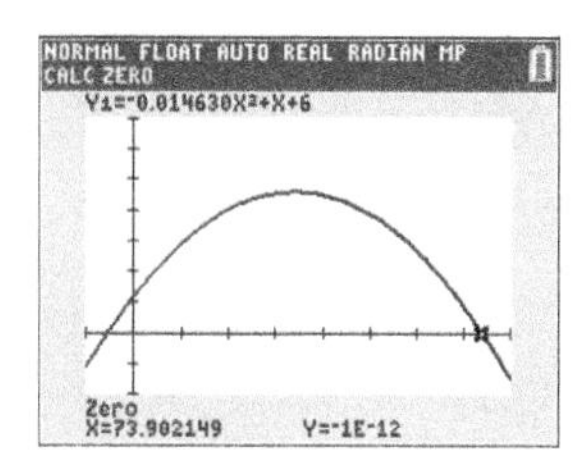

(73.902149, 0); (−5.549449, 0)

b. Is either x-intercept determined in part a significant to the problem situation? Explain.

Yes, the rightmost x-intercept represents the landing point of the shot. But the leftmost x-intercept, being negative, has no significance.

11. a. Use your result from Problem 10 to determine the practical domain of the shot put function. How does this compare with your answer in Problem 3a?

$0 \leq x \leq 73.9$. They are the same.

b. Sketch the path of the winning throw of the shot put. Be sure to label all key points, including the vertex and intercepts.

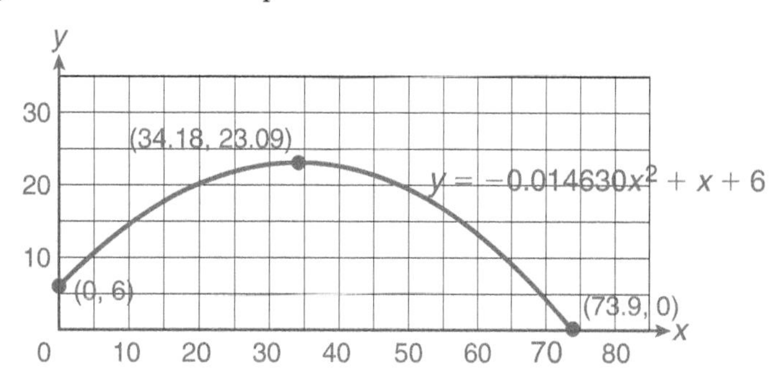

c. From the graph of the winning throw, over what horizontal distance (x-interval) is the height of the shot put increasing?

$0 < x < 34.18$ ft.

d. Determine the x-interval over which the height of the shot put is decreasing.

$34.18 < x < 73.90$ ft.

e. What is the practical range?

$0 \le y \le 23.09$ ft.

The graph of $H(x) = -0.014630x^2 + x + 6$ has two x-intercepts. Does the graph of every parabola have x-intercepts? Problems 12 and 13 will help answer this question.

LC LEARNING CATALYTICS

Use the values of a, b, and c to determine the coordinates of the vertex of the graph of $y = x^2 - 2x + 2$.

12. a. Use the values of a, b, and c to determine the coordinates of the vertex of the graph of $y = x^2 + 6x + 12$.

$a = 1, b = 6, c = 12; \left(-\frac{b}{2a}, f\left(-\frac{b}{2a}\right)\right) = (-3, f(-3)) = (-3, 3)$

b. Use the y-coordinate of the vertex to determine whether the vertex is above or below the x-axis.

The y-coordinate of the vertex is 3, so it is above the x-axis.

c. Use the value of a in $y = x^2 + 6x + 12$ to determine whether the parabola opens upward or downward.

$a = 1$. The parabola opens upward because a is positive.

12. e.

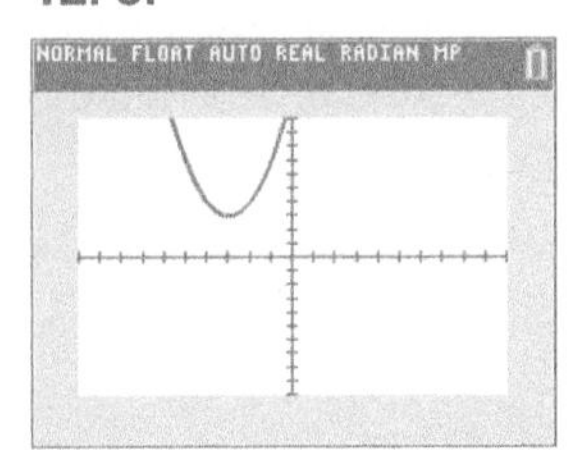

d. Use the results from parts b and c to determine whether the parabola has x-intercepts.

The vertex is above the x-axis, and the parabola opens upward. Therefore, there are no x-intercepts.

e. Use your graphing calculator to verify your answer to part d.

13. a. Use the values of a, b, and c to determine the coordinates of the vertex of the graph of $y = -x^2 + 8x - 21$.

$a = -1, b = 8, c = -21; \left(-\frac{b}{2a}, f\left(-\frac{b}{2a}\right)\right) = (4, f(4)) = (4, -5)$

b. Use the y-coordinate of the vertex to determine whether the vertex is above or below the x-axis.

The y-coordinate of the vertex is -5, so it is below the x-axis.

c. Use the value of a in $y = -x^2 + 8x - 21$ to determine whether the parabola opens upward or downward.

$a = -1$. The parabola opens downward.

d. Use the results from parts b and c to determine whether the parabola has x-intercepts.

The vertex is below the x-axis, and the parabola opens downward. Therefore, there are no x-intercepts.

13. e.

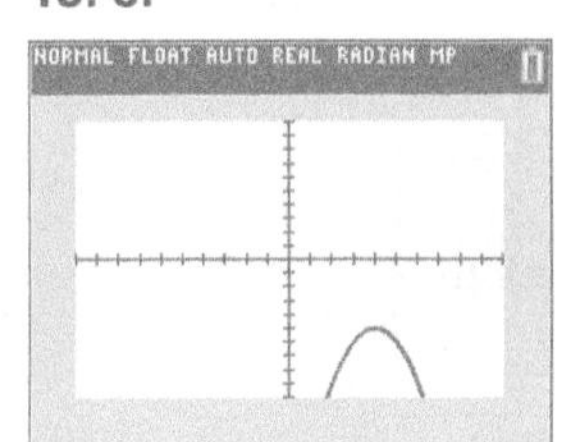

e. Use your graphing calculator to verify your answer to part d.

If a parabola opens upward and the vertex is above the x-axis, there are no x-intercepts. If a parabola opens downward and the vertex is below the x-axis, there are no x-intercepts.

SUMMARY Activity 4.2

The following characteristics are commonly used in analyzing the quadratic function defined by $f(x) = ax^2 + bx + c, a \neq 0$, and its graph.

1. The **axis of symmetry** is a vertical line that separates the parabola into two mirror images. The equation of the vertical axis of symmetry is given by $x = -\frac{b}{2a}$.

2. The **vertex** (turning point) always falls on the axis of symmetry. The x-coordinate of the vertex is given by $-\frac{b}{2a}$. Its y-coordinate is determined by evaluating the function at this x-value.

   In other words, the y-coordinate of the vertex is given by $f\left(-\frac{b}{2a}\right)$.

3. If the parabola changes from increasing to decreasing at the turning point, then the vertex is a maximum point. If the parabola changes from decreasing to increasing at the turning point, then the vertex is a minimum point.

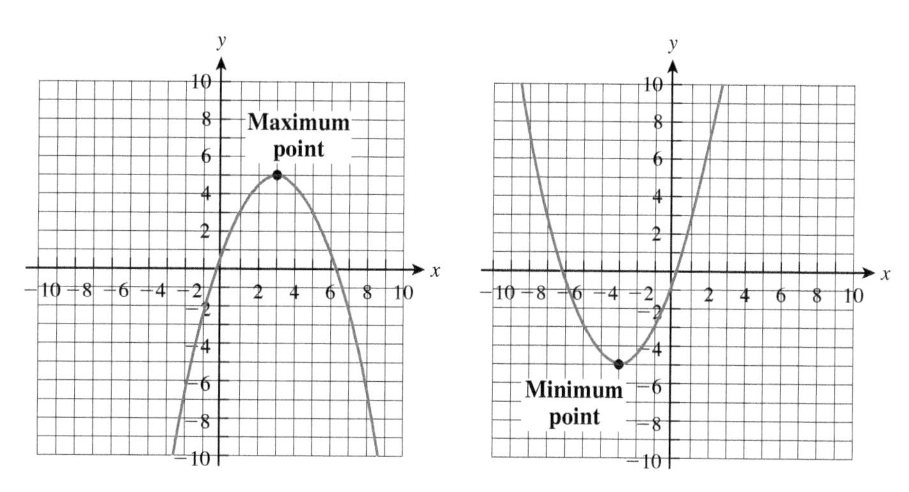

4. The **vertical intercept**, the point where the parabola crosses the y-axis (i.e., where its x-coordinate is zero), is always given by $(0, c)$.

5. A **horizontal intercept** is a point or points (if any) where the parabola crosses the x-axis (i.e., where its y-coordinate is zero).

6. If a parabola opens upward and the vertex is above the x-axis, there are no x-intercepts. If a parabola opens downward and the vertex is below the x-axis, there are no x-intercepts.

7. The **domain** of a general quadratic function is the set of all real numbers.

8. If the parabola opens upward, the **range** is all real numbers greater than or equal to the output value of the vertex. If the parabola opens downward, the *range* is all real numbers less than or equal to the output value of the vertex.

EXERCISES Activity 4.2

For Exercises 1–8, determine the following characteristics of each quadratic function:

a. *The direction in which the graph opens*

b. *The axis of symmetry*

c. *The turning point (vertex), determining whether it is the maximum or minimum*

d. *The y-intercept*

1. $f(x) = x^2 - 3$
   a. upward
   b. $x = 0$
   c. $(0, -3)$; minimum
   d. $(0, -3)$

2. $g(x) = x^2 + 2x - 8$
   a. upward
   b. $x = -1$
   c. $(-1, -9)$; minimum
   d. $(0, -8)$

3. $y = x^2 + 4x - 3$
   a. upward
   b. $x = -2$
   c. $(-2, -7)$; minimum
   d. $(0, -3)$

4. $f(x) = 3x^2 - 2x$
   a. upward
   b. $x = \frac{1}{3}$
   c. $\left(\frac{1}{3}, -\frac{1}{3}\right)$; minimum
   d. $(0, 0)$

5. $h(x) = x^2 + 3x + 4$
   a. upward
   b. $x = -1.5$
   c. $(-1.5, 1.75)$; minimum
   d. $(0, 4)$

6. $g(x) = -x^2 + 7x - 6$
   a. downward
   b. $x = 3.5$
   c. $(3.5, 6.25)$; maximum
   d. $(0, -6)$

7. $y = 2x^2 - x - 3$
   a. upward
   b. $x = 0.25$
   c. $(0.25, -3.125)$; minimum
   d. $(0, -3)$

8. $f(x) = x^2 + x + 3$
   a. upward
   b. $x = -0.5$
   c. $(-0.5, 2.75)$; minimum
   d. $(0, 3)$

For Exercises 9–16, use your graphing calculator to sketch the graphs of the functions; then determine each of the following:

a. *The coordinates of the x-intercepts for each function, if they exist*

b. *The domain and range for each function*

c. *The horizontal interval over which each function is increasing*

d. *The horizontal interval over which each function is decreasing*

9. $g(x) = -x^2 + 7x - 6$
   a. $(1, 0)$, $(6, 0)$
   b. D: all real numbers;
   R: $g(x) \leq 6.25$
   c. $x < 3.5$
   d. $x > 3.5$

10. $h(x) = 3x^2 + 6x + 4$
    a. none
    b. D: all real numbers;
    R: $h(x) \geq 1$
    c. $x > -1$
    d. $x < -1$

11. $y = x^2 - 12$

a. $(3.46, 0), (-3.46, 0)$

b. D: all real numbers; R: $y \geq -12$

c. $x > 0$

d. $x < 0$

12. $f(x) = x^2 + 4x - 5$

a. $(1, 0), (-5, 0)$

b. D: all real numbers; R: $f(x) \geq -9$

c. $x > -2$

d. $x < -2$

13. $g(x) = -x^2 + 2x + 3$

a. $(-1, 0), (3, 0)$

b. D: all real numbers; R: $g(x) \leq 4$

c. $x < 1$

d. $x > 1$

14. $h(x) = x^2 + 2x - 8$

a. $(2, 0), (-4, 0)$

b. D: all real numbers; R: $h(x) \geq -9$

c. $x > -1$

d. $x < -1$

15. $y = -5x^2 + 6x - 1$

a. $(0.2, 0), (1, 0)$

b. D: all real numbers; R: $y \leq 0.8$

c. $x < 0.6$

d. $x > 0.6$

16. $f(x) = 3x^2 - 2x + 1$

a. none

b. D: all real numbers; R: $f(x) \geq \frac{2}{3}$

c. $x > \frac{1}{3}$

d. $x < \frac{1}{3}$

17. You shoot an arrow vertically into the air from a height of 5 feet with an initial velocity of 96 feet per second. The height, h, in feet above the ground, at any time, t (in seconds), is modeled by

$$h(t) = 5 + 96t - 16t^2.$$

a. Determine the maximum height the arrow will attain.

149 ft.

b. Approximately when will the arrow reach the ground?

6.05 sec.

c. What is the significance of the vertical intercept?

It indicates the height of the arrow when it is shot.

d. What are the practical domain and practical range in this situation?

The practical domain is 0 second $\leq x \leq$ 6.05 seconds. The practical range is 0 feet $\leq h(x) \leq$ 149 feet.

e. Use your graphing calculator to determine the horizontal intercepts. Determine the practical meaning of these intercepts in this situation.

$(-0.05, 0), (6.05, 0)$. The first has no meaning. The second indicates the time in seconds it takes the arrow to hit the ground.

18. As part of a recreational waterfront grant, the city council plans to enclose a rectangular area along the waterfront of Lake Erie and create a park and swimming area. The budget calls for the purchase of 3000 feet of fencing. (**Note:** There is no fencing along the lake.)

a. Draw a picture of the planned recreational area. Let x represent the length of one of the two equal sides that are perpendicular to the water.

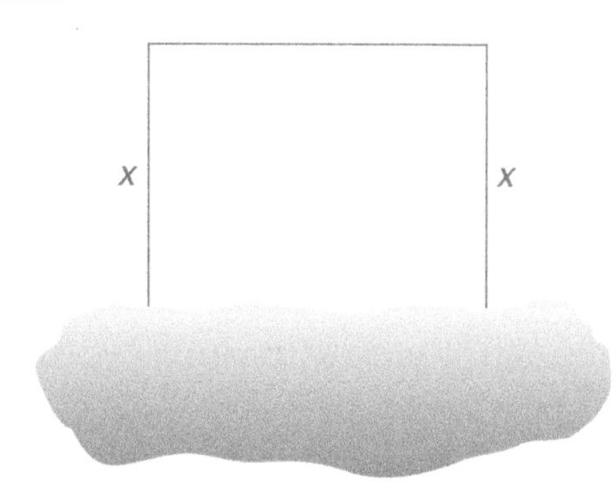

b. Write an expression that represents the width (side opposite the water) in terms of x. (**Note:** You have 3000 feet of fencing.)

$3000 - 2x$

c. Write an equation that expresses the area, $A(x)$, of this rectangular site as a quadratic function of x.

$A(x) = (3000 - 2x)x = -2x^2 + 3000x$

d. Determine the value of x for which $A(x)$ is a maximum.

$$x = -\frac{b}{2a} = \frac{-3000}{-4} = 750$$

e. What is the maximum area that can be enclosed?

$A(750) = 1{,}125{,}000$ sq. ft.

f. What are the dimensions of the maximum enclosed area?

750 ft. × 1500 ft.

g. Use your graphing calculator to graph the area function. What point on the graph represents the maximum area?

The vertex is (750, 1,125,000). This shows the maximum area 1,125,000, and the x-dimension, 750.

h. What is the vertical intercept? Does this point have any practical meaning in this situation?

(0, 0). It has no practical meaning; $x = 0$ is not in the practical domain.

i. From the graph, determine the horizontal intercepts. Do they have any practical meaning in this situation? Explain.

(0, 0), (1500, 0). No, neither $x = 0$ nor $x = 1500$ is in the practical domain of this function because they produce an area of zero.

19. a.

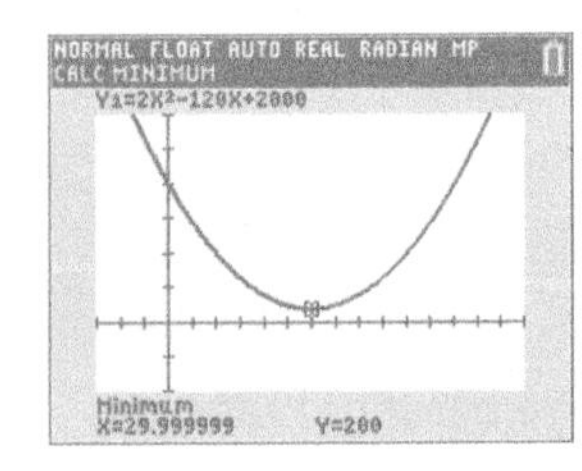

19. The average cost to produce pewter oil lamps is given by

$$\overline{C}(x) = 2x^2 - 120x + 2000,$$

where x represents the number of oil lamps produced and $\overline{C}(x)$ is the average cost of producing x oil lamps.

a. Use your graphing calculator to graph the average cost function and determine the coordinates of the turning point.

(30, 200)

b. Determine the vertex algebraically.

$$x = -\frac{b}{2a} = \frac{120}{4} = 30, \overline{C}(30) = 200.$$ The vertex is (30, 200).

c. How do your answers in parts a and b compare?

They are the same.

d. Is the vertex a minimum or maximum point?

minimum point

e. What is the practical meaning of the vertex in this situation?

The average cost of production is minimized when 30 oil lamps are produced.

f. What is the vertical intercept? What is the practical meaning of this intercept?

(0, 2000). It costs $2000 even when no oil lamps are produced.

20. You are selling ceramic lawn ornaments. After several months, your accountant tells you that your profit, $P(n)$, can be modeled by

$$P(n) = -0.002n^2 + 5.5n - 1200,$$

where n is the number of ornaments sold each month.

20. a.

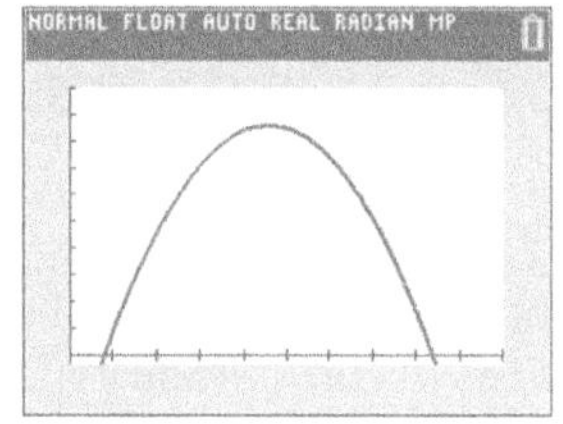

a. Use your graphing calculator to produce a graph of this function. Use the table feature set at TblStart $= 0$ and ΔTbl $= 500$ to help you set your window. Include the x-intercepts and the vertex.

b. Determine the x-intercepts of the graph of the profit function.

(239, 0) and (2511, 0) rounded to the nearest whole number

c. Determine the practical domain of the profit function.

$0 \leq n \leq 2511$ (Answers will vary.)

d. Determine the practical range of the profit function.

$-1200 \leq P(n) \leq 2581.25$ (Answers will vary.)

e. How many ornaments must be sold to maximize the profit?

1375

f. Write the equation that must be solved to determine the number of ornaments that need to be sold to produce a profit of $2300.

$-0.002n^2 + 5.5n - 1200 = \$2300$

20. g.

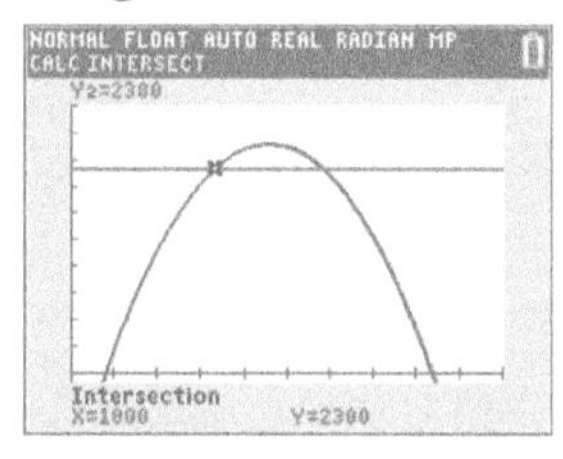

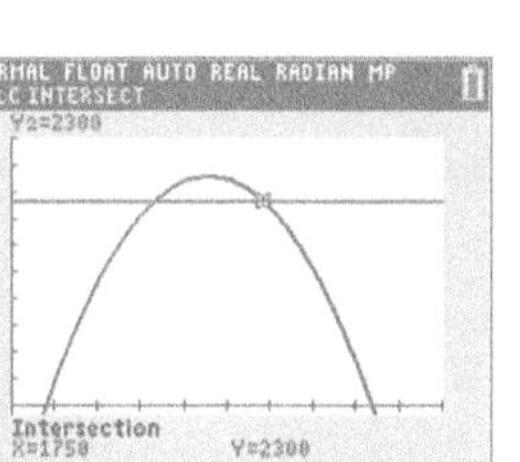

g. Solve the equation in part f graphically.

1000 or 1750

ACTIVITY 4.3

Spotify
Solving Quadratic Equations Numerically and Graphically

OBJECTIVES

1. Solve quadratic equations numerically.
2. Solve quadratic equations graphically.
3. Solve quadratic inequalities graphically.

Spotify is a music, podcast, and video streaming service. It was developed in Stockholm, Sweden, and was officially launched on October 7, 2008. Since then it has posted very impressive revenue reports. From 2009 through 2016, Spotify's annual revenue can be modeled by

$$R(t) = 76.9t^2 - 148.9t + 99.7$$

where t represents the number of years since 2009 and R is Spotify's revenue measured in millions of euros.

1. What is the practical domain for the model represented by the function R?

   The practical domain is values of t with $0 \le t \le 7$.

2. Use your graphing calculator to complete the following table of values for t, the number of years from 2009, and $R(t)$, Spotify's annual income in millions of euros. Round your results to the nearest whole number.

YEAR	2009	2010	2011	2012	2013	2014	2015	2016
t	0	1	2	3	4	5	6	7
$R(t)$	100	28	110	345	735	1278	1975	2826

3.

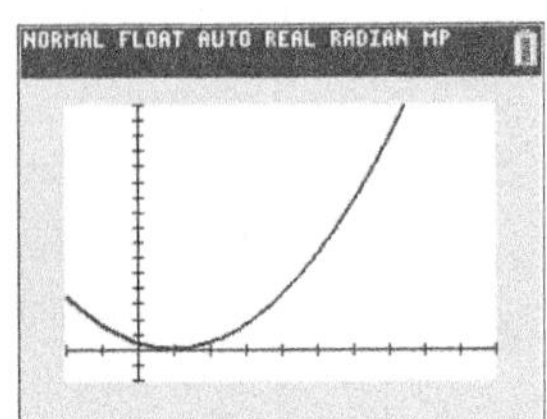

3. Sketch a graph of the function using your graphing calculator and the window Xmin = –2, Xmax = 10, Ymin = –400, Ymax = 3200. The graph appears to the left.

4. Use the model to estimate Spotify's revenue in 2019 ($t = 10$).

   $R(10) = 76.9 \cdot 10^2 - 148.9 \cdot 10 + 99.7 = 6301$ million euros

5. You want to determine the year that Spotify's Revenue will first reach 10 billion euros. In other words, for what value of t will $R(t) = 10{,}000$? Write an equation to determine the value of t when $R(t) = 10{,}000$.

   $76.9t^2 - 148.9t + 99.7 = 10{,}000$

The equation in Problem 5 is a quadratic equation because it involves a polynomial expression of degree 2.

The general form of a quadratic equation is $ax^2 + bx + c = 0, a \neq 0$.

EXAMPLE 1 ***Solve the quadratic equation $x^2 + 3x - 1 = 9$ numerically (using tables of data).***

SOLUTION

Create a table in which x is the input and $y = x^2 + 3x - 1$ is the output. The solution to the equation is the x-value corresponding to a y-value of 9. Using the graphing calculator, the solution is $x = 2$.

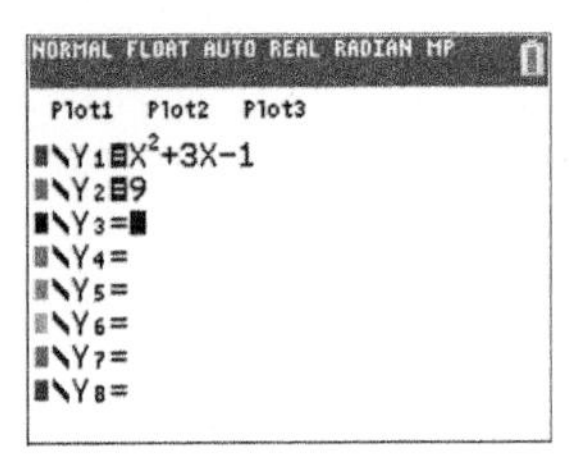

NORMAL FLOAT AUTO REAL RADIAN MP
PRESS + FOR △Tbl

X	Y1	Y2
1	3	9
2	9	9
3	17	9
4	27	9
5	39	9
6	53	9
7	69	9
8	87	9
9	107	9
10	129	9
11	153	9

X=2

A second solution is $x = -5$; try it yourself.

6. Determine the solution to $76.9t^2 - 148.9t + 99.7 = 10{,}000$ numerically using a table of appropriate data points (see Problem 2). What is your approximation using this approach?

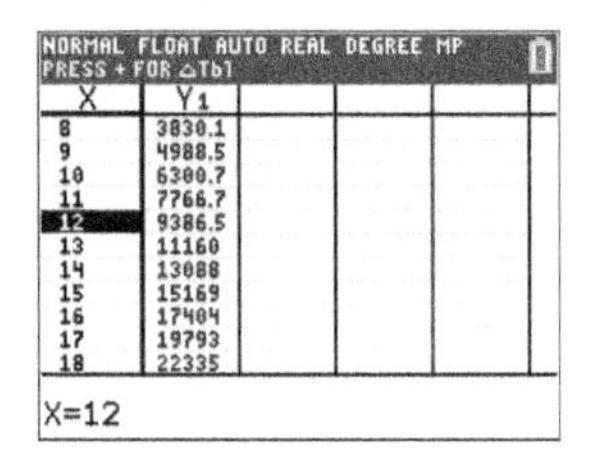
NORMAL FLOAT AUTO REAL DEGREE MP
PRESS + FOR △Tbl

X	Y1
8	3830.1
9	4988.5
10	6300.7
11	7766.7
12	9386.5
13	11160
14	13088
15	15169
16	17404
17	19793
18	22335

X=12

$t \approx 12(2021)$

Solving Quadratic Equations Graphically

A second method of solving the quadratic equation in Problem 5 is a graphical approach using your graphing calculator. Recall from Chapter 1 that you can solve the equation $76.9t^2 - 148.9t + 99.7 = 10{,}000$ by solving the following system of equations graphically:

$$Y_1 = 76.9t^2 - 148.9t + 99.7$$
$$Y_2 = 10{,}000$$

The expression for Y_1 gives the revenue in any given year. The value Y_2 is the specific revenue in which you are interested. The solution to the equation is the t-value for which $Y_1 = Y_2$. To do this, determine the point of intersection of these two graphs. Using the intersect option under the CALC menu, the graph should appear as follows:

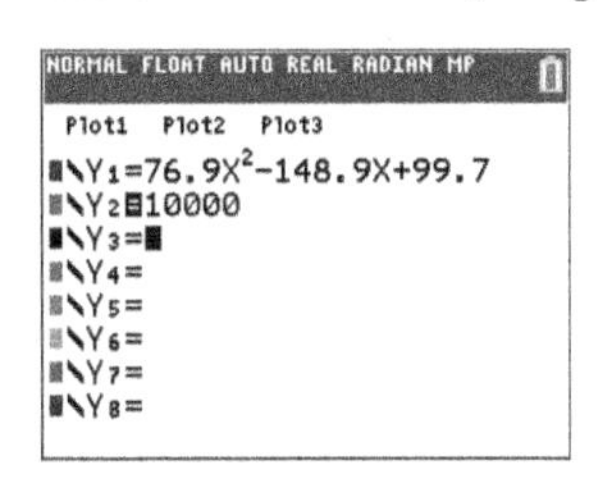

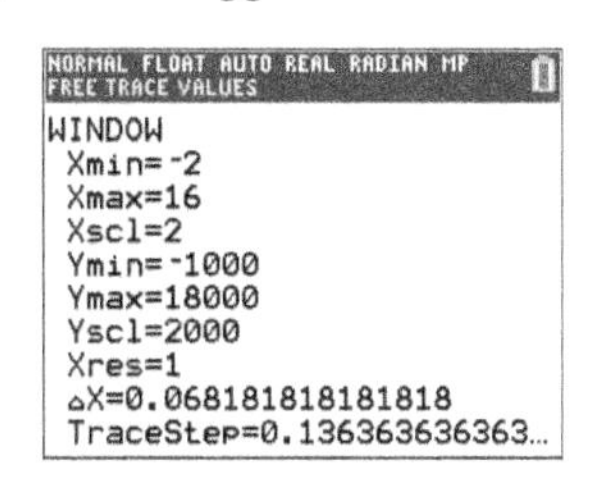

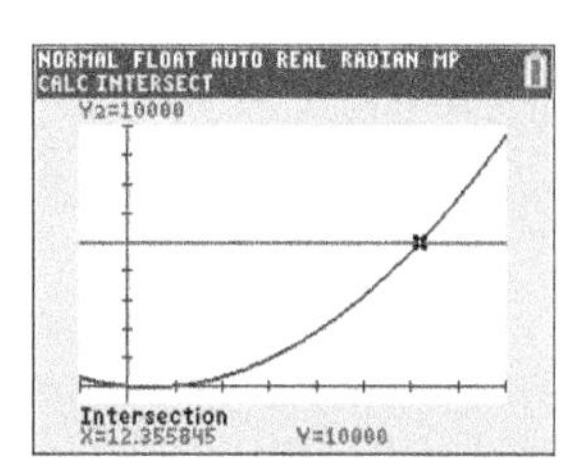

Another graphical method for solving the problem is to rearrange the quadratic equation $76.9t^2 - 148.9t + 99.7 = 10{,}000$ (1) so that the right-hand side is equal to zero. Subtracting 10,000 from each side, you have

$$76.9t^2 - 148.9t - 9900.3 = 0.$$

If you let $y = 76.9t^2 - 148.9t - 9900.3$, then the solution to equation (1) is the t-value for which $y = 0$, if it exists. This is the t-value of the t-intercept of the graph, also called the **zero** of the function.

7. a. Use your graphing calculator to sketch a graph of $y = 10.4t^2 + 263.0t - 19{,}152.2$. Use the window Xmin $= -20$, Xmax $= 20$, Ymin $= -16{,}000$, Ymax $= 20{,}000$. The screen should appear as follows:

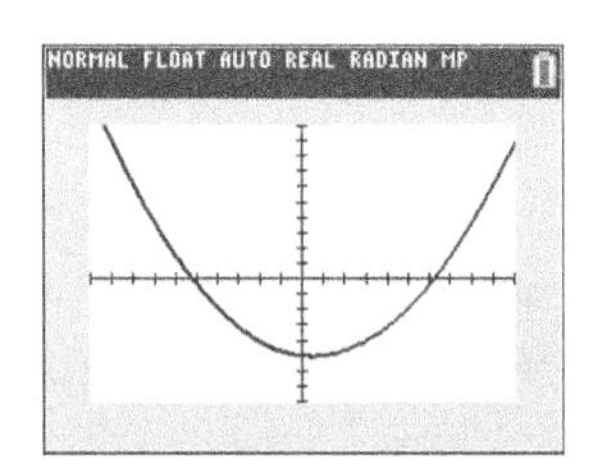

b. Use the zero option of the CALC menu to determine the t-intercepts of the new function defined by $y = 76.9t^2 - 148.9t - 9900.3$.

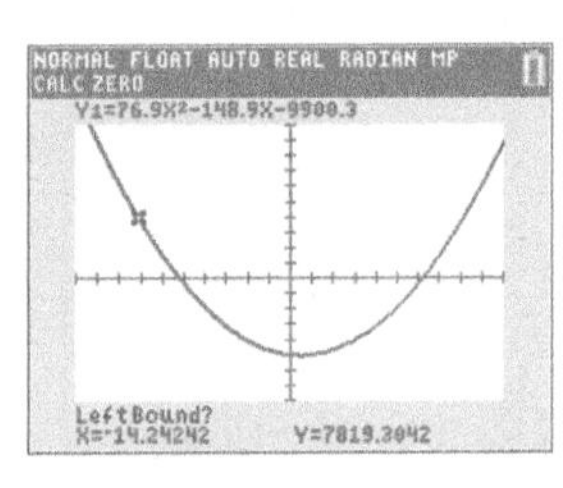

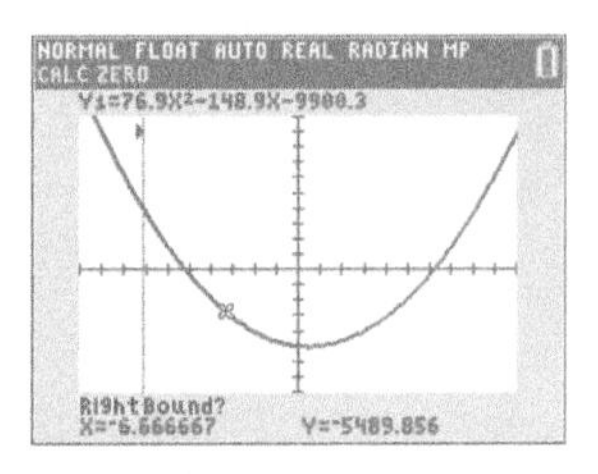

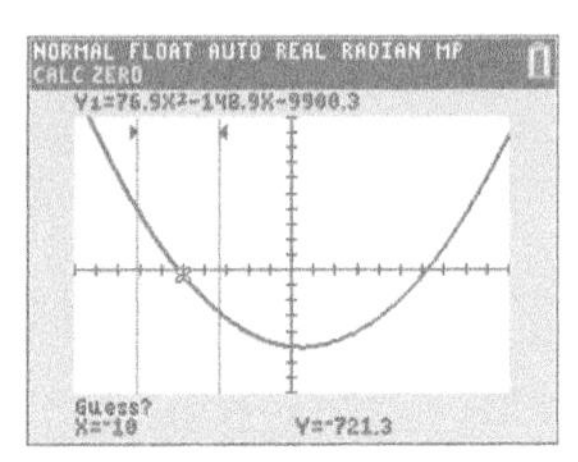

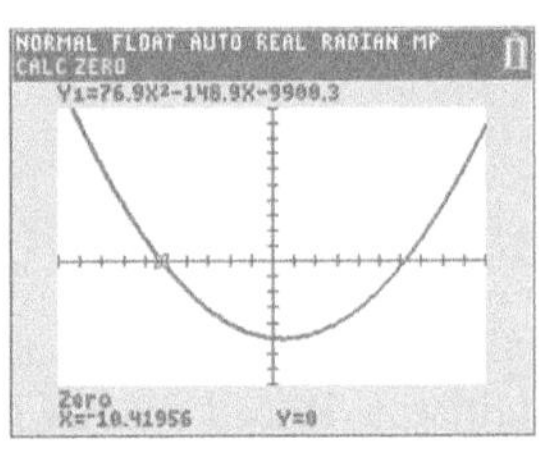

One solution is approximately (–10.4, 0).

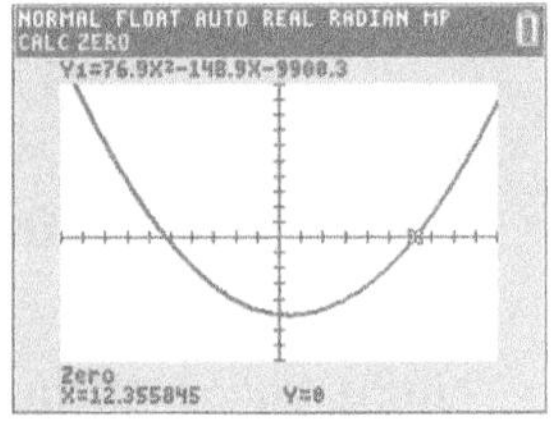

Another solution is approximately (12.4, 0).

c. Using the results from part b, determine whether the solutions are both of the values relevant to our problem? Explain.

The solution $t \approx 12.4$ is relevant and corresponds to the result from Problem 6. The solution $t \approx -10.4$ represents 10.4 years before 2009. This is before Spotify launched, so this solution is not relevant.

8. Describe two different ways to solve the equation $2x^2 - 4x + 3 = 2$ using a graphing approach. Solve the equation using each graphing method. How do your answers compare?

a. Graph $Y_1 = 2x^2 - 4x + 3$ and $Y_2 = 2$. Use the intersection feature of my graphing calculator.

b. Graph $Y_1 = 2x^2 - 4x + 1$. Use the zero feature of my graphing calculator to locate the x-intercepts.

c. Solution: $x = 0.29$ or $x = 1.71$. The solutions are the same using either method.

Solving Quadratic Inequalities Graphically

You are interested in determining in which years the U.S. per capita personal income was more than \$15,000. To answer this problem, you need to solve the inequality

$$10.4t^2 + 263.0t + 1347.8 > 15{,}000,$$

where t equals the number of years since 1960.

The following example demonstrates a procedure for solving an inequality similar to this one.

EXAMPLE 2 ***Solve the inequality $2x^2 - 4x + 3 > 7$ using a graphing approach.***

SOLUTION

Form the following system of equations:

$$Y_1 = 2x^2 - 4x + 3$$
$$Y_2 = 7$$

and graph each equation. The screen should resemble the following:

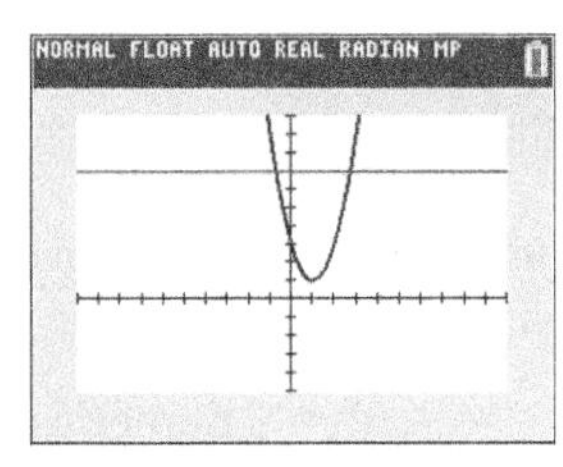

Use the intersection option on the CALC menu to determine where $Y_1 = Y_2$.

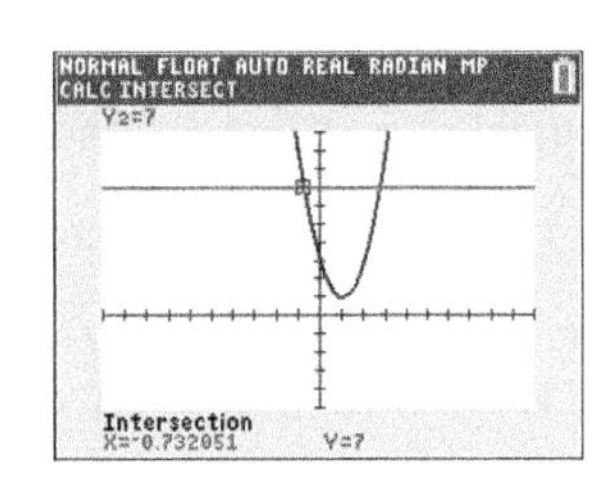

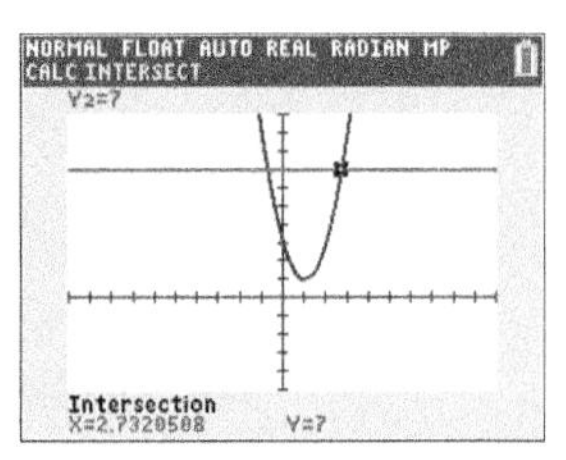

The solutions to the inequality are the values of x where $Y_1 > Y_2$ (i.e., the x-values of points where the graph of Y_1 is above the graph of Y_2). Therefore, the solutions are $x < -0.732$ or $x > 2.732$.

If the problem had been the reverse inequality $2x^2 - 4x + 3 < 7$, then the solution would be x-values of points where the graph of Y_1 is below the graph of Y_2. The solution would be $-0.732 < x < 2.732$.

9. a. Solve the inequality $10.4t^2 + 263.0t + 1347.8 > 15{,}000$ using a graphing approach. Be careful. The practical domain is $t \geq 0$.

The parabola is above the line if $t > 25.73$ or 26.

b. In which years was the per capita personal income more than \$15,000?

The per capita income was more than \$15,000 from sometime during 1986 until the present time.

c. There are negative values of t for which $Y_1 > 15{,}000$. Determine them graphically.

$Y_1 > 15{,}000$ if $t < -51.02$.

d. Explain why the values you determined in part c are not relevant to the original problem situation.

The numbers would not be relevant because they indicate that per capita income was greater than \$15,000 for all years before 1909, which we know is not true.

LC LEARNING CATALYTICS

Which of the following represents the solution of $x^2 - 2x - 3 < 0$?

a. $-3 < x < 1$
b. $x < -1$ or $x > 3$
c. $-1 < x < 3$
d. none of these

10. Solve the following quadratic inequalities using a graphing approach:

a. $x^2 - x - 6 < 0$

The graph of $y = x^2 - x - 6$ is below the x-axis for $-2 < x < 3$.

b. $x^2 - x - 6 > 0$

The graph of $y = x^2 - x - 6$ is above the x-axis if $x < -2$ or $x > 3$.

SUMMARY Activity 4.3

1. A **quadratic equation** is an equation involving polynomial expressions of degree 2. The standard form of a quadratic equation is $ax^2 + bx + c = 0, a \neq 0$.
2. To solve $f(x) = c$ **numerically**, construct a table and determine the x-values that produce c as an output.
3. To solve $f(x) = c$ **graphically**,
   a. graph $y = f(x)$, graph $y = c$, and determine the x-values of the points of intersection,
   b. or graph $y = f(x) - c$ and determine the x-intercepts.
4. To solve $f(x) > c$ graphically, graph $y = f(x)$, graph $y = c$, and determine all x-values for which the graph of f is above the graph of $y = c$.
5. To solve $f(x) < c$ graphically, graph $y = f(x)$, graph $y = c$, and determine all x-values for which the graph of f is below the graph of $y = c$.

EXERCISES Activity 4.3

In Exercises 1–4, solve the quadratic equation numerically (using tables of x- and y-values). Verify your solutions graphically.

1. $-4x = -x^2 + 12$
   $x = 6$ or $x = -2$
2. $x^2 + 9x + 18 = 0$
   $x = -3$ or $x = -6$
3. $2x^2 = 8x + 90$
   $x = 9$ or $x = -5$
4. $x^2 - x - 3 = 0$
   $x \approx 2.3$ or $x \approx -1.3$

In Exercises 5–8, solve the quadratic equation graphically using at least two different approaches. When necessary, give your solutions to the nearest hundredth.

5. $x^2 + 12x + 11 = 0$
   $x = -11$ or $x = -1$
6. $2x^2 - 3 = 2x$
   $x \approx 1.82$ or $x \approx -0.82$
7. $16x^2 - 400 = 0$
   $x = \pm 5$
8. $4x^2 + 12x = -4$
   $x \approx -0.38$ or $x \approx -2.62$

In Exercises 9–12, solve the equation by using either a numerical or graphical approach.

9. $x^2 + 2x - 3 = 0$
   $x = -3$ or $x = 1$
10. $x^2 + 11x + 24 = 0$
    $x = -8$ or $x = -3$
11. $x^2 - 2x - 8 = x + 20$
    $x = 7$ or $x = -4$
12. $x^2 - 10x + 6 = 5x - 50$
    $x = 7$ or $x = 8$

In Exercises 13–14, solve the given inequality using a graphing approach.

13. **a.** $x^2 - 4x - 1 < 11$
    $-2 < x < 6$
    b. $x^2 - 4x - 1 > 11$
    $x < -2$ or $x > 6$
14. **a.** $2x^2 + 5x - 3 < 0$
    $-3 < x < \frac{1}{2}$
    b. $2x^2 + 5x - 3 \geq 0$
    $x \leq -3$ or $x \geq \frac{1}{2}$

Exercise numbers appearing in color are answered in the Selected Answers appendix.

15. The stopping distance, d (in feet), for a car moving at a velocity (speed) v miles per hour is modeled by the equation

$$d(v) = 0.04v^2 + 1.1v.$$

a. What is the stopping distance for a velocity of 55 miles per hour?

$d(55) = 181.5$ ft.

b. What is the speed of the car if it takes 200 feet to stop?

$0.04v^2 + 1.1v = 200$; $v \approx 58$ mph

16. An international rule for determining the number, n, of board feet (usable finished lumber) in a 16-foot log is modeled by the equation

$$n(d) = 0.22d^2 - 0.71d,$$

where d is the diameter of the log in inches.

a. How many board feet can be obtained from a 16-foot log with a 14-inch diameter?

$n(14) = 0.22(14)^2 - 0.71(14) \approx 33$ board feet

b. Sketch a graph of this function. What is the practical domain of this function?

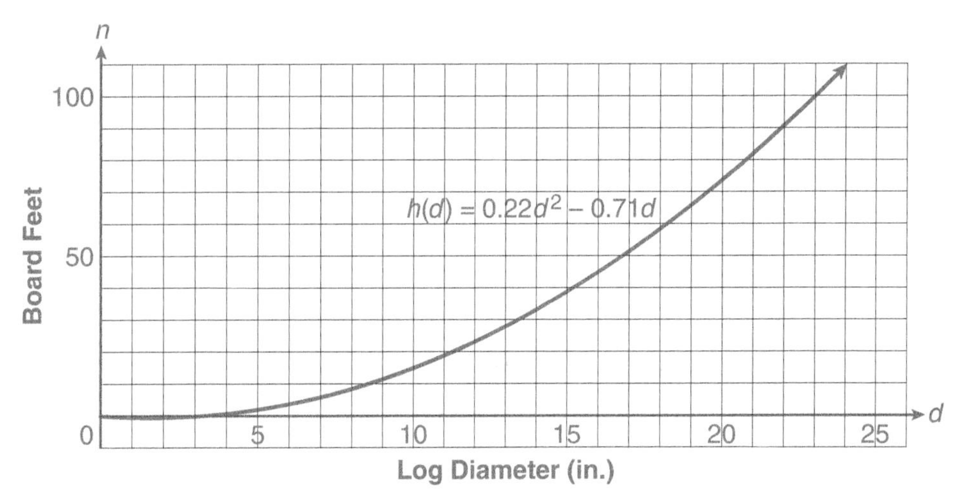

The practical domain is $3.2 < d < D$, where D is the maximum diameter of a log.

c. Use the graph to approximate the horizontal intercept(s). What is the practical meaning in this situation?

$d = 0$ and $d \approx 3.2$

The rule $n(d) = 0.22d^2 - 0.71d$ is valid only for logs whose diameters are greater than 3.2 inches.

d. What is the diameter of a 16-foot log that has 200 board feet?

$0.22d^2 - 0.71d = 200$; d is approximately 32 inches.

e. What inequality would you solve to determine the diameter when the board feet are, at most, 200?

$0.22d^2 - 0.71d \leq 200$

f. Solve the inequality by using the graph of the function.

3.2 in. $< d \leq 32$ in.

ACTIVITY 4.4

Sir Isaac Newton

Solving Quadratic Equations by Factoring

OBJECTIVES

1. Use the zero-product property to solve equations.
2. Factor expressions by removing the greatest common factor.
3. Factor trinomials using trial and error.
4. Solve quadratic equations by factoring.

Sir Isaac Newton XIV, a descendant of the famous physicist and mathematician, takes you to the top of a building to demonstrate a physics property discovered by his famous ancestor. He throws your math book straight up into the air. The book's distance, s, above the ground as a function of time, t, is modeled by

$$s = -16t^2 + 16t + 32.$$

1. When the book strikes the ground, what is the value of s?

   $s = 0$

2. Write the equation that you must solve to determine when the book strikes the ground.

   $-16t^2 + 16t + 32 = 0$

The quadratic equation in Problem 2 can be solved using a numerical or graphical approach. However, an algebraic technique is efficient in this case and will give an exact answer. The algorithm is based on the algebraic principle known as the **zero-product property**.

Zero-Product Principle

If a and b are any numbers and $a \cdot b = 0$, then either a or b or both must be equal to zero.

EXAMPLE 1 ***Solve the equation $x(x + 5) = 0$.***

SOLUTION

The two factors in this equation are x and $x + 5$. The zero-product property tells us that at least one of these factors must equal zero. That is,

$$x = 0 \quad \text{or} \quad x + 5 = 0.$$

The first equation tells you that $x = 0$ is a solution. To determine a second solution, solve $x + 5 = 0$.

$$\begin{aligned} x + 5 &= 0 \\ -5 \ &\ -5 \\ \hline x &= -5 \end{aligned}$$

There are two solutions, $x = 0$ and $x = -5$.

Your graphing calculator verifies the solutions as follows:

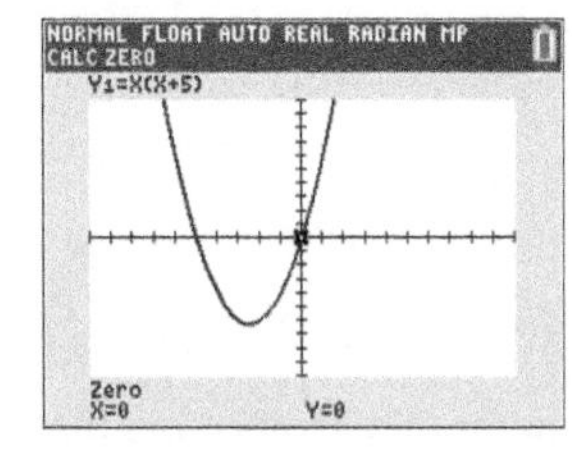

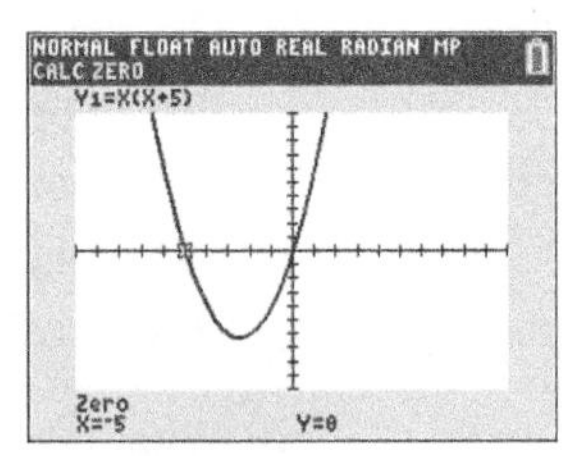

3. Solve each of the following equations using the zero-product property:

   a. $3x(x - 2) = 0$

   $3x = 0$ or $x - 2 = 0$

   $x = 0$ or $x = 2$

   b. $(2x - 3)(x + 2) = 0$

   $2x - 3 = 0$ or $x + 2 = 0$

   $x = \frac{3}{2}$ or $x = -2$

   c. $(x + 2)(x + 3) = 0$

   $x + 2 = 0$ or $x + 3 = 0$

   $x = -2$ or $x = -3$

For the zero-product property to be applied, one side of the equation must be zero. Therefore, at first glance, the zero-product property can be used to solve the quadratic equation $3x^2 - 6x = 0$. However, a second condition must be satisfied. The nonzero side of the equation must be written as a product.

The process of writing an expression such as $3x^2 - 6x$ as a product is called factoring.

DEFINITION

Rewriting an expression as a product is called **factoring**.

Factoring Common Factors

A **common factor** is a number or an expression that is a factor of each term of the entire expression. Whenever you want to factor a polynomial, look first for a common factor.

PROCEDURE

Removing a Common Factor from a Polynomial: First, identify the common factor, and then apply the distributive property in reverse.

EXAMPLE 2 ***Given the binomial $3x + 6$, 3 is a common factor because 3 is a factor of both terms $3x$ and 6. Applying the distributive property in reverse, you write***

$$3x + 6 \text{ as } 3(x + 2).$$

You can always check the factored binomial by multiplying.

$$3(x + 2) = 3(x) + 3(2) = 3x + 6$$

When you look for a common factor, determine the largest or **greatest common factor** (or GCF). You can see that 3 is a common factor of $6x + 24$ because 3 is a factor of both 6 and 24. However, there is a larger common factor, 6. Therefore,

$$6x + 24 = 6(x + 4).$$

EXAMPLE 3 ***Given $6x^2 + 14x - 30$, you can see that 2 is a common factor. Is 2 the greatest common factor? Yes, because no larger number is a factor of every term.***

If you divide each term by 2, you obtain $3x^2 + 7x - 15$. The expression $6x^2 + 14x - 30$ can now be written in factored form as $2(3x^2 + 7x - 15)$. Check the factored trinomial by multiplying.

EXAMPLE 4 ***Factor $4x^3 - 8x^2 + 28x$.***

SOLUTION

Four is a factor of each term, but x is as well. Therefore, the greatest common factor is $4x$. You remove the GCF by dividing each term by $4x$. This leads to the factored form $4x(x^2 - 2x + 7)$.

You can check your factoring by applying the distributive property.

4. Factor the following polynomials by removing the greatest common factor.

a. $9a^6 + 18a^2$

$9a^2(a^4 + 2)$

b. $21xy^3 + 7xy$

$7xy(3y^2 + 1)$

c. $3x^2 - 21x + 33$

$3(x^2 - 7x + 11)$

d. $4x^3 - 16x^2 - 24x$

$4x(x^2 - 4x - 6)$

Factoring Trinomials

With patience, you can factor trinomials of the form $ax^2 + bx + c$ by trial and error, using the FOIL Method in reverse.

PROCEDURE

Factoring Trinomials by Trial and Error

1. Remove the greatest common factor, GCF, if any.
2. To factor the resulting trinomial into the product of two binomials, try combinations of factors for the first and last terms in the two binomials.
3. Check the outer and inner products of the FOIL method to match the middle term of the original trinomial.
   a. If the constant term, c, is positive, both of its factors are positive or both are negative.
   b. If the constant term is negative, one factor is positive and one is negative.
4. If the check fails, repeat steps 2 and 3.

EXAMPLE 5 *Factor* $6x^2 - 7x - 3$.

SOLUTION

Step 1. There is no common factor, so go to step 2.

Step 2. You could factor the first term, $6x^2$, as $6x(x)$ or as $2x(3x)$. The last term, -3, has factors $3(-1)$ or $-3(1)$. Try $(2x + 1)(3x - 3)$.

Step 3. The outer product is $-6x$. The inner product is $3x$. The sum is $-3x$, not $7x$. The check fails.

Step 4. Try $(2x - 3)(3x + 1)$. The outer product is $2x$. The inner product is $-9x$. The sum is $-7x$. It checks.

5. Factor the following trinomials:

a. $x^2 - 7x + 12$

$(x - 3)(x - 4)$

b. $x^2 - 8x - 9$

$(x - 9)(x + 1)$

c. $x^2 + 14x + 49$

$(x + 7)(x + 7) = (x + 7)^2$

d. $25 + 10w + w^2$

$(5 + w)(5 + w) = (5 + w)^2$

Solving Quadratic Equations by Factoring

The following example demonstrates the procedure for solving quadratic equations written in standard form, $ax^2 + bx + c = 0$, by factoring.

LC LEARNING CATALYTICS

Solve the equation $2x^2 + 9x = 5$.

a. $x = \frac{1}{2}, -5$
b. $x = 1, -5$
c. $x = -\frac{1}{2}, 5$
d. none of these

EXAMPLE 6 ***Solve the equation $3x^2 - 2 = -x$ by factoring.***

Step 1. Rewrite the equation in the form $ax^2 + bx + c = 0$ (called general form).

$$\begin{array}{r} 3x^2 - 2 = -x \\ +x \qquad\quad +x \\ \hline 3x^2 + x - 2 = 0 \end{array}$$

Step 2. Factor the expression on the nonzero side of the equation.

$$(x + 1)(3x - 2) = 0$$

Step 3. Use the zero-product property to set each factor equal to zero, and then solve each equation.

$$(x + 1)(3x - 2) = 0$$

$$\begin{array}{r|r} x + 1 = 0 & 3x - 2 = 0 \\ x = -1 & 3x = 2 \\ & x = \frac{2}{3} \end{array}$$

Therefore, the solutions are $x = -1$ and $x = \frac{2}{3}$.

These solutions can be verified graphically as follows:

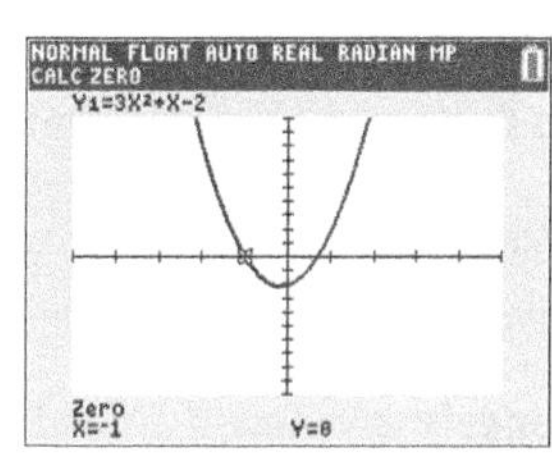

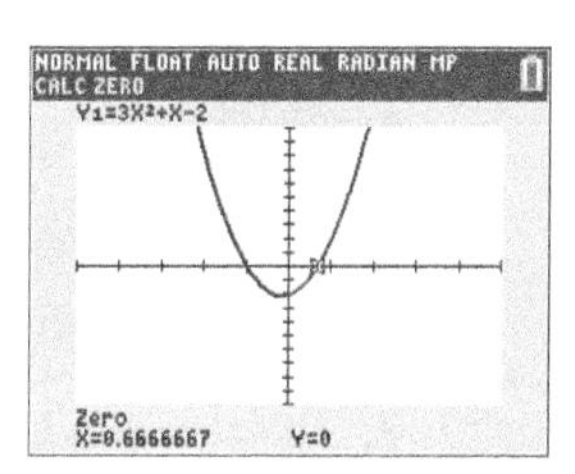

6. a. Returning to the math book problem from the beginning of this activity, solve the equation from Problem 2 by factoring.

$-16t^2 + 16t + 32 = 0$

$-16(t^2 - t - 2) = 0$

$-16(t - 2)(t + 1) = 0$

$$\begin{array}{r|r} t - 2 = 0 & t + 1 = 0 \\ t = 2 & t = -1 \end{array}$$

b. Are both solutions to the equation ($t = 2$ and $t = -1$) also solutions to the question, "At what time does the book strike the ground"? Explain.

Only $t = 2$ lies in the practical domain that starts at $t = 0$.

The book strikes the ground after 2 seconds.

7. a. You want to know at what time the book is 32 feet above the ground. Write a quadratic equation that represents this situation.

$-16t^2 + 16t + 32 = 32$

b. Solve the quadratic equation in part a by factoring.

$$-16t^2 + 16t + 32 = 32$$
$$-16t^2 + 16t = 0$$
$$-16t(t - 1) = 0$$

$-16t = 0$	$t - 1 = 0$
$t = 0$	$t = 1$

8. Solve each of the following quadratic equations by factoring:

a. $2x^2 - x - 6 = 0$

$(2x + 3)(x - 2) = 0$

$x = -\frac{3}{2}$ or $x = 2$

b. $3x^2 - 6x = 0$

$3x(x - 2) = 0$

$x = 0$ or $x = 2$

c. $x^2 + 4x = -x - 6$

$x^2 + 5x + 6 = 0$

$(x + 3)(x + 2) = 0$

$x = -3$ or $x = -2$

9. Determine the zeros of the function defined by $f(x) = 2x^2 - 3x + 1$.

$$2x^2 - 3x + 1 = 0$$
$$(x - 1)(2x - 1) = 0$$

$x - 1 = 0$	$2x - 1 = 0$
$x = 1$	$x = \frac{1}{2}$

SUMMARY Activity 4.4

1. To remove a **common factor** from a polynomial,

a. identify the common factor.

b. apply the distributive property in reverse.

2. The **zero-product property** says that if $ab = 0$ is a true statement, then either $a = 0$ or $b = 0$ or both.

3. To factor trinomials of the form $ax^2 + bx + c$ by **trial and error**,

a. remove the greatest common factor.

b. try combinations of factors for the first and last terms in two binomials.

c. check the outer and inner products of the FOIL method to match the middle term, bx, of the original trinomial:

- If the constant term, c, is positive, both factors of c are positive or both are negative.
- If the constant term is negative, one factor is positive and one is negative.

d. if the check fails, repeat steps 3b and 3c.

4. To solve equations by **factoring**,

a. use the addition principle to remove all terms from one side of the equation; this results in a polynomial being set equal to zero.

b. combine like terms and then factor the nonzero side of the equation.

c. use the zero-product property to set each factor containing a variable equal to zero and then solve the equations.

d. check your solutions in the original equation.

EXERCISES Activity 4.4

In Exercises 1–4, factor the polynomials by removing the GCF (greatest common factor).

1. $12x^5 - 18x^8$

$6x^5(2 - 3x^3)$

2. $14x^6y^3 - 6x^2y^4$

$2x^2y^3(7x^4 - 3y)$

3. $2x^3 - 14x^2 + 26x$

$2x(x^2 - 7x + 13)$

4. $5x^3 - 20x^2 - 35x$

$5x(x^2 - 4x - 7)$

In Exercises 5–13, completely factor the polynomials. Remember to look for the GCF first.

5. $x^2 + x - 6$

$(x + 3)(x - 2)$

6. $p^2 - 16p + 48$

$(p - 12)(p - 4)$

7. $x^2 + 7xy + 10y^2$

$(x + 5y)(x + 2y)$

8. $x^2 - 4x - 32$

$(x - 8)(x + 4)$

9. $12 + 8x + x^2$

$(6 + x)(2 + x)$

10. $2x^2 + 7x - 15$

$(2x - 3)(x + 5)$

11. $3x^2 + 19x - 14$

$(3x - 2)(x + 7)$

12. $8x^4 - 47x^3 - 6x^2$

$x^2(8x^2 - 47x - 6)$

$x^2(8x + 1)(x - 6)$

13. $20b^4 - 65b^3 - 60b^2$

$5b^2(4b^2 - 13b - 12)$

$5b^2(4b + 3)(b - 4)$

In Exercises 14–21, solve each quadratic equation by factoring.

14. $x^2 - 5x + 6 = 0$

$(x - 3)(x - 2) = 0$

$x - 3 = 0$	$x - 2 = 0$
$x = 3$	$x = 2$

15. $x^2 + 2x - 3 = 0$

$(x + 3)(x - 1) = 0$

$x + 3 = 0$	$x - 1 = 0$
$x = -3$	$x = 1$

16. $x^2 - x = 6$

$x^2 - x - 6 = 0$

$(x - 3)(x + 2) = 0$

$x = 3$ or $x = -2$

17. $x^2 - 5x = 14$

$x^2 - 5x - 14 = 0$

$(x - 7)(x + 2) = 0$

$x = 7$ or $x = -2$

18. $3x^2 + 11x - 4 = 0$

$(3x - 1)(x + 4) = 0$

$x = \frac{1}{3}$ or $x = -4$

19. $3x^2 - 12x = 0$

$3x(x - 4) = 0$

$x = 0$ or $x = 4$

20. $x^2 - 7x = 18$

$x^2 - 7x - 18 = 0$

$(x - 9)(x + 2) = 0$

$x = 9$ or $x = -2$

21. $3x(x - 6) - 5(x - 6) = 0$

$(x - 6)(3x - 5) = 0$

$x - 6 = 0$	$3x - 5 = 0$
$x = 6$	$x = \frac{5}{3}$

22. Your neighbors just finished installing a new swimming pool at their home. The pool measures 15 feet by 20 feet. They would like to plant a strip of grass of uniform width around three sides of the pool, the two short sides and one of the longer sides.

a. Sketch a diagram of the pool and the strip of lawn, using x to represent the width of the uniform strip.

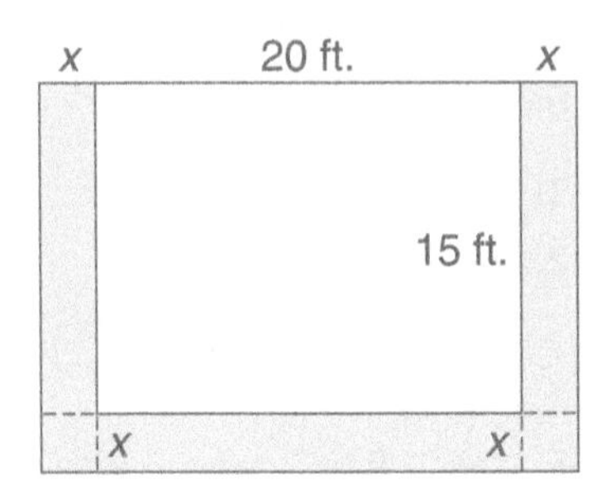

b. Write an equation for the area, A, in terms of x, that represents the lawn area around the pool.

$A = (20 + 2x)(15 + x) - 15(20) = 300 + 20x + 30x + 2x^2 - 300 = 2x^2 + 50x$

c. Your neighbors have enough seed for 168 square feet of lawn. Write an equation that relates the quantity of seed to the area of the uniform strip of lawn.

$2x^2 + 50x = 168$

d. Solve the equation in part c to determine the width of the uniform strip that can be seeded.

$$2x^2 + 50x = 168$$
$$2x^2 + 50x - 168 = 0$$
$$2(x^2 + 25x - 84) = 0$$
$$2(x + 28)(x - 3) = 0$$

$x + 28 = 0$ | $x - 3 = 0$

$x = -28$ | $x = 3$

The solution is 3 feet.

-28 feet makes no sense in this situation.

ACTIVITY 4.5

Drones

Solving Quadratic Equations Using the Quadratic Formula

OBJECTIVE

Solve quadratic equations by the quadratic formula.

One of the hottest consumer markets in the technology arena during the mid-2010s was the market for drones. The following table details sales of consumer drones to dealers in the United States between 2013 and 2017.

YEAR	DRONE SALES TO DEALERS (millions of U.S. dollars)
2013	44
2014	204
2015	443
2016	799
2017	1296

Source: The source for these numbers was the Consumer Technology Association.

1. a.

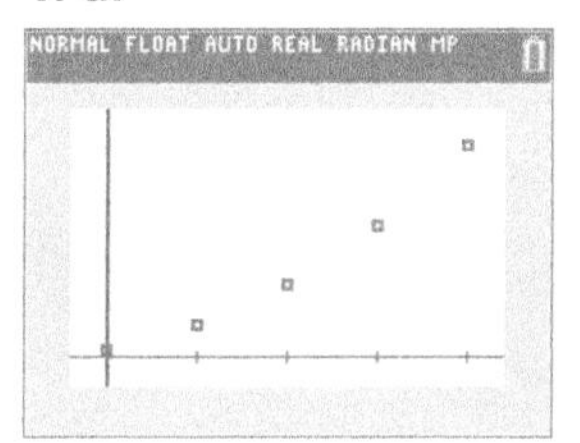

1. b.

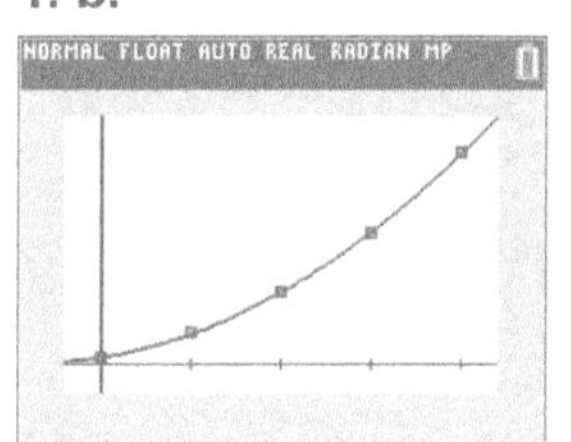

1. **a.** Use your graphing calculator to sketch a scatterplot of the given data. Let the input, t, represent the number of years since 2013.

   b. This data can be modeled by the function

$$S(t) = 56.5t^2 + 83.9t + 50.4$$

where $S(t)$ is the sales of drones in millions of dollars. Graph this function on the same coordinate axis as the scatterplot. Is this a good model for the data? Explain.

Yes, the function is a good model because the curve is very close to all of the data points.

   c. You want to determine when the sales will first reach 3 billion dollars. That is, you want to determine the value of t that yields $S = 3000$. Write the equation that must be solved.

$3000 = 56.5t^2 + 83.9t + 50.4$

The Quadratic Formula

The quadratic equation in Problem 1c can be solved using a numerical or graphical approach. However, the algebraic technique of factoring cannot be applied.

The technique of solving quadratic equations by factoring is limited. In most real-world applications involving quadratic equations, the quadratic is not factorable. In those cases, you can use a formula to solve the quadratic equation.

Beginning with the standard quadratic function defined by $y = ax^2 + bx + c, a \neq 0$, set $y = 0$ to obtain the equation

$$0 = ax^2 + bx + c.$$

The quadratic equation $ax^2 + bx + c = 0$ has two solutions,

$$x_1 = \frac{-b + \sqrt{b^2 - 4ac}}{2a} \quad \text{and} \quad x_2 = \frac{-b - \sqrt{b^2 - 4ac}}{2a}.$$

These solutions are often written as a single expression,

$$x = \frac{-b \pm \sqrt{b^2 - 4ac}}{2a}.$$

This formula is known as the **quadratic formula**.

For the details showing that the equation $ax^2 + bx + c = 0$ can be solved using the quadratic formula, see Appendix A. The section is called "Derivation of the Quadratic Formula."

The following example demonstrates the procedure for using the quadratic formula to solve an equation of the form $ax^2 + bx + c = 0$.

LC LEARNING CATALYTICS

Use the quadratic formula to approximate the solutions to $2x^2 + 3x = 6$.

a. $x = 4.55, -10.55$

b. $x = 2.27, -5.27$

c. $x = 1, -\frac{5}{2}$

d. none of these

EXAMPLE 1 ***Solve $x(3x + 4) = 5$ using the quadratic formula.***

SOLUTION

Step 1. Write the equation in general form, $ax^2 + bx + c = 0$.

$$x(3x + 4) = 5 \quad \text{Apply the distributive property on the left side.}$$
$$3x^2 + 4x = 5$$
$$-5 \quad -5 \quad \text{Subtract 5 from both sides.}$$
$$3x^2 + 4x - 5 = 0$$

Step 2. Identify the coefficients a and b and the constant term c.

$$a = 3, b = 4, c = -5$$

Step 3. Substitute the values a, b, and c into the quadratic formula and simplify.

$$x = \frac{-b \pm \sqrt{b^2 - 4ac}}{2a}$$

$$x = \frac{-4 \pm \sqrt{4^2 - 4(3)(-5)}}{2(3)}$$

$$x = \frac{-4 \pm \sqrt{16 - (-60)}}{6} = \frac{-4 \pm \sqrt{76}}{6}$$

$$x \approx -2.1196 \quad \text{or} \quad 0.7863$$

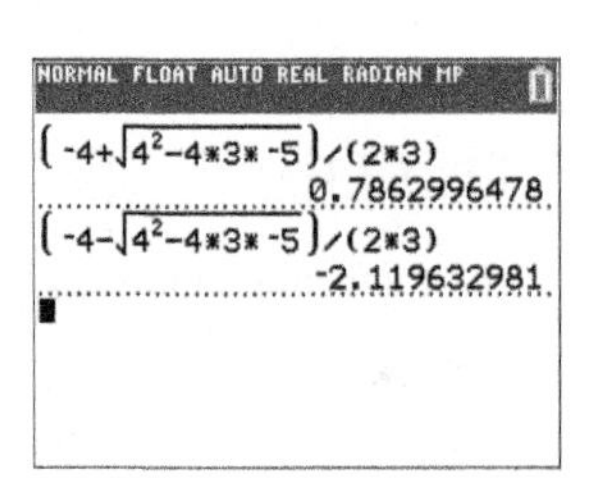

Step 4. Check your solutions. The following graphs verify the solutions:

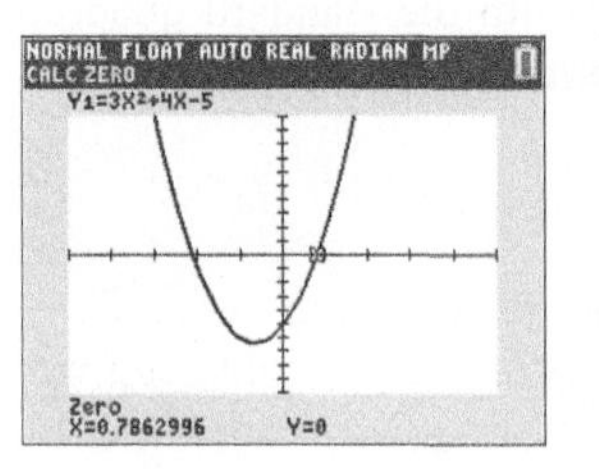

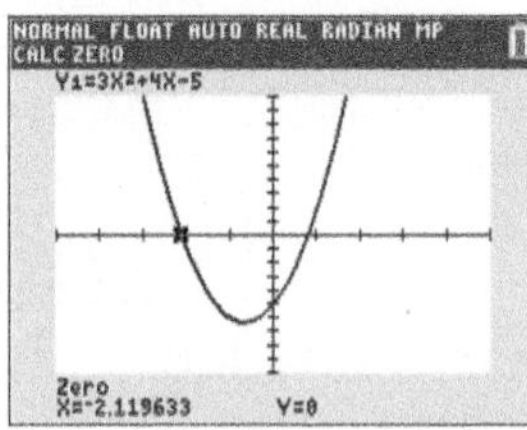

2. a.

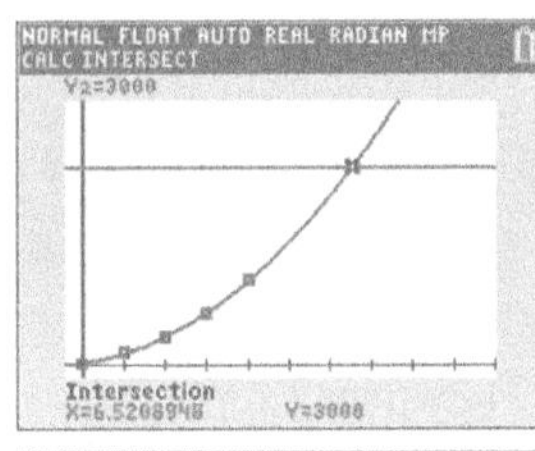

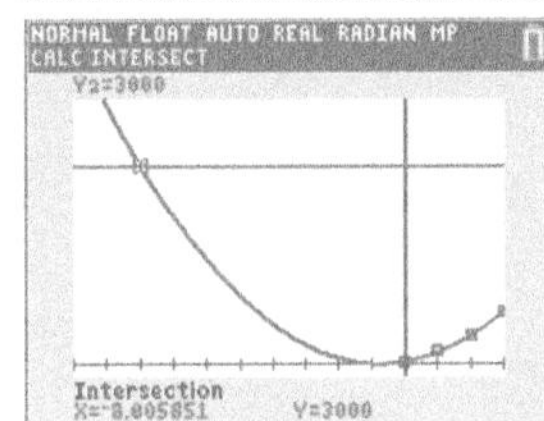

2. a. Solve the equation from Problem 1c, $56.5t^2 + 83.9t + 50.4 = 3000$, using the quadratic formula. Check your solution graphically.

$$56.5t^2 + 83.9t + 50.4 = 3000$$

$$56.5t^2 + 83.9t - 2949.6 = 0$$

$$t = \frac{-83.9 \pm \sqrt{83.9^2 - 4(56.5)(-2949.6)}}{2(56.5)}$$

$$t = \frac{-83.9 \pm \sqrt{673648.81}}{113}$$

$t \approx 6.52$ and $t \approx -8.01$

b. Use the results from part a to predict when drone sales will first reach 3 billion dollars.

You will reject the negative answer since that relates to a year clearly outside of our domain. We round 6.52 up to 7 and that is 7 years after 2013. Thus, the model predicts that the first year drone sales will reach 3 billion dollars is 2020.

3. a. Sketch a graph of $y = 2x^2 + 9x - 5$.

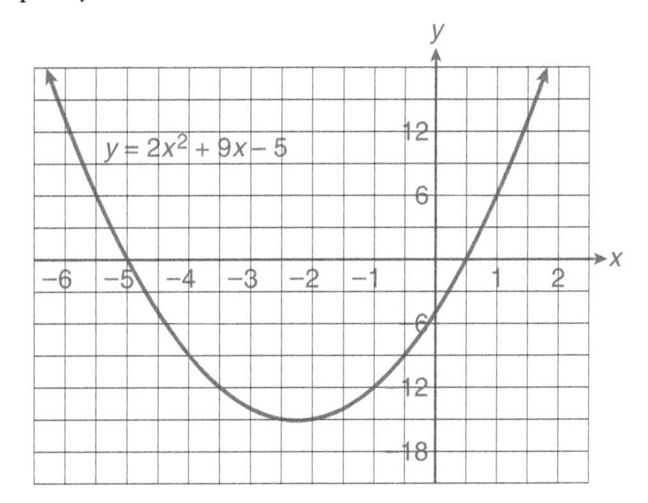

b. Write the equation that you need to solve to determine the x-intercepts of the graph.

$2x^2 + 9x - 5 = 0$

c. Solve the equation from part b using the quadratic formula.

$$x = \frac{-9 \pm \sqrt{81 - 4(2)(-5)}}{2(2)}$$

$$x = \frac{-9 \pm \sqrt{121}}{4}$$

$x = \frac{1}{2}$ or $x = -5$

d. Determine the x-intercepts of the graph using your graphing calculator.

$\left(\frac{1}{2}, 0\right), (-5, 0)$

e. Compare the solution using the quadratic formula (part c) with the x-intercepts determined from the graph.

The x-intercepts are the same using either process.

4. According to the Interactive Advertising Bureau (IAB) online advertising revenue in the United States from 2009 to 2016 was as follows:

Years since 2009, t	0	1	2	3	4	5	6	7
Total Online Advertising Revenue, $R(t)$ (billions \$)	22.7	26	31.7	36.6	42.8	49.5	59.6	72.5

This data can be modeled by $R(t) = 0.66t^2 + 2.22t + 23.31$, where t represents the number of years since 2009 and $R(t)$ is the total online advertising revenue in billions of dollars.

4. a.

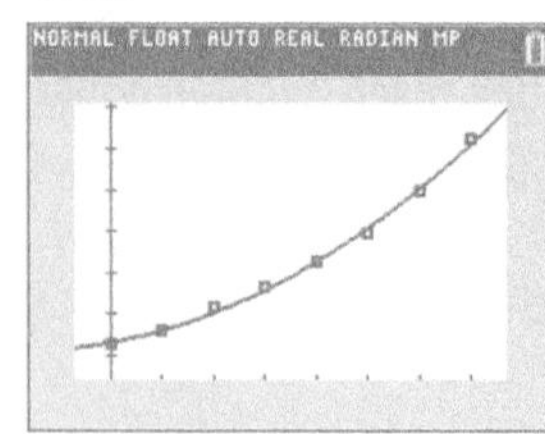

a. Use your graphing calculator to create a scatterplot of the data and a graph of the model P.

b. Does the model appear to be a good fit for the data? Explain.

Yes, the graph is very close to most of the data points.

c. Using the quadratic formula, determine the year in which online advertising revenue will first exceed 100 billion dollars.

$0.66t^2 + 2.22t + 23.31 = 100$

$0.66t^2 + 2.22t - 76.69 = 0$

$$x = \frac{-b \pm \sqrt{b^2 - 4ac}}{2a}$$

$$x = \frac{-2.22 \pm \sqrt{2.22^2 - 4(0.66)(-76.69)}}{2(0.66)}$$

$x \approx 9.23 \quad x \approx -12.59$

The negative value makes no sense because it is too far removed from the data points, but 9.23 corresponds to 2018.

d. Using your graphing calculator, graph $Y_2 = 100$ on the same axis as the function in part a. Determine the solution from the graph. How does the solution determined from the graph compare with the solution you determined in part c?

The x-intercept of (9.23, 0) determined from the graph indicates the same solution as determined by the quadratic formula.

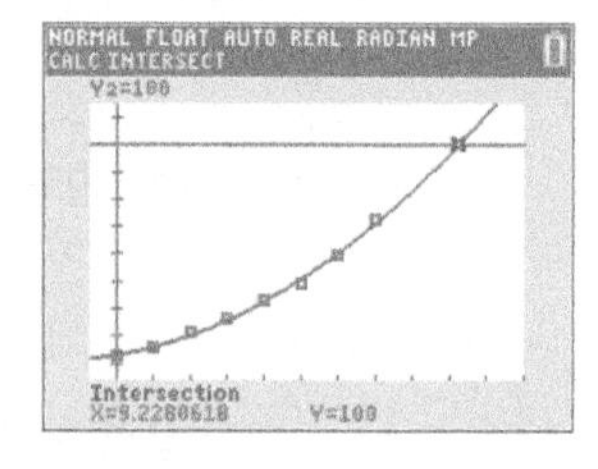

Axis of Symmetry Revisited

Recall that the axis of symmetry of a parabola defined by $f(x) = ax^2 + bx + c, a \neq 0$, is given by the formula $x = -\frac{b}{2a}$.

If you write the quadratic formula in a slightly different form, you obtain

$$x = -\frac{b}{2a} \pm \frac{\sqrt{b^2 - 4ac}}{2a} \quad \text{or} \quad x_1 = \frac{-b}{2a} + \frac{\sqrt{b^2 - 4ac}}{2a}, \quad x_2 = \frac{-b}{2a} - \frac{\sqrt{b^2 - 4ac}}{2a}.$$

The next problem uses the rewritten form of the quadratic formula to help identify a relationship between the x-intercepts and the axis of symmetry of the graph of f.

5. Consider the function defined by $f(x) = 2x^2 + 9x - 5$.

a. Determine the equation of the axis of symmetry of the graph of f.

$$x = \frac{-b}{2a} = -\frac{9}{4} = -2.25$$

b. What is the value of $\frac{\sqrt{b^2 - 4ac}}{2a}$ for function f?

$$\frac{\sqrt{9^2 - 4(2)(-5)}}{2 \cdot 2} = \frac{\sqrt{121}}{4} = \frac{11}{4}, \text{ or } 2.75$$

c. Sketch a graph of the function f, and label the axis of symmetry. Show where the value computed in part b is located graphically. What are the x-intercepts of the graph?

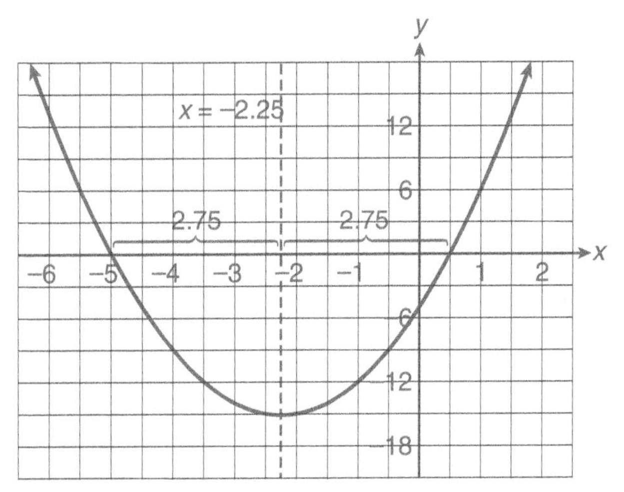

$(-5, 0)$ and $\left(\frac{1}{2}, 0\right)$

d. What is the relationship between the axis of symmetry and the x-intercepts of the parabola?

The axis of symmetry lies exactly midway between the x-intercepts.

6. For each of the following quadratic functions, determine the x-intercepts and the axis of symmetry of the graph. Solve the appropriate equation using the quadratic formula. Round your answers to the nearest hundredth.

a. $f(x) = 2x^2 - 6x - 3$

$0 = 2x^2 - 6x - 3$

$$x = \frac{6 \pm \sqrt{36 - 4(2)(-3)}}{4}$$

$$x = \frac{6 \pm \sqrt{60}}{4} \approx 3.44 \text{ or } -0.44$$

The intercepts are $(3.44, 0)$ and $(-0.44, 0)$.

The axis of symmetry is

$$x = \frac{3.44 + (-0.44)}{2}$$

$$x = \frac{3}{2}.$$

b. $h(x) = x^2 - 8x + 16$

$0 = x^2 - 8x + 16$

$$x = \frac{8 \pm \sqrt{64 - 4(1)16}}{2}$$

$x = 4$

The only intercept is $(4, 0)$.

The axis of symmetry is $x = 4$.

SUMMARY Activity 4.5

1. To solve a quadratic equation of the form $ax^2 + bx + c = 0$, $a \neq 0$, using the **quadratic formula** $x = \dfrac{-b \pm \sqrt{b^2 - 4ac}}{2a}$,

   a. rewrite the quadratic equation (if necessary) so that one side is equal to zero.

   b. identify the coefficients a and b and the constant term c.

   c. substitute these values into the formula and simplify.

   d. check your solutions graphically.

2. For a parabola with x-intercepts, the axis of symmetry is always midway between its x-intercepts.

3. The distance from the axis of symmetry to either x-intercept is $\dfrac{\sqrt{b^2 - 4ac}}{2a}$.

EXERCISES Activity 4.5

1. The height of a bridge arch located in the Thousand Islands is modeled by the function $f(x) = -0.04x^2 + 28$, where x is the distance, in feet, from the center of the arch and $f(x)$ is the height of the arch above the water of the river.

   a. Sketch a picture of this arch on a grid using the vertical axis as the center of the arch.

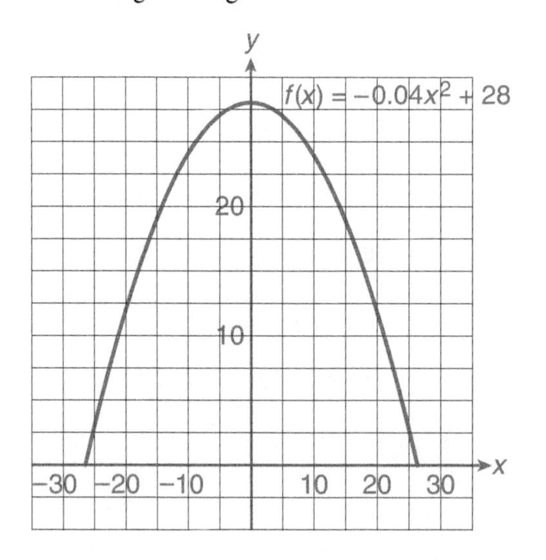

   b. Determine the vertical intercept. What is the practical meaning of this intercept in this situation?

   (0, 28). This point represents the vertex, or turning point, of the arch.

   c. Estimate the x-intercepts algebraically using the quadratic formula.

   $0 = -0.04x^2 + 28$, $a = -0.04$, $b = 0$, $c = 28$

   $$x = \frac{0 \pm \sqrt{0 - 4(-0.04)(28)}}{-0.08}$$

   $x \approx \pm 26.5$. The intercepts are approximately $(26.5, 0)$ and $(-26.5, 0)$.

Exercise numbers appearing in color are answered in the Selected Answers appendix.

d. Graph the function on your graphing calculator, and check the accuracy of the intercepts you determined in part c.

The intercepts are the same.

e. If the arch straddles the river exactly, how wide is the river?

The river is approximately 2(26.5), or 53 feet wide.

f. A sailboat is approaching the bridge. The top of the mast is 30 feet above the water. Will the boat clear the bridge? Explain.

No, the highest point of the arch is 28 feet above the water.

g. You want to install a flagpole on the bridge at an arch height of 20 feet. Write the equation you must solve to determine how far to the right or left of center the arch height is 20 feet.

$-0.04x^2 + 28 = 20$

h. Solve the equation in part g using the quadratic formula. Use your graphing calculator to check your result.

$-0.04x^2 + 8 = 0$

$$x = \frac{0 \pm \sqrt{0 - 4(-0.04)(8)}}{-0.08}$$

$x \approx \pm 14.14$ ft. Place the pole approximately 14.14 feet to the right or left of the center of the arch.

In Exercises 2–8, identify the values of a, b, and c, and then solve the equations using the quadratic formula. Round your answers to the nearest hundredth. Verify your solutions graphically.

2. $x^2 + 6x - 3 = 0$

$a = 1, b = 6, c = -3$

$$x = \frac{-6 \pm \sqrt{36 - 4(1)(-3)}}{2}$$

$$x = \frac{-6 \pm \sqrt{48}}{2}$$

$x \approx 0.46$ or $x \approx -6.46$

3. $4x^2 + 4x + 1 = 0$

$a = 4, b = 4, c = 1$

$$x = \frac{-4 \pm \sqrt{16 - 4(4)(1)}}{2(4)}$$

$$= -\frac{1}{2}$$

4. $x^2 + 5x = 13$

$x^2 + 5x - 13 = 0$

$a = 1, b = 5, c = -13$

$$x = \frac{-5 \pm \sqrt{77}}{2} \approx 1.89 \text{ or } -6.89$$

5. $2x^2 - 6x + 3 = 0$

$a = 2, b = -6, c = 3$

$$x = \frac{6 \pm \sqrt{12}}{4} \approx 2.37 \text{ or } 0.63$$

6. $2x^2 - 3x = 5$

$2x^2 - 3x - 5 = 0$

$a = 2, b = -3, c = -5$

$$x = \frac{3 \pm \sqrt{49}}{4} = 2.5 \text{ or } -1$$

7. $(2x - 1)(x + 2) = 1$

$2x^2 + 3x - 3 = 0$

$a = 2, b = 3, c = -3$

$$x = \frac{-3 \pm \sqrt{33}}{4} \approx 0.69 \text{ or } -2.19$$

8. $(x + 2)^2 + x^2 = 44$

$2x^2 + 4x - 40 = 0$

$a = 2, b = 4, c = -40$

$$x = \frac{-4 \pm \sqrt{336}}{4} = \frac{-2 \pm \sqrt{84}}{2} \approx 3.58 \text{ or } -5.58$$

In Exercises 9–11, determine the x-intercept(s) of the graph algebraically. Then check your results graphically.

9. $y = 3x^2 + 6x$

(0, 0) and (−2, 0)

10. $y = x^2 - x - 6$

(3, 0) and (−2, 0)

11. $f(x) = 2x^2 - x - 5$

$\left(\frac{1+\sqrt{41}}{4}, 0\right)$ and $\left(\frac{1-\sqrt{41}}{4}, 0\right)$, or approximately (1.851, 0) and (−1.351, 0)

12. The following data from the National Health and Nutrition Examination Survey indicates that the number of American adults who are overweight or obese is increasing:

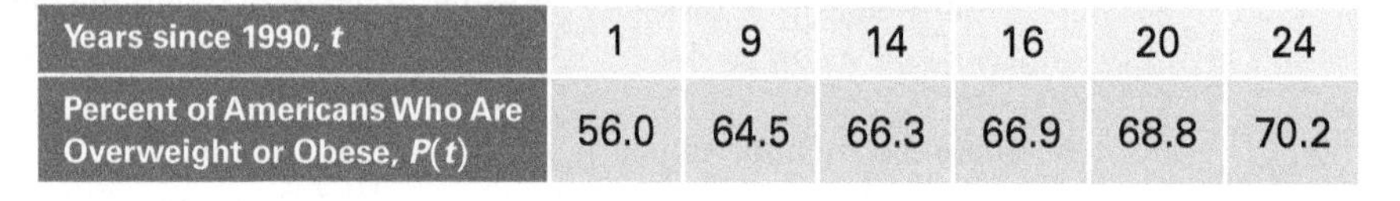

Years since 1990, t	1	9	14	16	20	24
Percent of Americans Who Are Overweight or Obese, $P(t)$	56.0	64.5	66.3	66.9	68.8	70.2

This data can be modeled by the equation $P(t) = -0.020t^2 + 1.090t + 55.249$, where $t = 0$ corresponds to the year 1990.

12. a.

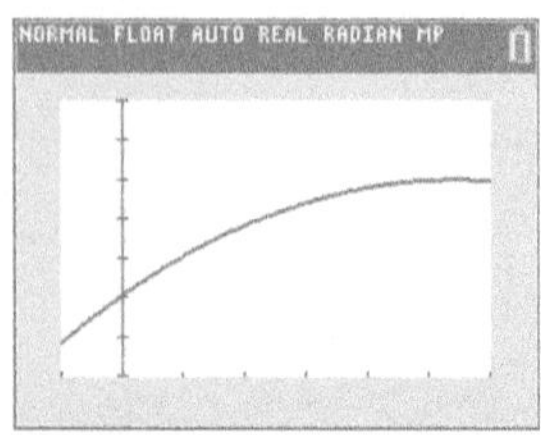

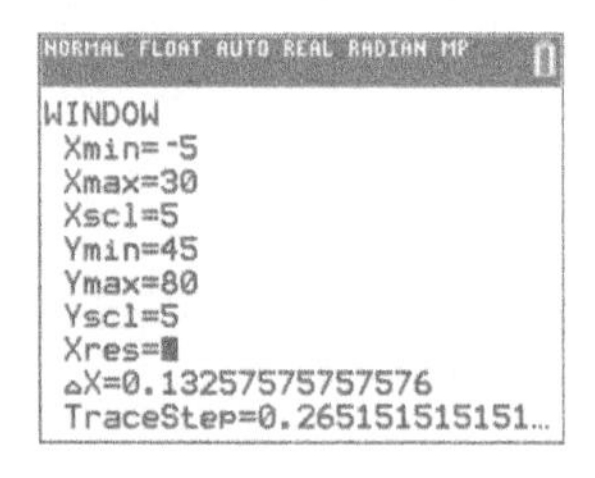

a. Use your graphing calculator to sketch a graph of the function.

b. What does the graph tell you that the model predicts will happen to the number of Americans who are overweight by 2020? Is it still increasing? Explain.

By 2020, the graph is starting to decrease. The model predicts that the number of Americans who are overweight or obese will be decreasing by the year 2020. The graph is a parabola opening downward and is past its vertex by the year 2020.

12. c.

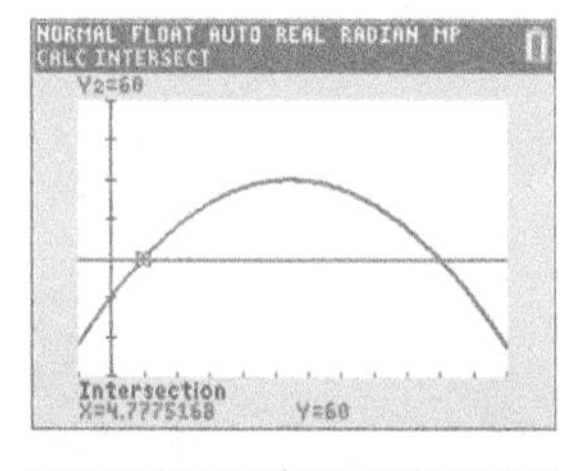

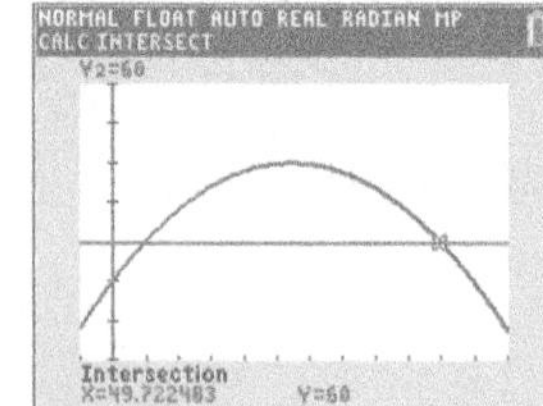

c. Use your graphing calculator to estimate the years in which 60% of Americans are overweight or obese.

$x \approx 49.7$ or $x \approx 4.78$. This means the model indicates that 60% of Americans were overweight or obese in 1995 and will be overweight or obese in 2040.

d. Use the quadratic formula to answer part c. How does your answer compare with the estimate you obtained using the graphical approach?

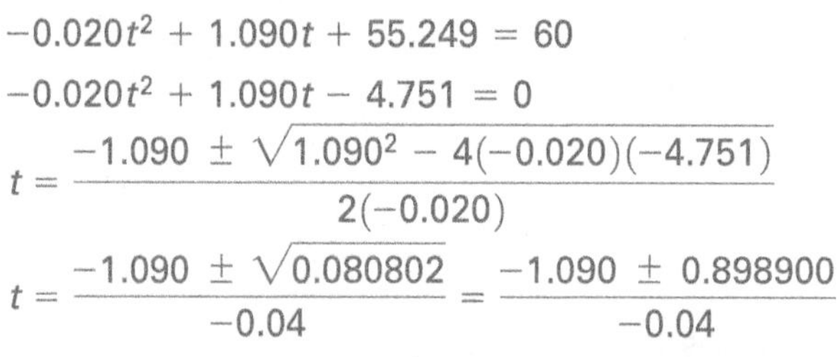

$$-0.020t^2 + 1.090t + 55.249 = 60$$

$$-0.020t^2 + 1.090t - 4.751 = 0$$

$$t = \frac{-1.090 \pm \sqrt{1.090^2 - 4(-0.020)(-4.751)}}{2(-0.020)}$$

$$t = \frac{-1.090 \pm \sqrt{0.080802}}{-0.04} = \frac{-1.090 \pm 0.898900}{-0.04}$$

$t \approx 4.78$. $t \approx 49.72$. The estimates are the same.

e. How confident are you in the solutions given in part d?

The 4.78 solution is fairly reliable since it is inside the data, but the 49.72 solution is extrapolating a long way from the data, so my confidence in that solution is not very high.

13. The quadratic function defined by the equation

$$d = 2r^2 - 16r + 34$$

gives the density of smoke, d, in millions of particles per cubic foot for a certain type of diesel engine. The input variable, r, represents the speed of the engine in hundreds of revolutions per minute.

a. Determine the density of smoke when $r = 3.5$ (350 revolutions per minute).

$d = 2(3.5)^2 - 16(3.5) + 34 = 2.5$ million particles per cubic foot

b. Determine the number of revolutions per minute for minimum smoke. What is the minimum output?

The minimum occurs at the vertex; $r = \frac{-b}{2a} = \frac{16}{4} = 4$, or 400 revolutions per minute.

$d = 2(4)^2 - 16(4) + 34 = 2$ million particles per cubic foot

c. If the density of smoke is determined to be 100 million particles per cubic foot, determine the speed of the engine.

$$2r^2 - 16r + 34 = 100$$

$$r^2 - 8r - 33 = 0$$

$$(r - 11)(r + 3) = 0$$

$r = 11$; 1100 revolutions per minute is the speed of the engine.

$r = -3$ is not in the practical domain $r > 0$.

ACTIVITY 4.6

Heat Index

Modeling Data with Quadratic Regression Equations

OBJECTIVES

1. Determine quadratic regression models using a graphing calculator.
2. Solve problems using quadratic regression models.

On very hot and humid summer days, it is common for the National Weather Service to issue warnings due to a very high heat index. The heat index is a measurement that combines air temperature and relative humidity to determine a relative temperature. The heat index is the temperature you perceive (how it feels) on a hot and humid day. Heat index is similar to windchill, the temperature you perceive on a cold and windy day. Heat indices are not usually calculated until the air temperature reaches 80°F.

The following table gives the heat index for various temperatures when the relative humidity is 90%. Both air temperature and the heat index (perceived temperature) are measured in degrees Fahrenheit.

Temperature, t (°F)	80	82	84	86	88	90	92
Heat Index, h (°F)	86	91	98	105	113	122	131

Note: Relative humidity is 90%.

1. Sketch a scatterplot of the data. Let the air temperature, t, represent the input variable, and the heat index, h, represent the output variable. Does the data appear to be quadratic? Explain.

   Yes, the data appear to be quadratic. There is a slight curve indicated by the plotted points.

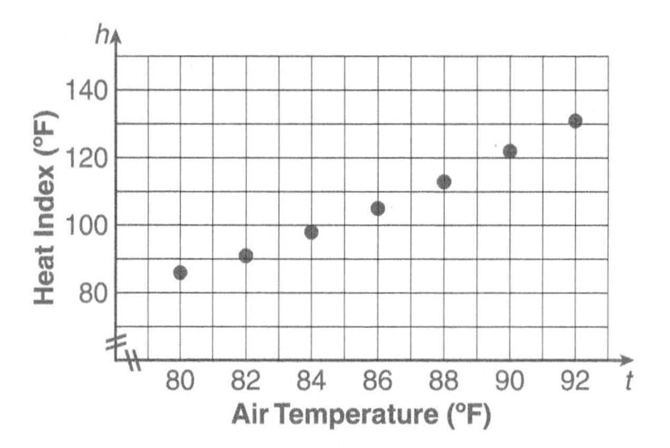

2. Use the regression feature of your graphing calculator to determine and plot a quadratic function that best fits this data. Your graph should appear as follows:

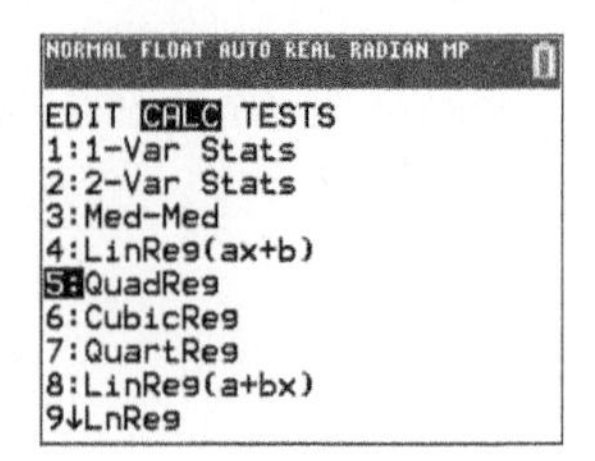

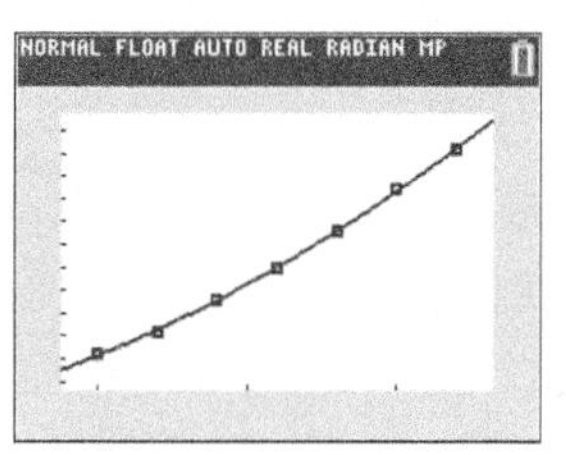

   The regression equation is $h = 0.09524t^2 - 12.59524t + 483.85714$.

3. **a.** How does the plot of the quadratic regression equation compare with your scatterplot of the data?

   The graph matches the scatterplot very well.

   b. Do you believe that the quadratic regression model is a good model for the heat index when the relative humidity is 90%?

   Based on the graph and the scatterplot, it is a very good model.

4. What is the practical domain of this function?

The practical domain is the set of integers from about 80 to 100 or so because heat indices are not calculated until the air temperature reaches 80°F.

5. a. Assuming that the humidity is 90%, use the quadratic regression equation to estimate the heat index for each of the following temperatures:

i. 45° **ii.** 85° **iii.** 100°

i. 110° ii. 101° iii. 177°

b. Which, if any, of these estimates do you think is most reliable? Explain.

The most reliable is the 85°F data. It makes sense that the temperature would feel like 101°F on a 85°F day with 90% humidity. However, the other two are not reasonable. The air temperatures for parts i and iii are outside the original data.

6. a. Assuming that the relative humidity is 90%, estimate the temperatures for which the heat index is equal to 100° using each of the following methods:

i. The given table (numerical)

If the humidity is 90%, the table on the calculator indicates that the heat index is equal to 100° for 48° and 85°.

ii. The graph of the quadratic regression equation (graphical)

If the humidity is 90%, the graph indicates that the heat index is equal to 100° for 48° and 85°.

b. Use the quadratic formula (algebraic) to estimate the temperature for which the heat index is 100° when the relative humidity is 90%.

$100 = 0.09524t^2 - 12.59524t + 483.85714$

$0 = 0.09524t^2 - 12.59524t + 383.85714$

The equation indicates that the temperatures are 85°F and 48°F.

c. How do the results in parts a and b compare?

The results are the same.

7. a. In Problem 6b, the stated relative humidity was 90% and you wrote an equation to determine the temperature for which the heat index (perceived temperature) was equal to 100°F. Write an equation to determine the temperature for which the air temperature and the heat index are equal.

$h = t$ or $0.09524t^2 - 12.59524t + 483.85714 = t$

b. Solve the equation in part a graphically to estimate the temperatures for which the air temperature and the heat index are equal.

The intersections occur near $t = 67.7$ and $t = 75.1$. Therefore, when the relative humidity is 90%, the temperatures for which the air temperature and the heat index are equal are about 68°F and 75°F.

8. a. The National Weather Service does not calculate the heat index unless the air temperature is at least 80°F and the relative humidity is at least 40%. Based on your results in Problems 5, 6, and 7, does this policy seem reasonable? Explain.

Yes, it is reasonable. The model gives some unrealistic results for temperatures below 80°F.

b. According to this policy, will the heat index reported by the National Weather Service ever be equal to the actual air temperature if the relative humidity is 90%? Explain.

No. According to the results in Problem 7b, the heat index and the air temperature will never be the same.

LC LEARNING CATALYTICS

Determine the quadratic regression equation for the following data.

x	−3	2	5	7	11
y	38	13	70	137	345

a. $y = 21.2x + 27.1$
b. $y = 2.99x^2 + 2.03x + 5.01$
c. $y = 2.99x^2 - 2.03x + 5.01$
d. none of these

9. The following data from the National Health and Nutrition Examination Survey appeared in Exercise 12 of Activity 4.5. The data indicate that the number of American adults who are overweight or obese is increasing.

Years since 1990, t	1	9	14	16	20	24
Percent of Americans Who Are Overweight or Obese, $P(t)$	56.0	64.5	66.3	66.9	68.8	70.2

Use the regression feature of your graphing calculator to verify that these data can be modeled by the equation $P(t) = -0.020t^2 + 1.090t + 55.249$, where $t = 0$ corresponds to the year 1990.

The result from the calculator verifies the equation.

SUMMARY Activity 4.6

1. Parabolic data can be modeled by a **quadratic regression equation**.

EXERCISES Activity 4.6

1. During one game, the Buffalo Bills punter was called upon to punt the ball 8 times. On one of these punts, the punter struck the ball at his own 30-yard line. The height, h, of the ball above the field in feet as a function of time, t, in seconds can be partially modeled by the following table:

t (sec.)	0	0.6	1.2	1.8	2.4	3.0
$h(t)$ (ft.)	2.50	28.56	43.10	46.12	37.12	17.60

1. a.

a. Sketch a scatterplot of the data using your graphing calculator.

b. Use your graphing calculator to obtain a quadratic regression function for this data. Round the values of a, b, and c to four decimal places.

$h(t) = -15.9752t^2 + 52.8875t + 2.5536$

c. Graph the equation from part b on the same coordinate axes as the data points. Does the curve appear to be a good fit for the data? Explain.

Yes, the curve touches nearly every data point.

1. c.

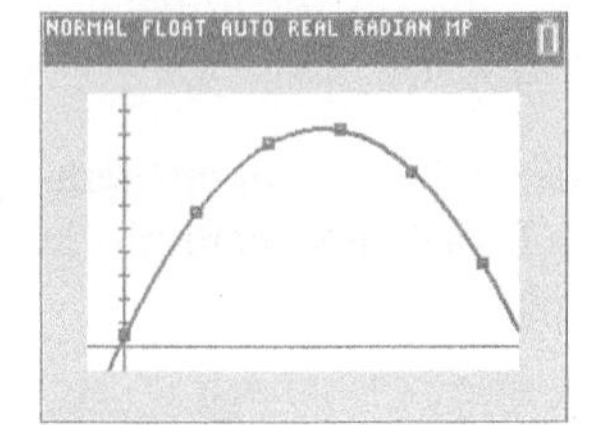

d. In this model, what is the practical domain of the quadratic regression function?

all real numbers from 0 to 3.36 seconds

e. Estimate the practical range of this model.

real numbers from 0 to 46.33 feet

f. How long after the ball was struck did the ball reach 35 feet above the field? Explain.

The ball reaches 35 feet on the way up after about 0.81 second. It reaches 35 feet again on the way down approximately 2.50 seconds after it was struck.

g. How many results did you obtain for part f? Do you think you have all of the solutions? Explain.

There are only at most two solutions to a quadratic equation, so I got them all.

2. Use the following data set to perform the tasks in parts a–e:

x	0	3	6	9	12
y	5	28	86	180	310

a. Determine an appropriate scale, and plot these points.

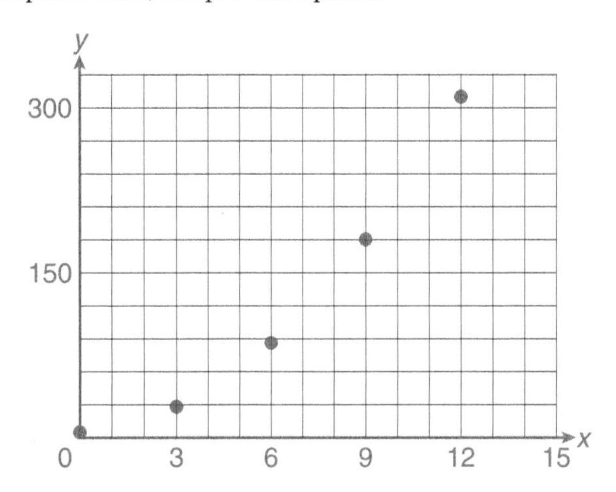

b. Use your graphing calculator to determine the quadratic regression equation for this data set.

$y = 1.984x^2 + 1.590x + 5.114$

2. c.

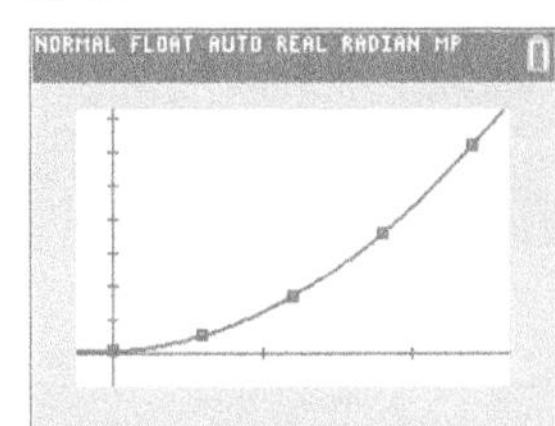

c. Graph the regression equation using your graphing calculator.

d. Use the table feature of your graphing calculator to compare the predicted outputs with the outputs given in the table.

If you round the model to the nearest whole number, all of the values are the same.

e. Predict the output for $x = 7$ and for $x = 15$.

113.46 and 475.364

3. The following table shows the stopping distance for a car at various speeds on dry pavement:

Speed (mph)	25	35	45	55	65	75
Distance (ft.)	65	108	167	245	340	450

a. Use your graphing calculator to determine a quadratic regression equation that represents this data.

$y = 0.086x^2 - 0.842x + 32.487$

b. Use the regression equation to predict the stopping distance at 90 mph.

approximately 653 feet

c. What speed would produce a stopping distance of 280 feet? (Round to the nearest tenth.) Explain how you arrived at your conclusion.

$280 = 0.086x^2 - 0.842x + 32.487$

$0 = 0.086x^2 - 0.842x - 247.513$

Using the quadratic formula, a speed of about 58.8 miles per hour requires a stopping distance of 280 feet.

4. According to Facebook, its revenues continued to increase from 2012 through 2016. The following table indicates Facebook's revenue for those years. Note that the revenue, R, is given in billions of U.S. dollars.

Year	2012	2013	2014	2015	2016
Years from 2012, t	0	1	2	3	4
Revenue, R (in billions)	5.09	7.87	12.47	17.93	27.64

a. Create a scatterplot of the data using your graphing calculator.

4.a.

NORMAL FLOAT AUTO REAL RADIAN MP
WINDOW
Xmin=-1
Xmax=5
Xscl=1
Ymin=-5
Ymax=40
Yscl=5
Xres=■
ΔX=0.022727272727273
TraceStep=0.045454545454...

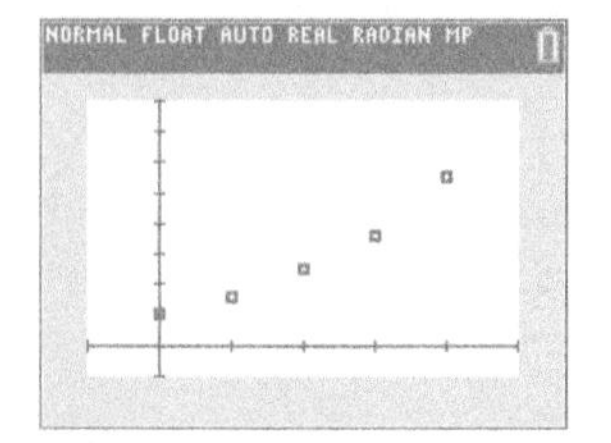

b. Use your graphing calculator to obtain a quadratic regression function for these data. Round the values of a, b, and c to three decimal places.

$R = 1.051t^2 + 1.320t + 5.271$

c. Graph the equation from part b on the same coordinate axes as the data points. Does the curve appear to be a good fit for the data? Explain.

Yes, a graph of the regression equation fits the data well. It is very close to the scatterplot points.

4. c.

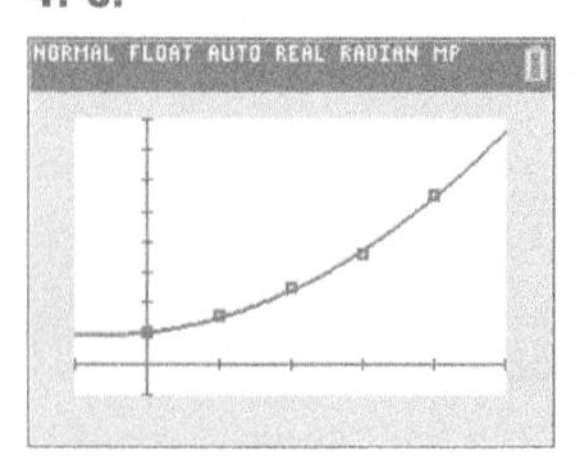

d. Use the model to predict Facebook's revenue in 2027. Do you have confidence in this prediction?

2027 is 15 years from 2012, so $t = 15$. $R(15) = 1.051 \cdot 15^2 + 1.320 \cdot 15 + 5.271 = 261.5$. The model predict that Facebook's 2027 revenue will be 261.5 billion dollars. $t = 15$ is well outside of the given data; therefore, the result is extrapolation. I do not have confidence in this prediction.

e. In what year does the model predict Facebook revenue will reach $100 billion?

Solve the equation $R = 1.051t^2 + 1.320t + 5.271 = 100$.

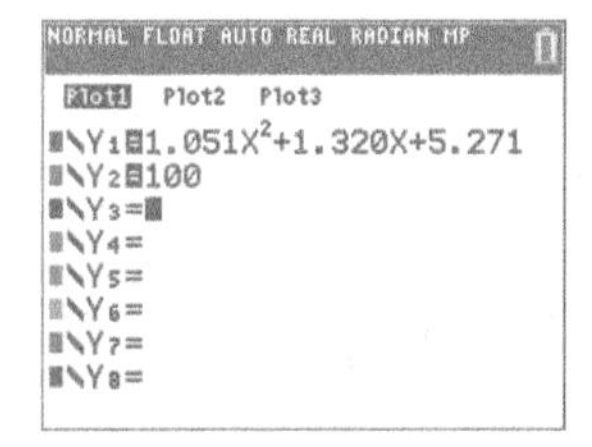

NORMAL FLOAT AUTO REAL RADIAN MP
PRESS + FOR ΔTbl

X	Y1	Y2
0	5.271	100
1	7.642	100
2	12.115	100
3	18.69	100
4	27.367	100
5	38.146	100
6	51.027	100
7	66.01	100
8	83.095	100
9	102.28	100
10	123.57	100

X=9

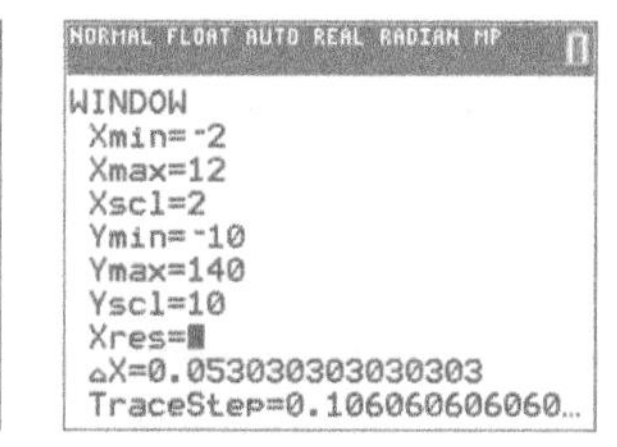

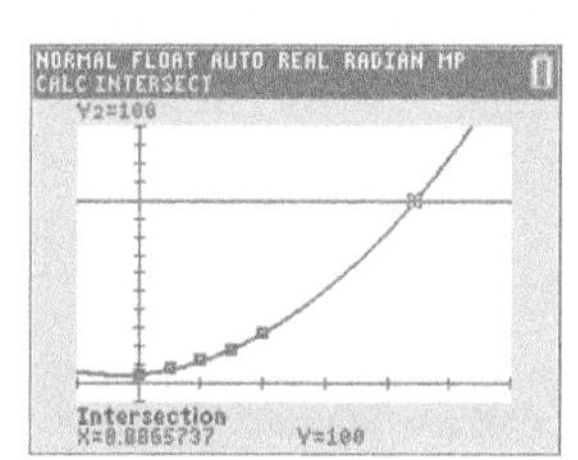

The solution to the equation is $t \approx 8.9$. The revenue will reach 100 billion dollars in 2020.

5. The shape of the main support cable in a suspension bridge is a parabola.

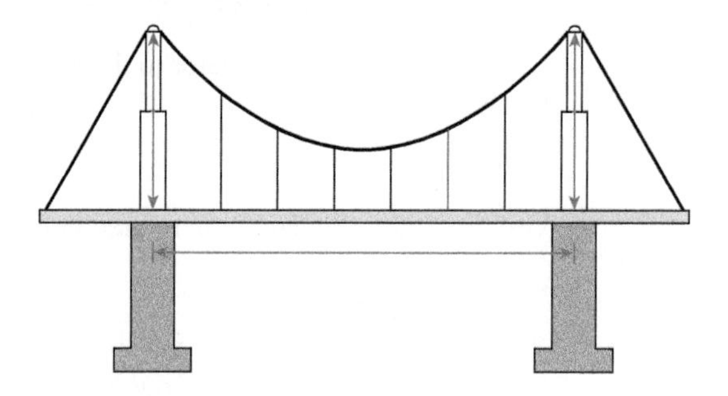

Use resources from the library or the Internet to determine specific dimensions, such as the distance between the main support columns, for one of the following bridges: Golden Gate Bridge, George Washington Bridge, or Verrazano Bridge.

Use the dimensions to develop points that lie on the graph of the main support cable of the bridge you selected. Draw a sketch of the support cable on graph paper. The orientation of the coordinate system, especially the origin, is very important. You will need at least three points that lie on the graph of the support cable in your drawing.

Use the points to determine a quadratic regression equation to represent the main support cable of the bridge. Use the model to determine the minimum distance from the cable to the highway.

Be prepared to give a presentation of your findings.

ACTIVITY
4.7

Complex Numbers

OBJECTIVES

1. Identify the imaginary unit $i = \sqrt{-1}$.
2. Identify complex numbers, $a + bi$.
3. Determine the value of the discriminant $b^2 - 4ac$ used in the quadratic formula.
4. Solve quadratic equations in the complex number system.
5. Determine the types of solutions to quadratic equations.

Recall that the solutions to the quadratic equation $ax^2 + bx + c = 0$ correspond to the x-intercepts of the parabola having equation $y = ax^2 + bx + c$.

Do all parabolas possess x-intercepts? Consider the graph of $y = 2x^2 + x + 5$. If you graph the function in the window Xmin $= -5$, Xmax $= 5$, Ymin $= -3$, and Ymax $= 15$, the graph will resemble the following:

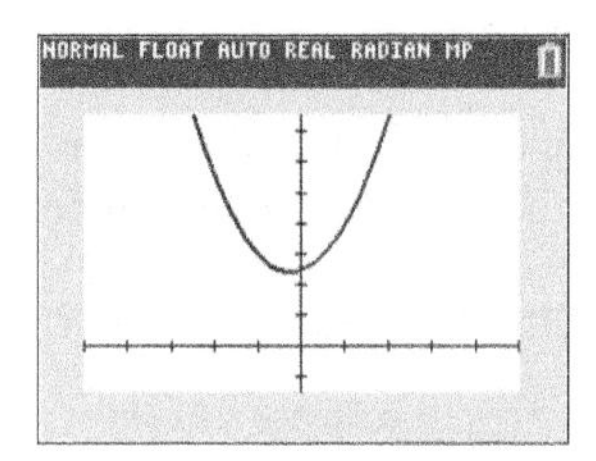

1. **a.** Based on what you know about parabolas, will the graph of $y = 2x^2 + x + 5$ have any x-intercepts? Explain.

   No, the parabola opens upward. The vertex is above the x-axis, so it will not have any x-intercepts.

   b. What can you say about the solutions to $2x^2 + x + 5 = 0$?

   There are no real number solutions.

Your response to Problem 1b should have been "There are no real number solutions." This is consistent with the graph. Because there are no x-intercepts for the graph of $y = 2x^2 + x + 5$, there are no real-valued solutions to the equation $2x^2 + x + 5 = 0$. Would you have discovered this if you had tried to solve the equation $2x^2 + x + 5 = 0$ algebraically using the quadratic formula? Problem 2 addresses this question.

2. Use the quadratic formula to solve $2x^2 + x + 5 = 0$. Where does the solution process break down? Explain.

   $b^2 - 4ac = 1 - 4(2)(5) = -39$ leads to a negative radicand; $\sqrt{-39}$ is not a real number.

Complex Numbers

Problem 2 illustrates that the breakdown with the quadratic formula occurs when you are asked to evaluate a radical with a negative radicand. If you try to evaluate $\sqrt{-39}$ using your TI-84 Plus C, the following screens will appear:

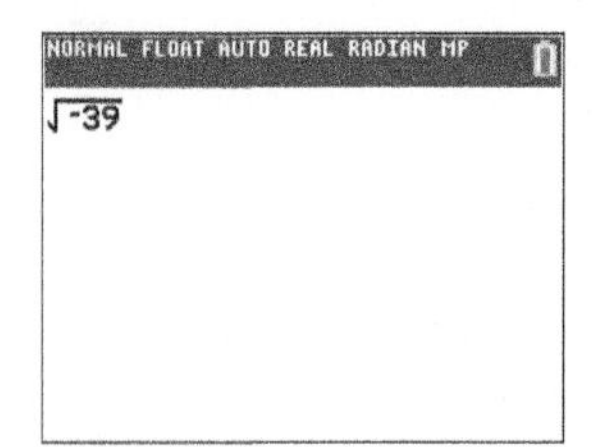

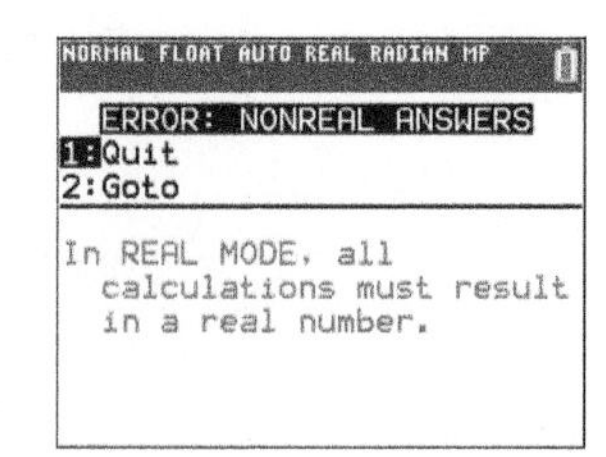

This tells you that there is no real number that is the square root of -39. Now change the MODE on the calculator from *real* to $a + bi$.

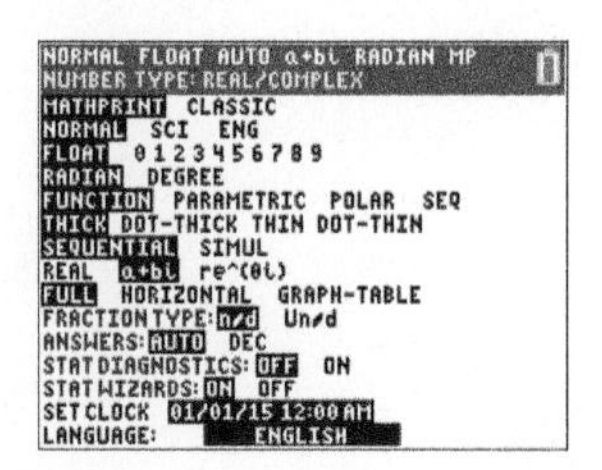

Now try to evaluate $\sqrt{-39}$ again. This time your calculator returns a value.

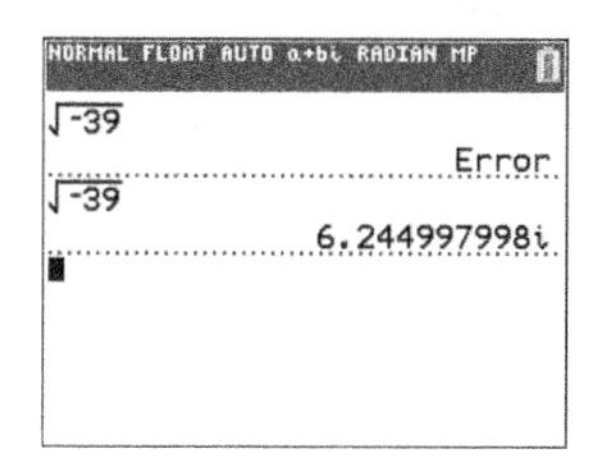

This is not a real number. Such a number is a **complex number** (an extension of the real numbers). The distinguishing characteristic of the complex numbers is the **imaginary unit**, $i = \sqrt{-1}$.

The quadratic formula solution to $2x^2 + x + 5 = 0$ uses the values $a = 2$, $b = 1$, and $c = 5$. The solution is

$$x = \frac{-1 \pm \sqrt{1^2 - 4(2)(5)}}{2(2)} = \frac{-1 \pm \sqrt{1 - 40}}{4} = -\frac{1}{4} \pm \frac{\sqrt{-39}}{4}.$$

The problem is that you cannot evaluate $\sqrt{-39}$ in the real number system because any real number multiplied by itself is nonnegative. Therefore, you need to introduce the imaginary unit, $i = \sqrt{-1}$, and interpret $\sqrt{-39}$ as

$$\sqrt{39(-1)} = \sqrt{39}\sqrt{-1} = \sqrt{39} \cdot i = i\sqrt{39}.$$

This approach can be used to rewrite any radical expression with a negative radicand.

EXAMPLE 1

a. $\sqrt{-16} = \sqrt{16}\sqrt{-1} = 4i$

b. $\sqrt{-25} = \sqrt{25}\sqrt{-1} = 5i$

c. $\sqrt{-53} = \sqrt{53}\sqrt{-1} = \sqrt{53}i$, or $i\sqrt{53}$

3. Rewrite each of the following in the form bi, where $i = \sqrt{-1}$:

a. $\sqrt{-26}$

$i\sqrt{26}$

b. $\sqrt{-5}$

$i\sqrt{5}$

c. $\sqrt{-64}$

$8i$

d. $\sqrt{3 \cdot (-7)}$

$i\sqrt{21}$

e. $\sqrt{-18}$

$3i\sqrt{2}$

f. $\sqrt{-27}$

$3i\sqrt{3}$

g. $\sqrt{-\frac{3}{4}}$

$\frac{\sqrt{3}}{2}i$

h. $\sqrt{-\frac{15}{27}}$

$\frac{\sqrt{5}}{3}i$

DEFINITION

Numbers of the form bi, where b is a real number and $i = \sqrt{-1}$, are called **pure imaginary numbers**. Numbers of the form $a \pm bi$, where a and b are real and $i = \sqrt{-1}$, are called **complex numbers**. The term a is called the *real part*. The term bi is called the *imaginary part*. Imaginary numbers are complex numbers of the form $0 + bi$. Real numbers are complex numbers of the form $a + 0i$.

EXAMPLE 2

a. ***The numbers*** $-3i, \frac{2}{3}i,$ ***and*** $7.4i$ ***are pure imaginary numbers.***

b. ***The numbers*** $-4 + 3i, \frac{1}{2} - \frac{2}{3}i, 4i, -2i,$ ***and*** $5 - 6i$ ***are complex numbers.***

Note that the set of real numbers is contained within the set of complex numbers. A real number a may be thought of as the complex number $a + 0i$.

In the sixteenth century, complex numbers were first used as solutions to polynomial equations. The notation $\sqrt{-1}$ was used during this time. Such numbers were called imaginary because their existence was not clearly understood. In 1777, Leonhard Euler introduced the notation i and wrote complex numbers in the form $a + bi$. Caspar Wessel in 1797 and Carl Friedrich Gauss in 1799 used the geometric interpretation of complex numbers as points in a plane. This made such numbers more concrete and less mysterious. Finally, in 1833, Sir William Hamilton showed that if the number i is defined to have the property

$$i^2 = -1,$$

then the set of real numbers can be extended to include numbers such as $\sqrt{-1}$.

Today, complex numbers are used in a variety of applications, including chaos theory (fractals) and engineering.

Operations with Complex Numbers

The operations of addition, subtraction, and multiplication of complex numbers are demonstrated in the following example.

EXAMPLE 3

a. ***To add complex numbers, add the real parts and the imaginary parts.***

$$(3 + 2i) + (5 - 7i) = 8 - 5i$$

b. ***To subtract complex numbers, add the opposite.***

$$(2 - 2i) - (-6 + i) = 2 - 2i + 6 - i = 8 - 3i$$

c. ***To multiply complex numbers, multiply each term of the first by each term of the second and simplify.***

$$\begin{aligned}(3 - 5i)(-1 + 8i) &= -3 + 24i + 5i - 40i^2 \\ &= -3 + 29i + 40 \\ &= 37 + 29i\end{aligned}$$

Remember that $i \cdot i = i^2 = -1$.

LC LEARNING CATALYTICS

Determine the product of $(2 + 2i)(3 - 4i)$.

The TI-84 Plus C is capable of operations with complex numbers. You first need to change the mode of the calculator from Real to $a + bi$. Note that the i key is second ⊡ period.

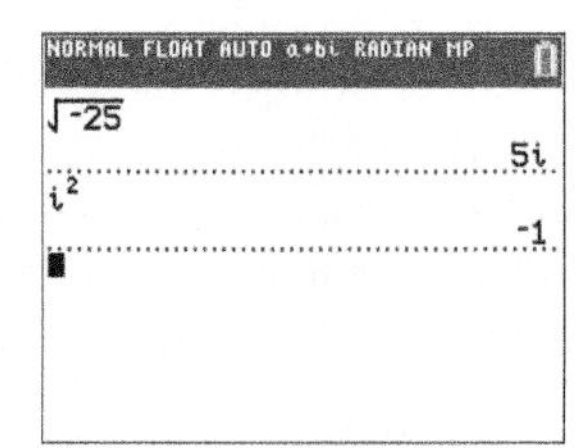

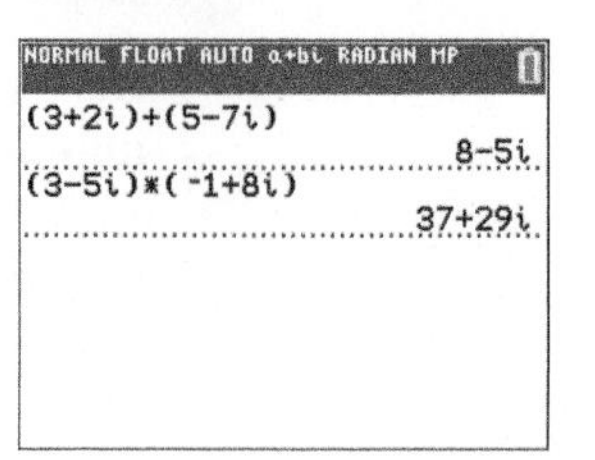

4. Perform the following operations with complex numbers. Use your graphing calculator to check your results.

a. $(3 + 4i) + (-5 + 6i)$

$-2 + 10i$

b. $(5 - 7i) - (-2 + 5i)$

$7 - 12i$

c. $5i(2 - 4i)$

$20 + 10i$

d. $(3 - 2i)(4 + 5i)$

$22 + 7i$

Discriminant

In the complex number system, every quadratic equation has at least one solution. In the quadratic formula

$$x = \frac{-b \pm \sqrt{b^2 - 4ac}}{2a},$$

the expression $b^2 - 4ac$ is called the **discriminant** because its value determines the number and type of solutions of a quadratic equation $ax^2 + bx + c = 0$. There are three possible cases, depending on whether the value of the discriminant is positive, zero, or negative. Problems 5, 6, and 7 investigate this relationship.

5. For each of the following quadratic functions, determine the sign of the discriminant. Then sketch a graph using your graphing calculator, and determine the number of x-intercepts.

a. $y = 2x^2 - 7x - 4$

$a = 2$

$b = -7$

$c = -4$

$b^2 - 4ac = 81$

two x-intercepts

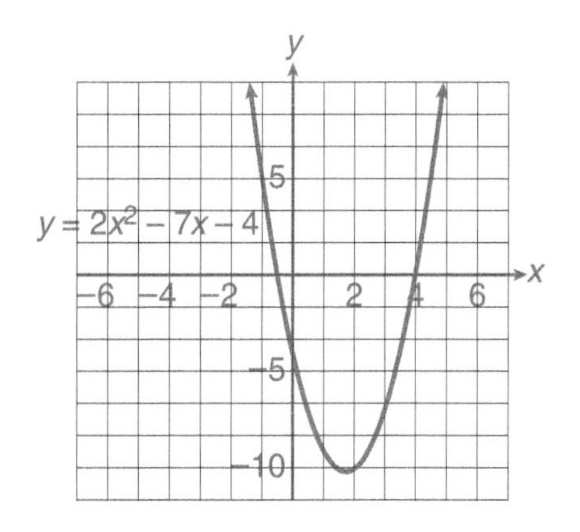

b. $y = 3x^2 + x + 1$

$a = 3$

$b = 1$

$c = 1$

$b^2 - 4ac = -11$

no x-intercepts

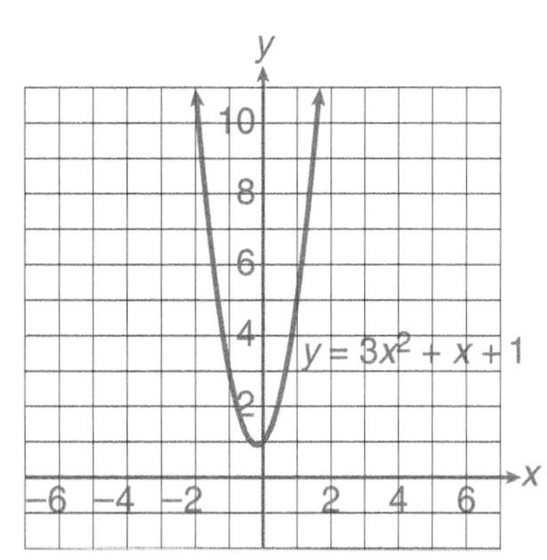

c. $y = x^2 + 2x + 1$

$a = 1$

$b = 2$

$c = 1$

$b^2 - 4ac = 0$

one x-intercept

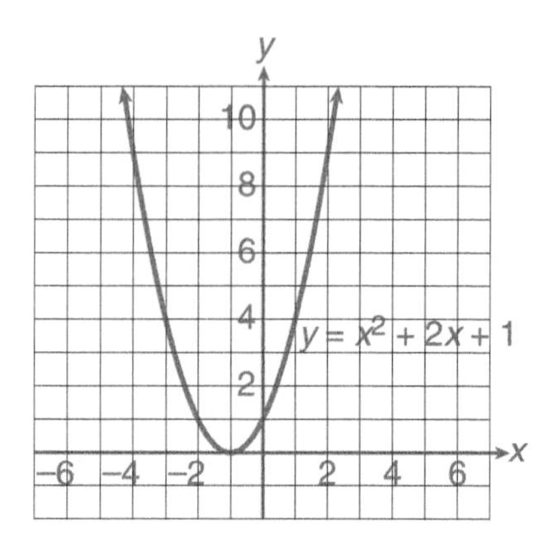

Recall that the real solutions to the equation $ax^2 + bx + c = 0$ and the x-intercepts of the graph of $y = ax^2 + bx + c$ are the same. Because the graph of $y = 2x^2 + x + 5$ (see Problem 1) has no x-intercept, it has no real solutions.

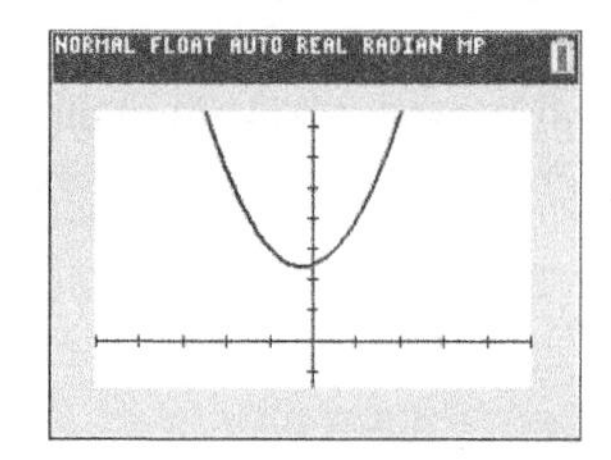

But the quadratic equation $2x^2 + x + 5 = 0$ does have exactly two solutions in the complex number system. The two solutions must be complex and not real. Similarly, if the graph of a quadratic function has two x-intercepts, the solutions to the equation $ax^2 + bx + c = 0$ must be two real numbers.

6. If the graph of $y = ax^2 + bx + c$ has exactly one x-intercept, what are the number and type (real or complex) of solutions to the equation $ax^2 + bx + c = 0$?

There is one real solution.

7. Return to Problem 5. Use the value of the discriminant $b^2 - 4ac$ and the number of x-intercepts in parts a, b, and c to complete the following table:

SOLUTIONS TO $ax^2 + bx + c = 0$ IN THE COMPLEX NUMBER SYSTEM	
$b^2 - 4ac$	NUMBER AND TYPE OF SOLUTIONS
Positive	2 real solutions
Zero	1 real solution
Negative	2 complex solutions

8. a. Evaluate the discriminant for each of the following equations, and indicate the number and type of solutions to the equations:

i. $2x^2 - 7x - 4 = 0$

81, two real solutions

ii. $3x^2 + x + 1 = 0$

−11, two complex solutions

iii. $x^2 - 2x + 1 = 0$

0, one real solution

iv. $3x^2 + 2x = -1$

−8, two complex solutions

b. Determine the solutions to each of the equations in part a to verify your results from part a.

i. $x = 4, -\frac{1}{2}$

ii. $x = \frac{-1}{6} \pm \frac{\sqrt{11}}{6}i$

iii. $x = 1$

iv. $x = \frac{-1}{3} \pm \frac{\sqrt{2}}{3}i$

SUMMARY Activity 4.7

1. The **imaginary unit** denoted by i is the number $\sqrt{-1}$. The number i is defined to have the property $i^2 = -1$.

2. Any number that can be written in the form $a + bi$, where a and b are real numbers and i is the imaginary unit, is called a **complex number**. The term a is called the **real part**. The term bi is called the **imaginary part**.

3. In the quadratic formula

$$x = \frac{-b \pm \sqrt{b^2 - 4ac}}{2a},$$

the expression $b^2 - 4ac$ is called the **discriminant**. Its value determines the number and type of solutions of a quadratic equation $ax^2 + bx + c = 0$.

4. Solutions of the quadratic equation $ax^2 + bx + c = 0$ in the complex number system are summarized in the following table:

$b^2 - 4ac$	NUMBER AND TYPE OF SOLUTIONS
Positive	2 real solutions
Zero	1 real solution
Negative	2 complex solutions

EXERCISES Activity 4.7

In Exercises 1–8, write each of the following in the form bi, where $i = \sqrt{-1}$:

1. $\sqrt{-25}$

   $5i$

2. $\sqrt{-20}$

   $2i\sqrt{5}$

3. $\sqrt{-36}$

   $6i$

4. $\sqrt{-10}$

   $i\sqrt{10}$

5. $\sqrt{-48}$

   $4i\sqrt{3}$

6. $\sqrt{-80}$

   $4i\sqrt{5}$

7. $\sqrt{-\frac{9}{16}}$

   $\frac{3}{4}i$

8. $\sqrt{\frac{-20}{75}}$

   $\sqrt{\frac{4}{15}}i = \frac{2}{\sqrt{15}}i$

In Exercises 9–13, perform the operations and express your answer in the form $a + bi$. Use your graphing calculator to verify the results.

9. $(2 + 8i) + (-7 + 2i)$

$-5 + 10i$

10. $(5 - 3i) - (2 - 6i)$

$3 + 3i$

11. $5i + (3 - 7i)$

$3 - 2i$

12. $3i(-2 + 4i)$

$-12 - 6i$

13. $(4 - 3i)(1 + 2i)$

$10 + 5i$

14. Complex numbers are used in electronics to describe the current in an electric circuit. In an alternating current, the resistance, R, in ohms, is the measure of how much the circuit resists (or impedes) the flow of current through it. The resistance, R, is related to the voltage, V, and current, I, by Ohm's Law:

$$V = IR$$

a. If $I = 0.3 + 2i$ amperes and $R = 0.5 - 3i$ ohms, determine the voltage, V.

$V = 6.15 + 0.1i$

b. If $I = 2 - 3i$ amperes and $R = 3 + 5i$ ohms, determine the voltage, V.

$V = 21 + i$

In Exercises 15–18, solve the quadratic equations in the complex number system using the quadratic formula. Verify your real solutions graphically. Verify that no real solutions mean no x-intercepts.

15. $3x^2 - 2x + 7 = 0$

$x = \frac{1}{3} \pm \frac{\sqrt{80}}{6}i$, or $\frac{1}{3} \pm \frac{2\sqrt{5}}{3}i$

no x-intercepts

16. $x^2 + x = 3$

$x = \frac{-1 \pm \sqrt{13}}{2}$, or $x \approx 1.303$ and $x \approx -2.303$

17. $2x^2 + 5x = 7$

$x = 1, -3.5$

18. $0.5x^2 - x + 3 = 0$

$x = 1 \pm i\sqrt{5}$

no x-intercepts

In Exercises 19–24, determine the number and type of solutions of each equation by examining the discriminant.

19. $2x^2 + 3x - 5 = 0$

$b^2 - 4ac = 49$

2 real solutions

20. $6x^2 + x + 5 = 0$

$b^2 - 4ac = -119$

2 complex solutions

21. $4x^2 - 4x + 1 = 0$

$b^2 - 4ac = 0$

1 real solution

22. $9x^2 + 6x + 1 = 0$

$b^2 - 4ac = 0$

1 real solution

23. $12x^2 = 4x - 3$

$12x^2 - 4x + 3 = 0$

$b^2 - 4ac = -128$

2 complex solutions

24. $3x^2 = 5x + 7$

$3x^2 - 5x - 7 = 0$

$b^2 - 4ac = 109$

2 real solutions

CLUSTER 1 What Have I Learned?

1. a. For the graph of the equation $y = ax^2 + bx + c$ to be a parabola, the value of the coefficient of x^2 cannot be zero. Explain.

   If $a = 0$, the equation would be $y = bx + c$, the equation of a line.

   b. What is the vertex of a parabola having an equation of the form $y = ax^2$?

   (0, 0)

   c. Describe the relationship between the vertex and the vertical intercept of the graph of $y = ax^2 + c$.

   The vertex and the y-intercept are the same, $(0, c)$.

2. Determine whether the vertex is a minimum point or a maximum point of $y = ax^2 + bx + c$ in each of the following situations:

   a. $a < 0$

   maximum

   b. $a > 0$

   minimum

3. a. What are the possibilities for the number of vertical intercepts of the graph of a quadratic function?

   There is always one vertical intercept.

   b. What are the possibilities for the number of horizontal intercepts of the graph of a parabola?

   There may be 0, 1, or 2 horizontal intercepts.

4. What is the relationship between the vertex and the x-intercept of the graph of $y = x^2 - 4x + 4$?

   They are the same, (2, 0).

5. a. The vertex of a parabola is (3, 1). Using this information, complete the following table:

x	1	2	3	4	5
y	5	2	1	2	5

   b. Assuming that the vertex of a parabola is (2, 4), complete the following table:

x	−2	0	2	4	6
y	0	3	4	3	0

6. a. Given the following graph, explain why choices i, ii, and iii do not fit the curve:

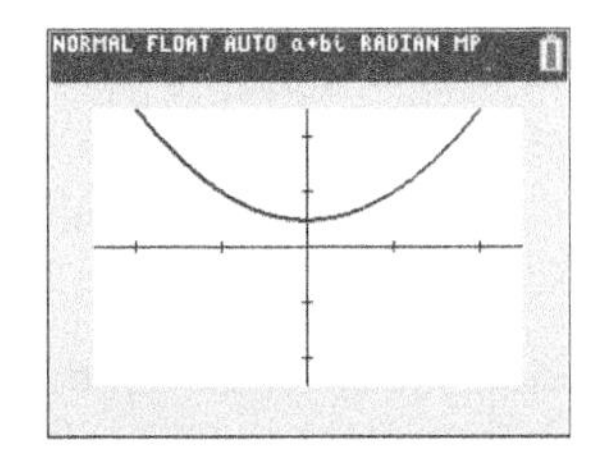

   i. $f(x) = ax^2 + bx$ with $a > 0, b < 0$

   $b < 0$ indicates that the vertex would not be on the y-axis.

ii. $g(x) = ax^2 + c$ with $a < 0, c > 0$

$a < 0$ indicates that the parabola would open downward.

iii. $h(x) = ax^2 + bx + c$ with $a < 0, b > 0, c < 0$

$c < 0$ indicates that the y-intercept is negative, and $a < 0$ indicates that the parabola would open downward.

b. What restrictions on a, b, and c are necessary to fit $y = ax^2 + bx + c$ to the given graph?

$a > 0$, the parabola opens upward; $b = 0$, the vertex is on the y-axis; and $c > 0$, the y-intercept is positive.

7. Review the steps in the following solution. Is the solution correct? Explain.

$$x^2 - 3x - 4 = 6$$
$$(x - 4)(x + 1) = 6$$
$$x - 4 = 6 \quad \Big| \quad x + 1 = 6$$
$$x = 10 \quad \Big| \quad x = 5$$

No. In the proposed solution, the equation is not set equal to zero.

When the product of two numbers is 6, there is no guarantee that either factor is equal to 6.

8. Describe how you would determine the solutions to $ax^2 + bx + c > 5$ graphically.

Graph $y = ax^2 + bx + c - 5$. Determine the x-intercepts. Determine all x-values for which the graph is above the x-axis.

9. Which of the following statements are true? In each case, justify your decision.

a. $3 + 2i$ is a pure imaginary number.

False; a pure imaginary number has a real part of 0, not 3.

b. $\sqrt{-7}$ is a complex number.

True; it can be written as $0 + i\sqrt{7}$.

c. 0 is a complex number.

True; it can be written as $0 + 0i$.

10. a. Describe the relationship between the x-intercepts (if they exist) of the graph of $y = ax^2 + bx + c$ and the solutions to the equation $ax^2 + bx + c = 0$.

The solutions (real-valued) of $ax^2 + bx + c = 0$ are precisely the x-coordinates of the x-intercepts of the associated graphs.

b. Describe the relationship between the x-intercepts (if they exist) of the graph of $y = ax^2 + bx + c$ and the discriminant $b^2 - 4ac$.

If the x-intercepts exist, you can calculate $\sqrt{\textit{discriminant}}$ and divide this value by $2a$ to obtain a real number. This number represents the number of units on either side of the axis of symmetry that must be moved to locate the x-intercepts.

11. Consider the quadratic equation $ax^2 + bx + c = 0$. If the quadratic expression $ax^2 + bx + c$ is factorable over the real numbers, what can you say about the sign of the discriminant, $b^2 - 4ac$? Is it positive, negative, or zero? Explain.

$b^2 - 4ac$ must be positive or 0. If it is negative, there are no real solutions.

There would be two complex solutions. If the discriminant was negative, the quadratic expression would not be factorable.

12. For what values of c are the solutions to $2x^2 - 5x + c = 0$ imaginary?

The discriminant would be negative. So $b^2 - 4ac = 25 - 8c < 0$; $c > \frac{25}{8}$.

13. For what values of k does $x^2 - kx + k = 0$ have only one solution? (*Hint:* Examine the discriminant.)

The discriminant must be 0; $a = 1$, $b = -k$, $c = k$. Substituting, you have the following:

$$b^2 - 4ac = k^2 - 4k = 0$$
$$k(k - 4) = 0$$
$$k = 0, 4$$

CLUSTER 1 How Can I Practice?

1. Complete the following table:

EQUATION OF THE FORM $y = ax^2 + bx + c$	VALUE OF a	VALUE OF b	VALUE OF c
$y = 5x^2$	5	0	0
$y = \frac{1}{3}x^2 + 3x - 1$	$\frac{1}{3}$	3	−1
$y = -2x^2 + x$	−2	1	0

For Exercises 2–7, determine the following characteristics for each graph:

a. *The direction in which the parabola opens*

b. *The equation of the axis of symmetry*

c. *The vertex; determine whether maximum or minimum*

d. *The y-intercept*

2. $y = -2x^2 + 4$

a. downward

b. $x = 0$

c. (0, 4); maximum

d. (0, 4)

3. $y = \frac{2}{3}x^2$

a. upward

b. $x = 0$

c. (0, 0); minimum

d. (0, 0)

4. $f(x) = -3x^2 + 6x + 7$

a. downward

b. $x = 1$

c. (1, 10); maximum

d. (0, 7)

5. $f(x) = 4x^2 - 4x$

a. upward

b. $x = \frac{1}{2}$

c. $\left(\frac{1}{2}, -1\right)$; minimum

d. (0, 0)

6. $y = x^2 + 6x + 9$

a. upward

b. $x = -3$

c. (−3, 0); minimum

d. (0, 9)

7. $y = x^2 - x + 1$

a. upward

b. $x = \frac{1}{2}$

c. $\left(\frac{1}{2}, \frac{3}{4}\right)$; minimum

d. (0, 1)

For Exercises 8–11, use your graphing calculator to sketch the graph of each quadratic function; then determine the following for each function:

a. *The coordinates of the x-intercepts (if they exist)*

b. *The domain and range*

c. *The horizontal interval over which the function is increasing*

d. *The horizontal interval over which the function is decreasing*

All exercises in How Can I Practice are answered in the Selected Answers appendix.

8. $y = -x^2 + 4$

a. $(-2, 0), (2, 0)$

b. D: all real numbers; R: $y \leq 4$

c. $x < 0$

d. $x > 0$

9. $y = x^2 - 5x + 6$

a. (2, 0), (3, 0)

b. D: all real numbers; R: $y \geq -0.25$

c. $x > 2.5$

d. $x < 2.5$

10. $y = -3x^2 - 6x + 8$

a. $(0.91, 0), (-2.91, 0)$

b. domain: all real numbers; range: $y \leq 11$

c. $x < -1$

d. $x > -1$

11. $y = 0.22x^2 - 0.71x + 2$

a. none

b. domain: all real numbers; range: $y \geq 1.427$

c. $x > 1.61$

d. $x < 1.61$

12. Use your graphing calculator to approximate the vertex of the graph of the parabola defined by the equation $y = -2x^2 + 3x + 25$.

(0.75, 26.125)

13. Completely factor the following polynomials if possible:

a. $9a^5 - 27a^2$

$9a^2(a^3 - 3)$

b. $24x^3 - 6x^2$

$6x^2(4x - 1)$

c. $4x^3 - 16x^2 - 20x$

$4x(x^2 - 4x - 5)$

$4x(x - 5)(x + 1)$

d. $5x^2 - 16x + 6$

cannot be factored using real numbers

e. $x^2 - 5x - 24$

$(x - 8)(x + 3)$

f. $y^2 + 10y + 25$

$(y + 5)^2$

14. Determine one solution of the following quadratic equations numerically. That is, construct a table of (x, y) ordered pairs, where $y = f(x)$, and estimate the value of x (input) that results in the required y-value (output).

a. $5x^2 = 7$

b. $x^2 - 7x + 10 = 5$

c. $3x^2 - 5x = 2$

a.

x	1	1.1	1.2	1.3	1.4	1.5
y	5.00	6.05	7.20	8.45	9.80	11.25

$x \approx 1.2$

b.

x	0.5	0.6	0.7	0.8	0.9	1
y	6.75	6.16	5.59	5.04	4.51	4.00

$x \approx 0.8$

c.

x	0	1	2	3	4	5
y	0	−2	2	12	28	50

$x = 2$

15. Solve each of the equations from Exercise 14 using the quadratic formula. When necessary, round your solutions to the nearest tenth. Check your solutions by graphing.

a. $x = 1.2, -1.2$

b. $x = 0.8, 6.2$

c. $x = -\frac{1}{3}, 2$

16. Solve each of the following equations by factoring:

a. $4x^2 - 8x = 0$

$4x(x - 2) = 0$

$x = 0, 2$

b. $x^2 - 6 = 7x + 12$

$x^2 - 7x - 18 = 0$

$(x - 9)(x + 2) = 0$

$x = 9, -2$

c. $2x(x - 4) = -6$

$2x^2 - 8x = -6$

$2x^2 - 8x + 6 = 0$

$2(x^2 - 4x + 3) = 0$

$2(x - 3)(x - 1) = 0$

$x = 3, 1$

d. $x^2 - 8x + 16 = 0$

$(x - 4)(x - 4) = 0$

$x = 4, 4$

e. $x^2 - 2x - 24 = 0$

$(x - 6)(x + 4) = 0$

$x = 6, -4$

f. $y^2 - 2y - 35 = -20$

$y^2 - 2y - 15 = 0$

$(y - 5)(y + 3) = 0$

$y = 5, -3$

g. $a^2 + 2a + 1 = 3a + 7$

$a^2 - a - 6 = 0$

$(a - 3)(a + 2) = 0$

$a = 3, -2$

h. $4x^2 + 4x - 3 = -3x - 1$

$4x^2 + 7x - 2 = 0$

$(4x - 1)(x + 2) = 0$

$x = \frac{1}{4}, -2$

17. Solve the following inequalities using a graphing approach:

a. $x^2 + 6x - 16 < 0$

$-8 < x < 2$

b. $x^2 + 6x - 16 > 0$

$x < -8$ or $x > 2$

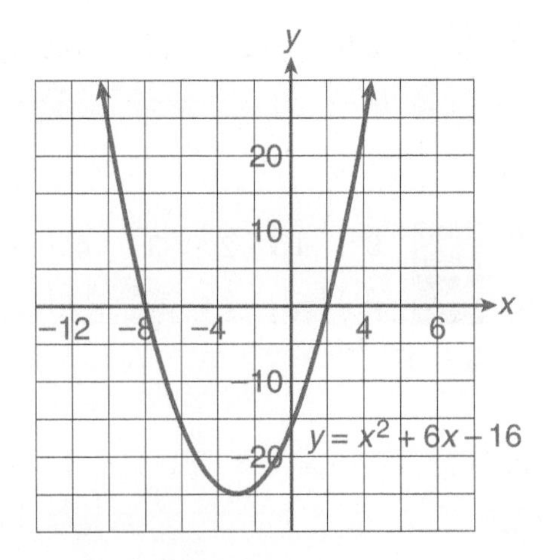

18. A fastball is hit straight up over home plate. The ball's height, h (in feet), from the ground is modeled by

$$h(t) = -16t^2 + 80t + 5,$$

where t is measured in seconds.

a. What is the maximum height of the ball above the ground?

$$t = \frac{-b}{2a} = \frac{-80}{-32} = 2.5$$

$h(2.5) = -16(2.5)^2 + 80(2.5) + 5 = 105$

The maximum height is 105 feet.

b. Write an equation to determine how long it will take the ball to reach the ground. Solve the equation using the quadratic formula. Check your solution by graphing.

Using the calculator to solve $-16t^2 + 80t + 5 = 0$, $t \approx 5.06$ seconds.

c. Write the equation you would need to determine when the ball is 101 feet above the ground.

$-16t^2 + 80t + 5 = 101$

d. Solve the equation you determined in part c algebraically to determine the time it will take the ball to reach a height of 101 feet. Verify your results graphically.

$$-16t^2 + 80t + 5 = 101$$
$$-16t^2 + 80t - 96 = 0$$
$$-16(t - 2)(t - 3) = 0$$
$$t = 2, 3$$

The ball reaches a height of 101 feet after 2 seconds on the way up and 1 second later on the way down.

19. A suspension bridge (shown in the accompanying figure) is 100 meters long. The bridge is supported by cables attached to the tops of towers 35 meters high at each end of the bridge. The cables hang from the towers approximately in the shape of a parabola. The height, $h(x)$ (in meters), of the cables above the surface of the roadway is modeled by

$$h(x) = 0.01x^2 - x + 35,$$

where x is the horizontal distance measured from the point where the left tower and roadway meet.

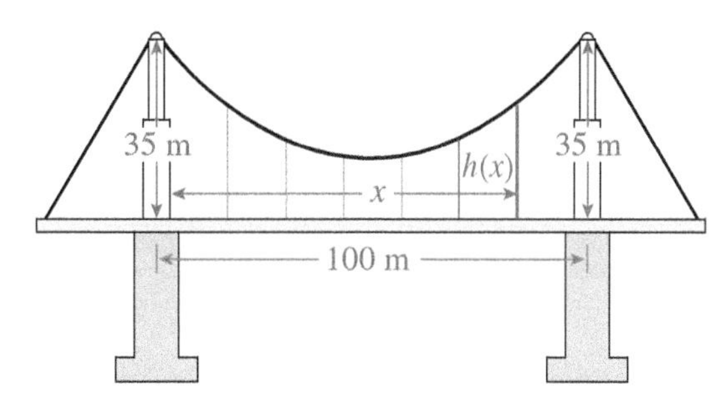

a. Use your graphing calculator to examine the height function. What is the practical domain of this function?

$0 \le x \le 100$

b. What is the minimum distance of the cables from the roadway?

$h(50) = 0.01(50)^2 - 50 + 35 = 10$ m

20. Use the following data set to perform the tasks in parts a–f:

x	0	1	3	5	7	8
y	10	4	−18	−54	−107	−145

a. Determine an appropriate scale, and plot these points.

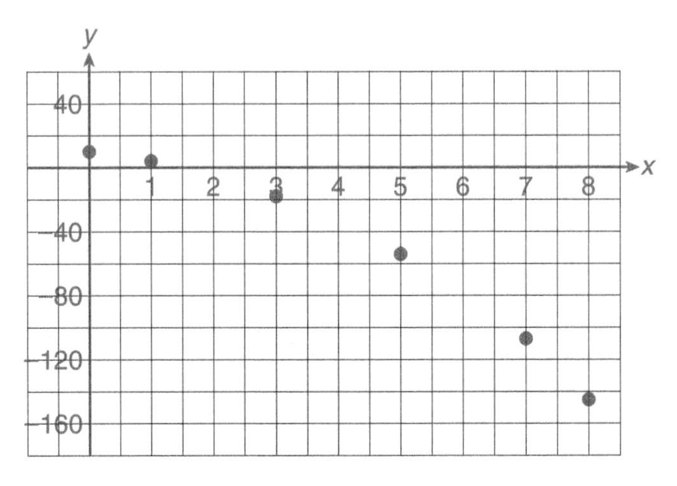

b. Use your graphing calculator to determine the quadratic regression equation for this data set.

$y = -2.096x^2 - 2.25x + 9.038$

c. Graph the regression equation on the same coordinate axes as the data points.

20. c.

d. Compare the predicted outputs with the given outputs in the table.

Predicted values are very close to the actual values.

e. What is the predicted output for $x = 4$ and for $x = 9$?

Approximately −33.5, −181

f. For what value(s) of x is $y = -40$? Use the quadratic formula.

$-40 = -2.096x^2 - 2.25x + 9.038$

$2.096x^2 + 2.25x - 49.038 = 0$

$$x = \frac{-2.25 \pm \sqrt{2.25^2 - 4(2.096)(-49.038)}}{2(2.096)}$$

$x \approx 4.330, -5.403$

Write each of the following in the form bi, where $i = \sqrt{-1}$:

21. $\sqrt{-49}$

$7i$

22. $\sqrt{-45}$

$3i\sqrt{5}$

23. $\sqrt{-121}$

$11i$

24. $\sqrt{-15}$

$i\sqrt{15}$

25. $\sqrt{-112}$

$4i\sqrt{7}$

26. $\sqrt{-125}$

$5i\sqrt{5}$

27. $\sqrt{-\frac{16}{25}}$

$\frac{4}{5}i$

28. $\sqrt{\frac{-24}{42}}$

$\sqrt{\frac{4}{7}}i = \frac{2}{\sqrt{7}}i$

Perform the following operations, and express your answer in the form $a + bi$. Use your graphing calculator to verify the results.

29. $(7 + 5i) + (-3 - 2i)$

$4 + 3i$

30. $(3 - 3i) - (8 - 9i)$

$-5 + 6i$

31. $3i + (6 - 7i)$

$6 - 4i$

32. $-3i(8 - 4i)$

$-12 - 24i$

33. $(3 - 4i)(-1 + 2i)$

$5 + 10i$

34. In parts a–d, perform the following tasks:

i. Identify the values of a, b, and c in $ax^2 + bx + c = 0$.

ii. Determine the type of solution by examining the sign of the discriminant.

iii. Solve the given equation using the quadratic formula. If necessary, round your solutions to the nearest hundredth.

iv. Check your real solutions by graphing as well as by substitution.

a. $3x^2 - x = 7$

$a = 3, b = -1, c = -7$

$b^2 - 4ac = 85$

two real solutions

$x = 1.70, -1.37$

b. $x^2 - 4x + 10 = 0$

$a = 1, b = -4, c = 10$

$b^2 - 4ac = -24$

two complex solutions

$x = 2 \pm i\sqrt{6}$

c. $2x^2 - 3x = 2x + 3$

$a = 2, b = -5, c = -3$

$b^2 - 4ac = 49$

two real solutions

$x = 3, -0.5$

d. $3x(3x - 2) + 1 = 0$

$a = 9, b = -6, c = 1$

$b^2 - 4ac = 0$

one real solution

$x = \frac{1}{3}$

35. Each of the following graphs represents a quadratic function. For each graph, determine whether the discriminant is positive, negative, or zero. Explain your decision.

a. 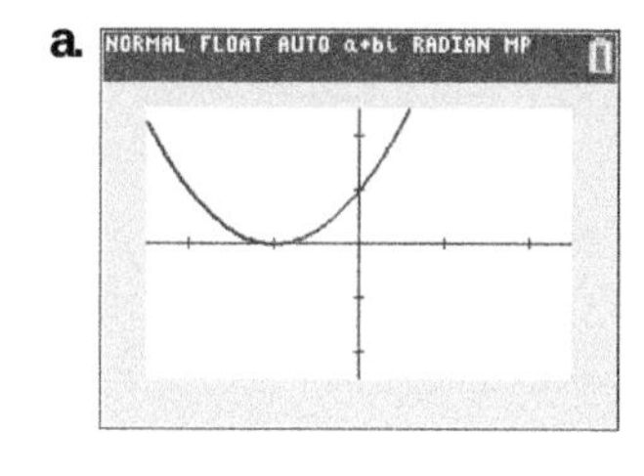

b. 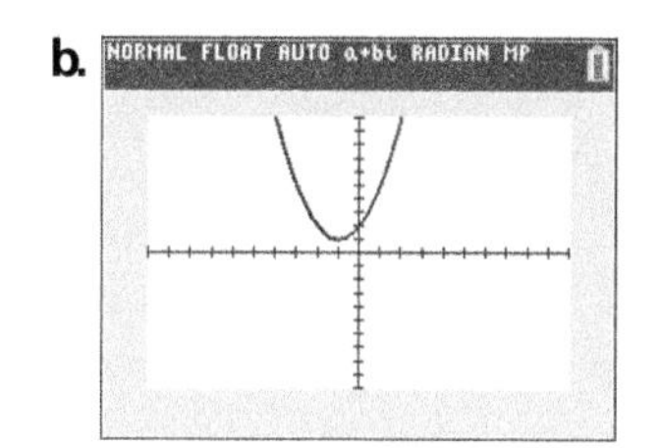

c. 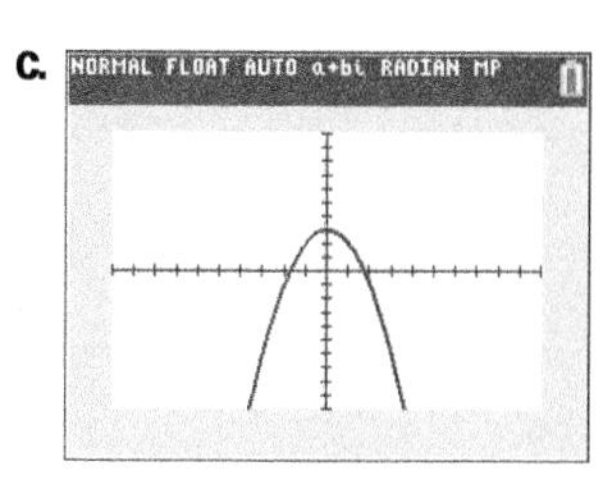

i. graph a

The discriminant is 0. The graph only touches the x-axis, indicating that there is one real solution.

ii. graph b

The discriminant is negative. The graph does not intersect the x-axis, indicating that there is no real solution.

iii. graph c

The discriminant is positive. The graph intersects the x-axis twice, indicating that there are two real solutions.

36. The following table shows the amount of online advertising revenue in the United States from 2009 to 2016, as reported by the Internet Advertising Bureau:

Year	2009	2010	2011	2012	2013	2014	2015	2016
Online Advertising Revenue, R (billions of dollars)	22.7	26.0	31.7	36.6	42.8	49.5	59.6	72.5

36. a.

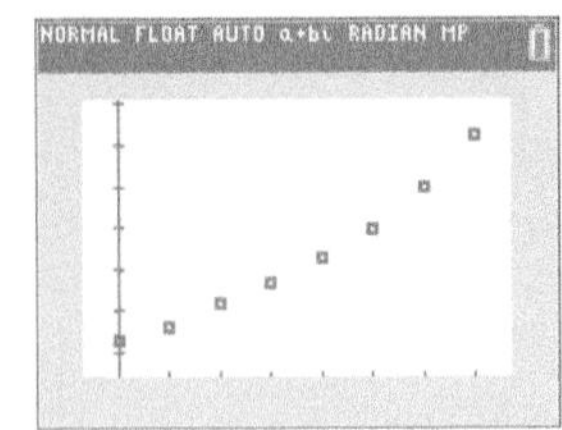

a. Use your graphing calculator to create a scatterplot of the data from the table. Let your input, t, represent the number of years since 2009, and your output, $R(t)$, be the online advertising revenue in billions of dollars.

b. Determine the quadratic regression model that best fits this data. Round your coefficients to two decimal places.

$R(t) = 0.66t^2 + 2.22t + 23.31$

c. Use the regression model to predict the online advertising revenue in 2020.

$2020 - 2009 = 11$. $R(11) = 0.66(11)^2 + 2.22(11) + 23.31 = 127.59$ billion dollars

36. d.

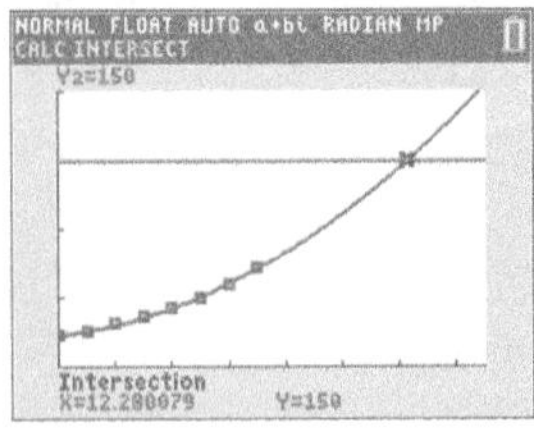

d. Use the regression model to predict the year in which the online advertising revenue will first reach 150 billion dollars.

The model shows 12.27 years. Rounding up to 13 years after 2009 or in the year 2022, the model predicts that online advertising revenue will first reach 150 billion dollars.

e. How confident are you in your answer to part d? Explain.

The model is projecting 6 years past the last data point, but a lot can happen, so I am not terribly confident.

CLUSTER 2 Curve Fitting and Higher-Order Polynomial Functions

ACTIVITY 4.8

The Power of Power Functions

Direct Variation Functions and Their Graphs

OBJECTIVES

1. Identify a direct variation function.
2. Determine the constant of variation.
3. Identify the properties of graphs of power functions defined by $y = kx^n$, where n is a positive integer, $k \neq 0$.

You are traveling in a hot air balloon when suddenly your binoculars drop from the edge of the balloon's basket. At that moment, the balloon is maintaining a constant height of 500 feet. The distance of the binoculars from the edge of the basket is modeled by

$$s = 16t^2.$$

The following table gives the distance, s (in feet), from the drop point at various times, t (in seconds):

Time, t (sec.)	0	1	2	3	4
Distance, s (ft.)	0	16	64	144	256

As the input values (units of time) increase, the corresponding output values (units of distance) increase. Let us look more closely at how this increase takes place.

Because $s = 16t^2$, you can say that the output, s, **varies directly** as the square of the input, t. Therefore, as t doubles in value from 1 to 2 or from 2 to 4, the corresponding output values become 4 times as large: increasing from 16 to 64 or 64 to 256.

1. a. As t triples from 1 to 3, the corresponding s-values become __9__ times as large.

   b. In general, if y varies directly as the square of x, then when x becomes n times as large, the corresponding y-values become __n^2__ times as large.

The volume, V, of a sphere is given by $V = \frac{4}{3}\pi r^3$. In this situation, you can say that the output, V, varies directly as the cube of the radius, r.

2. a. Complete the following table. Leave your answers for V in terms of π.

r	1	2	3	4	8
V	$\frac{4}{3}\pi$	$\frac{32}{3}\pi$	36π	$\frac{256}{3}\pi$	$\frac{2048}{3}\pi$

   Note that as r-values double from 2 to 4, the corresponding V-values increase from $\frac{32}{3}\pi$ to $\frac{256}{3}\pi$.

   b. As r doubles from 2 to 4 or from 4 to 8, the corresponding V-values become __8__ times as large.

   c. In general, if y varies directly as the cube of x, then when x becomes n times as large, the corresponding y-values become __n^3__ times as large.

d. Sketch a graph of the volume function. What is the practical domain of this function?

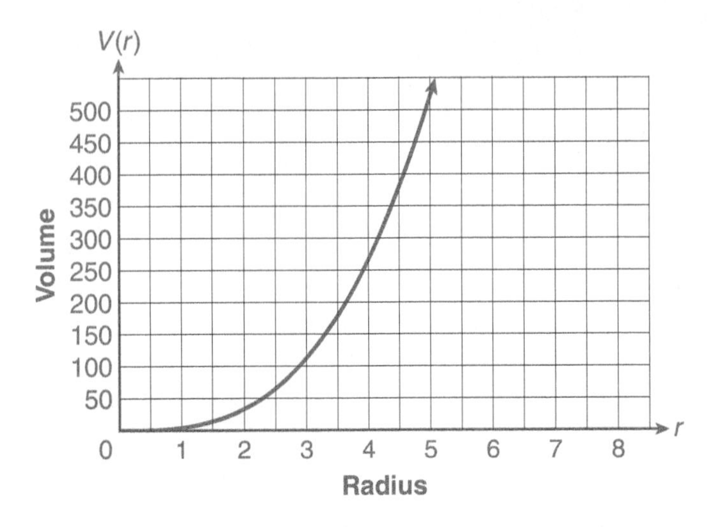

The value of r should be greater than 0 (a radius of 0 gives a point).

DEFINITION

The equation

$$y = kx^n,$$

where $k \neq 0$ and n is a positive integer, defines a **direct variation** function in which y varies directly as x^n. The constant, k, is called the **constant of variation**.

EXAMPLE 1 ***The constant of variation, k, in the free-falling object situation defined by $s = 16t^2$ is 16.***

3. What is the constant of variation for the direct variation function defined by $V = \frac{4}{3}\pi r^3$?

$\frac{4}{3}\pi$

In the falling binocular situation, you are given the direct variation equation. Suppose you only know that the distance, s, varies directly as the square of t and one data pair. Are you able to determine the direct variation equation? Example 2 demonstrates the process.

EXAMPLE 2 ***Let s vary directly as the square of t. If $s = 64$ when $t = 2$, determine the direct variation equation.***

LC LEARNING CATALYTICS

If s varies directly as the square of t, and if $s = 54$ when $t = 3$, determine the value of the constant of variation, k.

SOLUTION

Because s varies directly as the square of t, you have

$$s = kt^2,$$

where k is the constant of variation. Substituting 64 for s and 2 for t, you have

$$64 = k(2)^2, \quad \text{or} \quad 64 = 4k, \quad \text{or} \quad k = 16.$$

Therefore, the direct variation equation is

$$s = 16t^2.$$

4. For each table, determine the pattern and complete the table. Then write a direct variation equation for each table.

a. y varies directly as x.

x	1	2	4	8	12
y	6	12	24	48	72

$12 = 2k$, so $k = 6$ and $y = 6x$.

b. y varies directly as x^3.

x	1	2	3
y	4	32	108

$32 = k(2)^3$, so $k = 4$ and $y = 4x^3$.

c. y varies directly as x.

x	1	2	3	4	5
y	1	2	3	4	5

$3 = 3k$, so $k = 1$ and $y = x$.

5. The length, L (in feet), of the skid distance left by a car varies directly as the square of the initial velocity, v (in miles per hour), of the car.

a. Write a general equation for L as a function of v. Let k represent the constant of variation. $L = kv^2$

b. Suppose a car traveling 40 miles per hour leaves a skid distance of 60 feet. Use this information to determine the value of k. $60 = k(40)^2$, $k = 0.0375$

c. Use the function to determine the length of the skid distance left by the car traveling 60 miles per hour. $L = 0.0375(60)^2$, $L = 135$ feet

Power Functions

The direct variation functions that have equations of the form $y = kx^n$, where n is a positive integer and $k \neq 0$, are also called **power functions**. The graphs of this family of functions are very interesting and are useful in problem solving.

6. Sketch a graph of each of the following power functions. Use your graphing calculator to verify the graph.

a. $y = x$

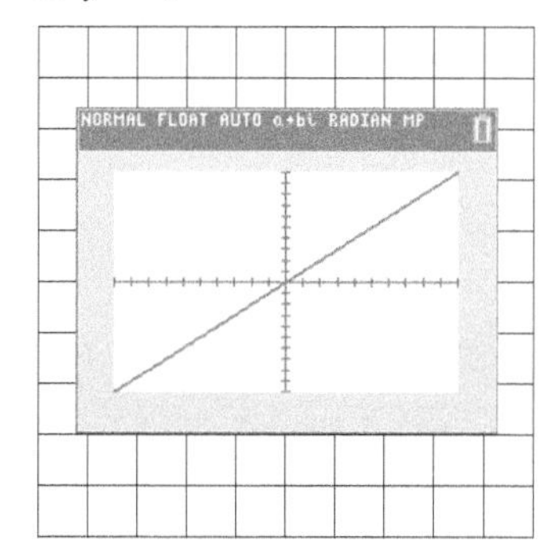

b. $y = x^2$

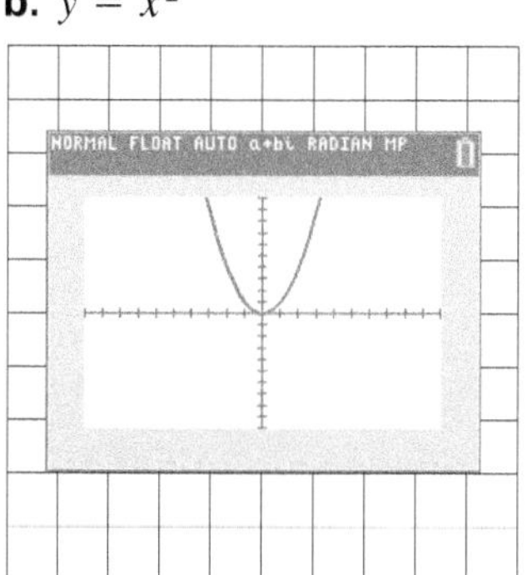

c. $y = x^3$

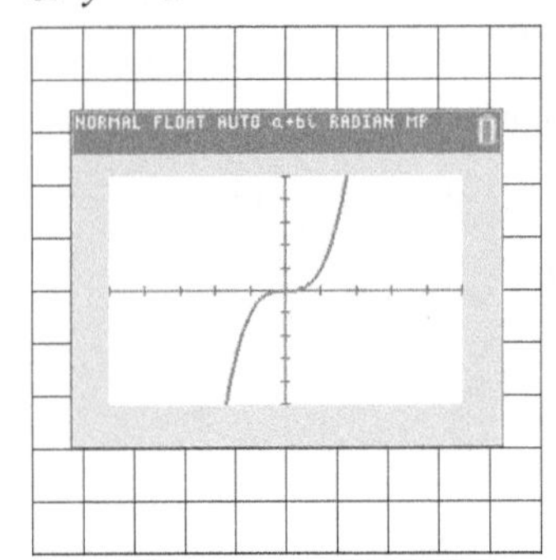

d. $y = x^4$

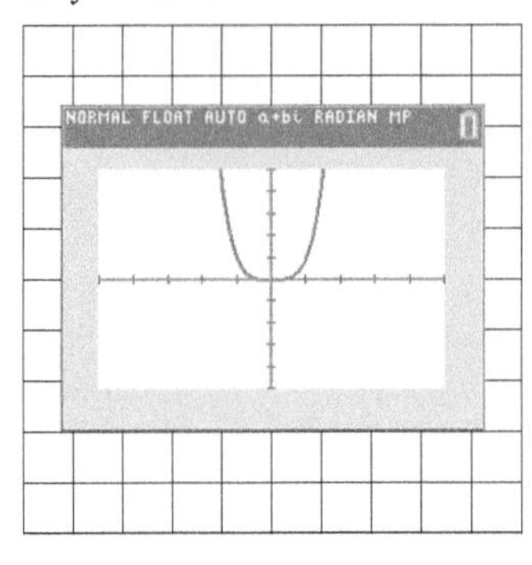

e. $y = x^5$

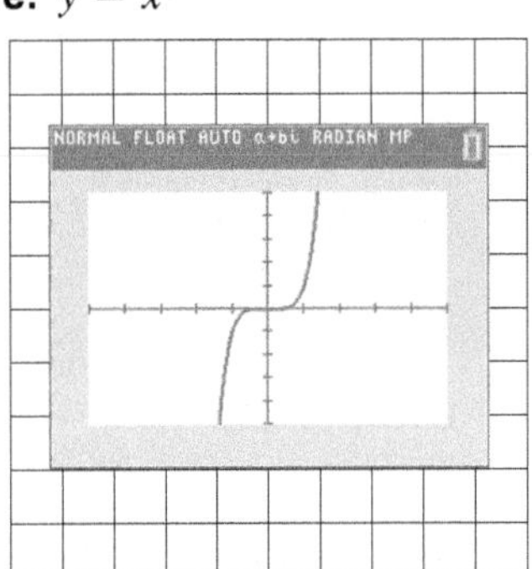

f. $y = x^6$

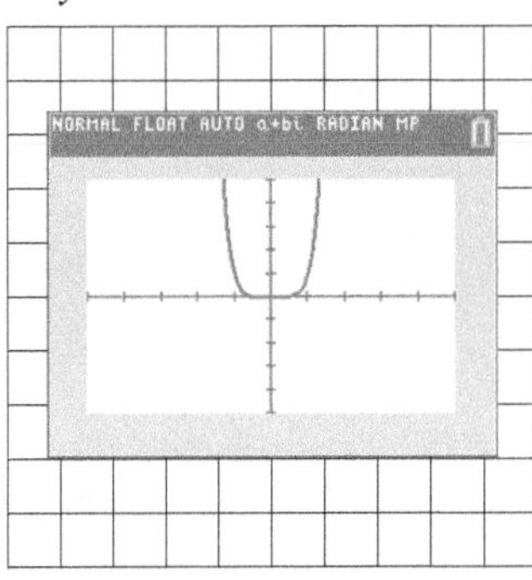

7. Each graph in Problem 6 has an equation of the form $y = x^n$, where n is a positive integer.

a. What is the basic shape of the graph if

i. n is even?

It is U-shaped.

ii. n is odd?

It rises from left to right.

b. If n is even, what happens to the graph as n gets larger in value?

The bottom of the U gets flatter, and the sides get steeper.

c. If n is odd, is the function increasing or decreasing?

increasing

8. Use the patterns from Problem 7 in combination with graphing techniques you learned previously to sketch a graph of each of the following without using a graphing calculator:

a. $y = x^2 + 1$

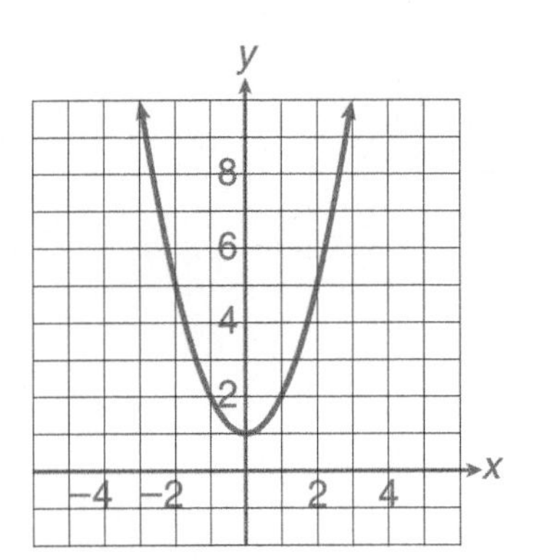

b. $y = -2x^4$

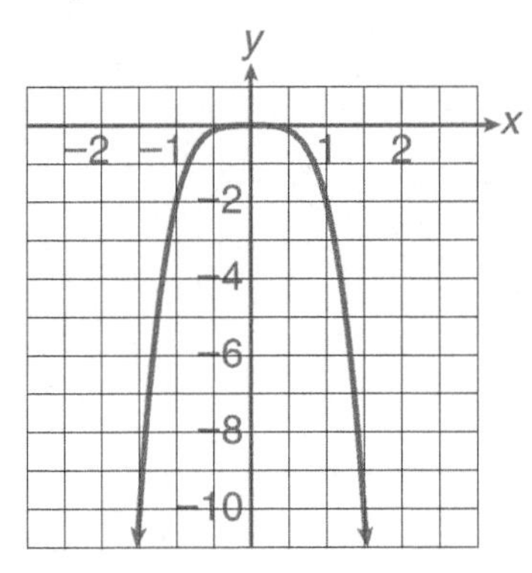

c. $y = 3x^8 + 1$

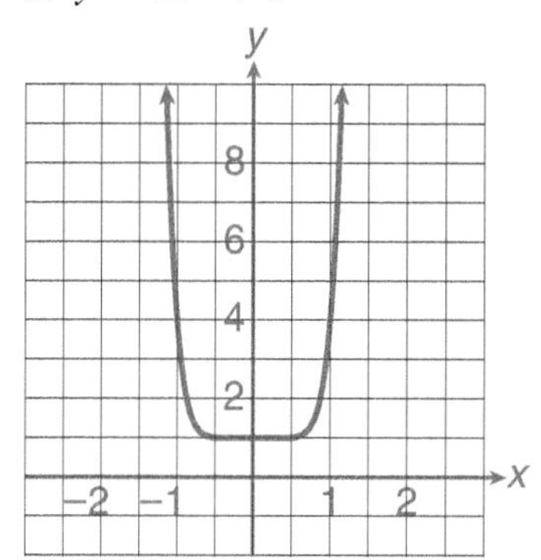

d. $y = -2x^5$

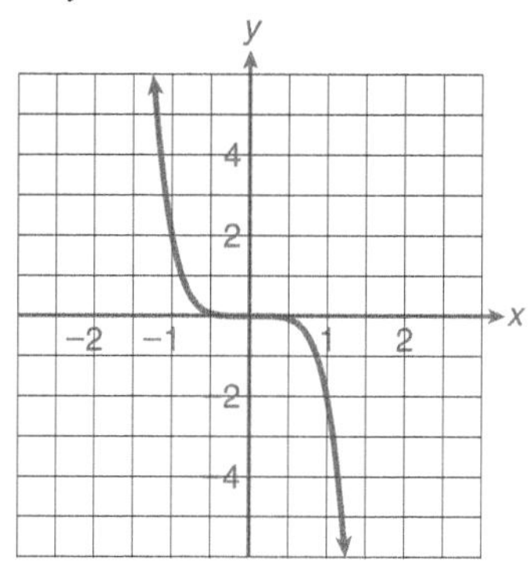

e. $y = x^{10}$

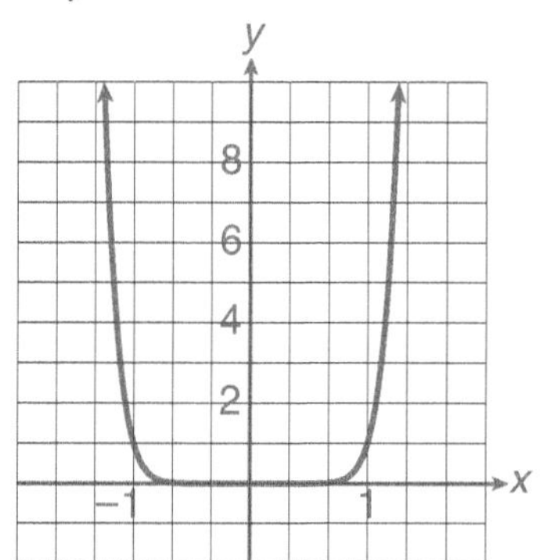

f. $y = 5x^3 + 2$

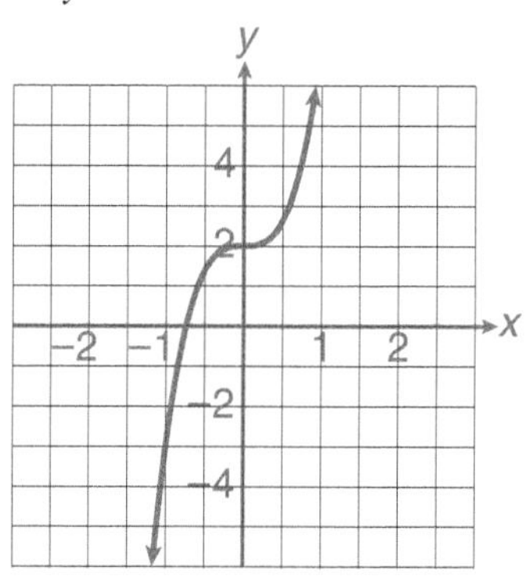

SUMMARY Activity 4.8

1. The equation $y = kx^n$, where $k \neq 0$ and n is a positive integer, defines a **direct variation function**. The constant, k, is called the **constant of variation**.

2. The direct variation functions that have equations of the form $y = kx^n$, where n is a positive integer, are also called **power functions**.

   a. Power functions in which n is even resemble parabolas. As n increases in value, the graph flattens near the vertex.

   b. Power functions in which n is odd resemble the graph of $y = kx^3$. If k is positive, the graph is increasing. If k is negative, the graph is decreasing.

EXERCISES Activity 4.8

1. For each table, determine the pattern and complete the table. Then write a direct variation equation for each table.

   a. y varies directly as x.

x	$\frac{1}{4}$	1	4	8
y	2	8	32	64

$y = kx$ and $8 = k(1)$, so $k = 8$ or $y = 8x$.

Exercise numbers appearing in color are answered in the Selected Answers appendix.

b. y varies directly as x^3.

x	$\frac{1}{2}$	1	3	6
y	$\frac{1}{8}$	1	27	216

$y = kx^3$ and $1 = k(1)^3$, so $k = 1$ or $y = x^3$.

2. The area, A, of a circle is given by the function $A = \pi r^2$, where r is the radius of the circle.

a. Does the area vary directly as the radius? Explain.

No, area varies directly as the square of the radius.

b. What is the constant of variation k?

$k = \pi$

3. Assume that y varies directly as the square of x and that when $x = 2$, $y = 12$. Determine y when $x = 8$.

$y = kx^2$ and $12 = k(2)^2$, so $k = 3$. Therefore, $y = 3x^2$. When $x = 8$, $y = 3(8)^2 = 192$.

4. The distance, d, that you drive at a constant speed varies directly as the time, t, that you drive. If you can drive 150 miles in 3 hours, how far can you drive in 6 hours?

$d = kt$ and $150 = k(3)$, so $k = 50$.

Now $d = 50t$. So in 6 hours, I drive $d = 50(6) = 300$ miles.

5. The number of meters, d, that a skydiver falls before her parachute opens varies directly as the square of the time, t, that she is in the air. A skydiver falls 20 meters in 2 seconds. How far will she fall in 2.5 seconds?

$d = kt^2$ and $20 = k(2)^2$, so $k = 5$.

Now $d = 5t^2$. So in 2.5 seconds, the skydiver travels $d = 5(2.5)^2 = 31.25$ meters.

In Exercises 6–10, sketch a graph of the given power function. Verify your graphs using your graphing calculator.

6. $y = -3x^2$

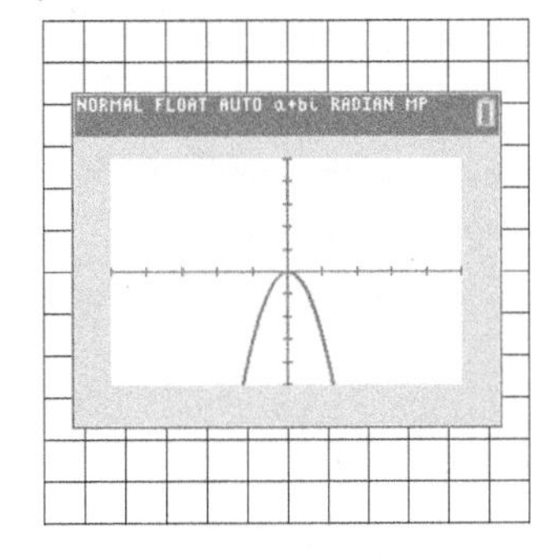

7. $y = x^4 + 1$

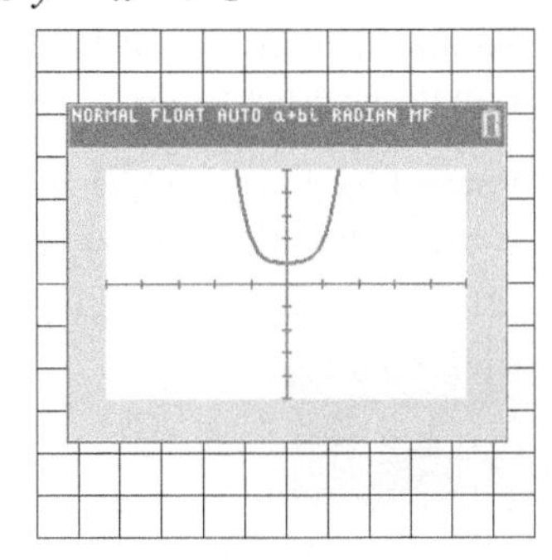

8. $y = -2x^5$

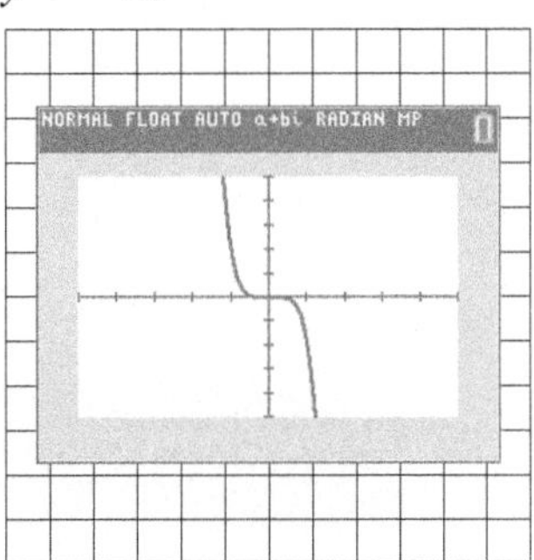

9. $f(x) = x^6$

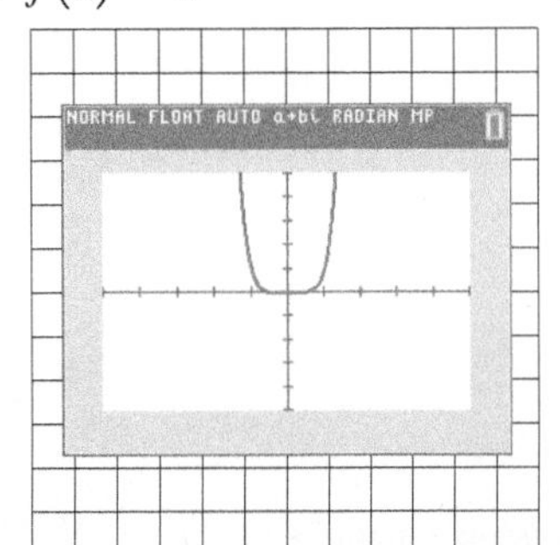

10. $g(x) = 3x^3 - 3$

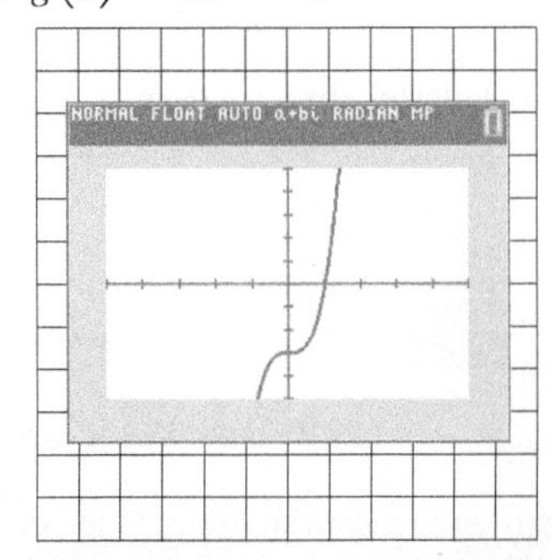

11. Determine the x-interval over which the function $f(x) = \frac{1}{2}x^4$ is increasing.

$f(x)$ is increasing for $x > 0$.

12. Does the function $g(x) = -\frac{1}{2}x^6$ have a maximum or a minimum point? Explain.

$g(x)$ has a maximum point. The graph opens downward, with an upside-down U shape, reaching the maximum at (0, 0).

13. For $x > 1$, is the graph of $y = x^2$ rising more quickly or more slowly than the graph of $y = x^3$? Explain.

The graph of $y = x^2$ is rising more slowly than $y = x^3$ for $x > 1$. Multiplying x^2 by x gives x^3, and this makes a larger output where $x > 1$.

14. Is the graph of $y = \frac{3}{2}x^4$ wider or narrower than the graph of $y = x^4$?

narrower

15. How are the graphs of $y = -2x^3$ and $y = 2x^3 + 1$ different? How are the graphs similar?

The graph of $y = -2x^3$ is decreasing and goes through (0, 0), whereas $y = 2x^3 + 1$ is increasing and does not pass through the origin. Both have a similar S-like shape.

16. a. For $x > 0$, is the graph of $y = x^2$ rising more quickly or more slowly than the graph of $y = 2^x$? Explain.

The graphs are fairly close until $x = 4$, when $y = 2^x$ begins to grow more quickly than $y = x^2$. Eventually, exponential growth is faster than power growth.

b. For $x > 0$, is the graph of $y = x^5$ rising more quickly or more slowly than the graph of $y = 2^x$? Explain.

The graph of $y = x^5$ is steeper than $y = 2^x$ until $x = 22$, when $y = 2^x$ begins to be steeper than $y = x^5$. This is similar to part a, but it took longer to surpass the fifth power.

ACTIVITY 4.9

Volume of a Storage Tank

Polynomial Functions and Their Graphs

OBJECTIVES

1. Identify equations that define polynomial functions.
2. Determine the degree of a polynomial function.
3. Determine the intercepts of the graph of a polynomial function.
4. Identify the properties of the graphs of polynomial functions.

The volume, V (in cubic feet), of a partially cylindrical storage tank of liquid fertilizer is represented by the formula

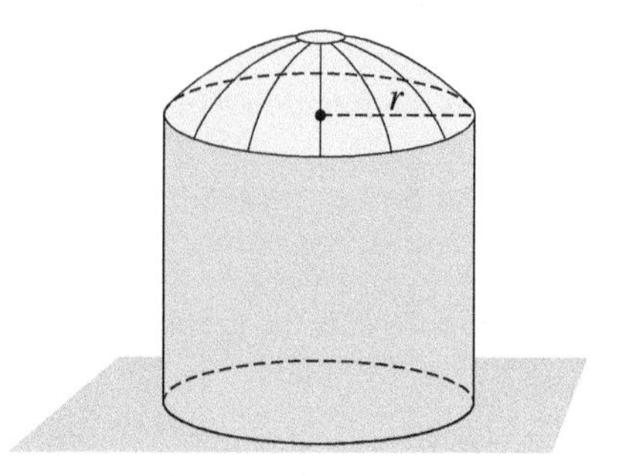

$$V = 2.1r^3 + 37.7r^2,$$

where r is the radius (in feet) of the cylindrical part of the tank.

1. Determine the volume of the tank if its radius is 3 feet.

   $V(3) = 2.1(3)^3 + 37.7(3)^2 = 396$ cubic feet

Polynomial Functions

Recall from Chapter 2 that **polynomial functions** are defined by equations of the form

output = polynomial expression involving the input.

If x and y represent the input and output, respectively, then y must equal sums and differences of terms of the form ax^n, where n is a whole number. The largest exponent on the input variable, n, is called the **degree** of the function. The following example gives several types of polynomial functions.

EXAMPLE 1 ***Examples of polynomial functions are listed in the following table:***

POLYNOMIAL FUNCTION	DEGREE OF THE POLYNOMIAL	NAME
$y = 3x - 2$	1	linear
$y = 2x^2 + 3x - 4$	2	quadratic
$y = 3x^3 - x - 4$	3	cubic
$y = 0.2x^4 - 2x^2 + 7x - 1$	4	quartic
$y = -2x^5 + 3x^4 + 2x - 6$	5	quintic

Note that the cubic function defined by $y = 3x^3 - x - 4$ can be written as $y = 3x^3 + 0x^2 - x - 4$. Also, the degree of a constant function defined by $y = c$, is zero.

Because the expression $2.1r^3 + 37.7r^2$ is a polynomial, the volume function defined by $V = 2.1r^3 + 37.7r^2$ is a polynomial function. Because the largest exponent on the input variable r is 3, this function is a **third-degree polynomial function**, or a **cubic function**.

Polynomial Functions of Degree 3 or Greater

You already studied polynomial functions of degree 1 (linear) and degree 2 (quadratic). The remainder of this activity explores some of the properties and shapes of the graphs of polynomial functions having degree 3 or greater.

2. a. What is the domain of the cubic function defined by $y = 2x^3 - 8x^2 - 10x$?

all real numbers

b. Determine the y-intercept of the graph of the cubic function in part a.

(0, 0)

c. Use your graphing calculator to sketch a graph. Use the window Xmin $= -5$, Xmax $= 8$, Ymin $= -50$, Ymax $= 10$, Yscl $= 5$. The graph should appear as follows:

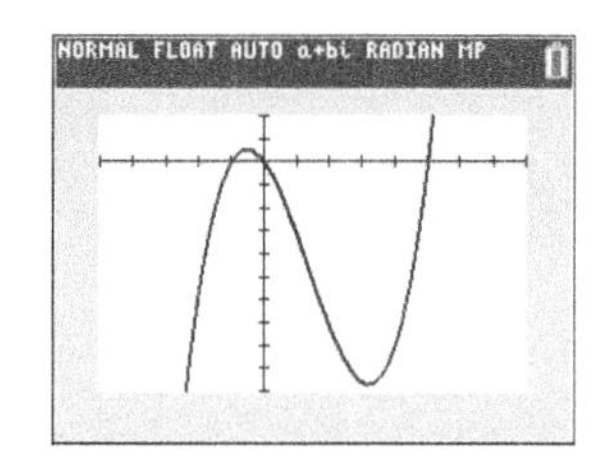

d. Write an equation to determine the x-intercepts of the graph.

$2x^3 - 8x^2 - 10x = 0$

e. Solve the equation in part d using a graphing approach. Use the zero option in the CALC menu of your graphing calculator.

$x = -1, 0,$ or 5

Can the equation $2x^3 - 8x^2 - 10x = 0$ be solved using an algebraic approach? Yes! The solution process is demonstrated in the following example.

LC LEARNING CATALYTICS

Solve $2x^3 + 7x^2 - 15x = 0$ algebraically.

a. $x = 0, \frac{2}{3}, -5$

b. $x = 1, \frac{3}{2}, -5$

c. $x = 0, \frac{3}{2}, -5$

d. $x = 0, -\frac{3}{2}, 5$

EXAMPLE 2 ***Solve the equation $2x^3 - 8x^2 - 10x = 0$ using an algebraic approach.***

SOLUTION

Step 1. The equation is already written in the form *polynomial expression* $= 0$.

Step 2. Completely factor the cubic expression on the left side of the equation.

$2x^3 - 8x^2 - 10x = 0$ Factor out the GCF.

$2x(x^2 - 4x - 5) = 0$ Factor the trinomial.

$2x(x + 1)(x - 5) = 0$

Step 3. Apply the zero-product property, and set each factor equal to zero. Solve each resulting equation.

$2x(x + 1)(x - 5) = 0$

$2x = 0$	$x + 1 = 0$	$x - 5 = 0$
$x = 0$	$x = -1$	$x = 5$

3. Using an algebraic approach (factoring), determine the x-intercepts of each of the following polynomial functions. Verify your results using a graphing approach.

a. $y = 2x^3 + 5x^2 - 12x$

$(0, 0)$, $(-4, 0)$, and $\left(\frac{3}{2}, 0\right)$

b. $f(x) = x^2(x^2 - 5) + 4$

$(-1, 0)$, $(1, 0)$, $(-2, 0)$, and $(2, 0)$

c. $g(x) = 2x^5 - 18x^3$

$(0, 0)$, $(-3, 0)$, and $(3, 0)$

4. a. Returning to the cubic function defined by $y = 2x^3 - 8x^2 - 10x$, the graph shows a high point (maximum point) in quadrant II and a low point (minimum point) in quadrant IV. Note that the maximum and minimum points occur at turning points of the graph. Use the maximum and minimum options in the CALC menu of your graphing calculator to approximate the coordinates of each of these points. Your screens should appear as follows:

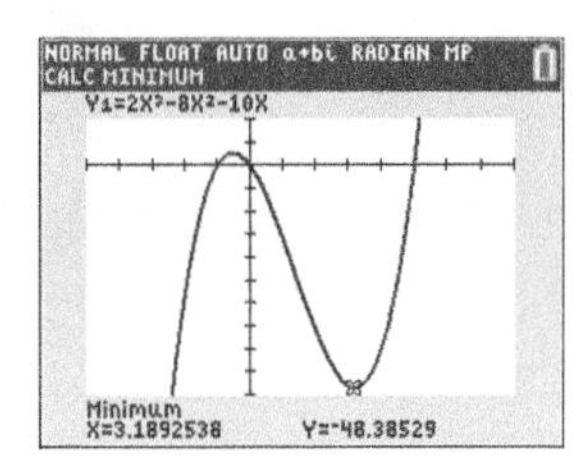

Min $(3.189, -48.385)$

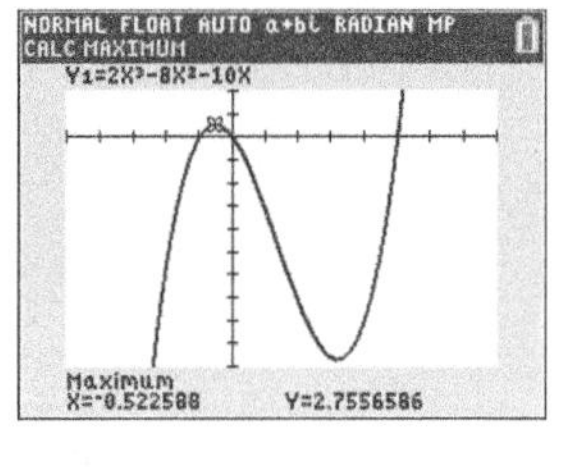

Max $(-0.522, 2.756)$

b. Using the results from part a, determine the interval or intervals along the x-axis where the function is

i. increasing

$x < -0.522$ or $x > 3.189$

ii. decreasing

$-0.522 < x < 3.189$

5. Use your graphing calculator to graph the following third-degree polynomials. Be careful of your choice of windows.

a. $f(x) = x^3$

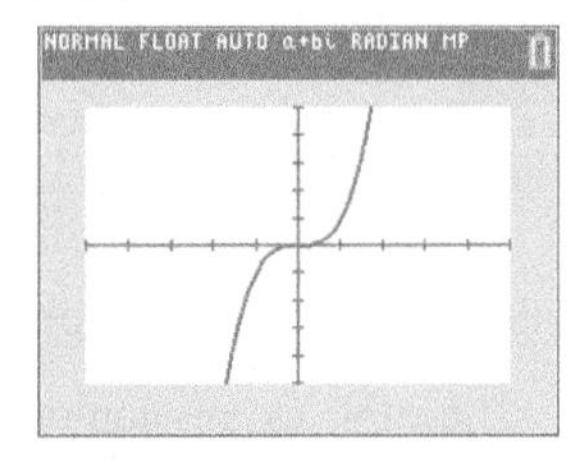

b. $i(x) = 3x^3 - x - 4$

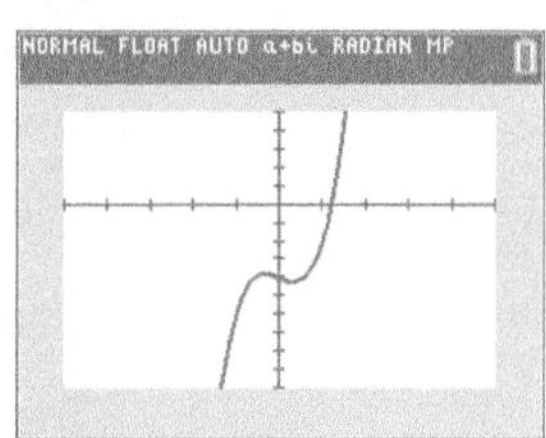

c. $g(x) = 0.2x^3 - 2x + 7$

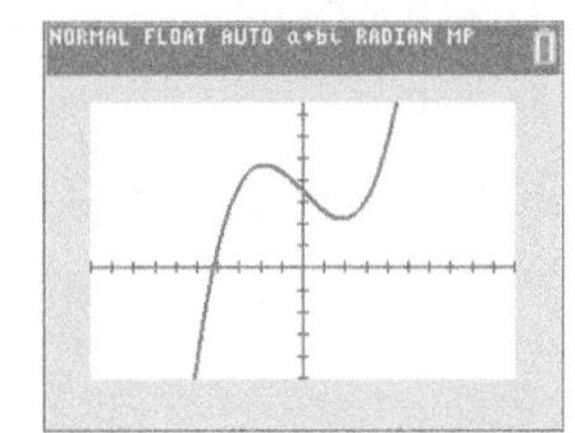

d. $j(x) = -5x^3 + 1$

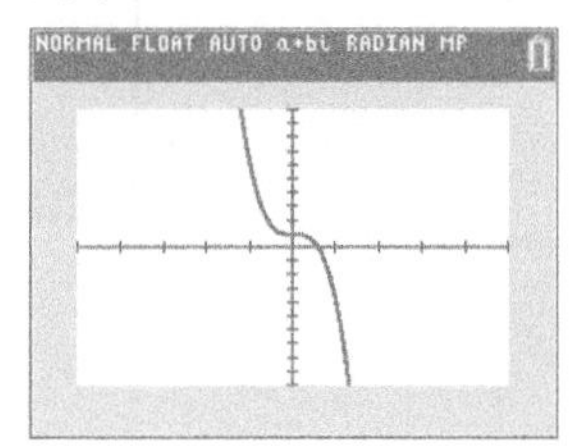

e. $h(x) = -0.6x^3 + 2x^2 - 1$

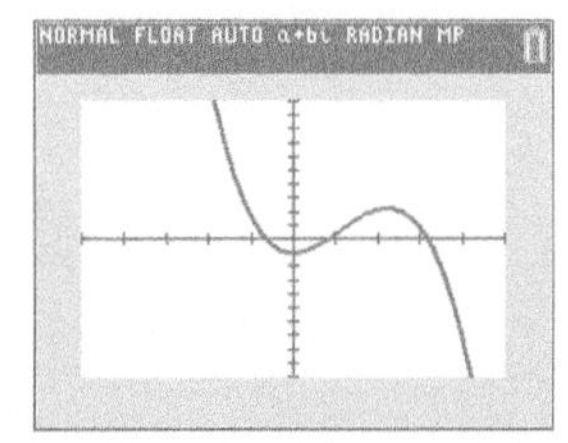

f. Use the graphs of cubic functions in parts a–e to write a few sentences comparing and contrasting the graph of the general quadratic equation, $y = ax^2 + bx + c, a \neq 0$, and the general cubic equation, $y = ax^3 + bx^2 + cx + d, a \neq 0$. Include comments on turning points and general trends, such as increasing and decreasing intervals.

The graphs of cubics and quadratics, like those of all polynomial functions, are continuous. The graph of any quadratic has exactly one turning point, whereas the graph of a cubic may have zero or two turning points. A quadratic with $a > 0$ will decrease until it reaches its turning point; then it will increase (with $a < 0$, it will increase until the turning point, then decrease). A quadratic will always have a line of symmetry through the turning point. A quadratic will have a vertical intercept and zero, one, or two horizontal intercepts. A cubic may be increasing everywhere or decreasing everywhere with no turning points. Any cubic with $a > 0$ will be increasing at the far right and far left of the graph. A cubic with $a < 0$ will decrease at the far right and far left. Cubics do not have any line of symmetry. Any cubic will have a vertical intercept and one, two, or three horizontal intercepts.

6. Use your graphing calculator to graph the following fourth-degree polynomials. Be careful of your choice of windows.

a. $f(x) = x^4$

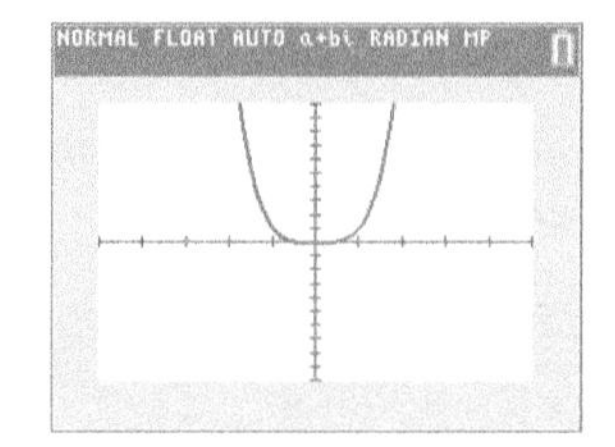

b. $i(x) = 3x^4 - x - 4$

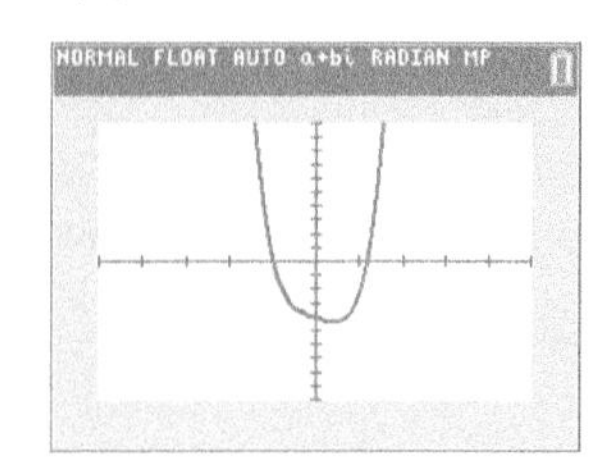

c. $g(x) = 0.2x^4 - 2x^2 + 7x - 1$

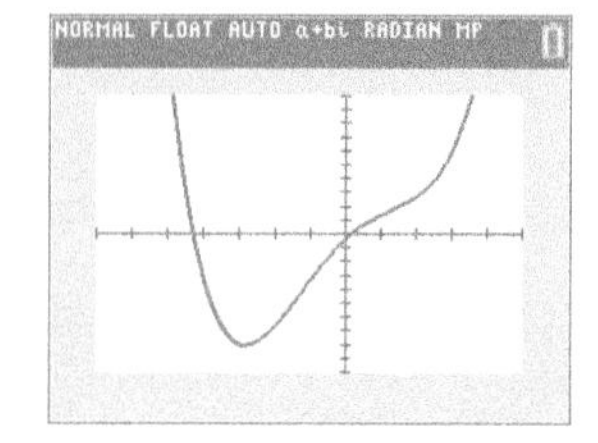

d. $j(x) = -5x^4 + 1$

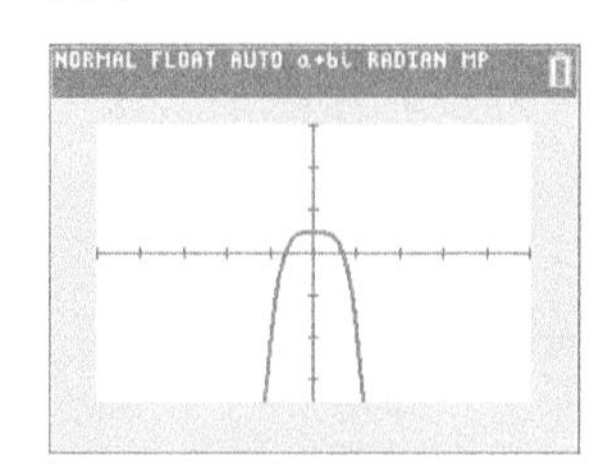

e. $h(x) = -0.6x^4 + 2x^3 - x + 1$

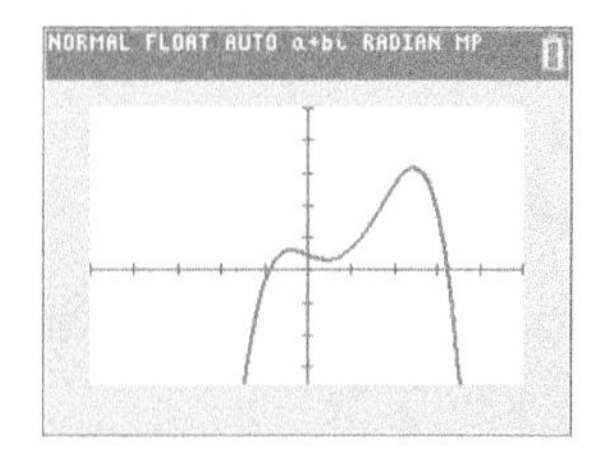

f. Use the graphs of quartic functions in parts a–e to write a few sentences comparing and contrasting the graph of the general quadratic equation, $y = ax^2 + bx + c, a \neq 0$, and the general quartic equation, $y = ax^4 + bx^3 + cx^2 + dx + e, a \neq 0$. Include comments on turning points and general trends, such as increasing and decreasing intervals.

Quartics and quadratics are continuous functions. The most basic quartic, $y = x^4$, and the most basic quadratic, $y = x^2$, are similar in their U shape, passing through the origin with symmetry on the y-axis. Any quartic will have a vertical intercept and from zero to four horizontal intercepts, whereas any quadratic will have a vertical intercept and from zero to two horizontal intercepts. A quadratic has exactly one turning point, but a quartic may have one or three turning points. For both quartics and quadratics, when $a > 0$, the graphs turn upward on both ends. When $a < 0$, the graphs turn downward on the ends.

SUMMARY Activity 4.9

1. **Polynomial functions** are defined by equations of the form

   output = polynomial expression involving the input.

2. The largest exponent, n, on the input variable is called the **degree** of the function.
3. Polynomial functions are continuous, with domain consisting of of all real numbers.
4. **Polynomial equations** of the form

   polynomial expression = 0

   can be solved graphically by locating the x-intercepts of the graph of the function defined by

   $y =$ polynomial expression

   or algebraically using factoring (if possible) and the zero-product property.

EXERCISES Activity 4.9

In Exercises 1–3, determine the x-intercept(s) of the graphs of each polynomial function using an algebraic approach (factoring). Verify your answer using your graphing calculator.

1. $f(x) = x^3 + 3x^2 + 2x$

   $f(x) = x(x + 2)(x + 1)$
   $(0, 0), (-1, 0), (-2, 0)$

2. $g(x) = 2x^2(x^2 - 4)$

   $g(x) = 2x^2(x + 2)(x - 2)$
   $(0, 0), (2, 0), (-2, 0)$

3. $h(x) = x^4 - 13x^2 + 36$

   $h(x) = (x^2 - 4)(x^2 - 9)$
   $h(x) = (x + 2)(x - 2)(x + 3)(x - 3)$
   $(2, 0), (-2, 0), (3, 0), (-3, 0)$

4. Determine the vertical intercept of each of the functions in Exercises 1–3.

   1. (0, 0) 2. (0, 0) 3. (0, 36)

5. Consider the function defined by $f(x) = x^4 - 6x^3 + 8x^2 + 1$.

   a. Use your graphing calculator to sketch a graph of the function.

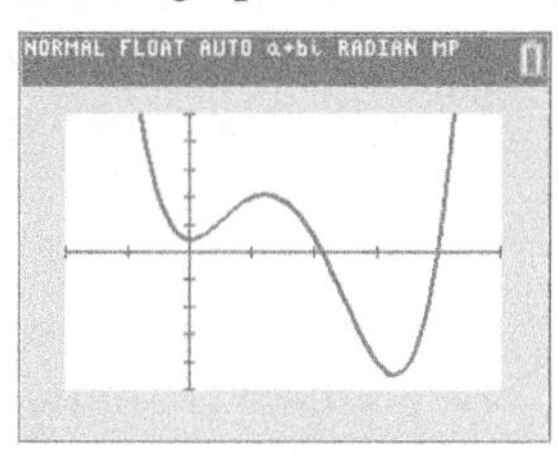

   b. Determine the domain of the function, f.

   The domain is all real numbers.

   c. Determine the range of f.

   The range is all y-values greater than or equal to -8.91.

   d. Use your graphing calculator to estimate the x-intercepts of the function, f.

   (2.12, 0) and (3.97, 0)

Exercise numbers appearing in color are answered in the Selected Answers appendix.

e. Use your graphing calculator to estimate any maximum and/or minimum points.

There are two minimum points, (0, 1) and (3.28, −8.91), and one maximum point, (1.22, 4.23).

6. Describe any symmetry of the graph of $y = x^4 - 4x^2 - 2$.

There is symmetry about the y-axis.

7. As the value of the input variable x increases without bound (e.g., 10 to 100 to 1000 etc.), do the output values decrease without bound for the function $y = x^3 + 3x^2 - x - 4$? Use a graph of the function to help answer the question.

No. As x increases without bound, y increases without bound.

8. Is the graph of $y = -x^3 - x + 3$ increasing or decreasing?

The graph is decreasing everywhere.

9. Consider the following graph of $y = f(x)$:

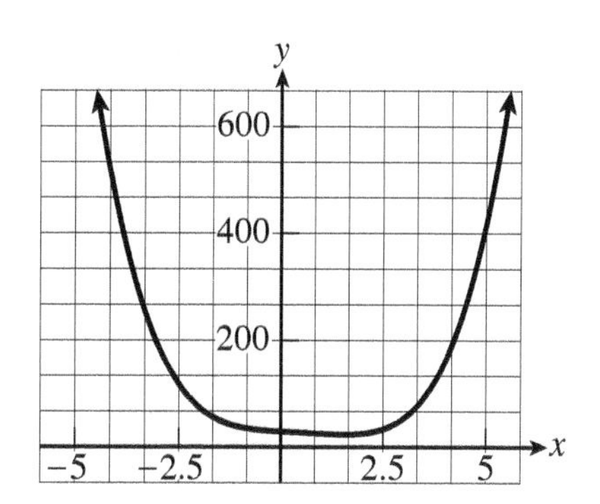

a. As x decreases without bound, the corresponding y-values increase.

b. Is the function f increasing or decreasing for $-2 < x < 2$? decreasing

c. How many turning points does the curve have? 1

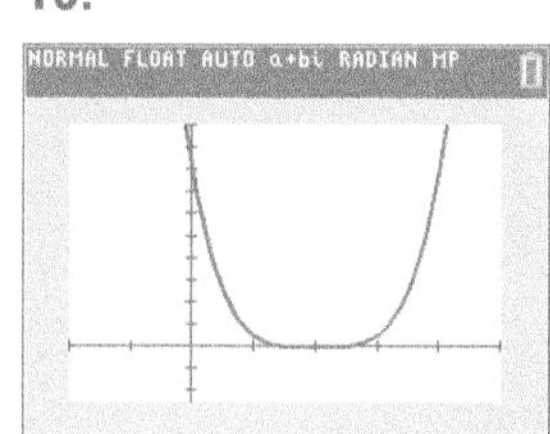

10. Using your graphing calculator, sketch a graph of $y = (x - 2)^4$. What is the relationship between the minimum point of the graph and its horizontal intercept?

The minimum point is the same as the horizontal intercept.

ACTIVITY 4.10

Federal Prison Population

Modeling Data with Polynomial Regression Equations

OBJECTIVE

Determine the regression equation of a polynomial function that best fits the data.

The federal prison population has been growing since the 1980s. The following table shows the total federal prison population for various years from 1980 to 2016:

Year	1980	1985	1990	1995	2000	2005	2010	2016
Total Federal Prison Population (thousands)	24.6	40.3	64.9	100.9	145.1	187.4	210.2	192.2

1. Let x represent the number of years since 1980. Let y represent the total federal prison population measured in thousands of inmates. Plot these points on your graphing calculator. The scatterplot should resemble the following:

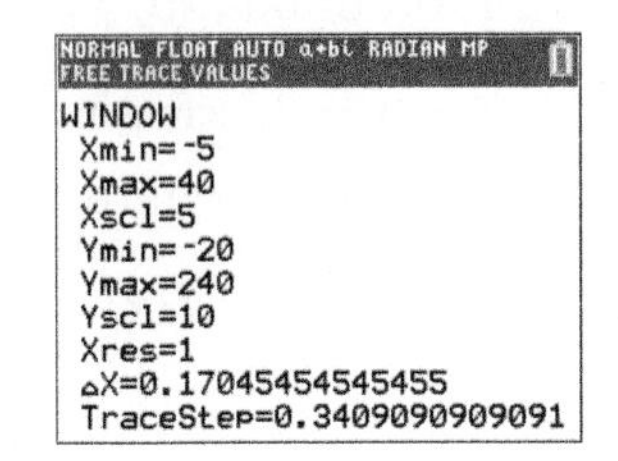

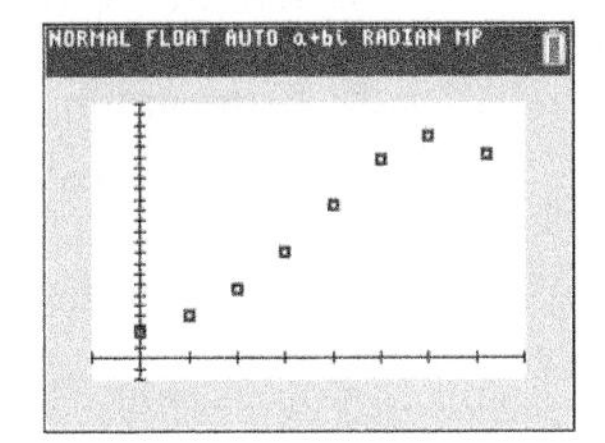

2. **a.** Using your graphing calculator, determine the regression equations of the first-, second-, and third-degree curves of best fit.

first degree (linear): $y = 5.671x + 20.757$

second degree (quadratic): $y = -0.065x^2 + 7.985x + 8.848$

third degree (cubic): $y = -0.013x^3 + 0.633x^2 - 1.440x + 27.636$

b. Use your calculator to fit each of your models from part a to your scatterplot. Your graphs should resemble the following:

Linear model

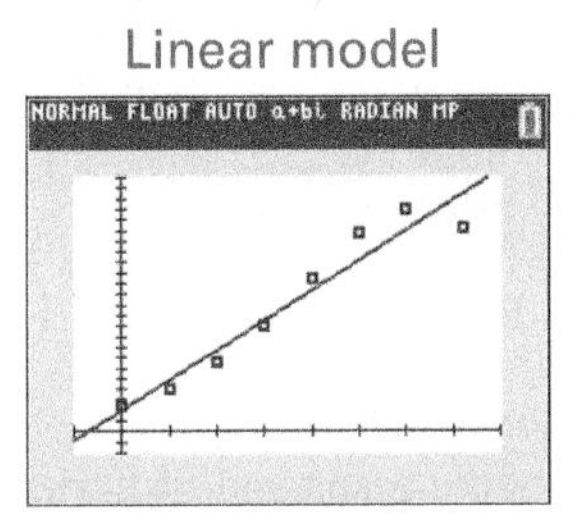

Quadratic model

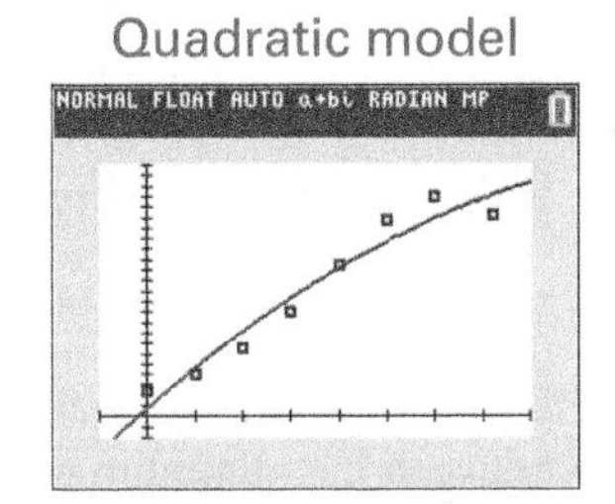

Cubic model

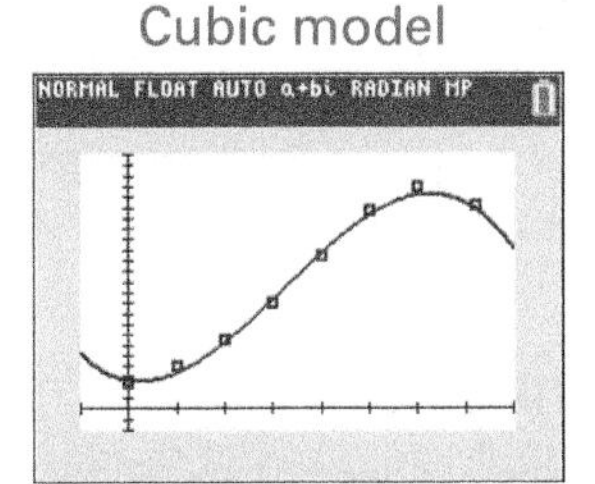

c. Which of these curves best represents the data? Explain.

All three are fairly good, but compared with the other two models, the cubic seems to better show the growth slowing at the end.

3. **a.** What is the practical domain of this prison population situation?

x can take on values from 0 to about 36. Answers may vary.

b. What is the practical range of this prison population situation?

The practical range is values from about 24 to 210 thousand inmates.

LC LEARNING CATALYTICS

For the following data, calculate the linear, quadratic, and cubic regression equations. Determine the curve that best fits the data.

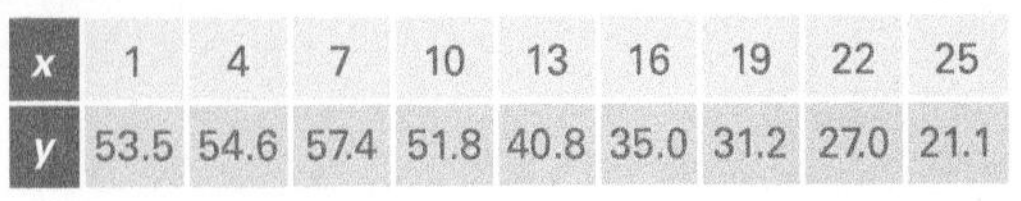

x	1	4	7	10	13	16	19	22	25
y	53.5	54.6	57.4	51.8	40.8	35.0	31.2	27.0	21.1

a. linear
b. quadratic
c. cubic
d. They all fit the about the same.

A quick-service restaurant is defined as a restaurant where patrons pay before eating and purchases may be consumed on-site, taken out, or delivered. This definition further excludes coffee and snack shops. The revenue and projected revenue for quick-service restaurants are given by the following table for various years from 2002 to 2020:

Years since 2000, t	2	4	6	8	10	12	14	16	18	20
Revenue of Quick-Service Restaurants in the U.S., $R(t)$ (billions of dollars)	159.2	173.7	180.8	185.9	190.5	194.8	198.9	206.3	214.8	223.9

4. a.

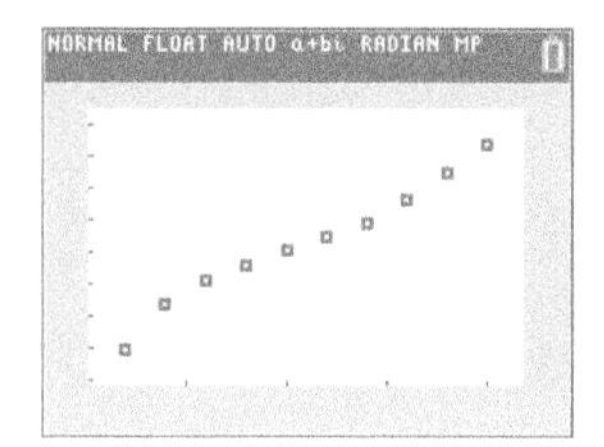

4. a. Use your graphing calculator to determine a scatterplot of this data.

b. Do you believe that a linear, quadratic, or cubic model would be the best fit for this data? Explain.

I believe cubic would be the best model. The data trends up, but it is a bit wavy rather than straight.

c. Use your graphing calculator to determine all three models, and record the best-fitting model. Round your coefficients to three decimal places.

$R(t) = 0.019t^3 - 0.644t^2 + 9.177t + 144.347$

d. Use your model to predict the revenue of quick-service restaurants in the United States in 2025 and 2035.

2025: $t = 25$; $R(25) = 0.019(25)^3 - 0.644(25)^2 + 9.177(25) + 144.347 \approx$ 268.1 billion dollars

2035: $t = 35$; $R(35) = 0.019(35)^3 - 0.644(35)^2 + 9.177(35) + 144.347 \approx$ 491.3 billion dollars

e. In which prediction in part d do you have the most confidence? Explain.

I have more confidence in the 2025 prediction. The 2035 prediction is extrapolating quite far from the original data.

SUMMARY Activity 4.10

The graphing calculator can be used to model data with cubic and quartic polynomial functions as well as linear and quadratic polynomial functions.

EXERCISES Activity 4.10

1. According to the National Institutes of Health, cigarette usage for 8th, 10th, and 12th graders combined for various years from 1991 to 2016 is as given in the following table:

Years since 1990, t	1	4	7	10	13	16	19	22	26
Percentage of 8th, 10th, and 12th Graders Who Admitted to Using Cigarettes, g	53.5	54.6	57.4	51.8	40.8	35.0	31.2	27.0	18.2

a. Sketch a scatterplot of the data. Let t represent the number of years since 1991. Does the data appear to be linear? Explain.

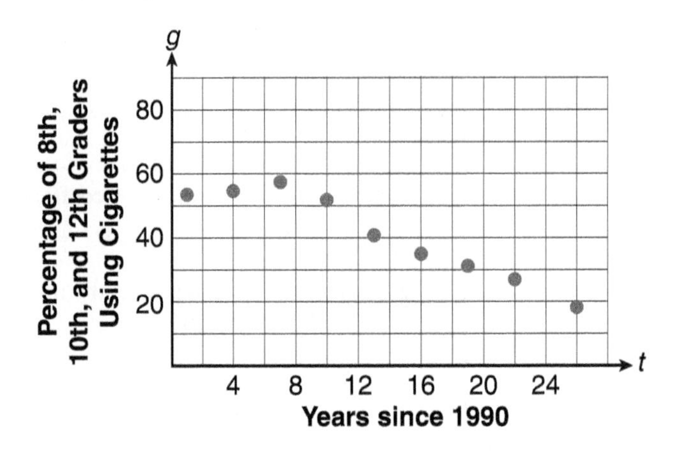

The data does not appear to be linear because as the input increases, the output increases, then decreases. No line would be close to all of the points.

b. Use your graphing calculator to create a STATPLOT; then determine the quadratic and cubic regression equations for the data.

Quadratic: $g = -0.04x^2 - 0.52x + 57.33$

Cubic: $g = 0.006x^3 - 0.276x^2 + 2.004x + 51.98$

2. a. Using the results from 1b, select the equation that best models the percent of 8th, 10th, and 12th graders who admitted to using cigarettes, g, as a function of the number of years since 1990, t.

Cubic: $g = 0.006x^3 - 0.276x^2 + 2.004x + 51.98$

b. What is the practical domain of this function?

The practical domain is the integers from 0 to 26.

c. What is the practical range of this function? Explain.

The practical range is real numbers from 15 to 60. (Answers will vary slightly.)

d. Use the equation in part a to estimate the percent of 8th, 10th, and 12th graders graders who admitted to using cigarettes in 1987, 2003 and 2020. In which of these estimates do you have the most confidence? Explain.

In 1987, $t = -3$; $g(-3) \approx 43.32\%$

In 2003, $t = 13$; $g(13) \approx 44.57\%$

In 2020, $t = 30$; $g(30) \approx 25.7\%$

I have the most confidence in 2003 because that is within the range of the data.

3. According to the U.S. Census Bureau, the consumption of cigarettes by adults in the United States is decreasing. The following table represents the per capita consumption of cigarettes in the United States, C, for various years between 2000 and 2015:

Year	2000	2003	2006	2009	2012	2015
Years since 2000, t	0	3	6	9	12	15
Per Capita Consumption, C	2076	1844	1695	1367	1196	1078

a. Use t as the input and sketch a scatterplot of the data.

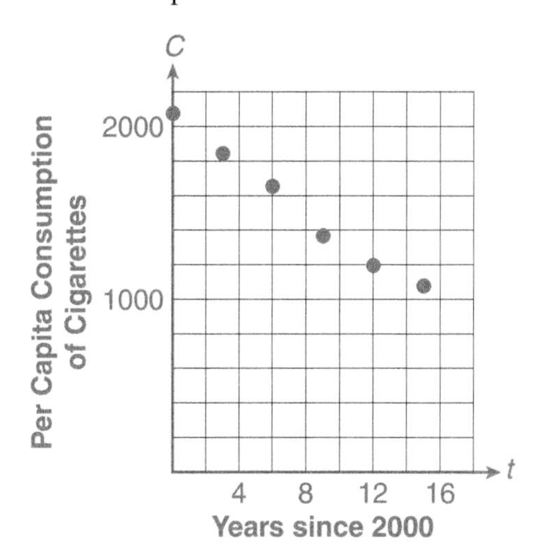

b. Use your graphing calculator to determine the linear regression equation for the data. Round the values of a and b to two decimal places.

$C = -69.16t + 2061.38$

c. Determine the quadratic model for the data. Round the values of a, b, and c to three decimal places.

$C = 0.956t^2 - 83.507t + 2090.071$

d. Use each model to predict the per capita consumption in 2020.

2020 is 20 years from 2000, so $t = 20$.

Linear model: $C = -69.16 \cdot 20 + 2061.38 = 678.18$

Quadratic model: $C = 0.956 \cdot 20^2 - 83.507 \cdot 20 + 2090.071 = 802.331$

e. How confident are you in the predictions determined in part d?

$t = 20$ is well outside of the given data; therefore, the result is extrapolation. I do not have confidence in either prediction.

CLUSTER 2 What Have I Learned?

1. In a hurricane, the wind pressure varies directly as the square of the wind velocity (speed). If the wind speed doubles in value, what change in the wind pressure do you experience?

   The wind pressure is multiplied by 4.

2. Is the graph of $y = 3x^4$ narrower or wider than the graph of $y = x^2$? Explain.

   $y = 3x^4$ is narrower than $y = x^2$ because it grows more quickly.

3. The graph of any cubic (third-degree polynomial) function must have one of the four following general shapes:

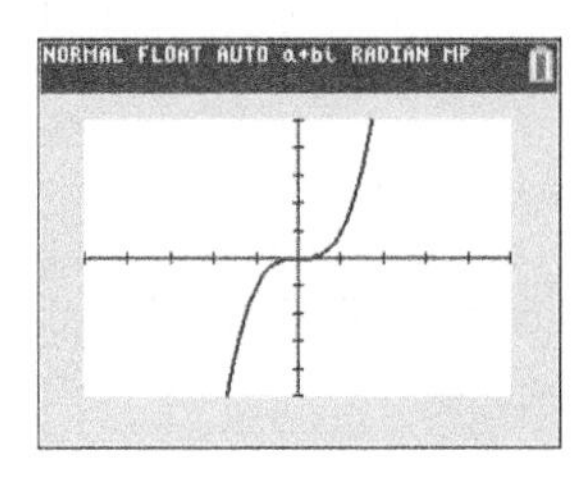

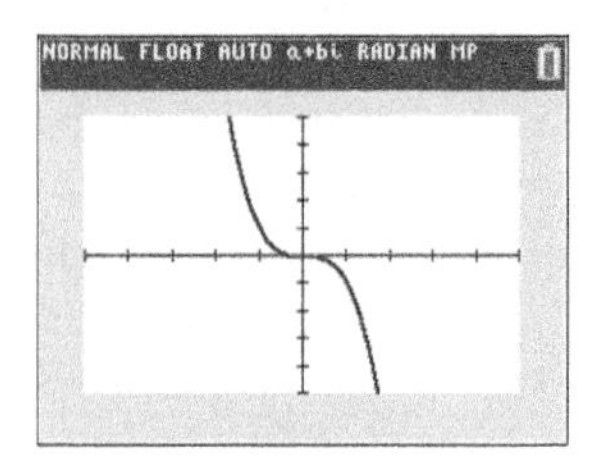

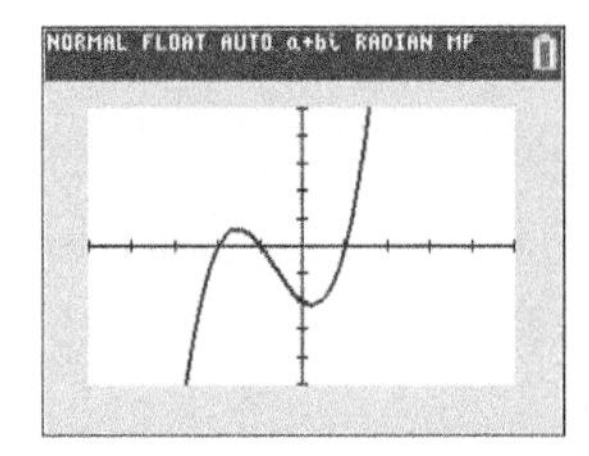

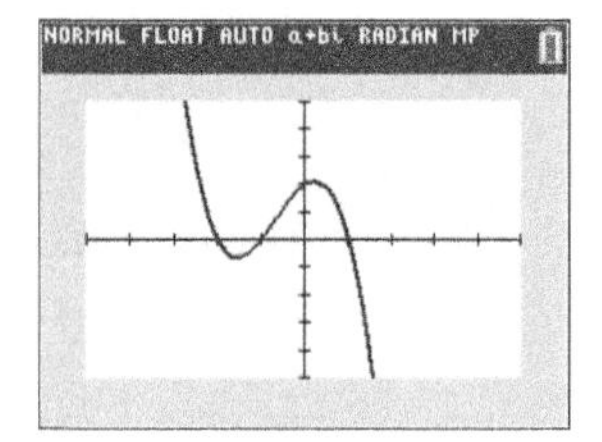

   a. Complete the following table, which gives the maximum number of turning points for a given family of polynomial functions:

DEGREE OF POLYNOMIAL FUNCTION	MAXIMUM NUMBER OF TURNING POINTS
1 (linear)	0
2 (quadratic)	1
3 (cubic)	2
4 (quartic)	3

   b. If n represents the degree, then write an expression that represents the maximum number of turning points.

   $n - 1$ will be the maximum number of turning points for a polynomial of degree n.

4. **a.** Sketch a graph of $y = x^4 - 4x^2$. Describe any symmetry that you observe.

   There is symmetry about the y-axis.

   b. Do all graphs of quartic (fourth-degree) functions have symmetry? Explain.

   No, a quartic will have symmetry only if all its terms have even powers of the input variable.

5. a. Does the graph of any cubic function have a horizontal intercept? Can the graph have more than one horizontal intercept? Explain.

Any cubic will have at least one horizontal intercept, possibly two or three, as can be seen from the shapes in Problem 3.

b. Does the graph of any cubic function have at least one vertical intercept? Explain.

Any cubic has one vertical intercept. Zero is always an element of the domain of a cubic function.

6. Given the following graph, determine whether any of the three functions in parts a–c fit the curve. Explain.

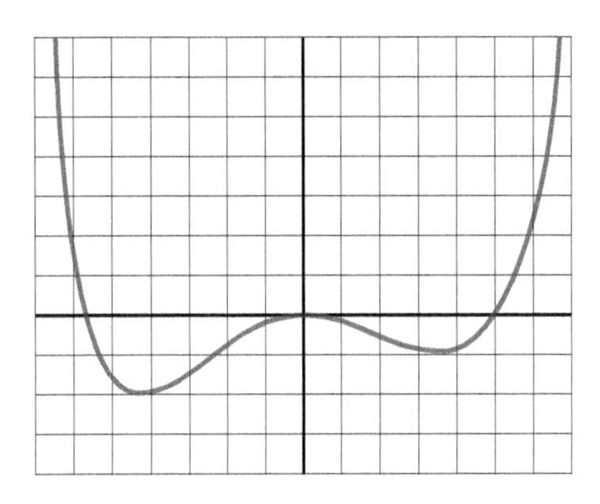

a. $f(x) = ax^4 + bx^3 + cx^2 + dx + e, \quad a > 0, e = 0$

This could be the equation. There are three turning points, and the graph of a quartic equation has at most three turning points. Because $a > 0$, the ends turn up, and $e = 0$ means that the graph passes through the origin.

b. $g(x) = ax^3 + bx^2 + cx + d, \quad a > 0, d = 0$

No. The graph has three turning points, whereas the graph of a cubic equation has at most two turning points; also the ends of the graph of a cubic function would go in opposite directions.

c. $h(x) = ax^4 + bx^3 + cx^2 + dx + e, \quad a < 0, e = 0$

This is not the equation of the curve. There are three turning points, but with $a < 0$, the graph of h would turn downward at both ends.

CLUSTER 2 How Can I Practice?

1. y varies directly as x^2. When $x = 3, y = 45$. Determine y when $x = 6$.

$y = kx^2$ and $45 = k(3)^2$, so $k = 5$.

$y = 5(6)^2 = 180$

2. Have you ever noticed that during a thunderstorm, you see lightning before you hear the thunder? This is true because light travels faster than sound. If d represents the distance (in feet) of the lightning from the observer, then d varies directly as the time, t (in seconds), it takes to hear the thunder. The relationship is modeled by

$$d = 1080t.$$

a. As the time t doubles (e.g., from 3 to 6), the corresponding d-values double.

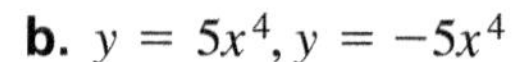

b. What is the value of k, the constant of variation, in this situation? What significance does k have in this problem?

$k = 1080$. k represents the speed at which the sound of thunder travels in feet per second.

3. The velocity, v, of a falling object varies directly as the time, t, of the fall. After 3 seconds, the velocity of the object is 60 feet per second. What will be its velocity after 4 seconds?

$v = kt$ and $60 = k3$, so $k = 20$.

$v = 20(4) = 80$ feet per second

4. Sketch a graph of each of the following pairs of functions. Describe the differences and the similarities in the graphs.

a. $y = 3x^2, y = 3x^2 + 5$

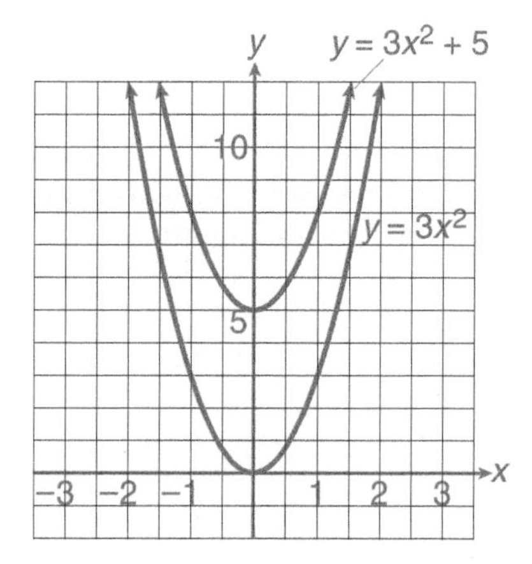

These are the same shape and size; however, $y = 3x^2 + 5$ is shifted up 5 units.

b. $y = 5x^4, y = -5x^4$

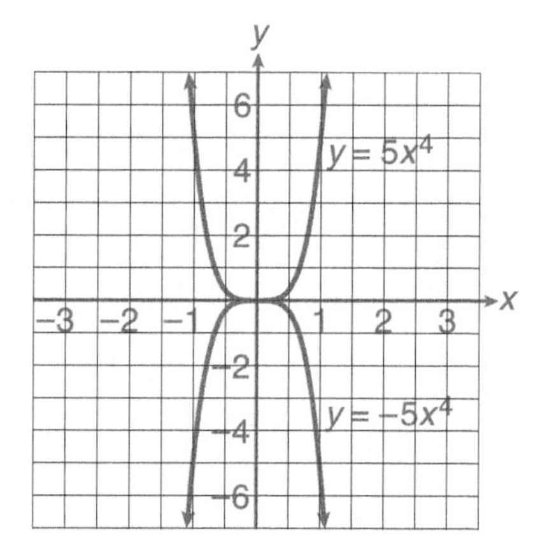

These are the same shape but are reflections of each other in the x-axis.

c. $y = 2x^3 + 1, y = 2x^3 - 4$

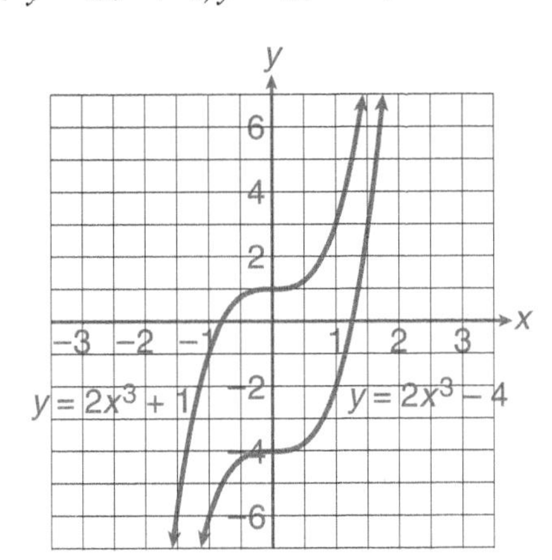

These are the same shape and size, but $y = 2x^3 - 4$ is shifted vertically 5 units below $y = 2x^3 + 1$.

d. $y = 4x^2, y = 4(x - 1)^2$

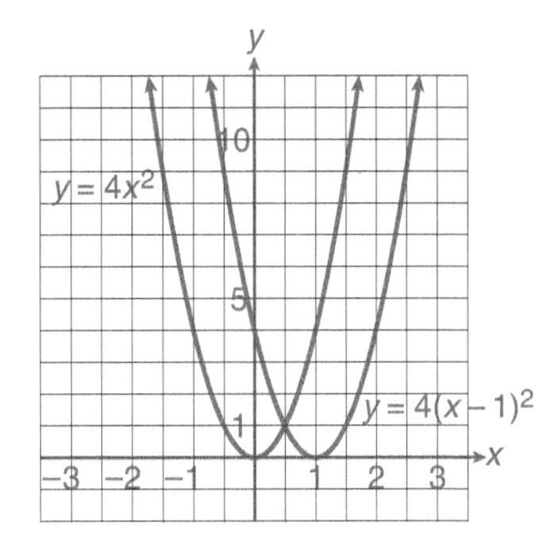

These are the same shape and size, but $y = 4(x - 1)^2$ is shifted horizontally 1 unit to the right of $y = 4x^2$.

5. Using your graphing calculator, graph each of the following polynomial functions. For each graph,

i. determine the vertical intercepts.

ii. approximate the horizontal intercepts (if they exist).

iii. determine the coordinates of any turning points.

a. $y = x^3 + 2x^2 - 8x$

i. (0, 0)

ii. (0, 0), (−4, 0), (2, 0)

iii. (−2.4, 16.9), (1.1, −5,0)

b. $y = -x^4 + 2x + 3$

i. (0, 3)

ii. (−1, 0), (1.57, 0)

iii. (0.79, 4.19)

6. In general, the average fuel economy of U.S.-manufactured cars has increased over the past several decades. The following table gives the average miles per gallon (mpg) for U.S.-made cars for selected years:

AVERAGE MILES PER GALLON FOR PASSENGER CARS IN THE UNITED STATES FROM 1970 TO 2015								
Year	1980	1990	1995	2000	2004	2009	2012	2015
Average mpg	16.0	20.3	21.1	21.9	22.5	22.4	23.2	24.8

a. Draw a scatterplot of the data points. Let $x =$ the number of years since 1980.

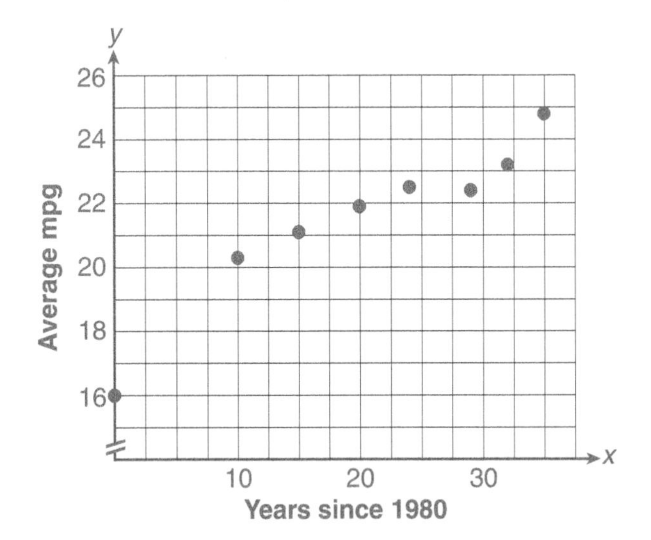

b. Determine the equations of the cubic and quartic functions that best fit the data.

cubic: $y = 0.00050186x^3 - 0.0305x^2 + 0.6982x + 15.9612$

quartic: $y = 0.000023738x^4 - 0.0012x^3 + 0.0084x^2 + 0.4248x + 16.0164$

c. Use the regression equations to predict the average miles per gallon in 2020.

cubic: 27.2 mpg quartic: 29.5 mpg

d. In which model, if either, do you have the most confidence? Explain.

In both cases, I am extrapolating. The cubic goes up moderately. The quartic goes up significantly. The cubic is better.

7. Your bathtub is partially filled. You finish filling the tub and settle in for a nice hot bath. The drain plug is broken, and water is running out. The amount of water (in gallons) in the bathtub is given by

$$W(t) = 10 + 7t^2 - t^3 \quad (t \geq 0),$$

where t is time in minutes and $W(t)$ represents the amount of water in gallons.

a. With the aid of your graphing calculator, sketch the graph of $W(t)$. Don't go beyond $t = 10$. Why?

$W(t)$ becomes negative after about 7 minutes.

7. a.

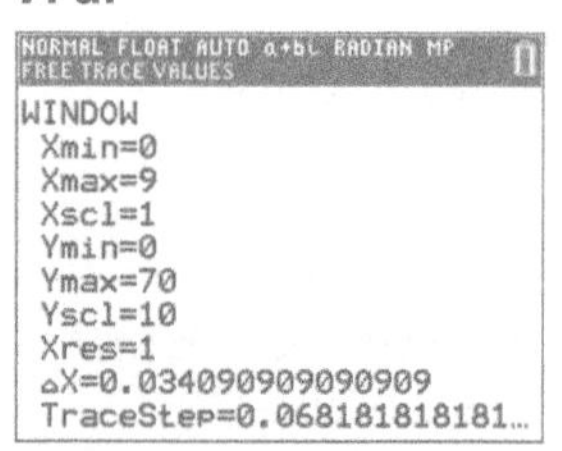

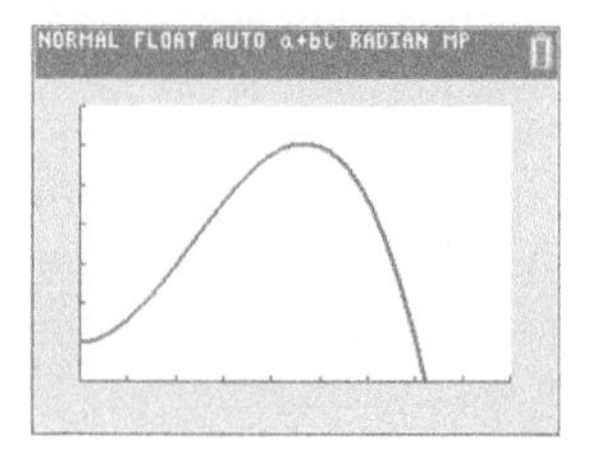

b. How much water was in the bathtub when you began to fill it?

$W(0) = 10$ gal.

c. Determine the maximum amount of water in the tub from the graph. Explain your result.

60.8 gallons, which I found by using the CALC menu on the graphing calculator. This is the output value at the highest point on the graph, after about 4.7 minutes.

d. Use the zero features of your graphing calculator to determine when the tub will be completely empty, to the nearest 0.01 minute. 7.19 min.

CHAPTER 4 Summary

The bracketed numbers following each concept indicate the activity in which the concept is discussed.

CONCEPT/SKILL	DESCRIPTION	EXAMPLE
Quadratic function [4.1]	The quadratic function with the input variable x has the standard form $y = ax^2 + bx + c$, where a, b, and c represent real numbers and $a \neq 0$.	$y = 2x^2 - 3x - 2$
Graph of a quadratic function (a parabola) [4.1]	For the quadratic function defined by $f(x) = ax^2 + bx + c$, if $a > 0$, the parabola opens upward; if $a < 0$, the parabola opens downward.	The graph of $y = 2x^2 - 3x - 2$ is a parabola that opens upward.
Vertical intercept of the graph of a quadratic function [4.1]	The constant term, c, of a quadratic function $f(x) = ax^2 + bx + c$ always indicates the vertical intercept of the parabola. The vertical intercept is $(0, c)$.	The vertical intercept of the graph $y = 2x^2 - 3x - 2$ is $(0, -2)$.
Axis of symmetry [4.2]	The axis of symmetry of a parabola is a vertical line that separates the parabola into two mirror images. The equation of the vertical axis of symmetry is given by $x = \frac{-b}{2a}$.	The axis of symmetry of the parabola defined by $y = 2x^2 - 3x - 2$ is $x = \frac{3}{4}$.
Vertex (turning point) [4.2]	The vertex of a parabola defined by $f(x) = ax^2 + bx + c$ is the point where the graph changes direction. It is given by $\left(\frac{-b}{2a}, f\left(-\frac{b}{2a}\right)\right)$.	The vertex of the parabola defined by $y = 2x^2 - 3x - 2$ is $\left(\frac{3}{4}, -\frac{25}{8}\right)$.
x-intercept(s) [4.2]	An x-intercept is the point or points (if any) where the graph crosses the x-axis (i.e., where its y-coordinate is zero).	The x-intercepts of the parabola defined by $y = 2x^2 - 3x - 2$ are $(2, 0)$ and $(-0.5, 0)$.
Domain of quadratic functions [4.2]	The domain of any quadratic function is all real numbers.	The domain of $y = 2x^2 - 3x - 2$ is all real numbers.
Range of quadratic functions [4.2]	Let k be the output value of the turning point. If the parabola opens upward, the range is $k \leq y < \infty$. If the parabola opens downward, the range is $-\infty < y \leq k$.	The range of the parabola defined by $y = 2x^2 - 3x - 2$ is $y \geq -\frac{25}{8}$.
Solving $f(x) = c$ graphically [4.3]	Graph $y = f(x)$, graph $y = c$, and determine the x-values of the points of intersection. Or graph $y = f(x) - c$ and determine the x-intercepts.	Problem 7, Activity 4.3.

CONCEPT/SKILL	DESCRIPTION	EXAMPLE
Solving $f(x) > c$ graphically [4.3]	Graph $y = f(x)$, graph $y = c$, and determine all x-values for which the graph of f is above the graph of $y = c$. Or graph $y = f(x) - c$ and determine all x-values for which the graph of $f(x) - c$ is above the x-axis.	Example 2, Activity 4.3
Greatest common factor (or GCF) [4.4]	The GCF is the largest factor common to all terms in an expression.	The GCF of $3x^4 - 6x^3 + 18x^2$ is $3x^2$.
Zero-product property [4.4]	If a and b are any real numbers and $a \cdot b = 0$, then either a or b or both must be equal to zero.	Example 1, Activity 4.4
Factoring trinomials by trial and error [4.4]	To factor trinomials by trial and error, 1. remove the GCF. 2. try combinations of factors for the first and last terms in two binomials. 3. check that the sum of the outer and inner products matches the middle term of the original trinomial. 4. if the check fails, repeat steps 2 and 3.	Example 5, Activity 4.4
Solving quadratic equations by factoring [4.4]	To solve a quadratic equation by factoring, 1. use the addition principle to remove all terms from one side of the equation. This results in a quadratic polynomial being set equal to zero. 2. combine like terms and then factor the nonzero side of the equation. 3. use the zero-product property to set each factor containing a variable equal to zero and then solve the equations. 4. check your solutions in the original equation.	Example 6, Activity 4.4
Quadratic formula [4.5]	$x = \dfrac{-b \pm \sqrt{b^2 - 4ac}}{2a}$	Example 1, Activity 4.5
Solving a quadratic equation of the form $ax^2 + bx + c = 0, a \neq 0$, using the quadratic formula [4.5]	To solve a quadratic equation, using the quadratic formula $x = \dfrac{-b \pm \sqrt{b^2 - 4ac}}{2a}$, 1. rewrite the quadratic equation with one side equal to zero. 2. identify the coefficients a and b and the constant term c. 3. substitute these values into the formula and simplify. 4. check your solutions.	Example 1, Activity 4.5

CONCEPT/SKILL	DESCRIPTION	EXAMPLE
Imaginary unit [4.7]	The imaginary unit is the number $\sqrt{-1}$. The notation for the imaginary unit is i.	$i = \sqrt{-1}, i^2 = -1$
Complex number [4.7]	Any number that can be written in the form $a + bi$, where a and b are real numbers and i is the imaginary unit, is called a complex number.	$2 + 6i$
Discriminant [4.7]	In the quadratic formula $x = \dfrac{-b \pm \sqrt{b^2 - 4ac}}{2a}$, the expression $b^2 - 4ac$ is called the discriminant. Its value determines the number and type of solutions of a quadratic equation $ax^2 + bx + c = 0$.	For the quadratic equation $2x^2 - 7x - 4 = 0$, the discriminant is $49 - 4(2)(-4) = 81$. The equation has two real solutions.
Direct variation function [4.8]	The equation $y = kx^n$, where $k \neq 0$ and n is a positive integer, defines a direct variation function in which y varies directly as x^n.	$y = 4x^3$
Constant of variation [4.8]	In the direct variation equation $y = kx^n$, the constant, k, is called the constant of variation.	In $y = 4x^3$, the constant of variation is 4.
Power functions [4.8]	The direct variation function having an equation of the form $y = kx^n$, where n is a positive integer, is also called a power function.	$y = 4x^3$ is a third-degree power function.
Polynomial functions [4.9]	Polynomial functions are defined by equations of the form output = polynomial expression involving the input.	$y = 5x^4 + 7x^2 - 3x + 1$
Degree of a polynomial [4.9]	The largest exponent on the input variable, n, is called the degree of the function.	$y = 5x^4 + 7x^2 - 3x + 1$ is a fourth-degree polynomial function.

CHAPTER 4 Gateway Review

In Exercises 1–8, determine the following characteristics of each quadratic function by inspecting its equation:

a. *the direction in which the graph opens*

b. *the equation of the axis of symmetry*

c. *the vertex*

d. *the y-intercept*

1. $f(x) = x^2 + 2$

a. up

b. $x = 0$

c. $(0, 2)$

d. $(0, 2)$

2. $F(x) = -3x^2$

a. down

b. $x = 0$

c. $(0, 0)$

d. $(0, 0)$

3. $g(x) = -3x^2 + 4$

a. down

b. $x = 0$

c. $(0, 4)$

d. $(0, 4)$

4. $f(x) = 2x^2 - x$

a. up

b. $x = \frac{1}{4}$

c. $\left(\frac{1}{4}, -\frac{1}{8}\right)$

d. $(0, 0)$

5. $h(x) = x^2 + 5x + 6$

a. up

b. $x = -2.5$

c. $(-2.5, -0.25)$

d. $(0, 6)$

6. $F(x) = x^2 - 3x + 4$

a. up

b. $x = 1.5$

c. $(1.5, 1.75)$

d. $(0, 4)$

7. $f(x) = x^2 - 2x + 1$

a. up

b. $x = 1$

c. $(1, 0)$

d. $(0, 1)$

8. $g(x) = -x^2 + 5x - 6$

a. down

b. $x = 2.5$

c. $(2.5, 0.25)$

d. $(0, -6)$

In Exercises 9–15, sketch the graph of each quadratic function using your graphing calculator. Then determine each of the following using the graph:

a. *the coordinates of the x-intercepts, if they exist*

b. *the domain and the range of the function*

c. *the horizontal interval in which the function is increasing*

d. *the horizontal interval in which the function is decreasing*

9. $g(x) = x^2 + 4x + 3$

a. $(-3, 0), (-1, 0)$

b. domain: all real numbers; range: $g(x) \geq -1$

c. $x > -2$

d. $x < -2$

10. $f(x) = x^2 + 2x - 3$

a. $(-3, 0), (1, 0)$

b. domain: all real numbers; range: $f(x) \geq -4$

c. $x > -1$

d. $x < -1$

9.

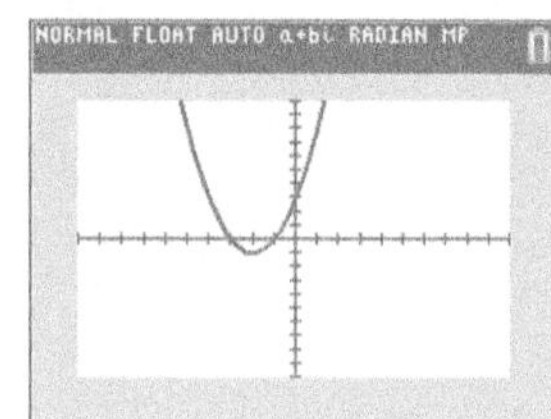

10.

NORMAL FLOAT AUTO a+bi RADIAN MP

Answers to all Gateway exercises are included in the Selected Answers appendix.

11. $F(x) = x^2 - 3x + 1$

a. $(0.382, 0), (2.62, 0)$

b. domain: all real numbers; range: $y \geq -1.25$

c. $x > 1.5$

d. $x < 1.5$

12. $h(x) = 2x^2 + 8x + 5$

a. $(-3.22, 0), (-0.775, 0)$

b. domain: all real numbers; range: $h(x) \geq -3$

c. $x > -2$

d. $x < -2$

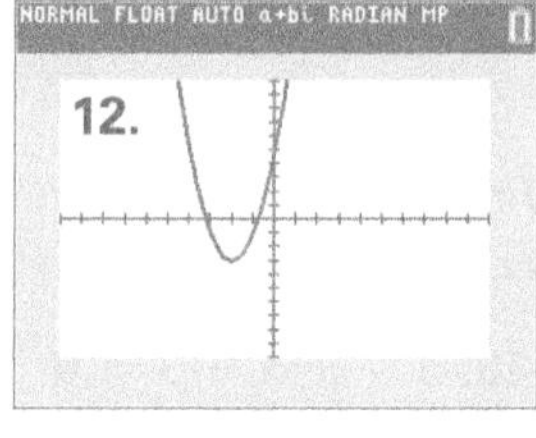

13. $F(x) = -2x^2 + 8$

a. $(2, 0), (-2, 0)$

b. domain: all real numbers; range: $y(x) \leq 8$

c. $x < 0$

d. $x > 0$

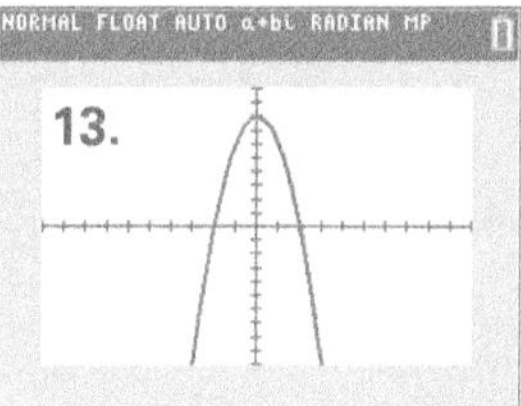

14. $f(x) = -3x^2 + 4x - 1$

a. $\left(\frac{1}{3}, 0\right), (1, 0)$

b. domain: all real numbers; range: $f(x) \leq \frac{1}{3}$

c. $x < \frac{2}{3}$

d. $x > \frac{2}{3}$

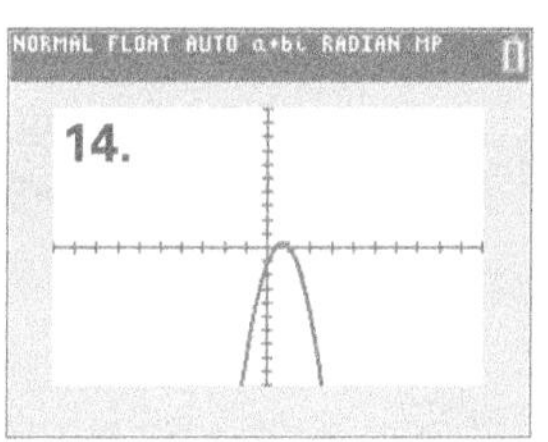

15. $g(x) = 4x^2 + 5$

a. none

b. domain: all real numbers; range: $g(x) \geq 5$

c. $x > 0$

d. $x < 0$

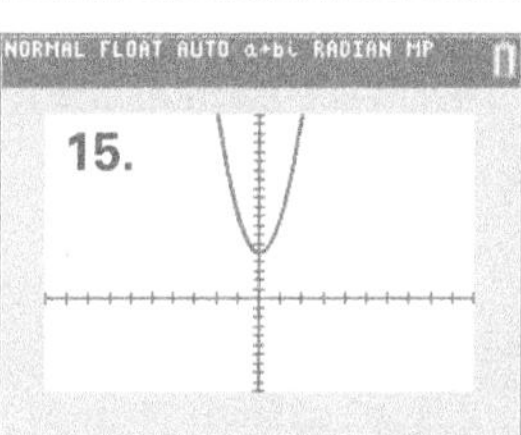

In Exercises 16–19, solve the quadratic equation numerically (using tables). Verify your solutions graphically.

16. $x^2 + 4x + 4 = 0$

$x = -2$

17. $x^2 - 5x + 6 = 0$

$x = 2, 3$

18. $3x^2 = 18x + 10$

$x \approx -0.51, 6.51$

19. $-x^2 = 3x - 10$

$x = -5, 2$

In Exercises 20 and 21, solve the equation using two different approaches. Round your answer to the nearest tenth when necessary.

20. $8x^2 = 10$

$x \approx \pm 1.1$

21. $5x^2 + 25x = -5$

$x \approx -0.2, -4.8$

22. Completely factor the following polynomials:

a. $9a^5 - 27a^2$

$9a^2(a^3 - 3)$

b. $24x^3 - 6x^2$

$6x^2(4x - 1)$

c. $4x^3 - 16x^2 - 20x$

$4x(x - 5)(x + 1)$

d. $5x^2 - 16x + 6$

cannot be factored using rational numbers

e. $x^2 - 5x - 24$

$(x - 8)(x + 3)$

f. $t^2 + 10t + 25$

$(t + 5)^2$

In Exercises 23–27, solve each equation by factoring. Verify your answer graphically or by substitution of the solutions in the equations.

23. $x^2 - 9 = 0$

$x = \pm 3$

24. $-x^2 + 36 = 0$

$x = \pm 6$

25. $x^2 - 7x + 12 = 0$

$x = 3, 4$

26. $x^2 - 6x = 27$

$x = -3, 9$

27. $x^2 = -x$

$x = 0, -1$

In Exercises 28–32, write each of the equations in the form $ax^2 + bx + c = 0$. Then identify a, b, and c, and solve the equation using the quadratic formula.

28. $x^2 + 5x + 3 = 0$

$a = 1, b = 5, c = 3$

$x \approx -0.7, -4.3$

29. $2x^2 - x = -3$

$a = 2, b = -1, c = 3$

$x \approx 0.25 \pm 1.2i$

30. $x^2 = 81$

$a = 1, b = 0, c = -81$

$x = \pm 9$

31. $3x^2 + 5x = 12$

$a = 3, b = 5, c = -12$

$x = -3, \frac{4}{3}$

32. $2x^2 = 3x + 5$

$a = 2, b = -3, c = -5$

$x = -1, 2.5$

33. For the quadratic function $f(x) = 2x^2 - 8x + 3$, determine the x-intercepts of the graph, if they exist. First, approximate the intercepts using your graphing calculator. Second, solve the equation using the quadratic formula. Approximate your answers to the nearest hundredth.

from the graphing calculator: (0.42, 0), (3.58, 0)

$$x = \frac{-(-8) \pm \sqrt{(-8)^2 - 4(2)(3)}}{2(2)} = \frac{8 \pm \sqrt{40}}{4} \approx 3.58, 0.42$$

34. Write each of the following using the imaginary unit, i:

a. $\sqrt{-49}$

$7i$

b. $\sqrt{-48}$

$4i\sqrt{3}$

c. $\sqrt{-9}$

$3i$

d. $\sqrt{-23}$

$i\sqrt{23}$

e. $\sqrt{-\frac{5}{9}}$

$\frac{\sqrt{5}}{3}i$

f. $\sqrt{-\frac{17}{16}}$

$\frac{\sqrt{17}}{4}i$

35. Perform the following operations with complex numbers. Use your graphing calculator to check your results.

a. $(2 + 7i) + (-7 + 10i)$

$-5 + 17i$

b. $(4 - 9i) - (-1 + 7i)$

$5 - 16i$

c. $4i(3 - 8i)$

$32 + 12i$

d. $(4 - i)(6 + 3i)$

$27 + 6i$

In Exercises 36–39, determine the type of solution to each of the equations, considering only its discriminant.

36. $2x^2 - 3x + 1 = 0$

$b^2 - 4ac = 1$; two real solutions

37. $4x^2 + 16x = 0$

$b^2 - 4ac = 256$; two real solutions

38. $x^2 - 9 = 0$

$b^2 - 4ac = 36$; two real solutions

39. $3x^2 + 2x + 2 = 0$

$b^2 - 4ac = -20$; two complex solutions

40. Solve the equation in Problem 39 in the complex number system using the quadratic formula. Verify your solution graphically.

$$x = \frac{-2 \pm \sqrt{2^2 - 4(3)(2)}}{2(3)} = \frac{-2 \pm \sqrt{-20}}{6} = \frac{-2 \pm 2i\sqrt{5}}{2(3)} = \frac{-1 \pm i\sqrt{5}}{3}$$

The graph has no x-intercepts, confirming complex solutions.

41. Solve the following inequalities using a graphing approach:

a. $x^2 - x - 6 < 0$

$-2 < x < 3$

b. $x^2 - x - 6 > 0$

$x < -2$ or $x > 3$

42. **a.** Suppose y varies directly as x. When $x = 3, y = 12$. Determine y when $x = 5$.

$y = 20$

b. Suppose y varies directly as x^2. When $x = 4, y = 8$. Determine y when $x = 8$.

$y = 32$

c. Suppose y varies directly as x^3. When $x = 1, y = 5$. Determine y when $x = 2$.

$y = 40$

In Exercises 43–47, graph the function using your graphing calculator. Then answer the following questions, referring to the graphing calculator.

a. *Determine the x-intercepts of the function, if it has any.*

b. *Determine the domain and range of the function.*

c. *Determine the values of x for which the function is increasing and the values of x for which the function is decreasing.*

43. $y = x^3 - 8$

a. $(2, 0)$

b. domain: all real numbers;
range: all real numbers

c. increasing for all real numbers

44. $y = -2x^3 - 2$

a. $(-1, 0)$

b. domain: all real numbers;
range: all real numbers

c. decreasing for all real numbers

45. $y = x^4 - 8$

a. $(-1.68, 0), (1.68, 0)$

b. domain: all real numbers;
range: $y \geq -8$

c. increasing: $x > 0$;
decreasing: $x < 0$

46. $y = x^4 + 2x$

a. $(0, 0), (-1.26, 0)$

b. domain: all real numbers;
range: $y \geq -1.19$

c. increasing: $x > -0.8$;
decreasing: $x < -0.8$

47. $y = x^4 + 5$

a. none

b. domain: all real numbers;
range: $y \geq 5$

c. increasing: $x > 0$;
decreasing: $x < 0$

48. The height, h (in feet), of a golf ball is a function of the time, t (in seconds), it has been in flight. A golfer strikes a golf ball with an initial upward velocity of 80 feet per second. The flight path of the ball is a parabola. The approximate height of the ball above the ground is modeled by

$$h(t) = -16t^2 + 80t.$$

a. Sketch a graph of the function. What is the practical domain in this situation?

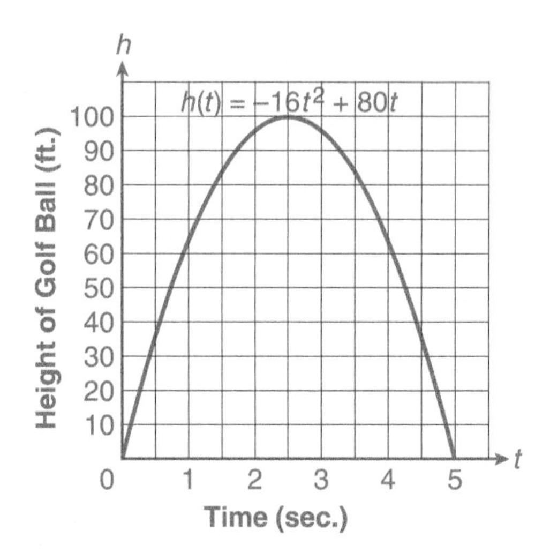

The practical domain is $0 \le t \le 5$.

b. Determine the vertex of the parabola. What is the practical meaning of this point?

(2.5, 100). The ball reaches its highest level, 100 feet, 2.5 seconds after being struck.

c. What is the vertical intercept, and what is its practical meaning in this situation?

(0, 0). The ball is on the ground when the club makes contact with it.

d. Determine the horizontal intercepts. What is the significance of these intercepts?

(0, 0), (5, 0). The ball is on the ground when the club makes contact, $t = 0$, and returns to the ground 5 seconds later.

e. What assumption are you making in this situation about the elevation of the spot where the ball is struck and the point where the ball lands?

I am assuming that the elevations are the same.

49. To use the regression feature of your calculator to determine the equation of a parabola, you need three distinct points. The stream of water flowing out of a water fountain is in the shape of a parabola. Suppose you let the origin of a coordinate system correspond to the point where the water begins to flow out of the nozzle (see figure).

6 in.
Origin
5 in.

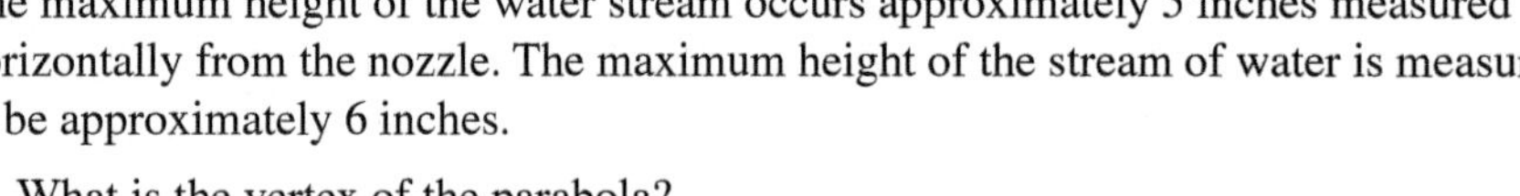
The maximum height of the water stream occurs approximately 5 inches measured horizontally from the nozzle. The maximum height of the stream of water is measured to be approximately 6 inches.

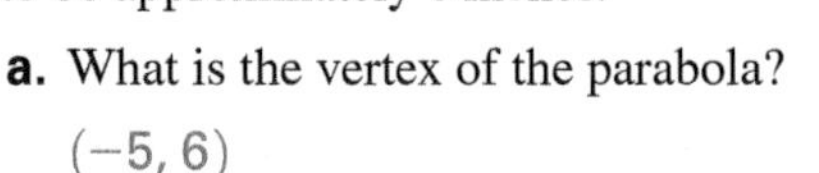
a. What is the vertex of the parabola?

(−5, 6)

b. You already have two points that lie on the parabola. What are they? Use symmetry to obtain a third point.

(0, 0), (−5, 6), (−10, 0)

c. Using these three points and the regression feature of your graphing calculator, determine the equation of the stream of water.

$y = -0.24x^2 - 2.4x$

50. A fastball is hit straight up over home plate. The ball's height, h (in feet), from the ground is modeled by

$$h(t) = -16t^2 + 80t + 5,$$

where t is measured in seconds.

a. What is the maximum height of the ball above the ground?

vertex: $(2.5, h(2.5))$ or $(2.5, 105)$; the maximum height is 105 feet.

b. How long will it take the ball to reach the ground?

Set $h(t) = 0$; $t \approx 5.06$ seconds.

51. Safe automobile spacing, S (in feet), is modeled by

$$S(v) = 0.03125v^2 + v + 18,$$

where v is average velocity in feet per second.

a. Suppose a car is traveling 44 feet per second. To be safe, how far should it be from the car in front of it?

$s(44) = 122.5$ feet away

b. If the car is following 50 feet behind a van, what is a safe speed for the car to be traveling? How fast is this in miles per hour (60 miles per hour $\approx$ 88 feet per second)?

$0.03125v^2 + v + 18 = 50$

$0.03125v^2 + v - 32 = 0$

$v \approx -51.78$ or 19.78

Reject the negative. 19.78 feet per second $\approx$ 13.5 miles per hour

52. The amount of monetary damage caused by reported cyber-crime to the Internet Crime Complaint Center (IC3) from 2011 to 2016 (in millions of U.S. dollars) is given in the following table:

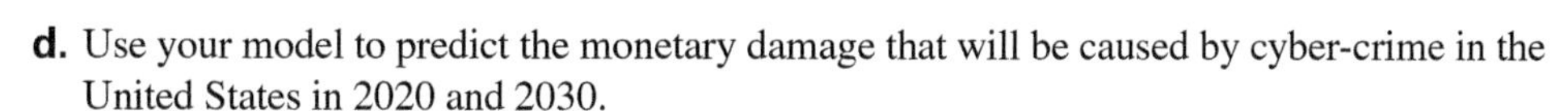

Years since 2010, t	1	2	3	4	5	6
Amount of Monetary Damage Caused by Cyber-crime, $D(t)$ (millions of dollars)	485.3	581.4	781.8	800.5	1070.7	1330.0

a. Use your graphing calculator to determine a scatterplot of this data.

52. a.

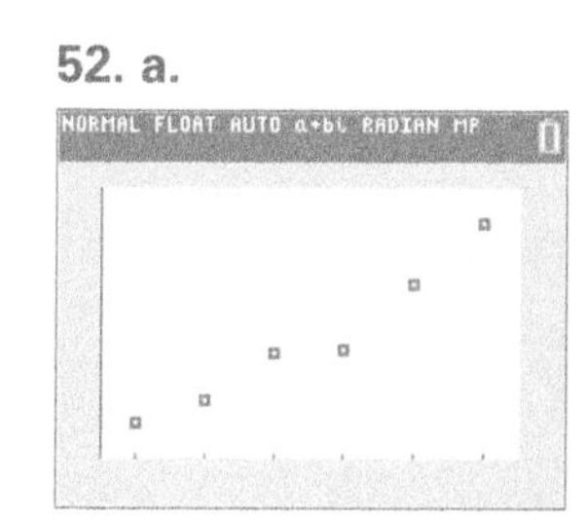

b. Use your graphing calculator to determine the quadratic regression model that best fits the data. Round your coefficients to two decimal places.

$D(t) = 19.56t^2 + 26.25t + 453.14$

c. Do you believe this is a good model for the data? Explain.

The model appears to fit the data quite well, passing very close to 4 of the 6 data points.

d. Use your model to predict the monetary damage that will be caused by cyber-crime in the United States in 2020 and 2030.

2020: $t = 10$; $D(10) = 19.56(10)^2 + 26.25(10) + 453.14 = 2671.64$ million dollars

2030: $t = 20$; $D(20) = 19.56(20)^2 + 26.25(20) + 453.14 = 8802.14$ million dollars

e. In which prediction in part d do you have the most confidence? Explain.

I have more confidence in the 2020 prediction. The 2030 prediction is extrapolating quite far from the original data.

CHAPTER 5

Rational and Radical Functions

CLUSTER 1

Rational Functions

ACTIVITY 5.1

Speed Limits

Properties and Graphs of Functions Defined by $y = \frac{k}{x}, k \neq 0$

OBJECTIVES

1. Determine the domain and range of functions defined by $y = \frac{k}{x}$, where k is a nonzero real number.
2. Determine the vertical and horizontal asymptotes of graphs of $y = \frac{k}{x}$.
3. Sketch graphs of functions of the form $y = \frac{k}{x}$.
4. Determine the properties of graphs having equation $y = \frac{k}{x}$.

The speed limit on the New York State Thruway is 65 miles per hour.

1. If you maintain an average speed of 65 miles per hour, how long will it take to make a 200-mile trip on the Thruway? Recall that *distance* = *rate* · *time*, *time* = $\frac{distance}{rate}$.

time at 65 mph $= \frac{200 \text{ mi.}}{65 \text{ mph}} \approx 3.08$ hr.

2. Complete the following table, in which the input variable r represents the average speed in miles per hour and the output variable t represents the time in hours to complete a 200-mile trip:

r (mph)	20	30	40	50	60	70	80
t (hr.)	10	6.67	5	4	3.33	2.86	2.5

3. Write an equation that defines travel time, t, as a function of the average speed, r.

$t = f(r) = \frac{200}{r}$

4. As the average speed, r, increases, what happens to the travel time, t? What does this mean in practical terms?

As the values of r increase, the values of t decrease but at a slower rate. Increasing the speed lowers the time less at higher speeds than at slower speeds.

5. During a winter storm, a combination of drifting snow and icy conditions reduces your average speed to almost a standstill. Complete the following table for a 200-mile trip on the New York State Thruway:

r (mph)	10	7	5	3	2	1
t (hr.)	20	28.6	40	66.7	100	200

6. As the average speed r gets closer to zero, what happens to the travel time t? Explain what this means in practical terms.

The values of t increase without bound. At very slow rates, the time of the trip is very, very long.

7. Can zero be used as an input value? Explain.

No, division by zero is undefined. If you average 0 miles per hour, you will never get there.

8. a. What is the practical domain of the function given in Problem 3?

The practical domain is $0 < r \leq 65$.

b. Sketch a graph of this function using the table values in Problems 2 and 5.

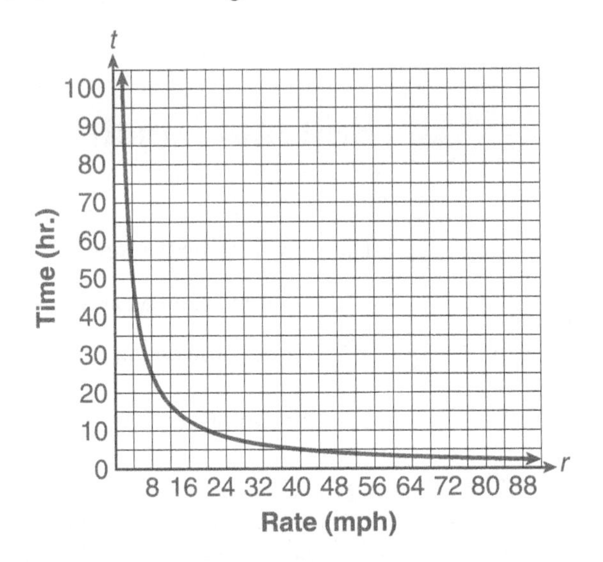

9. a. What are the horizontal and vertical intercepts of the graph?

There are no intercepts.

b. Describe the relationship between the horizontal axis $(t = 0)$ and the graph of the function as the values of r get very large.

As the input increases, the graph gets closer to the horizontal axis but never touches it.

In this situation, the horizontal line $t = 0$ is called a **horizontal asymptote**. Recall that a horizontal asymptote is a horizontal line $(y = b)$ that a graph approaches as the input values get very large in a positive direction (or get very large in a negative direction).

c. Describe the relationship between the vertical axis $(r = 0)$ and the graph of the function as the values of r get close to zero.

As the input gets close to zero, the graph increases without bound, approaching the vertical axis but never touching it.

In this situation, the vertical line $r = 0$ is called a **vertical asymptote**.

In general, a **vertical asymptote** is a vertical line, $x = a$, such that the function is undefined at $x = a$, and as the x-values get closer and closer to $x = a$, the y-values get larger and larger in magnitude. That is, the y-values become very large positive or very large negative values.

10. Using the graph in Problem 8, approximate your average speed if the 200-mile trip takes 5 hours.

(Answers will vary.) $r \approx 40$ mph

Functions Defined by $y = \frac{k}{x}$, Where k Is a Nonzero Constant

The function rule $t = \frac{200}{r}$ gives the relationship between the average speed, r, the time, t, and the given value for distance (200). This function belongs to a family of functions having a general equation of the form $f(x) = \frac{k}{x}$, where k represents some nonzero constant.

EXAMPLE 1 *Examples of this type of function are* $f(x) = \frac{1}{x}$, $g(x) = \frac{5}{x}$, *and* $h(x) = \frac{10}{x}$.

11. a. What is the domain of functions f, g, and h defined in Example 1?

all real numbers except zero

b. Complete the following table:

x	−20	−10	−5	−1	−0.5	−0.1	0	0.1	0.5	1	5	10	20
$f(x)$	−0.05	−0.1	−0.2	−1	−2	−10	undef.	10	2	1	0.2	0.1	0.05
$g(x)$	−0.25	−0.5	−1	−5	−10	−50	undef.	50	10	5	1	0.5	0.25
$h(x)$	−0.5	−1	−2	−10	−20	−100	undef.	100	20	10	2	1	0.5

c. Sketch graphs of f, g, and h on the same coordinate system. Verify using your graphing calculator with window Xmin = −5, Xmax = 5, Ymin = −25, and Ymax = 25.

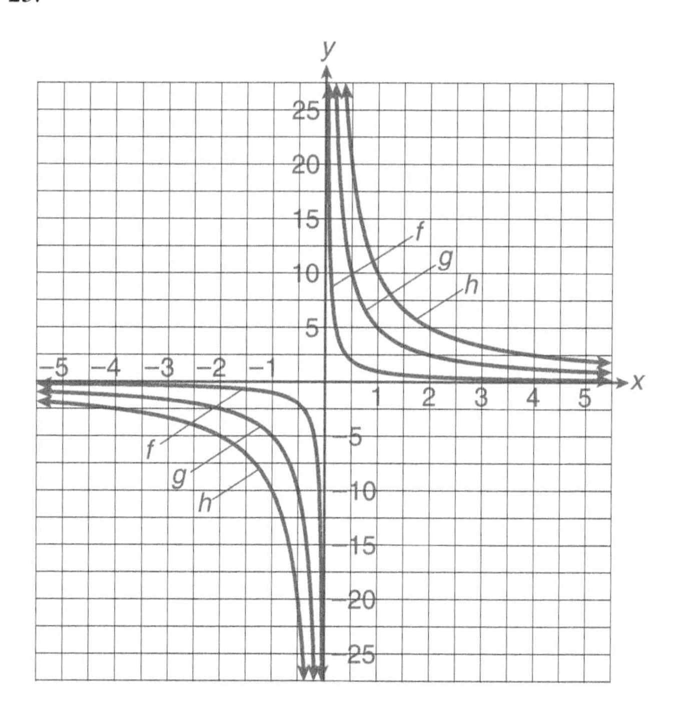

12. Using the table and graphs in Problem 11, answer each of the following questions:

a. What happens to the y-values as the x-values increase in magnitude infinitely (without bound) in both the positive and negative directions?

The y-values get closer and closer to zero.

b. What is the horizontal asymptote for each graph?

$y = 0$

c. What happens to the y-values as the positive x-values get closer to zero?

The y-values get very large (toward positive infinity).

d. What happens to the y-values as the negative x-values get closer to zero?

The y-values become very large negative numbers (toward negative infinity).

e. What is the vertical asymptote for each graph?

$x = 0$

13. a. Do the graphs of f, g, or h in Problem 11 have x- or y-intercepts?

None of the graphs has an x- or a y-intercept.

b. Do the functions f, g, and h have a maximum function value or a minimum function value? Explain.

No, the functions have neither a minimum nor maximum value.

They keep increasing or decreasing without turning.

14. a. Complete the following table, where $Q(x) = \frac{-1}{x}$:

x	−10	−5	−1	−0.5	−0.1	0	0.1	0.5	1	5	10
$Q(x)$	0.1	0.2	1	2	10	undef.	−10	−2	−1	−0.2	−0.1

b. Sketch a graph of Q on the first grid below. Verify using your graphing calculator with window Xmin = −4, Xmax = 4, Ymin = −4, and Ymax = 4.

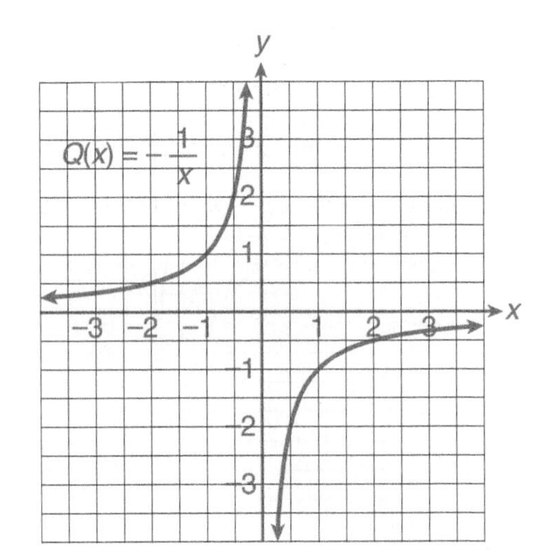

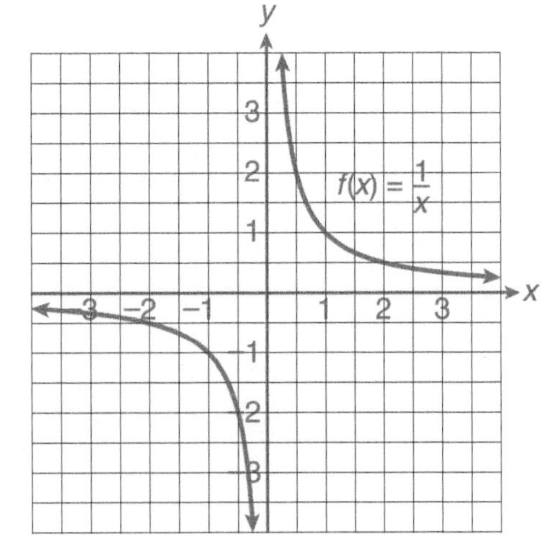

LC LEARNING CATALYTICS

Without graphing, determine the quadrants in which the graph of $f(x) = -\frac{1}{x}$ lies.

a. quadrants I and III
b. quadrants II and IV
c. quadrants I and II
d. quadrants III and IV

c. Sketch the graph of $f(x) = \dfrac{1}{x}$ on the second grid in part b.

d. Describe how the graph of $Q(x) = \dfrac{-1}{x}$ can be obtained from the graph of $f(x) = \dfrac{1}{x}$.

The change in sign results in opposite y-values. The graph of one is the reflection of the other about the x-axis.

SUMMARY Activity 5.1

Functions defined by $f(x) = \dfrac{k}{x}$, where k represents some nonzero constant, have the following properties:

1. The domain and the range consist of all real numbers except zero.
2. If $k > 0$, the graph of f has the general shape below.

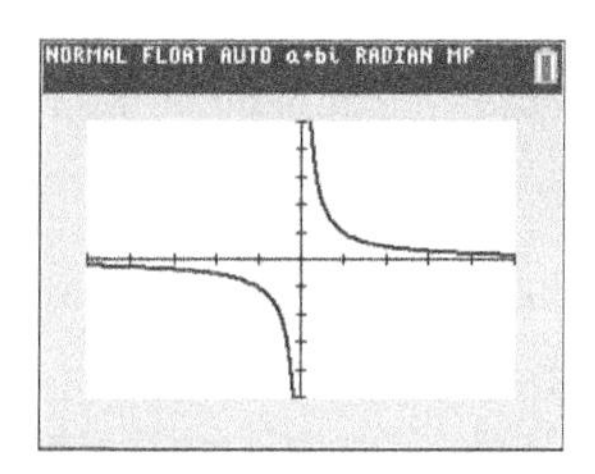

3. If $k < 0$, the graph of f has the general shape below.

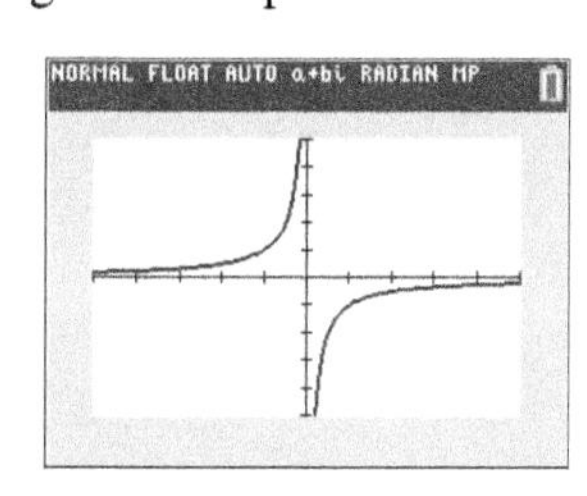

4. The vertical line $x = 0$ is the vertical asymptote.
5. The horizontal line $y = 0$ is the horizontal asymptote.
6. The graph does not intersect either axis (there are no intercepts).
7. There is no maximum or minimum y-value.
8. The function is not continuous at $x = 0$.

EXERCISES Activity 5.1

1. You are a member of a group of distance runners who compete in races ranging in length from 5 to 25 kilometers. In these races, each runner who finishes is told his or her time. Given the time and the length of the race, you can calculate your average running speed.

   a. If you finish a 20-kilometer race in 1 hour 15 minutes, what is your average speed?

   The average speed $= \dfrac{20 \text{ km}}{1 \text{ hr. } 15 \text{ min.}} = \dfrac{20 \text{ km}}{1.25 \text{ hr.}} = 16$ km/hr.

   b. Complete the following table for a 20-kilometer race:

t (hr.)	1.00	1.25	1.50	1.75	2.00	2.25	2.50
s (km/hr.)	20	16	13.33	11.43	10	8.89	8

   c. Write an equation that expresses the average speed, s, as a function of time t in a 20-kilometer race.

   $s = f(t) = \dfrac{20}{t}$

   d. i. What is the domain of the function?

   the set of all nonzero real numbers

   ii. What is the practical domain?

   (Answers will vary.) $1 \leq t \leq 5$. Because 20 kilometers per hour (when $t = 1$) is fast for a distance runner and 4 kilometers per hour (when $t = 5$) is slow for a distance runner, most times will fall between these values.

   iii. Sketch a graph of the function using the practical domain.

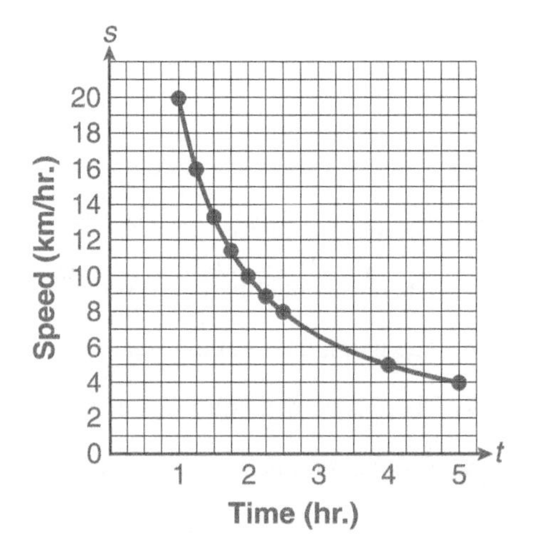

   e. As the running time, t, gets longer, what happens to the average speed, s?

   The average speed decreases, approaching zero.

   f. As the running time, t, is reduced (gets closer to zero), what happens to the average speed, s?

   The average speed increases without bound.

Exercise numbers appearing in color are answered in the Selected Answers appendix.

2. a. Sketch the graphs of the following pair of functions on separate coordinate systems. Use labels or colors to differentiate the graphs. Verify your sketches using your graphing calculator.

$$f(x) = \frac{5}{x}, \qquad g(x) = \frac{-5}{x}$$

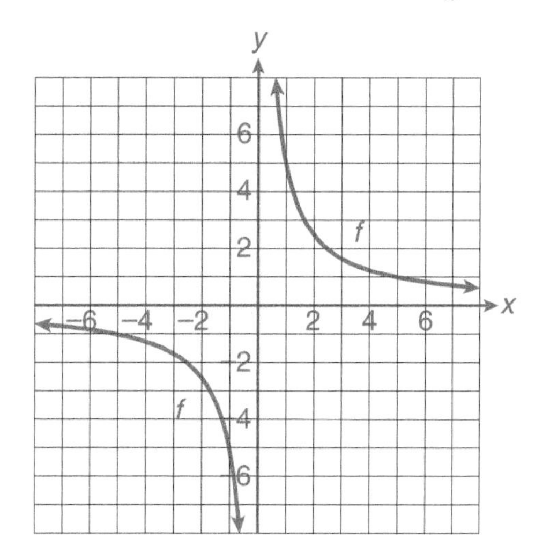

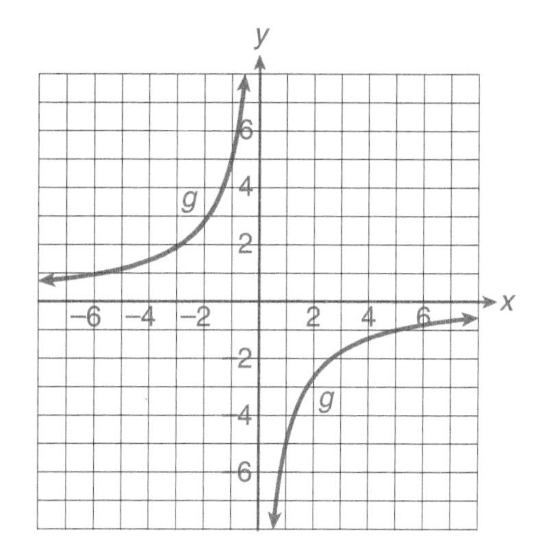

b. Describe how the graph of g can be obtained from the graph of function f.

The change in sign results in opposite output values. The graph of g is the reflection of f about the x-axis or the y-axis.

3. A commercial refrigerator has an initial cost, C, and a scrap value, V. If the life of the refrigerator is N years, then the amount, D, that can be depreciated each year is modeled by the formula

$$D = \frac{C - V}{N}.$$

a. If the initial cost is \$1400 and the scrap value is \$200, write an equation for D as a function of N.

$$D = f(N) = \frac{1400 - 200}{N} = \frac{1200}{N}$$

b. Complete the following table using the equation in part a:

N	1	2	3	6	12	24
D(\$)	1200	600	400	200	100	50

c. Sketch a graph of the function.

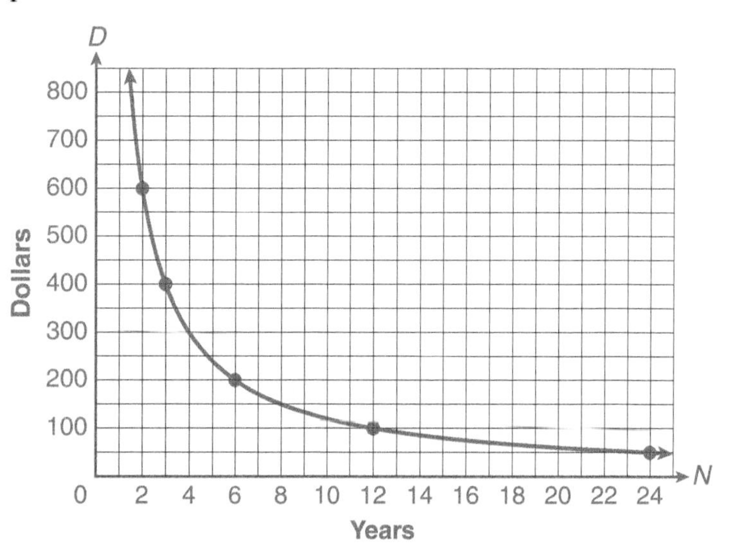

d. If the refrigerator is well constructed, it should have a long, useful life. Will an increase in the useful life of the refrigerator increase or decrease the amount, D, that can be depreciated each year? Explain.

decrease; as N gets larger, D gets smaller.

4. The speed limit on Route 66 in parts of Arizona is 75 miles per hour.

a. If you maintain an average speed of 75 miles per hour, how long will it take to make a 350-mile trip on Route 66?

distance = *rate* · *time* or $t = \frac{d}{r}$, where t is the time, d is the distance traveled, and r is the average rate. In this case, $t = \frac{350}{75} = 4\frac{2}{3}$ hours, or 4 hours and 40 minutes.

b. Write an equation that defines t as a function of r, in which r represents the average speed in miles per hour and t represents the time in hours to complete the 350-mile trip.

$$t = \frac{350}{r}$$

c. Complete the following table for the equation you determined in part b:

r (mph)	25	35	45	55	65	75	85
t (hr.)	14	10	7.8	6.4	5.4	4.7	4.1

d. As your average speed increases, what happens to the time it takes to complete the trip?

It decreases, but it decreases more slowly as r increases.

e. As your average speed for the trip gets closer to zero, what happens to the time it takes to complete the 350-mile trip?

The time increases very rapidly as the rate approaches zero.

f. What is the practical domain?

The practical domain is $0 < r \leq 85$. (Answers will vary slightly.)

g. Sketch a graph.

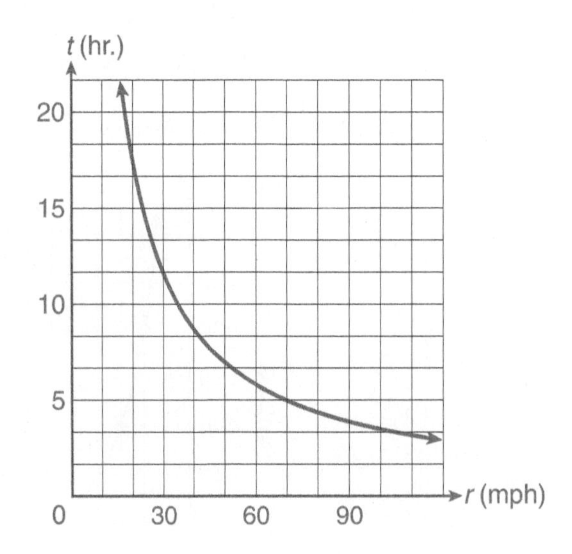

h. What are the vertical and horizontal asymptotes of the graph of the function?

The horizontal asymptote is $t = 0$; the vertical asymptote is $r = 0$.

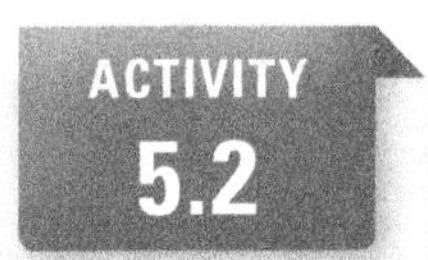

Loudness of a Sound

Inverse Variation Functions

OBJECTIVES

1. Graph functions defined by an equation of the form $y = \frac{k}{x^n}$, where n is a positive integer and k is a nonzero real number, $x \neq 0$.
2. Describe the properties of graphs having equation $y = \frac{k}{x^n}$, $x \neq 0$.
3. Determine k, the constant of proportionality (also called the constant of variation).

The loudness (or intensity) of any sound is a function of the listener's distance from the source of the sound. In general, the relationship between the intensity, I, and the distance, d, can be modeled by an equation of the form

$$I = \frac{k}{d^2},$$

where I is measured in microwatts per square meter, d is measured in meters, and k is a constant determined by the source of the sound and the nature of the surroundings.

1. The intensity, I, of a typical iPod at maximum setting can be given by the formula $I = \frac{64}{d^2}$.

Complete the following table:

d (m)	0.1	0.5	1	2	5	10	20	30
I(mW/m²)	6400	256	64	16	2.56	0.64	0.16	0.07

2. a. What is the practical domain of the function?

$0.01 < d < 20$ (Answers will vary slightly.)

b. Sketch a graph that shows the relationship between intensity of sound and distance from the source of the sound. Use the table in Problem 1 to help you determine the appropriate scale.

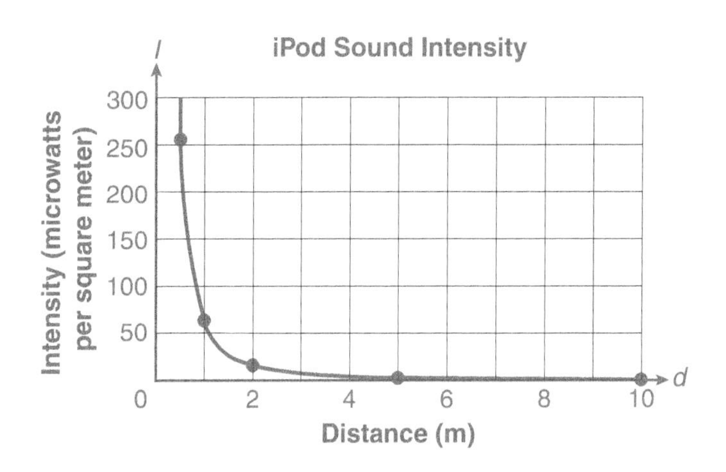

3. As you move closer to the iPod, what happens to the intensity of the sound?

It increases more and more rapidly as d approaches zero.

4. As you move away from the iPod, what happens to the intensity of the sound?

It decreases less and less rapidly as d gets larger.

Functions Defined by $y = \frac{k}{x^2}$, Where k Is a Nonzero Constant

The function defined by $I = \frac{64}{d^2}$ belongs to a family of functions having an equation of the form $y = \frac{k}{x^2}$, where k represents some nonzero constant.

constant of proportionality depends on the source of the sound and the nature of the surroundings. If the source of the sound changes, the value of the constant of proportionality will also change.

a. The intensity of the sound made by a heavy truck 20 meters away is 1000 microwatts per square meter. Determine the constant of proportionality.

Assume that $I = \frac{k}{d^2} \Rightarrow k = 1000 \cdot 20^2 = 400{,}000$.

b. Write a formula for the intensity, I, of the sound made by the truck when it is d meters away.

$$I = \frac{400{,}000}{d^2}$$

c. Use the formula from part b to determine the intensity of the sound made by the truck when it is 100 meters away.

$$I = \frac{400{,}000}{100^2} = 40 \text{ microwatts per square meter}$$

SUMMARY Activity 5.2

- Functions defined by equations of the form $f(x) = \frac{k}{x^n}$, where k is a nonzero real number, have the following properties:

1. The domain consists of all real numbers except zero.

2. The graph of f has the following general shape:

a. Where $k > 0$ and n is an even integer

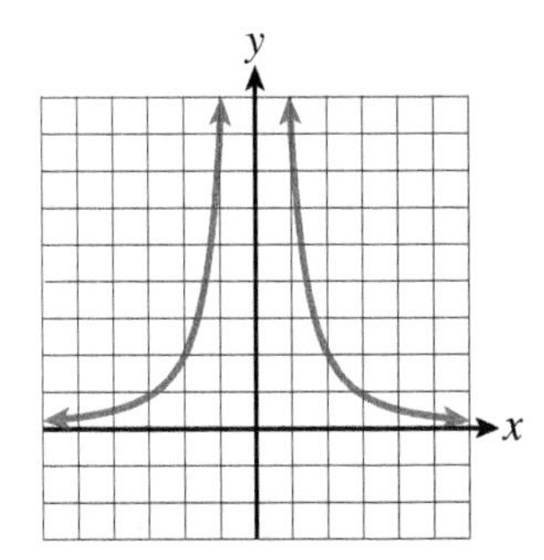

b. Where $k > 0$ and n is an odd integer

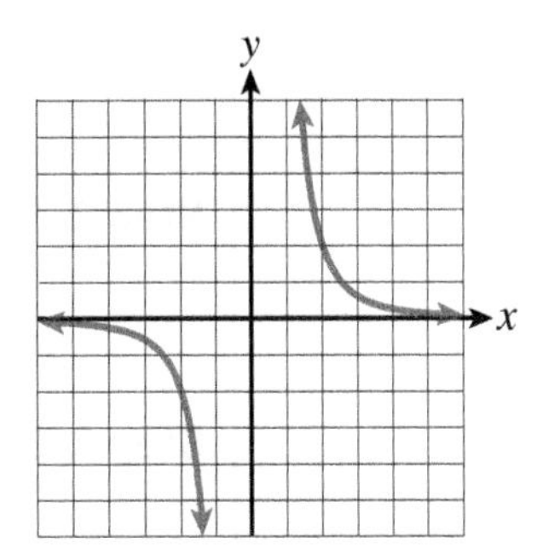

c. Where $k < 0$ and n is an even integer

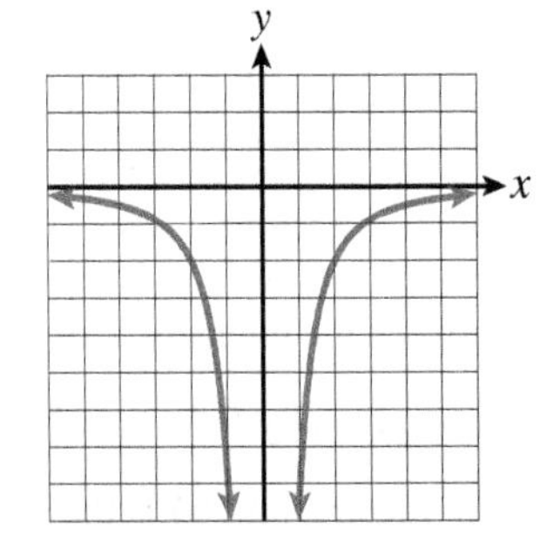

d. Where $k < 0$ and n is an odd integer

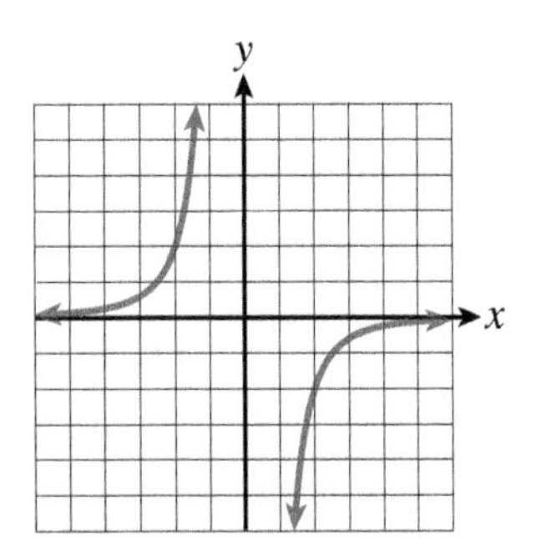

3. The vertical asymptote is the vertical line $x = 0$.

4. The horizontal asymptote is the horizontal line $y = 0$.

5. There are no vertical or horizontal intercepts.

6. There is no maximum or minimum function value.

7. The function f is not continuous at $x = 0$.

- Functions defined by $y = \dfrac{k}{x^n}$ are called **inverse variation functions** in which

1. y is said to vary inversely as the nth power of x.

2. k is called the constant of variation or constant of proportionality.

EXERCISES Activity 5.2

1. Doctors sometimes use a patient's body mass index to determine whether the patient should lose weight. The model for the body mass index, B, is the formula

$$B = \frac{703w}{h^2},$$

where w is weight in pounds and h is height in inches.

a. What is your body mass index?

(Answers will vary.) If a person is 6 feet = 72 inches tall and weighs 200 pounds, $B = \dfrac{703(200)}{72^2} \approx 27.12$.

b. Suppose your friend weighs 170 pounds. Substitute this value into the body mass index formula to obtain an equation for B in terms of height.

$B = \dfrac{119{,}510}{h^2}$

c. What is the practical domain of the body mass index function in part b?

$36 < h < 84$. This works unless the person is under 3 feet tall or over 7 feet tall.

d. Complete the following table using the formula for the body mass index of a 170-pound person:

Height in Inches, h	60	64	68	72	76	80
Body Mass Index, B	33.2	29.2	25.8	23.1	20.7	18.7

e. Sketch a graph of the function defined by $B = \frac{119{,}510}{h^2}$. Use the data values in part d to help you determine an appropriately scaled axis.

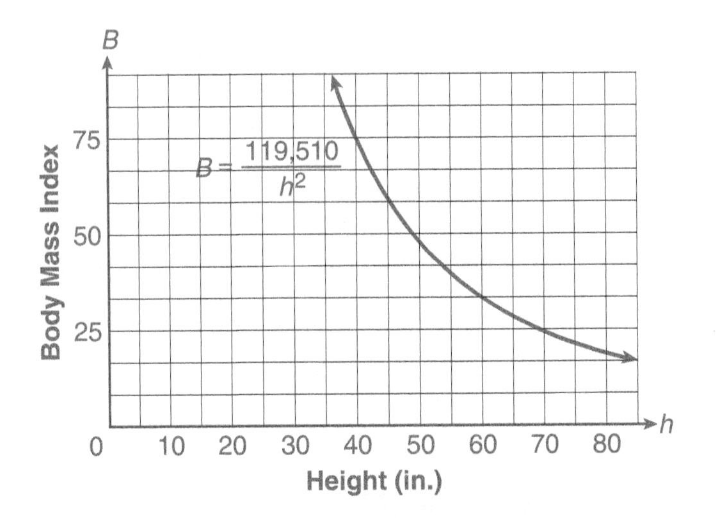

f. What happens to the body mass index as height increases? Does this make sense in the context of the situation? Explain.

Body mass index, B, gets smaller, and it actually approaches zero. Yes, this makes sense in this case because taller people with the same weight should be skinnier.

g. It is recommended that a person's body mass index be between 19 and 25. Use the graph and the trace key on your calculator to approximate the values of h for which a 170-lb. person has a BMI between 19 and 25.

$69.1 < h < 79.3$, measured in inches

2. Sketch a graph of the functions $f(x) = \frac{3}{x^2}$ on the first grid and $g(x) = \frac{-3}{x^2}$ on the second grid. Describe how the graph of g is related to the graph of f.

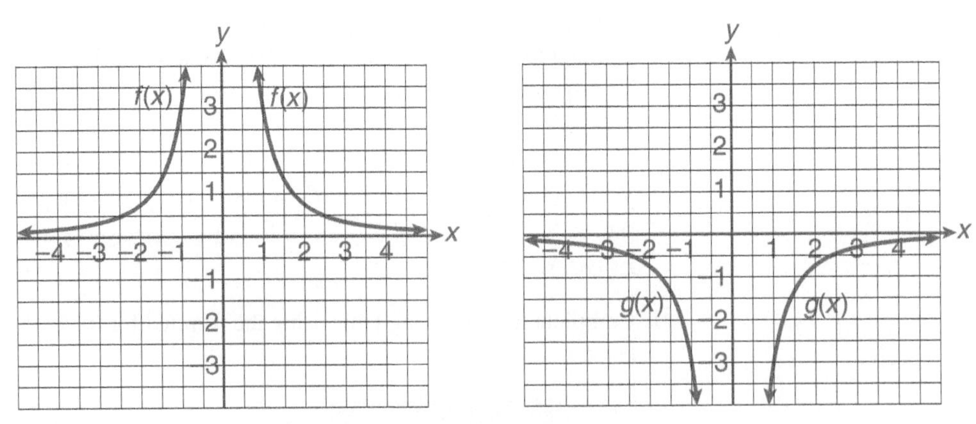

The graph of g is the reflection of f about the x-axis.

3. Match the following functions with the accompanying graphs:

i. $f(x) = \dfrac{10}{x^4}$

ii. $g(x) = \dfrac{100}{x^5}$

iii. $h(x) = \dfrac{-10}{x^3}$

iv. $F(x) = \dfrac{-1}{x^2}$

a.

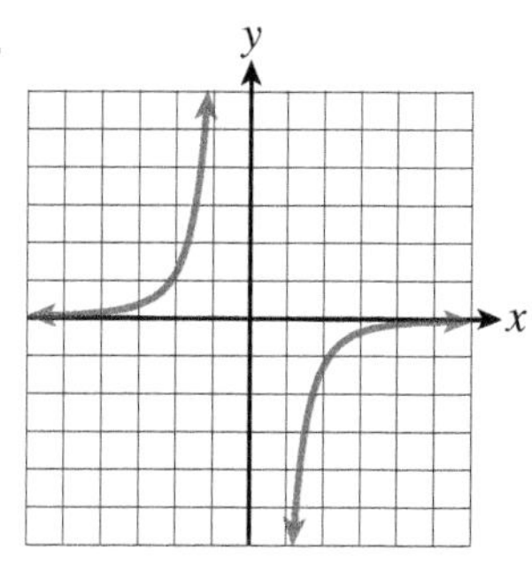

b.

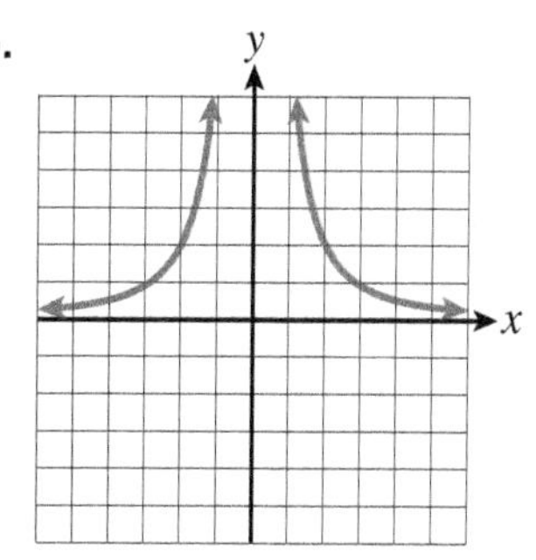

c.

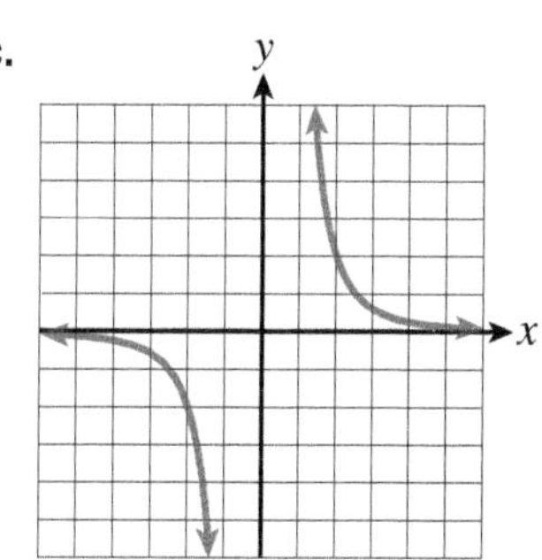

d.

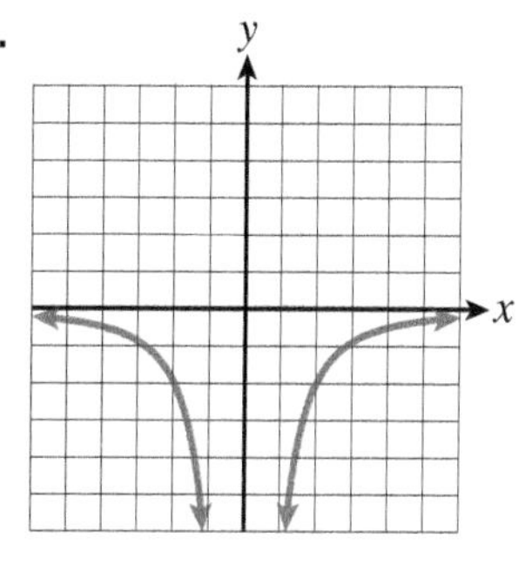

i. graph b

ii. graph c

iii. graph a

iv. graph d

4. Describe how the graphs of $y = \dfrac{1}{x^2}$ and $y = \dfrac{1}{x^3}$ are similar and how they are different.

The graphs have the same asymptotes and the same domain. Neither has a maximum or a minimum value. The graph of $\dfrac{1}{x^2}$ is symmetric with respect to the y-axis. The graph of $\dfrac{1}{x^3}$ is symmetric with respect to the origin. The output values for $y = \dfrac{1}{x^2}$ are always positive. The output values for $y = \dfrac{1}{x^3}$ are both positive and negative.

5. Consider the family of functions of the form $f(x) = \dfrac{k}{x^n}$, where k is a nonzero constant and n is a positive integer.

a. What is the domain of f?

all nonzero real numbers

b. Use several different values of k and n, where $k > 0$ and n is an odd positive integer, to determine the general shape of the graph of f.

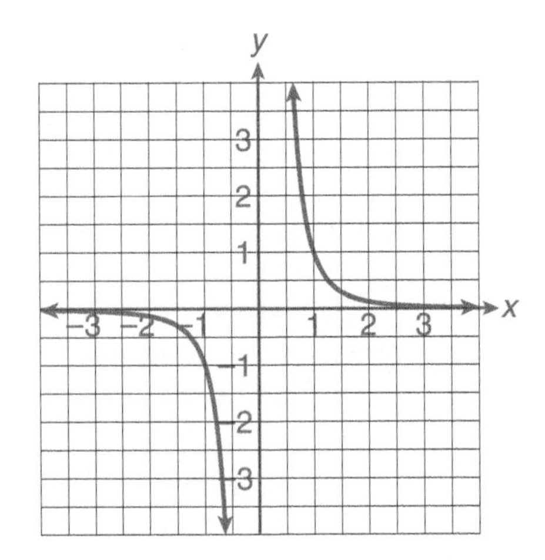

c. Use several different values of k and n, where $k > 0$ and n is an even positive integer, to determine the general shape of the graph of f.

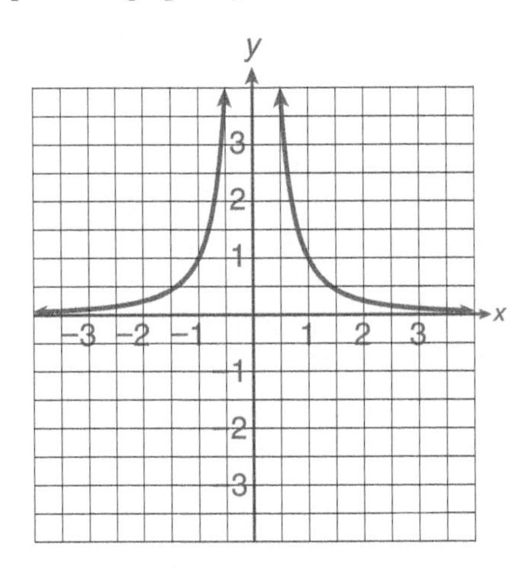

6. How will the general shapes of the graphs in Exercise 5 change if $k < 0$?

The output corresponding to a given input will be the opposite in sign.

The graphs will be reflected in the x-axis.

7. Complete the following tables of ordered pairs for the given inverse variations. Determine the constant of variation, and write the equation that is represented by the table.

a. y varies inversely as x. $y = \frac{2}{x}$

x	y
$\frac{1}{2}$	4
1	2
2	1
6	$\frac{1}{3}$

b. y varies inversely as x^3. $y = \frac{8}{x^3}$

x	y
$\frac{1}{2}$	64
1	8
2	1
6	$\frac{1}{27}$

8. If y varies inversely as the cube of x, determine the constant of proportionality if $y = 16$ when $x = 2$.

$y = \frac{k}{x^3}$; $\quad 16 = \frac{k}{2^3}$ or $k = 16 \cdot 2^3 = 128$

9. The amount of current, I (in amps), in a circuit varies inversely as the resistance, R (in ohms). A circuit containing a resistance of 10 ohms has a current of 12 amperes. Determine the current in a circuit containing a resistance of 15 ohms.

$I = \frac{k}{R}$; $\quad 12 = \frac{k}{10}$ or $k = 12 \cdot 10 = 120$

$I = \frac{120}{R}$; $\quad I = \frac{120}{15} = 8$ amps

10. The intensity, I, of light varies inversely as the square of the distance, d, between the source of light and the object being illuminated. A light meter reads 0.25 unit at a distance of 2 meters from a light source. What will the meter read at a distance of 3 meters from the source?

$I = \frac{k}{d^2}$; $\quad 0.25 = \frac{k}{2^2}$ or $k = (0.25) \cdot 4 = 1$

$I = \frac{1}{d^2} = \frac{1}{3^2} \approx 0.11$ unit

11. You are investigating the relationship between the volume, V, and pressure, P, of a gas. In a laboratory, you conduct the following experiment: While holding the temperature of a gas constant, you vary the pressure and measure the corresponding volume. The data that you collect appear in the following table:

p(psi)	20	30	40	50	60	70	80
V(ft.^{3})	82	54	41	32	27	23	20

a. Sketch a graph of the data.

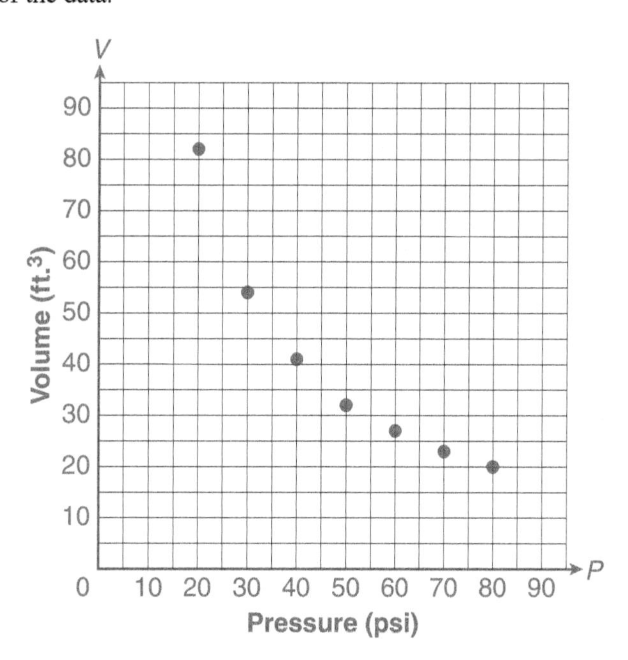

b. One possible model for the data is that V varies inversely as the square of P. Does the data fit the model $V = \frac{k}{P^2}$? Explain.

No. Using $P = 20$, $V = 82$, $k = 20^2(82) = 32{,}800$.

If $V = \frac{32{,}800}{P^2}$, then $P = 30$ would yield $V = \frac{32{,}800}{30^2} \approx 36.44$ (not very close).

c. Another possible model for the data is that V varies inversely as P. Does $V = \frac{k}{P}$ model the data? Explain.

Yes. Using $P = 20$, $V = 82$, $k = 20(82) = 1640$. Trying other P-values gives $V = \frac{1640}{30} \approx 54.67$, $V = \frac{1640}{40} = 41$, which are very close.

d. Predict the volume of the gas if the pressure is 65 pounds per square inch, using the graph in part a.

(Answers will vary if the graph is used.) $V \approx 25$ cubic feet

e. Verify your answer in part d using the model from part c.

$V = \frac{1640}{P}$, $V = \frac{1640}{65} \approx 25.2$ cubic feet

ACTIVITY 5.3

Percent Markup
Rational Functions and Their Graphs

OBJECTIVES

1. Determine the domains of rational functions defined by an equation of the form $y = \frac{k}{g(x)}$, where k is a nonzero constant and $g(x)$ is a first-degree polynomial.
2. Identify the vertical and horizontal asymptotes of $y = \frac{k}{g(x)}$.
3. Sketch graphs of rational functions defined by $y = \frac{k}{g(x)}$.

You are a buyer for a national chain of retail stores. You purchase merchandise at a wholesale cost. The merchandise is then sold at a retail price (called the selling price). The retailer's markup is the difference between the selling price (what the consumer pays) and the wholesale cost.

1. a. You acquire a line of sports jackets at a wholesale cost of $80 per jacket. If the jackets sell for $120 each at the retail level, what is the amount of the markup?

markup = $120 − $80 = $40

b. The markup is what percent of the selling price? (This percent is called the percent markup of the selling price.)

$\frac{40}{120} = \frac{1}{3} = 33\frac{1}{3}\%$

2. The relationship among the selling price, S, the wholesale cost, C, and the percent markup, P, of the selling price (expressed as a decimal) is modeled by

$$S = \frac{C}{1 - P}.$$

If the wholesale cost of a sports jacket is $80, write an equation for S in terms of P.

$S = \frac{80}{1 - P}$

3. a. Complete the following table for $S = \frac{80}{1 - P}$:

Percent Markup, P	0	0.01	0.05	0.10	0.25	0.50	0.75	0.95
Selling Price, S ($)	80	80.81	84.21	88.89	106.67	160	320	1600

b. As the values of P approach 1, what happens to the values of S? What does this mean in practical terms?

The values of S increase. As the percent markup increases, the selling price increases.

4. a. Can the percent markup of the selling price be 100% (i.e., can $P = 1$)? Explain.

No, P cannot equal 1 because the denominator would become $1 - 1 = 0$.

b. What is the practical domain of this function?

$0 \le P < 1$ because percent markup as a decimal is a number from 0 up to but not including 1. This formula will not permit you to mark up the selling price 100% or higher.

5. Sketch a graph of the function defined by $S = \frac{80}{1 - P}$. Use the table of data pairs in Problem 3 to help you determine an appropriate scale. Verify your sketch using your graphing calculator.

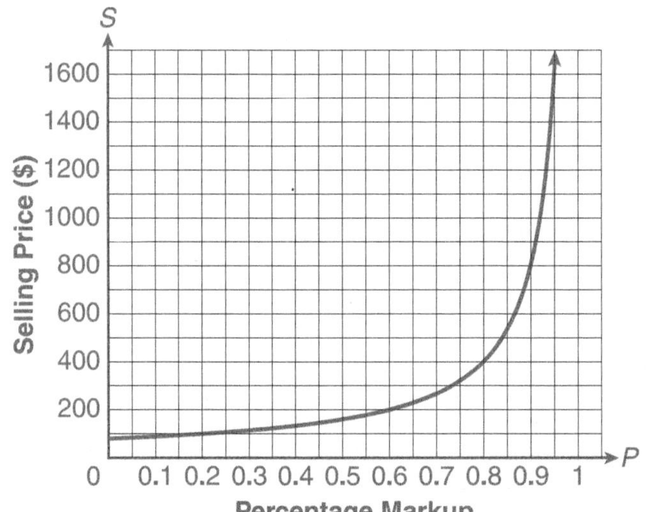

Graphs of $f(x) = \frac{k}{g(x)}$

Because S and P represent real-world quantities, the practical domain limits our investigation of the function defined by $S = \frac{80}{1 - P}$.

6. Consider the general function f defined by $f(x) = \frac{80}{1 - x}$.

a. What is the domain of function f?

all real numbers except $x = 1$

b. Complete the following table:

x	−10	−5	0	0.50	0.75	0.90	1	1.10	1.25	1.50	2	5	10
$f(x)$	7.27	13.33	80	160	320	800	undef.	−800	−320	−160	−80	−20	−8.89

c. Sketch a graph of the function f. Use the table in part b to determine an appropriate scale.

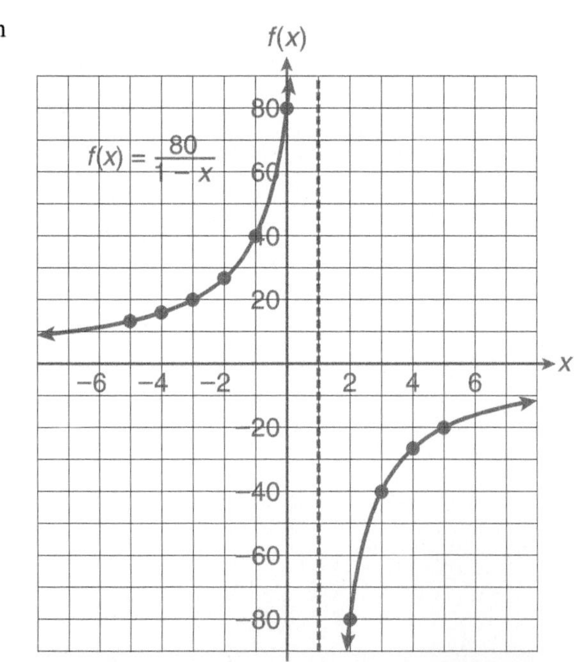

d. Does the graph of f have a horizontal asymptote? Explain. If you answer yes, what is the equation of the horizontal asymptote?

Yes, the equation of the horizontal asymptote is $y = 0$. As x gets larger in magnitude, $f(x)$ gets smaller (closer to zero).

e. Does the graph of f have a vertical asymptote? Explain. If you answer yes, what is the equation of the vertical asymptote?

Yes, the equation of the vertical asymptote is $x = 1$. f is not defined for $x = 1$, but as x approaches 1, $|f(x)|$ gets very large.

f. For what value of x is $f(x)$ maximum?

f has no maximum.

g. Does the graph have any intercepts?

It has a y-intercept at (0, 80). It has no x-intercept.

Note: If you have an older calculator, graphing functions of the form $f(x) = \frac{80}{1 - x}$ may not be accurate. Depending on your calculator, your choice of window, and the operating system, the calculator may attempt to draw the vertical asymptote ($x = 1$) as part of the graph. This problem can be alleviated by changing the mode to DOT mode.

7. Consider the function defined by $g(x) = \frac{80}{x + 1}$.

a. What is the domain of g?

all real numbers except $x = -1$

b. Construct a table of data points for g.

x	−4	−3	−2	−1	0	1	2
$g(x)$	−26.7	−40	−80	undef.	80	40	26.7

c. Sketch a graph of g using an appropriate scale.

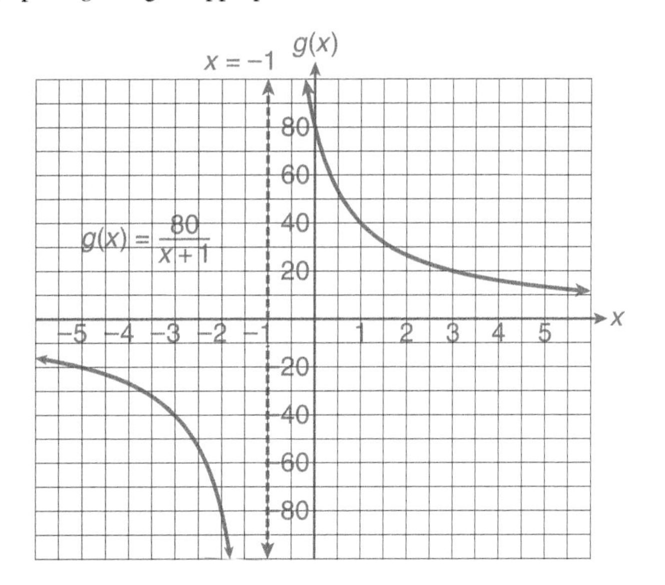

d. Determine the equation of the vertical asymptote. As a graphing aid, if the vertical asymptote is not the y-axis $(x = 0)$, the asymptote is drawn as a dotted vertical line. If you have not done so, draw the vertical asymptote in the graph in part c.

$x = -1$

e. Determine the equation of the horizontal asymptote.

$y = 0$

f. Does the graph of g have any intercepts?

It has no x-intercepts, but it does have a y-intercept at (0, 80).

g. Sketch a graph of $g(x) = \dfrac{80}{1 + x}$ using your graphing calculator. Your screens should appear as follows:

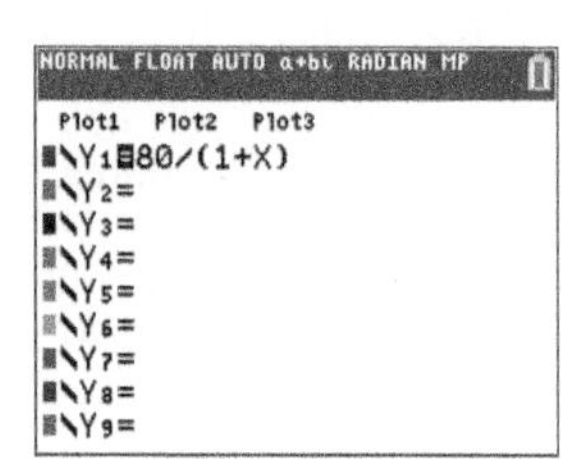

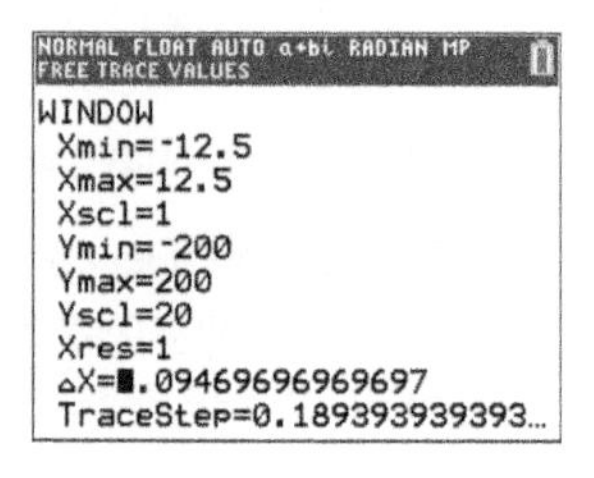

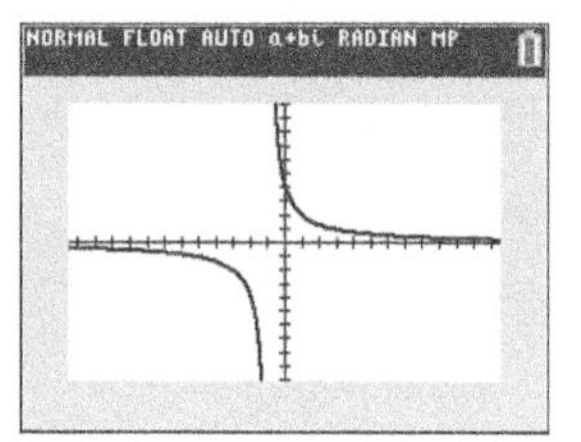

8. How are the graphs of $f(x) = \dfrac{80}{1 - x}$ and $g(x) = \dfrac{80}{x + 1}$ similar? How are they different?

The graphs have the same horizontal asymptote and the same y-intercept and basic shape. The vertical asymptotes are different. The graphs are mirror images with respect to the y-axis.

Rational Functions

A function Q, defined by an equation of the form

$$Q(x) = \frac{k}{g(x)},$$

where k is a nonzero constant and $g(x)$ is a polynomial with degree ≥ 1 and $g(x) \neq 0$, belongs to the family of functions known as **rational functions**. The inverse variation function in Activity 5.2 is a special case of a rational function, where $g(x) = x^n$. The only values at which the function is not defined are any values for which the denominator is zero.

If $g(a) = 0$, then $x = a$ is a vertical asymptote of the graph of $Q(x) = \dfrac{k}{g(x)}$.

The horizontal asymptote of the graph of function Q is the x-axis $(y = 0)$.

EXAMPLE 1 ***Determine the domain and the vertical and horizontal asymptotes for each of the following rational funtions:***

a. $f(x) = \dfrac{3}{x}$ **b.** $g(x) = \dfrac{10}{x - 4}$ **c.** $h(x) = \dfrac{-5}{2x + 6}$

SOLUTION

FUNCTION	DOMAIN	VERTICAL ASYMPTOTE	HORIZONTAL ASYMPTOTE
a. $f(x) = \frac{3}{x}$	all real numbers except $x = 0$	$x = 0$	$y = 0$
b. $g(x) = \frac{10}{x - 4}$	all real numbers except $x = 4$	$x = 4$	$y = 0$
c. $h(x) = \frac{-5}{2x + 6}$	all real numbers except $x = -3$	$x = -3$	$y = 0$

9. Consider the function defined by $f(x) = \frac{5}{3x - 6}$.

a. Determine the domain of f.

all real numbers except $x = 2$

b. Complete the following table:

x	−10	−5	0	1.5	1.9	2	2.1	2.5	3	8	13
$f(x)$	−0.14	−0.24	−0.83	−3.33	−16.67	undef.	16.67	3.33	1.67	0.28	0.15

LC LEARNING CATALYTICS

Determine the vertical asymptote(s), if any for the function defined by $f(x) = \frac{3}{2x + 4}$.

a. $x = 2$
b. $x = -2$
c. $x = \frac{3}{2}$
d. There are no vertical asymptotes.

c. Determine the vertical and horizontal asymptotes of the graph of f.

vertical: $x = 2$; horizontal: $y = 0$

d. Sketch a graph of f.

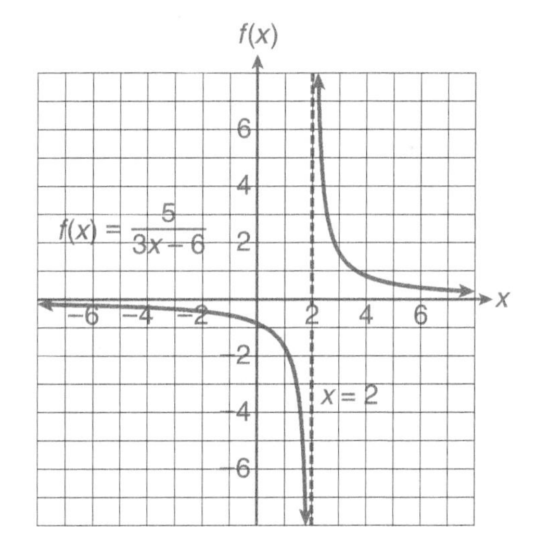

e. Verify the graph using your graphing calculator.

It is verified.

10. Without using your graphing calculator, match the following functions with the accompanying graphs. Use your graphing calculator to verify your matches.

i. $f(x) = \frac{5}{x}$ **ii.** $g(x) = \frac{5}{x^2}$ **iii.** $h(x) = \frac{-5}{x}$

iv. $F(x) = \frac{5}{x + 2}$ **v.** $G(x) = \frac{5}{2x - 4}$

a. 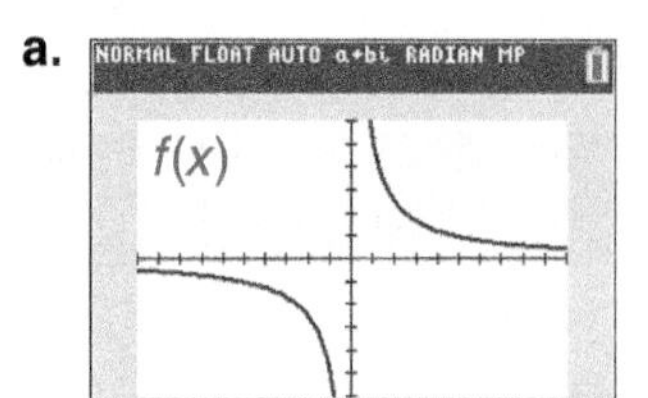

b. 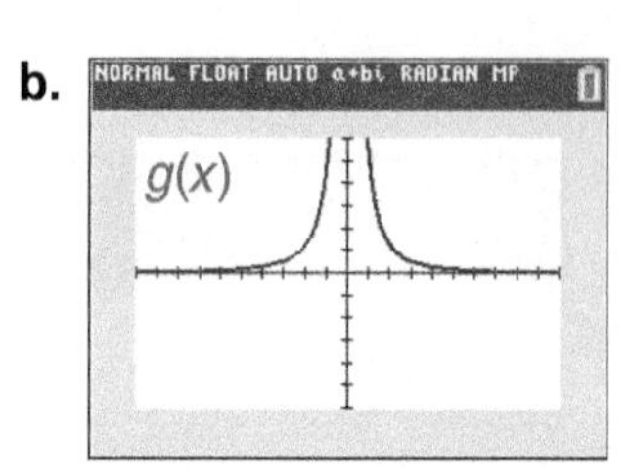

c. 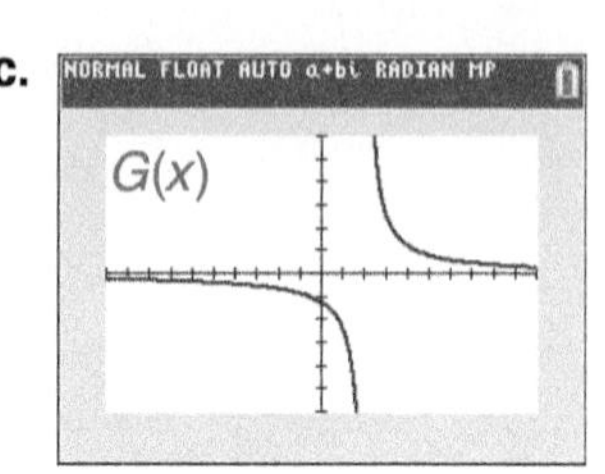

d. 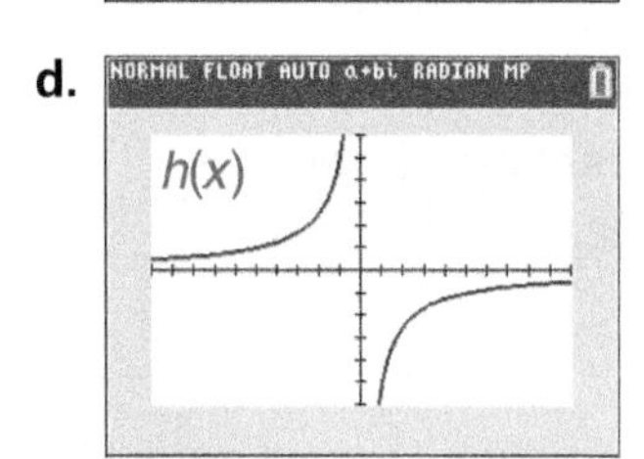

e. 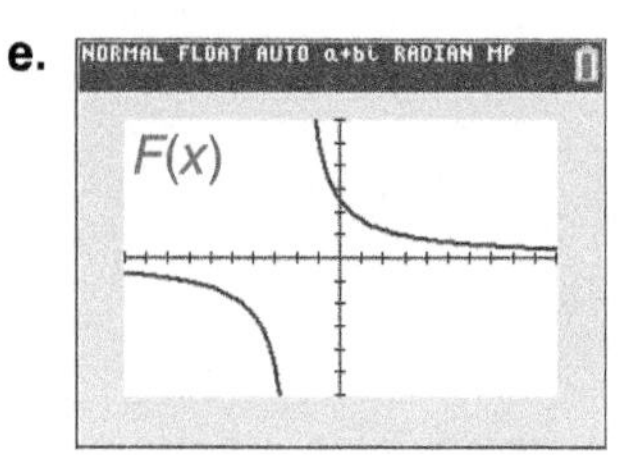

SUMMARY Activity 5.3

1. A function Q, defined by an equation of the form

$$Q(x) = \frac{k}{g(x)},$$

where k is any nonzero constant and $g(x)$ is a polynomial with degree ≥ 1 and $g(x) \neq 0$, belongs to a family of functions known as **rational functions**.

Examples include $f(x) = \frac{10}{x^2}$, $g(x) = \frac{10}{x - 4}$, and $h(x) = \frac{-5}{2x + 6}$.

2. The domain of Q is the set of all real numbers except those values of the input x such that $g(x) = 0$.

3. If $g(a) = 0$, then $x = a$ is a vertical asymptote of $Q(x) = \frac{k}{g(x)}$, $k \neq 0$.

4. The horizontal asymptote is the x-axis $(y = 0)$.

EXERCISES Activity 5.3

1. To obtain an estimate of the required volume, V, of timber that must be harvested for a logging company to break even, use the model

$$V = \frac{Y + L}{P - S - F - T},$$

where V is the required annual logging volume (in cubic meters),
Y is the yard cost (in dollars),
L is the loading cost (in dollars),
P is the selling price (in dollars per cubic meter),
S is the skidding cost (in dollars per cubic meter),
F is the felling cost (in dollars per cubic meter), and
T is the transportation cost (in dollars per cubic meter).

Exercise numbers appearing in color are answered in the Selected Answers appendix.

a. Suppose a logging company estimates that the yard cost will be \$25,000, the loading cost will be \$55,000, the skidding cost will be \$1.50 per cubic meter, the felling cost will be \$0.40 per cubic meter, and the transportation cost will be \$0.60 per cubic meter. Write a rule for V as a function of P.

$$V = \frac{25{,}000 + 55{,}000}{P - 1.5 - 0.40 - 0.60} = \frac{80{,}000}{P - 2.5}$$

b. Complete the following table using the equation in part a:

$P(\$)$	2.50	3.00	5.00	10.00	25.00
$V(m^3)$	undef.	160,000	32,000	10,667	3555.56

c. As the selling price per cubic meter increases, what happens to the corresponding required logging volume, V?

V decreases.

d. Determine the value of V when $P = 2$. What is the practical meaning of the negative value of V?

$V(2) = -160{,}000$

A price of \$2 per cubic meter is not practical.

e. What is the practical domain of this function?

$P > 2.5$

f. Sketch a graph of the function. Use the table in part b to determine an appropriate scale.

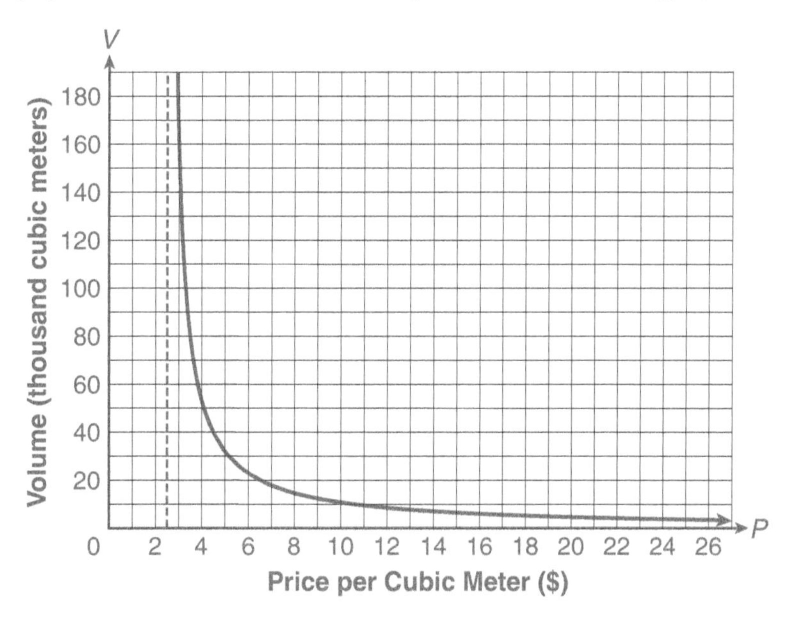

2. Two functions are defined by $f(x) = \dfrac{10}{x - 5}$ and $g(x) = \dfrac{10}{5 - x}$.

a. Describe how you can determine the vertical asymptote without graphing.

Set the denominator equal to 0, and solve for x.

b. Determine the vertical and horizontal asymptote for the graph of each function.

For both, vertical asymptote: $x = 5$; horizontal asymptote: $y = 0$.

c. Verify your answers by graphing each function on your graphing calculator.

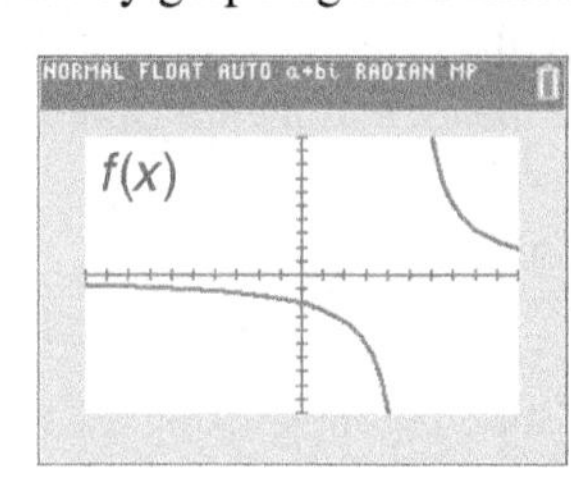

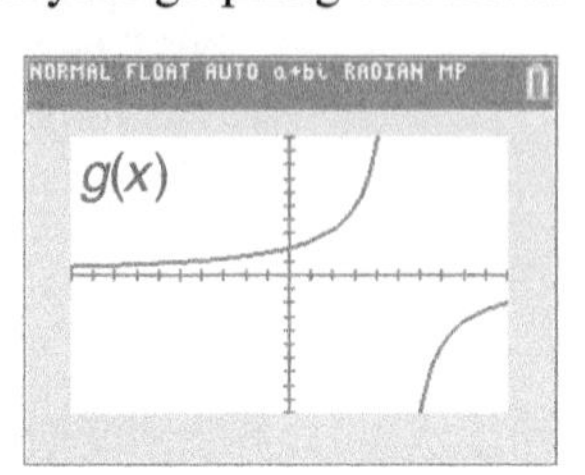

3. a.

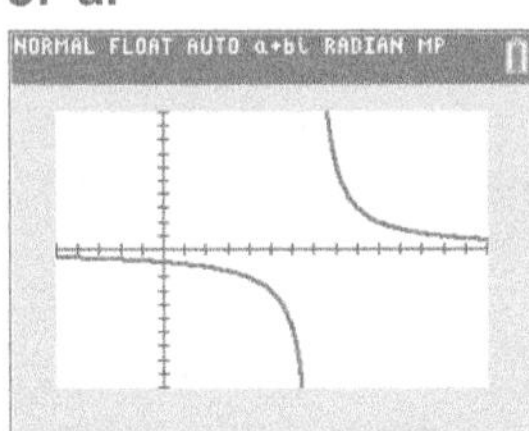

b.

c.

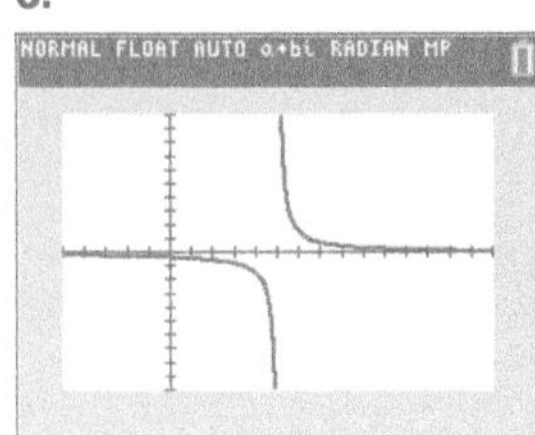

d.

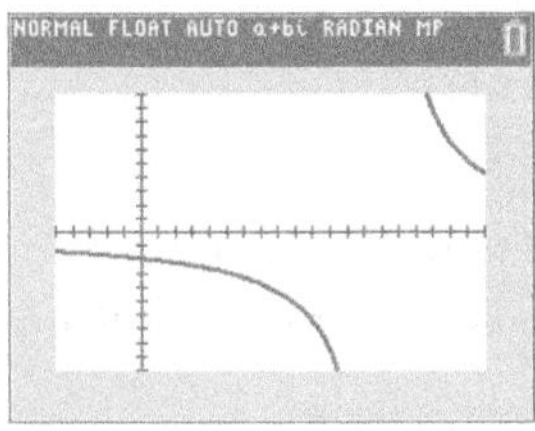

e.

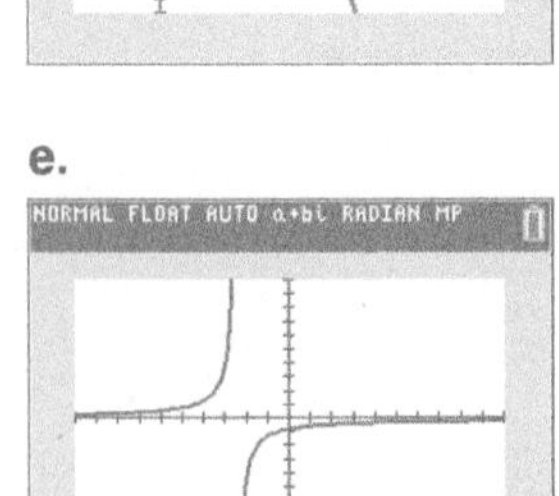

3. Without graphing, determine the domain of each of the following functions. Then determine the equation of the vertical asymptote for each function. Verify your answers using your graphing calculator.

a. 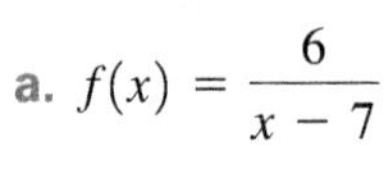 $f(x) = \dfrac{6}{x - 7}$

domain: all real numbers except $x = 7$

vertical asymptote: $x = 7$

b. $g(x) = \dfrac{20}{25 - x}$

domain: all real numbers except $x = 25$

vertical asymptote: $x = 25$

c. 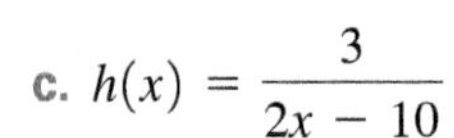 $h(x) = \dfrac{3}{2x - 10}$

domain: all real numbers except $x = 5$

vertical asymptote: $x = 5$

d. $F(x) = \dfrac{13}{0.5x - 7}$

domain: all real numbers except $x = 14$

vertical asymptote: $x = 14$

e. $G(x) = \dfrac{-4}{2x + 5}$

domain: all real numbers except $x = -2.5$

vertical asymptote: $x = -2.5$

4. Give examples of two different rational functions that have a vertical asymptote at $x = 10$.

$y = \dfrac{k}{x - 10}$ and $y = \dfrac{k}{10 - x}$ for any nonzero k

5. As the input value of a rational function gets closer to a vertical asymptote, the output becomes larger in magnitude, approaching either positive or negative infinity. Consider the functions $f(x) = \dfrac{10}{x - 5}$ and $g(x) = \dfrac{10}{5 - x}$.

a. Determine the equations of the vertical asymptotes for functions f and g.

$x = 5$

b. Describe what happens to the output value when x is near the vertical asymptote but to the right of it.

$f(x)$ gets large, approaching infinity. $g(x)$ gets large in magnitude in a negative direction, approaching negative infinity.

c. Describe what happens to the output value when x is near the vertical asymptote but to the left of it.

$f(x)$ gets large in magnitude in a negative direction, approaching negative infinity. $g(x)$ gets large, approaching infinity.

ACTIVITY
5.4

Blood-Alcohol Levels
Solving Rational Equations

OBJECTIVES

1. Solve an equation involving a rational expression using an algebraic approach.
2. Solve an equation involving a rational expression using a graphing approach.
3. Determine horizontal asymptotes of the graph of $y = \frac{f(x)}{g(x)}$, where $f(x)$ and $g(x)$ are first-degree polynomials.

As of 2013, every state in the United States has adopted a 0.08% blood-alcohol concentration as the legal measure of drunk driving. If you assume that a regular 12-ounce beer is 5% alcohol by volume and that the normal bloodstream contains 5 liters (or 169 ounces) of fluid, your blood-alcohol concentration, B, can be approximately modeled by the function having the equation

$$B = \frac{600n}{w(169 + 0.6n)},$$

where n is the number of beers consumed in 1 hour and w is your body weight in pounds.

1. a. Replace w with your body weight. Write an equation for B in terms of n.

(Answers will vary.) If $w = 200$, $B = \frac{600n}{200(169 + 0.6n)}$.

b. Complete the following table using your equation from part a. Round your results to three decimal places.

NUMBER OF BEERS CONSUMED IN ONE HOUR, n	BLOOD-ALCOHOL CONCENTRATION, B
1	0.018
2	0.035
3	0.053
4	0.070
5	0.087
6	0.104
7	0.121
8	0.138
9	0.155
10	0.171

(Answers will vary.) The table above uses $w = 200$ pounds.

2. According to this model, how many beers can you consume in 1 hour without exceeding the recommended legal measure of drunk driving?

(Answers will vary.) For $w = 200$, the answer is 4.

3. a. A football player friend of yours weighs 232 pounds. Rewrite the equation for B in terms of n. What is his maximum blood-alcohol level if he drinks four beers in 1 hour?

$B = \frac{600n}{232(169 + 0.6n)}$; $B \approx 0.06\%$ blood-alcohol concentration

b. Complete the following table using your equation from part a:

Number of Beers Consumed in 1 Hour, n	1	2	3	4	5	6	7	8	9	10
Blood-Alcohol Concentration, B	0.015	0.030	0.045	0.060	0.075	0.090	0.105	0.119	0.133	0.148

c. What is the practical domain of the blood-alcohol function of part a?

Generally, the practical domain is between 0 and 5 beers.

d. Does the weight of a person have any impact on the practical domain? Explain.

Yes, because the lower a person's weight, the lower the upper limit of beers that can be consumed.

e. What is the vertical intercept? Does this seem reasonable within the context of the problem?

When $n = 0$, $B = 0$. Yes. If you do not drink any beer, your blood-alcohol concentration is zero.

f. Sketch the graph of your blood-alcohol function over the practical domain identified in part c. Use the indicated scale.

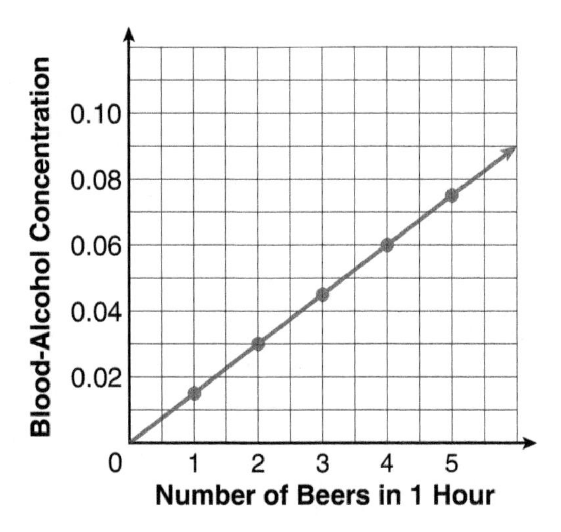

Note that the graph of the blood-alcohol function in Problem 3f looks like a line when drawn in the practical domain. The graph of the general function defined by

$$f(x) = \frac{600x}{232(169 + 0.6x)}$$

appears as follows:

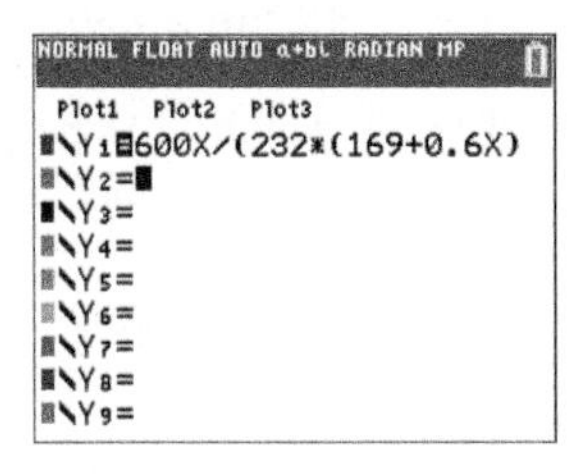

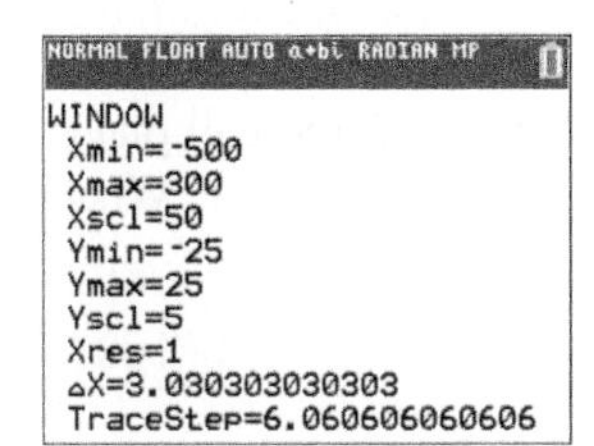

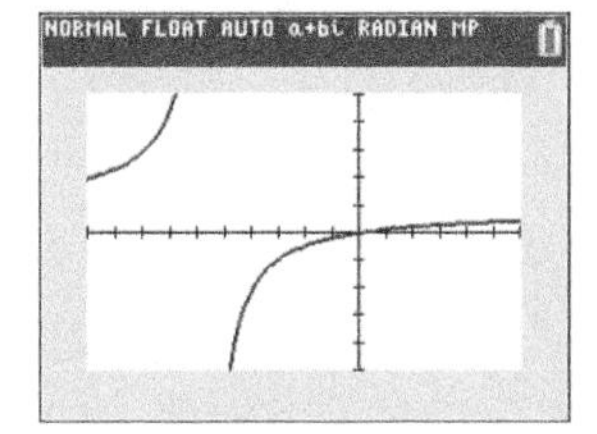

Solving Equations Involving a Rational Expression

4. a. Your 232-pound football player friend is given a breathalyzer test. The result is a blood-alcohol concentration of 0.05%. Using the blood-alcohol concentration function, write an equation that can be solved to determine the number of beers your friend consumed in the previous hour.

$$0.05 = \frac{600n}{232(169 + 0.6n)}$$

b. Solve the equation in part a graphically on the following grid. $n \approx 3.3$

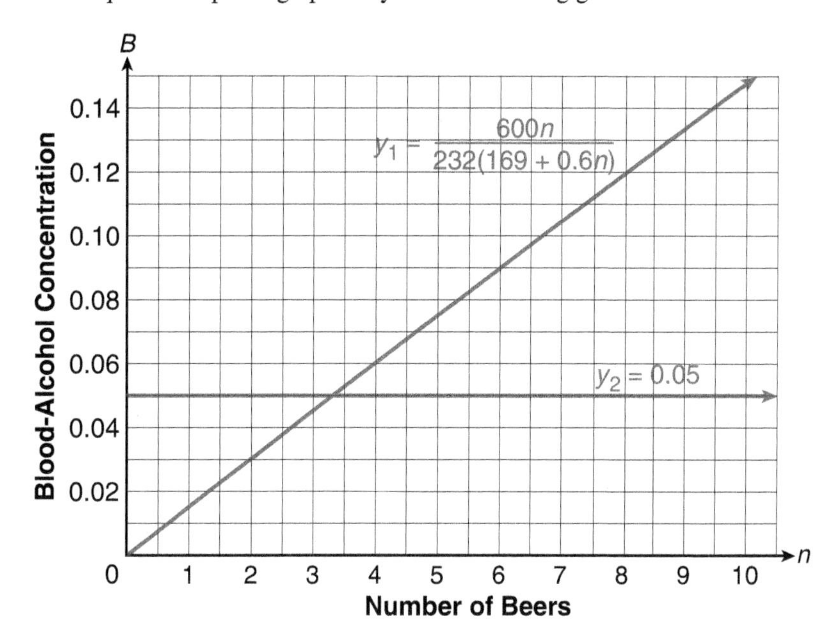

c. Use your graphing calculator to check the answer in part b. Your screen(s) should appear as follows:

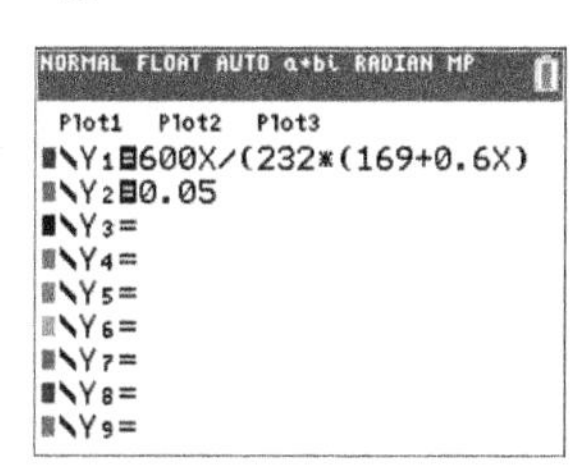

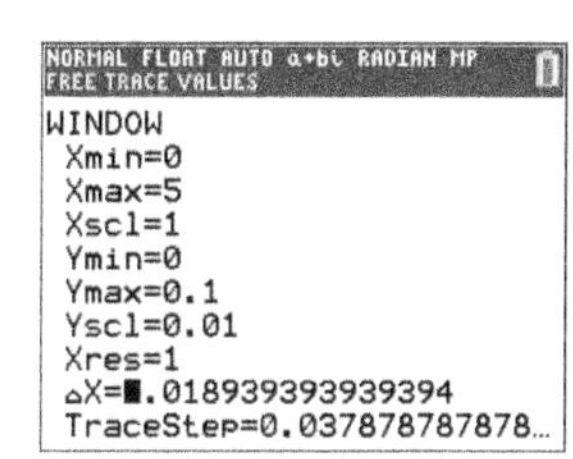

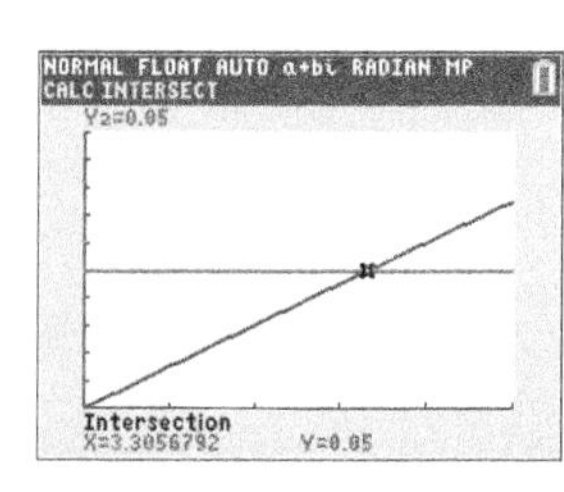

The equation in Problem 4 can be solved using an algebraic approach. If a variable appears in the denominator of a fraction, a general approach is to multiply both sides of the equation by the denominator and solve the resulting equation. Example 1 demonstrates this approach as well as two other methods.

LC LEARNING CATALYTICS

Solve $\frac{x+2}{2x-5} = 2$.

a. $x = 4$
b. $x = \frac{5}{2}$
c. $x = -2$
d. There is no solution.

EXAMPLE 1 ***Solve the equation*** $\mathbf{\frac{16}{x+3} = 2}$.

SOLUTION

Method 1. General Case

To solve the equation $\frac{16}{x+3} = 2$, multiply both sides of the equation by the denominator $x + 3$, as follows:

$$(x+3)\frac{16}{x+3} = 2(x+3)$$
$$16 = 2x + 6$$

Solving for x, you have

$$10 = 2x$$
$$5 = x.$$

Solutions to an equation involving rational functions should always be checked:

$$\frac{16}{5+3} = \frac{16}{8} = 2.$$

Therefore, 5 is a solution.

Method 2. Cross Multiplication

You can also solve the equation $\frac{16}{x+3} = 2$ by applying the following property:

If two ratios $\frac{a}{b}$ and $\frac{c}{d}$ represent the same value, then $\frac{a}{b} = \frac{c}{d}$ is equivalent to $ad = bc$.

This process is called cross multiplication. Therefore,

$$\begin{aligned} \frac{16}{x+3} &= \frac{2}{1} \qquad \text{Cross multiply.} \\ 2(x+3) &= 16(1) \\ 2x + 6 &= 16 \\ -6 &\quad -6 \\ 2x &= 10 \\ x &= 5 \end{aligned}$$

Method 3. Graphical Approach

You can verify that 5 is a solution to the given equation by graphing $y_1 = \frac{16}{x+3}$ and $y_2 = 2$ and determining the x-value of the point of intersection of the two graphs.

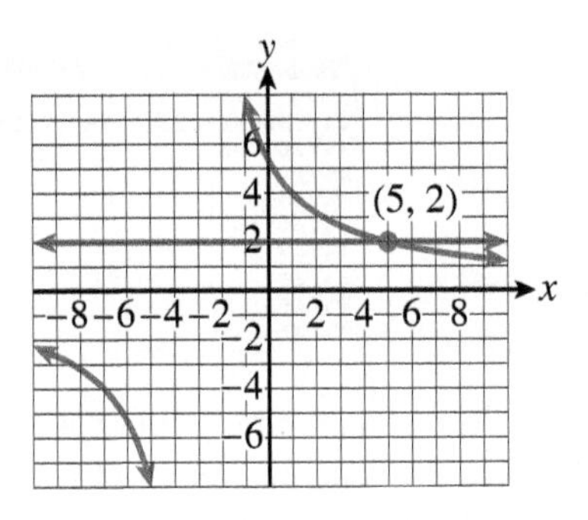

5. Solve each of the following equations using an algebraic approach. Verify your answer graphically.

a. $\frac{45}{x} = 9$

$x = 5$

b. $\frac{23}{x+2} = 15$

$x = -\frac{7}{15} \approx -0.467$

c. $\frac{13}{x} = \frac{2}{5}$

$x = 32.5$

d. $\frac{16}{x^2} = \frac{1}{4}$

$x = \pm 8$

6. a. Solve the equation in Problem 4a using an algebraic approach.

$$\begin{aligned} 0.05 &= \frac{600n}{232(169 + 0.6n)} \\ 0.05(232)(169 + 0.6n) &= 600n \\ 0.05(232)(169) + 0.05(232)(0.6n) &= 600n \\ 1960.4 + 6.96n &= 600n \\ 1960.4 &= 593.04n \\ 3.306 &\approx n \end{aligned}$$

b. How does your solution compare with the result in Problem 4c using a graphical approach?

They are the same.

Horizontal Asymptotes

The graphs you have studied so far in this chapter have at least one feature in common. The horizontal asymptote is the horizontal axis. As the input values increase infinitely in the positive direction or decrease infinitely in the negative direction, the output values have always approached zero.

7. Consider the function defined by $y = \frac{2x}{x+5}$. This equation is in the form $y = \frac{f(x)}{g(x)}$, where $f(x) = 2x$ and $g(x) = x + 5$.

a. What is the domain of this function?

all real numbers not equal to -5

b. What is the equation of the vertical asymptote?

$x = -5$

c. Complete the following table:

x	−100	−50	−10	−5.5	−5.1	−5	−4.9	−4.5	0	50	100
y	2.1	2.2	4	22	102	undef.	−98	−18	0	1.8	1.9

d. What appears to be happening to the y-values as the x-values increase infinitely to the right or decrease infinitely to the left?

The y-values appear to approach 2.

e. What is the horizontal asymptote?

$y = 2$

7. g.

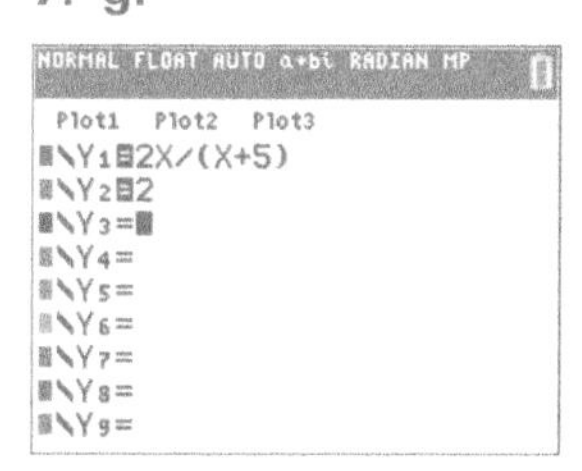

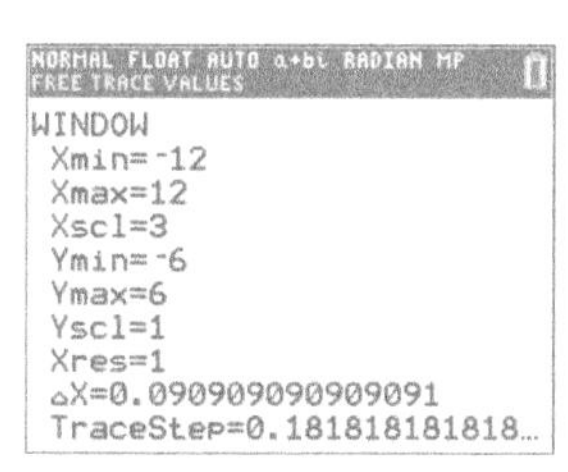

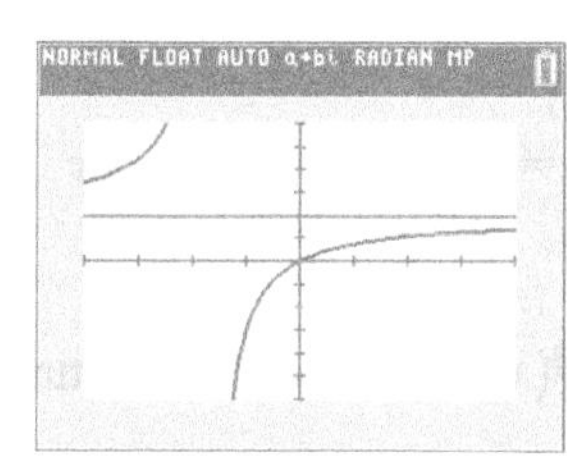

f. Sketch a graph of the function; include horizontal and vertical asymptotes as dotted lines.

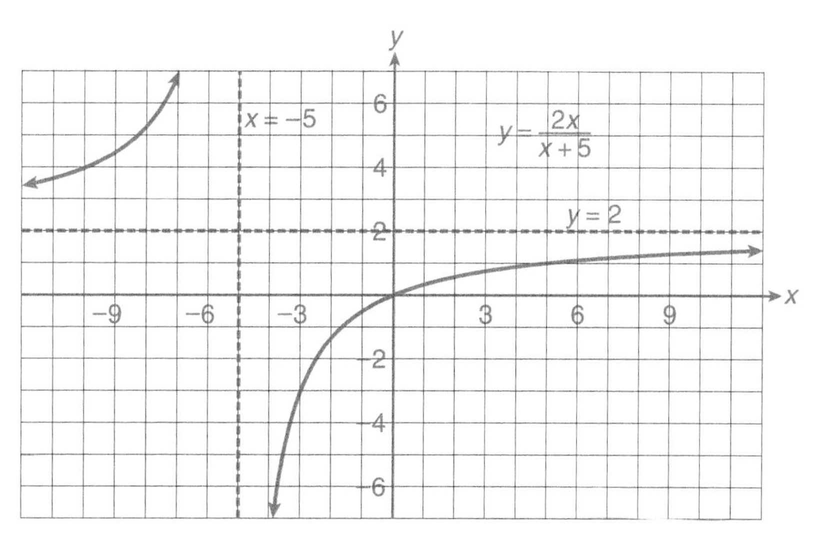

g. Verify the graph in part f using your graphing calculator. As a graphing aid, also sketch the graph of $y = 2$.

f. Complete the following table:

Number of People Attending, n	50	100	150	200	250
Mean Cost per Person, c (\$)	90	70	63.33	60	58

g. Does the graph of the cost function have a horizontal asymptote? Does it make sense in this situation? Explain.

Yes, $c = 50$ is the horizontal asymptote. It makes sense because as the number of attendees increases, the fixed costs attributed to each person get smaller so the average cost approaches the price of the meal, \$50.

h. Sketch a graph of the cost function over its practical domain.

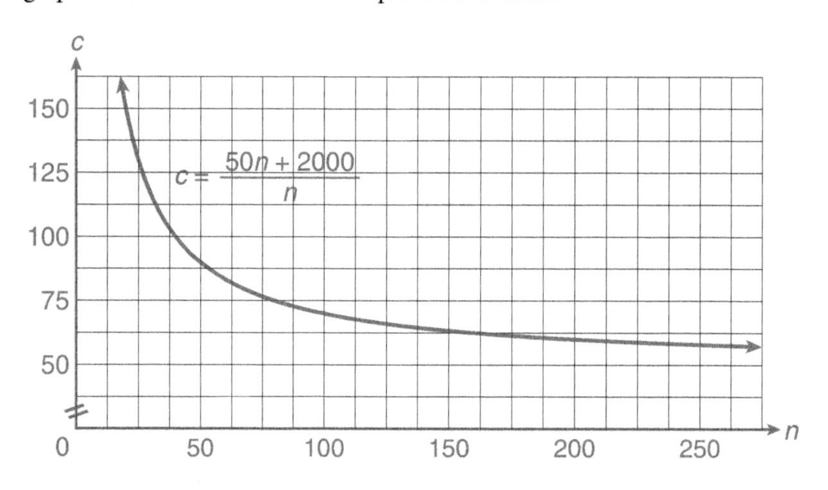

2. For each rational function,
   i. determine the domain.
   ii. determine the vertical asymptotes.
   iii. graph the function using your graphing calculator.
   iv. determine the horizontal asymptote by inspecting the graph of your function.

a. $y = \dfrac{4x}{x + 2}$

i. all real numbers except -2

ii. $x = -2$

iii.

iv. $y = 4$

b. $y = \dfrac{1 - x}{x + 1}$

i. all real numbers except -1

ii. $x = -1$

iii.

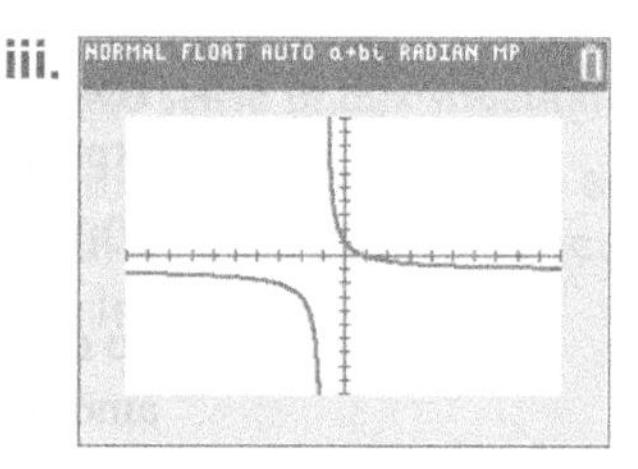

iv. $y = -1$

c. $y = \dfrac{3x}{x - 4}$

i. all real numbers except 4

ii. $x = 4$

iii.

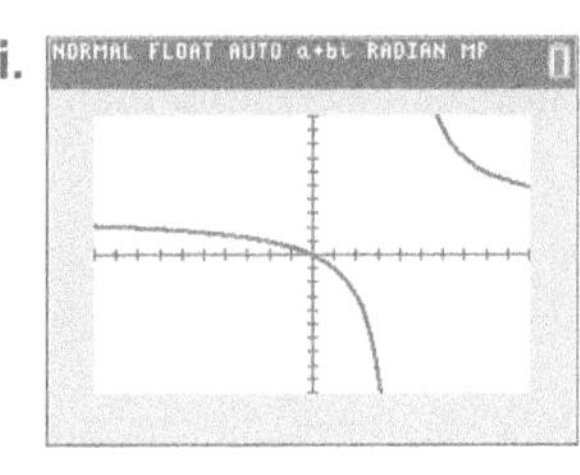
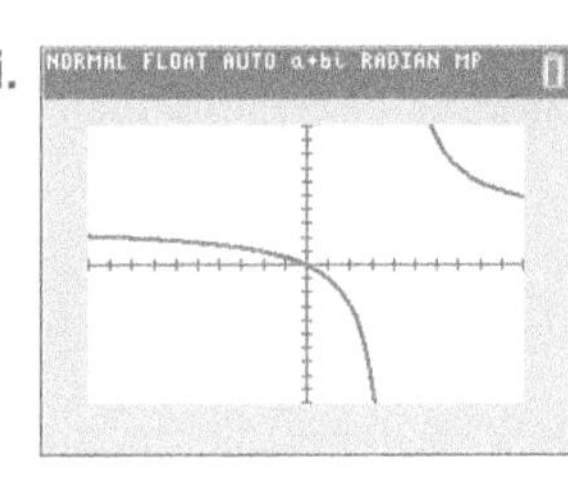

iv. $y = 3$

d. $y = 12 - \dfrac{6x}{1 - 2x}$

i. all real numbers except $\frac{1}{2}$

ii. $x = \frac{1}{2}$

iii.

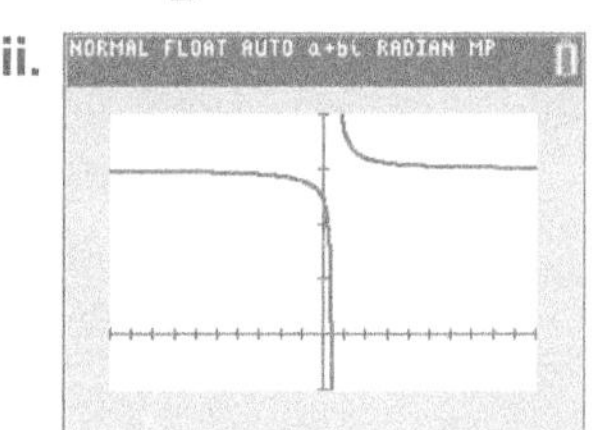

iv. $y = 15$

3. Solve the following equations algebraically and check your results by graphing:

a. $\dfrac{3x}{2x - 1} = 3$

$3x = 3(2x - 1)$

$3x = 6x - 3$

$-3x = -3$

$x = 1$

b. $\dfrac{x + 1}{5x - 3} = 2$

$x + 1 = 2(5x - 3)$

$x + 1 = 10x - 6$

$-9x = -7$

$x = \frac{7}{9} \approx 0.778$

c. $\dfrac{-7x}{2.8 + x} = 3.1$

$-7x = 3.1(2.8 + x)$

$-7x = 8.68 + 3.1x$

$-10.1x = 8.68$

$x \approx -0.859$

d. $\dfrac{x + 10}{3x + 2} = 3$

$3(3x + 2) = x + 10$

$9x + 6 = x + 10$

$8x = 4$

$x = 0.5$

3. a.

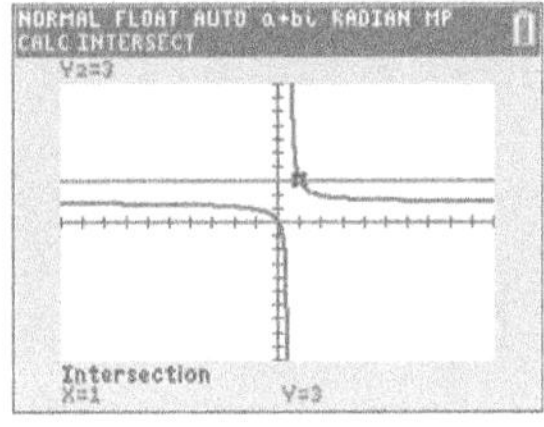

b.

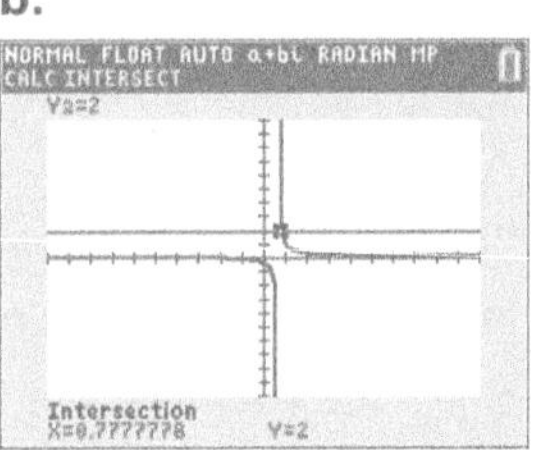

c.

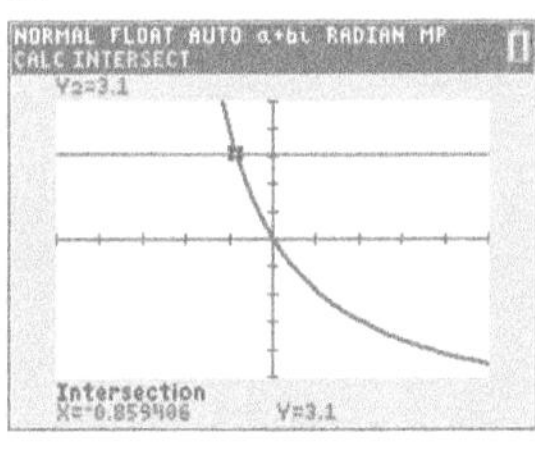

d.

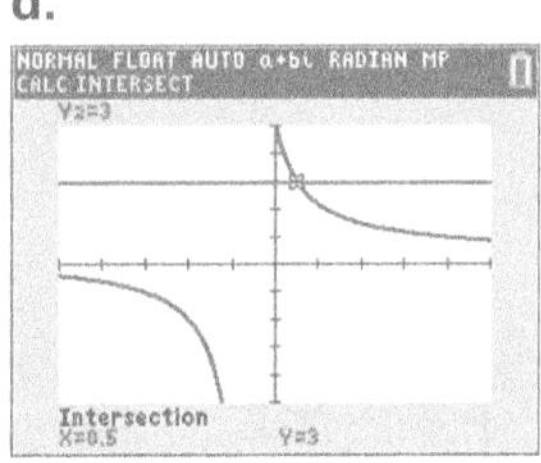

4. In a 20-kilometer race, a runner's average rate (in kilometers per hour) can be expressed as a function of time to run the race (in hours) by the equation $r = \dfrac{20}{t}$.

a. Determine your time to complete the race if you average 16 kilometers per hour.

$16 = \dfrac{20}{t}$; $16t = 20$, $t = \dfrac{20}{16} = 1.25$ hr.

b. What is your time if you average 18 kilometers per hour?

$18 = \dfrac{20}{t}$; $18t = 20$, $t = \dfrac{20}{18} \approx 1.11$ hr.

5. The intensity of the human voice to the listener varies inversely as the square of the distance from the source. This is given by a formula similar to the ones from Activity 5.2. Loudness of a voice could be given by $I = \dfrac{1500}{d^2}$, where I is in decibels and d is the distance in feet.

a. Determine the distance from the source when the intensity of the sound is 15 decibels.

$15 = \dfrac{1500}{d^2}$

$15d^2 = 1500$; $d^2 = 100$, $d = \pm 10$ ft., but only 10 ft. makes sense

b. What is the distance from the source when the intensity is 8000 decibels?

$8000 = \dfrac{1500}{d^2}$; $8000d^2 = 1500$, $d^2 = 0.1875$, $d \approx 0.433$ ft.

6. As a fund-raising project, the international club at your college decides to publish and sell a calendar. The cost of photographs and typesetting is \$450. It costs \$3 to print and assemble each calendar.

a. What is the total cost of printing 200 calendars?

$450 + 3(200) = \$1050$

b. What is the average cost per calendar of printing the 200 calendars?

$\frac{1050}{200} = \$5.25$

c. Write an expression for the total cost of printing n calendars.

$450 + 3n$

d. Let A represent the average cost per calendar. Write an equation that gives A as a function of n.

$$A = \frac{450 + 3n}{n}$$

e. Complete the following table using the equation in part d:

Number, n	50	75	100	500	750	1000
Average Cost, A(\$)	12	9	7.50	3.90	3.60	3.45

f. As the input n increases, what happens to the output A?

A gets closer to 3.

g. What is the horizontal asymptote of this function?

$A = 3$

h. Verify your answer in part g graphically.

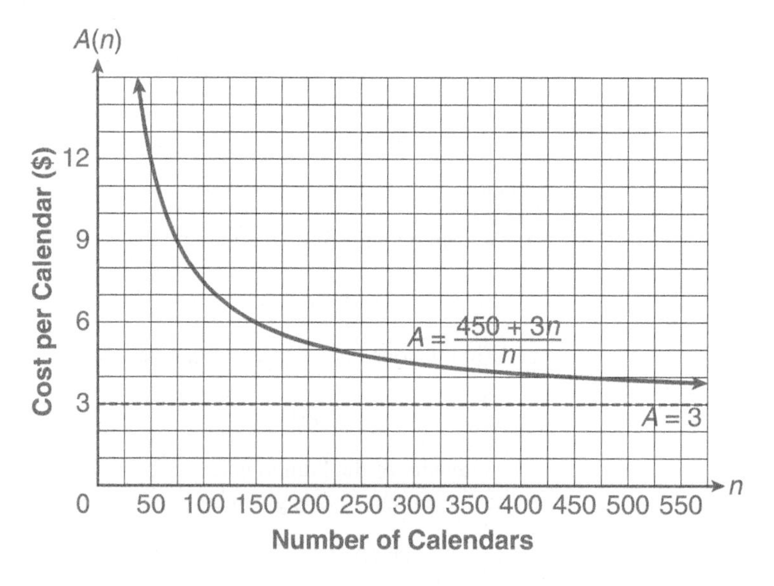

i. Interpret what the horizontal asymptote means in the context of the problem.

The average cost will always be at least \$3 per calendar. For larger numbers of calendars, the average cost gets closer to \$3.

6. j.

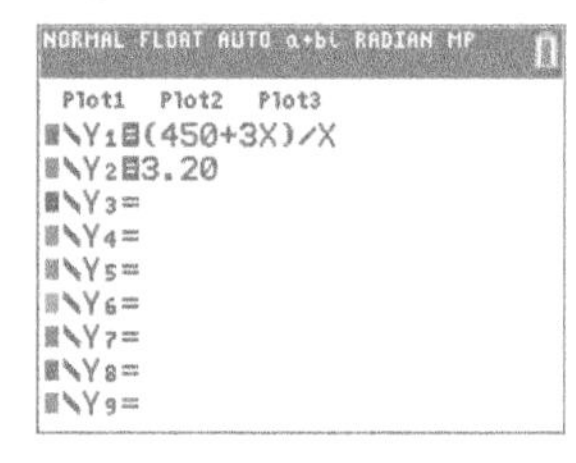

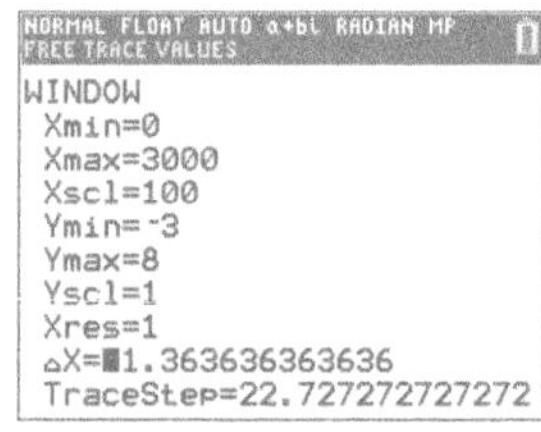

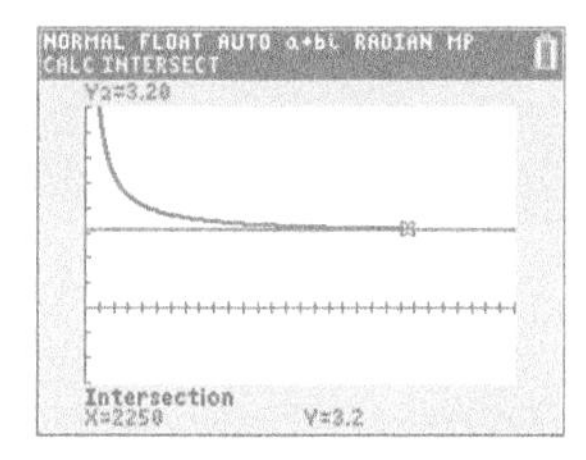

j. Suppose you want the average cost to be less than \$3.20. Model this problem with an inequality, solve it algebraically for n, and verify it graphically.

$$\frac{450 + 3n}{n} < 3.20$$

$$450 + 3n < 3.20n$$

$$450 < 0.20n$$

$$2250 < n$$

7. The following formula is used by the NCAA to calculate the quarterback passing efficiency rating:

$$R = \frac{8.4Y + 330T + 100C - 200I}{A},$$

where R = quarterback rating
A = passes attempted
C = passes completed
Y = passing yardage
T = touchdown passes
I = number of interceptions

In the 2017 regular season, Mason Rudolph, quarterback for Oklahoma State University, and Baker Mayfield, quarterback for the University of Oklahoma, had the following player statistics:

PLAYER	PASSES ATTEMPTED	PASSES COMPLETED	PASSING YARDAGE	NUMBER OF TOUCHDOWN PASSES	NUMBER OF INTERCEPTIONS
Mason Rudolph	489	318	4904	37	9
Baker Mayfield	404	285	4627	43	6

a. Determine the quarterback rating for Mason Rudolph for the 2017 NCAA football season.

$$R = \frac{8.4(4904) + 330(37) + 100(318) - 200(9)}{489} = 170.6 \text{ (to the nearest tenth)}$$

b. Determine the quarterback rating for Baker Mayfield.

$$R = \frac{8.4(4627) + 330(43) + 100(285) - 200(6)}{404} = 198.9 \text{ (to the nearest tenth)}$$

c. Visit the NCAA website, and select stats to determine the passing efficiency rating of your favorite quarterback.

8. In a predator–prey model from wildlife biology, the rate, R, at which prey are consumed by one predator is approximated by the function

$$R = \frac{0.623n}{1 + 0.046n},$$

measured in prey per week, where n is the number of prey available per square mile.

a. If the number of prey available per square mile is 30, what is R?

$$R = \frac{0.623(30)}{1 + 0.046(30)}$$

$R \approx 7.85$ prey per week

b. Approximately how many prey must be available per square mile for the predator to consume 10 prey per week?

$$10 = \frac{0.623(n)}{1 + 0.046(n)}$$

$$10(1 + 0.046n) = 0.623n$$

$$10 + 0.46n = 0.623n$$

$$10 = 0.163n$$

$$61.3 \approx n$$

Therefore, approximately 61 prey/sq. mi. must be available.

c. Suppose you want to maintain the prey population to ensure that a predator can obtain between 6 and 10 prey per week. One of the necessary equations for this situation was solved in part b. Write the other equation, and solve it.

$$6 = \frac{0.623n}{1 + 0.046n}$$

$$6(1 + 0.046n) = 0.623n$$

$$6 + 0.276n = 0.623n$$

$$6 = 0.347n$$

$$17.3 \approx n$$

Putting the two together, $18 \leq n \leq 61$.

d. Use the graph of this function to illustrate your solutions to parts b and c.

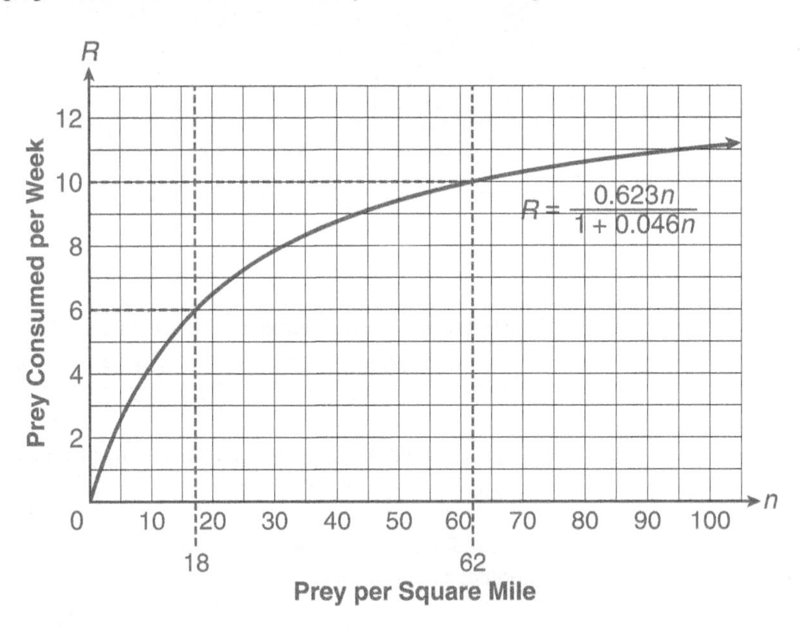

e. How many prey are required for a predator to consume 20 prey per week? Comment on the practical significance and how this shows up on the graph.

The result is negative, so discard it. It is not possible for the predator to consume 20 prey per week. Under these conditions, 20 is above the horizontal asymptote.

ACTIVITY 5.5

Traffic Flow

Solving Rational Equations Using Algebraic Methods

OBJECTIVES

1. Determine the least common denominator (LCD) of two or more rational expressions.
2. Solve an equation involving rational expressions using an algebraic approach.
3. Solve a formula for a specific variable.

You are an intern at an architecture firm. The company is designing an auditorium that will be annexed to the local high school building. The rate of traffic flow through the exits is an important consideration. The auditorium will have three exit doors. Two exits are single doors of slightly different sizes. The first exit, by itself, can be used to empty the auditorium in 10 minutes. The second exit can be used to empty the room in 8 minutes. The third exit is a double-wide door that, by itself, can be used to empty the auditorium in 5 minutes.

1. If only the first door is open, what fraction of the auditorium can be emptied in 5 minutes? 2 minutes? 1 minute?

   One-half of the auditorium can be emptied in 5 minutes. One-fifth of the auditorium can be emptied in 2 minutes. One-tenth of the auditorium can be emptied in 1 minute.

2. The rate at which a door can be used to empty the auditorium is the fraction of the job that can be completed in 1 minute. In this case, the units of measurement are auditoriums per minute. Determine the rate of emptying for each of the three exits, and record your answers in the following table:

EXIT	RATE OF EMPTYING
First	$\frac{1}{10}$
Second	$\frac{1}{8}$
Third	$\frac{1}{5}$

Your task is to determine the time, T, it takes to empty the auditorium if all three exit doors are open. The relationship between this time, T, and the individual emptying times is given by the formula

$$\frac{1}{t_1} + \frac{1}{t_2} + \frac{1}{t_3} = \frac{1}{T},$$

where t_1, t_2, and t_3 represent the times that each exit door can be used by itself to empty the auditorium.

Note that three single door rates are added to determine the rate, $\frac{1}{T}$, at which the three doors working together can be used to empty the auditorium.

3. a. Write the equation that can be used to determine the time, T, that it takes the auditorium to be emptied if all three exits are open.

   $$\frac{1}{T} = \frac{1}{10} + \frac{1}{8} + \frac{1}{5}$$

   b. Solve this equation graphically. Use the window Xmin = 0, Xmax = 10, Ymin = 0, Ymax = 2, and Yscl = 0.5.

   $T \approx 2.353$ min.

3. b.

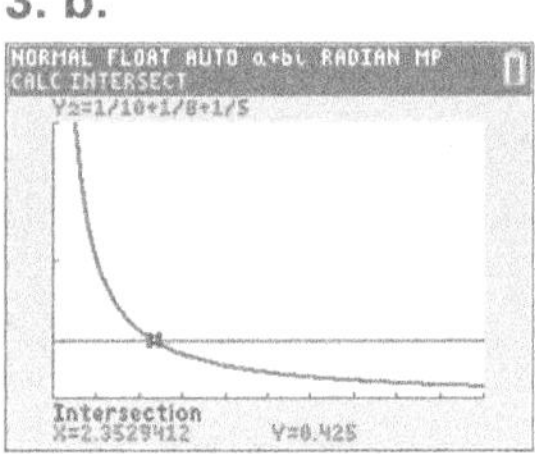

Solution Using an Algebraic Approach

An algebraic approach to solving the equation $\frac{1}{10} + \frac{1}{8} + \frac{1}{5} = \frac{1}{T}$ is to eliminate the fractions from the equation. This can be accomplished by first determining the least common denominator (LCD) of the fractions involved in the equation. The following example demonstrates the procedure for determining the LCD.

EXAMPLE 1 **a.** ***Determine the LCD for*** $\frac{5}{12}$ ***and*** $\frac{7}{45}$.

SOLUTION

Step 1. Write the prime factorization of each denominator. Express repeated factors as powers.

$$12 = 2 \cdot 2 \cdot 3 = 2^2 \cdot 3$$
$$45 = 3 \cdot 3 \cdot 5 = 3^2 \cdot 5$$

Step 2. Identify the different bases (factors) in step 1.

$$2, 3, 5$$

Step 3. Write the LCD as the product of the highest power of each of the different factors from step 2.

$$\text{LCD} = 2^2 \cdot 3^2 \cdot 5 = 4 \cdot 9 \cdot 5 = 180$$

The smallest number that both 12 and 45 will divide evenly is 180.

b. ***Determine the LCD for*** $\frac{11}{6xy^3}$ ***and*** $\frac{5a}{9x^2y}$.

SOLUTION

Step 1. $6xy^3 = 2 \cdot 3 \cdot x^1 \cdot y^3$
$9x^2y = 3^2 \cdot x^2 \cdot y$

Step 2. $2, 3, x, y$

Step 3. $\text{LCD} = 2 \cdot 3^2 \cdot x^2 \cdot y^3 = 18x^2y^3$

$18x^2y^3$ is evenly divided by both $6xy^3$ and $9x^2y$.

You are now ready to solve the equation $\frac{1}{10} + \frac{1}{8} + \frac{1}{5} = \frac{1}{T}$ using an algebraic approach.

4. a. Determine the LCD of the rational expressions in the equation $\frac{1}{10} + \frac{1}{8} + \frac{1}{5} = \frac{1}{T}$.

$10 = 2 \cdot 5$
$8 = 2^3$
$5 = 5$
$T = T$
$\text{LCD} = 2^3 \cdot 5T = 40T$

b. Multiply each side of the equation by the LCD, and solve the resulting equation.

$$40T\left(\frac{1}{10} + \frac{1}{8} + \frac{1}{5}\right) = 40T\left(\frac{1}{T}\right)$$
$$\frac{40T}{10} + \frac{40T}{8} + \frac{40T}{5} = \frac{40T}{T}$$
$$4T + 5T + 8T = 40$$
$$17T = 40$$
$$T = \frac{40}{17} \approx 2.3529 \text{ min.}$$

c. How does this solution compare with the solution you determined graphically in Problem 3?

They are the same.

5. A domed stadium is to be equipped with two ventilation fans. The first fan can exchange the air in the stadium in 4 hours. The building code requires a complete exchange of air in the stadium every 3 hours. To model this situation, use the equation

$$\frac{1}{t_1} + \frac{1}{t_2} = \frac{1}{T},$$

where t_1 and t_2 are the exchange times for the fans working alone and T is the exchange time for the two fans working together.

a. Let x represent the exchange time for the second ventilation fan. Write an equation that can be used to determine x so that the fans working together will satisfy the building code.

$$\frac{1}{4} + \frac{1}{x} = \frac{1}{3}$$

b. Solve this equation graphically.

$x = 12$ hr.

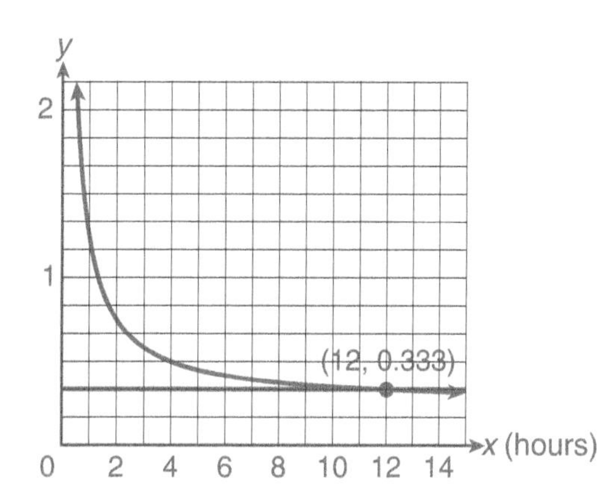

c. Solve the equation in part a algebraically by multiplying both sides by the LCD.

$$12x\left(\frac{1}{4} + \frac{1}{x}\right) = 12x\left(\frac{1}{3}\right)$$

$$\frac{12x}{4} + \frac{12x}{x} = \frac{12x}{3}$$

$$3x + 12 = 4x$$

$$12 = x$$

6. Switching to different fans, suppose the first fan can exchange the air in the stadium twice as fast as the second fan.

a. If x represents the exchange time for the first fan, write an expression that represents the exchange time for the second fan.

$2x$

b. Working together, the two fans can exchange the air in the stadium in 4 hours. Using the formula $\frac{1}{t_1} + \frac{1}{t_2} = \frac{1}{T}$, write an equation that can be solved to determine x.

$$\frac{1}{x} + \frac{1}{2x} = \frac{1}{4}$$

c. Solve this equation algebraically. Verify your solution graphically.

$$4x\left(\frac{1}{x} + \frac{1}{2x}\right) = 4x\left(\frac{1}{4}\right)$$

$$\frac{4x}{x} + \frac{4x}{2x} = \frac{4x}{4}$$

$$4 + 2 = x$$

$$6 = x$$

The exchange time for the first fan is 6 hours.

6. c.

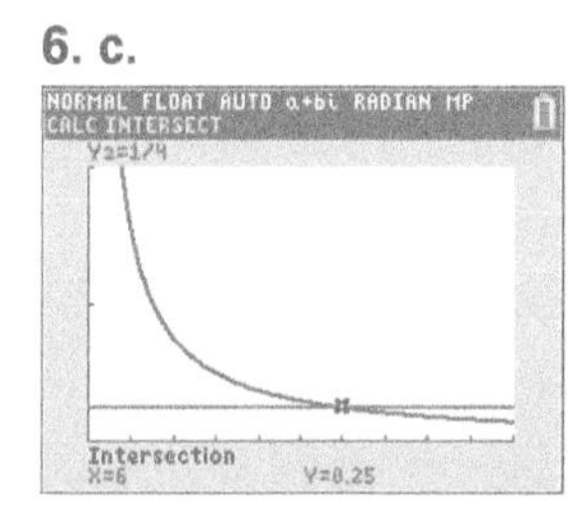

d. Determine the rate for each fan.

The exchange rate for the second fan is $\frac{1}{12}$ of the stadium per hour.

The exchange rate for the first fan is $\frac{1}{6}$ of the stadium per hour.

Solving a Formula for a Specified Letter

An alternative approach to solving problems involving formulas of the form $\frac{1}{t_1} + \frac{1}{t_2} = \frac{1}{T}$ is to use an equivalent formula that has been solved for the variable T.

EXAMPLE 2 *Solve* $\frac{1}{t_1} + \frac{1}{t_2} = \frac{1}{T}$ *for T.*

SOLUTION

Step 1. Determine the LCD.

$$\text{LCD} = t_1 t_2 T$$

Step 2. Multiply each side of the equation by the LCD.

$$\frac{t_1 t_2 T}{1}\left(\frac{1}{t_1} + \frac{1}{t_2}\right) = \frac{1}{T} \cdot \frac{t_1 t_2 T}{1}$$

$$\frac{t_1 t_2 T}{t_1} + \frac{t_1 t_2 T}{t_2} = \frac{t_1 t_2 T}{T}$$

Simplifying, you have $t_2 T + t_1 T = t_1 t_2$.

Step 3. Solve the resulting equation for T.

$T(t_2 + t_1) = t_1 t_2$ — T is the common factor on the left side.

$T = \frac{t_1 t_2}{t_2 + t_1}$ — Divide both sides by $t_2 + t_1$.

LC LEARNING CATALYTICS

Solve $\frac{1}{x} + \frac{1}{4x} = \frac{1}{4}$.

a. $x = 0$
b. $x = 4$
c. $x = 5$
d. $x = \frac{1}{4}$

7. a. If t_1 and t_2 are the exchange times for the fans working alone and T is the exchange time for the two fans working together, determine T if $t_1 = 4$ hours and $t_2 = 12$ hours.

Use the formula $\frac{1}{t_1} + \frac{1}{t_2} = \frac{1}{T}$.

$$\frac{1}{4} + \frac{1}{12} = \frac{1}{T}$$

$$3T + T = 12$$

$$4T = 12$$

$$T = 3$$

b. Repeat part a using the formula $T = \frac{t_1 t_2}{t_2 + t_1}$.

$$T = \frac{4 \cdot 12}{4 + 12} = \frac{48}{16} = 3$$

c. Compare the results in parts a and b.

They are the same.

8. The following formula is used in work with lenses and mirrors: $\frac{1}{p} + \frac{1}{q} = \frac{1}{f}$. Solve the formula for f.

$$pqf\left(\frac{1}{p} + \frac{1}{q}\right) = pqf\left(\frac{1}{f}\right)$$

$$\frac{pqf}{p} + \frac{pqf}{q} = \frac{pqf}{f}$$

$$qf + pf = pq$$

$$f(q + p) = pq$$

$$f = \frac{pq}{p + q}$$

SUMMARY Activity 5.5

1. To determine an LCD of two or more rational expressions,

Step 1. Write the prime factorization of each denominator. Express repeated factors as powers.

Step 2. Identify the different bases (factors) in step 1.

Step 3. Write the LCD as the product of the highest power of each of the different factors from step 2.

2. To solve an equation involving rational expressions,

Step 1. Determine the LCD of all denominators in the equation.

Step 2. Multiply each side of the equation by the LCD and simplify the resulting equation.

Step 3. Solve the resulting equation for the desired variable.

3. Many real-life applications involve equations of the form

$$\frac{1}{a} + \frac{1}{b} = \frac{1}{c},$$

where a, b, and $c \neq 0$. For example,

$\frac{1}{R_1} + \frac{1}{R_2} = \frac{1}{R}$ is used for resistance of electrical circuits,

$\frac{1}{p} + \frac{1}{q} = \frac{1}{f}$ is used in work with lenses and mirrors, and

$\frac{1}{t_1} + \frac{1}{t_2} = \frac{1}{T}$ is used to calculate the time it takes to complete a task when two machines or people are working together.

These formulas can also be extended to three or more resistors, lenses, or machines by simply adding additional fractions in each case.

EXERCISES Activity 5.5

1. Two pumps are working together to empty a gasoline tank.

   a. The emptying times for pump 1 and pump 2 are 30 minutes and 45 minutes, respectively. Determine the time required to empty the tank if both pumps are working. Use the formula $\frac{1}{t_1} + \frac{1}{t_2} = \frac{1}{T}$, where t_1 and t_2 are the emptying times for pump 1 and pump 2, respectively, and T is the total time required to empty the tank.

$$\frac{1}{30} + \frac{1}{45} = \frac{1}{T}$$
$$\frac{90T}{30} + \frac{90T}{45} = \frac{90T}{T}$$
$$3T + 2T = 90$$
$$5T = 90$$
$$T = 18 \text{ min.}$$

   b. Solve the equation $\frac{1}{t_1} + \frac{1}{t_2} = \frac{1}{T}$ for T.

$$\frac{t_1 t_2 T}{t_1} + \frac{t_1 t_2 T}{t_2} = \frac{t_1 t_2 T}{T}$$
$$t_2 T + t_1 T = t_1 t_2$$
$$T = \frac{t_1 t_2}{t_1 + t_2}$$

   c. Using the equation developed in part b, determine T if $t_1 = 20$ minutes and $t_2 = 15$ minutes.

$$T = \frac{20(15)}{20 + 15} = \frac{300}{35} \approx 8.57 \text{ min.}$$

   d. If one pump can empty the tank in 40 minutes, how fast must a second pump work for the pumps working together to empty the tank in 10 minutes? Use $\frac{1}{t_1} + \frac{1}{t_2} = \frac{1}{T}$.

$$\frac{1}{40} + \frac{1}{t_2} = \frac{1}{10}$$
$$\frac{40t_2}{40} + \frac{40t_2}{t_2} = \frac{40t_2}{10}$$
$$t_2 + 40 = 4t_2$$
$$3t_2 = 40$$
$$t_2 \approx 13.3 \text{ min.}$$

   e. Suppose three pumps are working together to empty the tank, with emptying times of 25 minutes, 30 minutes, and 50 minutes. How long will it take to empty the tank if all three pumps are working simultaneously? Use the formula $\frac{1}{t_1} + \frac{1}{t_2} + \frac{1}{t_3} = \frac{1}{T}$.

$$\frac{1}{25} + \frac{1}{30} + \frac{1}{50} = \frac{1}{T}$$
$$\frac{150T}{25} + \frac{150T}{30} + \frac{150T}{50} = \frac{150T}{T}$$
$$6T + 5T + 3T = 150$$
$$14T = 150$$
$$T \approx 10.7 \text{ min.}$$

Exercise numbers appearing in color are answered in the Selected Answers appendix.

2. Solve each of the following equations using an algebraic approach. Verify your answers using a graphing approach.

a. $\dfrac{10}{x+1} = 4$

$$\frac{10(x+1)}{x+1} = 4(x+1)$$
$$10 = 4x + 4$$
$$6 = 4x$$
$$x = 1.5$$

2. a.

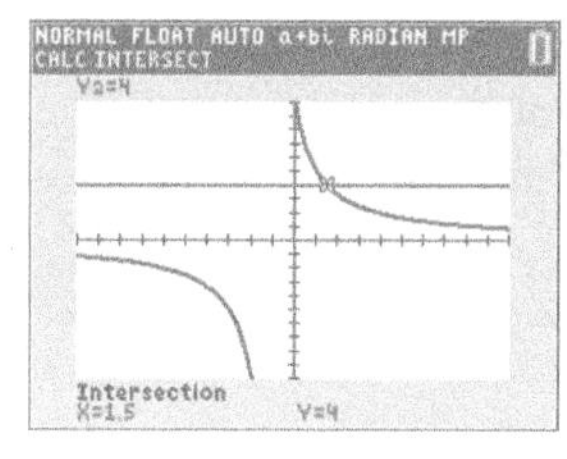

b. $\dfrac{2}{x} + \dfrac{3}{x} = 1$

$$\frac{2x}{x} + \frac{3x}{x} = 1 \cdot x$$
$$2 + 3 = x$$
$$x = 5$$

b.

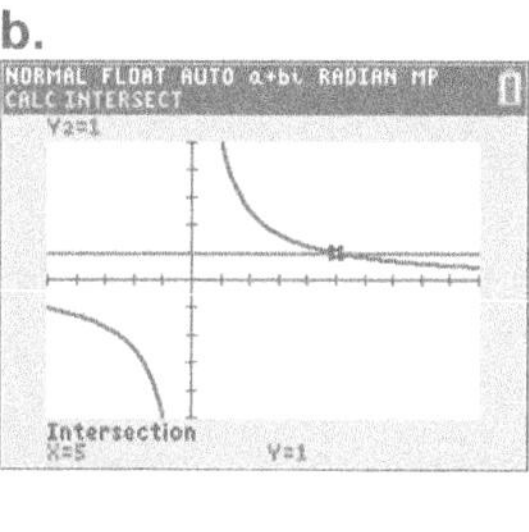

c. $\dfrac{1}{x} + \dfrac{1}{3x} = \dfrac{1}{5}$

$$\frac{15x}{x} + \frac{15x}{3x} = \frac{15x}{5}$$
$$15 + 5 = 3x$$
$$20 = 3x$$
$$x \approx 6.67$$

c.

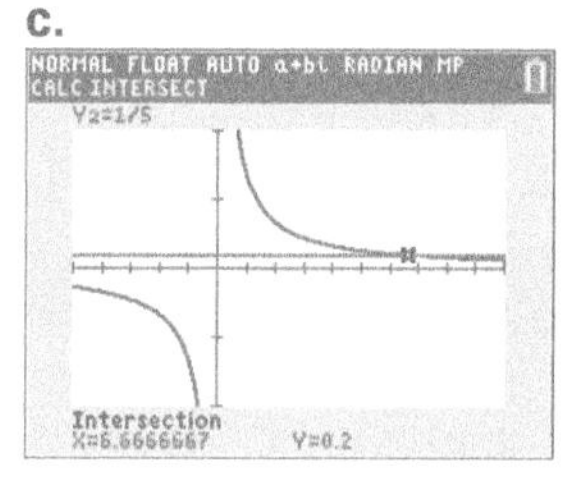

d. $\dfrac{3}{x} + \dfrac{2}{x} = \dfrac{4}{x}$

$$\frac{3x}{x} + \frac{2x}{x} = \frac{4x}{x}$$
$$3 + 2 = 4$$
$$5 = 4$$

5 = 4 is a false statement.

Therefore, there is no solution.

d.

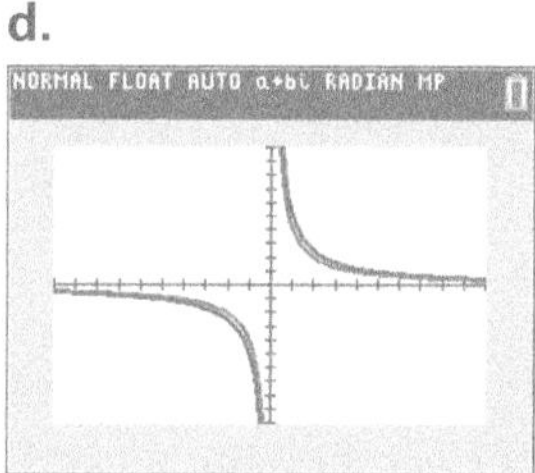

In Exercises 3–5, use the formula $\dfrac{1}{t_1} + \dfrac{1}{t_2} = \dfrac{1}{T}$.

3. The custodian in the mathematics building can buff the main floor 2 minutes faster than his supervisor can. If working together they can buff the floor in 35 minutes, how long does it take the supervisor working alone to buff the main floor of the mathematics building? (*Hint:* Let t represent the supervisor's time and $t - 2$ represent the custodian's time.)

$$\frac{1}{t} + \frac{1}{t-2} = \frac{1}{35}$$
$$\frac{35t(t-2)}{t} + \frac{35t(t-2)}{t-2} = \frac{35t(t-2)}{35}$$
$$35t - 70 + 35t = t^2 - 2t$$
$$0 = t^2 - 72t + 70$$
$$t = \frac{72 \pm \sqrt{72^2 - 4(70)}}{2}$$
$$t \approx 71 \text{ or } 1$$

The only value that makes sense in this situation is $t = 71$ minutes.

4. It takes you 4 hours working alone to clean your apartment, and your roommate takes 5 hours and 15 minutes. If you begin at noon and work together, will you complete the cleaning in time to leave for the game at 2:30? How late or early will you be?

$$\frac{1}{4} + \frac{1}{5.25} = \frac{1}{T}$$
$$5.25T + 4T = 4(5.25)$$
$$9.25T = 21$$
$$T \approx 2.27 \text{ hours}$$

T is approximately 2 hours and 16 minutes. We have approximately 14 minutes to spare.

5. One of your jobs as a work-study student at your college is sending out mailings (stuffing, sealing, and stamping envelopes). You can work twice as fast as your supervisor. If working together you complete a job in 7 hours, how long would it have taken you to complete the job by yourself? (*Hint:* Let t represent your total time working alone and $2t$ the time of your supervisor working alone.)

$$\frac{1}{t} + \frac{1}{2t} = \frac{1}{7}$$

$$14 + 7 = 2t$$

$$21 = 2t$$

$$t = 10.5 \text{ hr.}$$

6. Solve each of the following formulas for the indicated variable. Express your answer as a single fraction.

a. $\frac{1}{a} + \frac{2}{b} = \frac{3}{c}$, solve for a.

$$bc + 2ac = 3ab$$

$$2ac - 3ab = -bc$$

$$a(2c - 3b) = -bc$$

$$a = \frac{-bc}{2c - 3b}$$

b. $\frac{1}{x + y} = \frac{1}{z}$, solve for x.

$$\frac{z(x + y)}{x + y} = \frac{z(x + y)}{z}$$

$$z = x + y$$

$$x = z - y$$

c. $\frac{1}{a} + \frac{2}{b} = \frac{3}{c}$, solve for c.

$$bc + 2ac = 3ab$$

$$c(b + 2a) = 3ab$$

$$c = \frac{3ab}{b + 2a}$$

ACTIVITY 5.6

Electrical Circuits

Operations of Rational Expressions

OBJECTIVES

1. Multiply and divide rational expressions.
2. Add and subtract rational expressions.
3. Simplify complex fractions.

In performing some technical work in your new job at a local electronics firm, you need to be familiar with resistors combined in a circuit. The total resistance, R, of two resistors in a parallel circuit is modeled by the formula

$$R = \frac{1}{\frac{1}{R_1} + \frac{1}{R_2}},$$

where R_1 and R_2 are the resistances of the two resistors in the circuit, measured in ohms.

1. Calculate the total resistance for each pair of resistors.

R_1 (OHMS)	R_2 (OHMS)	R (OHMS)
10	10	5
10	5	3.333
15	5	3.75
20	10	6.667

If one resistor must be 10 ohms, then the formula

$$R = \frac{1}{\frac{1}{10} + \frac{1}{R_2}}$$

expresses the total resistance as a function of the second resistor's value.

Now the total resistance, R, is a function of R_2. The right side of the equation is a fraction in which fractions also appear in the denominator.

DEFINITION

A fraction that contains fractions in its numerator or denominator or both is called a **complex fraction**.

EXAMPLE 1 ***The following are examples of complex fractions:***

$$\frac{1}{\frac{1}{10} + \frac{1}{R_2}}; \quad \frac{\frac{50}{x}}{\frac{100}{x^2 + 5x}}; \quad \frac{4 + \frac{1}{x}}{\frac{10}{x^2} - \frac{2}{x}}$$

2. To make the equation $R = \frac{1}{\frac{1}{10} + \frac{1}{R_2}}$ less cumbersome to work with, simplify the right side of the equation so that it is written as a single fraction (with only one dividing line). This can be accomplished as follows.

   a. Determine the LCD of the fractions $\frac{1}{10}$ and $\frac{1}{R_2}$.

   $LCD = 10R_2$

   b. Add the fractions $\frac{1}{10}$ and $\frac{1}{R_2}$ by writing each fraction as an equivalent fraction that has the LCD as the denominator.

   $$\frac{R_2}{10R_2} + \frac{10}{10R_2} = \frac{R_2 + 10}{10R_2}$$

c. Divide the numerator by the denominator and simplify.

$$1 \div \frac{10 + R_2}{10R_2} = 1 \cdot \frac{10R_2}{10 + R_2} = \frac{10R_2}{10 + R_2}$$

3. a. Using your graphing calculator, sketch a graph of the function defined by

$$R = \frac{10R_2}{R_2 + 10},$$

where R is the total resistance of two resistors in a parallel circuit, one having resistance 10 ohms and the second having a resistance represented by R_2. Your screens should resemble the following:

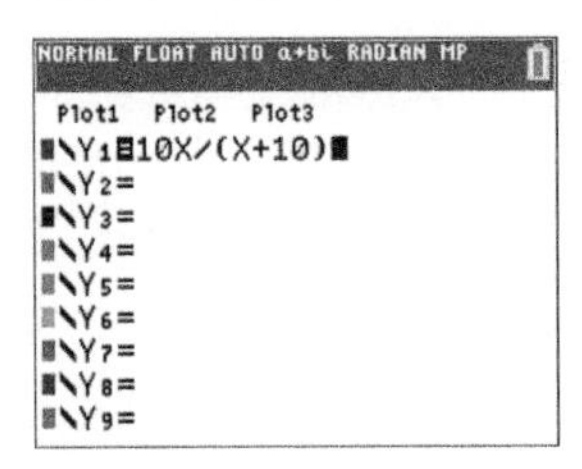

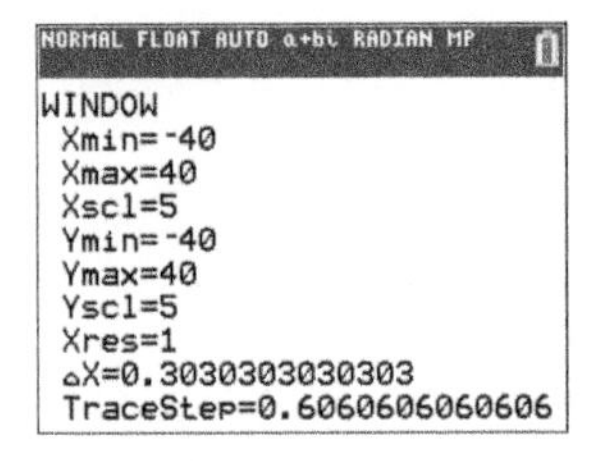

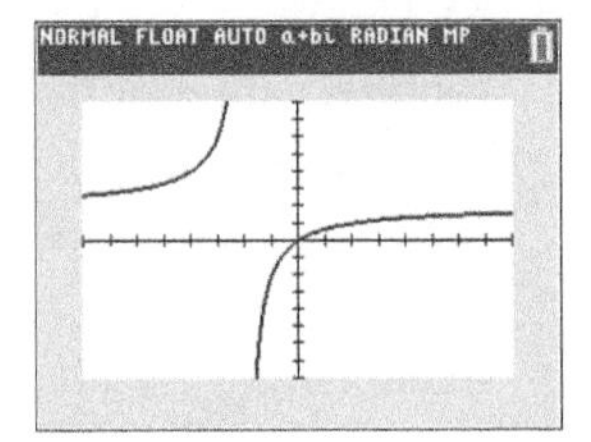

b. What is the domain of this function?

The domain is all real numbers except $R_2 = -10$.

c. What is the practical domain?

Because resistance is never negative, the domain is real numbers > 0.

d. What is the horizontal asymptote of this function?

$R = 10$

e. Interpret the practical meaning of the horizontal asymptote.

No matter how big the second resistor is, the total resistance of the circuit will never exceed 10 ohms.

4. a. Write an equation that you can use to determine the size of the second resistor you would need to add to the circuit to make a total resistance of 7 ohms.

$$7 = \frac{10R_2}{R_2 + 10}$$

b. Solve this equation graphically.

$R_2 = 23.33$ ohms

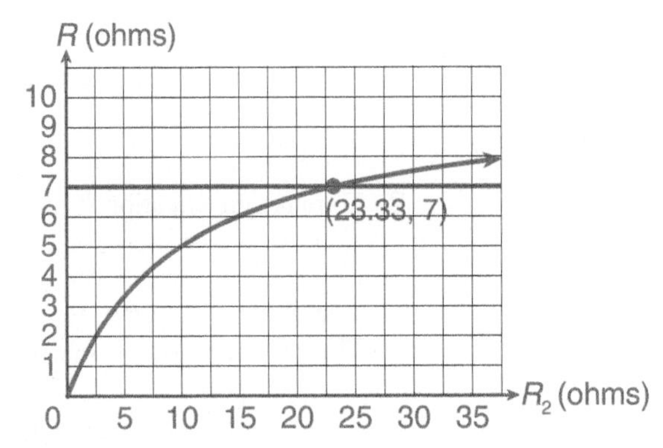

c. Solve the equation in part a algebraically.

$$7 = \frac{10R_2}{R_2 + 10}$$

$$7(R_2 + 10) = 10R_2$$

$$7R_2 + 70 = 10R_2$$

$$70 = 3R_2$$

$$R_2 = 70/3 \approx 23.3 \text{ ohms}$$

5. If resistors are available only in increments of 0.1 ohm, what size would you use to get as close as possible to but still have a total resistance of at least 7 ohms?

$R_2 = 23.4$ ohms

Operations with Rational Expressions

As you discovered in Problem 2, simplifying complex fractions involves a great deal of work with rational expressions. The following examples illustrate how to perform operations with algebraic fractions. Appendix A contains several additional examples and practice exercises involving operations with rational expressions.

EXAMPLE 2 ***Simplify the following complex fractions.***

a. $\dfrac{\frac{50}{x}}{\frac{100}{x^2 + 5x}}$

Both the numerator and denominator of the complex fraction contain single rational expressions. Therefore, write the complex fraction as a division problem and divide.

$$\frac{\frac{50}{x}}{\frac{100}{x^2 + 5x}} = \frac{50}{x} \div \frac{100}{x^2 + 5x}$$ Divide using the division rule $\frac{a}{b} \div \frac{c}{d} = \frac{a}{b} \cdot \frac{d}{c}$.

$$= \frac{50}{x} \cdot \frac{x^2 + 5x}{100}$$ Simplify if possible.

$$= \frac{\cancel{50}}{\cancel{x}} \cdot \frac{\cancel{x}(x + 5)}{2 \cdot \cancel{50}}$$

$$= \frac{x + 5}{2}$$

b. $\dfrac{4 + \frac{1}{x}}{\frac{10}{x^2} - \frac{2}{x}}$

The rational expressions in the numerator and denominator of the complex fraction can each be combined into a single rational expression.

Step 1. $\frac{4}{1} + \frac{1}{x} = \frac{4}{1} \cdot \frac{x}{x} + \frac{1}{x} = \frac{4x}{x} + \frac{1}{x} = \frac{4x + 1}{x}$

Step 2. $\frac{10}{x^2} - \frac{2}{x} = \frac{10}{x^2} - \frac{2x}{x \cdot x} = \frac{10}{x^2} - \frac{2x}{x^2} = \frac{10 - 2x}{x^2}$

Step 3. Now divide the numerator of the complex fraction by the denominator.

$$\frac{4 + \frac{1}{x}}{\frac{10}{x^2} - \frac{2}{x}} = \frac{4x + 1}{x} \div \frac{10 - 2x}{x^2} = \frac{(4x + 1)}{\cancel{x}} \cdot \frac{\cancel{x} \cdot x}{(10 - 2x)} = \frac{4x^2 + x}{10 - 2x}$$

LC LEARNING CATALYTICS

Simplify $\dfrac{\frac{x}{5} - \frac{5}{x}}{\frac{1}{5} + \frac{1}{x}}$.

a. $x + 5$
b. $x - 5$
c. $\frac{5x}{x + 5}$
d. -1

6. Simplify the following complex fraction:

$$\frac{\frac{4}{x+3}}{\frac{1}{x+2}+\frac{3}{x}}$$

$$\frac{\frac{4}{x+3}}{\frac{x}{x(x+2)}+\frac{3(x+2)}{x(x+2)}} = \frac{\frac{4}{x+3}}{\frac{x+3x+6}{x(x+2)}} = \frac{\frac{4}{x+3}}{\frac{4x+6}{x(x+2)}}$$

$$\frac{4}{x+3}\cdot\frac{x(x+2)}{4x+6} = \frac{4x^2+8x}{4x^2+18x+18} = \frac{4(x^2+2x)}{2(2x^2+9x+9)} = \frac{2(x^2+2x)}{2x^2+9x+9}$$

SUMMARY Activity 5.6

1. To **multiply or divide** rational expressions,

a. factor the numerator and denominator of each fraction completely.

b. divide out the common factors (cancel).

c. multiply remaining factors.

d. in division, proceed as above after inverting the divisor (the fraction after the division sign).

2. To **add or subtract** rational expressions,

a. determine the LCD (least common denominator).

b. build each fraction to have the LCD.

c. add or subtract numerators.

d. place the numerator over the LCD, and simplify if possible.

3. To simplify a complex fraction by simplifying the numerator and denominator,

a. express the numerator as a single fraction.

b. express the denominator as a single fraction.

c. divide the numerator by the denominator.

d. simplify if possible.

EXERCISES Activity 5.6

1. An emergency medical services (EMS) vehicle has a single-tone siren with a pitch of approximately 330 hertz (Hz). You are standing on the street corner as the vehicle approaches you at 40 miles per hour. Due to the Doppler effect, the pitch you hear is not 330 hertz. The pitch, h, that you hear is modeled by

$$h = \frac{a}{1-\frac{s}{770}},$$

where a is the actual pitch and s is the speed of the source of the sound in miles per hour.

a. Is the pitch you hear due to the Doppler effect lower or higher than the actual pitch?

$$h = \frac{330}{1 - \frac{40}{770}} \approx 348.08 \text{ hertz}$$

The pitch I hear is higher than the actual pitch.

b. Simplify the original equation by rewriting the complex fraction on the right side as a single fraction (using only one dividing line).

$$h = \frac{a}{1 - \frac{s}{770}} = \frac{a}{\frac{770}{770} - \frac{s}{770}} = \frac{a}{\frac{770 - s}{770}}$$

$$h = a \div \frac{770 - s}{770}$$

$$h = \frac{770a}{770 - s}$$

c. Redo part a using the new equation obtained from part b. How do the results compare?

$$h = \frac{770(330)}{770 - 40} \approx 348.08 \text{ hertz}$$

The results are the same.

d. What pitch sound would you hear if the EMS vehicle were traveling at 60 miles per hour?

$$h = \frac{770(330)}{770 - 60} \approx 357.89 \text{ hertz}$$

2. If three resistors having resistances R_1, R_2, and R_3 are connected in parallel, their combined resistance, R, is modeled by the formula

$$R = \frac{1}{\frac{1}{R_1} + \frac{1}{R_2} + \frac{1}{R_3}}.$$

a. Determine R if R_1 is 4 ohms, R_2 is 8 ohms, and R_3 is 12 ohms.

$$R = \frac{1}{\frac{1}{4} + \frac{1}{8} + \frac{1}{12}} \approx 2.18 \text{ ohms}$$

b. Simplify the complex fraction on the right side of the original formula.

$$\frac{1}{\frac{1}{R_1} + \frac{1}{R_2} + \frac{1}{R_3}} = \frac{1}{\frac{R_2R_3 + R_1R_3 + R_1R_2}{R_1R_2R_3}}$$

$$= 1 \div \frac{R_2R_3 + R_1R_3 + R_1R_2}{R_1R_2R_3}$$

$$= \frac{R_1R_2R_3}{R_2R_3 + R_1R_3 + R_1R_2}$$

c. Redo part a using the new formula from part b. How do the answers compare?

$$R = \frac{R_1R_2R_3}{R_2R_3 + R_1R_3 + R_1R_2} = \frac{4(8)(12)}{8(12) + 4(12) + 4(8)} \approx 2.18 \text{ ohms}$$

The answers are the same.

3. Simplify the following complex fractions:

a. $\dfrac{\frac{1}{2} - \frac{2}{x}}{x - 4}$

$$\frac{\frac{x}{2x} - \frac{4}{2x}}{x - 4} = \frac{x - 4}{2x} \div (x - 4)$$
$$= \frac{x - 4}{2x} \cdot \frac{1}{x - 4} = \frac{1}{2x}$$

b. $\dfrac{\frac{1}{x} - \frac{1}{2}}{\frac{1}{x^2} - \frac{1}{4}}$

$$\frac{\frac{2}{2x} - \frac{x}{2x}}{\frac{4 - x^2}{4x^2}} = \frac{\frac{2 - x}{2x}}{\frac{(2 - x)(2 + x)}{4x^2}}$$
$$= \frac{2 - x}{2x} \cdot \frac{4x^2}{(2 - x)(2 + x)} = \frac{2x}{2 + x}$$

c. $\dfrac{x + \frac{2x - 6}{x - 1}}{\frac{x}{3} - \frac{3}{x}}$

$$\frac{\frac{x(x - 1)}{x - 1} + \frac{2x - 6}{x - 1}}{\frac{x^2 - 9}{3x}} = \frac{\frac{x^2 + x - 6}{x - 1}}{\frac{x^2 - 9}{3x}}$$
$$= \frac{(x + 3)(x - 2)}{(x - 1)} \cdot \frac{3x}{(x - 3)(x + 3)}$$
$$= \frac{3x^2 - 6x}{x^2 - 4x + 3}$$

d. $\dfrac{\frac{x}{2} - 1}{x - \frac{4}{x}}$

$$\frac{\frac{x - 2}{2}}{\frac{x^2 - 4}{x}} = \frac{x - 2}{2} \cdot \frac{x}{(x + 2)(x - 2)}$$
$$= \frac{x}{2x + 4}$$

4. You decide to buy a new car, but you are concerned about the amount of the monthly payments. The amount, A, of each monthly payment is modeled by the formula

$$A = \frac{Pi}{1 - \frac{1}{(1 + i)^n}},$$

where P represents the principal or the amount borrowed,

i is the monthly interest rate, and

n is the number of monthly payments.

a. The car you are looking at costs \$16,000 at 1% monthly interest. If you want to pay for the car over 60 months, how much will your monthly payment be?

$$A = \frac{16{,}000 \cdot 0.01}{1 - \frac{1}{1.01^{60}}} \approx \$355.91$$

b. Simplify the complex fraction on the right side of the original formula.

$$\frac{Pi}{1 - \frac{1}{(1 + i)^n}} = \frac{Pi}{\frac{(1 + i)^n - 1}{(1 + i)^n}} = \frac{Pi(1 + i)^n}{(1 + i)^n - 1}$$

c. Use the simplified formula to determine the monthly payments. How does your answer compare with the answer in part a?

$$A = \frac{16{,}000 \cdot 0.01 \cdot (1.01)^{60}}{(1.01)^{60} - 1} \approx \$355.91$$

The answers are the same.

CLUSTER 1 What Have I Learned?

1. Make a list of some of the special features of the graphs of rational functions.

Graphs often have vertical and horizontal asymptotes. A vertical asymptote arises where a denominator is 0. A horizontal asymptote arises when the graph levels off as x gets larger (to positive infinity) or as x gets larger in the negative direction (to negative infinity).

2. Describe the connection, if any, between the domain of a rational function and the equation of its vertical asymptote.

First simplify the rational formula. If an input value of $x = a$ results in a value of 0 in a denominator of the simplified equation that defines the rational function, then the value a is not in the domain of the function. The vertical line having equation $x = a$ is a vertical asymptote of the graph.

3. Describe the algebraic steps required to determine the vertical asymptote(s) of the graph of a rational function.

Express the rational function as a single simplified fraction. Any input number, a, that makes the denominator equal to zero will then be the location of a vertical asymptote.

4. For what values of k will the graph of $y = \frac{k}{x}$ be in the second and fourth quadrants?

When k is negative, y will have the opposite sign of x. This will place the graph in the second and fourth quadrants.

5. a. Describe the algebraic steps required to solve $10 = \frac{35}{1 + 5x}$.

Multiply both sides of the equation by $1 + 5x$.

Distribute 10 to get $10 + 50x$ on one side of the equation.

Subtract 10 from both sides to get $50x = 25$.

Divide both sides by 50 to get the solution: $x = 0.5$.

Check by substituting $x = 0.5$ into the original equation.

b. Explain the technique for solving the equation in part a graphically.

To solve graphically, enter two functions on the calculator—one for the left side and one for the right side of the equation. Graph both, adjusting the window to see any points of intersection. Use the intersect routine to approximate the coordinates of the point of intersection. The x-coordinate is a solution to the equation.

6. Explain how you would determine the horizontal asymptote of the rational function $f(x) = \frac{6x + 1}{2 - 3x}$.

As positive values of x get larger, which you can observe in a graph or a table, the values of the function appear to get closer to -2. The horizontal asymptote is $y = -2$.

CLUSTER 1 How Can I Practice?

1. Describe the relationship between the graphs of $f(x) = \frac{1}{x}$ and $g(x) = \frac{-1}{x}$.

   The graphs of f and g are reflections of each other about the x-axis (or the y-axis).

2. Describe the relationship between the graphs of $f(x) = \frac{1}{x}$ and $g(x) = \frac{1}{x^3}$.

   They are similar, but the graph of g is closer to the x-axis, and the graph of f is closer to the y-axis.

3. **a.** Suppose you are taking a trip of 145 miles. Assume that you drive the entire distance at a constant speed. Express your time to take this trip as a function of your speed.

   T = time in hours; s = speed in mph

   $T = \frac{145}{s}$

   b. What is the practical domain of this function?

   $0 < s < 80$

   c. Using the equation for the function, determine the domain.

   The domain is all real numbers except 0.

4. Determine the domain of each of the following functions. Then give the equation of the vertical asymptote of each function.

   a. $g(x) = \frac{10}{x + 5}$

   domain: all real numbers except $x = -5$

   vertical asymptote: $x = -5$

   b. $f(x) = \frac{5}{13 - 2x}$

   domain: all real numbers except $x = \frac{13}{2}$

   vertical asymptote: $x = \frac{13}{2}$

   c. $g(x) = \frac{-3}{5x - 8}$

   domain: all real numbers except $x = \frac{8}{5}$

   vertical asymptote: $x = \frac{8}{5}$

   d. $h(x) = \frac{0.02}{5.7x - 3.2}$

   domain: all real numbers except $x \approx 0.5614$

   vertical asymptote: $x \approx 0.5614$

5. The weight of a body above the surface of Earth varies inversely with the square of the distance from the center of Earth. If an object weighs 100 pounds when it is 4000 miles from the center of Earth, how much will it weigh when it is 4500 miles from the center?

   $w = \frac{k}{d^2}$; $100 = \frac{k}{4000^2}$

   $4000^2 \cdot 100 = k$ or $k = 1.6 \cdot 10^9$

   $w = \frac{1.6 \cdot 10^9}{d^2} = \frac{1.6 \cdot 10^9}{(4500)^2} \approx 79.01$ lb.

Answers to all How Can I Practice exercises are included in the Selected Answers appendix.

6. A manufacturer of lawn mowers uses the function $C(x) = \dfrac{132x + 75{,}250}{x}$ to model the average cost per lawn mower, in dollars, where x is the number of lawn mowers produced.

a. What is the practical domain of this function?

The practical domain is all positive integers, with some realistic upper limit, depending on the specific situation.

b. What is the minimum number of lawn mowers that must be manufactured to bring the average cost per lawn mower down to \$199? Solve algebraically and verify graphically with your calculator.

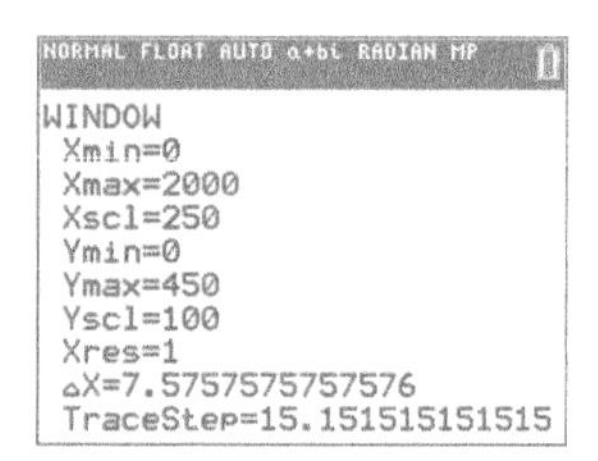

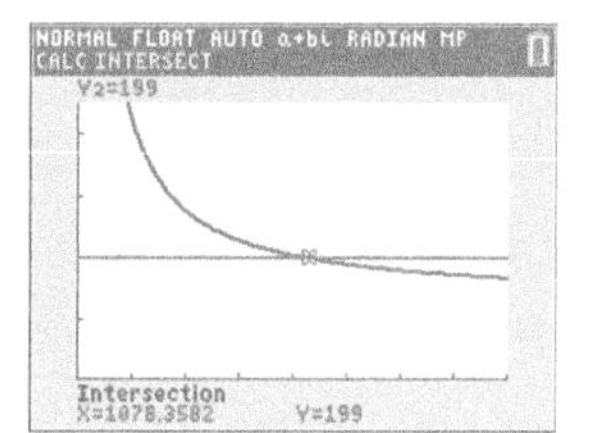

$$199 = \frac{132x + 75{,}250}{x}$$

$$199x = 132x + 75{,}250$$

$x \approx 1123.13$; $x = 1124$ mowers

7. The concentration of a certain drug in the bloodstream, measured in milligrams per liter, can be modeled by the function

$$C = \frac{14t}{3t^2 + 2.5},$$

where t is the number of minutes after injection of the drug.

a. How long after injection will it take the concentration to equal 0.05 milligram per liter? Solve algebraically and check graphically.

$$\frac{14t}{3t^2 + 2.5} = 0.05$$

$$14t = 0.05(3t^2 + 2.5)$$

$$14t = 0.15t^2 + 0.125$$

$0 = 0.15t^2 - 14t + 0.125.$ Determine the roots.

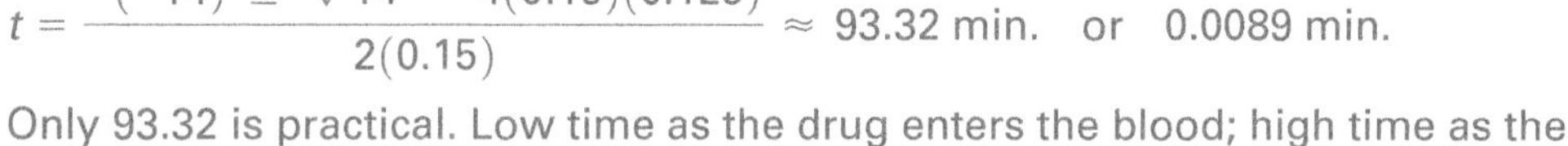

$$t = \frac{-(-14) \pm \sqrt{14^2 - 4(0.15)(0.125)}}{2(0.15)} \approx 93.32 \text{ min. or } 0.0089 \text{ min.}$$

Only 93.32 is practical. Low time as the drug enters the blood; high time as the drug dissipates.

b. Use the graph to determine when the drug will be at its highest concentration.

The drug will be at its highest concentration approximately 0.913 minute after injection.

7. a.

NORMAL FLOAT AUTO a+bi RADIAN MP
CALC INTERSECT
Y2=0.05
Intersection X=93.324404 Y=0.05

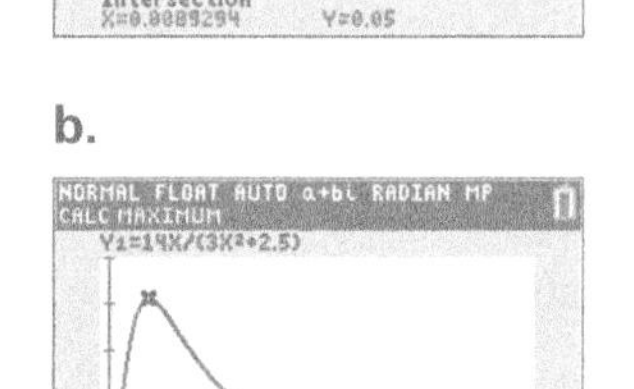

NORMAL FLOAT AUTO a+bi RADIAN MP
CALC INTERSECT
Y2=0.05
Intersection X=0.0089294 Y=0.05

b.

NORMAL FLOAT AUTO a+bi RADIAN MP
CALC MAXIMUM
Y1=14X/(3X²+2.5)
Maximum X=0.9128699 Y=2.5560386

8. Solve each equation algebraically. Verify your answer graphically.

a. $\dfrac{3}{x+1} = 4$

$x = -0.25$

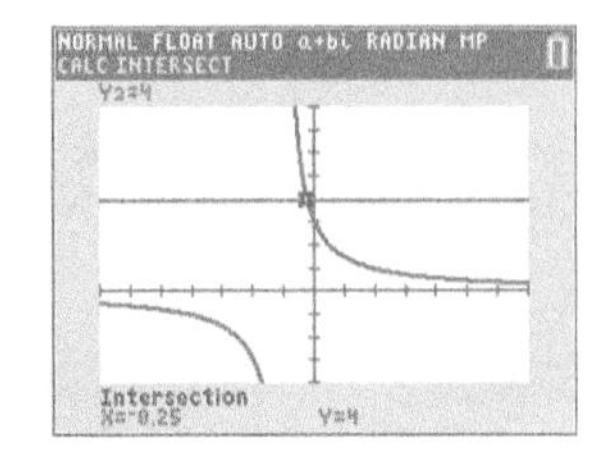

b. $\dfrac{3x}{2x-5} = 10$

$x = \dfrac{50}{17} \approx 2.94$

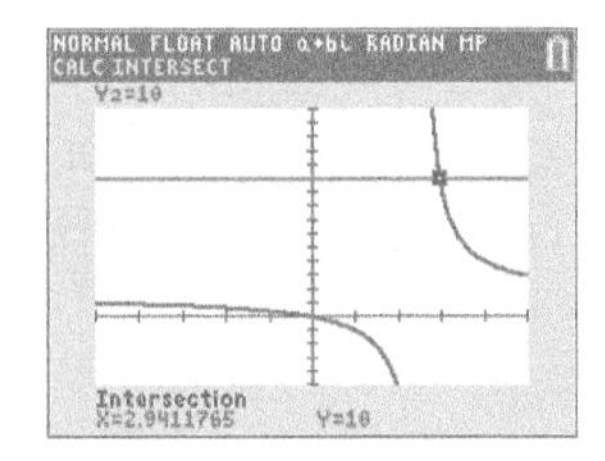

c. $\dfrac{4}{x+3} + 12 = 52$

$x = \dfrac{-116}{40} = -2.9$

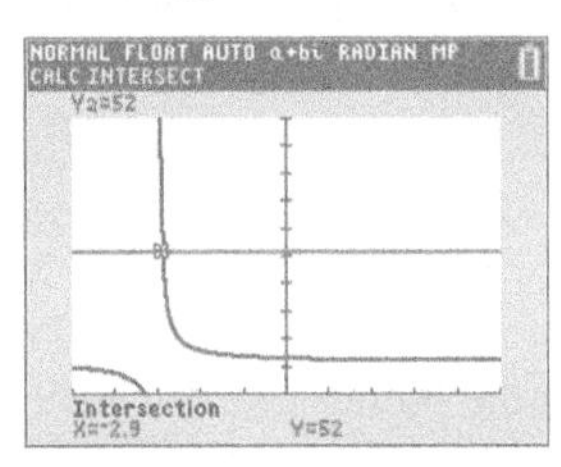

d. $\dfrac{2.4x}{1+0.3x} = 5.8$

$x = \dfrac{5.8}{2.4-(0.3)5.8} = \dfrac{290}{33} \approx 8.788$

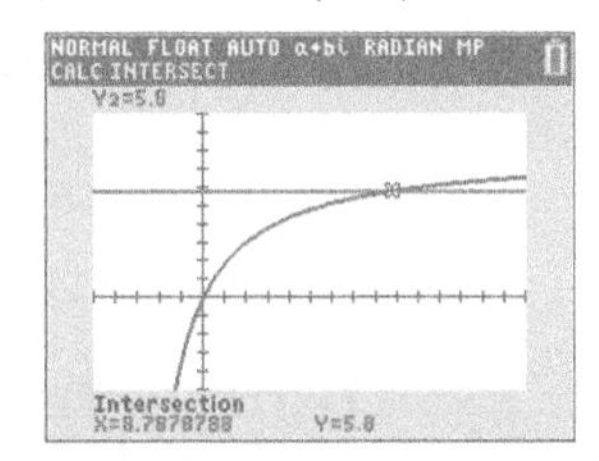

9. As an object rises, the effect of Earth's gravitational pull on the object is reduced.

If an object weighs E kilograms at sea level, then the weight, W (also in kilograms), of the object at a distance of h kilometers above sea level is modeled by the function

$$W = \frac{E}{\left(1 + \frac{h}{6400}\right)^2}.$$

a. Suppose you are flying in a commercial jetliner 15 kilometers above sea level. Replace E with your body weight, measured in kilograms (1 kilogram = 2.2 pounds), and calculate your weight at 15 kilometers above sea level.

(Answers will vary.) 200-pound man ≈ 90.9 kg

$$W = \frac{90.9}{\left(1 + \frac{15}{6400}\right)^2} \approx 90.475 \text{ kilograms}$$

b. If an astronaut weighs 70 kilograms at sea level, write a function equation that expresses the astronaut's weight as a function of his or her distance above sea level.

$$W = \frac{70}{\left(1 + \frac{h}{6400}\right)^2}$$

c. Complete the following table using the function equation from part b:

h (km)	0	10	100	1000	1500	2000	10,000	20,000
W (kg)	70	69.78	67.86	52.36	45.94	40.63	10.66	4.11

d. As the height, h, of the space shuttle increases, what happens to the corresponding weight of the astronaut?

The weight decreases.

e. As the space shuttle reaches its orbiting altitude of 650 kilometers above sea level, what is the weight of the astronaut?

57.69 kg

f. What is the practical domain of the weight function?

(Answers will vary at the upper end.)

The domain is $0 \leq h \leq 40{,}000$.

9. g.

NORMAL FLOAT AUTO a+bi RADIAN MP

g. Use your graphing calculator to sketch a graph of the weight function. Use the window Xmin = 0, Xmax = 40000, Ymin = 0, and Ymax = 80.

h. At what altitude does the astronaut's weight equal one-half of what it is at sea level?

Determine the zero of $W = \dfrac{70}{\left(1 + \dfrac{h}{6400}\right)^2} - 35$, or solve $35 = \dfrac{70}{\left(1 + \dfrac{h}{6400}\right)^2}$.

$h \approx 2650.9668$ km

10. An electrical circuit has three resistors. The total resistance of the circuit, R, is related to the individual resistances R_1, R_2, and R_3 by the equation

$$\frac{1}{R_1} + \frac{1}{R_2} + \frac{1}{R_3} = \frac{1}{R}.$$

a. You know that $R_1 = 4$ ohms, $R_2 = 6$ ohms, and the total resistance of the circuit is 2 ohms. Determine R_3.

$$\frac{1}{4} + \frac{1}{6} + \frac{1}{R_3} = \frac{1}{2}$$

$$\frac{12R_3}{4} + \frac{12R_3}{6} + \frac{12R_3}{R_3} = \frac{12R_3}{2}$$

$$3R_3 + 2R_3 + 12 = 6R_3$$

$$5R_3 + 12 = 6R_3$$

$$R_3 = 12 \text{ ohms}$$

b. Solve the equation $\dfrac{1}{R_1} + \dfrac{1}{R_2} + \dfrac{1}{R_3} = \dfrac{1}{R}$ for R.

$$\frac{R_1R_2R_3R}{R_1} + \frac{R_1R_2R_3R}{R_2} + \frac{R_1R_2R_3R}{R_3} = \frac{R_1R_2R_3R}{R}$$

$$R_2R_3R + R_1R_3R + R_1R_2R = R_1R_2R_3$$

$$R = \frac{R_1R_2R_3}{R_1R_2 + R_2R_3 + R_1R_3}$$

11. a.

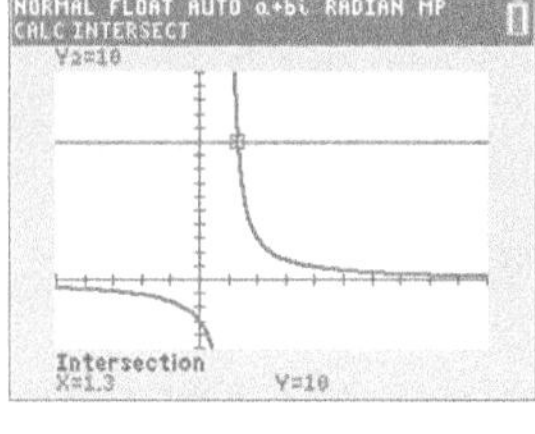

c. Using the equation developed in part b, determine R if $R_1 = 4$ ohms, $R_2 = 6$ ohms, and $R_3 = 12$ ohms.

$$R = \frac{4(6)(12)}{4(6) + 4(12) + 6(12)} = \frac{288}{24 + 48 + 72} = \frac{288}{144} = 2 \text{ ohms}$$

b.

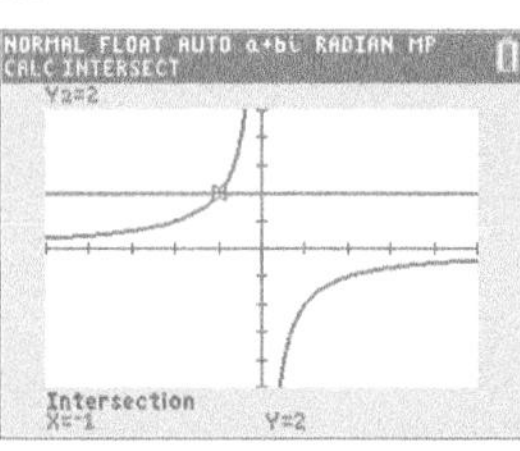

11. Solve each of the following equations using an algebraic approach. Verify your answers using a graphing approach.

a. $\dfrac{3}{x - 1} = 10$

$$\frac{3(x-1)}{x-1} = 10(x-1)$$

$$3 = 10x - 10$$

$$13 = 10x$$

$$x = 1.3$$

b. $\dfrac{2}{x} - \dfrac{4}{x} = 2$

$$\frac{2x}{x} - \frac{4x}{x} = 2 \cdot x$$

$$2 - 4 = 2x$$

$$-2 = 2x$$

$$x = -1$$

11. c.

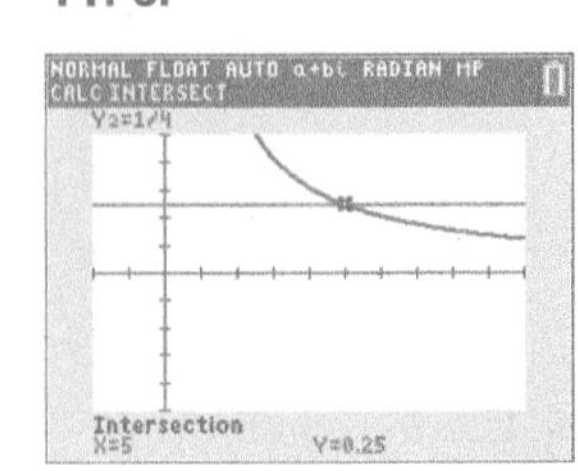

c. $\frac{1}{x} + \frac{1}{4x} = \frac{1}{4}$

$$\frac{4x}{x} + \frac{4x}{4x} = \frac{4x}{4}$$

$$4 + 1 = x$$

$$x = 5$$

d. $\frac{3}{x} - 4 = \frac{2}{x}$

$$\frac{3x}{x} - 4x = \frac{2x}{x}$$

$$3 - 4x = 2$$

$$1 = 4x$$

$$x = \frac{1}{4}$$

d.

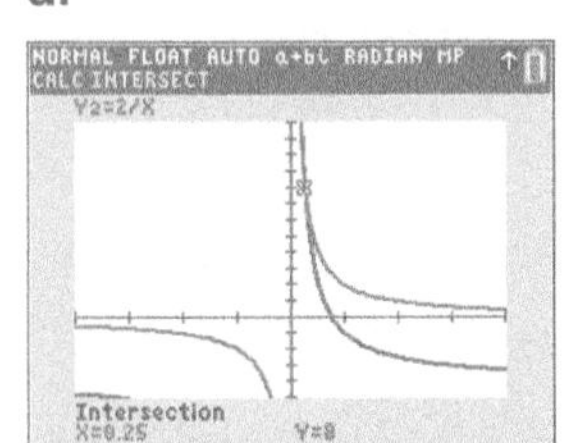

12. The average speed, s, of your round-trip commute from home to campus is modeled by

$$s = \frac{2d}{\frac{d}{r_1} + \frac{d}{r_2}},$$

where d is the one-way distance from home, r_1 is your average morning commute speed, and r_2 is your average afternoon commute speed.

a. Your one-way commute to campus is 15.3 miles. Your average morning commute speed is 45 miles per hour. Your average afternoon commute speed is 40 miles per hour. What is your average speed, s?

$$s = \frac{2(15.3)}{\frac{15.3}{45} + \frac{15.3}{40}} = \frac{30.6}{0.34 + 0.3825} \approx 42.4 \text{ mph}$$

b. Simplify the original equation by rewriting the complex fraction on the right side as a single fraction.

$$s = \frac{2d}{\frac{d}{r_1} + \frac{d}{r_2}} = \frac{2d}{\frac{dr_2}{r_1r_2} + \frac{dr_1}{r_1r_2}} = \frac{2d}{\frac{dr_2 + dr_1}{r_1r_2}}$$

$$s = 2d \div \frac{dr_2 + dr_1}{r_1r_2}$$

$$s = \frac{2dr_1r_2}{d(r_1 + r_2)} = \frac{2r_1r_2}{r_1 + r_2}$$

c. Redo part a using the new equation obtained in part b. How do the results compare?

$$s = \frac{2(45)(40)}{(40 + 45)} \approx 42.4 \text{ mph}$$

The results are the same.

13. Simplify the following complex fractions:

a. $\dfrac{4 + \frac{2}{x}}{1 - \frac{3}{x}}$

$$\frac{\frac{4x}{x} + \frac{2}{x}}{\frac{x}{x} - \frac{3}{x}} = \frac{\frac{4x + 2}{x}}{\frac{x - 3}{x}}$$

$$= \frac{4x + 2}{x} \cdot \frac{x}{x - 3}$$

$$= \frac{4x + 2}{x - 3}$$

b. $\dfrac{\frac{x}{5} - \frac{5}{x}}{\frac{1}{5} + \frac{1}{x}}$

$$\frac{\frac{x^2}{5x} - \frac{25}{5x}}{\frac{x}{5x} + \frac{5}{5x}} = \frac{\frac{x^2 - 25}{5x}}{\frac{x + 5}{5x}}$$

$$= \frac{x^2 - 25}{5x} \cdot \frac{5x}{x + 5}$$

$$= \frac{(x + 5)(x - 5)}{(x + 5)} = x - 5$$

c. $\dfrac{\frac{1}{x + 2}}{1 + \frac{1}{x + 2}}$

$$\frac{\frac{1}{x + 2}}{\frac{x + 2}{x + 2} + \frac{1}{x + 2}} = \frac{\frac{1}{x + 2}}{\frac{x + 3}{x + 2}}$$

$$= \frac{1}{x + 2} \cdot \frac{x + 2}{x + 3} = \frac{1}{x + 3}$$

CLUSTER 2 Radical Functions

ACTIVITY 5.7

Skydiving
Radical Functions and Their Graphs

OBJECTIVES

1. Determine the domain of a radical function defined by $y = \sqrt{g(x)}$, where $g(x)$ is a polynomial.
2. Graph functions having equation $y = \sqrt{g(x)}$ and $y = -\sqrt{g(x)}$.
3. Identify the properties of the graph of $y = \sqrt{g(x)}$ and $y = -\sqrt{g(x)}$.

The first recorded successful parachute jump was made from a hot air balloon in 1797. Parachute technology and equipment were significantly refined by the use of parachutes in the military. Modern parachuting, also known as skydiving, has developed into a popular and exciting recreational activity and competitive sport.

Typically, individuals jump out of an aircraft at approximately 14,000 feet and freefall for a period of time before activating a parachute. When first exiting the aircraft, a skydiver will have a slight feeling of falling. However, during their period of freefall, skydivers generally do not experience a "falling" sensation because the resistance of the air to their body provides some feeling of weight and direction.

The distance, s (in feet), that a skydiver travels in freefall is a function of time, t (in seconds). This function is defined by $s = 16t^2$. (**Note:** This model neglects air resistance.)

1. a. Use the formula $s = 16t^2$ to complete the following table:

t(sec.)	0	5	10	15	20	25
s(ft.)	0	400	1600	3600	6400	10,000

b. Use the table values to sketch a graph of the freefall distance function.

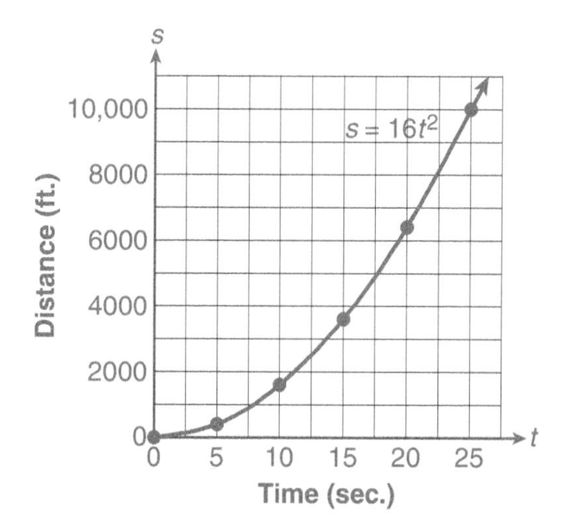

A skydiver needs to know how much time she has in freefall before having to activate the parachute. The freefall time will depend on the distance in freefall.

2. a. Suppose the skydiver jumps from a plane at 12,000 feet and must activate the parachute at 3000 feet. Determine the freefall distance.

$12{,}000 - 3000 = 9000$

b. Use the formula $s = 16t^2$ to determine how much time she will have in freefall.

$9000 = 16t^2,\ t^2 = \dfrac{9000}{16},\ t = \sqrt{\dfrac{9000}{16}},\ t \approx 23.7$ sec.

The process of determining freefall time for a given freefall distance would be simplified if you had a formula that expresses time t as a function of distance s.

3. a. Solve the equation $s = 16t^2$ for t, where $t > 0$, to determine an equation for the time in freefall, t, as a function of the distance, s.

$$s = 16t^2,\ t^2 = \frac{s}{16},\ t = \sqrt{\frac{s}{16}},\ \text{or } t = \frac{1}{4}\sqrt{s} = 0.25\sqrt{s}$$

b. Use the function from part a to complete the following table:

s(ft.)	0	400	1600	3600	6400	10,000
t(sec.)	0	5	10	15	20	25

c. Use the table values to sketch a graph of the function defined by $t = 0.25\sqrt{s}$.

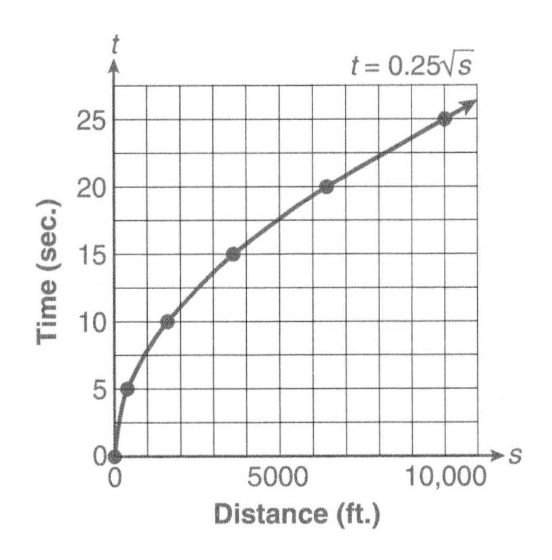

4. a. How are the ordered pairs in the tables in Problems 1a and 3b related?

The inputs and outputs are reversed.

b. Complete the following sentence. Functions that have their input and output values reversed are <u>inverse</u> functions.

Radical Functions

The investigation of the properties of the function defined by $t = 0.25\sqrt{s}$ is somewhat limited because of the restrictions placed on the variables t and s, which represent real-world quantities.

5. a. Consider the general function defined by $F(x) = 0.25\sqrt{x}$. What is the domain of F?

The domain is all nonnegative real numbers.

b. Sketch a graph of F.

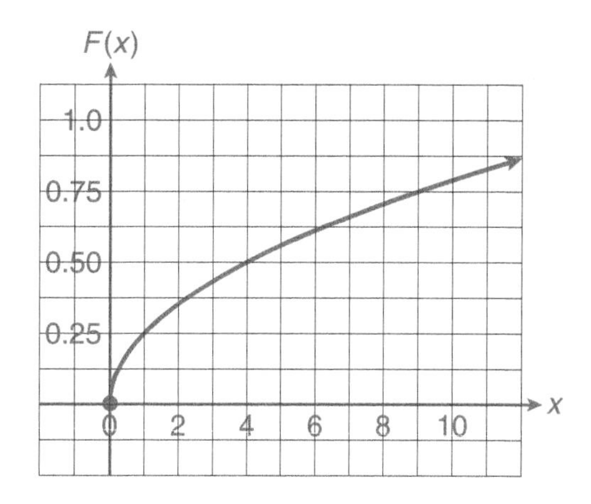

c. What is the range of F?

The range is all nonnegative real numbers.

Recall that the square root of a negative number is not a real number. Therefore, the domain of a function defined by an equation of the form $y = \sqrt{g(x)}$, where $g(x)$ is a polynomial, is the set of all real numbers for x such that $g(x) \geq 0$.

EXAMPLE 1 ***Determine the domain of the function defined by*** $f(x) = \sqrt{2x - 10}$.

SOLUTION

You need to determine all values of x such that $2x - 10 \geq 0$. Therefore,

$$2x - 10 \geq 0$$
$$2x \geq 10$$
$$x \geq 5.$$

The domain is all real numbers greater than or equal to 5.

LC LEARNING CATALYTICS

Determine the domain of $f(x) = \sqrt{3x - 6}$.

a. $x \geq 6$
b. $x > 2$
c. $x \leq 2$
d. $x \geq 2$

6. Consider the functions defined by the following equations:

$$f(x) = \sqrt{x} \qquad g(x) = \sqrt{x + 2} \qquad h(x) = \sqrt{x - 3}$$

a. Determine the domain of each function.

The domain of f is $x \geq 0$; the domain of g is $x \geq -2$; the domain of h is $x \geq 3$.

b. Sketch graphs of the functions f, g, and h on the same coordinate system.

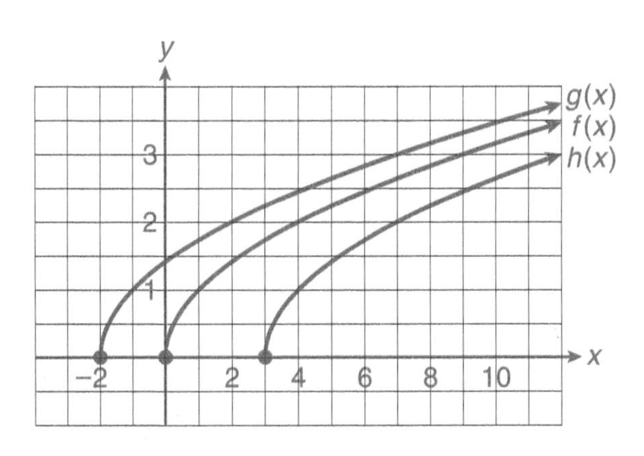

c. Are the functions f, g, and h increasing or decreasing?

All three are increasing.

7. a. Determine the domain and range of the functions defined by $f(x) = \sqrt{x}$ and $g(x) = -\sqrt{x}$.

The domain of f is $x \geq 0$, and the range of f is $f(x) \geq 0$. The domain of g is $x \geq 0$, and the range of g is $g(x) \leq 0$.

b. Sketch graphs of f and g on the same coordinate system.

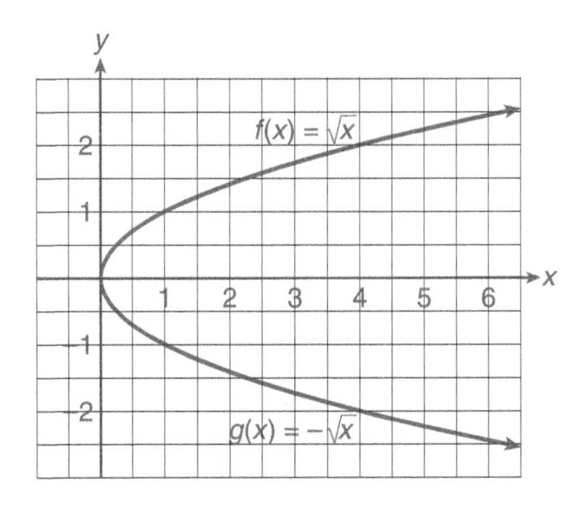

c. How would you obtain the graph of g from the graph of f?

Reflect the graph of f through the x-axis.

8. a. Determine the domains of the functions defined by $h(x) = \sqrt{x - 3}$ and $F(x) = \sqrt{3 - x}$.

The domain of h is $x - 3 \geq 0$ or $x \geq 3$. The domain of F is $3 - x \geq 0$ or $x \leq 3$.

b. Complete the following tables:

x	3	4	7	12	19	28
$h(x)$	0	1	2	3	4	5

x	3	2	−1	−6	−13	−22
$F(x)$	0	1	2	3	4	5

c. Sketch the graphs of h and F on the same coordinate system.

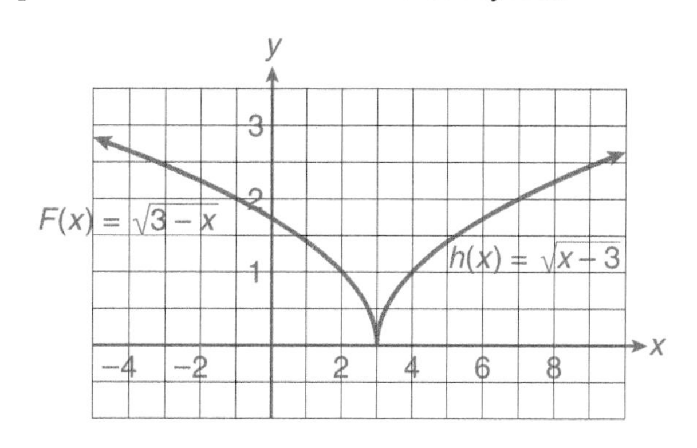

d. Determine whether the functions h and F are increasing or decreasing.

h is an increasing function. F is a decreasing function.

9. Consider the function defined by $f(x) = -\sqrt{3x - 6}$.

a. Determine the domain of f.

$3x - 6 \geq 0$, so $3x \geq 6$ or $x \geq 2$.

b. Determine the x-intercept of f.

(2, 0)

c. Complete the following table. If necessary, approximate $f(x)$ to the nearest tenth.

x	2	3	4	5	8	10
$f(x)$	0	$-\sqrt{3} \approx -1.7$	$-\sqrt{6} \approx -2.4$	-3	$-3\sqrt{2} \approx -4.2$	$-2\sqrt{6} \approx -4.9$

d. Sketch a graph of f.

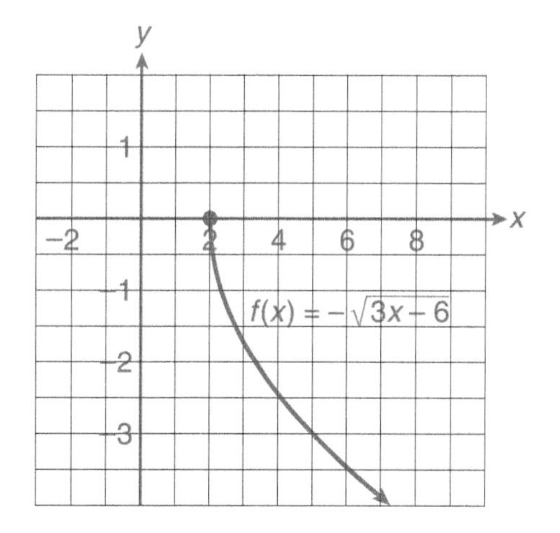

e. Use your graphing calculator to sketch a graph of f. Your screens should appear as follows:

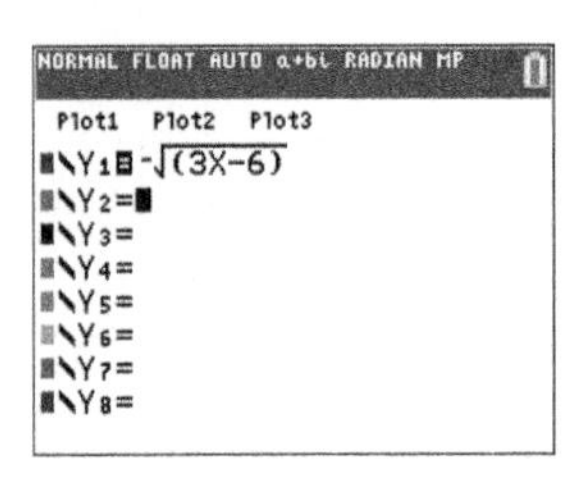

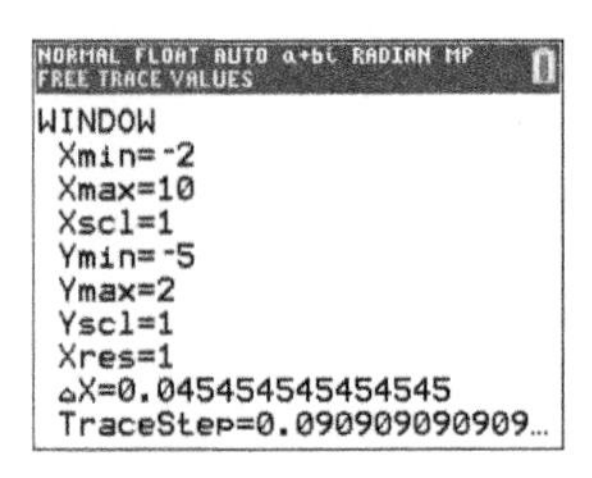

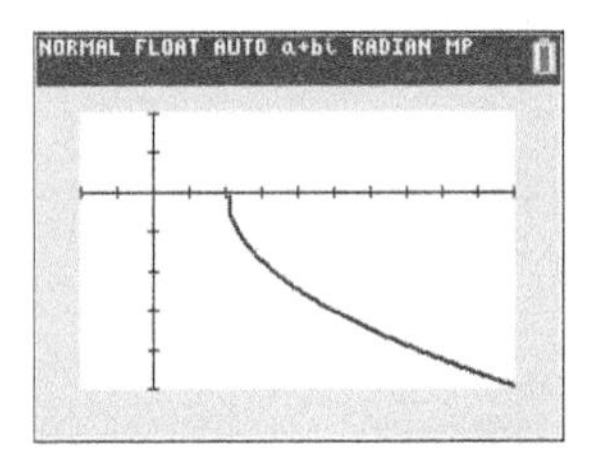

f. Determine the range of f.

The range is all real numbers less than or equal to 0.

g. Does the graph of f have a maximum value? If so, what is it?

Yes, the maximum value is 0.

10. Consider the function defined by $g(x) = \sqrt{x^2 + 4}$.

a. Determine the domain of g.

The domain is all real numbers because $x^2 + 4$ is always greater than 0.

b. Complete the following table. If necessary, approximate $g(x)$ to the nearest hundredth.

x	−4	−2	0	2	4	6
$g(x)$	4.47	2.83	2	2.83	4.47	6.32

c. Sketch a graph of g.

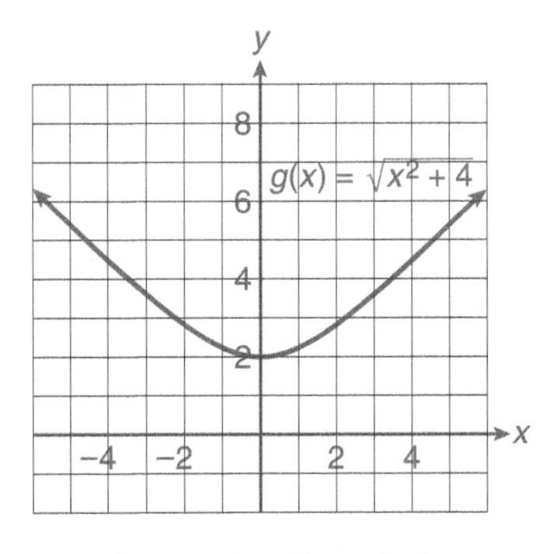

10. d.

NORMAL FLOAT AUTO a+bi RADIAN MP
FREE TRACE VALUES
WINDOW
Xmin=-6
Xmax=6
Xscl=1
Ymin=-2
Ymax=10
Yscl=1
Xres=1
ΔX=0.045454545454545
TraceStep=0.090909090909...

NORMAL FLOAT AUTO a+bi RADIAN MP

d. Verify your graph in part c using your graphing calculator.

e. Determine the range of g.

The range is all real numbers greater than or equal to 2.

f. Does the graph of g have a maximum or minimum value? If so, what is it?

There is no maximum value. 2 is the minimum value.

11. a. Complete the following table for $f(x) = \sqrt{x^2} + \sqrt{4}$:

x	−4	−2	0	2	4	6
$f(x)$	6	4	2	4	6	8

b. How do the outputs in the table in part a compare with the outputs for $g(x) = \sqrt{x^2 + 4}$ in Problem 10b?

They are different.

11. c.

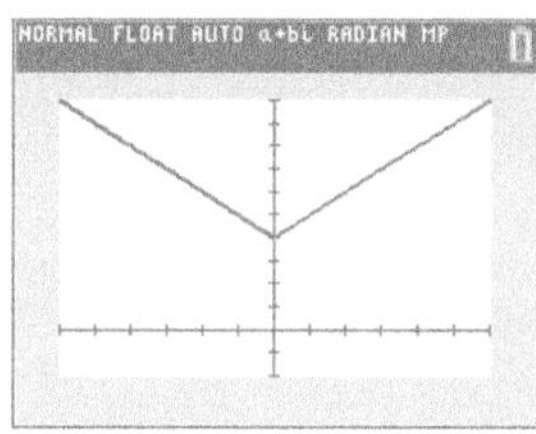

c. Use your graphing calculator to graph $f(x) = \sqrt{x^2} + \sqrt{4}$.

d. How does the graph of $f(x) = \sqrt{x^2} + \sqrt{4}$ compare with the graph of $g(x) = \sqrt{x^2 + 4}$ in Problem 10c?

They are different.

e. Is the expression $\sqrt{x^2 + 4}$ equivalent to $\sqrt{x^2} + \sqrt{4}$? Explain.

No, the graphs and tables indicate that there are different outputs for the same input.

Problem 11 demonstrates the following important fact about radicals:

$$\sqrt{a + b} \neq \sqrt{a} + \sqrt{b}$$

Recall from Chapter 2 that the expression $\sqrt{x}$ can also be written as $x^{1/2}$. The fractional exponent means that you are taking the positive square root of x. Therefore, the expression $\sqrt{3x - 2}$ can also be written as $(3x - 2)^{1/2}$.

12. Sketch the graph of $y = (3x - 2)^{1/2}$. Use your graphing calculator to verify that this is the same as $y = \sqrt{3x - 2}$. What is the domain?

The domain is $x \geq \frac{2}{3}$.

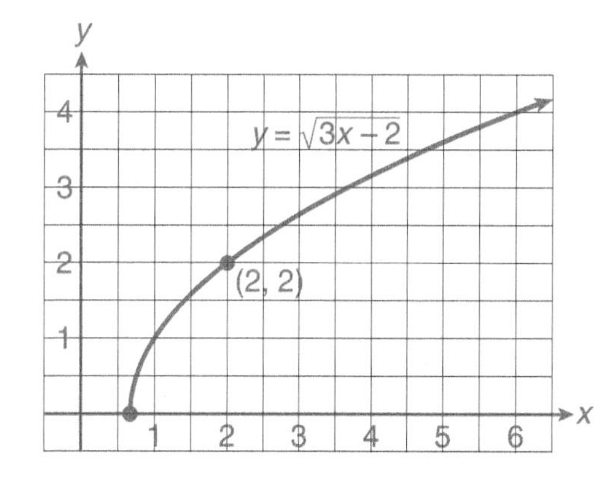

Space and Radicals

13. It is not unreasonable to imagine that someday travel in space will be a common occurrence. According to Einstein's theory of relativity, time would pass more quickly on Earth than it would for someone who is traveling in a spacecraft at a velocity close to the speed of light. As a result, a person on Earth would age more rapidly than a space traveler. The formula

$$A = F\sqrt{1 - \frac{v^2}{c^2}}$$

models the relationship between the aging rate, A, of an astronaut and the aging rate, F, of a person on Earth. The variable v represents the astronaut's velocity in miles per second, and c represents the speed of light (approximately 186,000 miles per second).

a. Suppose you are on a spaceship that is traveling at 80% of the speed of light—that is, $v = 0.8c$. What is your aging rate compared with that of a person on Earth?

$$A = F\sqrt{1 - \frac{0.64c^2}{c^2}} = F\sqrt{1 - 0.64} = F\sqrt{0.36} = 0.6F$$

My aging rate is 60% of the aging rate on Earth.

b. If you travel at 80% of the speed of light for 1 year (as you perceive it), approximately how much time has passed for a person on Earth?

1 year $= 0.6F$, so $F = \frac{1}{0.6}$ or $1\frac{2}{3}$ years or 20 months.

c. Suppose you are traveling at a velocity very close to the speed of light. Substitute c (speed of light) for v in the formula and simplify. Interpret your results.

$$A = F\sqrt{1 - \frac{c^2}{c^2}} = F\sqrt{1 - 1} = 0$$

This says that if you travel at the speed of light, you never age.

14. Escape velocity is the minimum speed that an object must attain to escape a planet's pull of gravity. Escape velocity, V, is modeled by the formula

$$V = \sqrt{\frac{2Gm}{r}},$$

where G is the universal gravitational constant,
m is the mass of the planet, and
r is the radius of the planet.

If Earth has mass 5.97×10^{24} kilograms and radius 6.37×10^6 meters, then determine the escape velocity for Earth. Round your answer to the nearest whole number. Use $G = 6.67 \times 10^{-11}\ \text{m}^3/\text{kg} \cdot \text{sec.}^2$

$$V = \sqrt{\frac{2 \cdot 6.67 \cdot 10^{-11} \cdot 5.97 \cdot 10^{24}}{6.37 \cdot 10^6}} \approx \sqrt{12.50 \cdot 10^7} = \sqrt{1.25 \cdot 10^8}$$

$$\approx 1.118 \cdot 10^4 = 11{,}180 \text{ m/sec.}$$

SUMMARY Activity 5.7

Square Root Notation and Terminology

1. $\sqrt[2]{n}$, or simply $\sqrt{n}$, represents the square root of a nonnegative number n. The 2 is called the **index**. In general, when you are working with square roots, the 2 is omitted.

2. The symbol $\sqrt{\ }$ is called the **radical sign**. The expression under the radical is called the **radicand**.

3. $\sqrt{n} \geq 0$, where $n \geq 0$

4. $\sqrt{a \cdot b} = \sqrt{a} \cdot \sqrt{b}$, where $a \geq 0, b \geq 0$

5. $\sqrt{\frac{a}{b}} = \frac{\sqrt{a}}{\sqrt{b}}, b \neq 0$

6. $\sqrt{a + b} \neq \sqrt{a} + \sqrt{b}$

Properties of Radical Functions

1. The function defined by $y = \sqrt{g(x)}$ has domain all real numbers x with $g(x) \geq 0$.
2. The function defined by $y = -\sqrt{g(x)}$ has domain all real numbers x with $g(x) \geq 0$. The graph of $y = -\sqrt{g(x)}$ is the reflection of the graph of $y = \sqrt{g(x)}$ about the x-axis.

EXERCISES Activity 5.7

1. Using your calculator, determine the value of each number to the nearest hundredth, if necessary.

   a. $\sqrt{30}$ — 5.48
   b. $6^{1/2}$ — 2.45
   c. $(\sqrt{13})^4$ — 169
   d. $(9^{1/2})^3$ — 27

2. Determine the domain of each function.

   a. $f(x) = \sqrt{x-5}$
   $x \geq 5$

   b. $g(x) = \sqrt{3x+2}$
   $x \geq -\frac{2}{3}$

   c. $h(x) = \sqrt{6-2x}$
   $x \leq 3$

   d. $R(x) = -\sqrt{2x}$
   $x \geq 0$

3. The following table gives the number of undergraduate students receiving federal Stafford loans over a period of 7 consecutive years:

Year, x	0	1	2	3	4	5	6
Number of Undergrads Receiving Stafford Loans, $N(x)$ (millions)	4.2	4.5	5.0	5.5	5.9	6.0	6.1

This data can be modeled by the function

$N(x) = 1.222\sqrt{x + 0.24} + 3.442.$

a. Determine the N-intercept. What does this intercept represent in this situation?

(0, 4.04). In the first year the data was collected, approximately 4.04 million undergraduate students received Stafford loans.

b. Complete the following table using the given equation. How well does the equation represent the actual data?

x	0	1	2	3	4	5	6
$N(x)$	4.04	4.80	5.27	5.64	5.96	6.24	6.49

The equation matches reasonably well, although it is not perfect.

c. Sketch a graph of the student loan function represented by $N(x) = 1.222\sqrt{x + 0.24} + 3.442$.

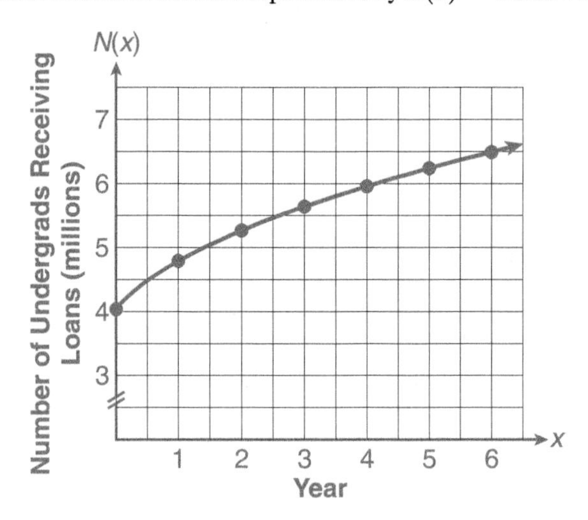

d. Use the model to predict the number of undergraduates who will receive Stafford loans in the year $x = 12$.

$x = 12$, so $N(12) = 1.222\sqrt{12 + 0.24} + 3.442 \approx 7.72$ million.

4. Describe how to obtain the graph of the second function from the graph of the first.

a. $g(x) = \sqrt{x}$, $h(x) = \sqrt{x} + 1$

The graph of h is obtained by shifting the graph of g up 1 unit.

b. $f(x) = \sqrt{x}$, $g(x) = -\sqrt{x}$

The graph of g is obtained from the graph of f by reflecting the graph of f through the x-axis.

5. a. Sketch a graph of $f(x) = x^2$ and $g(x) = \sqrt{x}$ on the same coordinate system. Use the graphs to answer parts b and c.

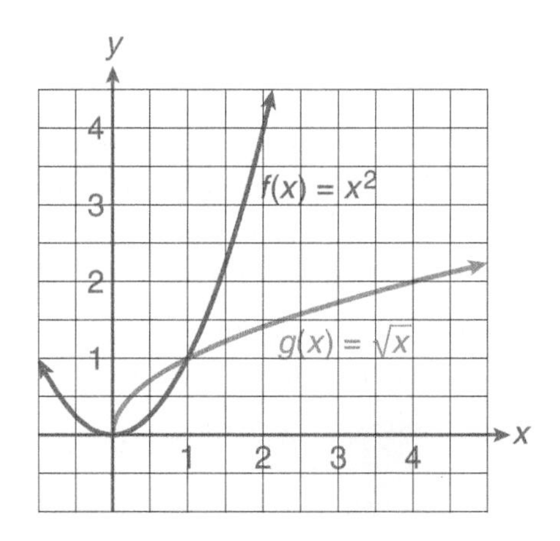

b. Is $x^2 > \sqrt{x}$ for $0 < x < 1$? Explain.

No, the graph of f is below the graph of g for $0 < x < 1$.

c. Is $x^2 > \sqrt{x}$ for $x > 1$? Explain.

Yes, the graph of f is above the graph of g for $x > 1$.

6. Which of the following functions increases more rapidly for $x > 4$: $f(x) = \sqrt{x}$ or $g(x) = \ln(x)$?

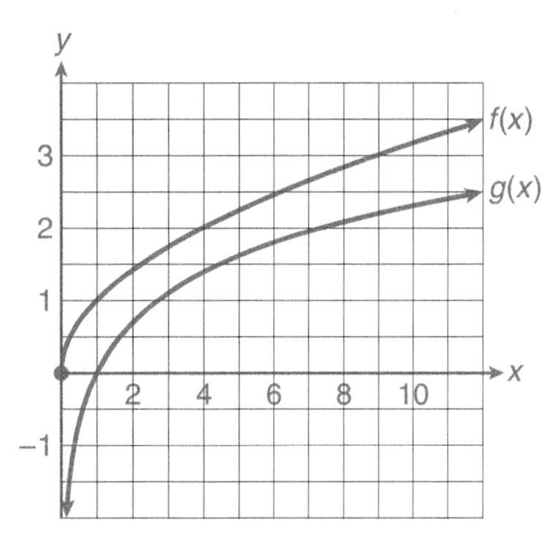

From viewing the graph, it appears that *f* grows more rapidly than *g*.

7. For each of the given functions, do the following:

i. Determine the domain.

ii. Determine the *x*- and *y*-intercepts.

iii. Sketch a graph.

a. $h(x) = \sqrt{2x + 3}$

i. The domain is all real numbers such that $2x + 3 \geq 0$, or $x \geq \frac{-3}{2}$.

ii. The *x*-intercept is $\left(\frac{-3}{2}, 0\right)$. The *y*-intercept is $(0, \sqrt{3})$.

iii.

$h(x) = \sqrt{2x + 3}$

b. $f(x) = -\sqrt{4x + 8}$

i. The domain is all real numbers such that $4x + 8 \geq 0$, or $x \geq -2$.

ii. The *x*-intercept is $(-2, 0)$; the *y*-intercept is $(0, -\sqrt{8})$.

iii.

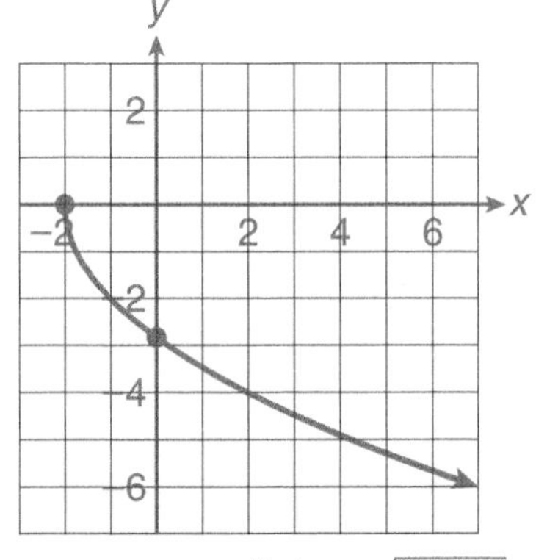

$f(x) = -\sqrt{4x + 8}$

c. $g(x) = \sqrt{5 - x}$

i. The domain is all real numbers such that $5 - x \geq 0$, or $x \leq 5$.

ii. The x-intercept is (5, 0). The y-intercept is $(0, \sqrt{5})$.

iii.

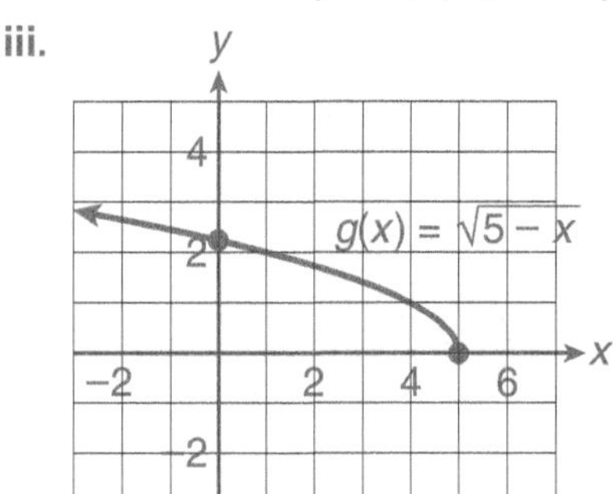

8. The surface area, S, of a cone is modeled by the formula

$$S = \pi r\sqrt{r^2 + h^2},$$

where r is the radius of the base and h is the height.

An umbrella is in the shape of a cone having radius 4 feet and height 2 feet. Determine the amount of material needed to make the umbrella.

$S = \pi(4)\sqrt{4^2 + 2^2} = 4\pi\sqrt{20} \approx 56.20$ sq. ft.

9. A shipping carton in the shape of a rectangular box has dimensions 12 inches × 24 inches × 17 inches. The diagonal, d, of a rectangular box is modeled by

$$d = \sqrt{w^2 + l^2 + h^2},$$

where w is the width, l is the length, and h is the height.

Will an umbrella measuring 34 inches long fit in the carton?

$d = \sqrt{12^2 + 24^2 + 17^2} = \sqrt{1009} \approx 31.8$ in. It will not fit.

10. Law enforcement officers investigating a car accident often use the formula $s = \sqrt{30fl}$ to estimate a car's speed, s, in miles per hour based on the length, l (in feet), of the skid marks. The f in the formula represents the road condition at the time of the accident.

a. On dry pavement, the f-value is 0.85. Write a function equation for speed, s, as a function of the length of the skid marks, l, on dry pavement.

$s = \sqrt{30(0.85)l} = \sqrt{25.5l}$

b. Estimate the speed of a car if the skid marks on dry pavement are 90 feet long.

$s = \sqrt{25.5(90)} \approx 47.9$ mph

c. What is the practical domain for the function in part a?

(Answers will vary.) 0 feet $\leq l \leq$ 300 feet is possible.

d. Graph this function using your graphing calculator.

10. d.

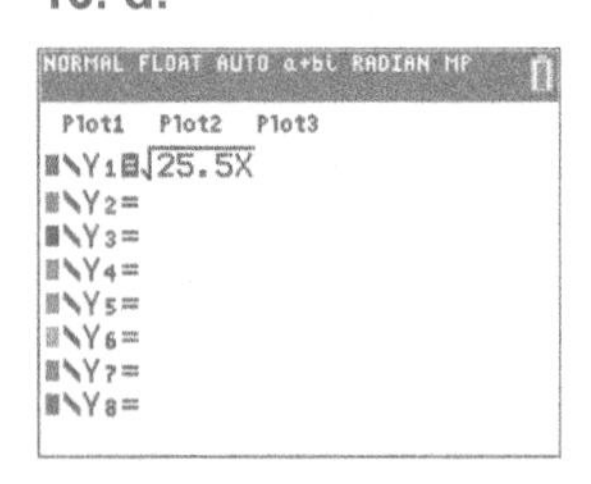

NORMAL FLOAT AUTO a+bi RADIAN MP
WINDOW
Xmin=0
Xmax=300
Xscl=50
Ymin=0
Ymax=90
Yscl=10
Xres=1
ΔX=1.1363636363636
TraceStep=2.2727272727272

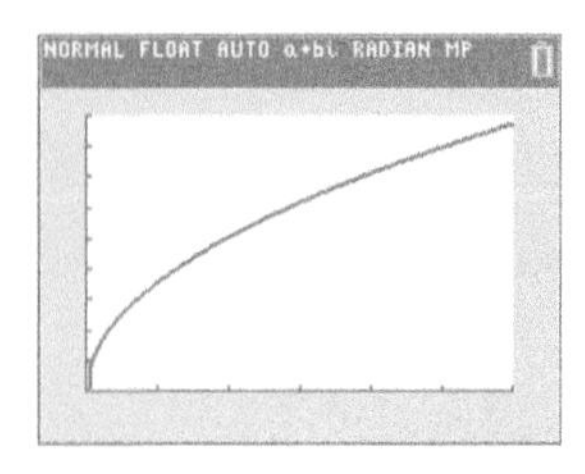

e. Use the graph to determine the length of the skid marks on dry pavement if the car was traveling 70 miles per hour when the brakes were applied.

The length of the skid marks is approximately 192 feet.

10. e.

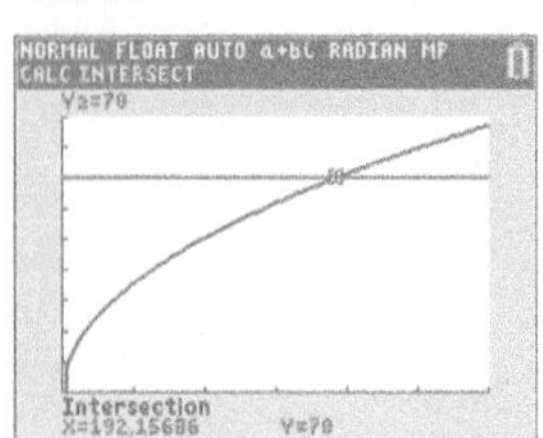

ACTIVITY 5.8

Falling Objects

Solving Equations Involving a Radical Expression

OBJECTIVE

Solve an equation involving a radical expression using graphical and algebraic approaches.

If an object is dropped from a tall building, the time, t, in seconds, it takes for the object to strike the ground is modeled by

$$t = \frac{\sqrt{d}}{4} = \frac{1}{4}\sqrt{d},$$

where the input, d, is the distance traveled in feet. The time it takes the object to hit the ground varies directly to the square root of the distance traveled. The number $\frac{1}{4}$ or 0.25 is the constant of proportionality or constant of variation.

1. a. How long will it take an object to fall from the top of the Willis Tower in Chicago, a distance of 1450 feet? Round to the nearest hundredth of a second.

$t = \frac{\sqrt{1450}}{4} \approx 9.52$ sec.

b. Complete the following table. Round to the nearest hundredth of a second.

d (ft.)	0	100	200	300	500	750	1000
t (sec.)	0	2.5	3.54	4.33	5.59	6.85	7.91

c. Sketch a graph of the given function. Use the table in part b to determine an appropriate scale.

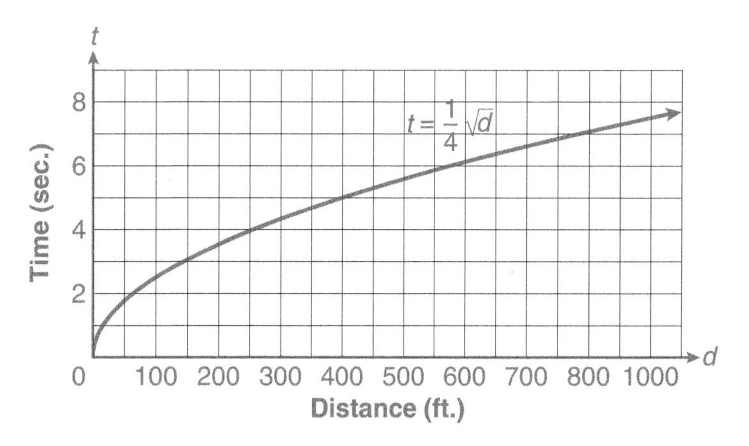

d. How tall must a building be for an object to take 6 seconds to fall to the ground? Use the graph from part c to approximate your answer.

(Answers will vary.) The building must be about 575 feet tall.

Solving Equations Involving Radical Expressions

Suppose you are interested in determining the value of d for many different values of t. In such a situation, the process could be simplified if you had a rule for d as a function of t.

The equation $t = \frac{\sqrt{d}}{4}$ gives t as a function of d. You need to solve this equation for d.

DEFINITION

An equation in which at least one side contains a radical with a variable in the radicand is called a **radical equation**.

EXAMPLE 1 ***Examples of radical equations include*** $t = \frac{\sqrt{d}}{4}$, $\sqrt{2x+1} = 5$, ***and*** $2\sqrt{3x} = \sqrt{5x-7}$.

Solving an equation algebraically when the variable appears under a radical involves using the following property of equations.

If a and b are two quantities such that $a = b$, then $a^n = b^n$, where n is a positive integer.

EXAMPLE 2 ***If*** $t = \sqrt{s}$, ***then apply the preceding property by squaring both sides of the equation.***

$t^2 = (\sqrt{s})^2$ Rewrite $\sqrt{s}$ as $s^{1/2}$.

$t^2 = (s^{1/2})^2$ Apply the property of exponents $(a^m)^n = a^{mn}$.

$t^2 = s^{1/2 \cdot 2}$ Simplify.

$t^2 = s^1$

Therefore, if $t = \sqrt{s}$, then $t^2 = s$.

2. a. Now solve the equation $t = \frac{\sqrt{d}}{4}$ for d by first squaring both sides of the equation.

$t^2 = \frac{d}{16}$

$d = 16t^2$

b. Using the new formula from part a, determine how tall a building must be for an object to take 6 seconds to fall to the ground.

$d = 16(6)^2 = 16(36) = 576$ ft.

c. How does your answer compare with the result in Problem 1d?

The answers are close.

d. You could also answer part b by solving the equation $6 = \frac{\sqrt{d}}{4}$. Solve the equation.

How does your answer compare with the result in part b?

$6 \cdot 4 = \sqrt{d}$

$24 = \sqrt{d}$

$576 = d$. The answers are the same.

The example that follows demonstrates a general algebraic procedure for solving radical equations.

LC LEARNING CATALYTICS

Solve $\sqrt{3x-2} - 6 = -2$.

a. $x = 6$
b. $x = 18$
c. $x = -6$
d. no solution

EXAMPLE 3 ***Solve for x:*** $2\sqrt{3x} - \sqrt{5x+7} = 0$.

SOLUTION

Step 1. If the equation involves more than one radical term, isolate one radical term on one side of the equation.

$$2\sqrt{3x} - \sqrt{5x+7} = 0$$
$$+\sqrt{5x+7} \quad +\sqrt{5x+7}$$
$$2\sqrt{3x} = \sqrt{5x+7}$$

Step 2. Square both sides of the equation.

$$(2\sqrt{3x})^2 = (\sqrt{5x + 7})^2$$

$$4 \cdot 3x = 5x + 7$$

Step 3. If a radical remains, repeat steps 1 and 2. Solve the resulting equation.

$$4 \cdot 3x = 5x + 7$$

$$12x = 5x + 7$$

$$7x = 7$$

$$x = 1$$

Step 4. Check all solutions in the original equation.

$$2\sqrt{3(1)} - \sqrt{5(1) + 7} = 2\sqrt{3} - \sqrt{12}$$

$$= 2\sqrt{3} - \sqrt{4 \cdot 3} = 2\sqrt{3} - 2\sqrt{3} = 0$$

You can also check your answer by solving the equation graphically.

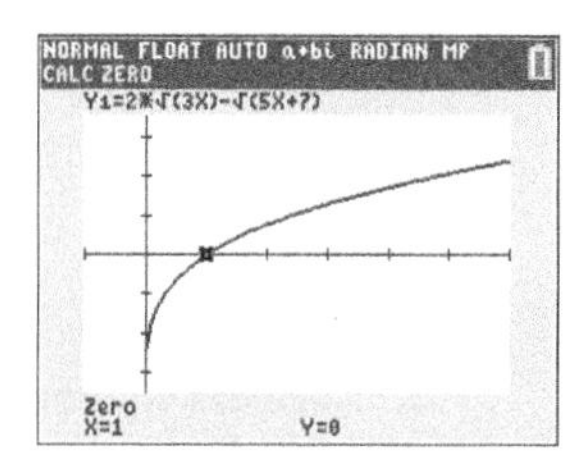

3. Suppose two different objects are dropped: a marble and a large beach ball. Because of air resistance, the beach ball will take longer than the marble to fall the same distance. Assume that the marble falls according to $t = 0.25\sqrt{d}$, as in Problem 1. The time for the beach ball to hit the ground is modeled by $t = k\sqrt{d}$, where the positive constant, k, is determined by experiment.

The beach ball is dropped from a height of 250 feet, and it takes 4.11 seconds to hit the ground. Determine the constant, k, accurate to the hundredths place. Remember that $(ab)^2 = a^2b^2$.

$$4.11 = k\sqrt{250}$$

$$16.8921 = 250k^2$$

$$0.0676 \approx k^2$$

$$k \approx 0.26$$

4. Now suppose the beach ball in Problem 3 is dropped from a height 50 feet lower than the marble. Then $t = 0.26\sqrt{d - 50}$ is the time for the beach ball to drop $d - 50$ feet, where d is the height the marble falls.

a. Write an equation that can be used to determine from what height the marble must be dropped so the beach ball and marble will hit the ground at the same time.

$$0.25\sqrt{d} = 0.26\sqrt{d - 50}$$

b. Solve this equation using an algebraic approach.

$$0.25\sqrt{d} = 0.26\sqrt{d - 50}$$

$$0.0625d = 0.0676(d - 50)$$

$$0 = 0.0051d - 3.38$$

$$d \approx 662.75 \text{ ft.}$$

c. Verify your solution in part b using your graphing calculator.

4. c.

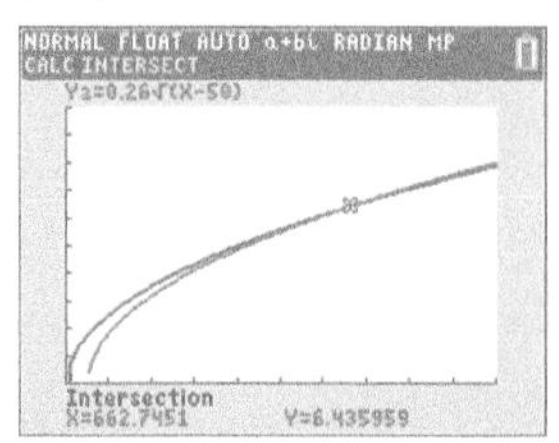

5. Consider the following algebraic solution of the equation $\sqrt{x + 3} + 5 = 0$:

$$\sqrt{x+3} + 5 = 0$$
$$\sqrt{x+3} = -5$$
$$(\sqrt{x+3})^2 = (-5)^2$$
$$x + 3 = 25$$
$$x = 22$$

a. It appears that $x = 22$ is a solution to the given equation. Check the solution by substituting 22 for x in the original equation. Does it check?

$$\sqrt{22 + 3} + 5 = 0$$
$$\sqrt{25} + 5 = 0$$
$$5 + 5 = 0$$
$$10 = 0.$$ No, it does not check.

b. What happened in the solution process to cause an **extraneous solution** (an apparent solution that does not check) to appear?

The second line of the solution has a positive radical on the left and a negative on the right. These cannot be equal. However, in the next line, squaring both sides produces two positive expressions.

5. c.

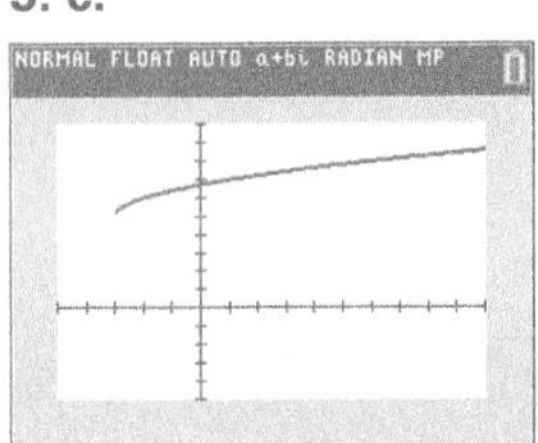

c. Does the equation $\sqrt{x + 3} + 5 = 0$ have a solution? Include a graph to help support your answer.

No, there is no solution. The graph never touches the x-axis.

6. In Exercise 3 on page 559, you were provided the following table giving the number of undergraduate students receiving federal Stafford loans over a period of 7 consecutive years:

Year, x	0	1	2	3	4	5	6
$N(x)$, Number of Undergraduates Receiving Stafford Loans (in millions)	4.2	4.5	5.0	5.5	5.9	6.0	6.1

This data was modeled by the function

$$N(x) = 1.222\sqrt{x + 0.24} + 3.442.$$

Determine the year this model would predict that the number of undergraduate students receiving Stafford loans will be approximately 8 million.

$$8 = 1.222\sqrt{x + 0.24} + 3.442$$
$$4.558 = 1.222\sqrt{x + 0.24}$$
$$\frac{4.558}{1.222} = \sqrt{x + 0.24}$$
$$\left(\frac{4.558}{1.222}\right)^2 = x + 0.24$$
$$13.7 \approx x$$

Stafford loans would be received by approximately 8 million students in year 14.

SUMMARY Activity 5.8

1. If a and b are two quantities such that $a = b$, then $a^n = b^n$, where n is a positive integer.
2. To solve an equation involving one radical expression,
   a. isolate the radical term on one side of the equation.
   b. square both sides of the equation.
   c. solve the resulting equation.
   d. check all solutions in the original equation.
3. To solve an equation involving more than one radical expression,
   a. if the equation involves more than one radical term, isolate one radical term on one side of the equation.
   b. square both sides of the equation.
   c. if a radical remains, repeat steps a and b solve the resulting equation.
   d. check all solutions in the original equation.
4. When you raise both sides of an equation to an even power, it is possible to introduce **extraneous solutions** into the process. These are values of the variable that appear by the process to be solutions but do not make the original equation true. It is important to check all potential solutions to determine whether or not they are real solutions or extraneous solutions.

EXERCISES Activity 5.8

1. Solve the following equations, or determine which does not have a solution. Explain your reasoning, or verify your solution.

   a. $\sqrt{2x+1} = 3$

   $2x + 1 = 9$

   $2x = 8$

   $x = 4$

   $\sqrt{2 \cdot 4 + 1} = 3$

   $\sqrt{9} = 3$

   It checks.

   b. $\sqrt{x+1} + 5 = 1$

   $\sqrt{x+1} = -4$

   This can't happen; a positive radical can't equal -4.

   There is no solution.

   Equation b has no solution. The left side of equation b will always be greater than 1, so no solution is possible.

   c. $\sqrt{2-x} = -x$

   $2 - x = x^2$

   $0 = x^2 + x - 2$

   $0 = (x + 2)(x - 1)$

   $x = -2$ or $x = 1$

   $\sqrt{2-(-2)} = -(-2)$

   $\sqrt{4} = 2$

   $x = -2$ checks.

   $\sqrt{2-(1)} = -(1)$

   $\sqrt{1} = -1$

   $x = 1$ does not check.

Exercise numbers appearing in color are answered in the Selected Answers appendix.

2. Solve each of the following equations algebraically. Then verify your answers graphically.

a. $\sqrt{x} = 2.5$

$x = 6.25$

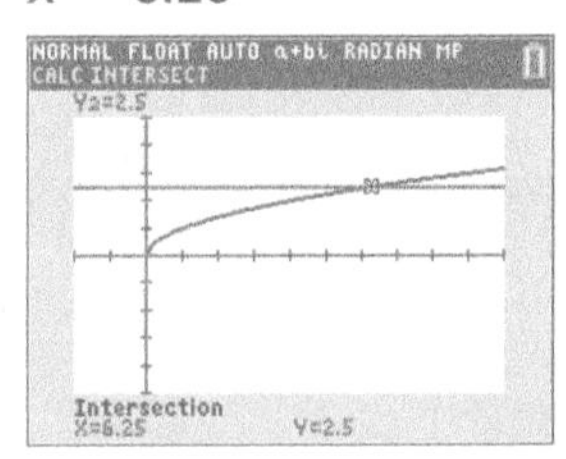

b. $\sqrt{x} - 3 = 0$

$x = 9$

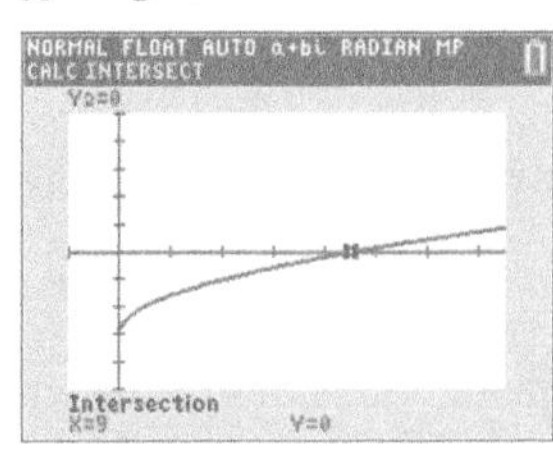

c. $\sqrt{2x} = 14$

$x = 98$

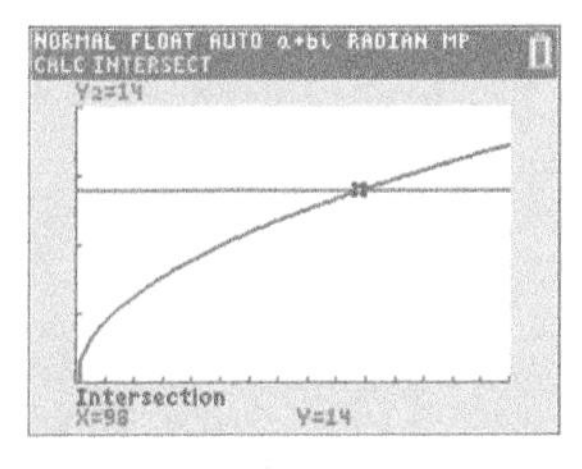

d. $3\sqrt{x} = 243$

$x = 6561$

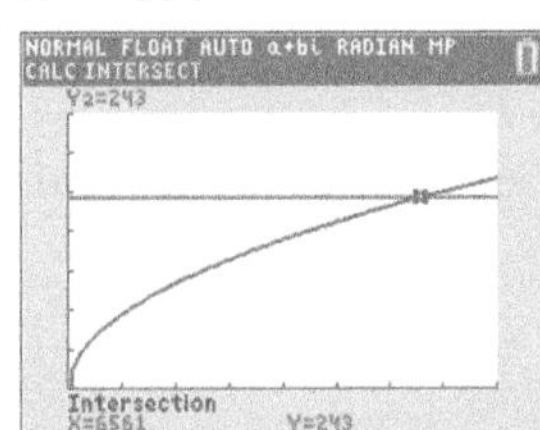

e. $4 - 5\sqrt{3x} = 1$

$x = 0.12$

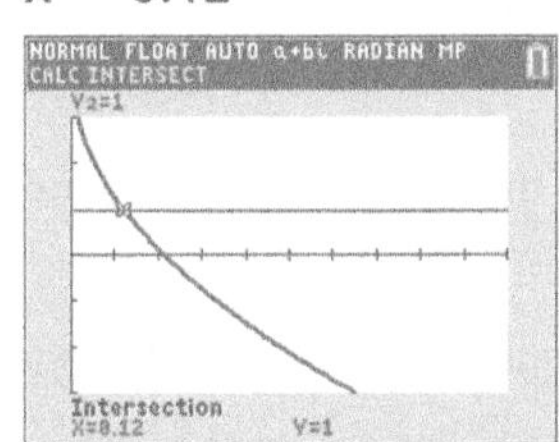

f. $\sqrt{x + 1} = 9$

$x = 80$

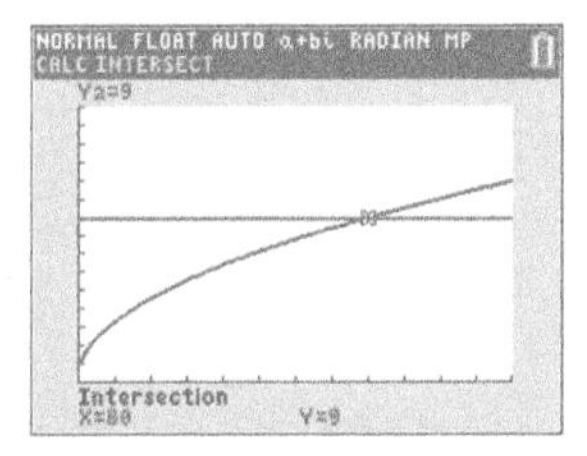

3. Solve each of the following equations algebraically and by graphing. Be aware of any extraneous roots.

a. $\sqrt{x + 5} = 1$

$x + 5 = 1$

$x = -4$

$\sqrt{-4 + 5} = 1$

$x = -4$ checks.

3. a.

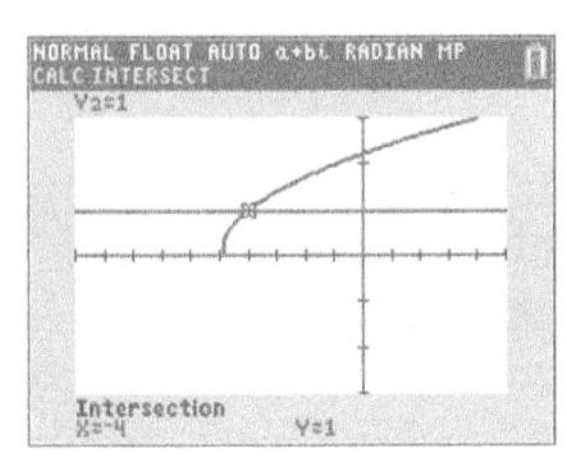

b. $10\sqrt{x + 2} = 20$

$\sqrt{x + 2} = 2$

$x + 2 = 4$

$x = 2$

$10\sqrt{2 + 2} = 20$

$x = 2$ checks.

3. b.

c. $\sqrt{5 - x} = x + 1$

$5 - x = x^2 + 2x + 1$

$0 = x^2 + 3x - 4$

$0 = (x + 4)(x - 1)$

$x = -4$ or $x = 1$

$\sqrt{5 - (-4)} = -4 + 1$

$\sqrt{9} = -3$, $x = -4$ does not check.

$\sqrt{5 - 1} = 1 + 1$

$\sqrt{4} = 2$ $x = 1$ checks.

3. c.

4. Solve algebraically and graphically: $\sqrt{1.4x + 3.2} = \sqrt{3.8x - 1}$.

$1.4x + 3.2 = 3.8x - 1$

$4.2 = 2.4x$

$1.75 = x$

4.

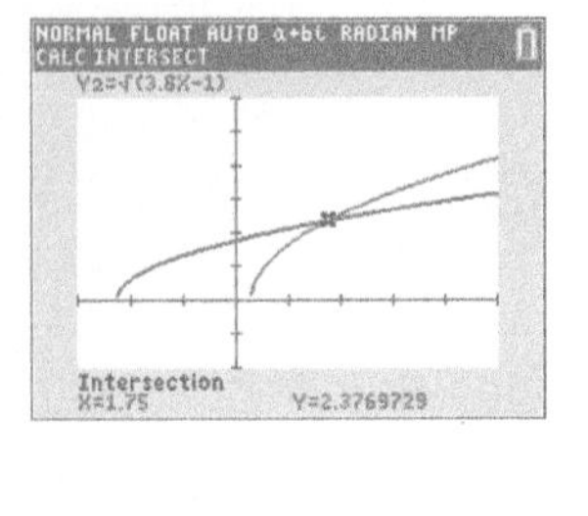

5. The time, t, in seconds, that it takes for a pendulum to complete one complete period (to swing back and forth one time) is modeled by

$$t = 2\pi\sqrt{\frac{L}{32}},$$

where L is the length of the pendulum, in feet (see diagram).

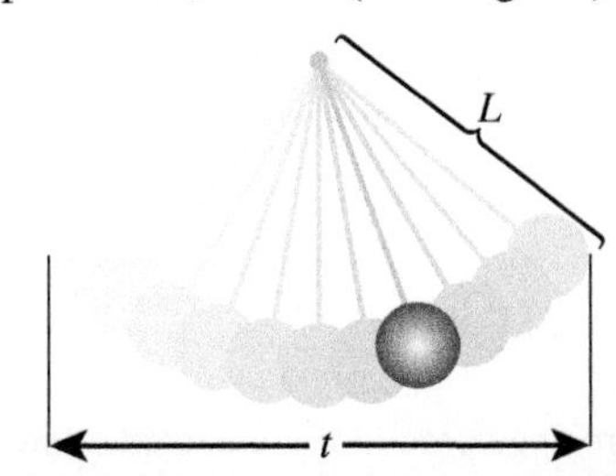

How long is the pendulum of a clock with a period of 1.95 seconds?

$$1.95 = 2\pi\sqrt{\frac{L}{32}}$$

$$\frac{1.95}{2\pi} = \sqrt{\frac{L}{32}}$$

$$\left(\frac{1.95}{2\pi}\right)^2 = \frac{L}{32}$$

$$L = 32\left(\frac{1.95}{2\pi}\right)^2$$

$$L \approx 3.08 \text{ ft.}$$

6. In a certain population, there are 28,520 births on a particular day. The number, N, of these people surviving to age x can be modeled by the function $N = 2850\sqrt{100 - x}$.

a. According to this model, how many of the 28,520 babies will survive to age 5?

$$N = 2850\sqrt{100 - 5} = 2850\sqrt{95} \approx 27{,}778$$

b. What is the practical domain of this function?

Because x and $100 - x$ must be greater than or equal to zero, the practical domain is integers from 0 to 100.

c. When only 5000 of this group are still alive, how old do you expect them to be?

$$5000 = 2850\sqrt{100 - x}$$

$$\frac{5000}{2850} = \sqrt{100 - x}$$

$$\left(\frac{5000}{2850}\right)^2 = 100 - x$$

$$x = 100 - \left(\frac{5000}{2850}\right)^2 \approx 97 \text{ years old}$$

d. After how many years will half of the original population of 28,520 people remain alive?

Half of 28,520 is 14,260.

$$14{,}260 = 2850\sqrt{100 - x}$$

$$\frac{14{,}260}{2850} = \sqrt{100 - x}$$

$$\left(\frac{14{,}260}{2850}\right)^2 = 100 - x$$

$$x = 100 - \left(\frac{14{,}260}{2850}\right)^2 \approx 75$$

In 75 years, half of the group will be alive.

7. a. A pressure gauge on a bridge indicates a wind pressure, P, of 10 pounds per square foot. What is the velocity, V, of the wind if

$$V = \sqrt{\frac{1000P}{3}},$$

where velocity is measured in miles per hour?

$$V = \sqrt{\frac{1000(10)}{3}} \approx 57.7 \text{ mph}$$

b. What is the wind pressure if the wind is blowing 70 miles per hour?

$$\sqrt{\frac{1000}{3}P} = 70$$

$$\frac{1000}{3}P = 4900$$

$$P = \frac{14{,}700}{1000} = 14.7 \text{ pounds per square foot}$$

8. Artificial gravity can be created in a space station by revolving the station. The number of revolutions required can be modeled by

$$N = \frac{1}{2\pi}\sqrt{\frac{a}{r}},$$

where N is measured in revolutions per second, a is the artificial gravity produced (measured in meters per second squared), and r is the radius of the space station in meters.

a. To produce an artificial acceleration simulating gravity on Earth, a must equal 9.8 meters per second squared. If the space station must revolve at the rate of one revolution every 5 minutes, what must its radius be? Solve both algebraically and graphically. Be careful; N is measured in revolutions per second.

1 revolution per 5 minutes is 1 revolution per 5(60) seconds, or 1 revolution per 300 seconds.

$$\frac{1}{300} = \frac{1}{2\pi}\sqrt{\frac{9.8}{r}}$$

$$\frac{2\pi}{300} = \sqrt{\frac{9.8}{r}}$$

$$\frac{4\pi^2}{90{,}000} = \frac{9.8}{r}$$

$$4\pi^2 r = 9.8 \cdot 90{,}000$$

$$r = \frac{9.8 \cdot 90{,}000}{4\pi^2} \approx 22{,}341 \text{ m}$$

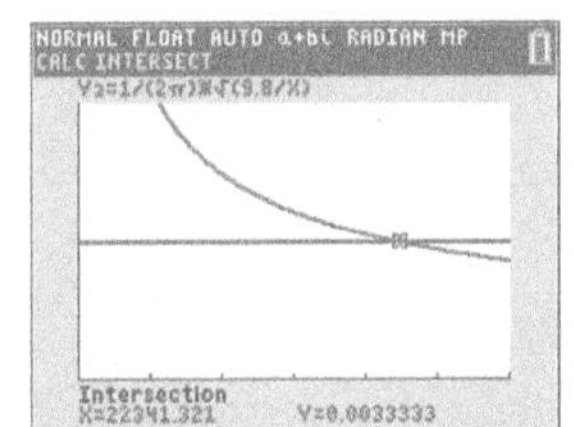

b. Solve the original formula for r.

$$2\pi N = \sqrt{\frac{a}{r}}$$

$$4\pi^2 N^2 = \frac{a}{r}$$

$$r = \frac{a}{4\pi^2 N^2}$$

c. Use the formula in part b to answer part a again. How do the answers compare?

$$r = \frac{9.8}{4\pi^2\left(\frac{1}{300}\right)^2}$$

$$= \frac{9.8 \cdot (300)^2}{4\pi^2}$$

$$\approx 22{,}341$$

The answers are the same.

9. The Masteller formula for calculating the adult body surface area, A, is

$$A = \sqrt{\frac{hw}{3131}},$$

where h is the person's height in inches and w is the adult's weight in pounds. A is the surface area in square meters.

a. Determine the body surface area, A, of an adult who is 70 inches tall and weighs 200 pounds.

$$A = \sqrt{\frac{70 \cdot 200}{3131}} \approx 2.11 \text{ sq. m.}$$

b. Solve the formula for w.

$$A^2 = \frac{hw}{3131}$$

$$3131A^2 = hw$$

$$w = \frac{3131A^2}{h}$$

ACTIVITY 5.9

Propane Tank

More Radical Functions and Their Graphs

OBJECTIVES

1. Determine the domain of a function defined by an equation of the form $y = \sqrt[n]{g(x)}$, where n is a positive integer and $g(x)$ is a polynomial.
2. Graph $y = \sqrt[n]{g(x)}$.
3. Identify the properties of graphs of $y = \sqrt[n]{g(x)}$.
4. Solve radical equations that contain radical expressions with an index other than 2.

A propane tank is in the shape of a sphere. The radius, r, of the spherical tank is modeled by the formula

$$r = \sqrt[3]{\frac{3V}{4\pi}},$$

where V is the volume of the tank.

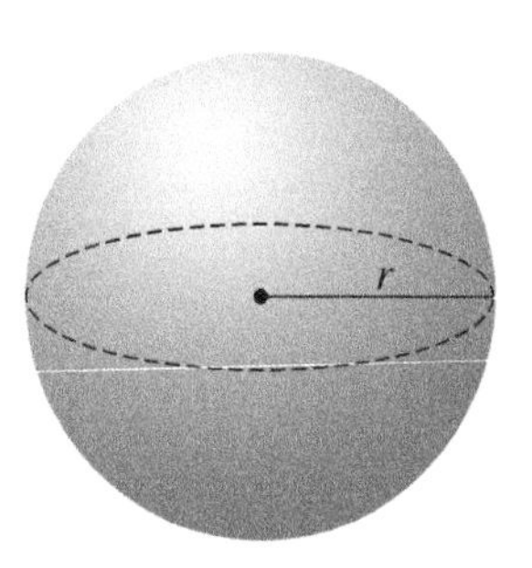

1. a. What is the radius of a propane tank having volume 50 cubic feet (round the answer to the nearest tenth)?

$$r = \sqrt[3]{\frac{3 \cdot 50}{4\pi}} \approx 2.3 \text{ ft.}$$

b. Complete the following table:

V (ft.^{3})	0	5	10	15	20	25	100
r (ft.)	0	1.06	1.34	1.53	1.68	1.81	2.88

c. Sketch a graph of the radius function over its practical domain.

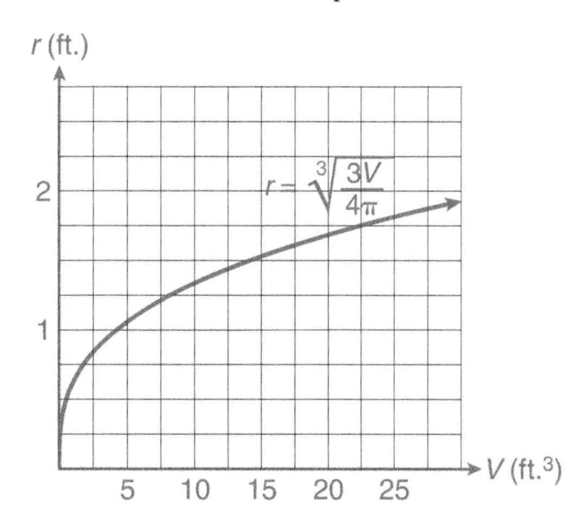

Graphs of $y = \sqrt[n]{g(x)}$, $n = 3, 4, 5$

The investigation of the properties of the radius function defined by $r = \sqrt[3]{\frac{3V}{4\pi}}$ is somewhat limited by the restrictions placed on the variables V and r.

2. a. Consider the cube root function defined by $f(x) = \sqrt[3]{x}$. What is the domain of f?

all real numbers

b. Complete the following table. Round $f(x)$ to nearest hundredth, if necessary.

x	−10	−7	−4	−1	0	1	4	7	10
$f(x)$	−2.15	−1.91	−1.59	−1	0	1	1.59	1.91	2.15

c. Sketch a graph of f. Verify your sketch using your graphing calculator.

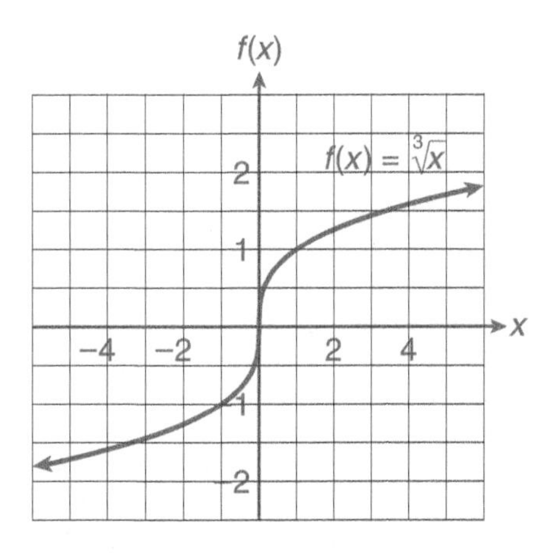

d. Is the function f increasing or decreasing?

increasing

There are two different ways to enter the cube root function on your calculator.

Method 1: Using fractional exponents, enter the cube root of x as x ^ (1/3) in the Y = editor.

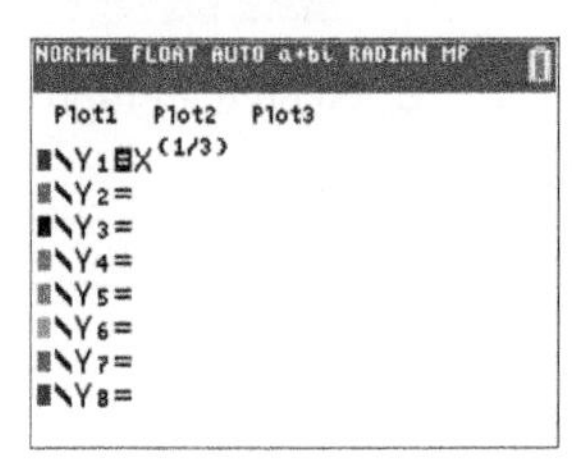

Method 2: Select the Y = editor, highlight the Y_n you want, and then select the MATH menu. Option 4 is the cube root.

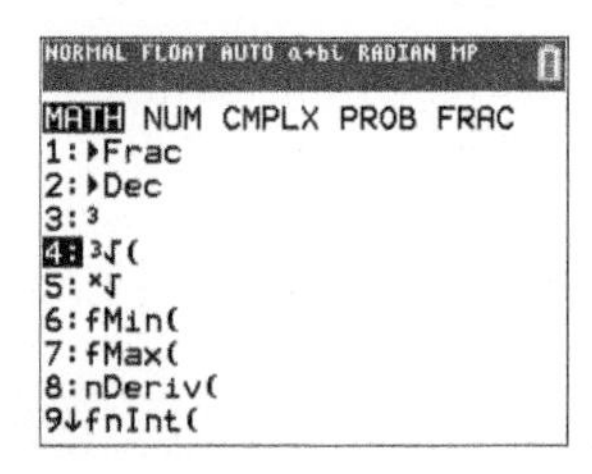

Select option 4, and press (ENTER). Insert the argument x and the right parenthesis to complete the function.

3. Use your graphing calculator to verify that the graphs of $y = x^{1/3}$ and $y = \sqrt[3]{x}$ are identical. Your screens should appear as follows:

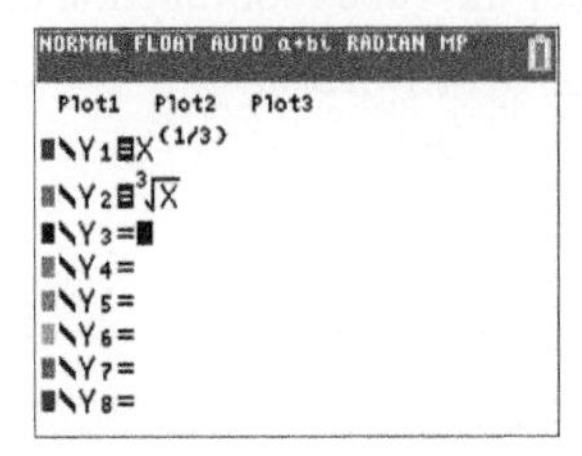

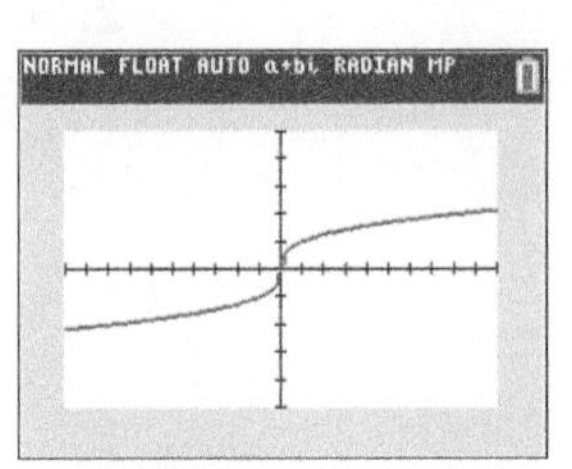

4. Consider functions defined by $f(x) = x^3$ and $g(x) = \sqrt[3]{x}$.

a. Determine the domain of each function.

The domain of f is all real numbers. The domain of g is all real numbers.

b. Complete the following table:

x	−10	−7	−4	−1	0	1	4	7	10
$f(x)$	−1000	−343	−64	−1	0	1	64	343	1000
$g(x)$	−2.15	−1.91	−1.59	−1	0	1	1.59	1.91	2.15

c. Sketch a graph of f and g on the same coordinate system. Verify your sketch using your graphing calculator.

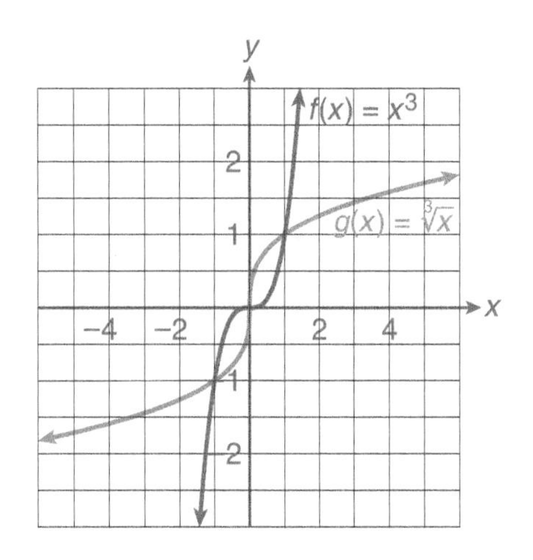

d. Are f and g inverse functions? Explain.

Yes, the graphs of f and g are symmetrical in the line $y = x$.

e. Determine the composition of f and g. That is, determine $f(g(x))$ and $g(f(x))$.

$f(g(x)) = f(\sqrt[3]{x}) = (\sqrt[3]{x})^3 = x$

$g(f(x)) = g(x^3) = \sqrt[3]{x^3} = x$

5. You are given the functions f and g defined by

$$f(x) = \sqrt[4]{x - 1} \quad \text{and} \quad g(x) = \sqrt[5]{x}.$$

a. Complete the following table:

x	−10	−5	−1	0	1	5	10
$f(x)$	undef.	undef.	undef.	undef.	0	1.41	1.73
$g(x)$	−1.58	−1.38	−1	0	1	1.38	1.58

b. Determine the domains of f and g.

The domain of f is $x \geq 1$. The domain of g is all real numbers.

c. Sketch a graph of f and g on the same coordinate axes.

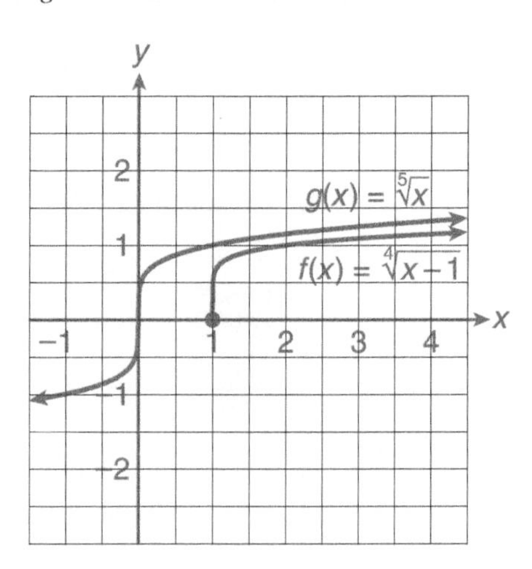

Solving Equations Involving Radical Expressions

Solving an equation such as $\sqrt[3]{x+1} + 5 = 9$ is similar to solving equations involving square roots.

EXAMPLE 1 ***Solve*** $\sqrt[3]{x+1} + 5 = 9$.

SOLUTION

Step 1. Isolate the radical term on one side of the equation.

$$\sqrt[3]{x+1} + 5 = 9$$
$$\sqrt[3]{x+1} = 4$$

Step 2. Raise each side of the equation to the power that matches the index of the radical. In this situation, cube each side and simplify.

$$(\sqrt[3]{x+1})^3 = 4^3$$

Step 3. If a radical remains, repeat steps 1 and 2. Solve the resulting equation.

$$x + 1 = 64$$
$$x = 63$$

Step 4. Check all solutions in the original equation.

$$\sqrt[3]{63+1} + 5 = 9$$
$$\sqrt[3]{64} + 5 = 9$$
$$4 + 5 = 9$$
$$9 = 9$$

You can also verify your solutions graphically.

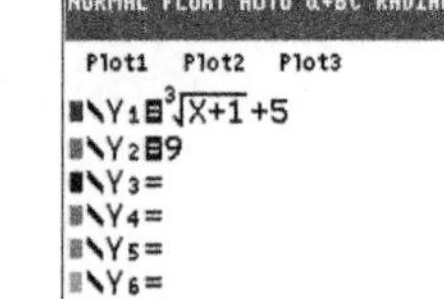

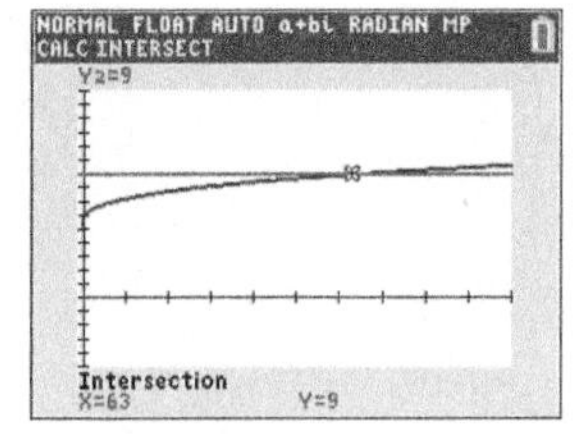

LC LEARNING CATALYTICS

Solve $\sqrt[3]{5x+4} = 3$.

a. $x = -\frac{4}{5}$

b. $x = \frac{23}{5}$

c. $x = \frac{5}{23}$

d. no solution

6. a. Returning to the propane tank situation, suppose the amount of space available in all directions for a propane tank is 10 feet. Write an equation to determine the maximum volume of a spherical tank that can fit into the given space.

Because the space is 10 feet, the maximum radius is 5 feet. Therefore, the equation is $5 = \sqrt[3]{\frac{3V}{4\pi}}$.

b. Solve this equation using an algebraic approach.

$$5 = \sqrt[3]{\frac{3V}{4\pi}}$$

$$5^3 = \left(\sqrt[3]{\frac{3V}{4\pi}}\right)^3$$

$$125 = \frac{3V}{4\pi}$$

$$500\pi = 3V$$

$$V = \frac{500\pi}{3} \approx 523.6 \text{ cubic feet}$$

c. Verify the solution using your graphing calculator. The screen should appear as follows:

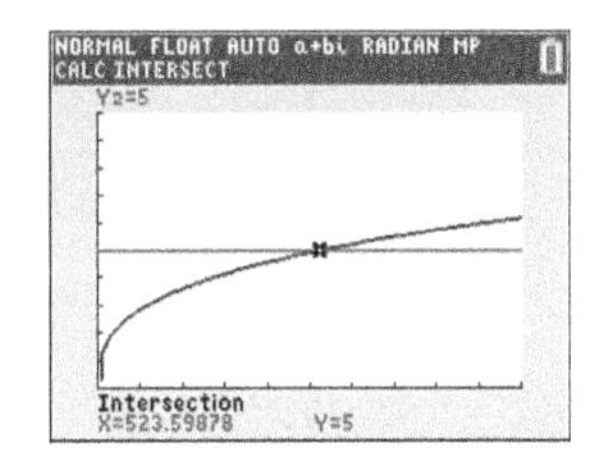

7. Solve the formula $V = l^3$ for l.

$l = \sqrt[3]{V}$

8. The basal metabolic rate (BMR) is the number of calories per day a person needs to maintain life. A person's basal metabolic rate is a function of his or her weight and is modeled by

$$B(w) = 70\sqrt[4]{w^3},$$

where $B(w)$ represents the basal metabolic rate measured in calories per day and w is the person's weight in kilograms.

a. Write the expression $70\sqrt[4]{w^3}$ using fractional exponents.

$70w^{3/4}$

b. If your friend weighs 50 kilograms (approximately 110 pounds), determine her basal metabolic rate. Round your answer to the nearest calorie.

$B(50) = 70 \cdot 50^{3/4} \approx 1316$ calories

c. Determine your basal metabolic rate. Be sure to convert your weight to kilograms.

(Answers will vary.) 1 kilogram is approximately 2.2 pounds.

A person weighing 180 pounds translates to 81.8 kilograms.
$B = 70 \cdot 81.8^{3/4} \approx 1904$ calories

d. Suppose a person is on a 2000-calorie-per-day diet. If the number of calories represents the person's basal metabolic rate, write an equation to determine the weight that is associated with this number of calories per day.

$2000 = 70w^{3/4}$

e. Solve the equation in part d. To help determine to what power you need to raise each side, consider how you simplify the expression $(x^{3/4})^{4/3}$.

$$2000 = 70w^{3/4}$$

$$\frac{2000}{70} = w^{3/4}$$

$$\left(\frac{200}{7}\right)^{4/3} = \left(w^{3/4}\right)^{4/3}$$

$$w \approx 87.3 \text{ kg}$$

$$w \approx 87.3(2.2) = 192 \text{ lb.}$$

f. If the person weighs 210 pounds, is the 2000-calorie diet healthy?

No, a 210-pound person requires 2138 calories per day.

Functions with Fractional/Decimal Exponents

Many growth and decay situations can be modeled by functions with fractional exponents.

9. The number of cell phone subscribers in the United States continues to increase every year. The following table gives the number of cell phone subscribers for various years from 2012 and projections through 2022:

Number of Years since 2010, t	2	4	6	8	10	12
Number of Cell Phone Subscribers in the U.S. (millions), $N(t)$	122.0	171.0	208.6	237.6	257.8	270.7

a. Plot the data from the table on the following grid:

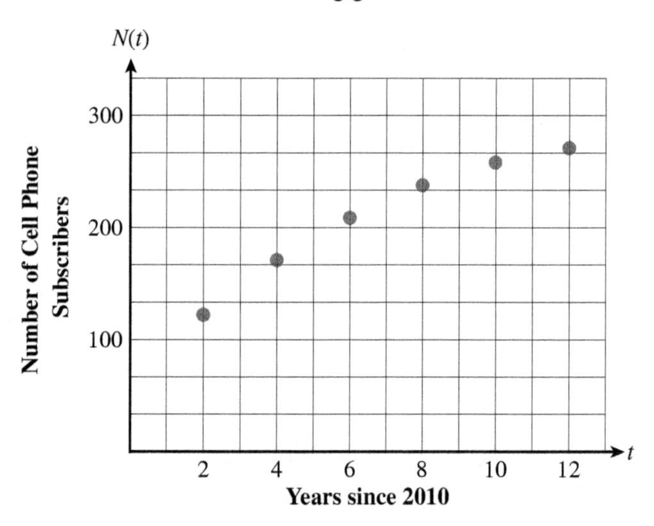

b. Use your calculator to create a statplot of the data, and determine the power regression $y = a \times x^b$ for the data. Use option A from the Stat Calc menu.

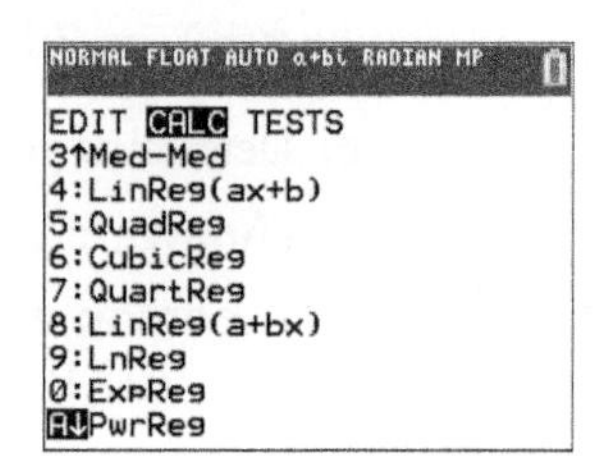

Round the values of a and b to two decimal places. Record the model below.

$N(t) = 90.61t^{0.454}$

c. According to the model, how many cell phone subscribers will there be in the United States in 2030?

$t = 20; N(20) = 90.61(20)^{0.454} \approx 353.1$ million subscribers

9. e.

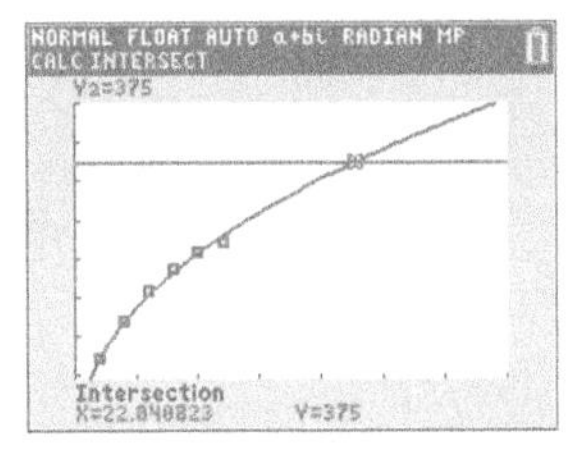

d. What equation do you need to solve to determine when the model predicts that there will be 375 million cell phone subscribers in the United States?

$90.61(t)^{0.454} = 375$

e. Solve the equation in part d graphically.

The graph indicates there will be approximately 375 million cell phone subscribers in the United States 23 years after 2010, or 2033.

SUMMARY Activity 5.9

1. To solve an equation when the variable appears under a radical, use the following two properties:

i. If a and b are two quantities such that $a = b$, then $a^n = b^n$, for any positive integer.

ii. $(b^{1/n})^n = b^1$ and $(b^{m/n})^{n/m} = b^1$

2. **i.** The domain of $f(x) = \sqrt[n]{x}$, where n is a positive odd integer, is all real numbers.

ii. The domain of $f(x) = \sqrt[n]{x}$, where n is a positive even integer, is $x \geq 0$.

3. To solve equations involving radicals,

Step 1. Isolate a radical term on one side of the equation.

Step 2. Raise each side of the equation to the power that matches the index of the radical.

Step 3. If a radical remains, repeat steps 1 and 2. Solve the resulting equation.

Step 4. Check all solutions in the original equation.

EXERCISES Activity 5.9

1. If possible, determine the exact value of each of the following:

a. $\sqrt[3]{64}$

4

b. $\sqrt[4]{16}$

2

c. $(-27)^{1/3}$

−3

d. $(625)^{1/4}$

5

e. $\sqrt{\frac{1}{36}}$

$\frac{1}{6}$

f. $(-81)^{1/4}$

not real

g. $(100{,}000)^{1/5}$

10

h. $(-1)^{1/6}$

not real

Exercise numbers appearing in color are answered in the Selected Answers appendix.

2. If the volume of a cube is 728 cubic centimeters, what is the length of one edge to the nearest tenth of a centimeter?

$e = (728)^{1/3} \approx 9.0$ cm

3. If the volume of a cube is decreased from 1450 cubic inches to 1280 cubic inches (and still remains a cube), by how much has the length of one edge decreased?

The difference is $\sqrt[3]{1450} - \sqrt[3]{1280} \approx 0.46$ in.

4. The volume of a sphere is 520 cubic meters. What is the diameter of the sphere?

The volume of a sphere is given by $V = \frac{4}{3}\pi r^3$, where r is the radius. So

$$520 = \frac{4}{3}\pi r^3$$

$$520 \cdot \frac{3}{4} = \pi r^3$$

$$390 = \pi r^3$$

$$\frac{390}{\pi} = r^3$$

$$r = \sqrt[3]{\frac{390}{\pi}} \approx 4.99 \text{ m.}$$

The diameter is $2(4.99) = 9.98$ meters.

5. What is the domain of each function?

a. $y = \sqrt[3]{x + 6}$

all real numbers

b. $f(x) = \sqrt[4]{x - 3}$

$x - 3 \geq 0$

$x \geq 3$

c. $g(x) = \sqrt[5]{2 - x}$

all real numbers

d. $f(x) = (2 - x)^{1/6}$

$2 - x \geq 0$

$x \leq 2$

6. Solve each of the following algebraically and graphically:

a. $\sqrt[3]{x + 4} = 3$

$x + 4 = 3^3$

$x + 4 = 27$

$x = 23$

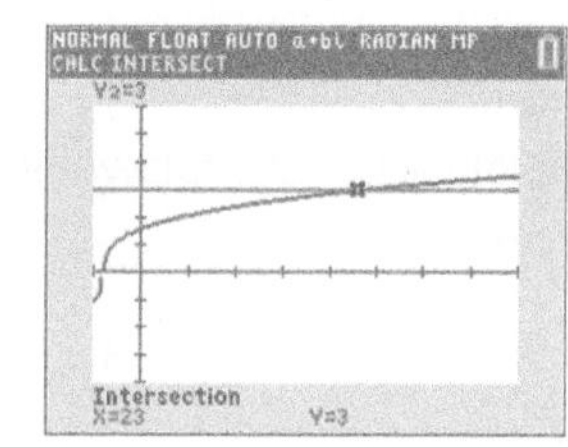

b. $\sqrt[4]{x + 5} = 2$

$x + 5 = 2^4$

$x + 5 = 16$

$x = 11$

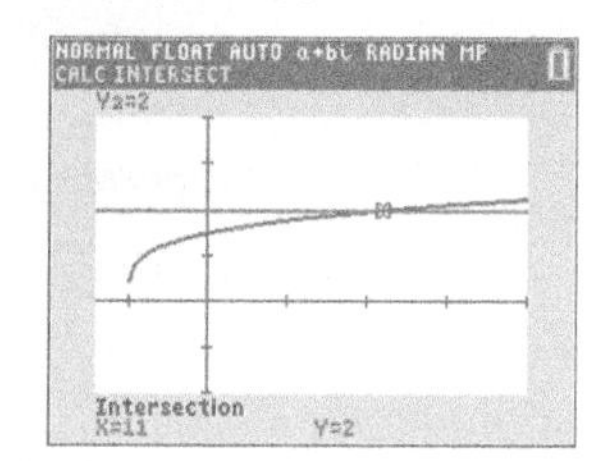

c. $\sqrt[3]{2x - 3} + 4 = 3$

$\sqrt[3]{2x - 3} = -1$

$2x - 3 = (-1)^3$

$2x - 3 = -1$

$2x = 2$

$x = 1$

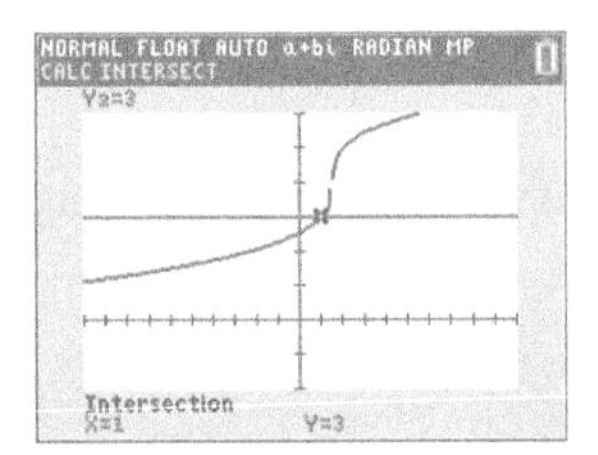

d. $\sqrt[4]{2 - x} = 5$

$2 - x = 5^4$

$2 - x = 625$

$2 - 625 = x$

$x = -623$

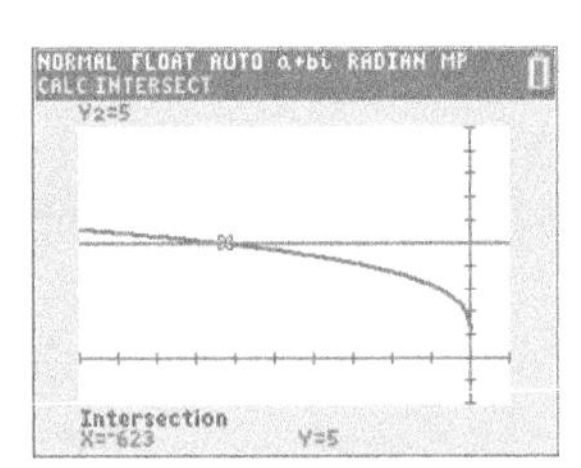

7. Solve each of the following algebraically, and verify your results graphically:

a. $x^{2/3} = 16$

$(x^{2/3})^{3/2} = 16^{3/2}$

$x = 64$

b. $2x^{3/4} = 54$

$x^{3/4} = 27$

$(x^{3/4})^{4/3} = 27^{4/3}$

$x = 81$

8. a. The diameter, d, of a sphere is modeled by the formula

$$d = \sqrt[3]{\frac{6V}{\pi}},$$

where V represents the volume of a sphere. Approximate the diameter of a sphere having a volume of 10 cubic inches.

$d = \sqrt[3]{\frac{6 \cdot 10}{\pi}} \approx 2.67$ in.

b. Determine the volume of a sphere having diameter 5 feet.

$5 = \sqrt[3]{\frac{6V}{\pi}}$

$125 = \frac{6V}{\pi}$

$\frac{125\pi}{6} = V$

$V \approx 65.45$ cubic feet

9. The radius, r, of a sphere is modeled by

$$r = \sqrt[3]{\frac{3V}{4\pi}},$$

where V is the volume of the sphere.

a. Determine the radius of a sphere that has a volume equal to 40 cubic centimeters.

$r = \sqrt[3]{\frac{3 \cdot 40}{4\pi}} \approx 2.12$ cm

b. Determine the volume of a sphere that has a radius equal to 3.5 feet.

$3.5 = \sqrt[3]{\frac{3V}{4\pi}}$

$(3.5)^3 = \frac{3V}{4\pi}$

$4\pi(3.5)^3 = 3V$

$V = \frac{4\pi(3.5)^3}{3} \approx 179.6$ cubic feet

c. Solve the formula for V, expressing volume as a function of the radius.

$$r = \sqrt[3]{\frac{3V}{4\pi}}$$

$$r^3 = \frac{3V}{4\pi}$$

$$4\pi r^3 = 3V$$

$$V = \frac{4\pi r^3}{3}$$

10. The number of flat-screen TVs sold are given in the following table:

Number of Years since 2010, t	1	2	3	4	5
Number of Flat-Screen TVs Sold in U.S. (millions), N	39.1	37.1	35.1	34.5	34.2

a. Plot the data from the table on the following grid:

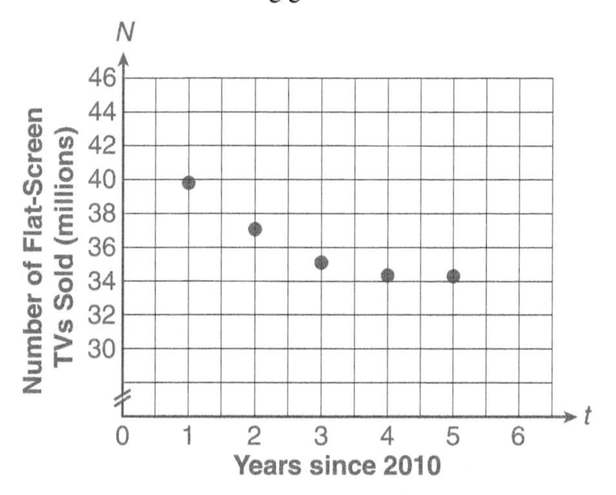

b. Use your calculator to create a statplot of the data, and determine the power regression $y = a \times x^b$ for the data. Round the values of a and b to two decimal places. Record the model below.

$N = 39.12 \cdot x^{-0.09}$

c. According to the model, how many flat-screen TVs will be sold in the United States in 2020?

$N = 39.12 \cdot 10^{-0.09} \approx 31.8$ million flat-screen TVs

d. What equation do you need to solve to determine when the model predicts that there will be 28 million flat-screen TVs sold in the United States?

$39.12 \cdot x^{-0.09} = 28$

e. Solve the equation in part d graphically.

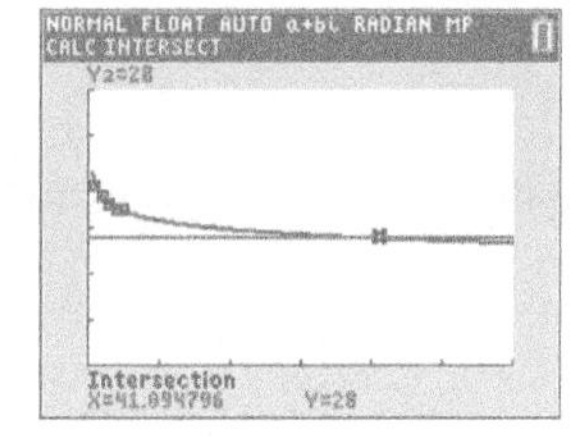

Approximately 28 million flat-screen TVs will be sold in 2051.

f. How confident are you in the results of part e?

(Answers will vary.) Not very confident. I am extrapolating too far into the future.

CLUSTER 2 What Have I Learned?

1. Explain the steps involved in solving the equation $\sqrt{2x + 3} = 5$ using an algebraic approach. Why must you be sure to check your solution?

Square both sides: $2x + 3 = 25$

Solve for x: $x = 11$

Check in the original: $\sqrt{2 \cdot 11 + 3} = \sqrt{25} = 5$

I must check for extraneous solutions.

2. What is the domain for the variable b in each of the following?

a. $\sqrt[n]{b}$, where n is an even positive integer

$b \geq 0$

b. $\sqrt[n]{b}$, where n is an odd positive integer

b can be any real number.

3. Is it possible for an extraneous solution to appear when the equation $\sqrt[3]{2x + 1} = -3$ is being solved? Explain.

No, if you cube a negative number the result is still negative.

4. Describe two ways to check for extraneous solutions when you are solving an equation by squaring both sides.

Check algebraically or check graphically.

5. Determine whether the following statements are true or false. Explain your answer.

a. If two numbers are equal, then their squares are equal.

True; multiplying equals by equals preserves the equality.

b. If the squares of two numbers are equal, then the two numbers are equal.

False; the numbers may be opposites; that is, they may have opposite signs.

6. a. For a given value of x, which is greater, $\sqrt[5]{x}$ or $x^{1/3}$? Explain how you determine your answer.

I look at the graphs. I must consider different cases. The answer depends on whether $x < -1$, $x = -1$, $-1 < x < 0$, $x = 0$, $0 < x < 1$, $x = 1$, or $x > 1$. For example, if $x > 1$, the cube root graph is above the 5th root graph and $x^{1/3} > \sqrt[5]{x}$.

b. In part a, did you assume that $x > 0$? Does your answer change if $x < 0$?

The two graphs are reflected in the x-axis for negative x, so the answer changes.

CLUSTER 2 How Can I Practice?

1. Solve each of the following equations algebraically, and check graphically:

a. $\sqrt{x+2} = 10$
$x = 98$

b. $(x-5)^{1/2} = 6$
$x = 41$

c. $\sqrt{2x+1} - 5 = 0$
$x = 12$

d. $\sqrt[3]{x^2+3} = 4$
$x = \pm\sqrt{61}$

e. $\sqrt{x} = \sqrt{x+2}$
no solution

f. $(2-x)^{1/3} = -2$
$x = 10$

g. $\sqrt[4]{2x-5} = 2$
$x = 10.5$

h. $(2.3x + 1.9)^{1/3} = 1.6$
$x \approx 0.95$

2. Identify the domain of each of the following functions:

a. $f(x) = \sqrt{6-x}$
$x \leq 6$

b. $g(x) = (2x-9)^{1/3}$
all real numbers

c. $h(x) = (x^2-4)^{1/4}$
$x \geq 2$ or $x \leq -2$

3. If the volume of a cube is 458 cubic inches, what is the length of one edge? Determine the value to the nearest hundredth of an inch.

Length is approximately 7.71 inches.

4. If the volume of a sphere is 620 cubic centimeters, what is its radius?

$r = \sqrt[3]{\frac{3V}{4\pi}}$, $V = 620$, $r \approx 5.29$ cm

5. When a stone is dropped to the ground, its velocity is modeled by the function $v = \sqrt{64d}$, where d is the distance the stone has fallen, in feet, and v is its velocity in feet per second. If the stone hits the ground at 100 feet per second, from what height was it dropped?

$v = 100$, $100 = \sqrt{64d}$; $d = 156.25$ ft.

6. A cardboard box with a square bottom has a height of 10 inches and a volume of 422.5 cubic inches. What are the dimensions of the bottom of the box?

$10x^2 = 422.5$

$x = 6.5$ in.

The dimensions of the bottom of the box are 6.5 inches × 6.5 inches.

7. Describe the similarities and differences between the graphs of $y = \sqrt{2-x}$ and $y = \sqrt{x-2}$.

The graphs are reflections about the line $x = 2$.

8. The sound intensity of a speaker varies inversely as the square of the distance from the speaker. The intensity is 20 microwatts per square meter when you are 10 meters from the speaker. Determine the intensity when you are 5 meters from the speaker.

$I = k/d^2$ $20 = k/10^2$ $20(100) = k$ $k = 2000$ $I = 2000/d^2$

$I = 2000/5^2$; $I = 80$ microwatts per square meter

Answers to all How Can I Practice exercises are included in the Selected Answers appendix.

CHAPTER 5 Summary

The bracketed numbers following each concept indicate the activity in which the concept is discussed.

CONCEPT/SKILL	DESCRIPTION	EXAMPLE
Domain and range of $y = \frac{k}{x}$ [5.1]	The domain and range consist of all real numbers except zero.	$y = \frac{3}{x}$
Graph of $y = \frac{k}{x}$ [5.1]	The graph is in the first and third quadrants if $k > 0$ and in the second and fourth quadrants if $k < 0$.	$y = \frac{3}{x}$ 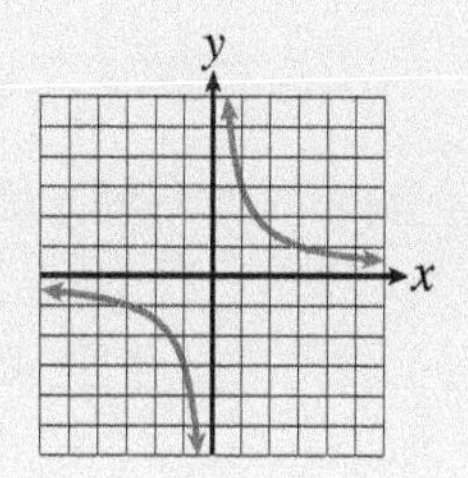
Asymptotes of the graph of $y = \frac{k}{x}$ [5.1]	The y-axis, $x = 0$, is the vertical asymptote. The x-axis, $y = 0$, is the horizontal asymptote.	$y = \frac{3}{x}$ 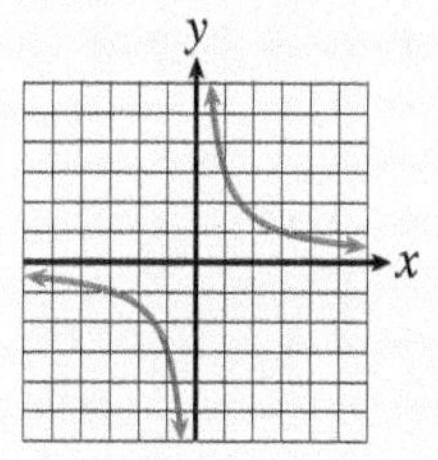
Domain of $y = \frac{k}{x^n}$ [5.2]	The domain consists of all real numbers except zero.	$y = \frac{4}{x^3}$
Graph of $y = \frac{k}{x^n}$ [5.2]	The graphs will vary depending on the values of k and n.	See the summary at the end of Activity 5.2.
Asymptotes of the graph of $y = \frac{k}{x^n}$ [5.2]	The y-axis, $x = 0$, is the vertical asymptote. The x-axis, $y = 0$, is the horizontal asymptote.	$y = \frac{4}{x^3}$ 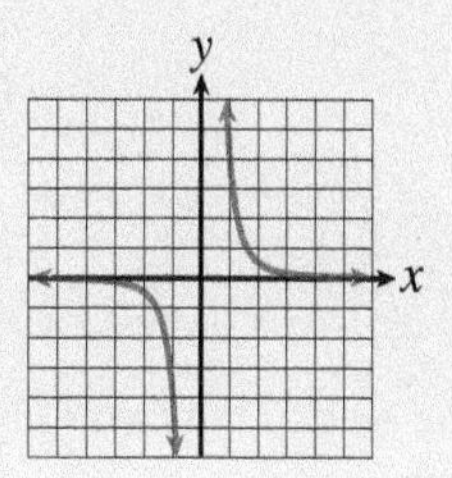
Inverse variation functions [5.2]	Functions defined by $y = \frac{k}{x^n}$ are called inverse variation functions in which y is said to vary inversely as the nth power of x; k is called the constant of variation.	For the function $y = \frac{4}{x^3}$, y varies inversely as the cube of x and 4 is the constant of variation.

CONCEPT/SKILL	DESCRIPTION	EXAMPLE
Rational function [5.3]	A function Q, defined by an equation of the form $Q(x) = \frac{k}{g(x)}$, where k is a nonzero constant and $g(x)$ is a polynomial, belongs to the family of functions known as rational functions.	$f(x) = \frac{10}{x - 3}$
Domain of a rational function [5.3]	The domain of $y = \frac{k}{g(x)}$ is the set of all real numbers except those values of the input x such that $g(x) = 0$.	The domain of $f(x) = \frac{10}{x - 3}$ is all real numbers except 3.
Vertical asymptote of a rational function [5.3]	The vertical asymptote is the vertical line that passes through the x-value for which $g(x) = 0$.	The vertical asymptote of $f(x) = \frac{10}{x - 3}$ is $x = 3$.
Horizontal asymptote of a rational function $y = \frac{k}{g(x)}$ [5.3]	The horizontal asymptote of $f(x) = \frac{k}{g(x)}$ is the x-axis $(y = 0)$.	The horizontal asymptote of $f(x) = \frac{10}{x - 3}$ is $y = 0$.
Solving rational equations [5.4]	Method 1. To solve an equation of the form $\frac{f(x)}{g(x)} = \frac{a}{b}$, where $g(x) \neq 0$ and $b \neq 0$, multiply both sides of the equation by the product $b \cdot g(x)$ and solve the resulting equation for x.	See Example 1, Activity 5.4, page 521.
Solving rational equations [5.4]	Method 2. To solve an equation of the form $\frac{f(x)}{g(x)} = \frac{a}{b}$, where $g(x) \neq 0$ and $b \neq 0$, cross multiply to obtain $b \cdot f(x) = a \cdot g(x)$ and solve the resulting equation for x.	See Example 1, Activity 5.4, page 522.
Horizontal asymptotes of rational functions [5.4]	Suppose the output values of a rational function R get closer to a number a as the input values increase infinitely in both the positive and negative directions. Then the graph of the function R has a horizontal asymptote. The equation of the horizontal asymptote is $y = a$.	The horizontal asymptote of $R(x) = \frac{3x + 1}{x - 2}$ is $y = 3$.
Determine an LCD of two or more expressions [5.5]	1. Write the prime factorization of each denominator. Express repeated factors as powers. 2. Identify the different bases (factors) in step 1. 3. Write the LCD as the product of the highest power of each of the different factors from step 2.	See Example 1, Activity 5.5, page 532.

CONCEPT/SKILL	DESCRIPTION	EXAMPLE
Solving an equation involving rational expressions [5.5]	1. Determine the LCD of all denominators in the equation. 2. Multiply each side of the equation by the LCD and simplify the resulting equation. 3. Solve the resulting equation for the desired variable.	See Problem 4, Activity 5.5, page 532.
Simplifying rational expressions [5.6]	1. Factor the numerator and the denominator. 2. Divide the numerator and the denominator by the common factors.	See Appendix A.
Multiplying or dividing rational expressions [5.6]	1. Factor the numerator and denominator of each fraction completely. 2. Divide out the common factors (cancel). 3. Multiply remaining factors. 4. In division, proceed as above after inverting the divisor (the fraction after the division sign).	See Appendix A.
Adding or subtracting rational expressions [5.6]	1. Determine the LCD. 2. Build each fraction to have the LCD. 3. Add or subtract numerators. 4. Place the numerator over the LCD, and simplify if necessary.	See Appendix A.
Simplifying a complex fraction [5.6]	1. Express the numerator as a single fraction. 2. Express the denominator as a single fraction. 3. Divide the numerator by the denominator. 4. Simplify if possible.	See Example 2, Activity 5.6, page 541.
Radical functions [5.7]	The function defined by $y = \sqrt{g(x)}$ has domain all real numbers x such that $g(x) \geq 0$.	$y = \sqrt{2x + 1}$. The domain is $x \geq -\frac{1}{2}$. 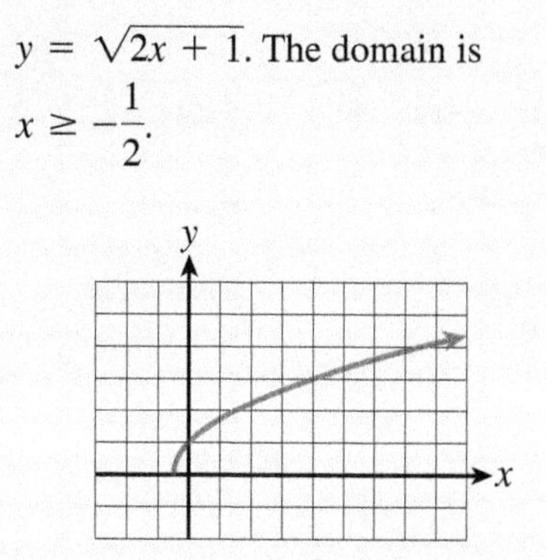

CONCEPT/SKILL	DESCRIPTION	EXAMPLE
Radical functions [5.7]	The function defined by $y = -\sqrt{g(x)}$ has domain all real numbers x such that $g(x) \geq 0$.	$y = -\sqrt{2x + 1}$. The domain is $x \geq -\frac{1}{2}$. 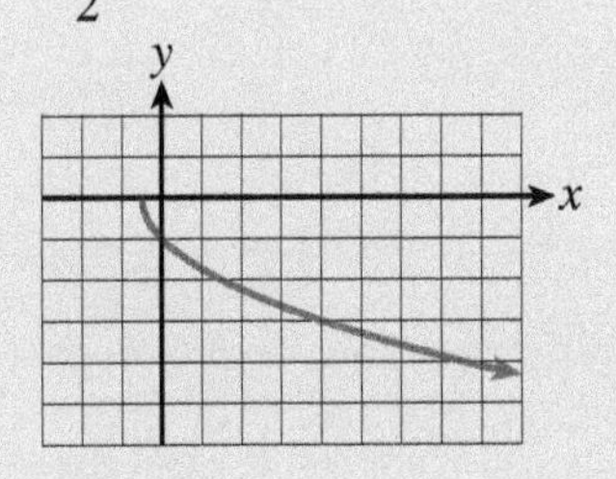
Solving an equation involving one radical expression [5.8]	1. Isolate the radical term on one side of the equation. 2. Square both sides of the equation. 3. Solve the resulting equation. 4. Check all solutions in the original equation.	See Appendix A.
Solving an equation involving more than one radical expression [5.8]	1. Isolate one radical term on one side of the equation. 2. Square both sides of the equation. 3. Solve the resulting equation. If a radical remains, repeat steps 1 and 2. 4. Check all solutions in the original equation.	See Example 3, Activity 5.8, pages 564–565.
Domain of a function defined by an equation of the form $y = \sqrt[n]{g(x)}$, where n is a positive integer and $g(x)$ is a polynomial [5.9]	The domain of $y = \sqrt[n]{g(x)}$ is all real numbers if n is an odd positive integer. The domain of $y = \sqrt[n]{g(x)}$ is all real numbers for which $g(x) \geq 0$ if n is an even positive integer.	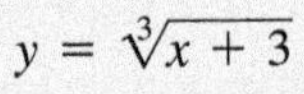 $y = \sqrt[3]{x + 3}$ 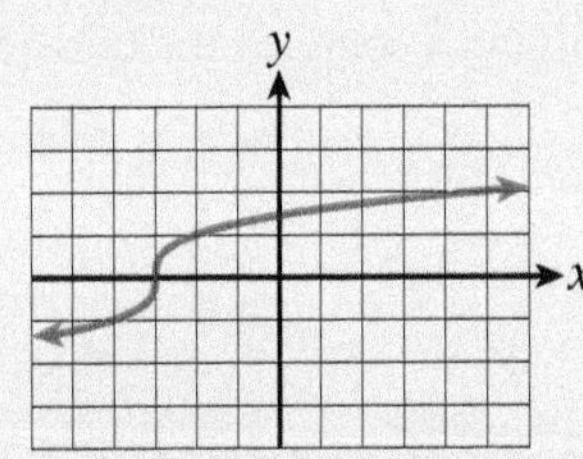
Solving radical equations that contain radical expressions with an index other than 2 [5.9]	1. Isolate the radical term on one side of the equation. 2. Raise each side of the equation to the power that matches the index of the radical. 3. If a radical remains, repeat steps 1 and 2. Solve the resulting equation. 4. Check all solutions in the original equation.	See Example 1, Activity 5.9, page 574.

CHAPTER 5 Gateway Review

1. According to the blueprint, the floor area of the stage in the new auditorium at your college must be rectangular and equal to 1200 square feet. The width of the stage is key to all of the theater productions. Therefore, in this situation, the stage's depth is a function of its width.

a. Let d represent the depth (in feet) and w represent the width (in feet). Write an equation that expresses d as a function of w.

$d = \frac{1200}{w}$

b. Complete the following table using the equation from part a:

w (ft.)	30	35	40	50	60
d (ft.)	40	34.286	30	24	20

c. What happens to the depth as the width increases?

As the width increases, the depth decreases.

d. What happens if the width is 100 feet? Is this realistic? Explain.

The depth is 12 feet, not enough room for most theater sets, so it is not realistic.

e. Can the width be zero? Explain.

No. Division by zero is undefined.

f. What do you think is the practical domain for this function?

(Answers will vary.) $30 \leq w \leq 60$

g. What type of a function do you have in this situation? a rational function

h. What is the domain of the general function? all real numbers except zero

i. What is the vertical asymptote? $w = 0$

j. What is the horizontal asymptote? Explain in words how you determined it.

$d = 0$

As w increases, d approaches 0.

2. Sketch the following graphs without using your graphing calculator:

a. $f(x) = \frac{1}{x^2}$

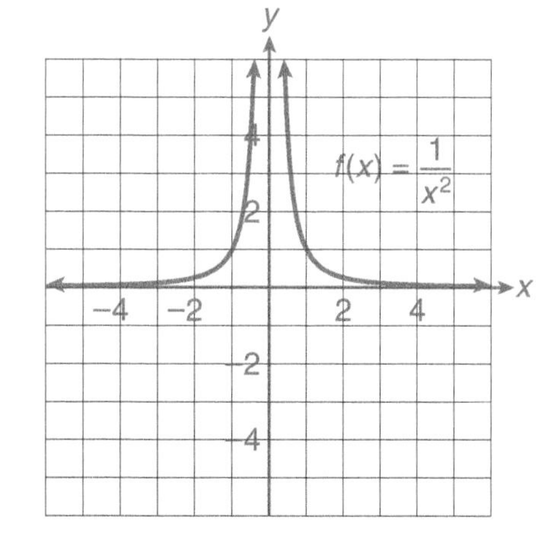

b. $g(x) = \frac{-1}{x^3}$

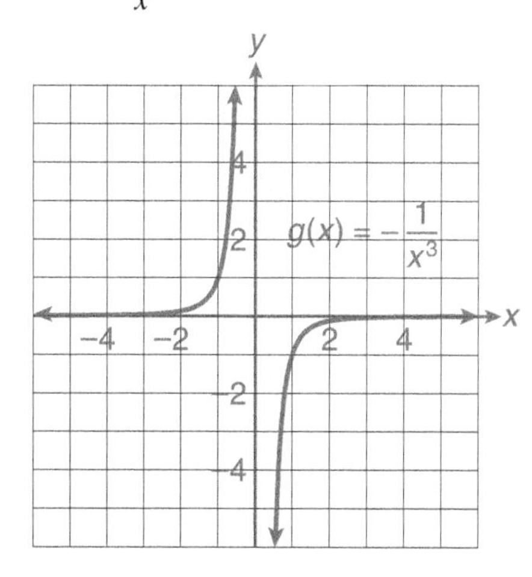

Answers to all Gateway exercises are included in the Selected Answers appendix.

c. Describe how the graphs are similar and how they are different.

The graphs have the same horizontal and vertical asymptotes.

$f(x) = \frac{1}{x^2}$ is symmetrical with respect to the y-axis.

$g(x) = -\frac{1}{x^3}$ is symmetrical with respect to the origin in quadrants II and IV. $f(x)$ is always positive.

$g(x)$ is both positive and negative.

3. a. If y varies inversely as x and $x = 10$ when $y = 12$, determine the value of y when $x = 30$.

$y = \frac{k}{x}$; $12 = \frac{k}{10}$; $k = 120$; $y = \frac{120}{30} = 4$

b. The loudness, in decibels, of a stereo varies inversely as the square of the distance from the speaker to the person listening. If the loudness is 32 decibels at a distance of 4 feet, what is the loudness when the listener is 10 feet from the speaker?

$I = \frac{k}{d^2}$; $32 = \frac{k}{16}$; $k = 512$; $I = \frac{512}{100} = 5.12$ decibels

c. When the volume of a circular cylinder is constant, the height varies inversely as the square of the radius. If the radius is 2 inches when the height is 8 inches, determine the height when the radius is 5 inches.

$h = \frac{k}{r^2}$; $8 = \frac{k}{4}$; $32 = k$; $h = \frac{32}{25} = 1.28$ in.

4. Determine the horizontal and vertical asymptotes and the intercepts of each of the following. Then sketch a graph of each function. Verify using a graphing calculator.

a. $f(x) = \frac{10}{x}$

H: $y = 0$

V: $x = 0$

no y-intercept

no x-intercept

b. $g(x) = \frac{4}{x - 3}$

H: $y = 0$

V: $x = 3$

$\left(0, -\frac{4}{3}\right)$

no x-intercept

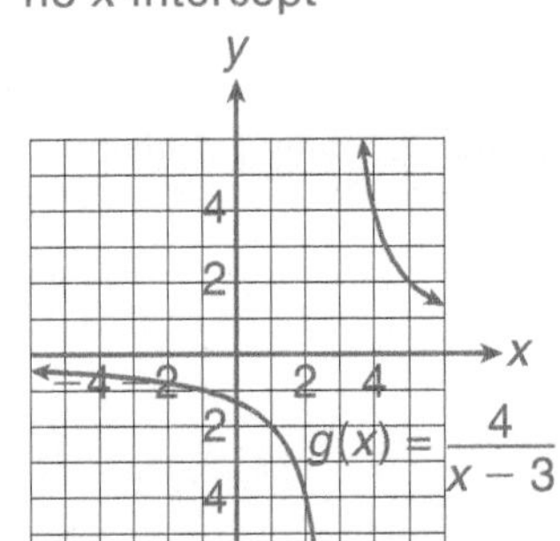

c. $f(x) = \frac{2x}{x + 2}$

H: $y = 2$

V: $x = -2$

(0, 0)

(0, 0)

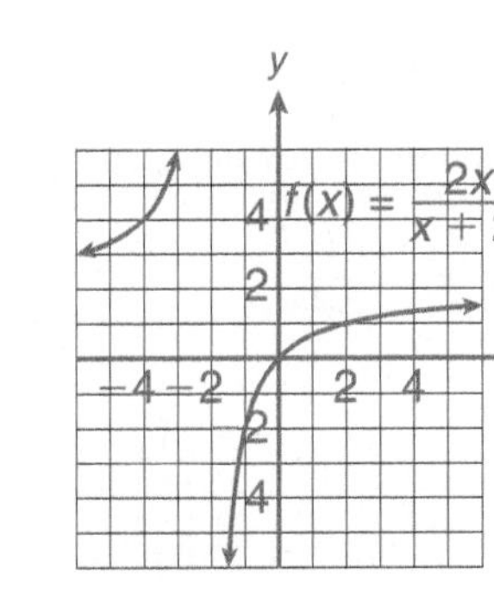

5. Students from the local community college plan to celebrate their 10-year reunion. They reserve a restaurant for an evening of entertainment. The fee for the band is $600, and food will cost each person $45.

a. If $f(n)$ represents the total cost for n people to participate in the reunion, write an equation for the total cost.

$f(n) = 45n + 600$

b. If 100 people attend, what will be the total cost of the event?

$f(100) = 45(100) + 600$

$f(100) = \$5100$

c. Determine a function, A in terms of n, that will represent the average cost per person to attend the event.

$$A(n) = \frac{45n + 600}{n}$$

d. If 100 people attend the reunion, how much should each person pay?

$$A(100) = \frac{45(100) + 600}{100} = \$51$$

e. Use your graphing calculator to complete the following table:

Number of People Attending, n	50	100	150	200	250
Cost per Person, $A(n)$ (\$)	57	51	49	48	47.40

f. If the committee thinks that each person should pay at most \$50, how many would have to attend for the cost per person to be \$50? Show your answer algebraically, and check it using your graphing calculator.

$$50 = \frac{45n + 600}{n};\ 50n = 45n + 600;\ 5n = 600;\ n = 120 \text{ people}$$

5. f.

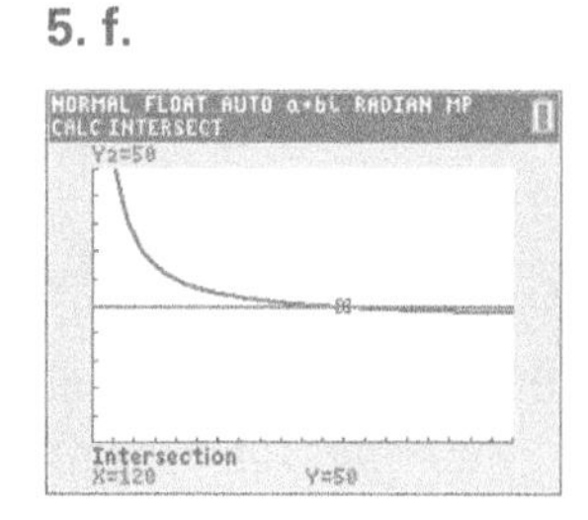

g. Determine the practical domain of the function.

$0 < n <$ seating capacity of restaurant

h. From the graph, determine the vertical asymptote. Is there a practical meaning of this asymptote in this situation?

The vertical asymptote is $n = 0$. You cannot calculate an average value if no one attends.

i. From the graph of function A, determine the horizontal asymptote. Is there a practical meaning of this asymptote in this situation?

The horizontal asymptote is $A(n) = 45$. As the number of people attending increases, the average cost approaches \$45 per person.

6. a. Solve the equation $\frac{4}{x - 2} + 3 = 9$ using an algebraic approach. Verify your answer graphically.

$$\frac{4}{x - 2} = 6;\ 4 = 6x - 12;\ 16 = 6x;\ x = \frac{8}{3}$$

b. When you graph the function $f(x) = \frac{4}{x - 2} - 6$, what do you discover about the solution to the equation in part a and the x-intercept of the graph of the function f? Explain.

The solution is the x-coordinate of the x-intercept.

7. Solve each of the following equations using an algebraic approach. Verify your solutions graphically.

a. $\dfrac{3}{x+2} = 5$

$3 = 5(x + 2)$

$3 = 5x + 10$

$-7 = 5x$

$x = -\dfrac{7}{5} = -1.4$

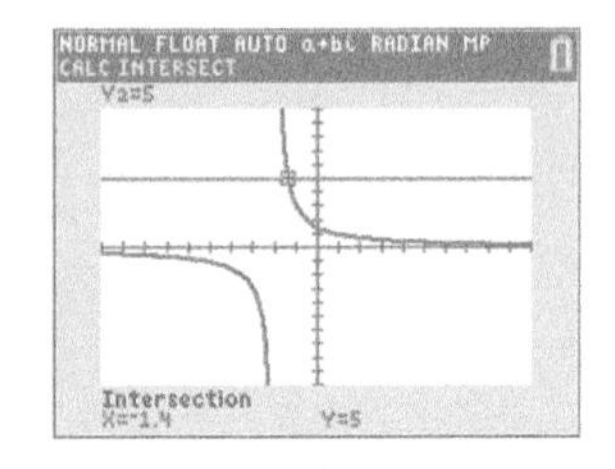

b. $\dfrac{-2x}{3x-4} = 2$

$-2x = 2(3x - 4)$

$-2x = 6x - 8$

$8 = 8x$

$x = 1$

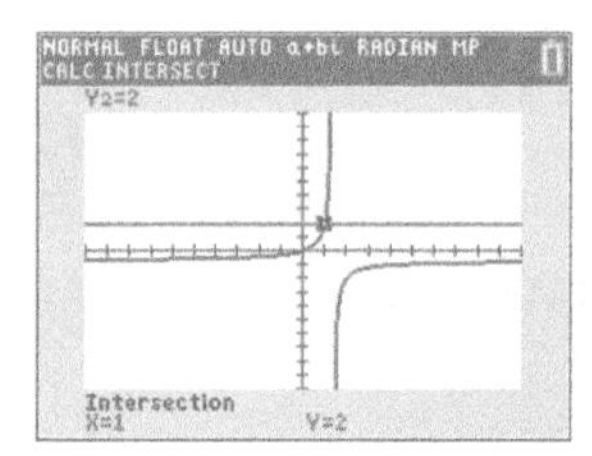

8. The local grocery store has just hired you and your friend. You can stock a shelf in 15 minutes. Your friend will take 20 minutes to stock the shelf. How long will it take to stock the shelf if you work together?

$\dfrac{1}{20} + \dfrac{1}{15} = \dfrac{1}{x}$; $60x\left(\dfrac{1}{20} + \dfrac{1}{15}\right) = 60x\left(\dfrac{1}{x}\right)$; $3x + 4x = 60$; $7x = 60$; $x \approx 8.57$ min.

9. You work in the admissions office at your community college. You must assemble all of the packets for the placement test sessions. You work with a friend who takes twice as long as you do to assemble the packets. If you work together, the packets can be completed in 45 minutes. On the day you must assemble the packets, you have a big exam. How many hours does it take your friend to do the job alone?

$\dfrac{1}{x} + \dfrac{1}{2x} = \dfrac{1}{45}$

$90x\left(\dfrac{1}{x} + \dfrac{1}{2x}\right) = 90x\left(\dfrac{1}{45}\right)$

$90 + 45 = 2x$

$2x = 135$

$x = 67.5$

$2x = 135$ min., or $2\frac{1}{4}$ hr.

10. Solve each of the following equations algebraically. Verify your answers graphically.

a. $\dfrac{1}{6} - \dfrac{3}{2x} = \dfrac{1}{5x}$

$30x\left(\dfrac{1}{6} - \dfrac{3}{2x}\right) = 30x\left(\dfrac{1}{5x}\right)$

$5x - 45 = 6$

$5x = 51$

$x = 10.2$

b. $\dfrac{2}{x} + \dfrac{3}{4x} = \dfrac{1}{12}$

$12x\left(\dfrac{2}{x} + \dfrac{3}{4x}\right) = 12x\left(\dfrac{1}{12}\right)$

$24 + 9 = x$

$x = 33$

11. Solve each of the following equations for the indicated variable. Express your answer as a single fraction.

a. Solve $S = \dfrac{C}{1 - r}$ for r.

$S(1 - r) = C;\ S - Sr = C;\ S - C = Sr;\ r = \dfrac{S - C}{S}$

b. Solve $\dfrac{1}{a} + \dfrac{3}{b} = \dfrac{4}{c}$ for b.

$abc\left(\dfrac{1}{a} + \dfrac{3}{b}\right) = abc\left(\dfrac{4}{c}\right);\ bc + 3ac = 4ab$

$bc - 4ab = -3ac;\ b(c - 4a) = -3ac;\ b = \dfrac{-3ac}{c - 4a}$

12. Simplify each of the following complex expressions:

a. $\dfrac{\frac{1}{a} + \frac{2}{b}}{\frac{2}{a} + \frac{1}{b}}$

$\dfrac{b + 2a}{2b + a}$

b. $\dfrac{1 + \frac{1}{x - 2}}{1 - \frac{3}{x + 2}}$

$\dfrac{x + 2}{x - 2}$

13. A camera lens possesses a measurement called the focal length, f. When an object is in focus, the focal length is related to the distance of the object from the lens, p, and the image distance from the lens, q, by the formula

$$f = \frac{1}{\frac{1}{p} + \frac{1}{q}}.$$

a. Determine f if p is 4 meters and q is 3 meters.

$f = \dfrac{1}{\frac{1}{4} + \frac{1}{3}} = \dfrac{1}{\frac{3}{12} + \frac{4}{12}} = 1 \div \dfrac{7}{12} = \dfrac{12}{7} \approx 1.71$ m

b. Simplify the complex fraction on the right side of the original formula.

$f = \dfrac{1}{\frac{1}{p} + \frac{1}{q}} = \dfrac{1}{\frac{q}{pq} + \frac{p}{pq}} = \dfrac{1}{\frac{p + q}{pq}} = \dfrac{pq}{p + q}$

c. Redo part a using the new formula from part b. How do the answers compare?

$f = \dfrac{4(3)}{4 + 3} = \dfrac{12}{7} \approx 1.71$ m

The values are the same.

14. a. What is the domain of the function defined by $f(x) = \sqrt{x + 4}$.

$x \geq -4$

b. Sketch the graph of the function f.

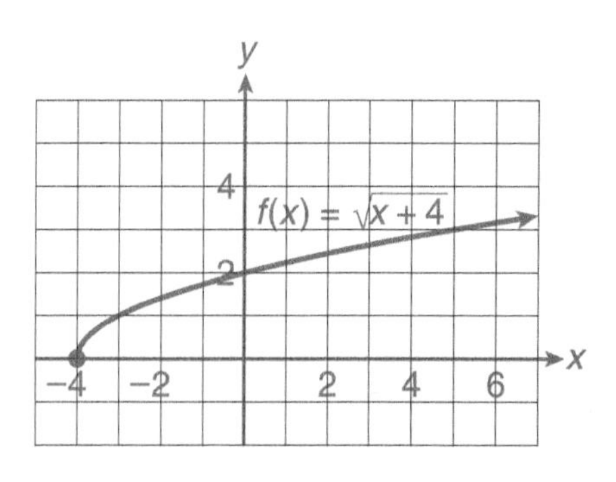

c. As the input increases, what happens to the output values?

The output is increasing.

d. What is the range of this function?

$y \geq 0$

e. Are there any intercepts? If so, what are they?

The x-intercept is $(-4, 0)$. The y-intercept is $(0, 2)$.

f. How is the function f similar to the function $g(x) = \sqrt{x - 4}$?

g has the same shape but is shifted 8 units to the right of f.

g. How is the graph of the function f similar to $h(x) = -\sqrt{x + 4}$?

The graphs are reflected through the x-axis.

15. a. Draw the graph of $f(x) = \sqrt{x}$.

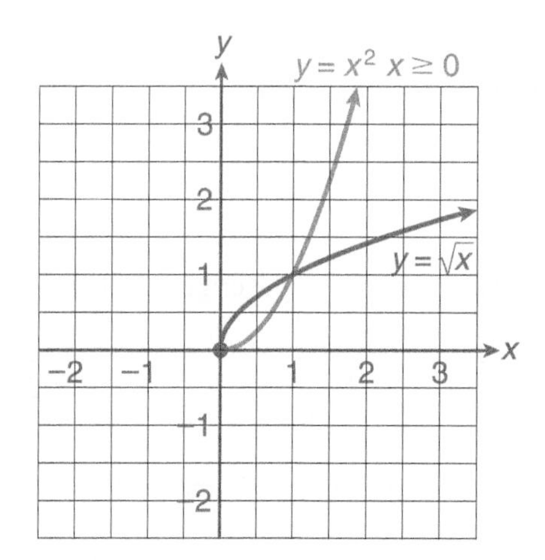

b. Determine the equation of the inverse of the function f.

$y = \sqrt{x}$; $x = \sqrt{y}$; $y = x^2$; $f^{-1}(x) = x^2$; $x \geq 0$

c. Sketch the graph of the inverse on the same axes as the graph of f in part a from $x \geq 0$.

d. From the graphs, describe how you know that they are inverses.

The graphs are reflections in $y = x$.

e. Show that f and f^{-1} are inverses using composition of functions.

$f(f^{-1}(x)) = f(x^2) = \sqrt{x^2} = x$

$(f^{-1}(f(x)) = f^{-1}(\sqrt{x}) = (\sqrt{x})^2 = x$

16. For each of the given functions,

i. determine the domain and range.

ii. determine the x- and y-intercepts.

iii. sketch a graph.

a. $f(x) = \sqrt{x} + 4$

i. $x \geq 0, y \geq 4$

ii. (0, 4) only

iii.

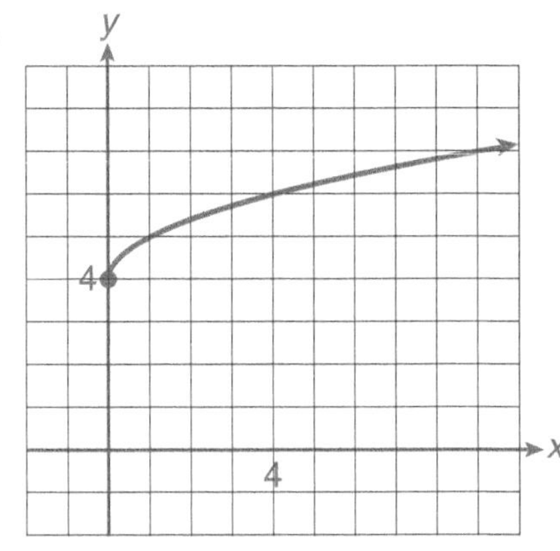

b. $f(x) = \sqrt{x + 4}$

i. $x \geq -4, y \geq 0$

ii. (0, 2) and (−4, 0)

iii.

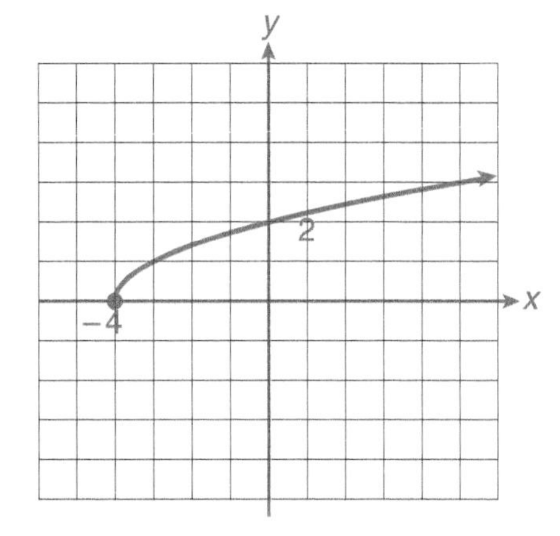

17. Solve each of the following equations using an algebraic approach. Verify your solutions graphically.

a. $\sqrt{3x - 2} - 6 = -2$

$\sqrt{3x - 2} = 4; 3x - 2 = 16$

$3x = 18$

$x = 6$

b. $\sqrt{2x + 1} - \sqrt{x + 7} = 0$

$2x + 1 = x + 7$

$x = 6$

c. $\sqrt[3]{5x + 4} = 3$

$5x + 4 = 27$

$5x = 23$

$x = \frac{23}{5} = 4.6$

d. $\sqrt{4x + 8} - 3 = -5$

$\sqrt{4x + 8} = -2$

$4x + 8 = 4$

$4x = -4$

$x = -1$

−1 does not check.

There is no solution.

e. $x^{4/3} = 81$

$(x^{4/3})^{3/4} = 81^{3/4}$

$x = 3^3$

$x = 27$

18. What is the domain of each function?

a. $y = \sqrt[3]{x + 8}$

all real numbers

b. $y = \sqrt[4]{x - 6}$

$x \geq 6$

c. $y = (x + 1)^{1/6}$

$x \geq -1$

19. A submarine periscope must be a certain distance above the water for it to be used to locate a ship a certain number of miles away. The model for the distance (in miles) that the submarine periscope can see is the formula $d = \sqrt{1.5h}$, where h represents the height (in feet) above the surface of the water.

How far above the surface of the water must the periscope be to see a ship that is 6 miles away?

$6 = \sqrt{1.5h}$

$36 = 1.5h$

$h = 24$ ft.

20. A ring is dropped from the American span of the Thousand Island Bridge and hits the water 3.1 seconds later. What is the height of the bridge? Use the formula $T = \sqrt{\frac{d}{16}}$, where d represents the distance in feet and T represents the time in seconds.

$3.1 = \sqrt{\frac{d}{16}}$

$9.61 = \frac{d}{16}$

$d = 153.76$ ft.

CHAPTER 6

Introduction to the Trigonometric Functions

CLUSTER 1

Introducing the Sine, Cosine, and Tangent Functions

ACTIVITY 6.1

The Leaning Tower of Pisa

Sine, Cosine, and Tangent of an Angle in a Right Triangle

OBJECTIVES

1. Identify the sides and corresponding angles of a right triangle.
2. Determine the length of the sides of similar right triangles using proportions.
3. Determine the sine, cosine, and tangent of an angle within a right triangle.
4. Determine the sine, cosine, and tangent of an acute angle by use of the graphing calculator.

You are a structural engineer studying the Leaning Tower of Pisa, which is located in Pisa, Italy. Originally the tower was 179 feet high. Although the tower has lost none of its height, it now makes an 85° angle with the ground. You want to know how high the top of the tower is above the ground today.

To answer this question accurately, you need a branch of mathematics called **trigonometry**. Although the development of trigonometry is generally credited to the ancient Greeks, there is evidence that the ancient Egyptian cultures used trigonometry in constructing the pyramids.

Right Triangles

As you will see, the accurate answer to the Leaning Tower question requires some knowledge of **right triangles**. Consider the following right triangle with angles A, B, and C and sides a, b, and c.

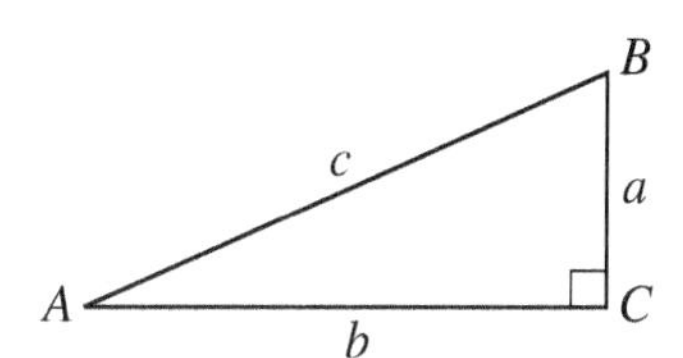

DEFINITION

Angle C is the right angle, the angle measuring 90°. The side opposite the right angle, c, is called the **hypotenuse**. Side a is said to be **opposite** angle A because it is not part of angle A. Side b is said to be **adjacent** to angle A because it and the hypotenuse form angle A. Similarly, side b is the side opposite angle B and side a is the side adjacent to angle B. The sides a and b are called **legs**.

Note that in any right triangle, the lengths of the sides are related by the Pythagorean theorem.

$$c^2 = a^2 + b^2$$

In words, the square of the hypotenuse is equal to the sum of the squares of the other two sides.

EXAMPLE 1 ***Consider the following right triangle.***

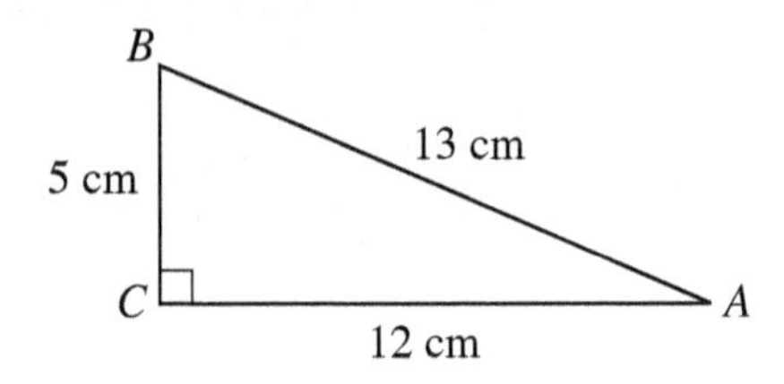

The side opposite angle B is 12 centimeters long. The side opposite angle A is 5 centimeters long. The side adjacent to angle B is 5 centimeters long. The side adjacent to angle A is 12 centimeters long. The hypotenuse is 13 centimeters long. Note that the lengths of the sides of the triangle satisfy the Pythagorean theorem: $13^2 = 5^2 + 12^2$.

1. Consider the right triangle below.

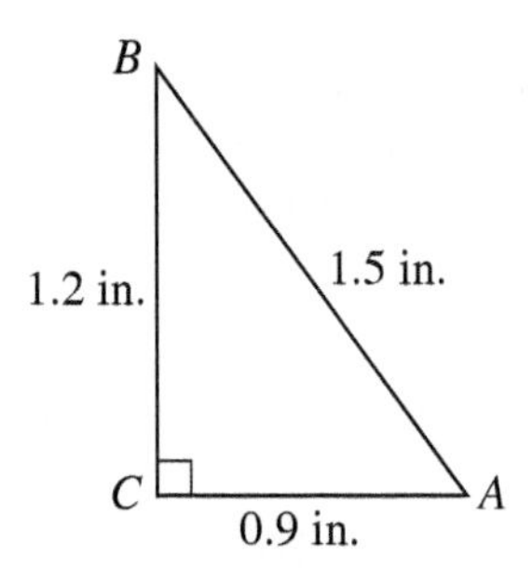

Determine the length of each of the following:

a. the side opposite angle B

0.9 in.

b. the side adjacent to angle B

1.2 in.

c. the side opposite angle A

1.2 in.

d. the side adjacent to angle A

0.9 in.

e. the hypotenuse

1.5 in.

f. Demonstrate that the length of the sides satisfies the Pythagorean theorem.

$1.5^2 = 1.2^2 + 0.9^2$

$2.25 = 1.44 + 0.81$

$2.25 = 2.25$

Similar Triangles

The measure of angle A in each of the right triangles below is 37°. The lengths of the sides are given in centimeters.

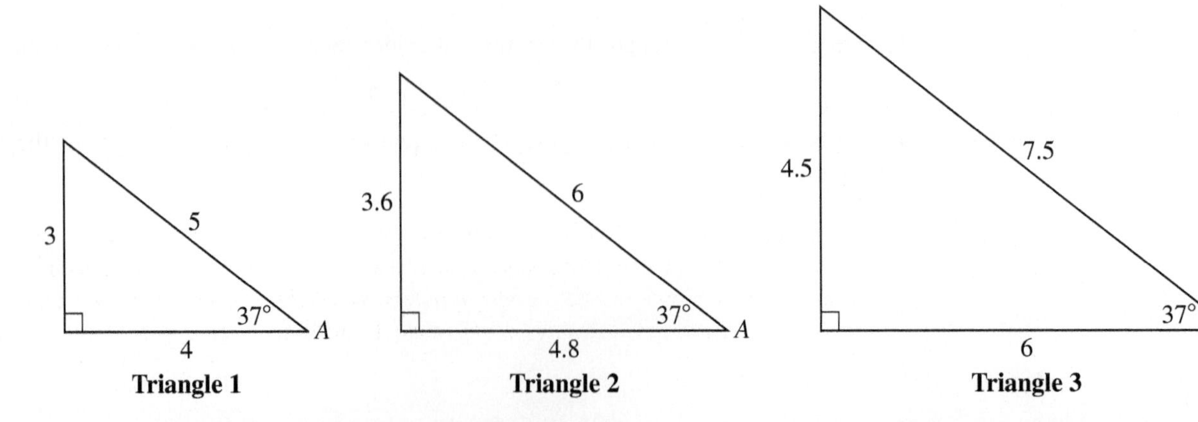

2. The following ratios for triangle 1 are given in the table below: $\frac{\textit{measure of side opposite A}}{\textit{measure of the hypotenuse}}$, $\frac{\textit{measure of side adjacent to A}}{\textit{hypotenuse}}$, $\frac{\textit{measure of side opposite A}}{\textit{measure of side adjacent to A}}$. Determine the same ratios for triangle 2 and triangle 3 and enter the results in the table.

	TRIANGLE 1	TRIANGLE 2	TRIANGLE 3
$\frac{\text{opposite}}{\text{hypotenuse}}$	$\frac{3}{5} = 0.6$	$\frac{3.6}{6} = 0.6$	$\frac{4.5}{7.5} = 0.6$
$\frac{\text{adjacent}}{\text{hypotenuse}}$	$\frac{4}{5} = 0.8$	$\frac{4.8}{6} = 0.8$	$\frac{6}{7.5} = 0.8$
$\frac{\text{opposite}}{\text{adjacent}}$	$\frac{3}{4} = 0.75$	$\frac{3.6}{4.8} = 0.75$	$\frac{4.5}{6} = 0.75$

3. What do you notice about each of the ratios for each of the three triangles?

The ratios of corresponding sides for each of the triangles are equal.

Triangles whose angles are equal are called **similar triangles**. For similar right triangles, these three ratios are always equal, and we say the corresponding sides are proportional.

Any right triangle with an angle of 37° will be similar to the triangles in Problem 1. Any right triangle with an angle of 37° will have the same ratios given in the table above. The value of the ratio is a function of the measure of the angle.

Sine, Cosine, and Tangent Functions

The ratios given in the table above have special names and are defined as follows.

DEFINITION

Let A be an acute angle (less than 90°) of a right triangle. The **sine**, **cosine**, and **tangent** of angle A are defined by

$$\text{sine of } A = \sin A = \frac{\textit{length of the side opposite A}}{\textit{length of the hypotenuse}},$$

$$\text{cosine of } A = \cos A = \frac{\textit{length of the side adjacent to A}}{\textit{length of the hypotenuse}},$$

$$\text{tangent of } A = \tan A = \frac{\textit{length of the side opposite A}}{\textit{length of the side adjacent to A}},$$

where sin, cos, and tan are the standard abbreviations for sine, cosine, and tangent, respectively.

- Sine, cosine, and tangent are called **trigonometric functions**.
- Note that the input values for the trigonometric functions are angles and the output values are ratios.
- A mnemonic device often used to help remember the trigonometric relationships is SOH CAH TOA, where SOH indicates that the ***s***ine function is the length of the side ***o***pposite divided by the length of the ***h***ypotenuse, CAH indicates the cosine definition, and TOA indicates the definition of the tangent function.

LC LEARNING CATALYTICS

Consider the right triangle ABC with sides a, b, and c, respectively. Angle C is the right angle. Side c is the hypotenuse, side a is the side opposite angle A, and side b is the side opposite angle B. If side a has length 12, side b has length 5, and the hypotenuse has length 13, what is the value of $\sin A$?

a. $\frac{5}{12}$.
b. $\frac{5}{13}$.
c. $\frac{12}{13}$.
d. $\frac{12}{5}$.

The sine, cosine, and tangent are function buttons labeled sin, cos, and tan on your graphing calculator. The values of the ratios in the table on page 597 can be found by finding sin(37°), cos(37°), *and* tan(37°). To get the correct results, your calculator needs to be in degree mode. Press the MODE button, move the cursor to degree, and hit ENTER. The screen should look like the one given below.

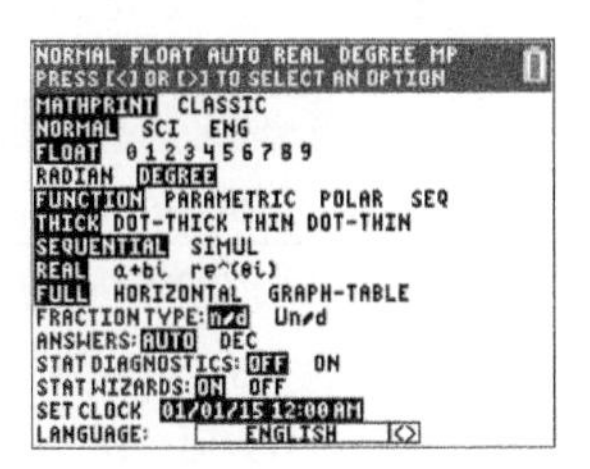

4. Now hit CLEAR to return to the home screen and find the values of sin(37°), cos(37°), *and* tan(37°). **Note:** To enter sin(37°) into the calculator, press SIN, then 3 7) ENTER. How do the results compare with the values of the ratios in the table above?

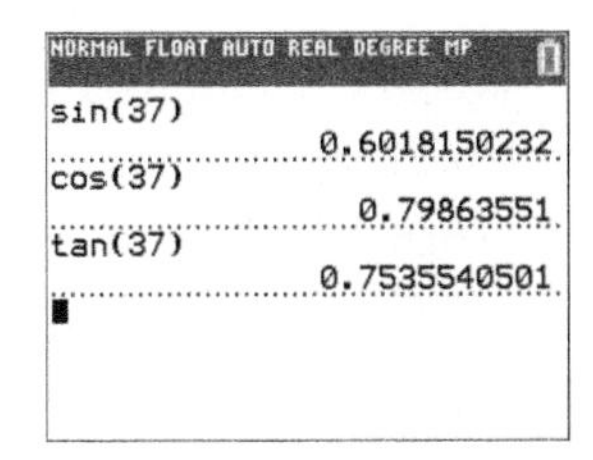

The rounded results are the same.

5. The sum of the angles in any triangle is 180°. Since a right triangle has a 90° angle, the sum of the other two angles must be 90°. Triangle 1, in the table contains a 37° angle.

a. What is the measure of the other acute angle in triangle 1?

$90° - 37° = 53°$

b. Use the angle from part a to calculate the ratios used in the table.

	TRIANGLE 1
$\frac{\text{opposite}}{\text{hypotenuse}}$	$\frac{4}{5} = 0.8$
$\frac{\text{adjacent}}{\text{hypotenuse}}$	$\frac{3}{5} = 0.6$
$\frac{\text{opposite}}{\text{adjacent}}$	$\frac{4}{3} = 1.33$

c. Verify the results in part b by using your calculator to determine the values of sin(53°), cos(53°), *and* tan(53°).

The rounded answers are the same as the ratios found in part b.

EXAMPLE 2 ***Consider the following right triangle with the given approximate dimensions.***

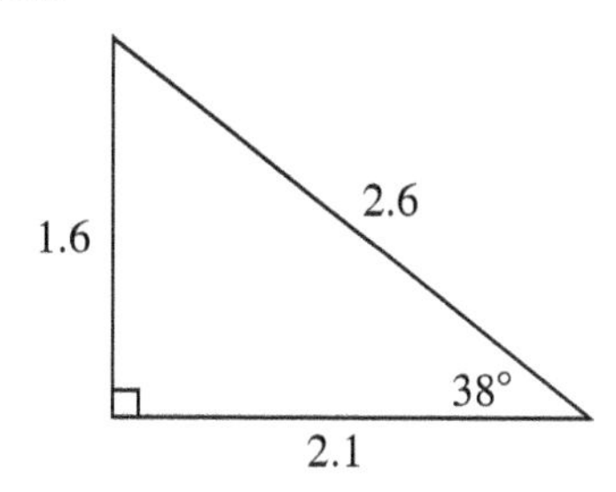

$$\sin 38° = \frac{1.6}{2.6} \approx 0.615, \text{ or } 0.6 \text{ (nearest tenth)}$$

$$\cos 38° = \frac{2.1}{2.6} \approx 0.808, \text{ or } 0.8 \text{ (nearest tenth)}$$

$$\tan 38° = \frac{1.6}{2.1} \approx 0.762, \text{ or } 0.8 \text{ (nearest tenth)}$$

Therefore, for any size right triangle with a 38° angle as one of its acute angles, the sine of 38° is always approximately 0.615, the cosine of 38° is always approximately 0.808, and the tangent of 38° is always approximately 0.762.

You can use your graphing calculator to evaluate sin 38°. Make sure the calculator is in degree mode. Press the (SIN) key, followed by (3), (8), ( )) and (ENTER).

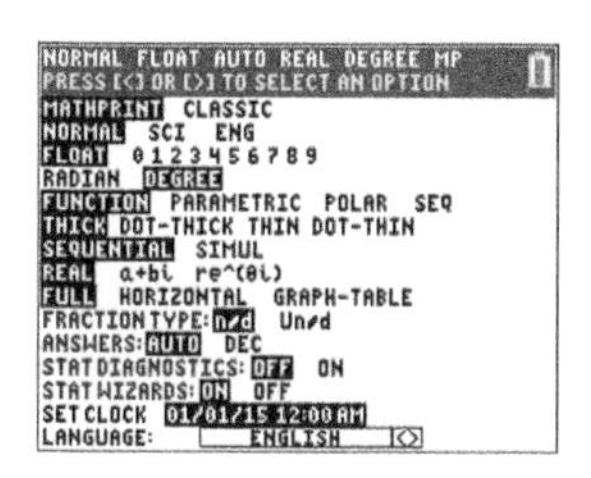

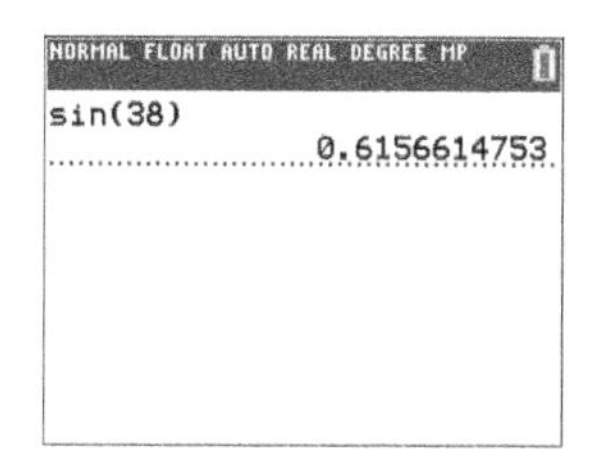

6. a. The sum of the three angles in a triangle is 180°. If one angle measures 90°, the sum of the measures of the other two angles must be 90°. If one of the acute angles measures 38°, the other acute angle is 52°. Use the accompanying figure to determine each of the following:

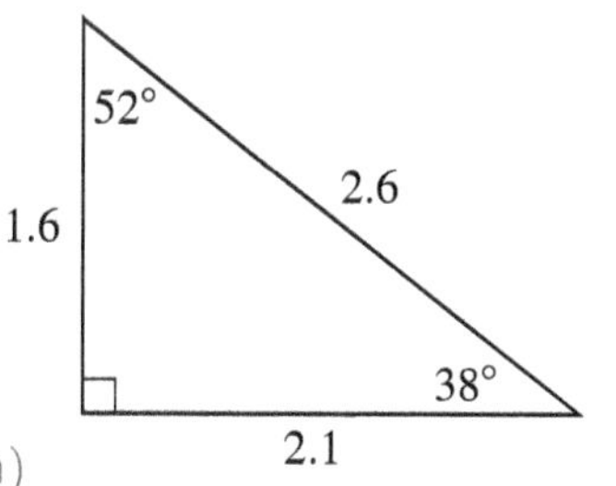

i. $\sin 52° = \frac{2.1}{2.6} \approx 0.808$, or 0.8 (nearest tenth)

ii. $\cos 52° = \frac{1.6}{2.6} \approx 0.615$, or 0.6 (nearest tenth)

iii. $\tan 52° = \frac{2.1}{1.6} \approx 1.313$, or 1.3 (nearest tenth)

b. Verify your rounded answers in part a using your graphing calculator.

NORMAL FLOAT AUTO REAL DEGREE MP
sin(52)
0.7880107536
cos(52)
0.6156614753
tan(52)
1.279941632

EXAMPLE 3 ***Consider the following right triangle, where A and B are acute angles.***

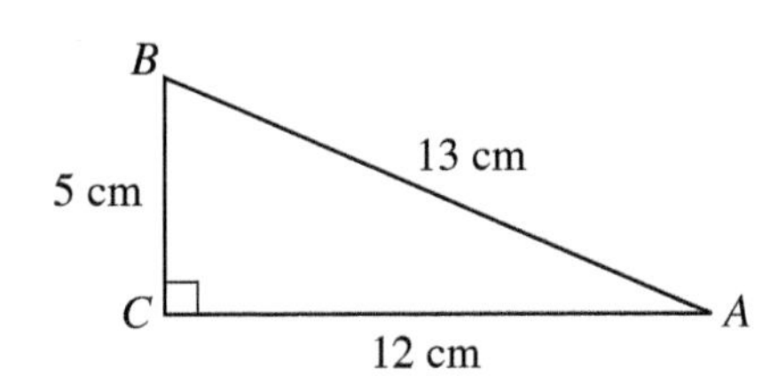

$$\sin A = \frac{5}{13}, \cos A = \frac{12}{13}, \tan A = \frac{5}{12}, \sin B = \frac{12}{13}, \cos B = \frac{5}{13}, \tan B = \frac{12}{5}$$

7. Consider this right triangle.

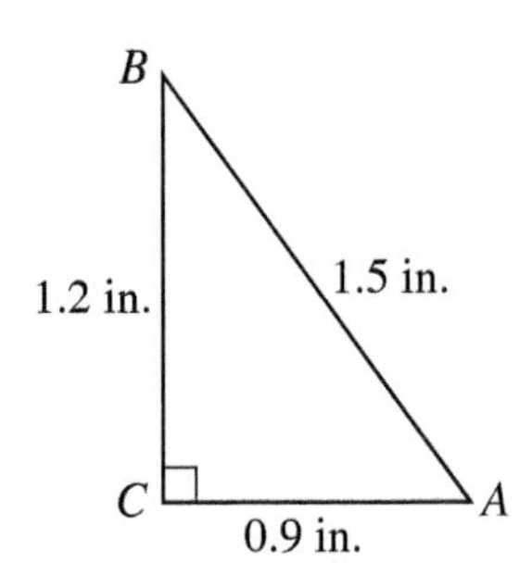

Calculate the following (round to four decimal places):

a. $\sin A$ 0.8000 **b.** $\cos A$ 0.6000 **c.** $\tan A$ 1.3333

d. $\sin B$ 0.6000 **e.** $\cos B$ 0.8000 **f.** $\tan B$ 0.7500

EXAMPLE 4 ***Solve the following equations. Round your answers to the nearest tenth.***

a. $\sin 56° = \frac{x}{15}$ **b.** $\cos 13° = \frac{24}{x}$ **c.** $\tan 72° = \frac{x}{24.7}$

SOLUTION

Use the calculator in degree mode to evaluate the function values,

a.
$$\begin{aligned} 15 \sin 56 &= x \\ 15(0.829) &= x \\ 12.4 &\approx x \end{aligned}$$

b.
$$\begin{aligned} x \cos 13 &= 24 \\ x(0.974) &= 24 \\ x &= \frac{24}{0.974} \\ x &\approx 24.6 \end{aligned}$$

c.
$$\begin{aligned} 24.7 \tan 72 &= x \\ (24.7)(3.078) &= x \\ 76.0 &\approx x \end{aligned}$$

8. Solve the following equations. Round your answers to the nearest tenth.

a. $\sin 24° = \frac{x}{10}$ **b.** $\cos 63° = \frac{x}{23.5}$ **c.** $\tan 48° = \frac{16}{x}$

$x \approx 4.1$ $x \approx 10.7$ $x \approx 14.4$

Trigonometric Values of Special Angles

To determine the trigonometric function values for 30° and 60°, start with an equilateral triangle, a triangle with three equal sides. Assume for now that all three sides are 2 units in length. Note that all three angles must measure 60°.

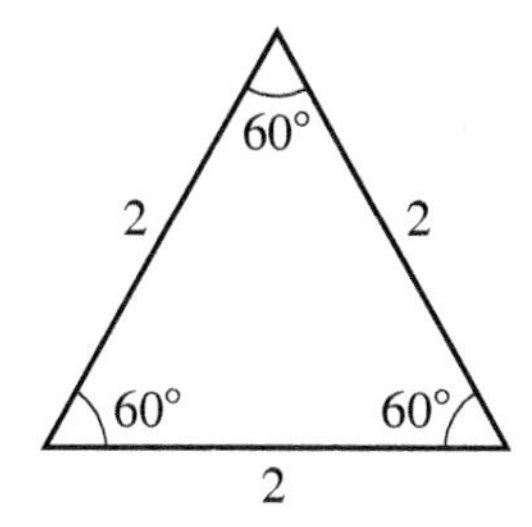

Now bisect one of the angles to form two congruent right triangles with angles measuring 30°, 60°, and 90°.

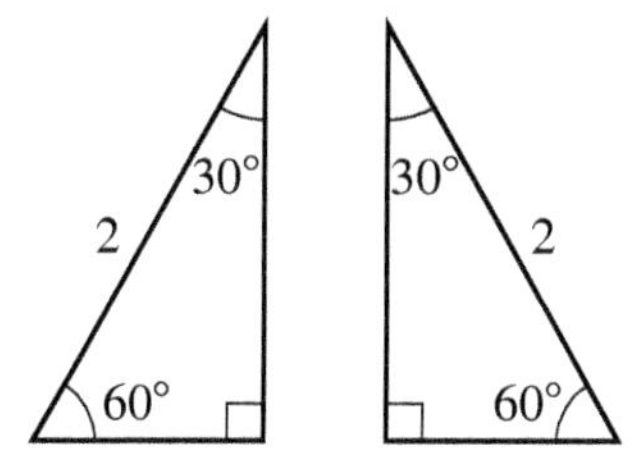

9. a. Considering one of these two new triangles, what are the lengths of the two legs of the right triangle?

The shorter leg has length 1, and the longer leg has length $\sqrt{3}$.

b. With respect to the 30° angle, what are the lengths of the opposite side, the adjacent side, and the hypotenuse, respectively?

The opposite side has length 1, the adjacent side has length $\sqrt{3}$, and the hypotenuse has length 2.

c. With respect to the 60° angle, what are the lengths of the opposite side, the adjacent side, and the hypotenuse, respectively?

The opposite side has length $\sqrt{3}$, the adjacent side has length 1, and the hypotenuse has length 2.

d. Use the results of parts b and c to complete the following table:

θ	$\sin\theta$	$\cos\theta$	$\tan\theta$
30°	$\frac{1}{2}$	$\frac{\sqrt{3}}{2} \approx 0.866$	$\frac{1}{\sqrt{3}} \approx 0.577$
60°	$\frac{\sqrt{3}}{2} \approx 0.866$	$\frac{1}{2}$	$\sqrt{3} \approx 1.732$

To determine the trigonometric function values of 45°, start with a 45-45-90 isosceles right triangle.

10. For convenience, now assume that the equal legs have length 1.

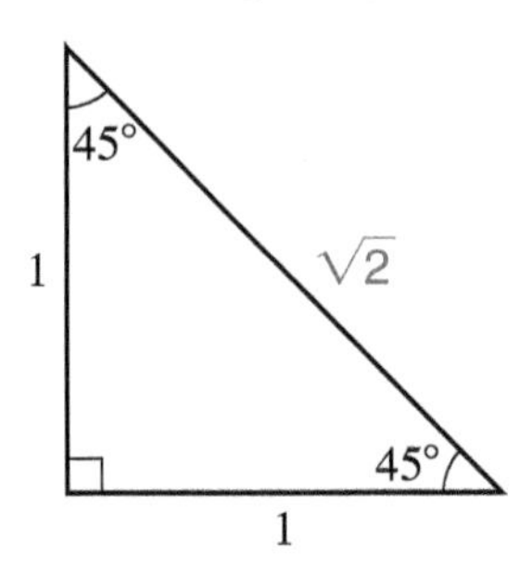

a. Determine the length of the hypotenuse.

$h = \sqrt{1^2 + 1^2} = \sqrt{2}$

b. With respect to the 45° angle, what are the lengths of the opposite side, the adjacent side, and the hypotenuse, respectively?

The opposite side has length 1, the adjacent side has length 1, and the hypotenuse has length $\sqrt{2}$.

c. Use the results of part b to complete the following table:

θ	$\sin\theta$	$\cos\theta$	$\tan\theta$
45°	$\frac{1}{\sqrt{2}} \approx 0.707$	$\frac{1}{\sqrt{2}} \approx 0.707$	1

Knowing the trigonometric function values for 30°, 45°, and 60° can be very helpful in understanding the behavior of these functions. Keeping these values handy or memorizing them is a good idea.

EXAMPLE 5 ***Consider the following two right triangles:***

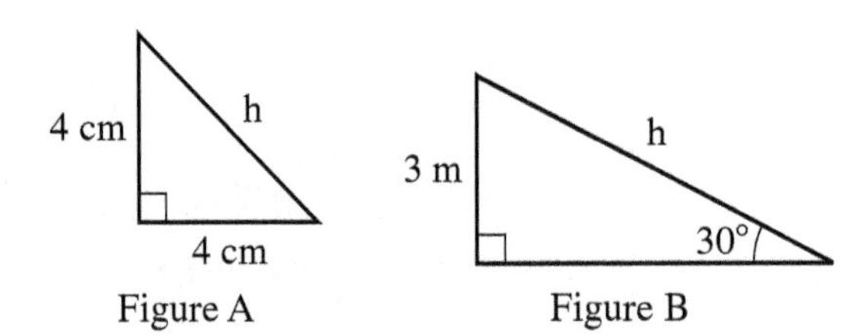

Figure A Figure B

a. Given that the length of one leg of a 45-45-90 triangle is 4 centimeters, determine the exact length of the other two sides. (See Figure A.)

SOLUTION

In Figure A, the two acute angles both measure 45° and the two legs are the same length. So the other leg is also 4 centimeters long. To determine the length of the hypotenuse, use the sine function.

$$\sin 45° = \frac{4}{h}; \quad \frac{1}{\sqrt{2}} = \frac{4}{h}; \quad h = 4\sqrt{2}\text{ cm}$$

b. Given that the length of the shortest side of a 30-60-90 triangle is 3 meters, determine the lengths of the other two sides. (See Figure B.)

SOLUTION

In Figure B, the smallest angle measures 30°. With respect to the 30° angle, the opposite side is given. To determine the length of the adjacent side, the tangent function could be used.

$$\tan 30° = \frac{3}{a}; \quad \frac{1}{\sqrt{3}} = \frac{3}{a}; \quad a = 3\sqrt{3}\text{ m}$$

To determine the length of the hypotenuse, the sine function could be used.

$$\sin 30° = \frac{3}{h}; \quad \frac{1}{2} = \frac{3}{h}; \quad h = 6\text{ m}$$

Tower of Pisa Problem Revisited

You are now ready to answer the original Tower of Pisa problem.

11. a. Recall that the tower was originally 179 feet high and that it now makes an angle of 85° with the ground. Construct a right triangle that models these conditions.

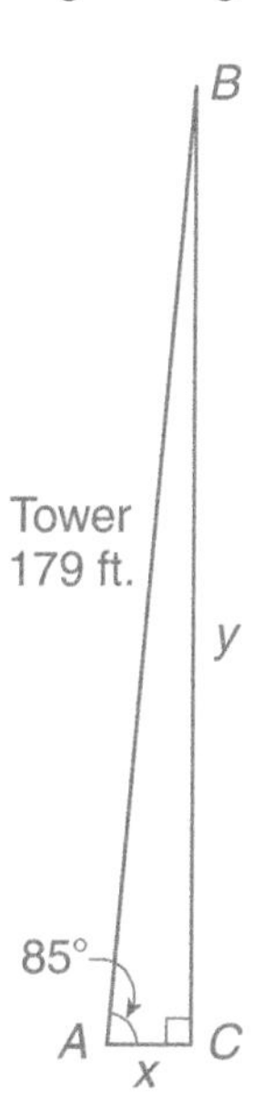

b. With respect to the 85° angle, is the vertical height of the tower represented by the length of an opposite side, the length of an adjacent side, or the length of the hypotenuse of your triangle?

The length of the tower is represented by the hypotenuse with respect to the 85° angle. The actual height of the tower (distance from top of tower to base measured vertically) is measured by the length of the opposite side, *y*.

c. You want to determine the vertical height of the top of the tower above the ground. Therefore, you want to determine the length of which side of the triangle with respect to the 85° angle?

I want to determine the side opposite the 85° angle.

d. Which trigonometric function relates the side with the length you know and the side with the length you want to know?

The sine function relates the 179-foot side (hypotenuse) to the vertical distance, *y*, the top of the tower is above the ground.

e. Write an equation using the information in parts a–d.

$\sin 85° = \frac{y}{179}$, where y is vertical distance the top of the tower is above the ground.

f. Using your calculator to evaluate sin 85°, solve the equation in part e.

$y \approx 178.3$ ft. y is approximately equal to 178.3 feet. Thus, the top of the tower is now 178.3 vertical feet from the ground.

SUMMARY Activity 6.1

1. The **trigonometric functions** are functions whose inputs are measures of the acute angles of a right triangle and whose outputs are ratios of the lengths of the sides of the right triangle.

2. The three sides of a right triangle are the **adjacent** side, the **opposite** side, and the **hypotenuse**. The hypotenuse is always the side opposite the right (90°) angle. The other two sides vary depending on which angle is used as the input.

3. The **sine, cosine**, and **tangent** of the acute angle A of a right triangle are defined by

$$\sin A = \frac{\textit{length of the side opposite A}}{\textit{length of the hypotenuse}}$$

$$\cos A = \frac{\textit{length of the side adjacent to A}}{\textit{length of the hypotenuse}}$$

$$\tan A = \frac{\textit{length of the side opposite A}}{\textit{length of the side adjacent to A}}.$$

For additional practice working with the trigonometric functions in right triangles and special triangles, see Appendix B.

4. Trigonometric values of special angles

θ	$\sin \theta$	$\cos \theta$	$\tan \theta$
30°	$\frac{1}{2}$	$\frac{\sqrt{3}}{2} \approx 0.866$	$\frac{1}{\sqrt{3}} \approx 0.577$
45°	$\frac{1}{\sqrt{2}} = \frac{\sqrt{2}}{2} \approx 0.707$	$\frac{1}{\sqrt{2}} = \frac{\sqrt{2}}{2} \approx 0.707$	1
60°	$\frac{\sqrt{3}}{2} \approx 0.866$	$\frac{1}{2}$	$\sqrt{3} \approx 1.732$

EXERCISES Activity 6.1

1. Triangle ABC is a right triangle.

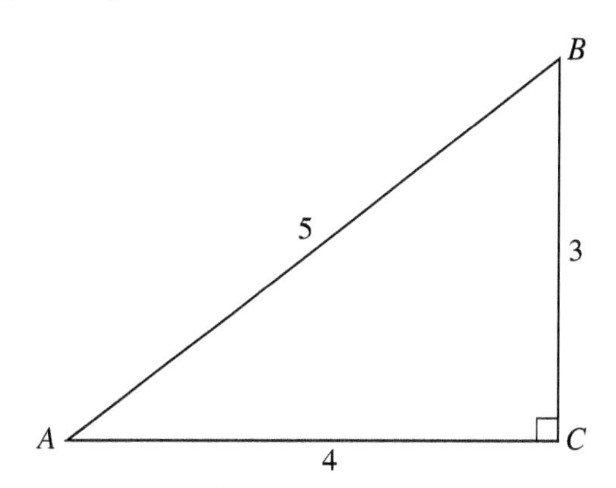

Determine each of the following:

a. $\sin A$	b. $\sin B$	c. $\cos A$
0.6000	0.8000	0.8000
d. $\cos B$	e. $\tan A$	f. $\tan B$
0.6000	0.7500	1.3333

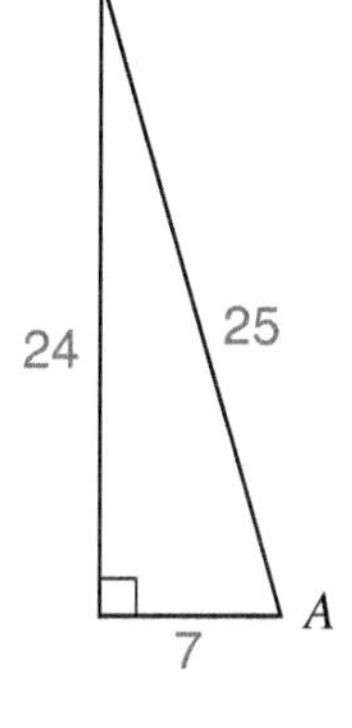

2. In a certain right triangle, $\sin A = \frac{24}{25}$.

   a. Determine possible lengths of the three sides of the right triangle.

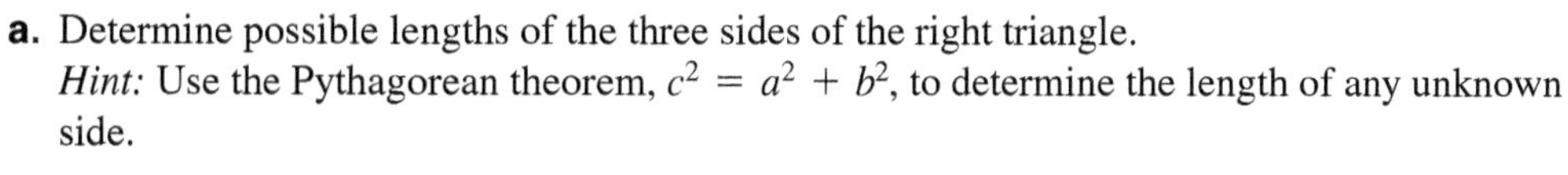

   Hint: Use the Pythagorean theorem, $c^2 = a^2 + b^2$, to determine the length of any unknown side.

   The length of the side opposite angle A is 24; the hypotenuse is 25. Using the Pythagorean theorem, I determine that the length of the side adjacent to angle A is 7.

   b. Determine $\cos A$.

   $\cos A = \frac{7}{25} = 0.2800$

   c. Determine $\tan A$.

   $\tan A = \frac{24}{7} \approx 3.4286$

3. In a certain right triangle, $\tan B = \frac{7}{4}$.

   a. Determine possible lengths of the three sides of the right triangle.

   Given $\tan B = \frac{7}{4}$, if the length of the side opposite angle B is 7, then the length of the side adjacent to angle B is 4. Using the Pythagorean theorem, I determine that the length of the hypotenuse is $\sqrt{65}$.

Exercise numbers appearing in color are answered in the Selected Answers appendix.

b. Determine sin *B*.

$$\sin B = \frac{7}{\sqrt{65}} \approx 0.8682$$

c. Determine cos *B*.

$$\cos B = \frac{4}{\sqrt{65}} \approx 0.4961$$

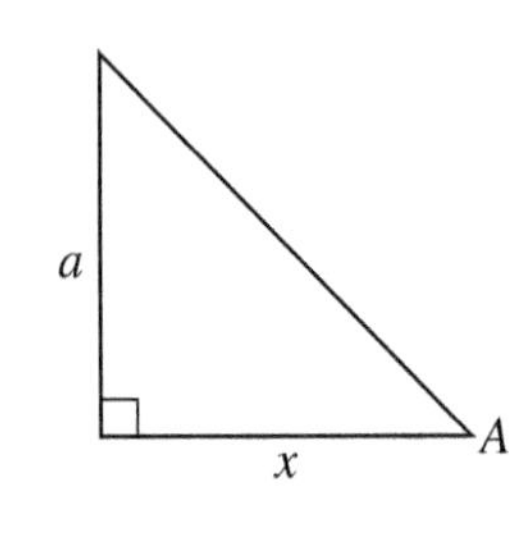

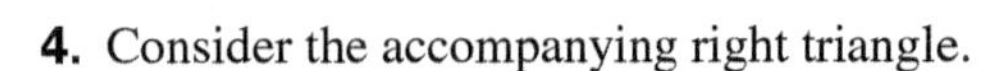

4. Consider the accompanying right triangle.

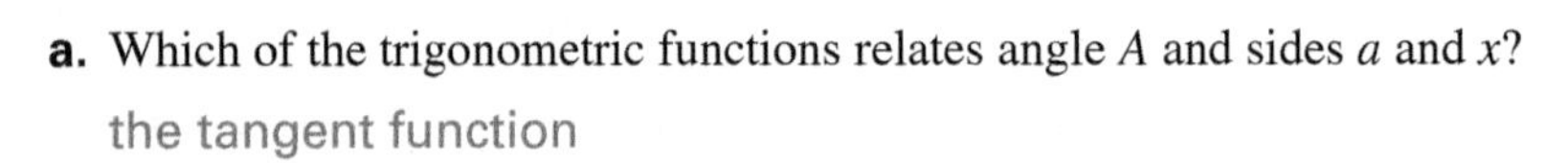

a. Which of the trigonometric functions relates angle *A* and sides *a* and *x*?

the tangent function

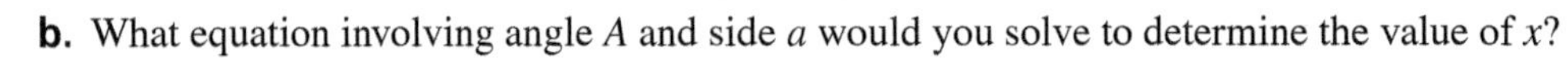

b. What equation involving angle *A* and side *a* would you solve to determine the value of *x*?

$$\tan A = \frac{a}{x}$$

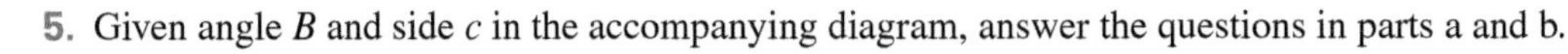

5. Given angle *B* and side *c* in the accompanying diagram, answer the questions in parts a and b.

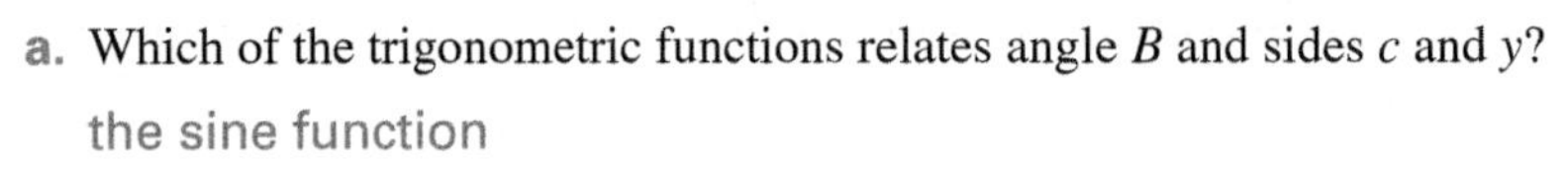

a. Which of the trigonometric functions relates angle *B* and sides *c* and *y*?

the sine function

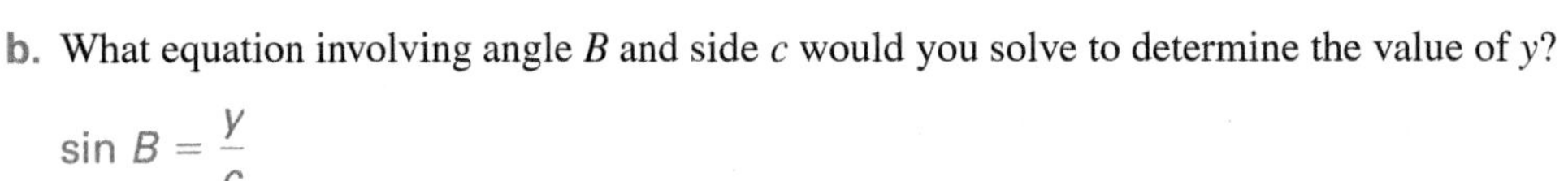

b. What equation involving angle *B* and side *c* would you solve to determine the value of *y*?

$$\sin B = \frac{y}{c}$$

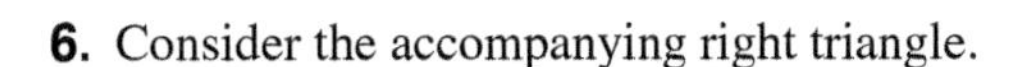

6. Consider the accompanying right triangle.

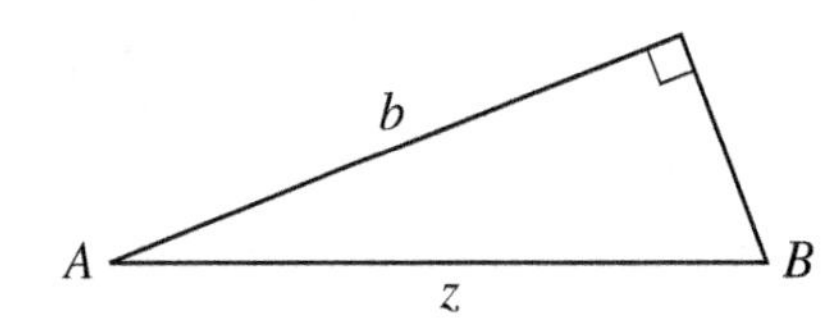

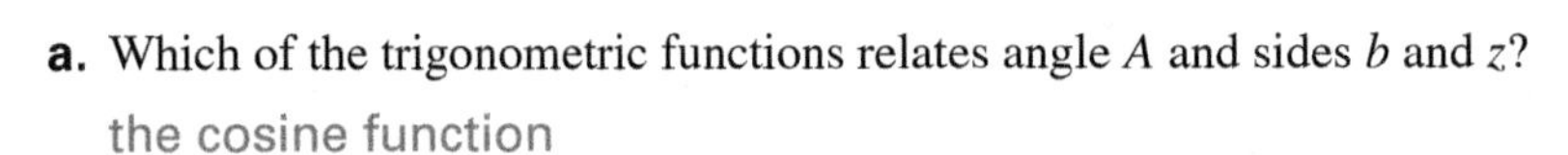

a. Which of the trigonometric functions relates angle *A* and sides *b* and *z*?

the cosine function

b. What equation involving angle *A* and side *b* would you solve to determine the value of *z*?

$$\cos A = \frac{b}{z}$$

7. Solve the following equations. Round your answers to the nearest tenth.

a. $\sin 49° = \dfrac{x}{12}$

$12 \sin 49° = x$

$x \approx 9.1$

b. $\tan 84° = \dfrac{x}{9}$

$x = 9 \tan 84°$

$x \approx 85.6$

c. $\sin 22° = \dfrac{23}{x}$

$x \sin 22° = 23$

$x = \dfrac{23}{\sin 22°}$

$x \approx 61.4$

8. a.

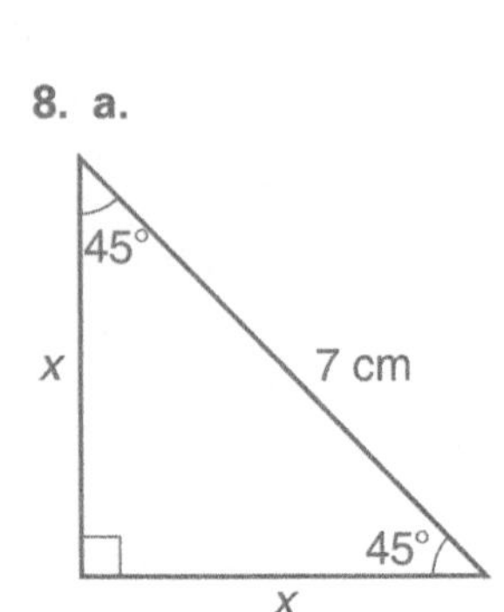

8. a. Given that the length of the hypotenuse of a 45-45-90 triangle is 7 centimeters, determine the exact length of the two legs. Sketch and label a diagram before calculating.

$\sin 45° = \dfrac{x}{7}$; $\dfrac{1}{\sqrt{2}} = \dfrac{x}{7}$; $7 = \sqrt{2}x$; $x = \dfrac{7}{\sqrt{2}}$. The third side is the same.

8. b.

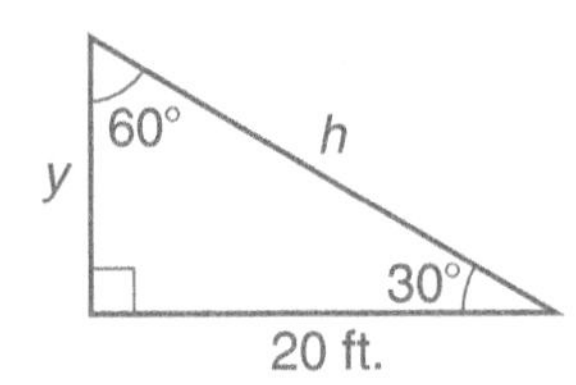

b. Given that the length of the longer leg of 30-60-90 triangle is 20 feet, determine the exact length of the short leg and the hypotenuse. Sketch and label a diagram before calculating.

$\tan 30° = \frac{y}{20}$; $\frac{1}{\sqrt{3}} = \frac{y}{20}$; $20 = \sqrt{3}y$; $y = \frac{20}{\sqrt{3}}$ ft.

$\cos 30° = \frac{20}{h}$; $\frac{\sqrt{3}}{2} = \frac{20}{h}$; $\sqrt{3}h = 40$; $h = \frac{40}{\sqrt{3}}$ ft.

9. A friend asks you to help build a ramp at his mother-in-law's house. Three 7-inch-high steps lead to the front door. Another friend donates a 15-foot ramp. The building inspector informs you that any access ramp for people with disabilities can have an inclination no greater than 5°.

a. Sketch a diagram assuming that the land in front of the steps is level and that the ramp makes a 5° angle with the top of the steps.

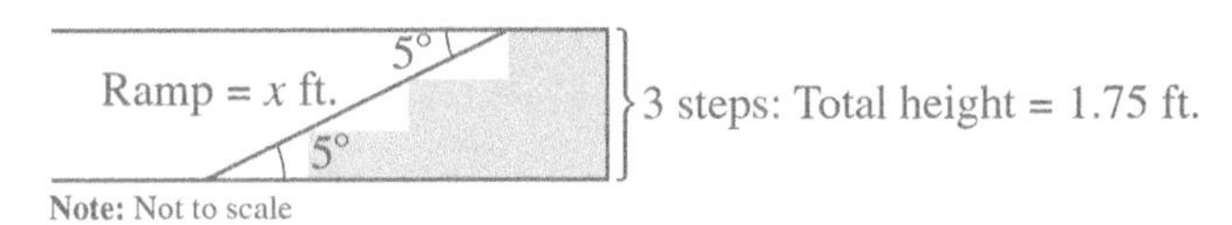

b. What is the increase in height from one end of the ramp to the other?

$3 \cdot 7 = 21$ in. The increase in height from one end of the ramp to the top of the stairs is $\frac{21}{12} = 1.75$ feet.

c. Would the donated ramp be long enough to meet the code? Explain.

$$\sin 5° = \frac{1.75}{x}$$

$$x = \frac{1.75}{\sin 5°}$$

$$x \approx 20.1 \text{ ft.}$$

The ramp needs to be at least 20.1 feet long. Therefore, the donated ramp will not be long enough to meet the code.

Alternative approach: $15 \sin 5° \approx 1.3$ feet. The three steps must measure at most 1.3 feet high for the 15-foot ramp to satisfy the code.

Each solution suggests ways to think about modifications to either the ramp or the steps (or both) that could be used to meet the code.

ACTIVITY

6.2

A Gasoline Problem

Cofunctions of Complementary Angles

OBJECTIVES

1. Identify complementary angles.
2. Demonstrate that the sine of one of the complementary angles equals the cosine of the other.

You and a friend take a camping trip to the American Southwest. On the way home, you realize that your Jeep is running low on gas, so you stop at a filling station. Unfortunately, the attendant informs you that his station has been without gas for several days. However, he is certain that a station on a side road up ahead has gas to sell.

A quick phone call confirms this information, and you begin to ask about the exact location of the station. The attendant's map of the area indicates that the station is about 10 miles away at an angle of 35° with the road you are currently traveling.

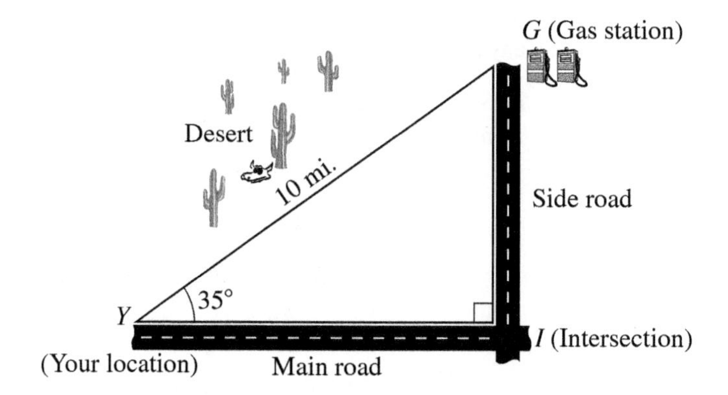

Your Jeep is equipped with a GPS, so driving through the desert is not a problem, although you would prefer to stay on the road. You estimate that you have enough gas for 15 miles. Do you have enough gas to make it to the other station without going through the desert?

This problem can be solved using the triangle above and some trigonometry to determine the distance from you to the intersection, side *YI*, and from the intersection to the gas station, side *IG*.

1. With respect to angle *Y*, select the correct response in parts a and b.

   a. The side *YI* is known as

   i. the side opposite angle *Y*.

   ii. the side adjacent to angle *Y*.

   iii. the hypotenuse.

   b. The side *YG* is known as

   i. the side opposite angle *Y*.

   ii. the side adjacent to angle *Y*.

   iii. the hypotenuse.

   c. The function that relates *YI*, *YG*, and angle *Y* is the cosine function.

2. **a.** Set up an equation indicated by Problem 1c.

   $$\cos Y = \frac{YI}{YG}$$

   b. Solve the equation for the length of *YI*.

   $$\cos 35° = \frac{YI}{10}$$

   $$YI = 10 \cos 35°$$

   $$YI \approx 8.2 \text{ mi.}$$

3. With respect to angle Y, select the correct response in parts a and b.

a. The side IG is known as

i. the side opposite angle Y.

ii. the side adjacent to angle Y.

iii. the hypotenuse.

b. The side YG is known as

i. the side opposite angle Y.

ii. the side adjacent to angle Y.

iii. the hypotenuse.

c. The function that relates IG, YG, and angle Y is the sine function.

4. a. Set up an equation indicated by Problem 3c.

$$\sin Y = \frac{IG}{YG}$$

b. Solve the equation for the length of IG.

$$\sin 35^\circ = \frac{IG}{10}$$

$$IG = 10 \sin 35^\circ$$

$$IG \approx 5.7 \text{ mi.}$$

5. Do you have enough gas to make it without journeying through the desert? Explain.

To avoid the desert, I would have to travel 8.2 + 5.7 = 13.9 miles on the two roads to the other station. I should be able to make it.

Meanwhile at the second service station, the attendant is also making some calculations. He is aware of your situation and is trying to anticipate which option you will choose in case he has to go look for you.

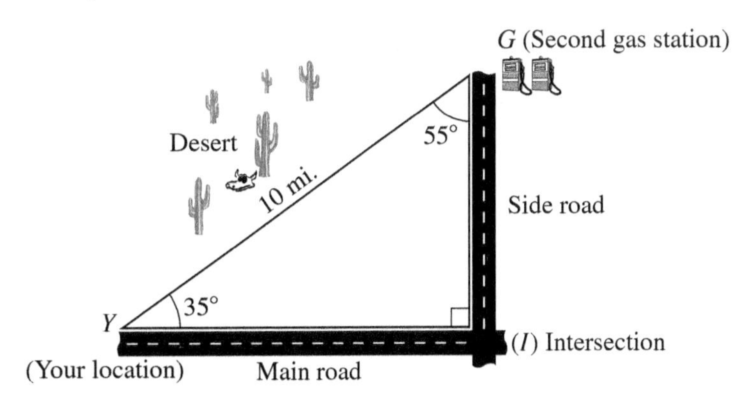

He knows that the line to the first station makes a 55° angle with his road.

6. With respect to angle G, select the correct response in parts a and b.

a. The side YI is known as

i. the side opposite angle G.

ii. the side adjacent to angle G.

iii. the hypotenuse.

b. The side YG is known as

i. the side opposite angle G.

ii. the side adjacent to angle G.

iii. the hypotenuse.

c. The function that relates YI, YG, and angle G is the sine function.

7. a. Set up an equation indicated by Problem 6c.

$$\sin G = \frac{YI}{YG}$$

b. Solve the equation for the length of YI.

$$\sin 55^\circ = \frac{YI}{10}$$

$$YI = 10 \sin 55^\circ$$

$$YI \approx 8.2 \text{ mi.}$$

c. How does the answer in part b compare with the answer in Problem 2b?

They are the same.

8. With respect to angle G, select the correct response in parts a and b.

a. The side IG is known as

i. the side opposite angle G.

ii. the side adjacent to angle G.

iii. the hypotenuse.

b. The side YG is known as

i. the side opposite angle G.

ii. the side adjacent to angle G.

iii. the hypotenuse.

c. The function that relates IG, YG, and angle G is the cosine function.

9. a. Set up an equation indicated by Problem 8c. $\cos G = \frac{IG}{YG}$

b. Solve the equation for the length of IG.

$$IG = 10 \cos 55^\circ$$

$$IG \approx 5.7 \text{ mi.}$$

c. How does the answer in part b compare with the answer in Problem 4b?

They are the same.

Complementary Angles

Although you used different angles of the triangle YGI in Problems 1–9, your results for the lengths of YI and IG should have been the same. The angles involved in these calculations were 35° and 55°. Because their sum is 90°, they are called **complementary angles**.

DEFINITION

Two acute angles A and B whose measures sum to 90° are called **complementary angles**. If x represents the measure of an acute angle, then $90 - x$ represents the measure of its complementary angle.

Because the measures of the two acute angles in a right triangle sum to 90°, they are complementary.

EXAMPLE 1

ACUTE ANGLE, x	COMPLEMENTARY ANGLE, $90° - x$
35°	55°
55°	35°
72°	18°
18°	72°

LC LEARNING CATALYTICS

Which of the following is equivalent to sin 57°?

a. cos 57°.

b. tan 33°.

c. cos 33°.

d. None of these.

10. a. The solution to Problem 2b involved cos 35°, and the solution to Problem 7b involved sin 55°. Compare the values of cos 35° and sin 55°.

cos 35° = sin 55° in the respective equations; thus, the resulting values are equal.

b. The solution to Problem 4b involved sin 35°, and the solution to Problem 9b involved cos 55°. Compare the values of sin 35° and cos 55°.

sin 35° = cos 55° in the respective equations; thus, the resulting values are equal.

This situation demonstrates another fundamental principle of trigonometry.

Cofunctions of complementary angles are equal. Symbolically, if x is an acute angle, then

$$\sin x = \cos(90° - x)$$

and

$$\cos x = \sin(90° - x).$$

In fact, the name *cosine* is derived from the words *sine* and *complement.*

EXAMPLE 2

a. $\sin 35° = \cos 55°$

b. $\sin 55° = \cos 35°$

c. $\sin 72° = \cos 18°$

d. $\sin 18° = \cos 72°$

11. Complete the following table, where x represents the measure of an angle in degrees. Use your graphing calculator to determine values of $\sin x$ and $\cos(90 - x)$ to four decimal places.

x	$90° - x$	$\sin x$	$\cos(90° - x)$
0°	90°	0.0000	0.0000
15°	75°	0.2588	0.2588
30°	60°	0.5000	0.5000
45°	45°	0.7071	0.7071
60°	30°	0.8660	0.8660
75°	15°	0.9659	0.9659
90°	0°	1.0000	1.0000

SUMMARY Activity 6.2

1. **Complementary** angles are two acute angles whose measures sum to 90°.
2. The two acute angles in a right triangle are complementary.
3. The *co* in cosine is from the word *complement.*
4. **Cofunctions** of complementary angles are equal.

EXERCISES Activity 6.2

1. You are in a rowboat on Devil Lake in Ontario, Canada. Your lakeside cabin has no running water, so you sometimes go to a fresh spring at a different point on the lakeshore. You row in a direction 60° north of east for half a mile. You are not yet tired, and it is a lovely day.

   a. How far would you now have to row if you were to return to your cabin by rowing directly south and then directly west? Note that the resulting figure is a 30-60-90 triangle. Round your answer to the nearest hundredth.

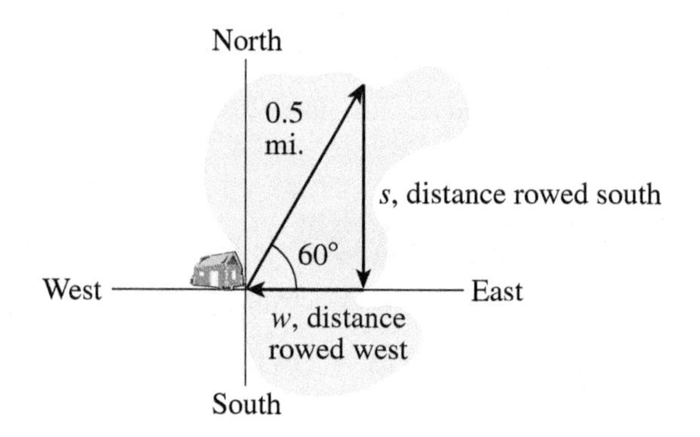

$s = 0.5 \sin 60° \approx 0.43$ mi.

$w = 0.5 \cos 60° = 0.25$ mi.

$s + w \approx 0.68$ mile

I would travel 0.68 mile, rowing back directly south and then directly west.

Exercise numbers appearing in color are answered in the Selected Answers appendix.

b. Check your solution to part a, using the complementary angle of the angle used in part a.

$s = 0.5 \cos 30° \approx 0.43$ mi.

$w = 0.5 \sin 30° = 0.25$ mi.

These calculations confirm the result in part a.

2. One afternoon you don't pay enough attention while rowing. You row 20° off course, too far north, instead of directly west as you had intended. You row off course for 300 meters.

a. Draw a diagram of this situation.

2. a.

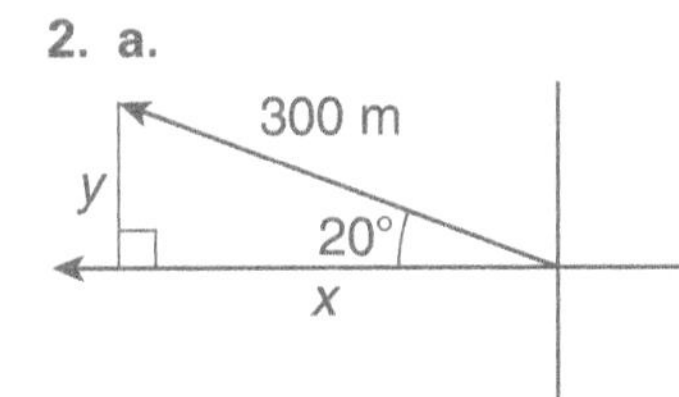

b. How far west have you gone? Round your answer to the nearest meter.

Let x be the distance to the west when I realize that I am off course.

$x = 300 \cos 20° \approx 282$ m

I have rowed 282 meters directly west.

c. How far north are you of where you had originally planned to be?

Let y be the distance I have rowed to the north when I realize that I am off course.

$y = 300 \sin 20° \approx 103$ m

I am 103 meters to the north when I realize that I am off course.

d. Check your solutions to parts b and c, using the complementary angle of the angle used in parts b and c.

$x = 300 \sin (90° - 20°) = 300 \sin (70°) \approx 282$ m

$y = 300 \cos (90° - 20°) = 300 \cos (70°) \approx 103$ m

These calculations verify the results in b and c.

3. a. Complete the following table, where x represents the measure of an angle in degrees. Use your graphing calculator to determine values of $\sin x$ and $\cos (90 - x)$ to four decimal places.

x	$90° - x$	$\sin x$	$\cos (90° - x)$
7°	83°	0.1219	0.1219
17°	73°	0.2924	0.2924
24°	66°	0.4067	0.4067
33°	57°	0.5446	0.5446
48°	42°	0.7431	0.7431
67°	23°	0.9205	0.9205
77°	13°	0.9744	0.9744

b. What trigonometric property does this table illustrate?

The table in part a illustrates the property that cofunctions of complementary angles are equal.

4. a. Complete the following table, where x represents the measure of an angle in degrees. For the special values in the table, determine the exact values of $\cos x$ and $\sin(90° - x)$.

x	$90° - x$	$\cos(x)$	$\sin(90° - x)$
30°	60°	$\frac{\sqrt{3}}{2}$	$\frac{\sqrt{3}}{2}$
45°	45°	$\frac{1}{\sqrt{2}}$	$\frac{1}{\sqrt{2}}$
60°	30°	$\frac{1}{2}$	$\frac{1}{2}$

b. What trigonometric property does this table illustrate?

The table in part a illustrates the property that cofunctions of complementary angles are equal.

ACTIVITY
6.3

The Sidewalks of New York
Inverse Sine, Cosine, and Tangent Functions

OBJECTIVES

1. Determine the inverse tangent of a number.
2. Determine the inverse sine and cosine of a number using the graphing calculator.
3. Identify the domain and range of the inverse sine, cosine, and tangent functions.

A friend of yours is having a party in her Manhattan apartment, which borders Central Park. She gives you the following directions from your place, which also borders the park:

i. If you are coming after dark, head east for three blocks, going around the park, and then go north for two blocks.

ii. If you can come early, you can cut through the park, a shorter route.

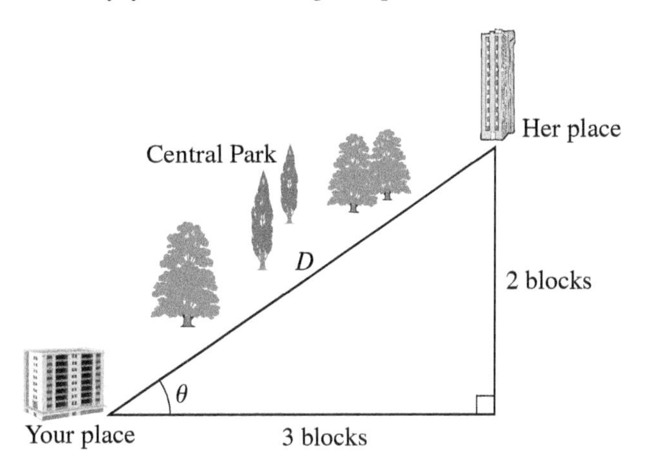

You leave your place early and decide to cut through the park. You want to compute the shortest distance, D (in blocks), between your apartments.

1. Using the Pythagorean theorem, $c^2 = a^2 + b^2$, determine the distance, D, in blocks, across the park.

$\sqrt{2^2 + 3^2} = \sqrt{13} \approx 3.6$ blocks

The shortest distance between the two apartments is 3.6 blocks.

Now you want to determine the direction (angle) you need to go to get to your friend's place. You can represent the angle by θ, the Greek letter *theta*. Because you now have the lengths of all three sides of the triangle, you can use any of the trigonometric functions to help determine θ. Begin with the tangent function.

Inverse Tangent Function

Recall that the input for the tangent function is an angle in a right triangle and that the output is the ratio of the length of the side opposite the length of the side adjacent.

$$\textit{tangent of an angle} = \frac{\textit{length of the opposite side}}{\textit{length of the adjacent side}}$$

2. Determine the value of the tangent of θ, written $\tan\theta$, for the Central Park triangle.

$$\tan\theta = \frac{2}{3} \approx 0.6667$$

If you want to determine the tangent of a known angle, you can use the TAN key on your calculator. From Problem 2, you know the value of the tangent of the angle θ, but not the value of θ. That is, you know the output for the tangent function, but not the input.

3. Use the table feature on your calculator to approximate θ from Problem 2. You should compute the tangent of several possible values of θ to get as close as possible to the desired answer. Complete the following table. Round your answers to five decimal places.

θ	$\tan \theta$
30°	0.57735
40°	0.83910
35°	0.70021
34°	0.67451
33.7°	0.66692
33.69°	0.66666

You don't have to experiment every time to determine θ when you know the tangent of θ. There is a more direct method using the inverse tangent function. Recall that with inverse functions, the inputs and outputs are interchanged.

DEFINITION

The input of the **inverse tangent** function is the ratio $\dfrac{\textit{length of the opposite side}}{\textit{length of the adjacent side}}$ for the angle. The output is the acute angle. The inverse tangent function is denoted by $\mathbf{tan^{-1}}$, or **arctan**, and defined by

$$\tan^{-1} x = \theta, \text{ where } \theta \text{ is the acute angle whose tangent is } x.$$

Thus, the input, x, represents the ratio of the length of the opposite side to the length of the adjacent side for the angle θ.

Just as $\log x = y$ has the equivalent exponential form $10^y = x$, $\tan^{-1} x = \theta$ is equivalent to $\tan \theta = x$.

EXAMPLE 1 ***Consider the following triangle:***

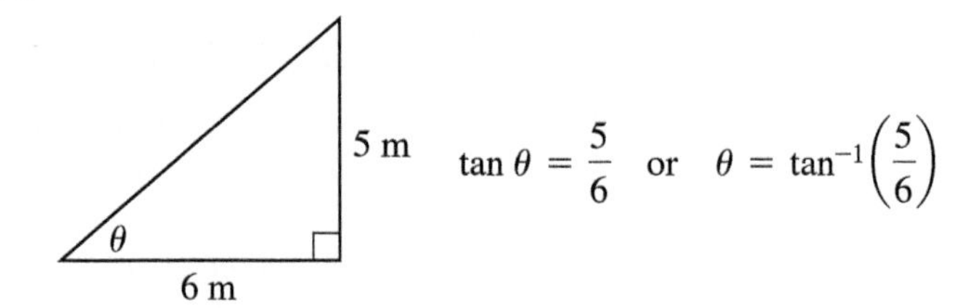

Using your calculator, you can determine that $\theta \approx 39.8°$. See the following TI-84 Plus C screen. Be careful; your calculator must be in degree mode.

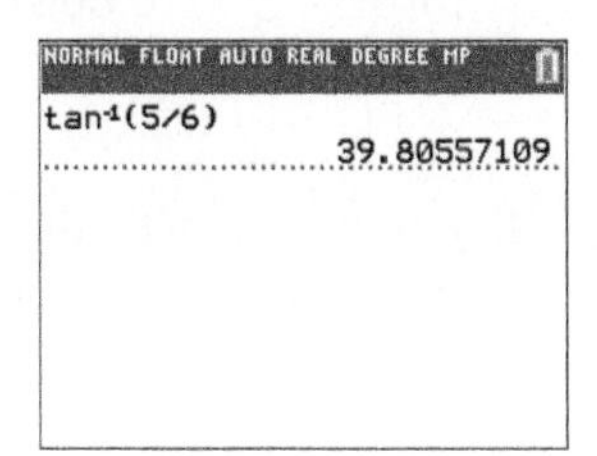

The inverse tangent function is located on your calculator as second function to tangent. That is, you will need to press the 2nd key before you press the tangent key. Note that your calculator uses the notation $\tan^{-1}$ rather than arctan.

4. Use your calculator to determine the inverse tangent of the answer to Problem 2. (Make sure your calculator is in degree mode.) Compare this answer with your approximation from Problem 3.

$$\textit{Remember}: \tan^{-1}(\text{ratio}) = \text{angle}$$
$$\tan(\text{angle}) = \text{ratio}$$

$\tan^{-1}\left(\frac{2}{3}\right) \approx 33.7°$

The value is very close to the approximation.

Inverse Sine and Cosine Function

There are similar definitions for the inverse sine and inverse cosine functions.

DEFINITION

1. The input of the **inverse sine** function is the ratio $\frac{\textit{length of the opposite side}}{\textit{length of the hypotenuse}}$ for the angle. The output is the acute angle. The inverse sine function is denoted by $\mathbf{sin^{-1}}$, or **arcsin**, and defined by

$$\sin^{-1}x = \theta, \text{ where } \theta \text{ is the acute angle whose sine is } x.$$

Thus, the input, x, represents the ratio of the length of the opposite side to the length of the hypotenuse for the angle θ.

2. The input of the **inverse cosine** function is the ratio $\frac{\textit{length of the adjacent side}}{\textit{length of the hypotenuse}}$ for the angle. The output is the acute angle. The inverse cosine function is denoted by $\mathbf{cos^{-1}}$, or **arccos**, and defined by

$$\cos^{-1}x = \theta, \text{ where } \theta \text{ is the acute angle whose cosine is } x.$$

Thus, the input, x, represents the ratio of the length of the adjacent side to the length of the hypotenuse for the angle θ.

Again, $\sin^{-1}x = \theta$ has the equivalent form $\sin\theta = x$ and $\cos^{-1}x = \theta$ has the equivalent form $\cos\theta = x$.

EXAMPLE 2 ***Consider the following triangle:***

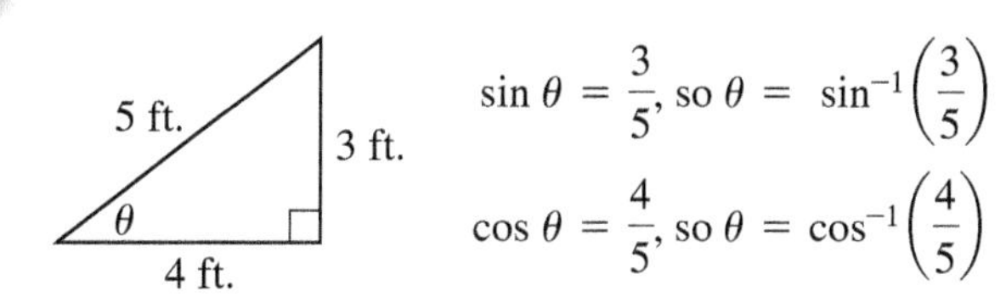

$$\sin\theta = \frac{3}{5}, \text{ so } \theta = \sin^{-1}\left(\frac{3}{5}\right)$$
$$\cos\theta = \frac{4}{5}, \text{ so } \theta = \cos^{-1}\left(\frac{4}{5}\right)$$

Using either the inverse sine or the inverse cosine function, you determine that $\theta \approx 36.9°$. See the following TI-84 C Plus screen.

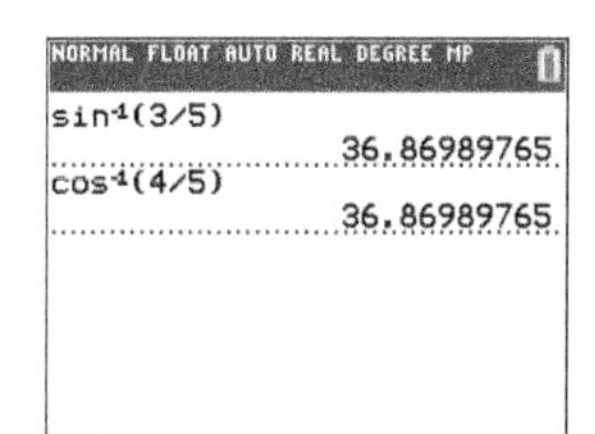

LC LEARNING CATALYTICS

Consider the right triangle ABC with sides a, b, and c, respectively. Angle C is the right angle. Side c is the hypotenuse, side a is the side opposite angle A, and side b is the side opposite angle B. If side a has length 10 and side b has length 24, what is the measure of angle A?

a. 30°.
b. 60°.
c. 22.6°.
d. 67.4°.

5. a. In Problems 2 and 4, you determined the values of θ in the Central Park situation using the inverse tangent function. Now use the inverse sine function to determine the value of θ in the Central Park situation.

$$\theta = \sin^{-1}\left(\frac{2}{\sqrt{13}}\right) \approx 33.6901°$$

b. Use the inverse cosine function to determine the value of θ in the Central Park situation.

$$\theta = \cos^{-1}\left(\frac{3}{\sqrt{13}}\right) \approx 33.6901°$$

6. The term used by highway departments when describing the steepness of a hill is **percent grade**. For example, a hill with a 5% grade possesses a slope of $\frac{5}{100}$, or $\frac{1}{20}$. This means there will be a 5-foot vertical change for every 100 feet of horizontal change or 1 foot of vertical change for every 20 feet of horizontal change as a car ascends or descends the hill.

a. You are driving along Route 17B in the Catskill Mountains of New York State. Just before coming to the top of a hill, you spot a sign that reads "7% Grade Next 3 Miles Trucks Use Lower Gear." Draw a triangle, and label the appropriate parts to model this situation.

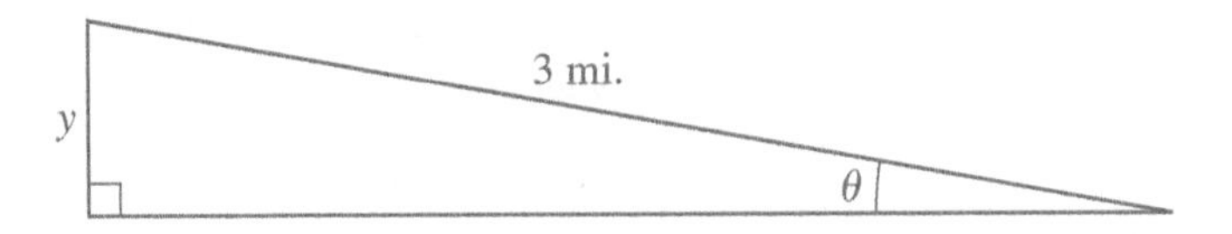

b. Use an inverse trigonometric function to determine the angle that the road makes with the horizontal.

$$\theta = \tan^{-1}\left(\frac{7}{100}\right) \approx 4.004°$$

c. Use trigonometry to determine how many feet of elevation you will lose from the top of the hill to the bottom.

$y = 3 \sin(4.004°) \approx 0.21$ mi.

In 3 miles, the elevation drops 0.21 mile, or approximately 1109 feet.

SUMMARY Activity 6.3

1. The domain (set of inputs) of each **inverse trigonometric function** is a set of ratios of the lengths of the sides of a right triangle.
2. The range (set of outputs) of each inverse trigonometric function is a set of angles.
3. For inverse trigonometric functions,

$\sin^{-1} x = \theta$ *is equivalent to* $\sin\theta = x$,

$\cos^{-1} x = \theta$ *is equivalent to* $\cos\theta = x$, and

$\tan^{-1} x = \theta$ *is equivalent to* $\tan\theta = x$.

EXERCISES Activity 6.3

1. For each of the following, use your calculator to determine θ to the nearest 0.01°:

a. $\theta = \arcsin\left(\frac{1}{2}\right)$

$\theta = 30°$

b. $\theta = \cos^{-1}\left(\frac{3}{7}\right)$

$\theta \approx 64.62°$

c. $\theta = \arctan(2.36)$

$\theta \approx 67.04°$

d. $\theta = \sin^{-1}(0.8974)$

$\theta \approx 63.82°$

e. $\tan\theta = \frac{7}{3}$

$\theta \approx 66.80°$

f. $\cos\theta = \frac{3}{7}$

$\theta \approx 64.62°$

g. $\sin\theta = 0.3791$

$\theta \approx 22.28°$

h. $\tan\theta = 0.3791$

$\theta \approx 20.76°$

2. For each of the following, determine θ without using your calculator:

a. $\tan\theta = 1$

$\theta = 45°$

b. $\sin\theta = 0.5$

$\theta = 30°$

c. $\cos\theta = \frac{\sqrt{3}}{2}$

$\theta = 30°$

3. Complete the accompanying table, which refers to right triangles, labeled as in the figure below.

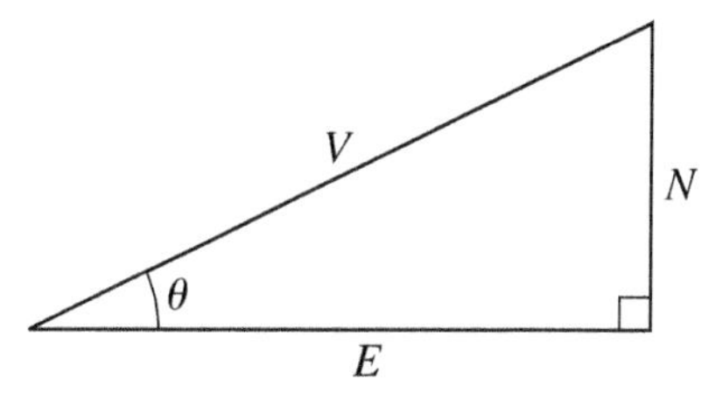

V	θ	E	N
32	65°	13.5	29.0
23.3	59°	12	20
4.1	54°	2.4	3.3
26	43.8°	18.8	18
4.5	45°	3.2	3.2

4. While hiking, you see an interesting rock formation on the side of a vertical cliff. You want to describe for a friend how he might see it when he walks down the path. If you stand on the path at a certain place, 50 feet from the base of the cliff, the rock formation is visible about 30 feet up the cliff. At what angle should you tell your friend to look?

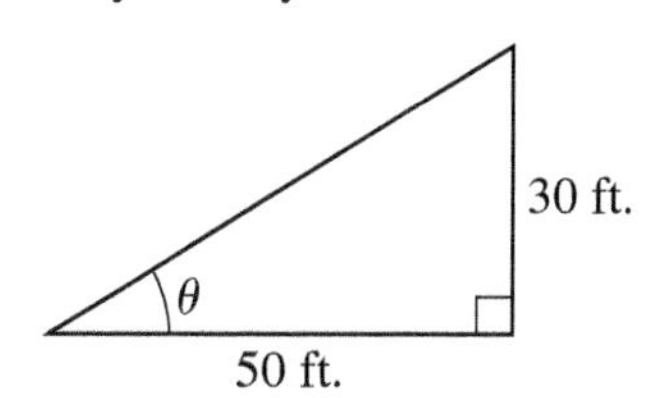

$\theta = \tan^{-1}\left(\frac{30}{50}\right) \approx 31°$

My friend should look up at an angle of approximately 31°.

Exercise numbers appearing in color are answered in the Selected Answers appendix.

5. A warehouse access ramp claims to have a 10% grade. The ramp is 15 feet long.

   a. Draw a diagram of this situation.

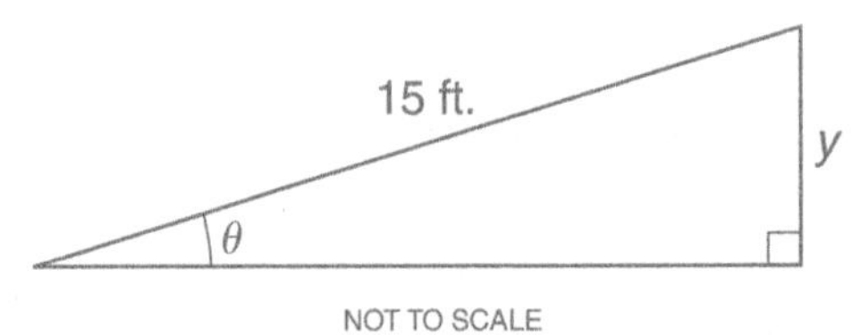

   b. What angle does the ramp make with the horizontal?

   Because grade is rise over run, $\tan\theta = 0.1$.

   $\theta = \tan^{-1}(0.1) \approx 5.7°$

   The ramp makes an angle of 5.7° with the horizontal.

   c. How much does the elevation change from one end of the ramp to the other?

   $y = 15\sin(5.7°) \approx 1.5$ ft.

   The elevation changes 1.5 feet from one end of the ramp to the other.

6. You are at a hot air balloon festival and enjoy watching the balloons inflate and rise straight up into the atmosphere. You are 200 feet from one particular balloon when it begins its ascent.

   a. Draw a diagram of this situation. Let θ represent the angle formed by the balloon, you, and the balloon's lift-off point.

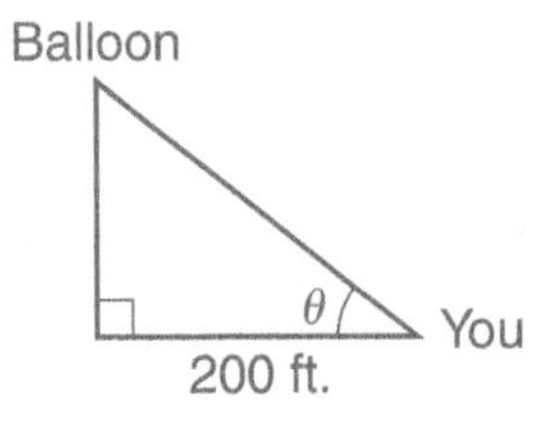

   b. Determine the value of θ when the balloon is 200 feet in the air.

   $\theta = 45°$ because $\theta = \tan^{-1}\left(\frac{200}{200}\right) = \tan^{-1}(1) = 45°$

   c. When the balloon is 400 feet from you, what is the value of θ?

   $\theta = \cos^{-1}\left(\frac{200}{400}\right) = \cos^{-1}\left(\frac{1}{2}\right) = 60°$

   d. Did you recognize the angles in parts b and c as special angles?

   Answers will vary; hopefully, most are yes.

7. You are an engineer studying bridges. You are standing at the edge of the river $\frac{1}{2}$ mile from the base of a particular bridge. You are told that the middle of the bridge is 185 feet above the water. What is the measure of the angle formed by the center of the bridge, you, and the water directly below the center of the bridge?

   $\frac{1}{2}$ mile is 2640 feet. So the situation resembles the following:

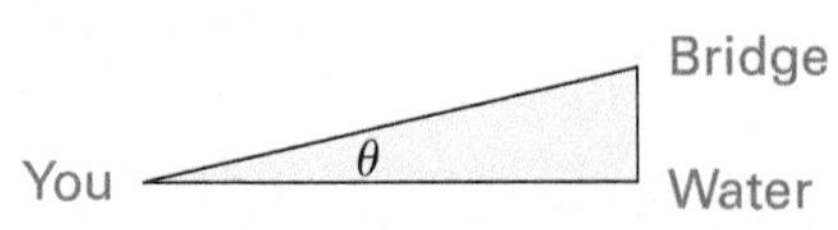

   The opposite side is 185 feet, and the adjacent side is 2640 feet, so $\tan\theta = \frac{185}{2640}$ or $\theta = \tan^{-1}\left(\frac{185}{2640}\right) \approx 4°$.

ACTIVITY
6.4

Solving a Murder
Trigonometric Problem-Solving Strategy

OBJECTIVE

Determine the measure of all sides and all angles of a right triangle.

There has been a fatal shooting 110 feet from the base of a 25-story building. Each story measures approximately 12 feet. Two suspects live in the building: one on the 7th floor and the other on the 20th. Both suspects were in their apartments at the time of the murder. Forensic specialists report that the bullet was fired from somewhere in the building and entered the body at an angle of approximately 58° with the ground.

1. Draw a diagram of this situation. Let α represent the angle formed by the base of the building, the victim, and the 20th floor of the building. Let θ represent the angle formed by the base of the building, the victim, and the 7th floor of the building.

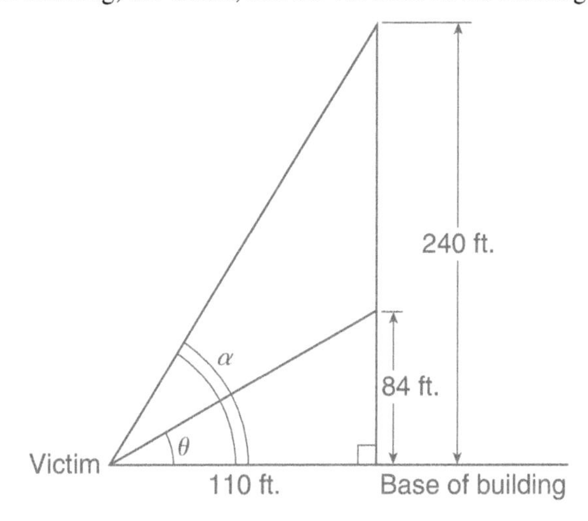

Based on the diagram, the question is now: Is α or is θ equal to 58°?

2. Use the appropriate trigonometric function to determine which of the two suspects could not have committed the murder.

$$\alpha = \tan^{-1}\left(\frac{84}{110}\right) \approx 37°$$

$$\theta = \tan^{-1}\left(\frac{240}{110}\right) \approx 65°$$

These calculations show that the suspect on the 7th floor probably could not have committed the murder.

3. On what other floors of the building should the police question additional possible suspects? Explain.

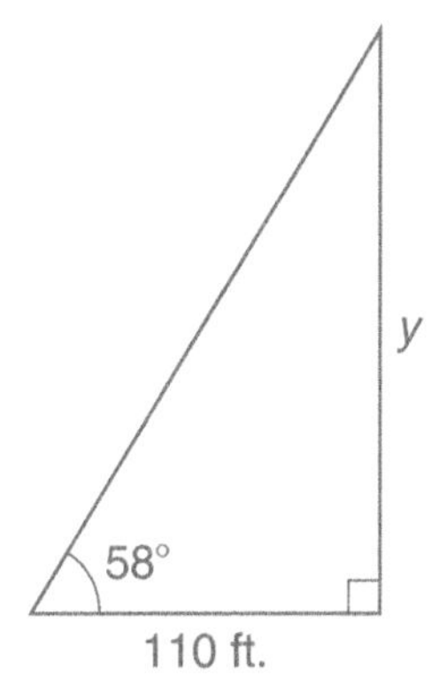

$y = 110 \tan 58° \approx 176$ ft., or about 14.7 floors

The police should question possible suspects on the 14th and 15th floors, but one or two floors above and below would be within the margin of error.

As demonstrated in Problems 2 and 3, determining the measure of the sides and angles of a right triangle can be useful.

DEFINITION

To **solve a triangle** means to determine the measure of all sides and all angles. This process can be especially useful for architects, surveyors, and navigators.

EXAMPLE 1 ***Solve the right triangle ABC, with $A = 33.0°$, $C = 90.0°$, and $c = 12.2$ inches.***

SOLUTION

Consider the following diagram:

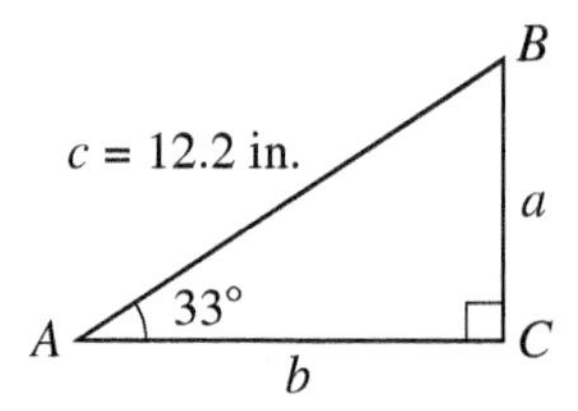

You need to determine the measurements of the remaining sides and angles: angle B, side a, and side b. Because $A + B = 90°$ and $A = 33°$,

$$33 + B = 90, \text{ or}$$
$$B = 90° - 33° = 57°.$$

To determine the length of side a, you can use angle A, side c, and the sine function.

$$\sin A = \frac{a}{c}$$
$$\sin 33° = \frac{a}{12.2}$$

So $a = 12.2 \sin 33° \approx 6.6$ inches (nearest tenth). See the following calculator screen:

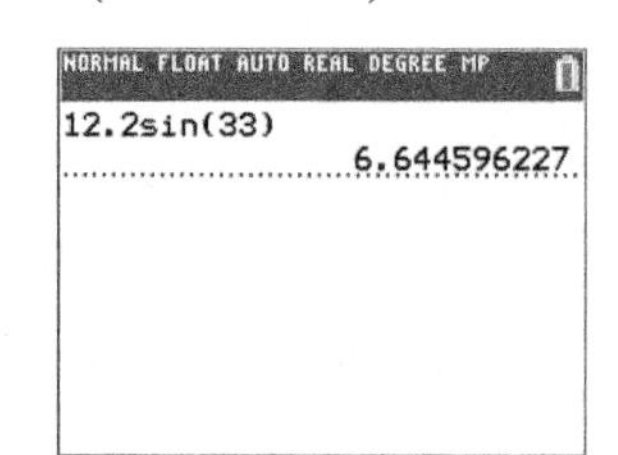

To determine the length of side b, you can use angle A, side c, and the cosine function.

$$\cos A = \frac{b}{c}$$
$$\cos 33° = \frac{b}{12.2}$$
$$b = 12.2 \cos 33° \approx 12.2(0.8387) \approx 10.2 \text{ in.}$$

LC LEARNING CATALYTICS

Consider the right triangle ABC with sides a, b, and c, respectively. Angle C is the right angle. Side c is the hypotenuse, side a is the side opposite angle A, and side b is the side opposite angle B. If side a has length 6.2 cm. and angle B measures 39°, what is the length of the hypotenuse?

a. 8.0 cm.
b. 9.9 cm.
c. 51°.
d. None of these.

4. Use the given information to solve each of the following right triangles:

a.

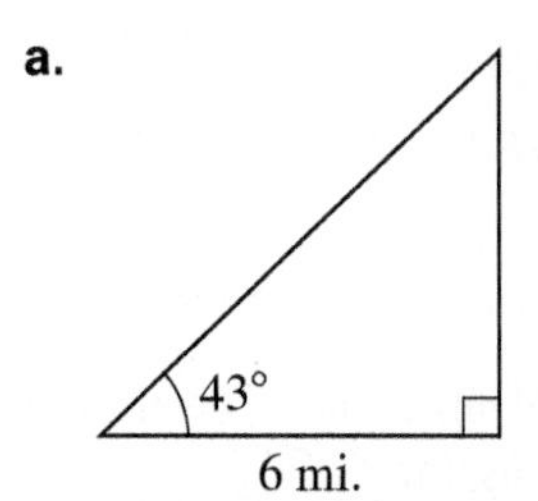

the side opposite the 43° angle: $6 \cdot \tan 43° \approx 5.6$ mi.

hypotenuse: $\dfrac{6}{\cos 43°} \approx 8.2$ mi.

the other acute angle: $90° - 43° = 47°$

b.

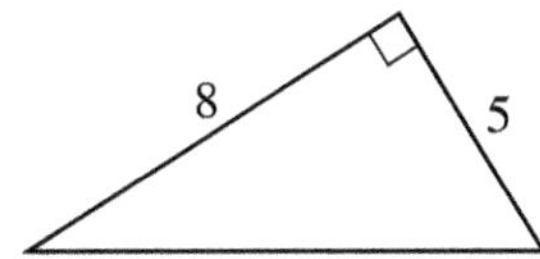

hypotenuse: $\sqrt{89} \approx 9.4$ using the Pythagorean theorem

the angle adjacent to side 8: $\tan^{-1}\left(\frac{5}{8}\right) \approx 32^\circ$

the angle adjacent to side 5: 58°

c.

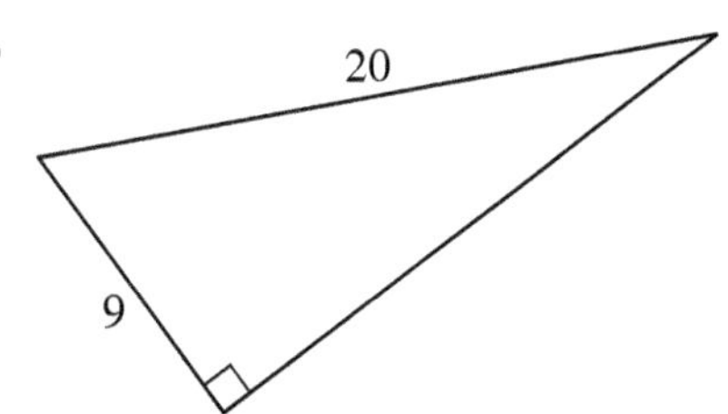

the other leg: $\sqrt{319} \approx 17.9$
the angle adjacent to side 9:

$\cos^{-1}\left(\frac{9}{20}\right) \approx 63^\circ$

the other acute angle: 27°

SUMMARY Activity 6.4

1. Many trigonometric problems involve solving right triangles; that is, determining the measures of all sides and angles.
2. The following is a trigonometric problem-solving strategy:
   - **a.** Draw a diagram of the situation using right triangles.
   - **b.** Identify all known sides and angles.
   - **c.** Identify sides and/or angles you want to know.
   - **d.** Identify functions that relate the known and unknown.
   - **e.** Write and solve the appropriate trigonometric equation(s).

EXERCISES Activity 6.4

1. Use the given information to solve each of the following right triangles:

   a.

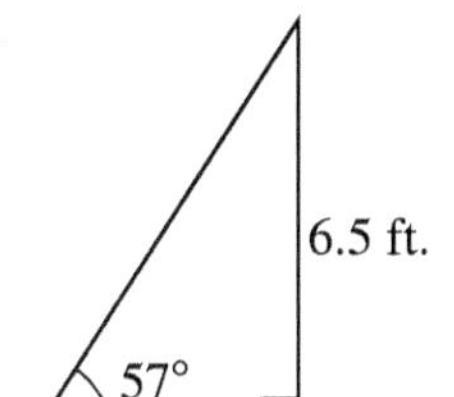

The side adjacent to the 57° angle is 4.2 feet.

The hypotenuse is 7.8 feet.

The other acute angle is 33°.

   b.

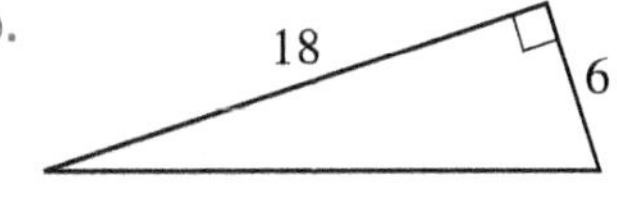

The hypotenuse is 19.0.

The angle adjacent to side 18 is 18.4°.

The other acute angle is 71.6°.

Exercise numbers appearing in color are answered in the Selected Answers appendix.

c.

12 in.

9 in.

The other leg is 7.9 inches.

The angle adjacent to side 9 inches is 41.4°.

The other angle is 48.6°.

2. You need to construct new steps for your deck and read that stringers are on sale at the local lumber company. Stringers are precut side supports to which you nail the steps; they are made in three-, four-, five-, six-, or seven-step sizes. Each step on the stringers is 7 inches high and 12 inches deep.

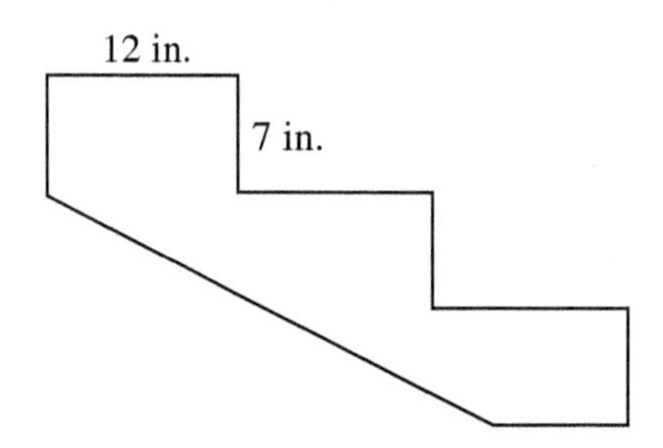

a. If the vertical rise of your deck measures 3.5 feet, which size stringer should you buy? Explain.

The height of the staircase is $3.5 \cdot 12 = 42$ inches. The number of steps $= \dfrac{42 \text{ inches}}{7 \text{ inches}} = 6$. Therefore, I need the six-step stringer.

b. How far out will your steps extend from the porch? Explain.

The steps will extend 6 steps · 12 inches = 72 inches, or 6 feet, from the porch.

c. What angle will a line from the lower right end of the stringer to the edge of the deck make with the ground? Explain.

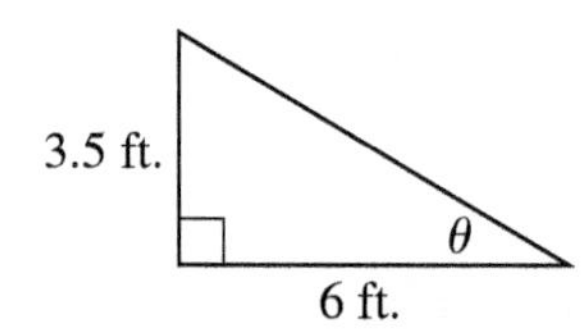

The height of the staircase is 3.5 feet, and the length is 6 feet.

$$\theta = \tan^{-1}\left(\frac{3.5}{6}\right) \approx 30.3°$$

The line makes an angle of about 30° with the ground.

d. What angle will the line in part c make with the vertical? Explain.

59.7° because the two acute angles of a right triangle are complementary.

3. Some application problems involve a horizontal line of sight, which is used as a reference line. An angle measured *above* the sight line is called an **angle of elevation**. An angle measured *below* the sight line is called an **angle of depression**.

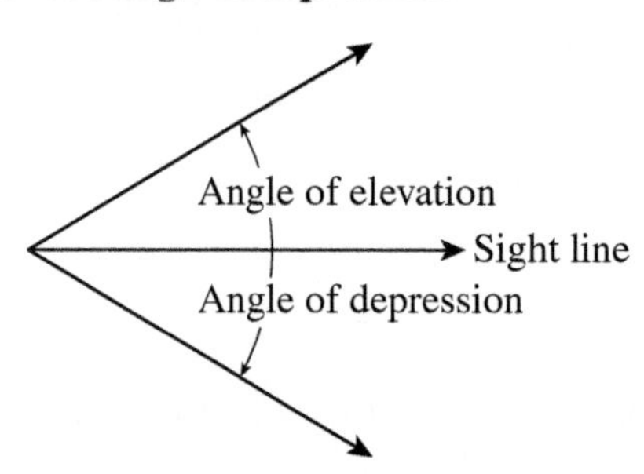

You and some friends take a trip to Colton Point State Park in the Grand Canyon of Pennsylvania. Some of your group go white-water rafting, while some of your friends join you for a hike. You reach the observation deck in time to see the rest of your party battling the white water. Someone in the group asks you how close the rafts actually get to the observation deck as they float by. You have no idea, but you ask a nearby park ranger.

She doesn't know either, but she does tell you that the canyon is approximately 800 feet deep at Colton Point and that the angle of depression to the creek is about 22°.

a. Draw a diagram of this situation.

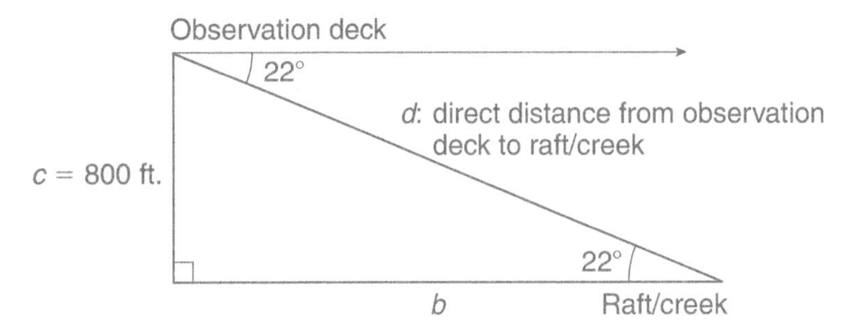

b. Use trigonometry to estimate how close the rafters get to you on the observation deck.

The direct distance, d, from the observation deck to the raft is approximately $\frac{800}{\sin 22^\circ}$ or 2136 feet.

If you could walk straight down the cliff and straight across at the base of the cliff to the creek, the distance would be approximately $b + c = 2780$ feet, where $b = \frac{800}{\tan 22^\circ}$, approximately equal to 1980.1 feet.

PROJECT ACTIVITY 6.5

How Stable Is That Tower?

Problem Solving Using Right Triangle Trigonometry

OBJECTIVES

1. Solve problems using right triangle trigonometry.
2. Solve optimization problems using right triangle trigonometry with a graphing approach.

Situation 1: Stabilizing a Tower

You are considering buying property near a cell phone tower and are concerned about your property values. You decide to read about issues involving towers, such as aesthetics, safety, and stability. Of course, anyone living near the tower would like a guarantee that it could not blow down. Guy wires are part of that guarantee.

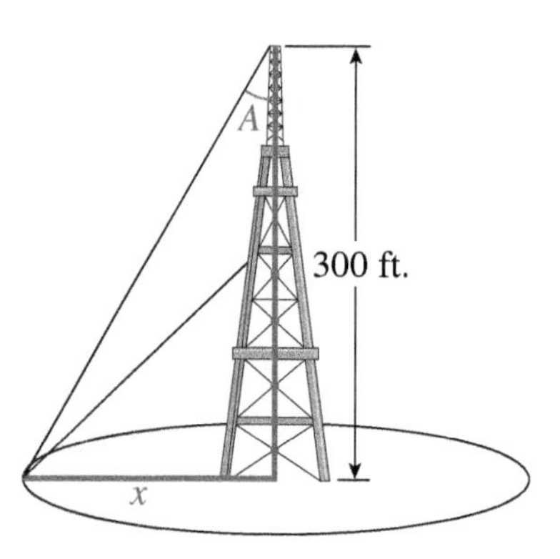

The tower rises 300 feet and is supported by several pairs of guy wires all attached on the ground at the same distance from the base of the tower. In each pair, one guy wire extends from the ground to the top of the tower, and the other attaches halfway up. The diagram above illustrates one pair of wires.

1. New guidelines for stability recommend that the angle (A) the guy wires make with the line through the center of the tower (not the angle of elevation) must be at least 40°. Therefore, the existing guy wires may need to be replaced. Because you are concerned about how close the wires will come to your property, you need to compute the shortest distance from the tower at which the guy wires may be attached. Use trigonometry to compute this distance.

   Let x represent the distance of the guy wire from the center of the base of the tower

   $x = 300 \cdot \tan 40° \approx 252$ ft.

   Because the angle must be at least 40°, the minimum distance from the tower at which the guy wires may be attached is 252 feet.

2. To improve stability, the authors of the guidelines propose increasing the minimum angle the guy wire makes with the tower from 40° to 50°. What is the effect on the shortest distance from the tower at which the guy wires may be attached?

   The shortest distance increases from about 252 feet to about 358 feet because $300 \cdot \tan 50° \approx 357.53$ feet.

Situation 2: Climbing a Mountain

You are camping in the Adirondacks and decide to climb a mountain that dominates the local area. You are curious about the vertical rise of the mountain and recall from your mathematics course that surveyors can measure the angle of elevation of a mountain summit with a theodolite. You borrow a theodolite from the local community college to gather some pertinent data. You find a level field and take a first reading of an angle of elevation of 23°. Then you walk 100 feet toward the mountain summit and take a second reading of 24° angle of elevation (see the illustration on the next page).

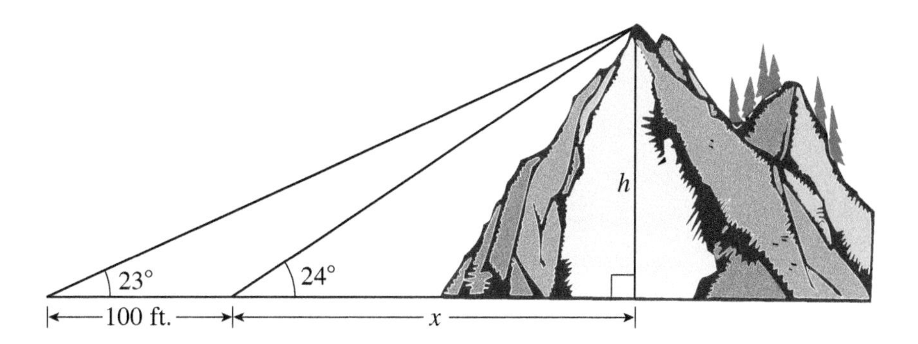

3. Use h to represent the vertical rise of the mountain, and write expressions for tan 23° and tan 24° in terms of h and x.

a. $\tan 23° = \dfrac{h}{100 + x}$ **b.** $\tan 24° = \dfrac{h}{x}$

4. a. Solve the system of equations in Problem 3 to determine x. Round your answer to the nearest tenth.

From 3a, $h = (\tan 23°)(100 + x)$.

From 3b, $h = (\tan 24°)(x)$.

$100 \tan 23° + x \tan 23° = x \tan 24°$

$x \tan 24° - x \tan 23° = 100 \tan 23°$

$x(\tan 24° - \tan 23°) = 100 \tan 23°$

$x = \dfrac{100 \tan 23°}{\tan 24° - \tan 23°}$

$x \approx 2045.3$ ft.

b. Use the value of x from part a to determine the vertical rise, h, of the mountain.

$h = 2045.3 \tan 24° \approx 910.6$

The vertical rise of the mountain is about 911 feet. To check, $h = 2145.3 \tan 23° \approx 910.6$.

Situation 3: Seeing Abraham Lincoln

You are traveling to South Dakota and plan to see Mount Rushmore. In preparation for your trip, you do some research and discover that from the observation center, the vertical rise of the mountain is approximately 500 feet and the height of Abraham Lincoln's face is 60 feet (see the diagram below). To get the best view, you want to position yourself so that your viewing angle of Lincoln's face is as large as possible.

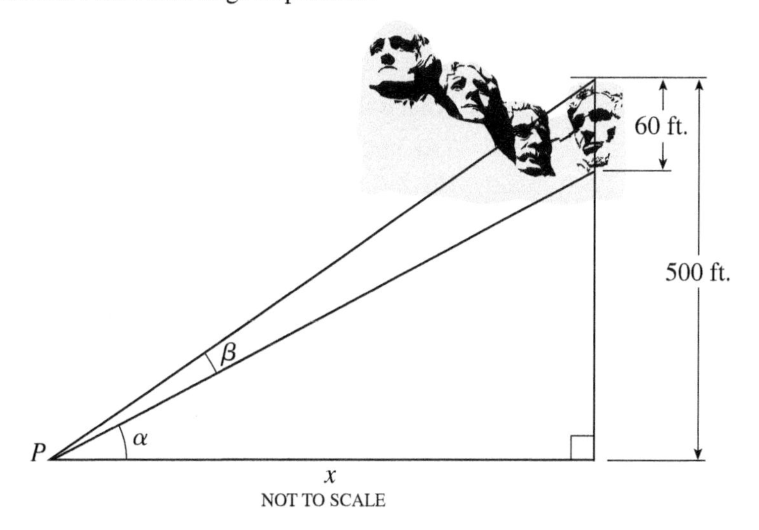

5. a. Determine from the diagram which angle is the viewing angle.

β

b. Write an appropriate trigonometric equation for the angle $\alpha + \beta$ in terms of x.

$$\tan(\alpha + \beta) = \frac{500}{x}$$

c. Write an appropriate trigonometric equation for α in terms of x.

$$\tan \alpha = \frac{440}{x}$$

d. Write the equivalent inverse function expression for the equations in parts b and c.

$$\alpha + \beta = \tan^{-1}\left(\frac{500}{x}\right)$$

$$\alpha = \tan^{-1}\left(\frac{440}{x}\right)$$

e. Write an equation that defines the viewing angle as a function of the distance, x, that you are standing from the base.

$$\beta = \tan^{-1}\left(\frac{500}{x}\right) - \tan^{-1}\left(\frac{440}{x}\right)$$

6. Enter the equation you determined in Problem 5e into Y1 of your graphing calculator. Choose a window that corresponds to the graph below. How does your graph compare with the graph below?

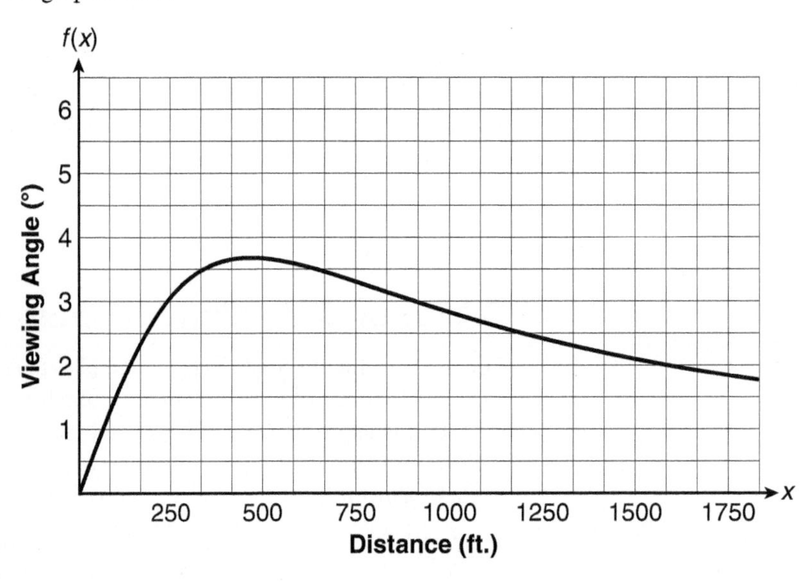

7. What is the largest value of the viewing angle? Justify your conclusion.

The largest viewing angle is approximately 3.7°. This is the output value of the highest point on the graph.

8. How far should you stand from the mountain to obtain this maximum value of the viewing angle? Explain.

I should stand approximately 470 feet from the mountain. This is the input value for the largest viewing angle as shown on the graph.

EXERCISES Activity 6.5

1. You are driving on a straight highway at sea level and begin to climb a hill with a 5% grade. (Recall that a grade is given in a percent but may be expressed as a fraction. In that way, you can view grade as a slope.)

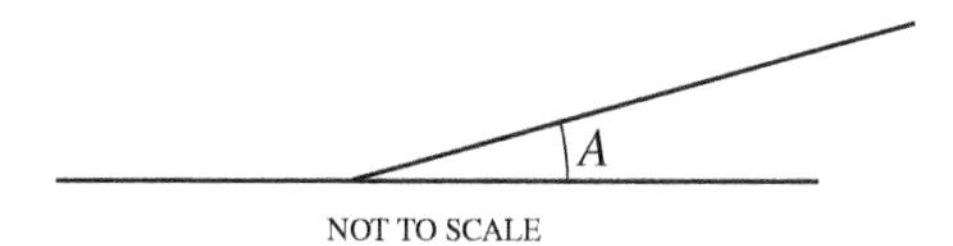

a. What is the slope of the highway with a grade of 5%?

The slope is $\frac{5}{100}$ or $\frac{1}{20}$ or 0.05.

b. What is the **angle of elevation**, A? That is, what angle does the highway make with the horizontal? Explain.

$A = \tan^{-1}(0.05) \approx 2.86°$. The highway makes an angle of 2.86° with the horizontal. This angle is called the angle of elevation.

c. If you walk along the highway for 1 mile, how many feet above sea level are you? Explain.

1 mile is equivalent to 5280 feet. If x represents the number of feet above sea level after walking 1 mile, then $x = 5280 \cdot \sin(2.86°)$ = approximately 263 feet. I would be 263 feet above sea level after 1 mile.

2. You are standing 92 meters from the base of the CN Tower in Toronto, Canada. You are able to measure the angle of elevation to the top of the tower as 80.6°.

a. Draw a diagram, and indicate the angle of elevation.

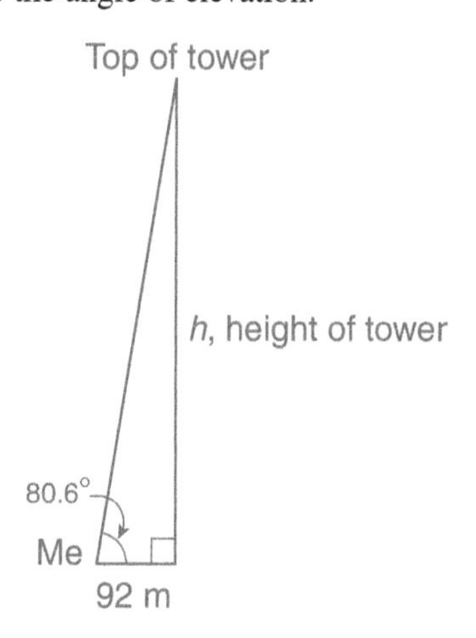

b. What is the height of the tower? Round your answer to the nearest meter.

556 meters because $h = 92 \tan 80.6° \approx 555.7$ m

Exercise numbers appearing in color are answered in the Selected Answers appendix.

3. You are in a spy satellite equipped with a measuring device like a theodolite, orbiting 5 miles above Earth. Your mission is to discover the length of a secret airport runway. You measure the angles of depression to each end of the runway as 30° and 25°, respectively. What is the length of the runway?

Using the following diagram:
The two equations are

(a) $\tan 25° = \dfrac{5}{x + y}$ and

(b) $\tan 30° = \dfrac{5}{y}$. Solving

equation (b), $y \approx 8.7$ miles.
Then equation (a) becomes

$\tan 25° = \dfrac{5}{x + 8.7}$.

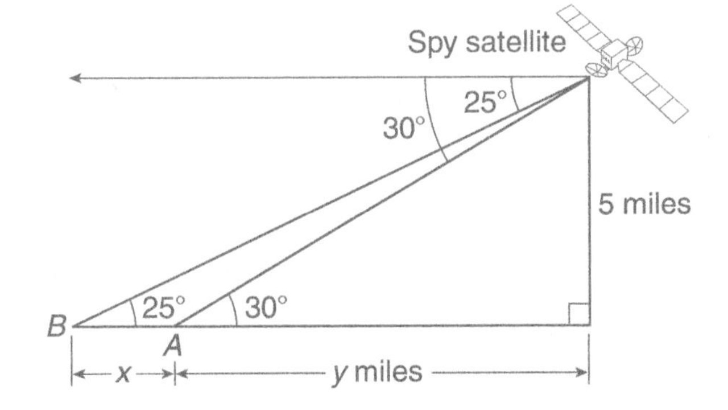

Solve this equation for x.

$(x + 8.7)\tan 25° = 5$

$x \tan 25° = 5 - 8.7 \tan 25°$

$x = \dfrac{5 - 8.7 \tan 25°}{\tan 25°}$

$x \approx 2.0$ mi. The runway is approximately 2 miles long.

4. The Empire State Building rises 1414 feet above the ground, and you are standing across 34th Street, approximately 80 feet from the base of the building.

a. If you look up to the top of the building, what angle of elevation does your line of sight make with the ground? Round your answer to the nearest degree.

The angle of elevation is $\theta = \tan^{-1}\left(\dfrac{1414}{80}\right) \approx 87°$.

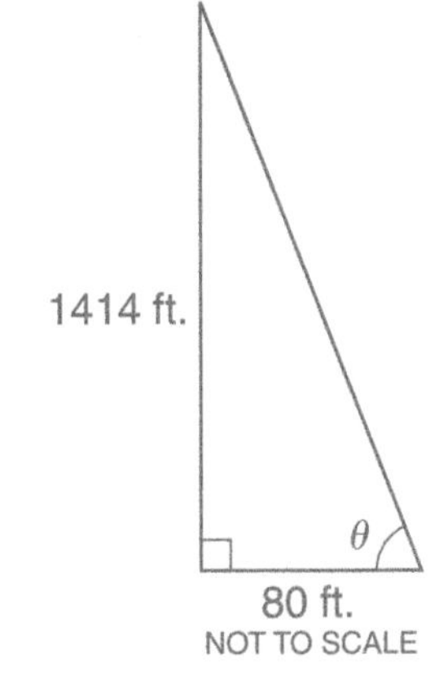

b. How far from the building must you be for the angle of elevation formed by your line of sight and the ground to be 85°? Round your answer to the nearest foot.

Let d represent my distance from the Empire State Building. $d = \dfrac{1414}{\tan 85°} \approx 124$ ft.

I must stand 124 feet from the base of the building for the angle of elevation to be 85°.

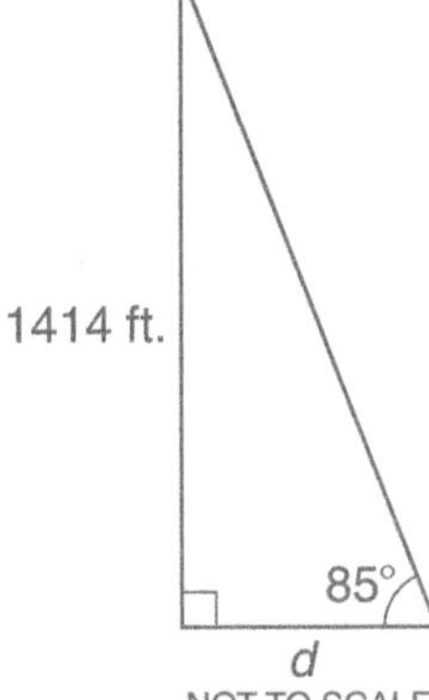

5. Consider the function defined by $f(x) = \arctan\frac{20}{x} - \arctan\frac{10}{x}$ defined for $0.01 \le x \le 100$.

a. Use your graphing calculator to sketch a graph of this function.

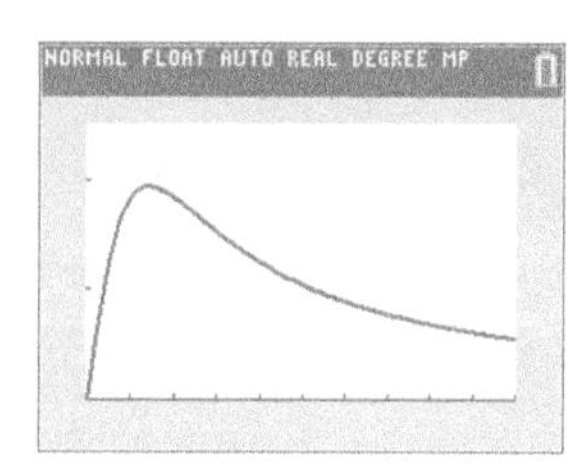

b. Over the given domain, what is the maximum value of the function? Where does it occur?

Over the given domain, the maximum value of the function is approximately 19.5. It occurs at $x \approx 14$.

c. Over the given domain, what is the minimum value? Where does it occur?

The minimum value is difficult to see on the graph. I used the trace or the table feature on my graphing calculator to determine that the minimum value in the given domain is 0.02865, and it occurs at $x = 0.01$.

6. You are interested in constructing a feeding trough for your cattle that can hold the largest amount of feed. You buy a 15-by-50-foot piece of aluminum to construct a 50-foot-long trapezoidal trough with a base of 5 feet. You bend up the two 5-foot sides through an angle of $t°$ with the horizontal. Each cross section is a trapezoid (see the accompanying diagram). You need to determine the angle, t, that produces the largest volume for the trough.

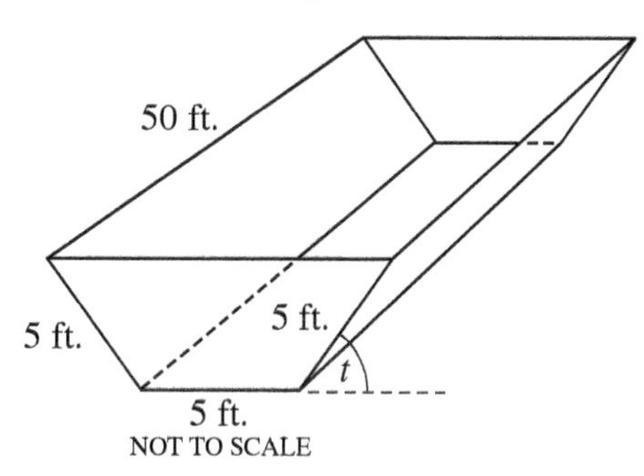

a. Recall the formula for the area of a trapezoid (see the inside back cover of the textbook, if necessary), and write the area of the trapezoidal cross section in terms of h and x (see the accompanying diagram).

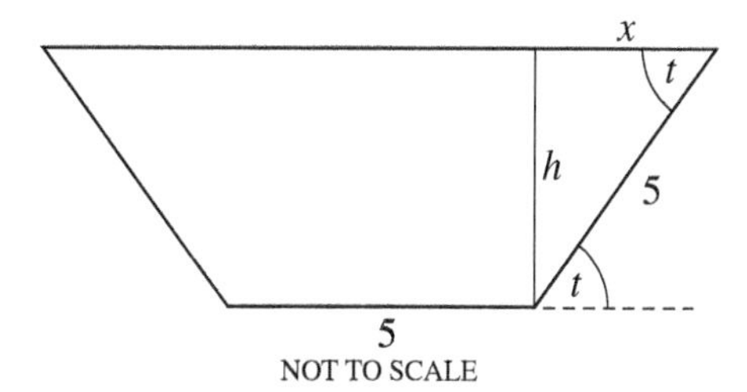

Let A represent the area of the trapezoidal cross section. The height of the cross section is h, and the two bases are 5 and $5 + 2x$.

The area is then determined by the formula $A = \frac{1}{2}h(10 + 2x)$, which, after simplifying, is $A = h(5 + x)$, or $A = 5h + hx$.

b. Using right triangle trigonometry, write h in terms of t. In a similar way, write x in terms of t. Using this information, write an equation for the area of the trapezoid as a function of t.

$h = 5 \sin t$

$x = 5 \cos t$

$A = 5(\sin t)(5 + 5 \cos t)$ or $A = 25 \sin t\,(1 + \cos t)$ or

$A = 25 \sin t + 25 \sin t\,(\cos t)$

c. Use your graphing calculator to graph the area function you constructed in part b with $0° < t < 90°$. Label your axes, and remember to indicate units.

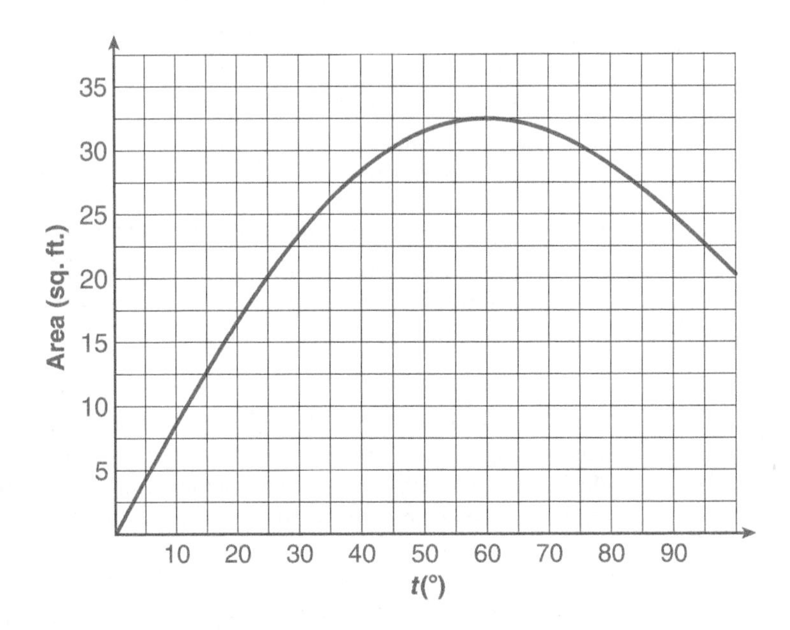

d. Use the graph to determine what angle produces the largest area for the trapezoid? Explain.

The graph in part c indicates that the area of the trapezoidal cross section (output) is greatest when the angle t is 60°.

e. What is the maximum area? Explain.

The area is approximately 32.5 square feet as read from the graph in part c.

f. Write the volume of the trough as a function of the angle t.

Let V represent the volume. Then
$V = 50 \cdot A$, where A is the cross-sectional area.
In terms of t, the volume is $V = 1250 \sin t\,(1 + \cos t)$.

g. Use your graphing calculator to graph the volume function you constructed in part f with $0° < t < 90°$. Label axes and show your units.

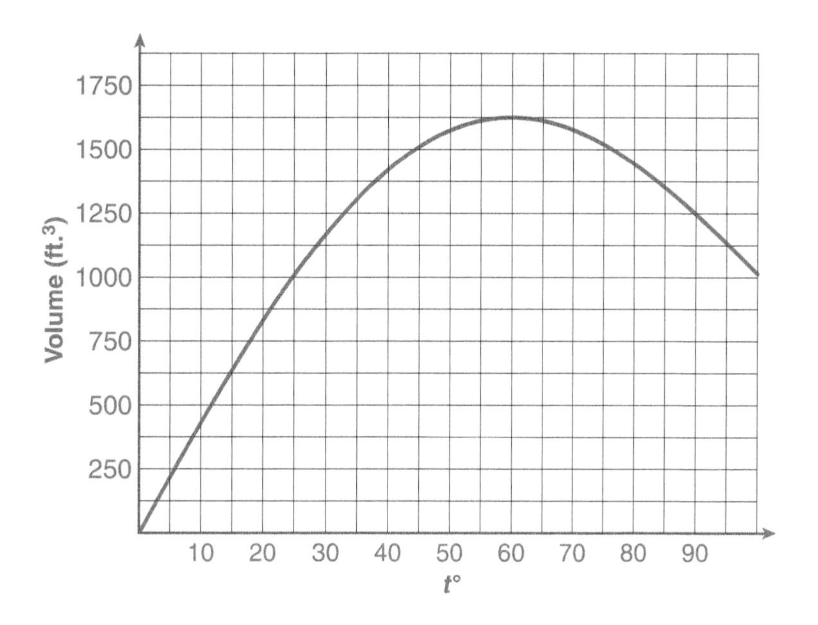

h. Use the graph to determine what angle produces the largest volume. What is that volume? Justify your conclusion.

The graph indicates that the greatest value for the volume between 0° and 90° is approximately 1625 cubic feet when the angle t is 60°.

i. What do you conclude about the angle that produces the largest cross-sectional area and the angle that produces the largest volume for the trough?

The angle is the same—namely, 60° in this scenario.

CLUSTER 1 What Have I Learned?

1. Classical right triangle trigonometry was developed by the ancient Greeks to solve problems in surveying, astronomy, and navigation. For purposes of computation, the side opposite the angle θ, side O, is called the opposite; the side opposite the right angle, side H, is called the hypotenuse; and the third side, side A, is called the adjacent side to angle θ.

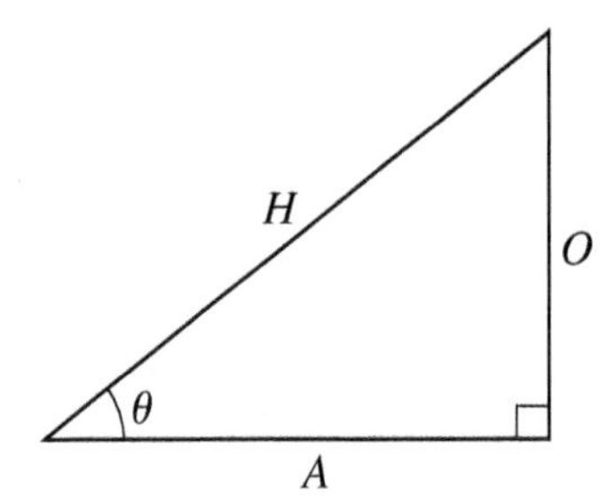

Define the three major trigonometric functions—sin θ, cos θ, and tan θ—in terms of H, A, and O.

$\sin\theta = \frac{O}{H}$; $\cos\theta = \frac{A}{H}$; $\tan\theta = \frac{O}{A}$

2. **a.** Given any right triangle, which trigonometric function would you use to determine the length of the opposite side if you knew the angle measure and the length of the hypotenuse?

   I would use the sine function.

   b. Which trigonometric function would you use to determine the length of the adjacent side if you knew the angle measure and the length of the hypotenuse?

   I would use the cosine function.

   c. Which trigonometric function would you use to determine the length of the adjacent side if you knew the angle measure and the length of the opposite side?

   I would use the tangent function.

   d. Which trigonometric function would you use to determine the length of the opposite side if you knew the angle measure and the length of the adjacent side?

   I would use the tangent function.

3. **a.** Suppose for a right triangle you know the length of the side opposite an angle and you know the length of the hypotenuse. How can you determine the angle?

   I would find the angle by using the arcsin function ($\sin^{-1}$).

   b. There is another way to solve part a. Describe this alternative technique.

   I could use the Pythagorean theorem to determine the length of the side adjacent and then determine the angle by using the arccos function.

4. If you know the lengths of two sides of a right triangle, how can you determine all the angles in the triangle? Make up an example, and determine all the angles. Remember that all the interior angles of a triangle add up to 180°.

Example:

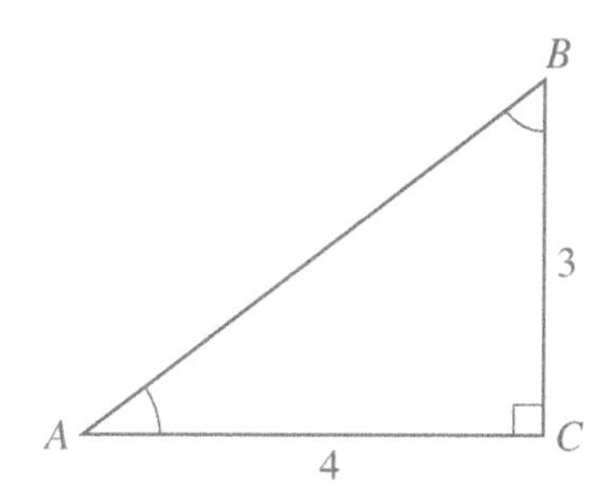

Angle C is 90° by definition of a right triangle.

Then $A = \tan^{-1}\left(\frac{3}{4}\right) \approx 36.9°$ and $B = \tan^{-1}\left(\frac{4}{3}\right) \approx 53.1°$.

Alternatively, $B = 90° - 36.9° = 53.1°$.

5. Consider the following right triangle:

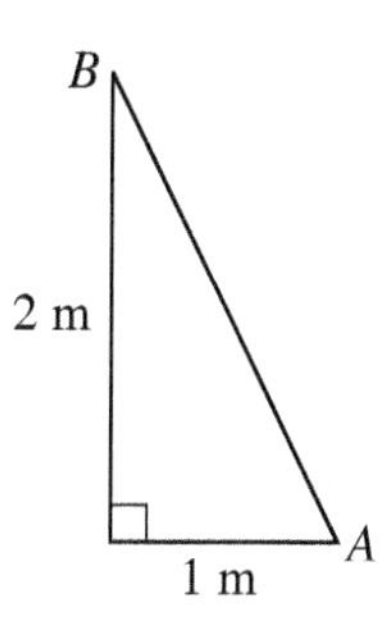

Determine the exact value of each of the following:

a. $\sin A$

$\frac{2}{\sqrt{5}}$

b. $\cos A$

$\frac{1}{\sqrt{5}}$

c. $\tan A$

$\frac{2}{1} = 2$

d. $\sin B$

$\frac{1}{\sqrt{5}}$

e. $\cos B$

$\frac{2}{\sqrt{5}}$

f. $\tan B$

$\frac{1}{2}$

g. What trigonometric property is illustrated by parts a–f? Explain.

For complementary angles in a given right triangle, the cofunction values of those angles are equal. Specifically, if A and B are complementary angles, then $\sin A = \cos B$ and $\sin B = \cos A$.

6. Consider the following two calculator screens:

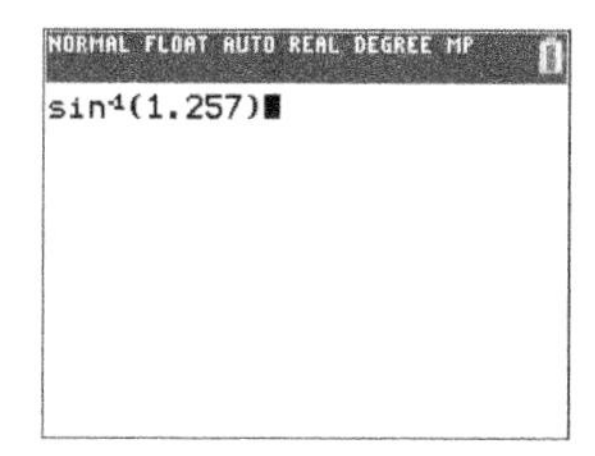

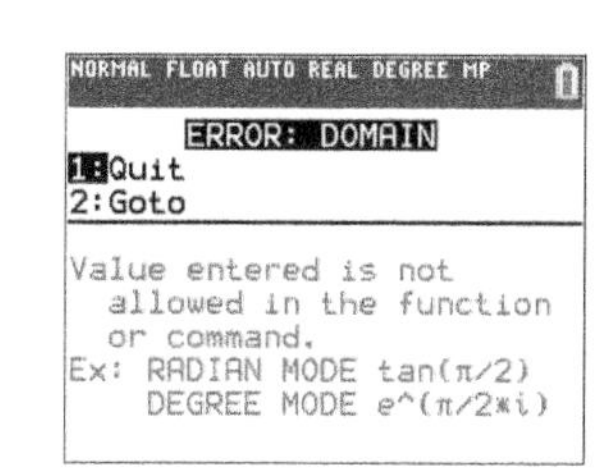

a. The second screen indicates that 1.257 is not in the domain of the inverse sine function. Do you agree? Explain.

Yes, I agree. Inputs are sine ratios, and the sine ratio is the length of the opposite side over the length of the hypotenuse. Because the opposite is never longer than the hypotenuse, the ratio cannot exceed 1.

b. The following screen indicates that you do not have the same problem for the inverse tangent function. Why not? Explain.

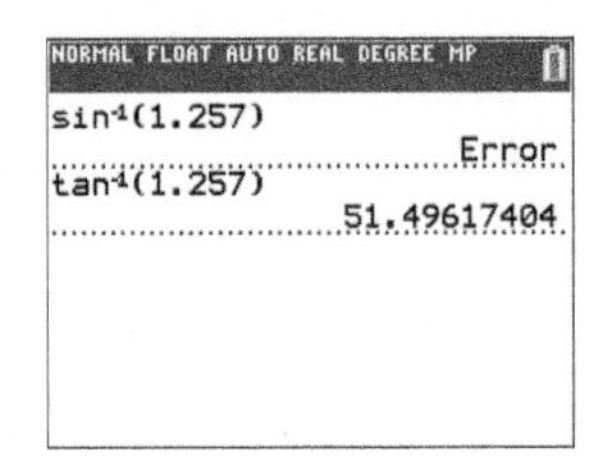

The ratios of $\frac{\textit{length of the opposite side}}{\textit{length of the adjacent side}}$ can be much larger than 1 because the length of the opposite can be larger than the length of the adjacent; therefore, $y = 1.257$ presents no problem.

7. Using diagrams, explain the difference between angle of depression and angle of elevation.

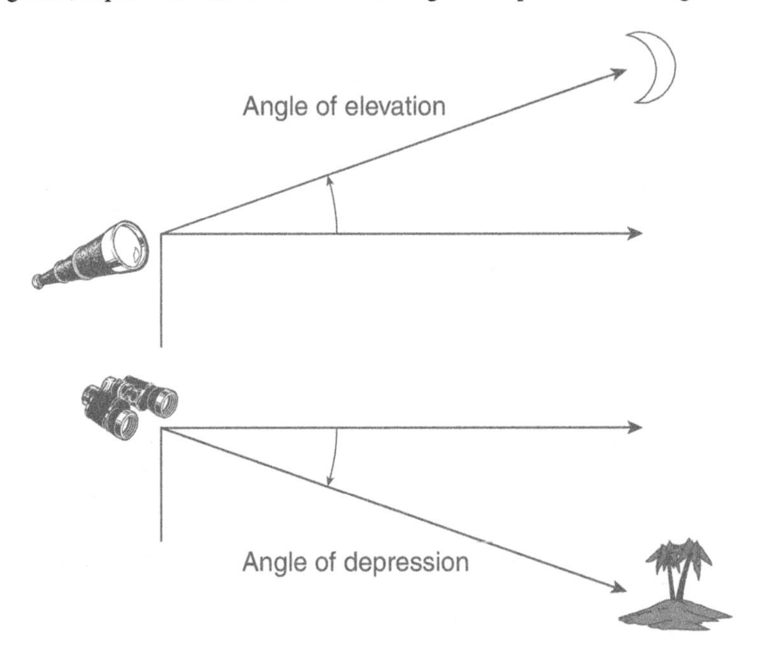

CLUSTER 1 How Can I Practice?

1. Triangle ABC is a right triangle.

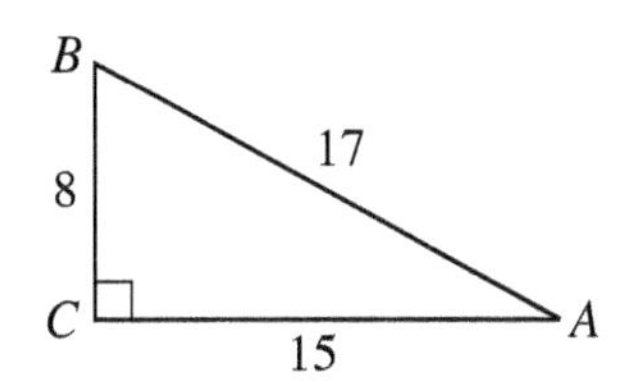

Determine each of the following. Write your answer as a ratio.

a. $\tan A$

$\frac{8}{15}$

b. $\tan B$

$\frac{15}{8}$

c. $\cos A$

$\frac{15}{17}$

d. $\cos B$

$\frac{8}{17}$

e. $\sin A$

$\frac{8}{17}$

f. $\sin B$

$\frac{15}{17}$

2. Use your graphing calculator to determine the values of each of the following. Round your answers to the nearest thousandth.

a. $\sin 47° = 0.731$

b. $\cos 55° = 0.574$

c. $\tan 31° = 0.601$

d. $\tan 80° = 5.671$

3. Given $\sin A = \frac{5}{13}$, determine $\cos A$ and $\tan A$ exactly.

$\cos A = \frac{12}{13}$ and $\tan A = \frac{5}{12}$

4. Given $\tan B = \frac{7}{4}$, determine $\sin B$ and $\cos B$ exactly.

$\sin B = \frac{7}{\sqrt{65}}$ and $\cos B = \frac{4}{\sqrt{65}}$

5. Use your calculator to determine θ, where $0° \le \theta \le 90°$. Round to the nearest tenth of a degree.

a. $\sin \theta = \frac{3}{4}$

$\theta \approx 48.6°$

b. $\cos \theta = 0.9172$

$\theta \approx 23.5°$

c. $\theta = \arctan \frac{7}{2}$

$\theta \approx 74.1°$

d. $\theta = \sin^{-1} \frac{2}{7}$

$\theta \approx 16.6°$

e. $\theta = \tan^{-1} 0.9714$

$\theta \approx 44.2°$

f. $\theta = \arccos 0.9714$

$\theta \approx 13.7°$

Answers to all How Can I Practice exercises are included in the Selected Answers appendix.

6. Solve the following right triangle. That is, determine all the missing sides and angles.

$\overline{BC} = 4.8$ cm

$\overline{AC} = 3.6$ cm

$\angle B = 37°$

$\angle C = 90°$

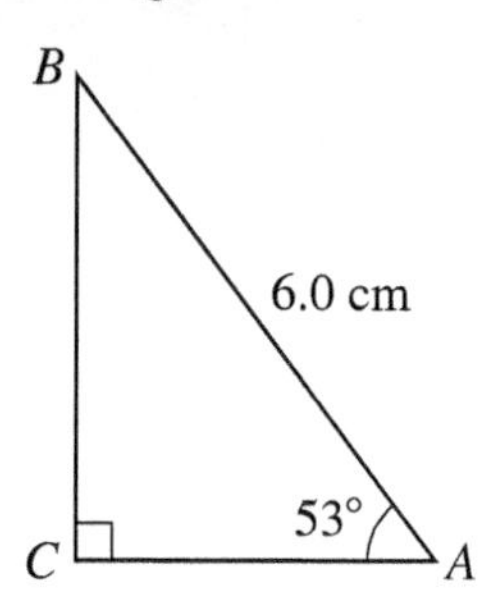

7. You are building a new garage that will be attached to your home and investigate several 30-foot-wide trusses to support the roof. You narrow down your choices to three: One has an angle of 45° with the horizontal, and the others have angles of 35° and 25° with the horizontal. You determine that the walls of the garage must be 10 feet high. To match the height of the rest of the house, the peak of the garage should be approximately 20 feet high. Which truss should you buy?

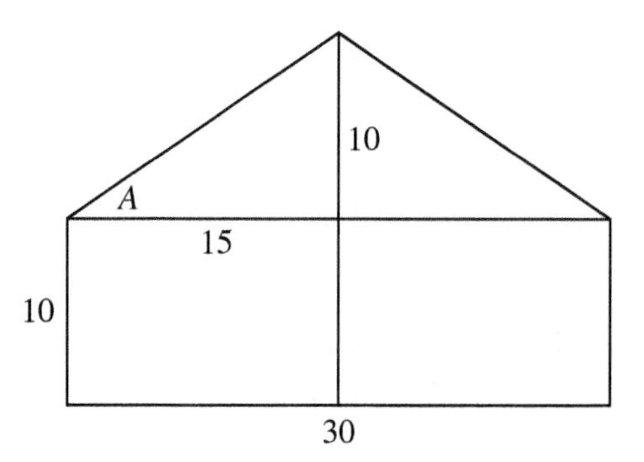

$A = \arctan\left(\frac{10}{15}\right) \approx 33.7°$

Therefore, I should buy the 35° trusses.

8. The side view of your swimming pool is shown here. The dimensions are in feet.

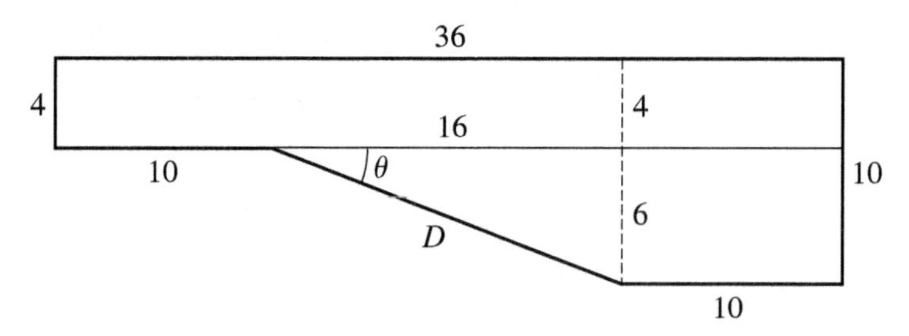

a. What is the angle of depression, θ?

$\theta = \tan^{-1}\left(\frac{6}{16}\right) \approx 20.6°$

b. What is the length of the inclined side, D?

$D = \sqrt{16^2 + 6^2} \approx 17.1$ ft.

9. a. As part of your summer vacation, you rent a cottage on a large lake. One day, you decide to visit a small island that is 6 miles east and $2\frac{1}{2}$ miles north of your cottage. Draw a diagram for this situation. How far from your cottage is the island?

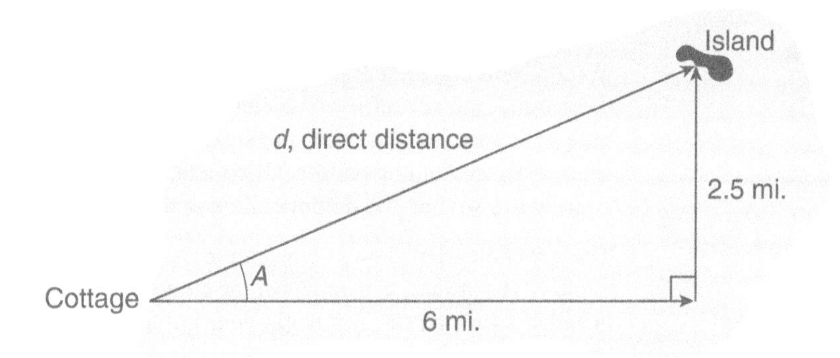

The direct distance, d, from the cottage to the island is $d = \sqrt{2.5^2 + 6^2} =$ 6.5 miles.

b. At what angle with respect to due east should you direct your boat to make the trip from your cottage to the island as short as possible?

$A = \arctan\left(\frac{2.5}{6}\right) \approx 22.6°$

I should direct my boat 22.6° north of east to get from the cottage to the island in the shortest distance.

CLUSTER 2 Why Are the Trigonometric Functions Called Circular Functions?

ACTIVITY 6.6

Learn Trig or Crash!
Graphs of Sine and Cosine Functions

OBJECTIVES

1. Determine the coordinates of points on a unit circle using sine and cosine functions.
2. Sketch the graph of $y = \sin x$ and $y = \cos x$.
3. Identify the properties of the graphs of the sine and cosine functions.

You are piloting a small plane and want to land at the local airport. Due to an emergency on the ground, the air traffic controller places you in a circular holding pattern at a constant altitude with a radius of 1 mile. Because your fuel is low, you are concerned about the distance traveled in the holding pattern. Of course, you communicate with air traffic control about your coordinates so that you do not collide with another airplane.

The given diagram shows your path in the air. The airport is located at the center $(0, 0)$ of the circle on the ground. The radius of the circle is 1 mile. A circle centered at $(0, 0)$ having radius 1 is called a **unit circle**. You begin your holding pattern at $(1, 0)$.

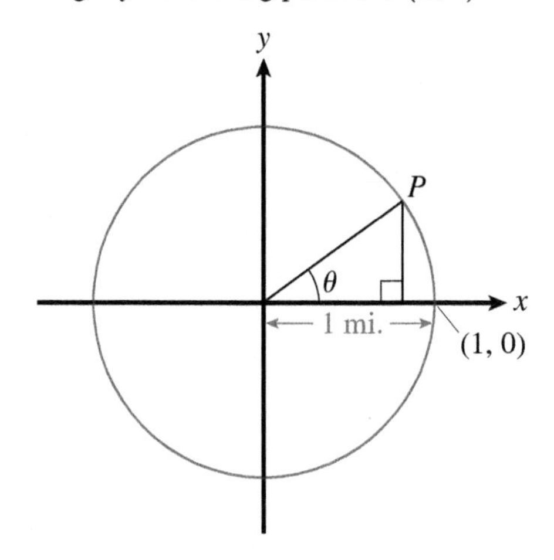

Let's examine the beginning of your first circular loop.

1. What distance (in miles) do you fly in one loop?

   One loop is the circumference of the circle:
   $C = 2\pi \cdot 1 \text{ mi.} = 2\pi \text{ mi.} \approx 6.28 \text{ mi.}$

2. Let P represent your position after flying only one-tenth of a loop. (See the preceding diagram.)

   a. Compute the distance traveled from $(1, 0)$ to P.

   $\frac{1}{10}(2\pi) \text{ mi.} = \frac{\pi}{5} \text{ mi.} \approx 0.628 \text{ mi.}$

   b. Determine the number of degrees of the central angle, θ, when you fly one-tenth of a loop (see the preceding diagram). Recall that there are 360 degrees in a circle.

   $\frac{1}{10}(360°) = 36°$

The angle in Problem 2b is called a **central angle**.

DEFINITION

A **central angle** is an angle with its vertex at the center of a circle.

3. Now refer back to the right triangle in the preceding diagram. The measure of the central angle, θ, is 36°.

   a. What is the length of the hypotenuse of the right triangle?

   The hypotenuse is 1 mile long.

b. If (x, y) represents the coordinates of point P, then which coordinate, x or y, represents the length of the side opposite the 36° angle? Which letter represents the length of the side adjacent to the 36° angle?

y represents the length of the opposite side, and x represents the length of the adjacent side.

c. Use the appropriate trigonometric function to determine the value of x.

$\cos 36° = \frac{x}{1}$; $x \approx 0.81$

d. Use the appropriate trigonometric function to determine the value of y.

$\sin 36° = \frac{y}{1}$; $y \approx 0.59$

e. What are the coordinates of point P?

$P(0.81, 0.59)$

Problem 3 demonstrates that if

i. an object is moving a distance d counterclockwise on the unit circle from the starting point (1, 0) and

ii. $P(x, y)$ represents the position of the object on the unit circle after it has moved distance d and

iii. θ represents the corresponding central angle, then the coordinates of P are given by $(\cos\theta, \sin\theta)$.

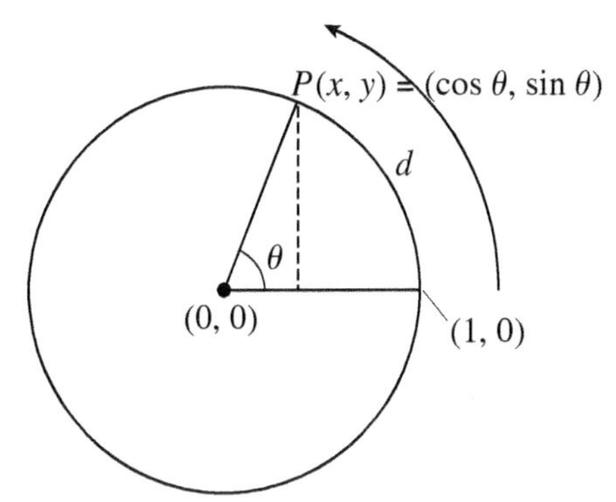

4. Repeat the procedure demonstrated in Problem 3 for the following fractions of a loop, and record your results in the table:

FRACTION OF LOOP	CENTRAL ANGLE, θ	DISTANCE TRAVELED	POSITION ON THE CIRCLE $(\cos\theta, \sin\theta)$
$\frac{1}{5}$	72°	$\frac{2\pi}{5}$ mi., or 1.26 mi.	(0.31, 0.95)
$\frac{1}{8}$	45°	$\frac{2\pi}{8}$ mi., or 0.79 mi.	(0.71, 0.71)
$\frac{1}{20}$	18°	$\frac{2\pi}{20}$ mi., or 0.31 mi.	(0.95, 0.31)

Trigonometric Functions of Angles Greater Than 90°

5. Your instruments tell you that you have traveled 2 miles in the holding pattern. What is the central angle, θ? (*Hint:* What fraction of the loop have you traveled?)

2 miles means $\frac{2}{2\pi} = \frac{1}{\pi} \approx 0.32$ of a loop. The angle is $\frac{1}{\pi}(360°) \approx 114.6°$.

For $\theta > 90°$, the coordinates of P are now *defined* to be $(\cos\theta, \sin\theta)$. This extends the idea of Problem 3 to larger angles, θ. Depending on where you are on the circle, these coordinates may be positive or negative.

6. a. From Problem 5, you know that the central angle is 114.6° (see the accompanying diagram). What is the measure of the central angle, θ', contained within the right triangle?

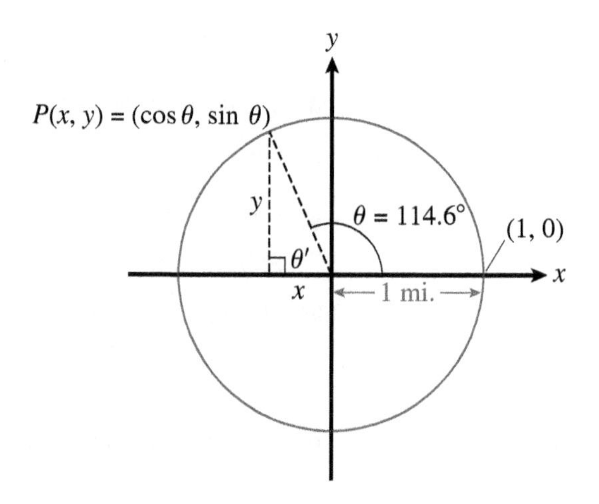

The measure of θ' is $180° - 114.6° = 65.4°$.

b. What is the length of the hypotenuse?

The length of the hypotenuse is 1 mile.

c. Using the sine and cosine functions, determine the lengths of the remaining two sides of the right triangle.

$\cos 65.4° = \frac{x}{1}$; the length of $x \approx 0.42$; $\sin 65.4° = \frac{y}{1}$; the length of $y \approx 0.91$

d. Using the results from part c, what are the coordinates of point P? Remember, the point is in quadrant II.

Because x is negative in quadrant II, the coordinates are $(-0.42, 0.91)$.

7. Your graphing calculator can be used to determine the coordinates of a point in a direct manner. Doing so involves calculating the sine and cosine of the central angle θ.

a. Determine the value of each of the following:

$\cos 114.6° \approx -0.42$

$\sin 114.6° \approx 0.91$

b. How do the results in part a compare with the coordinates of point P in Problem 6d?

The results are the same.

In general, the position $P(x, y)$ of an object moving on a unit circle is given by $x = \cos\theta$ and $y = \sin\theta$, where θ is a central angle with its initial side the positive x-axis and terminal side OP, where O is the origin, the center of the circle.

8. Give an example of where on the unit circle (circle of radius 1) both coordinates of P are negative. Place θ and the coordinates of P on the following diagram.

(Answers will vary.)

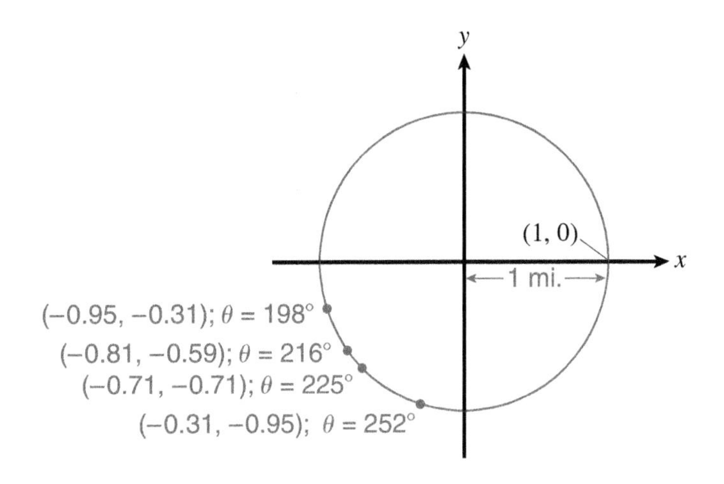

9. a. By the time you have traveled 5 miles in the holding pattern, what is the central angle, θ?

5 miles means $\frac{5}{2\pi}$ of a loop; $\theta = \frac{5}{2\pi}(360°) \approx 286.5°$.

b. What are the coordinates of point P?

$P(\cos 286.5°, \sin 286.5°)$, or approximately $(0.28, -0.96)$

Graphs of Sine and Cosine Functions

10. Complete the following table. Be sure your calculator is in degree mode. Round the first column to the nearest hundredth and the third and fourth columns to the nearest thousandth.

DISTANCE TRAVELED	CENTRAL ANGLE θ	$\cos\theta$	$\sin\theta$
0	0°	1	0
0.52	30°	0.866	0.5
0.79	45°	0.707	0.707
1.05	60°	0.5	0.866
1.57	90°	0	1
2.09	120°	−0.5	0.866
3.14	180°	−1	0
3.84	220°	−0.766	−0.643
6	344°	0.961	−0.276
6.28	360°	1	0

LC LEARNING CATALYTICS

An object is at the point (1, 0) of a unit circle of radius 1 m. If that object rotates on the circle for 170°, approximately how many meters has the object traveled?

a. 3.14 m.
b. 2.97 m.
c. 3.31 m.
d. 2.52 m.

11. Plot the data pairs $(\theta, \sin\theta)$ on the following grid. Draw a smooth curve through these points. Verify the results with your graphing calculator by sketching a graph of $y = \sin x$. Use the window Xmin = 0, Xmax = 360, Ymin = −1, and Ymax = 1 and degree mode.

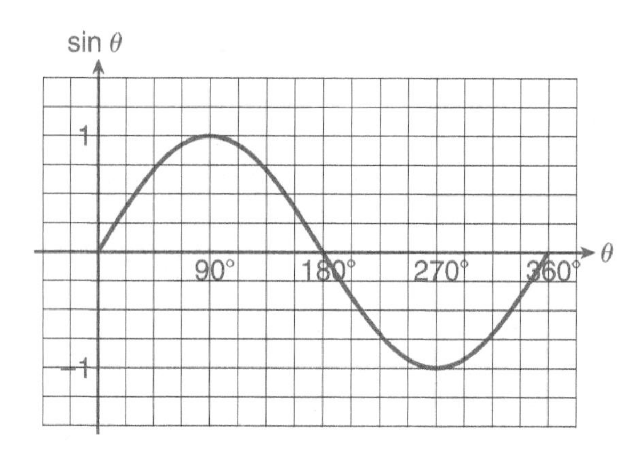

12. Repeat Problem 11 for data pairs $(\theta, \cos\theta)$.

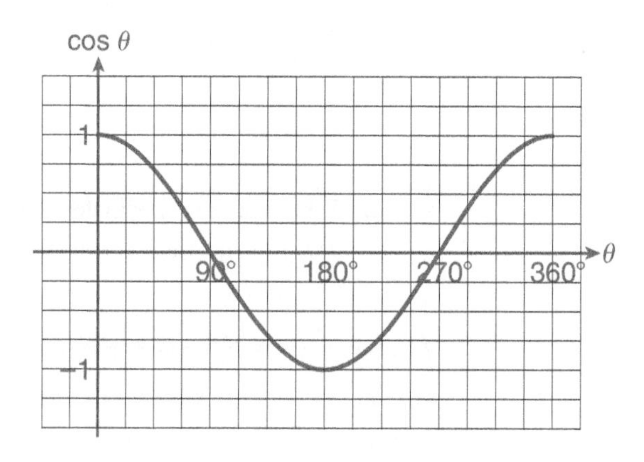

13. You notice your instruments read 11.28 miles from the start of the holding pattern. What is the central angle, θ, and what are the coordinates of point P?

$$\theta = \frac{11.28}{2\pi} \cdot 360° \approx 646.3°$$

$P(\cos 646.3, \sin 646.3)$, or approximately $(0.28, -0.96)$

14. Suppose you are at a particular point in the holding pattern. How many miles will you travel in the loop to return to the same coordinates?

2π, or approximately 6.28 mi.

15. a. Complete the following table:

θ (degrees)	0°	90°	180°	270°	360°	450°	540°	630°	720°
$\sin\theta$	0	1	0	−1	0	1	0	−1	0
$\cos\theta$	1	0	−1	0	1	0	−1	0	1

b. Sketch a graph of $y = \sin\theta$ for $0° \le \theta \le 720°$. Verify using your graphing calculator.

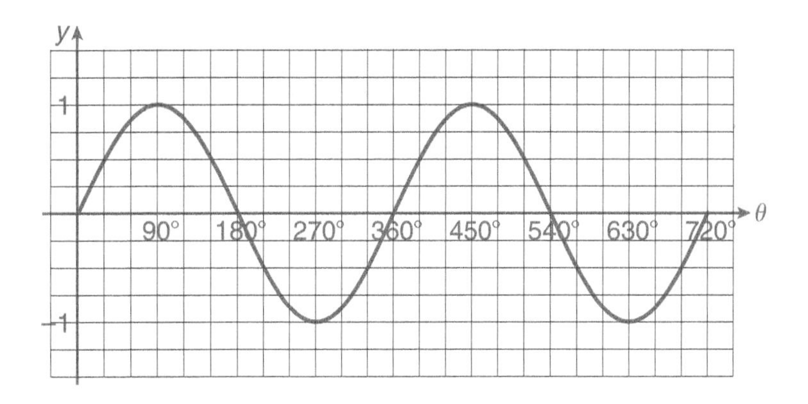

c. Sketch a graph of $y = \cos\theta$ for $0° \le \theta \le 720°$.

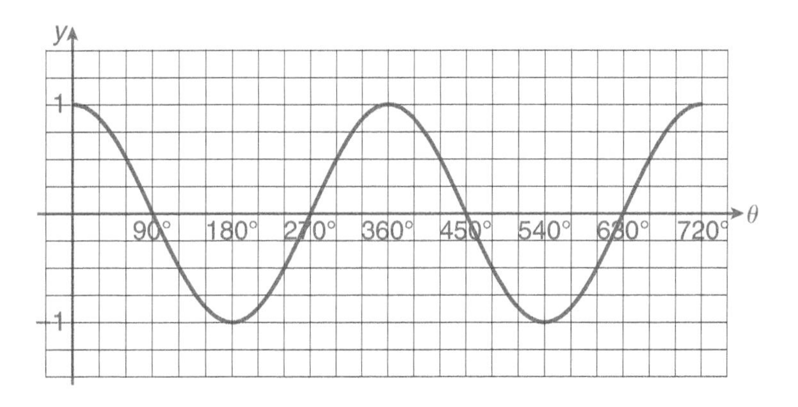

d. What pattern do you observe in each of the graphs?

(Answers will vary.) The patterns are wavelike, and they repeat.

Note that these repeating graphs in Problem 15 show the **periodic** or **cyclic** behavior of the trigonometric functions. Because many real-world phenomena involve this repeating behavior, the trigonometric functions are very useful in modeling these phenomena.

DEFINITION

A cycle is a portion of a graph of $y = f(x)$ that repeats over and over. The shortest change in x it takes for one cycle to be completed is called the **period**. The period is 360° for $y = \sin x$ and $y = \cos x$.

16. A negative angle is used to indicate that the object is moving along the circumference of a unit circle in a *clockwise* direction.

a. Locate the point P that has a central angle of $-30°$ on the following diagram:

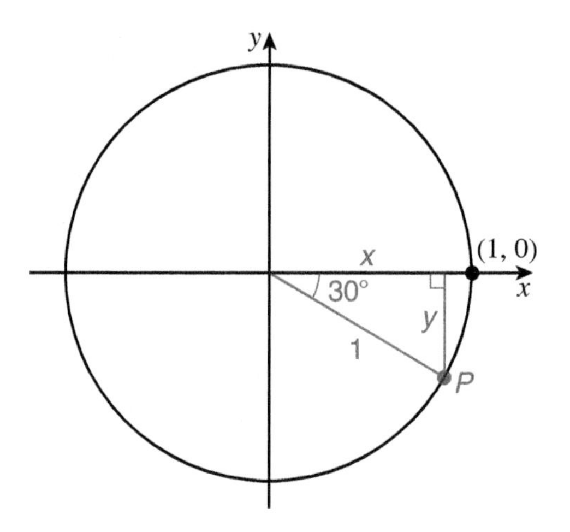

b. Construct the appropriate right triangle. What is the measure of the central angle within this right triangle? What is the length of the hypotenuse?

The measure of the central angle is 30°. The length of the hypotenuse is 1.

c. Use the sine and cosine functions to determine the lengths of the other two sides.

$\cos 30° = \frac{x}{1}$; the length of $x \approx 0.87$; $\sin 30° = \frac{y}{1}$; the length of $y = 0.5$

d. What are the coordinates of the point P?

Because P is located in quadrant IV, the coordinates of P are $(0.87, -0.5)$.

e. Use your graphing calculator to determine $\cos(-30°)$ and $\sin(-30°)$. How do these results compare with the coordinates of point P in part d?

The coordinates are the same.

17. a. Complete the following table:

θ (degrees)	$\cos\theta$	$\sin\theta$
0°	1	0
−30°	0.87	−0.5
−90°	0	−1
−180°	−1	0
−270°	0	1
−360°	1	0

b. Graph the points of the form $(\theta, \sin\theta)$ using the values from the table in part a.

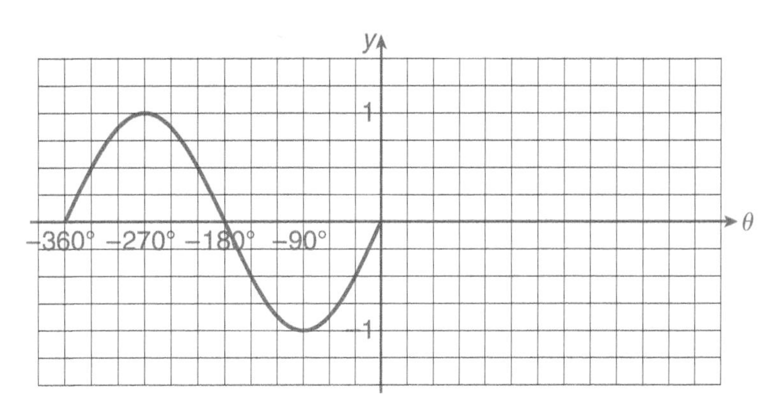

c. Graph the points of the form $(\theta, \cos\theta)$ using the values from the table in part a.

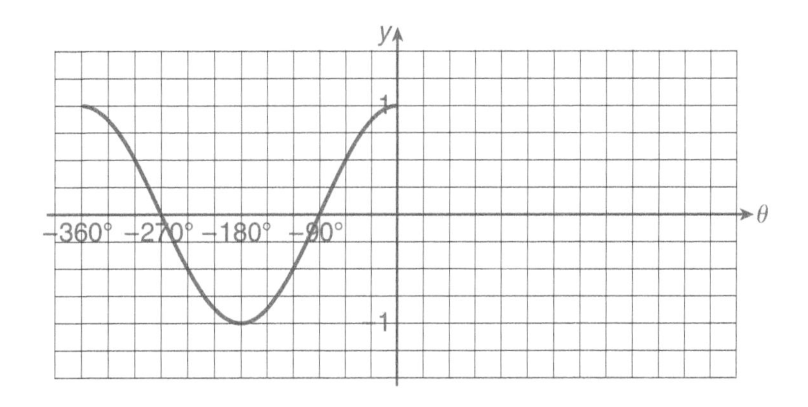

17. d.

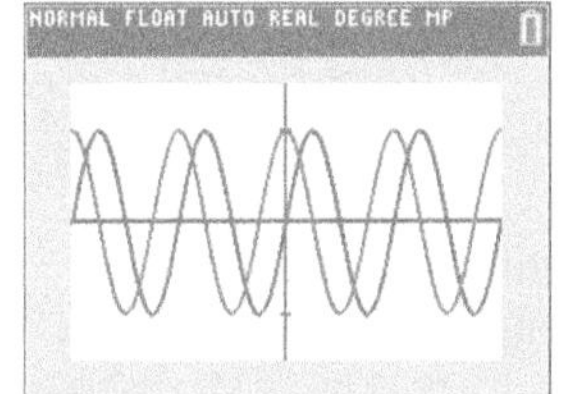

d. Use your graphing calculator to graph $y = \sin x$ and $y = \cos x$ for $-720° \le \theta \le 720°$.

18. a. What is the domain of the sine and cosine functions?

The domain is all possible degree values.

b. What is the range of the sine and cosine functions?

The range is all real numbers between −1 and 1 including −1 and 1.

SUMMARY Activity 6.6

1. A **central angle** is an angle whose vertex is the center of a circle.
2. The position $P(x, y)$ of an object moving on the unit circle from the point $(1, 0)$ defines the sine and cosine functions by the rules $x = \cos\theta$, $y = \sin\theta$, where θ is a central angle formed by the positive x-axis and the line segment OP. Because of this connection to the unit circle, the sine and cosine functions are often called **circular functions**.
3. The **domain** of both the sine and cosine functions consists of all angles, both positive and negative.
4. The **range** of both the sine and cosine functions consists of all values of N such that $-1 \le N \le 1$.

5. The graphs (one cycle) of $y = \sin x$ and $y = \cos x$ look like the following:

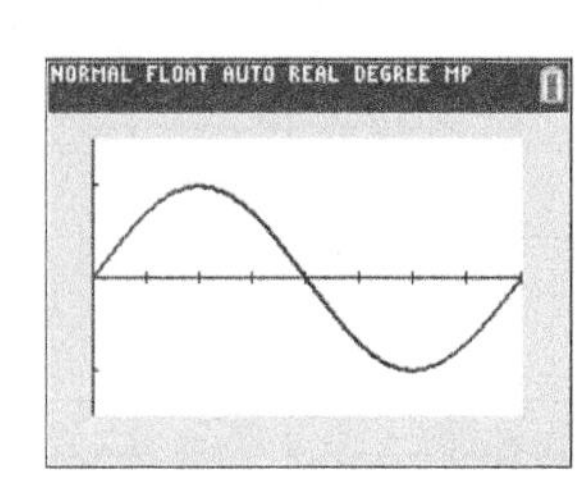

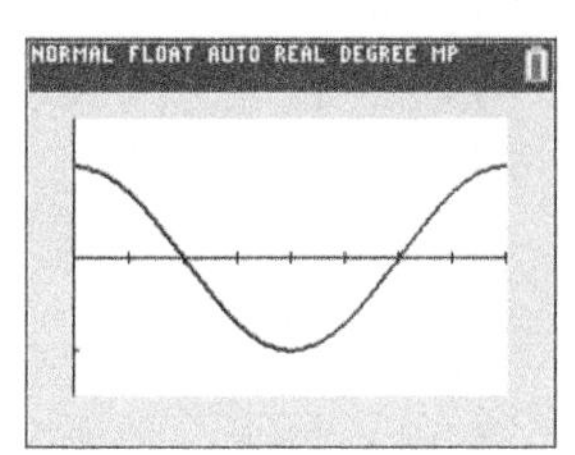

6. The **period** is 360° for $y = \sin x$ and $y = \cos x$.

EXERCISES Activity 6.6

1. Determine the coordinates of the point on the unit circle corresponding to the following central angles. Round to the nearest hundredth.

a. 72°
(0.31, 0.95)

b. 310°
(0.64, −0.77)

c. 270°
(0, −1)

d. 111°
(−0.36, 0.93)

e. 212°
(−0.85, −0.53)

f. 435°
(0.26, 0.97)

g. −70°
(0.34, −0.94)

2. For each of the points on the unit circle determined in Exercise 1, determine the distance traveled from (1, 0) to the point along the circle. Round your answers to the nearest hundredth.

a. $\frac{72}{360} \cdot 2\pi \approx 1.26$

b. $\frac{310}{360} \cdot 2\pi \approx 5.41$

c. $\frac{270}{360} \cdot 2\pi \approx 4.71$

d. $\frac{111}{360} \cdot 2\pi \approx 1.94$

e. $\frac{212}{360} \cdot 2\pi \approx 3.70$

f. $\frac{435}{360} \cdot 2\pi \approx 7.59$

g. $\frac{-70}{360} \cdot 2\pi \approx -1.22$

Therefore, the distance is 1.22.

3. Consider the graph of the following function:

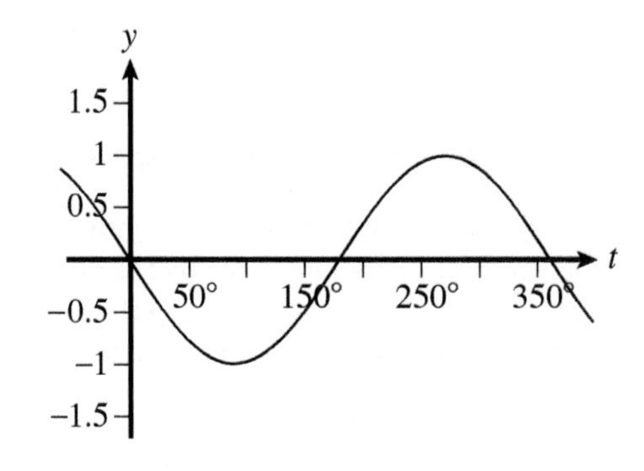

a. Compare this graph with graphs studied in this activity.
The graph looks like the sine curve reflected in the t-axis.

b. What is the motion along the unit circle modeled by the graph?

The motion is clockwise rather than counterclockwise.

4. Consider the graph of the following function:

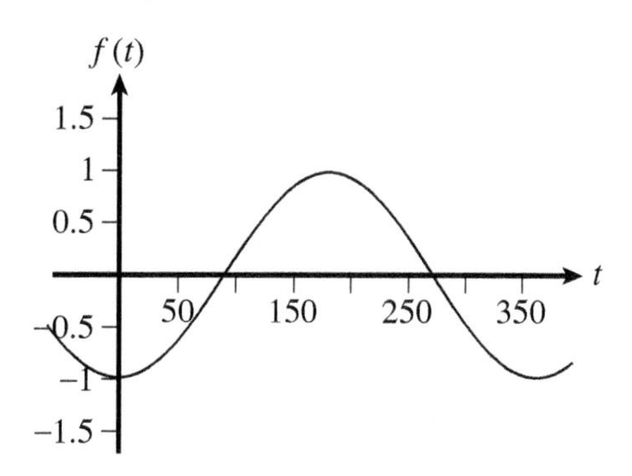

a. How does this graph compare with the graphs studied in this activity?

The graph looks like the cosine function reflected in the t-axis.

b. What is the motion around the unit circle modeled by the graph?

The graph indicates clockwise or counterclockwise rotation starting at $(-1, 0)$.

5. The following table represents the number of daylight hours for a certain city in the Western Hemisphere on the dates indicated:

MAR 21	APR 21	MAY 21	JUNE 21	JULY 21	AUG 21	SEPT 21	OCT 21	NOV 21	DEC 21	JAN 21	FEB 21	MAR 21
11.9	10.5	9.6	8.7	9.7	10.6	12.1	13.5	14.7	15.7	14.7	13.4	11.9

a. Plot the data on the following grid:

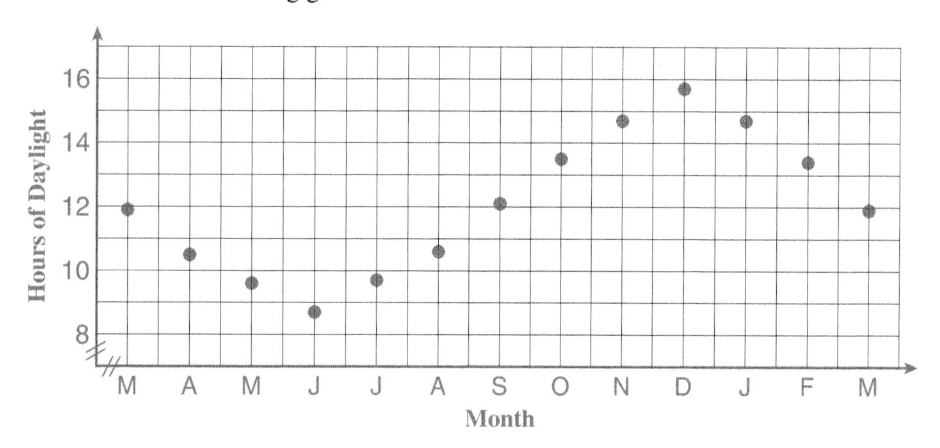

b. Does the data indicate any similarities to a circular function (a function defined by points on the unit circle)? Explain.

Yes, the number of hours of daylight is cyclical; the next year would look just like the first. The graph looks like a shifted and stretched sine graph reflected in the x-axis.

c. How does this function compare with the others in this activity?

The graph has the same wavelike shape.

d. Is the city in question north or south of the equator? Explain.

South; the number of hours of daylight is greater than 12 from October to February, summer in the Southern Hemisphere.

ACTIVITY
6.7

It Won't Hertz
Radians, Frequency and Periodic Behavior

OBJECTIVES

1. Convert between degree and radian measure.
2. Identify the period and frequency of a function defined by $y = a\sin(bx)$ or $y = a\cos(bx)$ using the graph.

Household electric current is called alternating current, or AC, because it changes magnitude and direction with time. The household current through a 60-watt lightbulb is given by the equation

$$A = 2\sin(120\pi t),$$

where A is the current in amperes and t is time in seconds.

Note that the input of the sine function in this activity is time measured by real numbers (seconds) and not angles measured in degrees. For this to make sense, an alternate real number method for measuring angles must be introduced. This method is called **radian measure**.

Degree and radian measure correspond in the following way:

180° is the same as 1π radians, or about 3.14 radians.

Therefore, $180° = 1\pi$ radians. Dividing each side by 180 gives you $1° = \frac{\pi}{180}$ radians.

Dividing each side by π gives you 1 radian $= \frac{180°}{\pi}$.

Whenever you see an angle measure without a degree symbol, assume that the angle is measured in radians.

EXAMPLE 1

Dividing each side of the equality 180° = π radians by 2 shows that 90 degrees is the same as $\frac{\pi}{2}$ radians. You can also convert 90 degrees to radians by multiplying 1° by 90.

$$90° = 90 \cdot 1° = 90 \cdot \frac{\pi}{180} \text{ radians} = \frac{90\pi}{180} \text{ radians} = \frac{\pi}{2} \text{ radians}$$

Similarly, $10° = 10 \cdot \frac{\pi}{180}$ radians $= \frac{\pi}{18}$ radians.

1. In the following table, convert degree measures to radian measure. Round to the nearest thousandth.

DEGREE MEASURE	RADIAN MEASURE
10°	$\frac{\pi}{18} \approx 0.175$
20°	$\frac{\pi}{9} \approx 0.349$
30°	$\frac{\pi}{6} \approx 0.524$
60°	$\frac{\pi}{3} \approx 1.047$
120°	$\frac{2\pi}{3} \approx 2.094$
360°	$2\pi \approx 6.283$

2. a. If you divide the equality $180° = \pi$ radians by π, you find that 1 radian $= \frac{180°}{\pi}$. Then 2 radians would be $2\left(\frac{180°}{\pi}\right)$ and so on.

   Generalize this to describe a procedure to convert radians to degrees.

   Multiply the number of radians by $\frac{180°}{\pi}$.

b. How many degrees are in 1.5π radians?

$$1.5\pi = 1.5\pi \cdot \frac{180^\circ}{\pi} = 270^\circ$$

c. How many degrees are in 2π radians?

$$2\pi \cdot \frac{180^\circ}{\pi} = 360^\circ$$

d. How many degrees are in $\frac{\pi}{10}$ radians?

$$\frac{\pi}{10} \cdot \frac{180^\circ}{\pi} = 18^\circ$$

For more practice converting degree measure to radian measure and vice versa, see Appendix B.

Periodic Behavior of Graphs of Sine and Cosine Functions

As stated, the function defined by $A = 2 \sin(120\pi t)$, where A is the current in amperes and t is time in seconds, gives the household current through a 60-watt lightbulb.

3. Graph the equation $A = 2 \sin(120\pi t)$ using your calculator. Your calculator must be in radian mode instead of degree mode. Using the indicated window, the graph should appear as follows:

$\text{Xmin} = 0$ $\qquad$ $\text{Ymin} = -3$

$\text{Xmax} = \frac{1}{20}$ $\qquad$ $\text{Ymax} = 3$

$\text{Xscl} = \frac{1}{240}$ $\qquad$ $\text{Yscl} = 1$

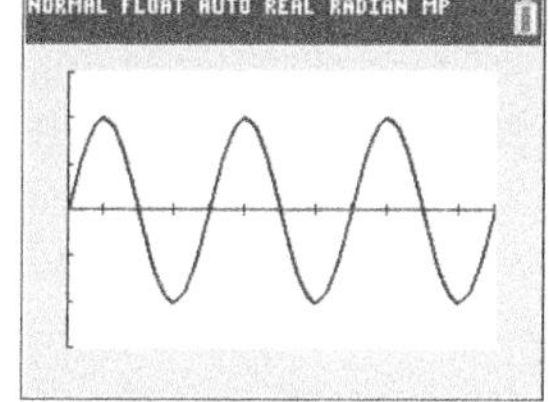

LC LEARNING CATALYTICS

Convert $\frac{5\pi}{4}$ radians to degrees.
a. 135°.
b. 45°.
c. 270°.
d. 225°.

4. Using the graph, determine the maximum current. Explain.

2 amps, the highest point represented on the graph; the lowest point, -2, represents 2 amperes in the opposite direction.

5. What do you think is happening to the current when the graph drops below the horizontal axis? (Reread the description of alternating current.)

It changes direction.

There is a pattern on the graph of $A = 2 \sin(120\pi t)$ starting at the 0 level, going up to $+2$, down to -2, and returning to the 0 level. The pattern repeats. This pattern is called a **cycle**.

Recall from Activity 6.6 that the shortest change in the input (on the input axis) it takes for one cycle to be completed is called the **period**.

EXAMPLE 2 *Determine the period of each of the following using its graph:*

a. $y = \sin(2x)$ $\qquad$ **b.** $y = \sin\left(\frac{1}{2}x\right)$

SOLUTION

a.

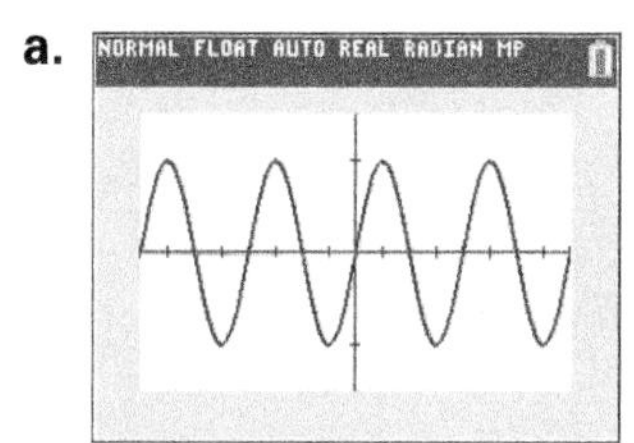

b.

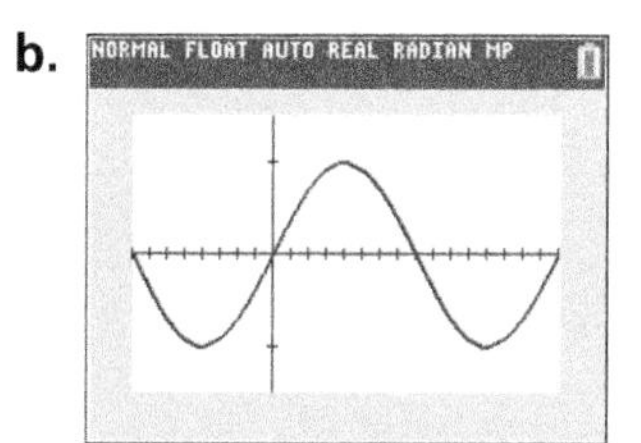

The Xscl in both graphs is $\frac{\pi}{4}$. Thus, the graph of $y = \sin(2x)$ completes one cycle from $(0, 0)$ to $(\pi, 0)$. Therefore, the period is π. The period in part b is 4π.

6. What is the period of the electric current function from Problem 3? (The tick marks on the horizontal axis are at $\frac{1}{240}$-second intervals.)

$\frac{1}{60}$th second

7. Use the graph of each of the following functions to determine its period. Use the window Xmin $= 0$, Xmax $= 6\pi$, Xscl $= \frac{\pi}{4}$, Ymin $= -2$, Ymax $= 2$, and Yscl $= 1$.

a. $y = \sin x$

2π

b. $y = \sin\left(\frac{1}{2}x\right)$

4π

c. $y = \cos(3x)$

$\frac{2\pi}{3}$

d. $y = \cos\left(\frac{2}{3}x\right)$

3π

DEFINITION

The **frequency** of $y = a\sin(bx)$ or $y = a\cos(bx)$ is the number of cycles completed when the input has completed a given interval. In trigonometry, the given interval is 2π because that is the distance required for one rotation around the unit circle. However, if the input value is time, t, the given interval is often 1 second. In this case, the frequency is $b/2\pi$.

EXAMPLE 3 ***Determine the number of cycles completed in 2π units for each of the following using its graph with Xscl $= \frac{\pi}{4}$:***

a. $y = \sin(2x)$

b. $y = \sin\left(\frac{1}{2}x\right)$

SOLUTION

a.

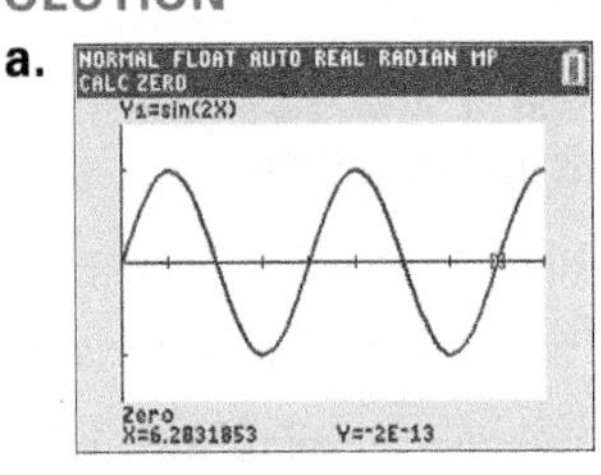

The number of cycles is 2.

b.

The number of cycles is 0.5.

8. Determine the number of cycles completed in 2π units for each of the following functions:

a. $y = \sin x$

1

b. $y = \cos\frac{1}{2}x$

$\frac{1}{2}$

c. $y = \cos 2x$

2

9. For normal household current described by $A = 2\sin(120\pi t)$, how many cycles occur in 1 second?

The equation indicates that the current completes 120π cycles in 2π seconds. So the current completes $\frac{120\pi}{2\pi} = 60$ cycles in 1 second.

SUMMARY Activity 6.7

1. **Radian measure** is used when the input of a repeating function is better defined by real numbers than angles measured in degrees.

2. Because 1° corresponds to $\frac{\pi}{180}$ radians, to convert degree measure to radian measure, multiply the number of degrees by $\frac{\pi \text{ radians}}{180°}$.

3. To convert radian measure to degree measure, multiply the number of radians by $\frac{180°}{\pi}$.

4. Radian and degree equivalences:

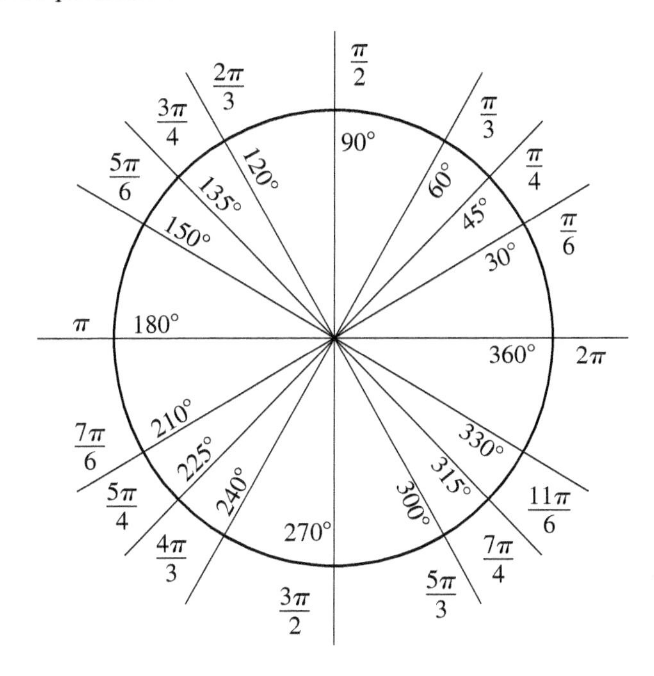

5. The pattern of a graph that covers all y-values once and is repeated is called a **cycle**.

6. The smallest interval of input necessary for the graph of a function to complete one cycle is called the **period**.

   Note: A formula for period will be developed in Activity 6.8.

7. The **frequency** of $y = a\sin(bx)$ or $y = a\cos(bx)$ is the number of cycles completed when the input has completed a given interval. In trigonometry, the given interval is 2π because that is the distance required for one rotation around the unit circle. However, if the input value is time, t, the given interval is often 1 second. In this case, the frequency is $b/2\pi$.

EXERCISES Activity 6.7

1. Convert the following degree measures to radian measures:

   a. 45°

   $45 \cdot \frac{\pi}{180} = \frac{\pi}{4} \approx 0.785$ radian

   b. 140°

   $140 \cdot \frac{\pi}{180} = \frac{7\pi}{9} \approx 2.443$ radians

c. 330°

$330 \cdot \frac{\pi}{180} = \frac{11\pi}{6} \approx 5.760$ radians

d. −36°

$-36 \cdot \frac{\pi}{180} = \frac{-\pi}{5} \approx -0.628$ radian

2. Convert the following radian measures to degree measures:

a. $\frac{3\pi}{4}$

$\frac{3\pi}{4} \cdot \frac{180}{\pi} = 135°$

b. 2.5π

$\frac{2.5\pi}{1} \cdot \frac{180}{\pi} = 450°$

c. 6π

$6\pi \cdot \frac{180}{\pi} = 1080°$

d. 1.8π

$1.8\pi \cdot \frac{180}{\pi} = 324°$

3. Complete the following table:

Degree Measure	0°	30°	45°	60°	90°	135°	180°	210°	270°	360°
Radian Measure	0	$\frac{\pi}{6}$	$\frac{\pi}{4}$	$\frac{\pi}{3}$	$\frac{\pi}{2}$	$\frac{3\pi}{4}$	π	$\frac{7\pi}{6}$	$\frac{3\pi}{2}$	2π

For Exercises 4–9, be sure your calculator is in radian mode.

4. a. Complete the following table:

Radian Measure, x	0	$\frac{\pi}{4}$	$\frac{\pi}{2}$	$\frac{3\pi}{4}$	π	$\frac{5\pi}{4}$	$\frac{3\pi}{2}$	$\frac{7\pi}{4}$	2π
$\sin x$	0	0.707	1	0.707	0	−0.707	−1	−0.707	0

b. Sketch a graph of $f(x) = \sin x$ using the table in part a.

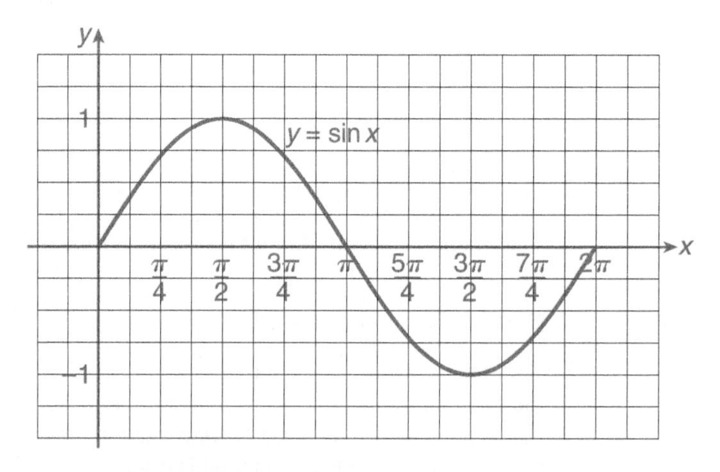

4. c.

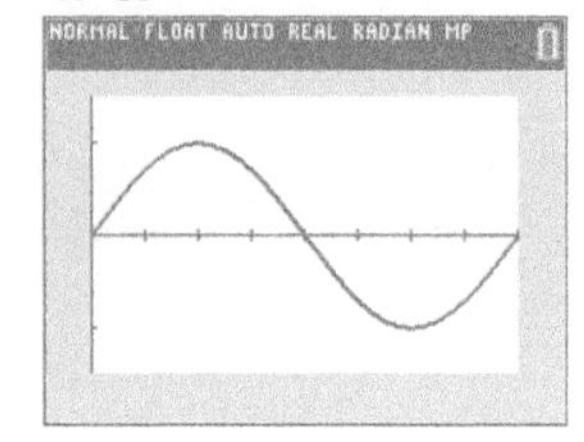

c. Use your graphing calculator to sketch a graph of $f(x) = \sin x$ for $0 \le x \le 2\pi$.

5. How do the graphs of each pair of the following functions compare? Use your graphing calculator.

a. $y = 2\cos x$, $y = \cos 2x$

$y = 2\cos x$ is $y = \cos x$ stretched vertically by a factor of 2. $y = \cos(2x)$ is $y = \cos x$ compressed horizontally by a factor of 2.

b. $y = \cos\frac{1}{3}x$, $y = \cos 3x$

$y = \cos\left(\frac{1}{3}x\right)$ is $y = \cos x$ stretched horizontally by a factor of 3.

$y = \cos(3x)$ is $y = \cos x$ compressed horizontally by a factor of 3.

For Exercises 6–9, graph each function and then determine the following for each function:

a. *The largest (maximum) value of the function*

b. *The smallest (minimum) value of the function*

c. *The period (the shortest interval for which the graph completes one cycle)*

6. $y = 0.5 \sin 2x$

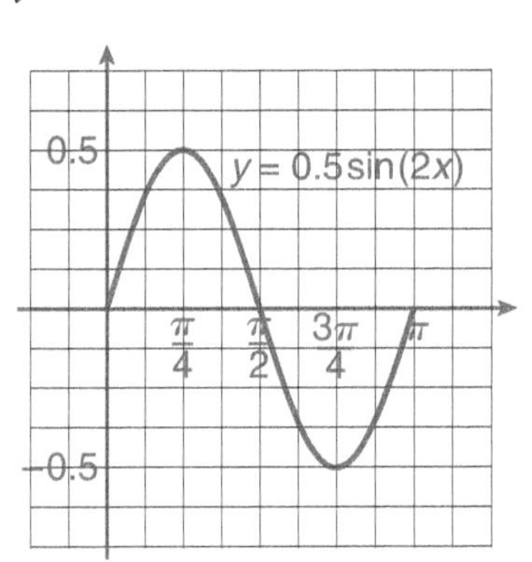

max: 0.5

min: −0.5

period: π

7. $y = -3 \sin 3x$

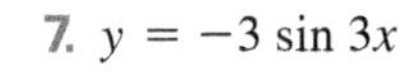

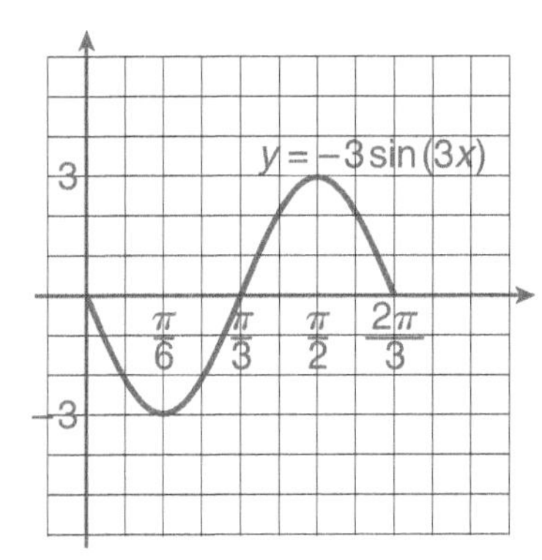

max: 3

min: −3

period: $\frac{2\pi}{3}$

8. $f(x) = 2.3 \cos (0.5x)$

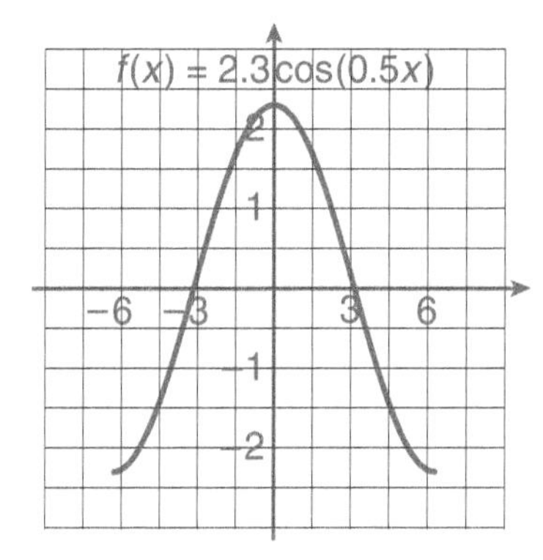

max: 2.3

min: −2.3

period: 4π

9. You are setting up a budget for the new year. Your utility bill for natural gas and electric usage is a large part of your budget. To help you determine the amount you might need to spend on gas and electricity, you examine your bills from the previous 3 years. Because you live in a rural area, you are billed every 2 months instead of every month. The data you obtain appears in the following graph:

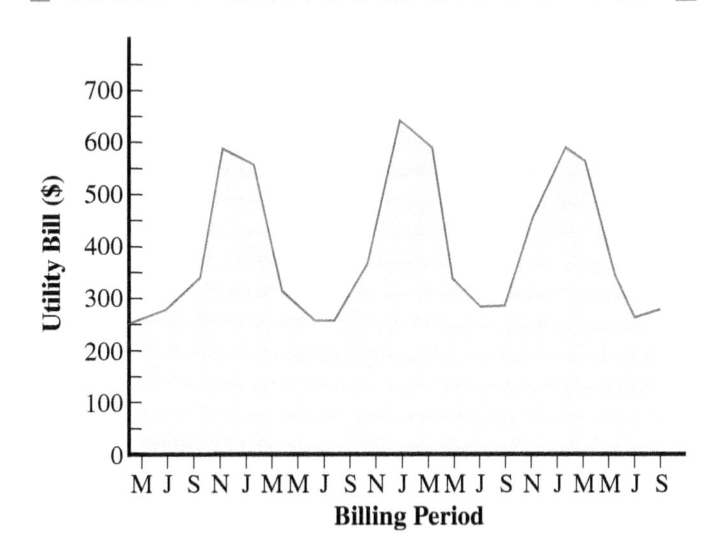

You notice that a pattern develops, which is repeated. This function, although not exactly periodic, models a periodic function for practical purposes. Label the horizontal axis with bimonthly periods, beginning with the June and July bill from 3 years ago.

a. For what months of the year is the utility bill highest? How much is this bill?

   The bill is highest for December and January. The amount of the bill is approximately $600.

b. What months of the year is the utility bill the lowest? How much is this bill?

   The bill is lowest for May and June. The amount of the bill is approximately $250.

c. What is the largest value for the function whose graph is given here?

   The largest value is $650.

d. What is the period of the graph?

   The period is 6 billing periods, or 12 months.

e. Your power company announces that its rates will increase 5% beginning in April of the coming year. How will this change affect the graph of this function? Will it affect the periodic nature of the function?

   The graph will be stretched vertically by a factor of 1.05. This will not affect the period of the function.

f. You heat your house and water with natural gas and use electricity for all other purposes. You do not currently have air-conditioning in your house. If you were to install a central air-conditioning unit next summer, what changes might occur in the shape of the graph?

   The amount of the bills for the summer months would increase. The graph would flatten out as the monthly charges become more equal.

ACTIVITY
6.8

Get in Shape

Amplitude and Period of the Sine and Cosine Functions

OBJECTIVES

1. Determine the amplitude of the graph of $y = a\sin(bx)$ and $y = a\cos(bx)$.
2. Determine the period of the graph of $y = a\sin(bx)$ and $y = a\cos(bx)$ using a formula.

You decide to try jogging to shape up. You are fortunate to have a large neighborhood park nearby that has a circular track with a radius of 100 meters.

1. If you run one lap around the track, how many meters have you traveled? Explain.

   1 lap = the circumference of the circle

   $C = 2\pi r = 2\pi(100 \text{ m}) \approx 628 \text{ m}$

2. You start off averaging a relatively slow rate of approximately 100 meters per minute. How long does it take you to complete one lap?

   $t \approx \dfrac{628 \text{ m}}{100 \text{ m/min.}} = 6.28$ min. More exactly, 2π min.

You want to improve your speed. Gathering data describing your position on the track as a function of time may be useful. You sketch the track on a coordinate system with a center at the origin. Assume that the starting line has coordinates (100, 0) and that you run counterclockwise.

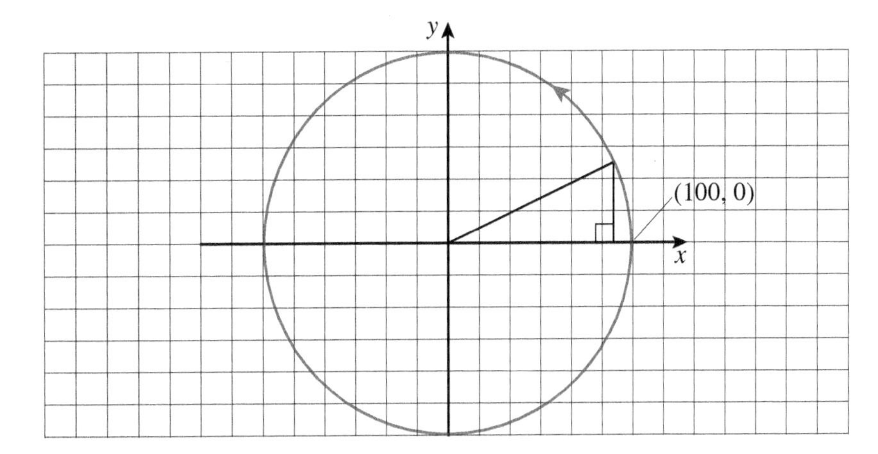

3. You begin to gather data about your position at various times. Using the preceding diagram and the results from Problem 2, complete the following table to locate your coordinates at selected times along your path:

t (min)	YOUR x-COORDINATE	YOUR y-COORDINATE
0	100	0
$\frac{\pi}{2}$	0	100
π	−100	0
$\frac{3\pi}{2}$	0	−100
2π	100	0
$\frac{5\pi}{2}$	0	100
3π	−100	0

4. What patterns do you notice about the numerical data in Problem 3? Predict your coordinates as your time increases.

   The data is periodic. For $t = \frac{7\pi}{2}$, $(x, y) = (0, -100)$;
   for $t = 4\pi$, $(x, y) = (100, 0)$.

5. To better analyze your position at times other than those listed in the previous table, you decide to make some educated guesses. Let's consider $t = \frac{\pi}{4}$ minutes. Use the graph to approximate the coordinates of your position when $t = \frac{\pi}{4}$, and label these coordinates on the graph.

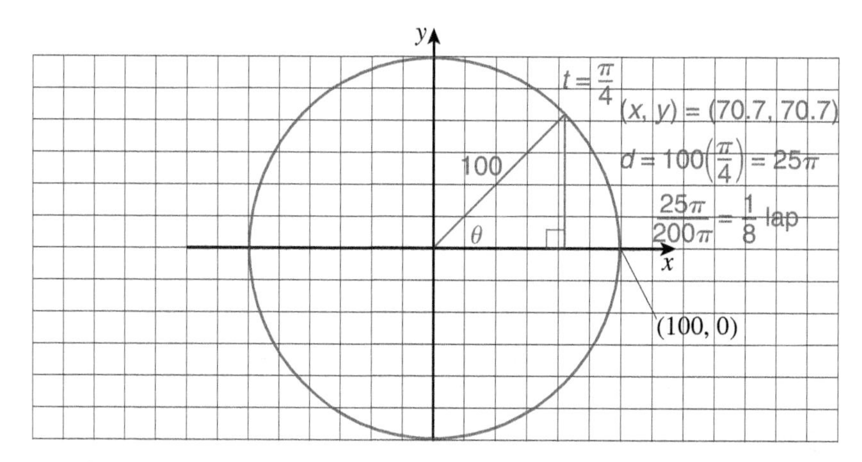

Appendix B

6. To check your guess, you recall that right triangle trigonometry gives some useful information about special right triangles. (If you are not familiar with special right triangles, see Appendix B.) First, however, you need to compute θ, given in the graph of Problem 5. Calculate θ and explain how you arrive at your answer.

$$\theta = \frac{\pi}{4} = 45°$$

$$\frac{1}{8}(2\pi) = \frac{\pi}{4}$$

$$\frac{1}{8}(360°) = 45°$$

There are two possible approaches: $\frac{\pi}{4}$ radians is equivalent to 45°, or a rotation of $\frac{\pi}{4}$ units is $\frac{1}{8}$ of a total rotation of 2π units; therefore, the angle is $\frac{1}{8}$ of 360°, or 45°.

7. On the following grid, use the data from Problems 3, 5, and 6 to plot (t, y), showing your y-coordinate as a function of t. Connect your data pairs to make a smooth graph, and predict what will happen to the graph for values of t before and after the values of t in the tables.

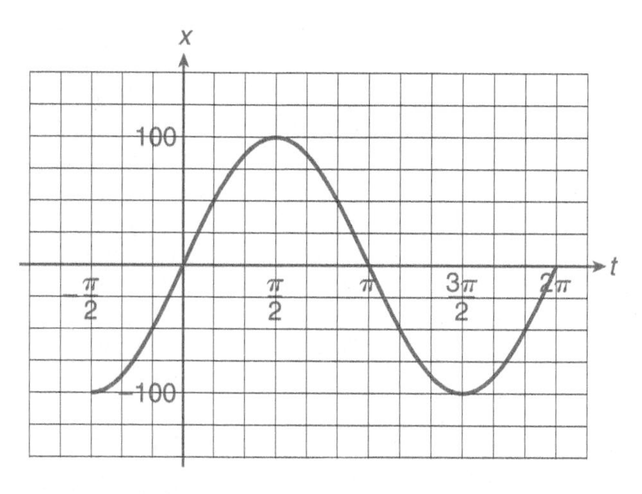

The pattern will continue for values of t before $-\frac{\pi}{2}$ and after 2π.

8.

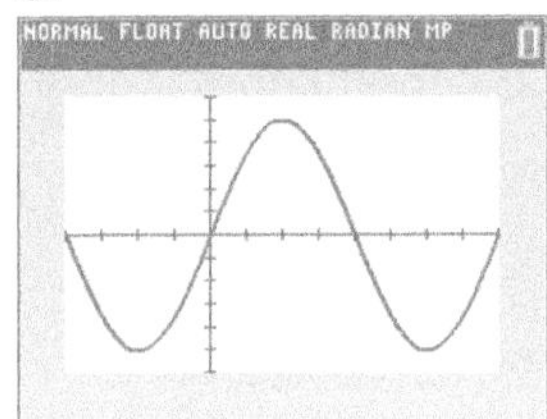

8. Use your graphing calculator to plot $y = 100 \sin(t)$. Make sure your graphing calculator is in radian mode. Note that the name of the input t must be changed to x to conform with the calculator.

9. Compare your answers to Problems 7 and 8.

The graphs are the same.

10. a. What is the maximum value of the sine function in Problem 8?

100 is the maximum value.

b. What is the minimum value of the sine function in Problem 8?

-100 is the minimum value.

Another important feature of the graphs of the sine and cosine functions is called the **amplitude**.

DEFINITION

The **amplitude** of a periodic function equals

$$\frac{1}{2}(M - m),$$

where M is the maximum output value of the function and m is the minimum output value of the function.

EXAMPLE 1

The amplitude of the function defined by $y = \sin x$ is 1. The maximum output value is 1. The minimum output value is -1. The amplitude is $\frac{1}{2}(1 - (-1)) = \frac{1}{2}(1 + 1) = \frac{1}{2}(2) = 1$.

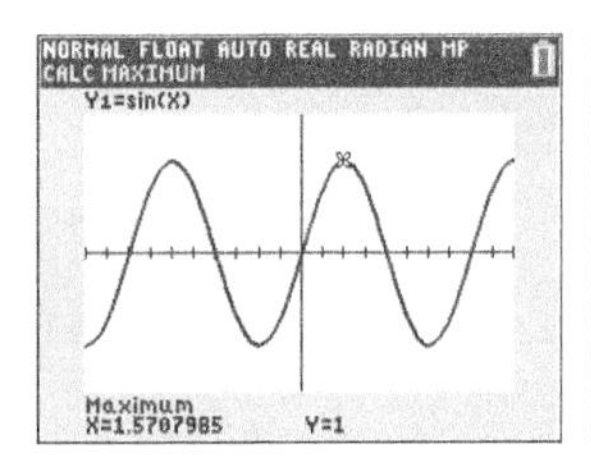

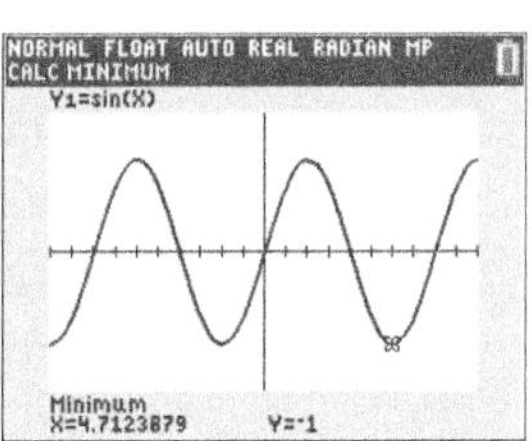

11. a. What is the amplitude of the sine function in the equation $y = 100 \sin x$?

100

b. Is there a relationship between the amplitude of $y = 100 \sin x$ and the coefficient 100? Explain.

Yes, the coefficient 100 is the amplitude.

12. Determine by inspection the amplitude of the following functions. Then verify your answers using your graphing calculator.

a. $y = 1.5 \sin x$

1.5

b. $f(x) = 15 \sin(2x)$

15

c. $y = 3 \cos\left(\frac{1}{3}x\right)$

3

13. a. Is -2 the amplitude of $y = -2 \sin x$? Explain.

No. Amplitude is one-half the maximum minus the minimum, so it can never be negative.

b. What is the amplitude of $y = -2 \sin x$?

2

13. c.

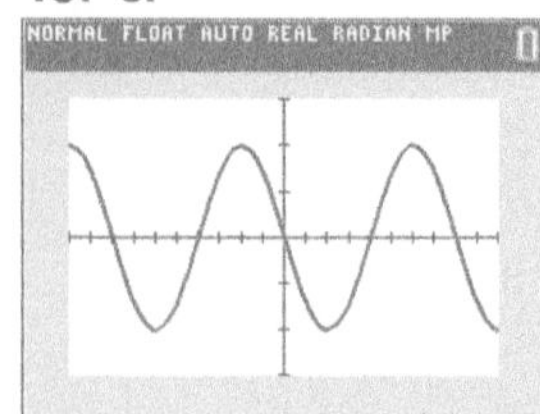

c. Use your graphing calculator to sketch a graph of $y = -2 \sin x$.

d. How does the graph of $y = -2 \sin x$ compare with the graph of $y = 2 \sin x$?

The graph of $y = -2 \sin x$ is the reflection of $y = 2 \sin x$ in the x-axis.

e. What is the general effect of the negative sign of the coefficient a in $y = a \sin x$?

The negative a will produce the graph of $y = |a| \sin x$ reflected in the x-axis.

In general, the amplitude of the functions defined by $y = a \sin (bx)$ and $y = a \cos (bx)$ is $|a|$.

Period of the Sine and Cosine Functions

After much practice, you begin to speed up on your circular track of radius 100 meters. You finally achieve your personal goal of 200 meters per minute.

14. a. If you run 200 meters per minute, how long does it take you to complete one lap? Note that this amount of time to complete one lap will be important in defining the key concept of period for the trigonometric functions in the following problems:

one lap $= 2\pi(100 \text{ m}) = 200\pi \text{ m}$

$$t = \frac{200\pi \text{ m}}{200 \text{ m/min.}} = \pi \text{ min.} \approx 3.14 \text{ min.}$$

b. Complete the following table to give your coordinates at selected special points along your path. Because one lap takes π min, each $\frac{\pi}{4}$ min. covers $\frac{1}{4}$ of the circular track.

t (min)	YOUR x-COORDINATE	YOUR y-COORDINATE
0	100	0
$\frac{\pi}{4}$	0	100
$\frac{\pi}{2}$	−100	0
$\frac{3\pi}{4}$	0	−100
π	100	0
$\frac{5\pi}{4}$	0	100
$\frac{3\pi}{2}$	−100	0
$\frac{7\pi}{4}$	0	−100
2π	100	0

c. On the following grid, use the data from Problem 14b to plot (t, y), your y-coordinate, as a function of t. Connect your data pairs to make a smooth graph, as in the previous activity, and predict what will happen to the graph for values of t before and after the values of t in the table.

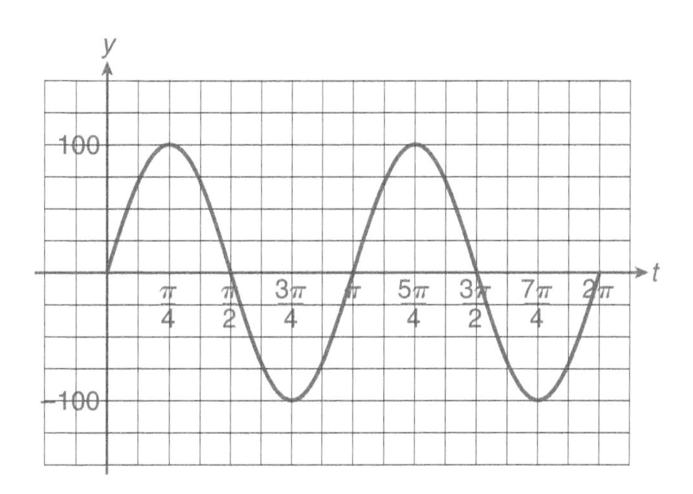

The graph continues in the same pattern in both directions.

d. What is the amount of time it takes for the graph to complete one full cycle (i.e., for you to complete one full lap)?

π min.

e. What effect does doubling your speed have on the amount of time to complete one cycle? Explain.

Doubling the speed cuts the time in half. Doubling the speed cuts the period and thus the time in half.

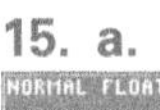

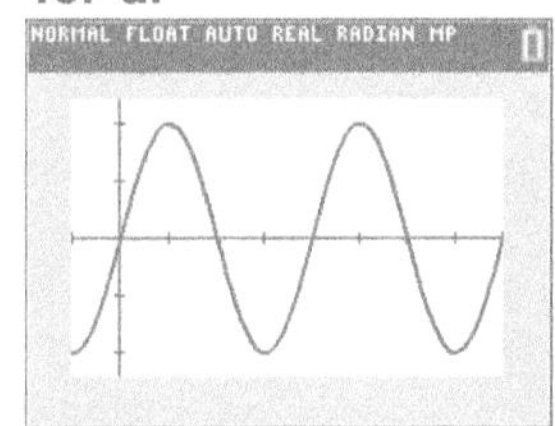

15. a. Use your graphing calculator to plot $y = 100 \sin(2t)$. Note that the name of the input t must be changed to x to conform with your calculator.

b. Compare the graph in part a with the graph in Problem 14c.

The graphs are the same.

16. Recall that the period of a trigonometric function is the shortest interval of inputs needed to complete one full cycle. Determine the period of each of the following using the graph:

a. $y = 100 \sin x$

period $= 2\pi$ units

b. $y = 100 \sin(2x)$

period $= \pi$ units

The period of the graph of a sine function can be determined directly from the equation that defines the function. For example, the period of $y = 100 \sin(2x)$ is π units, which is half of the period of $y = 100 \sin(x)$. It appears that the coefficient of x in the function affects the period of the function.

In general, if $y = a \sin(bx)$ and $b > 0$, then the period is $\frac{2\pi}{b}$.

17. On the following grid, repeat Problem 14c to plot the data pairs (t, x):

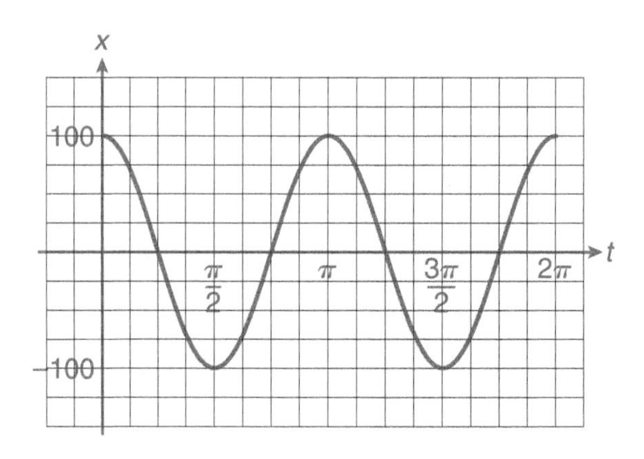

18. b.

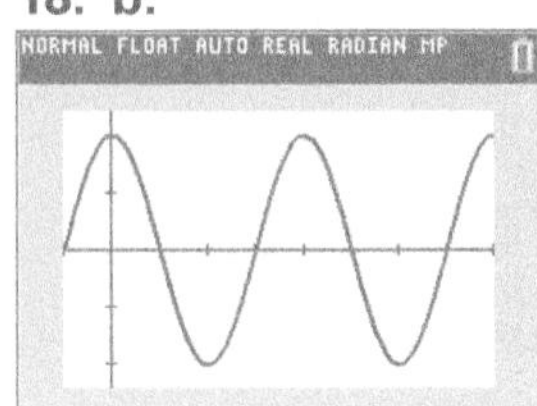

18. a. What function should you enter into your calculator to produce the graph from Problem 17?

$y = 100 \cos(2x)$

b. Enter the function from part a into your calculator, and obtain a graph.

19. a. If you were to triple your speed on the track from the original 100 meters per minute, what effect would this have on the amount of time to complete one lap?

The time would be reduced by a factor of 3, from 2π minutes (6.28 minutes) to $\frac{2\pi}{3}$ minutes (2.09 minutes).

b. What equations would describe the x- and y-coordinates of your position?

$x = 100 \cos(3t)$

$y = 100 \sin(3t)$

LC LEARNING CATALYTICS

Determine the period of the function defined by $y = 5\cos(\frac{3x}{4})$.

a. 2π.

b. π.

c. $\frac{8\pi}{3}$.

d. $\frac{3\pi}{2}$.

20. Determine the period of each of the following functions:

a. $y = 100 \cos x$

2π

b. $g(x) = 100 \cos(2x)$

π

c. $h(x) = 100 \cos(3x)$

$\frac{2\pi}{3}$

SUMMARY Activity 6.8

1. The **amplitude** of a periodic function is defined by

$$\frac{1}{2}(M - m),$$

where M represents the maximum function value and m represents the minimum function value. For trigonometric functions defined by $y = a\sin(bx)$ and $y = a\cos(bx)$, the amplitude is $|a|$ because

$$|a| = \frac{1}{2}(M - m).$$

2. For the trigonometric functions defined by $y = a\sin(bx)$ or $y = a\cos(bx)$, the **period** is $\frac{2\pi}{b}$, where $b > 0$.

EXERCISES Activity 6.8

1. On the following grid, use the data from Problem 3 to plot the data pairs (t, x):

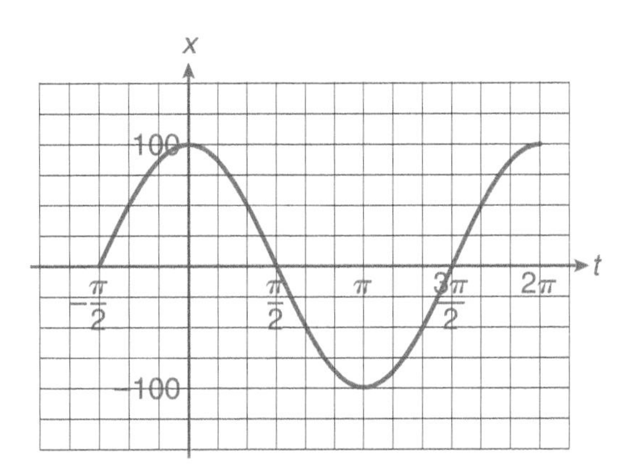

2. Use your graphing calculator to plot $x = 100 \cos(t)$. Make sure your graphing calculator is in radian mode. Note that the names of the input t and output (your x-coordinate) must be changed to x and y, respectively, to conform with your calculator.

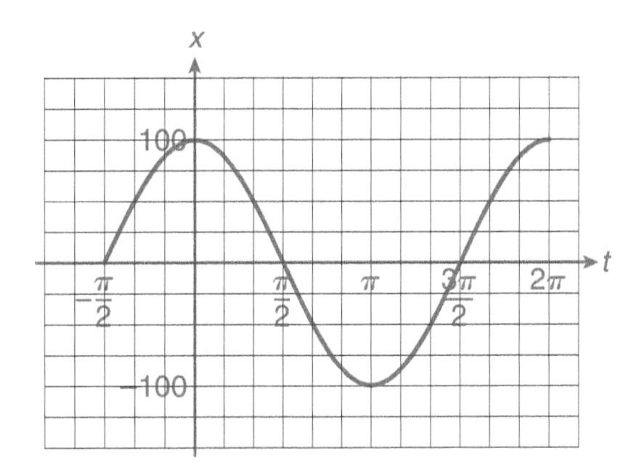

3. a. What is the maximum function value of the cosine function in Problem 2?

   The maximum value is 100.

   b. What is the minimum function value of the cosine function in Problem 2?

   The minimum value is −100.

4. What is the connection between the function values in Problem 3, the 100-meter radius of the circle, and the coefficient 100 of the function?

   100 is the amplitude of the function.

5. If you are forced to run on a larger circular track of radius 150 meters and you increase your speed to 150 meters per minute, predict the equations of the functions describing the x- and y-coordinates of your position. Explain.

   $x = 150 \cos \theta$; $y = 150 \sin \theta$

6. Determine by inspection the amplitude of the following functions. Then verify the results with your graphing calculator. (*Remember:* Amplitude *cannot* be negative.)

a. $y = 3 \sin x$

amplitude = 3

b. $y = 0.4 \cos x$

amplitude = 0.4

c. $f(x) = -2 \cos x$

amplitude = 2

d. $g(x) = -2.3 \sin x$

amplitude = 2.3

e. $y = 2 \sin(3x)$

amplitude = 2

f. $h(x) = -4 \cos x$

amplitude = 4

7. Is there any relationship between the amplitude and the period of a sine function?

No, they are independent.

8. For each of the following tables, identify a function of the form $y = a \sin bx$ or $y = a \cos bx$ that approximately satisfies the table.

a.

x	0	0.7854	1.5708	2.3562	3.1416
y	0	−15	0	15	0

$y = -15 \sin(2x)$

b.

x	0	2.244	4.488	6.732	8.976
y	1.3	0	−1.3	0	1.3

$y = 1.3 \cos(0.7x)$

9. For each of the following graphs, identify a function of the form $y = a \sin bx$ or $y = a \cos bx$ that the graph approximates.

a.

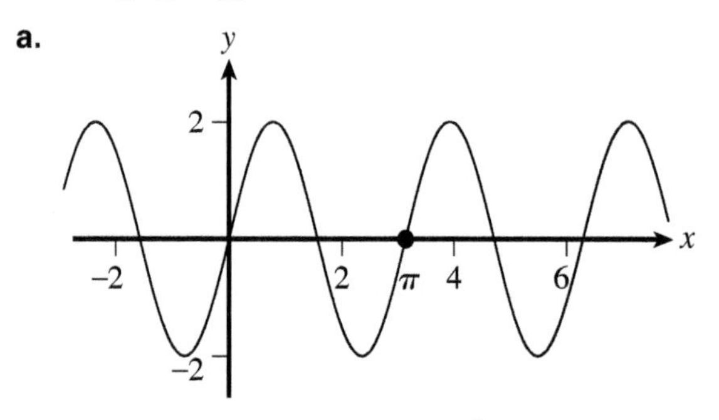

amplitude 2; period $= \pi = \frac{2\pi}{b}$ or $b = 2$

$y = 2 \sin(2x)$

b.

amplitude 1.5; period $= 2 = \frac{2\pi}{b}$, so that $b = \pi$

$y = -1.5 \cos(\pi x)$

10. Match the given function to one of the following graphs. Assume that Xscl = 1 and Yscl = 1.

a. $y = 2 \cos 0.5x$

Graph is iii.

b. $y = -0.5 \sin 2x$

Graph is iv.

c. $y = 0.5 \cos 2x$

Graph is i.

d. $y = 2 \sin 0.5x$

Graph is ii.

i.

NORMAL FLOAT AUTO REAL RADIAN MP

ii.

NORMAL FLOAT AUTO REAL RADIAN MP

iii.

NORMAL FLOAT AUTO REAL RADIAN MP

iv.

NORMAL FLOAT AUTO REAL RADIAN MP

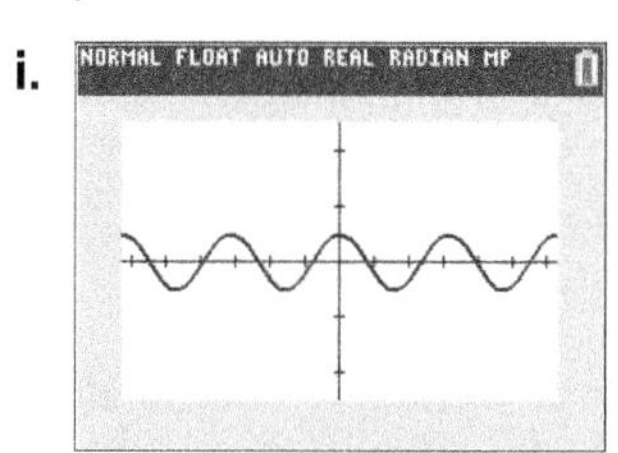

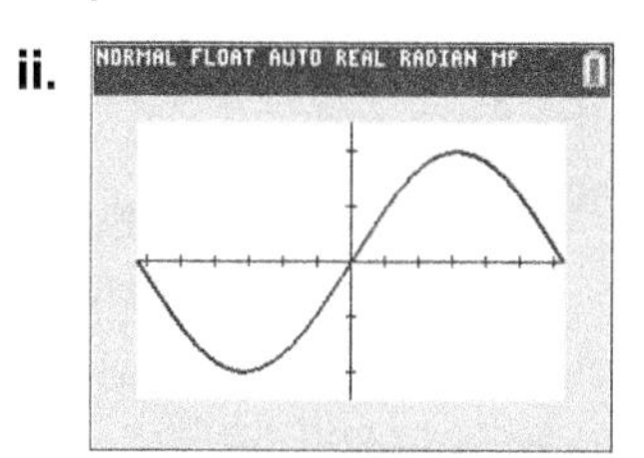

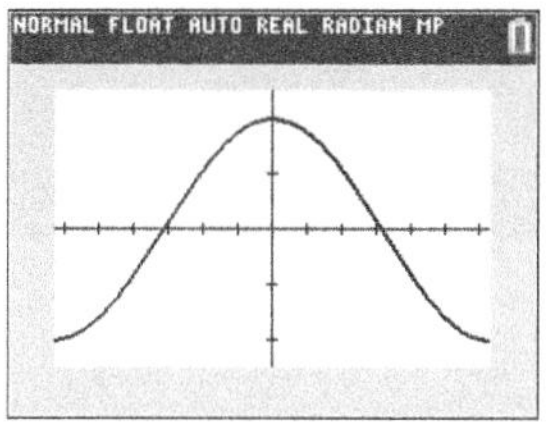

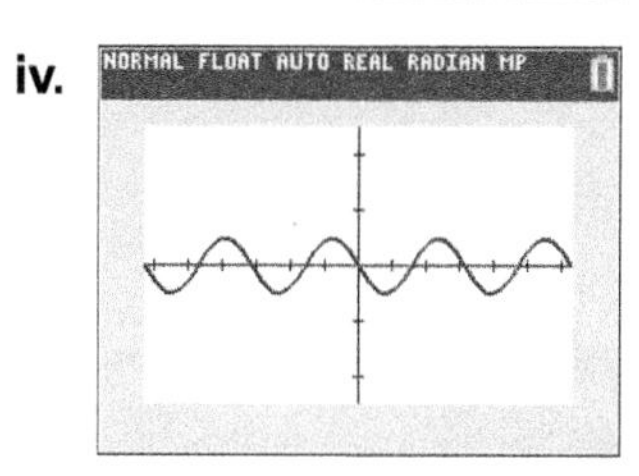

ACTIVITY

6.9

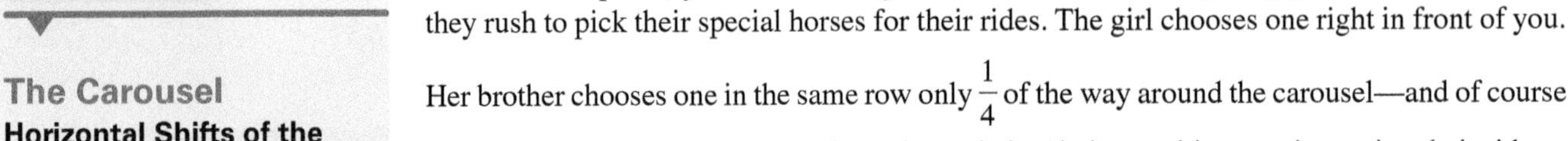

The Carousel

Horizontal Shifts of the Graphs of the Sine and Cosine Functions

OBJECTIVE

Determine the displacement of $y = a\sin(bx + c)$ and $y = a\cos(bx + c)$ using a formula.

You have developed a keen interest in carousels. You developed enough of an interest to visit six carousel parks on your vacation.

At one of the parks, you are watching a carousel and notice a young girl and her brother as they rush to pick their special horses for their rides. The girl chooses one right in front of you. Her brother chooses one in the same row only $\frac{1}{4}$ of the way around the carousel—and of course ahead of his sister. You decide to investigate their relative positions as they enjoy their rides.

Each child is 20 feet from the center of the carousel. Assume the girl starts at the point (20, 0) and her brother starts at (0, 20).

1. Complete the following table to give coordinates for both the girl and her brother at selected special points on their ride:

t (AMOUNT OF ROTATION IN RADIANS)	THE GIRL'S x-COORDINATE	THE GIRL'S y-COORDINATE	HER BROTHER'S x-COORDINATE	HER BROTHER'S y-COORDINATE
0	20	0	0	20
$\frac{\pi}{2}$	0	20	−20	0
π	−20	0	0	−20
$\frac{3\pi}{2}$	0	−20	20	0
2π	20	0	0	20
$\frac{5\pi}{2}$	0	20	−20	0

2. On the following grid, use the data from Problem 1 to plot (t, y) for both the girl and her brother. Connect the points to smooth out your graphs.

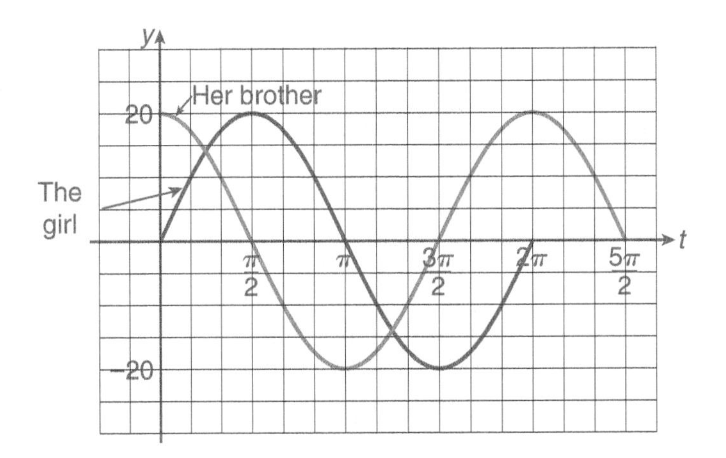

3. What is the relationship between the two graphs?

The graph for the boy is shifted $\frac{\pi}{2}$ units to the left of the graph for the girl.

DEFINITION

The **displacement**, or **horizontal shift**, of the graph of $y = a \sin(bx + c)$ is the smallest movement (left or right) necessary for the graph of $y = a \sin(bx)$ to match the graph of $y = a \sin(bx + c)$ exactly.

EXAMPLE 1 *Consider the graphs of* $y = \sin x$ *and* $y = \sin(x + 1)$.

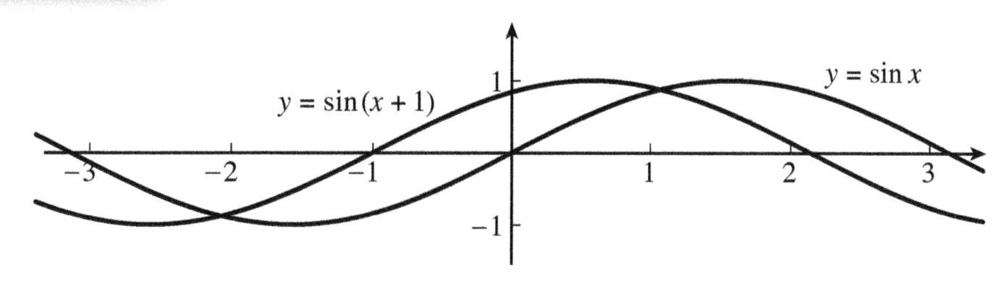

The graph of $y = \sin x$ must be moved 1 unit to the left to match the graph of $y = \sin(x + 1)$ exactly, so the displacement, or horizontal shift, is -1.

4. **a.** Which graph is displaced in Problem 2?

the graph for the boy

b. What is the displacement?

$\frac{\pi}{2}$ units to the left

5. Would you expect the same type of relationship between the two graphs representing the x-coordinates? Explain.

Yes, I can see from the table that the x-coordinates are the same for values of t that differ by $\frac{\pi}{2}$ units.

Displacement, or horizontal shift, is defined for the cosine function in the same manner it is defined for sine.

EXAMPLE 2 *Consider the graphs of* $y = 3 \cos 2x$ *and* $y = 3 \cos(2x - 1)$.

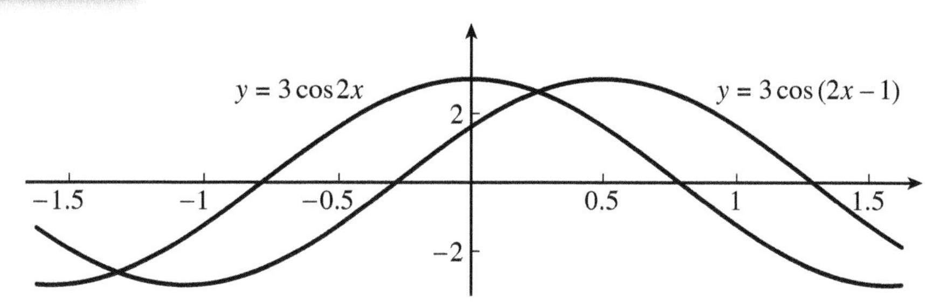

The graph of $y = 3 \cos(2x - 1)$ appears to be about $\frac{1}{2}$ unit to the right of the graph of $y = 3 \cos 2x$, so the displacement is approximately $\frac{1}{2}$.

6. a. If the girl's x-coordinate is given by 20 cos t, predict the defining equation of the boy's x-coordinate.

$$x = 20\cos\left(t + \frac{\pi}{2}\right)$$

b. Use your graphing calculator to test your prediction.

In general, in the functions defined by $y = a\sin(bx + c)$ and $y = a\cos(bx + c)$, the values for b and c affect the displacement, or horizontal shift, of the function. The horizontal shift is given by $-\frac{c}{b}$. Note that if $-\frac{c}{b}$ is negative, the shift is to the left. If $-\frac{c}{b}$ is positive, the shift is to the right.

LC LEARNING CATALYTICS

Determine the displacement of $y = 2\cos(3x - 0.5)$.

a. $\frac{\pi}{6}$.
b. $\frac{1}{6}$.
c. $\frac{1}{4}$.
d. None of these.

7. a. Using the expression $-\frac{c}{b}$, what is the displacement of $y = 3\cos(2x - 1)$?

$$-\frac{c}{b} = -\frac{(-1)}{2} = \frac{1}{2}$$

b. Is your result consistent with the graph in Example 2?

yes

8. In this activity, the boy's y-coordinate is given by $y = a\sin(bx + c)$, where $a = 20$ and $b = 1$. Calculate c.

$$-\frac{c}{b} = \frac{\pi}{-2}$$

$$\frac{-c}{1} = \frac{-\pi}{2}$$

$$c = +\frac{\pi}{2}$$

SUMMARY Activity 6.9

1. The **displacement**, or **horizontal shift**, of the graph of $y = a\sin(bx + c)$, $b > 0$, is the smallest movement (left or right) necessary for the graph of $y = a\sin bx$ to match the graph of $y = a\sin(bx + c)$ exactly.
2. The **displacement**, or **horizontal shift**, of the graph of $y = a\cos(bx + c)$, $b > 0$, is the smallest movement (left or right) necessary for the graph of $y = a\cos(bx)$ to match the graph of $y = a\cos(bx + c)$ exactly.
3. For the functions defined by $y = a\sin(bx + c)$ and $y = a\cos(bx + c)$, $b > 0$, the horizontal shift is given by $-\frac{c}{b}$.
4. If $-\frac{c}{b}$ is negative, the shift is to the left. If $-\frac{c}{b}$ is positive, the shift is to the right.

EXERCISES Activity 6.9

1. For each of the following equations, identify the amplitude, period, and displacement of its graph:

a. $y = 0.7 \cos\left(2x + \frac{\pi}{2}\right)$

amplitude: 0.7

period: π

displacement: $\frac{-\frac{\pi}{2}}{2} = -\frac{\pi}{4}$

b. $y = 3 \sin(x - 1)$

amplitude: 3

period: 2π

displacement: $\frac{-(-1)}{1} = 1$

c. $f(x) = -2.5 \sin\left(0.4x + \frac{\pi}{3}\right)$

amplitude: 2.5

period: 5π

displacement: $\frac{-\frac{\pi}{3}}{0.4} = -\frac{5\pi}{6}$

d. $g(x) = 15 \sin(2\pi x - 0.3)$

amplitude: 15

period: 1

displacement: $\frac{-(-0.3)}{2\pi} = \frac{3}{20\pi} \approx 0.0477$

2. For each of the following tables, identify a function of the form $y = a \sin(bx + c)$ or $y = a \cos(bx + c)$ that approximately satisfies the table.

a.

x	−0.7854	0.7854	2.3562	3.927	5.4978
y	0	3	0	−3	0

$y = 3 \sin\left(x + \frac{\pi}{4}\right)$

b.

x	1	2.5708	4.1416	5.7124	7.2832
y	−0.5	0	0.5	0	−0.5

$y = -0.5 \cos(x - 1)$

3. Sketch one cycle of the graph of the function defined by $f(x) = 2 \sin\left(x + \frac{\pi}{2}\right)$.

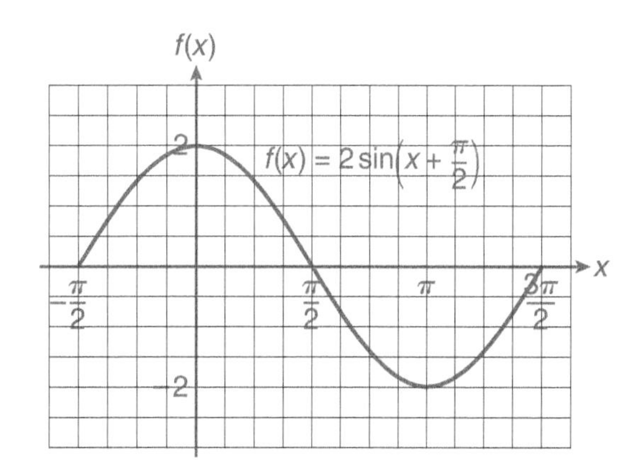

Exercise numbers appearing in color are answered in the Selected Answers appendix.

4. Determine an equation for the function defined by the following graph:

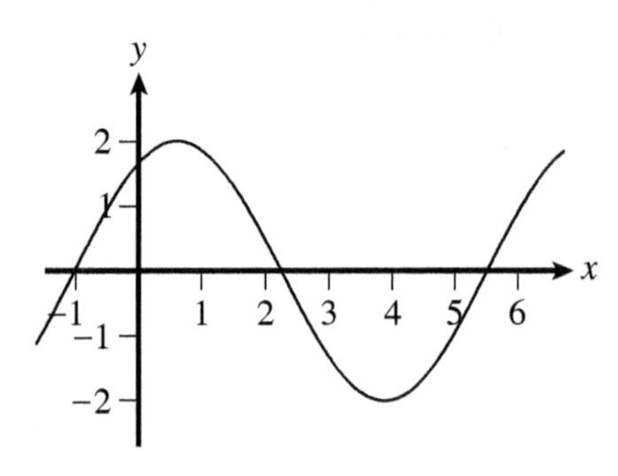

$y = 2 \sin (x + 1)$

5. Match the given equation to one of the graphs that follow. (Assume that Xscl = 1 and Yscl = 1.)

a. $y = 2 \cos (x - 1)$

Graph is iii.

b. $y = 2 \sin (x - 1)$

Graph is iv.

c. $y = 2 \cos (x + 2)$

Graph is i.

d. $y = 2 \sin (x + 2)$

Graph is ii.

i.
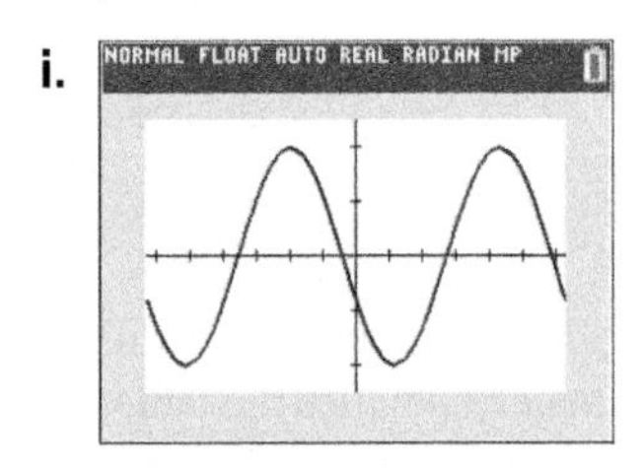

ii.
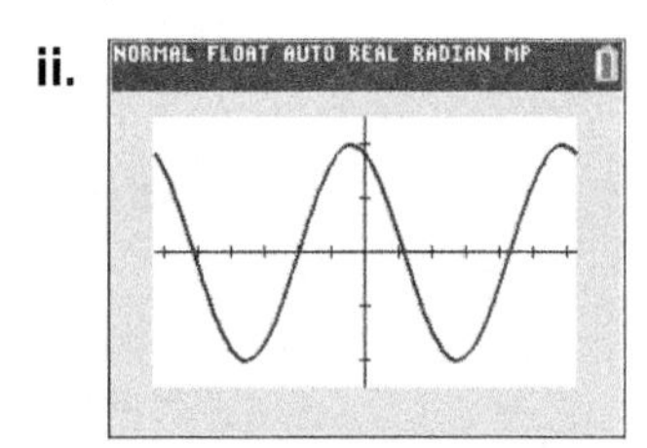

iii.
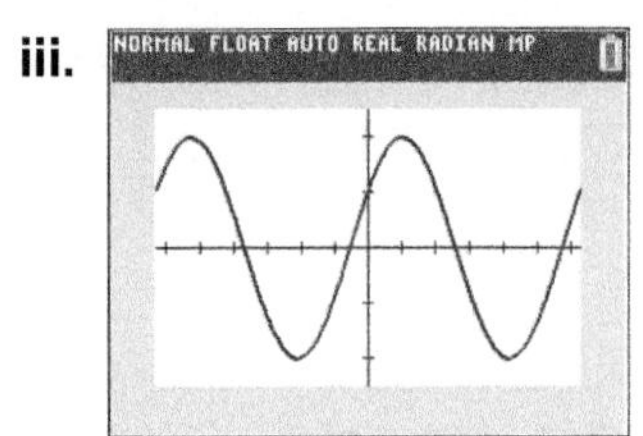

iv.
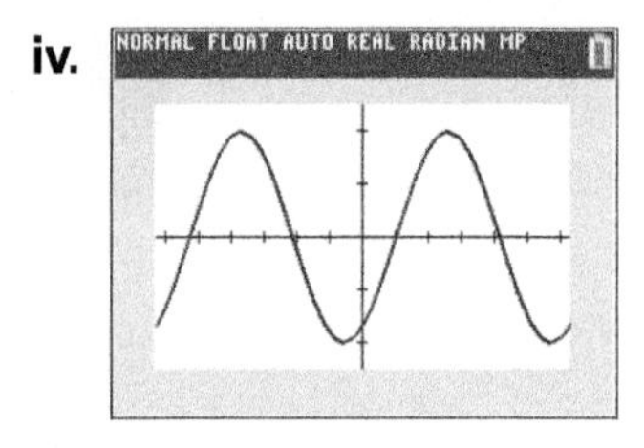

ACTIVITY 6.10

Texas Temperatures

Modeling Data with a Sine Regression Equation

OBJECTIVES

1. Determine the equation of a sine function that best fits the given data.
2. Make predictions using a sine regression equation.

Currently, several different websites provide average temperature data for many cities around the world. According to one website, 65-year average high temperatures for San Antonio, Texas, are as follows:

Month	Jan	Feb	Mar	Apr	May	June	July	Aug	Sept	Oct	Nov	Dec
Average High (°F)	62.2	66.5	73.5	80.4	86.4	92.1	94.8	95.3	89.8	82.1	71.6	64.5

1. To get a feel for the relationship between the month and the average high temperatures, let x represent the month, where $1 =$ Jan of the first year, $12 =$ Dec of the first year, $13 =$ Jan of the second year, and $24 =$ Dec of the second year. Plot the average maximum monthly temperatures for San Antonio over a 2-year period on the grid below. Use the same output data for the first and second year.

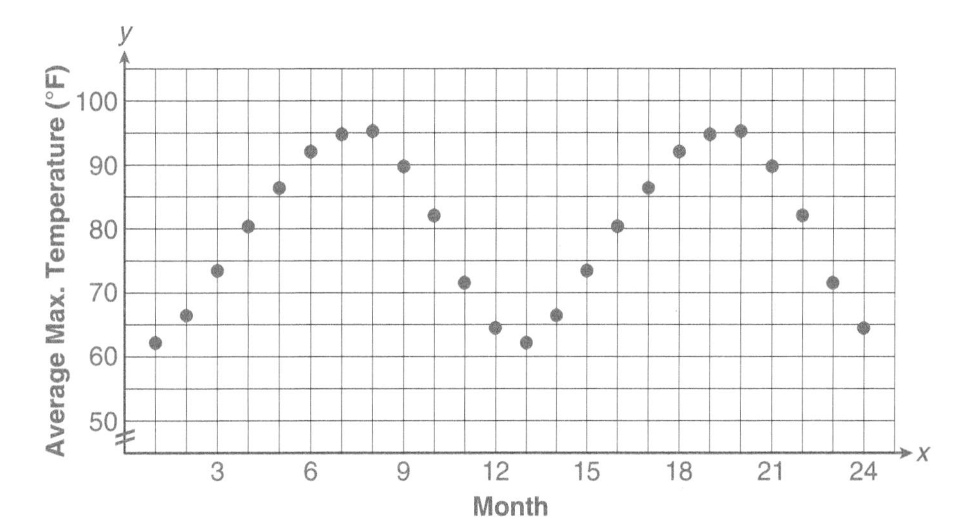

2. Enter the data plotted in Problem 1 into your calculator, and verify your scatterplot by creating a stat plot. Your plot should resemble the screen below.

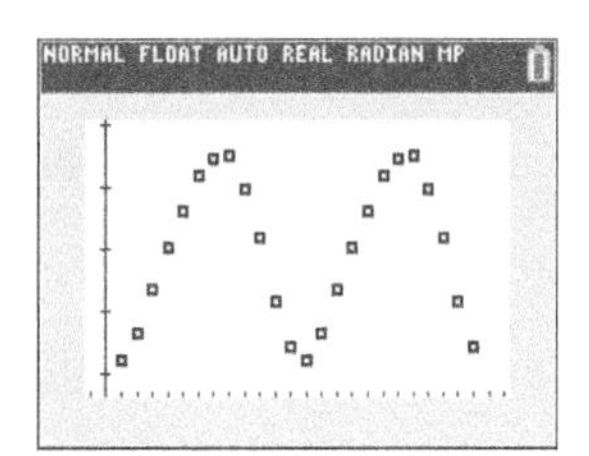

Notice the wavelike and repetitive nature of the data. This is typical of many natural periodic phenomena.

To produce a sine function that models the data, set your calculator to radian mode and use the Stat Calc menu and Option C: SinReg.

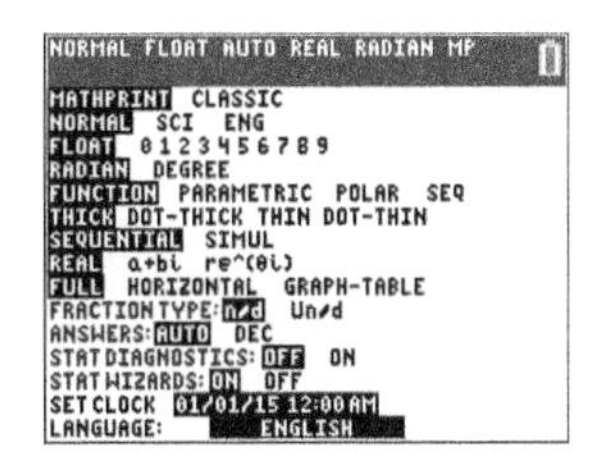

NORMAL FLOAT AUTO REAL RADIAN MP
EDIT CALC TESTS
5↑QuadReg
6:CubicReg
7:QuartReg
8:LinReg(a+bx)
9:LnReg
0:ExpReg
A:PwrReg
B:Logistic
C↓SinReg

Note that the sine regression model is of the form

$$y = a\sin(bx + c) + d.$$

LC LEARNING CATALYTICS

A trig function generates the following table:

x	0	$\frac{\pi}{4}$	$\frac{\pi}{2}$	$\frac{3\pi}{4}$	π
y	−2	0	2	0	−2

Which of the following could be the equation of the trig function that generated the table?

a. $y = 2\sin(2x)$.
b. $y = 2\cos(x)$.
c. $y = -2\sin(2x)$.
d. $y = -2\cos(2x)$.

3. Enter your regression model here, rounding the values of a, b, c, and d to the nearest 0.001.

   $y = 16.215\sin(0.521x - 2.097) + 79.854$

4. Add the graph of this function to your stat plot to verify that it is a good model. Your screen should appear as follows:

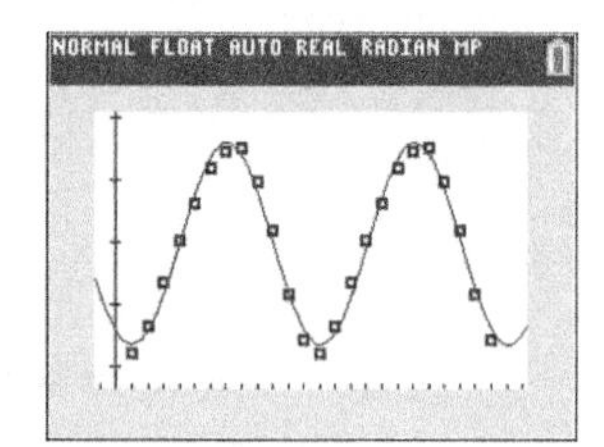

SUMMARY Activity 6.10

1. The **sine function** can be used to model periodic wavelike behavior.
2. To determine the sine regression using the TI-84 Plus C calculator,
   a. set your MODE to radian.
   b. use the SinReg option on the Stat Calc menu.

EXERCISES Activity 6.10

The phases of the moon are periodic and repetitive even though the names of the phases can be confusing.

i. A new moon occurs when no moon is visible to an observer on Earth.
ii. During the first quarter, half of the side of the moon facing Earth is visible.
iii. During a full moon, the entire side of the moon is visible.
iv. During the last quarter, the other half of the side of the moon facing Earth is visible.

The U.S. Naval Observatory lists the following dates for phases of the moon in a recent year:

Date	Jan 29	Feb 5	Feb 13	Feb 21	Feb 28	Mar 6	Mar 14	Mar 22	Mar 29
Moon Phase	New	1st quarter	Full	Last quarter	New	1st quarter	Full	Last quarter	New

1. a. Let x represent the number of days since Jan 29 (be careful here) and let y represent the amount of moon visible, where New = 0, the 1st quarter = 0.5, Full = 1, and the Last quarter = 0.5. Complete the following table:

Date	Jan 29	Feb 5	Feb 13	Feb 21	Feb 28	Mar 6	Mar 14	Mar 22	Mar 29
Days since Jan 29, x	0	7	15	23	30	36	44	52	59
The Amount of Moon Visible y	0	0.5	1	0.5	0	0.5	1	0.5	0

b. Plot the data points generated in the table in part a.

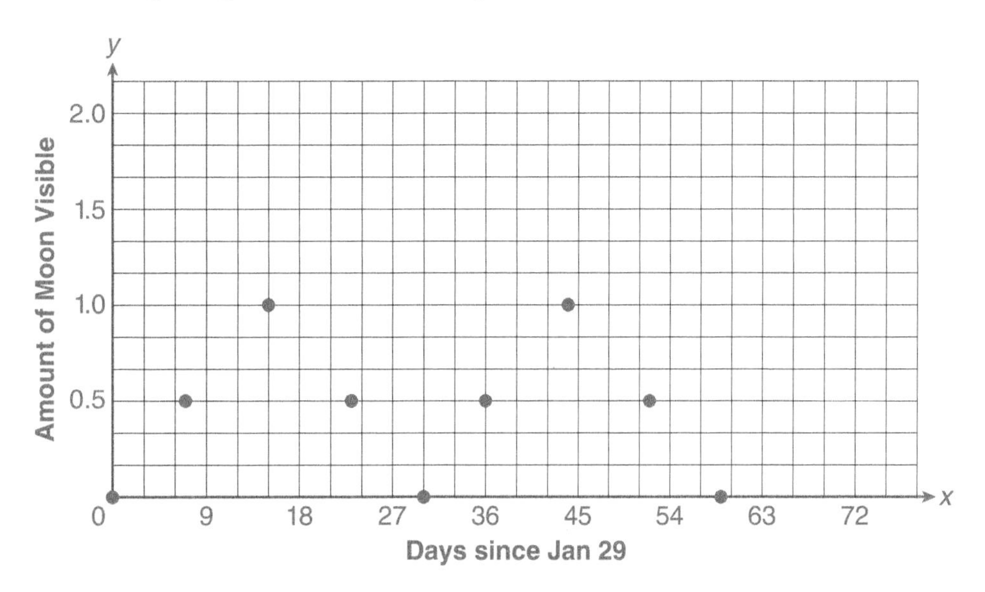

c. Does this data indicate that a sine regression may be appropriate in this case? Explain.

Yes, it does show repeated and periodic behavior.

d. Use your TI-84 Plus C calculator to produce a sine regression model for the data in part a. Round a, b, c, and d to the nearest 0.0001, and record the model below.

$y = 0.5019 \sin(0.2135x - 1.5888) + 0.5306$

e. Use your model to predict approximately how much of the moon will be visible 95 days after Jan 29, provided the sky is clear.

$y(95) \approx 0.45$, or about 45% of the side of the moon facing Earth will be visible 95 days after Jan 29.

2. Tides are the regular rising and falling of the ocean's surface. This is due in large part to the gravitational forces of the Moon. The following represents the water level of the tide off the coast of South Padre Island, Texas, for a 24-hour period in November of a recent year:

Hour #	0	2	4	6	8	10	12	14	16	18	20	22	24
Measurement above the Average Low Tide (ft.)	1.08	1.56	2.02	2.31	2.40	2.32	2.22	2.09	1.86	1.42	0.97	0.73	0.88

Source: NOAA tidesonline.noaa.gov.

a. Plot the data points given in the table above using hour as the independent variable.

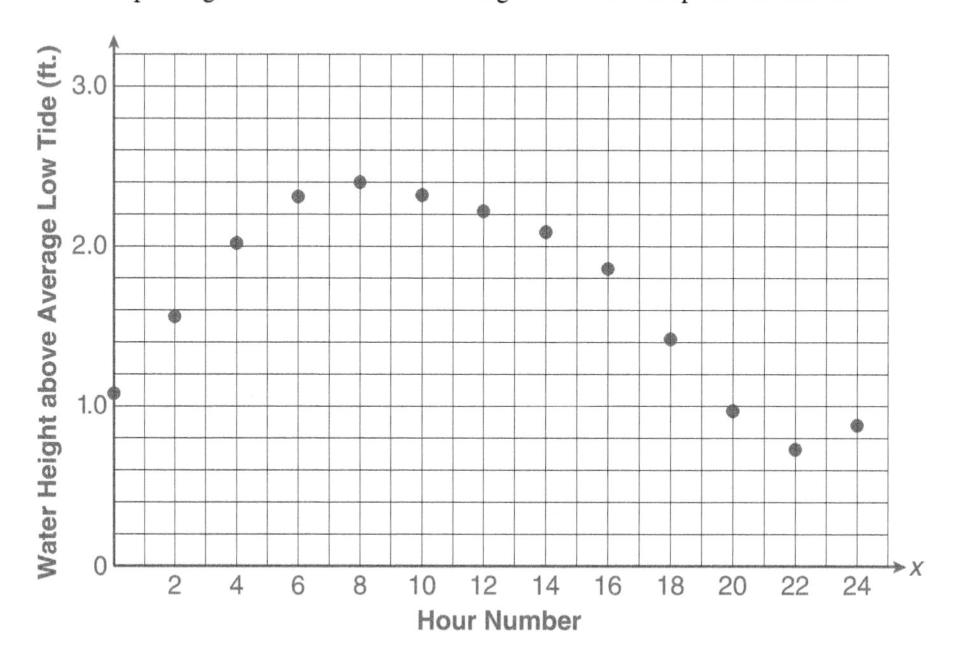

b. Does this data indicate that a sine regression may be appropriate in this case? Explain.

Yes, the wavelike behavior is apparent in the scatterplot.

c. Use your TI-84 Plus C calculator to produce a sine regression model for the data in the table. Round a, b, c, and d to the nearest 0.001, and record the model below.

$y = 0.839 \sin(0.220x - 0.506) + 1.611$

d. Use your model to predict the height of the tide 30 hours after the original observation.

$y(30) = 1.45$ feet above the average low tide for the period

e. Do you expect your model from part c to be a good predictor of tide height on a year-round basis?

No, the height of the tide is affected by weather, so the wavelike behavior should be there, but the heights will vary. (Students may also know that the tide is affected by the Moon, and its rotation period is about 29.5 days.)

3. The average rainfall in Chicago by season is given in the following table:

SEASON	SPRING	SUMMER	FALL	WINTER
Rainfall (in.)	10.49	12.06	9.97	6.57

a. Let t represent the seasons over a **two-year** period, where spring of year 1 is 1, summer of year 1 is 2, fall of year 1 is 3, etc. Plot these eight points on the following grid:

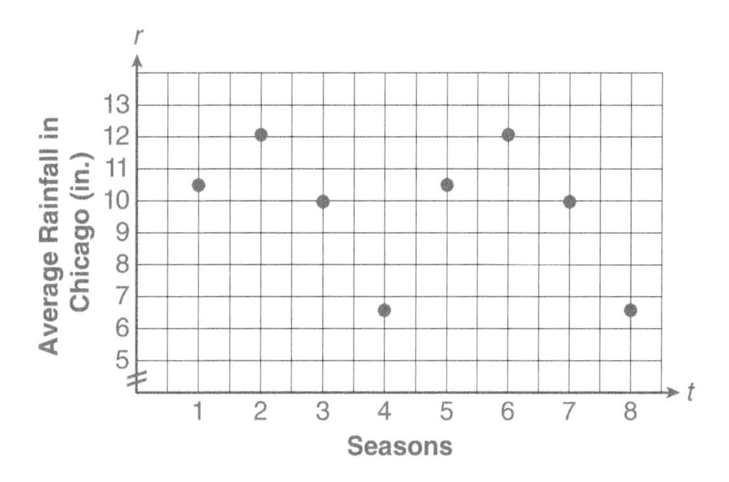

b. Does this data indicate that a sine regression may be appropriate in this case? Explain.

Yes, the data appears to be periodic.

c. Use your graphing calculator to produce a sine regression model for the data in part a. Round a, b, c, and d the nearest 0.001, and record the model here.

$y = 2.759 \sin(1.534x - 1.322) + 9.725$

d. What is the significance of the final constant in the model?

The 9.725 is the average output value. In this case, it represents the average seasonal rainfall in Chicago.

e. Determine the period of the sine regression model. Does the value make sense?

$\frac{2\pi}{1.534} \approx 4.10$ Yes, it makes sense. It should be close to 4 because there are four seasons in the input cycle.

Collecting and Analyzing Data

Visit www.weatherbase.com. Find your city or a city that is near you. Find the average low temperature for each month for that city, and repeat Problem 1–3 in the activity using this new data.

CLUSTER 2 What Have I Learned?

1. Explain why the trigonometric functions could be called circular functions.

Their output values are coordinates of points on a circle.

2. Sometimes the difference between the trigonometric functions and the circular functions is explained by the difference in inputs. The input values for the trigonometric functions are angle measurements, and the input values for the circular functions are real numbers. How does your knowledge of radian measure relate to this?

The radian measure of an angle is the distance traveled along an arc of a circle of radius 1 intercepted by the angle. This distance is a real number.

3. a. Estimate the amplitude of the function defined by the following graph:

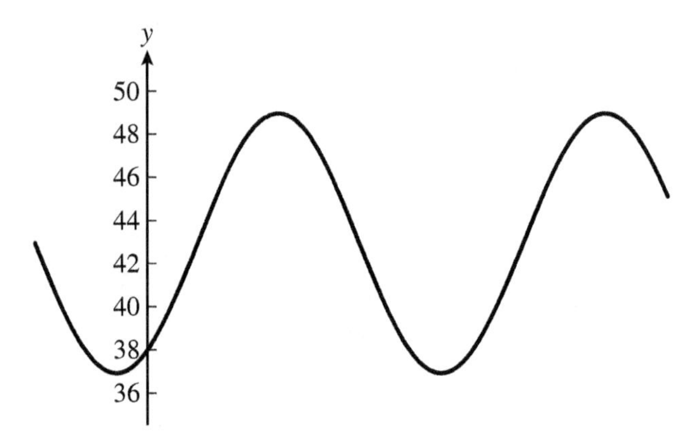

The amplitude is about 6.

b. Use the definition of amplitude to answer part a.

$$\text{amplitude} = \frac{1}{2}(M - m) = \frac{1}{2}(49 - 37) = \frac{1}{2}(12) = 6$$

4. The period of $y = \sin x$ is 2π, and the period of $y = \cos x$ is 2π. Explain why this makes sense when sine and cosine are viewed as circular functions.

The circumference of a circle is $2\pi r$. If the radius is 1, then the length of the intercepted arc is 2π units. So the x and y values (cosine and sine values) on the full-circle repeat every 2π units.

5. Given a function defined by $y = a\sin(bx + c)$ and $b > 0$ and given the fact that b and c possess opposite signs, determine whether the graph of the function is displaced to the right or to the left. Explain.

If b and c have opposite signs, then $\frac{c}{b}$ is negative and $-\frac{c}{b}$ is positive.

Therefore, the displacement is to the right.

CLUSTER 2 How Can I Practice?

1. Determine the coordinates of the point on the unit circle corresponding to the following central angles. If necessary, round your results to the nearest hundredth.

a. 36° (0.81, 0.59)

b. 210° (−0.87, −0.5)

c. −90° (0, −1)

d. 317° (0.73, −0.68)

e. −144° (−0.81, −0.59)

f. 450° (0, 1)

2. For each of the points on the unit circle you determined in Exercise 1, determine the distance traveled along the circle to the point from (1, 0).

a. 0.63 unit

b. 3.67 units

c. 1.57 units clockwise

d. 5.53 units

e. 2.51 units clockwise

f. 7.85 units

3. Convert the following degree measures to radian measures in terms of π:

a. 18° $18 \cdot \frac{\pi}{180} = \frac{\pi}{10}$

b. 150° $150 \cdot \frac{\pi}{180} = \frac{5\pi}{6}$

c. 390° $390 \cdot \frac{\pi}{180} = \frac{13\pi}{6}$

d. −72° $-72 \cdot \frac{\pi}{180} = \frac{-2\pi}{5}$

4. Convert the following radian measures to degree measures:

a. $\frac{5\pi}{6}$ $\frac{5\pi}{6} \cdot \frac{180}{\pi} = 150°$

b. 1.7π $1.7\pi \cdot \frac{180}{\pi} = 306°$

c. -3π $-3\pi \cdot \frac{180}{\pi} = -540°$

d. 0.9π $0.9\pi \cdot \frac{180}{\pi} = 162°$

Obtain the following information about each of the functions defined by the equations in Exercises 5–9.

a. *Use your graphing calculator to sketch a graph.*

b. *From the defining equation, determine the amplitude. Then use your graph to verify that your amplitude is correct.*

c. *From the defining equation, determine the period. Then use your graph to verify that your period is correct.*

d. *From the defining equation, determine the displacement. Then use your graph to verify that your displacement is correct.*

5. $y = 4 \cos 3x$

amplitude: 4

period: $\frac{2\pi}{3}$

displacement: 0

6. $y = -2 \sin (x - 1)$

amplitude: 2

period: $\frac{2\pi}{1} = 2\pi$

displacement: 1

5.

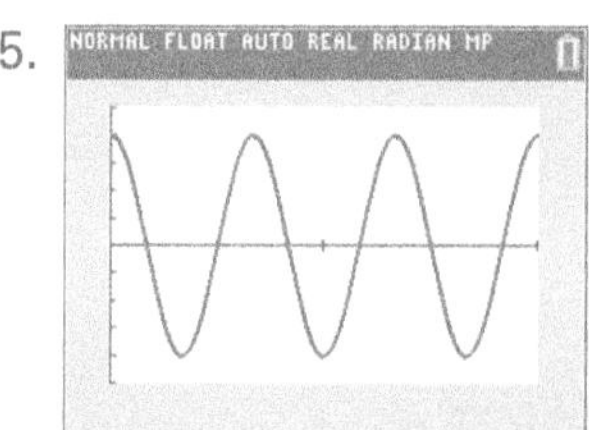

6.

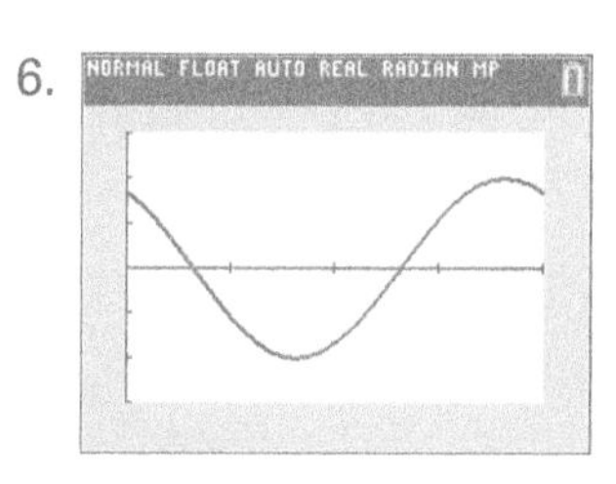

Answers to all How Can I Practice exercises are included in the Selected Answers appendix.

7. $s = -3.2 \sin(2x)$

amplitude: 3.2

period: $\frac{2\pi}{2} = \pi$

displacement: 0

8. $f(x) = -\cos\left(\frac{x}{2} + 1\right)$

amplitude: 1

period: $\frac{2\pi}{\frac{1}{2}} = 4\pi$

displacement: -2

7.

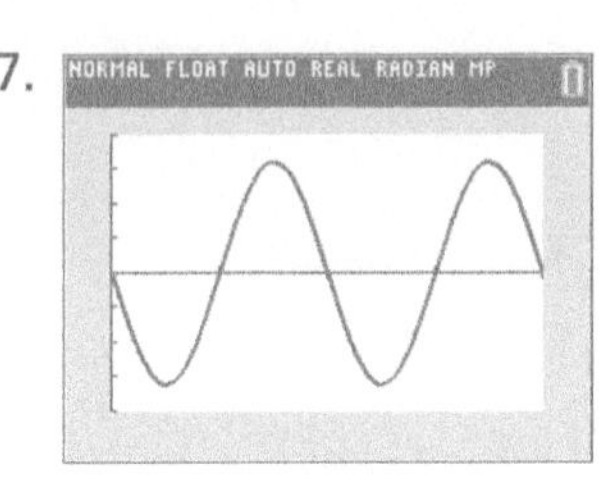

8.

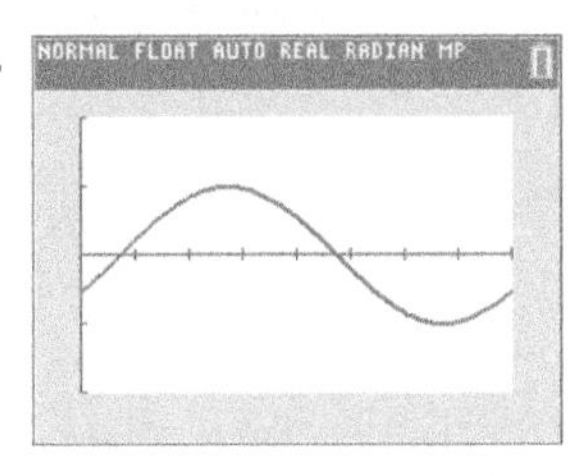

9. $g(x) = -3 \cos(4x - 1)$

amplitude: 3

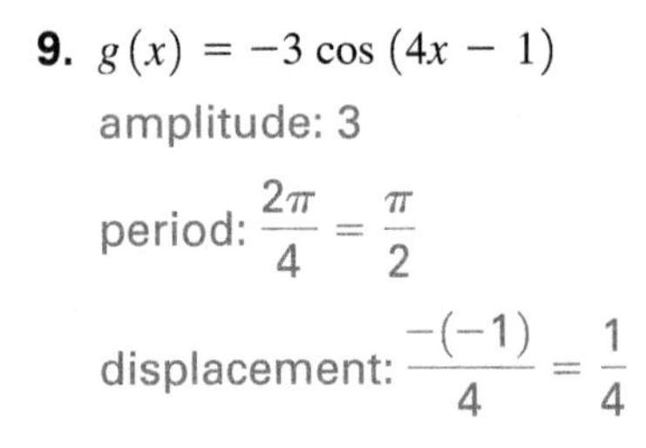

period: $\frac{2\pi}{4} = \frac{\pi}{2}$

displacement: $\frac{-(-1)}{4} = \frac{1}{4}$

9.

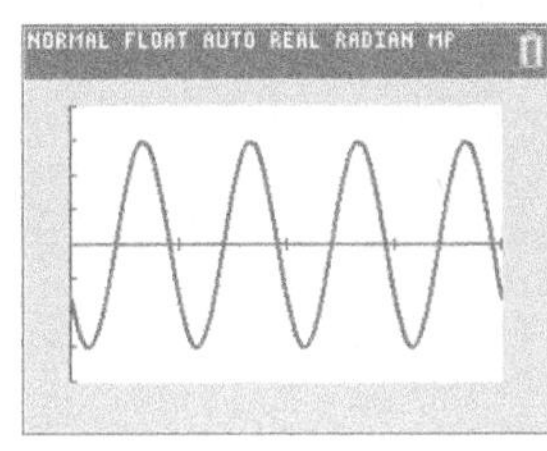

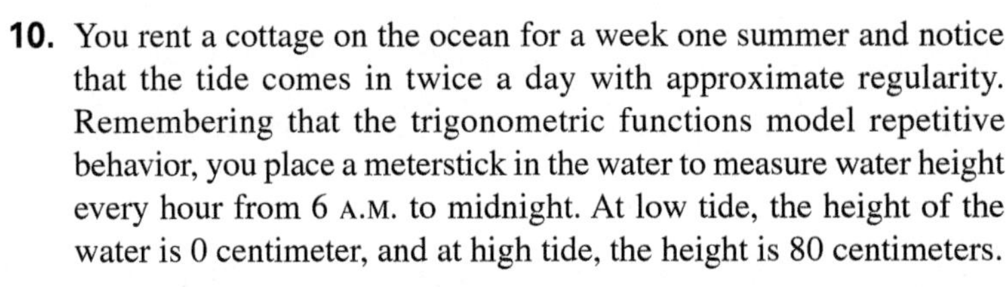

10. You rent a cottage on the ocean for a week one summer and notice that the tide comes in twice a day with approximate regularity. Remembering that the trigonometric functions model repetitive behavior, you place a meterstick in the water to measure water height every hour from 6 A.M. to midnight. At low tide, the height of the water is 0 centimeter, and at high tide, the height is 80 centimeters.

a. Explain why a sine or cosine function models this relationship between height of water in centimeters and time in hours.

A sine or cosine function models this relationship because of the repetitive nature of the height of the water as a function of time.

b. What is the amplitude of this function?

$$\text{amplitude} = \frac{80 - 0}{2} = 40$$

c. Approximate the period of this function. Explain.

The period is approximately 12 hours because high tide occurs twice a day.

d. Determine a reasonable defining equation for this function. Explain.

Let x represent the number of hours since midnight.

$y = a \sin(bx + c) + d$

$a = 40$, the amplitude

period: $\frac{2\pi}{b} = 12$

$b = \frac{\pi}{6}$

displacement: $3 = \frac{-c}{b} = \frac{-c}{\frac{\pi}{6}}$

$-c = 3 \cdot \left(\frac{\pi}{6}\right)$

$c = \frac{-\pi}{2}$

vertical shift: $d = 40$

$y = 40 \sin\left(\frac{\pi}{6}x - \frac{\pi}{2}\right) + 40$

Other equations are possible, depending on the choice of the displacement.

CHAPTER 6 Summary

The bracketed numbers following each concept indicate the activity in which the concept is discussed.

CONCEPT/SKILL	DESCRIPTION	EXAMPLE
The sine function of acute angles A of a right triangle [6.1]	$\sin A = \dfrac{\text{length of the side opposite } A}{\text{length of the hypotenuse}}$	Examples 2 and 3, Activity 6.1, pages 599–600
The cosine function of acute angles A of a right triangle [6.1]	$\cos A = \dfrac{\text{length of the side adjacent to } A}{\text{length of the hypotenuse}}$	Examples 2 and 3, Activity 6.1, pages 599–600
The tangent function of the acute angle A of a right triangle [6.1]	$\tan A = \dfrac{\text{length of the side opposite } A}{\text{length of the side adjacent to } A}$	Examples 2 and 3, Activity 6.1, pages 599–600
Complementary angles [6.2]	Complementary angles are two acute angles whose measures sum to 90°.	Angles of 30° and 60° are complementary angles.
Cofunctions related to complementary angles [6.2]	Cofunctions of complementary angles are equal.	$\sin 35° = \cos 55°$
The domain of the inverse trigonometric functions [6.3]	The domain (set of inputs) of an inverse trigonometric functions is the set of ratios of the lengths of the sides of a right triangle.	The domain of the inverse sine function is all real numbers from -1 to 1, including both -1 and 1.
The range of the inverse trigonometric functions [6.3]	The range (set of outputs) of an inverse trigonometric functions is the set of angles.	The range of the inverse tangent function is all angles from $-90°$ to 90° not including $-90°$ and 90°.
Inverse trigonometric functions [6.3]	The inverse trigonometric functions satisfy $\sin^{-1} x = \theta$ is equivalent to $\sin \theta = x$ $\cos^{-1} x = \theta$ is equivalent to $\cos \theta = x$ $\tan^{-1} x = \theta$ is equivalent to $\tan \theta = x$	Examples 1 and 2, Activity 6.3, pages 616–617
Solving right triangles [6.4]	When trigonometric problems involve solving right triangles, employ the following trigonometric problem-solving strategy: 1. Draw a diagram of the situation using right triangles. 2. Identify all known sides and angles. 3. Identify sides and/or angles you want to know. 4. Identify functions that relate the known and unknown. 5. Write and solve the appropriate trigonometric equation(s).	Example 1, Activity 6.4, page 622
Central angle [6.6]	A central angle is an angle whose vertex is the center of a circle.	Problem 2b, Activity 6.6, page 640

CONCEPT/SKILL	DESCRIPTION	EXAMPLE
The domain of the sine and cosine functions [6.6]	The domain of each of the sine and cosine functions is all angles, both positive and negative.	The domains are all real numbers.
The range of the sine and cosine functions [6.6]	The range of the sine and cosine functions is all values of N such that $-1 \leq N \leq 1$.	The range is all real numbers from -1 to 1 inclusive.
The graph of $y = \sin x$ [6.6]	The graph of $y = \sin x$ is a periodic wave. One cycle is shown in the Example.	
The graph of $y = \cos x$ [6.6]	The graph of $y = \cos x$ is a periodic wave. One cycle is shown in the Example.	
The period of the sine and cosine functions [6.6]	The period is the number of units of the input required to complete one cycle of the graph of a function.	The period is 360° for $y = \sin x$ and $y = \cos x$.
Radian measure [6.7]	Radian measure is used when the input of a repeating function is better defined by real numbers than angles measured in degrees.	360° is equivalent to 2π radians, the circumference of a unit circle.
Converting from radian measure to degree measure [6.7]	To convert radian measure to degree measure, multiply the radian measure by $\frac{180°}{\pi \text{ radians}}$.	$\frac{2\pi}{3}$ radians $= \frac{2\pi}{3} \cdot \frac{180°}{\pi} = \frac{360°}{3} = 120°$
Converting from degree measure to radian measure [6.7]	To convert degree measure to radian measure, multiply the degree measure by $\frac{\pi \text{ radians}}{180°}$.	$30° = 30° \cdot \frac{\pi}{180°} = \frac{\pi}{6}$ radians
The amplitude of trigonometric functions [6.8]	The amplitude of trigonometric functions defined by $y = a\sin(bx)$ or $y = a\cos(bx)$ is defined by $\lvert a \rvert = \frac{1}{2}(M - m)$, where M represents the maximum function value and m represents the minimum function value.	Example 1, Activity 6.8, page 659

CONCEPT/SKILL	DESCRIPTION	EXAMPLE
The period of sine and cosine functions [6.8]	For the trigonometric functions defined by $y = a \sin(bx)$ or $y = a \cos(bx)$, $b > 0$, the period is $\frac{2\pi}{b}$.	The period of $y = \sin(2x)$ is $\frac{2\pi}{2} = \pi$ units.
The displacement, or horizontal shift, of the graph of sine or cosine functions [6.9]	The displacement, or horizontal shift, of the graph of $y = a \sin(bx + c)$, $b > 0$, is the smallest movement (left or right) necessary for the graph of $y = a \sin bx$ to match the graph of $y = a \sin(bx + c)$ exactly. A similar scenario applies to the cosine.	Examples 1 and 2, Activity 6.9, page 667
The displacement, or horizontal shift, of the graph of sine or cosine functions [6.9]	For the functions $y = a \sin(bx + c)$ and $y = a \cos(bx + c)$, $b > 0$, the horizontal shift is given by $-\frac{c}{b}$.	The displacement of the function $y = 2 \sin\left(3x + \frac{\pi}{2}\right)$ is given by $-\frac{\frac{\pi}{2}}{3} = -\frac{\pi}{6}$. The shift is $\frac{\pi}{6}$ units to the left.
The sine function can be used to model periodic wavelike behavior [6.10]	Using the TI-84 Plus C calculator, set your MODE to radian and use the SinReg option on the Stat Calc menu.	Problem 1–4, Activity 6.10, pages 671–672

CHAPTER 6 Gateway Review

1. You walk 7 miles in a straight line 63° north of east.

a. Determine how far north you have traveled.

$N = 7 \sin(63°) \approx 6.24$ mi.

b. Determine how far east you have traveled.

$E = 7 \cos(63°) \approx 3.18$ mi.

2. Solve the following triangles:

a.

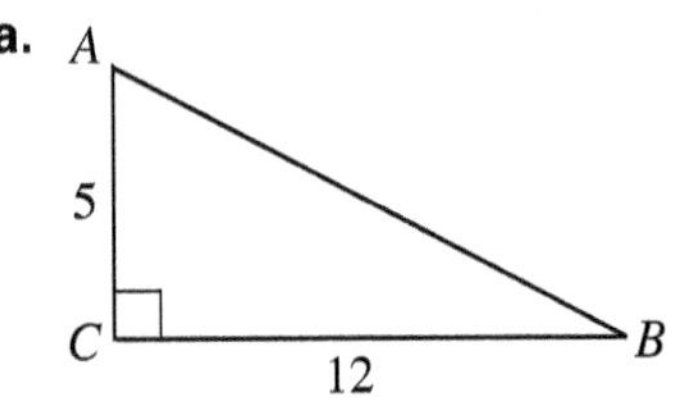

b.

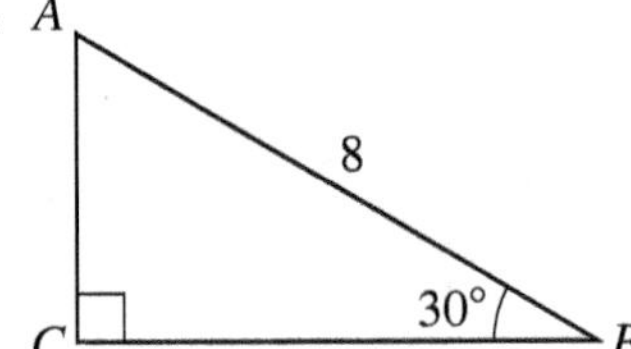

a. side $c = 13$, angle $A \approx 67.4°$, angle $B \approx 22.6°$

b. side $a \approx 6.93$, side $b = 4$, angle $A = 60°$

c.

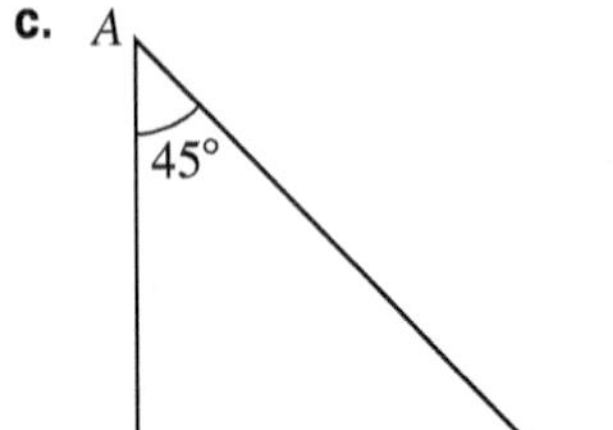
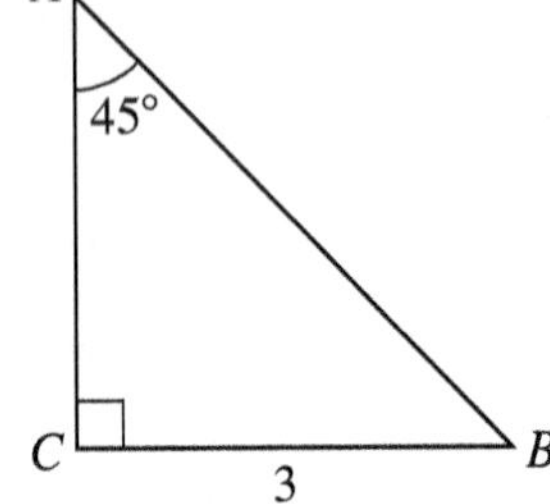

d.

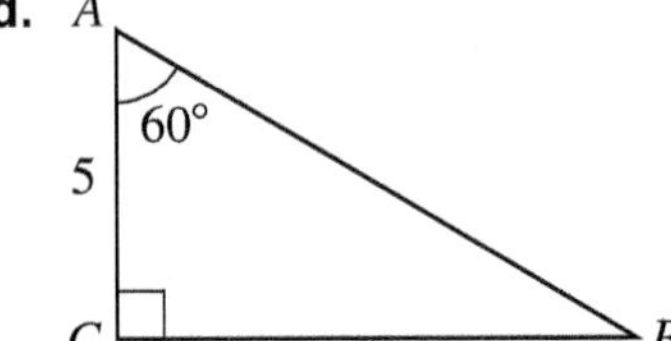

c. side $b = 3$, side $c \approx 4.24$, angle $B = 45°$

d. side $a \approx 8.66$, side $c = 10$, angle $B = 30°$

3. a. Given $\sin\theta = \frac{6}{10}$, determine $\cos\theta$ and $\tan\theta$ without using your calculator.

$\cos\theta = \frac{8}{10}$

$\tan\theta = \frac{6}{8}$

b. Given $\cos\theta = \frac{\sqrt{3}}{2}$, determine $\sin\theta$, $\tan\theta$, and θ without using your calculator.

$\sin\theta = \frac{1}{2}$

$\tan\theta = \frac{1}{\sqrt{3}}$

$\theta = 30°$

c. Given $\tan\theta = \frac{8}{5}$, determine $\sin\theta$ and $\cos\theta$ without using your calculator.

$c^2 = 5^2 + 8^2 = 89$

$c = \sqrt{89}$

$\sin\theta = \frac{8}{\sqrt{89}}$

$\cos\theta = \frac{5}{\sqrt{89}}$

Answers to all Gateway exercises are included in the Selected Answers appendix.

4. You are taking your nephew to see the Empire State Building. When you are 100 feet from the building, you and your nephew look up to see the top. You are 6 feet tall, your nephew is 3 feet tall, and the Empire State Building is 1414 feet high. You notice that even though he is only half your height, your nephew does not have to tilt his head any more than you do. Is your observation correct? Is the angle always independent of people's heights? Explain.

No. There is a difference, but it is so small that it is difficult to see.

for me: $\theta = \tan^{-1}\left(\frac{1408}{100}\right) \approx 85.9375°$

for my nephew: $\theta = \tan^{-1}\left(\frac{1411}{100}\right) \approx 85.9461°$

5. In the diagram below, determine the lengths of *a*, *b*, *c*, and *h* to the nearest tenth.

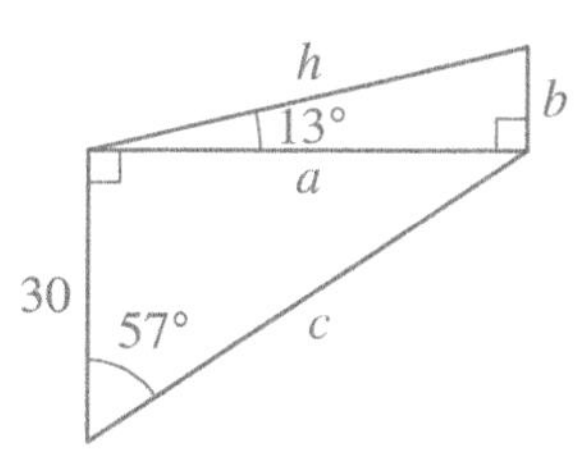

$$\tan 57° = \frac{a}{30}$$
$$30 \tan 57° = a$$
$$a \approx 46.2$$
$$\cos 57° = \frac{30}{c}$$
$$c = \frac{30}{\cos 57°}$$
$$c \approx 55.1$$

$$\tan 13° = \frac{b}{46.2}$$
$$b = 46.2 \tan 13°$$
$$b \approx 10.7$$
$$\cos 13° = \frac{46.2}{h}$$
$$h = \frac{46.2}{\cos 13°}$$
$$h \approx 47.4$$

6. In the following diagram, determine *x* and *h*. (*Hint:* See Exercises 2c and d.)

$$(2x)^2 = x^2 + (200 + x)^2$$
$$4x^2 = x^2 + 40{,}000 + 400x + x^2$$
$$2x^2 - 400x - 40{,}000 = 0$$
$$x = \frac{400 \pm \sqrt{400^2 + 4(2)(40{,}000)}}{2(2)}$$
$$x = \frac{400 \pm \sqrt{160{,}000 + 320{,}000}}{4}$$
$$x = \frac{400 \pm \sqrt{480{,}000}}{4}$$

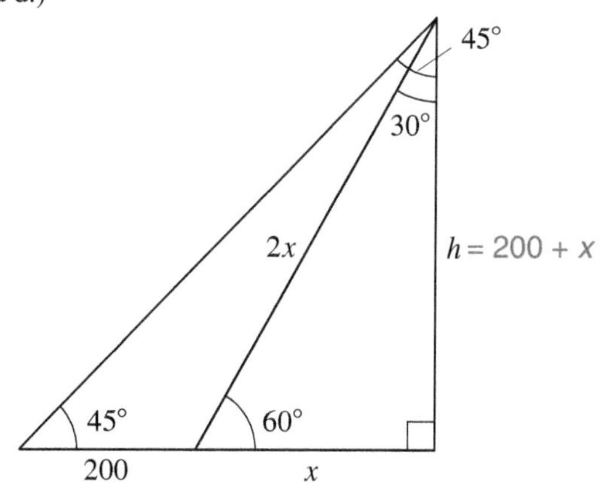

The negative does not make sense; thus, $x = \frac{400 + 400\sqrt{3}}{4}$.

$x = 100 + 100\sqrt{3}$; $h = 300 + 100\sqrt{3}$

$x \approx 273.2$; $h \approx 473.2$

7. Using the following triangles, complete the table without using your calculator:

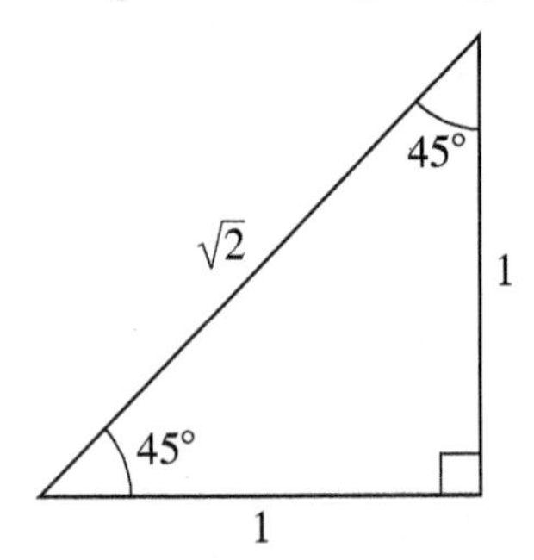

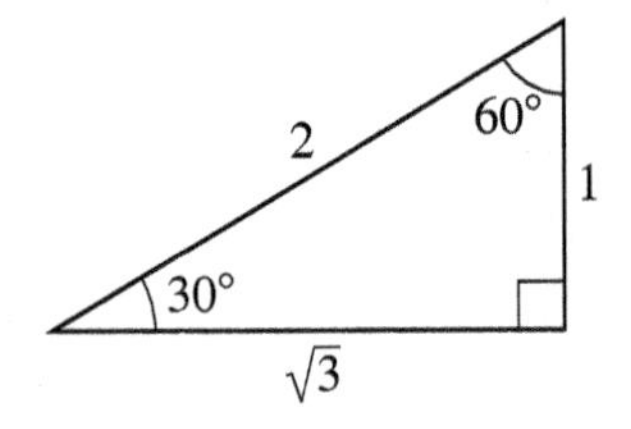

ANGLE θ	$\sin\theta$	$\cos\theta$	$\tan\theta$
120°	$\frac{\sqrt{3}}{2}$	$-\frac{1}{2}$	$-\sqrt{3}$
135°	$\frac{1}{\sqrt{2}}$	$-\frac{1}{\sqrt{2}}$	-1
150°	$\frac{1}{2}$	$-\frac{\sqrt{3}}{2}$	$-\frac{1}{\sqrt{3}}$
180°	0	-1	0
210°	$-\frac{1}{2}$	$-\frac{\sqrt{3}}{2}$	$\frac{1}{\sqrt{3}}$
225°	$-\frac{1}{\sqrt{2}}$	$-\frac{1}{\sqrt{2}}$	1
240°	$-\frac{\sqrt{3}}{2}$	$-\frac{1}{2}$	$\sqrt{3}$
270°	-1	0	undef.
300°	$-\frac{\sqrt{3}}{2}$	$\frac{1}{2}$	$-\sqrt{3}$
315°	$-\frac{1}{\sqrt{2}}$	$\frac{1}{\sqrt{2}}$	-1
330°	$-\frac{1}{2}$	$\frac{\sqrt{3}}{2}$	$-\frac{1}{\sqrt{3}}$
360°	0	1	0

8. Determine the amplitude and period of the given functions, and then sketch their graphs. Use your graphing calculator to verify your results:

a. $y = 2\sin x$

amplitude: 2

period: 2π

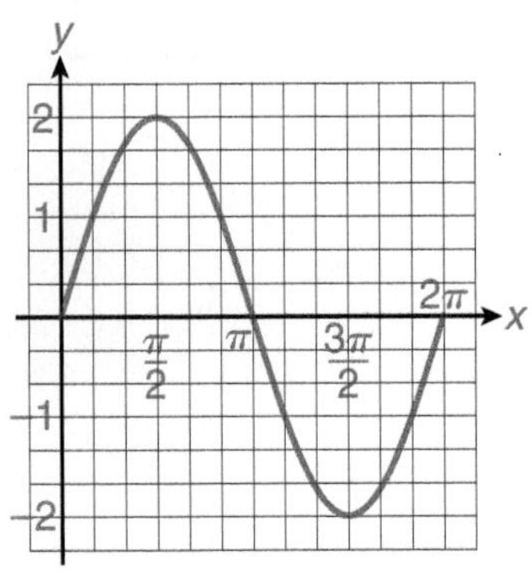

b. $y = -2\sin x$

amplitude: 2

period: 2π

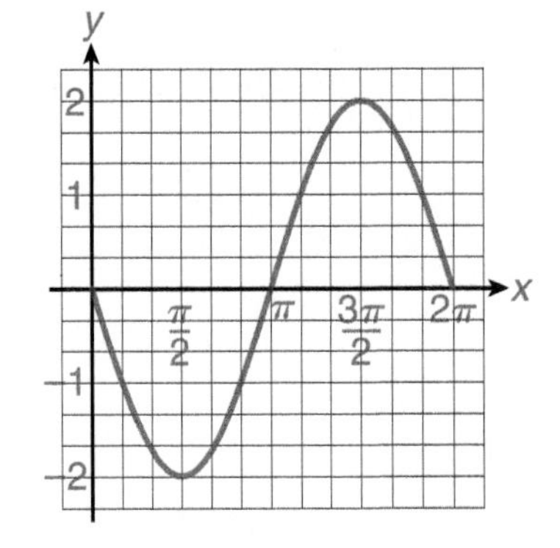

c. $y = \cos 2x$

amplitude: 1

period: π

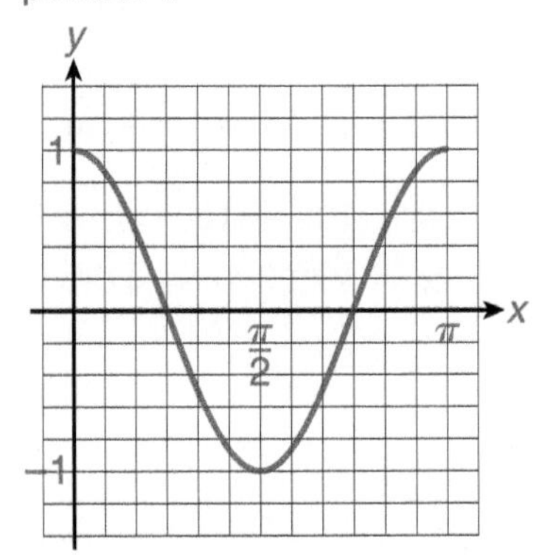

d. $y = \cos 2\pi x$

amplitude: 1

period: 1

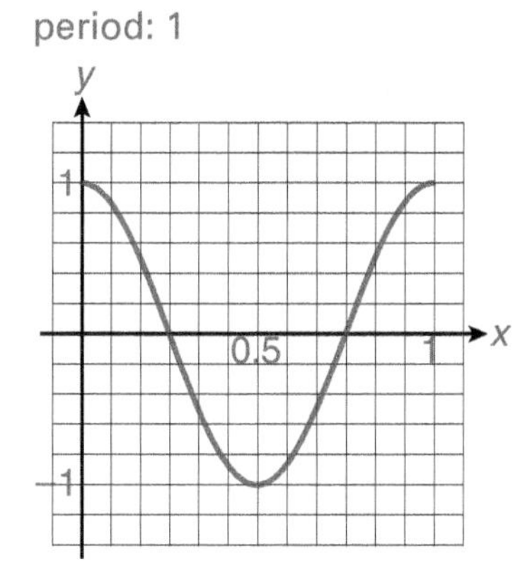

e. $y = \sin \frac{x}{2}$

amplitude: 1

period: 4π

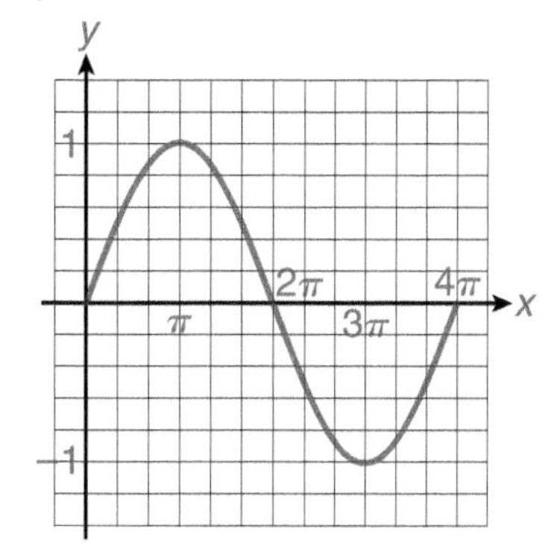

f. $y = \sin \frac{\pi x}{2}$

amplitude: 1

period: 4

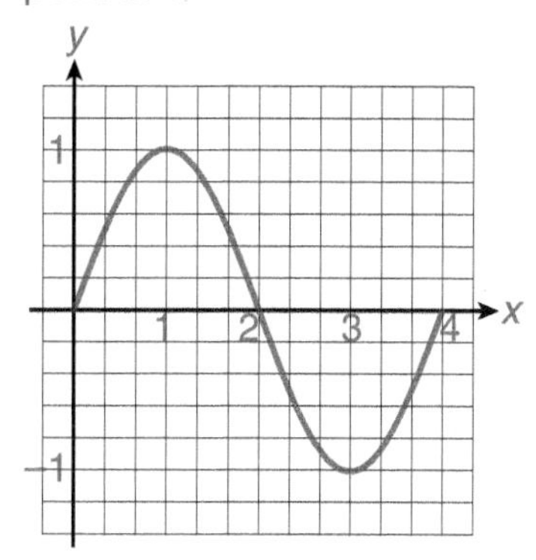

g. $y = \cos x$

amplitude: 1

period: 2π

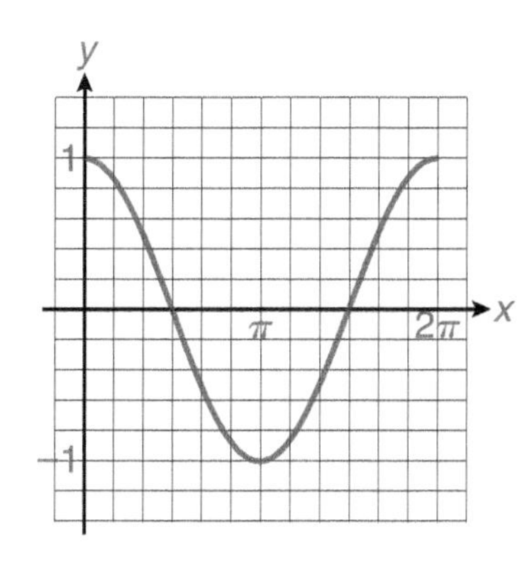

h. $y = \frac{2}{3} \cos (2x)$

amplitude: $\frac{2}{3}$

period: π

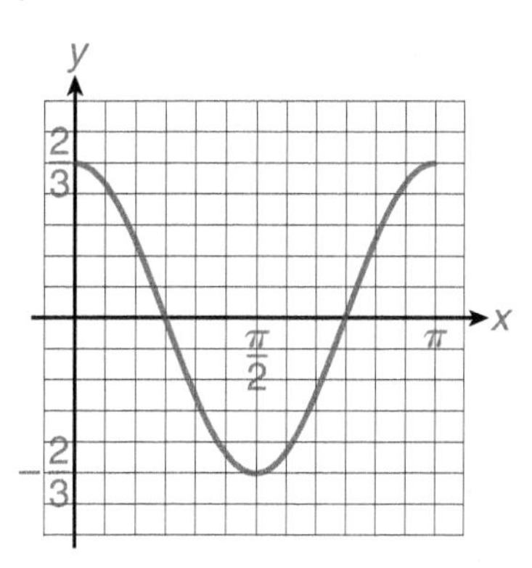

i. $y = \sin (2\pi x - 3\pi)$

amplitude: 1

period: 1

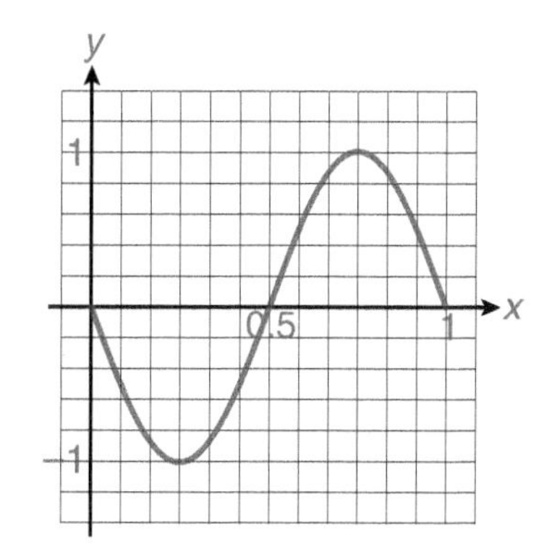

j. $y = 3 \sin (2\pi x - 3\pi)$

amplitude: 3

period: 1

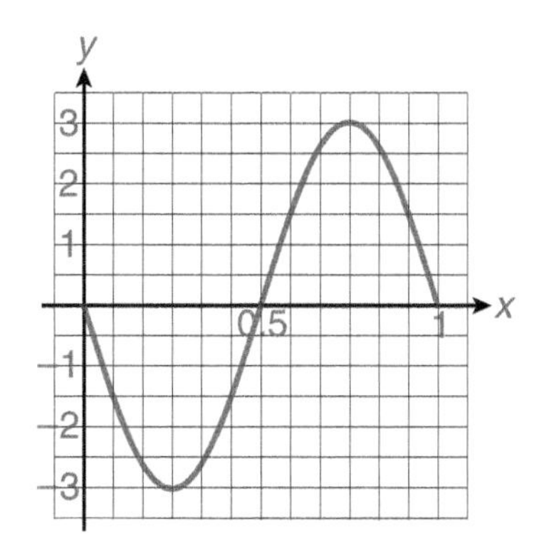

9. Match the given equation to one of the accompanying graphs. (Assume that Xscl = 1 and Yscl = 1.)

a. $y = -3 \sin x$

Graph is vi.

b. $y = 2 \sin\left(\dfrac{\pi x}{2} + \dfrac{\pi}{2}\right)$

Graph is ii.

c. $y = 2 \sin\left(\dfrac{\pi x}{2} - \dfrac{\pi}{2}\right)$

Graph is iv.

d. $y = -3 \cos x$

Graph is v.

e. $y = -2 \cos\left(\dfrac{\pi x}{2} + \dfrac{\pi}{2}\right)$

Graph is vii.

f. $y = -\cos\left(\pi x - \dfrac{\pi}{2}\right)$

Graph is viii.

i.

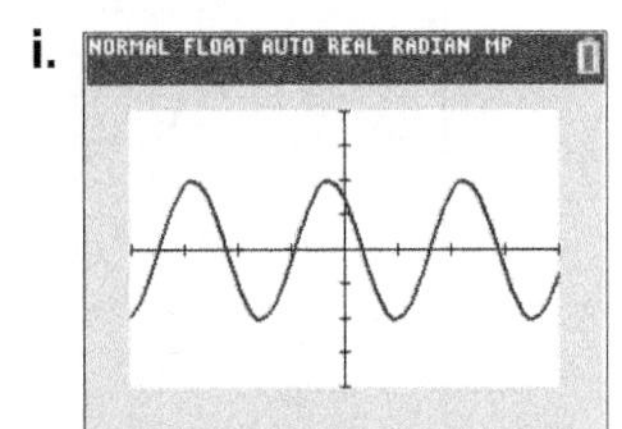

ii.

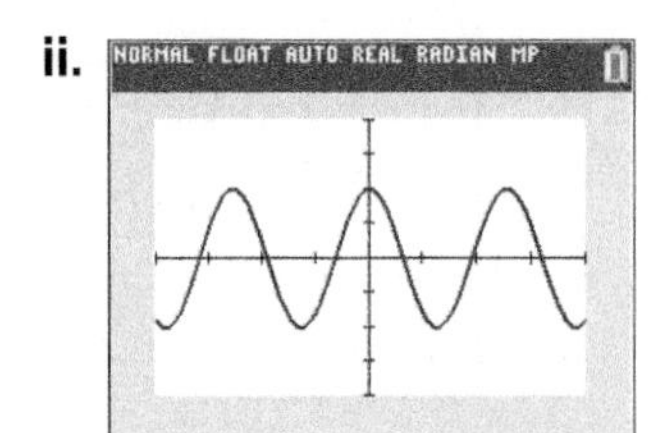

iii.

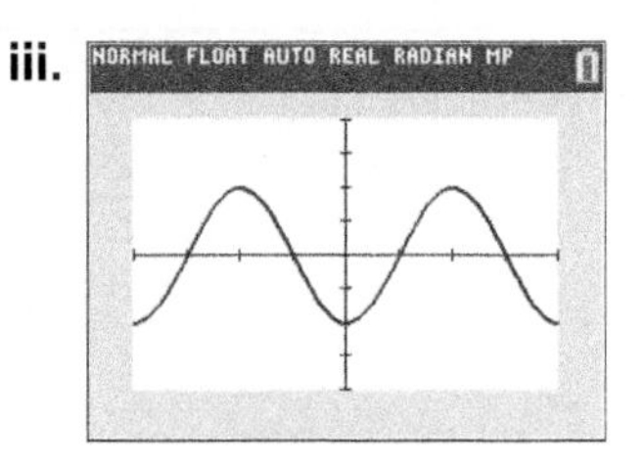

iv.

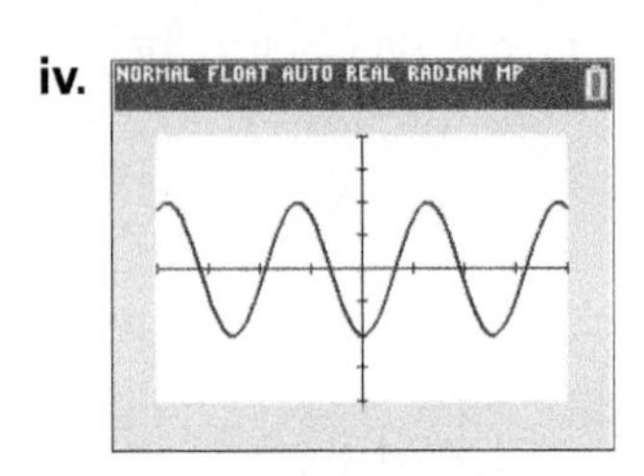

v.

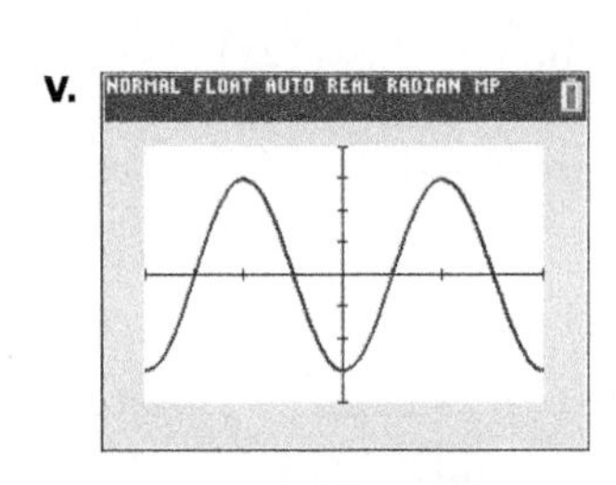

vi.

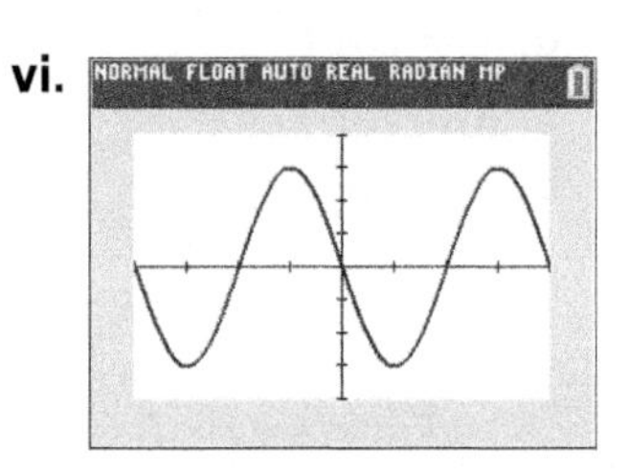

vii.

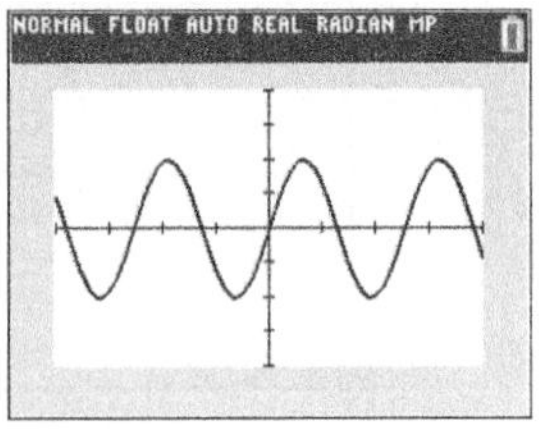

viii.

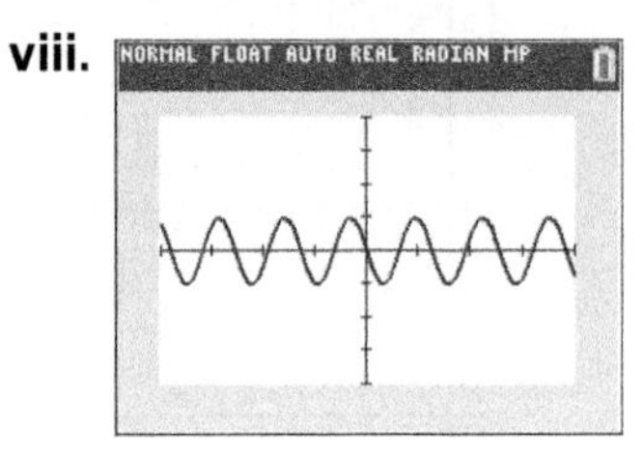

APPENDIX

A Concept Review

Properties of Exponents

The basic properties of exponents are summarized as follows:

If a is a nonzero real number and n and m are rational numbers, then

1. $a^n a^m = a^{n+m}$ **2.** $\dfrac{a^n}{a^m} = a^{n-m}$ **3.** $(a^n)^m = a^{nm}$

4. $(ab)^n = a^n b^n$ **5.** $a^0 = 1$ **6.** $a^{-n} = \left(\dfrac{1}{a}\right)^n = \dfrac{1}{a^n}$

Property 1: $a^n a^m = a^{n+m}$ If you are multiplying two powers of the same base, keep the base the same and add the exponents.

Example 1: $3^4 \cdot 3^7 = 3^{4+7} = 3^{11}$

Note that the exponents were added and the base did not change.

Property 2: $\dfrac{a^n}{a^m} = a^{n-m}$ If you are dividing two powers of the same base, keep the base the same and subtract the exponents.

Example 2: $\dfrac{6^6}{6^4} = 6^{6-4} = 6^2 = 36$

Note that the exponents were subtracted and the base did not change.

Property 3: $(a^n)^m = a^{nm}$ If an exponent is applied to a power, multiply the exponents.

Example 3: $(y^3)^4 = y^{12}$

The exponents were multiplied. The base did not change.

Property 4: $(ab)^n = a^n b^n$ If an exponent is applied to a product, the exponent is distributed to each factor.

Example 4: $(2x^2y^3)^3 = 2^3 \cdot (x^2)^3 \cdot (y^3)^3 = 8x^6y^9$

Note that because the base contained three factors, each of them was raised to the third power. The common mistake in an expansion such as this is not to raise the coefficient to the power.

Property 5: $a^0 = 1, a \neq 0$. Often presented as a definition, Property 5 states that any nonzero base raised to the zero power is 1. This property or definition is a result of Property 2 of exponents as follows:

Consider $\frac{x^5}{x^5}$. Using Property 2, $x^{5-5} = x^0$. However, you know that any fraction in which the numerator and the denominator are equal is equivalent to 1. Therefore, $x^0 = 1$.

Example 5: $\left(\frac{2x^3}{3yz^5}\right)^0 = 1$

Given a nonzero base, if the exponent is 0, the value is 1.

Property 6: $a^{-n} = \left(\frac{1}{a}\right)^n = \frac{1}{a^n}$. Sometimes presented as a definition, Property 6 states that any base raised to a negative power is equivalent to the reciprocal of the base raised to the positive power. Note that the negative exponent does not have any effect on the sign of the base. This property could also be viewed as a result of the second property of exponents as follows:

Consider $\frac{x^3}{x^5}$. Using Property 2, $x^{3-5} = x^{-2}$. If you view this expression algebraically, you have three factors of x in the numerator and five in the denominator. If you divide out the three common factors, you are left with $\frac{1}{x^2}$. Therefore, if Property 2 is true, then $x^{-2} = \frac{1}{x^2}$.

Example 6: Write each of the following without negative exponents.

a. 3^{-2}

b. $\frac{2}{x^{-3}}$

Solution:

a. $3^{-2} = \left(\frac{1}{3}\right)^2 = \frac{1}{3^2} = \frac{1}{9}$

b. $\frac{2}{x^{-3}} = \frac{2}{\frac{1}{x^3}} = 2 \div \frac{1}{x^3} = 2 \cdot x^3 = 2x^3$

A **factor** of the form b^x can be moved from a numerator to a denominator or from a denominator to a numerator by changing the *sign of the exponent.*

Example 7: Simplify and express your results with positive exponents only.

$$\left(\frac{x^3y^{-4}}{2x^{-3}y^{-2}z}\right) \cdot \left(\frac{4x^3y^2z}{x^5y^{-3}z^3}\right)$$

Solution: Simplify each factor by writing it with positive exponents only.

$$\left(\frac{x^6}{2y^2z}\right) \cdot \left(\frac{4y^5}{x^2z^2}\right)$$

Now multiply and simplify.

$$\frac{4x^6y^5}{2x^2y^2z^3} = \frac{2x^4y^3}{z^3}$$

Exercises

Simplify and express your results with positive exponents only. Assume all variables represent nonzero real numbers.

1. 5^{-3}

$\left(\frac{1}{5}\right)^3 = \frac{1}{5^3} = \frac{1}{125}$

2. $\frac{1}{x^{-5}}$

$\frac{1}{x^{-5}} = \frac{1}{\frac{1}{x^5}} = 1 \div \frac{1}{x^5} = 1 \cdot x^5 = x^5$

3. $\frac{3x}{y^{-2}}$

$\frac{3x}{y^{-2}} = 3x \div \frac{1}{y^2} = 3x \cdot y^2 = 3xy^2$

4. $\frac{10x^2y^5}{2x^{-3}}$

$5x^{2-(-3)}y^5 = 5x^5y^5$

5. $\frac{5^{-1}z}{x^{-1}z^{-2}}$

$\frac{5^{-1}z}{x^{-1}z^{-2}} = \frac{z}{5} \div \frac{1}{x^1z^2} = \frac{z}{5} \cdot xz^2 = \frac{xz^3}{5}$

6. $5x^0$

$5 \cdot 1 = 5$

7. $(a + b)^0$

1

8. $-3(x^0 - 4y^0)$

$-3(1 - 4(1)) = -3(-3) = 9$

9. $x^6 \cdot x^{-3}$

$x^{6+(-3)} = x^3$

10. $\frac{4^{-2}}{4^{-3}}$

$\frac{4^{-2}}{4^{-3}} = 4^{-2-(-3)} = 4^1 = 4$

11. $(4x^2y^3) \cdot (3x^{-3}y^{-2})$

$12x^{2+(-3)}y^{3+(-2)} = 12x^{-1}y^1 = \frac{12y}{x}$

12. $\frac{24x^{-2}y^3}{6x^3y^{-1}}$

$4x^{-2-3}y^{3-(-1)} = 4x^{-5}y^4 = \frac{4y^4}{x^5}$

13. $\frac{(4x^{-2}y^{-3}) \cdot (5x^3y^{-2})}{6x^2y^{-3}z^{-3}}$

$\frac{20x^1y^{-5}}{6x^2y^{-3}z^{-3}} = \frac{10x^{-1}y^{-2}z^3}{3} = \frac{10z^3}{3xy^2}$

14. $\left(\frac{2x^{-2}y^{-3}}{z^2}\right) \cdot \left(\frac{x^5y^3}{z^{-3}}\right)$

$\frac{2x^3y^0}{z^{-1}} = 2x^3z$

15. $\frac{(6x^4y^{-3}z^{-2})(3x^{-3}y^4)}{15x^{-3}y^{-3}z^2}$

$\frac{18x^1y^1z^{-2}}{15x^{-3}y^{-3}z^2} = \frac{6x^4y^4}{5z^4}$

Solving 3 × 3 Linear Systems Algebraically

Linear equations such as $3x + 2y - z = 4$ involve three variables, x, y, and z. A solution of such an equation is an ordered triple (x, y, z) such that if the values of x, y, and z in the ordered triple are substituted into the equation, the result is a true statement.

A system of three linear equations in three variables (a 3×3 system) such as

$$\begin{aligned} -x + y + z &= -3 \\ 3x + 9y + 5z &= 5 \\ x + 3y + 2z &= 4 \end{aligned}$$

has as its solution all ordered triples (x, y, z) that will make all three equations true.

Solving a 3 × 3 Linear System Algebraically

1. Eliminate one variable using any two of the given three equations to obtain an equation in two variables (or less).

2. Eliminate the same variable using the third equation not used in Step 1 and either of the other two equations to obtain a second equation in two variables (or less).

3. Solve the system consisting of the two equations found in Steps 1 and 2.

4. Substitute the values obtained in Step 3 into any equation involving all three variables to determine the value of the third variable.

5. Check your solution in all three equations.

Example 1: Determine all solutions of

$$\begin{aligned} -x + y + z &= -3 \\ 3x + 9y + 5z &= 5 \\ x + 3y + 2z &= 4. \end{aligned}$$

Solution: You may choose to eliminate x using the addition method.

Step 1. Multiply both sides of the first equation by 3, and add the results to the second equation.

$$\begin{aligned} 3(-x + y + z &= -3) \\ 3x + 9y + 5z &= 5 \end{aligned} \quad \text{or} \quad \begin{aligned} -3x + 3y + 3z &= -9 \\ 3x + 9y + 5z &= 5 \end{aligned}$$

The sum is $12y + 8z = -4$.

Step 2. Multiply the third equation by -3, and add the results to the second.

$$\begin{aligned} 3x + 9y + 5z &= 5 \\ -3(x + 3y + 2z &= 4) \end{aligned} \quad \text{or} \quad \begin{aligned} 3x + 9y + 5z &= 5 \\ -3x - 9y - 6z &= -12 \end{aligned}$$

The sum is $-z = -7$.

Step 3. The 2 × 2 system resulting from Steps 1 and 2 is

$$\begin{aligned} 12y + 8z &= -4 \\ -z &= -7. \end{aligned}$$

The second equation is equivalent to $z = 7$. Substituting this value into the first equation of the new system yields

$$\begin{aligned} 12y + 8(7) &= -4 \\ 12y + 56 &= -4 \\ 12y &= -60 \\ y &= -5. \end{aligned}$$

Step 4. Using the values of y and z in the third equation of the original system,

$$\begin{aligned} x + 3(-5) + 2(7) &= 4 \\ x - 15 + 14 &= 4 \\ x - 1 &= 4 \\ x &= 5. \end{aligned}$$

The potential solution is $(5, -5, 7)$ and should be checked in all three of the original equations.

Not every 3×3 linear system has unique solutions; some have multiple solutions. These are called **dependent systems**. Some systems have no solution and are called **inconsistent systems**.

Example 2: Solve the following system.

$$\begin{aligned} x + 2y + 3z &= 5 \\ -x + y - z &= -6 \\ 2x + y + 4z &= 4 \end{aligned}$$

Solution: Again, you may eliminate x using the addition method.

Step 1. Sum the first two equations to eliminate x.

$$\begin{aligned} x + 2y + 3z &= 5 \\ -x + y - z &= -6 \end{aligned}$$

The sum is $3y + 2z = -1$.

Step 2. Multiply the second equation by 2, and add the results to the third.

$$\begin{aligned} -2x + 2y - 2z &= -12 \\ 2x + y + 4z &= 4 \end{aligned}$$

The sum is $3y + 2z = -8$. The new system is

$$\begin{aligned} 3y + 2z &= -1 \\ 3y + 2z &= -8. \end{aligned}$$

Step 3. To solve the new system, multiply the first equation by -1 and add the results to the second.

$$\begin{aligned} -3y - 2z &= 1 \\ 3y + 2z &= -8 \end{aligned}$$

The sum is $0 = -7$.

Because $0 = -7$ is a false statement, the conclusion is there is no solution. The original system was an inconsistent system.

Had the sum of the equations in Step 3 resulted in a true statement such as $0 = 0$, the conclusion would have been that there were an infinite number of solutions. That is, the system would have been *dependent*.

Exercises

Solve the following systems algebraically. If the system is dependent or inconsistent, state this as the answer.

1. $x + y - z = 9$
$x + y + z = 5$
$x - y + 2z = 1$

Eliminate z variable.

Add equations 1 and 2.

$2x + 2y = 14$

Multiply equation 2 by -2, and add the result to equation 3.

$-x - 3y = -9$

Solve: $2x + 2y = 14$
$-x - 3y = -9$

Multiply equation 2 by 2 and add.

$2x + 2y = 14$
$-2x - 6y = -18$
$-4y = -4$
$y = 1$

Substitute $y = 1$, and solve for x.

$2x + 2(1) = 14$
$2x = 12; x = 6$

Substitute $x = 6$ and $y = 1$ in the first equation, and solve for z.

$6 + 1 - z = 9$
$7 - z = 9$
$z = -2$

$(6, 1, -2)$

2. $-2x + y + 4z = 3$
$x + y - 3z = 2$
$x - y + 2z = 1$

Eliminate y variable.

Multiply equation 2 by -1, and add to equation 1.

$-3x + 7z = 1$

Add equations 2 and 3.

$2x - z = 3$

Solve: $-3x + 7z = 1$
$2x - z = 3$

Multiply equation 2 by 7 and add.

$-3x + 7z = 1$
$14x - 7z = 21$
$11x = 22$
$x = 2$

Substitute $x = 2$ and solve for z.

$-3(2) + 7z = 1$
$-6 + 7z = 1$
$z = 1$

Substitute $x = 2, z = 1$ in the 2nd equation, and solve for y.

$2 + y - 3(1) = 2$
$y - 1 = 2$
$y = 3$

$(2, 3, 1)$

3. $x + 2y + 3z = 5$
$-x + y - z = -6$
$2x + y + 4z = 4$

Eliminate x variable.

Multiply equation 1 by -2, and add to equation 3.

$-3y - 2z = -6$

Add equations 1 and 2.

$3y + 2z = -1$

Solve: $-3y - 2z = -6$
$3y + 2z = -1$
$0 = -7$

inconsistent

4. $3x - 2y + 3z = 11$
$2x + 3y - 2z = -5$
$x + 4y - z = -5$

Eliminate x variable.

Multiply equation 3 by -3, and add to equation 1.

$-14y + 6z = 26$

Multiply equation 3 by -2 and add to equation 2.

$-5y = 5$
$y = -1$

Substitute $y = -1$, and solve for z.

$-14(-1) + 6z = 26$
$6z = 12$
$z = 2$

Solve for x: $3x - 2(-1) + 3(2) = 11$
$3x = 3$

$(1, -1, 2)$

5. $x - 4y + z = -5$

$3x - 12y + 3z = -15$

$-2x + 8y - 2z = 10$

Eliminate the x variable. Multiply equation 1 by -3, and add to equation 2.

$-3x + 12y - 3z = 15$

$3x - 12y + 3z = -15$

$0 = 0$

dependent

6. $2x + 3y + 4z = 3$

$6x - 6y + 8z = 3$

$4x + 3y - 4z = 2$

Eliminate the x variable. Multiply equation 1 by -3, and add to equation 2.

$-6x - 9y - 12z = -9$

$6x - 6y + 8z = 3$

$-15y - 4z = -6$

Multiply equation 1 by -2, and add to equation 3.

$-4x - 6y - 8z = -6$

$4x + 3y - 4z = 2$

$-3y - 12z = -4$

Solve: $-15y - 4z = -6$

$-3y - 12z = -4$

Multiply equation 1 by -3, and solve for y.

$45y + 12z = 18$

$-3y - 12z = -4$

$42y = 14$

$y = \frac{1}{3}$

Substitute $y = \frac{1}{3}$, and solve for z.

$-3\left(\frac{1}{3}\right) - 12z = -4$

$z = \frac{1}{4}$

Substitute $y = \frac{1}{3}$, $z = \frac{1}{4}$, and find x.

$2x + 3\left(\frac{1}{3}\right) + 4\left(\frac{1}{4}\right) = 3$

$x = \frac{1}{2}$

$\left(\frac{1}{2}, \frac{1}{3}, \frac{1}{4}\right)$

7. $x + 2y = 10$

$-x + 3z = -23$

$4y - z = 9$

Eliminate x variable.

Add equations 1 and 2.

$2y + 3z = -13$

$4y - z = 9$

Multiply equation 1 by -2, add the equations, and solve for z.

$-7z = 35$

$z = -5$

Substitute $z = -5$, and solve for y.

$2y + 3(-5) = -13$

$2y = 2$

$y = 1$

Substitute $y = 1$, and solve for x.

$x + 2(1) = 10$

$x = 8$

$(8, 1, -5)$

Inequalities Involving Absolute Value

The key to solving absolute value inequalities algebraically is to rewrite them using the following properties.

Absolute Value Properties

For any real number x and $a > 0$,

$$|x| < a \text{ is equivalent to } -a < x < a.$$

For any real number x and $a > 0$,

$$|x| > a \text{ is equivalent to the statement } x > a \text{ or } x < -a.$$

Solving Absolute Value Inequalities

1. Rewrite the inequality with the absolute value isolated.
2. Rewrite the inequality as a compound inequality or pair of inequalities.
3. Solve the resulting inequality or inequalities.

Example 1: Solve $|2x - 3| + 3 \leq 8$.

Solution: Subtract 3 from both sides.

$$|2x - 3| + 3 - 3 \leq 8 - 3 \quad \text{or} \quad |2x - 3| \leq 5$$

Using the first property,

$$-5 \leq 2x - 3 \leq 5.$$

Add 3 to each part.

$$-5 + 3 \leq 2x - 3 + 3 \leq 5 + 3$$
$$-2 \leq 2x \leq 8$$

Divide each part by 2.

$$\frac{-2}{2} \leq \frac{2x}{2} \leq \frac{8}{2} \quad \text{or} \quad -1 \leq x \leq 4$$

Example 2: Solve $|4x + 3| - 4 > 7$.

Solution: Add 4 to both sides to isolate the absolute value.

$$|4x + 3| - 4 + 4 > 7 + 4$$
$$|4x + 3| > 11$$

Using the second absolute value property,

$$4x + 3 > 11 \quad \text{or} \quad 4x + 3 < -11.$$

Solving these inequalities,

$$4x + 3 - 3 > 11 - 3 \quad \text{or} \quad 4x + 3 - 3 < -11 - 3$$
$$4x > 8 \qquad\qquad 4x < -14$$
$$x > 2 \quad \text{or} \quad x < -\frac{7}{2}.$$

Exercises

Solve the following inequalities.

1. $|3x - 5| < 5$

$-5 < 3x - 5 < 5$

$0 < 3x < 10$

$0 < x < \frac{10}{3}$

2. $|x - 3| - 2 \leq 3$

$|x - 3| \leq 5$

$-5 \leq x - 3 \leq 5$

$-2 \leq x \leq 8$

3. $|4x - 1| > 3$

$4x - 1 > 3$ or $4x - 1 < -3$

$4x > 4$ or $4x < -2$

$x > 1$ or $x < -\frac{1}{2}$

4. $|2x - 1| - 4 \geq 7$

$|2x - 1| \geq 11$

$2x - 1 \geq 11$ or $2x - 1 \leq -11$

$2x \geq 12$ or $2x \leq -10$

$x \geq 6$ or $x \leq -5$

Solving Equations by Factoring

Some quadratic and higher-order polynomial equations can be solved by using factoring and the zero-product property.

The process is as follows.

Solving an Equation by Factoring

1. Use the addition principle to remove all terms from one side of the equation. This results in the equation having one side equal to zero.
2. Combine like terms, and then factor.
3. Use the zero-product rule to set each factor containing a variable equal to zero and then solve the equations.
4. Check your solutions in the original equation.

Example 1: Solve the equation $x(x + 5) = 0$.

Solution: This equation already satisfies the first two steps in our process, so we simply start at Step 3.

$$x = 0 \quad x + 5 = 0$$

$$x = -5$$

Thus, we have two solutions: $x = 0$ and $x = -5$. The check is left to the reader.

Example 2: Solve the equation $6x^2 = 16x$.

Solution: Making one side of the equation equal to zero, $6x^2 - 16x = 0$. Because there are no like terms, factor the binomial.

$$2x(3x - 8) = 0$$

Using the zero-product principle,

$$2x = 0 \quad \text{or} \quad 3x - 8 = 0$$

$$x = 0 \quad \text{or} \quad 3x = 8$$

$$x = \frac{8}{3}.$$

The two potential solutions are $x = 0$ and $x = \frac{8}{3}$. The check is left to the reader.

Example 3: Solve $3x^2 - 2 = -x$.

Solution: Making one side of the equation equal to zero, $3x^2 + x - 2 = 0$. Factor the trinomial by trying factors of $3x^2$ and factors of -2 in two binomials.

$$(3x - 2)(x + 1) = 0$$

Using the zero-product principle,

$$\begin{aligned} 3x - 2 = 0 \quad &\text{or} \quad x + 1 = 0 \\ 3x = 2 \quad &\text{or} \quad x = -1 \\ x = \frac{2}{3}. \end{aligned}$$

The two potential solutions are $x = \frac{2}{3}$ and $x = -1$. The check is left to the reader.

Example 4: Solve $3x^3 - 8x^2 = 3x$.

Solution: Making one side of the equation equal to zero, $3x^3 - 8x^2 - 3x = 0$. Factor the trinomial by first removing the common monomial factor.

$$x(3x^2 - 8x - 3) = 0$$
$$x(3x + 1)(x - 3) = 0$$

Using the zero-product principle,

$$\begin{aligned} x = 0 \quad \text{or} \quad 3x + 1 = 0 \quad &\text{or} \quad x - 3 = 0 \\ 3x = -1 \quad &\text{or} \quad x = 3 \\ x = -\frac{1}{3}. \end{aligned}$$

The three potential solutions are $x = 0$, $x = -\frac{1}{3}$, and $x = 3$. The check is left to the reader.

Exercises

Solve each of the following equations.

1. $x(x + 7) = 0$

$x = 0$ or $x + 7 = 0$

$x = -7$

$x = 0, -7$

2. $3(x - 5)(2x + 1) = 0$

$x - 5 = 0$ or $2x + 1 = 0$

$x = 5$ $\quad 2x = -1$

$x = -\frac{1}{2}$

$x = 5, -\frac{1}{2}$

3. $12x = x^2$

$0 = x^2 - 12x$

$0 = x(x - 12)$

$x = 0$ or $x - 12 = 0$

$x = 12$

$x = 0, 12$

4. $x^2 + 5x = 0$

$x(x + 5) = 0$

$x = 0$ or $x + 5 = 0$

$x = -5$

$x = 0, -5$

5. $x^2 - 2x - 63 = 0$

$(x - 9)(x + 7) = 0$

$x - 9 = 0$ or $x + 7 = 0$

$x = 9$ or $x = -7$

$x = 9, -7$

6. $3x^2 - 9x - 30 = 0$

$3(x^2 - 3x - 10) = 0$

$3(x - 5)(x + 2) = 0$

$x - 5 = 0$ or $x + 2 = 0$

$x = 5$ or $x = -2$

$x = 5, -2$

7. $-7x + 6x^2 = 10$

$6x^2 - 7x - 10 = 0$

$(6x + 5)(x - 2) = 0$

$6x + 5 = 0$ or $x - 2 = 0$

$6x = -5$ or $x = 2$

$x = -\frac{5}{6}$

$x = -\frac{5}{6}, 2$

8. $3y^2 = 2 - y$

$3y^2 + y - 2 = 0$

$(3y - 2)(y + 1) = 0$

$3y - 2 = 0$ or $y + 1 = 0$

$3y = 2$ or $y = -1$

$y = \frac{2}{3}$

$y = \frac{2}{3}, -1$

9. $-28x^2 + 15x - 2 = 0$

$(7x - 2)(-4x + 1) = 0$

$7x - 2 = 0$ or $-4x + 1 = 0$

$7x = 2$ or $-4x = -1$

$x = \frac{2}{7}$ or $x = \frac{1}{4}$

$x = \frac{2}{7}, \frac{1}{4}$

10. $4x^2 - 25 = 0$

$(2x + 5)(2x - 5) = 0$

$2x + 5 = 0$ or $2x - 5 = 0$

$2x = -5$ or $2x = 5$

$x = -\frac{5}{2}$ or $x = \frac{5}{2}$

$x = \pm\frac{5}{2}$

11. $(x + 4)^2 - 16 = 0$

$x^2 + 8x + 16 - 16 = 0$

$x^2 + 8x = 0$

$x(x + 8) = 0$

$x = 0$ or $x + 8 = 0$

$x = -8$

$x = 0, -8$

12. $(x + 1)^2 - 3x = 7$

$x^2 + 2x + 1 - 3x - 7 = 0$

$x^2 - x - 6 = 0$

$(x - 3)(x + 2) = 0$

$x - 3 = 0$ or $x + 2 = 0$

$x = 3$ or $x = -2$

$x = 3, -2$

13. $2(x + 2)(x - 2) = (x - 2)(x + 3) - 2$

$2(x^2 - 4) = x^2 + x - 6 - 2$

$2x^2 - 8 = x^2 + x - 8$

$x^2 - x = 0$

$x(x - 1) = 0$

$x = 0$ or $x - 1 = 0$

$x = 1$

$x = 0, 1$

14. $18x^3 = 15x^2 + 12x$

$18x^3 - 15x^2 - 12x = 0$

$3x(6x^2 - 5x - 4) = 0$

$3x(3x - 4)(2x + 1) = 0$

$3x = 0$ or $3x - 4 = 0$ or $2x + 1 = 0$

$x = 0$ or $3x = 4$ or $2x = -1$

$x = \frac{4}{3}$ or $x = -\frac{1}{2}$

$x = 0, \frac{4}{3}, -\frac{1}{2}$

Solving Quadratic Equations by Completing the Square

The square root property can be used to solve equations of the form $B^2 = a$.

Square Root Property

If $B^2 = a$, where a is a real number, then $B = \pm\sqrt{a}$.

Example 1: Solve the equation $(x + 3)^2 = 9$.

Solution: This equation fits the form of the hypothesis of the square root property, where $B = x + 3$. Therefore,

$$x + 3 = \pm\sqrt{9} = \pm 3.$$

You now have two equations to solve, $x + 3 = 3$ and $x + 3 = -3$. The solutions are $x = 0, -6$. Both of these values make the original statement true. Hence, both are solutions.

The square root technique leads to an algebraic technique of solving quadratic equations known as *completing the square*. The strategy is to rewrite the quadratic equation $ax^2 + bx + c = 0$, $a \neq 0$, in the form $(x + h)^2 = k$ and solve as in Example 1. Rewriting requires an algebraic process known as completing the square.

Consider the binomial $x^2 + 6x$. What term must be added to the binomial to produce a trinomial that is a perfect square? The answer is one-half the coefficient of the linear term, squared. In this case, one-half of 6 is 3 and $3^2 = 9$ and

$$x^2 + 6x + 9 = (x + 3)^2.$$

This process can be helpful in solving quadratic equations as follows:

Completing the Square Method

1. Use the multiplication principle to make the coefficient of x^2 equal to 1.
2. Rewrite the equation with the constant term isolated on one side.
3. Use the addition principle to add the square of one-half the coefficient of the linear term to both sides of the equation.
4. Replace the trinomial with its factored form, a perfect square.
5. Apply the square root property.
6. Solve the resulting linear equations.
7. Check your solutions in the original equation.

Example 2: Solve $x^2 - 4x - 5 = 0$ by the completing the square method.

Because the coefficient of x^2 is 1, Step 1 is not necessary.

Step 2. Adding 5 to both sides to isolate the constant term yields

$$x^2 - 4x = 5.$$

Step 3. The value needed to complete the square is $\left(\frac{1}{2} \cdot (-4)\right)^2 = (-2)^2 = 4$. Adding this to both sides produces

$$x^2 - 4x + 4 = 5 + 4$$
$$x^2 - 4x + 4 = 9.$$

Step 4. Replacing the trinomial with its perfect square form,

$$(x - 2)^2 = 9.$$

Steps 5 and 6. Applying the square root principle and solving,

$$x - 2 = \pm 3$$
$$x = 2 \pm 3$$
$$x = -1, 5.$$

The check is left to the reader.

Example 3: Solve $6x + 6 = -x^2$ by the completing the square method.

Step 1. Multiply each term by -1 to make the coefficient of x^2 equal to 1.

$$-6x - 6 = x^2$$

Step 2. Using the addition principle to isolate the constant term,

$$-6 = x^2 + 6x.$$

Step 3. The value necessary to complete the square is $\left(\frac{1}{2}\cdot 6\right)^2 = 9$. Completing the square yields

$$-6 + 9 = x^2 + 6x + 9.$$

Step 4. Factoring and simplifying,

$$3 = (x + 3)^2.$$

Steps 5 and 6. Taking the square root of both sides and solving,

$$\pm\sqrt{3} = x + 3$$
$$x = -3 \pm \sqrt{3}.$$

The check is left to the reader. Note that the solutions in this case are real but not rational.

Exercises

Solve the following quadratic equations using the completing the square method.

1. $x^2 - 6x + 8 = 0$

$$x^2 - 6x = -8$$
$$x^2 - 6x + 9 = -8 + 9$$
$$(x - 3)^2 = 1$$
$$x - 3 = 1 \quad \text{or} \quad x - 3 = -1$$
$$x = 4 \quad \text{or} \quad x = 2$$

2. $x^2 - 9x + 14 = 0$

$$x^2 - 9x = -14$$
$$x^2 - 9x + \frac{81}{4} = \frac{81}{4} - 14$$
$$\left(x - \frac{9}{2}\right)^2 = \frac{81}{4} - \frac{56}{4} = \frac{25}{4}$$
$$x - \frac{9}{2} = \frac{5}{2} \quad \text{or} \quad x - \frac{9}{2} = -\frac{5}{2}$$
$$x = 7 \quad \text{or} \quad x = 2$$

3. $-4x = -x^2 + 12$

$$x^2 - 4x = 12$$
$$x^2 - 4x + 4 = 12 + 4$$
$$(x - 2)^2 = 16$$
$$x - 2 = 4 \quad \text{or} \quad x - 2 = -4$$
$$x = 6 \quad \text{or} \quad x = -2$$

4. $2x^2 + 2x - 24 = 0$

$$x^2 + x = 12$$
$$x^2 + x + \frac{1}{4} = 12 + \frac{1}{4}$$
$$\left(x + \frac{1}{2}\right)^2 = \frac{49}{4}$$
$$x + \frac{1}{2} = \frac{7}{2} \quad \text{or} \quad x + \frac{1}{2} = -\frac{7}{2}$$
$$x = 3 \quad \text{or} \quad x = -4$$

5. $3x^2 + 2x = 1$

$$x^2 + \frac{2}{3}x = \frac{1}{3}$$

$$x^2 + \frac{2}{3}x + \frac{1}{9} = \frac{1}{3} + \frac{1}{9}$$

$$\left(x + \frac{1}{3}\right)^2 = \frac{4}{9}$$

$$x + \frac{1}{3} = \frac{2}{3} \quad \text{or} \quad x + \frac{1}{3} = -\frac{2}{3}$$

$$x = \frac{1}{3} \quad \text{or} \quad x = -1$$

6. $-\frac{1}{2}x^2 - x + \frac{3}{2} = 0$

$$x^2 + 2x - 3 = 0$$

$$x^2 + 2x + 1 = 3 + 1$$

$$(x + 1)^2 = 4$$

$$x + 1 = 2 \quad \text{or} \quad x + 1 = -2$$

$$x = 1 \quad \text{or} \quad x = -3$$

7. $10x^2 + 6x = 5$

$$x^2 + \frac{3}{5}x = \frac{1}{2}$$

$$x^2 + \frac{3}{5}x + \frac{9}{100} = \frac{1}{2} + \frac{9}{100}$$

$$\left(x + \frac{3}{10}\right)^2 = \frac{59}{100}$$

$$x + \frac{3}{10} = \pm\frac{\sqrt{59}}{10}$$

$$x = -\frac{3}{10} \pm \frac{\sqrt{59}}{10}$$

8. $15x^2 - 10x - 3 = 0$

$$x^2 - \frac{2}{3}x = \frac{1}{5}$$

$$x^2 - \frac{2}{3}x + \frac{1}{9} = \frac{1}{5} + \frac{1}{9}$$

$$\left(x - \frac{1}{3}\right)^2 = \frac{9}{45} + \frac{5}{45} = \frac{14}{45}$$

$$x = \frac{1}{3} \pm \sqrt{\frac{14}{45}}$$

$$x = \frac{1}{3} \pm \frac{\sqrt{70}}{15}$$

Derivation of the Quadratic Formula

The quadratic formula results from applying the completing the square method to the general quadratic equation $ax^2 + bx + c = 0$, where $a > 0$.

Step 1. Make the coefficient of x^2 equal 1 by multiplying both sides of the equation by $\frac{1}{a}$.

$$x^2 + \frac{b}{a}x + \frac{c}{a} = 0$$

Step 2. Isolate the constant term on one side of the equal sign.

$$x^2 + \frac{b}{a}x = -\frac{c}{a}$$

Step 3. Complete the square of the binomial. The coefficient of the linear term is $\frac{b}{a}$. The term needed to complete the square is $\left(\frac{1}{2}\cdot\frac{b}{a}\right)^2 = \frac{b^2}{4a^2}$. Add this term to both sides.

$$x^2 + \frac{b}{a}x + \frac{b^2}{4a^2} = \frac{b^2}{4a^2} - \frac{c}{a}$$

Rewrite the right-hand side as a single fraction.

$$x^2 + \frac{b}{a}x + \frac{b^2}{4a^2} = \frac{b^2 - 4ac}{4a^2}$$

Step 4. Express the left-hand side in factored form.

$$\left(x + \frac{b}{2a}\right)^2 = \frac{b^2 - 4ac}{4a^2}$$

Step 5. Apply the square root property to the resulting equation.

$$x + \frac{b}{2a} = \pm\frac{\sqrt{b^2 - 4ac}}{2a}$$

Step 6. Solve for x.

$$x = -\frac{b}{2a} \pm \frac{\sqrt{b^2 - 4ac}}{2a} \quad \text{or} \quad x = \frac{-b \pm \sqrt{b^2 - 4ac}}{2a}$$

This formula can be used to solve any quadratic equation in standard form $ax^2 + bx + c = 0$, $a \neq 0$, and is called the **quadratic formula.**

Rational Expressions

Rational expressions are expressions of the form $\frac{f(x)}{g(x)}$, where $f(x)$ and $g(x)$ are polynomials.

Simplifying Rational Expressions

1. Factor the numerator and the denominator.
2. Divide the numerator and the denominator by the common factors, if any.

Example 1: Simplify $\frac{x^2 - 10x + 24}{x^2 - 5x + 4}$.

Solution:

Step 1. Factor the numerator and the denominator.

$$\frac{x^2 - 10x + 24}{x^2 - 5x + 4} = \frac{(x-4)(x-6)}{(x-1)(x-4)}$$

Step 2. Divide the numerator and denominator by the common factor, $x - 4$.

$$\frac{\frac{(x-4)(x-6)}{x-4}}{\frac{(x-1)(x-4)}{x-4}} = \frac{x-6}{x-1}$$

Multiplying or Dividing Rational Expressions

1. Factor the numerator and denominator of each expression completely.
2. Divide out the common factors (cancel).
3. Place remaining factors in a single rational expression.
4. In division, proceed as above after inverting the divisor (the fraction after the division sign).

Example 2: Divide and simplify $\frac{x^2 + 3x - 10}{2x} \div \frac{x^2 - 5x + 6}{x^2 - 3x}$.

Solution:

Step 1. Rewrite as multiplication.

$$\frac{x^2 + 3x - 10}{2x} \div \frac{x^2 - 5x + 6}{x^2 - 3x} = \frac{x^2 + 3x - 10}{2x} \cdot \frac{x^2 - 3x}{x^2 - 5x + 6}$$

Step 2. Factor each fraction completely.

$$\frac{x^2 + 3x - 10}{2x} \cdot \frac{x^2 - 3x}{x^2 - 5x + 6} = \frac{(x+5)(x-2)}{2 \cdot x} \cdot \frac{x \cdot (x-3)}{(x-2)(x-3)}$$

Step 3. Cancel common factors $x - 2$ and $x - 3$.

$$\frac{(x+5)(x-2)}{2 \cdot x} \cdot \frac{x \cdot (x-3)}{(x-2)(x-3)} = \frac{x+5}{2}$$

Step 4. The simplified expression is

$$\frac{x+5}{2}.$$

Adding or Subtracting Rational Expressions

1. Find the LCD (least common denominator).
2. Build each fraction to have the LCD.
3. Add or subtract numerators.
4. Place the resulting numerator over the LCD, and simplify if necessary.

Example 3: Add and simplify $\frac{x}{x+1} + \frac{3}{(x+1)^2}$.

Solution:

Step 1. Because the denominators are already factored, it is clear that the LCD is $(x+1)^2$.

Step 2. Build each fraction to have the LCD.

$$\frac{x}{x+1} + \frac{3}{(x+1)^2} = \frac{x(x+1)}{(x+1)(x+1)} + \frac{3}{(x+1)^2} = \frac{x^2+x}{(x+1)^2} + \frac{3}{(x+1)^2}$$

Step 3. Add or subtract the numerators.

$$\frac{x^2+x}{(x+1)^2} + \frac{3}{(x+1)^2} = \frac{x^2+x+3}{(x+1)^2}$$

Step 4. Because the numerator can't be factored, we are finished.

Solving Rational Equations

1. Find the LCD of all fractions in the equation.
2. Multiply both sides of the equation by $\frac{\text{LCD}}{1}$ (clear all denominators).
3. Solve the resulting equation.
4. Check for extraneous roots.

Example 4: Solve $3 - \frac{4}{x} = \frac{5}{2}$.

Solution:

Step 1. The LCD is $2x$.

Step 2. Multiply both sides of the equation by $\frac{2x}{1}$.

$$\frac{2x}{1}\left(3 - \frac{4}{x}\right) = \frac{2x}{1}\left(\frac{5}{2}\right)$$

This is equivalent to $6x - \frac{8x}{x} = \frac{10x}{2}$ or $6x - 8 = 5x$.

Step 3. Solving the resulting equation,

$$6x - 6x - 8 = 5x - 6x \quad \text{or} \quad -8 = -x, x = 8.$$

Step 4. The check is left to the reader.

Exercises

Simplify the following.

1. $\dfrac{3x^2 - 6x}{x^2 + x - 6}$

$\dfrac{3x}{x + 3}$

2. $\dfrac{2x^3 + 2x^2 - 4x}{2x + 4}$

$x(x - 1)$

3. $\dfrac{x^2 + 2x - 15}{3 - x}$

$-x - 5$

Perform the indicated operations and simplify.

4. $\dfrac{4x^2y}{5xz} \cdot \dfrac{15x^6}{8xy^2}$

$\dfrac{3x^6}{2yz}$

5. $\dfrac{x^2 + 2x - 15}{3x + 15} \div \dfrac{x - 3}{3}$

1

6. $\dfrac{3}{x^2} + \dfrac{5}{6x}$

$\dfrac{5x + 18}{6x^2}$

7. $\dfrac{2}{x - 5} - \dfrac{3}{x + 3}$

$\dfrac{-x + 21}{(x - 5)(x + 3)}$

8. $\dfrac{3}{x - 3} + \dfrac{x - 2}{x^2 - 9}$

$\dfrac{4x + 7}{(x - 3)(x + 3)}$

9. $\dfrac{5}{x^2 - x - 2} - \dfrac{2}{x^2 + 4x + 3}$

$\dfrac{3x + 19}{(x + 1)(x + 3)(x - 2)}$

10. $\dfrac{x - 3}{x^2 - 3x + 2} - \dfrac{x + 1}{x^2 - 4}$

$\dfrac{-x - 5}{(x - 2)(x + 2)(x - 1)}$

Solve the following equations.

11. $\dfrac{x}{3} + \dfrac{2x}{7} = 10$

$x = \dfrac{210}{13}$

12. $\dfrac{-2}{x} + \dfrac{8}{3} = \dfrac{2}{x}$

$x = \dfrac{3}{2}$

13. $\dfrac{x - 2}{x - 4} = \dfrac{x}{x - 1}$

$x^2 - 3x + 2 = x^2 - 4x$

$-3x + 2 = -4x$

$x = -2$

14. $\dfrac{1}{x - 4} + x = \dfrac{-3}{x - 4}$

$1 + x(x - 4) = -3$

$1 + x^2 - 4x = -3$

$x^2 - 4x + 4 = 0$

$x = 2$

15. Solve $\dfrac{1}{R_1} + \dfrac{1}{R_2} = \dfrac{1}{R}$ for R.

$R = \dfrac{R_1 R_2}{R_1 + R_2}$

Complex Fractions

Complex fractions are fractions with a fractional expression in the numerator, the denominator, or both. Examples include

$$\frac{\frac{2}{5}+\frac{1}{3}}{7}, \quad \frac{x+3}{\frac{x}{x+1}-2}, \quad \text{and} \quad \frac{x+\frac{1}{x}-3}{x^3-x-\frac{2}{x^2}}.$$

Two methods are commonly used to simplify complex fractions. The first is to express the numerator and denominator as single fractions and then divide.

Simplifying a Complex Fraction by Simplifying the Numerator and Denominator

1. Express the numerator as a single fraction.
2. Express the denominator as a single fraction.
3. Divide the numerator by the denominator.
4. Simplify, if possible.

Example 1: Simplify $\dfrac{1-\frac{7}{16}}{3-\frac{2}{5}}$.

Solution:

Step 1. Simplify the numerator: $1-\frac{7}{16}=\frac{16}{16}-\frac{7}{16}=\frac{9}{16}$.

Step 2. Simplify the denominator: $3-\frac{2}{5}=\frac{3}{1}-\frac{2}{5}=\frac{15}{5}-\frac{2}{5}=\frac{13}{5}$.

Step 3. Divide the numerator by the denominator: $\frac{9}{16}\div\frac{13}{5}=\frac{9}{16}\cdot\frac{5}{13}=\frac{45}{208}$.

Step 4. Because the fraction cannot be simplified, the simplified result is $\frac{45}{208}$.

Example 2: Simplify $\dfrac{\frac{1}{x}+\frac{2}{x^2}}{2+\frac{1}{x^2}}$.

Solution:

Step 1. Simplify the numerator: $\frac{1}{x}+\frac{2}{x^2}=\frac{x}{x^2}+\frac{2}{x^2}=\frac{x+2}{x^2}$.

Step 2. Simplify the denominator: $2+\frac{1}{x^2}=\frac{2}{1}+\frac{1}{x^2}=\frac{2x^2}{x^2}+\frac{1}{x^2}=\frac{2x^2+1}{x^2}$.

Step 3. Divide the numerator by the denominator.

$$\frac{x+2}{x^2}\div\frac{2x^2+1}{x^2}=\frac{x+2}{x^2}\cdot\frac{x^2}{2x^2+1}=\frac{x+2}{2x^2+1}$$

Step 4. The result in Step 3 is simplified.

The second method of simplifying a complex fraction is to multiply the numerator and denominator by the LCD of the entire fraction.

Simplifying a Complex Fraction by Multiplying by the LCD

1. Determine the LCD of the numerator fractions and denominator fractions.
2. Multiply the fraction by 1 in the form $\frac{\text{LCD}}{\text{LCD}}$.
3. Simplify, if possible.

Example 3: Simplify $\dfrac{1 - \dfrac{7}{16}}{3 - \dfrac{2}{5}}$.

Solution:

Step 1. The only denominators are 16 and 5. Because there are no common factors, the LCD is 80. Multiply by $\frac{80}{80}$.

Step 2.

$$\frac{80\left(1 - \dfrac{7}{16}\right)}{80\left(3 - \dfrac{2}{5}\right)} = \frac{80 - \dfrac{80 \cdot 7}{16}}{240 - \dfrac{80 \cdot 2}{5}} = \frac{80 - 5 \cdot 7}{240 - 16 \cdot 2} = \frac{80 - 35}{240 - 32} = \frac{45}{208}$$

Step 3. Because the fraction is simplified, $\frac{45}{208}$ is the desired result.

Example 4: Simplify $\dfrac{\dfrac{3}{n-5} - 2}{1 - \dfrac{4}{n-5}}$.

Solution:

Step 1. The LCD is $n - 5$.

Step 2. Multiply the original fraction by $\frac{n-5}{n-5}$.

$$\frac{(n-5) \cdot \left(\dfrac{3}{n-5} - 2\right)}{(n-5) \cdot \left(1 - \dfrac{4}{n-5}\right)} = \frac{\dfrac{3(n-5)}{n-5} - 2(n-5)}{(n-5) - \dfrac{4(n-5)}{n-5}} = \frac{3 - 2n + 10}{n - 5 - 4} = \frac{-2n + 13}{n - 9}$$

Step 3. Because the numerator and denominator have no common factors, the simplified result is $\frac{-2n + 13}{n - 9}$.

Exercises

Simplify the following complex fractions.

1. $\dfrac{\dfrac{1}{2} - \dfrac{1}{4}}{\dfrac{5}{8} + \dfrac{3}{4}}$

   $\frac{2}{11}$

2. $\dfrac{\dfrac{5}{6y}}{\dfrac{10}{3xy}}$

   $\frac{x}{4}$

3. $\dfrac{\dfrac{8x^2y}{3z^3}}{\dfrac{4xy}{9z^5}}$

$6xz^2$

4. $\dfrac{3 - \dfrac{1}{x}}{1 - \dfrac{1}{x}}$

$\dfrac{3x - 1}{x - 1}$

5. $\dfrac{\dfrac{x^2}{y} - y}{\dfrac{y^2}{x} - x}$

$-\dfrac{x}{y}$

6. $\dfrac{4 + \dfrac{6}{n+1}}{7 - \dfrac{4}{n+1}}$

$\dfrac{4n + 10}{7n + 3}$

7. $\dfrac{\dfrac{1}{y-2} + \dfrac{3}{x}}{\dfrac{5}{x} - \dfrac{4}{xy - 2x}}$

$\dfrac{x + 3y - 6}{5y - 14}$

8. $\dfrac{\dfrac{x}{x+1} - 1}{\dfrac{x+1}{x-1}}$

$\dfrac{-x + 1}{(x + 1)^2}$

9. $\dfrac{1 + \dfrac{x}{x+1}}{\dfrac{2x+1}{x-1}}$

$\dfrac{x - 1}{x + 1}$

10. $\dfrac{\dfrac{x+1}{x-1} + \dfrac{x-1}{x+1}}{\dfrac{x+1}{x-1} - \dfrac{x-1}{x+1}}$

$\dfrac{x^2 + 1}{2x}$

Radicals and Fractional Exponents

Radical expressions involve square roots, cube roots, etc., and powers of the roots. The keys to manipulating such expressions are fractional exponents.

Translating Radical Expressions to Expressions Using Rational Exponents

1. The integer power on the base, b, becomes the numerator of the exponent on b.
2. The index (root) becomes the denominator of the exponent.

As a mnemonic, in b to the p/r, p is ***power*** and r is ***root***.

For example, $\sqrt[3]{x^2} = x^{2/3}$. In reverse, $a^{5/4} = \sqrt[4]{a^5}$.

When a $\sqrt{\ }$ is written without an index, the index is assumed to be 2.

Fractional exponents also obey the laws of exponents as outlined on page A-1.

1. $a^n a^m = a^{n+m}$
2. $\dfrac{a^n}{a^m} = a^{n-m}$
3. $(a^n)^m = a^{nm}$
4. $(ab)^n = a^n b^n$
5. $a^0 = 1, a \neq 0$
6. $a^{-n} = \dfrac{1}{a^n}$

In Examples 1–6, the following steps are used.

1. Write each expression using rational exponents.
2. Apply the appropriate property of exponents.
3. Write the expression using radical notation.

Example 1: $\sqrt[3]{x^2} \cdot \sqrt[4]{x} = x^{2/3} \cdot x^{1/4} = x^{2/3+1/4}$
$= x^{(8/12)+(3/12)} = x^{11/12} = \sqrt[12]{x^{11}}$

Example 2: $\dfrac{\sqrt[5]{x^4}}{\sqrt[10]{x^7}} = \dfrac{x^{4/5}}{x^{7/10}} = x^{4/5-7/10} = x^{1/10} = \sqrt[10]{x}$

Example 3: $(\sqrt[3]{\sqrt{x^5}}) = (x^{5/2})^{1/3} = x^{(5/2)\cdot(1/3)} = x^{5/6} = \sqrt[6]{x^5}$

Example 4: $\sqrt[3]{ab^2} = (ab^2)^{1/3} = a^{1/3} \cdot b^{2/3} = \sqrt[3]{a} \cdot \sqrt[3]{b^2}$

Example 5: $(\sqrt[4]{x})^0 = (x^{1/4})^0 = x^0 = 1$

Example 6: $x^{-2/3} = \dfrac{1}{x^{2/3}} = \dfrac{1}{\sqrt[3]{x^2}}$

Exercises

Rewrite the following using exponents.

1. a. $\sqrt[5]{x^4}$
   $x^{4/5}$

   b. $\sqrt[6]{x^3}$
   $x^{1/2}$

   c. $\sqrt[3]{(x+y)^2}$
   $(x+y)^{2/3}$

   d. $\sqrt[3]{(a-b)^3}$
   $a - b$

Rewrite these expressions using a radical.

2. a. $9x^{3/2}$
   $9\sqrt{x^3}$

   b. $(9x)^{3/2}$
   $27\sqrt{x^3}$

   c. $(4x^2 - 9y^2)^{1/2}$
   $\sqrt{4x^2 - 9y^2}$

   d. $(x - y)^{4/5}$
   $\sqrt[5]{(x-y)^4}$

Simplify and express your results in radical form, if appropriate. Assume variables represent nonzero values.

3. a. $\sqrt{x^{12}}$
   x^6

   b. $\sqrt[5]{6.87^5}$
   6.87

   c. $\sqrt{\sqrt{a^2b}}$
   $\sqrt[4]{a^2b}$

   d. $(\sqrt[3]{a^4bc^3})^{30}$
   $(a^4bc^3)^{10} = a^{40}b^{10}c^{30}$

   e. $x^{1/4} \cdot x^{3/8}$
   $x^{5/8} = \sqrt[8]{x^5}$

   f. $\dfrac{x^{1/2}}{x^{1/3}}$
   $x^{1/6} = \sqrt[6]{x}$

   g. $(x^{-2/5})^{1/4}$
   $x^{-1/10} = \dfrac{1}{\sqrt[10]{x}}$

   h. $(2x^{1/3})^0$
   1

APPENDIX

B Trigonometry

In *degrees*, a protractor measures angles from 0 to 180. In *radians*, the angles range in value from 0 to π.

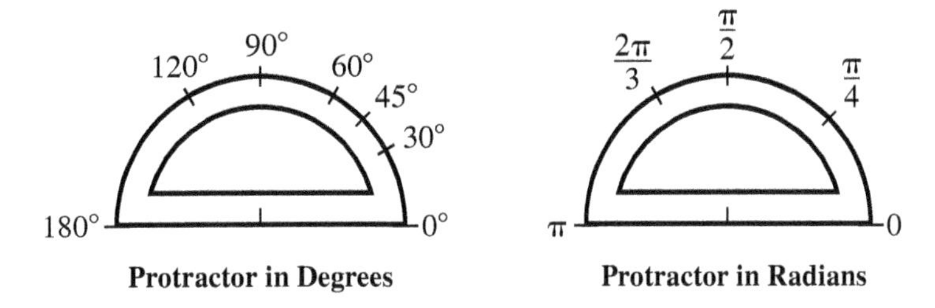

Protractor in Degrees **Protractor in Radians**

The fundamental idea is that the measure of a straight angle can be taken to be either 180 degrees or π radians.

$$180^\circ \equiv \pi \text{ radians} \qquad (1)$$

All other angles are done proportionately. The following table gives some examples. The last line of the table is useful for converting *any* angle from degrees to radians.

ANGLE (in degrees)	REASONING	CALCULATIONS	ANGLE (in radians)
90°	90 is *one-half* of 180	$\frac{1}{2} \cdot \pi$	$\frac{\pi}{2}$
60°	60 is *one-third* of 180	$\frac{1}{3} \cdot \pi$	$\frac{\pi}{3}$
45°	45 is *one-fourth* of 180	$\frac{1}{4} \cdot \pi$	$\frac{\pi}{4}$
30°	30 is *one-sixth* of 180	$\frac{1}{6} \cdot \pi$	$\frac{\pi}{6}$
120°	120 is *two-thirds* of 180	$\frac{2}{3} \cdot \pi$	$\frac{2\pi}{3}$
1°	1 is *one one-hundred-eightieth* of 180	$\frac{1}{180} \cdot \pi$	$\frac{\pi}{180}$

$$1^\circ \equiv \frac{\pi}{180} \text{ radians} \qquad (2)$$

The following table gives examples of the use of formula (2).

ANGLE (in degrees)	REASONING	CALCULATIONS	ANGLE (in radians)
12°	12 is *twelve* times 1	$12 \cdot \frac{\pi}{180}$	$\frac{\pi}{15}$
7°	7 is *seven* times 1	$7 \cdot \frac{\pi}{180}$	$\frac{7\pi}{180}$
345°	345 is *345* times 1	$345 \cdot \frac{\pi}{180}$	$\frac{23\pi}{12}$

The angle to wrap around a *full circle* is *twice* a straight angle of 180°; so it is 360°, or 2π radians. You can also have angles that wrap around a circle more than once! (Think of a fishing reel or spool of wire, with the string or wire wrapped around many times.)

Twice around the circle is 4π radians

π rad or 180° 0, 2π 4π

Equivalence (1) also enables us to convert angles from radians to degrees. Study these examples.

ANGLE (in radians)	REASONING	CALCULATIONS	ANGLE (in degrees)
$\frac{2\pi}{3}$	*Two-thirds* of π	$\frac{2}{3} \cdot 180$	120°
7π	*Seven* times π	$7 \cdot 180$	1260°
1	From (1), $\pi \equiv 180°$. Divide both sides of this equivalence by π.	$\frac{180}{\pi}$	$\frac{180}{\pi} \approx 57.3°$

The last line of the preceding table gives us an equivalence useful in converting from radians to degrees.

$$1 \text{ radian} = \frac{180°}{\pi} \qquad \textbf{(3)}$$

Because $\pi \approx 3.14$, equivalence (3) shows that 1 radian $\approx 57.3°$. This is worth seeing on a protractor.

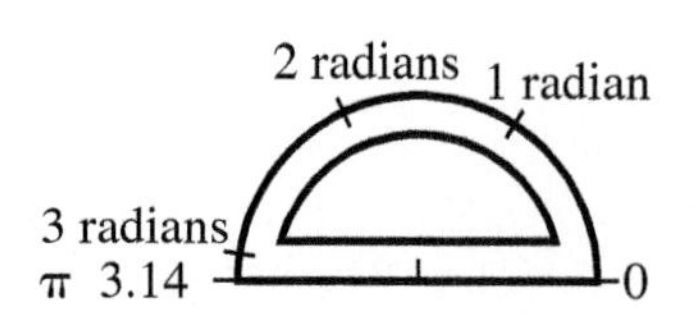

To convert 5 radians: $5 \cdot \frac{180}{\pi} = \frac{900}{\pi}$ degrees

To convert 0.6 radian: $0.6 \cdot \frac{180}{\pi} = \frac{108}{\pi}$ degrees

The answers just given can be written approximately: 5 radians $\approx$ 286.5° and 0.6 radian $\approx$ 34.4°.

Exercises

In Exercises 1–9, convert the given angle from degrees to radians or vice versa.

1. 30°

$30 \cdot \frac{\pi}{180} = \frac{\pi}{6}$ radians

2. 135°

$135 \cdot \frac{\pi}{180} = \frac{3\pi}{4}$ radians

3. $\frac{2\pi}{5}$ radians

$\frac{2\pi}{5} \cdot \frac{180}{\pi} = 72°$

4. 150°

$150 \cdot \frac{\pi}{180} = \frac{5\pi}{6}$ radians

5. $\frac{5\pi}{3}$ radians

$\frac{5\pi}{3} \cdot \frac{180}{\pi} = 5 \cdot 60 = 300°$

6. 1.5 radians

$1.5 \cdot \frac{180}{\pi} \approx \frac{1.5 \cdot 180}{3.14} \approx 86.0°$

7. 27°

$27 \cdot \frac{\pi}{180} = \frac{3\pi}{20}$ radians

8. $\frac{2}{3}$ radian

$\frac{2}{3} \cdot \frac{180}{\pi} \approx \frac{120}{3.14} \approx 38.2°$

9. 450°

$450 \cdot \frac{\pi}{180} = \frac{5\pi}{2}$ radians

10. How many times would you have to wrap a length of string around a circle to mark off an angle of 4π radians? 12π radians? 15π radians? 7 radians? 2000 radians?

$\frac{4\pi}{2\pi} = 2, \frac{12\pi}{2\pi} = 6, \frac{15\pi}{2\pi} = 7.5, \frac{7}{2\pi} \approx 1.11, \frac{2000}{2\pi} \approx 318.3$

11. Recall that the circumference of a circle, C, is given by the formula $C = 2\pi r$, where r is the radius of the circle. A **unit circle** is one whose radius is 1. Explain why the circumference of a unit circle equals the radian measure of the angle needed to wrap once around the circle.

The circumference of the circle is $C = 2\pi(1) = 2\pi$ units. The radian measure of the angle needed to wrap once around the circle is 2π.

12. a. Explain why a central angle of 1 radian in a unit circle subtends an arc whose length is 1 unit. (Hint: See Exercise 11.)

The ratio of $\frac{\text{arc length of the unit circle}}{\text{radian measure}}$ is $\frac{1}{1}$.

b. Explain why a central angle of t radians in a unit circle subtends an arc whose length is t units.

Same answer as part a.

c. Explain why a central angle of t radians in a circle of radius r subtends an arc whose length is tr units.

The arc length of a sector with central angle t radians is
$\frac{t}{2\pi} \cdot C = \frac{t}{2\pi} \cdot 2\pi r = tr.$

13. Label the following radian measures on the circle:
$\frac{\pi}{4}, \frac{\pi}{2}, \frac{5\pi}{4}, \frac{3\pi}{2}, 2\pi$

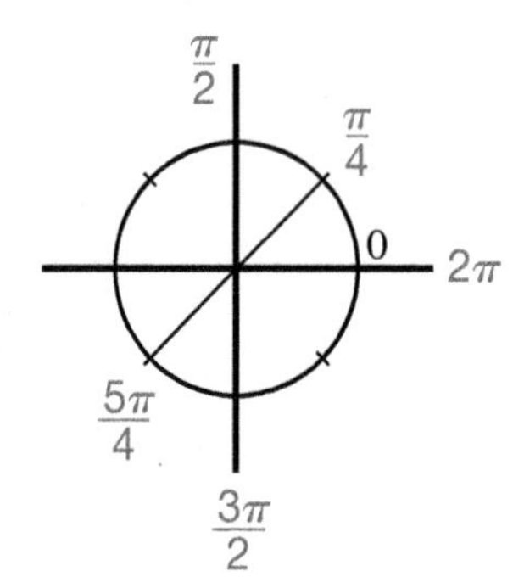

14. Label the following radian measures on the circle:
$\frac{\pi}{3}, \frac{\pi}{2}, \frac{2\pi}{3}, \frac{4\pi}{3}, \pi$

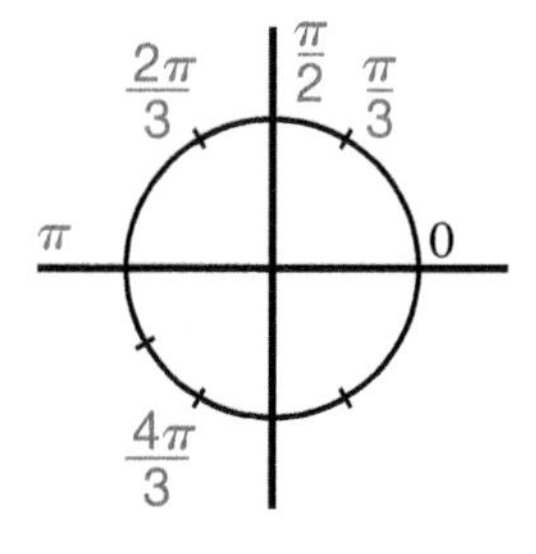

15. Label the following radian measures on the circle:
$\frac{\pi}{6}, \frac{\pi}{2}, \frac{7\pi}{6}, \frac{4\pi}{3}, \frac{11\pi}{6}$

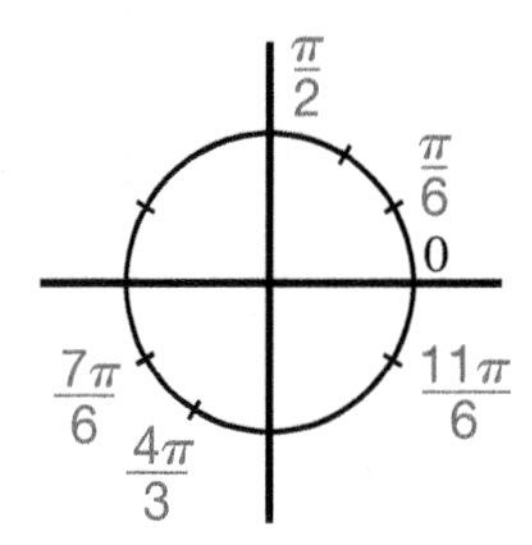

16. Locate approximately the following radian measures on the circle:
$1, \frac{\pi}{6}, 2, 0.6, 5, 3, \frac{3\pi}{4}, \frac{3\pi}{2}, 1.4$

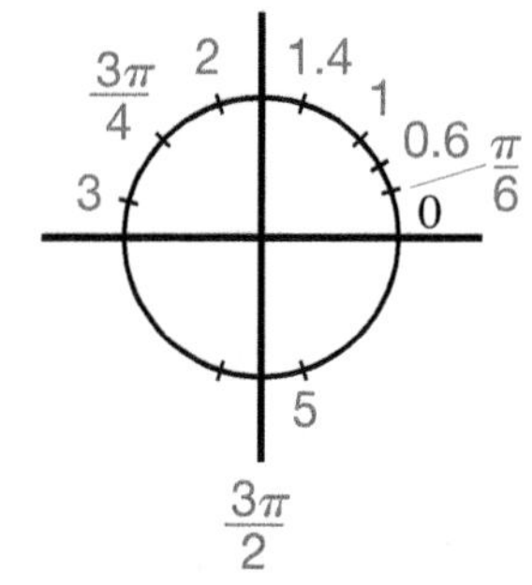

Trigonometric Functions in Right Triangles

For an angle θ in a right triangle (as pictured), the basic trigonometric functions (sine, cosine, tangent) are defined by the following.

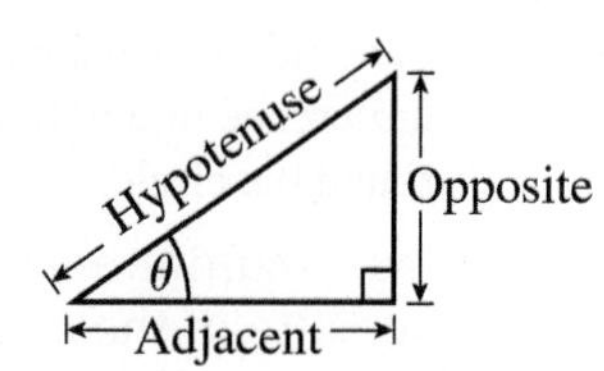

$$\sin\theta = \frac{\text{opposite}}{\text{hypotenuse}} \qquad \cos\theta = \frac{\text{adjacent}}{\text{hypotenuse}} \qquad \tan\theta = \frac{\text{opposite}}{\text{adjacent}}$$

The acronym **SOH CAH TOA** summarizes this; for example, **SOH** tells you that **s**ine equals **o**pposite over the **h**ypotenuse. Using the accompanying triangle,

$$\sin\theta = \frac{\text{opp}}{\text{hyp}} = \frac{6}{10} = \frac{3}{5}$$

$$\cos\theta = \frac{\text{adj}}{\text{hyp}} = \frac{8}{10} = \frac{4}{5}$$

$$\tan\theta = \frac{\text{opp}}{\text{adj}} = \frac{6}{8} = \frac{3}{4}.$$

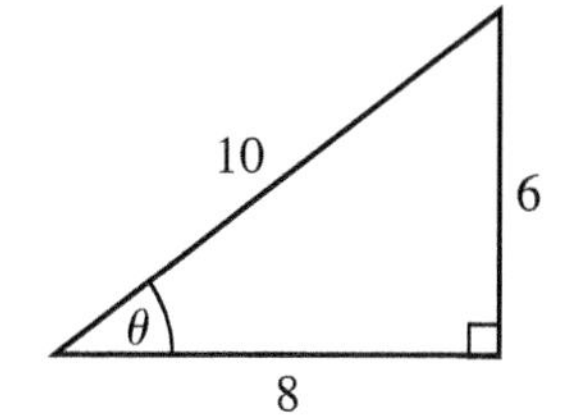

Frequently, you must use the **Pythagorean theorem** for right triangles.

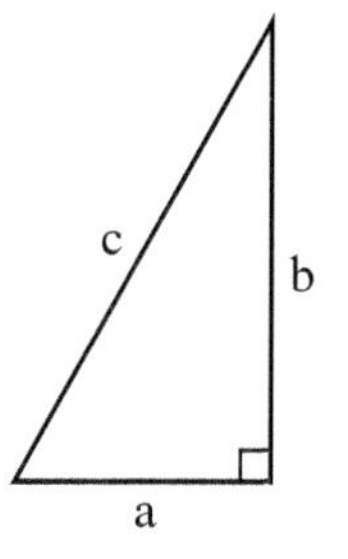

As you recall, the theorem says that in a right triangle, $c^2 = a^2 + b^2$. For example, to determine the values of trigonometric functions in the second triangle to the right, you first use the Pythagorean theorem to determine the missing side:

$$2^2 = 1^2 + x^2 \quad \text{or} \quad 4 = 1 + x^2 \quad \text{or} \quad x^2 = 3$$
$$x = \sqrt{3}$$

Note: Because x represents the length of the side of a triangle, we take the positive root.

Then, as before,

$$\sin\theta = \frac{\text{opp}}{\text{hyp}} = \frac{\sqrt{3}}{2}$$

$$\cos\theta = \frac{\text{adj}}{\text{hyp}} = \frac{1}{2}$$

$$\tan\theta = \frac{\text{opp}}{\text{adj}} = \frac{\sqrt{3}}{1} = \sqrt{3}.$$

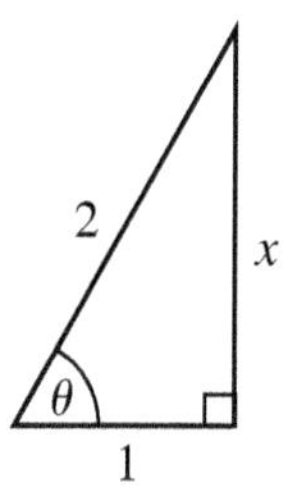

Exercises

In Exercises 1–6, find sin θ, cos θ, and tan θ.

1.

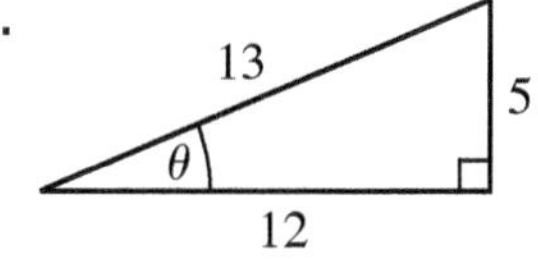

2.

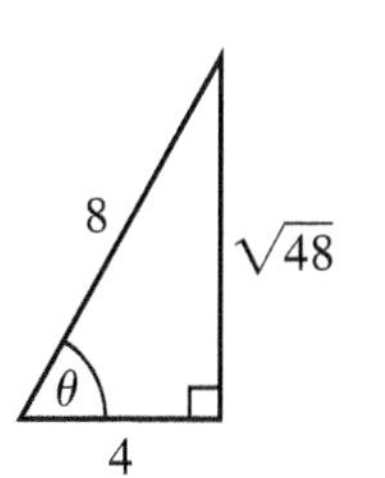

3.

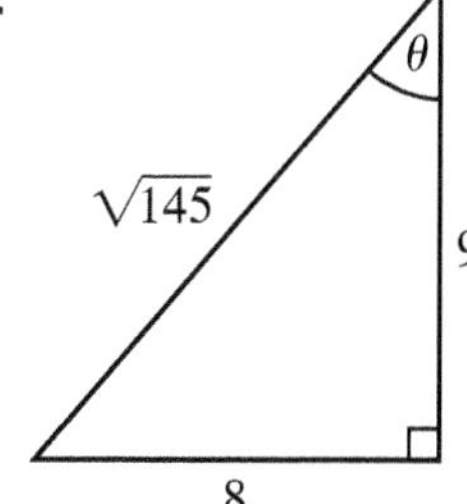

1. $\sin\theta = \frac{5}{13}$, $\cos\theta = \frac{12}{13}$, $\tan\theta = \frac{5}{12}$

2. $\sin\theta = \frac{\sqrt{48}}{8} = \frac{\sqrt{3}}{2}$, $\cos\theta = \frac{4}{8} = \frac{1}{2}$, $\tan\theta = \frac{\sqrt{48}}{4} = \sqrt{3}$

3. $\sin\theta = \frac{8}{\sqrt{145}}$, $\cos\theta = \frac{9}{\sqrt{145}}$, $\tan\theta = \frac{8}{9}$

4.

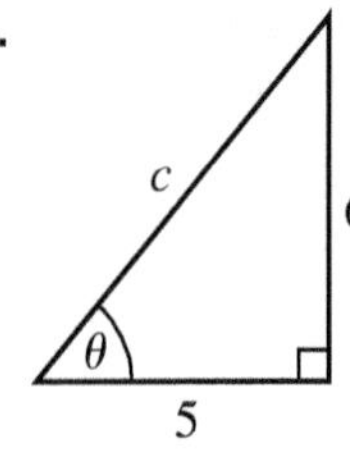

$5^2 + 6^2 = c^2$, so $c = \sqrt{61}$

$\sin\theta = \frac{6}{\sqrt{61}}$

$\cos\theta = \frac{5}{\sqrt{61}}$

$\tan\theta = \frac{6}{5}$

5.

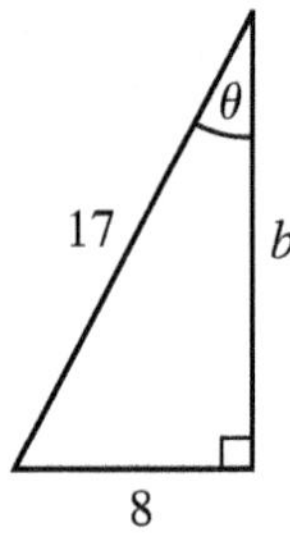

$17^2 = 8^2 + b^2$, so $b = 15$

$\sin\theta = \frac{8}{17}$

$\cos\theta = \frac{15}{17}$

$\tan\theta = \frac{8}{15}$

6.

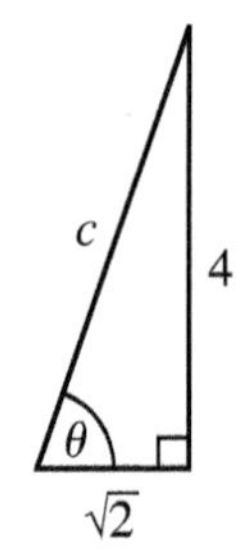

$\sqrt{2}^2 + 4^2 = c^2$ so $c = \sqrt{18}$

$\sin\theta = \frac{4}{\sqrt{18}}$

$\cos\theta = \frac{\sqrt{2}}{\sqrt{18}} = \frac{1}{3}$

$\tan\theta = \frac{4}{\sqrt{2}}$

7. Using the Pythagorean theorem and SOH CAH TOA, show that $\sin^2\theta + \cos^2\theta = 1$ for any angle θ in a right triangle.

$$\sin^2\theta + \cos^2\theta = \frac{O^2}{H^2} + \frac{A^2}{H^2} = \frac{O^2 + A^2}{H^2} = \frac{H^2}{H^2} = 1$$

Trigonometric Function Values of Special Angles

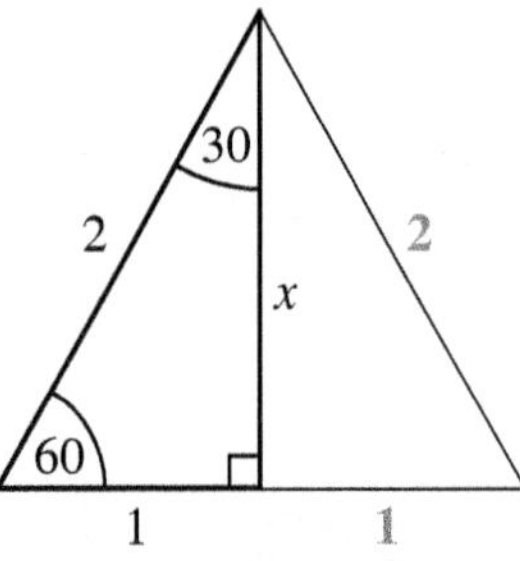

30-60-90: Half of an equilateral triangle

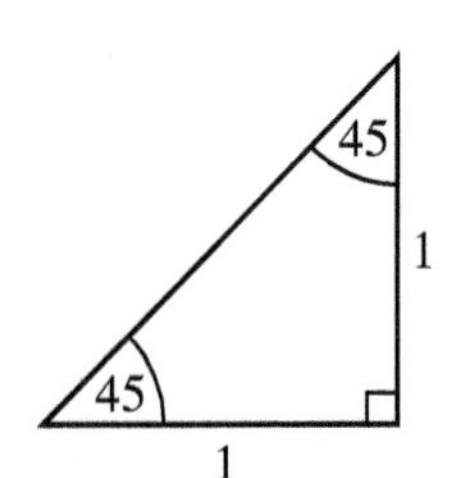

45-45-90: Isosceles; two equal legs

The third side in each triangle can be found using the Pythagorean theorem. The proportions are as follows:

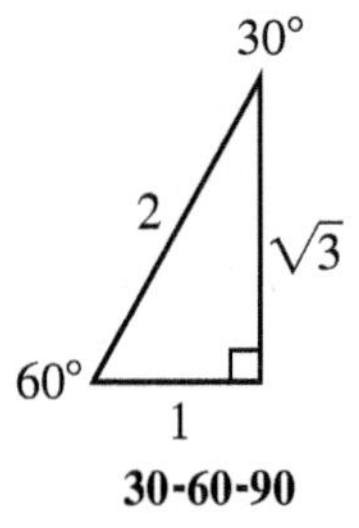

30-60-90

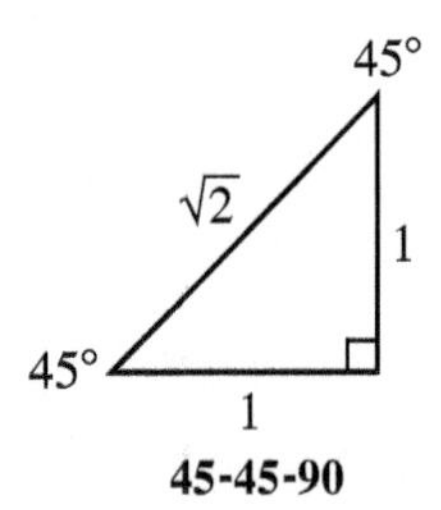

45-45-90

From the triangles, you obtain these important values for trigonometric functions.

ANGLE (in degrees)	sin θ	cos θ	tan θ
30°	$\frac{1}{2}$	$\frac{\sqrt{3}}{2}$	$\frac{1}{\sqrt{3}}$
45°	$\frac{\sqrt{2}}{2}$	$\frac{\sqrt{2}}{2}$	1
60°	$\frac{\sqrt{3}}{2}$	$\frac{1}{2}$	$\sqrt{3}$

Trigonometric Functions for More General Angles

When an angle θ is larger than 90° (in radians, $\theta > \frac{\pi}{2}$), you can still evaluate sine, cosine, and tangent. You work in an x-y plane, make the positive x-axis the initial side of the angle, and make a *reference triangle* by dropping a perpendicular to the x-axis from a point on the terminal side of the angle. (See the figure below.) For positive angles, you rotate *counterclockwise* to find the terminal side.

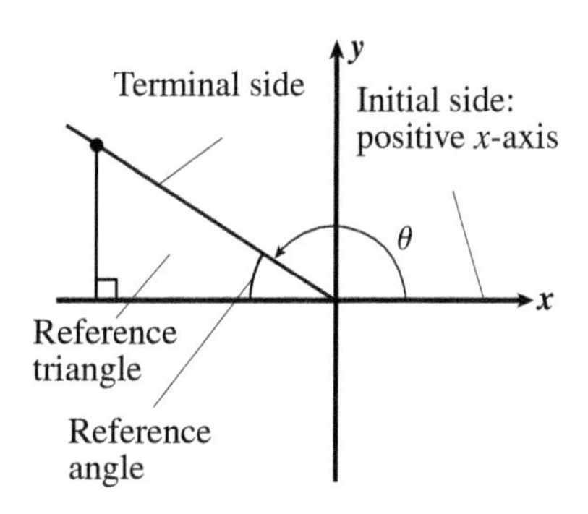

You then use SOH CAH TOA on the *reference triangle.* **Warning:** The adjacent and opposite sides may be *negative* in this situation, depending on the quadrant in which the terminal side of the angle lies.

For example, in the figure shown, note the negative sign for the adjacent side. (It lies on the *negative* x-axis.) You have

$$\sin\theta = \frac{\text{opp}}{\text{hyp}} = \frac{4}{5}$$

$$\cos\theta = \frac{\text{adj}}{\text{hyp}} = \frac{-3}{5}$$

$$\tan\theta = \frac{\text{opp}}{\text{adj}} = \frac{4}{-3} = \frac{-4}{3}.$$

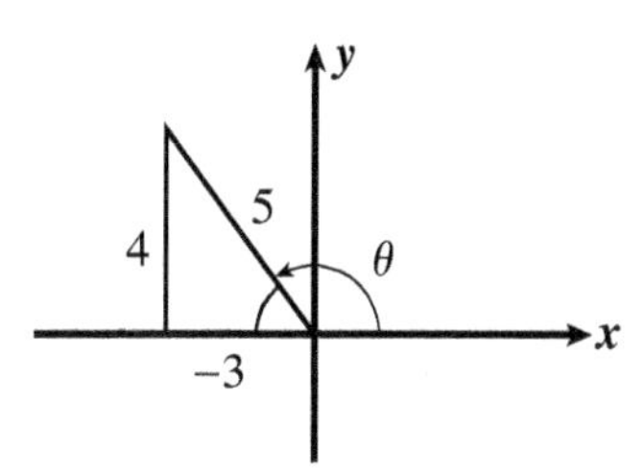

Note: The hypotenuse is ALWAYS positive.

You also need to use the Pythagorean theorem at times. In the next figure, you have $6^2 = (-2)^2 + y^2$ or $y^2 = 32$. Because y must be *negative* (do you see why?), $y = -\sqrt{32} = -4\sqrt{2}$. You now find

$$\sin\theta = \frac{\text{opp}}{\text{hyp}} = \frac{-4\sqrt{2}}{6} = \frac{-2\sqrt{2}}{3}$$

$$\cos\theta = \frac{\text{adj}}{\text{hyp}} = \frac{-2}{6} = \frac{-1}{3}$$

$$\tan\theta = \frac{\text{opp}}{\text{adj}} = \frac{-4\sqrt{2}}{-2} = 2\sqrt{2}.$$

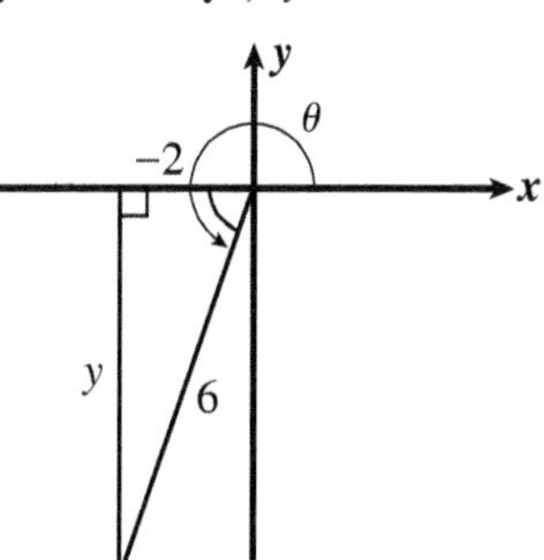

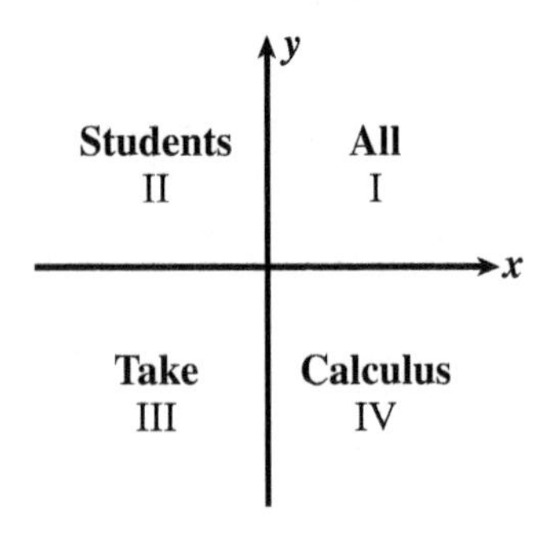

As noted previously, reference triangles with angles of 30°, 60°, or 45° show up frequently because of the symmetry involved. To find the values of the three trigonometric functions for $\theta = 5\pi/3$ radians, we convert to degrees and sketch the angle and the reference triangle.

$$\frac{5\pi}{3} = \frac{5}{3} \cdot 180° = 300°$$

The reference angle is $360° - 300° = 60°$

$$\sin 60° = \frac{\sqrt{3}}{2} \quad \sin 300° = \frac{-\sqrt{3}}{2}$$

$$\cos 60° = \frac{1}{2} \quad \cos 300° = \frac{1}{2}$$

$$\tan 60° = \sqrt{3} \quad \tan 300° = -\sqrt{3}$$

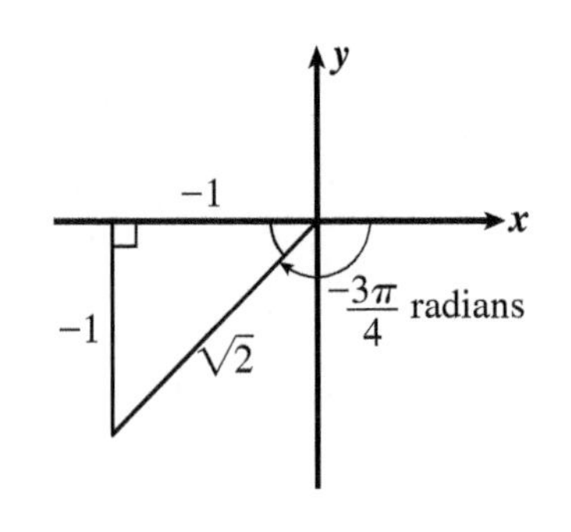

To help you remember the $\pm$ signs, the phrase *All Students Take Calculus* is useful. The four words go in the four quadrants of the x-y plane.

The All means *all* trigonometric functions are positive in quadrant I; **S** for **S**tudents means S*ine* is positive in quadrant II; **T** for **T**ake means T*angent* is positive in quadrant III; **C** for **C**alculus means C*osine* is positive in quadrant IV.

Note that for *negative* angles, we locate the terminal side by rotating *clockwise* from the positive x-axis. The figure at the left shows that

$$\sin\left(\frac{-3\pi}{4}\right) = \frac{-1}{\sqrt{2}} = \frac{-\sqrt{2}}{2},$$

and by similar calculations,

$$\cos\left(\frac{-3\pi}{4}\right) = \frac{-\sqrt{2}}{2} \quad \text{and} \quad \tan\left(\frac{-3\pi}{4}\right) = 1.$$

Exercises

In Exercises 1–4, find the remaining side of the reference triangle and evaluate $\sin\theta$, $\cos\theta$, *and* $\tan\theta$.

1.

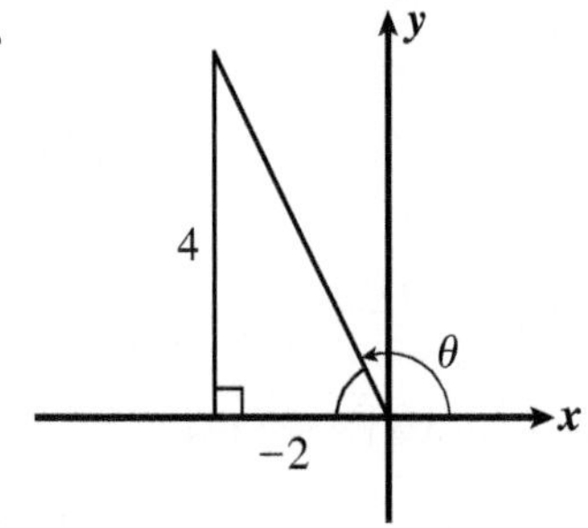

Hypotenuse is $\sqrt{20} = 2\sqrt{5}$.

$$\sin\theta = \frac{4}{2\sqrt{5}} = \frac{2}{\sqrt{5}}$$

$$\cos\theta = \frac{-2}{2\sqrt{5}} = \frac{-1}{\sqrt{5}}$$

$$\tan\theta = \frac{4}{-2} = -2$$

2.

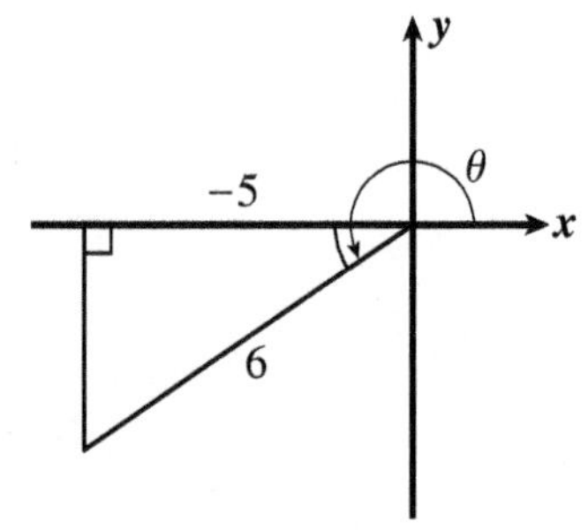

Third side is $-\sqrt{11}$.

$$\sin\theta = \frac{-\sqrt{11}}{6}$$

$$\cos\theta = -\frac{5}{6}$$

$$\tan\theta = \frac{\sqrt{11}}{5}$$

3.

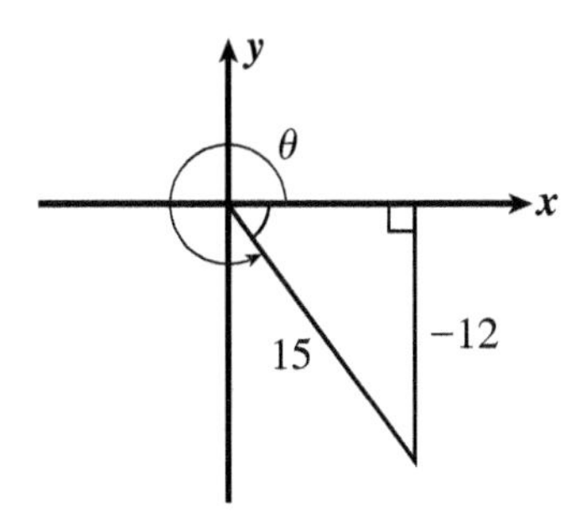

Third side is 9.

$$\sin\theta = \frac{-12}{15} = \frac{-4}{5}$$

$$\cos\theta = \frac{9}{15} = \frac{3}{5}$$

$$\tan\theta = \frac{-12}{9} = -\frac{4}{3}$$

4.

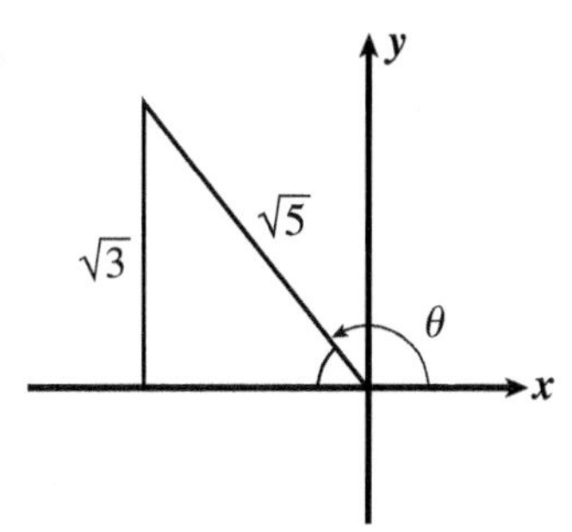

Third side is $-\sqrt{2}$.

$$\sin\theta = \frac{\sqrt{3}}{\sqrt{5}} = \frac{\sqrt{15}}{5}$$

$$\cos\theta = \frac{-\sqrt{2}}{\sqrt{5}} = \frac{-\sqrt{10}}{5}$$

$$\tan\theta = \frac{\sqrt{3}}{-\sqrt{2}} = -\frac{\sqrt{6}}{2}$$

In Exercises 5–10, sketch the angle and reference triangle and then work out the values of the three trigonometric functions for that angle.

5. $\theta = \frac{\pi}{2}$ radians

no triangle; opp = hyp and adj = 0

$$\sin\frac{\pi}{2} = 1$$

$$\cos\frac{\pi}{2} = 0$$

$$\tan\frac{\pi}{2} = \textit{undefined}$$

6. $\theta = 225°$

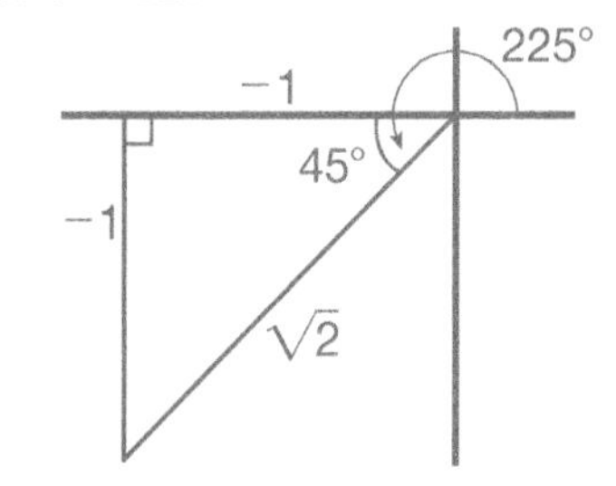

$$\sin 225° = \frac{-1}{\sqrt{2}}$$

$$\cos 225° = \frac{-1}{\sqrt{2}}$$

$$\tan 225° = 1$$

7. $\theta = -150°$

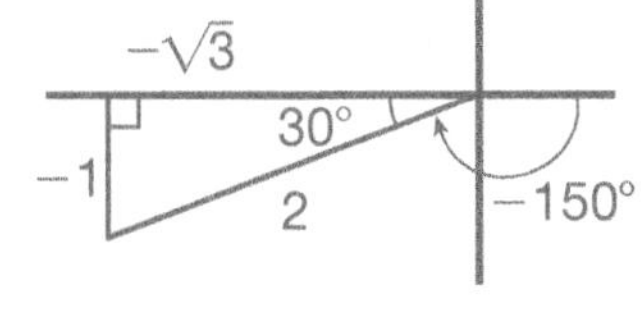

$$\sin(-150°) = \frac{-1}{2}$$

$$\cos(-150°) = \frac{-\sqrt{3}}{2}$$

$$\tan(-150°) = \frac{1}{\sqrt{3}}$$

8. $\theta = 330°$

8. 330° √3 30° 2 −1

$$\sin 330° = \frac{-1}{2}$$

$$\cos 330° = \frac{\sqrt{3}}{2}$$

$$\tan 330° = \frac{-1}{\sqrt{3}}$$

9.

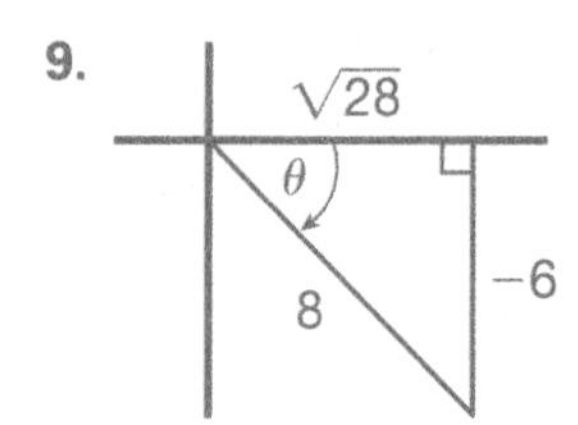

10.

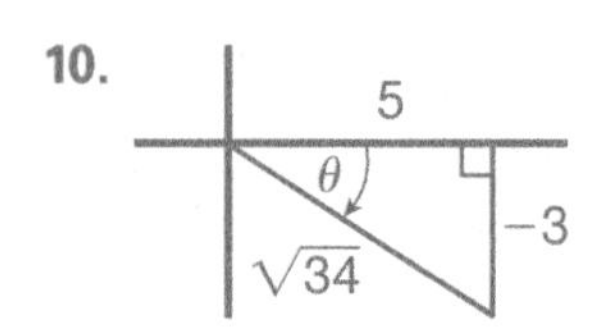

9. θ is fourth quadrant angle whose sine is $-\frac{6}{8}$.

$\cos\theta = \frac{\sqrt{28}}{8}$

$\tan\theta = \frac{-6}{\sqrt{28}}$

10. θ is fourth quadrant angle whose tangent is $-\frac{3}{5}$.

$\sin\theta = \frac{-3}{\sqrt{34}}$

$\cos\theta = \frac{5}{\sqrt{34}}$

11. Show that for θ in *any* quadrant, the relationship $\sin^2\theta + \cos^2\theta = 1$ holds.

Assume that α is the reference angle for θ. Then $\sin\alpha = |\sin\theta|$ and $\cos\alpha = |\cos\theta|$.

Because $\sin\alpha = \frac{O}{H}$ and $\cos\alpha = \frac{A}{H}$,

$$\sin^2\alpha + \cos^2\alpha = \sin^2\theta + \cos^2\theta = \frac{O^2}{H^2} + \frac{A^2}{H^2} = \frac{O^2 + A^2}{H^2} = \frac{H^2}{H^2} = 1.$$

Dealing with Special Angles: 0°, ±90°, ±180°

For multiples of 90° (equivalently, multiples of $\frac{\pi}{2}$ radians), the reference triangle degenerates to a straight line segment. Either the adjacent or opposite side degenerates to 0.

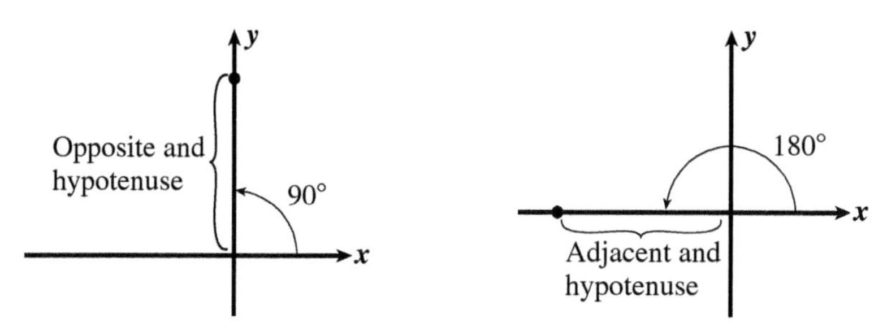

For simplicity, notice that for ordinary angles, when we mark a point on the terminal side of an angle θ, the *x-value* gives the value of the *adjacent* side and the *y*-value gives the *opposite* side. Let us also use r for the length of the hypotenuse. Therefore, we could have defined

$$\sin\theta = \frac{\text{opp}}{\text{hyp}} = \frac{y}{\text{hyp}} = \frac{y}{r} \qquad \cos\theta = \frac{\text{adj}}{\text{hyp}} = \frac{x}{\text{hyp}} = \frac{x}{r} \qquad \tan\theta = \frac{\text{opp}}{\text{adj}} = \frac{y}{x}. \tag{4}$$

We use $\frac{y}{r}, \frac{x}{r}$, and $\frac{y}{x}$ for the special angles to find the values of the trigonometric functions. When we have a multiple of 90°, we mark a point on the terminal side and label it with its x and y numbers. Use 0 and ±1 for simplicity; then use formula (4). Here are the figures for −90° and 180°. Remember that r is always positive.

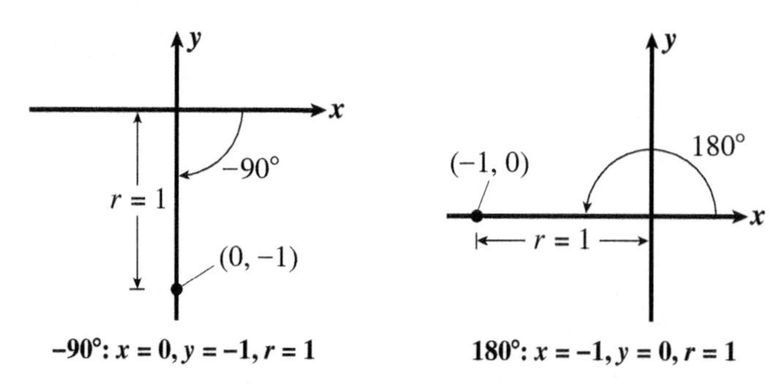

Now the trigonometric values are as follows:

$$\sin(-90^\circ) = \frac{\text{opp}}{\text{hyp}} = \frac{y}{r} = \frac{-1}{1} = -1 \qquad \sin 180^\circ = \frac{\text{opp}}{\text{hyp}} = \frac{y}{r} = \frac{0}{1} = 0$$

$$\cos(-90^\circ) = \frac{\text{adj}}{\text{hyp}} = \frac{x}{r} = \frac{0}{1} = 0 \qquad \cos 180^\circ = \frac{\text{adj}}{\text{hyp}} = \frac{x}{r} = -\frac{1}{1} = -1$$

$$\tan(-90^\circ) = \frac{\text{opp}}{\text{adj}} = \frac{y}{x} = \frac{-1}{0} = \textit{undefined} \qquad \tan 180^\circ = \frac{\text{opp}}{\text{adj}} = \frac{y}{x} = \frac{0}{-1} = 0$$

If you use only 0 and ± 1 for x- and y-values of these special angles, r always equals 1. Also, whenever the x value is 0, the tangent is *undefined*, because the formula then involves division by 0.

Exercises

In Exercises 1–6, for each given value of θ, draw the angle, label a point on the terminal side, and use formulas to get the values of $\sin\theta$, $\cos\theta$, and $\tan\theta$.

1.

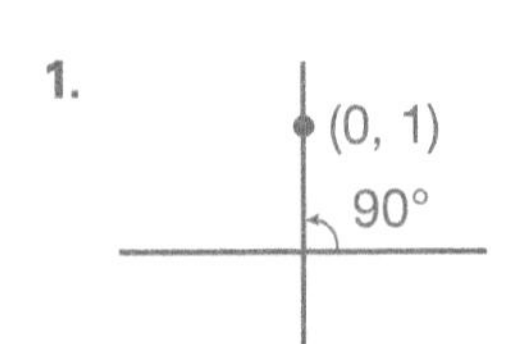

2.

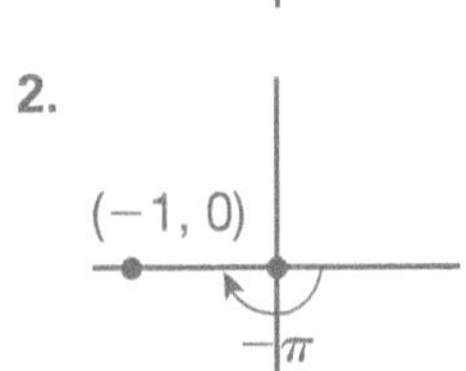

3.

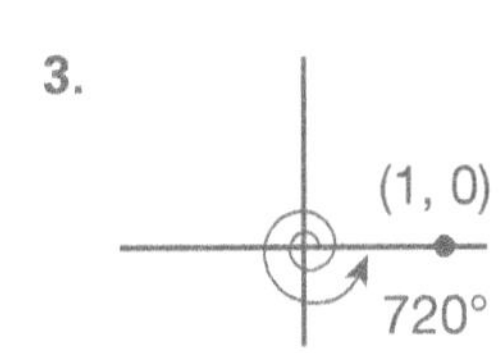

4.

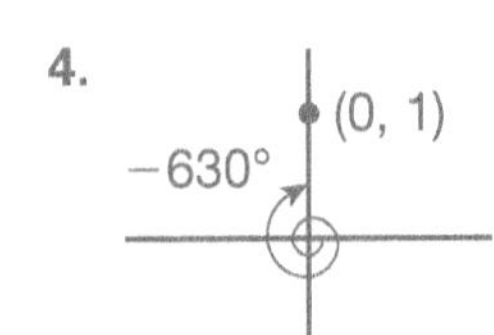

5. $\frac{7\pi}{2}$ (0, −1)

6. (−1, 0)

23π radians = 11.5 rotations; terminal side is $-x$-axis.

1. $\theta = 90^\circ$

$x = 0, y = 1, r = 1$

$\sin 90^\circ = \frac{1}{1} = 1$

$\cos 90^\circ = \frac{0}{1} = 0$

$\tan 90^\circ = \frac{1}{0} = \text{undef.}$

2. $\theta = -\pi$ radians

$x = -1, y = 0, r = 1$

$\sin(-\pi) = \frac{0}{1} = 0$

$\cos(-\pi) = \frac{-1}{1} = -1$

$\tan(-\pi) = \frac{0}{-1} = 0$

3. $\theta = 720^\circ$

$x = 1, y = 0, r = 1$

$\sin(720^\circ) = \frac{0}{1} = 0$

$\cos(720^\circ) = \frac{1}{1} = 1$

$\tan(720^\circ) = \frac{0}{1} = 0$

4. $\theta = -630^\circ$

$x = 0, y = 1, r = 1$

$\sin(-630^\circ) = \frac{1}{1} = 1$

$\cos(-630^\circ) = \frac{0}{1} = 0$

$\tan(-630^\circ) = \frac{1}{0} = \text{undef.}$

5. $\theta = \frac{7\pi}{2}$ radians

$x = 0, y = -1, r = 1$

$\sin\left(\frac{7\pi}{2}\right) = \frac{-1}{1} = -1$

$\cos\left(\frac{7\pi}{2}\right) = \frac{0}{1} = 0$

$\tan\left(\frac{7\pi}{2}\right) = \frac{-1}{0} = \text{undef.}$

6. $\theta = 23\pi$ radians

$x = -1, y = 0, r = 1$

$\sin(23\pi) = \frac{0}{1} = 0$

$\cos(23\pi) = \frac{-1}{1} = -1$

$\tan(23\pi) = \frac{0}{-1} = 0$

APPENDIX C

Getting Started with the TI-84 Plus Family of Calculators

The calculator screens in this appendix are from the TI-84 Plus CE calculator. The procedures given in this appendix apply to the TI-84 Plus with the latest operating system as well.

ON-OFF

To turn on the TI-84 Plus CE, press the ON key. To turn off the TI-84 Plus CE, press 2nd and ON in sequence.

In general, to access any of the white commands, simply press the black or gray key. To access the blue commands, press 2nd and then the black or gray key below the desired command. Similarly, to access any of the green commands or symbols, press ALPHA followed by the appropriate black or gray key.

MODE

The MODE key controls many calculator settings. The activated settings are highlighted. For most of your work in this course, the settings in the left-hand column should be highlighted.

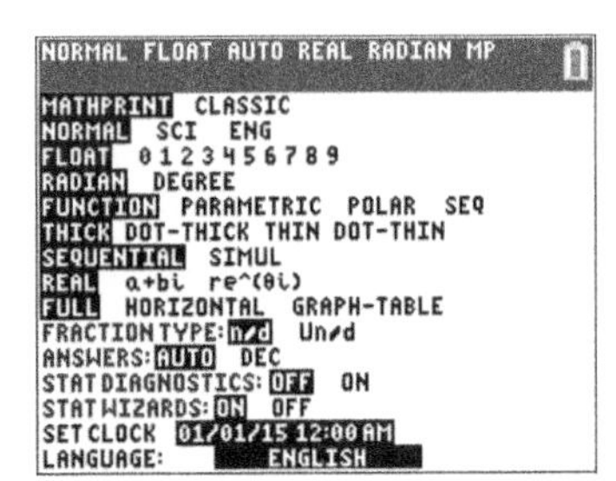

To change a setting, move the cursor to the desired setting and press ENTER.

The Home Screen

The home screen is used for calculations.

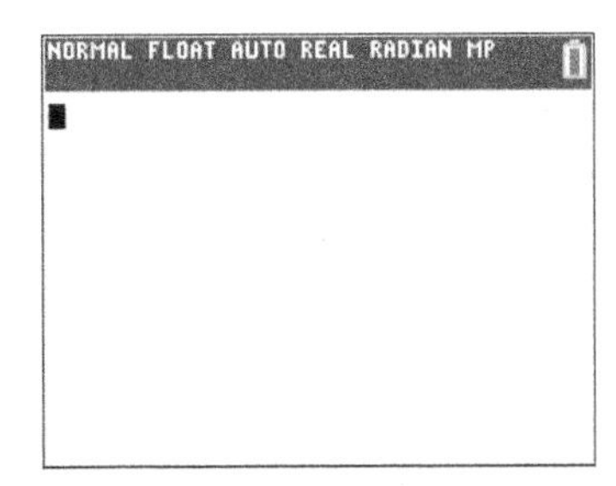

Note: The status at the top gives selected mode settings as well as the battery level.

You may return to the home screen at any time using the QUIT command. This command is accessed by pressing 2nd MODE. All calculations in the home screen are subject to the order of operations.

Enter all expressions as you would write them. Always observe the order of operations. Once you have typed the expression, press ENTER to obtain the simplified result. Before you hit ENTER, you may edit your expression by using the arrow keys, the delete command DEL, and the insert command 2nd DEL.

Three keys of special note are the reciprocal key X^{-1}, the caret key ^ and the negative key (−).

The reciprocal command X^{-1} reciprocates the number in the home screen.

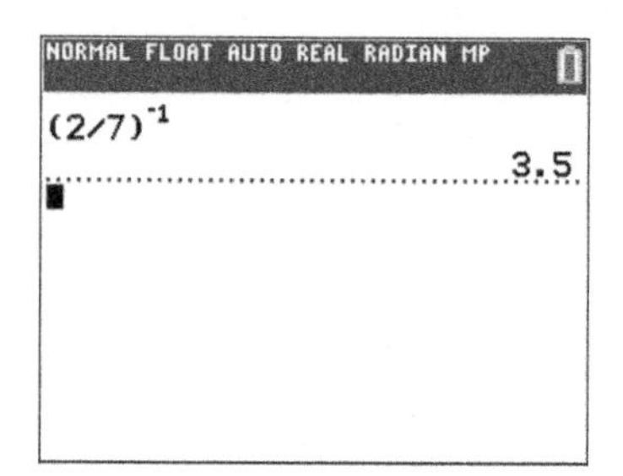

If a decimal result represents a rational number, it can be displayed as a fraction in lowest terms as follows. Press ALPHA Y= 4 ENTER.

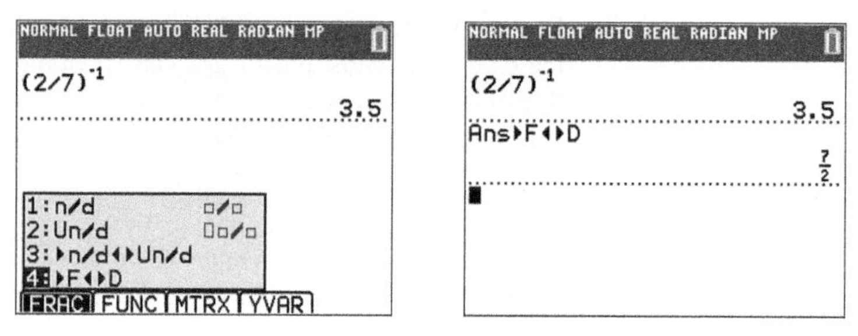

Pressing ALPHA Y= accesses the F1 menu which has functions used for working with fractions.

The carat key ^ is used to raise numbers to powers.

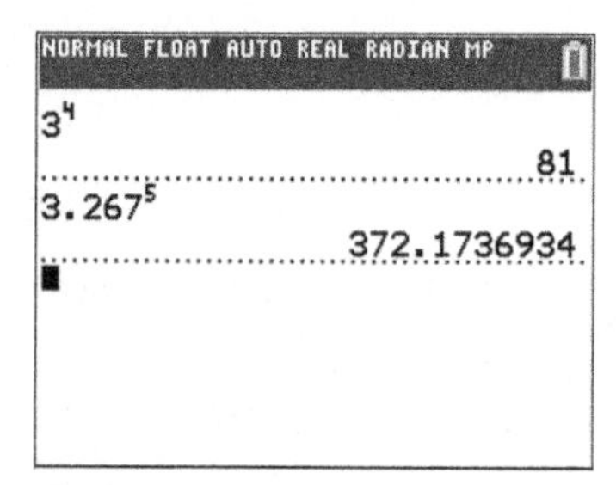

The square of a number can be evaluated two ways. Using the ^ key as above or using the x^2 key. To use the x^2 key first enter the value to be squared then press the x^2 key and ENTER.

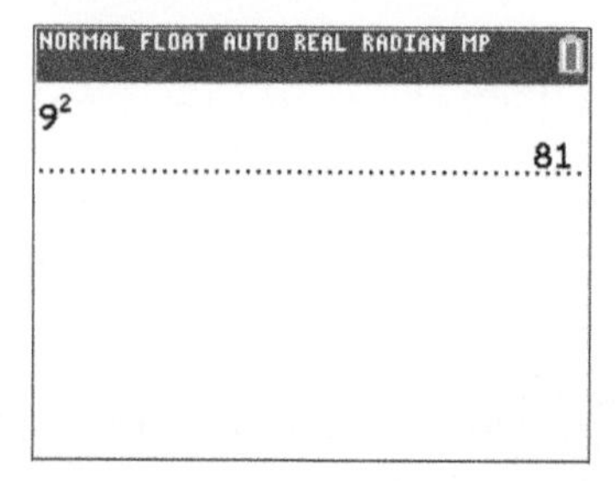

The key colors vary from calculator to calculator. The negative key is different from the minus key. To enter a negative number, use the (−) key on the bottom row, not the blue − key on the right side of the calculator.

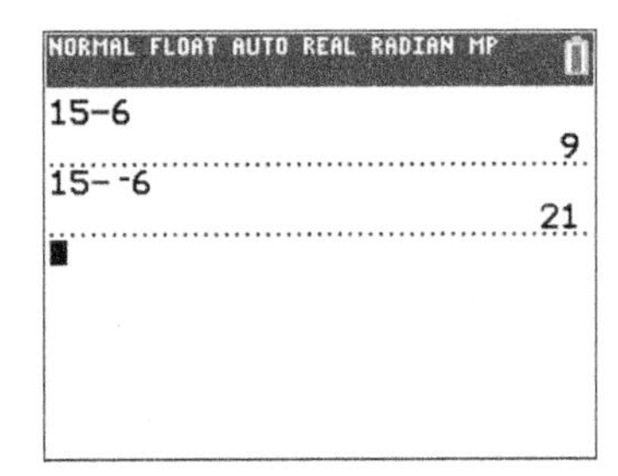

A table of keys and their functions follows.

KEY	FUNCTION DESCRIPTION
ON	Turns calculator on (or off)
Clear	Clears text screen
ENTER	Executes a command
(−)	Calculates the additive inverse
MODE	Displays current operating settings
DEL	Deletes the character at the cursor
^	Symbol used for exponentiation
ANS	Storage location of the last calculation
ENTRY	Retrieves a previously Executed expression

ANS and ENTRY

The last two commands in the above table can be real time-savers. The result of your last calculation is always stored in a memory location known as ANS. It is accessed by pressing 2nd (−), or it is automatically accessed by pressing any operation button.

Suppose you want to evaluate $12.5\sqrt{1 + 0.5 \cdot (0.55)^2}$. It could be evaluated in one expression and checked with a series of calculations using ANS.

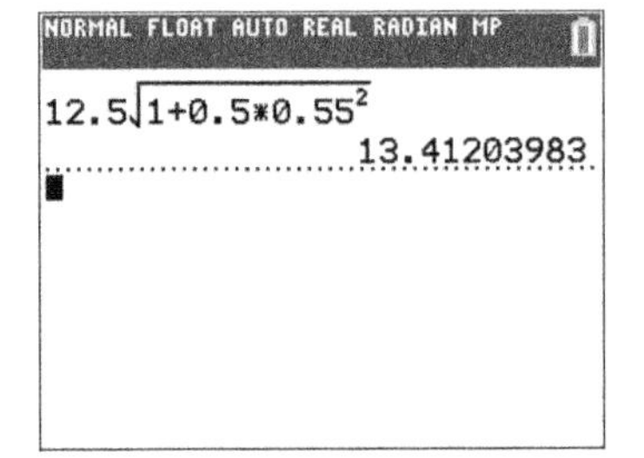

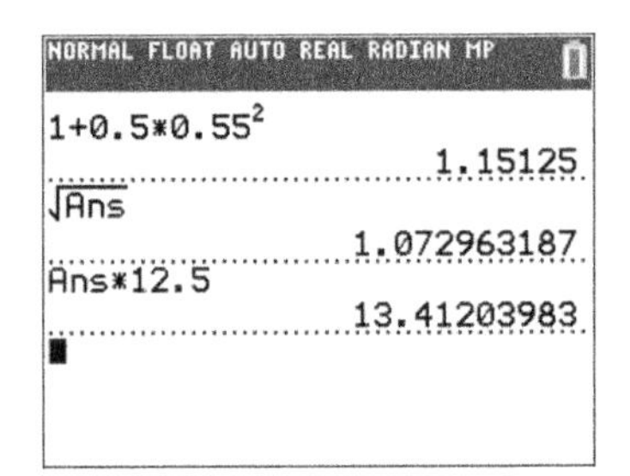

After you have keyed in an expression and pressed ENTER, you cannot move the cursor back up to edit or recalculate this expression. This is where the ENTRY 2nd ENTER command is used. The ENTRY command retrieves the previous expression and places the cursor at the end of the expression. You can use the left and right arrow keys to move the cursor to any location in the expression that you want to modify.

Suppose you want to evaluate the compound interest expression $P\left(1 + \frac{r}{n}\right)^{nt}$, where P is the principal, r is the interest rate, n is the number of compounding periods annually, and t is the number of years when $P = \$1000$, $r = 6.5\%$, $n = 1$, and $t = 2, 5,$ and 15 years. Using the ENTRY command, this expression would be entered once and edited twice.

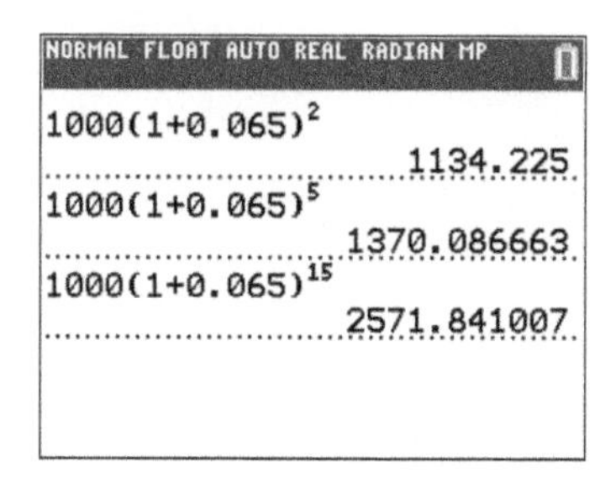

Note that many "last expressions" are stored in the ENTRY memory location. You can repeat the ENTRY command as many times as you want to retrieve a previously entered expression.

An alternative to ANS and ENTRY is simply to use the up arrow button to select previous answers or entries and then press ENTER to insert the selected expression at the cursor.

If you repeat the proceeding example using the up arrow button to retrieve the previous entries, the resulting screen will look exactly the same.

Functions and Graphing with the TI-84 Plus CE

Y= menu

Functions of the form $y = f(x)$ can be entered in the TI-84 Plus CE using the Y= menu. To access the Y= menu, press the Y= key. Type the expression $f(x)$ after Y1 using the X,T,θ,n key for the variable x, and press ENTER.

For example, enter the function $f(x) = 3x^5 - 4x + 1$. Remember to press the right arrow key after you input the exponent 5.

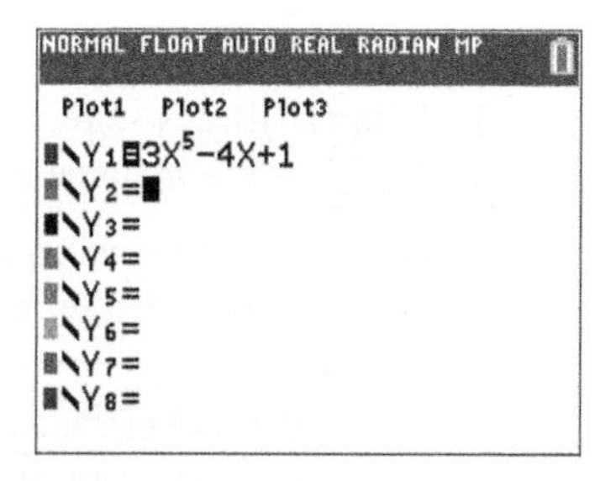

Note that the = sign after Y1 is highlighted. This indicates that the function Y1 is active and will be graphed when the graphing command is executed. The highlighting may be turned on or off by using the arrow keys to move the cursor to the = symbol and then pressing ENTER. Notice in the screen below that the Y1 has been deactivated and will not be graphed or appear in a table.

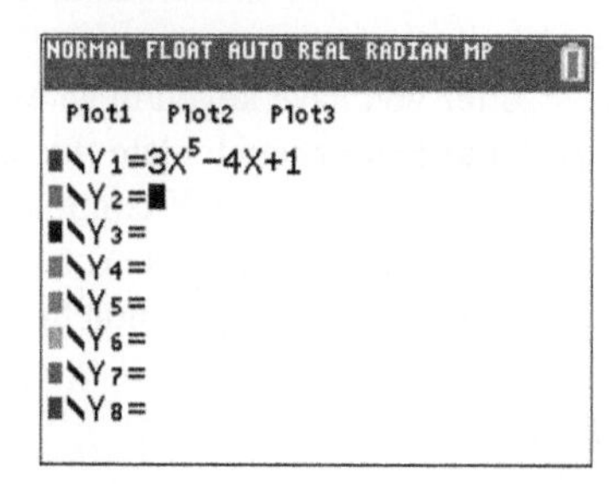

A function entered into Y_1 can be evaluated for specific values of x on the home screen. $Y_1(4)$ will evaluate the function in Y_1, $f(x) = 3x^5 - 4x + 1$, at $x = 4$, that is, $f(4)$.

With the cursor on the home screen, press the ALPHA TRACE buttons to access the F4 menu. Then press 1 to select Y_1 and place it at the cursor. Enter (4) ENTER to display the result.

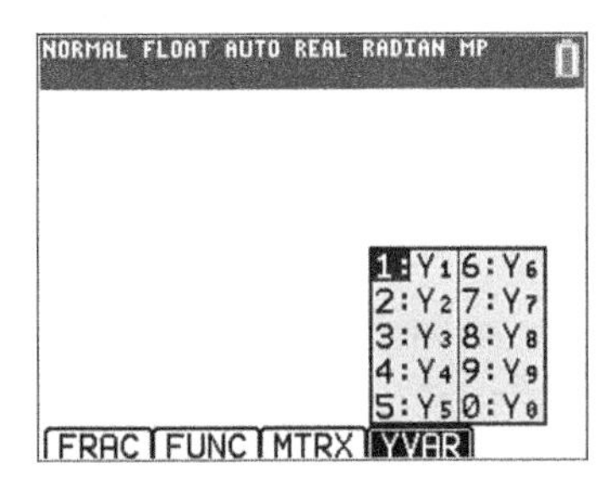

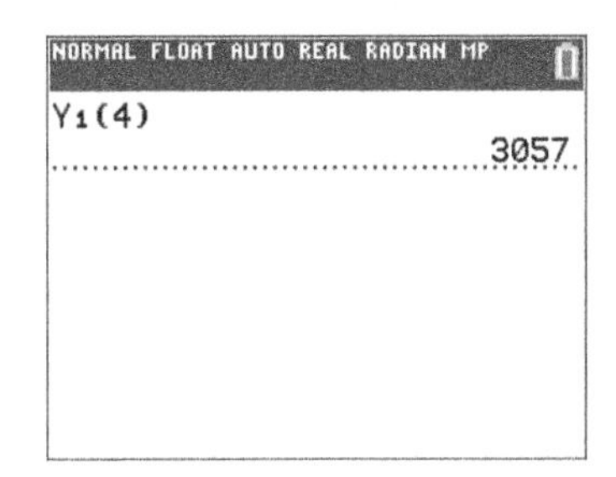

Tables of Values

If you are interested in viewing several function values for the same function, you may want to construct a table.

Before constructing the table, make sure the function appears in the "Y=" menu with its "=" highlighted. You may also want to deactivate or clear any functions that you do not need to see in your table. Next, you need to check the settings in the Table Setup menu. To do this, use the TBLSET command 2nd WINDOW.

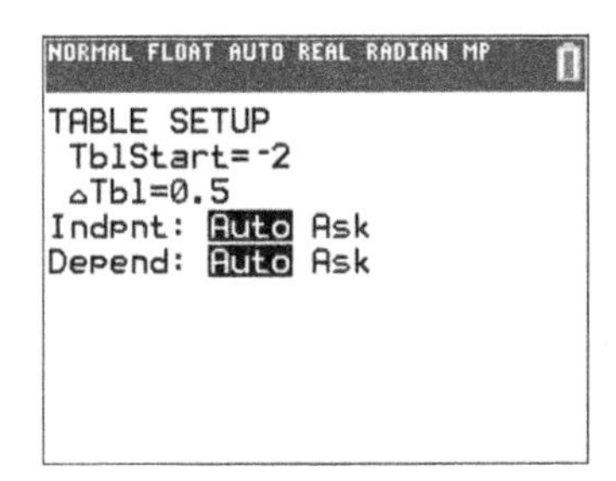

As shown in the screen above, the default setting for the table highlights the Auto options for both the independent (x) and the dependent (y) variables. Choosing this option will display ordered pairs of the function with equally spaced x-values. TblStart is the first x-value to be displayed and here is assigned the value -2. ΔTbl represents the equal spacing between consecutive x-values and here is assigned the value 0.5. The TABLE command 2nd GRAPH brings up the table displayed in the screen below.

NORMAL FLOAT AUTO REAL RADIAN MP
PRESS + FOR ΔTbl

X	Y1
-2	-87
-1.5	-15.78
-1	2
-0.5	2.9063
0	1
0.5	-0.906
1	0
1.5	17.781
2	89
2.5	283.97
3	718

X= -2

Use the up and down arrows to view additional ordered pairs of the function.

If the input values of interest are not evenly spaced, you may want to choose the Ask mode for the independent variable from the Table Setup menu.

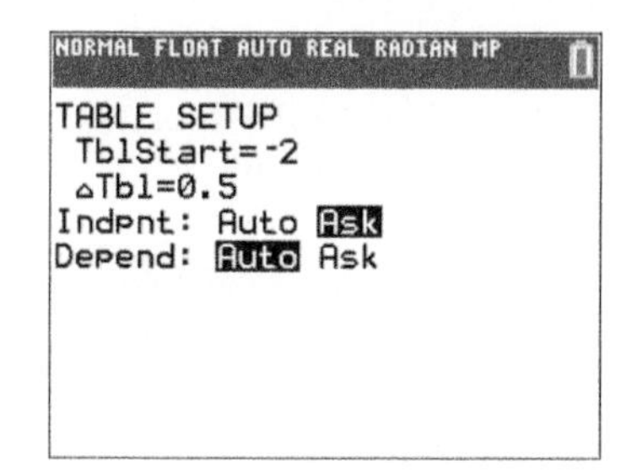

The resulting table is blank, but you can fill it by choosing any values for x that you like and pressing (ENTER) after each value.

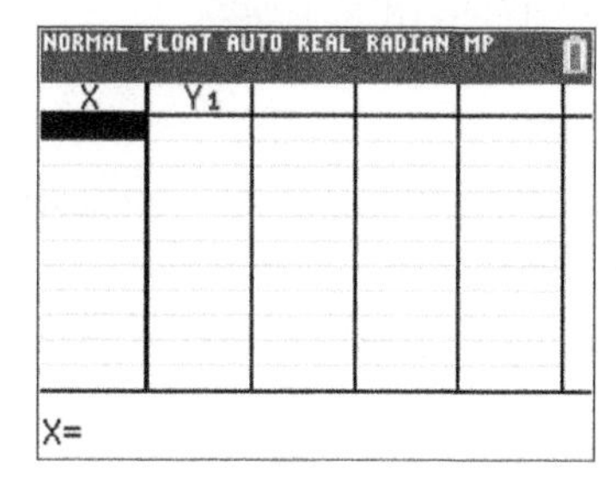

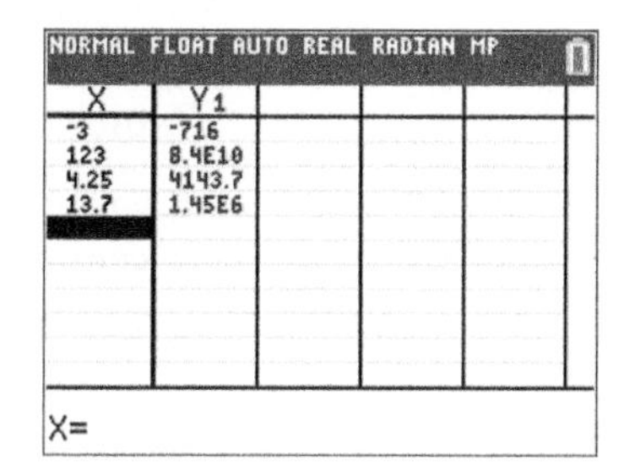

Note that the number of digits shown in the output is limited by the table width. But if you want more digits, move the cursor to the desired output, more digits then appear at the bottom of the screen.

Graphing a Function

Once a function is entered in the "Y=" menu and activated, it can be displayed and analyzed. For this discussion, we will use the function $f(x) = -x^2 + 10x + 12$. Enter this as Y1, making sure to use the negation key ((–)) and not the subtraction key (–).

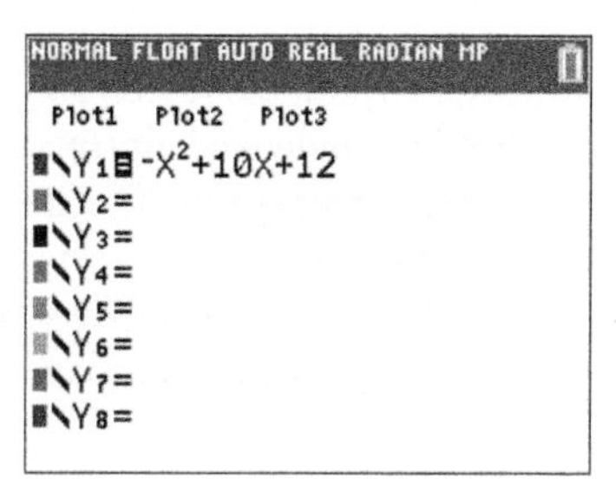

The Viewing Window

The viewing window is the portion of the rectangular coordinate system that is displayed when you graph a function.

Xmin defines the left edge of the window.

Xmax defines the right edge of the window.

Xscl defines the distance between horizontal tick marks.

Ymin defines the bottom edge of the window.

Ymax defines the top edge of the window.

Yscl defines the distance between vertical tick marks.

In the standard viewing window, Xmin $= -10$, Xmax $= 10$, Xscl $= 1$, Ymin $= -10$, Ymax $= 10$, and Yscl $= 1$.

To select the Standard Viewing Window, press ZOOM 6.

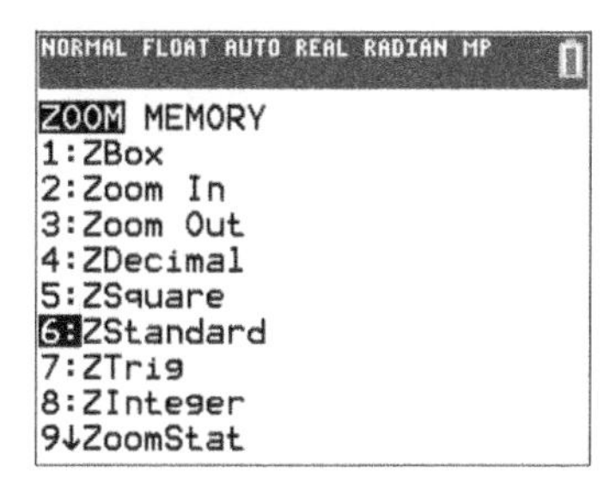

You will see the following.

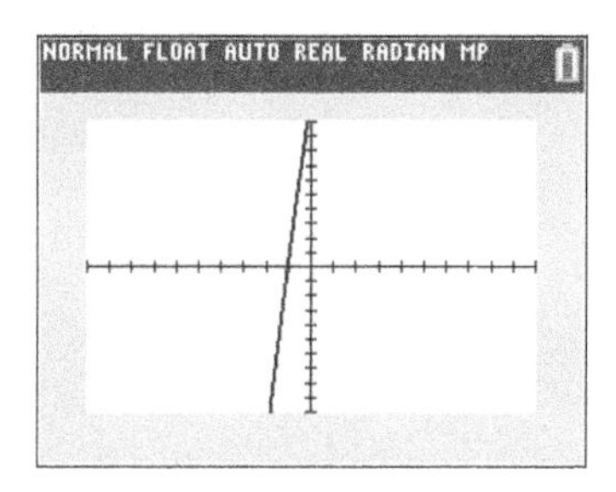

Is this an accurate and/or complete picture of your function, or is the window giving you a misleading impression? You may want to use your table function to view the output values from -10 to 10.

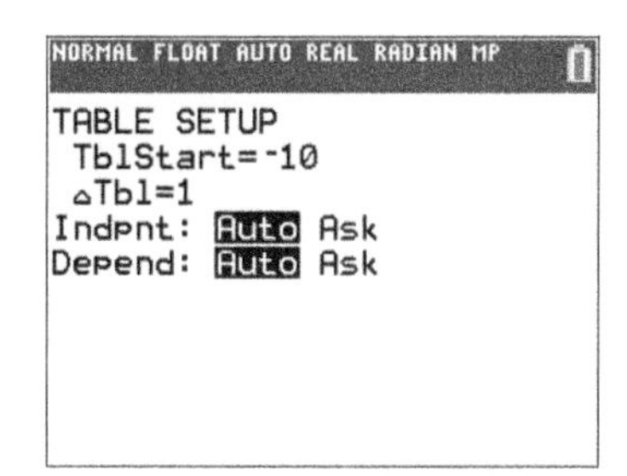

NORMAL FLOAT AUTO REAL RADIAN MP
PRESS + FOR ΔTbl

X	Y1
-10	-188
-9	-159
-8	-132
-7	-107
-6	-84
-5	-63
-4	-44
-3	-27
-2	-12
-1	1
0	12

X=-10

NORMAL FLOAT AUTO REAL RADIAN MP
PRESS + FOR ΔTbl

X	Y1
0	12
1	21
2	28
3	33
4	36
5	37
6	36
7	33
8	28
9	21
10	12

X=5

NORMAL FLOAT AUTO REAL RADIAN MP
PRESS + FOR ΔTbl

X	Y1
0	12
1	21
2	28
3	33
4	36
5	37
6	36
7	33
8	28
9	21
10	12

X=10

The table indicates that the minimum output value on the interval from $x = -10$ to $x = 10$ is -188, occurring at $x = -10$, and that the maximum output value is 37, occurring at $x = 5$. Press WINDOW and reset the settings to approximately the following.

$$\text{Xmin} = -10, \text{Xmax} = 10, \text{Xscl} = 1,$$

$$\text{Ymin} = -190, \text{Ymax} = 40, \text{Yscl} = 10$$

Press GRAPH to view the graph with these new settings.

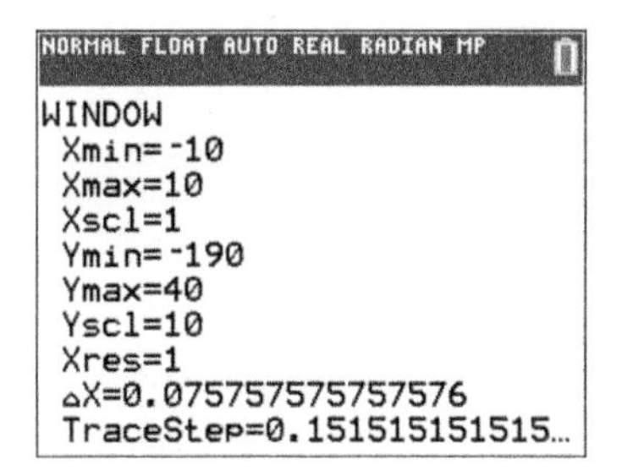

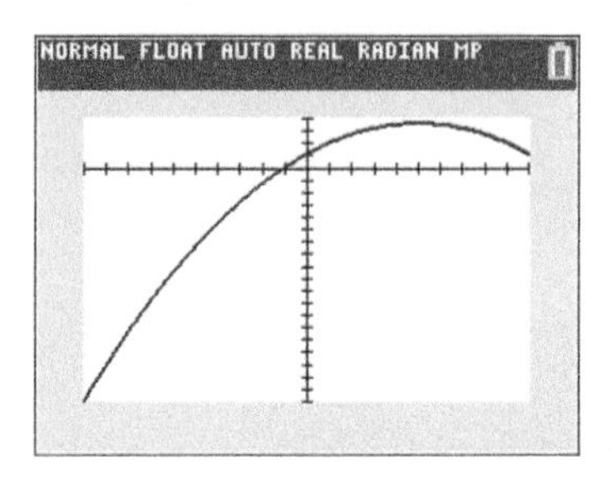

The new graph gives us a more complete picture of the behavior of the function on the interval $[-10, 10]$.

The coordinates of specific points on the curve can be viewed by activating the trace feature. While in the graph window, press TRACE. The function equation will be displayed at the top of the screen, a flashing cursor will appear on the curve at the middle of the screen, and the coordinates of the cursor will be displayed at the bottom of the screen.

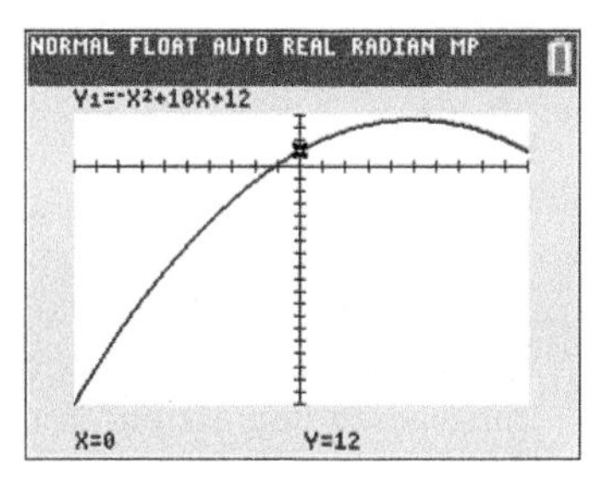

The left arrow key will move the cursor toward smaller input values. The right arrow key will move the cursor toward larger input values. If the cursor reaches the edge of the window and you continue to move the cursor, the window will adjust automatically.

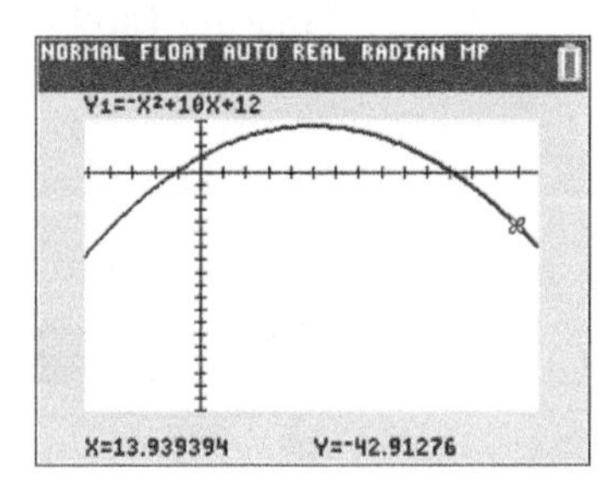

Zoom Menu

The Zoom menu offers several options for changing the window very quickly.

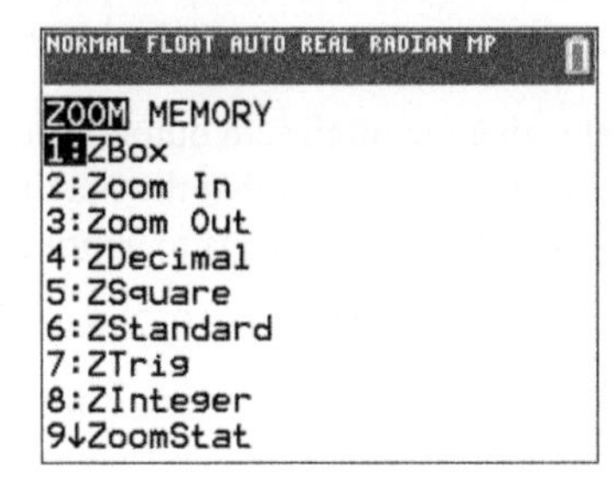

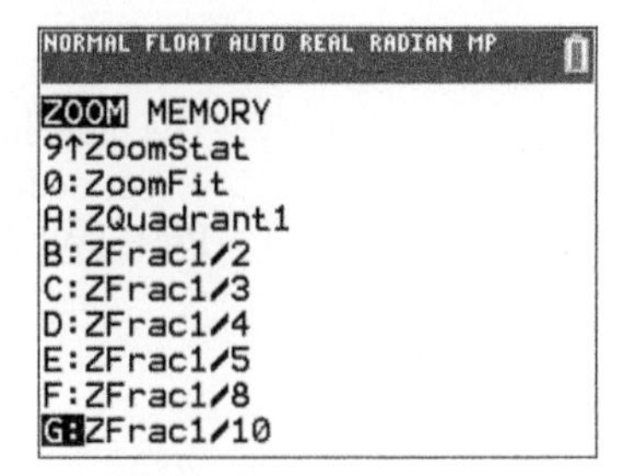

The features of many of the commands are summarized in the following table.

ZOOM COMMAND	DESCRIPTION
1:ZBox	Draws a box to define the viewing window
2:Zoom In	Magnifies the graph near the cursor
3:Zoom Out	Increases the viewing window around the cursor
4:ZDecimal	Sets a window so that Xscl and Yscl are 0.1
5:ZSquare	Sets equal-size pixels on the *x*- and *y*-axes
6:ZStandard	Sets the window to standard settings
7:ZTrig	Sets built-in trig window variables
8:ZInteger	Sets integer values on the *x*- and *y*-axes
9:Zoom Stat	Sets window based on the current values in the stat lists
0:ZoomFit	Replots graph to include the max and min output values for the current Xmin and Xmax

Solving Equations Graphically Using the TI-84 Plus CE

The Intersection Method

This method is based on the fact that solutions to the equation $f(x) = g(x)$ are input values of x that produce the same output for the functions f and g. Graphically, these are the x-coordinates of the intersection points of $y = f(x)$ and $y = g(x)$.

The following procedure illustrates how to use the intersection method to solve $x^3 + 3 = 3x$ graphically.

Step 1. Enter the left-hand side of the equation as Y_1 and the right-hand side as Y_2 in the "Y=" editor. Select the standard viewing window.

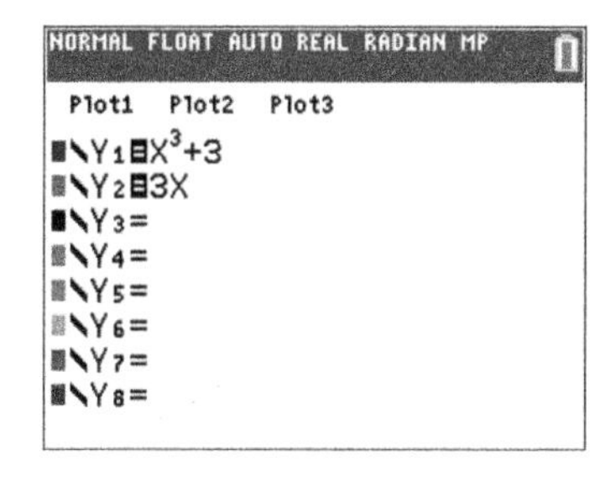

Step 2. Examine the graphs to determine the number of intersection points.

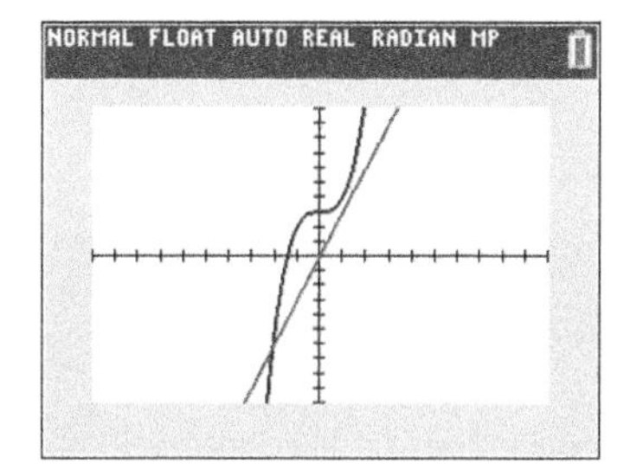

You may need a couple of windows to be certain of the number of intersection points.

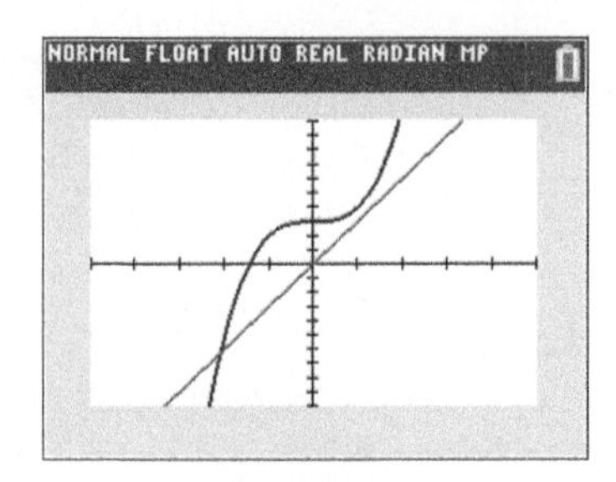

Step 3. Access the Calculate menu by pressing (2nd) (TRACE), and then choose option 5: **intersect**.

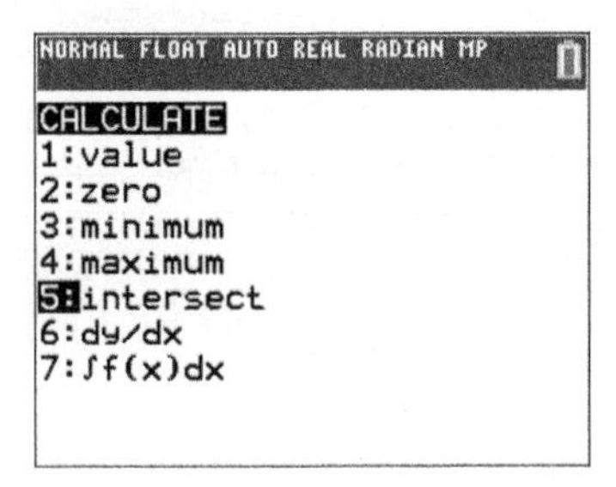

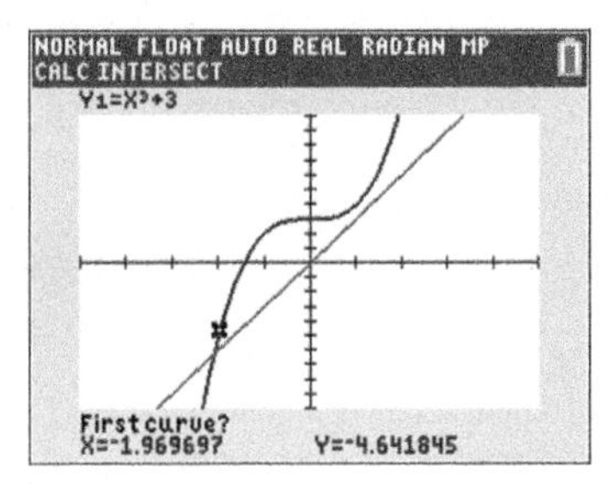

Step 4. Move the cursor close to an intersection point for the first curve and press (ENTER).

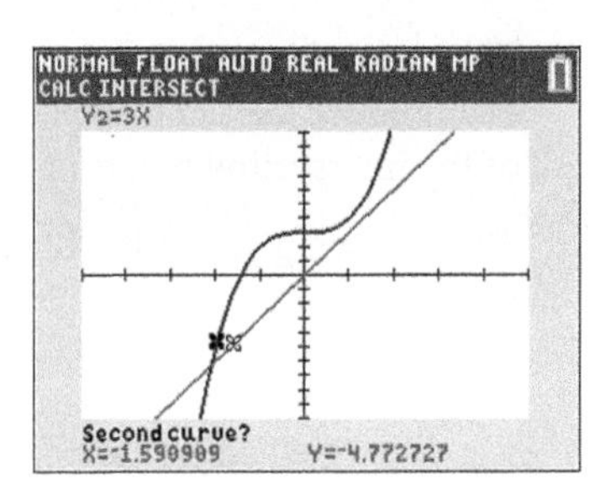

Step 5. Repeat Step 4 for the second curve.

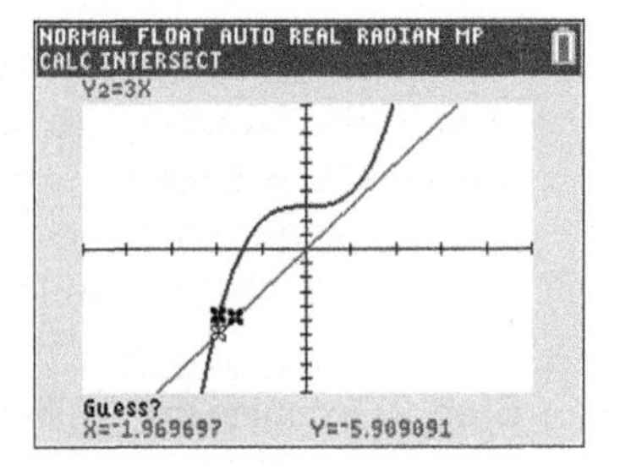

Step 6. To use the cursor's current location as your guess, press ENTER in response to the question on the screen that asks Guess? If you want to move to a better guess, do so before you press ENTER. You can also enter a Guess using the calculator keypad and pressing ENTER.

The coordinates of the intersection point appear below the word *Intersection*.

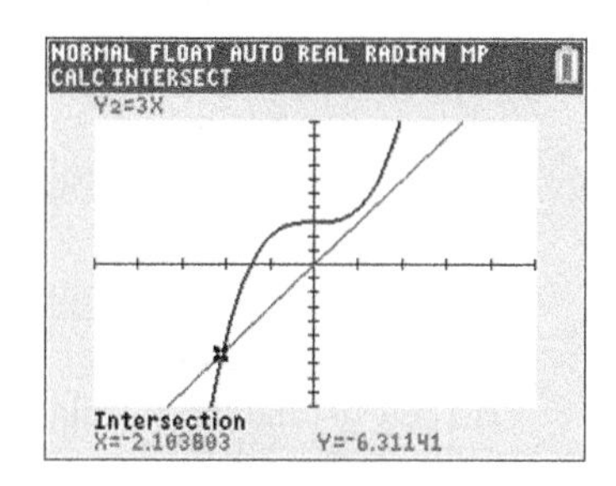

The x-coordinate is a solution of the equation.
If there are other intersection points, repeat the process as necessary.

Finding Zeros of a Function

A **zero** of a function, f, is the value, a, such that $f(a) = 0$. In words, a zero of a function is a value of the input that yields an output value of zero. $f(a) = 0$ corresponds to the ordered pair $(a, 0)$ which is on the horizontal axis so the value a is a horizontal intercept. **Note:** a function may have one or more zeros or no zero.

The graph of the function, $f(x) = x^2 - 4x - 5$, in the window $-5 \leq x \leq 10$ and $-15 \leq y \leq 10$, is given below.

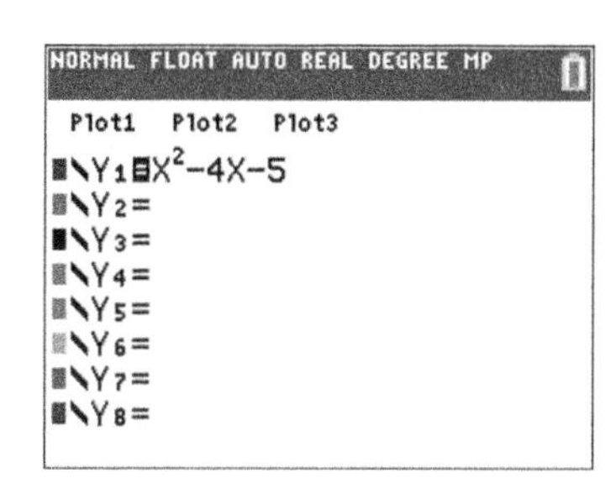

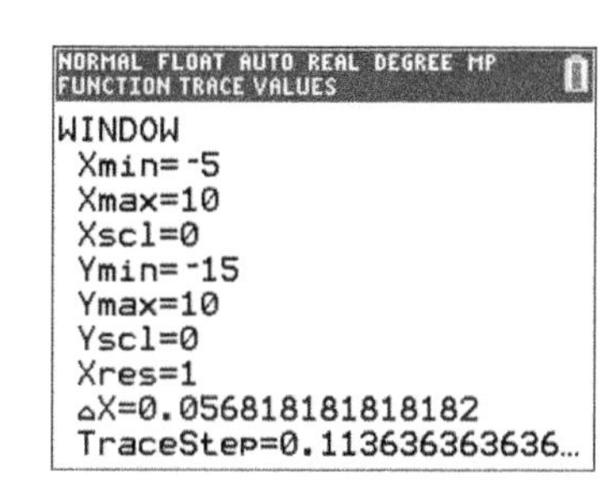

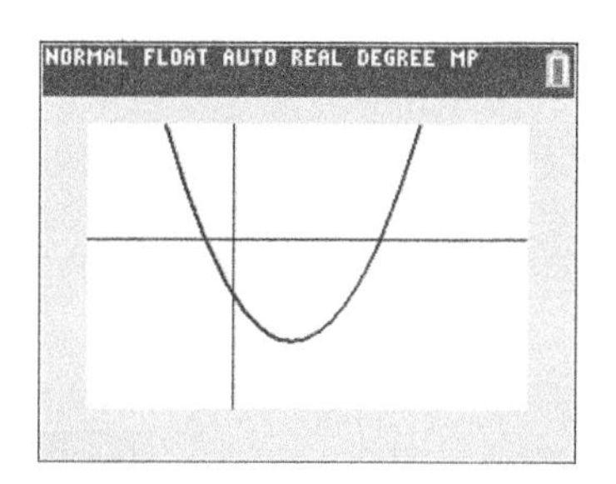

The graph shows 2 zeros. Use the following procedure to determine the zero to the left of the y-axis. The calculator will look for a zero between two x-values, one to the left or less than the zero (Left Bound) and one to the right or greater than the zero (Right Bound). With the graph displayed on the calculator go to the calc menu by pressing 2nd–TRACE. Zero is the second menu item so press 2.

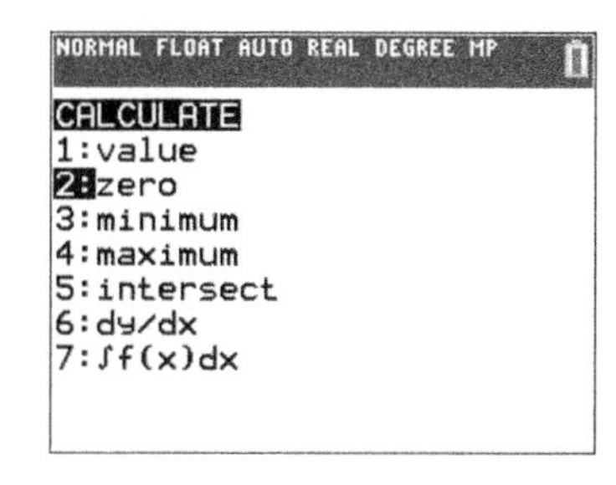

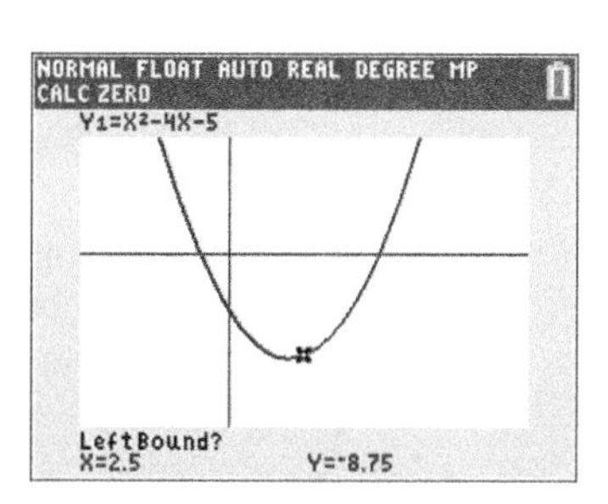

The calculator now asks for the Left Bound. Use the arrows to move the cursor to the left of the zero and hit (ENTER). The calculator now asks for the Right Bound. Use the arrows to move the cursor to the right of the zero.

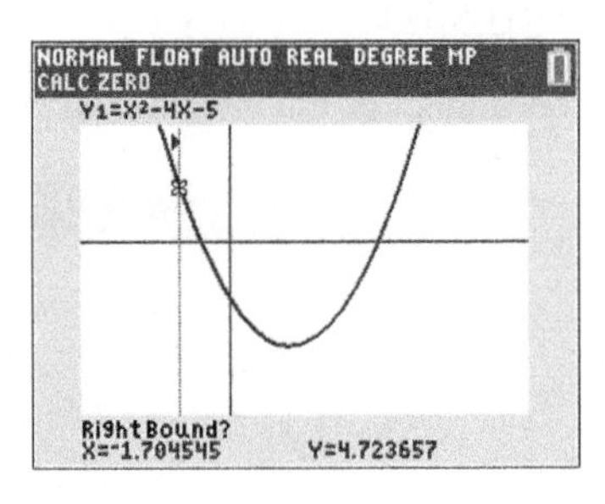

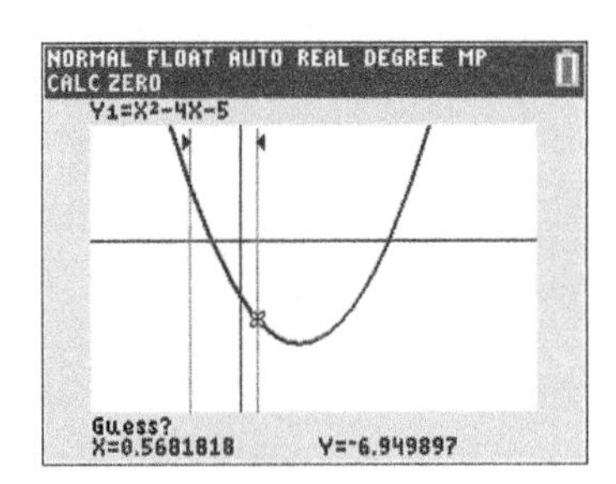

Hit (ENTER) and the calculator now asks for a guess. Move the cursor closer to the zero and hit (ENTER). The calculator now displays the zero, $x = -1$.

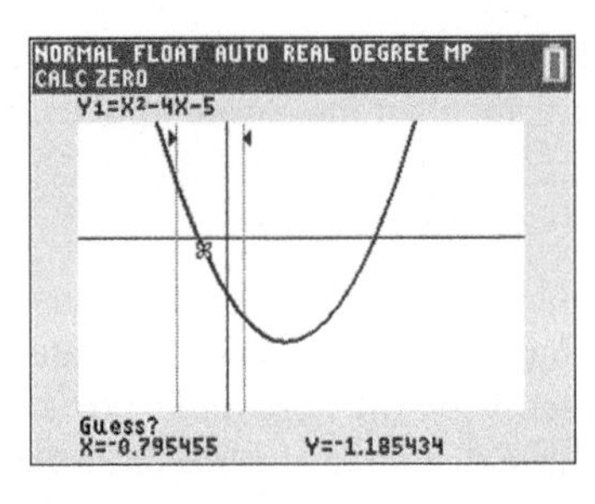

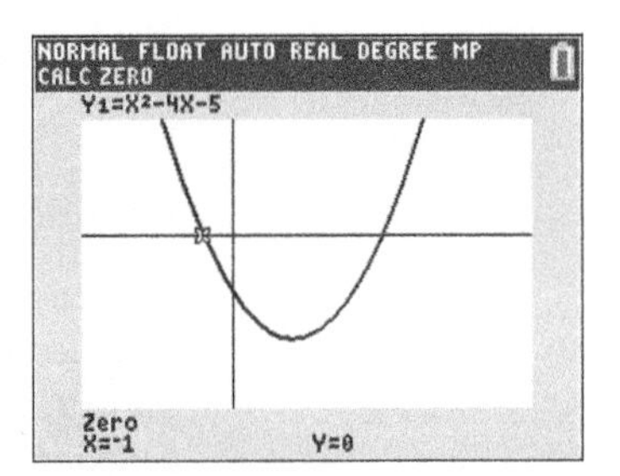

Use the same procedure to determine the zero to the right of the y-axis.

Finding Minimum/Maximum Values

The graph of the function $f(x) = x^2 - 4x - 5$ indicates that the function has a minimum value. The minimum value is the y-value (output) of the turning point. The procedure used to determine the minimum value is similar to the procedure for finding zeros.

The graph of the function, $f(x) = x^2 - 4x - 5$, in the window $-5 \le x \le 10$ and $-15 \le x \le 10$, is given below.

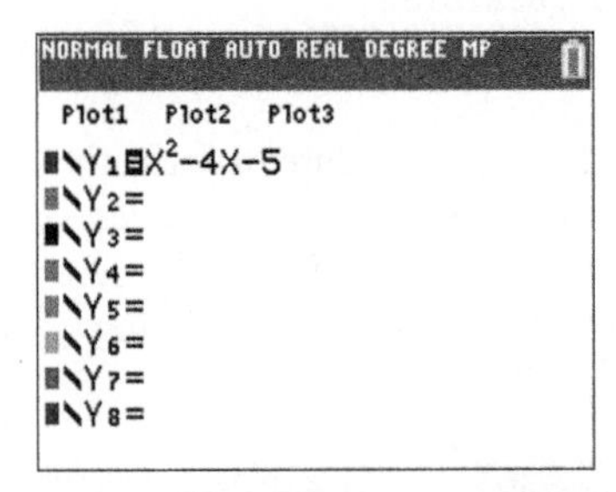

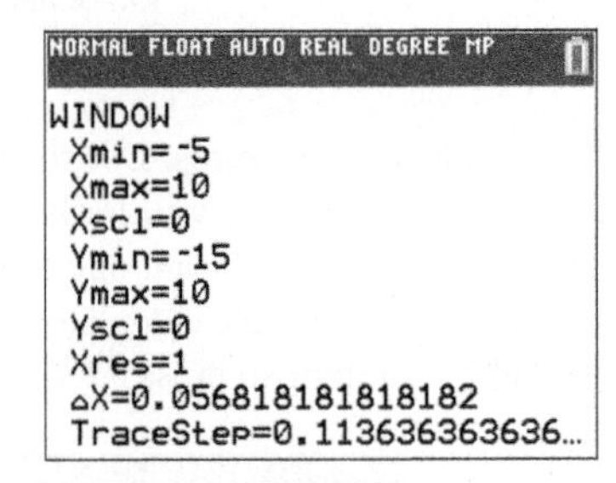

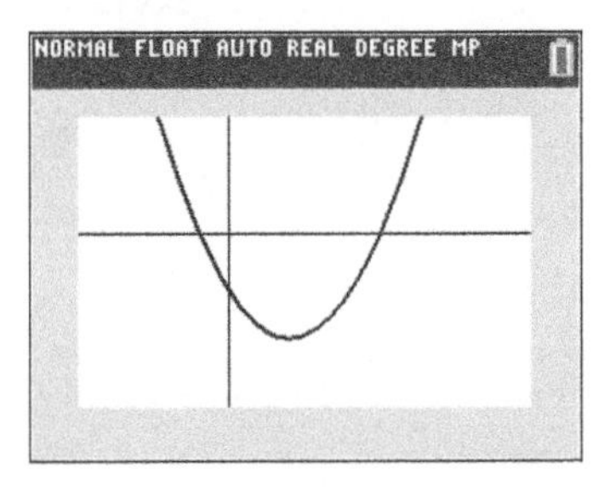

Use the following procedure to determine the minimum value of the function. The calculator will look for the minimum between two x-values, one to the left or less than the minimum (Left Bound) and one to the right or greater than the minimum (Right Bound). With the graph displayed on the calculator go to the calc menu by pressing (2nd)–(TRACE). Minimum is the third menu item so press (3).

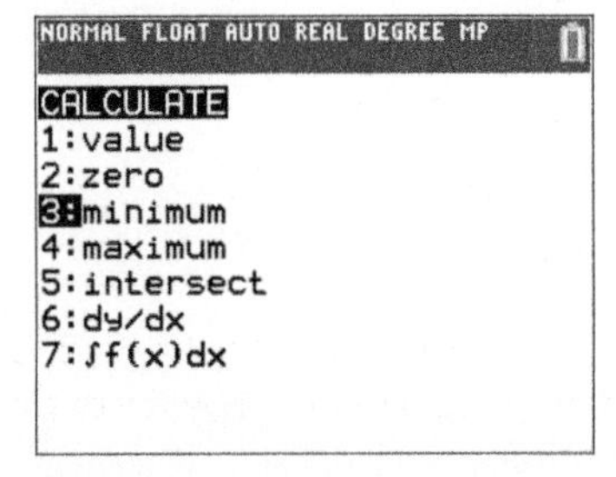

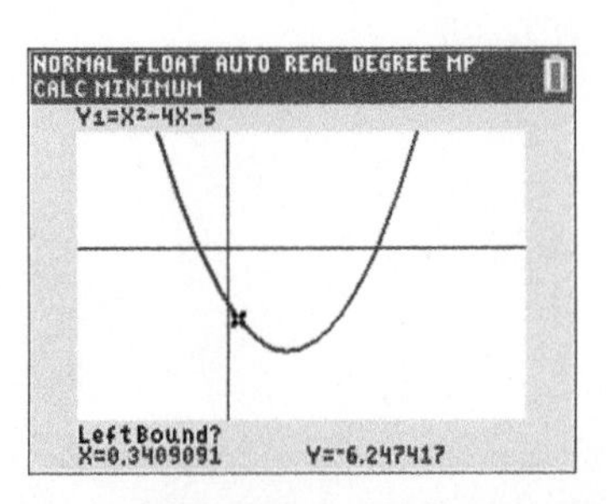

The calculator now asks for the Left Bound. Use the arrows to move the cursor to the left of the zero and hit (ENTER). The calculator now asks for the Right Bound. Use the arrows to move the cursor to the right of the zero.

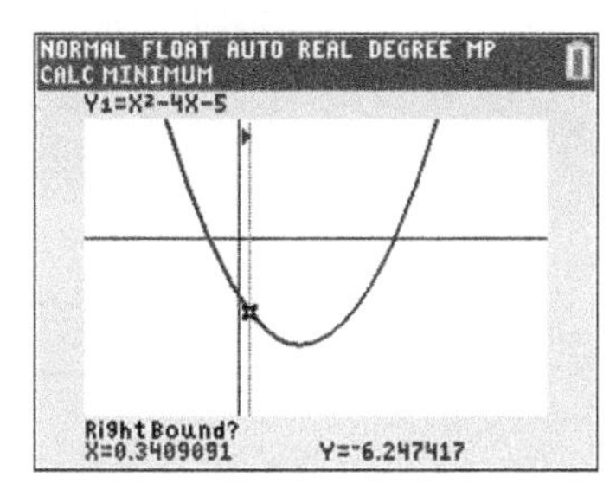

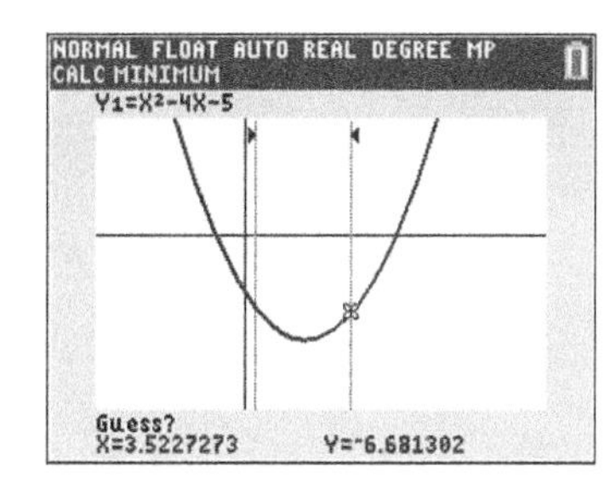

Hit (ENTER) and the calculator now asks for a guess. Move the cursor closer to the minimum and hit (ENTER). The calculator now displays the minimum value, $y = -9$ at the input value, $x = 2$.

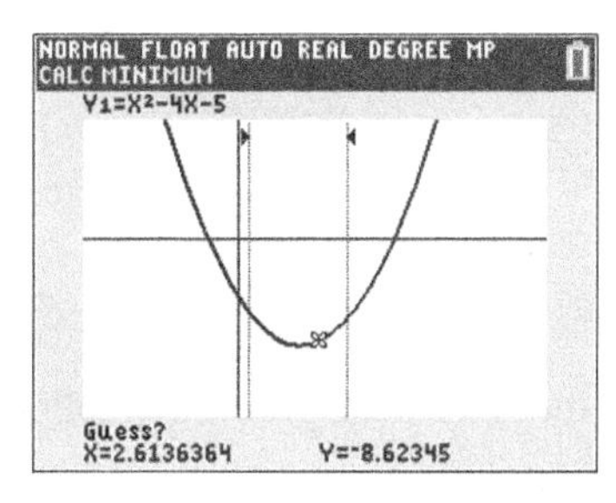

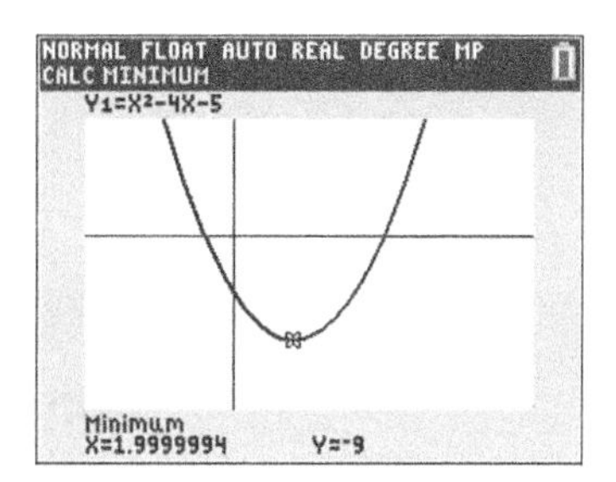

Note: The minimum is $(2, -9)$. The calculator displays $x = 1.99999$ which rounds to 2.

The procedure to determine a maximum value is the same as the procedure to determine the minimum value.

Using the TI-84 Plus CE to Determine the Linear Regression Equation for a Set of Paired Data Values

Example 1:

INPUT	OUTPUT
2	2
3	5
4	3
5	7
6	9

Enter the data into the calculator as follows:

1. Press the (STAT) button, and choose edit.

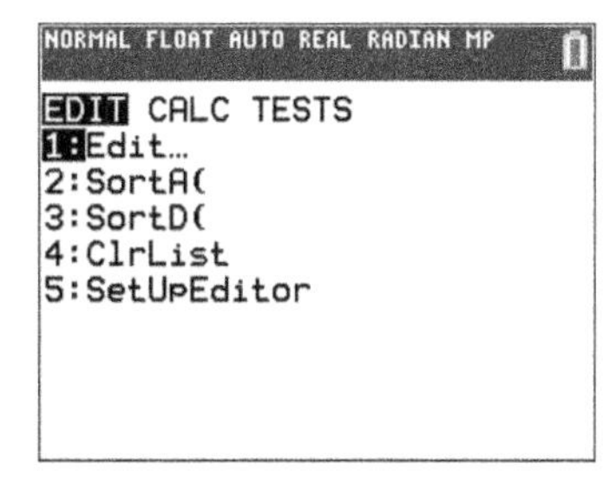

2. The TI-84 Plus CE has six built-in lists: L1, L2, . . . , L6. If there is data in L1, clear the list as follows:

   a. Use the arrows to place the cursor on L1 at the top of the list. Press (CLEAR) followed by (ENTER) followed by the down arrow.

b. Follow the same procedure to clear L2, if necessary.

c. Enter the input values into L1 and the corresponding output values into L2.

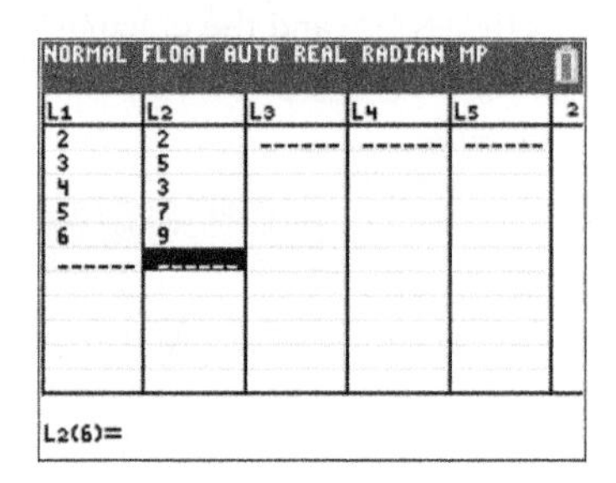

To see a scatterplot of the data, proceed as follows:

1. STAT PLOT is the second function of the (Y=) button. You must press the (2nd) button before pressing the (Y=) button to access the Stat Plot menu.

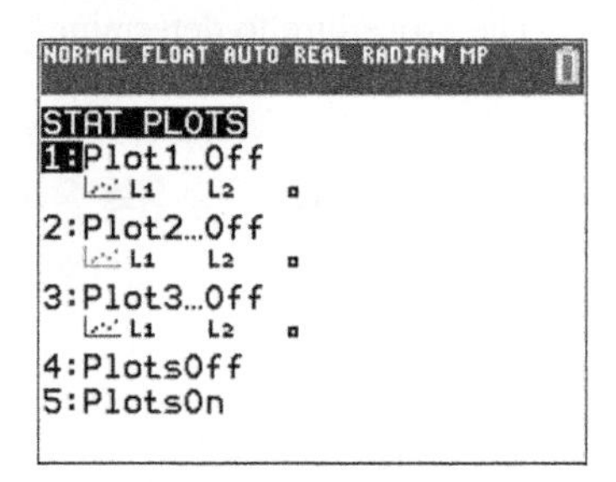

2. Select Plot 1, and make sure the other Plots are OFF. The screen below will appear. Select On and then choose the scatterplot option (first icon) on the Type line. Confirm that your x- and y-values are stored, respectively, in L1 and L2. The symbols L1 and L2 are second functions of the (1) and (2) keys, respectively. Finally, select the small square as the mark that will be used to plot each point.

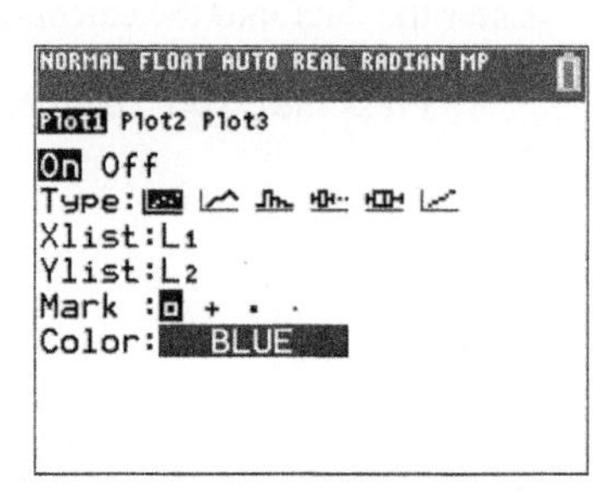

3. Press (Y=) and clear or turn off any functions entered.

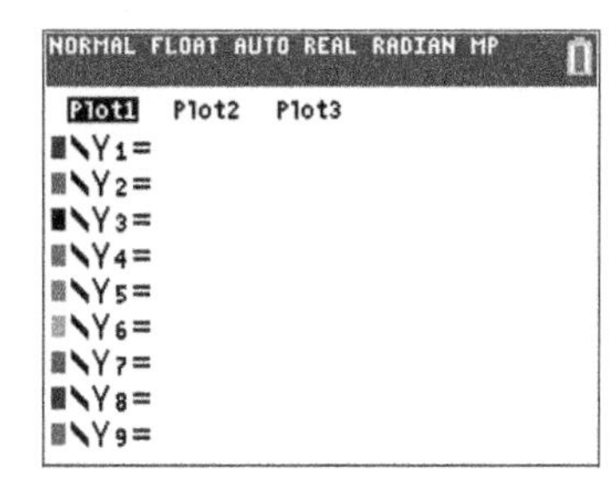

4. To display the scatterplot, have the calculator determine an appropriate window by pressing (ZOOM) and then (9) (ZoomStat).

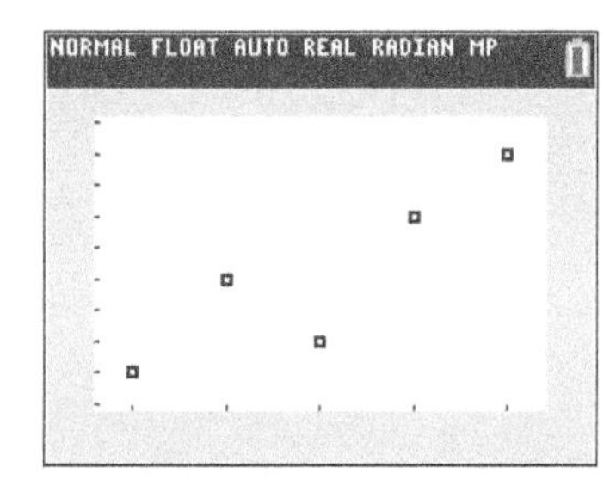

The following instructions will calculate the linear regression equation and store it in Y1.

1. Press (STAT) and right-arrow to CALC.

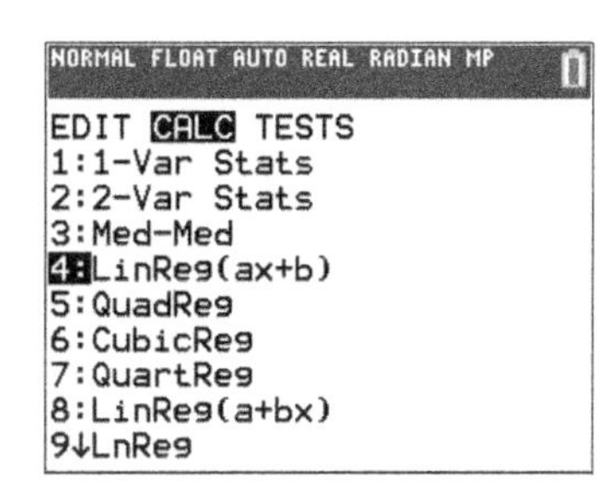

2. Choose 4 LinReg(ax+b). To tell the calculator where the data is, press (2nd) and (1) (for L1), then (,), and then (2nd) and (2) (for L2) because the Xlist and the Ylist were stored in L1 and L2, respectively. The display looks like this.

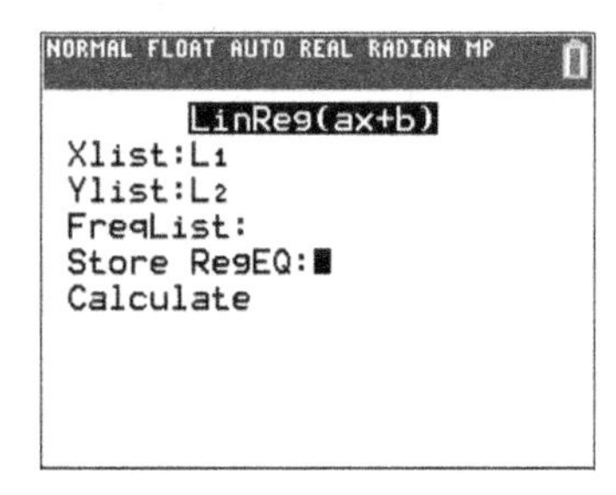

If you press (ENTER) until Calculate is highlighted, the calculator will calculate the linear regression equation from the data in L1 and L2. However, the TI-84 Plus C will automatically paste the equation into Y1 (or Y2, Y3, . . .) if you do the following:

3. Arrow down to Store RegEQ: Press the ALPHA TRACE buttons to access the F4 menu.

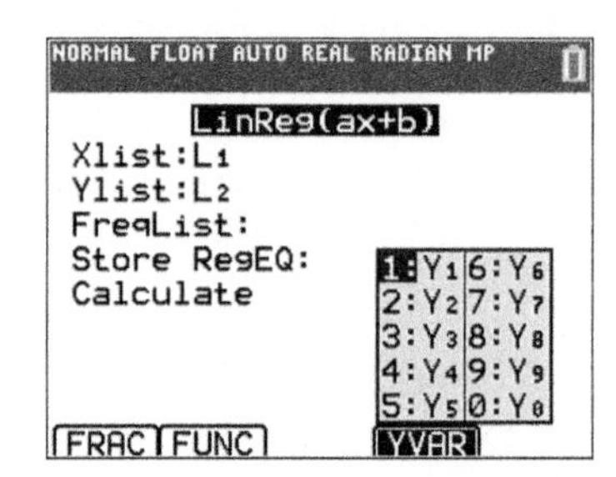

4. Then press 1 to select Y_1 (or 2 for Y_2, etc.) and place it at the cursor.

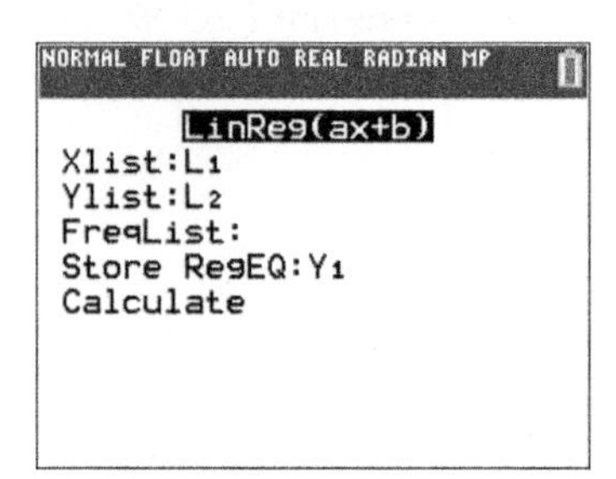

5. Move the cursor to Calculate and press ENTER.

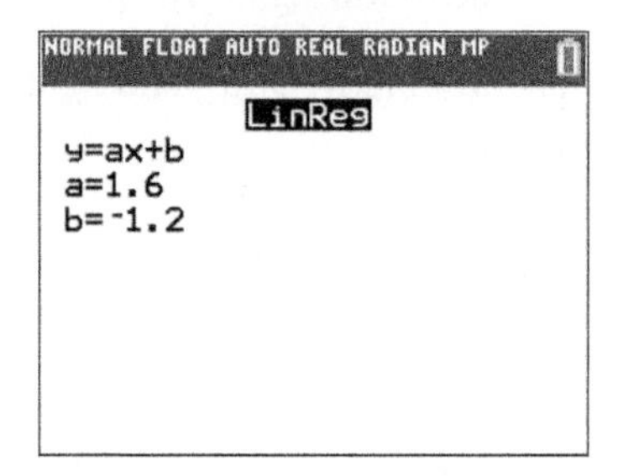

The linear regression equation for this data is $y = 1.6x - 1.2$.

6. To display the regression line, press GRAPH.

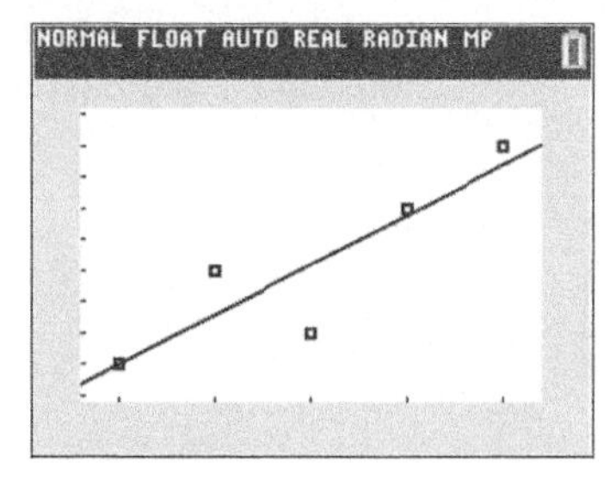

7. Press the Y= key to view the equation.

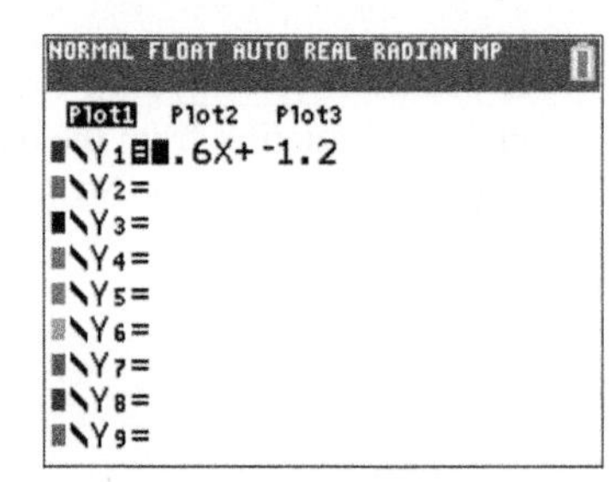

Selected Answers

Chapter 1

Activity 1.1 Exercises: 1. a. Weight is input, and height is output. **c.** Height is input, and weight is output. **3. a.** Yes, for each numerical grade there will correspond only one letter grade. **b.** No, for each letter grade there may correspond several different numerical grades. For example, an A may be based on 92% or 94%. **c.** This is not a function. For each number of hours studied for the exam, there may correspond several different scores. For example, a study time of 3 hours could result in a score of 71, 82, or 94. **d.** Yes, this is a function. For each date, there is just one corresponding number of tweets. **e.** This situation represents a function. For any given final selling price, there is just one corresponding amount of tax. **5. a.** Yes, in this table, each elevation is paired with only one amount of snowfall. **b.** Yes, in this table, each quantity of snow is paired with one elevation. **7. a.** Yes, each input value has only one output value. **b.** No, the one input 5 is paired with four different outputs. **9. a.** The input is x, the output is $g(x)$ or y. The function name is g, y equals g of x, **c.** The input is 6. The output is 3.527. The name is f, f of 6 equals 3.527. **e.** The input is price. The output is sales tax. The name is T. Sales tax is a function of price. **10. a.** 1600; 2400, **b.** $f(6) = 2400$

Activity 1.2 Exercises: 1. a. The independent (input) variable is the price of an item. The dependent (output) variable is the sales tax. **b.** $h(x) = 0.08x$; **3.** $f(2) = -1, f(-3.2) = -11.4, f(a) = 2a - 5$; **5.** $f(2) = 4, f(-3.2) = 4, f(a) = 4$;

7.

x	$h(x)$
10	0.1
20	0.05
30	0.033
40	0.025

9. a. The distance traveled is 2 times the number of hours I have hiked. **b.** The input is hours. The output is distance. **d.** $h(t)$ is the dependent variable because distance is the output. **f.** $h(7) = 2(7) = 14$. $(7, 14)$. If I hike for 7 hours, I expect to travel 14 miles. **h.** The practical domain depends on the individual and in this situation is probably real numbers from 0 to about 8. Using this domain, the range is real numbers from 0 to 16. **10. a.** domain $\{-2, 0, 5, 8\}$, range $\{3, 4, 8, 11\}$; **b.** domain $\{-6, -2, 0, 3\}$; range $\{5\}$ **11. a.** $0 \le C \le 100$, **b.** $32 \le F \le 212$

Activity 1.3 Exercises:

1. a.

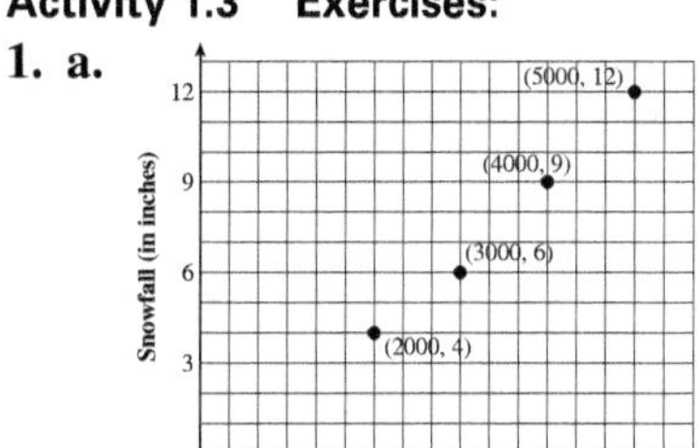

b. Continuous because the amount of snowfall is defined at any elevation. **3.** Graph d; **4.** Graph b; **5.** Graph a; **6.** Graph c; **7. a.** No graph appears on the screen. **b.** 24, 19, 16, 15, 16, 19, 24, **c.** The y-values are increasing from a minimum of 15. Therefore, you need to have Ymax of at least 30.

d.

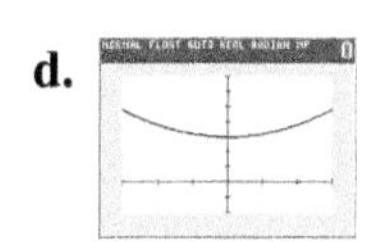

Activity 1.4 Exercises: 1. a. 3.7, 6.4, 9.1, 11.8 **b.** $f(2500) = 7.75$ inches of snow, At an elevation of 2500 feet, there were approximately 7.75 inches of snow. **c.** $f(-2000) = -4.4$ It has no meaning in this context. -2000 would mean 2000 feet below sea level, but -4.4 inches of snow is not possible. **d.** Xmin $= 0$, Xmax $= 5000$, Ymin $= 0$, Ymax $= 15$, **e.** Yes, any vertical line will intersect the graph no more than once. **f.** increasing, **g.** It is the same, 7.75. **3. a.** What is the value of the truck after a certain number of years? **b.** The value of the truck and the number of years of ownership, **c.** The independent variable is the number of years you own the truck. The dependent variable is the value of the truck. **d.** 28,000, 24,500, 21,000, 17,500, 14,000 **e.** The value of the truck is obtained by subtracting the product of 3500 and the number of years from 31,500. **f.** Let v represent the value of the truck and t represent the number of years. $v = 31{,}500 - 3500t$, **g.** Using the equation from part f, $v = 31{,}500 - 3500(7) = \$7000$. Using the table from part d, I could subtract 3500 twice from the value of the truck after 5 years. **h.** The rate of depreciation was a constant 3500 dollars per year. No, the truck would depreciate more when it's newer and has greater value.

4. a.

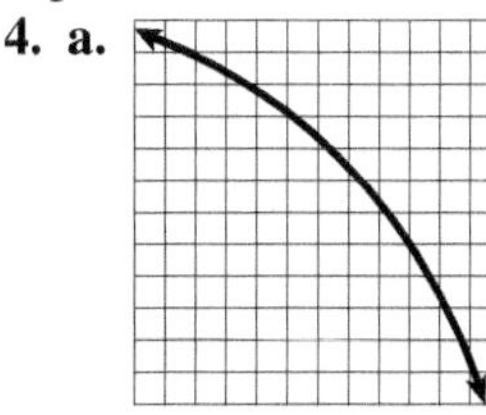

b. The graph is a horizontal line. **5. a.** This is a function. **b.** This is not a function. **6.** Xmin $= -9$, Xmax $= 9$, Ymin $= -54{,}000$, Ymax $= 54{,}000$, answers may vary.

Activity 1.5 Exercises: 2. a. Number of attempts, time in seconds, **b.** The time required to complete a task decreases as the number of attempts increases. As a person attempts a task more times, the task takes less time to complete. After five attempts, no further improvement is made. **3. a.** Selling price, number of units sold, **b.** As the selling price increases, the number of units sold increases slightly at first, reaches a maximum, and then declines until none are sold.

How Can I Practice? 1. a. Yes, because for each point total there is only one grade. **b.** Yes. For each numerical grade in the table, there is one value of total points. **c.** $f = \{(432, 86.4), (394, 78.8), (495, 99), (330, 66), (213, 42.6)\}$,

d.

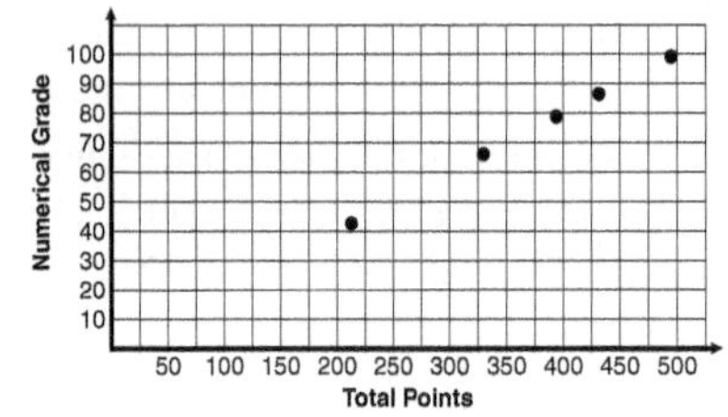

e. $f(394) = 78.8$,

f. It is the numerical grade that corresponds to 394 points. **g.** $f(213) = 42.6$, **h.** It is the numerical grade that corresponds to 213 points. **i.** $n = 330$; **2.** This is a function. **3.** This could be a function, depending on how activity level is measured. **4.** This is a function. **5.** This is not a function. **6.** This is a function. **7.** This is not a function. The input -3 has two different outputs. **8.** This is a function. **9.** This is a function. **10.** This is not a function. **11. a.** $c = 245h$, **b.** $f(h) = 245h$, **c.** 490, 980, 1715, 1960, 2695, **d.** \$735, (3,735), **e.** $h = 5$ hours, **f.** $f(h)$ or c is the output variable. This is the variable that depends on the number of credits taken. **g.** h is the independent variable. It is the input variable. **h.** For each value of input, there is one value of output. **i.** Assuming there care no half-credit courses, the practical domain is all whole numbers from 0 to 11. **j.** The horizontal axis represents the input.

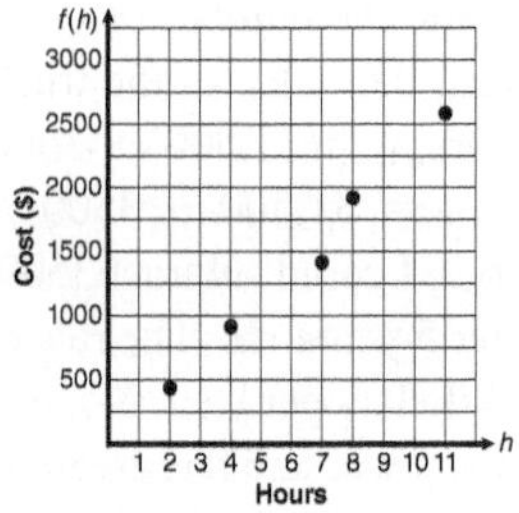

k. $f(h)$ is a function because the graph passes the vertical line test, **m.** \$2205; **12. a.** $p(3) = 13$, **b.** $p(-4) = -1$, **c.** $p\left(\frac{1}{2}\right) = 8$, **d.** $p(0) = 7$; **13. a.** $t(2) = -3$, **b.** $t(-3) = 18 + 9 - 5 = 22$; **14. a.** 65.0, 66.9, 68.8, 71.7, 74.5, 76.4, 78.3, **b.** The life expectancy for a male born in 1980 will be 70.7 years (answers may vary). **c.** Xmin = 0, Xmax = 75, Ymin = 60, Ymax = 80, **d.** increasing, **e.** They are the same. **15. a.** domain $\{3, 4, 5, 6\}$, range $\{5, 8, 10\}$, **b.** domain $\{2008, 2009, 2010, 2011, 2012, 2013, 2014\}$, range $\{4.0, 4.4, 4.7, 5.3, 6.0, 6.4, 6.9\}$, **c.** domain $-3 \le x \le 4$, range $-1 \le y \le 3$, **d.** domain $-3 \le x \le 3$, range $0 \le y \le 4$, **e.** Domain is all real numbers; range is all real numbers. **16. a.** The net profit increases during the first two quarters of 2009. The net profit then decreases for about two quarters, and then it increases through the final quarter of 2010. **b.** The annual income rises rather steadily for three years; in the fourth year, it rises sharply. Then it suffers a sharp decline during the next year. During the last year, the income recovers to about the point it was originally. **17.**

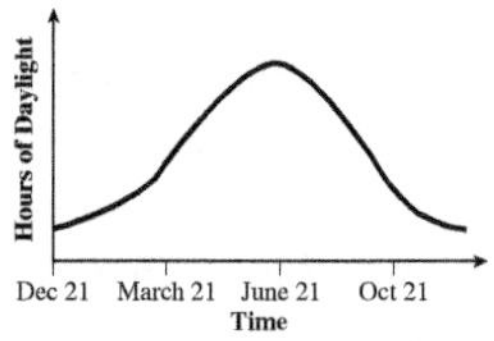

Activity 1.6 Exercises: 1. a. $\frac{28.2 - 22.8}{2010 - 1950} = \frac{5.4}{60} = 0.09$ years of age/year, **b.** The median age of a man at the time of his first marriage is increasing at an average rate of 0.09 years of age/year. **3.** $\frac{29.2 - 24.6}{95} = \approx 0.048$ years of age/year; **5. a.** It means that the median age of a man at the time of his first marriage is decreasing. **b.** 1920–1930 or 1940–1950, **c.** The graph would fall to the right.

7. a.

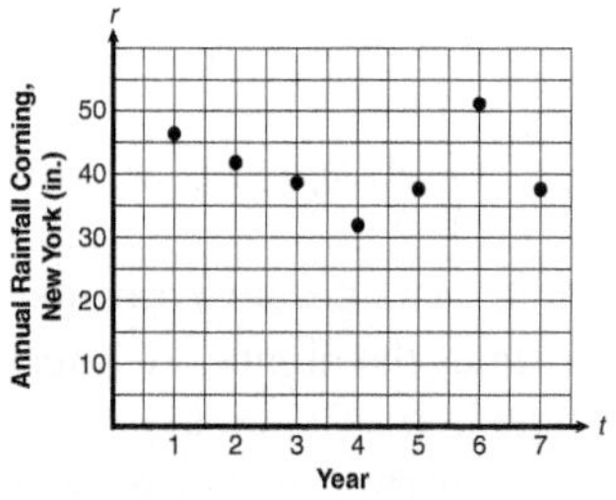

b. $\frac{37.54 - 32.91}{7 - 4} = \frac{4.63}{3} \approx 1.54$ inches/year,

c. $\frac{32.91 - 45.49}{4 - 1} = \frac{-12.58}{3} \approx -4.19$ inches/year;

8. a. $\frac{760 - 668}{1970 - 1960} = \frac{92}{10} = 9.2$ gallons/year,

b. $\frac{520 - 668}{1990 - 1960} = \frac{-148}{30} \approx -4.93$ gallons/year,

c. $\frac{567 - 530}{2005 - 1995} = \frac{37}{10} = 3.7$ gallons/year,

d. $\frac{480 - 668}{2015 - 1960} = \frac{-188}{55} \approx -3.4$ gallons/year,

e. It means that from 1960 to 2015, the average fuel consumption per year of a passenger car in the United States decreased by about 3.4 gallons/year.

Activity 1.7 Exercises: 1. a. yes, linear; the constant rate of change is 10. **b.** no, not linear, **c.** yes, linear; the constant rate of change is $\frac{-9}{4}$. **d.** yes, linear; the constant rate of change is 2, assuming the scales on each axis is 1 per tic mark **3. a.** Yes, the average rate of change is a constant -3. **b.** No, between weeks 1 and 2, the slope is -5. Between weeks 2 and 3, the slope is -4. **c.** Yes, the slope is 0 for all pairs of points.

5. a. $m = \frac{5 - (-7)}{0 - 2} = \frac{12}{-2} = -6$, **b.** $(0, 5)$, **c.** $f(x) = -6x + 5$, **d.** $\left(\frac{5}{6}, 0\right)$; **7. a.** Yes, the slope is a constant. **b.** $m = \frac{3000 - 3500}{20 - 0} = \frac{-500}{20} = -25$ ft/sec, **c.** The jet is losing altitude, **d.** $(0, 3500)$, **e.** $h = -25t + 3500$, **f.** $(140, 0)$; The jet lands in 140 seconds. **8. a.** $\left(\frac{-1}{2}, 0\right)$, **b.** $(6, 0)$

Activity 1.8 Exercises: 1. a. $y = \frac{1}{2}x - 1$, **b.** $y = -\frac{4}{3}x + 1$, **c.** $y = -3x - 5$, **d.** $m = \frac{6 - (-3)}{2 - (-4)} = \frac{6 + 3}{2 + 4} = \frac{9}{6} = \frac{3}{2}$, $y = \frac{3}{2}x + 3$, **e.** $y = 3x - 11$; **2. a.** $(0, 35)$; The vertical intercept occurs where the input $x = 0$. **b.** $m = \frac{40 - 35}{100 - 0} = \frac{5}{100} = 0.05$; The mileage charges are \$0.05 per mile. **c.** $c = 0.05x + 35$; **3. a.** $m = \frac{145 - 75}{4 - 2} = \frac{70}{2} = 35$ mph. This represents the average rate of change or the average speed of the boat from $t = 2$ to $t = 4$. **b.** $d = 35t + 5$; **4. a. i.** $m = 1$, **ii.** $(0, -2)$, **iii.** $y = x - 2$, **c. i.** $m = -2$, **ii.** $(0, 6)$, **iii.** $y = -2x + 6$; **7. a.** $m = \frac{31.42 - 0}{5 - 0} = 6.28$, **b.** $(0, 0)$, **c.** $C = 6.28r$, **d.** $C = 2\pi r$, **e.** Yes, π is approximately 3.14, so 2π is approximately 6.28. **9. a.** $(0, 5000)$; the vertical intercept occurs where the input value is 0. **b.** $\frac{\Delta c}{\Delta h} = \frac{380{,}000 - 5000}{3000} = \frac{375{,}000}{3000} = 125$, building costs are \$125 per square foot. **c.** $c = 125h + 5000$, **d.** $c = 125(2500) + 5000 = 317{,}500$; **11. a.** The slope $= \frac{2226 - 1952}{95 - 75} = \frac{274}{20} = 13.7$. **b.** For each additional 1 kilogram increase in weight, a 20-year-old, 190.5-centimeter-tall male has an increase of 13.7 additional calories in his basal energy requirement. **c.** $B = m(w - h) + k$, $B = 13.7(w - 75) + 1952$, $B = 13.7w - 1027.5 + 1952$, $B = 13.7w + 924.5$, the symbolic rule $B = 13.7w + 924.5$ expresses the basal energy rate B for a 20-year-old, 190.5-centimeter-tall male in terms of his weight, w. **d.** The B-intercept has no practical meaning in this situation because it would indicate a weight of 0 kilograms. A possible practical domain is a set of weights from 55 to 182 kilograms.

Activity 1.9 Exercises: 1. a. $y = 2x - 3$, $m = 2$, $(0, -3)$, **b.** $y = -x - 2$, $m = -1$, $(0, -2)$, **c.** $y = \frac{2}{3}x - \frac{7}{3}$, $m = \frac{2}{3}$, $\left(0, -\frac{7}{3}\right)$, **d.** $y = \frac{1}{2}x + 2$, $m = \frac{1}{2}$, $(0, 2)$, **e.** $y = 4$, $m = 0$, $(0, 4)$;

3. a.

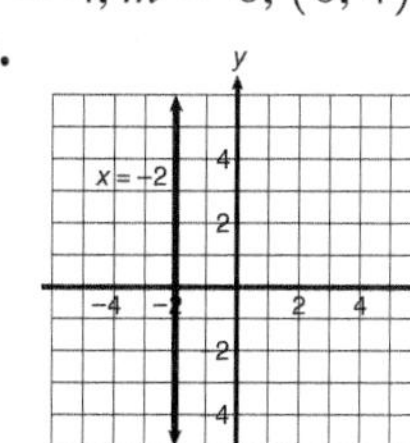

b. This is not a function. It does not pass the vertical line test. **c.** $x = -2$, **d.** The slope is undefined. **e.** vertical: none; horizontal: $(-2, 0)$; **5. a.** $f(x) = 2000$, **b.** 2000, 2000, 2000,

c.

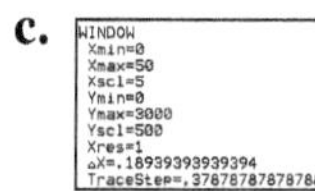
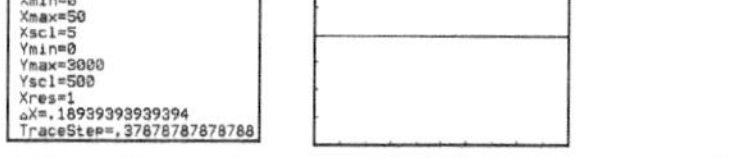

d. The slope is zero. This means the fee does not change. **e.** The graph is a horizontal line through $(0, 2000)$. **6. a.** $250w$, **b.** $200d$, **c.** $250w + 200d = 10{,}000$, **d.** $d = \frac{10{,}000 - 250w}{200} = 50 - \frac{5}{4}w$, **e.** $(40, 0)$; the maximum number of washers I can purchase is 40. **f.** $(0, 50)$; the maximum number of dryers I can purchase is 50.

Activity 1.10 Exercises:

1. a.

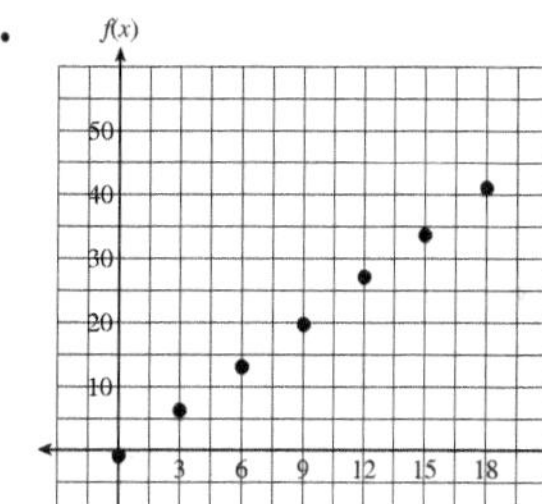

b. (Answers will vary.) Yes, the points are very close to a line. **c.** $f(x) = 2.299x - 0.761$, **d.** 22.229, **e.** 56.714,

f. $f(10)$ is more accurate; 10 is within the given data. 25 is not. $f(10)$ uses interpolation. $f(25)$ uses extrapolation; **2. b.** $c = f(t) = -24.06t + 368.84$ **c.** The slope of the line is -24.68. This means that the number of PCs shipped worldwide per year is declining at an average rate of 24,060,000 per year. **d.** The regression line predicts the number of PC shipments to be 320.72 million. This is 4.42 million unit above the actual. The error is about 1.4%. **e.** The regression equation predicts PC shipments in 2020 to be 152.3 million units. **f.** Extrapolation **g.** $r = -0.99$ There is a strong negative correlation.

4. a.

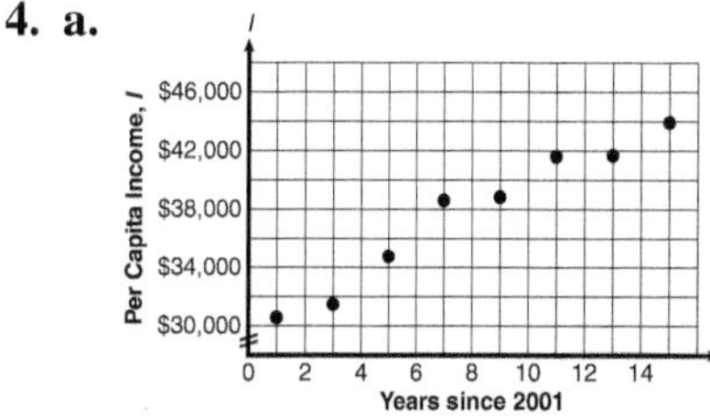

b. $I = 984.70x + 30799.08$

c. The slope is 984.70. This means that on average the per capita income in the United States increases by \$984.70 each year from 2001 through 2015. **d.** According to the model the per capita income will reach \$50,000 for the first time 20 years after 2001 or 2021. I am not very confident because this prediction is extrapolating into the future 6 years past my last data point.

How Can I Practice? 1. a. $C = 50 + 10t$, where the independent variable, t, represents the number of months and the dependent variable, C, represents the total cost in dollars. The vertical intercept is called the C-intercept and represents the \$50 down payment. The slope represents the month charge of \$10. **b.** $V = 16{,}000 - 1500x$, where the independent variable, x, represents the number of years since the car was purchased and the dependent variable, V, represents the value of the car in dollars. The vertical intercept is called the V-intercept and represents the value of the car on the day of purchase. The slope represents the car's yearly depreciation. **2. a.** $g(x) = 2x - 3$, **b.** $h(x) = -2x - 3$, **c.** $x = 2$, **d.** none, **e.** $f(x) = -2x + 3$, **f.** $y = -2x$, **g.** $y = 2$, **h.** none, **i.** none; **3. a.** 28, 40, 52, 60, 68, **b.** yes, **c.** 4, **d.** $c = f(m) = 4m + 20$,

e.

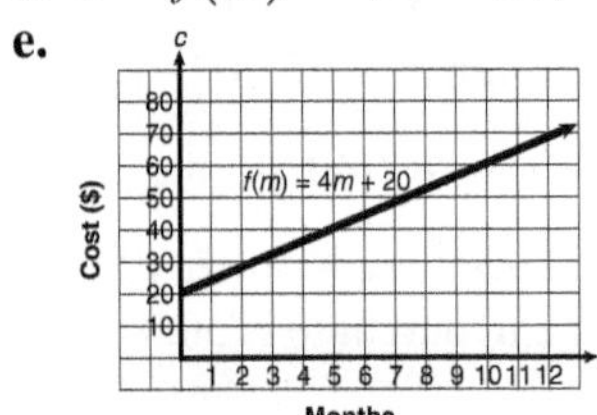

f. This represents the monthly charge. **g.** (0, 20); it indicates that the initial rental cost is \$20. **h.** $(-5, 0)$; it has no practical meaning in this case.

i. $65 = 4m + 20$ or $45 = 4m$ or $m = 11.25$; I can keep the graphing calculator for 11 months. **4. a.** 1.5, **b.** 1.5, **c.** $s(t)$ is a linear function because the rate of change is constant. **5.** $m = \frac{12 - 8}{-5 - 3} = -\frac{1}{2}$; **6.** $m = -4$; **7.** $m = \frac{2}{5}$; **8.** $y = -7x + 4$; **9.** $y = 2x + 10$; **10.** $y = 5$; **11.** $x = -3$; **12.** $y = -\frac{1}{2}x - 2$; **13.** $y = \frac{1}{3}x - 3$;

14.

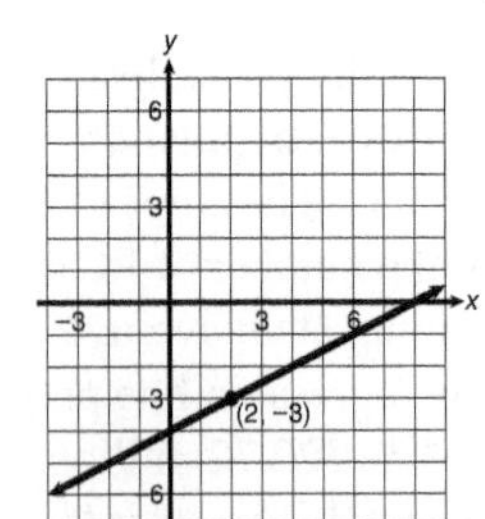

15.

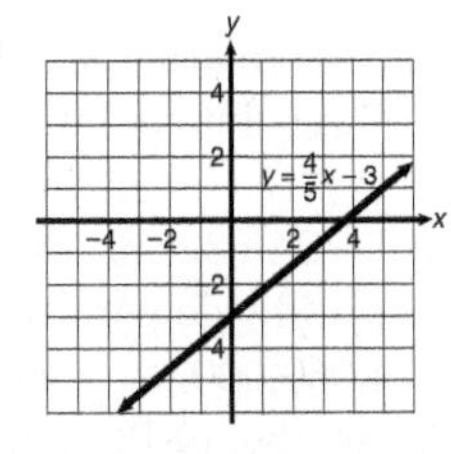

16.

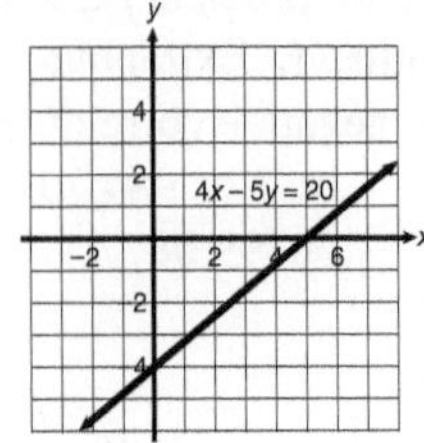

17.

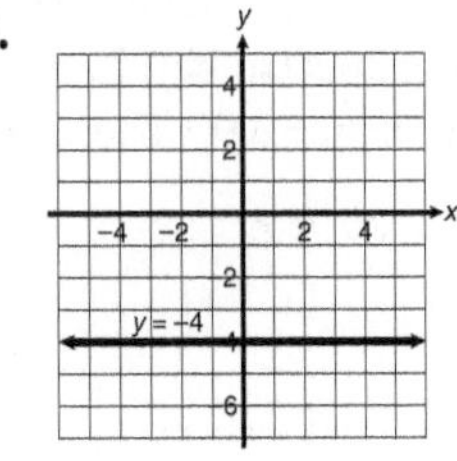

18. a.

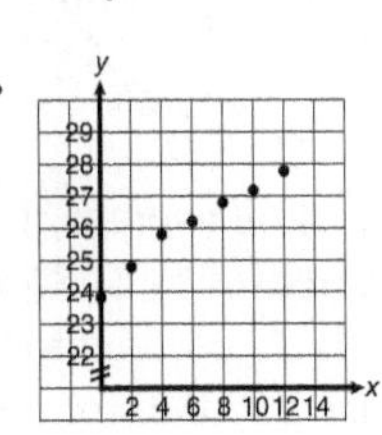

b. $y = 0.322x + 24.155$; $r = 0.976$, **c.** $y = 27.053$, **d.** $y = 30.595$;

19. a.

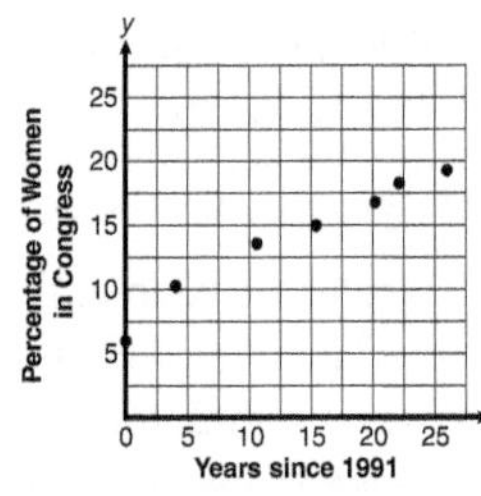

b. Regression equation $y = 0.479x + 7.552$, **c.** 2008, 15.69%; 2023, 23.83%, **d.** 2008 is within the given data; therefore, I have confidence in the result of 15.69%; 2025 is well outside the given data; therefore, I have little or no confidence in the result of 23.83%.

Activity 1.11 Exercises:

1. a. Numerically

x	y_1	y_2
−2	−1	8
−1	1	7
0	3	6
1	5	5
2	7	4
3	9	3

Graphically

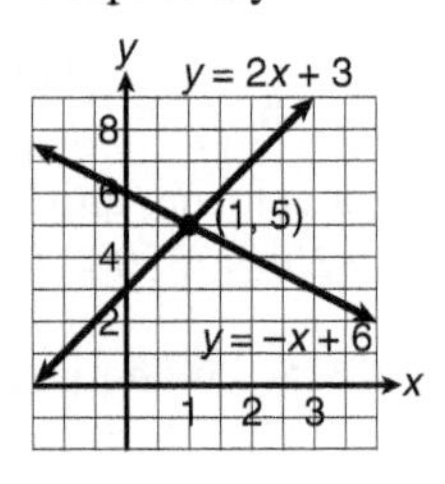

Algebraically (substitution method)

$y = 2x + 3 \quad y = -x + 6$

$$2x + 3 = -x + 6$$
$$+x \qquad +x$$
$$3x + 3 = 6$$
$$-3 \quad -3$$
$$\frac{3x}{3} = \frac{3}{3}$$
$$x = 1$$

$y = -1 + 6$

$y = 5$

The answer is $(1, 5)$.

c. Numerically

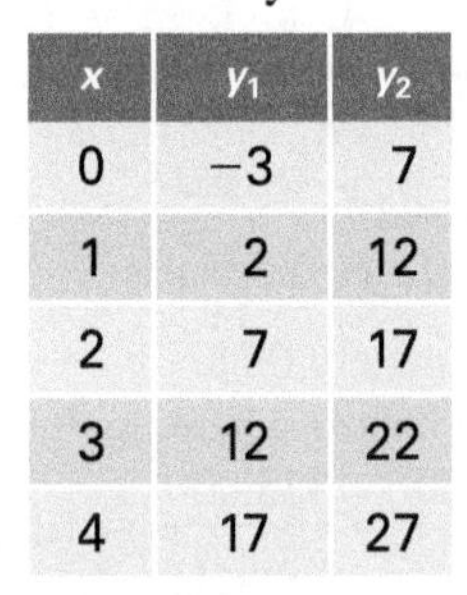

x	y_1	y_2
0	−3	7
1	2	12
2	7	17
3	12	22
4	17	27

Graphically

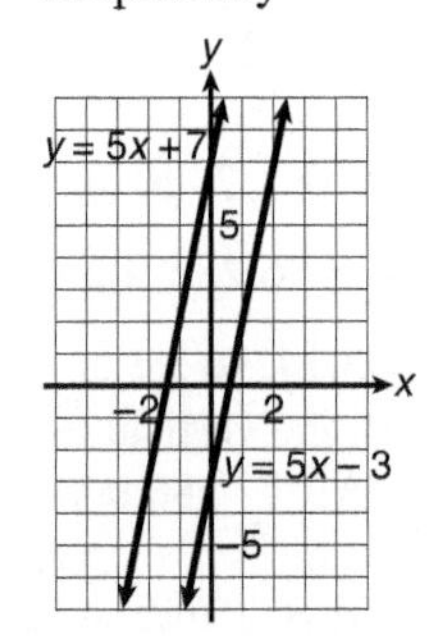

Algebraically (substitution method)

$y = 5x - 3 \quad y = 5x + 7$

$$5x - 3 = 5x + 7$$
$$-5x \qquad -5x$$
$$-3 = 7 \text{ false}$$

There is no solution.

2. a. $s = 17.2 + 1.5n$, **b.** $s = 9.6 + 2.3n$, **c.** in the year 2028; **3. a.** $c = 3560 + 15n$, **b.** $c = 2850 + 28n$, **c.** $n = 54.6$ or 55 months, **d.** dealer 1's system;

5. a.

t, NUMBER OF YEARS SINCE 1975	LIFE EXPECTANCY FOR WOMEN	LIFE EXPECTANCY FOR MEN
0	77.5	70.0
25	80.2	74.8
50	83.0	79.6
100	88.4	89.2
85	86.8	86.3
88	87.1	86.9
90	87.3	87.3

b. 90 years after 1980, in the year 2070, the life expectancy for both men and women will be 87.3 years. **c.** (90.4, 87.3), **d.** $t \approx 90.36, E = 87.34$

Activity 1.12 Exercises: 1. $x = 4, y = 2$;
3. $x = 3.5, y = -1.5$; **5.** $x = 1, y = -1$;
7. $x = 8, y = -5$; **9.** $x = 2, y = 0$; **11.** $x = -2, y = 1$;
13. $x = 3, x = -1$; **15.** $x = -2, y = 1$;
17. a. $8x + 5y = 106$
$x + 6y = 24$,
b. $x = -6y + 24$
$8(-6y + 24) + 5y = 106$
$-48y + 192 + 5y = 106$
$-43y = -86$
$y = 2$
$x = -6(2) + 24 = 12$
(12, 2) Each centerpiece costs \$12, and each glass costs \$2.
c. $8x + 5y = 106$ $\quad 8x + 5y = 106$
$-8(x + 6y) = -8(24)$ $\quad -8x - 48y = -192$
$-43y = -86$
$y = 2$
$x + 6(2) = 24$
$x + 12 = 24$
$x = 12$,

d. 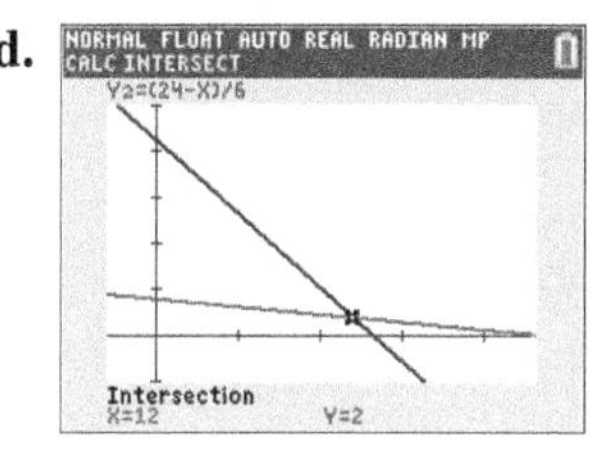

Activity 1.13 Exercises: 1. $(0, -3, 5)$; **3.** $(-5, 3, 1)$;
5. a. dependent, **b.** inconsistent.

Activity 1.14 Exercises:

1. $\begin{bmatrix} 4 & 3 & -1 \\ 2 & -1 & -13 \end{bmatrix}$; **3.** $\begin{bmatrix} 4 & -2 & 1 & 15 \\ 3 & 2 & -2 & -4 \\ 1 & 0 & 1 & 5 \end{bmatrix}$;

5. $4x + 3y = 15$
$2x - 5y = 1$;
7. $(x, y, z) = (1, -1, 2)$; **9.** Thirty-four 2-pointers and eleven 3-pointers were made for a total of 101 points.
11. 6 luxury sedans, 14 small hatchbacks, 8 hybrids

Activity 1.15 Exercises: 1. $l + w + d \le 61$;
3. $C(A) < C(B)$; **5.** $24{,}650 < i \le 59{,}750$;
7. $x > -2$

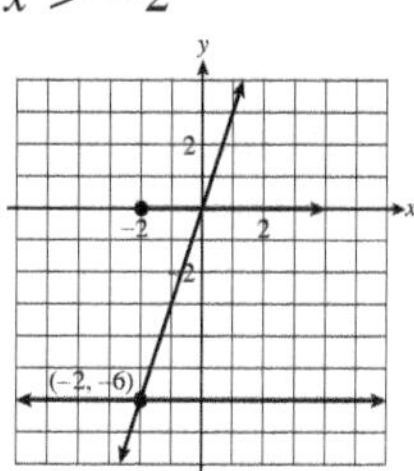

9. $x < 5$

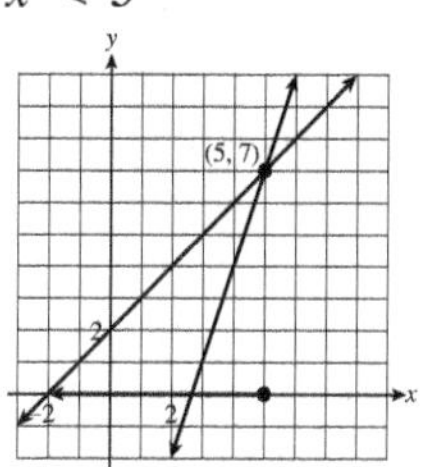

11. $x \ge 8$

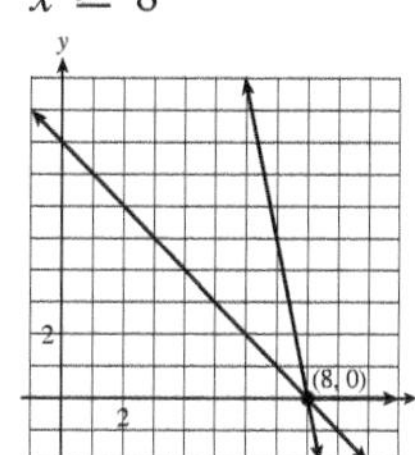

13. $x \ge -0.4$

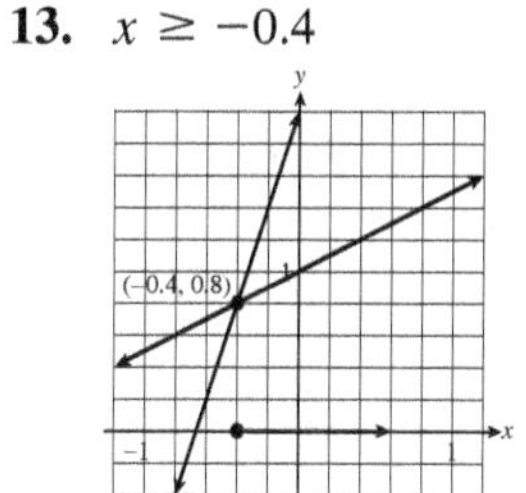

15. $1 < x < 2$; **17. a.** $-13.4t + 44.52 < 200$,
b. $t > 18.3$, 18 years from 2000, the year 2018;
18. a. $C = 60 + 75n$, **b.** $C = 30 + 1.00n$,
c. $30 + 1.00n < 60 + 0.75n$, **d.** $n < 120$;
19. a. $150 + 60n$, **b.** $150 + 60n \le 1200$, **c.** $n \le 17.5$
The maximum number of boxes that can be placed in the elevator is 17. **20.** $57.5 \le w \le 70$;
21. a. $-79.8 \le F \le 134$, **b.** $-79.8 \le 1.8C + 32 \le 134$,
c. $-62.1 \le C \le 56.7$

Activity 1.16 Exercises:
1. f. $5.00 + 0.90(11) = \$14.90$;
3. a.

$$f(x) = \begin{cases} 2.5x & x \le 15{,}000 \\ 37{,}500 + 3(x - 15{,}000) & 15{,}000 < x \le 21{,}000 \\ 55{,}500 + 4(x - 21{,}000) & x > 21{,}000 \end{cases}$$

c. 23,375 books; **5. a.** $f(x)$: 8, 7, 6, 5, 4, 3, 2, 1, 0, 1, 2;
$g(x)$: 2, 1, 0, 1, 2, 3, 4, 5, 6, 7, 8,
b.

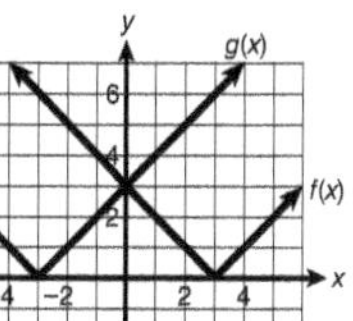

c. f is the same as $y = |x|$, only it is shifted 3 units to the right. g is the same as $y = |x|$, only it is shifted 3 units to the left.
d. $f(x) = \begin{cases} -x + 3 & x \le 3 \\ x - 3 & x > 3 \end{cases}$
e. The domains of f and g are all real numbers.
f. The ranges of f and g are $y \ge 0$.

How Can I Practice?
1. a. $(1, -4)$

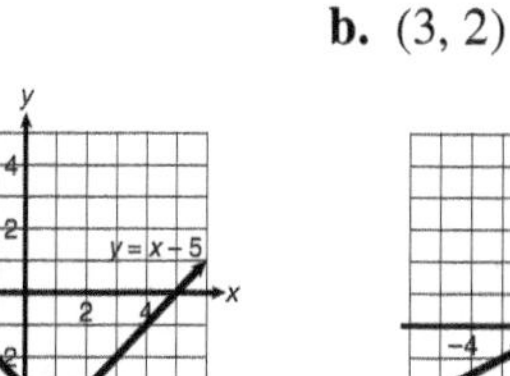

b. $(3, 2)$

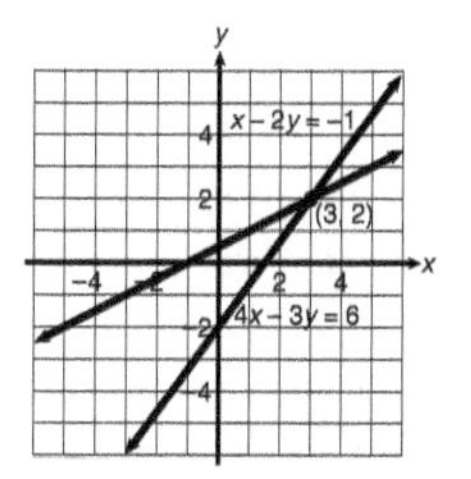

c. $(-4, -5)$

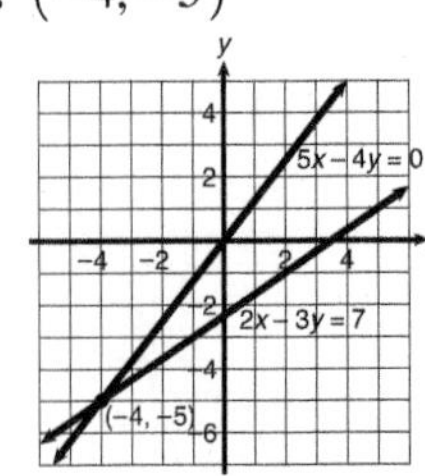

d. $-2 = 6$ inconsistent; no solution

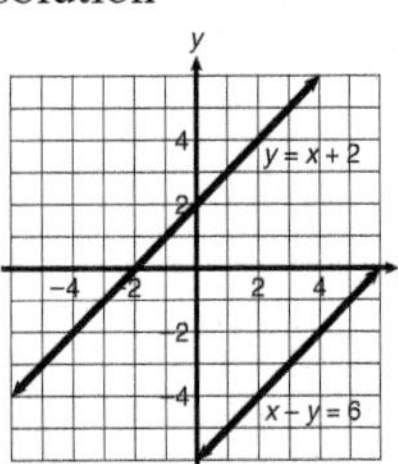

2. a. $(1, -4)$, **b.** $(3, 2)$, **c.** $(-4, -5)$, **d.** inconsistent; **3.** $(1, 5, -2)$; **4. a.** $x \geq 1.8$, **b.** $x > -6$, **c.** $1 \leq x < 5$; **5. a.** $t + d = 80, 0.50t + 0.75d = 52$, **b.** $d = 48, t = 32$, **c.**

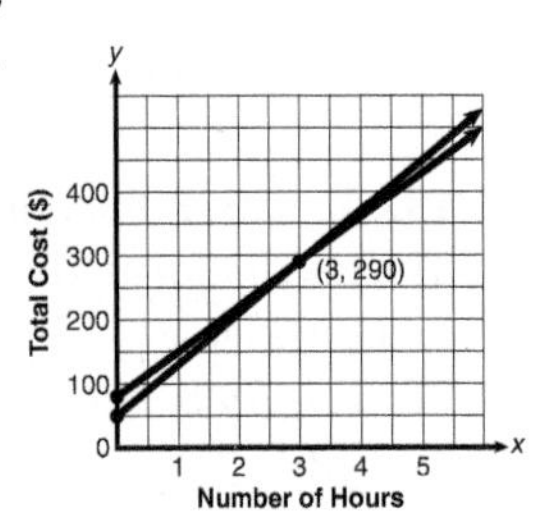

6. a. $y = 80 + 70x, y = 50 + 80x$,

b.

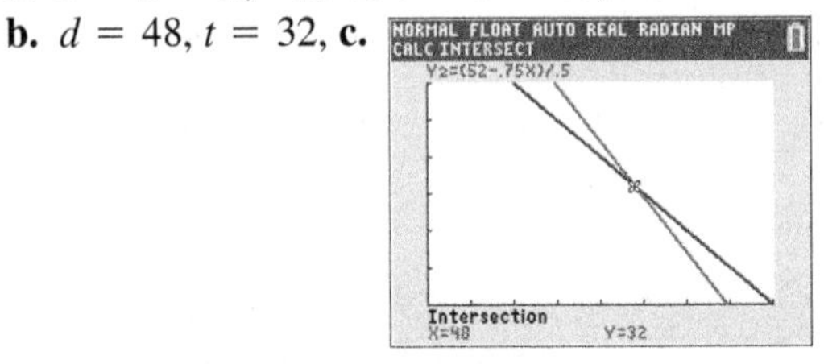

COLUMN 2	COLUMN 3
220	210
360	370
500	530
640	690

c.

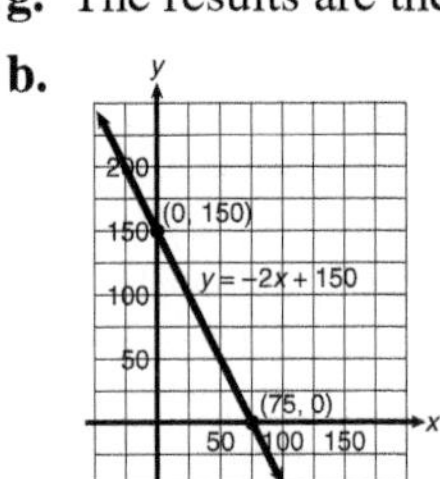

d. 3 hours; the total cost is \$290, **e.** $x = 3, y = 80 + 210 = 290$, **f.** I will use Towne Truck; its graph is below World Transport for $x = 6$. **7. a.** $y = 2x$, **b.** $x = 7, y = 14, z = 6$, **c.** It checks. **8. a.** $-5 < x \leq 6$, **b.** $-3 \leq x < 4$; **9. a.** $x \geq 2700$, **b.** $2367 \leq x \leq 3367$

10.

$$Tax = f(x) = \begin{cases} 0.04x, & 0 \leq x \leq 8500 \\ 340 + 0.45(x - 8500), & 8500 < x \leq 11{,}700 \\ 484 + 5.25(x - 11{,}700), & 11{,}700 < x \leq 13{,}900 \end{cases}$$

11.

$$f(x) = \begin{cases} -(x + 2), & x + 2 \leq 0 \\ x + 2 & x + 2 > 0 \end{cases} = \begin{cases} -x - 2, & x \leq -2 \\ x + 2, & x > -2 \end{cases}$$

Gateway Review 1. a. Yes, it is a function. **b.** No, it is not a function. There are two different outputs paired with 2. **c.** Yes, it is a function. **2.** 20, 36, 44, 60, 76, **a.** Yes, for each input there is one output. **b.** The input is x, the number of hours worked. **c.** The dependent variable is $f(x)$, the total cost. **d.** Negative values would not be realistic domain values. A negative number of hours worked does not make sense. **e.** The rate of change is \$8 per hour. **f.** The rate of change is \$8 per hour. **g.** The rate of change between any two points is \$8 per hour. **h.** The relationship is linear. **i.** $f(x) = 8x + 20$, **j.** The slope is the hourly rate I charge, \$8 per hour, **k.** (0, 20) is the vertical intercept. The 20 represents the fertilizer cost. **l.** $f(4) = 8(4) + 20 = 52$, **m.** $8x + 20 = 92$ or $8x = 72$ or $x = 9$; I need to work 9 hours for the cost to equal exactly \$92. **3. a.** $f(-2) = 14, g(-2) = 10$, **b.** $-6 + (-5) = -11$, **c.** $24 - 16 = 8$, **d.** $36(-2) = -72$; **4. a.** This represents a linear function. The slope is 4. **b.** This represents a linear function, $m = \frac{-2}{5}$. **c.** This does not represent a linear function. **d.** This represents a linear function, $m = 10.7$. **5. a.** $m = \frac{9 + 3}{-4 - 5} = \frac{12}{-9} = \frac{-4}{3}$, **b.** $m = \frac{3}{7}$, **c.** $m = \frac{1}{2}$; **6. a.** $y = 4$, **b.** $y = 2x + 5$, **c.** $-14 = -3(6) + b, b = 4, y = -3x + 4$, **d.** $-2 = 2(7) + b, b = -16, y = 2x - 16$, **e.** $x = 2$, **f.** $0 = -5(4) + b, b = 20, y = -5x + 20$, **g.** $16 = 4(2) + b, b = 8, y = 4x + 8$, **h.** $y = \frac{-1}{2}x + 5$; **7.** $y = \frac{-2}{5}x + 2$; **8. a.** $f(x) = 300{,}000 - 10{,}000x$, **b.** $m = -10{,}000$. The building depreciates \$10,000 per year. **c.** (0, 300,000); the original value is \$300,000. **d.** (30, 0); it takes 30 years for the building to fully depreciate. **9. a.** $(0, -3)$, **b.** $(0, -3)$, **c.** $(0, -3)$, **d.** The graphs all intersect at the point $(0, -3)$. **e.** The results are the same. **10. a.** $m = -2$; (0, 1), **b.** $m = -2$; $(0, -1)$, **c.** $m = -2$; $(0, -3)$, **d.** The graphs are parallel lines. **e.** The results are the same. **11. a.** $m = -3$; (0, 2), **b.** $m = -3$; (0, 2), **c.** $m = -3$; (0, 2), **d.** The graphs are all the same. **e.** the slopes, **f.** the slopes and the y-intercepts, **g.** The results are the same. **12. a.** (0, 150), (75, 0),

b.

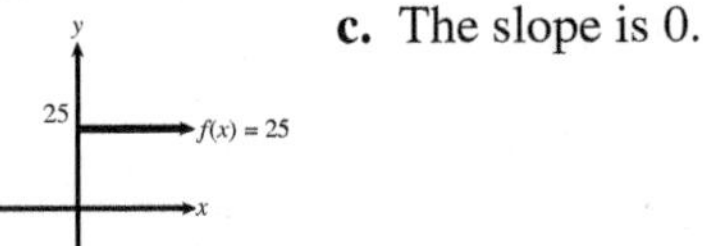

c. The domain and range are all real numbers. **d.** $w(t) = -2t + 150$, **e.** They are essentially the same. **f.** The vertical intercept is (0, 150). It indicates the person's initial weight of 150 pounds. The horizontal intercept (75, 0) indicates that after 75 weeks of weight loss, the person weighs nothing. **g.** The practical domain is $0 \leq t \leq 15$. The practical range is $120 \leq w(t) \leq 150$. **13. a.** $f(x) = 25$, horizontal line through (0, 25), **b.**

$f(x) = 25$

c. The slope is 0.

14. a.

$$f(x) = \begin{cases} 1500 & x \leq 10{,}000 \\ 1500 + 0.02(x - 10{,}000) & 10{,}000 < x \leq 40{,}000, \\ 2100 + 0.04(x - 40{,}000) & x > 40{,}000 \end{cases}$$

b. NORMAL FLOAT AUTO REAL RADIAN MP

c. $f(25{,}000) = 1500 + 0.02(15{,}000) = 1800$, **d.** $x = 66{,}250$; **15. a.** $y = 1040x + 7900$ or $t(n) = 1040n + 7900$, **b.** 1040; the number of finishers increased at a rate of 1040 per year, **c.** (0, 7900); the model

indicates that there were 7900 finishers in 2014, **d.** fairly well, **e.** 14,140, **f.** I used extrapolation because I am predicting outside the original data. **g.** No, 2016 is farther from the data than 2012. The farther removed we are from the data, the more likely our prediction is incorrect.
16. a. $(3, -1)$, **b.** $(-1, 6)$,
c.

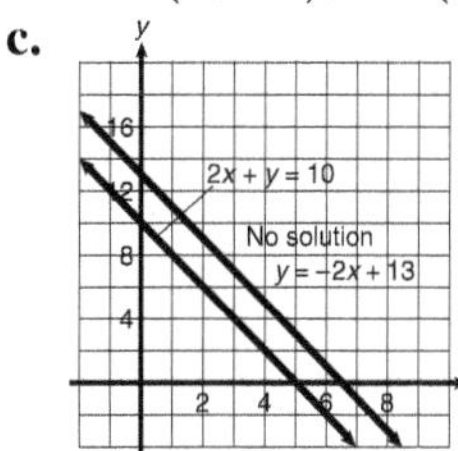

d. $4 = 4$; this is a dependent system. Any pair of numbers that satisfies one equation satisfies both equations.

17. a. $C = 5000 + 5400t$, **b.** $C = 8000 + 4500t$, **c.** $t = 3.33$ yr. $C = \$22{,}982$; **18.** $x = 4.50, y = 0.75$; **19. a.** $(0, 1, 2)$, **b.** $(-3, 1, 0)$, **c.** $(0.5, 0.25, -0.5)$, **d.** $(12, 7, 9)$; **20. a.** $x = \$0.50, y = \$1.00, z = \$2.00$ It is a good deal. **b.** Answers will vary. I would not take advantage of this. I don't give away many pictures.
21. a.

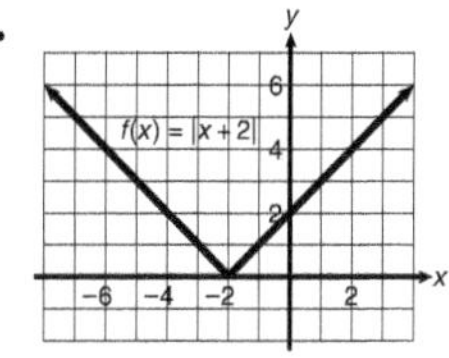

b. increasing $x > -2$, decreasing $x < -2$, **c.** The domain is all real numbers. **d.** The range is $y \geq 0$. **e.** g is the reflection through the x-axis. **f.** f shifts the graph of $y = |x|$ 2 units to the left. h shifts the graph of $y = |x|$ 2 units up.

Chapter 2

Activity 2.1 Exercises:

1. a.

COLUMN 2	COLUMN 3	COLUMN 4
100	250	350
100	500	600
100	750	850
100	1000	1100
100	1250	1350

b. $C(x) = 12.50x + 100$, **c.** $0.15(20) = 3$, **d.** $3(750) = \$2250$, **e.** $0.13(2250) = \$292.50$,

f.

COLUMN 2	COLUMN 3	COLUMN 4
3	2250	292.50
6	4500	585.00
9	6750	877.50
12	9000	1,170.00
15	11,250	1,462.50

g. $T(x) = 0.13(750)(0.15x) = 14.625x$,

h. $P(x) = 2.125x - 100$, **i.** $0 \leq x \leq$ the number of people the banquet room will accommodate.
j. Set $P(x) = 0$ and solve for x, $x \approx 47.06$; 48 people must attend. **k.** Set the profit equal to 500 and solve for x; $500 = 2.125x - 100$; $x \approx 282.35$; 283 must attend. **3.** $f(x) + g(x)$: $4, -6, 1, 6, 2, 8$; $f(x) - g(x)$: $2, -4, -1, 8, -4, 0$; **5. a.** $5x - 2$, **b.** $x^2 + 3x - 8$, **c.** $-x + 30$, **d.** $-3x^2 + 8x - 3$, **e.** $15x - 15$, **f.** 11, **g.** $5x^2 + 5x - 9$, **h.** $-6x^2 + 15x - 4$, **i.** $-5x + 25$, **j.** $33x - 10$; **7. a.** $4x^2 - x$, **b.** $4x^2 - 5x - 9$, **c.** $3 + 4x$, **d.** $4x^2 - 5x + 3$; **9. a.** $1, -9, -11, -5, 9, 31$, **b.** $f(x) - g(x) = x^2 + 5x - 5$, **c.** The answers check.

Activity 2.2 Exercises: 1. a. $(3x)(2x) = 6x^2$, **b.** $2x - 3$, **c.** $(2x - 3)(3x) = 6x^2 - 9x$; **3. a.** $P(x) = 40 - x$, **b.** $N(x) = 3000 + 100x$, **c.** $R(x) = (40 - x)(3000 + 100x)$, **d.** The domain is $0 \leq x \leq 10$.

e.

PRICE PER TICKET, $P(x)$	NUMBER OF TICKETS SOLD, $N(x)$	TOTAL REVENUE, $R(x)$
40	3000	120,000
38	3200	121,600
36	3400	122,400
34	3600	122,400
32	3800	121,600
30	4000	120,000

f. The values in the fourth row are the product of the values in the second and third rows. **g.**

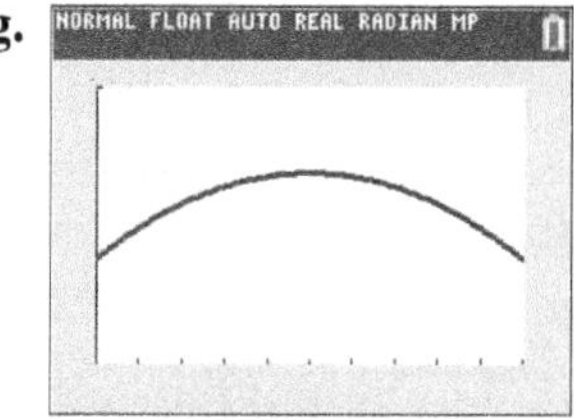

Xmin = 0, Xmax = 10, Ymin = 117,000, Ymax = 125,000, **h.** The maximum point of the graph is (5, 122,500). This means the price that produces the maximum revenue is $40 - 5$ or 35 dollars. The maximum revenue is \$122,500 **i.** $R(x) = -100x^2 + 1000x + 120{,}000$ **j.** There is only one graph. The graphs are the same.

5.

	x^2	$3x$	-5
x	x^3	$3x^2$	$-5x$
3	$3x^2$	$9x$	-15

$x^3 + 6x^2 + 4x - 15$; **7. a.** $6x^2 + 19x + 10$, **b.** $6x^2 - 19x + 10$, **c.** $4x^2 + 5x - 6$, **d.** $4x^2 - 5x - 6$; **9. a.** $4x^2 + 12x + 9$, **b.** $9x^2 - 12x + 4$, **c.** $25x^2 - 4$, **d.** $x^4 - 25$, **e.** $x^3 + 12x^2 + 48x + 64$, **f.** The outer product and inner product are opposites. Their sum is 0. **10. a.** $f(x) \cdot g(x) = 2x^2 - x - 3$,

b.

x	$f(x)$	$g(x)$	$f(x) \cdot g(x)$
0	1	−3	−3
1	2	−1	−2
2	3	1	3
3	4	3	12
4	5	5	25

c. Answers may vary depending on the choices of x.

Activity 2.3 Exercises: 1. 1.58×10^{12}, 1.58E12; **3. a.** 3×10^{21}, **b.** 4.5×10^{16}, **c.** 9,000,000,000,000,000,000,000,000,000, **d.** $\frac{9 \times 10^{27}}{4.5 \times 10^{16}} = 2 \times 10^{11}$; **5. a.** $\frac{6.35 \times 10^{8}}{2.27 \times 10^{9}}$, **b.** 2.797×10^{-1}; **7.** 1; **9.** 10; **11.** $\frac{4}{x^4}$; **13.** $2x^7$; **15.** $\frac{3y^4}{5x^4}$; **17.** $3x^{-3} = \frac{3}{x^3}$; **19.** $\frac{-1}{2a^2b^3}$

Activity 2.4 Exercises: 1. a. x^{18}, **b.** $4x^{10}$, **c.** $-27x^6$, **d.** $\frac{1}{x^{12}}$, **e.** $\frac{16}{a^{10}}$, **f.** $\frac{b^8}{a^8c^{12}}$, **g.** $-x^{18}$, **h.** x^{18}; **3. a.** $x^{13/12} = \sqrt[12]{x^{13}}$, **b.** $x^{1/6} = \sqrt[6]{x}$, **c.** $x^{7/15} = \sqrt[15]{x^7}$; **5. a.** 5, **b.** 0, **c.** not a real number; **7. a.** $t = f(L) = 2\pi\left(\frac{L}{32}\right)^{1/2}$, **b.** $t \approx 2.22$ seconds

How Can I Practice? 1. a. $x = 30$, **b.** $N = f(t) = 30 + t$, **c.** $C = g(t) = 20 - 0.5t$, **d.** $N = f(t)$: 30, 32, 34, 36, 38, 40; $C = g(t)$: 20, 19, 18, 17, 16, 15, **e.** $R(t)$: 600, 608, 612, 612, 608, 600, **f.** $R(t) = f(t) \cdot g(t) = (30 + t)(20 - 0.5t) = -0.5t^2 + 5t + 600$,

g.

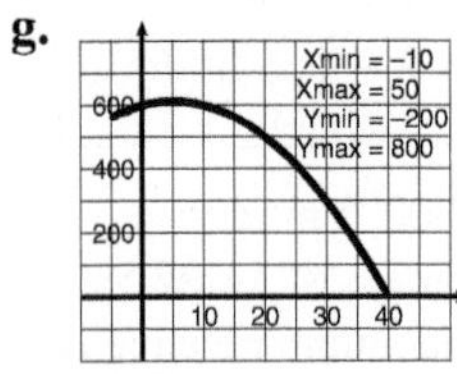

h. \$612.50 is the maximum revenue if 35 couples attend. **i.** 35 tickets must be sold to obtain the maximum revenue.

2. a. $3x - 1$, **b.** $-x + 5$, **c.** $2x^2 + x - 6$, **d.** 2, **e.** 0, **f.** $3x + 6$; **3. a.** $-3x + 5$, **b.** $x^4 - x^3 - 5x^2 + 9x - 4$, **c.** 0, **d.** $-x^2 - 7x + 14$; **4. a.** $5x - 2$, **b.** $2x^2 - 2x - 8$, **c.** $-2x + 12$, **d.** $2x^2 - 13x - 8$, **e.** $5x^2 - x + 2$; **5. a.** x^4, **b.** x^9, **c.** $6x^8$, **d.** x^5y^6z, **e.** $10x^6y^5z^8$, **f.** $-30a^5b^3$; **6. a.** $x^2 - 7x + 10$, **b.** $4x^2 + 25x - 21$, **c.** $4x^2 - 9$, **d.** $x^3 + x^2 - 11x + 10$, **e.** $2x^3 - x^2 + 3x + 2$, **f.** $-2x^2 - 5x - 21$, **g.** $11x^2 - 2x$, **h.** $-x^5 - x^3 + 3x^2 + 2x - 1$, **i.** $9x^2 + 30x + 25$, **j.** $4x^2 - 28x + 49$, **k.** $x^3 + 12x^2 + 48x + 64$, **l.** $25x^2 - 49$; **7. a.** 5636, 6436, 7236, 8036, 8836, **b.** $f(t) = 5636 + 160t$ **c.** 12,660, 14,810, 16,960, 19,110, 21,260 **d.** $g(t) = 12{,}660 + 430t$ **e.** 650, 650, 650, 650, 650 **f.** $h(t) = 650$ **g.** $k(t) = 18{,}946 + 590t$

h.

3	6116	13,950	650	20,716
12	7556	17,820	650	26,026
18	8516	20,400	650	29,566
25	9636	23,410	650	33,696

i. $t > 10.26$ Total cost will exceed \$25,000 approximately 11 years after 2016 or 2027. **8. a.** 17, −3, −7, 5, 33, 77, **b.** $f(x) - g(x) = 2x^2 + 2x - 7$, **c.** The answers check. **9. a.** $\frac{2}{x^3}$, **b.** $9x^2$, **c.** $\frac{1}{3^4}$, **d.** 1, **e.** $2x^7$, **f.** $-6x^4y^8$, **g.** 4, **h.** $\frac{2}{3x^4}$, **i.** $\frac{5y^2}{x^4}$, **j.** $\frac{-15}{x^5}$, **k.** $-2x^6$, **l.** $\frac{a^4}{b^4c^5}$; **10. a.** x^8, **b.** x^3y^3, **c.** $32x^{20}y^5$, **d.** $432x^{10}y^4$, **e.** 25, **f.** −25, **g.** $4x^2$, **h.** $-8x^3$, **i.** $-8x^{22}$, **j.** 125, **k.** 16, **l.** −3, **m.** $\frac{27}{64}$, **n.** 2, **o.** x^2y^4, **p.** $x^{7/6}$, **q.** $2x^{1/3}y^{1/3}$; **11. a.** 181,440,000 gal. **b.** I own approximately 0.2 of a square mile. **c.** 3.4339×10^{14} cubic miles; **12.** 3.26963×10^8, 7.591584×10^9; **13.** $\frac{2.27 \times 10^{10}}{7.57 \times 10^9} \approx 0.3 \times 10^1 = 3$. To divide powers with the same base, keep the base and subtract the exponents. **14. a.** $A = 35 + 12x + x^2$, **b.** $(7 + 4)(5 + 4) = 99$, or $35 + 12(4) + 4^2 = 99$; $99 - 35 = 64$ square feet, **c.** $A = \pi r^2$, for the umbrella $A = \pi(r + 2)^2$, or $A = \pi r^2 + 4\pi r + 4\pi$

Activity 2.5 Exercises: 1. a. $g(2) = 200$. The radius of the slick is 200 feet 2 hours after the spill. **b.** $f(g(2)) = f(200) = \pi(200)^2 \approx 125{,}664$. The area of the oil slick is 125,664 square feet 2 hours after the spill. **c.** $f(g(10)) = f(1000) = \pi(1000)^2 = 1{,}000{,}000\pi$. This is the area of the slick after 10 hours. **d.** $f(g(t)) = f(100t) = \pi(100t)^2 = 10{,}000\pi t^2$, **e.** $f(g(10)) = 10{,}000\pi(10)^2 = 1{,}000{,}000\pi$. The results are the same. **3. a.** $-18x^2 + 18x - 3$, **b.** $-6t^2 + 6t + 2$; **5. a.** $L(x) = 0.99x$, **b.** $D(x) = 0.90x$, **c.** $S(x) = L(D(x)) = L(0.90x) = 0.99(0.90x) = 0.891x$, **d.** $S(500) = 0.891(500) \approx 446$, **e.** $D(x)$ needs to be improved because the air bags fail 10% of the time, whereas the seat belts fail only 1% of the time. **6. a.** $f(x) = x - 6$, **b.** $g(x) = 0.99x$, **c.** $g(f(10{,}000)) = g(9994) = 0.99(9994) \approx 9894$. If 10,000 bottles are processed, 9894 of them should be properly labeled and capped.

Activity 2.6 Exercises: 1. a. The clerk was thinking 25% + 40% = 65%. **b.** Yes, because I am getting a larger discount than I should. **c.** No, because I should be getting a 55% discount. **3. a.** $f(x) = x - 1500$, **b.** $g(x) = 0.9x$, **c.** $g(f(20{,}000)) = g(20{,}000 - 1500) = 0.9(18{,}500) = 16{,}650$. The price of a \$20,000 car with a \$1500 rebate and a 10% discount is \$16,650; **4. a.** $g(f(t)) = g(0.5t) = \pi(0.5t)^2 = 0.25\pi t^2$, **b.** The input is the number of seconds after the pebble hits the water. The output is the area of the outer ripple in square feet.

Activity 2.7 Exercises: 1. a. 7, **b.** 4, **c.** x, **d.** x; **2. a.** $h^{-1} = \{(3, 2), (4, 3), (5, 4), (6, 5)\}$, **b.** $h(3) = 4 \quad h^{-1}(h(3)) = 3$, **c.** $h^{-1}(5) = 4 \quad h(h^{-1}(5)) = 5$; **3.** No, the interchange of the input and output does not result in a function.

5. a. $m = \dfrac{126.27 - 63.13}{100 - 50} = 1.2628$, **b.** You have 1.2628 Canadian dollars for every U.S. dollar. **c.** $f(x) = 1.2628x$, **d.** $f(3000) = 1.2628(3000) = 3788.40$. If I have 3000 U.S. dollars, then I can exchange them for \$3788.40 Canadian currency. **e.** $m = \dfrac{100 - 50}{126.27 - 63.13} \approx 0.7919$, **f.** You have 0.7919 U.S. dollar for every Canadian dollar. **g.** $g(x) = 0.7919x$, **h.** $g(6000) = 0.7919(6000) = 4751.40$. If I have 6000 Canadian dollars, then I can exchange them for 4751.40 U.S. dollars. **i.** $f(g(x)) = f(0.7919x) = 1.2628(0.7919x) = x$, $g(f(x)) = f(1.2628x) = 0.7919(1.2628x) = x$

Activity 2.8 Exercises: 1. a. $P = f(S) = 0.05S + 500$, **b.** $f(6000) = 0.05(6000) + 500 = 800$. The weekly salary for \$6000 worth of sales is \$800.

c. $S = g(P) = \dfrac{P - 500}{0.05}$, **d.** $g(600) = \dfrac{600 - 500}{0.05} = 2000$. A weekly salary of \$600 means I sold \$2000 worth of merchandise. **e.** $g(f(8000)) = 8000$; **3. a.** $y + 4 = 3x$ or $x = \dfrac{y + 4}{3}$ or $y = f^{-1}(x) = \dfrac{x + 4}{3}$, **b.** $2w = z - 4$ or $z = 2w + 4$ or $g^{-1}(z) = 2z + 4$ **c.** $t = \dfrac{5}{s}$ or $s = \dfrac{5}{t}$;

5.

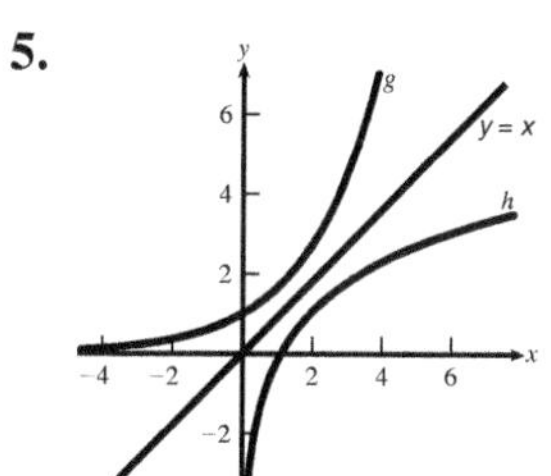

g and h are inverses. The graphs of g and h are symmetric with respect to the line $y = x$.

7. a. $g^{-1}(x) = \dfrac{3x - 6}{4}$, **b.**

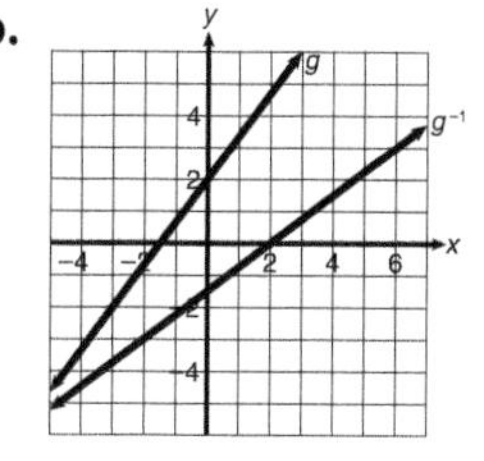

c. Yes, because the graphs are reflections in the line $y = x$.

d. $g^{-1}\left(\dfrac{6 + 4x}{3}\right) = \dfrac{3\left(\dfrac{6 + 4x}{3}\right) - 6}{4} = \dfrac{4x}{4} = x.$

Yes, because $g^{-1}(g(x)) = x$.

9. a.

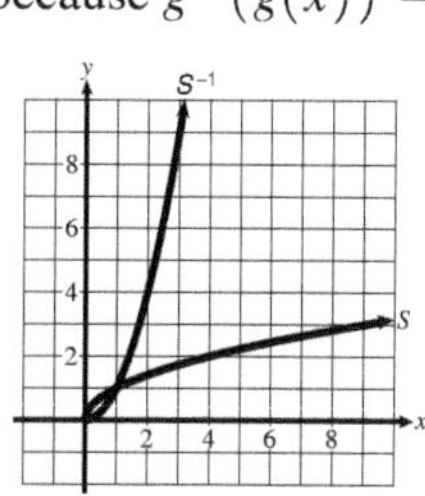

b.

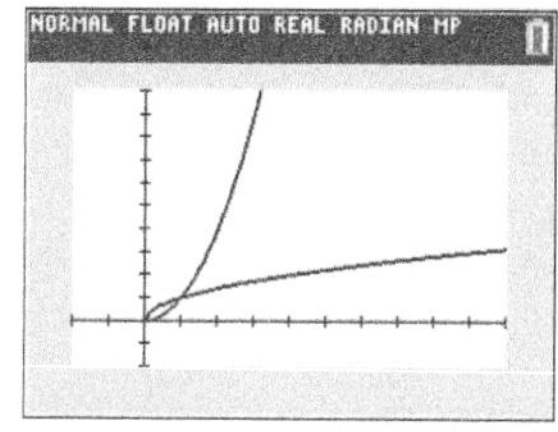

c. The area of the square is the input. The length of the side of the square is the output. **d.** The length of the side of the square is the input. The area of the square is the output. **e.** Given the length of the side, we can determine the area of the granite top.

How Can I Practice? 1. a. $f(-1) = -3$, **b.** $g(5) = 7$, **c.** $f(x + 2) = (x + 2)^2 - 4 = x^2 + 4x$, **d.** $g(x^2 - 4) = x^2 - 4 + 2 = x^2 - 2$, **e.** $f(x^2 - 4) = (x^2 - 4)^2 - 4 = x^4 - 8x^2 + 12$, **f.** $y = g^{-1}(x) = x - 2$; **2. a.** $f(-2) = -6$, **b.** $g(-1) = 2$, **c.** $g(-2) = -2$, **d.** $f(4 + x - x^2) = (4 + x - x^2) - 4 = x - x^2$, **e.** $= -x^2 + 9x - 16$, **f.** $y = f^{-1}(x) = x + 4$; **3.** $-1, 3, 2, 1, 0$; **4. a.** $g(3x^2) = -2(3x^2)^3 = -54x^6$, **b.** $f(-2x^3) = 3(-2x^3)^2 = 12x^6$, **c.** $g(48) = -2(48)^3 = -221{,}184$; **5. a.** $s(4x - 1) = (4x - 1)^2 + 4(4x - 1) - 1 = 16x^2 - 8x + 1 + 16x - 4 - 1 = 16x^2 + 8x - 4$, **b.** $t(x^2 + 4x - 1) = 4(x^2 + 4x - 1) - 1 = 4x^2 + 16x - 5$, **c.** $y = t^{-1}(x) = \dfrac{x + 1}{4}$;

6. a. $p(\sqrt{x + 2}) = \dfrac{1}{\sqrt{x + 2}}$, **b.** $c\left(\dfrac{1}{x}\right) = \sqrt{\dfrac{1}{x} + 2}$, **c.** $y = p^{-1}(x) = \dfrac{1}{x}$;

7. $\{(6, 4), (-9, 7), (1, -2), (0, 0)\}$;

8. $f(g(x)) = f\left(\dfrac{x + 3}{2}\right) = 2\left(\dfrac{x + 3}{2}\right) - 3 = x$

$g(f(x)) = g(2x - 3) = \dfrac{(2x - 3) + 3}{2} = x$. Because $f(g(x)) = g(f(x)) = x$, f and g are inverse functions.

9. a. $y = 4x + 3$; $\dfrac{y - 3}{4} = x$; $y = f^{-1}(x) = \dfrac{x - 3}{4}$,

b.

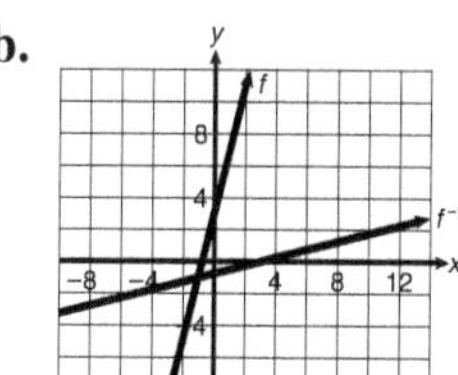

c. The intercepts of f are $(0, 3)$ and $\left(-\dfrac{3}{4}, 0\right)$. The intercepts of f^{-1} are $\left(0, -\dfrac{3}{4}\right)$ and $(3, 0)$.

d. The slope of the graph of f is 4. The slope of the graph of the inverse is $\dfrac{1}{4}$.

e.

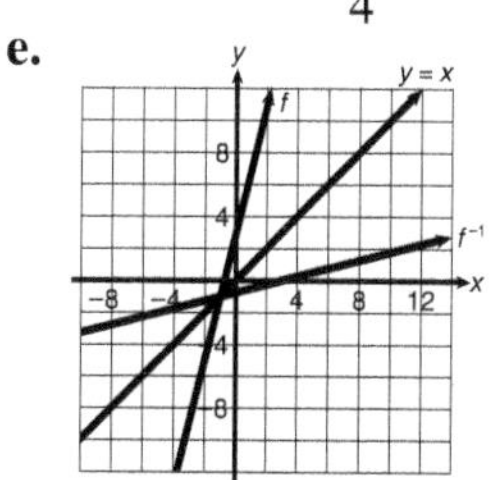

10. a. 26.4, 24.5, 22.5, 20.6, 18.6, 16.7, 14.7, 10.8

b.

Percent of The Adult Population That Smokes	26.4	24.5	22.5	20.6	18.6	16.7	14.7	10.8
Years since 1900	0	5	10	15	20	25	30	35

c. $f^{-1}(p) = \dfrac{26.4 - p}{0.39}$, **d.** 10% of the U.S. adult population will smoke 42 years after 1990, or in the year 2032. **e.** The graphs are symmetric in the line $y = x$. **f.** For f horizontal $(67.7, 0)$ vertical $(0, 26.4)$, for f^{-1} horizontal $(26.4, 0)$, vertical $(0, 67.7)$. The horizontal intercept of the function interchanged is the vertical intercept of the inverse. The vertical intercept of the function interchanged is the horizontal intercept of its inverse. **g.** Function: $m = -0.39$; inverse function: $m = \dfrac{-1}{0.39} \approx -2.56$. The slopes are reciprocals. **h.** 68 years after 1990, or 2058 **11. a.** $f(x) = 0.00740x$, **b.** 444 euros, **c.** \$421.40, **d.** $g(f(x)) = g(0.00740x) = 1.20400\,(0.00740x) = 0.00891x$, **e.** $g(f(60{,}000)) = 0.0891(60{,}000) = \$534.60$ **12.** $x = \dfrac{4}{3}$; **13. a.** $b = f(x) = x^2, V = g(b) = 10b$, **b.** $V = g(f(x)) = g(x^2) = 10x^2$.

Gateway Review 1. a. $2x^2 - 2x - 1$, **b.** $-x^2 + 5x - 4$, **c.** $4x^2 - 13x + 3$, **d.** $x^3 - 7x^2 + 13x - 15$, **e.** $-11x + 11$, **f.** $2x^4 - 5x^3 + 4x^2 + 7x - 4$; **2. a.** $6x^8$, **b.** $16x^6y^2$, **c.** $-2x^5y^3$, **d.** $-10x^5y^5z^4$, **e.** $9x^6y^2$, **f.** $-125x^3y^3$, **g.** $2x^3$, **h.** 2, **i.** 1, **j.** $\dfrac{3y^3}{2z^3}$, **k.** $-5x^{-8} = \dfrac{-5}{x^8}$, **l.** $-4x^2$, **m.** $(-5)^3(x^{-3})^3 = -125x^{-9} = \dfrac{-125}{x^9}$, **n.** $x^{4/5+1/2} = x^{8/10+5/10} = x^{13/10}$, **o.** $x^{(2/3)\cdot 3} = x^2$; **3. a.** -20, **b.** $4x + 1$, **c.** $f(3) - g(3) = 16 - (-3) = 19$, **d.** $(6x - 2)(-2x + 3) = -12x^2 + 22x - 6$, **e.** $-12x + 16$, **f.** $g(10) = -2(10) + 3 = -17$, **g.** $y = 6x - 2$ or $y = f^{-1}(x) = \dfrac{x + 2}{6}$; **4. a.** $x^2 - 4x + 5$, **b.** $= 3x^3 - 5x^2 + 11x - 6$, **c.** $= 9x^2 - 15x + 9$, **d.** $g(5) = 3(5) - 2 = 13$; **5. a.** 7, **b.** 4, **c.** 81, **d.** 3.214, **e.** 9, **f.** 32, **g.** $\dfrac{1}{16}$; **6. a.** $f(x) = 2(0.01)x^2 = 0.02x^2$, **b.** $g(x) = 4(0.004)(x)(3x) = 0.048x^2$, **c.** $(f + g)(x) = 0.02x^2 + 0.048x^2 = 0.068x^2$,

d.

$f(x)$	g	$t(x) = f(x) + g(x)$
0.08	0.19	0.27
0.32	0.77	1.09
0.72	1.73	2.45
1.28	3.07	4.35
2.00	4.80	6.80

7. a. $C(x) = 12x + 300$, **b.** $R(x) = 25.95x$, **c.** $p(x) = 25.95x - (12x + 300) = 13.95x - 300$, **d.** 22 hats must be sold because 21 hats is not quite enough. The solution was obtained graphically. **e.** $C(50) = 900$. The cost of producing 50 hats is \$900. $R(50) = 1297.50$. The revenue from 50 hats is \$1297.50. $p(50) = 397.50$. The profit from selling 50 hats is \$397.50. **f.** The profit is the difference between the revenue and cost functions. **8. a.** $f(x) = 60(110) + x(110 - 2x)$, **b.** $f(x) = 6600 + 110x - 2x^2$, **c.** integers $0 \le x \le 30$, **d.** $f(15) = 7800$. At regular price, the cost is \$8250; so the savings is \$450. **9. a.** -10, **b.** 41, **c.** 2, **d.** 5; **10. a.** integers $0 \le x \le 30$, **b.** $f(22) = \$2864.40$. If 22 snowboards are produced, the cost is \$2864.40. **c.** $f(3.75t) = 150(3.75t) - 0.9(3.75t)^2 = 562.50t - 12.65625t^2$, **d.** The input variable is t, **e.** $f(g(4)) = \$2047.50$, **f.** $3500 = 562.50t - 12.65625t^2$, t is about 7.5 hours (determined graphically);

11. a. $y = \dfrac{2x - 3}{5}$; $x = \dfrac{5y + 3}{2}$; or $f^{-1}(x) = \dfrac{5x + 3}{2}$

b. The slope of f is $\dfrac{2}{5}$. The slope of f^{-1} is $\dfrac{5}{2}$. The slopes are reciprocals.

12. a. $f(g(x)) = f\left(\dfrac{1 - x}{2}\right) = -2\left(\dfrac{1 - x}{2}\right) + 1 = x$;

$g(f(x)) = g(-2x + 1) = \dfrac{1 - (-2x + 1)}{2} = \dfrac{2x}{2} = x$

Because $f(g(x)) = g(f(x)) = x$, f and g are inverses.

b.

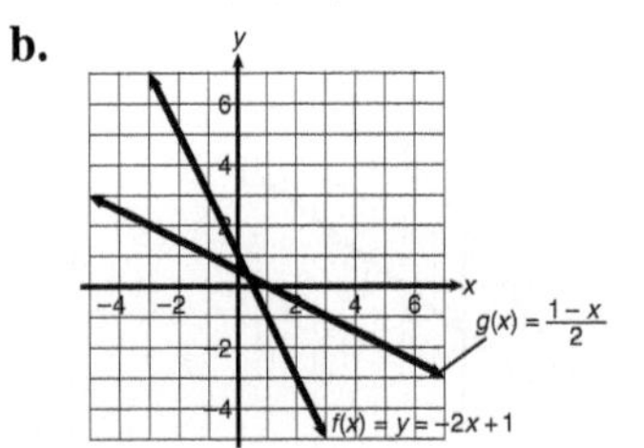

c. f and g are symmetric with respect to the line $y = x$;

13. a. Yes, the ratio (change in cost)/(change in number of tickets) is constant. **b.** $f(n) = 8.85n$, **c.** The cost of one ticket is \$8.85. This represents the slope.

d.

TOTAL COST	NUMBER OF TICKETS
\$17.70	2
\$44.25	5
\$61.95	7
\$106.00	12

e. $n = g(c) = \frac{1}{8.85}c = \left(\frac{20}{177}\right)c$,

f. The slope is $\frac{20}{177}$. The slopes are reciprocals.

g. $f(g(c)) = f\left(\frac{20}{177}\right)c = 8.85\left(\frac{20}{177}\right)c = c$; $g(f(n)) = g(8.85n) = \frac{20}{177}(8.85n) = n$. The functions are inverses because they undo one another.

Chapter 3

Activity 3.1 Exercises: 1. a. 2, 4, 8, 16, 32, 64, 128, 256, **b.** $2 = 2^1, 4 = 2^2, 8 = 2^3, 16 = 2^4, 32 = 2^5, 64 = 2^6, 128 = 2^7, 256 = 2^8$, **c.** $P(n) = 2^n$, **d.** $P(31) =$ 2,147,483,648 cents = \$21,474,836.48. No, I could not afford him. **e.** The function is not linear. The average rate of change is not constant. **f.** The input, n, is the July date; so the practical domain is whole numbers $n = 1$ to 31.

g.

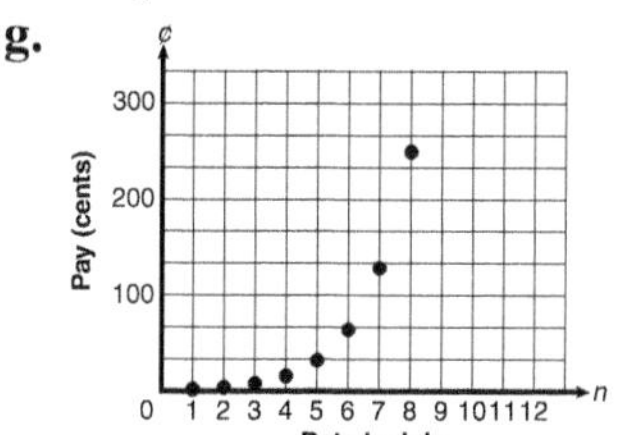

h. The function is discrete because the function is not defined for real numbers between 1 and 2, 2 and 3, and so on.

i. $2^{63} = 9.223372036854775808 \times 10^{18} =$ 9,223,372,036,854,775,808

9 quintillion 223 quadrillion 372 trillion 36 billion 854 million 775 thousand 808

3. a. top table: 0.008, 0.04, 0.2, 1, 5, 25, 125; bottom table: 0.0537, 0.1424, 0.3774, 1, 2.65, 7.0225, 18.61

b.

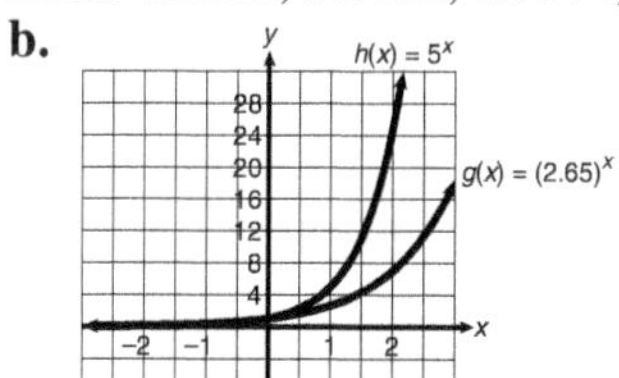

c.

BASE, b	GROWTH FACTOR	x-INTERCEPT	y-INTERCEPT	HORIZONTAL ASYMPTOTE	INCREASING OR DECREASING
5	5	none	(0, 1)	$y = 0$	increasing
2.65	2.65	none	(0, 1)	$y = 0$	increasing

5. a. This data is exponential with a growth factor of 3. **b.** This data is linear with a slope of 0.5. **c.** This data is exponential with a growth factor of 2. **7. a.** 2, 4, 8, 16, 32, **b.** 256, **c.** The data is exponential with a growth factor of 2. **d.** (Answers may vary.) The practical domain is the set of nonnegative integers from 0 to 10. The practical range is the set of whole number powers of 2 up to 2^{10}.

Activity 3.2 Exercises: 1. a. first table 23.32, 8.16, 2.86, 1, 0.35, 0.1225, 0.043, second table 125, 25, 5, 1, 0.2, 0.04, 0.008,

b.

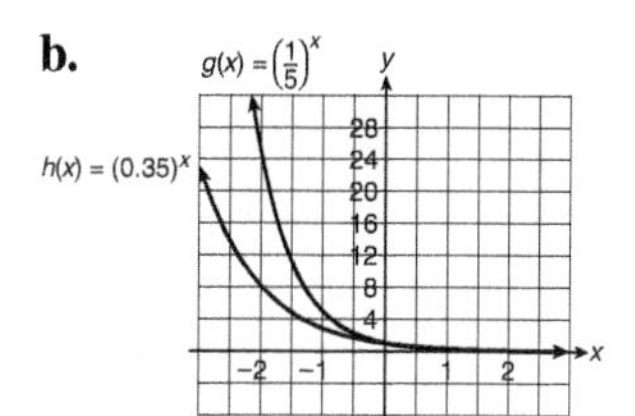

c. first row 0.35, decay, none, (0, 1) $y = 0$, decreasing, second row $\frac{1}{5}$, decay, none, (0, 1), $y = 0$, decreasing;

3. a. This data is linear with a slope of 0.5. **b.** This data is exponential with a growth factor of 4. **c.** This data is exponential with a decay factor of 0.4. **5. a.** The function f will increase slower than g because its growth factor is smaller, **b.** The function f will decrease faster than g because its decay factor is smaller.

Activity 3.3 Exercises: 1. a. $P = 328.1(0.991)^t$, **b.** Close; substituting 26 for t yields $P = 328.1(0.991)^{26} \approx 259.4$. **c.** $P = 328.1(0.991)^{35} \approx 239.1$ thousand; **d.** Not too much confidence in the projection. In 10 years, a lot of factors can enter into the increase or decrease of a population. **2. a.** ii, **b.** i; **4. a.** The function is increasing. The base 5 is greater than 1 and is a growth factor. **b.** The function is decreasing. The base $\frac{1}{2}$ is less than 1 and is a decay factor. **c.** The function is increasing. The base 1.5 is a growth factor (greater than 1). **d.** The function is decreasing. The base 0.2 is a decay factor (less than 1).

6. a. $f(-2) = \frac{3}{16}$, **b.** $f\left(\frac{1}{2}\right) = 6$, **c.** $f(2) = 48$, **d.** $f(1.3) \approx 18.19$; **8. a.** 2.5 ppm, **b.** 2.5, 1.75, 1.225, 0.8575, 0.6003, 0.4202,

c.

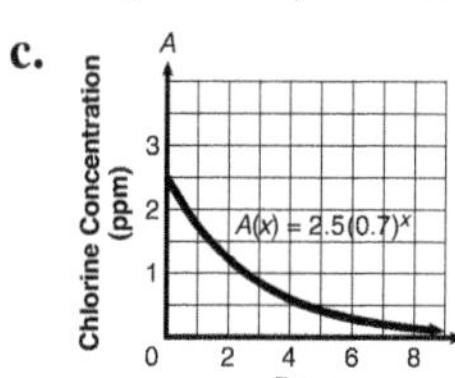

d. $A(3) = 0.8575$ ppm, **e.** Chlorine should be added in 1.4 days. **9. a.** Possible, the consecutive ratios are approximately constant. **b.** The growth factor is $b \approx 1.035$. **c.** $N = 62.4(1.035)^t$,

d.

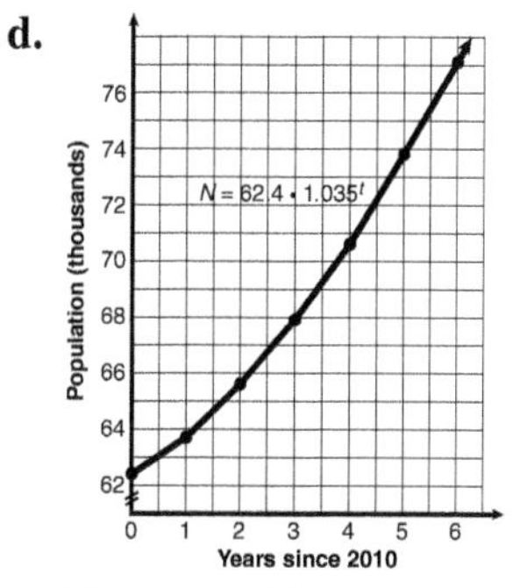

e. The vertical intercept is (0, 62.4). 62.4 is the population of Ft. Myers, in 2010 measured in thousands. **f.** In 2008, $t = 18$, so $N = 62.4 \cdot (1.035)^{18} \approx 115.9$ thousand. This is probably not a good estimate. 18 years is extrapolating too far into the future. **g.** According to the model, the population of Ft. Myers, will double in 20.1 years.

Activity 3.4 Exercises: 1.

GROWTH FACTOR	GROWTH RATE
1.02	2%
1.029	2.9%
2.23	123%
1.34	34%
1.0002	0.02%

DECAY FACTOR	DECAY RATE
0.77	23%
0.32	68%
0.953	4.7%
0.803	19.7%
0.9948	0.52%

2. a. Miami: $M(t) = 400.9(1.0136)^t$, Baltimore: $B(t) = 642.0(0.9948)^t$ **b.** 2020 is 10 years from 2010 so $t = 10$. Miami: $M(10) = 400.9(1.0136)^{10} =$ 458.9 thousand, Baltimore: $B(10) = 642.0(0.9948)^{10} =$ 609.4 thousand **c.** The population of Baltimore will have in 133 years from 2010. **d.** The populations will be equal in the 25th year after 2010 or 2035

4. a.

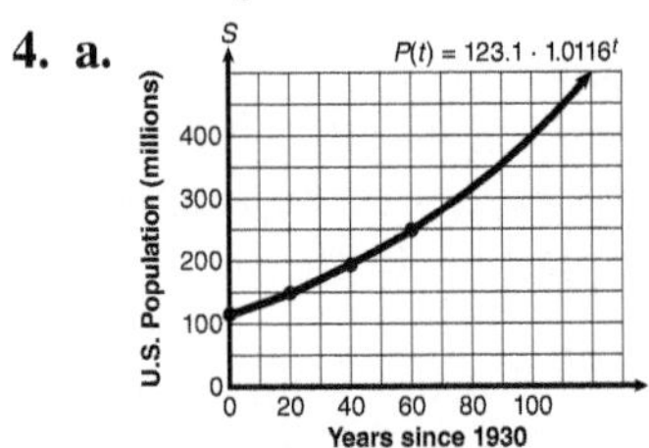

b. The growth factor is $b = 1.0116$; the growth rate is 0.0116, or 1.16%. **c.** $P(85) = 123.1 \cdot 1.0116^{85} \approx 328.1$ million. The model is a little higher than the actual. **5. a.** $V(t) = 25{,}000(0.85)^t$, **b.** The decay factor is $b = 0.85$ **c.** The decay rate is 15%. **d.** $V(5) = 25{,}000(0.85)^5 \approx \$11{,}092.63$,

e.

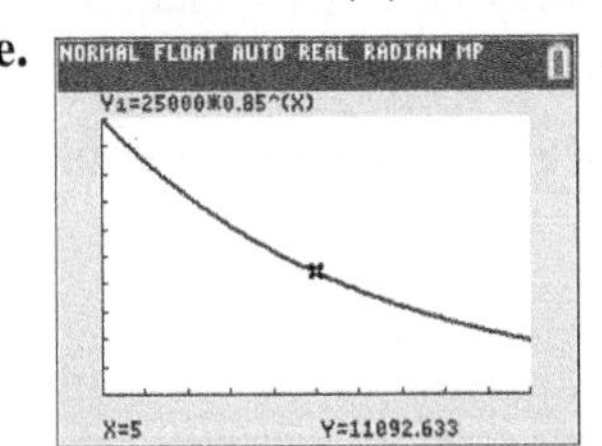

f. The value will be $15,000 when $t \approx 3.1$ years.

Activity 3.5 Exercises: 1. a. $A = 25{,}000\left(1 + \dfrac{0.045}{4}\right)^{4t}$, **c.** approximately 21.4 years, **e.** approximately 21.2 years; **3. a.** $A = 1900e^{0.03 \cdot 2} \approx \$2017.49$, **b.** approximately 23.1 years; **5. a.** $b = \left(1 + \dfrac{0.048}{12}\right)^{12}$, **b.** $b \approx 1.04907$, **c.** $r_e = 4.907\%$

Activity 3.6 Exercises: 1. a. Increasing. $b = 1.015$ is greater than 1 so it is a growth factor. **b.** The annual growth rate is $r = b - 1 = 1.015 - 1 = 0.015 = 1.5\%$ **c.** The initial value is $f(0) = 88.5(1.015)^0 = 88.5$. The 2010 ($t = 0$) population of Orem, Utah was 88.5 thousand (88,500) people. **d.** $k \approx 0.0149$ **e.** $f(t) = 88.5e^{0.0149t}$ **f.** 2025 corresponds to $t = 15$. $f(15) = 88.5e^{0.0149 \cdot 15} = 110.664$. If the population continues to grow at the same rate, the 2025 population will be 110,664 people. **3. a.** The initial value is 33 and R is increasing at the rate of 9.7%. **b.** The initial value is 97.8 and $f(x)$ is decreasing at the continuous rate of 23%. **c.** The initial value is 3250 and S is decreasing at the rate of 27%. **d.** The initial value is 0.987 and B is increasing at the continuous rate of 7.6%. **5. a.** $y = 20e^{-0.0244(20)} \approx 12.277$ grams, **b.** The half-life is approximately 28.4 years. **c.** The decay factor is $b = e^{-0.0244} \approx 0.9759$. The decay rate is 0.0241 or 2.41%. **7. a.** 2012 is 5 years from 2007 so $t = 5$. $f(5) = 15.75e^{0.247 \cdot 5} = 54.154$ billion. Amazon's 2012 revenue was $54,154,000,000.

b.

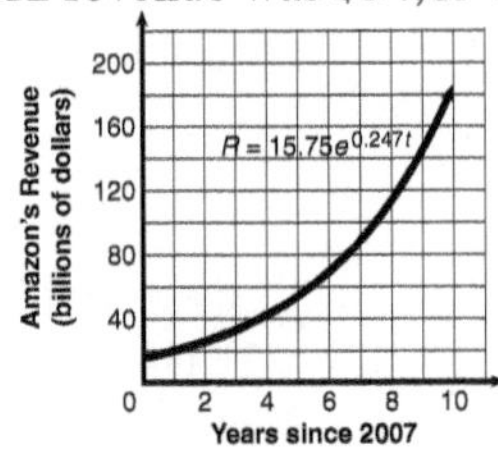

c. The vertical intercept of the graph is (0, 15.75). Amazon's 2007 revenue was $15,750,000,000. **d.** Amazon's revenue will reach $100 billion in about 7.5 years from 2007 or mid 2014. **e.** The doubling time is approximately 2.8 years.

Activity 3.7 Exercises:

1. a.

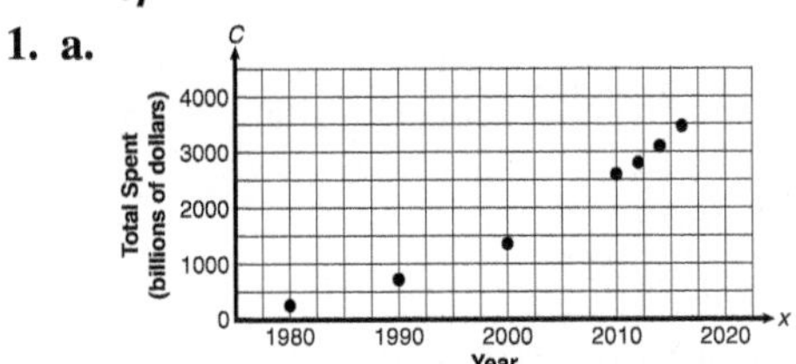

b. $C(t) = 299.73(1.0728)^t$ **c.** 2005 is 25 years from 1980; so $t = 25$. $C(25) = 299.73(1.0728)^{25} = 1736.598$. The 2005 estimated expenditure is $1,736,598,000,000. **d.** The growth factor is $b = 1.0728$. **e.** The growth rate is $r = 1.0728 - 1 = 0.0728 = 7.28\%$ **f.** the healthcare costs will exceed $1.5 trillion 23 years from 1980 or in 2003. **g.** The health costs will double in about 9.9 years. **3. a.** the set of all real numbers, **b.** the set of all positive real numbers, **c.** $y = a \cdot b^x$ is positive for all values of x. **d.** $y = a \cdot b^x$ is never negative. **e.** $(0, a)$

How Can I Practice? 1. a. $C = 23{,}000(1.04)^t$, **b.** The growth rate is 0.04; the growth factor is 1.04. **c.** $C = 23{,}000(1.04)^3 \approx \$25{,}871.87$, **d.** 10.7 years; **2. a.** Graph i is function g. **b.** Graph ii is function h. **3.** Graph i is function g because it is decreasing with a decay factor of 0.47, which is between 0 and 1. Graph ii is function h because it is increasing with the growth factor of 1.47. **4. a.** 13.01, 33.18, 84.61, **b.** 1.26, 0.76, 0.46, **c.** 216, 7776, 279,936; **5. a.** $y = 2(2.55)^x$, **b.** $y = 3.5(0.6)^x$, **c.** $y = \frac{1}{6}(36)^x$; **6. a.** $f(0) = 1.3$; decreasing, **b.** $f(0) = 0.6$; increasing, **c.** $f(0) = 3$; decreasing; **7. a.** Yes, **b.** The constant ratio is approximately 2.5 or $\frac{5}{2}$, **c.** $y = 2(2.5)^x$; **8. a.** Plan 1: $S = 30{,}000 + 1000x$; Plan 2: $S = 30{,}000(1.03)^x$, **b.** Plan 1: $30,000, $31,000, $33,000, $35,000, $40,000, $45,000; Plan 2: $30,000, $30,900, $32,782, $34,778, $40,317, $46,739,

c. It depends. If I plan to be with the company less than 8 years, I will take plan 1 because it takes plan 2 about 8 years to catch up. If I expect to be with the company for a long time, say 20 years, I will choose plan 2 because by then, I will be better off by more than \$4000 per year.
9. a. $N = f(t) = 2e^{0.075t}$,
b. $N = f(8) = 2e^{0.075(8)} \approx 3.6442$ thousand,
c.

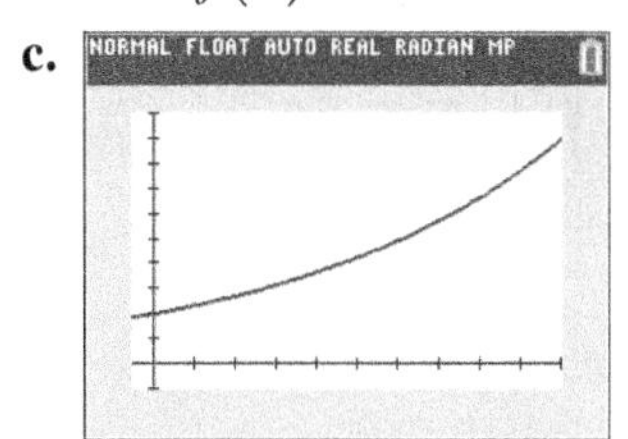

d. 14.6482 or about 15 weeks;

10. a. top table: $\frac{1}{64}, \frac{1}{16}, \frac{1}{4}$, 1, 4, 16, 64; bottom table: 64, 16, 4, 1, $\frac{1}{4}, \frac{1}{16}, \frac{1}{64}$, **b.**

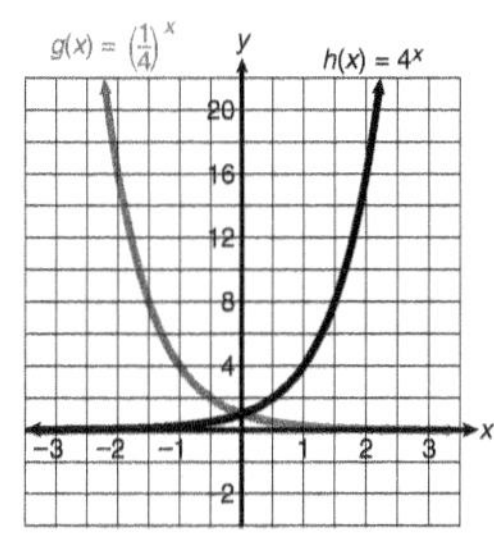

c. Row 1: 4, growth, none, (0, 1), x-axis, increasing; row 2: $\frac{1}{4}$, decay, none, (0, 1), x-axis, decreasing;
11. a. \$415, **b.** 0.0118 or 1.18% per month, **c.** 1.0118, **d.** $f(x) = 415(1.0118)^x$, **e.** (0, 415), **f.** This represents the initial balance on the card. **g.** \$466.65, **h.** With no payments, I exceed my credit limit during the sixteenth month; **12. a.** The ratios are all approximately 1.02, **b.** 1.02, **c.** $w(t) = 12.50(1.02)^t$, **d.** 2%, **e.** \$15.24, **f.** about 35 years; **13. a.** $A = 10{,}000(1.0025)^{12t}$, **b.** approximately \$11,616.17, **c.** approximately 23 years, **d.** $A = 10{,}000e^{0.03t}$, **e.** \$11,618.34, \$2.17 more than in part b;
14. a.

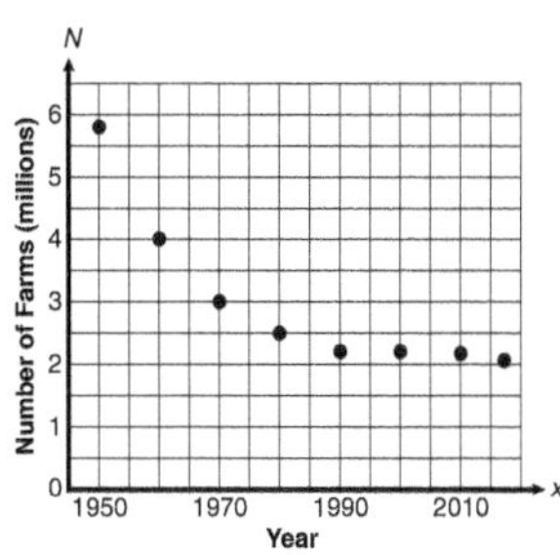

b. An exponential decay model would better model the data. The data is decreasing, but not at a constant rate.
c. $N = 4.5569(0.98597)^x$, **d.** $4.5569(0.9897)^{70} \approx 1.695$ million farms, **e.** 0.98597, **f.** $0.9897 = 1 - r$, $r = 1 - 0.98597 = 0.01403$ or 1.403%, **g.** The number of farms is decreasing at a rate of 1.403% per year, **h.** The half-life is approximately 49 years.

Activity 3.8 Exercises: 1. a. 5, **b.** 3, **c.** −1, **d.** −6, **e.** 0, **f.** 2, **g.** $\frac{1}{2}$, **h.** $\frac{1}{2}$, **i.** 0, **j.** 0, **k.** 5, **l.** −2, **m.** 0;
3. a. $\log_3 9 = 2$, **b.** $\log_{121} 11 = \frac{1}{2}$, **c.** $\log_4 27 = t$, **d.** $\log_b 19 = 3$; **5. a.** $x = 0.512$, **b.** $x = 2.771$, **c.** $x = -5.347$

Activity 3.9 Exercises: 1. a. $x > 0$, **b.** the set of all real numbers, **c.** $x > 1$, **d.** $0 < x < 1$, **e.** $x = 1$, **f.** $x = 10$; **3.** $y = \log_2 x$; **4. a.** $x > 0$, **b.** the set of all real numbers, **c.** $x > 1$, **d.** $0 < x < 1$, **e.** $x = 1$, **f.** $x = e$;
5. $n = \dfrac{\log(2500) - \log(45{,}000)}{\log(1 - 0.40)} \approx 5.7$ years

Activity 3.10 Exercises: 1. a.

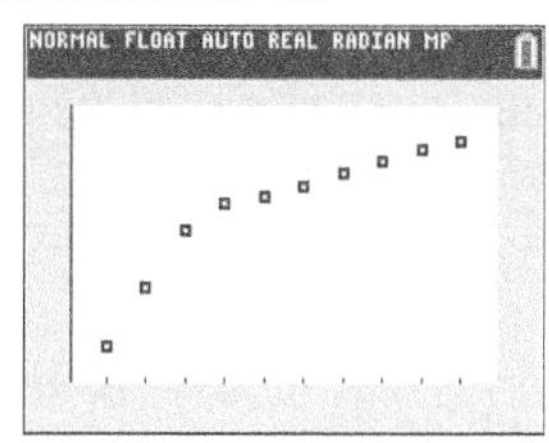

b. It could very well be logarithmic. It tends to increase more slowly as the input increases.
c. $g(t) = 757.43 + 434.68\ln(t)$, **d.** Yes, it is a very good fit, **e.** $T = 14$; $g(14) = 757.43 + 434.68\ln(14) \approx 1904.6$. The model predicts that in 2025, there will be 1904.6 million smartphones shipped globally.
3. a. $f(0) = 27$ inches. This is the pressure in the eye of the storm.
b.

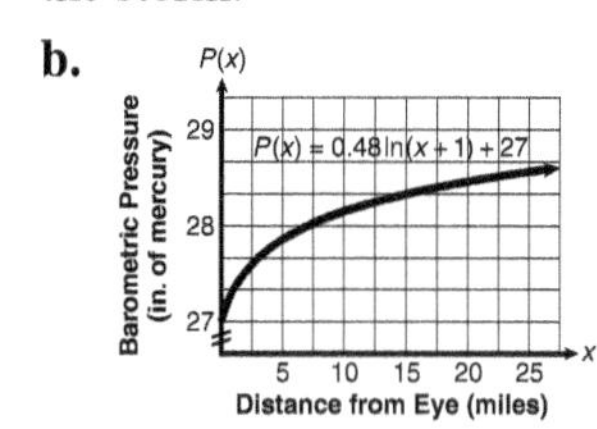

c. As you move away from the hurricane's eye, the pressure increases quickly at first and then more slowly.

5. a.

b. $R = 80.4 - 11\ln(300) \approx 17.66$ cubic feet;
7. a. $t = \dfrac{\ln 2}{4\ln\left(1 + \dfrac{0.055}{4}\right)} \approx 12.689$ years,
b. $t = \dfrac{\ln 3}{4\ln\left(1 + \dfrac{0.055}{4}\right)} \approx 20.11$ years,
c. $t = \dfrac{\ln 2}{12\ln\left(1 + \dfrac{0.055}{12}\right)} \approx 12.632$ years

Activity 3.11 Exercises: 1. a. $\log_b 3 + \log_b 7$, **b.** $\log_3 3 + \log_3 13 = 1 + \log_3 13$, **c.** $\log_7 13 - \log_7 17$, **d.** $\log_3 x + \log_3 y - \log_3 3 = \log_3 x + \log_3 y - 1$;

3. a.

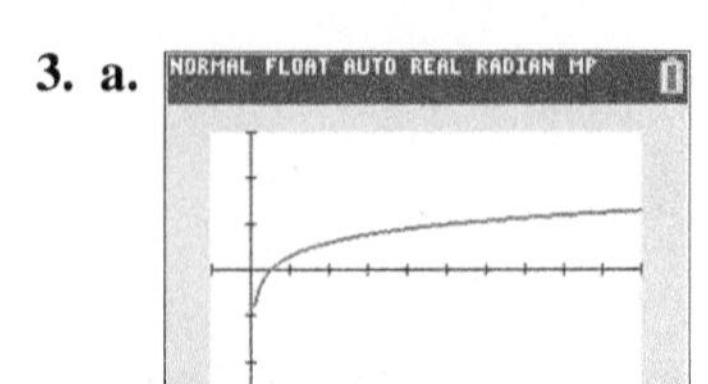

b. The graphs are the same. This is not surprising because the log of a product is the sum of the logs.

5. a. $7.4 \log(15) \approx 8.7$ or 9 cars, **b.** 9, 13, 21, **c.** The sum of the sales from the smaller ads exceeds the sales from the larger ad by 1. **d.** Fairly close. 15 times 50 equals 750, so I would have expected the sum of the sales from the smaller ads to equal the sales from the largest. The error is due to rounding. **e.** Forget about the giant ad. It is a waste of money. **7. a.** $\log_2 245$, **b.** $\log \sqrt[4]{\dfrac{x^3}{z^5}}$, **c.** $\ln \dfrac{2^2 5^2 z^4}{5^3} = \ln \dfrac{4z^4}{5}$, **d.** $\log_5 \dfrac{x^2 + 3x + 2}{x^2 + 6x + 9}$; **9. a.** 0.8271, **b.** 0.7557, **c.** 1.5011, **d.** 0.7112

Activity 3.12 Exercises: 1. a. $D = 23.3 \cdot 1.312^t$, **b.** $D = 23.3 \times 1.312^{13} \approx 795.2$ million dollars. **c.** $23.3 \cdot 1.312^t = 2000$

$$t \ln(1.312) = \ln\left(\frac{2000}{23.3}\right),\ t = \frac{\ln\left(\frac{2006}{23.3}\right)}{\ln(1.312)} \approx 16.4.$$

According to the model, the amount of cyber-crime will first reach 2 billion dollars 16.4 years after 2000, or in 2017.

3. a. Yes

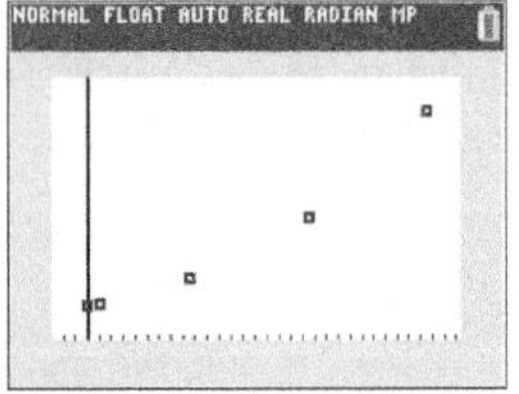

b. $A(t) = 352.65(1.006)^t$,

c. $t = \dfrac{\ln\left(\frac{705.3}{352.65}\right)}{\ln(1.006)} \approx 116$; 116 years after 1990 would be the year 2106.

d.

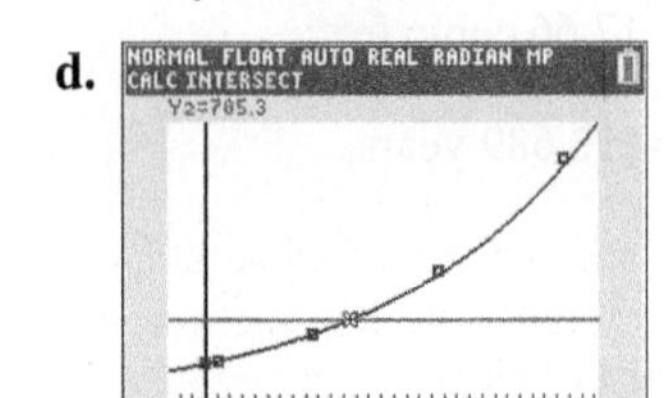

4. $x = \dfrac{\ln 14}{\ln 2} \approx 3.81$; **6.** $t = \dfrac{\ln 2}{\ln(1.04)} \approx 17.7$;

8. $x \approx 1.881$; **10. a.** $t \approx 8$ days, **b.** $\frac{1}{5}P_0 = P_0 e^{-0.086t}$ or $t = \dfrac{\ln(0.2)}{-0.086} \approx 19$ days

How Can I Practice? 1. a. $\log_4 16 = 2$, **b.** $\log_{10}(0.0001) = -4$, **c.** $\log_3\left(\frac{1}{81}\right) = -4$;

2. a. $2^5 = 32$, **b.** $5^0 = 1$, **c.** $10^{-3} = .001$, **d.** $e^1 = e$;

3. a. $x = 4^{-3} = \frac{1}{64}$, **b.** $b = 2$, **c.** $y = 3$;

4. a. $-1; -0.5; 0; 1; 2; 3$, **b.**

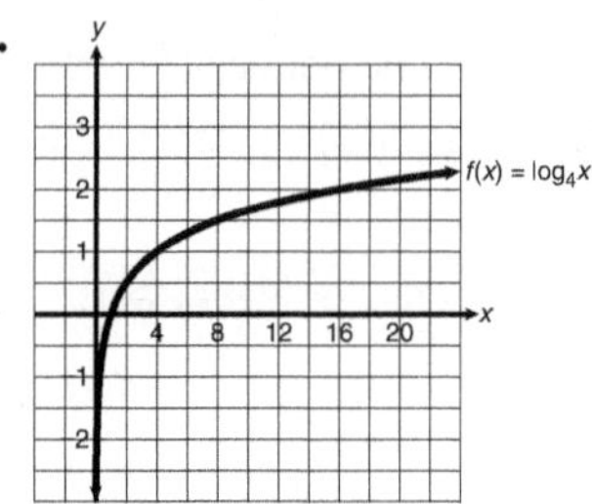

d. $(1, 0)$, **e.** $x > 0$, **f.** all real numbers, **g.** increasing, **h.** The y-axis ($x = 0$) is a vertical asymptote. **i.** $f(32) = 2.5$, **j.** $x \approx 90.5$;

5. a. $\log_b x + 2\log_b y - \log_b z$, **b.** $\frac{3}{2}\log_3 x + \frac{1}{2}\log_3 y - \log_3 z$, **c.** $\log_5 x + \frac{1}{2}\log_5(x^2 + 4)$, **d.** $\frac{1}{3}\log_4 x + \frac{2}{3}\log_4 y - \frac{2}{3}\log_4 z$; **6. a.** $\log \dfrac{x\sqrt[3]{y}}{\sqrt{z}}$, **b.** $\log_3 (x + 3)^3 z^2$, **c.** $\log_3 \sqrt[3]{\dfrac{x}{y^2 z^4}}$;

7. a. $\dfrac{\log 17}{\log 5} = 1.76$, **b.** $\dfrac{1}{3} \cdot \dfrac{\log 41}{\log 13} = 0.4826$;

8. a. $x \approx 0.0067$, **b.** $x = 646.08$;

9. a. $x = \dfrac{\log 17}{\log 3} = 2.5789$, **b.** $x = \dfrac{\ln 14}{1.7} \approx 1.55$;

10. a. $V = 10{,}000 \cdot e^{0.04t}$, **b.** $t \approx 17.3$. It will take 17.3 years for the account to grow to $20,000.

c.

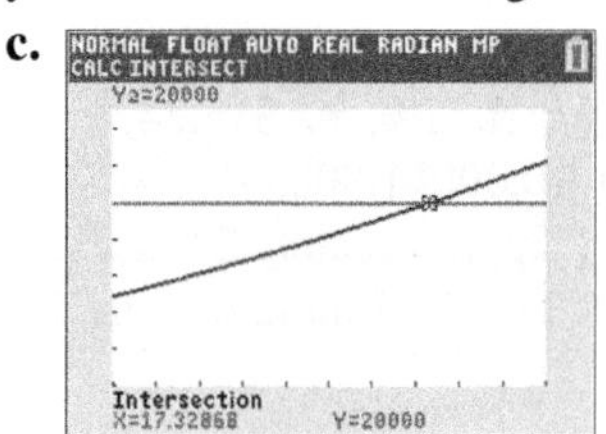

11. a.

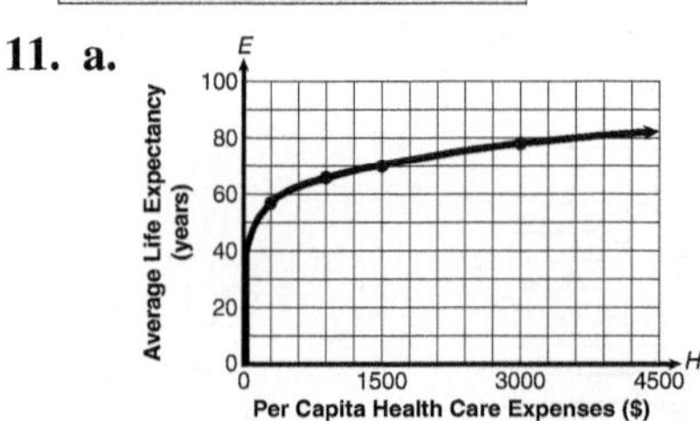

b. $E = 0.035 + 9.669 \ln(1500) \approx 70.7$ years, **c.** $H \approx \$2865$

Gateway Review 1. a. The growth factor is $100\% + 3\% = 103\% = 1.03$. Let T represent the tuition. $T = 300(1.03)^t$, $T = 300(1.03)^5 \approx 347.78$. In 5 years, the tuition will be $347.78 per credit. $T = 300(1.03)^{10} \approx 403.17$. In 10 years, the tuition will be $403.17 per credit.

b. $\dfrac{347.78 - 300}{5} = \dfrac{47.78}{5} \approx 9.56$; Tuition increases

\$9.56 per credit per year. **c.** $\frac{403.17 - 300}{10} = \frac{103.17}{10} \approx 10.32$, **c.** Tuition increases \$10.32 per credit per year. **d.** Tuition will double in approximately 23.4 years if the rate of increase stays at 3%. **2. a.** $\frac{1}{8}, \frac{1}{2}$; 1; 8; 16; 64; 512,

b.

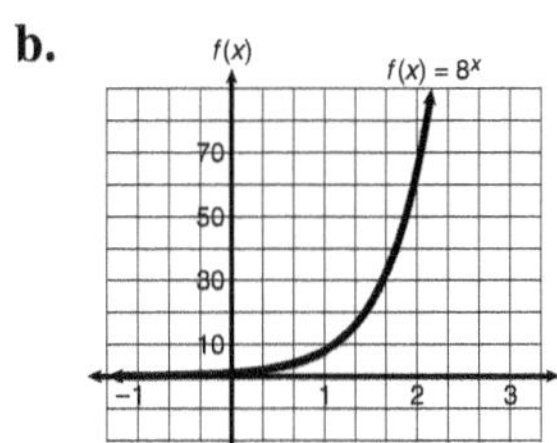

c. The function is increasing because $b = 8 > 1$. **d.** all real numbers, **e.** $y > 0$, **f.** There is no x-intercept. The y-intercept is (0, 1). **g.** There is one horizontal asymptote, the x-axis, $y = 0$,

h. The domain and range are the same. The graphs are reflections in the y-axis. f is increasing; g is decreasing. **i.** f is moved vertically upward 5 units to obtain h. **j.** Solve $y = 8^x$ for x: $x = \log_8 y$. Interchange x and y: $y = \log_8 x$.

3.

BASE, b	GROWTH OR DECAY FACTOR	x-INTERCEPT	y-INTERCEPT	HORIZONTAL ASYMPTOTE	INCREASING OR DECREASING
6	growth	none	(0, 1)	$y = 0$	increasing
$\frac{1}{3}$	decay	none	(0, 1)	$y = 0$	decreasing
2.34	growth	none	(0, 5)	$y = 0$	increasing
0.78	decay	none	(0, 3)	$y = 0$	decreasing
2	growth	(2, 0)	(0, −3)	$y = -4$	increasing

4.

all reals	all reals	all reals	$x > 0$	$x > 3$
$y > 0$	$y > 2$	$y > -5$	all reals	all reals

5. a. The table is approximately exponential. The growth factor is about 1.55. **b.** $y = 10 \cdot 1.55^x$; **6. a.** 12.48, 25.46, 51.94, **b.** 2.21, 1.55, 1.09, **c.** 64, 1024, 16,384, **d. i.** $y = 3.00(2.04)^x$, **ii.** $y = 4.50(0.7)^x$, **iii.** $y = 0.25(16)^x$; **7. a.** \$35,000, \$35,525, \$36,058, \$36,599, \$37,148, \$37,705, **b.** $y = 35{,}000(1.015)^x$, **c.** $x = 8$; $y = \$39{,}427$ This is realistic if you assume that \$35,000 is a reasonable starting salary and that the 1.5% salary increase per year remains constant. **d.** $70{,}000 = 35{,}000(1.015)^x$; $2 = 1.015^x$; $x = \frac{\ln 2}{\ln 1.015} \approx 46.6$ yr

8. a. $A = 10{,}000 \cdot e^{0.04(8)}$; $A = \$13{,}771.28$ **b.** $16{,}000 = 10{,}000 \cdot e^{(0.04 \cdot t)}$; $0.04t = \ln(8/5)$, $t \approx 11.75$ years c. **9. a.** 125, **b.** 27, **c.** $\frac{1}{32}$, **d.** 25, **e.** −2, **f.** 4, **g.** −3, **h.** 2; **10. a.** $\log_6 36 = 2$, **b.** $\log_{10} 0.000001 = -6$, **c.** $\log_2 \frac{1}{32} = -5$;

11. a. $3^4 = 81$, **b.** $7^0 = 1$, **c.** $10^{-4} = 0.0001$, **d.** $e^1 = e$, **e.** $q^b = y$; **12. a.** $x = \frac{1}{125}$, **b.** $b = 4$, **c.** $y = 6$, **d.** $x = 8$; **13. a.** −3, −2, −1, 0, 1, 2,

b. **c.**

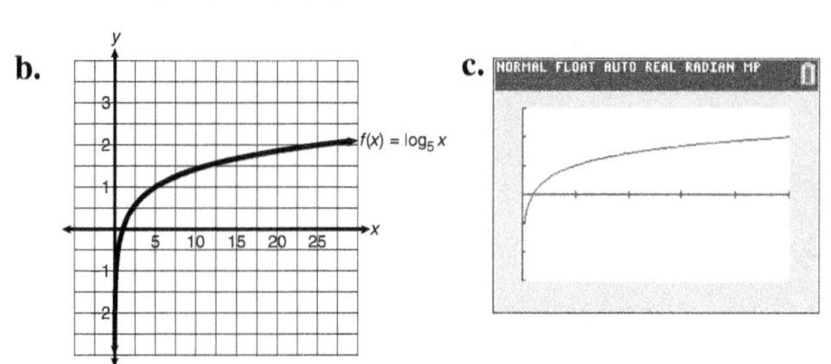

d. (1, 0), **e.** $x > 0$, **f.** all real numbers, **g.** It has a vertical asymptote at $x = 0$. The function gets closer and closer to the y-axis but does not cross it. **h.** $f(23) \approx 1.948$, **i.** $x \approx 52.416$; **14. a.** $\frac{\log 21}{\log 7} \approx 1.56$, **b.** $\frac{\log\left(\frac{8}{9}\right)}{\log 15} \approx -0.0435$;

15. a. $3\log_2 x + \log_2 y - \left(\frac{1}{2}\right)\log_2 z$,
b. $\left(\frac{1}{3}\right)(4\log x + 3\log y - \log z)$; **16. a.** $\log \frac{x\sqrt[4]{y}}{z^3}$,
b. $\log \sqrt[3]{\frac{x}{y^2 z}}$; **17. a.** $3 + x = \frac{\log 7}{\log 3}$; $x \approx -1.23$,
b. $4x + 9 = 2^4$; $x = 1.75$, **c.** $x \approx 341.5$;

18. a.

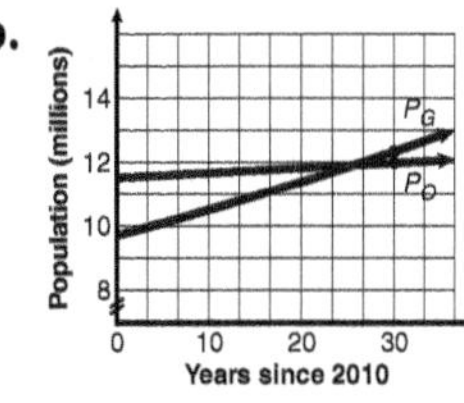

c. 2.87744,
d. $2.319 = \frac{\log x}{2\log 2}$; $x \approx 24.9$;

19. a. Ohio: 11.53 million; Georgia: 9.71 million

b.

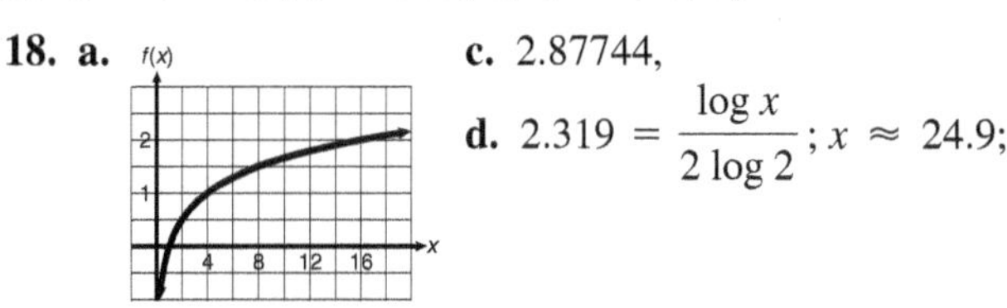

c. The models indicate that the populations of Ohio and Georgia will be equal 25.6 years after 2010 or sometime in the year 2035. **d.** $9.71 \cdot e^{0.00} > 13, e^{0.008x} > \frac{13}{9.71}$, $0.008x > \ln\left(\frac{13}{9.71}\right)$, $x > \frac{\ln(^{13}/_{9.71})}{0.008}$, $x > 36.5$
According to the model sometime during the year 2046 the population of Georgia will first exceed 13 million people
20. a. $y = 45.786 - 6.903\ln x$ **b.** $y = 45.786 - 6.903 \ln(500) \approx 2.887$ km.

Chapter 4

Activity 4.1 Exercises: 1. a. 9, 4, 1, 0, 1, 4, 9,

b.

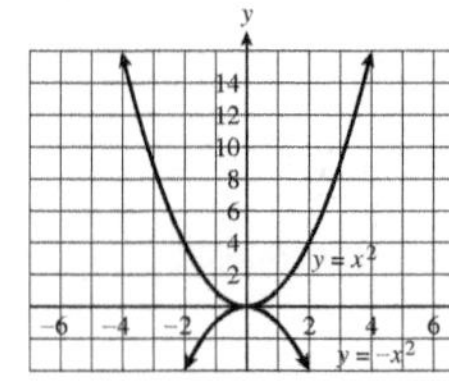

c. 1, **d.** The domain is all real numbers. The range is all real numbers greater than or equal to 0. **e.** $-9, -4, -1, 0, -1, -4, -9$, **g.** -1, **h.** The graph of $y = -x^2$ is a reflection of $y = x^2$ over the x-axis.

2. a. $a = -2, b = 0, c = 0$, **b.** $a = \frac{2}{5}, b = 0, c = 3$, **c.** $a = -1, b = 5, c = 0$, **d.** $a = 5, b = 2, c = -1$; **4. a.** f opens upward; g opens downward; both pass through $(0, 0)$. **b.** Both f and h open upward. h is wider than f. **c.** h is g shifted up 2 units; both open upward. **d.** Both f and g open upward. The low point of f is 3 units below the x-axis; the low point of g is 3 units above the x-axis. **e.** f opens upward with a vertical intercept at $(0, 1)$; h opens downward with a vertical intercept at $(0, -1)$; both are symmetric with respect to the y-axis. **6. a.** downward, **b.** $(0, -4)$; **8. a.** upward, **b.** $(0, 3)$; **10. a.** downward, **b.** $(0, -7)$; **12. a.** The graph of $y = \frac{3}{5}x^2$ is wider than the graph of $y = x^2$. **b.** The graph of $y = x^2$ would have a greater output value for any input x except $x = 0$.

Activity 4.2 Exercises: 1. a. upward, **b.** $x = 0$, **c.** $(0, -3)$, minimum, **d.** $(0, -3)$; **3. a.** upward, **b.** $x = -2$, **c.** $(-2, -7)$, minimum, **d.** $(0, -3)$; **5. a.** upward, **b.** $x = -1.5$, **c.** $(-1.5, 1.75)$, minimum, **d.** $(0, 4)$; **7. a.** upward, **b.** $x = 0.25$, **c.** $(0.25, -3.125)$, minimum, **d.** $(0, -3)$; **9. a.** (1, 0), (6, 0), **b.** D: all real numbers; R: $g(x) \leq 6.25$, **c.** $x < 3.5$, **d.** $x > 3.5$; **11. a.** $(3.46, 0), (-3.46, 0)$, **b.** D: all real numbers; R: $y \geq -12$, **c.** $x > 0$, **d.** $x < 0$; **13. a.** $(-1, 0), (3, 0)$, **b.** D: all real numbers; R: $g(x) \leq 4$, **c.** $x < 1$, **d.** $x > 1$; **15. a.** (0.2, 0), (1, 0), **b.** D: all real numbers; R: $y \leq 0.8$, **c.** $x < 0.6$, **d.** $x > 0.6$; **17. a.** 149 feet, **b.** 6.05 seconds, **c.** It indicates the height of the arrow when it is shot, **d.** The practical domain is 0 second $\leq x \leq$ 6.05 seconds. The practical range is 0 feet $\leq h(x) \leq$ 149 feet, **e.** $(-0.05, 0), (6.05, 0)$; the first has no meaning; the second indicates the time in seconds it takes for the arrow to hit the ground. **19. a.** (30, 200),

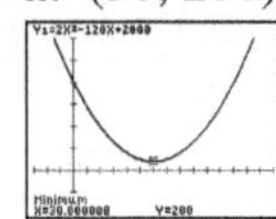

b. $x = -\frac{b}{2a} = \frac{120}{4} = 30, \overline{C}(30) = 200$; the vertex is (30, 200). **c.** They are the same. **d.** minimum point, **e.** The average cost of production is minimized when 30 oil lamps are produced. **f.** (0, 2000); it costs \$2000, even if no oil lamps are produced.

Activity 4.3 Exercises: 1. $x = 6$ or $x = -2$; **3.** $x = 9$ or $x = -5$; **5.** $x = -11$ or $x = -1$; **7.** $x = \pm 5$; **9.** $x = -3$ or $x = 1$; **11.** $x = 7$ or $x = -4$; **13. a.** $-2 < x < 6$, **b.** $x < -2$ or $x > 6$; **15. a.** $d(55) = 181.5$ ft., **b.** $0.04v^2 + 1.1v = 200$; $v \approx 58$ mph

Activity 4.4 Exercises: 1. $6x^5(2 - 3x^3)$; **3.** $2x(x^2 - 7x + 13)$; **5.** $(x + 3)(x - 2)$; **7.** $(x + 5y)(x + 2y)$; **9.** $(6 + x)(2 + x)$; **11.** $(3x - 2)(x + 7)$; **13.** $5b^2(4b + 3)(b - 4)$; **14.** $x = 3$ or $x = 2$; **16.** $x = 3$ or $x = -2$; **18.** $x = \frac{1}{3}$ or $x = -4$; **20.** $x = 9$ or $x = -2$;

22. a.

x 20 ft. x

15 ft.

x x

b. $A = (20 + 2x)(15 + x) - 15(20) = 300 + 20x + 30x + 2x^2 - 300 = 2x^2 + 50x$, **c.** $2x^2 + 50x = 168$, **d.** $x + 28 = 0$ or $x - 3 = 0$, $x = -28$ or $x = 3$; the solution is 3 ft. -28 ft. makes no sense in this situation.

Activity 4.5 Exercises:

1. a.

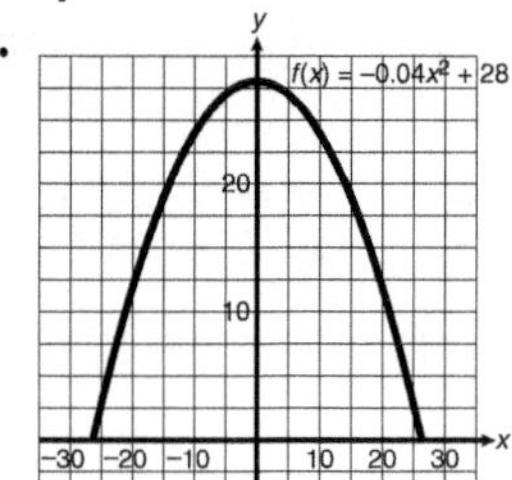

b. (0, 28) represents the vertex or turning point of the arch. **c.** $x \approx \pm 26.5$; the intercepts are (26.5, 0) and $(-26.5, 0)$. **d.** The intercepts are the same. **e.** The river is approximately 2(26.5) or 53 feet wide. **f.** No; the highest point of the arch is 28 feet above the water. **g.** $-0.04x^2 + 28 = 20$, **h.** $x = \pm 14.14$ ft. Place the pole approximately 14.14 feet to the right or left of the center of the arch.

3. $x = -\frac{1}{2}$; **5.** $x = \frac{6 \pm \sqrt{12}}{4} \approx 2.37$ or 0.63; **7.** $x = \frac{-3 \pm \sqrt{33}}{4} \approx 0.69$ or -2.19; **9.** (0, 0) and $(-2, 0)$;

11. $\left(\frac{1 + \sqrt{41}}{4}, 0\right)$ and $\left(\frac{1 - \sqrt{41}}{4}, 0\right)$ or approximately (1.851, 0) and (−1.351, 0); **13. a.** $d = 2.5$ million particles per ft^3, **b.** The minimum occurs at the vertex. $r = \frac{-b}{2a} = \frac{16}{4} = 4$ or 400 rpm; $d = 2(4)^2 - 16(4) + 34 = 2$ million particles per ft^3, **c.** $r = 11$; 1100 rpm is the speed of the engine. $r = -3$ is not in the practical domain $r > 0$.

Activity 4.6 **Exercises: 1. a.**

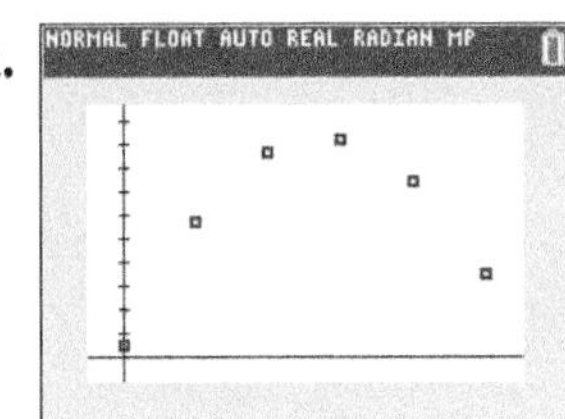

b. $h(t) = -15.9752t^2 + 52.8875t + 2.5536$,

c.

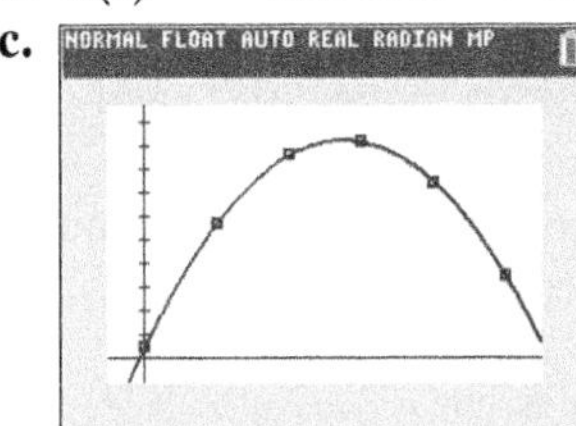

Yes, the curve touches nearly every data point. **d.** all real numbers from 0 to 3.36 seconds, **e.** real numbers from 0 to 46.33 feet, **f.** The ball reaches 35 feet on the way up after 0.81 second. It reaches 35 feet again on the way down, approximately 2.50 seconds after it was struck. **g.** There are only two solutions, so I got them all. **3. a.** $y = 0.086x^2 - 0.842x + 32.487$, **b.** approx. 653 feet, **c.** $0 = 0.086x^2 - 0.842x - 247.513$; using the quadratic formula, a speed of 58.8 mph requires a stopping distance of 280 feet.

Activity 4.7 **Exercises: 1.** $5i$; **3.** $6i$; **5.** $4i\sqrt{3}$; **7.** $\frac{3}{4}i$; **9.** $-5 + 10i$; **11.** $3 - 2i$; **13.** $10 + 5i$; **15.** $x = \frac{1}{3} \pm \frac{\sqrt{80}}{6}i$ or $\frac{1}{3} + \frac{2\sqrt{5}}{3}i$, no x-intercepts; **17.** $x = 1, -3.5$; **19.** 2 real solutions; **21.** 1 real solution; **23.** 2 complex solutions

How Can I Practice?

1.

VALUE OF a	VALUE OF b	VALUE OF c
5	0	0
$\frac{1}{3}$	3	−1
−2	1	0

2. a. downward, **b.** $x = 0$, **c.** $(0, 4)$, maximum, **d.** $(0, 4)$; **3. a.** upward, **b.** $x = 0$, **c.** $(0, 0)$, minimum **d.** $(0, 0)$; **4. a.** downward, **b.** $x = 1$, **c.** $(1, 10)$, maximum, **d.** $(0, 7)$; **5. a.** upward, **b.** $x = \frac{1}{2}$, **c.** $\left(\frac{1}{2}, -1\right)$, minimum, **d.** $(0, 0)$; **6. a.** upward, **b.** $x = -3$, **c.** $(-3, 0)$, minimum, **d.** $(0, 9)$; **7. a.** upward, **b.** $x = \frac{1}{2}$, **c.** $\left(\frac{1}{2}, \frac{3}{4}\right)$, minimum, **d.** $(0, 1)$; **8. a.** $(-2, 0), (2, 0)$, **b.** D: all real numbers; R: $y \leq 4$, **c.** $x < 0$, **d.** $x > 0$; **9. a.** $(2, 0), (3, 0)$, **b.** D: all real numbers; R: $y \geq -0.25$, **c.** $x > 2.5$, **d.** $x < 2.5$; **10. a.** $(0.91, 0), (-2.91, 0)$, **b.** D: all real numbers; R: $y \leq 11$, **c.** $x < -1$, **d.** $x > -1$; **11. a.** none, **b.** D: all real numbers; R: $y \geq 1.427$, **c.** $x > 1.61$, **d.** $x < 1.61$; **12.** $(0.75, 26.125)$; **13. a.** $9a^2(a^3 - 3)$, **b.** $6x^2(4x - 1)$, **c.** $4x(x - 5)(x + 1)$, **d.** cannot be factored, **e.** $(x - 8)(x + 3)$, **f.** $(y + 5)^2$;

14. a. 5.00, 6.05, 7.20, 8.45, 9.80, 11.25, $x \approx 1.2$, **b.** 6.75, 6.16, 5.59, 5.04, 4.51, 4.00, $x \approx 0.8$, **c.** 0, −2, 2, 12, 28, 50, $x = 2$; **15. a.** $x = 1.2, -1.2$, **b.** $x = 0.8, 6.2$, **c.** $x = -\frac{1}{3}, 2$; **16. a.** $x = 0, 2$, **b.** $x = 9, -2$, **c.** $x = 3, 1$, **d.** $x = 4, 4$, **e.** $x = 6, -4$, **f.** $y = 5, -3$, **g.** $a = 3, -2$, **h.** $x = \frac{1}{4}, -2$; **17. a.** $-8 < x < 2$, **b.** $x < -8$ or $x > 2$;

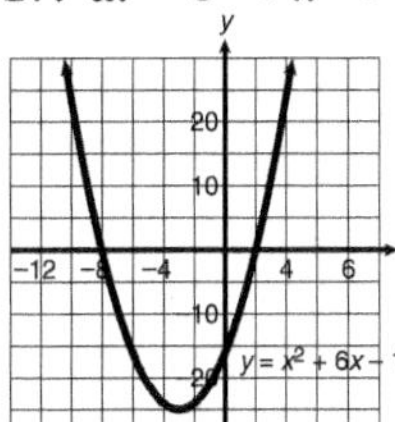

18. a. 105 feet, **b.** Using the calculator to solve $-16t^2 + 80t + 5 = 0$, $t \approx 5.06$ seconds. **c.** $-16t^2 + 80t + 5 = 101$, **d.** $t = 2, 3$; the ball reaches a height of 101 feet after 2 seconds on the way up and 1 second later on the way down.

19. a. $0 \leq x \leq 100$, **b.** $h(50) = 0.01(50)^2 - 50 + 35 = 10$ m; **20. a.**

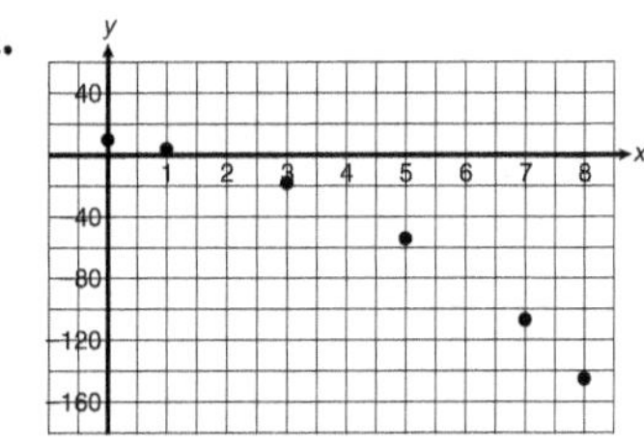

b. $y = -2.096x^2 - 2.25x + 9.038$,

c.

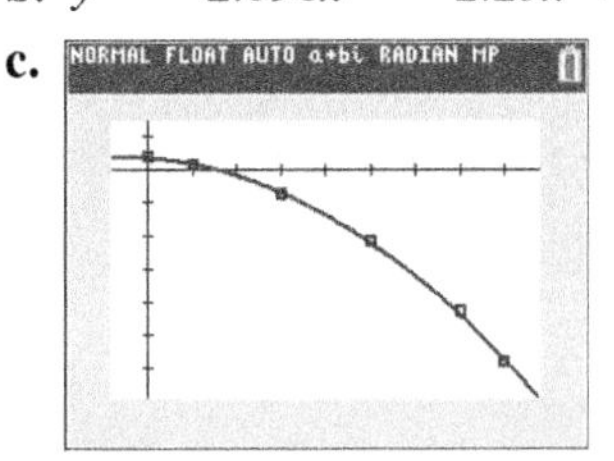

d. Predicted values are very close to the actual values. **e.** −33.5, −181,**f.** $x = 4.330, -5.403$; **21.** $7i$; **22.** $3i\sqrt{5}$; **23.** $11i$; **24.** $i\sqrt{15}$; **25.** $4i\sqrt{7}$; **26.** $5i\sqrt{5}$; **27.** $\frac{4}{5}i$; **28.** $\sqrt{\frac{4}{7}}i = \frac{2}{\sqrt{7}}i$; **29.** $4 + 3i$; **30.** $-5 + 6i$; **31.** $6 - 4i$; **32.** $-12 - 24i$; **33.** $5 + 10i$; **34. i.** $a = 3, b = -1, c = -7$; $b^2 - 4ac = 85$; two real solutions; $x = 1.70, -1.37$, **ii.** $a = 1, b = -4, c = 10$; $b^2 - 4ac = -24$; two complex solutions; $x = 2 \pm i\sqrt{6}$, **iii.** $a = 2, b = -5, c = -3$; $b^2 - 4ac = 49$; two real solutions; $x = 3, -0.5$, **iv.** $a = 9, b = -6, c = 1$; $b^2 - 4ac = 0$; one real solution; $x = \frac{1}{3}$; **35. i.** The discriminant is 0. The graph only touches the x-axis, indicating that there is one real solution, **ii.** The discriminant is negative. The graph does not intersect the x-axis, indicating that there is no real solution, **iii.** The discriminant is positive. The graph intersects the x-axis twice, indicating that there are two real solutions

36. a.

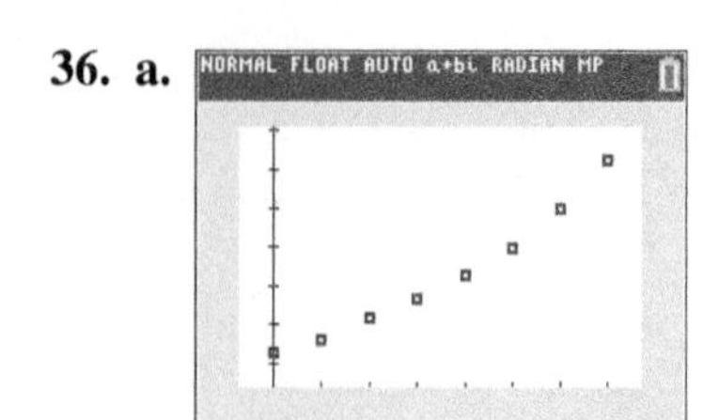

b. $R(t) = 0.66t^2 + 2.22t + 23.31$ **c.** $2020 - 2009 = 11$. $R(11) = 0.66(11)^2 + 2.22(11) + 23.31 = 127.59$ billion dollars **d.** The model shows 12.27 years. Rounding up to 13 years, after 2009 or in the year 2022, the model predicts that online advertising revenue will first reach 150 billion dollars. **e.** The model is projecting 6 years past the last data point, but a lot can happen, so I am not terribly confident.

Activity 4.8 Exercises: 1. a. 2, 32, 64, $y = kx$ and $8 = k1$, so $k = 8$ or $y = 8x$, **b.** $\frac{1}{8}$, 27, 216, $y = kx^3$ and $1 = k1^3$, so $k = 1$ or $y = x^3$; **3.** $y = kx^2$ and $12 = k2^2$, so $k = 3$. Therefore $y = 3x^2$. When $x = 8$, $y = 3(8)^2 = 192$. **5.** $d = kt^2$ and $20 = k(2)^2$; so $k = 5$. Now $d = 5t^2$, so in 2.5 seconds, the skydiver travels $d = 5(2.5)^2$ or 31.25 meters.

7.

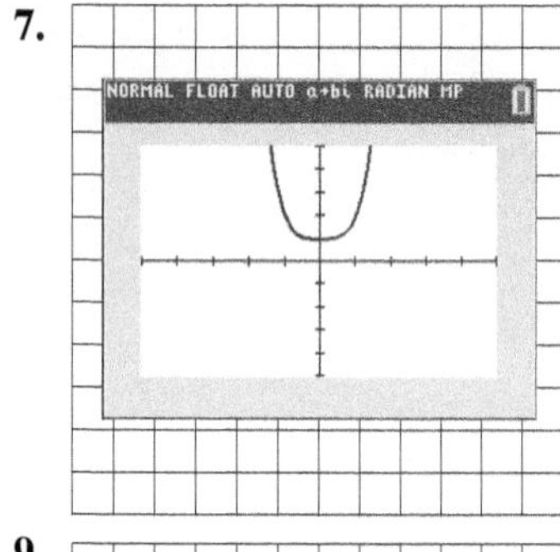

9.

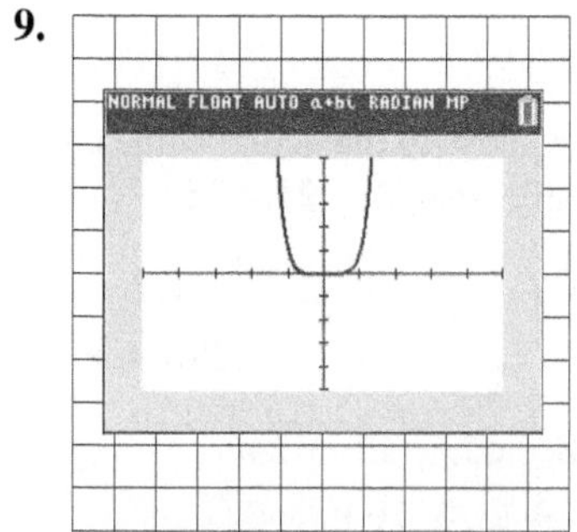

11. $f(x)$ is increasing for $x > 0$; **13.** The graph of $y = x^2$ is rising more slowly than $y = x^3$ for $x > 1$. Multiplying x^2 by x gives x^3, and this makes a larger output when $x > 1$. **15.** The graph of $y = -2x^3$ is decreasing and goes through (0, 0), whereas $y = 2x^3 + 1$ is increasing and does not pass through the origin. Both have a similar S-like shape.

Activity 4.9 Exercises: 1. $f(x) = x(x+2)(x+1)$, $(0, 0)$, $(-1, 0)$, $(-2, 0)$; **3.** $h(x) = (x^2 - 4)(x^2 - 9)$, $h(x) = (x + 2)(x - 2)(x + 3)(x - 3)$, $(2, 0)$, $(-2, 0)$, $(3, 0)$, $(-3, 0)$;

5. a.

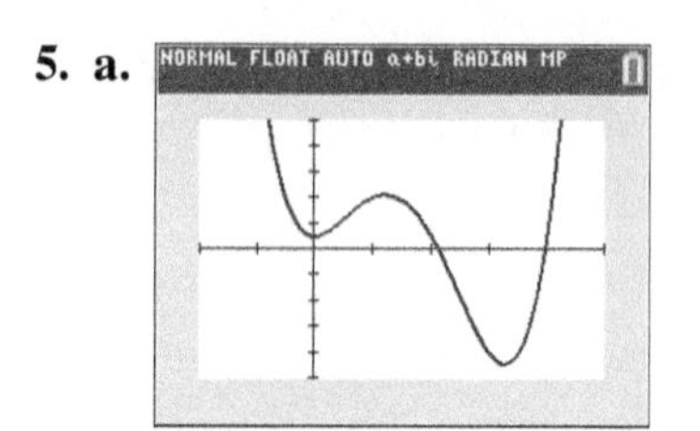

b. The domain is all real numbers. **c.** The range is y-values greater than or equal to -8.91. **d.** $(2.12, 0)$ and $(3.97, 0)$, **e.** There are two minimum points $(0, 1)$ and $(3.28, -8.91)$ and one maximum point $(1.22, 4.23)$. **7.** No, as x increases without bound, y increases without bound. **9. a.** increase, **b.** decreasing, **c.** 1

Activity 4.10 Exercises:

1. a.

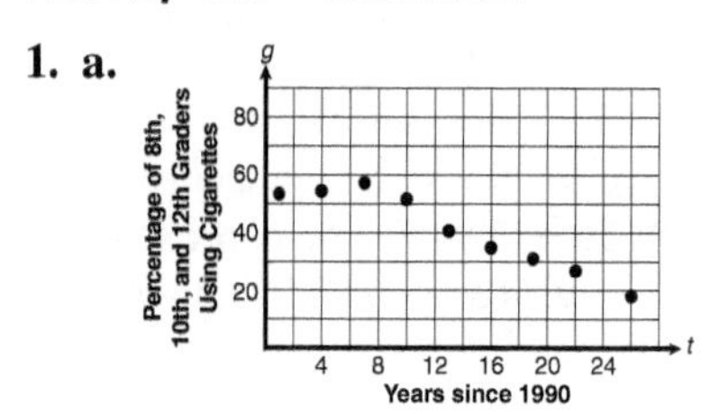

The data does not appear to be linear because as the input increases, the output increases and decreases. No line would be close to all of the points.
b. quadratic: $g = -0.04x^2 - 0.52x + 57.33$
cubic: $g = 0.006x^3 - 0.276x^2 + 2.004x + 51.98$

3. a.

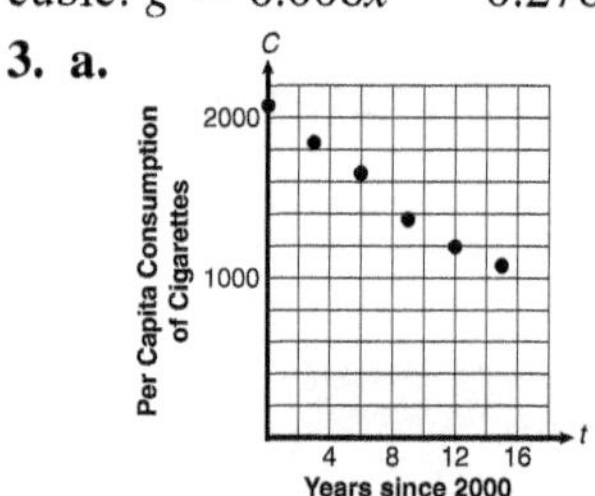

b. $C = -69.16t + 2061.38$, **c.** $C = 0.956t^2 - 83.507t + 2090.071$ **d.** 2020 is 20 years from 2000, so $t = 20$. Linear model: $C = -69.16(20) + 2061.38 = 678.18$; Quadratic model: $C = 0.956(20)^2 - 83.507(20) + 2090.071 = 802.331$ **e.** $t = 20$ is well outside of the given data; therefore, the values calculated are the result of extrapolation. I do not have confidence in either prediction.

How Can I Practice? 1. $y = kx^2$ and $45 = k(3)^2$, so $k = 5$ $y = 5(6)^2 = 180$; **2. a.** double, **b.** $k = 1080$; k represents the speed at which the sound of thunder travels in feet per second. **3.** $v = kt$ and $60 = k(3)$, so $k = 20$; $v = 20(4) = 80$ feet per second

4. a.

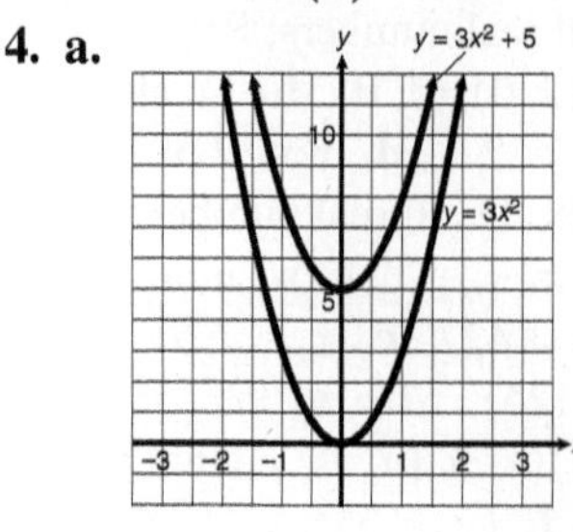

These are the same shape and size; however, $y = 3x^2 + 5$ is shifted up 5 units.

b.

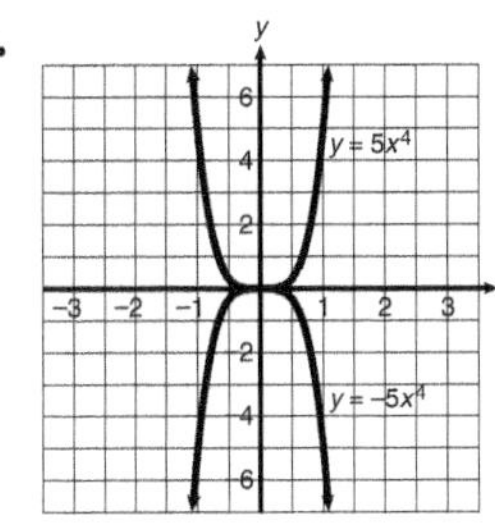

These are the same shape but are reflections of each other in the x-axis.

c.

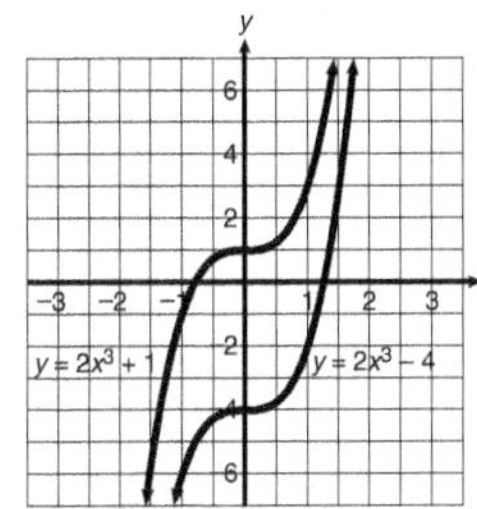

These are the same shape and size, but $y = 2x^3 - 4$ is shifted vertically 5 units below $y = 2x^3 + 1$.

d.

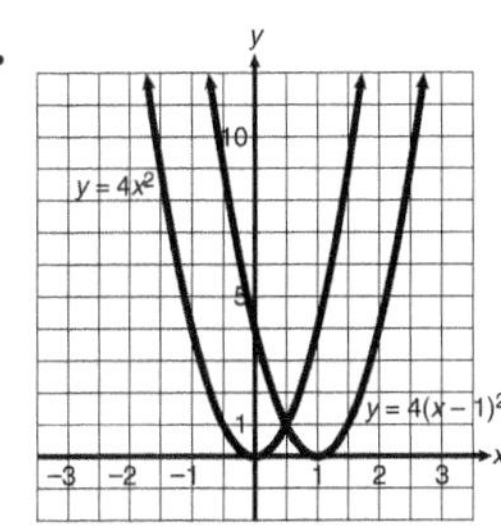

These are the same shape and size, but $y = 4(x - 1)^2$ is shifted horizontally 1 unit to the right of $y = 4x^2$.

5. a. i. $(0, 0)$, **ii.** $(0, 0), (-4, 0), (2, 0)$, **iii.** $(-2.4, 16.9), (1.1, -5.0)$, **b. i.** $(0, 3)$, **ii.** $(-1, 0), (1.57, 0)$, **iii.** $(0.79, 4.19)$

6. a.

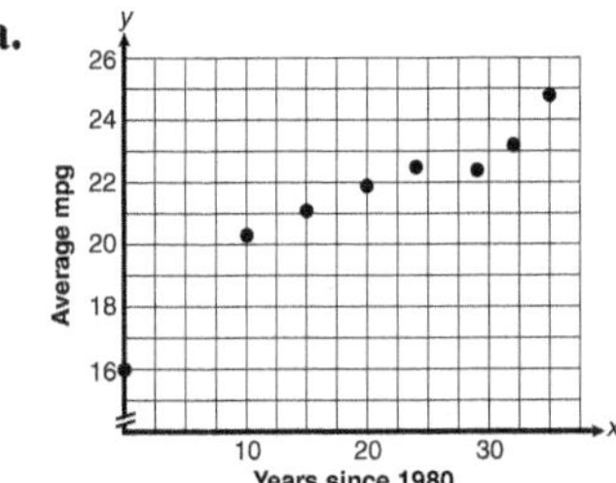

b. cubic: $y = 0.00050186x^3 - 0.0305x^2 + 0.6982x + 15.9612$; quartic: $y = 0.000023738x^4 - 0.0012x^3 + 0.0084x^2 + 0.4248x + 16.0164$ **c.** cubic: 27.2 mpg: quartic: 29.5 mpg **d.** In both cases, you are extrapolating. The cubic goes up moderately. The quartic goes up significantly. The cubic is better.

7. a.

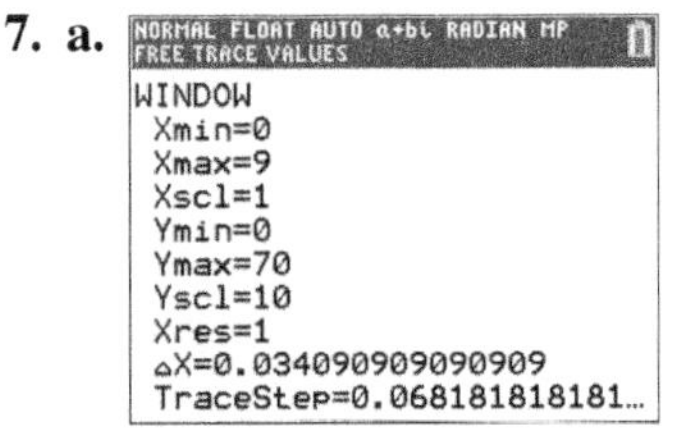

$W(t)$ becomes negative after about 7 minutes. **b.** $W(0) =$ 10 gallons, **c.** 60.8 gallons, found by using the CALC menu on the graphing calculator. This is the output value at the highest point on the graph. **d.** 7.19 minutes

Gateway Review 1. a. up, **b.** $x = 0$, **c.** $(0, 2)$, **d.** $(0, 2)$; **2. a.** down, **b.** $x = 0$, **c.** $(0, 0)$, **d.** $(0, 0)$; **3. a.** down, **b.** $x = 0$, **c.** $(0, 4)$, **d.** $(0, 4)$; **4. a.** up, **b.** $x = \frac{1}{4}$, **c.** $\left(\frac{1}{4}, -\frac{1}{8}\right)$, **d.** $(0, 0)$; **5. a.** up, **b.** $x = -2.5$, **c.** $(-2.5, -0.25)$, **d.** $(0, 6)$; **6. a.** up, **b.** $x = \frac{3}{2}$, **c.** $(1.5, 1.75)$, **d.** $(0, 4)$; **7. a.** up, **b.** $x = 1$, **c.** $(1, 0)$, **d.** $(0, 1)$; **8. a.** down, **b.** $x = 2.5$, **c.** $(2.5, 0.25)$, **d.** $(0, -6)$; **9. a.** $(-3, 0), (-1, 0)$, **b.** D: all real numbers; R: $g(x) \geq -1$, **c.** $x > -2$, **d.** $x < -2$; **10. a.** $(-3, 0), (1, 0)$ **b.** D: all real numbers; R: $f(x) \geq -4$, **c.** $x > -1$, **d.** $x < -1$; **11. a.** $(0.382, 0), (2.62, 0)$, **b.** D: all real numbers; R: $y \geq -1.25$, **c.** $x > 1.5$, **d.** $x < 1.5$; **12. a.** $(-3.22, 0), (-0.775, 0)$, **b.** D: all real numbers; R: $h(x) \geq -3$, **c.** $x > -2$, **d.** $x < -2$; **13. a.** $(2, 0), (-2, 0)$, **b.** D: all real numbers; R: $y(x) \leq 8$, **c.** $x < 0$, **d.** $x > 0$; **14. a.** $\left(\frac{1}{3}, 0\right)(1, 0)$, **b.** D: all real numbers; R: $f(x) \leq \frac{1}{3}$, **c.** $x < \frac{2}{3}$, **d.** $x > \frac{2}{3}$; **15. a.** none, **b.** D: all real numbers; R: $g(x) \geq 5$, **c.** $x > 0$, **d.** $x < 0$; **16.** $x = -2$; **17.** $x = 2, 3$; **18.** $x \approx -0.51, 6.51$; **19.** $x = -5, 2$; **20.** $x \approx \pm 1.1$; **21.** $x \approx -0.2, -4.8$; **22. a.** $9a^2(a^3 - 3)$, **b.** $6x^2(4x - 1)$, **c.** $4x(x - 5)(x + 1)$, **d.** cannot be factored, **e.** $(x - 8)(x + 3)$, **f.** $(t + 5)^2$; **23.** $x = \pm 3$; **24.** $x = \pm 6$; **25.** $x = 3, 4$; **26.** $x = -3, 9$; **27.** $x = 0, -1$; **28.** $a = 1, b = 5, c = 3; x \approx -0.7, -4.3$; **29.** $a = 2, b = -1, c = 3; x \approx 0.25 \pm 1.2i$; **30.** $a = 1, b = 0, c = -81; x = \pm 9$; **31.** $a = 3, b = 5, c = -12; x = -3, \frac{4}{3}$; **32.** $a = 2, b = -3, c = -5; x = -1, 2.5$; **33.** From the graphing calculator: $(0.42, 0), (3.58, 0)$

$$x = \frac{-(-8) \pm \sqrt{(-8)^2 - 4(2)(3)}}{2(2)} = \frac{8 \pm \sqrt{40}}{4} \approx$$

$3.58, 0.42$; **34. a.** $7i$, **b.** $4i\sqrt{3}$, **c.** $3i$, **d.** $i\sqrt{23}$, **e.** $\frac{\sqrt{5}}{3}i$, **f.** $\frac{\sqrt{17}}{4}i$; **35. a.** $-5 + 17i$, **b.** $5 - 16i$, **c.** $32 + 12i$, **d.** $27 + 6i$; **36.** $b^2 - 4ac = 1$; two real solutions; **37.** $b^2 - 4ac = 256$; two real solutions; **38.** $b^2 - 4ac = 36$; two real solutions; **39.** $b^2 - 4ac = -20$; two complex solutions;

40. $$x = \frac{-2 \pm \sqrt{2^2 - 4(3)(2)}}{2(3)} = \frac{-2 \pm \sqrt{-20}}{6} = \frac{-2 \pm 2i\sqrt{5}}{2(3)} = \frac{-1 \pm i\sqrt{5}}{3}$$; the graph has no x-intercepts, confirming complex solution; **41. a.** $-2 < x < 3$, **b.** $x < -2$ or $x > 3$; **42. a.** $y = 20$, **b.** $y = 32$, **c.** $y = 40$; **43. a.** $(2, 0)$, **b.** D: all real numbers; R: all real numbers, **c.** increasing for all real numbers; **44. a.** $(-1, 0)$, **b.** D: all real numbers; R: all real numbers,

c. decreasing for all real numbers; **45. a.** $(-1.68, 0)$, $(1.68, 0)$, **b.** D: all real numbers; R: $y \geq -8$, **c.** inc: $x > 0$; dec: $x < 0$; **46. a.** $(0, 0), (-1.26, 0)$, **b.** D: all real numbers; R: $y \geq -1.19$, **c.** inc: $x > -0.8$; dec: $x < -0.8$; **47. a.** none, **b.** D: all real numbers; R: $y \geq 5$, **c.** inc: $x > 0$; dec: $x < 0$;

48. a.

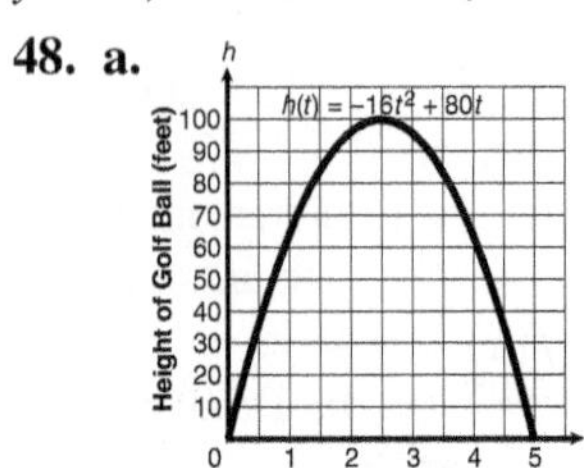

The practical domain is $0 \leq x \leq 5$.

b. (2.5, 100); the ball reaches its highest level, 100 feet, 2.5 seconds after being struck. **c.** (0, 0); the ball is on the ground when the club makes contact with it. **d.** (0, 0), (5, 0); the ball is on the ground when the club makes contact, $t = 0$, and returns to the ground 5 seconds later. **e.** I am assuming that the elevations are the same. **49. a.** $(-5, 6)$, **b.** $(0, 0), (-5, 6), (-10, 0)$, **c.** $y = -0.24x^2 - 2.4x$; **50. a.** vertex: $(2.5, h(2.5))$ or (2.5, 105); the maximum height is 105 feet, **b.** Set $h(t) = 0$; $t \approx 5.06$ seconds; **51. a.** $s(44) = 122.5$ feet away, **b.** $v \approx -51.78$ or 19.78; reject the negative; 19.78 feet per second $\approx$ 13.5 miles per hour

52. a.

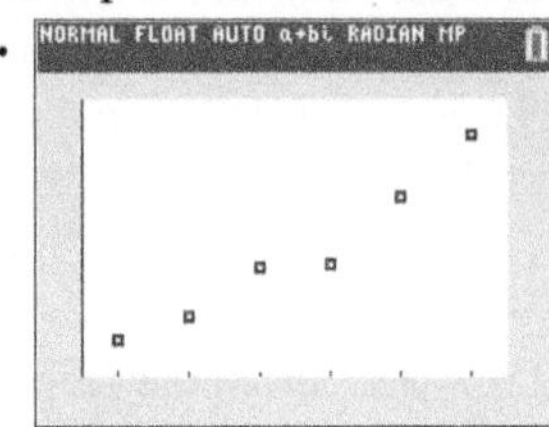

b. $D(t) = 19.56t^2 + 26.25t + 453.13$ **c.** The model appears to fit the data quite well, passing very close to 4 of the 6 data points. **d.** 2020$t = 10$; $D(10) = 19.56(10) + 26.25(10) + 453.13 = 2671.64$ million dollars; 2030$t = 20$; $D(20) = 19.56(20)^2 + 26.25(20) + 453.13 = 8802.14$ million dollars **e.** I have more confidence in the 2020 prediction. The 2030 prediction is extrapolating quite far from the original data.

Chapter 5

Activity 5.1 Exercises: 1. a. The average speed $= \dfrac{20 \text{ km}}{1 \text{ hr } 15 \text{ min}} = \dfrac{20 \text{ km}}{1.25 \text{ hr}} = 16 \text{ km/hr.}$, **b.** 20, 16, 13.33, 11.43, 10, 8.89, 8, **c.** $s = f(t) = \dfrac{20}{t}$, **d. i.** the set of all nonzero real numbers, **ii.** (Answers will vary.) $1 \leq t \leq 5$. Because 20 kilometers per hour (when $t = 1$) is fast for a distance runner and 4 kilometers per hour (when $t = 5$) is slow for a distance runner, most times will fall between these values.

iii.

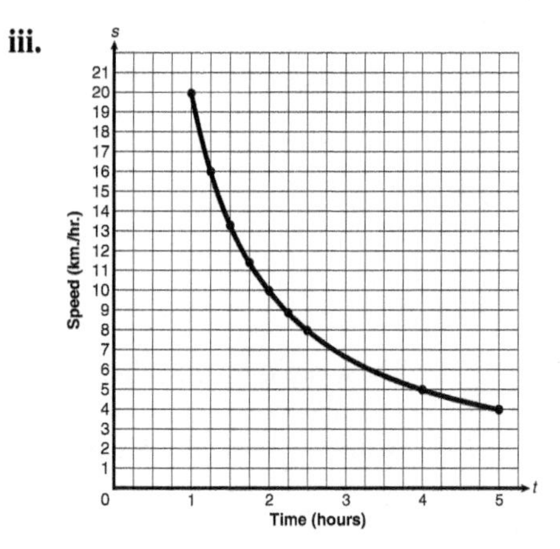

e. The average speed decreases, approaching 0. **f.** The average speed increases without bound.

3. a. $D = f(N) = \dfrac{1400 - 200}{N} = \dfrac{1200}{N}$, **b.** 1200, 600, 400, 200, 100, 50,

c.

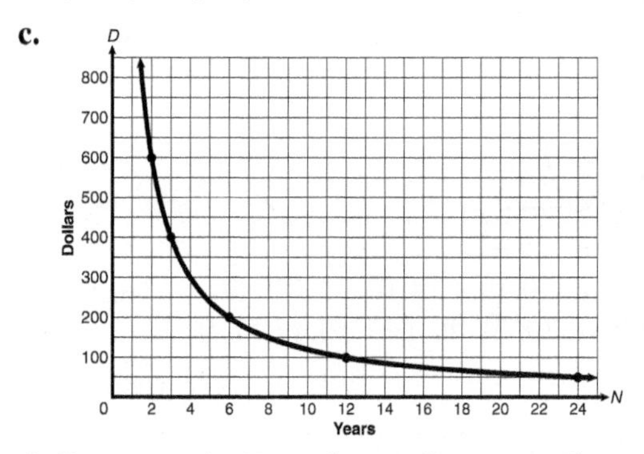

d. Decrease. As N gets larger, D gets smaller.

Activity 5.2 Exercises: 1. a. (Answers will vary.) If a person is 6 feet = 72 inches tall and weighs 200 pounds, $B = \dfrac{705(200)}{72^2} \approx 27.12$. **b.** $B = \dfrac{119{,}850}{h^2}$, **c.** $36 < h < 84$. This works unless the person is under 3 feet tall or over 7 feet tall. **d.** 33.2, 29.2, 25.8, 23.1, 20.7, 18.7,

e.

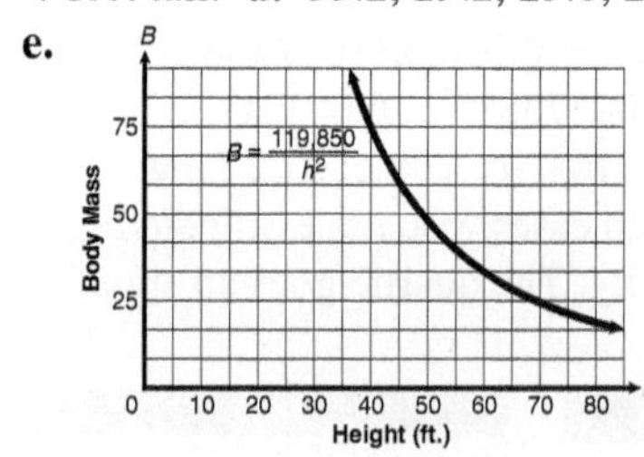

f. Body mass index B gets smaller, and it actually approaches 0. Yes, this makes sense in this case because taller persons with the same weight should be skinnier.

g. $69.1 < h < 79.3$ measured in inches; **3. i.** graph b, **ii.** graph c, **iii.** graph a, **iv.** graph d; **5. a.** all nonzero real numbers,

b. **c.**

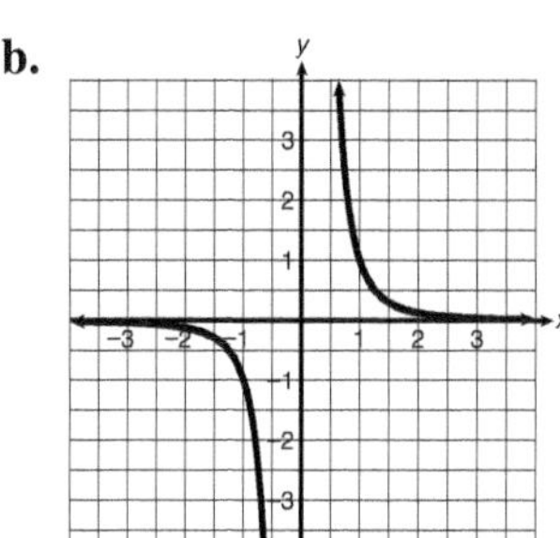

7. a. $y = \frac{2}{x}$; Table: 4, 1, $\frac{1}{3}$, **b.** $y = \frac{8}{x^3}$; Table: 64, 8, $\frac{1}{27}$;

9. $I = \frac{120}{R}$; $I = \frac{120}{15} = 8$ amps;

11. a.

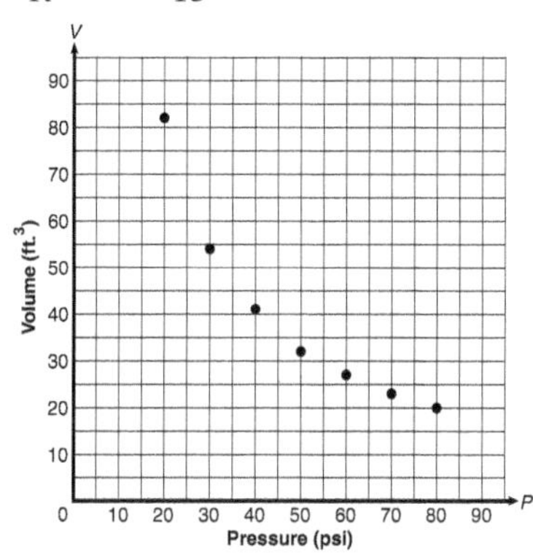

b. No, using $P = 20$, $V = 82$; $k = 20^2(82) = 32{,}800$. If $V = \frac{32{,}800}{P^2}$, then $P = 30$ would yield $V = \frac{32{,}800}{30^2} \approx 36.44$ (not very close). **c.** Yes, using $P = 20$, $V = 82$; $k = 20(82) = 1640$; $V = \frac{1640}{30} \approx 54.67$; $V = \frac{1640}{40} = 41$, **d.** Answers will vary. $V \approx 25$ cubic feet, **e.** $V = \frac{1640}{P}$; $V = \frac{1640}{65} \approx 25.2$ cubic feet

Activity 5.3 Exercises:

1. a. $V = \frac{25{,}000 + 55{,}000}{P - 1.5 - 0.40 - 0.60} = \frac{80{,}000}{P - 2.5}$,

b. undefined, 160,000, 32,000, 10,667, 3555.56, **c.** V decreases, **d.** $V(2) = -160{,}000$. A price of \$2 per cubic meter is not practical. **e.** $P > 2.5$,

f.

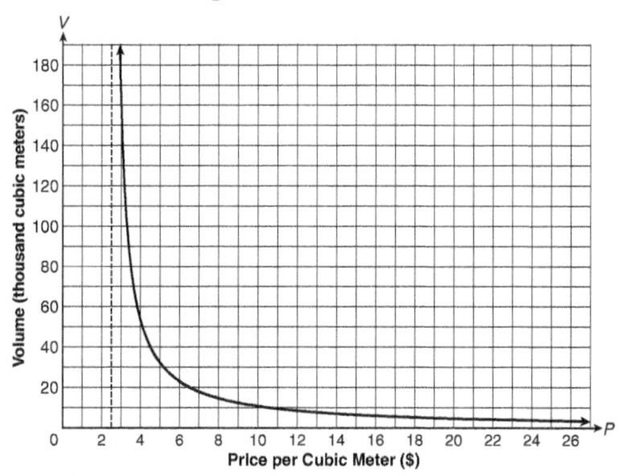

3. a. domain: all real numbers except $x = 7$; vertical asymptote: $x = 7$ **b.** domain: all real numbers except $x = 25$; vertical asymptote: $x = 25$

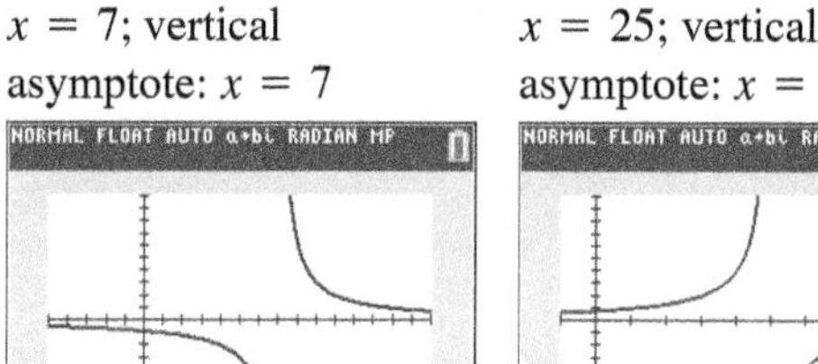

c. domain: all real numbers except $x = 5$; vertical asymptote: $x = 5$ **d.** domain: all real numbers except $x = 14$; vertical asymptote: $x = 14$

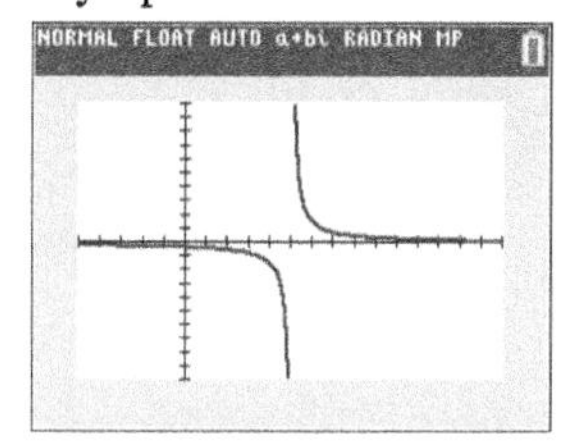

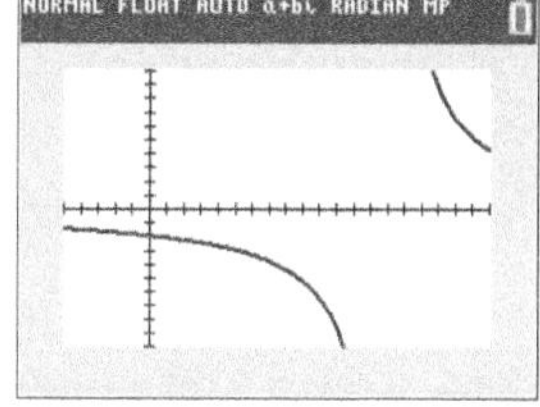

e. domain: all real numbers except $x = -2.5$; vertical asymptote: $x = -2.5$

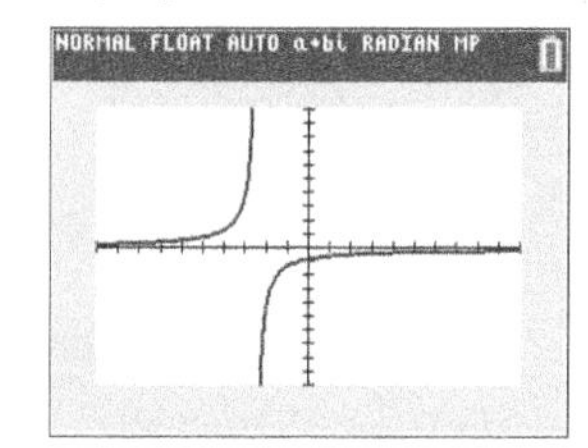

5. a. $x = 5$, **b.** $f(x)$ gets large, approaching infinity. $g(x)$ gets large in magnitude in a negative direction, approaching negative infinity. **c.** $f(x)$ gets large in magnitude in a negative direction, approaching negative infinity. $g(x)$ gets large, approaching infinity.

Activity 5.4 Exercises:

1. a. $500 + 600 + 500 + 400 = \$2000$, **b.** $2000 + 50n$, **c.** $c = \frac{50n + 2000}{n}$, **d.** $c = \frac{50(100) + 2000}{100} = \$70$, **e.** The practical domain is whole numbers from 1 to the size of your class, say 250. **f.** 90, 70, 63.33, 60, 58, **g.** Yes, $c = 50$ is the horizontal asymptote. It makes sense because as the number of attendees increases, the fixed costs attributed to each person get smaller and smaller.

h.

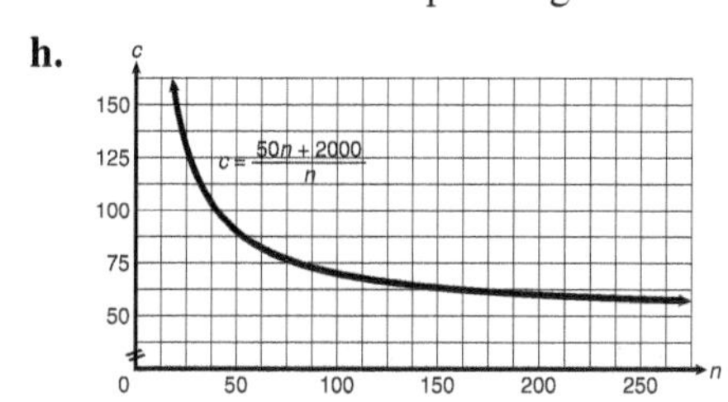

2. a. i. all real numbers except −2, **ii.** $x = -2$, **iii.**

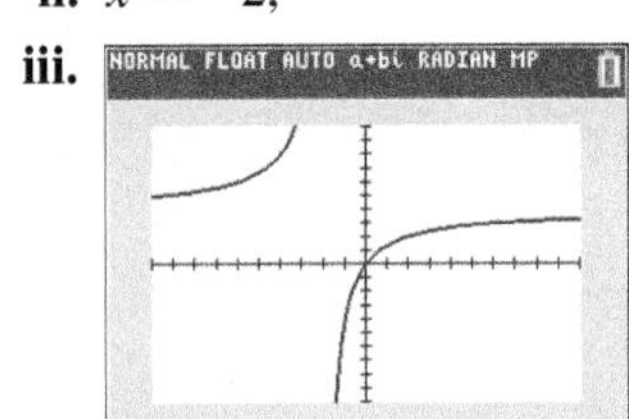

iv. $y = 4$,

b. i. all real numbers except −1, **ii.** $x = -1$, **iii.** **iv.** $y = -1$,

c. i. all real numbers except 4, **ii.** $x = 4$, **iii.**

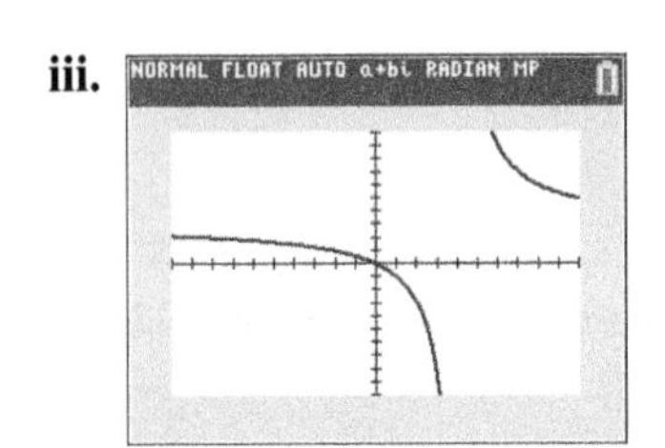

iv. $y = 3$,

d. i. all real numbers except $\frac{1}{2}$, **ii.** $x = \frac{1}{2}$, **iii.** **iv.** $y = 15$;

3. a. $x = 1$

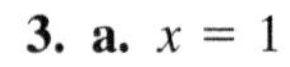

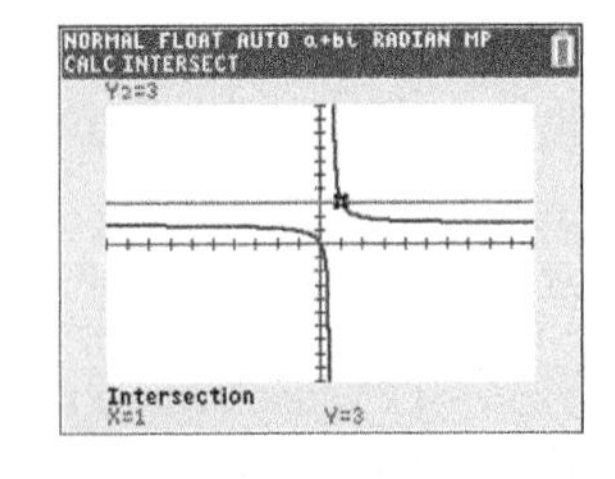

b. $x = \frac{7}{9} \approx 0.778$

c. $x \approx -0.859$

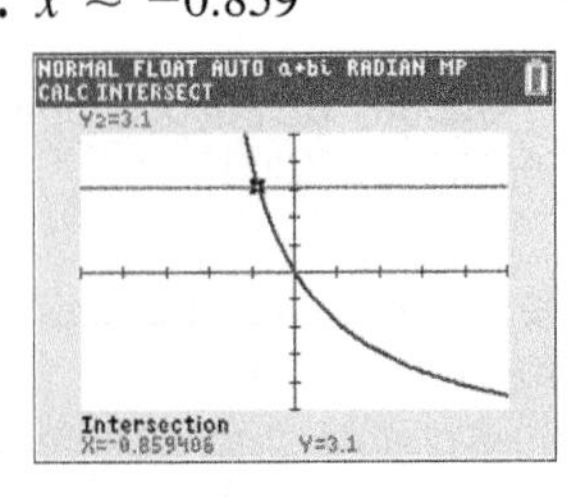

d. $x = 0.5$

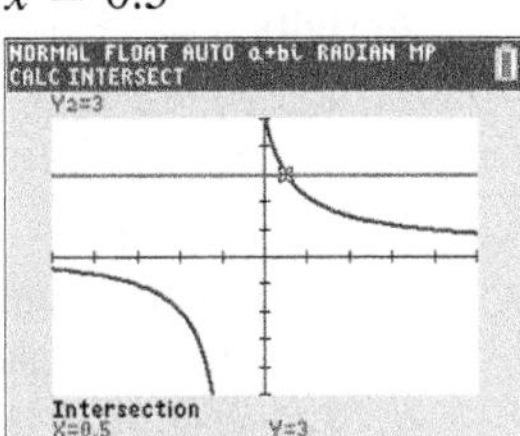

4. a. $16 = \frac{20}{t}$; $16t = 20$, $t = \frac{20}{16} = 1.25$ hours, **b.** $18 = \frac{20}{t}$; $18t = 20$, $t = \frac{20}{18} \approx 1.11$ hours;

5. a. $15d^2 = 1500$; $d^2 = 100$; $d = \pm 10$ feet, but only 10 makes sense, **b.** $8000 = \frac{1500}{d^2}$; $8000d^2 = 1500$; $d^2 = 0.1875$; $d \approx 0.433$ feet;

7. a. $R = \frac{8.4(4904) + 330(37) + 100(318) - 200(9)}{489}$
$= 170.6$ (to the nearest tenth)

b. $R = \frac{8.4(4627) + 330(43) + 100(285) - 200(6)}{404}$
$= 198.9$ (to the nearest tenth)

8. a. $R \approx 7.85$ prey per week, **b.** $61.3 \approx n$, $n =$ 61 prey/sq. mi., **c.** $17.3 \approx n$; putting the two together, $18 \le n \le 61$.

d.

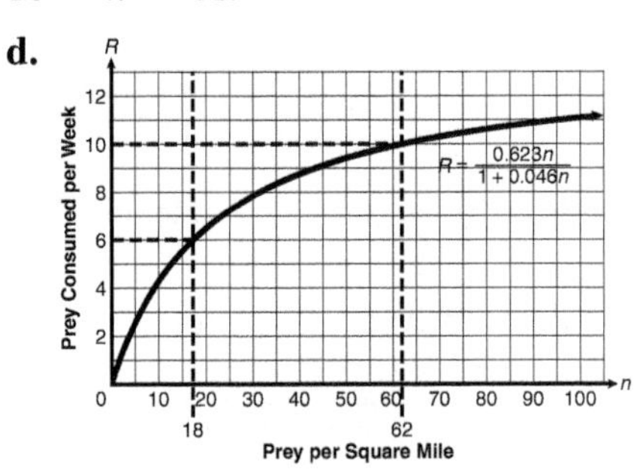

e. The result is negative, so discard it. It is not possible for the predator to consume 20 prey/week. Under these conditions, 20 is above the horizontal asymptote.

Activity 5.5 Exercises: 1. a. $5T = 90$, $T = 18$ minutes, **b.** $t_2T + t_1T = t_1t_2$, $T = \frac{t_1t_2}{t_1 + t_2}$, **c.** $T = \frac{20(15)}{20 + 15} = \frac{300}{35} \approx 8.57$ minutes, **d.** $3t_2 = 40$, $t_2 \approx 13.3$ minutes, **e.** $14T = 150$, $T \approx 10.7$ minutes;
3. $t = \frac{72 \pm \sqrt{72^2 - 4(70)}}{2}$, $t \approx 71$ or 1; the only value that makes sense in this situation is $t = 71$ minutes.
5. $21 = 2t$, $t = 10.5$ hours

Activity 5.6 Exercises:
1. a. $h = \frac{330}{1 - \frac{40}{770}} \approx 348.08$ hertz; the pitch I hear is higher than the actual pitch. **b.** $h = a \div \frac{770 - s}{770}$; $h = \frac{770a}{770 - s}$, **c.** $h = \frac{770(330)}{770 - 40} \approx 348.08$ hertz; the results are the same. **d.** $h = \frac{770(330)}{770 - 60} \approx 357.89$ hertz;
3. a. $\frac{x - 4}{2x} \cdot \frac{1}{x - 4} = \frac{1}{2x}$, **b.** $\frac{2 - x}{2x} \cdot \frac{4x^2}{(2 - x)(2 + x)} = \frac{2x}{2 + x}$, **c.** $\frac{3x^2 - 6x}{x^2 - 4x + 3}$, **d.** $\frac{x}{2x + 4}$

How Can I Practice? 1. The graphs of f and g are reflections of each other about the x-axis (or the y-axis). **2.** They are similar, but the graph of g is closer to the x-axis, and the graph of f is closer to the y-axis. **3. a.** T = time in hours, s = speed in mph, $T = \frac{145}{s}$, **b.** $0 < s < 80$,

c. all real numbers except 0; **4. a.** domain: all real numbers except $x = -5$; vertical asymptote: $x = -5$, **b.** domain: all real numbers except $x = \frac{13}{2}$; vertical asymptote: $x = \frac{13}{2}$, **c.** domain: all real numbers except $x = \frac{8}{5}$; vertical asymptote: $x = \frac{8}{5}$, **d.** domain: all real numbers except $x \approx 0.5614$; vertical asymptote: $x = 0.5614$; **5.** $4000^2 \cdot 100 = k = 1.6 \cdot 10^9$; $w = \frac{1.6 \cdot 10^9}{d^2} = \frac{1.6 \cdot 10^9}{(4500)^2} \approx 79.01$ lb.; **6. a.** The practical domain is all positive integers, with some realistic upper limit, depending on the specific situation.

b. 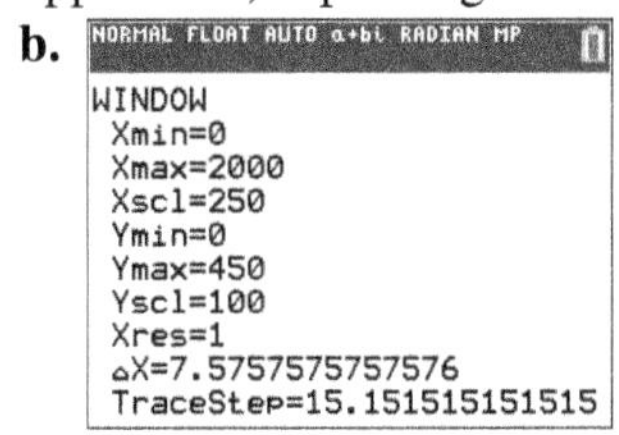

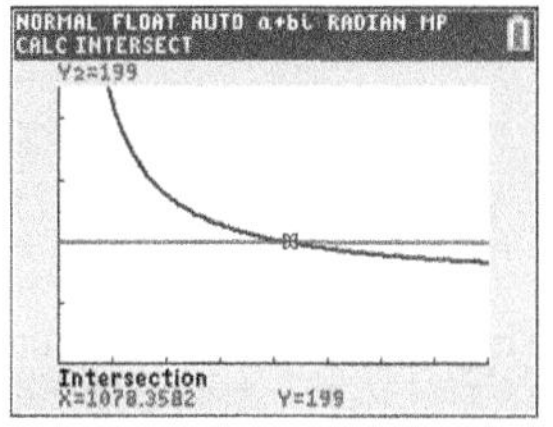

$199x = 132x + 75{,}250$, $x = 1124$ mowers

7. a. $t = \frac{-(-14) \pm \sqrt{14^2 - 4(0.15)(0.125)}}{2(0.15)} \approx$ 93.32 minutes or 0.0089 minutes. Only 93.32 is practical.

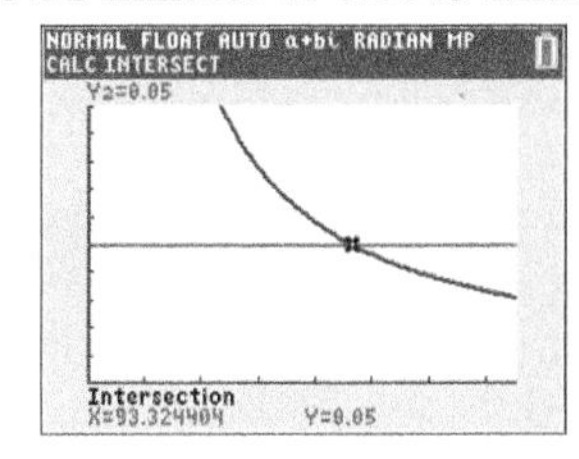

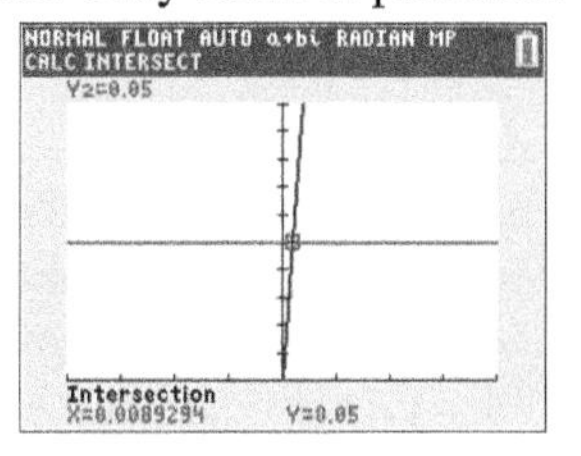

b. The drug will be at its highest concentration approximately 0.913 minute after injection.

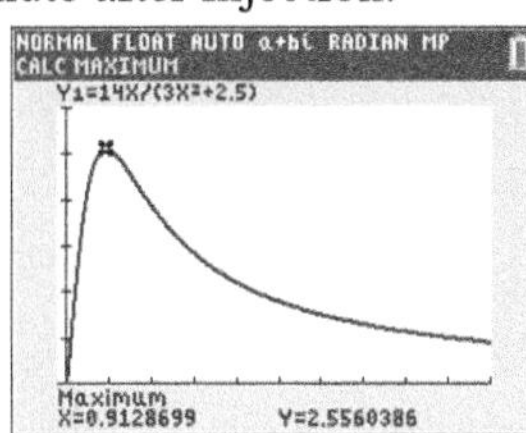

8. a. $x = -0.25$

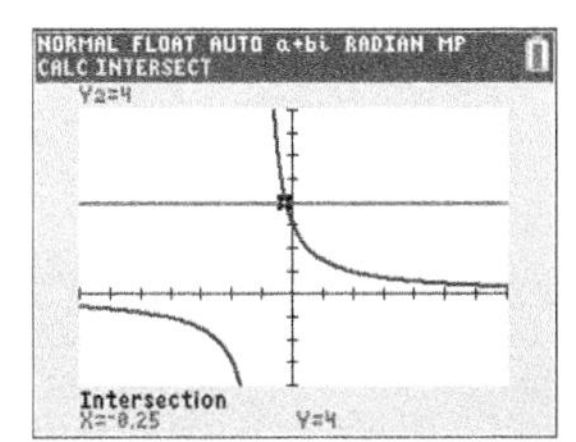

b. $x = \frac{50}{17} \approx 2.94$

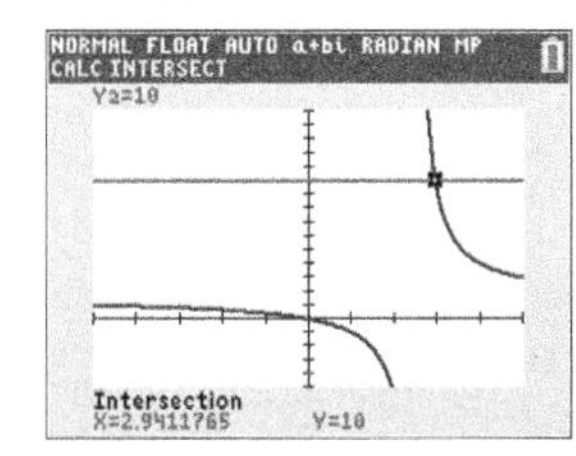

c. $x = \frac{-116}{40} = -2.9$

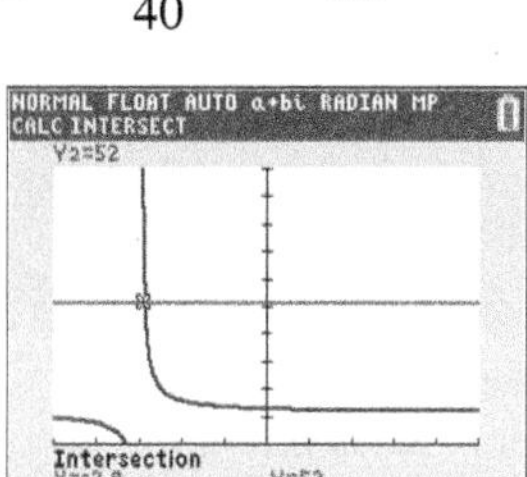

d. $x = \frac{5.8}{2.4 - (0.3)5.8} = \frac{290}{33} \approx 8.788$

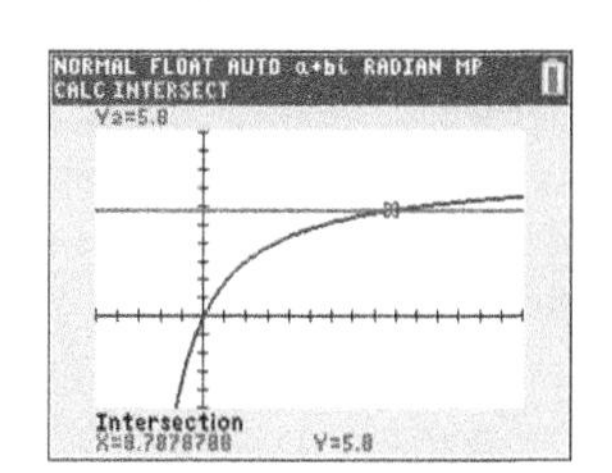

9. a. (Answers will vary.) 200-pound man ≈ 90.9 kilograms, $W = \frac{90.9}{\left(1 + \frac{15}{6400}\right)^2} \approx 90.475$ kilograms, **b.** $W = \frac{70}{\left(1 + \frac{h}{6400}\right)^2}$, **c.** 70, 69.78, 67.86, 52.36, 45.94, 40.64, 10.66, 4.11, **d.** The weight decreases. **e.** 57.69 kilograms, **f.** (Answers will vary at the upper end.) The domain is $0 \leq h \leq 40{,}000$.

g 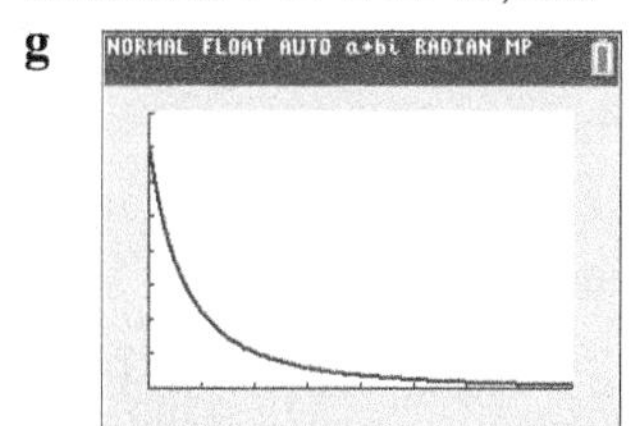

h. $h \approx 2650.9668$ kilometers

10. a. $R_3 = 12$ ohms, **b.** $R = \frac{R_1R_2R_3}{R_1R_2 + R_2R_3 + R_1R_3}$, **c.** $R = \frac{4(6)(12)}{4(6) + 4(12) + 6(12)} = \frac{288}{24 + 48 + 72} = \frac{288}{144} = 2$ ohms;

11. a. $x = 1.3$

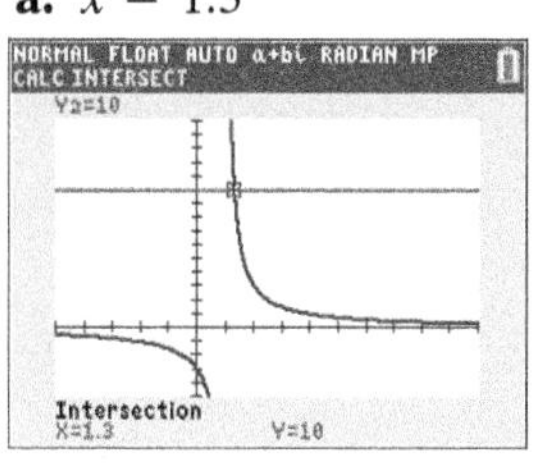

b. $x = -1$

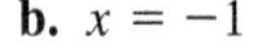

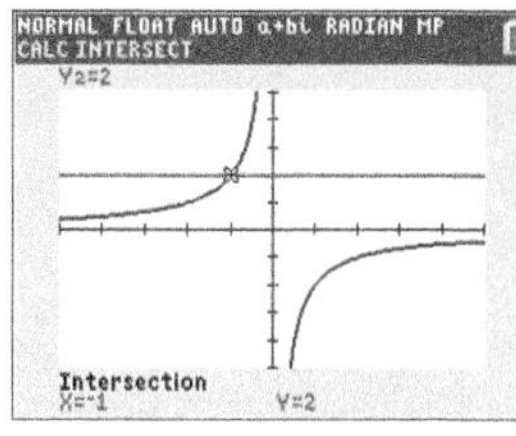

c. $x = 5$

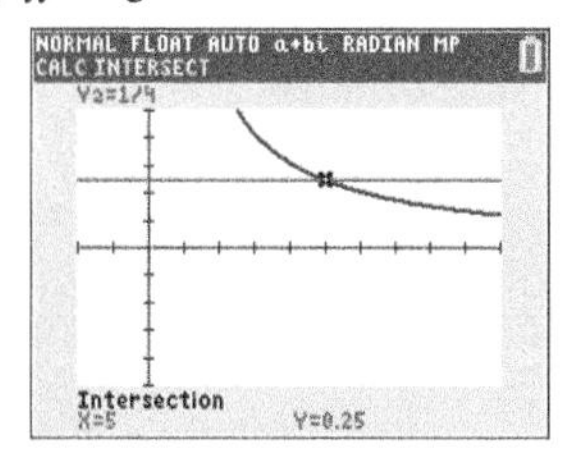

d. $x = \frac{1}{4}$

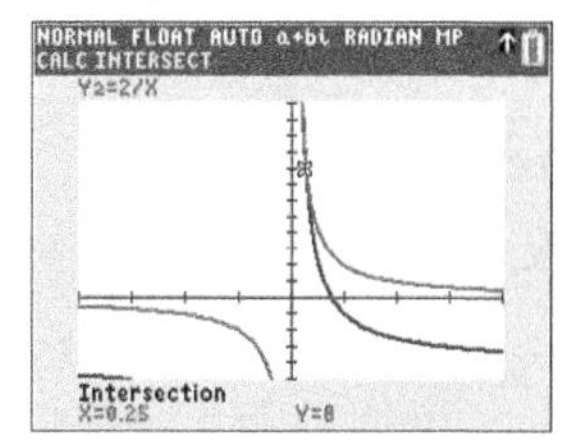

12. a. $s = \frac{2(15.3)}{\frac{15.3}{45} + \frac{15.3}{40}} = \frac{30.6}{0.34 + 0.3825} \approx 42.4$ mph,

b. $s = \dfrac{2dr_1r_2}{d(r_1 + r_2)} = \dfrac{2r_1r_2}{r_1 + r_2}$ **c.** $s = \dfrac{2(45)(40)}{(40 + 45)}$ ≈ 42.4 mph; the results are the same.

13. a. $\dfrac{4x + 2}{x - 3}$, **b.** $x - 5$, **c.** $\dfrac{1}{x + 3}$

Activity 5.7 Exercises: 1. a. 5.48, **b.** 2.45, **c.** 169, **d.** 27; **3. a.** (0, 4.04). In the first year the data was collected 4.04 million undergraduate students received Stafford loans. **b.** 4.04, 4.80, 5.27, 5.64, 5.96, 6.24, 6.49; the equation matches reasonably although it is not perfect.

c.

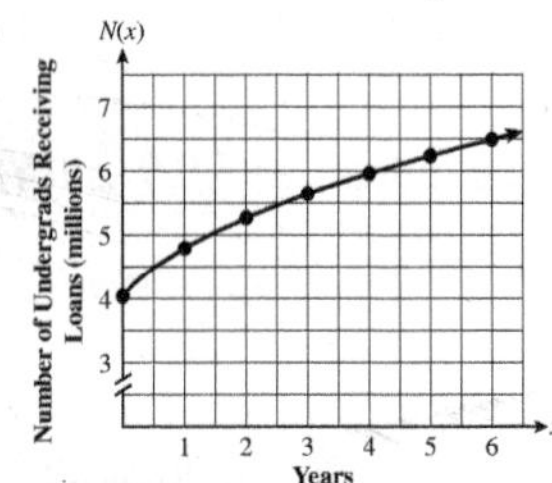

d. $x = 12$, so $N(12) = 1.222\sqrt{12 + 0.24} + 3.442 \approx$ 7.72 million.

5. a.

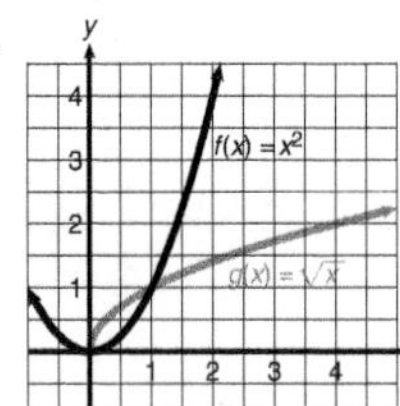

b. No, the graph of f is below the graph of g for $0 < x < 1$. **c.** Yes, the graph of f is above the graph of g for $x > 1$.

7. a. i. The domain is all real numbers such that $2x + 3 \geq 0$ or $x \geq -\dfrac{3}{2}$. **ii.** The x-intercept is $\left(-\dfrac{3}{2}, 0\right)$. The y-intercept is $(0, \sqrt{3})$.

iii.

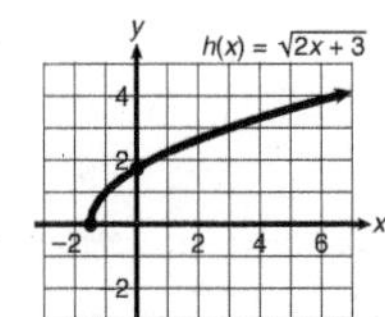

b. i. The domain is all real numbers such that $4x + 8 \geq 0$ or $x \geq -2$. **ii.** The x-intercept is $(-2, 0)$; the y-intercept is $(0, -\sqrt{8})$.

iii.

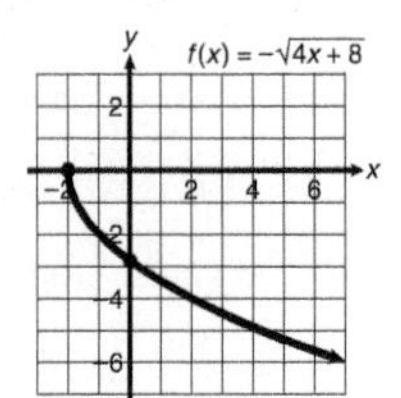

c. i. The domain is all real numbers such that $5 - x \geq 0$ or $x \leq 5$. **ii.** The x-intercept is (5, 0). The y-intercept is $(0, \sqrt{5})$.

iii.

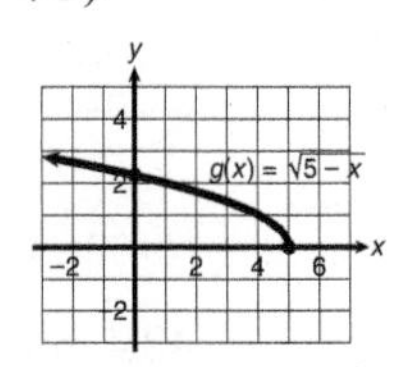

9. $d = \sqrt{12^2 + 24^2 + 17^2} = \sqrt{1009} \approx 31.8$ inches. It will not fit. **10. a.** $s = \sqrt{30(0.85)l} = \sqrt{25.5l}$, **b.** $s = \sqrt{25.5(90)} \approx 47.9$ mph, **c.** (answers may vary) 0 feet $\leq l \leq$ 300 feet is possible,

d.

```
NORMAL FLOAT AUTO a+bi RADIAN MP
Plot1 Plot2 Plot3
\Y1=√(25.5X)
\Y2=
\Y3=
\Y4=
\Y5=
\Y6=
\Y7=
\Y8=
```

```
NORMAL FLOAT AUTO a+bi RADIAN MP
WINDOW
Xmin=0
Xmax=300
Xscl=50
Ymin=0
Ymax=90
Yscl=10
Xres=1
ΔX=1.1363636363636
TraceStep=2.2727272727272
```

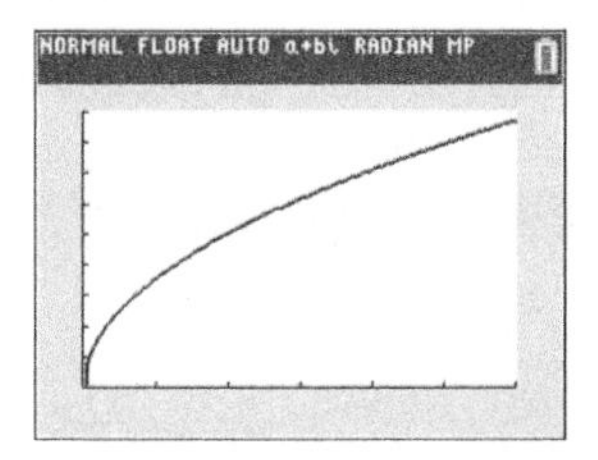

e. The length of the skid marks is approximately 192 feet.

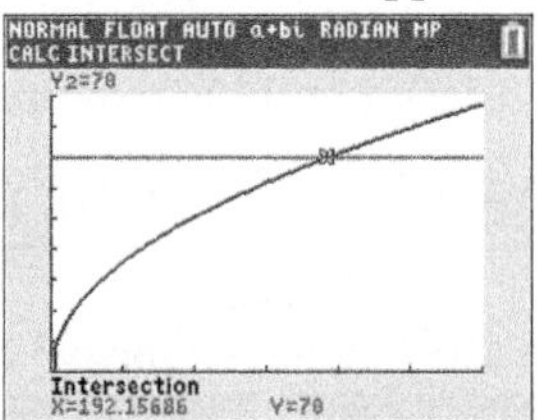

Activity 5.8 Exercises: 1. a. $x = 4$, **b.** $\sqrt{x + 1} = -4$; this can't happen; a positive radical can't equal -4. There is no solution. Equation b has no solution. The left side of equation b will always be greater than 1, so no solution is possible. **c.** $x = -2$, but $x = 1$ does not check.

3. a. $x = -4$ checks. **b.** $x = 2$ checks.

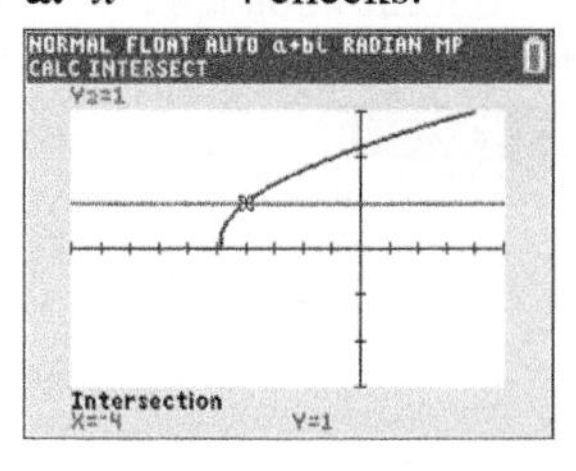

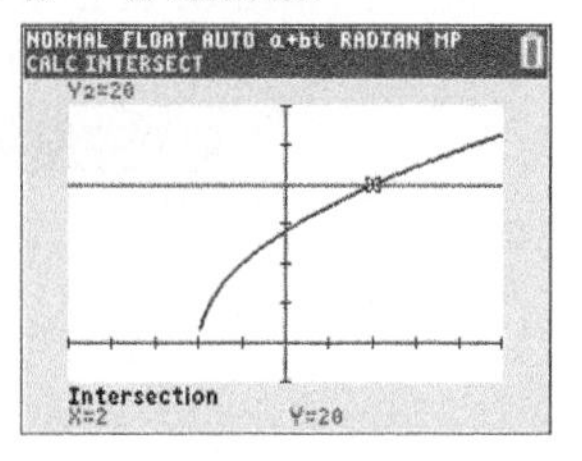

c. $x = 1$; $x = -4$ does not check.

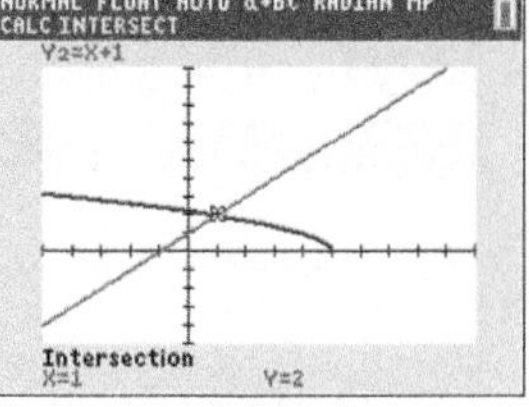

5. $L = 32\left(\dfrac{1.95}{2\pi}\right)^2$, $L \approx 3.08$ feet;

7. a. $V = \sqrt{\dfrac{1000(10)}{3}} \approx 57.7$ mph,

b. $P = \dfrac{14{,}700}{1000} = 14.7$ pound per square feet;

9. a. $A = \sqrt{\dfrac{70 \cdot 200}{3131}} \approx 2.11$ square meters, **b.** $w = \dfrac{3131A^2}{h}$

Activity 5.9 Exercises: 1. a. 4, **b.** 2, **c.** −3, **d.** 5, **e.** $\dfrac{1}{6}$, **f.** not real, **g.** 10, **h.** not real; **3.** The difference is $\sqrt[3]{1450} - \sqrt[3]{1280} \approx 0.46$ in.; **5. a.** all real numbers, **b.** $x \geq 3$, **c.** all real numbers, **d.** $x \leq 2$; **7. a.** $x = 64$, **b.** $x = 81$; **9. a.** $r = \sqrt[3]{\dfrac{3 \cdot 40}{4\pi}} \approx 2.12$ centimeters, **b.** $V = \dfrac{4\pi(3.5)^3}{3} \approx 179.6$ cubic feet, **c.** $V = \dfrac{4\pi r^3}{3}$

How Can I Practice? 1. a. $x = 98$, **b.** $x = 41$, **c.** $x = 12$, **d.** $x = \pm\sqrt{61}$, **e.** no solution, **f.** $x = 10$, **g.** $x = 10.5$, **h.** $x \approx 0.95$; **2. a.** $x \leq 6$, **b.** all real numbers, **c.** $x \geq 2$ or $x \leq -2$; **3.** Length is approximately 7.71 inches.

4. $r = \sqrt[3]{\dfrac{3V}{4\pi}}$, $V = 620$, $r \approx 5.29$ centimeters;
5. $v = 100$, $100 = \sqrt{64d}$, $d = 156.25$ feet;
6. $x = 6.5$ inches. The dimensions of the bottom of the box are 6.5 inches × 6.5 inches. **7.** The graphs are reflections about the line $x = 2$. **8.** $I = 80$ microwatts per square meter

Gateway Review 1. a. $d = \dfrac{1200}{w}$, **b.** 40, 34.286, 30, 24, 20, **c.** As width increases, the depth decreases. **d.** The depth is 12 feet, not enough room for most theater sets. **e.** No, division by 0 is undefined. **f.** (Answers may vary.) $30 \leq w \leq 60$, **g.** a rational function, **h.** all real numbers except 0, **i.** $w = 0$, **j.** $d = 0$; as w increases, d approaches 0.

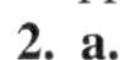

2. a.

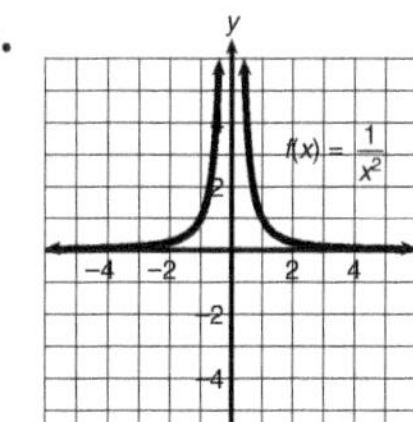

b.

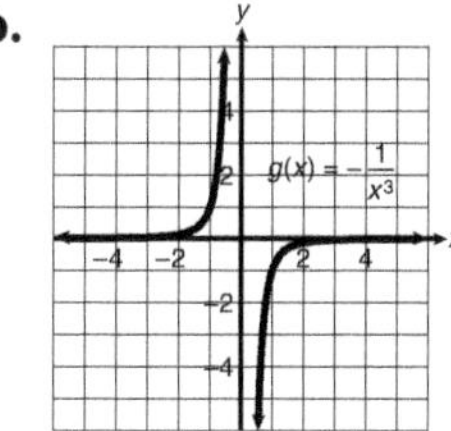

c. The graphs have the same horizontal and vertical asymptotes. $f(x) = \dfrac{1}{x^2}$ is symmetrical with respect to the y-axis. $g(x) = -\dfrac{1}{x^3}$ is symmetrical with respect to the origin in quadrants II and IV. $f(x)$ is always positive. $g(x)$ is both positive and negative. **3. a.** $y = \dfrac{k}{x}$; $12 = \dfrac{k}{10}$; $k = 120$; $y = \dfrac{120}{30} = 4$, **b.** $I = \dfrac{k}{d^2}$; $32 = \dfrac{k}{16}$; $k = 512$; $I = \dfrac{512}{100} = 5.12$ decibels, **c.** $h = \dfrac{k}{r^2}$; $8 = \dfrac{k}{4}$; $32 = k$; $h = \dfrac{32}{25} = 1.28$ inches; **4. a.** H: $y = 0$; V: $x = 0$; no y-intercept, no x-intercept, **b.** H: $y = 0$, V: $x = 3$, $\left(0, -\dfrac{4}{3}\right)$; no x-intercept, **c.** H: $y = 2$; V: $x = -2$, $(0, 0)$, $(0, 0)$;

4. a.

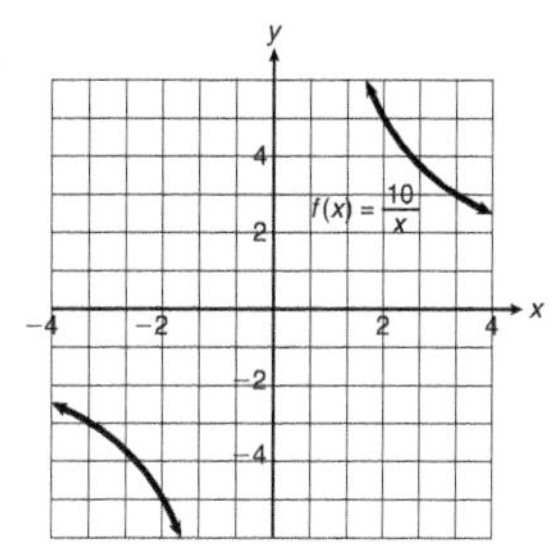

b.

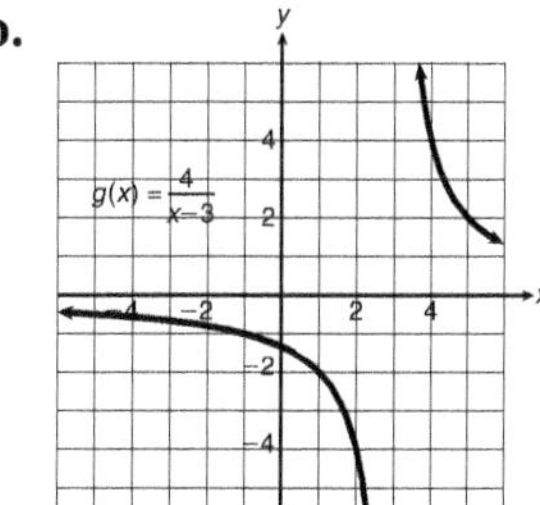

c.

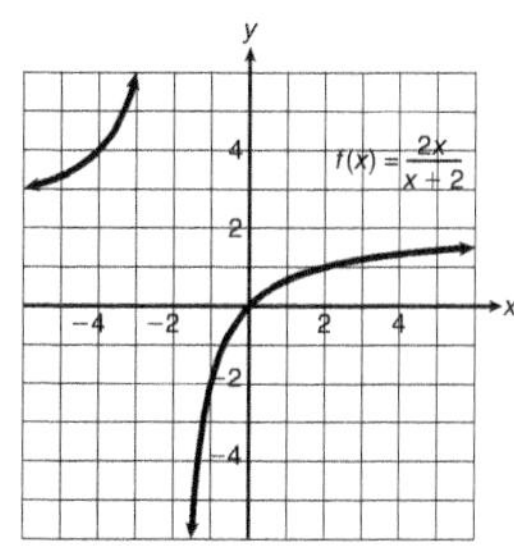

5. a. $f(n) = 45n + 600$, **b.** $f(100) = \$5100$, **c.** $A(n) = \dfrac{45n + 600}{n}$, **d.** $A(100) = \dfrac{45(100) + 600}{100} = \$51$, **e.** 57, 51, 49, 48, 47.40, **f.** $50 = \dfrac{45n + 600}{n}$; $50n = 45n + 600$; $5n = 600$; $n = 120$ people

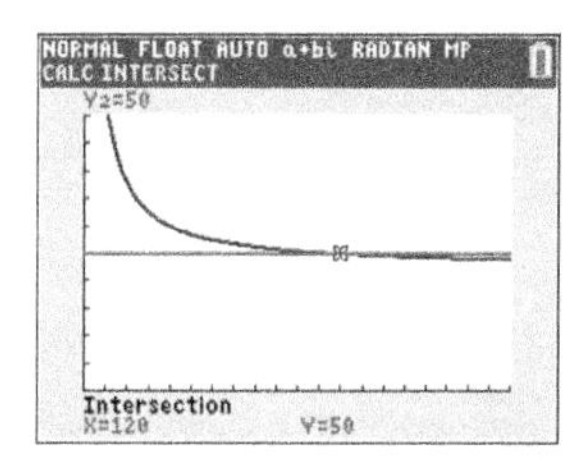

g. $0 < n <$ seating capacity of restaurant, **h.** The vertical asymptote is $n = 0$. You cannot calculate an average value if no people attend. **i.** The horizontal asymptote is $A(n) = 45$. As the number of people attending increases, the average cost approaches \$45 per person.

6. a. $\dfrac{4}{x - 2} = 6$; $4 = 6x - 12$; $16 = 6x$; $x = \dfrac{8}{3}$, **b.** The solution is the x-coordinate of the x-intercept.

7. a. $x = -\dfrac{7}{5} = -1.4$ **b.** $x = 1$

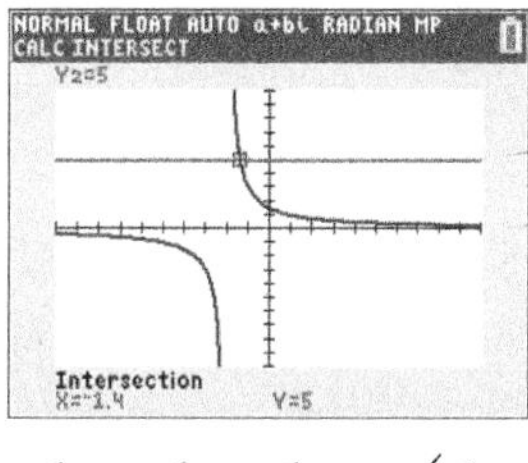

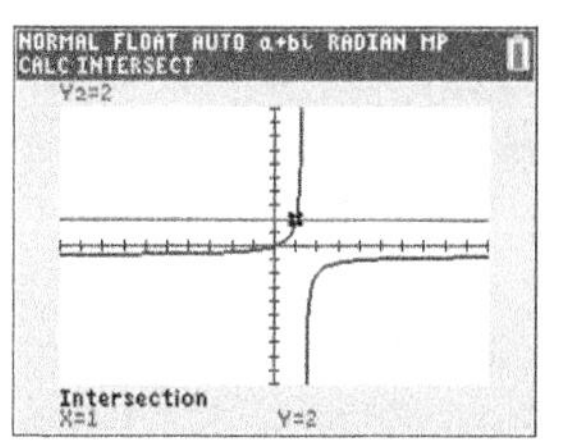

8. $\dfrac{1}{20} + \dfrac{1}{15} = \dfrac{1}{x}$; $60x\left(\dfrac{1}{20} + \dfrac{1}{15}\right) = 60x\left(\dfrac{1}{x}\right)$; $3x + 4x = 60$; $7x = 60$; $x \approx 8.57$ minutes; **9.** $x = 67.5$, $2x = 135$ minutes, or $2\dfrac{1}{4}$ hours; **10. a.** $x = 10.2$, **b.** $x = 33$;

11. a. $S = \dfrac{C}{1-r}$; $S(1-r) = C$; $S - Sr = C$; $S - C = Sr$; $r = \dfrac{S-C}{S}$, **b.** $bc - 4ab = -3ac$; $b(c - 4a) = -3ac$; $b = \dfrac{-3ac}{c-4a}$; **12. a.** $\dfrac{b+2a}{2b+a}$, **b.** $\dfrac{x+2}{x-2}$;

13. a. $f = \dfrac{1}{\frac{1}{4}+\frac{1}{3}} = \dfrac{1}{\frac{3}{12}+\frac{4}{12}} = 1 \div \dfrac{7}{12} = \dfrac{12}{7} \approx 1.71$ meters, **b.** $f = \dfrac{1}{\frac{1}{p}+\frac{1}{q}} = \dfrac{1}{\frac{q}{pq}+\frac{p}{pq}} = \dfrac{1}{\frac{p+q}{pq}} = \dfrac{pq}{p+q}$,

c. $f = \dfrac{4(3)}{4+3} = \dfrac{12}{7} \approx 1.71$ meters; the values are the same.

14. a. $x \geq -4$, **b.**

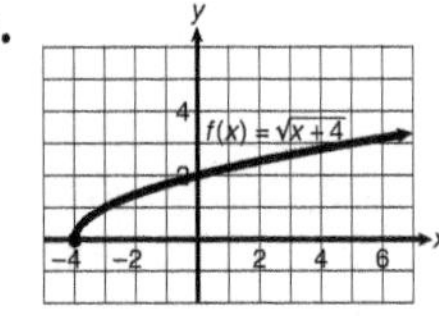

c. The output is increasing. **d.** $y \geq 0$, **e.** The x-intercept is $(-4, 0)$. The y-intercept is $(0, 2)$. **f.** g has the same shape but is shifted 8 units to the right. **g.** The graphs are reflected through the x-axis.

15. a.

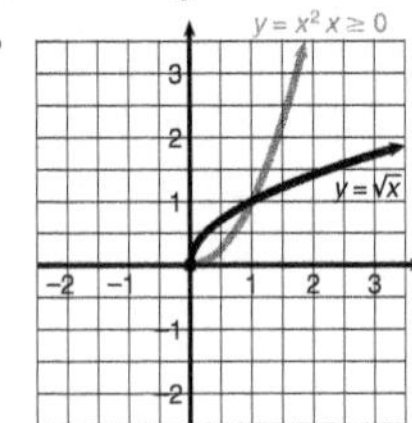

b. $y = \sqrt{x}$; $x = \sqrt{y}$; $y = x^2$; $f^{-1}(x) = x^2; x \geq 0$, **d.** The graphs are reflections in $y = x$. **e.** $f(f^{-1}(x)) = f(x^2) = \sqrt{x^2} = x$; $f^{-1}(f(x)) = f^{-1}(\sqrt{x}) = (\sqrt{x})^2 = x$;

16. a. i. $x \geq 0, y \geq 4$, **ii.** $(0, 4)$ only,

iii.

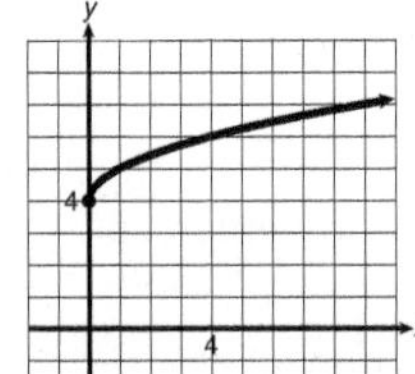

b. i. $x \geq -4, y \geq 0$, **ii.** $(0, 2)$ and $(-4, 0)$,
iii.

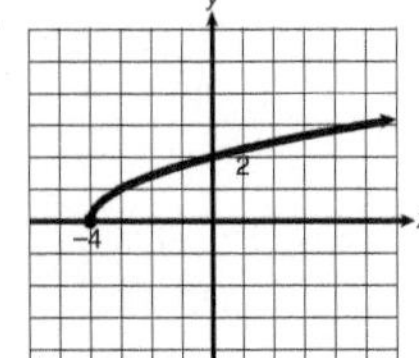

17. a. $x = 6$, **b.** $x = 6$, **c.** $x = \dfrac{23}{5} = 4.6$, **d.** $x = -1$, -1 does not check. There is no solution. **e.** $x = 27$;
18. a. all real numbers, **b.** $x \geq 6$, **c.** $x \geq -1$;
19. $36 = 1.5h$, $h = 24$ feet; **20.** $d = 153.76$ feet

Chapter 6

Activity 6.1 Exercises: 1. a. 0.6000, **b.** 0.8000, **c.** 0.8000, **d.** 0.6000, **e.** 0.7500, **f.** 1.3333;

3. a. Given $\tan B = \dfrac{7}{4}$, if the side opposite angle B is 7, then the side adjacent to angle B is 4. Using the Pythagorean theorem, I determine that the hypotenuse is $\sqrt{65}$.

b. $\sin B = \dfrac{7}{\sqrt{65}} \approx 0.8682$, **c.** $\cos B = \dfrac{4}{\sqrt{65}} \approx 0.4961$;

5. a. the sine function, **b.** $\sin B = \dfrac{y}{c}$; **7. a.** $x \approx 9.1$, **b.** $x \approx 85.6$, **c.** $x \approx 61.4$;

9. a.

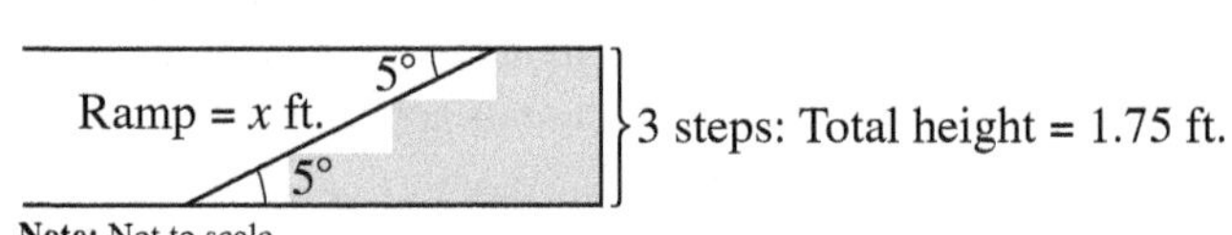

b. $3 \cdot 7 = 21$ inches. The increase in height from one end of the ramp to the top of the stairs is $\dfrac{21}{12} = 1.75$ feet.

c. $x \approx 20.1$ feet. The ramp needs to be at least 20.1 feet long. Therefore, the donated ramp will not be long enough to meet the code. Alternative approach: $15 \sin 5° \approx 1.3$ feet. The three steps must measure at most 1.3 feet high for the 15-foot ramp to satisfy the code. Each solution suggests ways to think about modifications to either the ramp or the steps (or both) that could be used to meet the code.

Activity 6.2 Exercises: 1. a. $s + w \approx 0.68$ miles, **b.** These calculations confirm the result in part a.

3. a.

$(90° - x)$	$\sin x$	$\cos(90° - x)$
83°	0.1219	0.1219
73°	0.2924	0.2924
66°	0.4067	0.4067
57°	0.5446	0.5446
42°	0.7431	0.7431
23°	0.9205	0.9205
13°	0.9744	0.9744

b. The table in part a illustrates the property that cofunctions of complementary angles are equal.

Activity 6.3 Exercises: 1. a. $\theta = 30°$, **b.** $\theta \approx 64.62°$, **c.** $\theta \approx 67.04°$, **d.** $\theta \approx 63.82°$, **e.** $\theta \approx 66.80°$, **f.** $\theta \approx 64.62°$, **g.** $\theta \approx 22.28°$, **h.** $\theta \approx 20.76°$;

3.

V	θ	E	N
32	65°	13.5	29.0
23.3	59°	12	20
4.1	54°	2.4	3.3
26	43.8°	18.8	18
4.5	45°	3.2	3.2

4. $\theta = \tan^{-1}\left(\frac{30}{50}\right) \approx 31°$. My friend should look up at an angle of approximately 31°.

5. a.

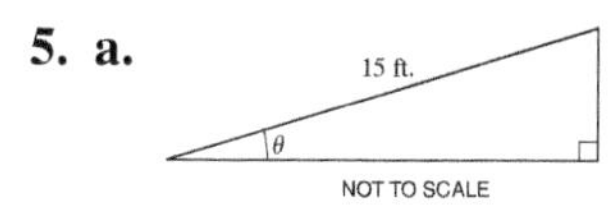

b. Because grade is rise over run, $\tan\theta = 0.1$. $\theta = \tan^{-1}(0.1) = 5.7°$.

The ramp makes an angle of 5.7° with the horizontal. **c.** $y = 15\sin(5.7°) \approx 1.5$ feet. The elevation changes 1.5 feet from one end of the ramp to the other.

7. $\tan\theta = \frac{185}{2640}$ or $\theta = \tan^{-1}\left(\frac{185}{2640}\right) \approx 4°$

Activity 6.4 Exercises: 1. a. The side adjacent to the 57° angle is 4.2 feet. The hypotenuse is 7.8 feet. The other acute angle is 33°. **b.** The hypotenuse is 19.0. The angle adjacent to side 18 is 18.4°. The other acute angle is 71.6°. **c.** The other leg is 7.9 inches. The angle adjacent to side 9 inches is 41.4°. The other angle is 48.6°.

3. a.

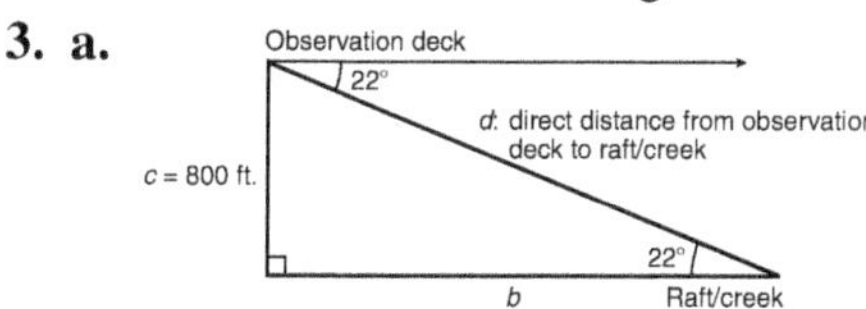

b. The direct distance, d, from the observation deck to the raft is approximately $\frac{800}{\sin(22°)}$ or 2136 feet. If you could walk straight down the cliff and straight across at the base of the cliff to the creek, the distance would be approximately $b + c = 2780$ feet, where $b = \frac{800}{\tan 22°}$ which is approximately equal to 1980.1 feet

Project Activity 6.5 Exercises: 1. a. The slope is $\frac{5}{100}$ or $\frac{1}{20}$ or 0.05. **b.** $A = \tan^{-1}(0.05) \approx 2.86°$. The highway makes an angle of 2.86° with the horizontal. This angle is called the angle of elevation. **c.** 1 mile is equivalent to 5280 feet. If x represents the number of feet above sea level after walking 1 mile, then $x = 5280 \cdot \sin(2.86°)$ which is approximately 263 feet. I would be 263 feet above sea level after 1 mile. **3.** Using the following diagram, the two equations are

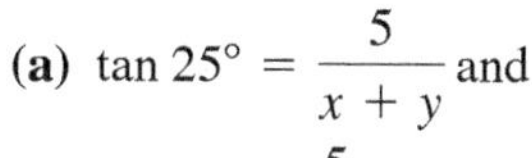

(a) $\tan 25° = \frac{5}{x + y}$ and

(b) $\tan 30° = \frac{5}{y}$.

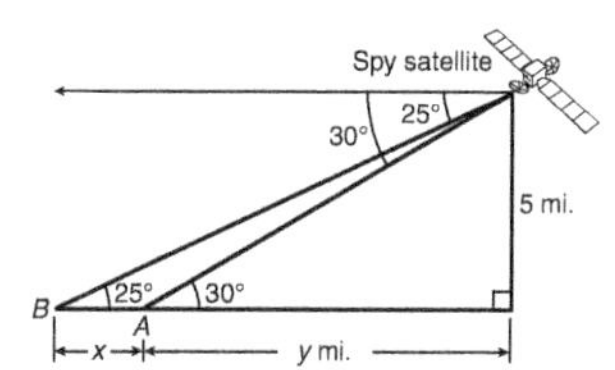

Solving equation (b), $y \approx 8.7$ miles. Then equation (a) becomes $\tan 25° = \frac{5}{x + 8.7}$. Solve this equation for x. $(x + 8.7)\tan 25° = 5$, $x\tan 25° = 5 - 8.7\tan 25°$,

$x = \frac{5 - 8.7\tan 25°}{\tan 25°}$, $x \approx 2.0$ miles. The runway is approximately 2 miles long. **6. a.** Let A represent the area of the trapezoidal cross section. The height of the cross section is h, and the two bases are 5 and $5 + 2x$, respectively. The area is then determined by the formula $A = \frac{1}{2}h(10 + 2x)$, which, after simplifying, is $A = h(5 + x)$ or $A = 5h + hx$. **b.** $h = 5\sin t$, $x = 5\cos t$, $A = 5(\sin t)(5 + 5\cos t)$ or $A = 25\sin t(1 + \cos t)$ or $A = 25\sin t + 25\sin t(\cos t)$,

c.

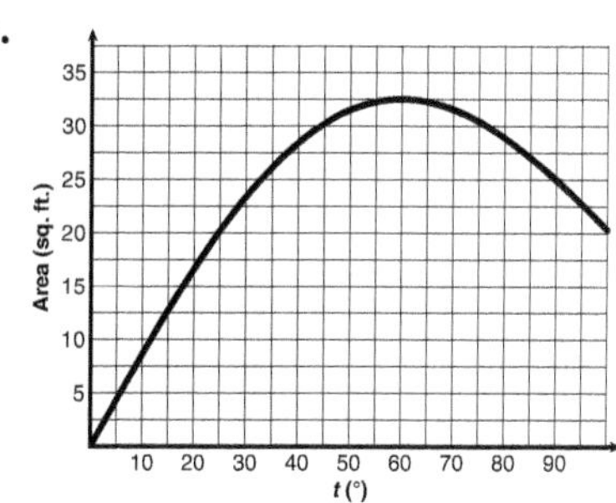

d. The graph in part c indicates that the area of the trapezoidal cross section (output) is greatest when the angle t is 60°.

e. The area is approximately 32.5 square feet as read from the graph in part c. **f.** Let V represent the volume. Then $V = 50 \cdot A$, where A is the cross-sectional area. In terms of t, the volume is $V = 1250\sin t(1 + \cos t)$.

g.

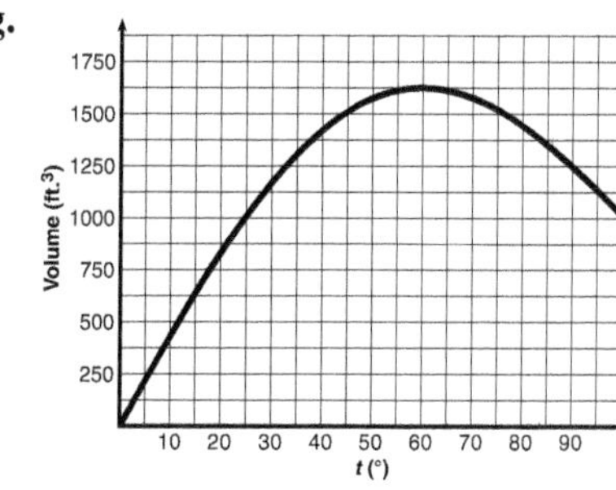

h. The graph indicates that the greatest value for the volume between 0° and 90° is approximately 1625 cubic feet when the angle t is 60°.

i. The angle is the same—namely, 60° in this scenario.

How Can I Practice? 1. a. $\frac{8}{15}$, **b.** $\frac{15}{8}$, **c.** $\frac{15}{17}$, **d.** $\frac{8}{17}$, **e.** $\frac{8}{17}$, **f.** $\frac{15}{17}$; **2. a.** 0.731, **b.** 0.574, **c.** 0.601, **d.** 5.671; **3.** $\cos A = \frac{12}{13}$ and $\tan A = \frac{5}{12}$; **4.** $\sin B = \frac{7}{\sqrt{65}}$ and $\cos B = \frac{4}{\sqrt{65}}$; **5. a** $\theta \approx 48.6°$, **b.** $\theta \approx 23.5°$, **c.** $\theta \approx 74.1°$, **d.** $\theta \approx 16.6°$, **e.** $\theta \approx 44.2°$, **f.** $\theta \approx 13.7°$; **6.** $\overline{BC} = 4.8$ centimeters, $\overline{AC} = 3.6$ centimeters, $\angle B = 37°$, $\angle C = 90°$; **7.** $A = \arctan\left(\frac{10}{15}\right) \approx 33.7°$; therefore, I should buy the 35° trusses. **8. a.** $\theta \approx 20.6°$, **b.** $D = \sqrt{16^2 + 6^2} \approx 17.1$ feet; **9. a.**

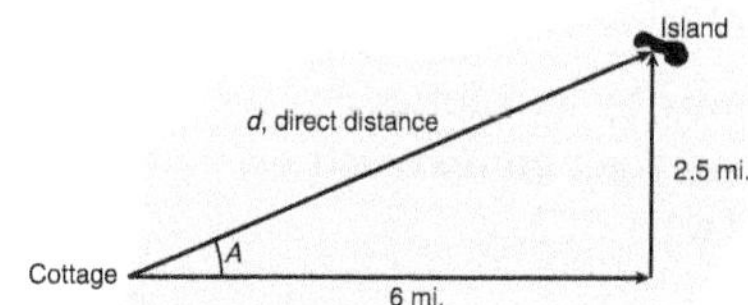

The direct distance, d, from the cottage to the island is $d = \sqrt{2.5^2 + 6^2} = 6.5$ miles. **b.** $A = \arctan\left(\frac{2.5}{6}\right) \approx 22.6°$; I should direct my boat 22.6° north of east to get from the cottage to the island in the shortest distance.

Activity 6.6 Exercises: 1. a. $(0.31, 0.95)$, **b.** $(0.64, -0.77)$, **c.** $(0, -1)$, **d.** $(-0.36, 0.93)$, **e.** $(-0.85, -0.53)$, **f.** $(0.26, 0.97)$, **g.** $(0.34, -0.94)$;
2. a. $\frac{72}{360} \cdot 2\pi = 1.26$, **b.** $\frac{310}{360} \cdot 2\pi = 5.41$, **c.** $\frac{270}{360} \cdot 2\pi = 4.71$, **d.** $\frac{111}{360} \cdot 2\pi = 1.94$, **e.** $\frac{212}{360} \cdot 2\pi = 3.70$, **f.** $\frac{435}{360} \cdot 2\pi = 7.59$, **g.** $\frac{-70}{360} \cdot 2\pi = -1.22$; distance is 1.22;
4. a. The graph looks like the cosine function reflected in the t-axis. **b.** The graph indicates clockwise rotation starting at $(-1, 0)$.
5. a.

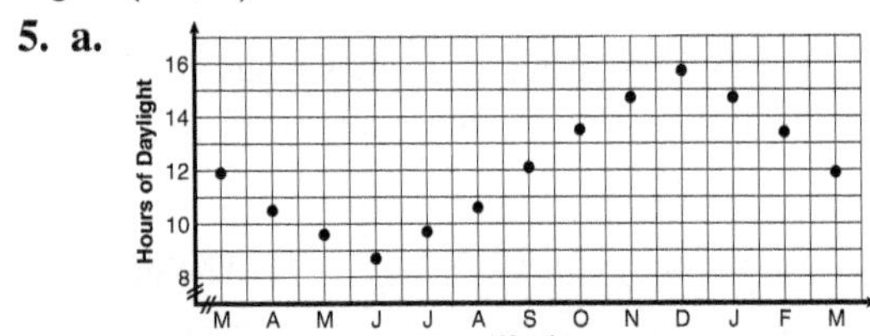

b. Yes, the number of hours of daylight is cyclical. The graph looks like a shifted and stretched sine graph reflected in the x-axis. **c.** The graph has the same wavelike shape. **d.** South; the number of hours of daylight is greater than 12 from October to February, summer in the Southern Hemisphere.

Activity 6.7 Exercises: 1. a. $45 \cdot \frac{\pi}{180} = \frac{\pi}{4} \approx 0.785$ radian, **b.** $140 \cdot \frac{\pi}{180} = \frac{7\pi}{9} \approx 2.443$ radians,
c. $330 \cdot \frac{\pi}{180} = \frac{11\pi}{6} \approx 5.760$ radians,
d. $-36 \cdot \frac{\pi}{180} = \frac{-\pi}{5} \approx -0.628$ radian;
3.

0°	30°	45°	60°	90°	135°	180°	210°	270°	360°
0	$\frac{\pi}{6}$	$\frac{\pi}{4}$	$\frac{\pi}{3}$	$\frac{\pi}{2}$	$\frac{3\pi}{4}$	π	$\frac{7\pi}{6}$	$\frac{3\pi}{2}$	2π

5. a. $y = 2\cos x$ is $y = \cos x$ stretched vertically by a factor of 2. $y = \cos(2x)$ is $y = \cos x$ compressed horizontally by a factor of 2. **b.** $y = \cos\left(\frac{1}{3}x\right)$ is $y = \cos x$ stretched horizontally by a factor of 3. $y = \cos(3x)$ is $y = \cos x$ compressed horizontally by a factor of 3.

7. max: 3, min: -3, period: $\frac{2\pi}{3}$;

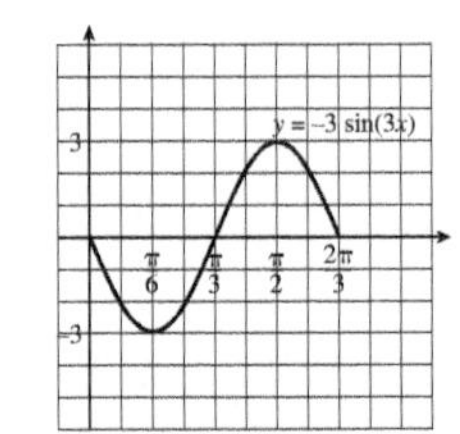

9. a. The bill is highest for December and January. The amount of the bill is approximately \$600. **b.** The bill is lowest for May and June. The amount of the bill is approximately \$250. **c.** The largest value is \$650. **d.** The period is 6 billing periods or 12 months. **e.** The graph will be stretched vertically by a factor of 1.05. This will not affect the period of the function. **f.** The amount of the bills for the summer months would increase. The graph would flatten out as the monthly charges become more equal.

Activity 6.8 Exercises:
1.

3. a. The maximum value is 100. **b.** The minimum value is -100. **5.** $x = 150\cos\theta$ $y = 150\sin\theta$;
8. a. $y = -15\sin(2x)$, **b.** $y = 1.3\cos(0.7x)$;
10. a. Graph is iii. **b.** Graph is iv. **c.** Graph is i. **d.** Graph is ii.

Activity 6.9 Exercises: 1. a. amplitude: 0.7, period: π, displacement: $\frac{-\frac{\pi}{2}}{2} = -\frac{\pi}{4}$, **b.** amplitude: 3, period: 2π, displacement: $\frac{-(-1)}{1} = 1$, **c.** amplitude: 2.5, period: 5π, displacement: $\frac{-\frac{\pi}{3}}{0.4} = -\frac{5\pi}{6}$, **d.** amplitude: 15, period: 1, displacement: $\frac{-(-0.3)}{2\pi} = \frac{3}{20\pi} \approx 0.0477$;
3.

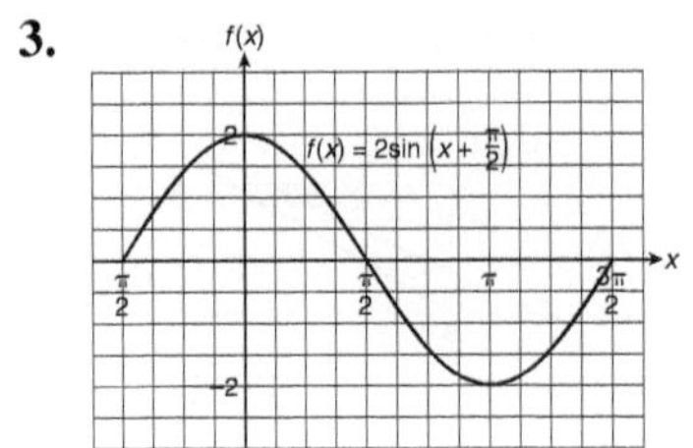

5. a. Graph is iii. **b.** Graph is iv. **c.** Graph is i. **d.** Graph is ii.

b.

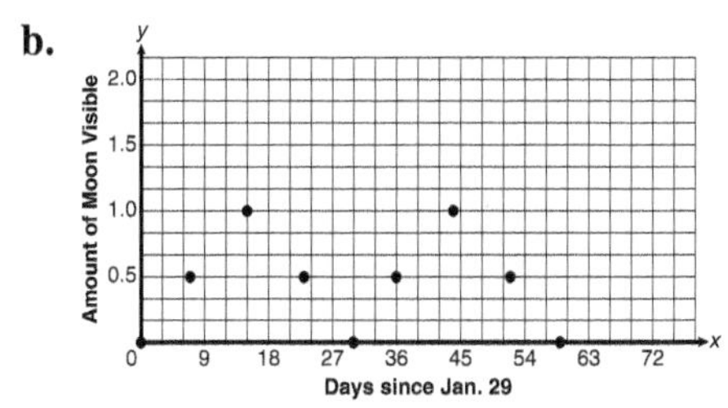

c. Yes, it does show repeated and periodic behavior. **d.** $y = 0.5019 \sin(0.2135x - 1.5888) + 0.5306$, **e.** $y(95) \approx 0.45$, or about 45% of the side of the moon facing Earth will be visible 95 days after January 29.

3. a.

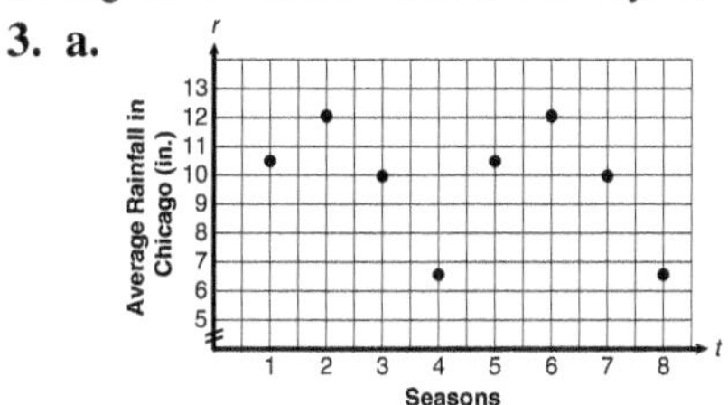

b. Yes, the data appears to be periodic. **c.** $y = 2.759 \sin(1.534x - 1.322) + 9.725$, **d.** The 9.725 is the average output value. In this case, it represents the average seasonal rainfall in Chicago. **e.** $\frac{2\pi}{1.534} \approx 4.10$. Yes, it makes sense. It should be close to 4 because there are four seasons in the input cycle.

How Can I Practice? 1. a. $(0.81, 0.59)$, **b.** $(-0.87, -0.5)$, **c.** $(0, -1)$, **d.** $(0.73, -0.68)$, **e.** $(-0.81, -0.59)$, **f.** $(0, 1)$; **2. a.** 0.63 units, **b.** 3.67 units, **c.** 1.57 units clockwise, **d.** 5.53 units, **e.** 2.51 units clockwise, **f.** 7.85 units; **3. a.** $18 \cdot \frac{\pi}{180} = \frac{\pi}{10}$, **b.** $150 \cdot \frac{\pi}{180} = \frac{5\pi}{6}$, **c.** $390 \cdot \frac{\pi}{180} = \frac{13\pi}{6}$, **d.** $-72 \cdot \frac{\pi}{180} = \frac{-2\pi}{5}$; **4. a.** $\frac{5\pi}{6} \cdot \frac{180}{\pi} = 150°$, **b.** $1.7\pi \cdot \frac{180}{\pi} = 306°$, **c.** $-3\pi \cdot \frac{180}{\pi} = -540°$, **d.** $0.9\pi \cdot \frac{180}{\pi} = 162°$;

5. amplitude: 4
period: $\frac{2\pi}{3}$
displacement: 0

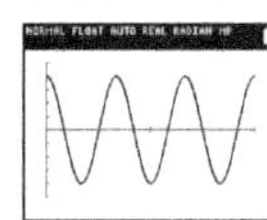

6. amplitude: 2
period: $\frac{2\pi}{1} = 2\pi$
displacement: 1

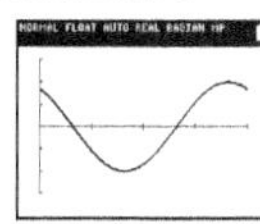

7. amplitude: 3.2
period: $\frac{2\pi}{2} = \pi$
displacement: 0

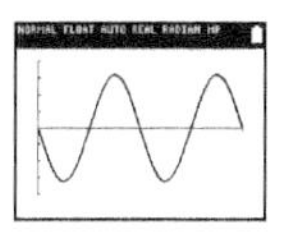

8. amplitude: 1
period: $\frac{2\pi}{\frac{1}{2}} = 4\pi$
displacement: -2

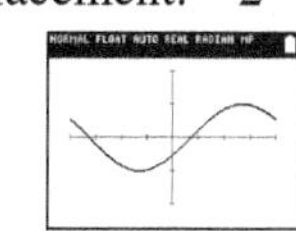

9. amplitude: 3
period: $\frac{2\pi}{4} = \frac{\pi}{2}$
displacement: $\frac{-(-1)}{4} = \frac{1}{4}$

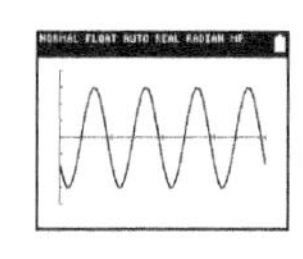

10. a. A sine or cosine function models this relationship because of the repetitive nature of the height of the water as a function of time, **b.** amplitude $= \frac{80 - 0}{2} = 40$, **c.** The period is approximately 12 hours because high tide occurs twice a day. **d.** Let x represent the number of hours since midnight. $y = a \sin(bx + c) + d$, $a = 40$, the amplitude, period: $\frac{2\pi}{b} = 12$, $b = \frac{\pi}{6}$, displacement: $3 = \frac{-c}{b} = \frac{-c}{\frac{\pi}{6}}$, $-c = 3 \cdot \left(\frac{\pi}{6}\right)$, $c = \frac{-\pi}{2}$, vertical shift: $d = 40$, $y = 40 \sin\left(\frac{\pi}{6}x - \frac{\pi}{2}\right) + 40$. Other equations are possible; it depends on the choice of displacement.

Gateway Review 1. a. $N = 7 \sin(63°) \approx 6.24$ miles, **b.** $E = 7 \cos(63°) \approx 3.18$ miles; **2. a.** side $c = 13$, angle $A \approx 67.4°$, angle $B \approx 22.6°$, **b.** side $a \approx 6.93$, side $b = 4$, angle $A = 60°$, **c.** side $b = 3$, side $c \approx 4.24$, angle $B = 45°$, **d.** side $a \approx 8.66$, side $c = 10$, angle $B = 30°$; **3. a.** $\cos\theta = \frac{8}{10}$, $\tan\theta = \frac{6}{8}$, **b.** $\sin\theta = \frac{1}{2}$, $\tan\theta = \frac{1}{\sqrt{3}}$, $\theta = 30°$, **c.** $c^2 = 5^2 + 8^2 = 89$, $c = \sqrt{89}$, $\sin\theta = \frac{8}{\sqrt{89}}$, $\cos\theta = \frac{5}{\sqrt{89}}$; **4.** No, there is a difference, but it is so small that it is difficult to see. For me: $\theta = \tan^{-1}\left(\frac{1408}{100}\right) \approx 85.9375°$. For my nephew: $\theta = \tan^{-1}\left(\frac{1411}{100}\right) \approx 85.9461°$.

Activity 6.10 Exercises: 1. a.

DATE	JAN 29	FEB 5	FEB 13	FEB 21	FEB 28	MAR 6	MAR 14	MAR 22	MAR 29
x; Days since Jan 29	0	7	15	23	30	36	44	52	59
y; The Amount of Moon Visible	0	0.5	1	0.5	0	0.5	1	0.5	0

5.

$$\tan 57^\circ = \frac{a}{30}$$
$$30 \tan 57^\circ = a$$
$$a \approx 46.2$$
$$\cos 57^\circ = \frac{30}{c}$$
$$c = \frac{30}{\cos 57^\circ}$$
$$c \approx 55.1$$

$$\tan 13^\circ = \frac{b}{46.2}$$
$$b = 46.2 \tan 13^\circ$$
$$b \approx 10.7$$
$$\cos 13^\circ = \frac{46.2}{h}$$
$$h = \frac{46.2}{\cos 13^\circ}$$
$$h \approx 47.4$$

6. So $(2x)^2 = x^2 + (200 + x)^2$; $x = \frac{400 \pm \sqrt{480{,}000}}{4}$.

The negative does not make sense; thus, $x = \frac{400 + 400\sqrt{3}}{4}$.

$x = 100 + 100\sqrt{3}$; $h = 300 + 100\sqrt{3}$

$x \approx 273.2$; $h \approx 473.2$

7.

$\sin\theta$	$\cos\theta$	$\tan\theta$
$\frac{\sqrt{3}}{2}$	$-\frac{1}{2}$	$-\sqrt{3}$
$\frac{1}{\sqrt{2}}$	$-\frac{1}{\sqrt{2}}$	-1
$\frac{1}{2}$	$-\frac{\sqrt{3}}{2}$	$-\frac{1}{\sqrt{3}}$
0	-1	0
$-\frac{1}{2}$	$-\frac{\sqrt{3}}{2}$	$\frac{1}{\sqrt{3}}$
$-\frac{1}{\sqrt{2}}$	$-\frac{1}{\sqrt{2}}$	1
$-\frac{\sqrt{3}}{2}$	$-\frac{1}{2}$	$\sqrt{3}$
-1	0	undef.
$-\frac{\sqrt{3}}{2}$	$\frac{1}{2}$	$-\sqrt{3}$
$-\frac{1}{\sqrt{2}}$	$\frac{1}{\sqrt{2}}$	-1
$-\frac{1}{2}$	$\frac{\sqrt{3}}{2}$	$-\frac{1}{\sqrt{3}}$
0	1	0

8. a. amplitude: 2
period: 2π

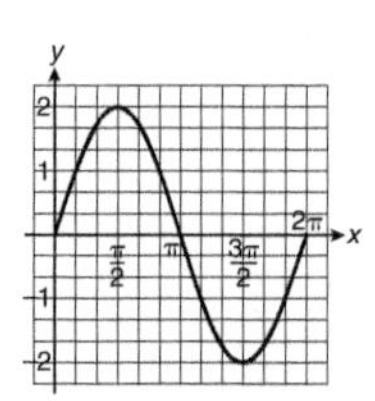

b. amplitude: 2
period: 2π

c. amplitude: 1
period: π

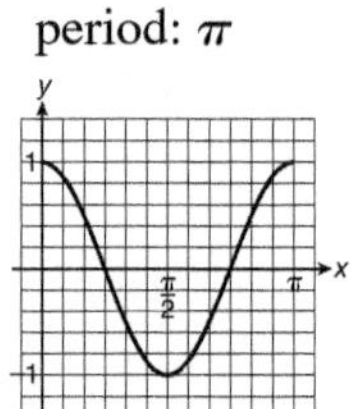

d. amplitude: 1
period: 1

e. amplitude: 1
period: 4π

f. amplitude: 1
period: 4

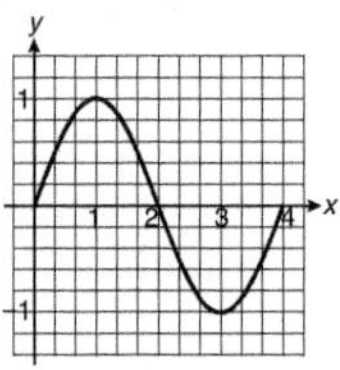

g. amplitude: 1
period: 2π

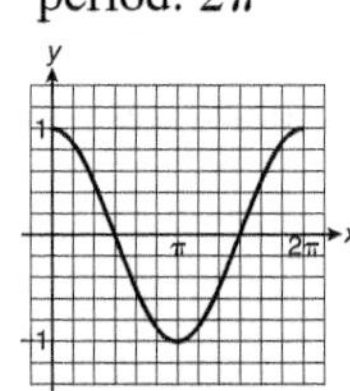

h. amplitude: $\frac{2}{3}$
period: π

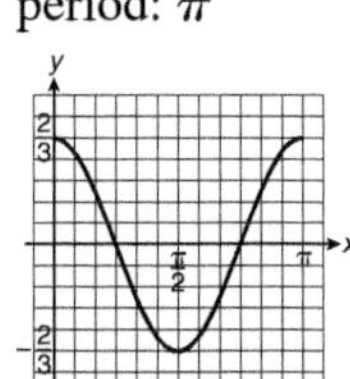

i. amplitude: 1
period: 1

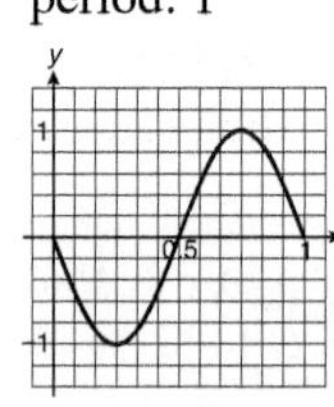

j. amplitude: 3
period: 1

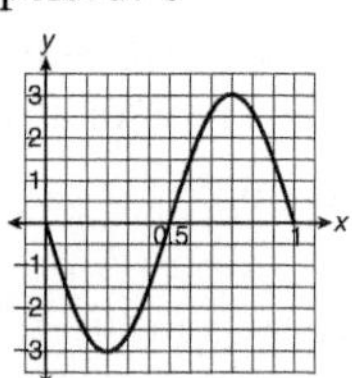

9. a. Graph is vi. **b.** Graph is ii. **c.** Graph is iv. **d.** Graph is v. **e.** Graph is vii. **f.** Graph is viii.

GLOSSARY

addition of functions See sum function.

argument Another name for the input of a function.

average rate of change, or, the rate of change The ratio $\frac{\Delta y}{\Delta t}$, where Δy represents the change in output and Δt represents the change in input. The average rate of change gives the change in the output for a one-unit increase in the input.

axis of symmetry A vertical line that separates the graph of a parabola into two mirror images.

change of base formula $\log_b x = \frac{\log_a x}{\log_a b}$, where $b > 0$, $b \neq 1$, is the formula used to change logarithms of one base to logarithms of another base.

coefficient The numerical multiplier of a variable.

common logarithms Base 10 logarithms.

complex numbers Numbers of the form $a + bi$, such that a and b are real numbers and $i = \sqrt{-1}$.

composition function The function that is created when the output of the function g becomes the input for the second function f. The rule is given by $y = f(g(x))$.

consistent system of linear equations A system with exactly one solution.

constant function A function in which there is no change in the output. The graph of a constant function is a horizontal line.

constant of proportionality A constant, k, that gives the rate of variation in the direct proportional relationship $y = kx^n$.

constant term A term that does not change in value.

continuous compounding of an investment Occurs when the compounding period is so short it is essentially an instant in time. The formula for continuous compounding is $A = Pe^{rt}$.

cubic function A third-degree polynomial function having the general equation $y = ax^3 + bx^2 + cx + d$, where a, b, c, and d are real numbers and $a \neq 0$.

decay factor of an exponential function The number, b, in the equation $y = ab^x$, where $0 < b < 1$ and a is the amount when $x = 0$.

decreasing function A function in which the output decreases in value as the input increases. The graph goes down to the right.

degree of a polynomial function The exponent of the term with the largest exponent.

dependent system system of linear equations A linear system with an infinite number of solutions. (Graphically both equations represent the same line.)

dependent variable of a function Another name for the output variable of a function.

difference function, $f - g$ The function that is created from two functions, f and g, by the rule $y = f(x) - g(x)$.

direct variation between two variables A relationship defined algebraically by $y = k \cdot x$, where k is a constant, in which whenever x *increases* by a multiplicative factor (e.g., doubles), y also *increases* by the same factor.

discriminant The expression $b^2 - 4ac$ under the radical of the quadratic formula. The value of the discriminant determines the type of solutions of the equation $ax^2 + bx + c = 0$.

domain of a function The set (collection) of all possible input values for a function.

exponential function A function of the form $y = a \cdot b^x$ with $b > 0$ and $b \neq 1$, where the independent variable, x, is the exponent.

extraneous solution A potential solution that is not really a solution to the original equation or problem.

function A relationship between the input and the output such that for each input value there is exactly one output value.

general form of a linear equation $Ax + By = C$, where A, B, and C are real numbers.

growth factor of the exponential function $y = a \cdot b^x$ The number b, where $b > 1$ and a is the amount when $x = 0$.

horizontal asymptote A horizontal line $y = b$ that the graph of a function approaches but not touches as the input variable increases in magnitude.

horizontal intercepts All points of the graph of a function whose y-coordinate is 0. (*See* x-intercept.)

identity function The function in which the output value is always identical to the input value.

imaginary unit $i = \sqrt{-1}$.

inconsistent system of linear equations A linear system of equations with no solution. (Graphically, two parallel lines represent the system.)

increasing function A function in which the output increases in value as the input increases. The graph goes up to the right.

input variable The independent variable.

inverse functions Two functions f and g related such that $f(g(x)) = x = g(f(x(x))$. Graphically, these functions are mirror images in the line $y = x$.

inverse variation between two variables A relationship in which as the independent variable (input) increases in value, the dependent variable (output) decreases. Also, the independent variable decreases as the dependent variable increases.

irrational number Any real number that cannot be written as a rational number.

linear function Any function in which the rate of change, or slope, is constant.

linear term The term of a polynomial function of the form bx, where b is a real number.

logarithm function A function of the form $y = \log_b x$ where the base $b > 0, b \neq 1$. (The logarithm function is the inverse of the exponential function.)

magnitude The relative size of a number or quantity, expressed as a distance or absolute value (and is therefore not negative).

mathematical model A function that best fits the actual data and can be used to predict output values for input values not in the table.

natural logarithm A logarithm to the base e. The logarithm is written as $\log_e x = \ln x$.

ordered pair A pair of values, separated by a comma and enclosed in a set of parentheses. The input value is written to the left of the output value.

output variable The dependent variable.

parabola The graph of a quadratic function (second-degree polynomial function). The graph is U-shaped, opening either upward or downward.

piecewise function A function in which the function rule for determining the output is given separately, or in pieces, for different values of the input.

point-slope form of the equation of a line The equation $y = m(x - x_0) + y_0$ where m represents the slope of the line and (x_0, y_0) is a fixed point on the line.

polynomial function Any function defined by a sum of a finite number of terms of the form ax^n, where a is a real number and n is a nonnegative integer.

practical domain The set of all input values that make sense in a problem situation.

practical range The set of all output values that make sense in a problem situation.

product function The function that is created from two functions f and g, by the rule $y = f(x) \cdot g(x)$

profit function A function that is common in the business world and is defined by profit = revenue − cost.

quadratic formula The formula $x = \dfrac{-b \pm \sqrt{b^2 - 4ac}}{2a}$ that gives the solutions to the quadratic equation $ax^2 + bx + c = 0$.

quadratic function A second-degree polynomial function defined by an equation of the form $f(x) = ax^2 + bx + c$, where a, b, and c are real numbers and $a \neq 0$.

quartic function A fourth-degree polynomial function defined by an equation of the form $f(x) = ax^4 + bx^3 + cx^2 + dx + e$, where a, b, c, d, and e are real numbers and a (is not equal to) 0.

radical function Any function involving a radical (square root, cube root, and so on).

radicand The expression under the radical.

range The collection of all values of the dependent variable.

rate of change See average rate of change.

rational equation An equation composed of fractions where the numerators and denominators are polynomials, with the variable appearing in the denominator.

rational function Any function that can be defined as the ratio of two polynomial functions.

real numbers All numbers that are either rational or irrational.

slope of a line The constant rate of change of output to input.

slope-intercept form of the equation of a line The equation $y = mx + b$, where m represents the slope of the line and $(0, b)$ is the vertical intercept.

sum function, $f + g$ The function that is created from two functions f and g by the rule $y = f(x) + g(x)$.

system of linear equations in two variables A pair of equations that can be written in the form $y = ax + b$ and $y = cx + d$, respectively where a, b, c, and d are real numbers.

variation How the dependent variable changes when the independent variable changes. (See direct variation or inverse variation.)

vertex The turning point of the graph of a parabola. It has coordinates $\left(\dfrac{-b}{2a}, f\left(\dfrac{-b}{2a}\right)\right)$, where a and b are determined from the equation $f(x) = ax^2 + bx + c$. The vertex is the highest or lowest point of a parabola.

vertical asymptote of the graph of $y = f(x)$ The vertical line, $x = c$, such that $f(c)$ is undefined and $f(x)$ becomes arbitrarily large in magnitude as x approaches c.

vertical intercept The point of the graph of the function whose x-coordinate is 0. (*See* y-intercept.)

x-intercepts All points of the graph of the function at which the y-coordinate is 0. (See horizontal intercept.)

Xmax The largest value of input visible in the window of a graphing calculator.

Xmin The smallest value of input visible in the window of a graphing calculator.

y-intercept All points of the graph of an equation whose x-coordinate is 0. (See vertical intercept.)

Ymax The largest value of output visible in the window of the graphing calculator.

Ymin The smallest value of output visible in the window of a graphing calculator.

zero-product principle The algebraic rule that says if a and b are real numbers such that a $b = 0$, then either a or b, or both must be equal to zero.

INDEX

G

H

I

L

M

N

O

P

Q

R

S

T

U

V

X

Y

Z

Geometric Formulas

Perimeter and Area of a Triangle, and Sum of Measures of the Angles

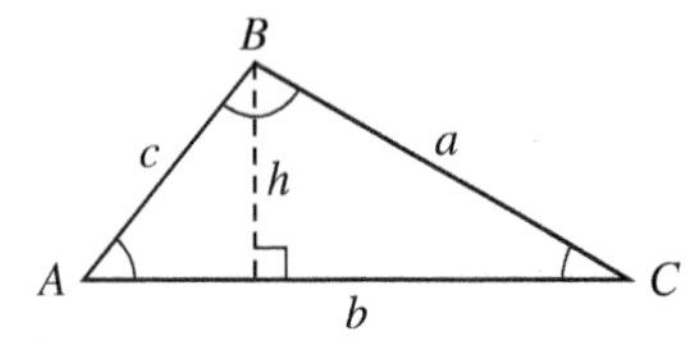

$$P = a + b + c$$
$$A = \frac{1}{2}bh$$
$$A + B + C = 180°$$

Pythagorean Theorem

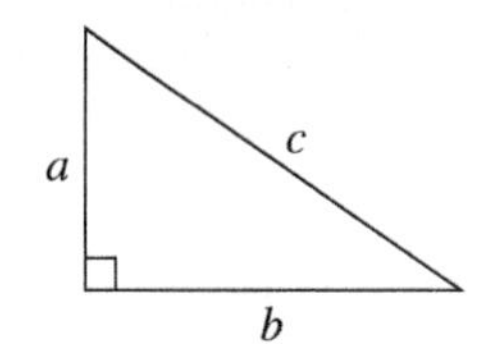

$$a^2 + b^2 = c^2$$

Perimeter and Area of a Rectangle

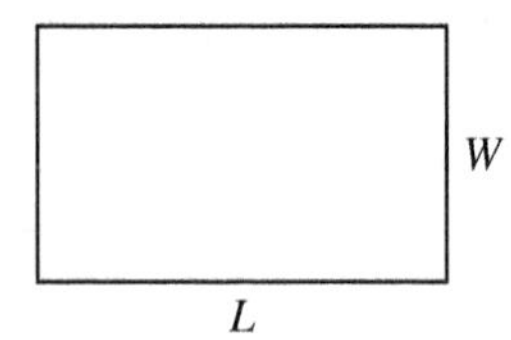

$$P = 2L + 2W$$
$$A = LW$$

Perimeter and Area of a Square

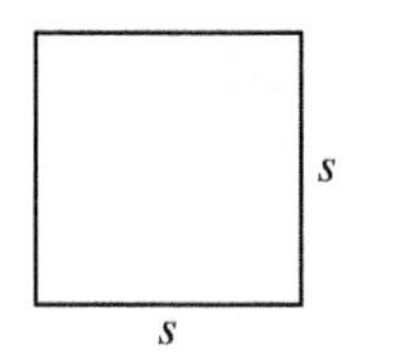

$$P = 4s$$
$$A = s^2$$

Area of a Trapezoid

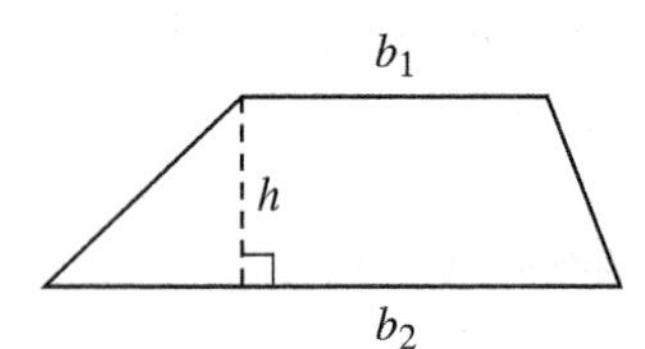

$$A = \frac{1}{2}h(b_1 + b_2)$$

Circumference and Area of a Circle

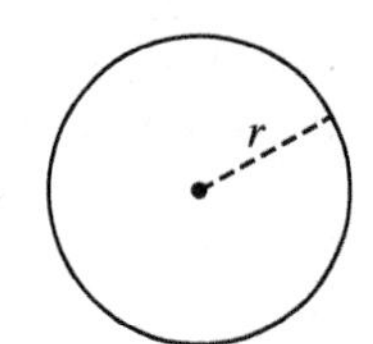

$$C = 2\pi r$$
$$A = \pi r^2$$

Volume and Surface Area of a Rectangular Solid

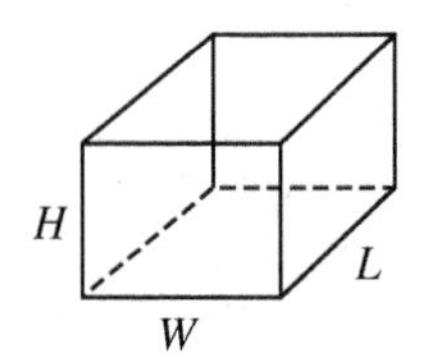

$$V = LWH$$
$$SA = 2LW + 2LH + 2WH$$

Volume and Surface Area of a Sphere

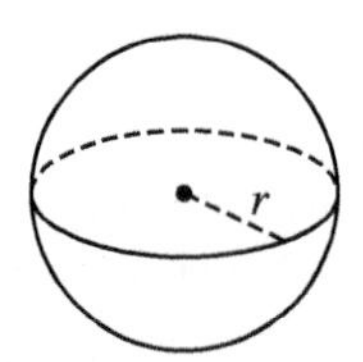

$$V = \frac{4}{3}\pi r^3$$
$$SA = 4\pi r^2$$

Volume and Surface Area of a Right Circular Cylinder

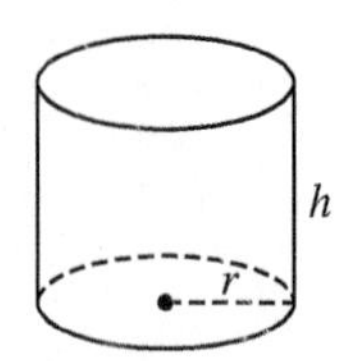

$$V = \pi r^2 h$$
$$SA = 2\pi r^2 + 2\pi rh$$

Volume and Surface Area of a Right Circular Cone

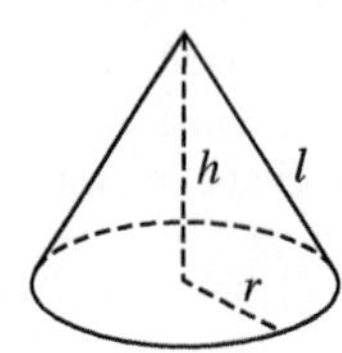

$$V = \frac{1}{3}\pi r^2 h$$
$$SA = \pi r^2 + \pi rl$$

Metric—U.S. Conversion

Length

1 meter = 3.28 feet

1 meter = 1.09 yards

1 centimeter = 0.39 inches

1 kilometer = 0.62 miles

Weight

1 gram = 0.035 ounces

1 kilometer = 2.20 pounds

1 gram = 0.0022 pounds

Volume

1 liter = 1.06 quarts

1 liter = 0.264 gallons

Temperature

Celsius to Fahrenheit

$F = \frac{9}{5}C + 32$

$C = \frac{5}{9}(F - 32)$